Signal Processing
and
Linear Systems

Signal Processing
and
Linear Systems

B. P. Lathi

Berkeley Cambridge Press

Carmichael, California

SIGNAL PROCESSING AND LINEAR SYSTEMS

Berkeley-Cambridge Press
P.O. Box 947, Carmichael, CA. 95609-0947

10   9   8   7   6   5   4   3   2

Library of congress cataloging in publication data
Lathi, B. P. (Bhagawandas Pannalal)
    Signal Processing and Linear Systems
    Includes bibliographical references and index.
    1. Signal processing—Mathematics.   2. System analysis
3. Linear time-invariant systems.   4. Digital filters (Mathematics)
I. Title.

TK5102.9.L38   1998   621.382'2—dc21   98-10609
ISBN 0-941413-35-7   CIP

Cover and graphic design by Jeff Clinton.
Printed in the United States of America

Preface

This book presents a comprehensive treatment of signals and linear systems suitable for juniors and seniors in electrical engineering. The book contains most of the material from my earlier popular book *Linear Systems and Signals* (1992) with added chapters on analog and digital filters and digital signal processing. There are also additional applications to communications and controls. The sequence of topics in this book is somewhat different from the earlier book. Here, the Laplace transform follows Fourier, whereas in the 1992 book, the sequence was the exact opposite. Moreover, the continuous-time and the discrete-time are treated sequentially, whereas in the 1992 book, both approaches were interwoven. The book contains enough material in discrete-time systems so that it can be used not only for a traditional course in *Signals and Systems*, but also for an introductory course in *Digital Signal processing*.

A perceptive author has said: "The function of a teacher is not so much to cover the topics of study as to uncover them for the students." The same can be said of a textbook. This book, as all my previous books, emphasizes the physical appreciation of concepts rather than mere mathematical manipulation of symbols. There is a temptation to treat an engineering subject, such as this, as a branch of applied mathematics. This view ignores the physical meaning behind various results and derivations, which deprives a student of intuitive understanding of the subject. I have used mathematics not so much to prove an axiomatic theory as to enhance the physical and intuitive understanding. Wherever possible, theoretical results are interpreted heuristically† and are supported by carefully chosen examples and analogies.‡

Notable Features

The notable features of the book include the following:

1. Emphasis on intuitive and heuristic understanding of the concepts and physical meaning of mathematical results leading to deeper appreciation and easier comprehension of the concepts. As one reviewer put it, "One thing I found very appealing about this book is great balance of mathematical and intuitive explanation." Most reviewers of the book have noted the reader friendly character of the book with unusual clarity of presentation.

2. The book provides extensive applications in the areas of communication, controls, and filtering.

3. For those who like to get students involved with computers, computer solutions of several examples are provided using MATLAB®, which is becoming a

---

†Heuristic [Greek *heuriskein*, to invent, discover]: a method of education in which the pupil is trained to find out things for himself. The word 'Eureka' (I have found it) is the 1st pers. perf. indic. act., of heuriskein.

‡If these lines appear familiar to you, there is a good reason. I have used them in the preface of some of my earlier books, including *Signals, Systems, and Communication* (Wiley, 1965). What is more interesting, many other authors also have borrowed them for their preface.

standard software package in an electrical engineering curriculum.

4. Many students are handicapped by an inadequate background in basic material such as complex numbers, sinusoids, sketching signals, Cramer's rule, partial fraction expansion, and matrix algebra. I have added a chapter that addresses these basic and pervasive topics in electrical engineering. Response by student has been unanimously enthusiastic.

5. There are over 200 worked examples along with exercises (with answers) for students to test their understanding. There are also about 400 selected problems of varying difficulty at the end of the chapters. Many problems are provided with hints to steer a student in the proper direction.

6. The discrete-time and continuous-time systems are covered sequentially, with flexibility to teach them concurrently if so desired.

7. The summary at the end of each chapter proves helpful to students in summing up essential developments in the chapter, and is an effective tool in the study for tests. Answers to selected problems are helpful in providing feedback to students trying to assess their knowledge.

8. There are several historical notes to enhance student's interest in the subject. These facts introduce students to historical background that influenced the development of electrical engineering.

Organization

The book opens with a chapter titled *Background*, which deals with the mathematical background material that a student taking this course is expected to have already mastered. It includes topics such as complex numbers, sinusoids, sketching signals, Cramer's rule, partial fraction expansion, matrix algebra. The next 7 chapters deal with continuous-time signals and systems followed by 5 chapters treating discrete-time signals and systems. The last chapter deals with state-space analysis. There are MATLAB examples dispersed throught the book. The book can be readily tailored for a variety of courses of 30 to 90 lecture hours. It can also be used as a text for a first undergraduate course in *Digital signal Processing* (DSP).

The organization of the book permits a great deal of flexibility in teaching the continuous-time and discrete-time concepts. The natural sequence of chapters is meant for a sequential approach in which all the continuous-time analysis is covered first, followed by discrete-time analysis. It is also possible to integrate (interweave) continuous-time and discrete-time analysis by using a appropriate sequence of chapters.

Credits

The photographs of Gauss (p. 3), Laplace (p. 380), Heaviside (p. 380), Fourier (p. 188), Michelson (p. 206) have been reprinted courtesy of the Smithsonian Institution. The photographs of Cardano (p. 3) and Gibbs (p. 206) have been reprinted courtesy of the Library of Congress. Most of the MATLAB examples were prepared by Dr. O. P. Mandhana of IBM, Austin, TX.

Acknowledgments

Several individuals have helped me in the preparation of this book. I am grateful for the helpful suggestions of the reviewers Professors. Dwight Day (Kansas State

University), Prof. Mark Herro (University of Notre Dame), Hua Lee (University of California, Santa Barbara), Tina Tracy (University of Missouri, Columbia), J.K. Tugnait (Auburn University), R.L. Tummala (Michigan State University). I owe Dr. O.P. Mandhana a debt of gratitude for his helpful suggestions and his painstaking solutions to most of the MATLAB problems. Special thanks go to Prof. James Simes for generous help with computer solution of several problems. I am much obliged to Ing Ming Chang for his enthusiastic and crucial help in solving MATLAB problems and using computer to prepare the manuscript. Finally I would like to mention the enormous but invisible sacrifices of my wife Rajani in this endeavor.

B. P. Lathi

MATLAB

Throughout this book, examples have been provided to familiarize the reader with computer tools for systems design and analysis using the powerful and versatile software package MATLAB. Much of the time and cost associated with the analysis and design of systems can be reduced by using computer software packages for simulation. Many corporations will no longer support the development systems without prior computer simulation and numerical results which suggest a design will work. The examples and problems in this book will assist the reader in learning the value of computer packages for systems design and simulation.

MATLAB is the software package used throughout this book. MATLAB is a powerful package developed to perform matrix manipulations for system designers. MATLAB is easily expandable and uses its own high level language. These factors make developing sophisticated systems easier. In addition, MATLAB has been carefully written to yield numerically stable results to produce reliable simulations.

All the computer examples in this book are verified to be compatible with the student edition of the MATLAB when used according to the instructions given in its manual. The reader should make sure that \MATLAB\BIN is added in the DOS search path. MATLAB can be invoked by executing the command MATLAB. The MATLAB banner will appear after a moment with the prompt '>>'. MATLAB has a useful on-line help. To get help on a specific command, type HELP COMMAND NAME and then press the ENTER key. DIARY FILE is a command to record all the important keyboard inputs to a file and the resulting output of your MATLAB session to be written on the named file. MATLAB can be used interactively, or by writing functions (subroutines) often called M files because of the .M extension used for these files. Once familiar with the basics of MATLAB, the reader can easily learn how to write functions and to use MATLAB's existing functions.

The MATLAB M-files have been created to supplement this text. This includes all the examples solved by MATLAB in the text. These M-files may be retrieved from the Mathworks anonymous FTP site at

ftp://ftp.mathworks.com/pub/books/lathi/.

O. P. Mandhana

Contents

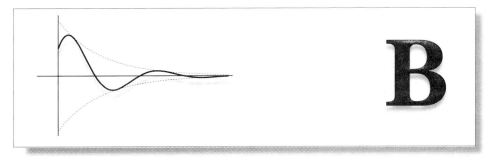

Background

The topics discussed in this chapter are not entirely new to students taking this course. You have already studied many of these topics in earlier courses or are expected to know them from your previous training. Even so, this background material deserves a review because it is so pervasive in the area of signals and systems. Investing a little time in such a review will pay big dividends later. Furthermore, this material is useful not only for this course but also for several courses that follow. It will also be helpful as reference material in your future professional career.

B.1 Complex Numbers

Complex numbers are an extension of ordinary numbers and are an integral part of the modern number system. Complex numbers, particularly **imaginary numbers**, sometimes seem mysterious and unreal. This feeling of unreality derives from their unfamiliarity and novelty rather than their supposed nonexistence! Mathematicians blundered in calling these numbers "imaginary," for the term immediately prejudices perception. Had these numbers been called by some other name, they would have become demystified long ago, just as irrational numbers or negative numbers were. Many futile attempts have been made to ascribe some physical meaning to imaginary numbers. However, this effort is needless. In mathematics we assign symbols and operations any meaning we wish as long as internal consistency is maintained. A healthier approach would have been to define a symbol i (with any term but "imaginary"), which has a property $i^2 = -1$. The history of mathematics is full of entities which were unfamiliar and held in abhorrence until familiarity made them acceptable. This fact will become clear from the following historical note.

B.1-1 A Historical Note

Among early people the number system consisted only of natural numbers (positive integers) needed to count the number of children, cattle, and quivers of arrows. These people had no need for fractions. Whoever heard of two and one-half children or three and one-fourth cows!

1

However, with the advent of agriculture, people needed to measure continuously varying quantities, such as the length of a field, the weight of a quantity of butter, and so on. The number system, therefore, was extended to include fractions. The ancient Egyptians and Babylonians knew how to handle fractions, but **Pythagoras** discovered that some numbers (like the diagonal of a unit square) could not be expressed as a whole number or a fraction. Pythagoras, a number mystic, who regarded numbers as the essence and principle of all things in the universe, was so appalled at his discovery that he swore his followers to secrecy and imposed a death penalty for divulging this secret.[1] These numbers, however, were included in the number system by the time of Descartes, and they are now known as **irrational numbers**.

Until recently, **negative numbers** were not a part of the number system. The concept of negative numbers must have appeared absurd to early man. However, the medieval Hindus had a clear understanding of the significance of positive and negative numbers.[2,3] They were also the first to recognize the existence of absolute negative quantities.[4] The works of **Bhaskar** (1114-1185) on arithmetic (*Līlāvatī*) and algebra (*Bījaganit*) not only use the decimal system but also give rules for dealing with negative quantities. Bhaskar recognized that positive numbers have two square roots.[5] Much later, in Europe, the banking system that arose in Florence and Venice during the late Renaissance (fifteenth century) is credited with developing a crude form of negative numbers. The seemingly absurd subtraction of 7 from 5 seemed reasonable when bankers began to allow their clients to draw seven gold ducats while their deposit stood at five. All that was necessary for this purpose was to write the difference, 2, on the debit side of a ledger.[6]

Thus the number system was once again broadened (generalized) to include negative numbers. The acceptance of negative numbers made it possible to solve equations such as $x + 5 = 0$, which had no solution before. Yet for equations such as $x^2 + 1 = 0$, leading to $x^2 = -1$, the solution could not be found in the real number system. It was therefore necessary to define a completely new kind of number with its square equal to -1. During the time of Descartes and Newton, imaginary (or complex) numbers came to be accepted as part of the number system, but they were still regarded as algebraic fiction. The Swiss mathematician **Leonhard Euler** introduced the notation i (for **imaginary**) around 1777 to represent $\sqrt{-1}$. Electrical engineers use the notation j instead of i to avoid confusion with the notation i often used for electrical current. Thus

$$j^2 = -1$$

and

$$\sqrt{-1} = \pm j$$

This notation allows us to determine the square root of any negative number. For example,

$$\sqrt{-4} = \sqrt{4} \times \sqrt{-1} = \pm 2j$$

When imaginary numbers are included in the number system, the resulting numbers are called **complex numbers**.

Origins of Complex Numbers

Ironically (and contrary to popular belief), it was not the solution of a quadratic equation, such as $x^2 + 1 = 0$, but a cubic equation with real roots that made

Gerolamo Cardano (left) and Karl Friedrich Gauss (right).

imaginary numbers plausible and acceptable to early mathematicians. They could dismiss $\sqrt{-1}$ as pure nonsense when it appeared as a solution to $x^2 + 1 = 0$ because this equation has no real solution. But in 1545, **Gerolamo Cardano** of Milan published *Ars Magna* (*The Great Art*), the most important algebraic work of the Renaissance. In this book he gave a method of solving a general cubic equation in which a root of a negative number appeared in an intermediate step. According to his method, the solution to a third-order equation†

$$x^3 + ax + b = 0$$

is given by

$$x = \sqrt[3]{-\frac{b}{2} + \sqrt{\frac{b^2}{4} + \frac{a^3}{27}}} + \sqrt[3]{-\frac{b}{2} - \sqrt{\frac{b^2}{4} + \frac{a^3}{27}}}$$

For example, to find a solution of $x^3 + 6x - 20 = 0$, we substitute $a = 6$, $b = -20$ in the above equation to obtain

$$x = \sqrt[3]{10 + \sqrt{108}} + \sqrt[3]{10 - \sqrt{108}} = \sqrt[3]{20.392} - \sqrt[3]{0.392} = 2$$

We can readily verify that 2 is indeed a solution of $x^3 + 6x - 20 = 0$. But when Cardano tried to solve the equation $x^3 - 15x - 4 = 0$ by this formula, his solution

---

†This equation is known as the *depressed cubic* equation. A general cubic equation

$$y^3 + py^2 + qy + r = 0$$

can always be reduced to a depressed cubic form by substituting $y = x - \frac{p}{3}$. Therefore any general cubic equation can be solved if we know the solution to the depressed cubic. The depressed cubic was independently solved, first by **Scipione del Ferro** (1465-1526) and then by **Niccolo Fontana** (1499-1557). The latter is better known in the history of mathematics as **Tartaglia** ("Stammerer"). Cardano learned the secret of the depressed cubic solution from Tartaglia. He then showed that by using the substitution $y = x - \frac{p}{3}$, a general cubic is reduced to a depressed cubic.

was

$$x = \sqrt[3]{2 + \sqrt{-121}} + \sqrt[3]{2 - \sqrt{-121}}$$

What was Cardano to make of this equation in the year 1545? In those days negative numbers were themselves suspect, and a square root of a negative number was doubly preposterous! Today we know that

$$(2 \pm j)^3 = 2 \pm j11 = 2 \pm \sqrt{-121}$$

Therefore, Cardano's formula gives

$$x = (2 + j) + (2 - j) = 4$$

We can readily verify that $x = 4$ is indeed a solution of $x^3 - 15x - 4 = 0$. Cardano tried to explain halfheartedly the presence of $\sqrt{-121}$ but ultimately dismissed the whole enterprise as being "as subtle as it is useless." A generation later, however, **Raphael Bombelli** (1526-1573), after examining Cardano's results, proposed acceptance of imaginary numbers as a necessary vehicle that would transport the mathematician from the *real* cubic equation to its *real* solution. In other words, while we begin and end with real numbers, we seem compelled to move into an unfamiliar world of imaginaries to complete our journey. To mathematicians of the day, this proposal seemed incredibly strange.[7] Yet they could not dismiss the idea of imaginary numbers so easily because this concept yielded the real solution of an equation. It took two more centuries for the full importance of complex numbers to become evident in the works of Euler, Gauss, and Cauchy. Still, Bombelli deserves credit for recognizing that such numbers have a role to play in algebra.[7]

In 1799, the German mathematician **Karl Friedrich Gauss**, at a ripe age of 22, proved the fundamental theorem of algebra, namely that every algebraic equation in one unknown has a root in the form of a complex number. He showed that every equation of the nth order has exactly n solutions (roots), no more and no less. Gauss was also one of the first to give a coherent account of complex numbers and to interpret them as points in a complex plane. It is he who introduced the term *complex numbers* and paved the way for general and systematic use of complex numbers. The number system was once again broadened or generalized to include imaginary numbers. Ordinary (or real) numbers became a special case of generalized (or complex) numbers.

The utility of complex numbers can be understood readily by an analogy with two neighboring countries X and Y, as illustrated in Fig. B.1. If we want to travel from City a to City b (both in Country X), the shortest route is through Country Y, although the journey begins and ends in Country X. We may, if we desire, perform this journey by an alternate route that lies exclusively in X, but this alternate route is longer. In mathematics we have a similar situation with real numbers (Country X) and complex numbers (Country Y). All real-world problems must start with real numbers, and all the final results must also be in real numbers. But the derivation of results is considerably simplified by using complex numbers as an intermediary. It is also possible to solve all real-world problems by an alternate method, using real numbers exclusively, but such procedure would increase the work needlessly.

B.1-2 Algebra of Complex Numbers

A complex number (a, b) or $a + jb$ can be represented graphically by a point whose Cartesian coordinates are (a, b) in a complex plane (Fig. B.2). Let us denote this complex number by z so that

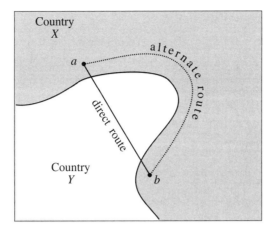

Fig. B.1 Use of complex numbers can reduce the work.

$$z = a + jb \tag{B.1}$$

The numbers a and b (the abscissa and the ordinate) of z are the **real part** and the **imaginary part**, respectively, of z. They are also expressed as

$$\text{Re } z = a$$

$$\text{Im } z = b$$

Note that in this plane all real numbers lie on the horizontal axis, and all imaginary numbers lie on the vertical axis.

Complex numbers may also be expressed in terms of polar coordinates. If (r, θ) are the polar coordinates of a point $z = a + jb$ (see Fig. B.2), then

$$a = r \cos \theta$$

$$b = r \sin \theta$$

and

$$z = a + jb = r \cos \theta + jr \sin \theta$$

$$= r(\cos \theta + j \sin \theta) \tag{B.2}$$

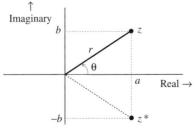

Fig. B.2 Representation of a number in the complex plane.

The **Euler formula** states that

$$e^{j\theta} = \cos\theta + j\sin\theta$$

To prove the Euler formula, we expand $e^{j\theta}$, $\cos\theta$, and $\sin\theta$ using a Maclaurin series

$$e^{j\theta} = 1 + j\theta + \frac{(j\theta)^2}{2!} + \frac{(j\theta)^3}{3!} + \frac{(j\theta)^4}{4!} + \frac{(j\theta)^5}{5!} + \frac{(j\theta)^6}{6!} + \cdots$$

$$= 1 + j\theta - \frac{\theta^2}{2!} - j\frac{\theta^3}{3!} + \frac{\theta^4}{4!} + j\frac{\theta^5}{5!} - \frac{\theta^6}{6!} - \cdots$$

$$\cos\theta = 1 - \frac{\theta^2}{2!} + \frac{\theta^4}{4!} - \frac{\theta^6}{6!} + \frac{\theta^8}{8!} \cdots$$

$$\sin\theta = \theta - \frac{\theta^3}{3!} + \frac{\theta^5}{5!} - \frac{\theta^7}{7!} + \cdots$$

Hence, it follows that

$$e^{j\theta} = \cos\theta + j\sin\theta \tag{B.3}$$

Using (B.3) in (B.2) yields

$$z = a + jb$$

$$= re^{j\theta} \tag{B.4}$$

Thus, a complex number can be expressed in Cartesian form $a + jb$ or polar form $re^{j\theta}$ with

$$a = r\cos\theta, \qquad b = r\sin\theta \tag{B.5}$$

and

$$r = \sqrt{a^2 + b^2}, \qquad \theta = \tan^{-1}\left(\frac{b}{a}\right) \tag{B.6}$$

Observe that r is the distance of the point z from the origin. For this reason, r is also called the **magnitude** (or **absolute value**) of z and is denoted by $|z|$. Similarly θ is called the angle of z and is denoted by $\angle z$. Therefore

$$|z| = r, \qquad \angle z = \theta$$

and

$$z = |z|e^{j\angle z} \tag{B.7}$$

Also

$$\frac{1}{z} = \frac{1}{re^{j\theta}} = \frac{1}{r}e^{-j\theta} = \frac{1}{|z|}e^{-j\angle z} \tag{B.8}$$

Conjugate of a Complex Number

We define z^*, the **conjugate** of $z = a + jb$, as

$$z^* = a - jb = re^{-j\theta} \tag{B.9a}$$

$$= |z|e^{-j\angle z} \tag{B.9b}$$

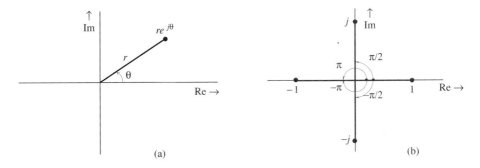

Fig. B.3 Understanding some useful identities in terms of $re^{j\theta}$.

The graphical representation of a number z and its conjugate z^* is depicted in Fig. B.2. Observe that z^* is a mirror image of z about the horizontal axis. *To find the conjugate of any number, we need only to replace j by $-j$ in that number* (which is the same as changing the sign of its angle).

The sum of a complex number and its conjugate is a real number equal to twice the real part of the number:

$$z + z^* = (a + jb) + (a - jb) = 2a = 2\operatorname{Re} z \qquad \text{(B.10a)}$$

The product of a complex number z and its conjugate is a real number $|z|^2$, the square of the magnitude of the number:

$$zz^* = (a + jb)(a - jb) = a^2 + b^2 = |z|^2 \qquad \text{(B.10b)}$$

Understanding Some Useful Identities

In a complex plane, $re^{j\theta}$ represents a point at a distance r from the origin and at an angle θ with the horizontal axis, as shown in Fig. B.3a. For example, the number -1 is at a unit distance from the origin and has an angle π or $-\pi$ (in fact, any odd multiple of $\pm\pi$), as seen from Fig. B.3b. Therefore,

$$1e^{\pm j\pi} = -1$$

In fact,

$$e^{\pm jn\pi} = -1 \qquad n \text{ odd integer} \qquad \text{(B.11)}$$

The number 1, on the other hand, is also at a unit distance from the origin, but has an angle 2π (in fact, $\pm 2n\pi$ for any integral value of n). Therefore,

$$e^{\pm j2n\pi} = 1 \qquad n \text{ integer} \qquad \text{(B.12)}$$

The number j is at unit distance from the origin and its angle is $\pi/2$ (see Fig. B.3b). Therefore,

$$e^{j\pi/2} = j$$

Similarly,

$$e^{-j\pi/2} = -j$$

Thus

$$e^{\pm j\pi/2} = \pm j \qquad \text{(B.13a)}$$

In fact,
$$e^{\pm jn\pi/2} = \pm j \qquad n = 1, 5, 9, 13, \cdots \tag{B.13b}$$
and
$$e^{\pm jn\pi/2} = \mp j \qquad n = 3, 7, 11, 15, \cdots \tag{B.13c}$$

These results are summarized in Table B.1.

TABLE B.1

r	θ	$re^{j\theta}$	
1	0	$e^{j0} = 1$	
1	$\pm\pi$	$e^{\pm j\pi} = -1$	
1	$\pm n\pi$	$e^{\pm jn\pi} = -1$	n odd integer
1	$\pm 2\pi$	$e^{\pm j2\pi} = 1$	
1	$\pm 2n\pi$	$e^{\pm j2n\pi} = 1$	n integer
1	$\pm\pi/2$	$e^{\pm j\pi/2} = \pm j$	
1	$\pm n\pi/2$	$e^{\pm jn\pi/2} = \pm j$	$n = 1, 5, 9, 13, \ldots$
1	$\pm n\pi/2$	$e^{\pm jn\pi/2} = \mp j$	$n = 3, 7, 11, 15, \ldots$

This discussion shows the usefulness of the graphic picture of $re^{j\theta}$. This picture is also helpful in several other applications. For example, to determine the limit of $e^{(\alpha+j\omega)t}$ as $t \to \infty$, we note that
$$e^{(\alpha+j\omega)t} = e^{\alpha t}e^{j\omega t}$$

Now the magnitude of $e^{j\omega t}$ is unity regardless of the value of ω or t because $e^{j\omega t} = re^{j\theta}$ with $r = 1$. Therefore, $e^{\alpha t}$ determines the behavior of $e^{(\alpha+j\omega)t}$ as $t \to \infty$ and
$$\lim_{t\to\infty} e^{(\alpha+j\omega)t} = \lim_{t\to\infty} e^{\alpha t}e^{j\omega t} = \begin{cases} 0 & \alpha < 0 \\ \infty & \alpha > 0 \end{cases} \tag{B.14}$$

In future discussions you will find it very useful to remember $re^{j\theta}$ as a number at a distance r from the origin and at an angle θ with the horizontal axis of the complex plane.

A Warning About Using Electronic Calculators in Computing Angles

From the Cartesian form $a + jb$ we can readily compute the polar form $re^{j\theta}$ [see Eq. (B.6)]. Electronic calculators provide ready conversion of rectangular into polar and vice versa. However, if a calculator computes an angle of a complex number using an inverse trigonometric function $\theta = \tan^{-1}(b/a)$, proper attention must be paid to the quadrant in which the number is located. For instance, θ corresponding to the number $-2 - j3$ is $\tan^{-1}(\frac{-3}{-2})$. This result is not the same as $\tan^{-1}(\frac{3}{2})$. The former is $-123.7°$, whereas the latter is $56.3°$. An electronic calculator cannot make this distinction and can give a correct answer only for angles in the first and

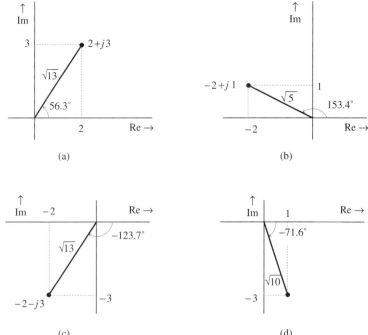

Fig. B.4 From Cartesian to polar form.

fourth quadrants. It will read $\tan^{-1}(\frac{-3}{-2})$ as $\tan^{-1}(\frac{3}{2})$, which is clearly wrong. In computing inverse trigonometric functions, if the angle appears in the second or third quadrant, the answer of the calculator is off by 180°. The correct answer is obtained by adding or subtracting 180° to the value found with the calculator (either adding or subtracting yields the correct answer). For this reason it is advisable to draw the point in the complex plane and determine the quadrant in which the point lies. This issue will be clarified by the following examples.

■ **Example B.1**

Express the following numbers in polar form:

(a) $2 + j3$ (b) $-2 + j1$ (c) $-2 - j3$ (d) $1 - j3$

(a)

$$|z| = \sqrt{2^2 + 3^2} = \sqrt{13} \qquad \angle z = \tan^{-1}\left(\tfrac{3}{2}\right) = 56.3°$$

In this case the number is in the first quadrant, and a calculator will give the correct value of 56.3°. Therefore, (see Fig. B.4a)

$$2 + j3 = \sqrt{13}\, e^{j56.3°}$$

(b)

$$|z| = \sqrt{(-2)^2 + 1^2} = \sqrt{5} \qquad \angle z = \tan^{-1}\left(\tfrac{1}{-2}\right) = 153.4°$$

In this case the angle is in the second quadrant (see Fig. B.4b), and therefore the answer given by the calculator $(\tan^{-1}(\frac{1}{-2}) = -26.6°)$ is off by 180°. The correct answer is $(-26.6 \pm 180)° = 153.4°$ or $-206.6°$. Both values are correct because they represent the same angle. As a matter of convenience, we choose an angle whose numerical value is less than 180°, which in this case is 153.4°. Therefore,

$$-2 + j1 = \sqrt{5}e^{j153.4°}$$

(c)
$$|z| = \sqrt{(-2)^2 + (-3)^2} = \sqrt{13} \qquad \angle z = \tan^{-1}(\tfrac{-3}{-2}) = -123.7°$$

In this case the angle appears in the third quadrant (see Fig. B.4c), and therefore the answer obtained by the calculator ($\tan^{-1}(\tfrac{-3}{-2}) = 56.3°$) is off by 180°. The correct answer is $(56.3 \pm 180)° = 236.3°$ or $-123.7°$. As a matter of convenience, we choose the latter and (see Fig. B.4c)

$$-2 - j3 = \sqrt{13}e^{-j123.7°}$$

(d)
$$|z| = \sqrt{1^2 + (-3)^2} = \sqrt{10} \qquad \angle z = \tan^{-1}(\tfrac{-3}{1}) = -71.6°$$

In this case the angle appears in the fourth quadrant (see Fig. B.4d), and therefore the answer given by the calculator ($\tan^{-1}(\tfrac{-3}{1}) = -71.6°$) is correct (see Fig. B.4d).

$$1 - j3 = \sqrt{10}e^{-j71.6°} \quad \blacksquare$$

⊙ **Computer Example CB.1**
Express the following numbers in polar form: **(a)** $2 + j3$ **(b)** $-2 + j1$
MATLAB function $cart2pol(a,b)$ can be used to convert the complex number $a + jb$ to its polar form.

(a)
[Zangle_in_rad,Zmag]=cart2pol(2,3)
Zangle_in_rad = 0.9828
Zmag =3.6056
Zangle_in_deg=Zangle_in_rad*(180/pi)
Zangle_in_deg=56.31
Therefore
$$z = 2 + j3 = 3.6056e^{j56.31°}$$

(b)
[Zangle_in_rad,Zmag]=cart2pol(-2,1)
Zangle_in_rad = 2.6779
Zmag =2.2361
Zangle_in_deg=Zangle_in_rad*(180/pi)
Zangle_in_deg=153.4349
Therefore
$$z = -2 + j1 = 2.2361e^{j153.4349°}$$

Note that MATLAB automatically takes care of the quadrant in which the complex number lies. ⊙

■ **Example B.2**
Represent the following numbers in the complex plane and express them in Cartesian form: **(a)** $2e^{j\pi/3}$ **(b)** $4e^{-j3\pi/4}$ **(c)** $2e^{j\pi/2}$ **(d)** $3e^{-j3\pi}$ **(e)** $2e^{j4\pi}$ **(f)** $2e^{-j4\pi}$.

(a) $2e^{j\pi/3} = 2\left(\cos\frac{\pi}{3} + j\sin\frac{\pi}{3}\right) = 1 + j\sqrt{3}$ (see Fig. B.5a)
(b) $4e^{-j3\pi/4} = 4\left(\cos\frac{3\pi}{4} - j\sin\frac{3\pi}{4}\right) = -2\sqrt{2} - j2\sqrt{2}$ (see Fig. B.5b)
(c) $2e^{j\pi/2} = 2\left(\cos\frac{\pi}{2} + j\sin\frac{\pi}{2}\right) = 2(0 + j1) = j2$ (see Fig. B.5c)
(d) $3e^{-j3\pi} = 3(\cos 3\pi - j\sin 3\pi) = 3(-1 + j0) = -3$ (see Fig. B.5d)
(e) $2e^{j4\pi} = 2(\cos 4\pi + j\sin 4\pi) = 2(1 + j0) = 2$ (see Fig. B.5e)
(f) $2e^{-j4\pi} = 2(\cos 4\pi - j\sin 4\pi) = 2(1 - j0) = 2$ (see Fig. B.5f)   ■

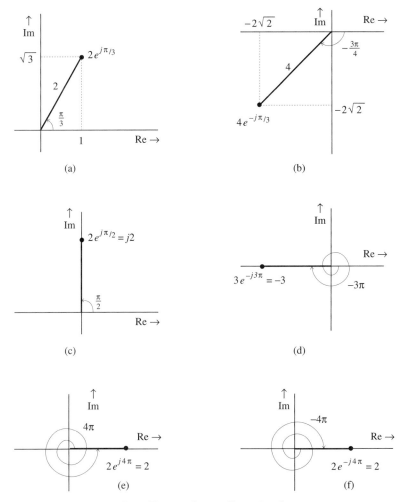

Fig. B.5 From polar to Cartesian form.

⊙ **Computer Example CB.2**

Represent $4e^{-j\frac{3\pi}{4}}$ in Cartesian form.
MATLAB function pol2cart(θ, r) converts the complex number $re^{j\theta}$ to Cartesian form.

```
[Zreal,Zimag]=pol2cart(-3*pi/4,4)
Zreal=-2.8284
Zimag=-2.8284
```

Therefore

$$4e^{-j\frac{3\pi}{4}} = -2.8284 - j2.8284 \qquad ⊙$$

Arithmetical Operations, Powers, and Roots of Complex Numbers

To perform addition and subtraction, complex numbers should be expressed in Cartesian form. Thus, if

$$z_1 = 3 + j4 = 5e^{j53.1°}$$

and

$$z_2 = 2 + j3 = \sqrt{13}e^{j56.3°}$$

then

$$z_1 + z_2 = (3 + j4) + (2 + j3) = 5 + j7$$

If z_1 and z_2 are given in polar form, we would need to convert them into Cartesian form for the purpose of adding (or subtracting). Multiplication and division, however, can be carried out in either Cartesian or polar form, although the latter proves to be much more convenient. This is because if z_1 and z_2 are expressed in polar form as

$$z_1 = r_1 e^{j\theta_1} \quad \text{and} \quad z_2 = r_2 e^{j\theta_2}$$

then

$$z_1 z_2 = \left(r_1 e^{j\theta_1}\right)\left(r_2 e^{j\theta_2}\right) = r_1 r_2 e^{j(\theta_1 + \theta_2)} \tag{B.15a}$$

and

$$\frac{z_1}{z_2} = \frac{r_1 e^{j\theta_1}}{r_2 e^{j\theta_2}} = \frac{r_1}{r_2} e^{j(\theta_1 - \theta_2)} \tag{B.15b}$$

Moreover,

$$z^n = \left(re^{j\theta}\right)^n = r^n e^{jn\theta} \tag{B.15c}$$

and

$$z^{1/n} = \left(re^{j\theta}\right)^{1/n} = r^{1/n} e^{j\theta/n} \tag{B.15d}$$

This shows that the operations of multiplication, division, powers, and roots can be carried out with remarkable ease when the numbers are in polar form.

■ Example B.3

Determine $z_1 z_2$ and z_1/z_2 for the numbers

$$z_1 = 3 + j4 = 5e^{j53.1°}$$

$$z_2 = 2 + j3 = \sqrt{13}e^{j56.3°}$$

We shall solve this problem in both polar and Cartesian forms.

Multiplication: Cartesian Form

$$z_1 z_2 = (3 + j4)(2 + j3) = (6 - 12) + j(8 + 9) = -6 + j17$$

Multiplication: Polar Form

$$z_1 z_2 = \left(5e^{j53.1°}\right)\left(\sqrt{13}e^{j56.3°}\right) = 5\sqrt{13}e^{j109.4°}$$

Division: Cartesian Form

$$\frac{z_1}{z_2} = \frac{3 + j4}{2 + j3}$$

In order to eliminate the complex number in the denominator, we multiply both the numerator and the denominator of the right-hand side by $2 - j3$, the denominator's conjugate. This yields

$$\frac{z_1}{z_2} = \frac{(3+j4)(2-j3)}{(2+j3)(2-j3)} = \frac{18-j1}{2^2+3^2} = \frac{18-j1}{13} = \frac{18}{13} - j\frac{1}{13}$$

Division: Polar Form

$$\frac{z_1}{z_2} = \frac{5e^{j53.1°}}{\sqrt{13}e^{j56.3°}} = \frac{5}{\sqrt{13}}e^{j(53.1°-56.3°)} = \frac{5}{\sqrt{13}}e^{-j3.2°} \quad \blacksquare$$

It is clear from this example that multiplication and division are easier to accomplish in polar form than in Cartesian form.

■ **Example B.4**

For $z_1 = 2e^{j\pi/4}$ and $z_2 = 8e^{j\pi/3}$, find **(a)** $2z_1 - z_2$ **(b)** $\frac{1}{z_1}$ **(c)** $\frac{z_1}{z_2^2}$ **(d)** $\sqrt[3]{z_2}$

(a) Since subtraction cannot be performed directly in polar form, we convert z_1 and z_2 to Cartesian form:

$$z_1 = 2e^{j\pi/4} = 2\left(\cos\tfrac{\pi}{4} + j\sin\tfrac{\pi}{4}\right) = \sqrt{2} + j\sqrt{2}$$

$$z_2 = 8e^{j\pi/3} = 8\left(\cos\tfrac{\pi}{3} + j\sin\tfrac{\pi}{3}\right) = 4 + j4\sqrt{3}$$

Therefore,

$$2z_1 - z_2 = 2(\sqrt{2} + j\sqrt{2}) - (4 + j4\sqrt{3})$$
$$= (2\sqrt{2} - 4) + j(2\sqrt{2} - 4\sqrt{3})$$
$$= -1.17 - j4.1$$

(b)

$$\frac{1}{z_1} = \frac{1}{2e^{j\pi/4}} = \frac{1}{2}e^{-j\pi/4}$$

(c)

$$\frac{z_1}{z_2^2} = \frac{2e^{j\pi/4}}{(8e^{j\pi/3})^2} = \frac{2e^{j\pi/4}}{64e^{j2\pi/3}} = \frac{1}{32}e^{j(\frac{\pi}{4} - \frac{2\pi}{3})} = \frac{1}{32}e^{-j\frac{5\pi}{12}}$$

(d)

$$\sqrt[3]{z_2} = z_2^{1/3} = \left(8e^{j\pi/3}\right)^{\frac{1}{3}} = 8^{\frac{1}{3}}\left(e^{j\pi/3}\right)^{1/3} = 2e^{j\pi/9} \quad \blacksquare$$

⊙ **Computer Example CB.3**

Determine $z_1 z_2$ and z_1/z_2 if $z_1 = 3 + j4$ and $z_2 = 2 + j3$

Multiplication and division: Cartesian Form

```
z1=3+j*4; z2=2+j*3;
z1z2=z1*z2
z1z2=-6.000+17.0000i
z1_over_z2=z1/z2
z1_over_z2=1.3486-0.0769i
```

Therefore

$$(3+j4)(2+j3) = -6+j17 \quad \text{and} \quad (3+j4)/(2+j3) = 1.3486 - 0.0769 \qquad \odot$$

◼ **Example B.5**

Consider $F(\omega)$, a complex function of a real variable ω:

$$F(\omega) = \frac{2 + j\omega}{3 + j4\omega} \tag{B.16a}$$

(a) Express $F(\omega)$ in Cartesian form, and find its real and imaginary parts. **(b)** Express $F(\omega)$ in polar form, and find its magnitude $|F(\omega)|$ and angle $\angle F(\omega)$.

(a) To obtain the real and imaginary parts of $F(\omega)$, we must eliminate imaginary terms in the denominator of $F(\omega)$. This is readily done by multiplying both the numerator and denominator of $F(\omega)$ by $3 - j4\omega$, the conjugate of the denominator $3 + j4\omega$ so that

$$F(\omega) = \frac{(2 + j\omega)(3 - j4\omega)}{(3 + j4\omega)(3 - j4\omega)} = \frac{(6 + 4\omega^2) - j5\omega}{9 + 16\omega^2} = \frac{6 + 4\omega^2}{9 + 16\omega^2} - j\frac{5\omega}{9 + \omega^2} \tag{B.16b}$$

This is the Cartesian form of $F(\omega)$. Clearly the real and imaginary parts $F_r(\omega)$ and $F_i(\omega)$ are given by

$$F_r(\omega) = \frac{6 + 4\omega^2}{9 + 16\omega^2}, \qquad F_i(\omega) = \frac{-5\omega}{9 + 16\omega^2}$$

(b)

$$F(\omega) = \frac{2 + j\omega}{3 + j4\omega} = \frac{\sqrt{4 + \omega^2}\, e^{j\tan^{-1}\left(\frac{\omega}{2}\right)}}{\sqrt{9 + 16\omega^2}\, e^{j\tan^{-1}\left(\frac{4\omega}{3}\right)}}$$

$$= \sqrt{\frac{4 + \omega^2}{9 + 16\omega^2}}\, e^{j\left[\tan^{-1}\left(\frac{\omega}{2}\right) - \tan^{-1}\left(\frac{4\omega}{3}\right)\right]} \tag{B.16c}$$

This is the polar representation of $F(\omega)$. Observe that

$$|F(\omega)| = \sqrt{\frac{4 + \omega^2}{9 + 16\omega^2}}, \qquad \angle F(\omega) = \tan^{-1}\left(\frac{\omega}{2}\right) - \tan^{-1}\left(\frac{4\omega}{3}\right) \tag{B.17}$$

◼

B.2 Sinusoids

Consider the sinusoid

$$f(t) = C\cos\left(2\pi\mathcal{F}_0 t + \theta\right) \tag{B.18}$$

We know that

$$\cos\varphi = \cos\left(\varphi + 2n\pi\right) \qquad n = 0, \pm1, \pm2, \pm3, \cdots$$

Therefore, $\cos\varphi$ repeats itself for every change of 2π in the angle φ. For the sinusoid in Eq. (B.18), the angle $2\pi\mathcal{F}_0 t + \theta$ changes by 2π when t changes by $1/\mathcal{F}_0$. Clearly, this sinusoid repeats every $1/\mathcal{F}_0$ seconds. As a result, there are $\mathcal{F}_0$ repetitions per second. This is the **frequency** of the sinusoid, and the repetition interval T_0 given by

$$T_0 = \frac{1}{\mathcal{F}_0} \tag{B.19}$$

is the **period**. For the sinusoid in Eq. (B.18), C is the **amplitude**, $\mathcal{F}_0$ is the **frequency** (in **Hertz**), and θ is the phase. Let us consider two special cases of this sinusoid when $\theta = 0$ and $\theta = -\pi/2$ as follows:

(a) $f(t) = C \cos 2\pi \mathcal{F}_0 t$ $(\theta = 0)$

(b) $f(t) = C \cos \left(2\pi \mathcal{F}_0 t - \frac{\pi}{2}\right) = C \sin 2\pi \mathcal{F}_0 t$ $(\theta = -\pi/2)$

The angle or phase can be expressed in units of degrees or radians. Although the radian is the proper unit, in this book we shall often use the degree unit because students generally have a better feel for the relative magnitudes of angles when expressed in degrees rather than in radians. For example, we relate better to the angle 24° than to 0.419 radians. Remember, however, when in doubt, use the radian unit and, above all, be consistent. In other words, in a given problem or an expression do not mix the two units.

It is convenient to use the variable ω_0 (*radian frequency*) to express $2\pi \mathcal{F}_0$:

$$\omega_0 = 2\pi \mathcal{F}_0 \tag{B.20}$$

With this notation, the sinusoid in Eq. (B.18) can be expressed as

$$f(t) = C \cos (\omega_0 t + \theta)$$

in which the period T_0 is given by [see Eqs. (B.19) and (B.20)]

$$T_0 = \frac{1}{\omega_0/2\pi} = \frac{2\pi}{\omega_0} \tag{B.21a}$$

and

$$\omega_0 = \frac{2\pi}{T_0} \tag{B.21b}$$

In future discussions, we shall often refer to ω_0 as the frequency of the signal $\cos (\omega_0 t + \theta)$, but it should be clearly understood that the frequency of this sinusoid is $\mathcal{F}_0$ Hz ($\mathcal{F}_0 = \omega_0/2\pi$), and ω_0 is actually the **radian frequency**.

The signals $C \cos \omega_0 t$ and $C \sin \omega_0 t$ are illustrated in Figs. B.6a and B.6b respectively. A general sinusoid $C \cos (\omega_0 t + \theta)$ can be readily sketched by shifting the signal $C \cos \omega_0 t$ in Fig. B.6a by the appropriate amount. Consider, for example,

$$f(t) = C \cos (\omega_0 t - 60°)$$

This signal can be obtained by shifting (delaying) the signal $C \cos \omega_0 t$ (Fig. B.6a) to the right by a phase (angle) of 60°. We know that a sinusoid undergoes a 360° change of phase (or angle) in one cycle. A quarter-cycle segment corresponds to a 90° change of angle. Therefore, an angle of 60° corresponds to two-thirds of a quarter-cycle segment. We therefore shift (delay) the signal in Fig. B.6a by two-thirds of a quarter-cycle segment to obtain $C \cos (\omega_0 t - 60°)$, as shown in Fig. B.6c.

Observe that if we delay $C \cos \omega_0 t$ in Fig. B.6a by a quarter-cycle (angle of 90° or $\pi/2$ radians), we obtain the signal $C \sin \omega_0 t$, depicted in Fig. B.6b. This verifies the well-known trigonometric identity

$$C \cos \left(\omega_0 t - \frac{\pi}{2}\right) = C \sin \omega_0 t \tag{B.22a}$$

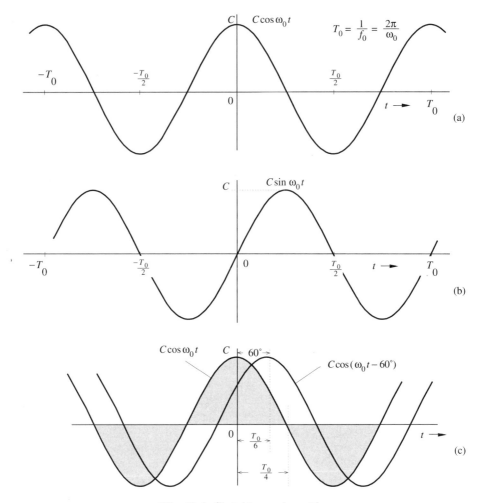

Fig. B.6 Sketching a sinusoid.

Alternatively, if we advance $C \sin \omega_0 t$ by a quarter-cycle, we obtain $C \cos \omega_0 t$. Therefore,

$$C \sin \left(\omega_0 t + \tfrac{\pi}{2}\right) = C \cos \omega_0 t \qquad \text{(B.22b)}$$

This observation means $\sin \omega_0 t$ lags $\cos \omega_0 t$ by $90°(\pi/2$ radians), or $\cos \omega_0 t$ leads $\sin \omega_0 t$ by $90°$.

B.2-1   Addition of Sinusoids

Two sinusoids having the same frequency but different phases add to form a single sinusoid of the same frequency. This fact is readily seen from the well-known trigonometric identity

$$C \cos \left(\omega_0 t + \theta\right) = C \cos \theta \cos \omega_0 t - C \sin \theta \sin \omega_0 t$$

$$= a \cos \omega_0 t + b \sin \omega_0 t \qquad \text{(B.23a)}$$

in which

$$a = C \cos \theta, \qquad\qquad b = -C \sin \theta$$

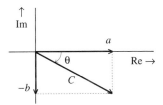

Fig. B.7 Phasor addition of sinusoids.

Therefore,

$$C = \sqrt{a^2 + b^2} \tag{B.23b}$$

$$\theta = \tan^{-1}\left(\frac{-b}{a}\right) \tag{B.23c}$$

Equations (B.23b) and (B.23c) show that C and θ are the magnitude and angle, respectively, of a complex number $a - jb$. In other words, $a - jb = Ce^{j\theta}$. Hence, to find C and θ, we convert $a - jb$ to polar form and the magnitude and the angle of the resulting polar number are C and θ, respectively.

To summarize,

$$a \cos \omega_0 t + b \sin \omega_0 t = C \cos (\omega_0 t + \theta)$$

in which C and θ are given by Eqs. (B.23b) and (B.23c), respectively. These happen to be the magnitude and angle, respectively, of $a - jb$.

The process of adding two sinusoids with the same frequency can be clarified by using **phasors** to represent sinusoids. We represent the sinusoid $C \cos (\omega_0 t + \theta)$ by a phasor of length C at an angle θ with the horizontal axis. Clearly, the sinusoid $a \cos \omega_0 t$ is represented by a horizontal phasor of length a ($\theta = 0$), while $b \sin \omega_0 t = b \cos (\omega_0 t - \frac{\pi}{2})$ is represented by a vertical phasor of length b at an angle $-\pi/2$ with the horizontal (Fig. B.7). Adding these two phasors results in a phasor of length C at an angle θ, as depicted in Fig. B.7. From this figure, we verify the values of C and θ found in Eqs. (B.23b) and (B.23c), respectively.

Proper care should be exercised in computing θ. Recall that $\tan^{-1}(\frac{-b}{a}) \neq \tan^{-1}(\frac{b}{-a})$. Similarly, $\tan^{-1}(\frac{-b}{-a}) \neq \tan^{-1}(\frac{b}{a})$. Electronic calculators cannot make this distinction. When calculating such an angle, it is advisable to note the quadrant where the angle lies and not to rely exclusively on an electronic calculator. A foolproof method is to convert the complex number $a - jb$ to polar form. The magnitude of the resulting polar number is C and the angle is θ. The following examples clarify this point.

■ **Example B.6**
In the following cases, express $f(t)$ as a single sinusoid:

(a) $f(t) = \cos \omega_0 t - \sqrt{3} \sin \omega_0 t$

(b) $f(t) = -3 \cos \omega_0 t + 4 \sin \omega_0 t$

(a) In this case, $a = 1, b = -\sqrt{3}$, and from Eqs. (B.23)

$$C = \sqrt{1^2 + (\sqrt{3})^2} = 2$$

$$\theta = \tan^{-1}\left(\frac{\sqrt{3}}{1}\right) = 60°$$

Therefore,

$$f(t) = 2\cos(\omega_0 t + 60°)$$

We can verify this result by drawing phasors corresponding to the two sinusoids. The sinusoid $\cos \omega_0 t$ is represented by a phasor of unit length at a zero angle with the horizontal. The phasor $\sin \omega_0 t$ is represented by a unit phasor at an angle of $-90°$ with the horizontal. Therefore, $-\sqrt{3}\sin \omega_0 t$ is represented by a phasor of length $\sqrt{3}$ at $90°$ with the horizontal, as depicted in Fig. B.8a. The two phasors added yield a phasor of length 2 at $60°$ with the horizontal (also shown in Fig. B.8a). Therefore,

$$f(t) = 2\cos(\omega_0 t + 60°)$$

Alternately, we note that $a - jb = 1 + j\sqrt{3} = 2e^{j\pi/3}$. Hence, $C = 2$ and $\theta = \pi/3$.

Observe that a phase shift of $\pm\pi$ amounts to multiplication by -1. Therefore, $f(t)$ can also be expressed alternatively as

$$f(t) = -2\cos(\omega_0 t + 60° \pm 180°)$$

$$= -2\cos(\omega_0 t - 120°)$$

$$= -2\cos(\omega_0 t + 240°)$$

In practice, an expression with an angle whose numerical value is less than $180°$ is preferred.

(b) In this case, $a = -3, b = 4$, and from Eqs. (B.23)

$$C = \sqrt{(-3)^2 + 4^2} = 5$$

$$\theta = \tan^{-1}\left(\frac{-4}{-3}\right) = -126.9°$$

Observe that

$$\tan^{-1}\left(\frac{-4}{-3}\right) \neq \tan^{-1}\left(\frac{4}{3}\right) = 53.1°$$

Therefore,

$$f(t) = 5\cos(\omega_0 t - 126.9°)$$

This result is readily verified in the phasor diagram in Fig. B.8b. Alternately, $a - jb = -3 - j4 = 5e^{-j126.9°}$. Hence, $C = 5$ and $\theta = -126.9°$. ■

⊙ **Computer Example CB.4**

Express $f(t) = -3\cos \omega_0 t + 4\sin \omega_0 t$ as a single sinusoid.

Recall that $a\cos \omega_0 t + b\sin \omega_0 t = C\cos[\omega_0 t + \tan^{-1}(-b/a)]$. Hence, the amplitude C and the angle θ of the resulting sinusoid are the magnitude and angle of a complex number $a - jb$. We use the 'cart2pol' function to convert it to the polar form to obtain C and θ.

```
a=-3;b=4;
[theta,C]=cart2pol(a,-b);
Theta_deg=(180/pi)*theta;
C,Theta_deg
C=5
Theta_deg=-126.8699
```

Therefore

$$-3\cos \omega_0 t + 4\sin \omega_0 t = 5\cos(\omega_0 t - 126.8699°) \qquad ⊙$$

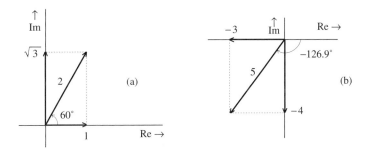

Fig. B.8 Phasor addition of sinusoids in Example B.6.

We can also perform the reverse operation, expressing

$$f(t) = C \cos (\omega_0 t + \theta)$$

in terms of $\cos \omega_0 t$ and $\sin \omega_0 t$ using the trigonometric identity

$$C \cos (\omega_0 t + \theta) = C \cos \theta \cos \omega_0 t - C \sin \theta \sin \omega_0 t$$

For example,

$$10 \cos (\omega_0 t - 60°) = 5 \cos \omega_0 t + 5\sqrt{3} \sin \omega_0 t$$

Sinusoids in Terms of Exponentials: Euler's Formula

Sinusoids can be expressed in terms of exponentials using Euler's formula [see Eq. (B.3)]

$$\cos \varphi = \frac{1}{2} \left(e^{j\varphi} + e^{-j\varphi} \right) \tag{B.24a}$$

$$\sin \varphi = \frac{1}{2j} \left(e^{j\varphi} - e^{-j\varphi} \right) \tag{B.24b}$$

Inversion of these equations yields

$$e^{j\varphi} = \cos \varphi + j \sin \varphi \tag{B.25a}$$

$$e^{-j\varphi} = \cos \varphi - j \sin \varphi \tag{B.25b}$$

B.3   Sketching Signals

In this section we discuss the sketching of a few useful signals, starting with exponentials.

B.3-1   Monotonic Exponentials

The signal e^{-at} decays monotonically, and the signal e^{at} grows monotonically with t (assuming $a > 0$) as depicted in Fig. B.9. For the sake of simplicity, we shall consider an exponential e^{-at} starting at $t = 0$, as shown in Fig. B.10a.

The signal e^{-at} has a unit value at $t = 0$. At $t = 1/a$, the value drops to $1/e$ (about 37% of its initial value), as illustrated in Fig. B.10a. This time interval over which the exponential reduces by a factor e (that is, drops to about 37% of its value) is known as the **time constant** of the exponential. Therefore, the time

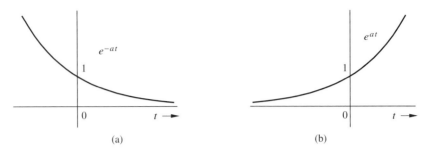

Fig. B.9 Monotonic exponentials.

constant of e^{-at} is $1/a$. Observe that the exponential is reduced to 37% of its initial value over any time interval of duration $1/a$. This can be shown by considering any set of instants t_1 and t_2 separated by one time constant so that

$$t_2 - t_1 = \tfrac{1}{a}$$

Now the ratio of e^{-at_2} to e^{-at_1} is given by

$$\frac{e^{-at_2}}{e^{-at_1}} = e^{-a(t_2-t_1)} = \tfrac{1}{e} \approx 0.37$$

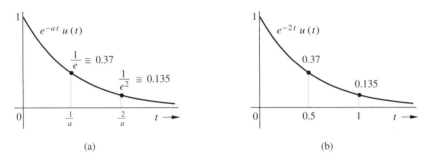

Fig. B.10 (a) Sketching e^{-at} (b) sketching e^{-2t}.

We can use this fact to sketch an exponential quickly. For example, consider

$$f(t) = e^{-2t}$$

The time constant in this case is $1/2$. The value of $f(t)$ at $t = 0$ is 1. At $t = 1/2$ (one time constant) it is $1/e$ (about 0.37). The value of $f(t)$ continues to drop further by the factor $1/e$ (37%) over the next half-second interval (one time constant). Thus $f(t)$ at $t = 1$ is $(1/e)^2$. Continuing in this manner, we see that $f(t) = (1/e)^3$ at $t = 3/2$ and so on. A knowledge of the values of $f(t)$ at $t = 0$, 0.5, 1, and 1.5 allows us to sketch the desired signal† as shown in Fig. B.10b. For a monotonically

---

†If we wish to refine the sketch further, we could consider intervals of half the time constant over which the signal decays by a factor $1/\sqrt{e}$. Thus, at $t = 0.25$, $f(t) = 1/\sqrt{e}$, and at $t = 0.75$, $f(t) = 1/e\sqrt{e}$, etc.

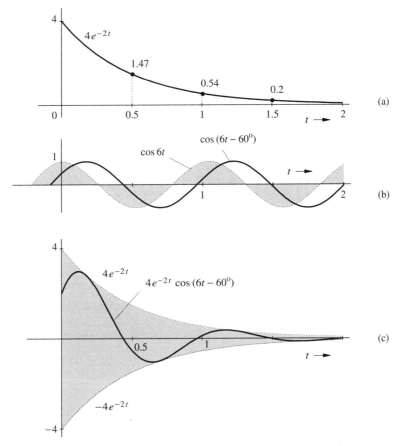

Fig. B.11 Sketching an exponentially varying sinusoid

growing exponential e^{at}, the waveform increases by a factor e over each interval of $1/a$ seconds.

B.3-2   The Exponentially Varying Sinusoid

We now discuss sketching an exponentially varying sinusoid

$$f(t) = Ae^{-at} \cos(\omega_0 t + \theta)$$

Let us consider a specific example

$$f(t) = 4e^{-2t} \cos(6t - 60°) \qquad (B.26)$$

We shall sketch $4e^{-2t}$ and $\cos(6t - 60°)$ separately and then multiply them.

(i)   Sketching $4e^{-2t}$

This monotonically decaying exponential has a time constant of $1/2$ second and an initial value of 4 at $t = 0$. Therefore, its values at $t = 0.5$, 1, 1.5, and 2 are $4/e$, $4/e^2$, $4/e^3$, and $4/e^4$, or about 1.47, 0.54, 0.2, and 0.07 respectively. Using these values as a guide, we sketch $4e^{-2t}$, as illustrated in Fig. B.11a.

(ii)   Sketching $\cos{(6t - 60°)}$

The procedure for sketching $\cos{(6t - 60°)}$ is discussed in Sec. B.2 (Fig. B.6c). Here the period of the sinusoid is $T_0 = 2\pi/6 \approx 1$, and there is a phase delay of $60°$, or two-thirds of a quarter-cycle, which is equivalent to about a $(60/360)(1) \approx 1/6$ second delay (see Fig. B.11b).

(iii)   Sketching $4e^{-2t}\cos{(6t - 60°)}$

We now multiply the waveforms in **(i)** and **(ii)**. The multiplication amounts to forcing the amplitude of the sinusoid $\cos{(6t - 60°)}$ to decrease exponentially with a time constant of 0.5. The initial amplitude (at $t = 0$) is 4, decreasing to $4/e$ (=1.47) at $t = 0.5$, to $1.47/e$ (=0.54) at $t = 1$, and so on. This is depicted in Fig. B.11c. Note that at the instants where $\cos{(6t - 60°)}$ has a value of unity (peak amplitude),

$$4e^{-2t}\cos{(6t - 60°)} = 4e^{-2t} \tag{B.27}$$

Therefore, $4e^{-2t}\cos{(6t - 60°)}$ touches $4e^{-2t}$ at those instants where the sinusoid $\cos{(6t - 60°)}$ is at its positive peaks. Clearly $4e^{-2t}$ is an envelope for positive amplitudes of $4e^{-2t}\cos{(6t - 60°)}$. Similarly, at those instants where the sinusoid $\cos{(6t - 60°)}$ has a value of -1 (negative peak amplitude),

$$4e^{-2t}\cos{(6t - 60°)} = -4e^{-2t} \tag{B.28}$$

and $4e^{-2t}\cos{(6t - 60°)}$ touches $-4e^{-2t}$ at its negative peaks. Therefore, $-4e^{-2t}$ is an envelope for negative amplitudes of $4e^{-2t}\cos{(6t - 60°)}$. Thus, to sketch $4e^{-2t}\cos{(6t - 60°)}$, we first draw the envelopes $4e^{-2t}$ and $-4e^{-2t}$ (the mirror image of $4e^{-2t}$ about the horizontal axis), and then sketch the sinusoid $\cos{(6t - 60°)}$, with these envelopes acting as constraints on the sinusoid's amplitude (see Fig. B.11c).

In general, $Ke^{-at}\cos{(\omega_0 t + \theta)}$ can be sketched in this manner, with Ke^{-at} and $-Ke^{-at}$ constraining the amplitude of $\cos{(\omega_0 t + \theta)}$.

B.4   Cramer's Rule

This is a very convenient rule used to solve simultaneous linear equations. Consider a set of n linear simultaneous equations in n unknowns x_1, x_2, ..., x_n:

$$a_{11}x_1 + a_{12}x_2 + \cdots + a_{1n}x_n = y_1$$

$$a_{21}x_1 + a_{22}x_2 + \cdots + a_{2n}x_n = y_2$$

$$\cdots\cdots\cdots\cdots\cdots\cdots\cdots\cdots\cdots\cdots \tag{B.29}$$

$$a_{n1}x_1 + a_{n2}x_2 + \cdots + a_{nn}x_n = y_n$$

These equations can be expressed in matrix form as

$$\begin{bmatrix} a_{11} & a_{12} & \cdots & a_{1n} \\ a_{21} & a_{22} & \cdots & a_{2n} \\ \cdots\cdots\cdots\cdots\cdots \\ a_{n1} & a_{n2} & \cdots & a_{nn} \end{bmatrix} \begin{bmatrix} x_1 \\ x_2 \\ \vdots \\ x_n \end{bmatrix} = \begin{bmatrix} y_1 \\ y_2 \\ \vdots \\ y_n \end{bmatrix} \tag{B.30}$$

We denote the matrix on the left-hand side formed by the elements a_{ij} as $\mathbf{A}$. The determinant of $\mathbf{A}$ is denoted by $|\mathbf{A}|$. If the determinant $|\mathbf{A}|$ is not zero, the set of equations (B.29) has a unique solution given by Cramer's formula

$$x_k = \frac{|\mathbf{D}_k|}{|\mathbf{A}|} \qquad k = 1, 2, \ldots, n \tag{B.31}$$

where $|\mathbf{D}_k|$ is obtained by replacing the kth column of $|\mathbf{A}|$ by the column on the right-hand side of Eq. (B.30) (with elements $y_1, y_2, \ldots, y_n$).

We shall demonstrate the use of this rule with an example.

■ **Example B.7**

Using Cramer's rule, solve the following simultaneous linear equations in three unknowns:

$$2x_1 + x_2 + x_3 = 3$$

$$x_1 + 3x_2 - x_3 = 7$$

$$x_1 + x_2 + x_3 = 1$$

In matrix form these equations can be expressed as

$$\begin{bmatrix} 2 & 1 & 1 \\ 1 & 3 & -1 \\ 1 & 1 & 1 \end{bmatrix} \begin{bmatrix} x_1 \\ x_2 \\ x_3 \end{bmatrix} = \begin{bmatrix} 3 \\ 7 \\ 1 \end{bmatrix}$$

Here,

$$|\mathbf{A}| = \begin{vmatrix} 2 & 1 & 1 \\ 1 & 3 & -1 \\ 1 & 1 & 1 \end{vmatrix} = 4$$

Since $|\mathbf{A}| = 4 \neq 0$, a unique solution exists for x_1, x_2, and x_3. This solution is provided by Cramer's rule (B.31) as follows:

$$x_1 = \frac{1}{|\mathbf{A}|} \begin{vmatrix} 3 & 1 & 1 \\ 7 & 3 & -1 \\ 1 & 1 & 1 \end{vmatrix} = \frac{8}{4} = 2$$

$$x_2 = \frac{1}{|\mathbf{A}|} \begin{vmatrix} 2 & 3 & 1 \\ 1 & 7 & -1 \\ 1 & 1 & 1 \end{vmatrix} = \frac{4}{4} = 1$$

$$x_3 = \frac{1}{|\mathbf{A}|} \begin{vmatrix} 2 & 1 & 3 \\ 1 & 3 & 7 \\ 1 & 1 & 1 \end{vmatrix} = \frac{-8}{4} = -2 \quad \blacksquare$$

⊙ **Example CB.5**

Using a Computer, solve Example B.7.

```
A = [2 1 1;1 3 -1;1 1 1];  b=[3 7 1]';
for k=1:3
A1=A;
A1(:,k)=b;
D=A1;
x(k)=det(D)/det(A);
end
x=x'
x =  2
     1
    -2   ⊙
```

B.5 Partial Fraction Expansion

In the analysis of linear time-invariant systems, we encounter functions that are ratios of two polynomials in a certain variable, say x. Such functions are known as **rational functions**. A rational function $F(x)$ can be expressed as

$$F(x) = \frac{b_m x^m + b_{m-1} x^{m-1} + \cdots + b_1 x + b_0}{x^n + a_{n-1} x^{n-1} + \cdots + a_1 x + a_0} \tag{B.32}$$

$$= \frac{P(x)}{Q(x)} \tag{B.33}$$

The function $F(x)$ is **improper** if $m \geq n$ and **proper** if $m < n$. An improper function can always be separated into the sum of a polynomial in x and a proper function. Consider, for example, the function

$$F(x) = \frac{2x^3 + 9x^2 + 11x + 2}{x^2 + 4x + 3} \tag{B.34a}$$

Because this is an improper function, we divide the numerator by the denominator until the remainder has a lower degree than the denominator.

$$
\begin{array}{r}
2x + 1 \\
x^2 + 4x + 3 \overline{)\, 2x^3 + 9x^2 + 11x + 2} \\
\underline{2x^3 + 8x^2 + 6x} \\
x^2 + 5x + 2 \\
\underline{x^2 + 4x + 3} \\
x - 1
\end{array}
$$

Therefore, $F(x)$ can be expressed as

$$F(x) = \frac{2x^3 + 9x^2 + 11x + 2}{x^2 + 4x + 3} = \underbrace{2x + 1}_{\text{polynomial in } x} + \underbrace{\frac{x - 1}{x^2 + 4x + 3}}_{\text{proper function}} \qquad \text{(B.34b)}$$

A proper function can be further expanded into partial fractions. The remaining discussion in this section is concerned with various ways of doing this.

B.5-1   Partial Fraction Expansion: Method of Clearing Fractions

This method consists of writing a rational function as a sum of appropriate partial fractions with unknown coefficients, which are determined by clearing fractions and equating the coefficients of similar powers on the two sides. This procedure is demonstrated by the following example.

■ **Example B.8**

Expand the following rational function $F(x)$ into partial fractions:

$$F(x) = \frac{x^3 + 3x^2 + 4x + 6}{(x + 1)(x + 2)(x + 3)^2}$$

This function can be expressed as a sum of partial fractions with denominators $(x + 1)$, $(x + 2)$, $(x + 3)$, and $(x + 3)^2$, as shown below.

$$F(x) = \frac{x^3 + 3x^2 + 4x + 6}{(x + 1)(x + 2)(x + 3)^2} = \frac{k_1}{x + 1} + \frac{k_2}{x + 2} + \frac{k_3}{x + 3} + \frac{k_4}{(x + 3)^2}$$

To determine the unknowns k_1, k_2, k_3, and k_4 we clear fractions by multiplying both sides by $(x + 1)(x + 2)(x + 3)^2$ to obtain

$$x^3 + 3x^2 + 4x + 6 = k_1(x^3 + 8x^2 + 21x + 18) + k_2(x^3 + 7x^2 + 15x + 9)$$
$$+ k_3(x^3 + 6x^2 + 11x + 6) + k_4(x^2 + 3x + 2)$$
$$= x^3(k_1 + k_2 + k_3) + x^2(8k_1 + 7k_2 + 6k_3 + k_4)$$
$$+ x(21k_1 + 15k_2 + 11k_3 + 3k_4) + (18k_1 + 9k_2 + 6k_3 + 2k_4)$$

Equating coefficients of similar powers on both sides yields

$$k_1 + k_2 + k_3 = 1$$
$$8k_1 + 7k_2 + 6k_3 + k_4 = 3$$
$$21k_1 + 15k_2 + 11k_3 + 3k_4 = 4$$
$$18k_1 + 9k_2 + 6k_3 + 2k_4 = 6$$

Solution of these four simultaneous equations yields

$$k_1 = 1, \qquad k_2 = -2, \qquad k_3 = 2, \qquad k_4 = -3$$

Therefore,

$$F(x) = \frac{1}{x + 1} - \frac{2}{x + 2} + \frac{2}{x + 3} - \frac{3}{(x + 3)^2} \qquad ■$$

Although this method is straightforward and applicable to all situations, it is not necessarily the most efficient. We now discuss other methods which can reduce numerical work considerably.

B.5-2   Partial Fractions: The Heaviside "Cover-Up" Method

1.   Unrepeated Factors of $Q(x)$

We shall first consider the partial fraction expansion of $F(x) = P(x)/Q(x)$, in which all the factors of $Q(x)$ are unrepeated. Consider the proper function

$$
\begin{aligned}
F(x) &= \frac{b_m x^m + b_{m-1} x^{m-1} + \cdots + b_1 x + b_0}{x^n + a_{n-1} x^{n-1} + \cdots + a_1 x + a_0} \qquad m < n \\[2mm]
&= \frac{P(x)}{(x - \lambda_1)(x - \lambda_2) \cdots (x - \lambda_n)}
\end{aligned}
\tag{B.35a}
$$

We can show that $F(x)$ in Eq. (B.35a) can be expressed as the sum of partial fractions

$$
F(x) = \frac{k_1}{x - \lambda_1} + \frac{k_2}{x - \lambda_2} + \cdots + \frac{k_n}{x - \lambda_n}
\tag{B.35b}
$$

To determine the coefficient k_1, we multiply both sides of Eq. (B.35b) by $x - \lambda_1$ and then let $x = \lambda_1$. This yields

$$
(x - \lambda_1) F(x)\big|_{x=\lambda_1} = k_1 + \frac{k_2(x - \lambda_1)}{(x - \lambda_2)} + \frac{k_3(x - \lambda_1)}{(x - \lambda_3)} + \cdots + \frac{k_n(x - \lambda_1)}{(x - \lambda_n)}\bigg|_{x=\lambda_1}
$$

On the right-hand side, all the terms except k_1 vanish. Therefore,

$$
k_1 = (x - \lambda_1) F(x)\big|_{x=\lambda_1}
\tag{B.36}
$$

Similarly, we can show that

$$
k_r = (x - \lambda_r) F(x)\big|_{x=\lambda_r} \qquad\qquad r = 1, 2, \cdots, n
\tag{B.37}
$$

■ **Example B.9**
Expand the following rational function $F(x)$ into partial fractions:

$$
F(x) = \frac{2x^2 + 9x - 11}{(x + 1)(x - 2)(x + 3)} = \frac{k_1}{x + 1} + \frac{k_2}{x - 2} + \frac{k_3}{x + 3}
$$

To determine k_1, we let $x = -1$ in $(x + 1)F(x)$. Note that $(x + 1)F(x)$ is obtained from $F(x)$ by omitting the term $(x + 1)$ from its denominator. Therefore, to compute k_1 corresponding to the factor $(x + 1)$, we cover up the term $(x + 1)$ in the denominator of $F(x)$ and then substitute $x = -1$ in the remaining expression. (Mentally conceal the term $(x + 1)$ in $F(x)$ with a finger and then let $x = -1$ in the remaining expression.) The procedure is explained step by step below.

$$
F(x) = \frac{2x^2 + 9x - 11}{(x + 1)(x - 2)(x + 3)}
$$

Step 1: Cover up (conceal) the factor $(x + 1)$ from $F(x)$:

$$\frac{2x^2 + 9x - 11}{\cancel{(x+1)}(x-2)(x+3)}$$

Step2: Substitute $x = -1$ in the remaining expression to obtain k_1:

$$k_1 = \frac{2 - 9 - 11}{(-1-2)(-1+3)} = \frac{-18}{-6} = 3$$

Similarly, to compute k_2, we cover up the factor $(x-2)$ in $F(x)$ and let $x = 2$ in the remaining function, as shown below.

$$k_2 = \frac{2x^2 + 9x - 11}{(x+1)\cancel{(x-2)}(x+3)}\bigg|_{x=2} = \frac{8 + 18 - 11}{(2+1)(2+3)} = \frac{15}{15} = 1$$

and

$$k_3 = \frac{2x^2 + 9x - 11}{(x+1)(x-2)\cancel{(x+3)}}\bigg|_{x=-3} = \frac{18 - 27 - 11}{(-3+1)(-3-2)} = \frac{-20}{10} = -2$$

Therefore,

$$F(x) = \frac{2x^2 + 9x - 11}{(x+1)(x-2)(x+3)} = \frac{3}{x+1} + \frac{1}{x-2} - \frac{2}{x+3} \quad \blacksquare$$

Complex Factors in $F(x)$

The procedure above works regardless of whether the factors of $Q(x)$ are real or complex. Consider, for example,

$$F(x) = \frac{4x^2 + 2x + 18}{(x+1)(x^2 + 4x + 13)} \tag{B.38}$$

$$= \frac{4x^2 + 2x + 18}{(x+1)(x+2-j3)(x+2+j3)}$$

$$= \frac{k_1}{x+1} + \frac{k_2}{x+2-j3} + \frac{k_3}{x+2+j3}$$

where

$$k_1 = \left[\frac{4x^2 + 2x + 18}{\cancel{(x+1)}(x^2 + 4x + 13)}\right]_{x=-1} = 2$$

Similarly,

$$k_2 = \left[\frac{4x^2 + 2x + 18}{(x+1)\cancel{(x+2-j3)}(x+2+j3)}\right]_{x=-2+j3} = 1 + j2 = \sqrt{5}e^{j63.43°}$$

$$k_3 = \left[\frac{4x^2 + 2x + 18}{(x+1)(x+2-j3)\cancel{(x+2+j3)}}\right]_{x=-2-j3} = 1 - j2 = \sqrt{5}e^{-j63.43°}$$

Therefore,

$$F(x) = \frac{2}{x+1} + \frac{\sqrt{5}e^{j63.43°}}{x+2-j3} + \frac{\sqrt{5}e^{-j63.43°}}{x+2+j3} \tag{B.39}$$

The coefficients k_2 and k_3 corresponding to the complex conjugate factors are also conjugates of each other. This is generally true when the coefficients of a rational function are real. In such a case, we need to compute only one of the coefficients.

2. Quadratic Factors

Often we are required to combine the two terms arising from complex conjugate factors into one quadratic factor. For example, $F(x)$ in Eq. (B.38) can be expressed as

$$F(x) = \frac{4x^2 + 2x + 18}{(x+1)(x^2+4x+13)} = \frac{k_1}{x+1} + \frac{c_1 x + c_2}{x^2 + 4x + 13}$$

The coefficient k_1 is found by the Heaviside method to be 2. Therefore,

$$\frac{4x^2 + 2x + 18}{(x+1)(x^2+4x+13)} = \frac{2}{x+1} + \frac{c_1 x + c_2}{x^2 + 4x + 13} \qquad (B.40)$$

The values of c_1 and c_2 are determined by clearing fractions and equating the coefficients of similar powers of x on both sides of the resulting equation. Clearing fractions on both sides of Eq. (B.40) yields

$$4x^2 + 2x + 18 = 2(x^2 + 4x + 13) + (c_1 x + c_2)(x+1)$$
$$= (2 + c_1)x^2 + (8 + c_1 + c_2)x + (26 + c_2) \qquad (B.41)$$

Equating terms of similar powers yields $c_1 = 2$, $c_2 = -8$, and

$$\frac{4x^2 + 2x + 18}{(x+1)(x^2+4x+13)} = \frac{2}{x+1} + \frac{2x - 8}{x^2 + 4x + 13} \qquad (B.42)$$

Short-Cuts

The values of c_1 and c_2 in Eq. (B.40) can also be determined by using short-cuts. After computing $k_1 = 2$ by the Heaviside method as before, we let $x = 0$ on both sides of Eq. (B.40) to eliminate c_1. This gives us

$$\frac{18}{13} = 2 + \frac{c_2}{13}$$

Therefore,

$$c_2 = -8$$

To determine c_1, we multiply both sides of Eq. (B.40) by x and then let $x \to \infty$. Remember that when $x \to \infty$, only the terms of the highest power are significant. Therefore,

$$4 = k_1 + c_1 = 2 + c_1$$

and

$$c_1 = 2$$

In the procedure discussed here, we let $x = 0$ to determine c_2 and then multiply both sides by x and let $x \to \infty$ to determine c_1. However, nothing is sacred about these values ($x = 0$ or $x = \infty$). We use them because they reduce the number of computations involved. We could just as well use other convenient values for x, such as $x = 1$. Consider the case

$$F(x) = \frac{2x^2 + 4x + 5}{x(x^2 + 2x + 5)}$$

$$= \frac{k}{x} + \frac{c_1 x + c_2}{x^2 + 2x + 5}$$

We find $k = 1$ by the Heaviside method in the usual manner. As a result,

$$\frac{2x^2 + 4x + 5}{x(x^2 + 2x + 5)} = \frac{1}{x} + \frac{c_1 x + c_2}{x^2 + 2x + 5} \tag{B.43}$$

To determine c_1 and c_2, if we try letting $x = 0$ in Eq. (B.43), we obtain ∞ on both sides. So let us choose $x = 1$. This yields

$$F(1) = \frac{11}{8} = 1 + \frac{c_1 + c_2}{8}$$

or

$$c_1 + c_2 = 3$$

We can now choose some other value for x, such as $x = 2$, to obtain one more relationship to use in determining c_1 and c_2. In this case, however, a simple method is to multiply both sides of Eq. (B.43) by x and then let $x \to \infty$. This yields

$$2 = 1 + c_1$$

so that

$$c_1 = 1 \quad \text{and} \quad c_2 = 2$$

Therefore,

$$F(x) = \frac{1}{x} + \frac{x + 2}{x^2 + 2x + 5}$$

B.5-3   Repeated Factors in $Q(x)$

If a function $F(x)$ has a repeated factor in its denominator, it has the form

$$F(x) = \frac{P(x)}{(x - \lambda)^r (x - \alpha_1)(x - \alpha_2) \cdots (x - \alpha_j)} \tag{B.44}$$

Its partial fraction expansion is given by

$$F(x) = \frac{a_0}{(x - \lambda)^r} + \frac{a_1}{(x - \lambda)^{r-1}} + \cdots + \frac{a_{r-1}}{(x - \lambda)}$$

$$+ \frac{k_1}{x - \alpha_1} + \frac{k_2}{x - \alpha_2} + \cdots + \frac{k_j}{x - \alpha_j} \tag{B.45}$$

The coefficients $k_1, k_2, \ldots, k_j$ corresponding to the unrepeated factors in this equation are determined by the Heaviside method, as before [Eq. (B.37)]. To find the coefficients a_0, a_1, a_2, ..., a_{r-1}, we multiply both sides of Eq. (B.45) by $(x - \lambda)^r$. This gives us

$$(x - \lambda)^r F(x) = a_0 + a_1(x - \lambda) + a_2(x - \lambda)^2 + \cdots + a_{r-1}(x - \lambda)^{r-1}$$

$$+ k_1 \frac{(x - \lambda)^r}{x - \alpha_1} + k_2 \frac{(x - \lambda)^r}{x - \alpha_2} + \cdots + k_n \frac{(x - \lambda)^r}{x - \alpha_n} \qquad \text{(B.46)}$$

If we let $x = \lambda$ on both sides of Eq. (B.46), we obtain

$$(x - \lambda)^r F(x)|_{x=\lambda} = a_0 \qquad \text{(B.47a)}$$

Therefore, a_0 is obtained by concealing the factor $(x - \lambda)^r$ in $F(x)$ and letting $x = \lambda$ in the remaining expression (the Heaviside "cover up" method). If we take the derivative (with respect to x) of both sides of Eq. (B.46), the right-hand side is a_1+ terms containing a factor $(x - \lambda)$ in their numerators. Letting $x = \lambda$ on both sides of this equation, we obtain

$$\frac{d}{dx}[(x - \lambda)^r F(x)]\Big|_{x=\lambda} = a_1$$

Thus, a_1 is obtained by concealing the factor $(x - \lambda)^r$ in $F(x)$, taking the derivative of the remaining expression, and then letting $x = \lambda$. Continuing in this manner, we find

$$a_j = \frac{1}{j!} \frac{d^j}{dx^j}[(x - \lambda)^r F(x)]\Big|_{x=\lambda} \qquad \text{(B.47b)}$$

Observe that $(x - \lambda)^r F(x)$ is obtained from $F(x)$ by omitting the factor $(x - \lambda)^r$ from its denominator. Therefore, the coefficient a_j is obtained by concealing the factor $(x - \lambda)^r$ in $F(x)$, taking the jth derivative of the remaining expression, and then letting $x = \lambda$ (while dividing by $j!$).

■ **Example B.10**

Expand $F(x)$ into partial fractions if

$$F(x) = \frac{4x^3 + 16x^2 + 23x + 13}{(x + 1)^3(x + 2)}$$

The partial fractions are

$$F(x) = \frac{a_0}{(x + 1)^3} + \frac{a_1}{(x + 1)^2} + \frac{a_2}{x + 1} + \frac{k}{x + 2}$$

The coefficient k is obtained by concealing the factor $(x+2)$ in $F(x)$ and then substituting $x = -2$ in the remaining expression:

$$k = \frac{4x^3 + 16x^2 + 23x + 13}{(x + 1)^3(x + 2)}\Big|_{x=-2} = 1$$

To find a_0, we conceal the factor $(x + 1)^3$ in $F(x)$ and let $x = -1$ in the remaining expression:

$$a_0 = \frac{4x^3 + 16x^2 + 23x + 13}{(x + 1)^3(x + 2)}\Big|_{x=-1} = 2$$

To find a_1, we conceal the factor $(x+1)^3$ in $F(x)$, take the derivative of the remaining expression, and then let $x = -1$:

$$a_1 = \frac{d}{dx}\left[\frac{4x^3 + 16x^2 + 23x + 13}{(x+1)^3(x+2)}\right]\Bigg|_{x=-1} = 1$$

Similarly,

$$a_2 = \frac{1}{2!}\frac{d^2}{dx^2}\left[\frac{4x^3 + 16x^2 + 23x + 13}{(x+1)^3(x+2)}\right]\Bigg|_{x=-1} = 3$$

Therefore,

$$F(x) = \frac{2}{(x+1)^3} + \frac{1}{(x+1)^2} + \frac{3}{x+1} + \frac{1}{x+2} \quad\blacksquare$$

B.5-4   A Hybrid Method: Mixture of the Heaviside "Cover-Up" and Clearing Fractions

For multiple roots, especially of higher order, the Heaviside expansion method, which requires repeated differentiation, can become cumbersome. For a function which contains several repeated and unrepeated roots, a hybrid of the two procedures proves the best. The simpler coefficients are determined by the Heaviside method, and the remaining coefficients are found by clearing fractions or short-cuts, thus incorporating the best of the two methods. We demonstrate this procedure by solving Example B.10 once again by this method.

In Example B.10, coefficients k and a_0 are relatively simple to determine by the Heaviside expansion method. These values were found to be $k_1 = 1$ and $a_0 = 2$. Therefore,

$$\frac{4x^3 + 16x^2 + 23x + 13}{(x+1)^3(x+2)} = \frac{2}{(x+1)^3} + \frac{a_1}{(x+1)^2} + \frac{a_2}{x+1} + \frac{1}{x+2}$$

We now multiply both sides of the above equation by $(x+1)^3(x+2)$ to clear the fractions. This yields

$$4x^3 + 16x^2 + 23x + 13$$
$$= 2(x+2) + a_1(x+1)(x+2) + a_2(x+1)^2(x+2) + (x+1)^3$$
$$= (1+a_2)x^3 + (a_1 + 4a_2 + 3)x^2 + (5 + 3a_1 + 5a_2)x + (4 + 2a_1 + 2a_2 + 1)$$

Equating coefficients of the third and second powers of x on both sides, we obtain

$$\left.\begin{array}{r} 1 + a_2 = 4 \\ a_1 + 4a_2 + 3 = 16 \end{array}\right\} \implies \begin{array}{l} a_1 = 1 \\ a_2 = 3 \end{array}$$

We may stop here if we wish because the two desired coefficients, a_1 and a_2, are now determined. However, equating the coefficients of the two remaining powers of x yields a convenient check on the answer. Equating the coefficients of the x^1 and x^0 terms, we obtain

$$23 = 5 + 3a_1 + 5a_2$$

$$13 = 4 + 2a_1 + 2a_2 + 1$$

These equations are satisfied by the values $a_1 = 1$ and $a_2 = 3$, found earlier, providing an additional check for our answers. Therefore,

$$F(x) = \frac{2}{(x+1)^3} + \frac{1}{(x+1)^2} + \frac{3}{x+1} + \frac{1}{x+2}$$

which agrees with the previous result.

A Mixture of the Heaviside "Cover-Up" and Short Cuts

In the above example, after determining the coefficients $a_0 = 2$ and $k = 1$ by the Heaviside method as before, we have

$$\frac{4x^3 + 16x^2 + 23x + 13}{(x+1)^3(x+2)} = \frac{2}{(x+1)^3} + \frac{a_1}{(x+1)^2} + \frac{a_2}{x+1} + \frac{1}{x+2}$$

There are only two unknown coefficients, a_1 and a_2. If we multiply both sides of the above equation by x and then let $x \to \infty$, we can eliminate a_1. This yields

$$4 = a_2 + 1 \quad \Longrightarrow \quad a_2 = 3$$

Therefore,

$$\frac{4x^3 + 16x^2 + 23x + 13}{(x+1)^3(x+2)} = \frac{2}{(x+1)^3} + \frac{a_1}{(x+1)^2} + \frac{3}{x+1} + \frac{1}{x+2}$$

There is now only one unknown a_1, which can be readily found by setting x equal to any convenient value, say $x = 0$. This yields

$$\tfrac{13}{2} = 2 + a_1 + 3 + \tfrac{1}{2} \quad \Longrightarrow \quad a_1 = 1$$

which agrees with our earlier answer.

B.5-5   Improper $F(x)$ with $m = n$

A general method of handling an improper function is indicated in the beginning of this section. However, for a special case where the numerator and denominator polynomials of $F(x)$ are of the same degree ($m = n$), the procedure is the same as that for a proper function. We can show that for

$$F(x) = \frac{b_n x^n + b_{n-1}x^{n-1} + \cdots + b_1 x + b_0}{x^n + a_{n-1}x^{n-1} + \cdots + a_1 x + a_0}$$

$$= b_n + \frac{k_1}{x - \lambda_1} + \frac{k_2}{x - \lambda_2} + \cdots + \frac{k_n}{x - \lambda_n}$$

the coefficients $k_1, k_2, \ldots, k_n$ are computed as if $F(x)$ were proper. Thus,

$$k_r = (x - \lambda_r)F(x)\big|_{x=\lambda_r}$$

For quadratic or repeated factors, the appropriate procedures discussed in Secs. B.5-2 or B.5-3 should be used as if $F(x)$ were proper. In other words, when $m = n$, the only difference between the proper and improper case is the appearance of an extra constant b_n in the latter. Otherwise the procedure remains the same. The proof is left as an exercise for the reader.

■ **Example B.11**

Expand $F(x)$ into partial fractions if

$$F(x) = \frac{3x^2 + 9x - 20}{x^2 + x - 6} = \frac{3x^2 + 9x - 20}{(x-2)(x+3)}$$

Here $m = n = 2$ with $b_n = b_2 = 3$. Therefore,

$$F(x) = \frac{3x^2 + x - 20}{(x-2)(x+3)} = 3 + \frac{k_1}{x-2} + \frac{k_2}{x+3}$$

in which

$$k_1 = \left. \frac{3x^2 + 9x - 20}{(x-2)(x+3)} \right|_{x=2} = \frac{12 + 18 - 20}{(2+3)} = \frac{10}{5} = 2$$

and

$$k_2 = \left. \frac{3x^2 + 9x - 20}{(x-2)(x+3)} \right|_{x=-3} = \frac{27 - 27 - 20}{(-3-2)} = \frac{-20}{-5} = 4$$

Therefore,

$$F(x) = \frac{3x^2 + 9x - 20}{(x-2)(x+3)} = 3 + \frac{2}{x-2} + \frac{4}{x+3} \qquad ■$$

B.5-6   Modified Partial Fractions

Often we require partial fractions of the form $\frac{kx}{(x-\lambda_i)^r}$ rather than $\frac{k}{(x-\lambda_i)^r}$. This can be achieved by expanding $F(x)/x$ into partial fractions. Consider, for example,

$$F(x) = \frac{5x^2 + 20x + 18}{(x+2)(x+3)^2}$$

Dividing both sides by x yields

$$\frac{F(x)}{x} = \frac{5x^2 + 20x + 18}{x(x+2)(x+3)^2}$$

Expansion of the right-hand side into partial fractions as usual yields

$$\frac{F(x)}{x} = \frac{5x^2 + 20x + 18}{x(x+2)(x+3)^2} = \frac{a_1}{x} + \frac{a_2}{x+2} + \frac{a_3}{(x+3)} + \frac{a_4}{(x+3)^2}$$

Using the procedure discussed earlier, we find $a_1 = 1$, $a_2 = 1$, $a_3 = -2$, and $a_4 = 1$. Therefore,

$$\frac{F(x)}{x} = \frac{1}{x} + \frac{1}{x+2} - \frac{2}{x+3} + \frac{1}{(x+3)^2}$$

Now multiplying both sides by x yields

$$F(x) = 1 + \frac{x}{x+2} - \frac{2x}{x+3} + \frac{x}{(x+3)^2}$$

This expresses $F(x)$ as the sum of partial fractions having the form $\frac{kx}{(x-\lambda_i)^r}$.

B.6 Vectors and Matrices

An entity specified by n numbers in a certain order (ordered n-tuple) is an n-dimensional **vector**. Thus, an ordered n-tuple $(x_1, x_2, \ldots, x_n)$ represents an n-dimensional vector $\mathbf{x}$. Vectors may be represented as a row (**row vector**):

$$\mathbf{x} = \begin{bmatrix} x_1 & x_2 & \cdots & x_n \end{bmatrix}$$

or as a column (**column vector**):

$$\mathbf{x} = \begin{bmatrix} x_1 \\ x_2 \\ \vdots \\ x_n \end{bmatrix}$$

Simultaneous linear equations can be viewed as the transformation of one vector into another. Consider, for example, the n simultaneous linear equations

$$y_1 = a_{11}x_1 + a_{12}x_2 + \cdots + a_{1n}x_n$$

$$y_2 = a_{21}x_1 + a_{22}x_2 + \cdots + a_{2n}x_n$$

$$\cdots\cdots\cdots\cdots\cdots\cdots\cdots\cdots\cdots\cdots\cdots\cdots\cdots \tag{B.48}$$

$$y_m = a_{m1}x_1 + a_{m2}x_2 + \cdots + a_{mn}x_n$$

If we define two column vectors $\mathbf{x}$ and $\mathbf{y}$ as

$$\mathbf{x} = \begin{bmatrix} x_1 \\ x_2 \\ \vdots \\ x_n \end{bmatrix}, \qquad \mathbf{y} = \begin{bmatrix} y_1 \\ y_2 \\ \vdots \\ y_m \end{bmatrix} \tag{B.49}$$

then Eqs. (B.48) may be viewed as the relationship or the function that transforms vector $\mathbf{x}$ into vector $\mathbf{y}$. Such a transformation is called the **linear transformation** of vectors. In order to perform a linear transformation, we need to define the array of coefficients a_{ij} appearing in Eqs. (B.48). This array is called a **matrix** and is denoted by $\mathbf{A}$ for convenience:

$$\mathbf{A} = \begin{bmatrix} a_{11} & a_{12} & \cdots & a_{1n} \\ a_{21} & a_{22} & \cdots & a_{2n} \\ \cdot & \cdot & & \cdot \\ a_{m1} & a_{m2} & \cdots & a_{mn} \end{bmatrix} \tag{B.50}$$

A matrix with m rows and n columns is called a matrix of the order (m, n) or an $(m \times n)$ matrix. For the special case where $m = n$, the matrix is called a **square matrix** of order n.

It should be stressed at this point that a matrix is not a number such as a determinant, but an array of numbers arranged in a particular order. It is convenient to abbreviate the representation of matrix $\mathbf{A}$ in Eq. (B.50) with the form $(a_{ij})_{m \times n}$, implying a matrix of order $m \times n$ with a_{ij} as its ijth element. In practice, when the order $m \times n$ is understood or need not be specified, the notation can be abbreviated to (a_{ij}). Note that the first index i of a_{ij} indicates the row and the second index j indicates the column of the element a_{ij} in matrix $\mathbf{A}$.

The simultaneous equations (B.48) may now be expressed in a symbolic form as

$$\mathbf{y} = \mathbf{A}\mathbf{x} \tag{B.51}$$

or

$$
\begin{bmatrix} y_1 \\ y_2 \\ \cdot \\ y_m \end{bmatrix} = \begin{bmatrix} a_{11} & a_{12} & \cdots & a_{1n} \\ a_{21} & a_{22} & \cdots & a_{2n} \\ \cdot & \cdot & & \cdot \\ a_{m1} & a_{m2} & \cdots & a_{mn} \end{bmatrix} \begin{bmatrix} x_1 \\ x_2 \\ \cdot \\ x_n \end{bmatrix} \tag{B.52}
$$

Equation (B.51) is the symbolic representation of Eq. (B.48). As yet, we have not defined the operation of the multiplication of a matrix by a vector. The quantity $\mathbf{A}\mathbf{x}$ is not meaningful until we define such an operation.

B.6-1   Some Definitions and Properties

A square matrix whose elements are zero everywhere except on the main diagonal is a **diagonal matrix**. An example of a diagonal matrix is

$$
\begin{bmatrix} 2 & 0 & 0 \\ 0 & 1 & 0 \\ 0 & 0 & 5 \end{bmatrix}
$$

A diagonal matrix with unity for all its diagonal elements is called an **identity matrix** or a **unit matrix**, denoted by $\mathbf{I}$. Note that this is a square matrix:

$$\mathbf{I} = \begin{bmatrix} 1 & 0 & 0 & \cdots & 0 \\ 0 & 1 & 0 & \cdots & 0 \\ 0 & 0 & 1 & \cdots & 0 \\ & & \cdots & & \\ 0 & 0 & 0 & \cdots & 1 \end{bmatrix} \tag{B.53}$$

The order of the unit matrix is sometimes indicated by a subscript. Thus, $\mathbf{I}_n$ represents the $n \times n$ unit matrix (or identity matrix). However, we shall omit the subscript. The order of the unit matrix will be understood from the context.

A matrix having all its elements zero is a **zero matrix**.

A square matrix **A** is a **symmetric matrix** if $a_{ij} = a_{ji}$ (symmetry about the main diagonal).

Two matrices of the same order are said to be **equal** if they are equal element by element. Thus, if

$$\mathbf{A} = (a_{ij})_{m \times n} \qquad \text{and} \qquad \mathbf{B} = (b_{ij})_{m \times n}$$

then $\mathbf{A} = \mathbf{B}$ only if $a_{ij} = b_{ij}$ for all i and j.

If the rows and columns of an $m \times n$ matrix **A** are interchanged so that the elements in the ith row now become the elements of the ith column (for $i = 1, 2, \ldots, m$), the resulting matrix is called the **transpose** of **A** and is denoted by $\mathbf{A}^T$. It is evident that $\mathbf{A}^T$ is an $n \times m$ matrix. For example, if

$$\mathbf{A} = \begin{bmatrix} 2 & 1 \\ 3 & 2 \\ 1 & 3 \end{bmatrix} \quad \text{then} \quad \mathbf{A}^T = \begin{bmatrix} 2 & 3 & 1 \\ 1 & 2 & 3 \end{bmatrix}$$

Thus, if

$$\mathbf{A} = (a_{ij})_{m \times n}$$

then

$$\mathbf{A}^T = (a_{ji})_{n \times m} \tag{B.54}$$

Note that

$$(\mathbf{A}^T)^T = \mathbf{A} \tag{B.55}$$

B.6-2   Matrix Algebra

We shall now define matrix operations, such as addition, subtraction, multiplication, and division of matrices. The definitions should be formulated so that they are useful in the manipulation of matrices.

1. Addition of Matrices

For two matrices **A** and **B**, both of the same order $(m \times n)$,

$$\mathbf{A} = \begin{bmatrix} a_{11} & a_{12} & \cdots & a_{1n} \\ a_{21} & a_{22} & \cdots & a_{2n} \\ \multicolumn{4}{c}{\dotfill} \\ a_{m1} & a_{m2} & \cdots & a_{mn} \end{bmatrix} \quad \text{and} \quad \mathbf{B} = \begin{bmatrix} b_{11} & b_{12} & \cdots & b_{1n} \\ b_{21} & b_{22} & \cdots & b_{2n} \\ \multicolumn{4}{c}{\dotfill} \\ b_{m1} & b_{m2} & \cdots & b_{mn} \end{bmatrix}$$

we define the sum $\mathbf{A} + \mathbf{B}$ as

$$\mathbf{A} + \mathbf{B} = \begin{bmatrix} (a_{11} + b_{11}) & (a_{12} + b_{12}) & \cdots & (a_{1n} + b_{1n}) \\ (a_{21} + b_{21}) & (a_{22} + b_{22}) & \cdots & (a_{2n} + b_{2n}) \\ \cdots\cdots\cdots\cdots\cdots\cdots\cdots\cdots\cdots\cdots\cdots\cdots\cdots \\ (a_{m1} + b_{m1}) & (a_{m2} + b_{m2}) & \cdots & (a_{mn} + b_{mn}) \end{bmatrix}$$

or

$$\mathbf{A} + \mathbf{B} = (a_{ij} + b_{ij})_{m \times n}$$

Note that two matrices can be added only if they are of the same order.

2. Multiplication of a Matrix by a Scalar

We define the multiplication of a matrix $\mathbf{A}$ by a scalar c as

$$c\mathbf{A} = c \begin{bmatrix} a_{11} & a_{12} & \cdots & a_{1n} \\ a_{21} & a_{22} & \cdots & a_{2n} \\ \cdots\cdots\cdots\cdots\cdots\cdots\cdots \\ a_{m1} & a_{m2} & \cdots & a_{mn} \end{bmatrix} = \begin{bmatrix} ca_{11} & ca_{12} & \cdots & ca_{1n} \\ ca_{21} & ca_{22} & \cdots & ca_{2n} \\ \cdots\cdots\cdots\cdots\cdots\cdots\cdots \\ ca_{m1} & ca_{m2} & \cdots & ca_{mn} \end{bmatrix}$$

3. Matrix Multiplication

We define the product

$$\mathbf{AB} = \mathbf{C}$$

in which c_{ij}, the element of $\mathbf{C}$ in the ith row and jth column, is found by adding the products of the elements of $\mathbf{A}$ in the ith row with the corresponding elements of $\mathbf{B}$ in the jth column. Thus,

$$c_{ij} = a_{i1}b_{1j} + a_{i2}b_{2j} + \cdots + a_{in}b_{nj}$$

$$= \sum_{k=1}^{n} a_{ik}b_{kj} \tag{B.56}$$

This result is shown below.

$$\begin{bmatrix} & & & \\ & & & \\ a_{i1} \ a_{i2} & \cdots a_{in} & \\ & & & \\ & & & \end{bmatrix} \begin{bmatrix} & b_{1j} & \\ & b_{2j} & \\ & \vdots & \\ \cdots & b_{ij} & \cdots \\ & \vdots & \\ & b_{nj} & \end{bmatrix} = \begin{bmatrix} & & \\ & & \\ \cdots & c_{ij} & \cdots \\ & & \\ & & \end{bmatrix}$$

$$\underbrace{}_{\mathbf{A}(m \times n)} \quad \underbrace{}_{\mathbf{B}(n \times p)} \quad \underbrace{}_{\mathbf{C}(m \times p)}$$

Note carefully that the number of columns of **A** must be equal to the number of rows of **B** if this procedure is to work. In other words, **AB**, the product of matrices **A** and **B**, is defined only if the number of columns of **A** is equal to the number of rows of **B**. If this condition is not satisfied, the product **AB** is not defined and is meaningless. When the number of columns of **A** is equal to the number of rows of **B**, matrix **A** is said to be **conformable** to matrix **B** for the product **AB**. Observe that if **A** is an $m \times n$ matrix and **B** is an $n \times p$ matrix, **A** and **B** are conformable for the product, and **C** is an $m \times p$ matrix.

We demonstrate the use of the rule in Eq. (B.56) with the following examples.

$$
\begin{bmatrix} 2 & 3 \\ 1 & 1 \\ 3 & 1 \end{bmatrix}
\begin{bmatrix} 1 & 3 & 1 & 2 \\ 2 & 1 & 1 & 1 \end{bmatrix}
=
\begin{bmatrix} 8 & 9 & 5 & 7 \\ 3 & 4 & 2 & 3 \\ 5 & 10 & 4 & 7 \end{bmatrix}
$$

$$
\begin{bmatrix} 2 & 1 & 3 \end{bmatrix}
\begin{bmatrix} 2 \\ 1 \\ 1 \end{bmatrix} = 8
$$

In both cases above, the two matrices are conformable. However, if we interchange the order of the matrices as follows,

$$
\begin{bmatrix} 1 & 3 & 1 & 2 \\ 2 & 1 & 1 & 1 \end{bmatrix}
\begin{bmatrix} 2 & 3 \\ 1 & 1 \\ 3 & 1 \end{bmatrix}
$$

the matrices are no longer conformable for the product. It is evident that in general,

$$
\mathbf{AB} \neq \mathbf{BA}
$$

Indeed, **AB** may exist and **BA** may not exist, or vice versa, as in the above examples. We shall see later that for some special matrices,

$$
\mathbf{AB} = \mathbf{BA} \tag{B.57}
$$

When Eq. (B.57) is true, matrices **A** and **B** are said to **commute**. We must stress here again that in general, matrices do not commute. Operation (B.57) is valid only for some special cases.

In the matrix product **AB**, matrix **A** is said to be **postmultiplied** by **B** or matrix **B** is said to be **premultiplied** by **A**. We may also verify the following relationships:

$$
(\mathbf{A} + \mathbf{B})\mathbf{C} = \mathbf{AC} + \mathbf{BC} \tag{B.58}
$$

$$
\mathbf{C}(\mathbf{A} + \mathbf{B}) = \mathbf{CA} + \mathbf{CB} \tag{B.59}
$$

We can verify that any matrix **A** premultiplied or postmultiplied by the identity matrix **I** remains unchanged:

$$\mathbf{AI} = \mathbf{IA} = \mathbf{A} \tag{B.60}$$

Of course, we must make sure that the order of $\mathbf{I}$ is such that the matrices are conformable for the corresponding product.

4. Multiplication of a Matrix by a Vector

Consider the matrix Eq. (B.52), which represents Eq. (B.48). The right-hand side of Eq. (B.52) is a product of the $m \times n$ matrix $\mathbf{A}$ and a vector $\mathbf{x}$. If, for the time being, we treat the vector $\mathbf{x}$ as if it were an $n \times 1$ matrix, then the product $\mathbf{Ax}$, according to the matrix multiplication rule, yields the right-hand side of Eq. (B.48). Thus, we may multiply a matrix by a vector by treating the vector as if it were an $n \times 1$ matrix. Note that the constraint of conformability still applies. Thus, in this case, $\mathbf{xA}$ is not defined and is meaningless.

5. Matrix Inversion

To define the inverse of a matrix, let us consider the set of equations

$$\begin{bmatrix} y_1 \\ y_2 \\ \cdots \\ y_n \end{bmatrix} = \begin{bmatrix} a_{11} & a_{12} & \cdots & a_{1n} \\ a_{21} & a_{22} & \cdots & a_{2n} \\ \cdots\cdots\cdots\cdots\cdots \\ a_{n1} & a_{n2} & \cdots & a_{nn} \end{bmatrix} \begin{bmatrix} x_1 \\ x_2 \\ \cdots \\ x_n \end{bmatrix} \tag{B.61}$$

We can solve this set of equations for $x_1, x_2, \ldots, x_n$ in terms of $y_1, y_2, \ldots, y_n$ by using Cramer's rule [see Eq. (B.31)]. This yields

$$\begin{bmatrix} x_1 \\ x_2 \\ \vdots \\ x_n \end{bmatrix} = \begin{bmatrix} \frac{|\mathbf{D}_{11}|}{|\mathbf{A}|} & \frac{|\mathbf{D}_{21}|}{|\mathbf{A}|} & \cdots & \frac{|\mathbf{D}_{n1}|}{|\mathbf{A}|} \\ \frac{|\mathbf{D}_{12}|}{|\mathbf{A}|} & \frac{|\mathbf{D}_{22}|}{|\mathbf{A}|} & \cdots & \frac{|\mathbf{D}_{n2}|}{|\mathbf{A}|} \\ \cdots\cdots\cdots\cdots\cdots\cdots \\ \frac{|\mathbf{D}_{1n}|}{|\mathbf{A}|} & \frac{|\mathbf{D}_{2n}|}{|\mathbf{A}|} & \cdots & \frac{|\mathbf{D}_{nn}|}{|\mathbf{A}|} \end{bmatrix} \begin{bmatrix} y_1 \\ y_2 \\ \vdots \\ y_n \end{bmatrix} \tag{B.62}$$

in which $|\mathbf{A}|$ is the determinant of the matrix $\mathbf{A}$ and $|\mathbf{D}_{ij}|$ is the cofactor of element a_{ij} in the matrix $\mathbf{A}$. The cofactor of element a_{ij} is given by $(-1)^{i+j}$ times the determinant of the $(n-1) \times (n-1)$ matrix that is obtained when the ith row and the jth column in matrix $\mathbf{A}$ are deleted.

We can express Eq. (B.61) in matrix form as

$$\mathbf{y} = \mathbf{Ax} \tag{B.63}$$

We can now define $\mathbf{A}^{-1}$, the inverse of a square matrix $\mathbf{A}$, with the property

$$\mathbf{A}^{-1}\mathbf{A} = \mathbf{I} \quad \text{(unit matrix)} \tag{B.64}$$

Then, premultiplying both sides of Eq. (B.63) by $\mathbf{A}^{-1}$, we obtain

$$\mathbf{A}^{-1}\mathbf{y} = \mathbf{A}^{-1}\mathbf{Ax} = \mathbf{Ix} = \mathbf{x}$$

or

$$\mathbf{x} = \mathbf{A}^{-1}\mathbf{y} \qquad (B.65)$$

A comparison of Eq. (B.65) with Eq. (B.62) shows that

$$\mathbf{A}^{-1} = \frac{1}{|\mathbf{A}|} \begin{bmatrix} |\mathbf{D}_{11}| & |\mathbf{D}_{21}| & \cdots & |\mathbf{D}_{n1}| \\ |\mathbf{D}_{12}| & |\mathbf{D}_{22}| & \cdots & |\mathbf{D}_{n2}| \\ \cdots\cdots\cdots\cdots\cdots\cdots\cdots \\ |\mathbf{D}_{1n}| & |\mathbf{D}_{2n}| & \cdots & |\mathbf{D}_{nn}| \end{bmatrix} \qquad (B.66)$$

One of the conditions necessary for a unique solution of Eq. (B.61) is that the number of equations must equal the number of unknowns. This implies that the matrix $\mathbf{A}$ must be a square matrix. In addition, we observe from the solution as given in Eq. (B.62) that if the solution is to exist, $|\mathbf{A}| \neq 0.$† Therefore, the inverse exists only for a square matrix and only under the condition that the determinant of the matrix be nonzero. A matrix whose determinant is nonzero is a **nonsingular** matrix. Thus, an inverse exists only for a nonsingular (square) matrix. By definition, we have

$$\mathbf{A}^{-1}\mathbf{A} = \mathbf{I} \qquad (B.67a)$$

Postmultiplying this equation by $\mathbf{A}^{-1}$ and then premultiplying by $\mathbf{A}$, we can show that

$$\mathbf{A}\mathbf{A}^{-1} = \mathbf{I} \qquad (B.67b)$$

Note that the matrices $\mathbf{A}$ and $\mathbf{A}^{-1}$ commute.

■ **Example B.12**

Let us find $\mathbf{A}^{-1}$ if

$$\mathbf{A} = \begin{bmatrix} 2 & 1 & 1 \\ 1 & 2 & 3 \\ 3 & 2 & 1 \end{bmatrix}$$

Here

$$|\mathbf{D}_{11}| = -4, \quad |\mathbf{D}_{12}| = 8, \quad |\mathbf{D}_{13}| = -4$$

$$|\mathbf{D}_{21}| = 1, \quad |\mathbf{D}_{22}| = -1, \quad |\mathbf{D}_{23}| = -1$$

$$|\mathbf{D}_{31}| = 1, \quad |\mathbf{D}_{32}| = -5, \quad |\mathbf{D}_{33}| = 3$$

and $|\mathbf{A}| = -4$. Therefore,

$$\mathbf{A}^{-1} = -\frac{1}{4} \begin{bmatrix} -4 & 1 & 1 \\ 8 & -1 & -5 \\ -4 & -1 & 3 \end{bmatrix} \qquad ■$$

---

†These two conditions imply that the number of equations is equal to the number of unknowns and that all the equations are independent.

B.6-3   Derivatives and Integrals of a Matrix

Elements of a matrix need not be constants; they may be functions of a variable. For example, if

$$\mathbf{A} = \begin{bmatrix} e^{-2t} & \sin t \\ e^{t} & e^{-t} + e^{-2t} \end{bmatrix} \tag{B.68}$$

then the matrix elements are functions of t. Here, it is helpful to denote $\mathbf{A}$ by $\mathbf{A}(t)$. Also, it would be helpful to define the derivative and integral of $\mathbf{A}(t)$.

The derivative of a matrix $\mathbf{A}(t)$ (with respect to t) is defined as a matrix whose ijth element is the derivative (with respect to t) of the ijth element of the matrix $\mathbf{A}$. Thus, if

$$\mathbf{A}(t) = [a_{ij}(t)]_{m \times n}$$

then

$$\frac{d}{dt}[\mathbf{A}(t)] = \left[\frac{d}{dt} a_{ij}(t)\right]_{m \times n} \tag{B.69a}$$

or

$$\dot{\mathbf{A}}(t) = [\dot{a}_{ij}(t)]_{m \times n} \tag{B.69b}$$

Thus, the derivative of the matrix in Eq. (B.68) is given by

$$\dot{\mathbf{A}}(t) = \begin{bmatrix} -2e^{-2t} & \cos t \\ e^{t} & -e^{-t} - 2e^{-2t} \end{bmatrix}$$

Similarly, we define the integral of $\mathbf{A}(t)$ (with respect to t) as a matrix whose ijth element is the integral (with respect to t) of the ijth element of the matrix $\mathbf{A}$:

$$\int \mathbf{A}(t)\, dt = \left(\int a_{ij}(t)\, dt\right)_{m \times n} \tag{B.70}$$

Thus, for the matrix $\mathbf{A}$ in Eq. (B.68), we have

$$\int \mathbf{A}(t)\, dt = \begin{bmatrix} \int e^{-2t}\, dt & \int \sin dt \\ \int e^{t}\, dt & \int (e^{-t} + 2e^{-2t})\, dt \end{bmatrix}$$

We can readily prove the following identities:

$$\frac{d}{dt}(\mathbf{A} + \mathbf{B}) = \frac{d\mathbf{A}}{dt} + \frac{d\mathbf{B}}{dt} \tag{B.71a}$$

$$\frac{d}{dt}(c\mathbf{A}) = c\frac{d\mathbf{A}}{dt} \tag{B.71b}$$

$$\frac{d}{dt}(\mathbf{A}\mathbf{B}) = \frac{d\mathbf{A}}{dt}\mathbf{B} + \mathbf{A}\frac{d\mathbf{B}}{dt} = \dot{\mathbf{A}}\mathbf{B} + \mathbf{A}\dot{\mathbf{B}} \tag{B.71c}$$

The proofs of identities (B.71a) and (B.71b) are trivial. We can prove Eq. (B.71c) as follows. Let $\mathbf{A}$ be an $m \times n$ matrix and $\mathbf{B}$ an $n \times p$ matrix. Then, if

$$\mathbf{C} = \mathbf{AB}$$

from Eq. (B.56), we have

$$c_{ik} = \sum_{j-1}^{n} a_{ij}b_{jk}$$

and

$$\dot{c}_{ik} = \underbrace{\sum_{j-1}^{n} \dot{a}_{ij}b_{jk}}_{d_{ik}} + \underbrace{\sum_{j-1}^{n} a_{ij}\dot{b}_{jk}}_{e_{ik}} \qquad (B.72)$$

or

$$\dot{c}_{ij} = d_{ij} + e_{ik}$$

Equation (B.72) along with the multiplication rule clearly indicate that d_{ik} is the ikth element of matrix $\dot{\mathbf{A}}\mathbf{B}$ and e_{ik} is the ikth element of matrix $\mathbf{A}\dot{\mathbf{B}}$. Equation (B.71c) then follows.

If we let $\mathbf{B} = \mathbf{A}^{-1}$ in Eq. (B.71c), we obtain

$$\frac{d}{dt}(\mathbf{A}\mathbf{A}^{-1}) = \frac{d\mathbf{A}}{dt}\mathbf{A}^{-1} + \mathbf{A}\frac{d}{dt}\mathbf{A}^{-1}$$

But since

$$\frac{d}{dt}(\mathbf{A}\mathbf{A}^{-1}) = \frac{d}{dt}\mathbf{I} = 0$$

we have

$$\frac{d}{dt}(\mathbf{A}^{-1}) = -\mathbf{A}^{-1}\frac{d\mathbf{A}}{dt}\mathbf{A}^{-1} \qquad (B.73)$$

B.6-4 The Characteristic Equation of a Matrix: The Cayley-Hamilton Theorem

For an $(n \times n)$ square matrix $\mathbf{A}$, any vector $\mathbf{x}$ $(\mathbf{x} \neq 0)$ that satisfies the equation

$$\mathbf{A}\mathbf{x} = \lambda\mathbf{x} \qquad (B.74)$$

is an **eigenvector** (or **characteristic vector**), and λ is the corresponding **eigenvalue** (or **characteristic value**) of $\mathbf{A}$. Equation (B.74) can be expressed as

$$(\mathbf{A} - \lambda\mathbf{I})\mathbf{x} = 0 \qquad (B.75)$$

The solution for this set of homogeneous equations exists if and only if

$$|\mathbf{A} - \lambda\mathbf{I}| = |\lambda\mathbf{I} - \mathbf{A}| = 0 \qquad (B.76a)$$

or

$$\begin{vmatrix} a_{11} - \lambda & a_{12} & \cdots & a_{1n} \\ a_{21} & a_{22} - \lambda & \cdots & a_{2n} \\ \hdotsfor{4} \\ a_{n1} & a_{n2} & \cdots & a_{nn} - \lambda \end{vmatrix} = 0 \qquad \text{(B.76b)}$$

Equation (B.76a) [or (B.76b)] is known as the **characteristic equation** of the matrix $\mathbf{A}$ and can be expressed as

$$Q(\lambda) = |\lambda \mathbf{I} - \mathbf{A}| = \lambda^n + a_{n-1}\lambda^{n-1} + \cdots + a_1\lambda + a_0\lambda^0 = 0 \qquad \text{(B.77)}$$

$Q(\lambda)$ is called the **characteristic polynomial** of the matrix $\mathbf{A}$. The n zeros of the characteristic polynomial are the eigenvalues of $\mathbf{A}$ and, corresponding to each eigenvalue, there is an eigenvector that satisfies Eq. (B.74).

The **Cayley-Hamilton theorem** states that every $n \times n$ matrix $\mathbf{A}$ satisfies its own characteristic equation. In other words, Eq. (B.77) is valid if λ is replaced by $\mathbf{A}$:

$$\mathbf{Q}(\mathbf{A}) = \mathbf{A}^n + a_{n-1}\mathbf{A}^{n-1} + \cdots + a_1\mathbf{A} + a_0\mathbf{A}^0 = 0 \qquad \text{(B.78)}$$

Functions of a Matrix

The Cayley-Hamilton theorem can be used to evaluate functions of a square matrix $\mathbf{A}$, as shown below.

Consider a function $f(\lambda)$ in the form of an infinite power series:

$$f(\lambda) = \alpha_0 + \alpha_1\lambda + \alpha_2\lambda_2^2 + \cdots + \cdots = \sum_{i=0}^{\infty} \alpha_i \lambda^i \qquad \text{(B.79)}$$

Because λ satisfies the characteristic Eq. (B.77), we can write

$$\lambda^n = -a_{n-1}\lambda^{n-1} - a_{n-2}\lambda^{n-2} - \cdots - a_1\lambda - a_0 \qquad \text{(B.80)}$$

If we multiply both sides by λ, the left-hand side is λ^{n+1}, and the right-hand side contains the terms λ^n, λ^{n-1}, ..., λ. Using Eq. (B.80), if we substitute λ^n in terms of λ^{n-1}, λ^{n-2}, ..., λ, the highest power on the right-hand side is reduced to $n-1$. Continuing in this way, we see that λ^{n+k} can be expressed in terms of λ^{n-1}, λ^{n-2}, ..., λ for any k. Hence, the infinite series on the right-hand side of Eq. (B.79) can always be expressed in terms of λ^{n-1}, λ^{n-2}, ..., λ as

$$f(\lambda) = \beta_0 + \beta_1\lambda + \beta_2\lambda^2 + \cdots + \beta_{n-1}\lambda^{n-1} \qquad \text{(B.81)}$$

If we assume that there are n distinct eigenvalues λ_1, λ_2, ..., λ_n, then Eq. (B.81) holds for these n values of λ. The substitution of these values in Eq. (B.81) yields n simultaneous equations

$$
\begin{bmatrix} f(\lambda_1) \\ f(\lambda_2) \\ \cdots \\ f(\lambda_n) \end{bmatrix} = \begin{bmatrix} 1 & \lambda_1 & \lambda_1^2 & \cdots & \lambda_1^{n-1} \\ 1 & \lambda_2 & \lambda_2^2 & \cdots & \lambda_2^{n-1} \\ \cdots\cdots\cdots\cdots\cdots\cdots\cdots \\ 1 & \lambda_n & \lambda_n^2 & \cdots & \lambda_n^{n-1} \end{bmatrix} \begin{bmatrix} \beta_0 \\ \beta_1 \\ \cdots \\ \beta_{n-1} \end{bmatrix} \tag{B.82a}
$$

and

$$
\begin{bmatrix} \beta_0 \\ \beta_1 \\ \cdots \\ \beta_{n-1} \end{bmatrix} = \begin{bmatrix} 1 & \lambda_1 & \lambda_1^2 & \cdots & \lambda_1^{n-1} \\ 1 & \lambda_2 & \lambda_2^2 & \cdots & \lambda_2^{n-1} \\ \cdots\cdots\cdots\cdots\cdots\cdots\cdots \\ 1 & \lambda_n & \lambda_n^2 & \cdots & \lambda_n^{n-1} \end{bmatrix}^{-1} \begin{bmatrix} f(\lambda_1) \\ f(\lambda_2) \\ \cdots \\ f(\lambda_n) \end{bmatrix} \tag{B.82b}
$$

Since $\mathbf{A}$ also satisfies Eq. (B.80), we may advance a similar argument to show that if $f(\mathbf{A})$ is a function of a square matrix $\mathbf{A}$ expressed as an infinite power series in $\mathbf{A}$, then

$$
f(\mathbf{A}) = \alpha_0\mathbf{I} + \alpha_1\mathbf{A} + \alpha_2\mathbf{A}^2 + \cdots + \cdots = \sum_{i=0}^{\infty} \alpha_i \mathbf{A}^i \tag{B.83a}
$$

and

$$
f(\mathbf{A}) = \beta_0\mathbf{I} + \beta_1\mathbf{A} + \beta_2\mathbf{A}^2 + \cdots + \beta_{n-1}\mathbf{A}^{n-1} \tag{B.83b}
$$

in which the coefficients β_is are found from Eq. (B.82b). If some of the eigenvalues are repeated (multiple roots), the results are somewhat modified.

We shall demonstrate the utility of this result with the following two examples.

B.6-5 Computation of an Exponential and a Power of a Matrix

Let us compute $e^{\mathbf{A}t}$ defined by

$$
e^{\mathbf{A}t} = \mathbf{I} + \mathbf{A}t + \frac{\mathbf{A}^2 t^2}{2!} + \cdots + \frac{\mathbf{A}^n t^n}{n!} + \cdots
$$

$$
= \sum_{k=0}^{\infty} \frac{\mathbf{A}^k t^k}{k!}
$$

From Eq. (B.83b), we can express

$$
e^{\mathbf{A}t} = \sum_{i=1}^{n-1} \beta_i (\mathbf{A})^i
$$

in which the β_is are given by Eq. (B.82b), with $f(\lambda_i) = e^{\lambda_i t}$.

■ **Example B.13**

Let us consider the case where

$$\mathbf{A} = \begin{bmatrix} 0 & 1 \\ -2 & -3 \end{bmatrix}$$

The eigenvalues are

$$|\lambda\mathbf{I} - \mathbf{A}| = \begin{vmatrix} \lambda & -1 \\ 2 & \lambda + 3 \end{vmatrix} = \lambda^2 + 3\lambda + 2 = (\lambda + 1)(\lambda + 2) = 0$$

Hence, $\lambda_1 = -1$, $\lambda_2 = -2$, and

$$e^{\mathbf{A}t} = \beta_0\mathbf{I} + \beta_1\mathbf{A}$$

in which

$$\begin{bmatrix} \beta_0 \\ \beta_1 \end{bmatrix} = \begin{bmatrix} 1 & -1 \\ 1 & -2 \end{bmatrix}^{-1} \begin{bmatrix} e^{-t} \\ e^{-2t} \end{bmatrix}$$

$$= \begin{bmatrix} 2 & -1 \\ 1 & -1 \end{bmatrix} \begin{bmatrix} e^{-t} \\ e^{-2t} \end{bmatrix} = \begin{bmatrix} 2e^{-t} - e^{-2t} \\ e^{-t} - e^{-2t} \end{bmatrix}$$

and

$$e^{\mathbf{A}t} = (2e^{-t} - e^{-2t}) \begin{bmatrix} 1 & 0 \\ 0 & 1 \end{bmatrix} + (e^{-t} - e^{-2t}) \begin{bmatrix} 0 & 1 \\ -2 & -3 \end{bmatrix}$$

$$= \begin{bmatrix} 2e^{-t} - e^{-2t} & (e^{-t} - e^{-2t}) \\ -2e^{-t} + 2e^{-2t} & -e^{-t} + 2e^{-2t} \end{bmatrix} \qquad (B.84)$$

■

Computation of $\mathbf{A}^k$

As Eq. (B.83b) indicates, we can express $\mathbf{A}^k$ as

$$\mathbf{A}^k = \beta_0\mathbf{I} + \beta_1\mathbf{A} + \cdots + \beta_{n-1}\mathbf{A}^{n-1}$$

in which the β_is are given by Eq. (B.82b) with $f(\lambda_i) = \lambda_i^k$. For a completed example of the computation of $\mathbf{A}^k$ by this method, see Example 13.12.

B.7 Miscellaneous

B.7-1 L'Hôpital's Rule

If $\lim f(x)/g(x)$ results in the indeterministic form $0/0$ or ∞/∞, then

$$\lim \frac{f(x)}{g(x)} = \lim \frac{\dot{f}(x)}{\dot{g}(x)}$$

B.7-2 The Taylor and Maclaurin Series

$$f(x) = f(a) + \frac{(x-a)}{1!}\dot{f}(a) + \frac{(x-a)^2}{2!}\ddot{f}(a) + \cdots$$

$$f(x) = f(0) + \frac{x}{1!}\dot{f}(0) + \frac{x^2}{2!}\ddot{f}(0) + \cdots$$

B.7-3 Power Series

$$e^x = 1 + x + \frac{x^2}{2!} + \frac{x^3}{3!} + \cdots + \frac{x^n}{n!} + \cdots$$

$$\sin x = x - \frac{x^3}{3!} + \frac{x^5}{5!} - \frac{x^7}{7!} + \cdots$$

$$\cos x = 1 - \frac{x^2}{2!} + \frac{x^4}{4!} - \frac{x^6}{6!} + \frac{x^8}{8!} - \cdots$$

$$\tan x = x + \frac{x^3}{3} + \frac{2x^5}{15} + \frac{17x^7}{315} + \cdots \qquad x^2 < \pi^2/4$$

$$\tanh x = x - \frac{x^3}{3} + \frac{2x^5}{15} - \frac{17x^7}{315} + \cdots \qquad x^2 < \pi^2/4$$

$$(1+x)^n = 1 + nx + \frac{n(n-1)}{2!}x^2 + \frac{n(n-1)(n-2)}{3!}x^3 + \cdots + \binom{n}{k}x^k + \cdots + x^n$$

$$\approx 1 + nx \qquad |x| \ll 1$$

$$\frac{1}{1-x} = 1 + x + x^2 + x^3 + \cdots \qquad |x| < 1$$

B.7-4 Sums

$$\sum_{m=0}^{k} r^m = \frac{r^{k+1}-1}{r-1} \qquad r \neq 1$$

$$\sum_{m=M}^{N} r^m = \frac{r^{N+1}-r^M}{r-1} \qquad r \neq 1$$

$$\sum_{m=0}^{k} \left(\frac{a}{b}\right)^m = \frac{a^{k+1}-b^{k+1}}{b^k(a-b)} \qquad a \neq b$$

B.7-5 Complex Numbers

$$e^{\pm j\pi/2} = \pm j$$

$$e^{\pm jn\pi} = \begin{cases} 1 & n \text{ even} \\ -1 & n \text{ odd} \end{cases}$$

$$e^{\pm j\theta} = \cos\theta \pm j\sin\theta$$

$$a + jb = re^{j\theta} \qquad r = \sqrt{a^2 + b^2}, \qquad \theta = \tan^{-1}\left(\frac{b}{a}\right)$$

$$(re^{j\theta})^k = r^k e^{jk\theta}$$

$$(r_1 e^{j\theta_1})(r_2 e^{j\theta_2}) = r_1 r_2 e^{j(\theta_1 + \theta_2)}$$

B.7-6 Trigonometric Identities

$$e^{\pm jx} = \cos x \pm j\sin x$$

$$\cos x = \tfrac{1}{2}[e^{jx} + e^{-jx}]$$

$$\sin x = \tfrac{1}{2j}[e^{jx} - e^{-jx}]$$

$$\cos\left(x \pm \tfrac{\pi}{2}\right) = \mp\sin x$$

$$\sin\left(x \pm \tfrac{\pi}{2}\right) = \pm\cos x$$

$$2\sin x\cos x = \sin 2x$$

$$\sin^2 x + \cos^2 x = 1$$

$$\cos^2 x - \sin^2 x = \cos 2x$$

$$\cos^2 x = \tfrac{1}{2}(1 + \cos 2x)$$

$$\sin^2 x = \tfrac{1}{2}(1 - \cos 2x)$$

$$\cos^3 x = \tfrac{1}{4}(3\cos x + \cos 3x)$$

$$\sin^3 x = \tfrac{1}{4}(3\sin x - \sin 3x)$$

$$\sin(x \pm y) = \sin x\cos y \pm \cos x\sin y$$

$$\cos(x \pm y) = \cos x\cos y \mp \sin x\sin y$$

$$\tan(x \pm y) = \frac{\tan x \pm \tan y}{1 \mp \tan x\tan y}$$

$$\sin x\sin y = \tfrac{1}{2}[\cos(x - y) - \cos(x + y)]$$

$$\cos x\cos y = \tfrac{1}{2}[\cos(x - y) + \cos(x + y)]$$

$$\sin x\cos y = \tfrac{1}{2}[\sin(x - y) + \sin(x + y)]$$

$$a\cos x + b\sin x = C\cos(x + \theta)$$

$$\text{in which } C = \sqrt{a^2 + b^2} \quad \text{and} \quad \theta = \tan^{-1}\left(\frac{-b}{a}\right)$$

B.7-7 Indefinite Integrals

$$\int u\, dv = uv - \int v\, du$$

$$\int f(x)\dot{g}(x)\, dx = f(x)g(x) - \int \dot{f}(x)g(x)\, dx$$

$$\int \sin ax\, dx = -\frac{1}{a}\cos ax \qquad\qquad \int \cos ax\, dx = \frac{1}{a}\sin ax$$

$$\int \sin^2 ax\, dx = \frac{x}{2} - \frac{\sin 2ax}{4a} \qquad\qquad \int \cos^2 ax\, dx = \frac{x}{2} + \frac{\sin 2ax}{4a}$$

$$\int x \sin ax\, dx = \frac{1}{a^2}(\sin ax - ax \cos ax)$$

$$\int x \cos ax\, dx = \frac{1}{a^2}(\cos ax + ax \sin ax)$$

$$\int x^2 \sin ax\, dx = \frac{1}{a^3}(2ax \sin ax + 2\cos ax - a^2 x^2 \cos ax)$$

$$\int x^2 \cos ax\, dx = \frac{1}{a^3}(2ax \cos ax - 2\sin ax + a^2 x^2 \sin ax)$$

$$\int \sin ax \sin bx\, dx = \frac{\sin (a-b)x}{2(a-b)} - \frac{\sin (a+b)x}{2(a+b)} \qquad a^2 \neq b^2$$

$$\int \sin ax \cos bx\, dx = -\left[\frac{\cos (a-b)x}{2(a-b)} + \frac{\cos (a+b)x}{2(a+b)}\right] \qquad a^2 \neq b^2$$

$$\int \cos ax \cos bx\, dx = \frac{\sin (a-b)x}{2(a-b)} + \frac{\sin (a+b)x}{2(a+b)} \qquad a^2 \neq b^2$$

$$\int e^{ax}\, dx = \frac{1}{a}e^{ax}$$

$$\int x e^{ax}\, dx = \frac{e^{ax}}{a^2}(ax - 1)$$

$$\int x^2 e^{ax}\, dx = \frac{e^{ax}}{a^3}(a^2 x^2 - 2ax + 2)$$

$$\int e^{ax} \sin bx\, dx = \frac{e^{ax}}{a^2 + b^2}(a \sin bx - b \cos bx)$$

$$\int e^{ax} \cos bx\, dx = \frac{e^{ax}}{a^2 + b^2}(a \cos bx + b \sin bx)$$

$$\int \frac{1}{x^2 + a^2}\, dx = \frac{1}{a}\tan^{-1}\frac{x}{a}$$

$$\int \frac{x}{x^2 + a^2}\, dx = \frac{1}{2}\ln(x^2 + a^2)$$

B.7-8   Differentiation Table

$$\frac{d}{dx}f(u) = \frac{d}{du}f(u)\frac{du}{dx} \qquad\qquad \frac{d}{dx}a^{bx} = b(\ln a)a^{bx}$$

$$\frac{d}{dx}(uv) = u\frac{dv}{dx} + v\frac{du}{dx} \qquad\qquad \frac{d}{dx}\sin ax = a\cos ax$$

$$\frac{d}{dx}\left(\frac{u}{v}\right) = \frac{v\frac{du}{dx} - u\frac{dv}{dx}}{v^2} \qquad\qquad \frac{d}{dx}\cos ax = -a\sin ax$$

$$\frac{dx^n}{dx} = nx^{n-1} \qquad\qquad \frac{d}{dx}\tan ax = \frac{a}{\cos^2 ax}$$

$$\frac{d}{dx}\ln(ax) = \frac{1}{x} \qquad\qquad \frac{d}{dx}(\sin^{-1}ax) = \frac{a}{\sqrt{1-a^2x^2}}$$

$$\frac{d}{dx}\log(ax) = \frac{\log e}{x} \qquad\qquad \frac{d}{dx}(\cos^{-1}ax) = \frac{-a}{\sqrt{1-a^2x^2}}$$

$$\frac{d}{dx}e^{bx} = be^{bx} \qquad\qquad \frac{d}{dx}(\tan^{-1}ax) = \frac{a}{1+a^2x^2}$$

B.7-9   Some Useful Constants

$\pi \approx 3.1415926535$

$e \approx 2.7182818284$

$\frac{1}{e} \approx 0.3678794411$

$\log_{10} 2 = 0.30103$

$\log_{10} 3 = 0.47712$

B.7-10   Solution of Quadratic and Cubic Equations

Any **quadratic** equation can be reduced to the form

$$ax^2 + bx + c = 0$$

The solution of this equation is provided by

$$x = \frac{-b \pm \sqrt{b^2 - 4ac}}{2a}$$

A general **cubic** equation

$$y^3 + py^2 + qy + r = 0$$

may be reduced to the **depressed cubic** form

$$x^3 + ax + b = 0$$

by substituting

$$y = x - \tfrac{p}{3}$$

This yields

$$a = \tfrac{1}{3}(3q - p^2) \qquad b = \tfrac{1}{27}(2p^3 - 9pq + 27r)$$

Now let

$$A = \sqrt[3]{-\tfrac{b}{2} + \sqrt{\tfrac{b^2}{4} + \tfrac{a^3}{27}}}, \qquad B = \sqrt[3]{-\tfrac{b}{2} - \sqrt{\tfrac{b^2}{4} + \tfrac{a^3}{27}}}$$

The solution of the depressed cubic is

$$x = A + B, \qquad x = -\tfrac{A+B}{2} + \tfrac{A-B}{2}\sqrt{-3}, \qquad x = -\tfrac{A+B}{2} - \tfrac{A-B}{2}\sqrt{-3}$$

and

$$y = x - \tfrac{p}{3}$$

References

1.   Asimov, Isaac, *Asimov on Numbers,* Bell Publishing Co., N.Y., 1982.

2.   Calinger, R., Ed., *Classics of Mathematics,* Moore Publishing Co.,Oak Park, IL., 1982.

3.   Hogben, Lancelot, *Mathematics in the Making,* Doubleday & Co. Inc., New York, 1960.

4.   Cajori, Florian, *A History of Mathematics*, 4th ed., Chelsea, New York, 1985.

5.   Encyclopaedia Britannica, 15th ed., *Micropaedia*, vol. 11, p. 1043, 1982.

6.   Singh, Jagjit, *Great Ideas of Modern Mathematics*, Dover, New York, 1959.

7.   Dunham, William, *Journey through Genius,* Wiley, New York, 1990.

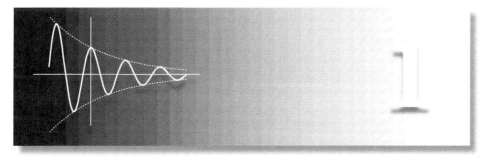

Introduction to Signals and Systems

In this chapter we shall discuss certain basic aspects of signals. We shall also introduce important basic concepts and qualitative explanations of the how's and why's of systems theory, thus building a solid foundation for understanding the quantitative analysis in the remainder of the book.

Signals

A **signal**, as the term implies, is a set of information or data. Examples include a telephone or a television signal, monthly sales of a corporation, or the daily closing prices of a stock market (e.g., the Dow Jones averages). In all these examples, the signals are functions of the independent variable *time*. This is not always the case, however. When an electrical charge is distributed over a body, for instance, the signal is the charge density, a function of *space* rather than time. In this book we deal almost exclusively with signals that are functions of time. The discussion, however, applies equally well to other independent variables.

Systems

Signals may be processed further by **systems**, which may modify them or extract additional information from them. For example, an antiaircraft gun operator may want to know the future location of a hostile moving target that is being tracked by his radar. Knowing the radar signal he knows the past location and velocity of the target. By properly processing the radar signal (the input) he can approximately estimate the future location of the target. Thus, a system is an entity that *processes* a set of signals (**inputs**) to yield another set of signals (**outputs**). A system may be made up of physical components, as in electrical, mechanical, or hydraulic systems (hardware realization), or it may be an algorithm that computes an output from an input signal (software realization).

1.1   Size of a Signal

The size of any entity is a number that indicates the largeness or strength of that entity. Generally speaking, the signal amplitude varies with time. How can a signal that exists over a certain time interval with varying amplitude be measured by one number that will indicate the signal size or signal strength? Such a measure must consider not only the signal amplitude, but also its duration. For instance, if we are to devise a single number V as a measure of the size of a human being, we must consider not only his or her width (girth), but also the height. If we make a simplifying assumption that the shape of a person is a cylinder of variable radius r (which varies with the height h) then a reasonable measure of the size of a person of height H is the person's volume V, given by

$$V = \pi \int_0^H r^2(h)\, dh$$

Signal Energy

Arguing in this manner, we may consider the area under a signal $f(t)$ as a possible measure of its size, because it takes account of not only the amplitude, but also the duration. However, this will be a defective measure because $f(t)$ could be a large signal, yet its positive and negative areas could cancel each other, indicating a signal of small size. This difficulty can be corrected by defining the signal size as the area under $f^2(t)$, which is always positive. We call this measure the **signal energy** E_f, defined (for a real signal) as

$$E_f = \int_{-\infty}^{\infty} f^2(t)\, dt \tag{1.1}$$

This definition can be generalized to a complex valued signal $f(t)$ as

$$E_f = \int_{-\infty}^{\infty} |f(t)|^2\, dt \tag{1.2}$$

There are also other possible measures of signal size, such as the area under $|f(t)|$. The energy measure, however, is not only more tractable mathematically, but is also more meaningful (as shown later) in the sense that it is indicative of the energy that can be extracted from the signal.

Signal Power

The signal energy must be finite for it to be a meaningful measure of the signal size. A necessary condition for the energy to be finite is that the signal amplitude $\to 0$ as $|t| \to \infty$ (Fig. 1.1a). Otherwise the integral in Eq. (1.1) will not converge.

In some cases, for instance, when the amplitude of $f(t)$ does not $\to 0$ as $|t| \to \infty$ (Fig. 1.1b), then, the signal energy is infinite. A more meaningful measure of the signal size in such a case would be the time average of the energy, if it exists. This measure is called the **power** of the signal. For a signal $f(t)$, we define its power P_f as

$$P_f = \lim_{T \to \infty} \frac{1}{T} \int_{-T/2}^{T/2} f^2(t)\, dt \tag{1.3}$$

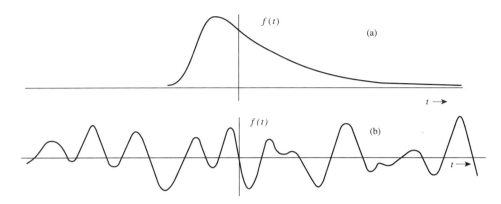

Fig. 1.1 Examples of Signals: (a) a signal with finite energy (b) a signal with finite power.

We can generalize this definition for a complex signal $f(t)$ as

$$P_f = \lim_{T \to \infty} \frac{1}{T} \int_{-T/2}^{T/2} |f(t)|^2 \, dt \qquad (1.4)$$

Observe that the signal power P_f is the time average (mean) of the signal amplitude squared, that is, the *mean-squared* value of $f(t)$. Indeed, the square root of P_f is the familiar **rms** (root mean square) value of $f(t)$.

The mean of an entity averaged over a large time interval approaching infinity exists if the entity is either periodic or has a statistical regularity. If such a condition is not satisfied, the average may not exist. For instance, a ramp signal $f(t) = t$ increases indefinitely as $|t| \to \infty$, and neither the energy nor the power exists for this signal.

Comments

The signal energy as defined in Eq. (1.1) or Eq. (1.2) does not indicate the actual energy of the signal because the signal energy depends not only on the signal, but also on the load. It can, however, be interpreted as the energy dissipated in a normalized load of a 1-ohm resistor if a voltage $f(t)$ were to be applied across the 1-ohm resistor (or if a current $f(t)$ were to be passed through the 1-ohm resistor). The measure of "energy" is, therefore indicative of the energy capability of the signal and not the actual energy. For this reason the concepts of conservation of energy should not be applied to this "signal energy". Parallel observation applies to "signal power" defined in Eq. (1.3) or (1.4). These measures are but convenient indicators of the signal size, which prove useful in many applications. For instance, if we approximate a signal $f(t)$ by another signal $g(t)$, the error in the approximation is $e(t) = f(t) - g(t)$. The energy (or power) of $e(t)$ is a convenient indicator of the goodness of the approximation. It provides us with a quantitative measure of determining the closeness of the approximation. In communication systems, during transmission over a channel, message signals are corrupted by unwanted signals (noise). The quality of the received signal is judged by the relative sizes of the

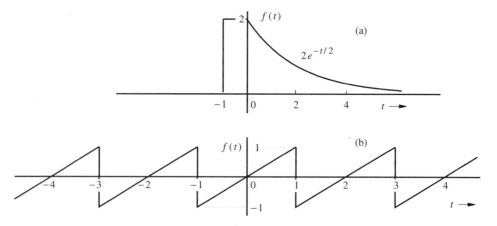

Fig. 1.2 Signals for Example 1.1.

desired signal and the unwanted signal (noise). In this case the ratio of the message signal and noise signal powers (signal to noise power ratio) is a good indication of the received signal quality.

Units of Energy and Power: Equations (1.1) and (1.2) are not correct dimensionally. This is because here we are using the term *energy* not in its conventional sense, but to indicate the signal size. The same observation applies to Eqs. (1.3) and (1.4) for power. The units of energy and power, as defined here, depend on the nature of the signal $f(t)$. If $f(t)$ is a voltage signal, its energy E_f has units of V^2s (volts squared-seconds) and its power P_f has units of V^2 (volts squared). If $f(t)$ is a current signal, these units will be A^2s (amperes squared-seconds) and A^2 (amperes squared), respectively.

■ **Example 1.1**

Determine the suitable measures of the signals in Fig 1.2.

In Fig. 1.2a, the signal amplitude $\to 0$ as $|t| \to \infty$. Therefore the suitable measure for this signal is its energy E_f given by

$$E_f = \int_{-\infty}^{\infty} f^2(t)\, dt = \int_{-1}^{0} (2)^2\, dt + \int_{0}^{\infty} 4e^{-t}\, dt = 4 + 4 = 8$$

In Fig. 1.2b, the signal amplitude does not $\to 0$ as $|t| \to \infty$. However, it is periodic, and therefore its power exists. We can use Eq. (1.3) to determine its power. We can simplify the procedure for periodic signals by observing that a periodic signal repeats regularly each period (2 seconds in this case). Therefore, averaging $f^2(t)$ over an infinitely large interval is identical to averaging this quantity over one period (2 seconds in this case). Thus

$$P_f = \frac{1}{2} \int_{-1}^{1} f^2(t)\, dt = \frac{1}{2} \int_{-1}^{1} t^2\, dt = \frac{1}{3}$$

Recall that the signal power is the square of its rms value. Therefore, the rms value of this signal is $1/\sqrt{3}$. ■

■ **Example 1.2**

Determine the power and the rms value of

(a) $f(t) = C \cos(\omega_0 t + \theta)$ (b) $f(t) = C_1 \cos(\omega_1 t + \theta_1) + C_2 \cos(\omega_2 t + \theta_2)$ $(\omega_1 \neq \omega_2)$
(c) $f(t) = De^{j\omega_0 t}$.

(a) This is a periodic signal with period $T_0 = 2\pi/\omega_0$. The suitable measure of this signal is its power. Because it is a periodic signal, we may compute its power by averaging its energy over one period $T_0 = 2\pi/\omega_0$. However, for the sake of demonstration, we shall solve this problem by averaging over an infinitely large time interval using Eq (1.3).

$$P_f = \lim_{T \to \infty} \frac{1}{T} \int_{-T/2}^{T/2} C^2 \cos^2(\omega_0 t + \theta) \, dt = \lim_{T \to \infty} \frac{C^2}{2T} \int_{-T/2}^{T/2} [1 + \cos(2\omega_0 t + 2\theta)] \, dt$$

$$= \lim_{T \to \infty} \frac{C^2}{2T} \int_{-T/2}^{T/2} dt + \lim_{T \to \infty} \frac{C^2}{2T} \int_{-T/2}^{T/2} \cos(2\omega_0 t + 2\theta) \, dt$$

The first term on the right-hand side is equal to $C^2/2$. Moreover, the second term is zero because the integral appearing in this term represents the area under a sinusoid over a very large time interval T with $T \to \infty$. This area is at most equal to the area of half the cycle because of cancellations of the positive and negative areas of a sinusoid. The second term is this area multiplied by $C^2/2T$ with $T \to \infty$. Clearly this term is zero, and

$$P_f = \frac{C^2}{2} \tag{1.5a}$$

This shows that a sinusoid of amplitude C has a power $C^2/2$ regardless of the value of its frequency ω_0 $(\omega_0 \neq 0)$ and phase θ. The rms value is $C/\sqrt{2}$. If the signal frequency is zero (dc or a constant signal of amplitude C), the reader can show that the power is C^2.

(b) In Chapter 4, we show that a sum of two sinusoids may or may not be periodic, depending on whether the ratio ω_1/ω_2 is a rational number or not. Therefore, the period of this signal is not known. Hence, its power will be determined by averaging its energy over T seconds with $T \to \infty$. Thus,

$$P_f = \lim_{T \to \infty} \frac{1}{T} \int_{-T/2}^{T/2} [C_1 \cos(\omega_1 t + \theta_1) + C_2 \cos(\omega_2 t + \theta_2)]^2 \, dt$$

$$= \lim_{T \to \infty} \frac{1}{T} \int_{-T/2}^{T/2} C_1^2 \cos^2(\omega_1 t + \theta_1) \, dt + \lim_{T \to \infty} \frac{1}{T} \int_{-T/2}^{T/2} C_2^2 \cos^2(\omega_2 t + \theta_2) \, dt$$

$$+ \lim_{T \to \infty} \frac{2C_1 C_2}{T} \int_{-T/2}^{T/2} \cos(\omega_1 t + \theta_1) \cos(\omega_2 t + \theta_2) \, dt$$

The first and second integrals on the right-hand side are the powers of the two sinusoids, which are $C_1^2/2$ and $C_2^2/2$ as found in part (a). Arguing as in part (a), we see that the third term is zero, and we have†

$$P_f = \frac{C_1^2}{2} + \frac{C_2^2}{2} \tag{1.5b}$$

and the rms value is $\sqrt{(C_1^2 + C_2^2)/2}$.

We can readily extend this result to a sum of any number of sinusoids with distinct frequencies. Thus, if

---

†This is true only if $\omega_1 \neq \omega_2$. If $\omega_1 = \omega_2$, the integrand of the third term contains a constant $\cos(\theta_1 - \theta_2)$, and the third term $\to 2C_1 C_2 \cos(\theta_1 - \theta_2)$ as $T \to \infty$.

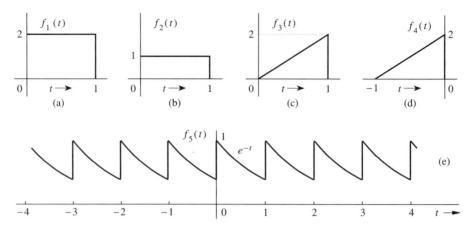

Fig. 1.3   Signals for Exercise E1.1.

$$f(t) = \sum_{n=1}^{\infty} C_n \cos(\omega_n t + \theta_n)$$

where none of the two sinusoids have identical frequencies, then

$$P_f = \frac{1}{2} \sum_{n=1}^{\infty} C_n{}^2 \tag{1.5c}$$

(c) In this case the signal is complex, and we use Eq. (1.4) to compute the power.

$$P_f = \lim_{T \to \infty} \frac{1}{T} \int_{-T/2}^{T/2} |D e^{j\omega_0 t}|^2 \, dt$$

Recall that $|e^{j\omega_0 t}| = 1$ so that $|D e^{j\omega_0 t}|^2 = |D|^2$, and

$$P_f = |D|^2 \tag{1.5d}$$

The rms value is $|D|$.   ∎

Comment: In part **(b)** we have shown that the power of the sum of two sinusoids is equal to the sum of the powers of the sinusoids. It appears that the power of $f_1(t) + f_2(t)$ is $P_{f_1} + P_{f_2}$. Unfortunately, this conclusion is not true in general. It is true only under a certain condition (orthogonality) discussed later in Sec. 3.1-3.

△ **Exercise E1.1**

    Show that the energies of the signals in Figs. 1.3a,b,c and d are 4, 1, 4/3, and 4/3, respectively. Observe that doubling a signal quadruples the energy, and time-shifting a signal has no effect on the energy. Show also that the power of the signal in Fig. 1.3e is 0.4323. What is the rms value of signal in Fig. 1.3e?   ▽

△ **Exercise E1.2**

    Redo Example 1.2a to find the power of a sinusoid $C \cos(\omega_0 t + \theta)$ by averaging the signal energy over one period $T_0 = 2\pi/\omega_0$ (rather than averaging over the infinitely large interval). Show also that the power of a constant signal $f(t) = C_0$ is C_0^2, and its rms value is C_0.   ▽

△ **Exercise E1.3**

    Show that if $\omega_1 = \omega_2$, the power of $f(t) = C_1 \cos(\omega_1 t + \theta_1) + C_2 \cos(\omega_2 t + \theta_2)$ is $[C_1{}^2 + C_2{}^2 + 2C_1 C_2 \cos(\theta_1 - \theta_2)]/2$, which is not equal to $(C_1{}^2 + C_2{}^2)/2$.   ▽

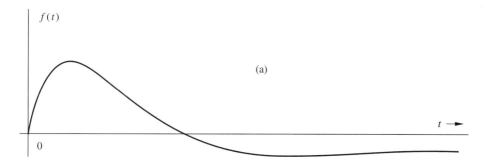

(a)

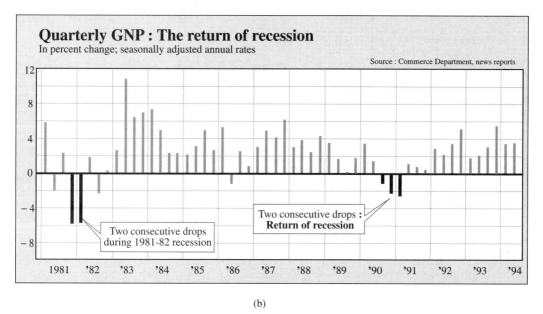

(b)

Fig. 1.4 Continuous-time and Discrete-time Signals.

1.2 Classification of Signals

There are several classes of signals. Here we shall consider only the following classes, which are suitable for the scope of this book:

1. Continuous-time and discrete-time signals
2. Analog and digital signals
3. Periodic and aperiodic signals
4. Energy and power signals
5. Deterministic and probabilistic signals

1.2-1 Continuous-Time and Discrete-Time Signals

A signal that is specified for every value of time t (Fig. 1.4a) is a **continuous-time signal**, and a signal that is specified only at discrete values of t (Fig. 1.4b) is a **discrete-time signal**. Telephone and video camera outputs are continuous-time

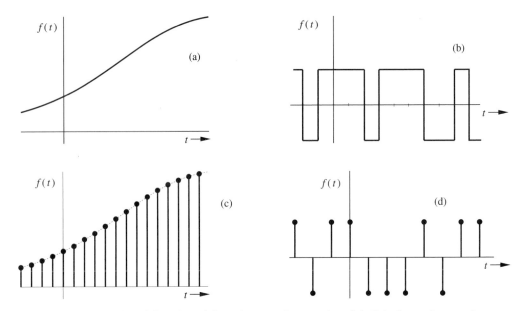

Fig. 1.5 Examples of Signals: (a) analog, continuous-time (b) digital, continuous-time (c) analog, discrete-time (d) digital, discrete-time.

signals, whereas the quarterly gross national product (GNP), monthly sales of a corporation, and stock market daily averages are discrete-time signals.

1.2-2 Analog and Digital Signals

The concept of continuous-time is often confused with that of analog. The two are not the same. The same is true of the concepts of discrete-time and digital. A signal whose amplitude can take on any value in a continuous range is an **analog signal**. This means that an analog signal amplitude can take on an infinite number of values. A **digital signal**, on the other hand, is one whose amplitude can take on only a finite number of values. Signals associated with a digital computer are digital because they take on only two values (binary signals). A digital signal whose amplitudes can take on M values is an **M-ary** signal of which binary ($M = 2$) is a special case. The terms *continuous-time* and *discrete-time* qualify the nature of a signal along the time (horizontal) axis. The terms *analog* and *digital*, on the other hand, qualify the nature of the signal amplitude (vertical axis). Figure 1.5 shows examples of various types of signals. It is clear that analog is not necessarily continuous-time and digital need not be discrete-time. Figure 1.5c shows an example of an analog discrete-time signal. An analog signal can be converted into a digital signal [analog-to-digital (A/D) conversion] through quantization (rounding off), as explained in Sec. 5.1-3.

1.2-3 Periodic and Aperiodic Signals

A signal $f(t)$ is said to be **periodic** if for some positive constant T_0

$$f(t) = f(t + T_0) \qquad \text{for all } t \qquad (1.6)$$

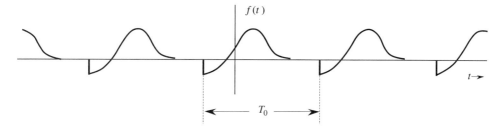

Fig. 1.6 A periodic signal of period T_0.

The *smallest* value of T_0 that satisfies the periodicity condition (1.6) is the **period** of $f(t)$. The signals in Figs. 1.2b and 1.3e are periodic signals with periods 2 and 1, respectively. A signal is **aperiodic** if it is not periodic. Signals in Figs. 1.2a, 1.3a, 1.3b, 1.3c, and 1.3d are all aperiodic.

By definition, a periodic signal $f(t)$ remains unchanged when time-shifted by one period. For this reason a periodic signal must start at $t = -\infty$ because if it starts at some finite instant, say $t = 0$, the time-shifted signal $f(t + T_0)$ will start at $t = -T_0$ and $f(t + T_0)$ would not be the same as $f(t)$. Therefore a *periodic signal, by definition, must start at $t = -\infty$ and continuing forever, as illustrated in Fig. 1.6.*

Another important property of a periodic signal $f(t)$ is that $f(t)$ can be generated by **periodic extension** of any segment of $f(t)$ of duration T_0 (the period). As a result we can generate $f(t)$ from any segment of $f(t)$ with a duration of one period by placing this segment and the reproduction thereof end to end ad infinitum on either side. Figure 1.7 shows a periodic signal $f(t)$ of period $T_0 = 6$. The shaded portion of Fig. 1.7a shows a segment of $f(t)$ starting at $t = -1$ and having a duration of one period (6 seconds). This segment, when repeated forever in either direction, results in the periodic signal $f(t)$. Figure 1.7b shows another shaded segment of $f(t)$ of duration T_0 starting at $t = 0$. Again we see that this segment, when repeated forever on either side, results in $f(t)$. The reader can verify that this construction is possible with any segment of $f(t)$ starting at any instant as long as the segment duration is one period.

It is helpful to label signals that start at $t = -\infty$ and continue for ever as everlasting signals. Thus, an everlasting signal exists over the entire interval $-\infty < t < \infty$. The signals in Figs. 1.1b and 1.2b are examples of everlasting signals. Clearly, a periodic signal, by definition, is an everlasting signal.

A signal that does not start before $t = 0$ is a **causal** signal. In other words, $f(t)$ is a causal signal if

$$f(t) = 0 \qquad t < 0 \qquad\qquad (1.7)$$

Signals in Figs. 1.3a, b, c, as well as in Figs. 1.9a and 1.9b are causal signals. A signal that starts before $t = 0$ is a **noncausal** signal. All the signals in Figs. 1.1 and 1.2 are noncausal. Observe that an everlasting signal is always noncausal but a noncausal signal is not necessarily everlasting. The everlasting signal in Fig. 1.2b is noncausal; however, the noncausal signal in Fig. 1.2a is not everlasting. A signal that is zero for all $t \geq 0$ is called an **anticausal** signal.

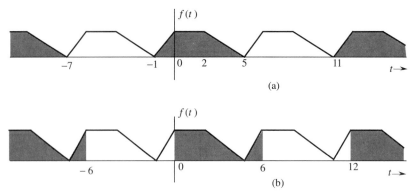

Fig. 1.7 Generation of a periodic signal by periodic extension of its segment of one-period duration.

Comment:

A true everlasting signal cannot be generated in practice for obvious reasons. Why should we bother to postulate such a signal? In later chapters we shall see that certain signals (including an everlasting sinusoid) which cannot be generated in practice *do* serve a very useful purpose in the study of signals and systems.

1.2-4 Energy and Power Signals

A signal with finite energy is an **energy signal**, and a signal with finite and nonzero power is a **power signal**. Signals in Fig. 1.2a and 1.2b are examples of energy and power signals, respectively. Observe that power is the time average of energy. Since the averaging is over an infinitely large interval, a signal with finite energy has zero power, and a signal with finite power has infinite energy. Therefore, a signal cannot both be an energy and a power signal. If it is one, it cannot be the other. On the other hand, there are signals that are neither energy nor power signals. The ramp signal is such an example.

Comments

All practical signals have finite energies and are therefore energy signals. A power signal must necessarily have infinite duration; otherwise its power, which is its energy averaged over an infinitely large interval, will not approach a (nonzero) limit. Clearly, it is impossible to generate a true power signal in practice because such a signal has infinite duration and infinite energy.

Also, because of periodic repetition, periodic signals for which the area under $|f(t)|^2$ over one period is finite are power signals; however, not all power signals are periodic.

△ **Exercise E1.4**

Show that an everlasting exponential e^{-at} is neither an energy nor a power signal for any real value of a. However, if a is imaginary, it is a power signal with power $P_f = 1$ regardless of the value of a. ▽

1.2-5 Deterministic and Random Signals

A signal whose physical description is known completely, either in a mathematical form or a graphical form, is a **deterministic signal**. A signal whose values

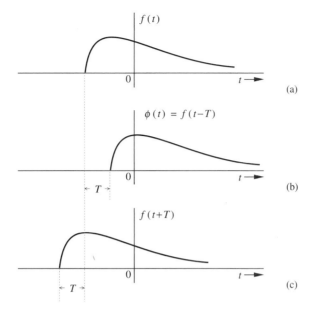

Fig. 1.8 Time shifting a signal.

cannot be predicted precisely but are known only in terms of probabilistic description, such as mean value, mean squared value, and so on is a **random signal**. In this book we shall exclusively deal with deterministic signals. Random signals are beyond the scope of this study.

1.3 Some Useful Signal Operations

We discuss here three useful signal operations: shifting, scaling, and inversion. Since the independent variable in our signal description is time, these operations are discussed as time shifting, time scaling, and time inversion (or folding). However, this discussion is valid for functions having independent variables other than time (e.g., frequency or distance).

1.3-1 Time Shifting

Consider a signal $f(t)$ (Fig. 1.8a) and the same signal delayed by T seconds (Fig. 1.8b), which we shall denote by $\phi(t)$. Whatever happens in $f(t)$ (Fig. 1.8a) at some instant t also happens in $\phi(t)$ (Fig. 1.8b) T seconds later at the instant $t+T$. Therefore

$$\phi(t+T) = f(t) \tag{1.8}$$

and

$$\phi(t) = f(t-T) \tag{1.9}$$

Therefore, to time-shift a signal by T, we replace t with $t - T$. Thus $f(t - T)$ represents $f(t)$ time-shifted by T seconds. If T is positive, the shift is to the right (delay). If T is negative, the shift is to the left (advance). Thus, $f(t - 2)$ is $f(t)$ delayed (right-shifted) by 2 seconds, and $f(t + 2)$ is $f(t)$ advanced (left-shifted) by 2 seconds.

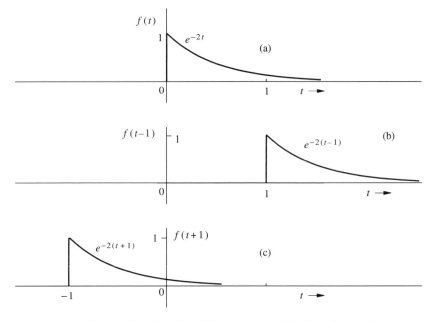

Fig. 1.9 (a) signal $f(t)$ (b) $f(t)$ delayed by 1 second (c) $f(t)$ advanced by 1 second.

■ **Example 1.3**

An exponential function $f(t) = e^{-2t}$ shown in Fig. 1.9a is delayed by 1 second. Sketch and mathematically describe the delayed function. Repeat the problem if $f(t)$ is advanced by 1 second.

The function $f(t)$ can be described mathematically as

$$f(t) = \begin{cases} e^{-2t} & t \geq 0 \\ 0 & t < 0 \end{cases} \tag{1.10}$$

Let $f_d(t)$ represent the function $f(t)$ delayed (right-shifted) by 1 second as illustrated in Fig. 1.9b. This function is $f(t - 1)$; its mathematical description can be obtained from $f(t)$ by replacing t with $t - 1$ in Eq. (1.10). Thus

$$f_d(t) = f(t-1) = \begin{cases} e^{-2(t-1)} & t-1 \geq 0 \quad \text{or} \quad t \geq 1 \\ 0 & t-1 < 0 \quad \text{or} \quad t < 1 \end{cases} \tag{1.11}$$

Let $f_a(t)$ represent the function $f(t)$ advanced (left-shifted) by 1 second as depicted in Fig. 1.9c. This function is $f(t + 1)$; its mathematical description can be obtained from $f(t)$ by replacing t with $t + 1$ in Eq. (1.10). Thus

$$f_a(t) = f(t+1) = \begin{cases} e^{-2(t+1)} & t+1 \geq 0 \quad \text{or} \quad t \geq -1 \\ 0 & t+1 < 0 \quad \text{or} \quad t < -1 \end{cases} \tag{1.12}$$

■

△ **Exercise E1.5**

Write a mathematical description of the signal $f_3(t)$ in Fig. 1.3c. This signal is delayed by 2 seconds. Sketch the delayed signal. Show that this delayed signal $f_d(t)$ can be described mathematically as $f_d(t) = 2(t - 2)$ for $2 \leq t \leq 3$, and equal to 0 otherwise. Now repeat the

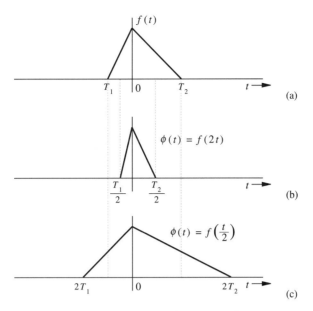

Fig. 1.10 Time scaling a signal.

procedure if the signal is advanced (left-shifted) by 1 second. Show that this advanced signal $f_a(t)$ can be described as $f_a(t) = 2(t+1)$ for $-1 \leq t \leq 0$, and equal to 0 otherwise. $\triangledown$

1.3-2   Time Scaling

The compression or expansion of a signal in time is known as **time scaling**. Consider the signal $f(t)$ of Fig. 1.10a. The signal $\phi(t)$ in Fig. 1.10b is $f(t)$ compressed in time by a factor of 2. Therefore, whatever happens in $f(t)$ at some instant t also happens to $\phi(t)$ at the instant $t/2$, so that

$$\phi\left(\tfrac{t}{2}\right) = f(t) \tag{1.13}$$

and

$$\phi(t) = f(2t) \tag{1.14}$$

Observe that because $f(t) = 0$ at $t = T_1$ and T_2, we must have $\phi(t) = 0$ at $t = T_1/2$ and $T_2/2$, as shown in Fig. 1.10b. If $f(t)$ were recorded on a tape and played back at twice the normal recording speed, we would obtain $f(2t)$. In general, if $f(t)$ is compressed in time by a factor a $(a > 1)$, the resulting signal $\phi(t)$ is given by

$$\phi(t) = f(at) \tag{1.15}$$

Using a similar argument, we can show that $f(t)$ expanded (slowed down) in time by a factor a $(a > 1)$ is given by

$$\phi(t) = f\left(\tfrac{t}{a}\right) \tag{1.16}$$

Figure 1.10c shows $f(\tfrac{t}{2})$, which is $f(t)$ expanded in time by a factor of 2. Observe that in time scaling operation, the origin $t = 0$ is the anchor point, which remains unchanged under scaling operation because at $t = 0$, $f(t) = f(at) = f(0)$.

In summary, to time-scale a signal by a factor a, we replace t with at. If $a > 1$, the scaling results in compression, and if $a < 1$, the scaling results in expansion.

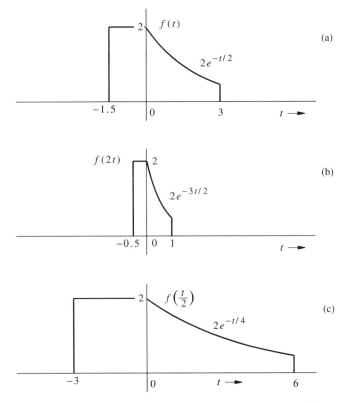

Fig. 1.11 (a) signal $f(t)$ (b) signal $f(3t)$ (c) signal $f(\frac{t}{2})$.

■ **Example 1.4**

Figure 1.11a shows a signal $f(t)$. Sketch and describe mathematically this signal time-compressed by factor 3. Repeat the problem for the same signal time-expanded by factor 2.

The signal $f(t)$ can be described as

$$f(t) = \begin{cases} 2 & -1.5 \leq t < 0 \\ 2\,e^{-t/2} & 0 \leq t < 3 \\ 0 & \text{otherwise} \end{cases} \tag{1.17}$$

Figure 1.11b shows $f_c(t)$, which is $f(t)$ time-compressed by factor 3; consequently, it can be described mathematically as $f(3t)$, which is obtained by replacing t with $3t$ in the right-hand side of Eq. 1.17. Thus

$$f_c(t) = f(3t) = \begin{cases} 2 & -1.5 \leq 3t < 0 \quad \text{or} \quad -0.5 \leq t < 0 \\ 2\,e^{-3t/2} & 0 \leq 3t < 3 \quad \text{or} \quad 0 \leq t < 1 \\ 0 & \text{otherwise} \end{cases} \tag{1.18a}$$

Observe that the instants $t = -1.5$ and 3 in $f(t)$ correspond to the instants $t = -0.5$, and 1 in the compressed signal $f(3t)$.

Figure 1.11c shows $f_e(t)$, which is $f(t)$ time-expanded by factor 2; consequently, it can be described mathematically as $f(t/2)$, which is obtained by replacing t with $t/2$ in $f(t)$. Thus

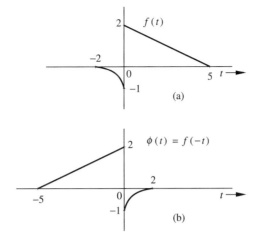

Fig. 1.12 Time inversion (reflection) of a signal.

$$f_e(t) = f\left(\frac{t}{2}\right) = \begin{cases} 2 & -1.5 \leq \frac{t}{2} < 0 \quad \text{or} \quad -3 \leq t < 0 \\ 2\,e^{-t/4} & 0 \leq \frac{t}{2} < 3 \quad \text{or} \quad 0 \leq t < 6 \\ 0 & \text{otherwise} \end{cases} \qquad (1.18b)$$

Observe that the instants $t = -1.5$ and 3 in $f(t)$ correspond to the instants $t = -3$ and 6 in the expanded signal $f(\frac{t}{2})$. ∎

△ **Exercise E1.6**

Show that the time-compression by a factor n $(n > 1)$ of a sinusoid results in a sinusoid of the same amplitude and phase, but with the frequency increased n-fold. Similarly the time expansion by a factor n $(n > 1)$ of a sinusoid results in a sinusoid of the same amplitude and phase, but with the frequency reduced by a factor n. Verify your conclusion by sketching a sinusoid $\sin 2t$ and the same sinusoid compressed by a factor 3 and expanded by a factor 2. ▽

1.3-3 Time Inversion (Time Reversal)

Consider the signal $f(t)$ in Fig. 1.12a. We can view $f(t)$ as a rigid wire frame hinged at the vertical axis. To time-invert $f(t)$, we rotate this frame 180° about the vertical axis. This time inversion or folding [the reflection of $f(t)$ about the vertical axis] gives us the signal $\phi(t)$ (Fig. 1.12b). Observe that whatever happens in Fig. 1.12a at some instant t also happens in Fig. 1.12b at the instant $-t$. Therefore

$$\phi(-t) = f(t)$$

and

$$\phi(t) = f(-t) \qquad (1.19)$$

Therefore, to time-invert a signal we replace t with $-t$. Thus, the time inversion of signal $f(t)$ yields $f(-t)$. Consequently, the mirror image of $f(t)$ about the vertical axis is $f(-t)$. Recall also that the mirror image of $f(t)$ about the horizontal axis is $-f(t)$.

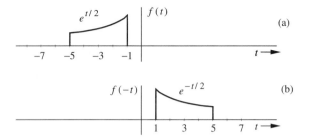

Fig. 1.13 An example of time inversion.

■ **Example 1.5**

For the signal $f(t)$ illustrated in Fig. 1.13a, sketch $f(-t)$, which is time inverted $f(t)$.

The instants -1 and -5 in $f(t)$ are mapped into instants 1 and 5 in $f(-t)$. Because $f(t) = e^{t/2}$, we have $f(-t) = e^{-t/2}$. The signal $f(-t)$ is depicted in Fig. 1.13b. We can describe $f(t)$ and $f(-t)$ as

$$f(t) = \begin{cases} e^{t/2} & -1 \geq t > -5 \\ 0 & \text{otherwise} \end{cases}$$

and its time inverted version $f(-t)$ is obtained by replacing t with $-t$ in $f(t)$ as

$$f(-t) = \begin{cases} e^{-t/2} & -1 \geq -t > -5 \quad \text{or} \quad 1 \leq t < 5 \\ 0 & \text{otherwise} \end{cases} \quad ■$$

1.3-4 Combined Operations

Certain complex operations require simultaneous use of more than one of the above operations. The most general operation involving all the three operations is $f(at - b)$, which is realized in two possible sequences of operation:

1. Time-shift $f(t)$ by b to obtain $f(t-b)$. Now time-scale the shifted signal $f(t-b)$ by a (that is, replace t with at) to obtain $f(at - b)$.

2. Time-scale $f(t)$ by a to obtain $f(at)$. Now time-shift $f(at)$ by $\frac{b}{a}$ (that is, replace t with $(t - \frac{b}{a})$ to obtain $f[a(t - \frac{b}{a})] = f(at - b)$. In either case, if a is negative, time scaling involves time inversion.

For instance, the signal $f(2t - 6)$ can be obtained in two ways: first, delay $f(t)$ by 6 to obtain $f(t - 6)$ and then time-compress this signal by factor 2 (replace t with $2t$) to obtain $f(2t - 6)$. Alternately, we first time-compress $f(t)$ by factor 2 to obtain $f(2t)$, then delay this signal by 3 (replace t with $t - 3$) to obtain $f(2t - 6)$.

1.4 Some Useful Signal Models

In the area of signals and systems, the step, the impulse, and the exponential functions are very useful. They not only serve as a basis for representing other signals, but their use can simplify many aspects of the signals and systems.

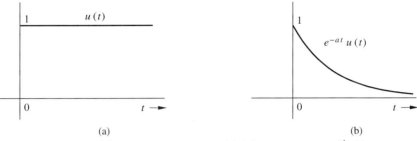

Fig. 1.14 (a) Unit step function $u(t)$ (b) exponential $e^{-at}u(t)$.

1. Unit Step Function u(t)

In much of our discussion, the signals begin at $t = 0$ (causal signals). Such signals can be conveniently described in terms of unit step function $u(t)$ shown in Fig. 1.14a. This function is defined by

$$u(t) = \begin{cases} 1 & t \geq 0 \\ 0 & t < 0 \end{cases} \tag{1.20}$$

If we want a signal to start at $t = 0$ (so that it has a value of zero for $t < 0$), we only need to multiply the signal with $u(t)$. For instance, the signal e^{-at} represents an everlasting exponential that starts at $t = -\infty$. The causal form of this exponential illustrated in Fig. 1.14b can be described as $e^{-at}u(t)$.

The unit step function also proves very useful in specifying a function with different mathematical descriptions over different intervals. Examples of such functions appear in Fig. 1.11. These functions have different mathematical descriptions over different segments of time as seen from Eqs. (1.17), (1.18a), and (1.18b). Such a description often proves clumsy and inconvenient in mathematical treatment. Using the unit step function, we can describe such functions by a single expression that is valid for all t.

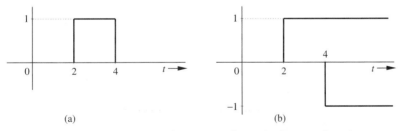

Fig. 1.15 Representation of a rectangular pulse by step functions.

Consider, for example, the rectangular pulse depicted in Fig. 1.15a. We can express such a pulse in terms of familiar step functions by observing that the pulse $f(t)$ can be expressed as the sum of the two delayed unit step functions as shown in Fig. 1.15b. The unit step function $u(t)$ delayed by T seconds is $u(t - T)$. From Fig. 1.15b, it is clear that

$$f(t) = u(t - 2) - u(t - 4)$$

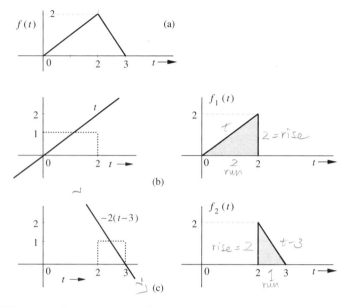

Fig. 1.16 Representation of a signal defined interval by interval.

■ **Example 1.6**

Describe the signal in Fig. 1.16a.

The signal illustrated in Fig. 1.16a can be conveniently handled by breaking it up into the two components $f_1(t)$ and $f_2(t)$, depicted in Figs. 1.16b and 1.16c respectively. Here, $f_1(t)$ can be obtained by multiplying the ramp t by the gate pulse $u(t) - u(t - 2)$, as shown in Fig. 1.16b. Therefore

$$f_1(t) = t\,[u(t) - u(t - 2)]$$

The signal $f_2(t)$ can be obtained by multiplying another ramp by the gate pulse illustrated in Fig. 1.16c. This ramp has a slope -2; hence it can be described by $-2t+c$. Now, because the ramp has a zero value at $t = 3$, the constant $c = 6$, and the ramp can be described by $-2(t - 3)$. Also, the gate pulse in Fig. 1.16c is $u(t - 2) - u(t - 3)$. Therefore

$$f_2(t) = -2(t - 3)\,[u(t - 2) - u(t - 3)]$$

and

$$f(t) = f_1(t) + f_2(t)$$
$$= t\,[u(t) - u(t - 2)] - 2(t - 3)\,[u(t - 2) - u(t - 3)]$$
$$= tu(t) - 3(t - 2)u(t - 2) + 2(t - 3)u(t - 3)\quad ■$$

decreasing

$-tu(t-2) - 2(t-3)u(t-2)$
$[-t - 2(t-3)]u(t-2)$
$[-t - 2t + 6]u(t-2)$
$(-3t+6)\,u(t-2)$
$-3(t-2)u(t-2)$

■ **Example 1.7**

Describe the signal in Fig. 1.11a by a single expression valid for all t.

Over the interval from -1.5 to 0, the signal can be described by a constant 2, and over the interval from 0 to 3, it can be described by $2\,e^{-t/2}$. Therefore

$$f(t) = \underbrace{2[u(t + 1.5) - u(t)]}_{f_1(t)} + \underbrace{2e^{-t/2}[u(t) - u(t - 3)]}_{f_2(t)}$$

$$= 2u(t + 1.5) - 2(1 - e^{-t/2})u(t) - 2e^{-t/2}u(t - 3)$$

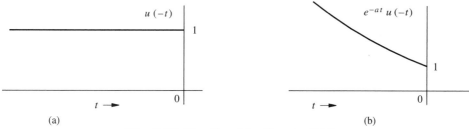

Fig. 1.17 The Signal for Exercise E1.7.

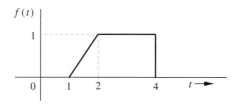

Fig. 1.18 The signal for Exercise E1.8.

Compare this expression with the expression for the same function found in Eq. 1.17.   ■

△ **Exercise E1.7**

Show that the signals depicted in Figs. 1.17a and 1.17b can be described as $u(-t)$, and $e^{-at}u(-t)$, respectively.   ▽

△ **Exercise E1.8**

Show that the signal shown in Fig. 1.18 can be described as

$$f(t) = (t-1)u(t-1) - (t-2)u(t-2) - u(t-4)$$ ▽

2.   The Unit Impulse Function $\delta(t)$

The unit impulse function $\delta(t)$ is one of the most important functions in the study of signals and systems. This function was first defined by P. A. M Dirac as

$$\delta(t) = 0 \qquad t \neq 0$$

$$\int_{-\infty}^{\infty} \delta(t)\, dt = 1 \tag{1.21}$$

We can visualize an impulse as a tall, narrow rectangular pulse of unit area, as illustrated in Fig. 1.19b. The width of this rectangular pulse is a very small value $\epsilon \to 0$. Consequently, its height is a very large value $1/\epsilon$. The unit impulse therefore can be regarded as a rectangular pulse with a width that has become infinitesimally small, a height that has become infinitely large, and an overall area that has been maintained at unity. Thus $\delta(t) = 0$ everywhere except at $t = 0$, where it is undefined. For this reason a unit impulse is represented by the spear-like symbol in Fig. 1.19a.

Other pulses, such as exponential pulse, triangular pulse, or Gaussian pulse may also be used in impulse approximation. The important feature of the unit impulse function is not its shape but the fact that its effective duration (pulse width)

Fig. 1.19 A unit impulse and its approximation.

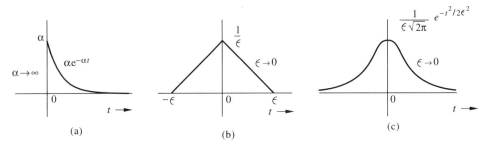

Fig. 1.20 Other possible approximations to a unit impulse.

approaches zero while its area remains at unity. For example, the exponential pulse $\alpha e^{-\alpha t} u(t)$ in Fig. 1.20a becomes taller and narrower as α increases. In the limit as $\alpha \to \infty$, the pulse height $\to \infty$, and its width or duration $\to 0$. Yet, the area under the pulse is unity regardless of the value of α because

$$\int_0^\infty \alpha e^{-\alpha t}\, dt = 1 \tag{1.22}$$

The pulses in Figs. 1.20b and 1.20c behave in a similar fashion.

From Eq. (1.21), it follows that the function $k\delta(t) = 0$ for all $t \neq 0$, and its area is k. Thus, $k\delta(t)$ is an impulse function whose area is k (in contrast to the unit impulse function, whose area is 1).

Multiplication of a Function by an Impulse

Let us now consider what happens when we multiply the unit impulse $\delta(t)$ by a function $\phi(t)$ that is known to be continuous at $t = 0$. Since the impulse exists only at $t = 0$, and the value of $\phi(t)$ at $t = 0$ is $\phi(0)$, we obtain

$$\phi(t)\delta(t) = \phi(0)\delta(t) \tag{1.23a}$$

Similarly, if $\phi(t)$ is multiplied by an impulse $\delta(t - T)$ (impulse located at $t = T$), then

$$\phi(t)\delta(t - T) = \phi(T)\delta(t - T) \tag{1.23b}$$

provided $\phi(t)$ is continuous at $t = T$.

Sampling Property of the Unit Impulse Function

From Eq. (1.23a) it follows that

$$\int_{-\infty}^{\infty} \phi(t)\delta(t)\,dt = \phi(0) \int_{-\infty}^{\infty} \delta(t)\,dt$$

$$= \phi(0) \tag{1.24a}$$

provided $\phi(t)$ is continuous at $t = 0$. This result means that *the area under the product of a function with an impulse $\delta(t)$ is equal to the value of that function at the instant where the unit impulse is located.* This property is very important and useful, and is known as the **sampling** or **sifting property** of the unit impulse.

From Eq. (1.23b) it follows that

$$\int_{-\infty}^{\infty} \phi(t)\delta(t - T)\,dt = \phi(T) \tag{1.24b}$$

Equation (1.24b) is just another form of sampling or sifting property. In the case of Eq. (1.24b), the impulse $\delta(t - T)$ is located at $t = T$. Therefore, the area under $\phi(t)\delta(t - T)$ is $\phi(T)$, the value of $\phi(t)$ at the instant where the impulse is located (at $t = T$). In these derivations we have assumed that the function is continuous at the instant where the impulse is located.

Unit Impulse as a Generalized Function

The definition of the unit impulse function given in Eq. (1.21) is not mathematically rigorous, which leads to serious difficulties. First, the impulse function does not define a unique function: for example, it can be shown that $\delta(t) + \dot{\delta}(t)$ also satisfies Eq. (1.21).[1] Moreover, $\delta(t)$ is not even a true function in the ordinary sense. An ordinary function is specified by its values for all time t. The impulse function is zero everywhere except at $t = 0$, and at this only interesting part of its range it is undefined. These difficulties are resolved by defining the impulse as a generalized function rather than an ordinary function. A **generalized function** is defined by its effect on other functions instead of by its value at every instant of time.

In this approach the impulse function is defined by the sampling property [Eq. (1.24)]. We say nothing about what the impulse function is or what it looks like. Instead, the impulse function is defined in terms of its effect on a test function $\phi(t)$. We define a unit impulse as a function for which the area under its product with a function $\phi(t)$ is equal to the value of the function $\phi(t)$ at the instant where the impulse is located. It is assumed that $\phi(t)$ is continuous at the location of the impulse. Therefore, either Eq. (1.24a) or (1.24b) can serve as a definition of the impulse function in this approach. Recall that the sampling property [Eq. (1.24)] is the consequence of the classical (Dirac) definition of impulse in Eq. (1.21). In contrast, *the sampling property [Eq. (1.24)] defines the impulse function in the generalized function approach.*

We now present an interesting application of the generalized function definition of an impulse. Because the unit step function $u(t)$ is discontinuous at $t = 0$, its derivative du/dt does not exist at $t = 0$ in the ordinary sense. We now show that this derivative *does* exist in the generalized sense, and it is, in fact, $\delta(t)$. As a proof, let us evaluate the integral of $(du/dt)\phi(t)$, using integration by parts:

$$\int_{-\infty}^{\infty} \frac{du}{dt}\phi(t)\,dt = u(t)\phi(t)\Big|_{-\infty}^{\infty} - \int_{-\infty}^{\infty} u(t)\dot{\phi}(t)\,dt \qquad (1.25)$$

$$= \phi(\infty) - 0 - \int_{0}^{\infty} \dot{\phi}(t)\,dt$$

$$= \phi(\infty) - \phi(t)\big|_{0}^{\infty}$$

$$= \phi(0) \qquad (1.26)$$

This result shows that du/dt satisfies the sampling property of $\delta(t)$. Therefore it is an impulse $\delta(t)$ in the generalized sense—that is,

$$\frac{du}{dt} = \delta(t) \qquad (1.27)$$

Consequently

$$\int_{-\infty}^{t} \delta(\tau)\,d\tau = u(t) \qquad (1.28)$$

These results can also be obtained graphically from Fig. 1.19b. We observe that the area from $-\infty$ to t under the limiting form of $\delta(t)$ in Fig. 1.19b is zero if $t < 0$ and unity if $t \geq 0$. Consequently

$$\int_{-\infty}^{t} \delta(\tau)\,d\tau = \begin{cases} 0 & t < 0 \\ 1 & t \geq 0 \end{cases}$$

$$= u(t) \qquad (1.29)$$

Derivatives of impulse function can also be defined as generalized functions (see Prob. 1.4-10).

△   **Exercise E1.9**
    Show that

(a) $(t^3 + 3)\delta(t) = 3\delta(t)$ (b) $\left[\sin\left(t^2 - \frac{\pi}{2}\right)\right]\delta(t) = -\delta(t)$

(c) $e^{-2t}\delta(t) = \delta(t)$ (d) $\dfrac{\omega^2 + 1}{\omega^2 + 9}\delta(\omega - 1) = \dfrac{1}{5}\delta(\omega - 1)$

Hint: Use Eqs. (1.23).   ▽

△   **Exercise E1.10**
    Show that

(a) $\displaystyle\int_{-\infty}^{\infty} \delta(t)e^{-j\omega t}\,dt = 1$

(b) $\displaystyle\int_{-\infty}^{\infty} \delta(t - 2)\cos\left(\frac{\pi t}{4}\right)\,dt = 0$

(c) $\displaystyle\int_{-\infty}^{\infty} e^{-2(x-t)}\delta(2 - t)\,dt = e^{-2(x-2)}$

Hint: In part **c** recall that $\delta(x)$ is located at $x = 0$. Therefore $\delta(2 - t)$ is located at $2 - t = 0$; that is at $t = 2$.   ▽

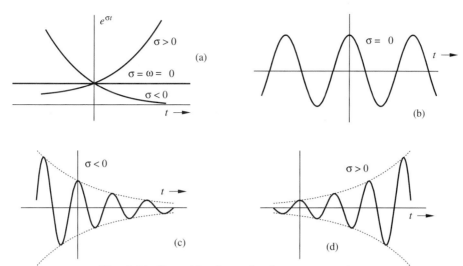

Fig. 1.21 Sinusoids of complex frequency $\sigma + j\omega$.

3. The Exponential Function e^{st}

One of the most important functions in the area of signals and systems is the exponential signal e^{st}, where s is complex in general, given by

$$s = \sigma + j\omega$$

Therefore

$$e^{st} = e^{(\sigma + j\omega)t} = e^{\sigma t}e^{j\omega t} = e^{\sigma t}(\cos \omega t + j \sin \omega t) \qquad (1.30a)$$

If $s^* = \sigma - j\omega$ (the conjugate of s), then

$$e^{s^* t} = e^{\sigma - j\omega} = e^{\sigma t}e^{-j\omega t} = e^{\sigma t}(\cos \omega t - j \sin \omega t) \qquad (1.30b)$$

and

$$e^{\sigma t} \cos \omega t = \frac{1}{2}(e^{st} + e^{s^* t}) \qquad (1.30c)$$

Comparison of this equation with Euler's formula shows that e^{st} is a generalization of the function $e^{j\omega t}$, where the frequency variable $j\omega$ is generalized to a complex variable $s = \sigma + j\omega$. For this reason we designate the variable s as the **complex frequency**. From Eqs. (1.30) it follows that the function e^{st} encompasses a large class of functions. The following functions are special cases of e^{st}:

1 A constant $k = ke^{0t}$ $(s = 0)$
2 A monotonic exponential $e^{\sigma t}$ $(\omega = 0,\ s = \sigma)$
3 A sinusoid $\cos \omega t$ $(\sigma = 0,\ s = \pm j\omega)$
4 An exponentially varying sinusoid $e^{\sigma t} \cos \omega t$ $(s = \sigma \pm j\omega)$

These functions are illustrated in Fig. 1.21.

The complex frequency s can be conveniently represented on a **complex frequency plane** (s plane) as depicted in Fig. 1.22. The horizontal axis is the real axis (σ axis), and the vertical axis is the imaginary axis ($j\omega$ axis). The absolute value of the imaginary part of s is $|\omega|$ (the *radian* frequency), which indicates the frequency

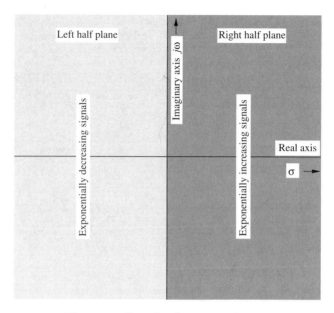

Fig. 1.22 Complex frequency plane.

of oscillation of e^{st}; the real part σ (the *neper* frequency) gives information about the rate of increase or decrease of the amplitude of e^{st}. For signals whose complex frequencies lie on the real axis (σ-axis, where $\omega = 0$), the frequency of oscillation is zero. Consequently these signals are monotonically increasing or decreasing exponentials (Fig. 1.21a). For signals whose frequencies lie on the imaginary axis ($j\omega$ axis where $\sigma = 0$), $e^{\sigma t} = 1$. Therefore, these signals are conventional sinusoids with constant amplitude (Fig. 1.21b). The case $s = 0$ ($\sigma = \omega = 0$) corresponds to a constant (dc) signal because $e^{0t} = 1$. For the signals illustrated in Figs. 1.21c and 1.21d, both σ and ω are nonzero; the frequency s is complex and does not lie on either axis. The signal in Fig. 1.21c decays exponentially. Therefore, σ is negative, and s lies to the left of the imaginary axis. In contrast, the signal in Fig. 1.21d *grows* exponentially. Therefore, σ is positive, and s lies to the right of the imaginary axis. Thus the s-plane (Fig. 1.21) can be differentiated into two parts: the **left half-plane** (LHP) corresponding to exponentially decaying signals and the **right half-plane** (RHP) corresponding to exponentially growing signals. The imaginary axis separates the two regions and corresponds to signals of constant amplitude.

An exponentially growing sinusoid $e^{2t}\cos{(5t + \theta)}$, for example, can be expressed as a sum of exponentials $e^{(2+j5)t}$ and $e^{(2-j5)t}$ with complex frequencies $2 + j5$ and $2 - j5$, respectively, which lie in the RHP. An exponentially decaying sinusoid $e^{-2t}\cos{(5t + \theta)}$ can be expressed as a sum of exponentials $e^{(-2+j5)t}$ and $e^{(-2-j5)t}$ with complex frequencies $-2 + j5$ and $-2 - j5$, respectively, which lie in the LHP. A constant amplitude sinusoid $\cos{(5t+\theta)}$ can be expressed as a sum of exponentials e^{j5t} and e^{-j5t} with complex frequencies $\pm j5$, which lie on the imaginary axis. Observe that the monotonic exponentials $e^{\pm 2t}$ are also generalized sinusoids with complex frequencies ± 2.

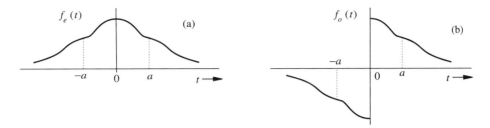

Fig. 1.23 An even and an odd function of t.

1.5 Even and Odd Functions

A function $f_e(t)$ is said to be an **even function** of t if

$$f_e(t) = f_e(-t) \tag{1.31}$$

and a function $f_o(t)$ is said to be an **odd function** of t if

$$f_o(t) = -f_o(-t) \tag{1.32}$$

An even function has the same value at the instants t and $-t$ for all values of t. Clearly, $f_e(t)$ is symmetrical about the vertical axis, as shown in Fig. 1.23a. On the other hand, the value of an odd function at the instant t is the negative of its value at the instant $-t$. Therefore, $f_o(t)$ is anti-symmetrical about the vertical axis, as depicted in Fig. 1.23b.

1.5-1 Some Properties of Even and Odd Functions

Even and odd functions have the following property:

even function $\times$ odd function = odd function

odd function $\times$ odd function = even function

even function $\times$ even function = even function

The proofs of these facts are trivial and follow directly from the definition of odd and even functions [Eqs. (1.31) and (1.32)].

Area

Because $f_e(t)$ is symmetrical about the vertical axis, it follows from Fig. 1.23a that

$$\int_{-a}^{a} f_e(t)\,dt = 2 \int_{0}^{a} f_e(t)\,dt \tag{1.33a}$$

It is also clear from Fig. 1.23b that

$$\int_{-a}^{a} f_o(t)\,dt = 0 \tag{1.33b}$$

These results can also be proved formally by using the definitions in Eqs. (1.31) and (1.32). We leave them as an exercise for the reader.

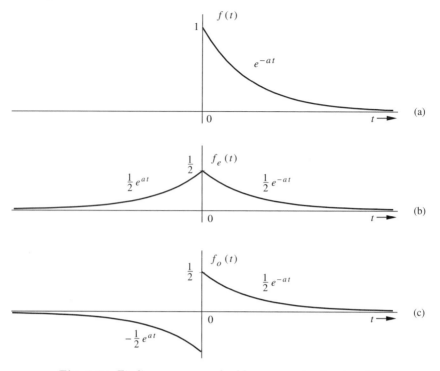

Fig. 1.24 Finding an even and odd components of a signal.

1.5-2   Even and Odd Components of a Signal

Every signal $f(t)$ can be expressed as a sum of even and odd components because

$$f(t) = \tfrac{1}{2}\underbrace{[f(t) + f(-t)]}_{\text{even}} + \tfrac{1}{2}\underbrace{[f(t) - f(-t)]}_{\text{odd}} \tag{1.34}$$

From the definitions in Eqs. (1.31) and (1.32), we can clearly see that the first component on the right-hand side is an even function, while the second component is odd. This is apparent from the fact that replacing t by $-t$ in the first component yields the same function. The same maneuver in the second component yields the negative of that component.

Consider the function

$$f(t) = e^{-at}u(t)$$

Expressing this function as a sum of the even and odd components $f_e(t)$ and $f_o(t)$, we obtain

$$f(t) = f_e(t) + f_o(t)$$

where [from Eq. (1.34)]

$$f_e(t) = \tfrac{1}{2}\left[e^{-at}u(t) + e^{at}u(-t)\right] \tag{1.35a}$$

and

$$f_o(t) = \tfrac{1}{2}\left[e^{-at}u(t) - e^{at}u(-t)\right] \tag{1.35b}$$

The function $e^{-at}u(t)$ and its even and odd components are illustrated in Fig. 1.24.

■ **Example 1.8**

Find the even and odd components of e^{jt}.

From Eq. (1.34)

$$e^{jt} = f_e(t) + f_o(t)$$

where

$$f_e(t) = \tfrac{1}{2}\left[e^{jt} + e^{-jt}\right] = \cos t$$

and

$$f_o(t) = \tfrac{1}{2}\left[e^{jt} - e^{-jt}\right] = j\sin t \quad ■$$

1.6 Systems

As mentioned in Sec. 1.1, systems are used to process signals in order to modify or to extract additional information from the signals. A system may consist of physical components (hardware realization) or may consist of an algorithm that computes the output signal from the input signal (software realization).

A system is characterized by its **inputs**, its **outputs** (or **responses**), and the **rules of operation** (or **laws**) adequate to describe its behavior. For example, in electrical systems, the laws of operation are the familiar voltage-current relationships for the resistors, capacitors, inductors, transformers, transistors, and so on, as well as the laws of interconnection (i.e., Kirchhoff's laws). Using these laws, we derive mathematical equations relating the outputs to the inputs. These equations then represent a **mathematical model** of the system. Thus a system is characterized by its inputs, its outputs, and its mathematical model.

A system can be conveniently illustrated by a "black box" with one set of accessible terminals where the input variables $f_1(t)$, $f_2(t)$, …, $f_j(t)$ are applied and another set of accessible terminals where the output variables $y_1(t)$, $y_2(t)$, …, $y_k(t)$ are observed. Note that the direction of the arrows for the variables in Fig. 1.25 is always from cause to effect.

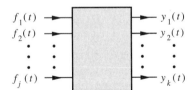

Fig. 1.25 Representation of a system.

The study of systems consists of three major areas: mathematical modeling, analysis, and design. Although we shall be dealing with mathematical modeling, our main concern is with analysis and design. The major portion of this book is devoted to the analysis problem—how to determine the system outputs for the given inputs and a given mathematical model of the system (or rules governing the system). To a lesser extent, we will also consider the problem of design or synthesis—how to construct a system which will produce a desired set of outputs for the given inputs.

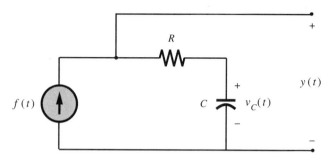

Fig. 1.26   An example of a simple electrical system.

Data Needed to Compute System Response

In order to understand what data we need to compute a system response, consider a simple RC circuit with a current source $f(t)$ as its input (Fig. 1.26). The output voltage $y(t)$ is given by

$$y(t) = Rf(t) + \frac{1}{C} \int_{-\infty}^{t} f(\tau)\, d\tau \qquad (1.36\text{a})$$

The limits of the integral on the right-hand side are from $-\infty$ to t because this integral represents the capacitor charge due to the current $f(t)$ flowing in the capacitor, and this charge is the result of the current flowing in the capacitor from $-\infty$. Now, Eq. (1.36a) can be expressed as

$$y(t) = Rf(t) + \frac{1}{C} \int_{-\infty}^{0} f(\tau)\, d\tau + \frac{1}{C} \int_{0}^{t} f(\tau)\, d\tau \qquad (1.36\text{b})$$

The middle term on the right-hand side is $v_C(0)$, the capacitor voltage at $t = 0$. Therefore

$$y(t) = v_C(0) + Rf(t) + \frac{1}{C} \int_{0}^{t} f(\tau)\, d\tau \qquad (1.36\text{c})$$

This equation can be readily generalized as

$$y(t) = v_C(t_0) + Rf(t) + \frac{1}{C} \int_{t_0}^{t} f(\tau)\, d\tau \qquad (1.36\text{d})$$

From Eq. (1.36a), the output voltage $y(t)$ at an instant t can be computed if we know the input current flowing in the capacitor throughout its entire past ($-\infty$ to t). Alternatively, if we know the input current $f(t)$ from some moment t_0 onward, then, using Eq. (1.36d), we can still calculate $y(t)$ for $t \geq t_0$ from a knowledge of the input current, provided we know $v_C(t_0)$, the initial capacitor voltage (voltage at t_0). Thus $v_C(t_0)$ contains all the relevant information about the circuit's entire past ($-\infty$ to t_0) that we need to compute $y(t)$ for $t \geq t_0$. Therefore, the response of a system at $t > t_0$ can be determined from its input(s) during the interval t_0 to t and from certain **initial conditions** at $t = t_0$.

In the preceding example, we needed only one initial condition. However, in more complex systems, several initial conditions may be necessary. We know, for example, that in passive RLC networks, the initial values of all inductor currents and all capacitor voltages† are needed to determine the outputs at any instant $t \geq 0$ if the inputs are given over the interval $[0, t]$.

1.7   Classification of Systems

Systems may be classified broadly in the following categories:‡

1. Linear and nonlinear systems;
2. Constant-parameter and time-varying-parameter systems;
3. Instantaneous (memoryless) and dynamic (with memory) systems;
4. Causal and noncausal systems;
5. Lumped-parameter and distributed-parameter systems;
6. Continuous-time and discrete-time systems;
7. Analog and Digital systems;

1.7-1   Linear and Nonlinear Systems

The Concept of Linearity

A system whose output is proportional to its input is an *example* of a linear system. But linearity implies more than this; it also implies **additivity property**, implying that if several causes are acting on a system, then the total effect on the system due to all these causes can be determined by considering each cause separately while assuming all the other causes to be zero. The total effect is then the sum of all the component effects. This property may be expressed as follows: for a linear system, if a cause c_1 acting alone has an effect e_1, and if another cause c_2, also acting alone, has an effect e_2, then, with both causes acting on the system, the total effect will be $e_1 + e_2$. Thus, if

$$c_1 \longrightarrow e_1 \qquad \text{and} \qquad c_2 \longrightarrow e_2 \tag{1.37}$$

then for all c_1 and c_2

$$c_1 + c_2 \longrightarrow e_1 + e_2 \tag{1.38}$$

In addition, a linear system must satisfy the **homogeneity** or scaling property, which states that for arbitrary real or imaginary number k, if a cause is increased k-fold, the effect also increases k-fold. Thus, if

$$c \longrightarrow e$$

then for all real or imaginary k

$$kc \longrightarrow ke \tag{1.39}$$

Thus, linearity implies two properties: homogeneity (scaling) and additivity$. Both these properties can be combined into one property (**superposition**), which is expressed as follows: If

$$c_1 \longrightarrow e_1 \qquad \text{and} \qquad c_2 \longrightarrow e_2$$

then for all values of constants k_1 and k_2,

---

† Strictly speaking, independent inductor currents and capacitor voltages.

‡ Other classifications, such as deterministic and probabilistic systems, are beyond the scope of this text and are not considered.

$ A linear system must also satisfy the additional condition of **smoothness**, where small changes in the system's inputs must result in small changes in its outputs.[2]

$$k_1 c_1 + k_2 c_2 \longrightarrow k_1 e_1 + k_2 e_2 \tag{1.40}$$

This is true for all c_1 and c_2.

It may appear that additivity implies homogeneity. Unfortunately, there are cases where homogeneity does not follow from additivity. See the case in Exercise E1.11 below.

△ **Exercise E1.11**

Show that a system with the input (cause) $c(t)$ and the output (effect) $e(t)$ related by $e(t) =$ Re$\{c(t)\}$ satisfies the additivity property but violates the homogeneity property. Hence, such a system is not linear.

Hint: show that Eq. (1.39) is not satisfied when k is complex.   ▽

Response of a Linear System

For the sake of simplicity, we discuss below only **single-input, single-output** (**SISO**) systems. But the discussion can be readily extended to **multiple-input, multiple-output** (**MIMO**) systems.

A system's output for $t \geq 0$ is the result of two independent causes: the initial conditions of the system (or the system state) at $t = 0$ and the input $f(t)$ for $t \geq 0$. If a system is to be linear, the output must be the sum of the two components resulting from these two causes: first, the **zero-input response** component that results only from the initial conditions at $t = 0$ with the input $f(t) = 0$ for $t \geq 0$, and then the **zero-state response** component that results only from the input $f(t)$ for $t \geq 0$ when the initial conditions (at $t = 0$) are assumed to be zero. When all the appropriate initial conditions are zero, the system is said to be in **zero state**. The system output is zero when the input is zero only if the system is in zero state.

In summary, a linear system response can be expressed as the sum of a zero-input and a zero-state component:

Total response = zero-input response + zero-state response (1.41)

This property of linear systems which permits the separation of an output into components resulting from the initial conditions and from the input is called the **decomposition property**.

For the RC circuit of Fig. 1.26, the response $y(t)$ was found to be [see Eq. (1.36c)]

$$y(t) = \underbrace{v_C(0)}_{z-i \text{ component}} + \underbrace{Rf(t) + \frac{1}{C} \int_0^t f(\tau)\, d\tau}_{z-s \text{ component}} \tag{1.42}$$

From Eq. (1.42), it is clear that if the input $f(t) = 0$ for $t \geq 0$, the output $y(t) = v_C(0)$. Hence $v_C(0)$ is the zero-input component of the response $y(t)$. Similarly, if the system state (the voltage v_C in this case) is zero at $t = 0$, the output is given by the second component on the right-hand side of Eq. (1.42). Clearly this is the zero-state component of the response $y(t)$.

In addition to the decomposition property, linearity implies that both the zero-input and zero-state components must obey the principle of superposition with respect to each of their respective causes. For example, if we increase the initial condition k-fold, the zero-input component must also increase k-fold. Similarly, if we increase the input k-fold, the zero-state component must also increase k-fold.

These facts can be readily verified from Eq. (1.42) for the RC circuit in Fig. 1.26. For instance, if we double the initial condition $v_C(0)$, the zero-input component doubles; if we double the input $f(t)$, the zero-state component doubles.

■ Example 1.9
Show that the system described by the equation

$$\frac{dy}{dt} + 3y(t) = f(t) \tag{1.43}$$

is linear.

Let the system response to the inputs $f_1(t)$ and $f_2(t)$ be $y_1(t)$ and $y_2(t)$, respectively. Then

$$\frac{dy_1}{dt} + 3y_1(t) = f_1(t)$$

and

$$\frac{dy_2}{dt} + 3y_2(t) = f_2(t)$$

Multiplying the first equation by k_1, the second with k_2, and adding them yields

$$\frac{d}{dt}\left[k_1 y_1(t) + k_2 y_2(t)\right] + 3\left[k_1 y_1(t) + k_2 y_2(t)\right] = k_1 f_1(t) + k_2 f_2(t)$$

But this equation is the system equation [Eq. (1.43)] with

$$f(t) = k_1 f_1(t) + k_2 f_2(t)$$

and

$$y(t) = k_1 y_1(t) + k_2 y_2(t)$$

Therefore, when the input is $k_1 f_1(t) + k_2 f_2(t)$, the system response is $k_1 y_1(t) + k_2 y_2(t)$. Consequently, the system is linear. Using this argument, we can readily generalize the result to show that a system described by a differential equation of the form

$$\frac{d^n y}{dt^n} + a_{n-1}\frac{d^{n-1}y}{dt^{n-1}} + \cdots + a_0 y = b_m\frac{d^m f}{dt^m} + \cdots + b_1\frac{df}{dt} + b_0 f \tag{1.44}$$

is a linear system. The coefficients a_i and b_i in this equation can be constants or functions of time. ■

△   **Exercise E1.12**
Show that the system described by the following equation is linear:

$$\frac{dy}{dt} + t^2 y(t) = (2t + 3)f(t) \quad \triangledown$$

△   **Exercise E1.13**
Show that a system described by the following equation is nonlinear:

$$y(t)\frac{dy}{dt} + 3y(t) = f(t) \quad \triangledown$$

More Comments on Linear Systems

Almost all systems observed in practice become nonlinear when large enough signals are applied to them. However, many systems show linear behavior for small signals. The analysis of nonlinear systems is generally difficult. Nonlinearities can arise in so many ways that describing them with a common mathematical form is impossible. Not only is each system a category in itself, but even for a given system, changes in initial conditions or input amplitudes may change the nature of the problem. On the other hand, the superposition property of linear systems is a powerful unifying principle which allows for a general solution. The superposition property (linearity) greatly simplifies the analysis of linear systems. Because of the decomposition property, we can evaluate separately the two components of the output. The zero-input component can be computed by assuming the input to be zero, and the zero-state component can be computed by assuming zero initial conditions. Moreover, if we express an input $f(t)$ as a sum of simpler functions,

$$f(t) = a_1 f_1(t) + a_2 f_2(t) + \cdots + a_m f_m(t)$$

then, by virtue of linearity, the response $y(t)$ is given by

$$y(t) = a_1 y_1(t) + a_2 y_2(t) + \cdots + a_m y_m(t) \qquad (1.45)$$

where $y_k(t)$ is the zero-state response to an input $f_k(t)$. This apparently trivial observation has profound implications. As we shall see repeatedly in later chapters, it proves extremely useful and opens new avenues for analyzing linear systems.

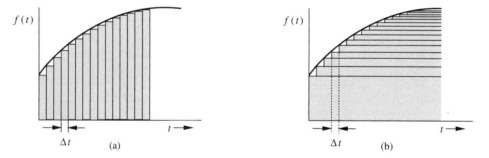

Fig. 1.27 Signal representation in terms of impulse and step components.

As an example, consider an arbitrary input $f(t)$ such as the one shown in Fig. 1.27a. We can approximate $f(t)$ with a sum of rectangular pulses of width Δt and of varying heights. The approximation improves as $\Delta t \to 0$, when the rectangular pulses become impulses spaced Δt seconds apart (with $\Delta t \to 0$). Thus, an arbitrary input can be replaced by a weighted sum of impulses spaced Δt ($\Delta t \to 0$) seconds apart. Therefore, if we know the system response to a unit impulse, we can immediately determine the system response to an arbitrary input $f(t)$ by adding the system response to each impulse component of $f(t)$. A similar situation is depicted in Fig. 1.27b, where $f(t)$ is approximated by a sum of step functions of varying magnitude and spaced Δt seconds apart. The approximation improves as Δt becomes smaller. Therefore, if we know the system response to a unit step input, we can compute the system response to any arbitrary input $f(t)$ with relative ease. Time-domain analysis of linear systems (discussed in Chapter 2) uses this approach.

In Chapters 4,6,10, and 11 we employ the same approach but instead use sinusoids or exponentials as our basic signal components. There, we show that any arbitrary input signal can be expressed as a weighted sum of sinusoids (or exponentials) having various frequencies. Thus a knowledge of the system response to a sinusoid enables us to determine the system response to an arbitrary input $f(t)$.

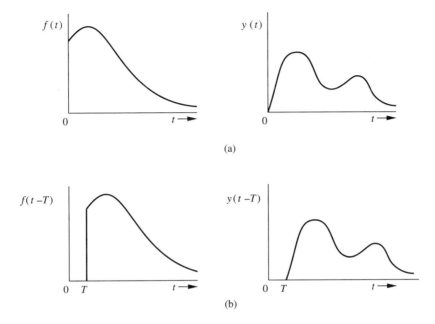

Fig. 1.28   Time-invariance property.

1.7-2   Time-Invariant and Time-Varying Parameter Systems

Systems whose parameters do not change with time are **time-invariant** (also **constant-parameter**) systems. For such a system, if the input is delayed by T seconds, the output is the same as before but delayed by T (assuming identical initial conditions). This property is expressed graphically in Fig. 1.28.

It is possible to verify that the system in Fig. 1.26 is a time-invariant system. Networks composed of RLC elements and other commonly used active elements such as transistors are time-invariant systems. A system with an input-output relationship described by a linear differential equation of the form (1.44) is a linear time-invariant (LTI) system when the coefficients a_i and b_i of such equation are constants. If these coefficients are functions of time, then the system is a linear **time-varying** system. The system described in exercise E1.12 is an example of a linear time-varying system. Another familiar example of a time-varying system is the carbon microphone, in which the resistance R is a function of the mechanical pressure generated by sound waves on the carbon granules of the microphone. An equivalent circuit for the microphone appears in Fig. 1.29. The response is the current $i(t)$, and the equation describing the circuit is

$$L\frac{di(t)}{dt} + R(t)i(t) = f(t)$$

One of the coefficients in this equation, $R(t)$, is time-varying.

△   **Exercise E1.14**
Show that a system described by the following equation is time-varying parameter system:

$$y(t) = (\sin t)\, f(t-2)$$

Hint: Show that the system fails to satisfy the time-invariance property.   ▽

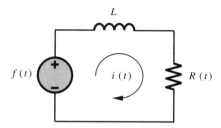

Fig. 1.29   An example of a linear time-varying system.

1.7-3 Instantaneous and Dynamic Systems

As observed earlier, a system's output at any instant t generally depends upon the entire past input. However, in a special class of systems, the output at any instant t depends only on its input at that instant. In resistive networks, for example, any output of the network at some instant t depends only on the input at the instant t. In these systems, past history is irrelevant in determining the response. Such systems are said to be **instantaneous** or **memoryless** systems. More precisely, a system is said to be instantaneous (or memoryless) if its output at any instant t depends, at most, on the strength of its input(s) at the same instant but not on any past or future values of the input(s). Otherwise, the system is said to be **dynamic** (or a system with memory). A system whose response at t is completely determined by the input signals over the past T seconds [interval from $(t - T)$ to t] is a **finite-memory system** with a memory of T seconds. Networks containing inductive and capacitive elements generally have infinite memory because the response of such networks at any instant t is determined by their inputs over the entire past $(-\infty, t)$. This is true for the RC circuit of Fig. 1.26.

In this book we will generally examine dynamic systems. Instantaneous systems are a special case of dynamic systems.

1.7-4 Causal and Noncausal Systems

A **causal** (also known as a **physical** or **non-anticipative**) system is one for which the output at any instant t_0 depends only on the value of the input $f(t)$ for $t \le t_0$. In other words, the value of the output at the present instant depends only on the past and present values of the input $f(t)$, not on its future values. To put it simply, in a causal system the output cannot start before the input is applied. If the response starts before the input, it means that the system knows the input in the future and acts on this knowledge before the input is applied. A system that violates the condition of causality is called a **noncausal** (or **anticipative**) system.

Any practical system that operates in real time† must necessarily be causal. We do not yet know how to build a system that can respond to future inputs (inputs not yet applied). A noncausal system is a prophetic system that knows the future input and acts on it in the present. Thus, if we apply an input starting at $t = 0$ to a noncausal system, the output would begin even before $t = 0$. As an example, consider the system specified by

$$y(t) = f(t - 2) + f(t + 2) \tag{1.46}$$

---

†In real-time operations, the response to an input is essentially simultaneous (contemporaneous) with the input itself.

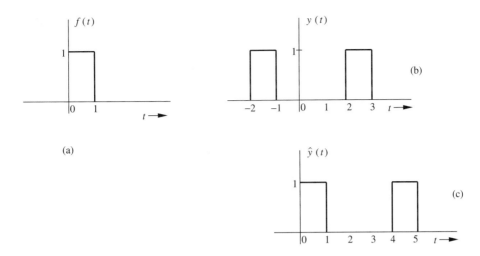

Fig. 1.30 A noncausal system and its realization by a delayed causal system.

For the input $f(t)$ illustrated in Fig. 1.30a, the output $y(t)$, as computed from Eq. (1.46) (shown in Fig. 1.30b), starts even before the input is applied. Equation (1.46) shows that $y(t)$, the output at t, is given by the sum of the input values two seconds before and two seconds after t (at $t - 2$ and $t + 2$ respectively). But if we are operating the system in real time at t, we do not know what the value of the input will be two seconds later. Thus it is impossible to implement this system in real time. For this reason, noncausal systems are unrealizable in **real time**.

Why Study Noncausal Systems?

From the above discussion it may seem that noncausal systems have no practical purpose. This is not the case; they are valuable in the study of systems for several reasons. First, noncausal systems are realizable when the independent variable is other than "time" (e.g., **space**). Consider, for example, an electric charge of density $q(x)$ placed along the x-axis for $x \geq 0$. This charge density produces an electric field $E(x)$ that is present at every point on the x-axis from $x = -\infty$ to ∞. In this case the input [i.e., the charge density $q(x)$] starts at $x = 0$, but its output [the electric field $E(x)$] begins before $x = 0$. Clearly, this space charge system is noncausal. This discussion shows that only temporal systems (systems with time as independent variable) must be causal in order to be realizable. The terms "before" and "after" have a special connection to causality only when the independent variable is time. This connection is lost for variables other than time. Nontemporal systems, such as those occurring in optics, can be noncausal and still realizable.

Moreover, even for temporal systems, such as those used for signal processing, the study of noncausal systems is important. In such systems we may have all input data prerecorded. (This often happens with speech, geophysical, and meteorological signals, and with space probes.) In such cases, the input's future values are available to us. For example, suppose we had a set of input signal records available for the system described by Eq. (1.46). We can then compute $y(t)$ since, for any t, we need only refer to the records to find the input's value two seconds before and two seconds after t. Thus, noncausal systems can be realized, although not in real time. We

Noncausal systems are realizable with time delay!

may therefore be able to realize a noncausal system, provided that we are willing to accept a time delay in the output. Consider a system whose output $\hat{y}(t)$ is the same as $y(t)$ in Eq. (1.46) delayed by two seconds (Fig 1.30c), so that

$$\hat{y}(t) = y(t-2)$$
$$= f(t-4) + f(t)$$

Here the value of the output $\hat{y}$ at any instant t is the sum of the values of the input f at t and at the instant four seconds earlier [at $(t-4)$]. In this case, the output at any instant t does not depend on future values of the input, and the system is causal. The output of this system, which is $\hat{y}(t)$, is identical to that in Eq. (1.46) or Fig. 1.30b except for a delay of two seconds. Thus, a noncausal system may be realized or satisfactorily approximated in real time by using a causal system with a delay.

A third reason for studying noncausal systems is that they provide an upper bound on the performance of causal systems. For example, if we wish to design a filter for separating a signal from noise, then the optimum filter is invariably a noncausal system. Although unrealizable, this noncausal system's performance acts as the upper limit on what can be achieved and gives us a standard for evaluating the performance of causal filters.

At first glance, noncausal systems may seem inscrutable. Actually, there is nothing mysterious about these systems and their approximate realization through using physical systems with delay. If we want to know what will happen one year from now, we have two choices: go to a prophet (an unrealizable person) who can give the answers immediately, or go to a wise man and allow him a delay of one year to give us the answer! If the wise man is truly wise, he may even be able to shrewdly guess the future very closely with a delay of less than a year by studying trends. Such is the case with noncausal systems—nothing more and nothing less.

△ **Exercise E1.15**

Show that a system described by the equation below is noncausal:

$$y(t) = \int_{t-5}^{t+5} f(\tau)\, d\tau$$

Show that this system can be realized physically if we accept a delay of 5 seconds in the output.

▽

1.7-5 Lumped-Parameter and Distributed-Parameter Systems

In the study of electrical systems, we make use of voltage-current relationships for various components (Ohm's law, for example). In doing so, we implicitly assume that the current in any system component (resistor, inductor, etc.) is the same at every point throughout that component. Thus, we assume that electrical signals are propagated instantaneously throughout the system. In reality, however, electrical signals are electromagnetic space waves requiring some finite propagation time. An electric current, for example, propagates through a component with a finite velocity and therefore may exhibit different values at different locations in the same component. Thus, an electric current is a function not only of time but also of space. However, if the physical dimensions of a component are small compared to the wavelength of the signal propagated, we may assume that the current is constant throughout the component. This is the assumption made in **lumped-parameter systems**, where each component is regarded as being lumped at one point in space. Such an assumption is justified at lower frequencies (higher wavelength). Therefore, in lumped-parameter models, signals can be assumed to be functions of time alone. For such systems, the system equations require only one independent variable (time) and therefore are ordinary differential equations.

In contrast, for **distributed-parameter systems** such as transmission lines, waveguides, antennas, and microwave tubes, the system dimensions cannot be assumed to be small compared to the wavelengths of the signals; thus the lumped-parameter assumption breaks down. The signals here are functions of space as well as of time, leading to mathematical models consisting of partial differential equations.[3] The discussion in this book will be restricted to lumped-parameter systems only.

1.7-6 Continuous-Time and Discrete-Time Systems

Distinction between discrete-time and continuous-time signals is discussed in Sec. 1.2-1. Systems whose inputs and outputs are continuous-time signals are **continuous-time systems**. On the other hand, systems whose inputs and outputs are discrete-time signals are **discrete-time systems**. If a continuous-time signal is sampled, the resulting signal is a discrete-time signal. We can process a continuous-time signal by processing its samples with a discrete-time system.

1.7-7 Analog and Digital Systems

Analog and digital signals are discussed in Sec. 1.2-2. A system whose input and output signals are analog is an **analog system**; a system whose input and output signals are digital is a **digital system**. A digital computer is an example of a digital (binary) system. Observe that a digital computer is an example of a system that is digital as well as discrete-time.

Additional Classification of Systems

There are additional classes of systems, such as **invertible** and **noninvertible** systems. A system S performs certain operation(s) on input signal(s). If we can obtain the input $f(t)$ back from the output $y(t)$ by some operation, the system

S is said to be invertible. For a noninvertible system, different inputs can result in the same output (as in a rectifier), and it is impossible to determine the input for a given output. Therefore, for an invertible system, it is essential that distinct inputs result in distinct outputs so that there is one-to-one mapping between an input and the corresponding output. This ensures that every output has a unique input. Consequently, the system is invertible. The system that achieves this inverse operation [of obtaining $f(t)$ from $y(t)$] is the **inverse system** of S. For instance, A system whose input and output are related by equation $y(t) = a\,f(t) + b$ is an invertible system. But a rectifier, specified by the equation $y(t) = |f(t)|$ is noninvertible because the rectification operation cannot be undone.

An ideal differentiator is noninvertible because integration of its output cannot restore the original signal unless we know one piece of information about the signal. For instance, if $f(t) = 3t + 5$, the output of the differentiator is $y(t) = 3$. If this output is applied to an integrator, the output is $3t + c$, where c is an arbitrary constant. If we know one piece of information about $f(t)$, such as $f(0) = 5$, we can determine the input to be $f(t) = 3t + 5$. Thus, a differentiator along with one piece of information (known as auxiliary condition) is an invertible system.† Similarly, a system consisting of a cascade of two differentiators is invertible, if we know two independent pieces of information (auxiliary conditions) about the input signal.

In addition, systems can also be classified as **stable** or **unstable** systems. The concept of stability is discussed in more depth in later chapters.

△ **Exercise E1.16**

Show that a system described by the equation $y(t) = f^2(t)$ is noninvertible.   ▽

1.8 System Model: Input-output Description

As mentioned earlier, systems theory encompasses a variety of systems, such as electrical, mechanical, hydraulic, acoustic, electromechanical, and chemical, as well as social, political, economic, and biological. The first step in analyzing any system is the construction of a system model, which is a mathematical expression or a rule that satisfactorily approximates the dynamical behavior of the system. In this chapter we shall consider only the continuous-time systems. (Modeling of discrete-time systems is discussed in Chapter 8.)

To construct a system model, we must study the relationships between different variables in the system. In electrical systems, for example, we must determine a satisfactory model for the voltage-current relationship of each element, such as Ohm's law for a resistor. In addition, we must determine the various constraints on voltages and currents when several electrical elements are interconnected. These are the laws of interconnection—the well-known Kirchhoff's voltage and current laws (KVL and KCL). From all these equations, we eliminate unwanted variables to obtain equation(s) relating the desired output variable(s) to the input(s). The following examples demonstrate the procedure of deriving input-output relationships for some LTI electrical systems.

---

†The additional piece of information cannot be just any information. For instance, in the above example, if we are given $\dot{f}(0) = 0$, it will not help in determining c, and the system is noninvertible.

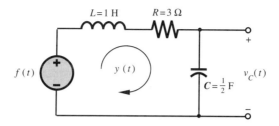

Fig. 1.31   Circuit for Example 1.10.

■ **Example 1.10**

For the series RLC circuit of Fig. 1.31, find the input-output equation relating the input voltage $f(t)$ to the output current (loop current) $y(t)$.

Application of the Kirchhoff's voltage law around the loop yields

$$v_L(t) + v_R(t) + v_C(t) = f(t) \qquad (1.47)$$

By using the voltage-current laws of each element (inductor, resistor, and capacitor), we can express this equation as

$$\frac{dy}{dt} + 3y(t) + 2\int_{-\infty}^{t} y(\tau)\,d\tau = f(t) \qquad (1.48)$$

Differentiating both sides of this equation obtains

$$\frac{d^2y}{dt^2} + 3\frac{dy}{dt} + 2y(t) = \frac{df}{dt} \qquad (1.49)$$

This differential equation is the input-output relationship between the output $y(t)$ and the input $f(t)$. ■

It proves convenient to use a compact notation D for the differential operator $\frac{d}{dt}$. Thus

$$\frac{dy}{dt} \equiv Dy(t) \qquad (1.50)$$

$$\frac{d^2y}{dt^2} \equiv D^2y(t) \qquad (1.51)$$

and so on. With this notation, Eq. (1.49) can be expressed as

$$\left(D^2 + 3D + 2\right)y(t) = Df(t) \qquad (1.52)$$

The differential operator is the inverse of the integral operator, so we can use the operator $1/D$ to represent integration†.

$$\int_{-\infty}^{t} y(\tau)\,d\tau \equiv \frac{1}{D}y(t) \qquad (1.53)$$

---

† Use of operator $1/D$ for integration generates some subtle mathematical difficulties because the operators D and $1/D$ do not commute. For instance, we know that $D(1/D) = 1$ because $\frac{d}{dt}[\int_{-\infty}^{t} y(\tau)\,d\tau] = y(t)$. However, $\frac{1}{D}D$ is not necessarily unity. Use of Cramer's rule in solving simultaneous integro-differential equations will always result in cancellation of operators $1/D$ and D. This procedure may yield erroneous results in those cases where the factor D occurs in the

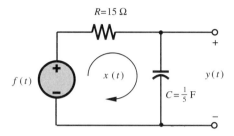

Fig. 1.32 Circuit for Example 1.11.

Consequently, the loop equation (1.48) can be expressed as

$$\left(D + 3 + \frac{2}{D}\right) y(t) = f(t) \tag{1.54}$$

Multiplying both sides by D, that is, differentiating Eq. (1.54), we obtain

$$\left(D^2 + 3D + 2\right) y(t) = D f(t) \tag{1.55}$$

which is identical to Eq. (1.52).

Recall that Eq. (1.55) is not an algebraic equation, and $D^2 + 3D + 2$ is not an algebraic term that multiplies $y(t)$; it is an operator that operates on $y(t)$. It means that we must perform the following operations on $y(t)$: take the second derivative of $y(t)$ and add to it 3 times the first derivative of $y(t)$ and 2 times $y(t)$. Clearly, a polynomial in D multiplied by $y(t)$ represents a certain differential operation on $y(t)$.

■ **Example 1.11**

Find the equation relating the input to output for the series RC circuit of Fig. 1.32 if the input is the voltage $f(t)$ and output is **(a)** the loop current $x(t)$ **(b)** the capacitor voltage $y(t)$.

The loop equation for the circuit is

$$Rx(t) + \frac{1}{C} \int_{-\infty}^{t} x(\tau)\, d\tau = f(t) \tag{1.56}$$

or

$$15x(t) + 5 \int_{-\infty}^{t} x(\tau)\, d\tau = f(t) \tag{1.57}$$

With operational notation, this equation can be expressed as

---

numerator as well as in the denominator. This happens, for instance, in circuits with all-inductor loops or all-capacitor cutsets. To eliminate this problem, avoid the integral operation in system equations so that the resulting equations are differential rather than integro-differential. In electrical circuits, this can be done by using charge (instead of current) variables in loops containing capacitors and using current variables for loops without capacitors. In the literature this problem of commutativity of D and $1/D$ is largely ignored. As mentioned earlier, such procedure gives erroneous results only in special systems, such as the circuits with all-inductor loops or all-capacitor cutsets. Fortunately such systems constitute a very small fraction of the systems we deal with. For further discussion of this topic and a correct method of handling problems involving integrals, see Ref. 4

$$15x(t) + \frac{5}{D}x(t) = f(t) \tag{1.58}$$

Multiplying both sides of the above equation by D (that is, differentiating the above equation), we obtain

$$(15D + 5)x(t) = Df(t) \tag{1.59a}$$

or

$$15\frac{dx}{dt} + 5x(t) = \frac{df}{dt} \tag{1.59b}$$

Moreover,

$$x(t) = C\frac{dy}{dt}$$

$$= \frac{1}{5}Dy(t)$$

Substitution of this result in Eq. (1.59a) yields

$$(3D + 1)y(t) = f(t) \tag{1.60}$$

or

$$3\frac{dy}{dt} + y(t) = f(t) \tag{1.61}$$

■

△ **Exercise E1.17**
For the RLC circuit in Fig. 1.31, find the input-output relationship if the output is the inductor voltage $v_L(t)$.
Hint: $v_L(t) = LDy(t) = Dy(t)$.   Answer: $\left(D^2 + 3D + 2\right)v_L(t) = D^2 f(t)$ ▽

△ **Exercise E1.18**
For the RLC circuit in Fig. 1.31, find the input-output relationship if the output is the capacitor voltage $v_C(t)$.
Hint: $v_C(t) = \frac{1}{CD}y(t) = \frac{2}{D}y(t)$.   Answer: $\left(D^2 + 3D + 2\right)v_C(t) = 2f(t)$ ▽

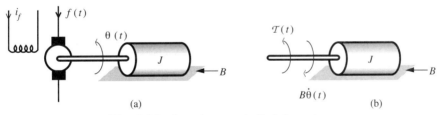

Fig. 1.33 Armature controlled dc motor.

■ **Example 1.12**
In rotational systems, the equations of motion are similar to those in translational systems. In place of force F, we have torque $\mathcal{T}$. In place of mass M, we have **moment of inertia** J (the rotational mass), and in place of linear acceleration $\ddot{x}$, we have angular acceleration $\ddot{\theta}$. The equation of motion for rotational systems is $\mathcal{T} = J\ddot{\theta}$ (in place of $F = M\ddot{x}$).

A wide variety of electromechanical systems convert electrical signals into mechanical motion (mechanical energy) and vice versa. Here we consider a rather simple example of an armature-controlled dc motor (with a constant field current i_f) driven by a current source $f(t)$, as depicted in Fig. 1.33a. Let $\theta(t)$ be the angular position of the rotor. The torque $T(t)$ generated in the rotor is proportional to the armature current $f(t)$. Therefore

$$T(t) = K_T f(t) \tag{1.62}$$

where K_T is a constant of the motor. This torque drives a mechanical load whose free-body diagram is illustrated in Fig. 1.33b. The viscous damping (with coefficient B), which is proportional to the angular velocity $\dot{\theta}$, dissipates a torque $B\dot{\theta}(t)$. If J is the moment of inertia of the load (including the rotor of the motor), then the net torque $T(t) - B\dot{\theta}(t)$ available must equal to $J\ddot{\theta}(t)$;

$$J\ddot{\theta}(t) = T(t) - B\dot{\theta}(t) \tag{1.63}$$

Thus

$$\left(JD^2 + BD\right)\theta(t) = T(t)$$

$$= K_T f(t) \tag{1.64}$$

which can be expressed as

$$D(D + a)\theta(t) = K_1 f(t) \tag{1.65}$$

where $a = B/J$ and $K_1 = K_T/J$.

■

1.8-1   Internal and External Descriptions of a System

With a knowledge of the internal structure of a system, we can write system equations yielding an **internal description** of the system. In contrast, the system description seen from the system's input and output terminals is the system's **external description**. To understand an external description, suppose that a system is enclosed in a "black box" with only its input(s) and output(s) terminals accessible. In order to describe or characterize such a system, we must perform some measurements at these terminals. For example, we might apply a known input, such as a unit impulse or a unit step, and then measure the system's output. The description provided by such a measurement is an external description of the system.

Suppose the circuit in Fig. 1.34a with the input $f(t)$ and the output $y(t)$ is enclosed inside a "black box" with only the input and output terminals accessible. Under these conditions the only way to describe or specify the system is with external measurements. We can, for example, apply a known voltage $f(t)$ at the input terminals and measure the resulting output voltage $y(t)$. From this information we can describe or characterize the system. This is the *external description*.

Assuming zero initial capacitor voltage, the input voltage $f(t)$ produces a current i (Fig. 1.34a), which divides equally between the two branches because of the balanced nature of the circuit. Thus, the voltage across the capacitor continues to remain zero. Therefore, for the purpose of computing the current i, the capacitor may be removed or replaced by a short. The resulting circuit is equivalent to that shown in Fig. 1.34b. It is clear from Fig. 1.34b that $f(t)$ sees a net resistance of $5\,\Omega$, and

$$i(t) = \frac{1}{5}f(t)$$

Also, because $y(t) = 2 \times (i/2) = i$,

$$y(t) = \frac{1}{5}f(t) \tag{1.66}$$

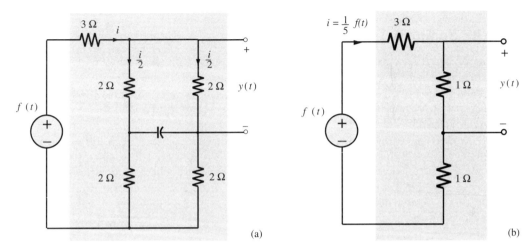

Fig. 1.34 A system that cannot be described by external measurements.

The equivalent system as seen from the system's external terminals is depicted in Fig. 1.34b. Clearly, for the external description, the capacitor does not exist. For most systems, the external and internal descriptions are identical, but there are a few exceptions, as in the present case, where the external description gives an inadequate picture of the systems. This happens when the system is **uncontrollable** and/or **unobservable**. Figures 1.35a and 1.35b show a structural representation of simple uncontrollable and unobservable systems respectively. In Fig. 1.35a we note that part of the system (subsystem S_2) inside the box cannot be controlled by the input $f(t)$. In Fig. 1.35b some of the system outputs (those in subsystem S_2) cannot be observed from the output terminals. If we try to describe either of these systems by applying an external input $f(t)$ and then measuring the output $y(t)$, the measurement will not characterize the complete system but only the part of the system (here S_1) that is both controllable and observable (linked to both the input and output). Such systems are undesirable in practice and should be avoided in any system design. The system in Fig. 1.35a can be shown to be neither controllable nor observable. It can be represented structurally as a combination of the systems in Figs. 1.35a and 1.35b.

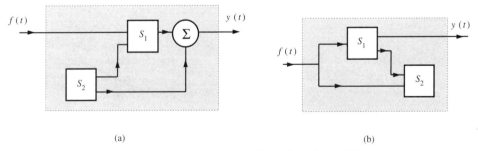

Fig. 1.35 Structures of uncontrollable and unobservable systems.

1.9 Summary

A *signal* is a set of information or data. A *system* processes input signals
to modify them or extract additional information from them to produce output
signals (response). A system may be made up of physical components (hardware
realization) or may be an algorithm that computes an output signal from an input
signal (software realization).

A convenient measure of the size of a signal is its energy if it is finite. If the
signal energy is infinite, the appropriate measure is its power, if it exists. The signal
power is the time average of its energy (averaged over the entire time interval from
$-\infty$ to ∞). For periodic signals the time averaging need be performed only over
one period in view of the periodic repetition of the signal. Signal power is also equal
to the mean squared value of the signal (averaged over the entire time interval from
$t = -\infty$ to ∞).

Signals can be classified in several ways as follows:

1. A *continuous-time signal* is specified for a continuum of values of the indepen-
    dent variable (such as time t). A *discrete-time signal* is specified only at a finite
    or a countable set of time instants.

2. An *analog signal* is a signal whose amplitude can take on any value over a con-
    tinuum. On the other hand, a signal whose amplitudes can take on only a finite
    number of values is a *digital signal*. The terms *discrete-time* and *continuous-
    time* qualify the nature of a signal along the time axis (horizontal axis). The
    terms *analog* and *digital*, on the other hand, qualify the nature of the signal
    amplitude (vertical axis).

3. A *periodic signal* $f(t)$ is defined by the fact that $f(t) = f(t + T_0)$ for some
    T_0. The smallest value of T_0 for which this relationship is satisfied is called the
    period. A periodic signal remains unchanged when shifted by an integral mul-
    tiple of its period. A periodic signal can be generated by a periodic extension
    of any segment of $f(t)$ of duration T_0. Finally, a periodic signal, by definition,
    must exist over the entire time interval $-\infty < t < \infty$. A signal is *aperiodic* if
    it is not periodic.

    An *everlasting signal* starts at $t = -\infty$ and continues forever to $t = \infty$. A
    causal signal is a signal that is zero for $t < 0$. Hence, periodic signals are
    everlasting signals.

4. A signal with finite energy is an *energy signal*. Similarly a signal with a finite
    and nonzero power (mean square value) is a *power signal*. A signal can either
    be an energy signal or a power signal, but not both. However, there are signals
    that are neither energy nor power signals.

5. A signal whose physical description is known completely in a mathematical or
    graphical form is a *deterministic signal*. A *random signal* is known only in
    terms of its probabilistic description such as mean value, mean square value,
    and so on, rather than its mathematical or graphical form.

    A signal $f(t)$ delayed by T seconds (right-shifted) is given by $f(t - T)$; on the
other hand, $f(t)$ advanced by T (left-shifted) is given by $f(t + T)$. A signal $f(t)$
time-compressed by a factor a ($a > 1$) is given by $f(at)$; on the other hand, the same

signal time-expanded by factor a is given by $f(\frac{t}{a})$. The same signal time-inverted is given by $f(-t)$.

The unit step function $u(t)$ is very useful in representing causal signals and signals with different mathematical descriptions over different intervals.

In the classical definition, the unit impulse function $\delta(t)$ is characterized by unit area, and the fact that it is concentrated at a single instant $t = 0$. The impulse function has a sampling (or sifting) property, which states that the area under the product of a function with a unit impulse is equal to the value of that function at the instant where the impulse is located (assuming the function to be continuous at the impulse location). In the modern approach, the impulse function is viewed as a generalized function and is defined by the sampling property.

The exponential function e^{st}, where s is complex, encompasses a large class of signals that includes a constant, a monotonic exponential, a sinusoid, and an exponentially varying sinusoid.

A signal that is symmetrical about the vertical axis $(t = 0)$ is an *even* function of time, and a signal that is antisymmetrical about the vertical axis is an *odd* function of time. The product of an even function with an odd function results in an odd function. However, the product of an even function with an even function or an odd function with an odd function results in an even function. The area under an odd function from $t = -a$ to a is always zero regardless of the value of a. On the other hand, the area under an even function from $t = -a$ to a is two times the area under the same function from $t = 0$ to a (or from $t = -a$ to 0). Every signal can be expressed as a sum of odd and even function of time.

A system processes input signals to produce output signals (response). The input is the cause and the output is its effect. In general, the output is affected by two causes: the internal conditions of the system (such as the initial conditions) and the external input.

Systems can be classified in several ways:

1. Linear systems are characterized by the linearity property, which implies superposition; if several causes (such as various inputs and initial conditions) are acting on a linear system, the total effect (response) is the sum of the responses from each cause, assuming that all the remaining causes are absent. A system is nonlinear if it is not linear.

2. Time-invariant systems are characterized by the fact that system parameters do not change with time. The parameters of time-varying parameter systems change with time.

3. For memoryless (or instantaneous) systems, the system response at any instant t depends only on the present value of the input (value at t). For systems with memory (also known as dynamic systems), the system response at any instant t depends not only on the present value of the input, but also on the past values of the input (values before t).

4. In contrast, if a system response at t also depends on the future values of the input (values of input beyond t), the system is noncausal. In causal systems, the response does not depend on the future values of the input. Because of the dependence of the response on the future values of input, the effect (response)

of noncausal systems occurs before cause. When the independent variable is time (temporal systems), the noncausal systems are prophetic systems, and therefore, unrealizable, although close approximation is possible with some time delay in the response. Noncausal systems with independent variables other than time (e.g., space) are realizable.

5. If the dimensions of system elements are small compared to the wavelengths of the signals, we may assume that each element is lumped at a single point in space, and the system may be considered as a lumped-parameter system. The signals under this assumption are functions of time only. If this assumption does not hold, the signals are functions of space and time; such a system is a distributed-parameter system.

6. Systems whose inputs and outputs are continuous-time signals are continuous-time systems; systems whose inputs and outputs are discrete-time signals are discrete-time systems. If a continuous-time signal is sampled, the resulting signal is a discrete-time signal. We can process a continuous-time signal by processing the samples of this signal with a discrete-time system.

7. Systems whose inputs and outputs are analog signals are analog systems; those whose inputs and outputs are digital signals are digital systems.

8. If we can obtain the input $f(t)$ back from the output $y(t)$ of a system $\mathcal{S}$ by some operation, the system $\mathcal{S}$ is said to be invertible. Otherwise the system is noninvertible.

The system model derived from a knowledge of the internal structure of the system is its internal description. In contrast, an external description of a system is its description as seen from the system's input and output terminals; it can be obtained by applying a known input and measuring the resulting output. In the majority of practical systems, an external description of a system so obtained is equivalent to its internal description. In some cases, however, the external description fails to give adequate information about the system. Such is the case with the so-called uncontrollable or unobservable systems.

References

1. Papoulis, A., *The Fourier Integral and Its Applications*, McGraw-Hill, New York, 1962.

2. Kailath, T., *Linear Systems*, Prentice-Hall, Englewood Cliffs, New Jersey, 1980.

3. Lathi, B.P., *Signals, Systems, and Communication*, Wiley, New York, 1965.

4. Lathi, B.P., *Signals and Systems*, Berkeley-Cambridge Press, Carmichael, California, 1987.

Problems

1.1-1 Find the energies of the signals illustrated in Fig. P1.1-1. Comment on the effect on energy of sign change, time shifting, or doubling of the signal. What is the effect on the energy if the signal is multiplied by k?

1.1-2 Repeat Prob. 1.1-1 for the signals in Fig. P1.1-2.

1.1-3 (a) Find the energies of the pair of signals $x(t)$ and $y(t)$ depicted in Figs. P1.1-3a and b. Sketch and find the energies of signals $x(t) + y(t)$ and $x(t) - y(t)$. Can you make any observation from these results?

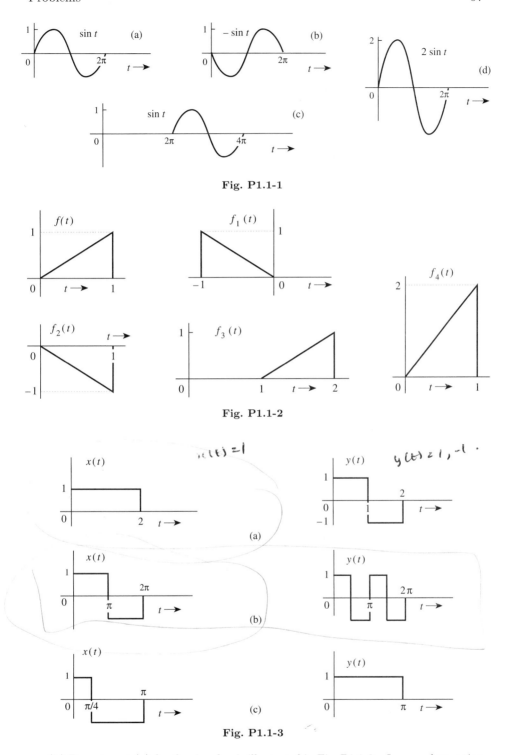

Fig. P1.1-1

Fig. P1.1-2

Fig. P1.1-3

(b) Repeat part (a) for the signal pair illustrated in Fig. P1.1-3c. Is your observation in part (a) still valid?

1.1-4   Find the power of the periodic signal $f(t)$ shown in Fig. P1.1-4. Find also the powers and the rms values of: **(a)** $-f(t)$ **(b)** $2f(t)$ **(c)** $cf(t)$. Comment.

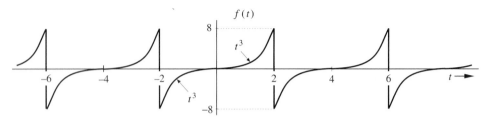

Fig. P1.1-4

1.1-5   Show that the power of a signal

$$f(t) = \sum_{k=m}^{n} D_k e^{j\omega_k t} \quad \text{is} \quad P_f = \sum_{k=m}^{n} |D_k|^2$$

assuming all frequencies to be distinct, that is, $\omega_i \neq \omega_k$ for all $i \neq k$

1.1-6   Determine the power and the rms value for each of the following signals:

(a) $10 \cos\left(100t + \dfrac{\pi}{3}\right)$ **(b)** $10 \cos\left(100t + \dfrac{\pi}{3}\right) + 16 \sin\left(150t + \dfrac{\pi}{5}\right)$

(c) $(10 + 2 \sin 3t)\cos 10t$ **(d)** $10 \cos 5t \cos 10t$

(e) $10 \sin 5t \cos 10t$ **(f)** $e^{j\alpha t} \cos \omega_0 t$

1.3-1   In Fig. P1.3-1, the signal $f_1(t) = f(-t)$. Express signals $f_2(t)$, $f_3(t)$, $f_4(t)$, and $f_5(t)$ in terms of signals $f(t)$, $f_1(t)$, and their time-shifted, time-scaled or time-inverted versions. For instance $f_2(t) = f(t - T) + f_1(t - T)$.

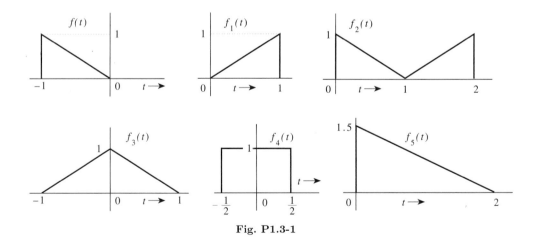

Fig. P1.3-1

1.3-2   For the signal $f(t)$ depicted in Fig. P1.3-2, sketch the signals: **(a)** $f(-t)$ **(b)** $f(t+6)$ **(c)** $f(3t)$ **(d)** $f(\frac{t}{2})$.

1.3-3   For the signal $f(t)$ illustrated in Fig. P1.3-3, sketch **(a)** $f(t-4)$ **(b)** $f(\frac{t}{1.5})$ **(c)** $f(-t)$ **(d)** $f(2t - 4)$ **(e)** $f(2 - t)$.

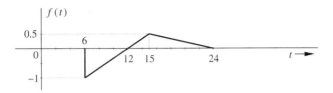

Fig. P1.3-2

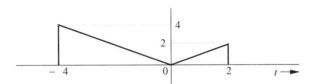

Fig. P1.3-3

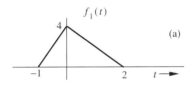

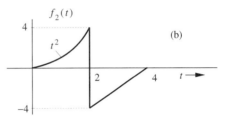

Fig. P1.4-2

1.4-1 Sketch the signals **(a)** $u(t-5)-u(t-7)$ **(b)** $u(t-5)+u(t-7)$ **(c)** $t^2[u(t-1)-u(t-2)]$
(d) $(t-4)[u(t-2)-u(t-4)]$

1.4-2 Express each of the signals in Fig. P1.4-2 by a single expression valid for all t.

1.4-3 For an energy signal $f(t)$ with energy E_f, show that the energy of any one of the signals $-f(t)$, $f(-t)$ and $f(t-T)$ is E_f. Show also that the energy of $f(at)$ as well as $f(at-b)$ is E_f/a. This shows that time-inversion and time-shifting does not affect signal energy. On the other hand, time compression of a signal $(a > 1)$ reduces the energy, and time expansion of a signal $(a < 1)$ increases the energy. What is the effect on signal energy if the signal is multiplied by a constant a?

1.4-4 Simplify the following expressions:

(a) $\left(\dfrac{\sin t}{t^2+2}\right)\delta(t)$ **(b)** $\left(\dfrac{j\omega+2}{\omega^2+9}\right)\delta(\omega)$

(c) $\left[e^{-t}\cos(3t-60°)\right]\delta(t)$ **(d)** $\left(\dfrac{\sin\left[\frac{\pi}{2}(t-2)\right]}{t^2+4}\right)\delta(t-1)$

(e) $\left(\dfrac{1}{j\omega+2}\right)\delta(\omega+3)$ **(f)** $\left(\dfrac{\sin k\omega}{\omega}\right)\delta(\omega)$

Hint: Use Eq. (1.23). For part **(f)** use L'Hôpital's rule.

1.4-5 Evaluate the following integrals:

(a) $\displaystyle\int_{-\infty}^{\infty} \delta(\tau)f(t-\tau)\,d\tau$

(e) $\displaystyle\int_{-\infty}^{\infty} \delta(t+3)e^{-t}\,dt$

(b) $\displaystyle\int_{-\infty}^{\infty} f(\tau)\delta(t-\tau)\,d\tau$

(f) $\displaystyle\int_{-\infty}^{\infty} (t^3+4)\delta(1-t)\,dt$

(c) $\displaystyle\int_{-\infty}^{\infty} \delta(t)e^{-j\omega t}\,dt$

(g) $\displaystyle\int_{-\infty}^{\infty} f(2-t)\delta(3-t)\,dt$

(d) $\displaystyle\int_{-\infty}^{\infty} \delta(t-2)\sin \pi t\,dt$

(h) $\displaystyle\int_{-\infty}^{\infty} e^{(x-1)}\cos\left[\tfrac{\pi}{2}(x-5)\right]\delta(x-3)\,dx$

Hint: $\delta(x)$ is located at $x=0$. For example, $\delta(1-t)$ is located at $1-t=0$, and so on.

1.4-6 **(a)** Find and sketch df/dt for the signal $f(t)$ shown in Fig. P1.3-3.
(b) Find and sketch d^2f/dt^2 for the signal $f(t)$ depicted in Fig. P1.4-2a.

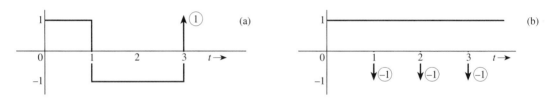

Fig. P1.4-7

1.4-7 Find and sketch $\int_{-\infty}^{t} f(x)\,dx$ for the signal $f(t)$ illustrated in Fig. P1.4-7.

1.4-8 Using the generalized function definition, show that $\delta(t)$ is an even function of t.

Hint: Start with Eq. (1.24a) as the definition of $\delta(t)$. Now change variable $t=-x$ to show that

$$\int_{-\infty}^{\infty} \phi(t)\delta(-t)\,dt = \phi(0)$$

1.4-9 Prove that

$$\delta(at) = \frac{1}{|a|}\delta(t)$$

Hint: Show that

$$\int_{-\infty}^{\infty} \phi(t)\delta(at)\,dt = \frac{1}{|a|}\phi(0)$$

1.4-10 Show that

$$\int_{-\infty}^{\infty} \dot{\delta}(t)\phi(t)\,dt = -\dot{\phi}(0)$$

where $\phi(t)$ and $\dot{\phi}(t)$ are continuous at $t=0$, and $\phi(t) \to 0$ as $t \to \pm\infty$. This integral defines $\dot{\delta}(t)$ as a generalized function. Hint: Use integration by parts.

1.4-11 A sinusoid $e^{\sigma t}\cos \omega t$ can be expressed as a sum of exponentials e^{st} and e^{-st} (Eq. (1.30c) with complex frequencies $s=\sigma+j\omega$ and $s=\sigma-j\omega$. Locate in the complex plane the frequencies of the following sinusoids: **(a)** $\cos 3t$ **(b)** $e^{-3t}\cos 3t$ **(c)** $e^{2t}\cos 3t$ **(d)** e^{-2t} **(e)** e^{2t} **(f)** 5.

1.5-1 Find and sketch the odd and the even components of **(a)** $u(t)$ **(b)** $tu(t)$ **(c)** $\sin \omega_0 t\, u(t)$
(d) $\cos \omega_0 t\, u(t)$ **(e)** $\sin \omega_0 t$ **(f)** $\cos \omega_0 t$.

1.6-1 Write the input-output relationship for an ideal integrator. Determine the zero-input
and zero-state components of the response.

1.7-1 For the systems described by the equations below, with the input $f(t)$ and output
$y(t)$, determine which of the systems are linear and which are nonlinear.

(a) $\dfrac{dy}{dt} + 2y(t) = f^2(t)$ $\qquad\qquad$ **(c)** $3y(t) + 2 = f(t)$

(b) $\dfrac{dy}{dt} + 3ty(t) = t^2 f(t)$ $\qquad$ **(f)** $\dfrac{dy}{dt} + (\sin t)y(t) = \dfrac{df}{dt} + 2f(t)$

(e) $\left(\dfrac{dy}{dt}\right)^2 + 2y(t) = f(t)$ $\qquad$ **(g)** $\dfrac{dy}{dt} + 2y(t) = f(t)\dfrac{df}{dt}$

(d) $\dfrac{dy}{dt} + y^2(t) = f(t)$ $\qquad\qquad$ **(h)** $y(t) = \displaystyle\int_{-\infty}^{t} f(\tau)\, d\tau$

1.7-2 For the systems described by the equations below, with the input $f(t)$ and output
$y(t)$, determine which of the systems are time-invariant parameter systems and which
are time-varying parameter systems.

(a) $y(t) = f(t-2)$ $\qquad\qquad$ **(d)** $y(t) = t\, f(t-2)$

(b) $y(t) = f(-t)$ $\qquad\qquad$ **(e)** $y(t) = \displaystyle\int_{-5}^{5} f(\tau)\, d\tau$

(c) $y(t) = f(at)$ $\qquad\qquad$ **(f)** $y(t) = \left(\dfrac{df}{dt}\right)^2$

1.7-3 For a certain LTI system with the input $f(t)$, the output $y(t)$ and the two initial
conditions $x_1(0)$ and $x_2(0)$, following observations were made:

$f(t)$	$x_1(0)$	$x_2(0)$	$y(t)$
0	1	-1	$e^{-t}u(t)$
0	2	1	$e^{-t}(3t+2)u(t)$
$u(t)$	-1	-1	$2u(t)$

Determine $y(t)$ when both the initial conditions are zero and the input $f(t)$ is as
shown in Fig. P1.7-3.

Hint: There are three causes: the input and each of the two initial conditions. Because
of linearity property, if a cause is increased by a factor k, the response to that cause
also increases by the same factor k. Moreover, if causes are added, the corresponding
responses add.

Fig. P1.7-3

1.7-4 A system is specified by its input-output relationship as

$$y(t) = f^2(t) \Big/ \left(\frac{df}{dt}\right)$$

Show that the system satisfies the homogeneity property but not the additivity property.

1.7-5 Show that the circuit in Fig. P1.7-5 is zero-state linear but is not zero-input linear. Assume all diodes to have identical (matched) characteristics.

Hint: In zero state (when the initial capacitor voltage $v_c(0) = 0$), the circuit is linear. If the input $f(t) = 0$, and $v_c(0)$ is nonzero, the current $y(t)$ does not exhibit linearity with respect to its cause $v_c(0)$.

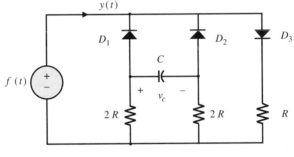

Fig. P1.7-5

1.7-6 The inductor L and the capacitor C in Fig. P1.7-6 are nonlinear, which makes the circuit nonlinear. The remaining 3 elements are linear. Show that the output $y(t)$ of this nonlinear circuit satisfies the linearity conditions with respect to the input $f(t)$ and the initial conditions (all the initial inductor currents and capacitor voltages). Recognize that a current source is an open circuit when the current is zero.

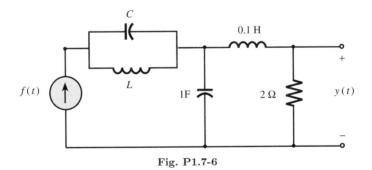

Fig. P1.7-6

1.7-7 For the systems described by the equations below, with the input $f(t)$ and output $y(t)$, determine which of the systems are causal and which are noncausal.

    (a) $y(t) = f(t-2)$ **(c)** $y(t) = f(at)$ $a > 1$

    (b) $y(t) = f(-t)$ **(d)** $y(t) = f(at)$ $a < 1$

1.7-8 For the systems described by the equations below, with the input $f(t)$ and output $y(t)$, determine which of the systems are invertible and which are noninvertible. For the invertible systems, find the input-output relationship of the inverse system.

(a) $y(t) = \displaystyle\int_{-\infty}^{t} f(\tau)\, d\tau$

(c) $y(t) = f^n(t)$ n, integer

(b) $y(t) = f(3t - 6)$

(d) $y(t) = \cos{[f(t)]}$

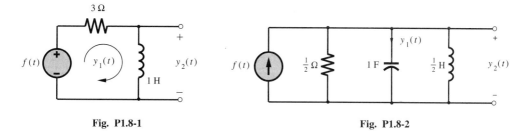

Fig. P1.8-1   **Fig. P1.8-2**

1.8-1 For the circuit depicted in Fig. P1.8-1, find the differential equations relating outputs $y_1(t)$ and $y_2(t)$ to the input $f(t)$.

1.8-2 Repeat Prob. 1.8-1 for the circuit in Fig. P1.8-2.

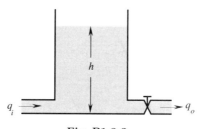

Fig. P1.8-3

1.8-3 Water flows into a tank at a rate of q_i units/s and flows out through the outflow valve at a rate of q_0 units/s (Fig. P1.8-3). Determine the equation relating the outflow q_0 to the input q_i. The outflow rate is proportional to the head h. Thus $q_0 = Rh$ where R is the valve resistance. Determine also the differential equation relating the head h to the input q_i.

(Hint: The net inflow of water in time $\triangle t$ is $(q_i - q_0)\triangle t$. This inflow is also $A\triangle h$ where A is the cross section of the tank.)

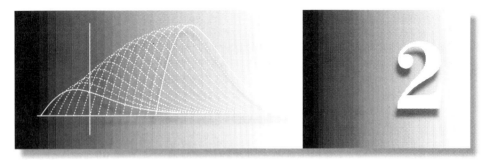

Time-Domain Analysis of Continuous-Time Systems

In this book we consider two methods of analysis of linear time-invariant (LTI) systems: the time-domain method and the frequency-domain method. In this chapter we discuss the **time-domain analysis** of linear, time-invariant, continuous-time (LTIC) systems.

2.1 Introduction

For the purpose of analysis, we shall consider **linear differential systems**. This is the class of LTIC systems discussed in Chapter 1, for which the input $f(t)$ and the output $y(t)$ are related by linear differential equations of the form

$$\frac{d^n y}{dt^n} + a_{n-1}\frac{d^{n-1}y}{dt^{n-1}} + \cdots + a_1\frac{dy}{dt} + a_0 y(t) =$$

$$b_m\frac{d^m f}{dt^m} + b_{m-1}\frac{d^{m-1}f}{dt^{m-1}} + \cdots + b_1\frac{df}{dt} + b_0 f(t) \qquad (2.1a)$$

where all the coefficients a_i and b_i are constants. Using operational notation D to represent d/dt, we can express this equation as

$$(D^n + a_{n-1}D^{n-1} + \cdots + a_1 D + a_0)\, y(t)$$

$$= (b_m D^m + b_{m-1}D^{m-1} + \cdots + b_1 D + b_0)\, f(t) \qquad (2.1b)$$

or

$$Q(D)y(t) = P(D)f(t) \qquad (2.1c)$$

where the polynomials $Q(D)$ and $P(D)$ are

$$Q(D) = D^n + a_{n-1}D^{n-1} + \cdots + a_1 D + a_0 \qquad (2.2a)$$

$$P(D) = b_m D^m + b_{m-1}D^{m-1} + \cdots + b_1 D + b_0 \qquad (2.2b)$$

Theoretically the powers m and n in the above equations can take on any value. Practical noise considerations, however, require $m \leq n$. Noise is any undesirable signal, natural or manmade, which interferes with the desired signals in the system. Some of the sources of noise are: the electromagnetic radiation from stars, random motion of electrons in system components, interference from nearby radio and television stations, transients produced by automobile ignition systems, fluorescent lighting, and so on. We show in Chapter 6 that a system specified by Eq. (2.1) behaves as an $(m - n)$th-order differentiator of high-frequency signals if $m > n$. Unfortunately, noise is a wideband signal containing components of all frequencies from 0 to ∞. For this reason noise contains a significant amount of rapidly varying components with derivatives that are, consequently, very large. Therefore, any system specified by Eq. (2.1) in which $m > n$ will magnify the high-frequency components of noise through differentiation. It is entirely possible for noise to be magnified so much that it swamps the desired system output even if the noise signal at the system's input is tolerably small. For the rest of this text we assume implicitly that $m \leq n$. For the sake of generality we shall assume $m = n$ in Eq. (2.1).

In Chapter 1, we demonstrated that a system described by Eq. (2.1) is linear.† Therefore, its response can be expressed as the sum of two components: the zero-input component and the zero-state component (decomposition property).‡ Therefore,

Total response = zero-input response + zero-state response   (2.3)

The zero-input component is the system response when the input $f(t) = 0$ so that it is the result of internal system conditions (such as energy storages, initial conditions) alone. It is independent of the external input $f(t)$. In contrast, the zero-state component is the system response to the external input $f(t)$ when the system is in zero state, meaning the absence of all internal energy storages; that is, all initial conditions are zero.

---

†To demonstrate that any system described by Eq. (2.1) is linear, let the input $f_1(t)$ to the system generate the output $y_1(t)$, and another input $f_2(t)$ generate the output $y_2(t)$. From Eq. (2.1c) it follows that

$$Q(D)y_1(t) = P(D)f_1(t) \qquad \text{and} \qquad Q(D)y_2(t) = P(D)f_2(t)$$

Multiplication of these equations by k_1 and k_2, respectively and then adding yields

$$Q(D)\left[k_1 y_1(t) + k_2 y_2(t)\right] = P(D)\left[k_1 f_1(t) + k_2 f_2(t)\right]$$

This equation shows that the input $k_1 f_1(t) + k_2 f_2(t)$ generates the response $k_1 y_1(t) + k_2 y_2(t)$. Therefore, the system is linear

‡ We can verify readily that the system described by Eq. (2.1) has the decomposition property. If $y_0(t)$ is the zero-input response, then, by definition,

$$Q(D)y_0(t) = 0$$

If $y(t)$ is the zero-state response, then $y(t)$ is the solution of

$$Q(D)y(t) = P(D)f(t)$$

subject to zero initial conditions (zero-state). The addition of these two equations yields

$$Q(D)\left[y_0(t) + y(t)\right] = P(D)f(t)$$

Clearly, $y_0(t) + y(t)$ is the general solution of Eq. (2.1).

2.2 System Response to Internal Conditions: Zero-Input Response

The zero-input response $y_0(t)$ is the solution of Eq. (2.1) when the input $f(t) = 0$ so that

$$Q(D)y_0(t) = 0 \tag{2.4a}$$

or

$$\left(D^n + a_{n-1}D^{n-1} + \cdots + a_1 D + a_0\right) y_0(t) = 0 \tag{2.4b}$$

A solution to this equation can be obtained systematically.[1] However, we will take a short cut by using heuristic reasoning. Equation (2.4b) shows that a linear combination of $y_0(t)$ and its n successive derivatives is zero, not at *some* values of t, but for all t. Such a result is possible *if and only if* $y_0(t)$ and all its n successive derivatives are of the same form. Otherwise their sum can never add to zero for all values of t. We know that only an exponential function $e^{\lambda t}$ has this property. So let us assume that

$$y_0(t) = ce^{\lambda t}$$

is a solution to Eq. (2.4b). Then

$$Dy_0(t) = \frac{dy_0}{dt} = c\lambda e^{\lambda t}$$

$$D^2 y_0(t) = \frac{d^2 y_0}{dt^2} = c\lambda^2 e^{\lambda t}$$

$$\cdots\cdots\cdots\cdots\cdots\cdots\cdots$$

$$D^n y_0(t) = \frac{d^n y_0}{dt^n} = c\lambda^n e^{\lambda t}$$

Substituting these results in Eq. (2.4b), we obtain

$$c\left(\lambda^n + a_{n-1}\lambda^{n-1} + \cdots + a_1\lambda + a_0\right)e^{\lambda t} = 0$$

For a nontrivial solution of this equation,

$$\lambda^n + a_{n-1}\lambda^{n-1} + \cdots + a_1\lambda + a_0 = 0 \tag{2.5a}$$

This result means that $ce^{\lambda t}$ is indeed a solution of Eq. (2.4), provided that λ satisfies Eq. (2.5a). Note that the polynomial in Eq. (2.5a) is identical to the polynomial $Q(D)$ in Eq. (2.4b), with λ replacing D. Therefore, Eq. (2.5a) can be expressed as

$$Q(\lambda) = 0 \tag{2.5b}$$

When $Q(\lambda)$ is expressed in factorized form, Eq. (2.5b) can be represented as

$$Q(\lambda) = (\lambda - \lambda_1)(\lambda - \lambda_2)\cdots(\lambda - \lambda_n) = 0 \tag{2.5c}$$

Clearly, λ has n solutions: $\lambda_1, \lambda_2, \ldots, \lambda_n$. Consequently, Eq. (2.4) has n possible solutions: $c_1 e^{\lambda_1 t}, c_2 e^{\lambda_2 t}, \ldots, c_n e^{\lambda_n t}$, with $c_1, c_2, \ldots, c_n$ as arbitrary constants. We can readily show that a general solution is given by the sum of these n solutions,† so that

$$y_0(t) = c_1 e^{\lambda_1 t} + c_2 e^{\lambda_2 t} + \cdots + c_n e^{\lambda_n t} \tag{2.6}$$

where $c_1, c_2, \ldots, c_n$ are arbitrary constants determined by n constraints (the auxiliary conditions) on the solution.

---

† To prove this assertion, assume that $y_1(t), y_2(t), \ldots, y_n(t)$ are all solutions of Eq. (2.4). Then

Observe that the polynomial $Q(\lambda)$, which is characteristic of the system, has nothing to do with the input. For this reason the polynomial $Q(\lambda)$ is called the **characteristic polynomial** of the system. The equation

$$Q(\lambda) = 0 \tag{2.7}$$

is called the **characteristic equation** of the system. Equation (2.5c) clearly indicates that $\lambda_1, \lambda_2, \ldots, \lambda_n$ are the roots of the characteristic equation; consequently, they are called the **characteristic roots** of the system. The terms **characteristic values**, **eigenvalues**, and **natural frequencies** are also used for characteristic roots.‡ The exponentials $e^{\lambda_i t}(i = 1, 2, \ldots, n)$ in the zero-input response are the **characteristic modes** (also known as **modes** or **natural modes**) of the system. There is a characteristic mode for each characteristic root of the system, and the *zero-input response is a linear combination of the characteristic modes of the system.*

The single most important attribute of an LTIC system is its characteristic modes. Characteristic modes not only determine the zero-input response but also play an important role in determining the zero-state response. In other words, the entire behavior of a system is dictated primarily by its characteristic modes. In the rest of this chapter we shall see the pervasive presence of characteristic modes in every aspect of system behavior.

Repeated Roots

The solution of Eq. (2.4) as given in Eq. (2.6) assumes that the n characteristic roots $\lambda_1, \lambda_2, \ldots, \lambda_n$ are distinct. If there are repeated roots (same root occurring more than once), the form of the solution is modified slightly. By direct substitution we can show that the solution of the equation

$$(D - \lambda)^2 y_0(t) = 0$$

is given by

$$y_0(t) = (c_1 + c_2 t)e^{\lambda t}$$

In this case the root λ repeats twice. Observe that the characteristic modes in this case are $e^{\lambda t}$ and $te^{\lambda t}$. Continuing this pattern, we can show that for the differential equation

$$(D - \lambda)^r y_0(t) = 0 \tag{2.8}$$

the characteristic modes are $e^{\lambda t}, te^{\lambda t}, t^2 e^{\lambda t}, \ldots, t^{r-1}e^{\lambda t}$, and that the solution is

$$y_0(t) = \left(c_1 + c_2 t + \cdots + c_r t^{r-1}\right) e^{\lambda t} \tag{2.9}$$

$$Q(D)y_1(t) = 0$$
$$Q(D)y_2(t) = 0$$
$$\cdots\cdots\cdots\cdots$$
$$Q(D)y_n(t) = 0$$

Multiplying these equations by $c_1, c_2, \ldots, c_n$, respectively, and adding them together yields
$$Q(D)\left[c_1 y_1(t) + c_2 y_2(t) + \cdots + c_n y_n(t)\right] = 0$$
This result shows that $c_1 y_1(t) + c_2 y_2(t) + \cdots + c_n y_n(t)$ is also a solution of the homogeneous equation (2.4).

‡ The term *eigenvalue* is German for characteristic value.

Consequently, for a system with the characteristic polynomial

$$Q(\lambda) = (\lambda - \lambda_1)^r (\lambda - \lambda_{r+1}) \cdots (\lambda - \lambda_n)$$

the characteristic modes are $e^{\lambda_1 t}$, $te^{\lambda_1 t}$, $\ldots$, $t^{r-1}e^{\lambda_1 t}$, $e^{\lambda_{r+1} t}$, $\ldots$, $e^{\lambda_n t}$ and the solution is

$$y_0(t) = (c_1 + c_2 t + \cdots + c_r t^{r-1})e^{\lambda_1 t} + c_{r+1}e^{\lambda_{r+1} t} + \cdots + c_n e^{\lambda_n t}$$

Complex Roots

The procedure for handling complex roots is the same as that for real roots. For complex roots the usual procedure leads to complex characteristic modes and the complex form of solution. However, it is possible to avoid the complex form altogether by selecting a real form of solution, as described below.

For a real system, complex roots must occur in pairs of conjugates if the coefficients of the characteristic polynomial $Q(\lambda)$ are to be real. Therefore, if $\alpha + j\beta$ is a characteristic root, $\alpha - j\beta$ must also be a characteristic root. The zero-input response corresponding to this pair of complex conjugate roots is

$$y_0(t) = c_1 e^{(\alpha+j\beta)t} + c_2 e^{(\alpha-j\beta)t} \tag{2.10a}$$

For a real system, the response $y_0(t)$ must also be real. This is possible only if c_1 and c_2 are conjugates. Let

$$c_1 = \frac{c}{2}e^{j\theta} \qquad \text{and} \qquad c_2 = \frac{c}{2}e^{-j\theta}$$

This yields

$$
\begin{aligned}
y_0(t) &= \frac{c}{2}e^{j\theta}e^{(\alpha+j\beta)t} + \frac{c}{2}e^{-j\theta}e^{(\alpha-j\beta)t} \\
&= \frac{c}{2}e^{\alpha t}\left[e^{j(\beta t+\theta)} + e^{-j(\beta t+\theta)}\right] \\
&= ce^{\alpha t}\cos(\beta t + \theta) \tag{2.10b}
\end{aligned}
$$

Therefore, the zero-input response corresponding to complex conjugate roots $\alpha \pm j\beta$ can be expressed in a complex form (2.10a) or a real form (2.10b). The latter is more convenient from a computational viewpoint because it avoids dealing with complex numbers.

■ **Example 2.1**

(a) Find $y_0(t)$, the zero-input component of the response for an LTI system described by the following differential equation:

$$\left(D^2 + 3D + 2\right) y(t) = Df(t)$$

when the initial conditions are $y_0(0) = 0$, $\dot{y}_0(0) = -5$. Note that $y_0(t)$, being the zero-input component ($f(t) = 0$), is the solution of $(D^2 + 3D + 2)y_0(t) = 0$.

The characteristic polynomial of the system is $\lambda^2 + 3\lambda + 2$. The characteristic equation of the system is therefore $\lambda^2 + 3\lambda + 2 = (\lambda + 1)(\lambda + 2) = 0$. The characteristic roots of the system are $\lambda_1 = -1$ and $\lambda_2 = -2$, and the characteristic modes of the system are e^{-t} and e^{-2t}. Consequently, the zero-input component of the loop current is

$$y_0(t) = c_1 e^{-t} + c_2 e^{-2t} \tag{2.11a}$$

To determine the arbitrary constants c_1 and c_2, we differentiate Eq. (2.11a) to obtain

$$\dot{y}_0(t) = -c_1 e^{-t} - 2c_2 e^{-2t} \tag{2.11b}$$

Setting $t = 0$ in Eqs. (2.11a) and (2.11b), and substituting the initial conditions $y_0(0) = 0$ and $\dot{y}_0(0) = -5$ we obtain

$$0 = c_1 + c_2$$

$$-5 = -c_1 - 2c_2$$

Solving these two simultaneous equations in two unknowns for c_1 and c_2 yields

$$c_1 = -5, \qquad c_2 = 5$$

Therefore

$$y_0(t) = -5e^{-t} + 5e^{-2t}$$

This is the zero-input component of $y(t)$ for $t \geq 0$.

(b) Similar procedure may be followed for repeated roots. For instance, for a system specified by

$$\left(D^2 + 6D + 9\right) y(t) = (3D + 5)f(t)$$

let us determine $y_0(t)$, the zero-input component of the response if the initial conditions are $y_0(0) = 3$ and $\dot{y}_0(0) = -7$.

The characteristic polynomial is $\lambda^2 + 6\lambda + 9 = (\lambda + 3)^2$, and its characteristic roots are $\lambda_1 = -3$, $\lambda_2 = -3$ (repeated roots). Consequently, the characteristic modes of the system are e^{-3t} and te^{-3t}. The zero-input response, being a linear combination of the characteristic modes, is given by

$$y_0(t) = (c_1 + c_2 t)e^{-3t}$$

We can find the arbitrary constants c_1 and c_2 from the initial conditions $y_0(0) = 3$ and $\dot{y}_0(0) = -7$ following the procedure in part **(a)**. The reader can show that $c_1 = 3$ and $c_2 = 2$. Hence,

$$y_0(t) = (3 + 2t)e^{-3t} \qquad t \geq 0$$

(c) For the case of complex roots, let us find the zero-input response of an LTI system described by the equation:

$$\left(D^2 + 4D + 40\right) y(t) = (D + 2)f(t)$$

with initial conditions $y_0(0) = 2$ and $\dot{y}_0(0) = 16.78$.

The characteristic polynomial is $\lambda^2 + 4\lambda + 40 = (\lambda + 2 - j6)(\lambda + 2 + j6)$. The characteristic roots are $-2 \pm j6$.† The solution can be written either in the complex form (2.10a) or in the real form (2.10b). The complex form is $y_0(t) = c_1 e^{\lambda_1 t} + c_2 e^{\lambda_2 t}$, where $\lambda_1 = -2 + j6$ and $\lambda_2 = -2 - j6$. Since $\alpha = -2$ and $\beta = 6$, the real form solution is [see Eq. (2.10b)]

$$y_0(t) = ce^{-2t} \cos{(6t + \theta)} \tag{2.12a}$$

---

†The complex conjugate roots of a second-order polynomial can be determined by using the formula in Sec. B.7-10 or by expressing the polynomial as a sum of two squares. The latter can be accomplished by completing the square with the first two terms, as shown below:

$$\lambda^2 + 4\lambda + 40 = (\lambda^2 + 4\lambda + 4) + 36 = (\lambda + 2)^2 + (6)^2 = (\lambda + 2 - j6)(\lambda + 2 + j6)$$

where c and θ are arbitrary constants to be determined from the initial conditions $y_0(0) = 2$ and $\dot{y}_0(0) = 16.78$. Differentiation of Eq. (2.12a) yields

$$\dot{y}_0(t) = -2ce^{-2t}\cos(6t + \theta) - 6ce^{-2t}\sin(6t + \theta) \qquad (2.12b)$$

Setting $t = 0$ in Eqs. (2.12a) and (2.12b), and then substituting initial conditions, we obtain

$$2 = c\cos\theta$$

$$16.78 = -2c\cos\theta - 6c\sin\theta$$

Solution of these two simultaneous equations in two unknowns $c\cos\theta$ and $c\sin\theta$ yields

$$c\cos\theta = 2 \qquad (2.13a)$$

$$c\sin\theta = -3.463 \qquad (2.13b)$$

Squaring and then adding the two sides of the above equations yields

$$c^2 = (2)^2 + (-3.464)^2 = 16 \Longrightarrow c = 4$$

Next, dividing (2.13b) by (2.13a); that is dividing $c\sin\theta$ by $c\cos\theta$ yields

$$\tan\theta = \frac{-3.463}{2}$$

and

$$\theta = \tan^{-1}\left(\frac{-3.463}{2}\right) = -\frac{\pi}{3}$$

Therefore

$$y_0(t) = 4e^{-2t}\cos\left(6t - \frac{\pi}{3}\right)$$

Figure B.11c shows the plot of $y_0(t)$. ∎

⊙ **Computer Example C2.1**
Find the roots of polynomial $\lambda^2 + 4\lambda + 40$

```
a=[1 4 40]; r= roots(a)
r=-2.0000 + 6.0000i
   -2.0000 - 6.0000i   ⊙
```

⊙ **Computer Example C2.2**
For an LTIC system specified by the differential equation

$$(D^2 + 4D + k)y(t) = (3D + 5)f(t)$$

determine the zero-input component of the response if the initial conditions are $y_0(0) = 3$, and $\dot{y}_0(0) = -7$ for two values of k: (a) 3 (b) 4 (c) 40.

```
(a)   y0=dsolve('D2y+4*Dy+3*y=0','y(0)=3','Dy(0)=-7','t')
      y0 = 2*exp(-3*t)+exp(-t)

(b)   y0=dsolve('D2y+4*Dy+4*y=0','y(0)=3','Dy(0)=-7','t')
      y0 = 3*exp(-2*t)-exp(-2t)*t

(c)   y0=dsolve('D2y+4*Dy+40*y=0','y(0)=3','Dy(0)=-7','t')
      y0 = 3*exp(-2*t)*cos(6*t)-1/6*exp(-2t)*sin(6*t)   ⊙
```

△ **Exercise E2.1**
Find the zero-input response of an LTIC system described by $(D+5)y(t) = f(t)$ if the initial condition is $y(0) = 5$.
Answer: $y_0(t) = 5e^{-5t} \qquad t \geq 0$ ▽

△ **Exercise E2.2**
Solve

$$\left(D^2 + 2D\right)y_0(t) = 0$$

if $y_0(0) = 1$ and $\dot{y}_0(0) = 4$. Hint: The characteristic roots are 0 and -2.
Answer: $y_0(t) = 3 - 2e^{-2t} \qquad t \geq 0$ ▽

Practical Initial Conditions and the meaning of 0^- and 0^+

In Example 2.1 the initial conditions $y_0(0)$ and $\dot{y}_0(0)$ were supplied. In practical problems, we must derive such conditions from the physical situation. For instance, in an RLC circuit, we may be given the conditions, such as initial capacitor voltages, and initial inductor currents, etc. From this information, we need to derive $y_0(0)$, $\dot{y}_0(0)$, $\cdots$ for the desired variable as demonstrated in the next example.

In much of our discussion, the input is assumed to start at $t = 0$, unless otherwise mentioned. Hence, $t = 0$ is the reference point. The conditions immediately before $t = 0$ (just before the input is applied) are the conditions at $t = 0^-$, and those immediately after $t = 0$ (just after the input is applied) are the conditions at $t = 0^+$ (compare this with the historical time frame B.C. and A.D.). In practice, we are likely to know the initial conditions at $t = 0^-$ rather than at $t = 0^+$. The two sets of conditions are generally different, although in some cases they may be identical.

We are dealing with the total response $y(t)$, which consists of two components; the zero-input component $y_0(t)$ (response due to the initial conditions alone with $f(t) = 0$) and the zero-state component resulting from the input alone with all initial conditions zero. At $t = 0^-$, the response $y(t)$ consists solely of the zero-input component $y_0(t)$ because the input has not started yet. Hence the initial conditions on $y(t)$ are identical to those of $y_0(t)$. Thus, $y(0^-) = y_0(0^-)$, $\dot{y}(0^-) = \dot{y}_0(0^-)$, and so on. Moreover, $y_0(t)$ is the response due to initial conditions alone and does not depend on the input $f(t)$. Hence, application of the input at $t = 0$ does not affect $y_0(t)$. This means the initial conditions on $y_0(t)$ at $t = 0^-$ and 0^+ are identical; that is $y_0(0^-)$, $\dot{y}_0(0^-)$, $\cdots$ are identical to $y_0(0^+)$, $\dot{y}_0(0^+)$, $\cdots$, respectively. It is clear that for $y_0(t)$, there is no distinction between the initial conditions at $t = 0^-$, 0 and 0^+. They are all the same. But this is not the case with the total response $y(t)$, which consists of both, the zero-input and the zero-state components. Thus, in general, $y(0^-) \neq y(0^+)$, $\dot{y}(0^-) \neq \dot{y}(0^+)$, and so on.

■ Example 2.2

A voltage $f(t) = 10e^{-3t}u(t)$ is applied at the input of the RLC circuit illustrated in Fig. 2.1a. Find the loop current $y(t)$ for $t \geq 0$ if the initial inductor current is zero; that is, $y(0^-) = 0$ and the initial capacitor voltage is 5 volts; that is, $v_C(0^-) = 5$.

The differential (loop) equation relating $y(t)$ to $f(t)$ was derived in Eq. (1.55) as

$$\left(D^2 + 3D + 2\right) y(t) = Df(t)$$

The zero-state component of $y(t)$ resulting from the input $f(t)$, assuming that all initial conditions are zero; that is, $y(0^-) = v_C(0^-) = 0$, will be obtained later in Example 2.5. In this example we shall find the zero-input component $y_0(t)$. For this purpose, we need two initial conditions $y_0(0)$ and $\dot{y}_0(0)$. These conditions can be derived from the given initial conditions, $y(0^-) = 0$ and $v_C(0^-) = 5$, as follows. Recall that $y_0(t)$ is the loop current when the input terminals are shorted at $t = 0$, so that the input $f(t) = 0$ (zero-input) as depicted in Fig. 2.1b. We now compute $y_0(0)$ and $\dot{y}_0(0)$, the values of the loop current and its derivative at $t = 0$, from the initial values of the inductor current and the capacitor voltage. Remember that the inductor current cannot change instantaneously in the absence of an impulsive voltage. Similarly, the capacitor voltage cannot change instantaneously in the absence of an impulsive current. Therefore, when the input terminals are shorted at $t = 0$, the inductor current is still zero and the capacitor voltage is still 5 volts. Thus,

$$y_0(0) = 0$$

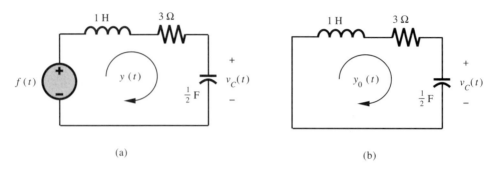

(a)   (b)

Fig. 2.1

To determine $\dot{y}_0(0)$, we use the loop equation for the circuit in Fig. 2.1b. Because the voltage across the inductor is $L(dy_0/dt)$ or $\dot{y}_0(t)$, this equation can be written as follows:

$$\dot{y}_0(t) + 3y_0(t) + v_C(t) = 0$$

Setting $t = 0$, we obtain

$$\dot{y}_0(0) + 3y_0(0) + v_C(0) = 0 \tag{2.14}$$

But $y_0(0) = 0$ and $v_C(0) = 5$. Consequently,

$$\dot{y}_0(0) = -5$$

Therefore, the desired initial conditions are

$$y_0(0) = 0 \qquad \text{and} \qquad \dot{y}_0(0) = -5$$

Thus, the problem reduces to finding $y_0(t)$, the zero-input component of $y(t)$ of the system specified by the equation $(D^2 + 3D + 2)y(t) = Df(t)$, when the initial conditions are $y_0(0) = 0$ and $\dot{y}_0(0) = -5$. We have already solved this problem in Example 2.1a, where we found

$$y_0(t) = -5e^{-t} + 5e^{-2t} \qquad t \geq 0 \tag{2.15}$$

This is the zero-input component of the loop current $y(t)$.

    It will be interesting to find the initial conditions at $t = 0^-$ and 0^+ for the total response $y(t)$. Let us compare $y(0^-)$ and $\dot{y}(0^-)$ with $y(0^+)$ and $\dot{y}(0^+)$. The two pairs can be compared by writing the loop equation for the circuit in Fig. 2.1a at $t = 0^-$ and $t = 0^+$. The only difference between the two situations is that at $t = 0^-$, the input $f(t) = 0$, whereas at $t = 0^+$, the input $f(t) = 10$ (because $f(t) = 10e^{-3t}$). Hence, the two loop equations are

$$\dot{y}(0^-) + 3y(0^-) + v_C(0^-) = 0$$

$$\dot{y}(0^+) + 3y(0^+) + v_C(0^+) = 10$$

The loop current $y(0^+) = y(0^-) = 0$ because it cannot change instantaneously in the absence of impulsive voltage. The same is true of the capacitor voltage. Hence, $v_C(0^+) = v_C(0^-) = 5$. Substituting these values in the above equations, we obtain $\dot{y}(0^-) = -5$ and $\dot{y}(0^+) = 5$. Thus

$$y(0^-) = 0, \ \dot{y}(0^-) = -5 \quad \text{and} \quad y(0^+) = 0, \ \dot{y}(0^+) = 5 \tag{2.16}$$

                                                                                            ■

△ **Exercise E2.3**

    In the circuit in Fig. 2.1a, the inductance $L = 0$ and the initial capacitor voltage $v_C(0) = 30$ volts. Show that the zero-input component of the loop current is given by $y_0(t) = -10e^{-2t/3}$ for $t \geq 0$. ▽

Independence of Zero-Input and Zero-State Response

In this example we computed the zero-input component without using the input $f(t)$. The zero-state component can be computed from the knowledge of the input $f(t)$ alone; the initial conditions are assumed to be zero (system in zero state). The two components of the system response (the zero-input and zero-state components) are independent of each other. *The two worlds of zero-input response and zero-state response coexist side by side, neither of them knowing or caring what the other is doing. For each component, the other is totally irrelevant.*

Role of Auxiliary Conditions in Solution of Differential Equations

Solution of a differential equation requires additional pieces of information (the **auxiliary conditions**). Why? We now show that a differential equation does not, in general, have a unique solution unless some additional constraints (or conditions) on the solution are known. The reason is that, as explained in the discussion on invertibility (Sec. 1.7), the differentiation operation is not invertible unless one piece of information about $y(t)$ is given. Thus, differentiation is an irreversible (noninvertible) operation during which certain information is lost. To invert this operation, one piece of information about $y(t)$ must be provided to restore the original $y(t)$. Using a similar argument, we can show that, given d^2y/dt^2, we can determine $y(t)$ uniquely only if two additional pieces of information (constraints) about $y(t)$ are given. In general, to determine $y(t)$ uniquely from its nth derivative, we need n additional pieces of information (constraints) about $y(t)$. These constraints are also called *auxiliary conditions*. When these conditions are given at $t = 0$, they are called *initial conditions*.

2.2-1 Some Insights into the Zero-Input Behavior of a System

By definition, the zero-input response is the system response to its internal conditions, assuming that its input is zero. Understanding this phenomenon provides interesting insight into system behavior. If a system is disturbed momentarily from its rest position and if the disturbance is then removed, the system will not come back to rest instantaneously. In general, it will come back to rest over a period of time and only through a special type of motion that is characteristic of the system.† For example, if we press on an automobile fender momentarily and then release it at $t = 0$, there is no external force on the automobile for $t > 0$‡. The auto body will eventually come back to its rest (equilibrium) position, but not through any arbitrary motion. It must do so using only a form of response which is sustainable by the system on its own without any external source, because the input is zero. Only characteristic modes satisfy this condition. *The system uses a proper combination of characteristic modes to come back to the rest position while satisfying appropriate boundary (or initial) conditions.*

If the shock absorbers of the automobile are in good condition (high damping coefficient), the characteristic modes will be monotonically decaying exponentials, and the auto body will come to rest rapidly without oscillation. In contrast, for

---

† This assumes that the system will eventually come back to its original rest (or equilibrium) position.

‡ We ignore the force of gravity, which merely causes a constant displacement of the auto body without affecting the other motion.

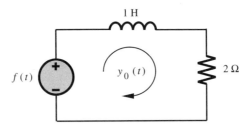

Fig. 2.2   Modes always get a free ride.

poor shock absorbers (low damping coefficients), the characteristic modes will be exponentially decaying sinusoids, and the body will come to rest through oscillatory motion. When a series RC circuit with an initial charge on the capacitor is shorted, the capacitor will start to discharge exponentially through the resistor. This response of the RC circuit is caused entirely by its internal conditions and is sustained by this system without the aid of any external input. The exponential current waveform is therefore the characteristic mode of this RC circuit.

Mathematically we know that *any combination of characteristic modes can be sustained by the system alone without requiring an external input.* This fact can be readily verified for the series RL circuit shown in Fig. 2.2. The loop equation for this system is

$$(D + 2)y(t) = f(t)$$

It has a single characteristic root $\lambda = -2$, and the characteristic mode is e^{-2t}. We now verify that a loop current $y(t) = ce^{-2t}$ can be sustained through this circuit without any input voltage. The input voltage $f(t)$ required to drive a loop current $y(t) = ce^{-2t}$ is given by

$$f(t) = L\frac{dy}{dt} + Ry(t)$$

$$= \frac{d}{dt}(ce^{-2t}) + 2ce^{-2t}$$

$$= -2ce^{-2t} + 2ce^{-2t}$$

$$= 0$$

Clearly, the loop current $y(t) = ce^{-2t}$ is sustained by the RL circuit on its own, without the necessity of an external input.

The Resonance Phenomenon

We have seen that any signal consisting of a system's characteristic modes is sustained by the system on its own; the system offers no obstacle to such signals. Imagine what would happen if we actually drove the system with an external input that is one of its characteristic modes. This would be like pouring gasoline on a fire in a dry forest or hiring an alcoholic to taste liquor. An alcoholic would gladly do the job without pay. Think what will happen if he were paid by the amount of liquor he tasted! The system response to characteristic modes would naturally be very high. We call this behavior the **resonance phenomenon**. An intelligent discussion of this important phenomenon requires an understanding of the zero-state response; for this reason we postpone this topic until Sec. 2.7-7.

2.3   The Unit Impulse Response h(t)

The impulse function $\delta(t)$ is also used in determining the response of a linear system to an arbitrary input $f(t)$. In Chapter 1 we explained how a system response to an input $f(t)$ may be found by breaking this input into narrow rectangular pulses, as illustrated in Fig. 1.27a, and then summing the system response to all the components. The rectangular pulses become impulses in the limit as their widths approach zero. Therefore, the system response is the sum of its responses to various impulse components. This discussion shows that if we know the system response to an impulse input, we can determine the system response to an arbitrary input $f(t)$. We now discuss a method of determining $h(t)$, the unit impulse response of an LTIC system described by the nth-order differential equation

$$Q(D)y(t) = P(D)f(t) \tag{2.17a}$$

where $Q(D)$ and $P(D)$ are the polynomials shown in Eq. (2.2). Recall that noise considerations restrict practical systems to $m \leq n$. Under this constraint, the most general case is $m = n$. Therefore, Eq. (2.17a) can be expressed as

$$(D^n + a_{n-1}D^{n-1} + \cdots + a_1 D + a_0)y(t) =$$
$$(b_n D^n + b_{n-1}D^{n-1} + \cdots + b_1 D + b_0)f(t) \tag{2.17b}$$

Before deriving the general expression for the unit impulse response $h(t)$, it is illuminating to understand qualitatively the nature of $h(t)$. The impulse response $h(t)$ is the system response to an impulse input $\delta(t)$ applied at $t = 0$ with all the initial conditions zero at $t = 0^-$. An impulse input $\delta(t)$ is like lightning, which strikes instantaneously and then vanishes. But in its wake, in that single moment, lightning rearranges things where it strikes. Similarly, an impulse input $\delta(t)$ appears momentarily at $t = 0$, and then it is gone forever. But in that moment it generates energy storages; that is, it creates nonzero initial conditions instantaneously within the system at $t = 0^+$. Although the impulse input $\delta(t)$ vanishes for $t > 0$ so that the system has no input after the impulse has been applied, the system will still have a response generated by these newly created initial conditions. The impulse response $h(t)$, therefore, must consist of the system's characteristic modes for $t \geq 0^+$. As a result

$$h(t) = \text{characteristic mode terms} \qquad t \geq 0^+$$

This response is valid for $t > 0$. But what happens at $t = 0$? At a single moment $t = 0$, there can at most be an impulse†, so the form of the complete response $h(t)$ is given by

---

† It might be possible for the derivatives of $\delta(t)$ to appear at the origin. However, if $m \leq n$, it is impossible for $h(t)$ to have any derivatives of $\delta(t)$. This conclusion follows from Eq. (2.17b) with $f(t) = \delta(t)$ and $y(t) = h(t)$. The coefficients of the impulse and all of its derivatives must be matched on both sides of this equation. If $h(t)$ contains $\delta^{(1)}(t)$, the first derivative of $\delta(t)$, the left-hand side of Eq. (2.17b) will contain a term $\delta^{(n+1)}(t)$. But the highest-order derivative term on the right-hand side is $\delta^{(n)}(t)$. Therefore, the two sides cannot match. Similar arguments can be made against the presence of the impulse's higher-order derivatives in $h(t)$.

$$h(t) = A_0\delta(t) + \text{characteristic mode terms} \qquad t \geq 0 \qquad (2.18)$$

The detailed derivation of $h(t)$ is neither illuminating nor necessary for our future development, so to prevent needless distraction, this derivation is placed in Appendix 2.1 at the end of the chapter. There, we show that for an LTIC system specified by Eq. (2.17), the unit impulse response $h(t)$ is given by

$$h(t) = b_n\delta(t) + [P(D)y_n(t)]u(t) \qquad (2.19)$$

where b_n is the coefficient of the nth-order term in $P(D)$ [see Eq. (2.17b)], and $y_n(t)$ is a linear combination of the characteristic modes of the system subject to the following initial conditions:

$$y_n^{(n-1)}(0) = 1, \quad \text{and} \quad y_n(0) = \dot{y}_n(0) = \ddot{y}_n(0) = \cdots = y_n^{(n-2)}(0) = 0 \qquad (2.20)$$

where $y_n^{(k)}(0)$ is the value of the kth derivative of $y_n(t)$ at $t = 0$. We can express this condition for various values of n (the system order) as follows:

$$n = 1: \ y_n(0) = 1$$

$$n = 2: \ y_n(0) = 0 \ \text{and} \ \dot{y}_n(0) = 1$$

$$n = 3: \ y_n(0) = \dot{y}_n(0) = 0 \ \ \text{and} \ \ddot{y}_n(0) = 1$$

$$n = 4: \ y_n(0) = \dot{y}_n(0) = \ddot{y}_n(0) = 0 \ \text{and} \ \dddot{y}_n(0) = 1 \qquad (2.21)$$

and so on.

If the order of $P(D)$ is less than the order of $Q(D)$, $b_n = 0$, and the impulse term $b_n\delta(t)$ in $h(t)$ is zero.

■ **Example 2.3**

Determine the unit impulse response $h(t)$ for a system specified by the equation

$$\left(D^2 + 3D + 2\right) y(t) = Df(t) \qquad (2.22)$$

This is a second-order system ($n = 2$) having the characteristic polynomial

$$\left(\lambda^2 + 3\lambda + 2\right) = (\lambda + 1)(\lambda + 2)$$

The characteristic roots of this system are $\lambda = -1$ and $\lambda = -2$. Therefore

$$y_n(t) = c_1 e^{-t} + c_2 e^{-2t} \qquad (2.23a)$$

Differentiation of this equation yields

$$\dot{y}_n(t) = -c_1 e^{-t} - 2c_2 e^{-2t} \qquad (2.23b)$$

The initial conditions are [see Eq. (2.21) for $n = 2$]

$$\dot{y}_n(0) = 1 \qquad \text{and} \qquad y_n(0) = 0$$

Setting $t = 0$ in Eqs. (2.23a) and (2.23b), and substituting the above initial conditions, we obtain

$$0 = c_1 + c_2$$

$$1 = -c_1 - 2c_2 \qquad (2.24)$$

Solution of these two simultaneous equations yields

$$c_1 = 1 \quad \text{and} \quad c_2 = -1$$

Therefore

$$y_n(t) = e^{-t} - e^{-2t} \tag{2.25}$$

Moreover, according to Eq. (2.22), $P(D) = D$, so that

$$P(D)y_n(t) = Dy_n(t) = \dot{y}_n(t) = -e^{-t} + 2e^{-2t}$$

Also in this case, $b_n = b_2 = 0$ [the second-order term is absent in $P(D)$]. Therefore

$$h(t) = b_n\delta(t) + [P(D)y_n(t)]u(t) = (-e^{-t} + 2e^{-2t})u(t) \quad \blacksquare \tag{2.26}$$

Comment

In the above discussion, we have assumed $m \leq n$, as specified by Eq. (2.17b). Appendix 2.1 shows that the expression for $h(t)$ applicable to all possible values of m and n is given by

$$h(t) = P(D)[y_n(t)u(t)]$$

where $y_n(t)$ is a linear combination of the characteristic modes of the system subject to initial conditions (2.20). This expression reduces to Eq. (2.19) when $m \leq n$.

Determination of the impulse response $h(t)$ using the procedure in this section is relatively simple. However, in Chapter 6 we shall discuss another, even simpler method using the Laplace transform.

△ **Exercise E2.4**

Determine the unit impulse response of LTIC systems described by the equations:

(a) $(D + 2)y(t) = (3D + 5)f(t)$
(b) $D(D + 2)y(t) = (D + 4)f(t)$
(c) $(D^2 + 2D + 1)y(t) = Df(t)$

Answers: (a) $3\delta(t) - e^{-2t}u(t)$ (b) $(2 - e^{-2t})u(t)$ (c) $(1 - t)e^{-t}u(t)$ ▽

⊙ **Computer Example C2.3**

Determine the impulse response $h(t)$ for an LTIC system specified by the differential equation

$$(D^2 + 3D + 2)y(t) = Df(t)$$

This is a second-order system with $b_n = b_2 = 0$. First we find the zero-input component for initial conditions $y(0^-) = 0$, and $\dot{y}(0^-) = 1$

Yzi=dsolve('D2y+3*Dy+2*y=0','y(0)=0','Dy(0)=1','t')
Yzi = -exp(-2*t)+exp(-t)
Since $P(D) = D$, we differentiate the zero-input response:
PYzi=symdiff(Yzi)
PYzi = 2*exp(-2*t)-exp(-t)

Therefore

$$h(t) = b_2\delta(t) + [Dy_0(t)]u(t) = (2e^{-2t} - e^{-t})u(t) \quad ⊙$$

System Response to Delayed Impulse

If $h(t)$ is the response of an LTIC system to the input $\delta(t)$, then $h(t-T)$ is the response of this same system to the input $\delta(t-T)$. This conclusion follows from the time-invariance property of LTIC systems. Thus, by knowing the unit impulse response $h(t)$, we can determine the system response to a delayed impulse $\delta(t-T)$.

2.4   System Response to External Input: Zero-state Response

This section is devoted to the determination of the zero-state response of an LTIC system. This is the system response $y(t)$ to an input $f(t)$ when the system is in zero state; that is, when all initial conditions are zero. **We shall assume that the systems discussed in this section are in zero state unless mentioned otherwise.** Under these conditions, the zero-state response will be the total response of the system.

We shall use the superposition principle here to derive a linear system's response to some arbitrary input $f(t)$. In this approach, we express $f(t)$ in terms of impulses. We begin by approximating $f(t)$ with narrow rectangular pulses, as depicted in Fig. 2.3a. This procedure gives us a staircase approximation of $f(t)$ that improves as pulse width is reduced. In the limit as pulse width approaches zero, this representation becomes exact, and the rectangular pulses become impulses delayed by various amounts. The system response to the input $f(t)$ is then given by the sum of the system's responses to each (delayed) impulse component of $f(t)$. In other words, we can determine $y(t)$, the system response to any arbitrary input $f(t)$, if we know the impulse response of the system.

For the sake of generality, we place no restriction on $f(t)$ as to where it starts and where it ends. It is therefore assumed to exist for all time, starting at $t = -\infty$. The system's total response to this input will then be given by the sum of its responses to each of these impulse components. This process is illustrated in Fig. 2.3.

Figure 2.3a shows $f(t)$ as a sum of rectangular pulses, each of width $\triangle\tau$. In the limit as $\triangle\tau \to 0$, each pulse approaches an impulse having a strength equal to the area under that pulse. For example, as $\triangle\tau \to 0$, the shaded rectangular pulse located at $t = n\triangle\tau$ in Fig. 2.3a will approach an impulse at the same location with strength $f(n\triangle\tau)\triangle\tau$ (the shaded area under the rectangular pulse). This impulse can therefore be represented by $[f(n\triangle\tau)\triangle\tau]\delta(t-n\triangle\tau)$, as shown in Fig. 2.3d.

If the system's response to a unit impulse $\delta(t)$ is $h(t)$ (Fig. 2.3b), its response to a delayed impulse $\delta(t-n\triangle\tau)$ will be $h(t-n\triangle\tau)$ (Fig. 2.3c). Consequently, the system's response to $[f(n\triangle\tau)\triangle\tau]\delta(t-n\triangle\tau)$ will be $[f(n\triangle\tau)\triangle\tau]h(t-n\triangle\tau)$, as illustrated in Fig. 2.3d. These results can be conveniently displayed as input-output pairs with an arrow directed from the input to the output as shown below. The left-hand side represents the input, and the right-hand side represents the corresponding system response:

$$\delta(t) \implies h(t)$$

$$\delta(t-n\triangle\tau) \implies h(t-n\triangle\tau)$$

$$\underbrace{[f(n\triangle\tau)\triangle\tau]\delta(t-n\triangle\tau)}_{\text{input}} \implies \underbrace{[f(n\triangle\tau)\triangle\tau]h(t-n\triangle\tau)}_{\text{output}} \qquad (2.27)$$

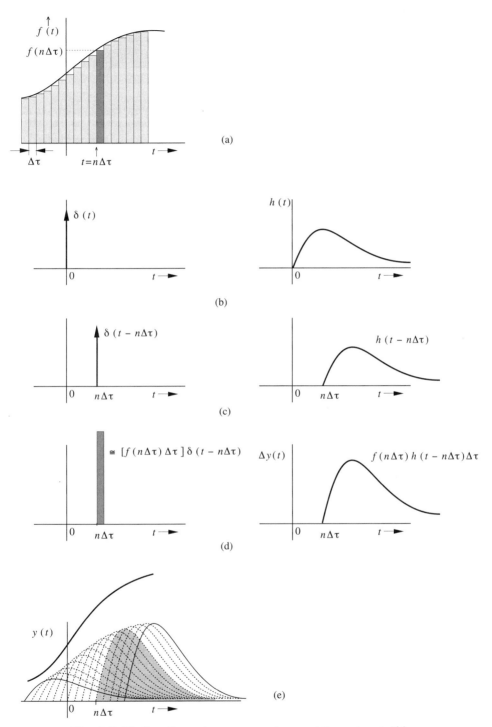

Fig. 2.3 Finding the system response to an arbitrary input $f(t)$.

This display shows the input-output pairs in Figs. 2.3b, c, and d, respectively. The last pair represents the system response to only one of the impulse components of $f(t)$. The total response $y(t)$ is obtained by summing all such components as depicted in Fig. 2.3e. Summing on both sides of the above display yields (with $\triangle\tau \to 0$)

$$\underbrace{\lim_{\triangle\tau \to 0} \sum_{n=-\infty}^{\infty} f(n\triangle\tau)\delta(t - n\triangle\tau)\triangle\tau}_{\text{The input } f(t)} \implies \underbrace{\lim_{\triangle\tau \to 0} \sum_{n=-\infty}^{\infty} f(n\triangle\tau)h(t - n\triangle\tau)\triangle\tau}_{\text{The output } y(t)}$$

The left-hand side is the input $f(t)$ represented as a sum of all the impulse components in a manner illustrated in Fig. 2.3a. The right-hand side is the output $y(t)$ represented as a sum of the output components as shown in Fig. 2.3e. Both the left-hand side and the right-hand side, by definition, are integrals given by†

$$\underbrace{\int_{-\infty}^{\infty} f(\tau)\delta(t - \tau)\,d\tau}_{f(t)} \implies \underbrace{\int_{-\infty}^{\infty} f(\tau)h(t - \tau)\,d\tau}_{y(t)} \qquad (2.28)$$

The left-hand side expresses the input $f(t)$ as made up of the impulse components in a manner depicted in Fig. 2.3a. The right-hand expresses the output as made up of the sum of the system responses to all the impulse components of the input as illustrated in Fig. 2.3e. To summarize, the (zero-state) response $y(t)$ to the input $f(t)$ is given by

$$y(t) = \int_{-\infty}^{\infty} f(\tau)h(t - \tau)\,d\tau \qquad (2.29)$$

This is the result we seek. We have obtained the system response $y(t)$ to input $f(t)$ in terms of the unit impulse response $h(t)$. Knowing $h(t)$, we can determine the response $y(t)$ to any input. *Observe once again the all-pervasive nature of the system's characteristic modes. The system response to any input is determined by the impulse response, which in turn is made up of characteristic modes of the system.*

It is important to keep in mind the assumptions used in deriving Eq. (2.29). We assumed a linear, time-invariant (LTI) system. Linearity allowed us to use the principle of superposition, and time-invariance made it possible to express the system's response to $\delta(t - n\triangle\tau)$ as $h(t - n\triangle\tau)$.

2.4-1 The Convolution Integral

The zero-state response $y(t)$ obtained in Eq. (2.29) is given by an integral that occurs frequently in the physical sciences, engineering, and mathematics. For

---

† In deriving this result we have assumed a time-invariant system. If the system is time-varying, then the system response to the input $\delta(t - n\triangle\tau)$ cannot be expressed as $h(t - n\triangle\tau)$, but instead has the form $h(t, n\triangle\tau)$. Using this form will modify Eq. (2.28) as

$$y(t) = \int_{-\infty}^{\infty} f(\tau)h(t, \tau)\,d\tau \qquad (2.28\text{n})$$

where $h(t, \tau)$ is the system response at instant t to a unit impulse input located at τ.

this reason this integral is given a special name: the **convolution integral**. The convolution integral of two functions $f_1(t)$ and $f_2(t)$ is denoted symbolically by $f_1(t) * f_2(t)$ and is defined as

$$f_1(t) * f_2(t) \equiv \int_{-\infty}^{\infty} f_1(\tau) f_2(t - \tau) \, d\tau \tag{2.30}$$

Some important properties of the convolution integral are given below.

1. **The Commutative Property**: Convolution operation is commutative; that is, $f_1(t) * f_2(t) = f_2(t) * f_1(t)$. This property can be proved by a change of variable. In Eq. (2.30), if we let $x = t - \tau$ so that $\tau = t - x$ and $d\tau = -dx$, we obtain

$$f_1(t) * f_2(t) = -\int_{\infty}^{-\infty} f_2(x) f_1(t - x) \, dx$$

$$= \int_{-\infty}^{\infty} f_2(x) f_1(t - x) \, dx$$

$$= f_2(t) * f_1(t) \tag{2.31}$$

2. **The Distributive Property**: According to this property:

$$f_1(t) * [f_2(t) + f_3(t)] = f_1(t) * f_2(t) + f_1(t) * f_3(t) \tag{2.32}$$

3. **The Associative Property**: According to this property:

$$f_1(t) * [f_2(t) * f_3(t)] = [f_1(t) * f_2(t)] * f_3(t) \tag{2.33}$$

The proofs of (2.32) and (2.33) follow directly from the definition of the convolution integral. They are left as an exercise for the reader.

4. **The Shift Property**: If

$$f_1(t) * f_2(t) = c(t)$$

then

$$f_1(t) * f_2(t - T) = c(t - T) \tag{2.34a}$$

$$f_1(t - T) * f_2(t) = c(t - T) \tag{2.34b}$$

and

$$f_1(t - T_1) * f_2(t - T_2) = c(t - T_1 - T_2) \tag{2.34c}$$

Proof: We are given

$$f_1(t) * f_2(t) = \int_{-\infty}^{\infty} f_1(\tau) f_2(t - \tau) \, d\tau = c(t)$$

Therefore

$$f_1(t) * f_2(t - T) \equiv \int_{-\infty}^{\infty} f_1(\tau) f_2(t - T - \tau) \, d\tau$$

$$= c(t - T)$$

Equation (2.34b) follows from (2.34a) and the commutative property of convolution; Eq. (2.34c) follows directly from (2.34a) and (2.34b).

5. **Convolution with an Impulse**: Convolution of a function $f(t)$ with a unit impulse results in the function $f(t)$ itself. By definition of convolution

$$f(t) * \delta(t) = \int_{-\infty}^{\infty} f(\tau)\delta(t - \tau)\, d\tau \qquad (2.35)$$

Because $\delta(t - \tau)$ is an impulse located at $\tau = t$, according to the sampling property of the impulse [Eq. (1.24)], the integral in the above equation is the value of $f(\tau)$ at $\tau = t$, that is, $f(t)$. Therefore

$$f(t) * \delta(t) = f(t) \qquad (2.36)$$

Actually this result has been derived earlier in Eq. (2.28).

6. **The Width Property**: If the durations (widths) of $f_1(t)$ and $f_2(t)$ are T_1 and T_2 respectively, then the duration (width) of $f_1(t) * f_2(t)$ is $T_1 + T_2$ (Fig. 2.4). The proof of this property follows readily from the graphical considerations discussed later in Sec. 2.4-2. This rule may superficially appear to be violated in some special cases discussed later.

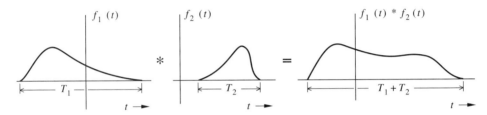

Fig. 2.4 Width property of convolution.

Zero-State Response and Causality

The (zero-state) response $y(t)$ of an LTIC system is

$$y(t) = f(t) * h(t) = \int_{-\infty}^{\infty} f(\tau)h(t - \tau)\, d\tau \qquad (2.37)$$

In deriving Eq. (2.37), we assumed the system to be linear and time-invariant. There were no other restrictions either on the system or on the input signal $f(t)$. In practice, most systems are causal, so that their response cannot begin before the input starts. Furthermore, most inputs are also causal, which means they start at $t = 0$.

Causality restrictions on both signals and systems further simplify the limits of integration in Eq. (2.37). By definition, the response of a causal system cannot begin before its input begins. Consequently, the causal system's response to a unit impulse $\delta(t)$ (which is located at $t = 0$) cannot begin before $t = 0$. Therefore, a *causal system's unit impulse response $h(t)$ is a causal signal.*

It is important to remember that the integration in Eq. (2.37) is performed with respect to τ (not t). If the input $f(t)$ is causal, $f(\tau) = 0$ for $\tau < 0$. Therefore, $f(\tau) = 0$ for $\tau < 0$, as illustrated in Fig. 2.5a. Similarly, if $h(t)$ is causal, $h(t-\tau) = 0$ for $t - \tau < 0$; that is, for $\tau > t$, as depicted in Fig. 2.5a. Therefore, the product $f(\tau)h(t - \tau) = 0$ everywhere except over the nonshaded interval $0 \leq \tau \leq t$ shown in Fig. 2.5a (assuming $t \geq 0$). Observe that if t is negative, $f(\tau)h(t - \tau) = 0$ for all τ as shown in Fig. 2.5b. Therefore, Eq. (2.37) reduces to

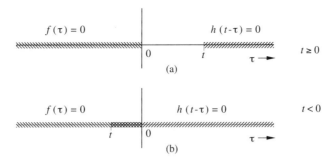

Fig. 2.5 Limits of convolution integral.

$$y(t) = f(t) * h(t) = \int_{0^-}^{t} f(\tau)h(t - \tau)\, d\tau \qquad t \geq 0 \qquad (2.38)$$

$$= 0 \qquad\qquad t < 0$$

The lower limit of integration in Eq. (2.38) is taken as 0^- to avoid the difficulty in integration that can arise if $f(t)$ contains an impulse at the origin. In subsequent discussion, the lower limit will be shown as 0 with the understanding that it means 0^-. This result shows that if $f(t)$ and $h(t)$ are both causal, the response $y(t)$ is also causal.

Because of the convolution's commutative property [Eq. (2.31)], we can also express Eq. (2.38) as [assuming causal $f(t)$ and $h(t)$]

$$y(t) = \int_{0}^{t} h(\tau)f(t - \tau)\, d\tau \qquad t \geq 0 \qquad (2.39)$$

As in Eq. (2.38), this result assumes that both the input and the system are causal.

■ **Example 2.4**

For an LTIC system with the unit impulse response $h(t) = e^{-2t}u(t)$, determine the response $y(t)$ for the input

$$f(t) = e^{-t}u(t) \qquad\qquad (2.40)$$

Here both $f(t)$ and $h(t)$ are causal (Fig. 2.6). Hence, we need only to perform the convolution's integration over the range $(0, t)$ [see Eq. (2.38)]. The system response is therefore given by

$$y(t) = \int_{0}^{t} f(\tau)h(t - \tau)\, d\tau \qquad t \geq 0$$

Because $f(t) = e^{-t}u(t)$ and $h(t) = e^{-2t}u(t)$

$$f(\tau) = e^{-\tau}u(\tau) \quad \text{and} \quad h(t - \tau) = e^{-2(t-\tau)}u(t - \tau)$$

Remember that the integration is performed with respect to τ (not t), and the region of integration is $0 \leq \tau \leq t$. In other words, τ lies between 0 and t. Therefore, if $t \geq 0$, then $\tau \geq 0$ and $t - \tau \geq 0$, so that $u(\tau) = 1$ and $u(t - \tau) = 1$; consequently

$$y(t) = \int_{0}^{t} e^{-\tau}e^{-2(t-\tau)}\, d\tau \qquad t \geq 0$$

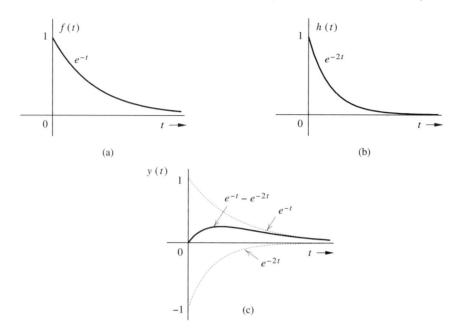

Fig. 2.6 Convolution of $f(t)$ and $h(t)$ in Example 2.4.

Because this integration is with respect to τ, we can pull e^{-2t} outside the integral, giving us

$$y(t) = e^{-2t} \int_0^t e^\tau \, d\tau = e^{-2t}(e^t - 1) = e^{-t} - e^{-2t} \qquad t \geq 0 \qquad (2.41)$$

Also, $y(t) = 0$ when $t < 0$ [see Eq. (2.38)]. This result, along with Eq. (2.41), yields

$$y(t) = (e^{-t} - e^{-2t})u(t) \qquad (2.42)$$

The response is depicted in Fig. 2.6c. ■

△ **Exercise E2.5**

For an LTIC system with the impulse response $h(t) = 6e^{-t}u(t)$, determine the system response to the input: **(a)** $2u(t)$ and **(b)** $3e^{-3t}u(t)$.

Answer: **(a)** $12(1 - e^{-t})u(t)$ **(b)** $9(e^{-t} - e^{-3t})u(t)$ ▽

△ **Exercise E2.6**

Repeat Exercise E2.5 if the input $f(t) = e^{-t}u(t)$.

Answer: $6te^{-t}u(t)$ ▽

The Convolution Table

The task of convolution is considerably simplified by a ready-made convolution table (Table 2.1). This table, which lists several pairs of signals and their resulting convolution, can conveniently determine $y(t)$, a system response to an input $f(t)$, without performing the tedious job of integration. For instance, we could have readily found the convolution in Example 2.4 using pair 4 (with $\lambda_1 = -1$ and $\lambda_2 = -2$) to be $(e^{-t} - e^{-2t})u(t)$. The following example demonstrates the utility of this table.

TABLE 2.1: Convolution Table

No	$f_1(t)$	$f_2(t)$	$f_1(t) * f_2(t) = f_2(t) * f_1(t)$
1	$f(t)$	$\delta(t - T)$	$f(t - T)$
2	$e^{\lambda t}u(t)$	$u(t)$	$\dfrac{1 - e^{\lambda t}}{-\lambda} u(t)$
3	$u(t)$	$u(t)$	$tu(t)$
4	$e^{\lambda_1 t}u(t)$	$e^{\lambda_2 t}u(t)$	$\dfrac{e^{\lambda_1 t} - e^{\lambda_2 t}}{\lambda_1 - \lambda_2} u(t) \qquad \lambda_1 \neq \lambda_2$
5	$e^{\lambda t}u(t)$	$e^{\lambda t}u(t)$	$te^{\lambda t}u(t)$
6	$te^{\lambda t}u(t)$	$e^{\lambda t}u(t)$	$\dfrac{1}{2}t^2 e^{\lambda t}u(t)$
7	$t^n u(t)$	$e^{\lambda t}u(t)$	$\dfrac{n!\,e^{\lambda t}}{\lambda^{n+1}} u(t) - \displaystyle\sum_{j=o}^{n} \dfrac{n!\,t^{n-j}}{\lambda^{j+1}(n-j)!} u(t)$
8	$t^m u(t)$	$t^n u(t)$	$\dfrac{m!\,n!}{(m+n+1)!} t^{m+n+1} u(t)$
9	$te^{\lambda_1 t}u(t)$	$e^{\lambda_2 t}u(t)$	$\dfrac{e^{\lambda_2 t} - e^{\lambda_1 t} + (\lambda_1 - \lambda_2)te^{\lambda_1 t}}{(\lambda_1 - \lambda_2)^2} u(t)$
10	$t^m e^{\lambda t}u(t)$	$t^n e^{\lambda t}u(t)$	$\dfrac{m!\,n!}{(n+m+1)!} t^{m+n+1} e^{\lambda t} u(t)$
11	$t^m e^{\lambda_1 t}u(t)$	$t^n e^{\lambda_2 t}u(t)$	$\displaystyle\sum_{j=0}^{m} \dfrac{(-1)^j m!(n+j)!\,t^{m-j}e^{\lambda_1 t}}{j!(m-j)!(\lambda_1 - \lambda_2)^{n+j+1}} u(t)$
	$\lambda_1 \neq \lambda_2$		$+ \displaystyle\sum_{k=0}^{n} \dfrac{(-1)^k n!(m+k)!\,t^{n-k}e^{\lambda_2 t}}{k!(n-k)!(\lambda_2 - \lambda_1)^{m+k+1}} u(t)$
12	$e^{-\alpha t}\cos(\beta t + \theta)u(t)$	$e^{\lambda t}u(t)$	$\dfrac{\cos(\theta - \phi)e^{\lambda t} - e^{-\alpha t}\cos(\beta t + \theta - \phi)}{\sqrt{(\alpha + \lambda)^2 + \beta^2}} u(t)$
			$\phi = \tan^{-1}[-\beta/(\alpha + \lambda)]$
13	$e^{\lambda_1 t}u(t)$	$e^{\lambda_2 t}u(-t)$	$\dfrac{e^{\lambda_1 t}u(t) + e^{\lambda_2 t}u(-t)}{\lambda_2 - \lambda_1} \quad \mathrm{Re}\,\lambda_2 > \mathrm{Re}\,\lambda_1$
14	$e^{\lambda_1 t}u(-t)$	$e^{\lambda_2 t}u(-t)$	$\dfrac{e^{\lambda_1 t} - e^{\lambda_2 t}}{\lambda_2 - \lambda_1} u(-t)$

■ **Example 2.5**

Find the loop current $y(t)$ of the RLC circuit in Example 2.2 for the input $f(t) = 10e^{-3t}u(t)$, when all the initial conditions are zero.

The loop equation for this circuit [see Example 1.11 or Eq. (1.55)] is

$$\left(D^2 + 3D + 2\right) y(t) = Df(t)$$

The impulse response $h(t)$ for this system, as obtained in Example 2.3, is

$$h(t) = \left(2e^{-2t} - e^{-t}\right) u(t)$$

The input is $f(t) = 10e^{-3t}u(t)$, and the response $y(t)$ is

$$y(t) = f(t) * h(t)$$
$$= 10e^{-3t}u(t) * \left[2e^{-2t} - e^{-t}\right] u(t)$$

Using the distributive property of the convolution [Eq. (2.32)], we obtain

$$y(t) = 10e^{-3t}u(t) * 2e^{-2t}u(t) - 10e^{-3t}u(t) * e^{-t}u(t)$$
$$= 20 \left[e^{-3t}u(t) * e^{-2t}u(t)\right] - 10 \left[e^{-3t}u(t) * e^{-t}u(t)\right]$$

Now the use of Pair 4 in Table 2.1 yields

$$y(t) = \frac{20}{-3 - (-2)} \left[e^{-3t} - e^{-2t}\right] u(t) - \frac{10}{-3 - (-1)} \left[e^{-3t} - e^{-t}\right] u(t)$$
$$= -20 \left(e^{-3t} - e^{-2t}\right) u(t) + 5 \left(e^{-3t} - e^{-t}\right) u(t)$$
$$= \left(-5e^{-t} + 20e^{-2t} - 15e^{-3t}\right) u(t) \qquad (2.43)$$

■

△ **Exercise E2.7**

Rework Probs. E2.5 and E2.6 using the convolution table.   ▽

△ **Exercise E2.8**

Using the convolution table, determine

$$e^{-2t}u(t) * \left(1 - e^{-t}\right) u(t)$$

Answer: $\left(\frac{1}{2} - e^{-t} + \frac{1}{2}e^{-2t}\right) u(t)$ ▽

△ **Exercise E2.9**

For an LTIC system with the unit impulse response $h(t) = e^{-2t}u(t)$, determine the zero-state response $y(t)$ if the input $f(t) = \sin 3t\, u(t)$. Hint: Use the convolution table Pair 12 with suitable values for α β θ, and λ.

Answer: $\frac{1}{13} \left[3e^{-2t} + \sqrt{13}\cos\left(3t - 146.32°\right)\right] u(t)$ or $\frac{1}{13} \left[3e^{-2t} - \sqrt{13}\cos\left(3t + 33.68°\right)\right] u(t)$ ▽

Multiple Inputs

Multiple inputs to LTI systems can be treated by applying the superposition principle. Each input is considered separately, with all other inputs assumed to be zero. The sum of all these individual system responses constitutes the total system output when all the inputs are applied simultaneously.

2.4-2 Graphical Understanding of Convolution

To have a proper grasp of convolution operation, we should understand the graphical interpretation of convolution. Such a comprehension also helps in evaluating the convolution integral of more complicated signals. In addition, graphical convolution allows us to grasp visually or mentally the convolution integral's result, which can be of great help in sampling, filtering, and many other problems. Finally, many signals have no exact mathematical description, so they can be described only graphically. If two such signals are to be convolved, we have no choice but to perform their convolution graphically.

We shall now explain the convolution operation by convolving the signals $f(t)$ and $g(t)$, illustrated in Figs. 2.7a and 2.7b respectively. If $c(t)$ is the convolution of $f(t)$ with $g(t)$, then

$$c(t) = \int_{-\infty}^{\infty} f(\tau)g(t-\tau)\,d\tau \tag{2.44}$$

One of the crucial points to remember here is that this integration is performed with respect to τ, so that t is just a parameter (like a constant). This consideration is especially important when we sketch the graphical representations of the functions $f(\tau)$ and $g(t-\tau)$ appearing in the integrand of Eq. (2.44). Both of these functions should be sketched as functions of τ, not of t.

The function $f(\tau)$ is identical to $f(t)$, with τ replacing t (Fig. 2.7c). Therefore, $f(t)$ and $f(\tau)$ will have the same graphical representations. Similar remarks apply to $g(t)$ and $g(\tau)$ (Fig. 2.7d).

The function $g(t-\tau)$ is not as easy to comprehend. To understand what this function looks like, let us start with the function $g(\tau)$ (Fig. 2.7d). Time-inversion of this function (reflection about the vertical axis $\tau = 0$) yields $g(-\tau)$ (Fig. 2.7e). Let us denote this function by $\phi(\tau)$

$$\phi(\tau) = g(-\tau)$$

Now $\phi(\tau)$ shifted by t seconds is $\phi(\tau - t)$, given by

$$\phi(\tau - t) = g[-(\tau - t)] = g(t - \tau)$$

Therefore, we first time-invert $g(\tau)$ to obtain $g(-\tau)$ and then time-shift $g(-\tau)$ by t to obtain $g(t-\tau)$. For positive t, the shift is to the right (Fig. 2.7f); for negative t, the shift is to the left (Fig. 2.7g).

The preceding discussion gives us a graphical interpretation of the functions $f(\tau)$ and $g(t-\tau)$. The convolution $c(t)$ is the area under the product of these two functions. Thus, to compute $c(t)$ at some positive instant $t = t_1$, we first obtain $g(-\tau)$ by inverting $g(\tau)$ about the vertical axis. Next, we right-shift or delay $g(-\tau)$ by t_1 to obtain $g(t_1 - \tau)$ (Fig. 2.7f), and then we multiply this function by $f(\tau)$, giving us the product $f(\tau)g(t_1 - \tau)$ (Fig. 2.7f). The area A_1 under this product

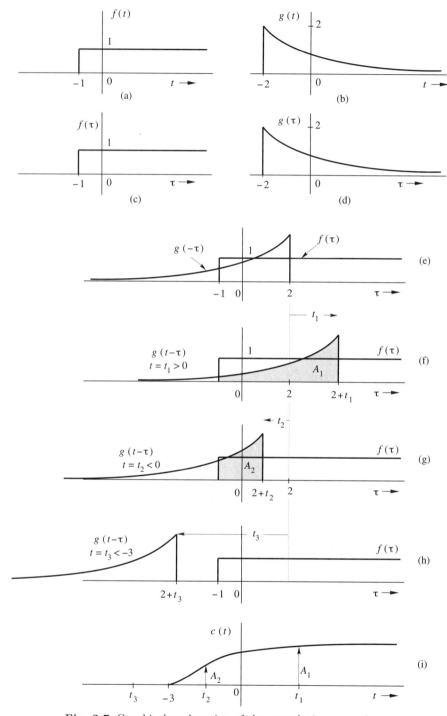

Fig. 2.7 Graphical explanation of the convolution operation.

is $c(t_1)$, the value of $c(t)$ at $t = t_1$. We can therefore plot $c(t_1) = A_1$ on a curve describing $c(t)$, as shown in Fig. 2.7i. Observe that the area under the product $f(\tau)g(-\tau)$ in Fig. 2.7e is $c(0)$, the value of the convolution for $t = 0$ (at the origin).

A similar procedure is followed in computing the value of $c(t)$ at $t = t_2$, where t_2 is negative (Fig. 2.7g). In this case, the function $g(-\tau)$ is shifted by a negative amount (that is, left-shifted) to obtain $g(t_2 - \tau)$. Multiplication of this function with $f(\tau)$ yields the product $f(\tau)g(t_2 - \tau)$. The area under this product is $c(t_2) = A_2$, giving us another point on the curve $c(t)$ at $t = t_2$ (Figure 2.7i). This procedure can be repeated for all values of t, from $-\infty$ to ∞. The result will be a curve describing $c(t)$ for all time t. Note that when $t \le -3$, $f(\tau)$ and $g(t - \tau)$ do not overlap (see Fig. 2.7h); therefore, $c(t) = 0$ for $t \le -3$.

Summary of the Graphical Procedure
The procedure for graphical convolution can be summarized as follows:

1. Keep the function $f(\tau)$ fixed.
2. Visualize the function $g(\tau)$ as a rigid wire frame, and rotate (or invert) this frame about the vertical axis ($\tau = 0$) to obtain $g(-\tau)$.
3. Shift the inverted frame along the τ axis by t_0 seconds. The shifted frame now represents $g(t_0 - \tau)$.
4. The area under the product of $f(\tau)$ and $g(t_0 - \tau)$ (the shifted frame) is $c(t_0)$, the value of the convolution at $t = t_0$.
5. Repeat this procedure, shifting the frame by different values (positive and negative) to obtain $c(t)$ for all values of t.

Convolution: its bark is worse than its bite!

The graphical procedure discussed here appears very complicated and discouraging at first reading. Indeed, some people claim that convolution has driven many electrical engineering undergraduates to contemplate theology either for salvation or as an alternative career (*IEEE Spectrum*, March 1991, p.60). Actually, the bark

of convolution is worse than its bite! In graphical convolution, we need to determine the area under the product $f(\tau)g(t-\tau)$ for all values of t from $-\infty$ to ∞.
However, a mathematical description of $f(\tau)g(t-\tau)$ is generally valid over a range
of t. Therefore, repeating the procedure for every value of t amounts to repeating
it only a few times for different ranges of t.

We can also use the commutative property of convolution to our advantage
by computing $f(t) * g(t)$ or $g(t) * f(t)$, whichever is simpler. As a rule of thumb,
*convolution computations are simplified if we choose to invert the simpler of the
two functions.* For example, if the mathematical description of $g(t)$ is simpler than
that of $f(t)$, then $f(t) * g(t)$ will be easier to compute than $g(t) * f(t)$. In contrast,
if the mathematical description of $f(t)$ is simpler, the reverse will be true.

We shall demonstrate graphical convolution with the following examples. Let
us start by reworking Example 2.4 using this graphical method.

■ **Example 2.6**

Determine graphically $y(t) = f(t) * h(t)$ for $f(t) = e^{-t}u(t)$ and $h(t) = e^{-2t}u(t)$.

Figures 2.8a and 2.8b show $f(t)$ and $h(t)$ respectively, and Fig. 2.8c shows $f(\tau)$ and
$h(-\tau)$ as functions of τ. The function $h(t-\tau)$ is now obtained by shifting $h(-\tau)$ by t. If t
is positive, the shift is to the right (delay); if t is negative, the shift is to the left (advance).
Figure 2.8d shows that for negative t, $h(t-\tau)$ [obtained by left-shifting $h(-\tau)$] does not
overlap $f(\tau)$, and the product $f(\tau)h(t-\tau)=0$, so that

$$y(t) = 0 \qquad t < 0$$

Figure 2.8e shows the situation for $t \geq 0$. Here $f(\tau)$ and $h(t-\tau)$ do overlap, but the
product is nonzero only over the interval $0 \leq \tau \leq t$ (shaded interval). Therefore

$$y(t) = \int_0^t f(\tau)h(t-\tau)\,d\tau \qquad t \geq 0$$

All we need to do now is substitute correct expressions for $f(\tau)$ and $h(t-\tau)$ in this integral.
Figures 2.8a and 2.8b clearly indicate that the segments of $f(t)$ and $g(t)$ to be used in this
convolution (Fig. 2.8e) are described by

$$f(t) = e^{-t} \qquad \text{and} \qquad h(t) = e^{-2t}$$

Therefore

$$f(\tau) = e^{-\tau} \qquad \text{and} \qquad h(t-\tau) = e^{-2(t-\tau)}$$

Consequently

$$y(t) = \int_0^t e^{-\tau} e^{-2(t-\tau)}\,d\tau$$

$$= e^{-2t} \int_0^t e^{\tau}\,d\tau$$

$$= e^{-t} - e^{-2t} \qquad t \geq 0$$

Moreover, $y(t) = 0$ for $t < 0$, so that

$$y(t) = \left(e^{-t} - e^{-2t}\right) u(t) \qquad ■$$

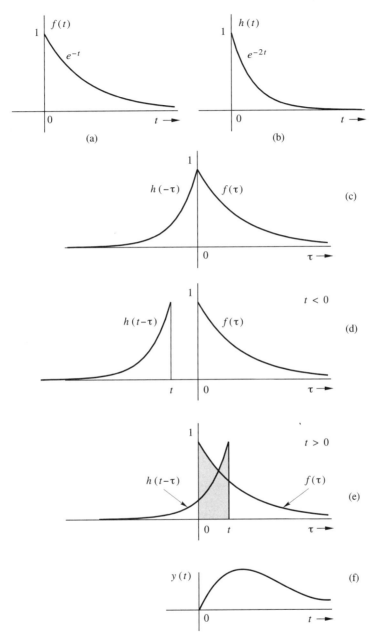

Fig. 2.8 Convolution of $f(t)$ and $h(t)$ in Example 2.6.

■ **Example 2.7**

Find $c(t) = f(t) * g(t)$ for the signals depicted in Figs. 2.9a and 2.9b.

Since $f(t)$ is simpler than $g(t)$, it is easier to evaluate $g(t) * f(t)$ than $f(t) * g(t)$. However, we shall intentionally take the more difficult route and evaluate $f(t) * g(t)$ to clarify some of the finer points of convolution.

Figures 2.9a and 2.9b show $f(t)$ and $g(t)$ respectively. Observe that $g(t)$ is composed of two segments. As a result, it can be described as

$$g(t) = \begin{cases} 2e^{-t} & \text{segment A} \\ -2e^{2t} & \text{segment B} \end{cases}$$

Therefore

$$g(t - \tau) = \begin{cases} 2e^{-(t-\tau)} & \text{segment A} \\ -2e^{2(t-\tau)} & \text{segment B} \end{cases} \tag{2.45}$$

The segment of $f(t)$ that is used in convolution is $f(t) = 1$, so that $f(\tau) = 1$. Figure 2.9c shows $f(\tau)$ and $g(-\tau)$.

To compute $c(t)$ for $t \geq 0$, we right-shift $g(-\tau)$ to obtain $g(t-\tau)$, as illustrated in Fig. 2.9d. Clearly, $g(t - \tau)$ overlaps with $f(\tau)$ over the shaded interval; that is, over the range $\tau \geq 0$; Segment A overlaps with $f(\tau)$ over the interval $(0, t)$, while Segment B overlaps with $f(\tau)$ over (t, ∞). Remembering that $f(\tau) = 1$, we have

$$c(t) = \int_0^\infty f(\tau)g(t-\tau)\, d\tau$$

$$= \int_0^t 2e^{-(t-\tau)}\, d\tau + \int_t^\infty -2e^{2(t-\tau)}\, d\tau$$

$$= 2\left(1 - e^{-t}\right) - 1$$

$$= 1 - 2e^{-t} \qquad\qquad t \geq 0$$

Figure 2.9e shows the situation for $t < 0$. Here the overlap is over the shaded interval; that is, over the range $\tau \geq 0$, where only the segment B of $g(t)$ is involved. Therefore

$$c(t) = \int_0^\infty f(\tau)g(t-\tau)\, d\tau$$

$$= \int_0^\infty g(t-\tau)\, d\tau$$

$$= \int_0^\infty -2e^{2(t-\tau)}\, d\tau$$

$$= -e^{2t} \qquad\qquad t < 0$$

Therefore

$$c(t) = \begin{cases} 1 - 2e^{-2t} & t \geq 0 \\ -e^{2t} & t < 0 \end{cases}$$

Figure 2.9f shows a plot of $c(t)$.   ■

■ Example 2.8

Find $f(t) * g(t)$ for the functions $f(t)$ and $g(t)$ shown in Figs. 2.10a and 2.10b.

Here, $f(t)$ has a simpler mathematical description than that of $g(t)$, so it is preferable to invert $f(t)$. Hence, we shall determine $g(t) * f(t)$ rather than $f(t) * g(t)$. Thus

$$c(t) = g(t) * f(t)$$

$$= \int_{-\infty}^\infty g(\tau)f(t-\tau)\, d\tau$$

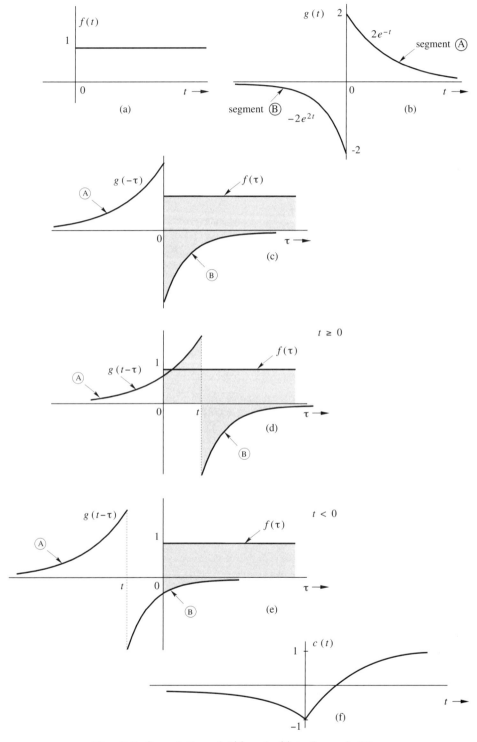

Fig. 2.9 Convolution of $f(t)$ and $g(t)$ in Example 2.7.

First, we determine the expressions for the segments of $f(t)$ and $g(t)$ used in finding $c(t)$. According to Figs. 2.10a and 2.10b, these segments can be expressed as

$$f(t) = 1 \quad\text{and}\quad g(t) = \tfrac{1}{3}t$$

Therefore

$$f(t - \tau) = 1 \quad\text{and}\quad g(\tau) = \tfrac{1}{3}\tau$$

Fig. 2.10c shows $g(\tau)$ and $f(-\tau)$, whereas Fig. 2.10d shows $g(\tau)$ and $f(t - \tau)$, which is $f(-\tau)$ shifted by t. Because the edges of $f(-\tau)$ are at $\tau = -1$ and 1, the edges of $f(t - \tau)$ are at $-1 + t$ and $1 + t$. The two functions overlap over the interval $(0, 1 + t)$ (shaded interval), so that

$$c(t) = \int_0^{1+t} g(\tau)f(t - \tau)\,d\tau$$

$$= \int_0^{1+t} \tfrac{1}{3}\tau\,d\tau$$

$$= \tfrac{1}{6}(t + 1)^2 \qquad -1 \le t \le 1 \qquad\qquad (2.46a)$$

This situation, depicted in Fig. 2.10d, is valid only for $-1 \le t \le 1$. For $t > 1$ but < 2, the situation is as illustrated in Fig. 2.10e. The two functions overlap only over the range $-1 + t$ to $1 + t$ (shaded interval). Note that the expressions for $g(\tau)$ and $f(t - \tau)$ do not change; only the range of integration changes. Therefore

$$c(t) = \int_{-1+t}^{1+t} \tfrac{1}{3}\tau\,d\tau$$

$$= \tfrac{2}{3}t \qquad 1 \le t \le 2 \qquad\qquad (2.46b)$$

Also note that the expressions in Eqs. (2.46a) and (2.46b) both apply at $t = 1$, the transition point between their respective ranges. We can readily verify that both expressions yield a value of $2/3$ at $t = 1$, so that $c(1) = 2/3$. The continuity of $c(t)$ at transition points indicates a high probability of a right answer.† For $t \ge 2$ but < 4 the situation is as shown in Fig. 2.10f. The functions $g(\tau)$ and $f(t - \tau)$ overlap over the interval from $-1 + t$ to 3 (shaded interval), so that

$$c(t) = \int_{-1+t}^{3} \tfrac{1}{3}\tau\,d\tau$$

$$= -\tfrac{1}{6}\left(t^2 - 2t - 8\right) \qquad\qquad (2.46c)$$

Again, both Eqs. (2.46b) and (2.46c) apply at the transition point $t = 2$. We can readily verify that $c(2) = 4/3$ when either of these expressions is used.

For $t \ge 4$, $f(t - \tau)$ has been shifted so far to the right that it no longer overlaps with $g(\tau)$ as depicted in Fig. 2.10g. Consequently

$$c(t) = 0 \qquad t \ge 4 \qquad\qquad (2.46d)$$

We now turn our attention to negative values of t. We have already determined $c(t)$ up to $t = -1$. For $t < -1$ there is no overlap between the two functions, as illustrated in Fig. 2.10h, so that

$$c(t) = 0 \qquad t < -1 \qquad\qquad (2.46e)$$

Figure 2.10i shows $c(t)$ plotted according to Eqs. (2.46a) through (2.46e). ∎

---

† Even if $c(t)$ is continuous at the transition, the answer could be wrong in the unlikely event of two or more errors canceling out their effects. Our discussion assumes that there are no impulses in $f(t - \tau)$ and $g(\tau)$ after the transition which were not present before.

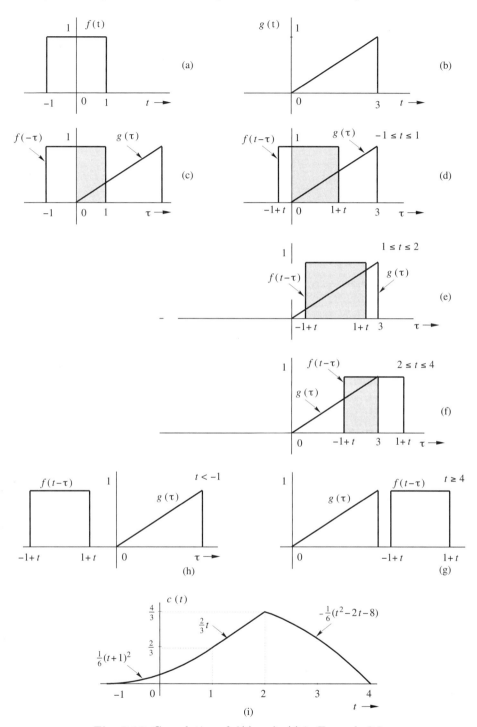

Fig. 2.10 Convolution of $f(t)$ and $g(t)$ in Example 2.8.

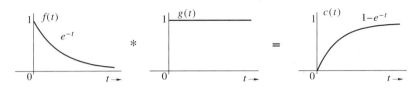

Fig. 2.11 Convolution of $f(t)$ and $g(t)$ in Exercise E2.11.

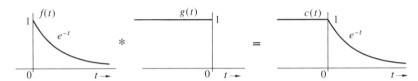

Fig. 2.12 Convolution of $f(t)$ and $g(t)$ in Exercise E2.12.

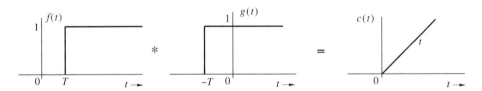

Fig. 2.13 Convolution of $f(t)$ and $g(t)$ in Exercise E2.13.

Width of the Convolved Function

The widths (durations) of $f(t)$, $g(t)$, and $c(t)$ in Example 2.8 (Fig. 2.10) are 2, 3, and 5 respectively. Note that the width of $c(t)$ in this case is the sum of the widths of $f(t)$ and $g(t)$. This observation is not a coincidence. Using the concept of graphical convolution, we can readily see that if $f(t)$ and $g(t)$ have the finite widths of T_1 and T_2 respectively, then the width of $c(t)$ is generally equal to $T_1 + T_2$. The reason is that the time it takes for a signal of width (duration) T_1 to completely pass another signal of width (duration) T_2 so that they become nonoverlapping is $T_1 + T_2$. When the two signals become nonoverlapping, the convolution goes to zero. However, there are cases where the two signals $f(\tau)$ and $g(t - \tau)$ overlap, yet the area under their product vanishes. Such is the case of signals in Fig. P2.4-14 for $\tau > 2\pi$. In this case the width property is superficially violated.†

△ **Exercise E2.10**

Rework Example 2.7 by evaluating $g(t) * f(t)$ ▽

△ **Exercise E2.11**

Use graphical convolution to show that $f(t) * g(t) = g(t) * f(t) = c(t)$ in Fig. 2.11.   ▽

△ **Exercise E2.12**

Repeat Prob. E2.11 for the functions in Fig. 2.12.   ▽

△ **Exercise E2.13**

Repeat Prob. E2.11 for the functions in Fig. 2.13.   ▽

---

†Even in such cases, the width property may be held valid if we consider that the region where the area under the product of two nonoverlapping signals vanishes to become a part of $c(t)$ (where $c(t)$ happens to be zero). Thus, $c(t)$ in this case has an infinite duration, but the value of $c(t) = 0$ for $t > 2\pi$.

⊙ **Computer Example C2.4**
Find $c(t) = f(t) * g(t)$ for the signals in Fig. 2.9.

```
t1=-10:.01:0;t1=t1';
g1=-2*exp(2*t1);
t2=0:.01:10;t2=t2';
g2=2*exp(-t2);
t=[t1;t2]; g=[g1;g2];
f=[zeros(size(g1));ones(size(g2))];
c=0.01*conv(f,g);
t=-20:.01:5;t=t';
plot(t,c(1:length(t)))      ⊙
```

Some Reflections on the Use of Impulse Function

In the study of signals and systems we often come across signals such as impulses, which cannot be generated in practice. One wonders why we even consider such signals. The answer should be clear from our discussion so far in this chapter. Even if the impulse function has no physical existence, we can compute the system response $h(t)$ to this phantom input according to the procedure in Sec. 2.3, and knowing $h(t)$, we can compute the system response to any arbitrary input. The concept of impulse response, therefore, provides an effective intermediary for computing system response to an arbitrary input. In addition, the impulse response $h(t)$ itself provides a great deal of information and insight about the system behavior. In Sec. 2.7 we show that the knowledge of impulse response provides much valuable information, such as the response time, pulse dispersion, and filtering properties of the system. Many other useful insights about the system behavior can be obtained by inspection of $h(t)$.

We have a similar situation in frequency-domain analysis (discussed in later chapters) where we use an *everlasting exponential* (or *sinusoid*) to determine system response. An everlasting exponential (or sinusoid) has no physical existence, but it provides another effective intermediary for computing the system response to an arbitrary input. Moreover, the system response to everlasting exponential (or sinusoid) provides valuable information and insight regarding the system's behavior.

2.4-3 A Very Special Function For LTI Systems: The Everlasting Exponential e^{st}

There is a very special connection of LTI systems with the everlasting exponential function e^{st}. We now show that the LTI system's (zero-state) response to everlasting exponential input e^{st} is also the same everlasting exponential (within a multiplicative constant). Moreover, no other function can make the same claim. Such an input for which the system response is also of the same form is called the **characteristic function** (also *eigenfunction*) of the system. Because a sinusoid is a form of exponential, everlasting sinusoid is also a characteristic function of an LTI system. Note that we are talking here of an everlasting exponential (or sinusoid), which starts at $t = -\infty$.

If $h(t)$ is the system's unit impulse response, then system response $y(t)$ to an everlasting exponential e^{st} is given by

$$y(t) = h(t) * e^{st} = \int_{-\infty}^{\infty} h(\tau)e^{s(t-\tau)}\, d\tau$$

$$= e^{st} \int_{-\infty}^{\infty} h(\tau)e^{-s\tau}\, d\tau$$

The integral on the right-hand side is a function of s. Let us denote it by $H(s)$. Thus†,

$$y(t) = H(s)e^{st} \tag{2.47}$$

where

$$H(s) = \int_{-\infty}^{\infty} h(\tau)e^{-s\tau}\, d\tau \tag{2.48}$$

Note that $H(s)$ is a constant for a given s. Thus, the input and the output are the same (within a multiplicative constant) for the everlasting exponential signal.

$H(s)$, which is called the **transfer function** of the system, is a function of complex variable s. We can define the transfer function $H(s)$ of an LTIC system from Eq. (2.47) as

$$H(s) = \left.\frac{\text{output signal}}{\text{input signal}}\right|_{\text{Input=everlasting exponential } e^{st}} \tag{2.49}$$

The transfer function is defined for, and is meaningful to, LTI systems only. It does not exist for nonlinear or time-varying systems in general.

In this discussion we are talking of the everlasting exponential, which starts at $t = -\infty$, not the causal exponential $e^{st}u(t)$, which starts at $t = 0$.

For a system specified by Eq. (2.1), the transfer function is given by

$$H(s) = \frac{P(s)}{Q(s)} \tag{2.50}$$

This follows readily by considering an everlasting input $f(t) = e^{st}$. According to Eq. (2.47), the output is $y(t) = H(s)e^{st}$. Substitution of this $f(t)$ and $y(t)$ in Eq. (2.1) yields $H(s)[Q(D)e^{st}] = P(D)e^{st}$. Moreover, $D^r e^{st} = d^r e^{st}/dt^r = s^r e^{st}$. Hence, $P(D)e^{st} = P(s)e^{st}$ and $Q(D)e^{st} = Q(s)e^{st}$. Consequently, $H(s) = P(s)/Q(s)$.

△ **Exercise E2.14**

Show that the transfer function of an ideal integrator is $H(s) = 1/s$ and that of an ideal differentiator is $H(s) = s$. Find the answer in two ways: using Eq. (2.49) and Eq. (2.50). ▽

2.4-4 Total Response

The total response of a linear system can be expressed as the sum of its zero-input and zero-state components:

$$\text{Total Response} = \underbrace{\sum_{j=1}^{n} c_j e^{\lambda_j t}}_{\text{zero-input component}} + \underbrace{f(t) * h(t)}_{\text{zero-state component}}$$

---

†This result is valid only for the values of s for which the $\int_{-\infty}^{\infty} h(\tau)e^{-s\tau}\, d\tau$ exists (or converges).

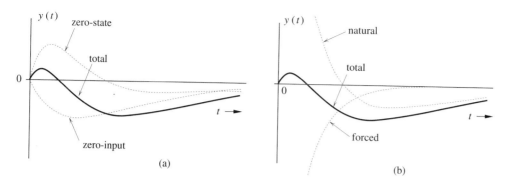

Fig. 2.14 Total response and its components.

For repeated roots, the zero-input component should be appropriately modified.

For the series RLC circuit in Example 2.2 with the input $f(t) = 10e^{-3t}u(t)$ and the initial conditions $y(0^-) = 0$, $v_C(0^-) = 5$, we determined the zero-input component in Example 2.2 [Eq. (2.15)]. We found the zero-state component in Example 2.5 Using the results in Examples 2.2 and 2.5, we obtain

$$\text{Total current} = \underbrace{\left(-5e^{-t} + 5e^{-2t}\right)}_{\text{zero-input current}} + \underbrace{\left(-5e^{-t} + 20e^{-2t} - 15e^{-3t}\right)}_{\text{zero-state current}} \qquad t \geq 0 \quad (2.51a)$$

Figure 2.14a shows the zero-input, the zero-state, and the total response.

Natural and Forced Response

For the RLC circuit in Example 2.2, the characteristic modes were found to be e^{-t} and e^{-2t}. As we expected, the zero-input response is composed exclusively of characteristic modes. Note, however, that even the zero-state response [Eq. (2.51a)] contains characteristic mode terms. This observation is generally true of LTI systems. We can now lump together all the characteristic mode terms in the total response, giving us a component known as the **natural response** $y_n(t)$. The remainder, consisting entirely of noncharacteristic mode terms, is known as the **forced response** $y_\phi(t)$. The total response of the RLC circuit in Example 2.2 can be expressed in terms of natural and forced components by regrouping the terms in Eq. (2.51a) as

$$\text{Total current} = \underbrace{\left(-10e^{-t} + 25e^{-2t}\right)}_{\text{natural response } y_n(t)} + \underbrace{\left(-15e^{-3t}\right)}_{\text{forced response } y_\phi(t)} \qquad t \geq 0 \qquad (2.51b)$$

Figure 2.14b shows the natural, forced, and total response.

2.5   Classical Solution of Differential Equations

In the classical method we solve differential equation to find the natural and forced components rather than the zero-input and zero-state components of the response. Although this method is relatively simple compared to the method discussed so far, as we shall see, it also has several glaring drawbacks.

As Sec. 2.4-4 shows, when all of the characteristic mode terms of the total system response are lumped together, they form the system's **natural response** $y_n(t)$ (also known as the **homogeneous solution** or **complementary solution**). The remaining portion of the response consists entirely of noncharacteristic mode terms and is called the system's **forced response** $y_\phi(t)$ (also known as the **particular solution**). Equation (2.51b) shows these two components for the loop current in the RLC circuit of Fig. 2.1a.

The total system response is $y(t) = y_n(t) + y_\phi(t)$. Since $y(t)$ must satisfy the system equation [Eq. (2.1)],

$$Q(D) [y_n(t) + y_\phi(t)] = P(D)f(t)$$

or

$$Q(D)y_n(t) + Q(D)y_\phi(t) = P(D)f(t)$$

But $y_n(t)$ is composed entirely of characteristic modes. Therefore

$$Q(D)y_n(t) = 0$$

so that

$$Q(D)y_\phi(t) = P(D)f(t) \qquad (2.52)$$

The natural response, being a linear combination of the system's characteristic modes, has the same form as that of the zero-input response; only its arbitrary constants are different. These constants are determined from auxiliary conditions, as explained later. We shall now discuss a method of determining the forced response.

2.5-1 Forced Response: The Method of Undetermined Coefficients

It is a relatively simple task to determine $y_\phi(t)$, the forced response of an LTIC system, when the input $f(t)$ is such that it yields only a finite number of independent derivatives. Inputs having the form $e^{\zeta t}$ or t^r fall into this category. For example, $e^{\zeta t}$ has only one independent derivative; the repeated differentiation of $e^{\zeta t}$ yields the same form as this input; that is, $e^{\zeta t}$. Similarly, the repeated differentiation of t^r yields only r independent derivatives. The forced response to such an input can be expressed as a linear combination of the input and its independent derivatives. Consider, for example, the input $at^2 + bt + c$. The successive derivatives of this input are $2at + b$ and $2a$. In this case, the input has only two independent derivatives. Therefore, the forced response can be assumed to be a linear combination of $f(t)$ and its two derivatives. The suitable form for $y_\phi(t)$ in this case is, therefore

$$y_\phi(t) = \beta_2 t^2 + \beta_1 t + \beta_0$$

The undetermined coefficients β_0, β_1, and β_2 are determined by substituting this expression for $y_\phi(t)$ in Eq. (2.52)

$$Q(D)y_\phi(t) = P(D)f(t)$$

and then equating coefficients of similar terms on both sides of the resulting expression.

TABLE 2.2

Input $f(t)$	Forced Response
1. $e^{\zeta t}$ $\zeta \neq \lambda_i \, (i = 1, 2, \cdots, n)$	$\beta e^{\zeta t}$
2. $e^{\zeta t}$ $\zeta = \lambda_i$	$\beta t e^{\zeta t}$
3. k	β
4. $\cos(\omega t + \theta)$	$\beta \cos(\omega t + \phi)$
5. $\left(t^r + \alpha_{r-1} t^{r-1} + \cdots + \alpha_1 t + \alpha_0\right) e^{\zeta t}$	$(\beta_r t^r + \beta_{r-1} t^{r-1} + \cdots + \beta_1 t$ $+ \beta_0) e^{\zeta t}$

Note: By definition, $y_\phi(t)$ cannot have any characteristic mode terms. If any term appearing in the right-hand column for the forced response is also a characteristic mode of the system, the correct form of the forced response must be modified to $t^i y_\phi(t)$, where i is the smallest possible integer that can be used and still can prevent $t^i y_\phi(t)$ from having a characteristic mode term. For example, when the input is $e^{\zeta t}$, the forced response (right-hand column) has the form $\beta e^{\zeta t}$. But if $e^{\zeta t}$ happens to be a characteristic mode of the system, the correct form of the forced response is $\beta t e^{\zeta t}$ (see Pair 2). If $t e^{\zeta t}$ also happens to be a characteristic mode of the system, the correct form of the forced response is $\beta t^2 e^{\zeta t}$, and so on.

Although this method can be used only for inputs with a finite number of derivatives, this class of inputs includes a wide variety of the most commonly encountered signals in practice. Table 2.2 shows a variety of such inputs and the form of the forced response corresponding to each input. We shall demonstrate this procedure with an example.

■ **Example 2.9**

Solve the differential equation

$$\left(D^2 + 3D + 2\right) y(t) = D f(t)$$

if the input

$$f(t) = t^2 + 5t + 3$$

and the initial conditions are $y(0^+) = 2$ and $\dot{y}(0^+) = 3$.

The characteristic polynomial of the system is

$$\lambda^2 + 3\lambda + 2 = (\lambda + 1)(\lambda + 2)$$

Therefore, the characteristic modes are e^{-t} and e^{-2t}. The natural response is then a linear combination of these modes, so that

$$y_n(t) = K_1 e^{-t} + K_2 e^{-2t} \qquad t \geq 0$$

Here the arbitrary constants K_1 and K_2 must be determined from the system's initial conditions.

The forced response to the input $t^2 + 5t + 3$, according to Table 2.2 (Pair 5 with $\zeta = 0$), is

$$y_\phi(t) = \beta_2 t^2 + \beta_1 t + \beta_0$$

Moreover, $y_\phi(t)$ satisfies the system equation [Eq. (2.52)]; that is,

Now
$$\left(D^2 + 3D + 2\right) y_\phi(t) = Df(t) \qquad\qquad (2.53)$$

$$Dy_\phi(t) = \frac{d}{dt}\left(\beta_2 t^2 + \beta_1 t + \beta_0\right) = 2\beta_2 t + \beta_1$$

$$D^2 y_\phi(t) = \frac{d^2}{dt^2}\left(\beta_2 t^2 + \beta_1 t + \beta_0\right) = 2\beta_2$$

and
$$Df(t) = \frac{d}{dt}\left[t^2 + 5t + 3\right] = 2t + 5$$

Substituting these results in Eq. (2.53) yields

or
$$2\beta_2 + 3(2\beta_2 t + \beta_1) + 2(\beta_2 t^2 + \beta_1 t + \beta_0) = 2t + 5$$

$$2\beta_2 t^2 + (2\beta_1 + 6\beta_2)t + (2\beta_0 + 3\beta_1 + 2\beta_2) = 2t + 5$$

Equating coefficients of similar powers on both sides of this expression yields

$$2\beta_2 = 0$$

$$2\beta_1 + 6\beta_2 = 2$$

$$2\beta_0 + 3\beta_1 + 2\beta_2 = 5$$

Solving these three equations for their unknowns, we obtain $\beta_0 = 1$, $\beta_1 = 1$, and $\beta_2 = 0$. Therefore

$$y_\phi(t) = t + 1 \qquad\qquad t > 0$$

The total system response $y(t)$ is the sum of the natural and forced solutions. Therefore

$$y(t) = y_n(t) + y_\phi(t)$$

$$= K_1 e^{-t} + K_2 e^{-2t} + t + 1 \qquad\qquad t > 0$$

so that

$$\dot{y}(t) = -K_1 e^{-t} - 2K_2 e^{-2t} + 1$$

Setting $t = 0$ and substituting $y(0) = 2$ and $\dot{y}(0) = 3$ in these equations, we have

$$2 = K_1 + K_2 + 1$$

$$3 = -K_1 - 2K_2 + 1$$

The solution of these two simultaneous equations is $K_1 = 4$ and $K_2 = -3$. Therefore

$$y(t) = 4e^{-t} - 3e^{-2t} + t + 1 \qquad\qquad t \geq 0 \quad \blacksquare$$

Comments on Initial Conditions

In the classical method, the initial conditions are required at $t = 0^+$. The reason is that because at $t = 0^-$, only the zero-input component exists, and the initial conditions at $t = 0^-$ can be applied to the zero-input component only. In the classical method, the zero-input and zero-state components cannot be separated. Consequently, the initial conditions must be applied to the total response, which begins at $t = 0^+$.

△ **Exercise E2.15**

An LTIC system is specified by the equation

$$\left(D^2 + 5D + 6\right) y(t) = (D + 1)f(t)$$

The input is $f(t) = 6t^2$. Find **(a)** the forced response $y_\phi(t)$ **(b)** the total response $y(t)$ if the initial conditions are $y(0^+) = \frac{25}{18}$ and $\dot{y}(0^+) = -\frac{2}{3}$.

Answers: **(a)** $y_\phi(t) = t^2 + \frac{1}{3}t - \frac{11}{18}$ **(b)** $y(t) = 5e^{-2t} - 3e^{-3t} + \left(t^2 + \frac{1}{3}t - \frac{11}{18}\right)$. ▽

The Exponential Input $e^{\zeta t}$

The exponential signal is the most important signal in the study of LTI systems. Interestingly, the forced response for an exponential input signal turns out to be very simple. From Table 2.2 we see that the forced response for the input $e^{\zeta t}$ has the form $\beta e^{\zeta t}$. We now show that $\beta = Q(\zeta)/P(\zeta)$.† To determine the constant β, we substitute $y_\phi(t) = \beta e^{\zeta t}$ in the system equation [Eq. (2.52)] to obtain

$$Q(D)\left[\beta e^{\zeta t}\right] = P(D)e^{\zeta t} \tag{2.54}$$

Now observe that

$$De^{\zeta t} = \frac{d}{dt}\left(e^{\zeta t}\right) = \zeta e^{\zeta t}$$

$$D^2 e^{\zeta t} = \frac{d^2}{dt^2}\left(e^{\zeta t}\right) = \zeta^2 e^{\zeta t}$$

$$\cdots\cdots\cdots\cdots\cdots\cdots\cdots$$

$$D^r e^{\zeta t} = \zeta^r e^{\zeta t}$$

Consequently

$$Q(D)e^{\zeta t} = Q(\zeta)e^{\zeta t} \quad \text{and} \quad P(D)e^{\zeta t} = P(\zeta)e^{\zeta t}$$

Therefore, Eq. (2.52) becomes

$$\beta Q(\zeta)e^{\zeta t} = P(\zeta)e^{\zeta t}$$

and

$$\beta = \frac{P(\zeta)}{Q(\zeta)}$$

Thus, for the input $f(t) = e^{\zeta t}u(t)$, the forced response is given by

$$y_\phi(t) = H(\zeta)e^{\zeta t} \quad t > 0 \tag{2.55}$$

where

$$H(\zeta) = \frac{P(\zeta)}{Q(\zeta)} \tag{2.56}$$

This is an interesting and significant result. It states that for an exponential input $e^{\zeta t}$ the forced response $y_\phi(t)$ is the same exponential multiplied by $H(\zeta) = P(\zeta)/Q(\zeta)$. The total system response $y(t)$ to an exponential input $e^{\zeta t}$ is then given by

$$y(t) = \sum_{j=1}^{n} K_j e^{\lambda_j t} + H(\zeta)e^{\zeta t} \tag{2.57}$$

---

† This result is valid only if ζ is not a characteristic root of the system.

where the arbitrary constants K_1, K_2, ..., K_n are determined from auxiliary conditions.

Recall that the exponential signal includes a large variety of signals, such as a constant ($\zeta = 0$), a sinusoid ($\zeta = \pm j\omega$), and an exponentially growing or decaying sinusoid ($\zeta = \sigma \pm j\omega$). Let us consider the forced response for some of these cases.

The Constant Input $f(t) = C$

Because $C = Ce^{0t}$, the constant input is a special case of the exponential input $Ce^{\zeta t}$ with $\zeta = 0$. The forced response to this input is then given by

$$y_\phi(t) = CH(\zeta)e^{\zeta t} \qquad \text{with} \qquad \zeta = 0$$

$$= CH(0) \tag{2.58}$$

The Exponential Input $e^{j\omega t}$

Here $\zeta = j\omega$ and

$$y_\phi(t) = H(j\omega)e^{j\omega t} \tag{2.59}$$

The Sinusoidal Input $f(t) = \cos \omega_0 t$

We know that the forced response for the input $e^{\pm j\omega t}$ is $H(\pm j\omega)e^{\pm j\omega t}$. Since $\cos \omega t = (e^{j\omega t} + e^{-j\omega t})/2$, the forced response to $\cos \omega t$ is

$$y_\phi(t) = \frac{1}{2}\left[H(j\omega)e^{j\omega t} + H(-j\omega)e^{-j\omega t}\right]$$

Because the two terms on the right-hand side are conjugates,

$$y_\phi(t) = \text{Re}\left[H(j\omega)e^{j\omega t}\right]$$

But

$$H(j\omega) = |H(j\omega)|e^{j\angle H(j\omega)}$$

so that

$$y_\phi(t) = \text{Re}\left\{|H(j\omega)|e^{j[\omega t + \angle H(j\omega)]}\right\}$$

$$= |H(j\omega)| \cos\left[\omega t + \angle H(j\omega)\right] \tag{2.60}$$

This result can be generalized for the input $f(t) = \cos(\omega t + \theta)$. The forced response in this case is

$$y_\phi(t) = |H(j\omega)| \cos\left[\omega t + \theta + \angle H(j\omega)\right] \tag{2.61}$$

■ **Example 2.10**
Solve the differential equation

$$\left(D^2 + 3D + 2\right)y(t) = Df(t)$$

if the initial conditions are $y(0^+) = 2$ and $\dot{y}(0^+) = 3$ and the input is

(a) $10e^{-3t}$ (b) 5 (c) e^{-2t} (d) $10\cos(3t + 30°)$.

According to Example 2.9, the natural response for this case is

$$y_n(t) = K_1 e^{-t} + K_2 e^{-2t}$$

For this case

$$H(\zeta) = \frac{P(\zeta)}{Q(\zeta)} = \frac{\zeta}{\zeta^2 + 3\zeta + 2}$$

(a) For input $f(t) = 10e^{-3t}$, $\zeta = -3$, and

$$y_\phi(t) = 10H(-3)e^{-3t} = 10\left[\frac{-3}{(-3)^2 + 3(-3) + 2}\right]e^{-3t} = -15e^{-3t} \qquad t > 0$$

The total solution (the sum of the forced and the natural response) is

$$y(t) = K_1 e^{-t} + K_2 e^{-2t} - 15e^{-3t} \qquad t > 0$$

and

$$\dot{y}(t) = -K_1 e^{-t} - 2K_2 e^{-2t} + 45e^{-3t} \qquad t > 0$$

The initial conditions are $y(0^+) = 2$ and $\dot{y}(0^+) = 3$. Setting $t = 0$ in the above equations and then substituting the initial conditions yields

$$K_1 + K_2 - 15 = 2 \qquad \text{and} \qquad -K_1 - 2K_2 + 45 = 3$$

Solution of these equations yields $K_1 = -8$ and $K_2 = 25$. Therefore

$$y(t) = -8e^{-t} + 25e^{-2t} - 15e^{-3t} \qquad t > 0$$

(b) For input $f(t) = 5 = 5e^{0t}$, $\zeta = 0$, and

$$y_\phi(t) = 5H(0) = 0 \qquad t > 0$$

The complete solution is $K_1 e^{-t} + K_2 e^{-2t}$. Using the initial conditions, we determine K_1 and K_2 as in part **(a)**.

(c) Here $\zeta = -2$, which is also a characteristic root of the system. Hence, (see Pair 2, Table 2.2, or the comment at the bottom of the table)

$$y_\phi(t) = \beta t e^{-2t}$$

To find β, we substitute $y_\phi(t)$ in the system equation to obtain

$$\left(D^2 + 3D + 2\right)y_\phi(t) = Df(t)$$

or

$$\left(D^2 + 3D + 2\right)\left[\beta t e^{-2t}\right] = De^{-2t}$$

But

$$D\left[\beta t e^{-2t}\right] = \beta(1 - 2t)e^{-2t}$$

$$D^2\left[\beta t e^{-2t}\right] = 4\beta(t - 1)e^{-2t}$$

$$De^{-2t} = -2e^{-2t}$$

Consequently

$$\beta(4t - 4 + 3 - 6t + 2t)e^{-2t} = -2e^{-2t}$$

or

$$-\beta e^{-2t} = -2e^{-2t}$$

Therefore, $\beta = 2$ so that

$$y_\phi(t) = 2te^{-2t}$$

The complete solution is $K_1e^{-t} + K_2e^{-2t} + 2te^{-2t}$. Using the initial conditions, we determine K_1 and K_2 as in part (a).

(d) For the input $f(t) = 10\cos(3t + 30°)$, the forced response [see Eq. (2.61)] is

$$y_\phi(t) = 10|H(j3)|\cos[3t + 30° + \angle H(j3)]$$

where

$$H(j3) = \frac{P(j3)}{Q(j3)} = \frac{j3}{(j3)^2 + 3(j3) + 2} = \frac{j3}{-7 + j9} = \frac{27 - j21}{130} = 0.263e^{-j37.9°}$$

Therefore

$$|H(j3)| = 0.263, \qquad \angle H(j3) = -37.9°$$

and

$$y_\phi(t) = 10(0.263)\cos(3t + 30° - 37.9°) = 2.63\cos(3t - 7.9°)$$

The complete solution is $K_1e^{-t} + K_2e^{-2t} + 2.63\cos(3t - 7.9°)$. Using the initial conditions, we determine K_1 and K_2 as in part (a). ■

■ **Example 2.11**

Using the classical method, find the loop current $y(t)$ in the RLC circuit of Fig. 2.1, Example 2.2 if the input voltage $f(t) = 10e^{-3t}$ and the initial conditions are $y(0^-) = 0$ and $v_C(0^-) = 5$.

The zero-input and zero-state responses for this problem are found in Examples 2.2 and 2.5, respectively. The natural and forced responses appear in Eq. 2.51b. Here we shall solve this problem by the classical method, which requires the initial conditions at $t = 0^+$. These conditions, already found in Eq. (2.16), are

$$y(0^+) = 0 \qquad \text{and} \qquad \dot{y}(0^+) = 5$$

The loop equation for this system [see Example 2.2 or Eq. (1.55)] is

$$\left(D^2 + 3D + 2\right)y(t) = Df(t)$$

The characteristic polynomial is $\lambda^2 + 3\lambda + 2 = (\lambda + 1)(\lambda + 2)$. Therefore, the natural response is

$$y_n(t) = K_1e^{-t} + K_2e^{-2t}$$

The forced response, already found in Example 2.10 (a), is

$$y_\phi(t) = -15e^{-3t}$$

The total response is

$$y(t) = K_1e^{-t} + K_2e^{-2t} - 15e^{-3t}$$

Differentiation of this equation yields

$$\dot{y}(t) = -K_1e^{-t} - 2K_2e^{-2t} + 45e^{-3t}$$

Setting $t = 0^+$ and substituting $y(0^+) = 0$, $\dot{y}(0^+) = 5$ in these equations yields

$$\left.\begin{array}{l} 0 = K_1 + K_2 - 15 \\ 5 = -K_1 - 2K_2 + 45 \end{array}\right\} \implies \begin{array}{l} K_1 = -10 \\ K_2 = 25 \end{array}$$

Therefore

$$y(t) = -10e^{-t} + 25e^{-2t} - 15e^{-3t}$$

which agrees with the solution found previously in Eq. 2.51b. ∎

⊙ **Computer Example C2.5**
Solve the differential equation

$$(D^2 + 3D + 2)y(t) = f(t)$$

for the input $f(t) = 5t + 3$.

```
f='5*t+3';
mpa('f',f)
yt=dsolve('D2y+3*Dy+2*y=f','y(0)=2','Dy(0)=3','t')
yt = -9/4+5/2*t-19/4*exp(-2*t)+9*exp(-t)     ⊙
```

Assessment of the Classical Method

The development in this section shows that the classical method is relatively simple compared to the method of finding the response as a sum of the zero-input and zero-state components. Unfortunately, the classical method has a serious drawback because it yields the total response, which cannot be separated into components arising from the internal conditions and the external input. In the study of systems it is important to be able to express the system response to an input $f(t)$ as an explicit function of $f(t)$. This is not possible in the classical method. Moreover, the classical method is restricted to a certain class of inputs; it cannot be applied to any input. Another minor problem is that because the classical method yields total response, the auxiliary conditions must be on the total response which exists only for $t \geq 0^+$. In practice we are most likely to know the conditions at $t = 0^-$ (before the input is applied). Therefore, we need to derive a new set of auxiliary conditions at $t = 0^+$ from the known conditions at $t = 0^-$.

If we must solve a particular linear differential equation or find a response of a particular LTI system, the classical method may be the best. In the theoretical study of linear systems, however, the classical method is practically useless.

2.6 System Stability

Because of the great variety of possible system behaviors, there are several definitions of stability in the literature. Here we shall consider a definition that is suitable for causal, linear, time-invariant (LTI) systems.

In order to understand system stability intuitively, let us examine the stability concept as applied to a right circular cone. Such a cone can be made to stand forever on its circular base, on its apex, or on its side. For this reason these three states of the cone are said to be **equilibrium states**. Qualitatively, however, the three states show very different behavior. If this cone, standing on its circular base, were to be disturbed slightly, it would eventually return to its original equilibrium state if left to itself. In this case, the cone is said to be in **stable equilibrium**. In contrast, if the cone stands on its apex, then the slightest disturbance will cause the cone to move farther and farther away from its equilibrium state. The cone in this case is said to be in an **unstable equilibrium**. The cone lying on its side, if disturbed, will neither go back to the original state nor continue to move farther away from the original state. The cone in this case is said to be in a **neutral equilibrium**.

Let us apply these observations to systems in general. If, in the absence of an external input, a system remains in a particular state (or condition) indefinitely, then that state is said to be an **equilibrium state of the system**. For an LTI system this equilibrium state is the zero state, in which all initial conditions are zero. Now suppose an LTI system is in equilibrium (zero state) and we change this state by creating some nonzero initial conditions. By analogy with the cone, if the system is stable it should eventually return to zero state. In other words, when left to itself, the system's output due to the nonzero initial conditions should approach 0 as $t \to \infty$. But the system output generated by initial conditions (zero-input response) is made up of its characteristic modes. For this reason we define stability as follows: a system is **(asymptotically) stable** if, and only if, all its characteristic modes $\to 0$ as $t \to \infty$. If any of the modes grows without bound as $t \to \infty$, the system is **unstable**. There is also a borderline situation in which the zero-input response remains bounded (approaches neither zero nor infinity), approaching a constant or oscillating with a constant amplitude as $t \to \infty$. For this borderline situation, the system is said to be **marginally stable** or just stable.

If an LTIC system has n distinct characteristic roots $\lambda_1, \lambda_2, \ldots, \lambda_n$, the zero-input response is given by

$$y_0(t) = \sum_{j=1}^{n} c_j e^{\lambda_j t} \tag{2.62}$$

We have shown elsewhere [see Eq. (B.14)]

$$\lim_{t \to \infty} e^{\lambda t} = \begin{cases} 0 & \text{Re } \lambda < 0 \\ \infty & \text{Re } \lambda > 0 \end{cases} \tag{2.63}$$

It is helpful to study system stability in terms of the location of the system's characteristic roots in the complex plane. Let us first assume that the system has distinct roots only. If a characteristic root λ is located in the left half of the complex plane (LHP), its real part is negative (Re $\lambda < 0$). Similarly, if a root λ is located in the right half of the complex plane (RHP), its real part is positive (Re $\lambda > 0$). Along the imaginary axis, the real part is zero (Re $\lambda = 0$). These regions are delineated in Fig. 2.15. Equation (2.63) clearly shows that the characteristic modes corresponding to roots in LHP vanish as $t \to \infty$, while the modes corresponding to roots in RHP grow without bound as $t \to \infty$. However, the modes corresponding to simple (unrepeated) roots on the imaginary axis are of the form $e^{j\beta t}$; these are bounded (neither vanish nor grow without limit) as $t \to \infty$.

From this discussion it follows that a system is asymptotically stable if, and only if, all of its characteristic roots lie in the left half of the complex plane. If any of the roots—even one—lies in RHP, the system is unstable. If none of the roots lie in RHP, but if some unrepeated (simple) roots lie on the imaginary axis, then the system is marginally stable (Fig. 2.15).

So far we have assumed all of the system's n roots to be distinct. The modes corresponding to a root λ repeated r times are $e^{\lambda t}, te^{\lambda t}, t^2 e^{\lambda t}, \cdots, t^{r-1} e^{\lambda t}$. But as

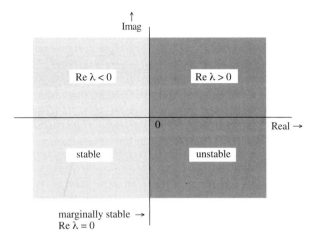

Fig. 2.15 Characteristic roots location and system stability.

$t \rightarrow \infty$, $t^k e^{\lambda t} \rightarrow 0$, if $\operatorname{Re}\lambda < 0$ (λ in LHP). Therefore, repeated roots in LHP do not cause instability. But when the repeated roots are on the imaginary axis ($\lambda = j\omega$), the corresponding modes $t^k e^{j\omega t}$ approach infinity as $t \rightarrow \infty$. Therefore, repeated roots on the imaginary axis cause instability. Figure 2.16 shows characteristic modes corresponding to characteristic roots at various location in the complex plane. Observe the central role played by the characteristic roots or characteristic modes in determining the system's stability.

To summarize:

1. An LTIC system is asymptotically stable if, and only if, all the characteristic roots are in the LHP. The roots may be simple (unrepeated) or repeated.

2. An LTIC system is unstable if, and only if, either one or both of the following conditions exist: (i) at least one root is in the RHP, (ii) there are repeated roots on the imaginary axis.

3. An LTIC system is marginally stable if, and only if, there are no roots in the RHP, and there are some unrepeated roots on the imaginary axis.

■ **Example 2.12**
Investigate the stability of LTIC system described by the following equations:

(a) $(D + 1)\left(D^2 + 4D + 8\right) y(t) = (D - 3)f(t)$

(b) $(D - 1)\left(D^2 + 4D + 8\right) y(t) = (D + 2)f(t)$

(c) $(D + 2)(D^2 + 4)y(t) = \left(D^2 + D + 1\right) f(t)$

(d) $(D + 1)(D^2 + 4)^2 y(t) = \left(D^2 + 2D + 8\right) f(t)$

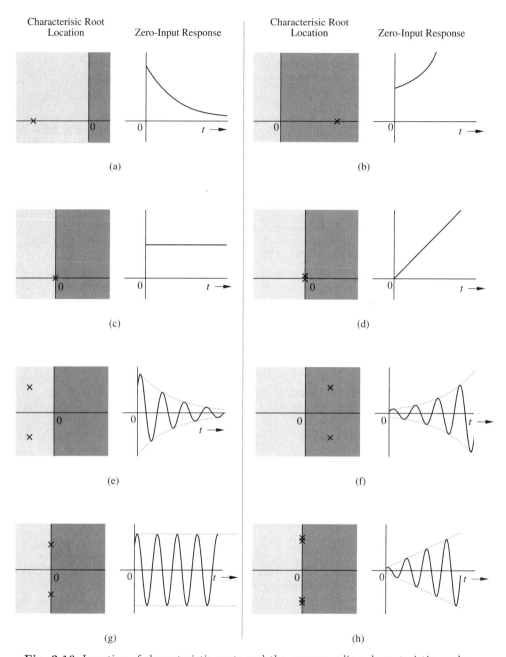

Fig. 2.16 Location of characteristic roots and the corresponding characteristic modes.

The characteristic polynomials of these systems are

(a) $(\lambda + 1)\left(\lambda^2 + 4\lambda + 8\right) = (\lambda + 1)(\lambda + 2 - j2)(\lambda + 2 + j2)$

(b) $(\lambda - 1)\left(\lambda^2 + 4\lambda + 8\right) = (\lambda - 1)(\lambda + 2 - j2)(\lambda + 2 + j2)$

(c) $(\lambda + 2)(\lambda^2 + 4) = (\lambda + 2)(\lambda - j2)(\lambda + j2)$

(d) $(\lambda + 1)(\lambda^2 + 4)^2 = (\lambda + 2)(\lambda - j2)^2(\lambda + j2)^2$

Consequently, the characteristic roots of the systems above are (see Fig. 2.17):

(a) -1, $-2 \pm j2$ (b) 1, $-2 \pm j2$ (c) -2, $\pm j2$ (d) -1, $\pm j2$, $\pm j2$.

System (a) is asymptotically stable (all roots in LHP), (b) is unstable (one root in RHP), (c) is marginally stable (unrepeated roots on imaginary axis) and no roots in RHP, and (d) is unstable (repeated roots on the imaginary axis).   ■

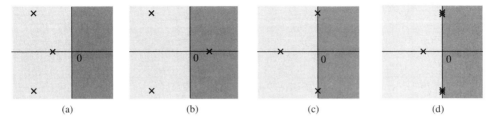

Fig. 2.17   Location of characteristic roots for systems in Example 2.12.

△ **Exercise E2.16**

For each of the systems specified by the equations below, plot its characteristic roots in the complex plane and determine whether it is asymptotically stable, marginally stable, or unstable.

(a) $D(D + 2)y(t) = 3f(t)$

(b) $D^2(D + 3)y(t) = (D + 5)f(t)$

(c) $(D + 1)(D + 2)y(t) = (2D + 3)f(t)$

(d) $(D^2 + 1)(D^2 + 9)y(t) = \left(D^2 + 2D + 4\right)f(t)$

(e) $(D + 1)\left(D^2 - 4D + 9\right)y(t) = (D + 7)f(t)$

Answer: (a) marginally stable (b) unstable (c) stable (d) marginally stable (e) unstable.   ▽

2.6-1 System Response to Bounded Inputs

From the example of the right circular cone, it appears that when a system is in stable equilibrium, application of a small force (input) produces a small response. In contrast, when the system is in unstable equilibrium, a small force (input) produces an unbounded response. Intuitively we feel that every bounded input should produce a bounded response in a stable system, whereas in an unstable system this would not be the case. We shall now verify this hunch and show that it is indeed true.

Recall that for an LTIC system

$$y(t) = h(t) * f(t)$$

$$= \int_{-\infty}^{\infty} h(\tau)f(t - \tau)\, d\tau \qquad (2.64)$$

Therefore

$$|y(t)| \le \int_{-\infty}^{\infty} |h(\tau)||f(t - \tau)|\, d\tau$$

Moreover, if $f(t)$ is bounded, then $|f(t - \tau)| < K_1 < \infty$, and

$$|y(t)| \le K_1 \int_{-\infty}^{\infty} |h(\tau)|\, d\tau$$

Because $h(t)$ contains terms of the form $e^{\lambda_j t}$ or $t^k e^{\lambda_j t}$, $h(t)$ decays exponentially with time if $\operatorname{Re} \lambda_j < 0$. Consequently, for an asymptotically stable system†

$$\int_{-\infty}^{\infty} |h(\tau)|\, d\tau < K_2 < \infty \qquad (2.65)$$

and

$$|y(t)| \leq K_1 K_2 < \infty$$

Thus, for an asymptotically stable system, a bounded input always produces a bounded output. Moreover, we can show that for an unstable or a marginally stable system, the output $y(t)$ is unbounded for some bounded input (see Problem 2.6-4). These results lead to the formulation of an alternative definition of stability known as **bounded-input, bounded-output (BIBO) stability**: a system is BIBO stable if, and only if, a bounded input produces a bounded output. Observe that an asymptotically stable system is always BIBO stable.‡ However, a marginally stable system is BIBO unstable.

Implications of Stability

All practical signal processing systems must be stable. Unstable systems are useless from the viewpoint of signal processing because any set of intended or unintended initial conditions leads to an unbounded response that either destroys the system or (more likely) leads it to some saturation conditions that change the nature of the system. Even if the discernible initial conditions are zero, stray voltages or thermal noise signals generated within the system will act as initial conditions. Because of exponential growth, a stray signal, no matter how small, will eventually cause an unbounded output in an unstable system.

Marginally stable systems do have one important application in the oscillator, which is a system that generates a signal on its own without the application of an external input. Consequently, the oscillator output is a zero-input response. If such a response is to be a sinusoid of frequency ω_0, the system should be marginally stable with characteristic roots at $\pm j\omega_0$. Thus, to design an oscillator of frequency ω_0, we should pick a system with the characteristic polynomial $(\lambda - j\omega_0)(\lambda + j\omega_0) = \lambda^2 + \omega_0{}^2$. A system described by the differential equation

$$\left(D^2 + \omega_0{}^2\right) y(t) = f(t)$$

will do the job.

---

† This can be shown as follows. If $\lambda_i = \alpha_i + j\beta_i$, then $e^{\lambda_i t} = e^{\alpha_i t} e^{j\beta_i t}$ and $\left|e^{\lambda_i t}\right| = e^{\alpha_i t}$. Therefore

$$\int_{-\infty}^{\infty} \left|e^{\lambda_i \tau} u(\tau)\right| d\tau = \int_{0}^{\infty} e^{\alpha_i \tau} d\tau = -\frac{1}{\alpha_i} \qquad \text{if } \operatorname{Re} \lambda_i = \alpha_i < 0$$

and Eq. (2.65) follows. This conclusion is also valid when the integrand is of the form $|t^k e^{\lambda_i t}| u(t)$.
‡ However, a BIBO stable system is not necessarily asymptotically stable because BIBO stability is determined from the system's impulse response, which is an external description of the system, while asymptotic stability is determined from the internal description of the system obtained from system equations. In certain systems (e.g., uncontrollable or unobservable systems), the two descriptions may not be the same. Remember that the external description describes only that part of the system which is coupled to both the input and the output. Hence, a system may be internally unstable while appearing stable from the system's external terminals (BIBO stable)[2].

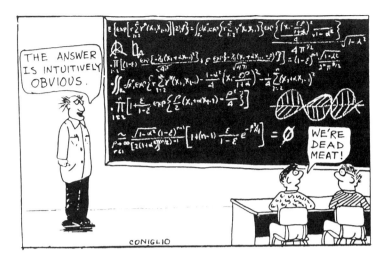

Intuition can cut the math jungle instantly!

2.7 Intuitive Insights into System Behavior

This section attempts to provide an understanding of what determines system behavior. Because of its intuitive nature, the following discussion will be more or less qualitative. We shall now show that the most important attributes of a system are its characteristic roots or characteristic modes because they determine not only the zero-input response but also the entire behavior of the system.

2.7-1 Dependence of System Behavior on Characteristic Modes

Recall that the zero-input response of a system consists of the system's characteristic modes. For a stable system, these characteristic modes decay exponentially and eventually vanish. This behavior may give the impression that these modes do not substantially affect system behavior in general and system response in particular. This impression is totally wrong! We shall now see that the system's characteristic modes leave their imprint on every aspect of the system behavior. *We may compare the system's characteristic modes (or roots) to a seed which eventually dissolves in the ground; however, the plant that springs from it is totally determined by the seed. The imprint of the seed exists on every cell of the plant.*

In order to understand this interesting phenomenon, recall that the characteristic modes of a system are very special to that system because it can sustain these signals without the application of an external input. In other words, the system offers a free ride and ready access to these signals. Now imagine what would happen if we actually drove the system with an input having the form of a characteristic mode! We would expect the system to respond strongly (this is, in fact, the resonance phenomenon discussed later in this section). If the input is not exactly a characteristic mode but is close to such a mode, we would still expect the system response to be strong. However, if the input is very different from any of the characteristic modes, we would expect the system to respond poorly. We shall now show that these intuitive deductions are indeed true.

Although we have devised a measure of similarity of signals (correlation) later in Chapter 3, we shall take a simpler approach here. Let us restrict the system's inputs only to exponentials of the form $e^{\zeta t}$, where ζ is generally a complex number. The similarity of two exponential signals $e^{\zeta t}$ and $e^{\lambda t}$ will then be measured by the closeness of ζ and λ. If the difference $\zeta - \lambda$ is small, the signals are similar; if $\zeta - \lambda$ is large, the signals are dissimilar.

Now consider a first-order system with a single characteristic mode $e^{\lambda t}$ and the input $e^{\zeta t}$. The impulse response of this system is then given by $Ae^{\lambda t}$, where the exact value of A is not important for this qualitative discussion. The system response $y(t)$ is given by

$$y(t) = h(t) * f(t)$$
$$= Ae^{\lambda t}u(t) * e^{\zeta t}u(t)$$

From the convolution table (Table 2.1), we obtain

$$y(t) = \frac{A}{\zeta - \lambda}\left[e^{\zeta t} - e^{\lambda t}\right]u(t) \tag{2.66}$$

Clearly, if the input $e^{\zeta t}$ is similar to $e^{\lambda t}$, $\zeta - \lambda$ is small, and the system response is large. *The closer the input $f(t)$ is to the characteristic mode, the stronger is the system response.* In contrast, if the input is very different from the natural mode, $\zeta - \lambda$ is large, and the system responds poorly. This is precisely what we set out to prove.

We have proved the above assertion for a single-mode (first-order) system. It can be generalized to an nth-order system, which has n characteristic modes. The impulse response $h(t)$ of such a system is a linear combination of its n modes. Therefore, if $f(t)$ is similar to any one of the modes, the corresponding response will be high; if it is similar to none of the modes, the response will be small. Clearly, the characteristic modes are very influential in determining system response to a given input.

It would be tempting to conclude on the basis of Eq. (2.66) that if the input is identical to the characteristic mode, so that $\zeta = \lambda$, then the response goes to infinity. Remember, however, that if $\zeta = \lambda$, the numerator on the right-hand side of Eq. (2.66) also goes to zero. We shall study this complex behavior (resonance phenomenon) later in this section.

We shall now show that *mere inspection of the impulse response $h(t)$ (which is composed of characteristic modes), reveals a great deal about the system behavior.*

2.7-2   Response Time of a System: The System Time Constant

Like human beings, systems have a certain response time. In other words, when an input (stimulus) is applied to a system, a certain amount of time elapses before the system fully responds to that input. This time lag or response time is called the system **time constant**. As we shall see, a system's time constant is equal to the width of its impulse response $h(t)$.

An input $\delta(t)$ to a system is instantaneous (zero duration), but its response $h(t)$ has a duration T_h. Therefore, the system requires a time T_h to respond fully to this input, and we are justified in viewing T_h as the system's response time or time constant. We arrive at the same conclusion via another argument. The output is a convolution of the input with $h(t)$. If an input is a pulse of width T_f, then the output pulse width is $T_f + T_h$ according to the width property of convolution. This conclusion shows that the system requires T_h seconds to respond fully to any input. *The system time constant indicates how fast the system is. A system with a smaller time constant is a faster system that responds quickly to an input. A system with a relatively large time constant is a sluggish system that cannot respond well to rapidly varying signals.*

Strictly speaking, the duration of the impulse response $h(t)$ is ∞ because the characteristic modes approach zero asymptotically as $t \to \infty$. However, beyond some value of t, $h(t)$ becomes negligible. It is therefore necessary to use some suitable measure of the impulse response's effective width.

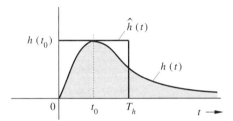

Fig. 2.18 Effective duration of an impulse response.

There is no single satisfactory definition of effective signal duration (or width) applicable to every situation. For the situation depicted in Fig. 2.18, a reasonable definition of the duration $h(t)$ would be T_h, the width of the rectangular pulse $\hat{h}(t)$. This rectangular pulse $\hat{h}(t)$ has an area identical to that of $h(t)$ and a height identical to that of $h(t)$ at some suitable instant $t = t_0$. In Fig. 2.18, t_0 is chosen as the instant at which $h(t)$ is maximum. According to this definition,†

$$T_h h(t_0) = \int_{-\infty}^{\infty} h(t)\, dt$$

or

$$T_h = \frac{\int_{-\infty}^{\infty} h(t)\, dt}{h(t_0)} \tag{2.67}$$

Now if a system has a single mode

$$h(t) = A e^{\lambda t} u(t)$$

with λ negative and real, then $h(t)$ is maximum at $t = 0$ with value $h(0) = A$. Therefore, according to Eq. (2.67)

$$T_h = \frac{1}{A} \int_0^{\infty} A e^{\lambda t}\, dt = -\frac{1}{\lambda} \tag{2.68}$$

---

† This definition is satisfactory when $h(t)$ is a single, mostly positive (or mostly negative) pulse. Such systems are lowpass systems. This definition should not be applied indiscriminately to all systems.

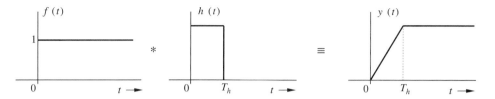

Fig. 2.19 Rise time of a system.

Thus, the time constant in this case is simply (the negative of the) reciprocal of the system's characteristic root. For the multimode case, $h(t)$ is a weighted sum of the system's characteristic modes, and T_h is a weighted average of the time constants associated with the n modes of the system.

2.7-3 Time Constant and Rise Time of a System

The system time constant may also be viewed from a different perspective. The unit step response $y(t)$ of a system is the convolution of $u(t)$ with $h(t)$. Let the impulse response $h(t)$ be a rectangular pulse of width T_h, as shown in Fig. 2.19. This assumption simplifies the discussion, yet gives satisfactory results for qualitative discussion. The result of this convolution is illustrated in Fig. 2.19. Note that the output does not rise from zero to a final value instantaneously as the input rises; instead, the output takes T_h seconds to accomplish this. Hence, the rise time T_r of the system is equal to the system time constant‡

$$T_r = T_h \tag{2.69}$$

This result and Fig. 2.19 show clearly that a system generally does not respond to an input instantaneously. Instead, it takes time T_h for the system to respond fully.

2.7-4 Time Constant and Filtering

A larger time constant implies a sluggish system because the system takes a longer time to respond fully to an input. Such a system cannot respond effectively to rapid variations in the input. In contrast, a smaller time constant indicates that a system is capable of responding to rapid variations in the input. Thus, there is a direct connection between a system's time constant and its filtering properties.

Consider a high-frequency sinusoid that varies rapidly with time. A system with a large time constant will not be able to respond well to this input. Therefore, such a system will suppress rapidly varying (high-frequency) sinusoids and other high-frequency signals, thereby acting as a lowpass filter (a filter allowing the transmission of low-frequency signals only). We shall now show that a system with a time constant T_h acts as a lowpass filter having a cutoff frequency of $\mathcal{F}_c = 1/T_h$ Hz, so that sinusoids with frequencies below $\mathcal{F}_c$ Hz are transmitted reasonably well, while those with frequencies above $\mathcal{F}_c$ Hz are suppressed.

To demonstrate this fact, let us determine the system response to a sinusoidal input $f(t)$ by convolving this input with the effective impulse response $h(t)$ in Fig. 2.20a. Figures 2.20b and 2.20c show the process of convolution of $h(t)$ with the

---

‡ Because of varying definitions of rise time, the reader may find different results in the literature. The qualitative and intuitive nature of this discussion should always be kept in mind.

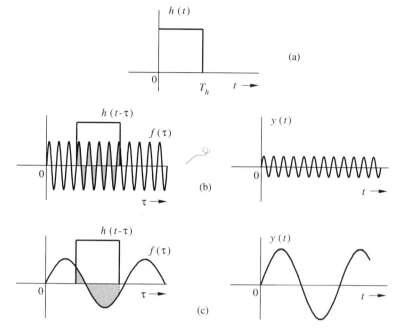

Fig. 2.20 Time constant and filtering.

sinusoidal inputs of two different frequencies. The sinusoid in Fig. 2.20b has a relatively high frequency, while the frequency of the sinusoid in Fig. 2.20c is low. Recall that the convolution of $f(t)$ and $h(t)$ is equal to the area under the product $f(\tau)h(t-\tau)$. This area is shown shaded in Figs. 2.20b and 2.20c for the two cases. For the high-frequency sinusoid, it is clear from Fig. 2.20b that the area under $f(\tau)h(t-\tau)$ is very small because its positive and negative areas nearly cancel each other out. In this case the output $y(t)$ remains periodic but has a rather small amplitude. This happens when the period of the sinusoid is much smaller than the system time-constant T_h. In contrast, for the low-frequency sinusoid, the period of the sinusoid is larger than T_h, so that the partial cancellation of area under $f(\tau)h(t-\tau)$ is less effective. Consequently, the output $y(t)$ is much larger, as depicted in Fig. 2.20c.

Between these two possible extremes in system behavior, a transition point occurs when the period of the sinusoid is equal to the system time constant T_h. The frequency at which this transition occurs is known as the **cutoff frequency** $\mathcal{F}_c$ of the system. Because T_h is the period of cutoff frequency $\mathcal{F}_c$,

$$\mathcal{F}_c = \frac{1}{T_h} \tag{2.70}$$

The frequency $\mathcal{F}_c$ is also known as the bandwidth of the system because the system transmits or passes sinusoidal components with frequencies below $\mathcal{F}_c$ while attenuating components with frequencies above $\mathcal{F}_c$. Of course, the transition in system behavior is gradual. There is no dramatic change in system behavior at $\mathcal{F}_c = 1/T_h$. Moreover, these results are based on an idealized (rectangular pulse) impulse response; in practice these results will vary somewhat, depending on the exact shape of $h(t)$. Remember that the "feel" of general system behavior is more important than exact system response for this qualitative discussion.

Since the system time constant is equal to its rise time, we have

$$T_r = \frac{1}{\mathcal{F}_c} \qquad \text{or} \qquad \mathcal{F}_c = \frac{1}{T_r} \qquad (2.71a)$$

Thus, a system's bandwidth is inversely proportional to its rise time. Although Eq. (2.71a) was derived for an idealized (rectangular) impulse response, its implications are valid for lowpass LTIC systems in general. For a general case, we can show that[1]

$$\mathcal{F}_c = \frac{k}{T_r} \qquad (2.71b)$$

where the exact value of k depends on the nature of $h(t)$. An experienced engineer often can estimate quickly the bandwidth of an unknown system by simply observing the system response to a step input on an oscilloscope.

2.7-5 Time Constant and Pulse Dispersion (Spreading)

In general, the transmission of a pulse through a system causes pulse dispersion (or spreading). Therefore, the output pulse is generally wider than the input pulse. This system behavior can have serious consequences in communication systems where information is transmitted by pulse amplitudes. Dispersion (or spreading) causes interference or overlap with neighboring pulses, thereby distorting pulse amplitudes and introducing errors in the received information.

Earlier we saw that if an input $f(t)$ is a pulse of width T_f, then T_y, the width of the output $y(t)$, is

$$T_y = T_f + T_h \qquad (2.72)$$

This result shows that an input pulse spreads out (disperses) as it passes through a system. Since T_h is also the system's time constant or rise time, the amount of spread in the pulse is equal to the time constant (or rise time) of the system.

2.7-6 Time Constant and Rate of Information Transmission

In pulse communications systems where information is conveyed through pulse amplitudes, the rate of information transmission is proportional to the rate of pulse transmission. We shall demonstrate that to avoid the destruction of information caused by dispersion of pulses during their transmission through the channel (transmission medium), the rate of information transmission should not exceed the bandwidth of the communications channel.

Since an input pulse spreads out by T_h seconds, the consecutive pulses should be spaced T_h seconds apart in order to avoid interference between pulses. Thus, the rate of pulse transmission should not exceed $1/T_h$ pulses/second. But $1/T_h = \mathcal{F}_c$, the channel's bandwidth, so that we can transmit pulses through a communications channel at a rate of $\mathcal{F}_c$ pulses/second and still avoid significant interference between the pulses. The rate of information transmission is therefore proportional to the channel's bandwidth (or to the reciprocal of its time constant).†.

---

† Theoretically, a channel of bandwidth $\mathcal{F}_c$ can transmit up to $2\mathcal{F}_c$ pulse amplitudes per second correctly.[3] Our derivation here, being very simple and qualitative, yields only half the theoretical limit. In practice it is difficult to attain the upper theoretical limit; transmission rates of $\mathcal{F}_c$ pulses per second are more common.

This discussion (Secs. 2.7-2 through 2.7-6) shows that the system time constant determines much of a system's behavior—its filtering characteristics, rise time, pulse dispersion, and so on. In turn, the time constant is determined by the system's characteristic roots. Clearly the characteristic roots and their relative amounts in the impulse response $h(t)$ determine the behavior of a system.

2.7-7 The Resonance Phenomenon

Finally, we come to the fascinating phenomenon of resonance. As we have mentioned already several times, this phenomenon is observed when the input signal is identical or is very similar to a characteristic mode of the system. For the sake of simplicity and clarity, we consider a first-order system which has only a single mode, $e^{\lambda t}$. Let the impulse response of this system be†

$$h(t) = Ae^{\lambda t} \tag{2.73}$$

and let the input be

$$f(t) = e^{(\lambda - \epsilon)t}$$

The system response $y(t)$ is then given by

$$y(t) = Ae^{\lambda t} * e^{(\lambda - \epsilon)t}$$

From the convolution table we obtain

$$y(t) = \frac{A}{\epsilon}\left[e^{\lambda t} - e^{(\lambda - \epsilon)t}\right]$$

$$= Ae^{\lambda t}\left(\frac{1 - e^{-\epsilon t}}{\epsilon}\right) \tag{2.74}$$

Now, as $\epsilon \to 0$, both the numerator and the denominator of the term in the parentheses approach zero. Applying L'Hôpital's rule to this term yields

$$\lim_{\epsilon \to 0} y(t) = Ate^{\lambda t} \tag{2.75}$$

Clearly, the response does not go to infinity as $\epsilon \to 0$, but it acquires a factor t, which approaches ∞ as $t \to \infty$. If λ has a negative real part (so that it lies in LHP), $e^{\lambda t}$ decays faster than t and $y(t) \to 0$ as $t \to \infty$. The resonance phenomenon in this case is present, but its manifestation is aborted by the signal's own exponential decay.

This discussion shows that *resonance is a cumulative phenomenon*, not instantaneous. It builds up linearly‡ with t. When the mode decays exponentially, the signal decays at a rate too fast for resonance to counteract the decay; as a result, the signal vanishes before resonance has a chance to build it up. However, if the mode were to decay at a rate less than $1/t$, we should see the resonance phenomenon clearly. This specific condition would be possible if Re $\lambda \geq 0$. For instance, when Re $\lambda = 0$, so that λ lies on the imaginary axis of the complex plane, and therefore

---

† For convenience we omit multiplying $f(t)$ and $h(t)$ by $u(t)$. Throughout this discussion, we assume that they are causal.
‡ If the characteristic root in question repeats r times, resonance effect increases as t^{r-1}. However, $t^{r-1}e^{\lambda t} \to 0$ as $t \to \infty$ for any value of r, provided Re $\lambda < 0$ (λ in LHP).

$$\lambda = j\omega$$

and Eq. (2.75) becomes

$$y(t) = Ate^{j\omega t} \tag{2.76}$$

Here, the response does go to infinity linearly with t.

For a real system, if $\lambda = j\omega$ is a root, $\lambda^* = -j\omega$ must also be a root; the impulse response is of the form $Ae^{j\omega t} + Ae^{-j\omega t} = 2A\cos\omega t$. The response of this system to input $A\cos\omega t$ is $2A\cos\omega t * \cos\omega t$. The reader can show that this convolution contains a term of the form $At\cos\omega t$. The resonance phenomenon is clearly visible. The system response to its characteristic mode increases linearly with time, eventually reaching ∞, as indicated in Fig. 2.21.

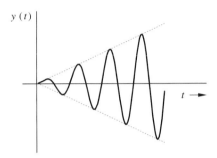

Fig. 2.21 Build up of system response in resonance.

Recall that when $\lambda = j\omega$, the system is marginally stable. As we have indicated, the full effect of resonance cannot be seen for an asymptotically stable system; only in a marginally stable system does the resonance phenomenon boost the system's response to infinity when the system's input is a characteristic mode. But even in an asymptotically stable system, we see a manifestation of resonance if its characteristic roots are close to the imaginary axis, so that Re λ is very small. We can show that when the characteristic roots of a system are $\sigma \pm j\omega_0$, then, the system response to the input $e^{j\omega_0 t}$ or the sinusoid $\cos\omega_0 t$ is very large for small σ.[†] The response drops off rapidly as the input signal frequency moves away from ω_0. This frequency-selective behavior can be studied more profitably after an understanding of frequency-domain analysis is acquired. For this reason we postpone full discussion of this subject until Chapter 7.

Importance of the Resonance Phenomenon

The resonance phenomenon is very important because it allows us to design frequency-selective systems by choosing their characteristic roots properly. Low-pass, bandpass, highpass, and bandstop filters are all examples of frequency selective networks. In mechanical systems, the inadvertent presence of resonance can cause signals of such tremendous magnitudes that the system may fall apart. A musical note (periodic vibrations) of proper frequency can shatter a glass if the frequency is matched to the characteristic root of the glass, which acts as a mechanical system. Similarly, a company of soldiers marching in step across a bridge

---

[†]This follows directly from Eq. (2.74) with $\lambda = \sigma + j\omega_0$ and $\epsilon = \sigma$

amounts to applying a periodic force to the bridge. If the frequency of this input force happens to be nearer to a characteristic root of the bridge, the bridge may respond (vibrate) violently and collapse, even though it would have been strong enough to carry many soldiers marching out of step. A case in point is the Tacoma Narrow Bridge failure of 1940. This bridge was opened to traffic in July 1940. Within four months of opening (on 7 November 1940), it collapsed in a mild gale, not because of the wind's brute force but because the frequency of wind-generated vortices, which matched the natural frequency (characteristic roots) of the bridge, causing resonance.

Because of the great damage which may occur, mechanical resonance is generally something to be avoided, especially in structures or vibrating mechanisms. If an engine with periodic force (such as piston motion) is mounted on a platform, the platform with its mass and springs should be designed so that their characteristic roots are not close to the engine's frequency of vibration. Proper design of this platform can not only avoid resonance, but also attenuate vibrations if the system roots are placed far away from the frequency of vibration.

2.8   Appendix 2.1: Determining the Impulse Response

We now derive the unit impulse response of an LTIC system S specified by the nth-order differential equation

or
$$Q(D)y(t) = P(D)f(t) \tag{2.77a}$$

$$(D^n + a_{n-1}D^{n-1} + \cdots + a_1 D + a_0)y(t)$$
$$= (b_n D^n + b_{n-1}D^{n-1} + \cdots + b_1 D + b_0)f(t) \tag{2.77b}$$

In Sec. 2.3 we showed that the impulse response $h(t)$ is given by

$$h(t) = A_0 \delta(t) + \text{characteristic modes} \tag{2.78}$$

We now show that in the above equation $A_0 = b_n$ where b_n is the coefficient of the nth-order term on the right-hand side of Eq. (2.77b). When the input $f(t) = \delta(t)$, the response $y(t) = h(t)$. Therefore, from Eq. (2.77b), we obtain

$$(D^n + a_{n-1}D^{n-1} + \cdots + a_1 D + a_0)h(t) = (b_n D^n + b_{n-1}D^{n-1} + \cdots + b_1 D + b_0)\delta(t)$$

In this equation we substitute $h(t)$ from Eq. (2.78) and compare the coefficients of similar impulsive terms on both sides. The highest order of the derivative of impulse on both sides is n, with its coefficient value as A_0 on the left-hand side and b_n on the right-hand side. The two values must be matched. Therefore, $A_0 = b_n$ and

$$h(t) = b_n \delta(t) + \text{characteristic modes} \tag{2.79}$$

To determine the characteristic mode terms in the above equation, let us consider a system S_0 whose input $f(t)$ and the corresponding output $x(t)$ are related by

$$Q(D)x(t) = f(t) \tag{2.80}$$

Observe that both the systems S and S_0 have the same characteristic polynomial; namely, $Q(\lambda)$, and, consequently, the same characteristic modes. Moreover, S_0 is the same as S with $P(D) = 1$, that is, $b_n = 0$. Therefore, according to Eq. (2.79), the impulse response of S_0 consists of characteristic mode terms only without an impulse at $t = 0$. Let us denote this impulse response of S_0 by $y_n(t)$. Observe that $y_n(t)$ consists of characteristic modes of S, and therefore may be viewed as a zero-input response of S. Now $y_n(t)$ is the response of S_0 to input $\delta(t)$. Therefore, according to Eq. (2.80)

$$Q(D)y_n(t) = \delta(t) \tag{2.81a}$$

or

$$(D^n + a_{n-1}D^{n-1} + \cdots + a_1 D + a_0)y_n(t) = \delta(t) \tag{2.81b}$$

or

$$y_n^{(n)}(t) + a_{n-1}y_n^{(n-1)}(t) + \cdots + a_1 y_n^{(1)}(t) + a_0 y_n(t) = \delta(t) \tag{2.81c}$$

where $y_n^{(k)}(t)$ represents the kth derivative of $y_n(t)$. The right-hand side contains a single impulse term $\delta(t)$. This is possible only if $y_n^{(n-1)}(t)$ has a unit jump discontinuity at $t = 0$, so that $y_n^{(n)}(t) = \delta(t)$. Moreover, the lower-order terms can not have any jump discontinuity because this would mean the presence of the derivatives of $\delta(t)$. Suppose, for instance, $y_n(t)$ has a jump discontinuity, then its derivative $\dot{y}_n(t)$ contains an impulse $\delta(t)$, and its second derivative $\ddot{y}_n(t)$ contains the first derivative of the impulse $\delta(t)$, and so on. But this is impossible because the right-hand side of Eq. (2.81c) consists of only $\delta(t)$. For this reason only $y_n^{(n-1)}(t)$ can have a unit jump discontinuity so that $y_n^{(n)}(t)$ is $\delta(t)$. There can be no jump discontinuities in any of the remaining variables because this would give rise to higher-order derivatives of $\delta(t)$ on the left-hand side. Therefore $y_n(0) = y_n^{(1)}(0) = \cdots = y_n^{(n-2)}(0) = 0$ (no discontinuity at $t = 0$). Therefore, the n initial conditions on $y_n(t)$ are

$$y_n^{(n-1)}(0) = 1$$
$$y_n(0) = y_n^{(1)}(0) = \cdots = y_n^{(n-2)}(0) = 0 \tag{2.82}$$

This discussion means that $y_n(t)$ is the zero-input response of the system S subject to initial conditions (2.82).

We now show that for the same input $f(t)$ to both systems, S and S_0, their respective outputs $y(t)$ and $x(t)$ are related by

$$y(t) = P(D)x(t) \tag{2.83}$$

To prove this result, we operate on both sides of Eq. (2.80) by $P(D)$ to obtain

$$Q(D)P(D)x(t) = P(D)f(t)$$

Comparison of this equation with Eq. (2.77a) leads immediately to Eq. (2.83).

Now if the input $f(t) = \delta(t)$, the output of S_0 is $y_n(t)$, and the output of S, according to Eq. (2.83), is $P(D)y_n(t)$. This output is $h(t)$, the unit impulse response of S. Note, however, that because it is an impulse response of a causal system S_0, the function $y_n(t)$ is causal. To incorporate this fact we must represent this function as $y_n(t)u(t)$. Now it follows that $h(t)$, the unit impulse response of the system S, is given by

$$h(t) = P(D)[y_n(t)u(t)] \tag{2.84}$$

where $y_n(t)$ is a linear combination of the characteristic modes of the system subject to initial conditions (2.82).

The right-hand side of Eq. (2.84) is a linear combination of the derivatives of $y_n(t)u(t)$. Evaluating these derivatives is clumsy and inconvenient because of the presence of $u(t)$. The derivatives will generate an impulse and its derivatives at the origin. Fortunately when $m \leq n$ [Eq. (2.77)], we can avoid this difficulty by using the observation in Eq. (2.79), which asserts that at $t = 0$ (the origin), $h(t) = b_n \delta(t)$. Therefore, we need not bother to find $h(t)$ at the origin. This simplification means that instead of deriving $P(D)[y_n(t)u(t)]$, we can derive $P(D)y_n(t)$ and add to it the term $b_n \delta(t)$, so that

$$h(t) = b_n \delta(t) + P(D)y_n(t) \qquad t \geq 0$$

$$= b_n \delta(t) + [P(D)y_n(t)]u(t) \tag{2.85}$$

This expression is valid when $m \leq n$ [the form given in Eq. (2.77b)]. When $m > n$, Eq. (2.84) should be used.

2.9 Summary

This chapter discusses time-domain analysis of LTIC systems. The total response of a linear system is a sum of the zero-input response and zero-state response. The zero-input response is the system response generated only by the internal conditions (initial conditions) of the system, assuming that the external input is zero; hence the term "zero-input." The zero-state response is the system response generated by the external input, assuming that all initial conditions are zero; that is, when the system is in zero state.

Every system can sustain certain forms of response on its own with no external input (zero input). These forms are intrinsic characteristics of the system; that is, they do not depend on any external input. For this reason they are called characteristic modes of the system. Needless to say, the zero-input response is made up of characteristic modes chosen in a suitable combination so as to satisfy the initial conditions of the system. For an nth-order system, there are n distinct modes.

The unit impulse function is an idealized mathematical model of a signal that cannot be generated in practice. Nevertheless, introduction of such a signal as an intermediary is very helpful in analysis of signals and systems. The unit impulse response of a system is a combination of the characteristic modes of the system† because the impulse $\delta(t) = 0$ for $t > 0$. Therefore, the system response for $t > 0$ must necessarily be a zero-input response, which, as seen earlier, is a combination of characteristic modes.

The zero-state response (response due to external input) of a linear system can be obtained by breaking the input into simpler components and then adding the responses to all the components. In this chapter we represent an arbitrary input $f(t)$ as a sum of narrow rectangular pulses [staircase approximation of $f(t)$]. In the limit as the pulse width $\rightarrow 0$, the rectangular pulse components approach impulses. Knowing the impulse response of the system, we can find the system response to all the impulse components and add them to yield the system response to the input

---

† There is the possibility of an impulse in addition to characteristic modes.

$f(t)$. The sum of the responses to the impulse components is in the form of an integral, known as the convolution integral. The system response is obtained as the convolution of the input $f(t)$ with the system's impulse response $h(t)$. Therefore, the knowledge of the system's impulse response allows us to determine the system response to any arbitrary input.

LTIC systems have a very special relationship to the everlasting exponential signal e^{st} because the response of an LTIC system to such an input signal is the same signal within a multiplicative constant. The response of an LTIC system to the everlasting exponential input e^{st} is $H(s)e^{st}$, where $H(s)$ is the transfer function of the system.

Differential equations of LTIC systems can also be solved by the classical method, where the response is obtained as a sum of natural and forced response. These are not the same as the zero-input and zero-state components, although they satisfy the same equations respectively. Although simple, this method suffers from the fact that it is applicable to a restricted class of input signals, and the system response cannot be expressed as an explicit function of the input. This limitation makes it useless in the theoretical study of systems.

A linear system is in a zero state if all initial conditions are zero. A system in a zero state is incapable of generating any response in the absence of an external input. When some initial conditions are applied to a system, then, if the system eventually goes to zero state in the absence of any external input, the system is said to be asymptotically stable. In contrast, if the system's response increases without bound, it is unstable. If neither the system goes to zero state nor the response increases indefinitely, the system is marginally stable. The stability criterion in terms of the location of a system's characteristic roots can be summarized as follows:

1. An LTIC system is asymptotically stable if, and only if, all the characteristic roots are in the LHP. The roots may be repeated or unrepeated.

2. An LTIC system is unstable if, and only if, either one or both of the following conditions exist: (i) at least one root is in the RHP; (ii) there are repeated roots on the imaginary axis.

3. An LTIC system is marginally stable if, and only if, there are no roots in the RHP, and there are some unrepeated roots on the imaginary axis.

According to an alternative definition of stability— bounded-input bounded-output (BIBO) stability—a system is stable if every bounded input produces a bounded output. Otherwise the system is (BIBO) unstable. Asymptotically stable system is always BIBO-stable. The converse is not necessarily true, however.

Characteristic behavior of a system is extremely important because it determines not only the system response to internal conditions (zero-input behavior), but also the system response to external inputs (zero-state behavior) and the system stability. The system response to external inputs is determined by the impulse response, which itself is made up of characteristic modes. The width of the impulse response is called the time constant of the system, which indicates how fast the system can respond to an input. The time constant plays an important role in determining such diverse system behavior as the response time and filtering properties of the system, dispersion of pulses, and the rate of pulse transmission through the system.

References

1. Lathi, B.P., *Signals and Systems*, Berkeley-Cambridge Press, Carmichael, Ca., 1987.
2. Kailath, T., *Linear System*, Prentice-Hall, Englewood Cliffs, N.J., 1980.
3. Lathi, B.P., *Modern Digital and Analog Communication Systems*, Third Ed., Oxford University Press, New York, 1998.

Problems

2.2-1 An LTIC system is specified by the equation

$$\left(D^2 + 5D + 6\right) y(t) = (D + 1)f(t)$$

(a) Find the characteristic polynomial, characteristic equation, characteristic roots, and characteristic modes of this system.
(b) Find $y_0(t)$, the zero-input component of the response $y(t)$ for $t \geq 0$, if the initial conditions are $y_0(0) = 2$ and $\dot{y}_0(0) = -1$.

2.2-2 Repeat Prob. 2.2-1 if

$$\left(D^2 + 4D + 4\right) y(t) = Df(t)$$

and $y_0(0) = 3$, $\dot{y}_0(0) = -4$.

2.2-3 Repeat Prob. 2.2-1 if

$$D(D + 1)y(t) = (D + 2)f(t)$$

and $y_0(0) = \dot{y}_0(0) = 1$.

2.2-4 Repeat Prob. 2.2-1 if

$$\left(D^2 + 9\right) y(t) = (3D + 2)f(t)$$

and $y_0(0) = 0$, $\dot{y}_0(0) = 6$.

2.2-5 Repeat Prob. 2.2-1 if

$$\left(D^2 + 4D + 13\right) y(t) = 4(D + 2)f(t)$$

with $y_0(0) = 5$, $\dot{y}_0(0) = 15.98$.

2.2-6 Repeat Prob. 2.2-1 if

$$D^2(D + 1)y(t) = (D^2 + 2)f(t)$$

with $y_0(0) = 4$, $\dot{y}_0(0) = 3$ and $\ddot{y}_0(0) = -1$.

2.2-7 Repeat Prob. 2.2-1 if

$$(D + 1)\left(D^2 + 5D + 6\right) y(t) = Df(t)$$

with $y_0(0) = 2$, $\dot{y}_0(0) = -1$ and $\ddot{y}_0(0) = 5$.

2.3-1 Find the unit impulse response of a system specified by the equation

$$\left(D^2 + 4D + 3\right) y(t) = (D + 5)f(t)$$

2.3-2 Repeat Prob. 2.3-1 if

$$\left(D^2 + 5D + 6\right) y(t) = \left(D^2 + 7D + 11\right) f(t)$$

2.3-3 Repeat Prob. 2.3-1 for the first-order allpass filter specified by the equation

$$(D + 1)y(t) = -(D - 1)f(t)$$

2.3-4 Find the unit impulse response of an LTIC system specified by the equation

$$\left(D^2 + 6D + 9\right) y(t) = (2D + 9) f(t)$$

2.4-1 If $c(t) = f(t) * g(t)$, then show that $A_c = A_f A_g$, where A_f, A_g, and A_c are the areas under $f(t)$, $g(t)$, and $c(t)$, respectively. Verify this **area property** of convolution in Examples 2.6 and 2.8.

2.4-2 If $f(t) * g(t) = c(t)$, then show that $f(at) * g(at) = |\frac{1}{a}|c(at)$. This **time-scaling property** of convolution states that if both $f(t)$ and $g(t)$ are time-scaled by a, their convolution is also time-scaled by a (and multiplied by $|1/a|$).

2.4-3 Show that the convolution of an odd and an even function is an odd function and the convolution of two odd or two even functions is an even function.
Hint: Use time-scaling property of convolution in Problem 2.4-2.

2.4-4 Using direct integration, find $e^{-at}u(t) * e^{-bt}u(t)$.

2.4-5 Using direct integration, find $u(t) * u(t)$, $e^{-at}u(t) * e^{-at}u(t)$, and $tu(t) * u(t)$.

2.4-6 Using direct integration, find $\sin t\, u(t) * u(t)$ and $\cos t\, u(t) * u(t)$.

2.4-7 The unit impulse response of an LTIC system is $h(t) = e^{-t}u(t)$. Find this system's (zero-state) response $y(t)$ if the input $f(t)$ is:

(a) $u(t)$ (b) $e^{-t}u(t)$ (c) $e^{-2t}u(t)$ (d) $\sin 3t\, u(t)$.

Use the convolution table to find your answers.

2.4-8 Repeat Prob. 2.4-7 if
$$h(t) = \left[2e^{-3t} - e^{-2t}\right]u(t)$$

and if the input $f(t)$ is: (a) $u(t)$ (b) $e^{-t}u(t)$ (c) $e^{-2t}u(t)$.

2.4-9 Repeat Prob. 2.4-7 if
$$h(t) = (1 - 2t)e^{-2t}u(t)$$

and if the input $f(t) = u(t)$.

2.4-10 Repeat Prob. 2.4-7 if $h(t) = 4e^{-2t}\cos 3t\, u(t)$ and if the input $f(t)$ is: (a) $u(t)$ (b) $e^{-t}u(t)$.

2.4-11 Repeat Prob. 2.4-7 if
$$h(t) = e^{-t}u(t)$$

and if the input $f(t)$ is: (a) $e^{-2t}u(t)$ (b) $e^{-2(t-3)}u(t)$ (c) $e^{-2t}u(t-3)$ (d) the gate pulse depicted in Fig. P2.4-11. For (d), sketch $y(t)$.

Hint: The input in (d) can be expressed as $u(t) - u(t - 1)$. For parts (c) and (d), use the shift property (2.34) of convolution. (Alternatively, you may want to invoke the system's time-invariance and superposition properties.)

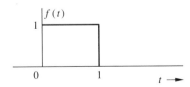

Fig. P2.4-11

Fig. P2.4-14

Fig. P2.4-15

2.4-12 A first-order allpass filter impulse response is given by

$$h(t) = -\delta(t) + 2e^{-t}u(t)$$

(a) Find the zero-state response of this filter for the input $e^t u(-t)$.

(b) Sketch the input and the corresponding zero-state response.

2.4-13 Sketch the functions $f(t) = \frac{1}{t^2+1}$ and $u(t)$. Now find $f(t) * u(t)$ and sketch the result.

2.4-14 Figure P2.4-14 shows $f(t)$ and $g(t)$. Find and sketch $c(t) = f(t) * g(t)$.

2.4-15 Find and sketch $c(t) = f(t) * g(t)$ for the functions depicted in Fig. P2.4-15.

2.4-16 Find and sketch $c(t) = f_1(t) * f_2(t)$ for the pairs of functions illustrated in Fig. P2.4-16.

2.4-17 For an LTIC system, if the (zero-state) response to an input $f(t)$ is $y(t)$, show that the (zero-state) response to the input $\dot{f}(t)$ is $\dot{y}(t)$ and that for the input $\int_{-\infty}^{t} f(\tau)\,d\tau$ is $\int_{-\infty}^{t} y(\tau)\,d\tau$.

Hint: Recognize that $\dot{f}(t) = \lim_{T \to 0} \frac{1}{T}[f(t) - f(t-T)]$. Now use linearity and time invariance to find the response to $\dot{f}(t)$. Also, recognize that $\int_{-\infty}^{t} f(\tau)\,d\tau = f(t) * u(t)$.

2.4-18 If $f(t) * g(t) = c(t)$, then show that

$$\dot{f}(t) * g(t) = f(t) * \dot{g}(t) = \dot{c}(t)$$

Extend this result to show that

$$f^{(m)}(t) * g^{(n)}(t) = c^{(m+n)}(t)$$

where $x^{(m)}(t)$ is the mth derivative of $x(t)$, and all the derivatives of $f(t)$ and $g(t)$ in this integral exist.

Hint: Use the first part of the hint in Prob. 2.4-17 and the time-shift property of convolution.

2.4-19 As mentioned in Chapter 1 (Fig. 1.27b), it is possible to express an input in terms of its step components, as shown in Fig. P2.4-19. If $g(t)$ is the unit step response of an LTIC system, show that the (zero-state) response $y(t)$ of the system to an input $f(t)$ can be expressed as

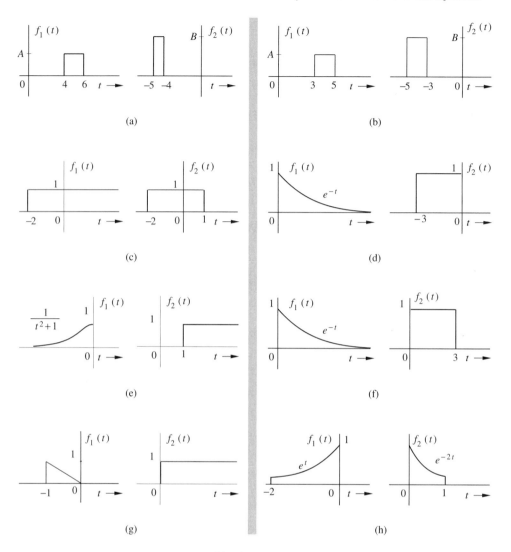

Fig. P2.4-16

$$y(t) = \int_{-\infty}^{\infty} \dot{f}(\tau)g(t-\tau)\,d\tau = \dot{f}(t) * g(t)$$

Hint: From Figure P2.4-19, the shaded step component of the input is given by $(\triangle f)u(t - n\triangle\tau) \simeq [\dot{f}(\tau)\triangle\tau]u(t - n\triangle\tau)$. Add the system responses to all such components.

2.4-20 A line charge is located along the x axis with a charge density $f(x)$. Show that the electric field $E(x)$ produced by this line charge at a point x is given by

$$E(x) = f(x) * h(x) \qquad \text{where} \qquad h(x) = \frac{1}{4\pi\epsilon x^2}$$

Hint: The charge over an interval $\triangle\tau$ located at $\tau = n\triangle\tau$ is $f(n\triangle\tau)\triangle\tau$. Also by

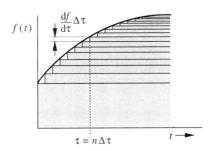

$$\tau = n\Delta\tau$$

Fig. P2.4-19

Coulomb's law, the electric field $E(r)$ at a distance r from a charge q is given by

$$E(r) = \frac{q}{4\pi\epsilon r^2}$$

2.4-21 Determine $H(s)$, the transfer function of an ideal time delay of T seconds. Find your answer by two methods; using Eq. (2.48) and using Eq. (2.49).

2.5-1 Using the classical method, solve

$$\left(D^2 + 7D + 12\right) y(t) = (D + 2)f(t)$$

if the initial conditions are $y(0^+) = 0$, $\dot{y}(0^+) = 1$, and if the input $f(t)$ is:

(a) $u(t)$ **(b)** $e^{-t}u(t)$ **(c)** $e^{-2t}u(t)$.

2.5-2 Using the classical method, solve

$$\left(D^2 + 6D + 25\right) y(t) = (D + 3)f(t)$$

if the initial conditions are $y(0^+) = 0$, $\dot{y}(0^+) = 2$, and if the input $f(t) = u(t)$.

2.5-3 Using the classical method, solve

$$\left(D^2 + 4D + 4\right) y(t) = (D + 1)f(t)$$

if the initial conditions are $y(0^+) = \frac{9}{4}$, $\dot{y}(0^+) = 5$, and if the input $f(t)$ is:

(a) $e^{-3t}u(t)$ **(b)** $e^{-t}u(t)$.

2.5-4 Using the classical method, solve

$$(D^2 + 2D)y(t) = (D + 1)f(t)$$

if the initial conditions are $y(0^+) = 2$, $\dot{y}(0^+) = 1$, and if the input is $f(t) = u(t)$.

2.5-5 Repeat Prob. 2.5-1 if the input

$$f(t) = e^{-3t}u(t)$$

2.6-1 Explain, with reasons, whether the LTIC systems described by the following equations are asymptotically stable, marginally stable, or unstable.

(a) $(D^2 + 8D + 12)y(t) = (D - 1)f(t)$
(b) $D(D^2 + 3D + 2)y(t) = (D + 5)f(t)$
(c) $D^2(D^2 + 2)y(t) = f(t)$
(d) $(D + 1)(D^2 - 6D + 5)y(t) = (3D + 1)f(t)$

2.6-2 Repeat Prob. 2.6-1 if
(a)$(D+1)(D^2+2D+5)^2 y(t) = f(t)$
(b)$(D+1)(D^2+9)y(t) = (2D+9)f(t)$
(c)$(D+1)(D^2+9)^2 y(t) = (2D+9)f(t)$
(d)$(D^2+1)(D^2+4)(D^2+9)y(t) = 3Df(t)$

2.6-3 For a certain LTIC system, the impulse response $h(t) = u(t)$.
(a) Determine the characteristic root(s) of this system.
(b) Is this system asymptotically or marginally stable, or is it unstable?
(c) Is this system BIBO stable?
(d) What can this system be used for?

2.6-4 In Sec. 2.6 we demonstrated that for an LTIC system, Condition (2.65) is sufficient for BIBO stability. Show that this is also a necessary condition for BIBO stability in such systems. In other words, show that if Eq. (2.65) is not satisfied, then there exists a bounded input that produces an unbounded output.

Hint: Assume that a system exists for which $h(t)$ violates Eq. (2.65) and yet produces an output that is bounded for every bounded input. Establish contradiction in this statement by considering an input $f(t)$ defined by $f(t_1 - \tau) = 1$ when $h(\tau) \geq 0$ and $f(t_1 - \tau) = -1$ when $h(\tau) < 0$, where t_1 is some fixed instant.

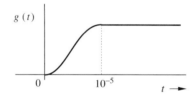

Fig. P2.7-1

2.7-1 Data at a rate of 1 million pulses per second are to be transmitted over a certain communications channel. The unit step response $g(t)$ for this channel is shown in Fig. P2.7-1.
(a) Explain if this channel can transmit data at the required rate.
(b) Can an audio signal consisting of components with frequencies up to 15kHz be transmitted over this channel with reasonable fidelity?

2.7-2 A certain communication channel has a bandwidth of 10 kHz. A pulse of 0.5 ms duration is transmitted over this channel.
(a) Determine the width (duration) of the received pulse.
(b) Find the maximum rate at which these pulses can be transmitted over this channel without interference between the successive pulses.

2.7-3 A first-order LTIC system has a characteristic root $\lambda = -10^4$.
(a) Determine T_r, the rise time of its unit step input response.
(b) Determine the bandwidth of this system.
(c) Determine the rate at which the information pulses can be transmitted through this system.

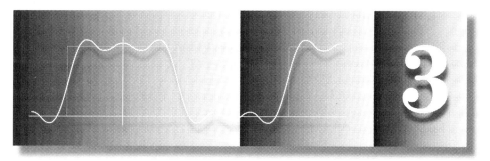

Signal Representation By Fourier Series

This chapter is important for basic understanding of signal representation and signal comparison. In Chapter 2 we expressed an arbitrary input $f(t)$ as a sum of its impulse components. The LTI system (zero-state) response to input $f(t)$ was obtained by summing the system's responses to all these components in the form of the convolution integral. There are infinite possible ways of expressing an input $f(t)$ in terms of other signals. For this reason, the problem of signal representation in terms of a set of signals is very important in the study of signals and systems. This chapter addresses the issue of representing a signal as a sum of its components. The problem is similar to that of representing a vector in terms of its components.

Signals and Vectors

There is a perfect analogy between signals and vectors; the analogy is so strong that the term *'analogy'* understates the reality. Signals are not just *like* vectors. Signals *are* vectors! A vector can be represented as a sum of its components in a variety of ways, depending upon the choice of coordinate system. A signal can also be represented as a sum of its components in a variety of ways. Let us begin with some basic vector concepts and then apply these concepts to signals.

3.1-1 Component of a Vector

A vector is specified by its magnitude and its direction. We shall denote all vectors by boldface. For example, $\mathbf{x}$ is a certain vector with magnitude or length $|\mathbf{x}|$. For the two vectors $\mathbf{f}$ and $\mathbf{x}$ shown in Fig. 3.1, we define their dot (inner or scalar) product as

$$\mathbf{f} \cdot \mathbf{x} = |\mathbf{f}||\mathbf{x}| \cos \theta \qquad (3.1)$$

where θ is the angle between these vectors. Using this definition we can express $|\mathbf{x}|$, the length of a vector $\mathbf{x}$ as

$$|\mathbf{x}|^2 = \mathbf{x} \cdot \mathbf{x} \tag{3.2}$$

Let the component of $\mathbf{f}$ along $\mathbf{x}$ be $c\mathbf{x}$ as depicted in Fig. 3.1. Geometrically the component of $\mathbf{f}$ along $\mathbf{x}$ is the projection of $\mathbf{f}$ on $\mathbf{x}$, and is obtained by drawing a perpendicular from the tip of $\mathbf{f}$ on the vector $\mathbf{x}$, as illustrated in Fig. 3.1. What is the mathematical significance of a component of a vector along another vector? As seen from Fig. 3.1, the vector $\mathbf{f}$ can be expressed in terms of vector $\mathbf{x}$ as

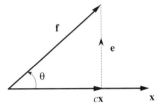

Fig. 3.1   Component (projection) of a vector along another vector.

$$\mathbf{f} = c\mathbf{x} + \mathbf{e} \tag{3.3}$$

However, this is not the only way to express $\mathbf{f}$ in terms of $\mathbf{x}$. Figure 3.2 shows two of the infinite other possibilities. From Figs. 3.2a and 3.2b, we have

$$\mathbf{f} = c_1\mathbf{x} + \mathbf{e}_1 = c_2\mathbf{x} + \mathbf{e}_2 \tag{3.4}$$

In each of these three representations $\mathbf{f}$ is represented in terms of $\mathbf{x}$ plus another vector called the **error vector**. If we approximate $\mathbf{f}$ by $c\mathbf{x}$,

$$\mathbf{f} \simeq c\mathbf{x} \tag{3.5}$$

the error in the approximation is the vector $\mathbf{e} = \mathbf{f} - c\mathbf{x}$. Similarly, the errors in approximations in Figs. 3.2a and 3.2b are $\mathbf{e}_1$ and $\mathbf{e}_2$. What is unique about the approximation in Fig. 3.1 is that the error vector is the smallest. We can now define mathematically the component of a vector $\mathbf{f}$ along vector $\mathbf{x}$ to be $c\mathbf{x}$ where c is chosen to minimize the length of the error vector $\mathbf{e} = \mathbf{f} - c\mathbf{x}$.

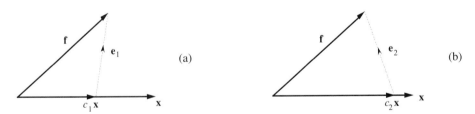

Fig. 3.2   Approximation of a vector in terms of another vector.

Now, the length of the component of $\mathbf{f}$ along $\mathbf{x}$ is $|\mathbf{f}| \cos \theta$. But it is also $c|\mathbf{x}|$ as seen from Fig. 3.1. Therefore

$$c|\mathbf{x}| = |\mathbf{f}| \cos \theta$$

Multiplying both sides by $|\mathbf{x}|$ yields

$$c|\mathbf{x}|^2 = |\mathbf{f}||\mathbf{x}|\cos\theta = \mathbf{f}\cdot\mathbf{x}$$

Therefore

$$c = \frac{\mathbf{f}\cdot\mathbf{x}}{\mathbf{x}\cdot\mathbf{x}} = \frac{1}{|\mathbf{x}|^2}\mathbf{f}\cdot\mathbf{x} \tag{3.6}$$

From Fig. 3.1, it is apparent that when $\mathbf{f}$ and $\mathbf{x}$ are perpendicular, or orthogonal, then $\mathbf{f}$ has a zero component along $\mathbf{x}$; consequently, $c = 0$. Keeping an eye on Eq. (3.6), we therefore define $\mathbf{f}$ and $\mathbf{x}$ to be **orthogonal** if the inner (scalar or dot) product of the two vectors is zero, that is, if

$$\mathbf{f}\cdot\mathbf{x} = 0 \tag{3.7}$$

3.1-2 Component of a Signal

The concept of a vector component and orthogonality can be extended to signals. Consider the problem of approximating a real signal $f(t)$ in terms of another real signal $x(t)$ over an interval $[t_1, t_2]$:

$$f(t) \simeq cx(t) \qquad t_1 \le t \le t_2 \tag{3.8}$$

The error $e(t)$ in this approximation is

$$e(t) = \begin{cases} f(t) - cx(t) & t_1 \le t \le t_2 \\ 0 & \text{otherwise} \end{cases} \tag{3.9}$$

We now select some criterion for the 'best approximation'. We know that the signal energy is one possible measure of a signal size. For best approximation, we need to minimize the error signal–that is, minimize its size, which is its energy E_e over the interval $[t_1, t_2]$ given by

$$E_e = \int_{t_1}^{t_2} e^2(t)\,dt$$

$$= \int_{t_1}^{t_2} [f(t) - cx(t)]^2\,dt$$

Note that the right-hand side is a definite integral with t as the dummy variable. Hence, E_e is a function of the parameter c (not t) and E_e is minimum for some choice of c. To minimize E_e, a necessary condition is

$$\frac{dE_e}{dc} = 0 \tag{3.10}$$

or

$$\frac{d}{dc}\left[\int_{t_1}^{t_2} [f(t) - cx(t)]^2\,dt\right] = 0$$

Expanding the squared term inside the integral, we obtain

$$\frac{d}{dc}\left[\int_{t_1}^{t_2} f^2(t)\,dt\right] - \frac{d}{dc}\left[2c\int_{t_1}^{t_2} f(t)x(t)\,dt\right] + \frac{d}{dc}\left[c^2\int_{t_1}^{t_2} x^2(t)\,dt\right] = 0$$

From which we obtain

$$-2 \int_{t_1}^{t_2} f(t)x(t)\, dt + 2c \int_{t_1}^{t_2} x^2(t)\, dt = 0$$

and

$$c = \frac{\displaystyle\int_{t_1}^{t_2} f(t)x(t)\, dt}{\displaystyle\int_{t_1}^{t_2} x^2(t)\, dt} = \frac{1}{E_x} \int_{t_1}^{t_2} f(t)x(t)\, dt \tag{3.11}$$

We observe a remarkable similarity between the behavior of vectors and signals, as indicated by Eqs. (3.6) and (3.11). It is evident from these two parallel expressions that *the area under the product of two signals corresponds to the inner (scalar or dot) product of two vectors.* In fact, the area under the product of $f(t)$ and $x(t)$ is called the *inner product* of $f(t)$ and $x(t)$, and is denoted by (f, x). The energy of a signal is the inner product of a signal with itself, and corresponds to the vector length square (which is the inner product of the vector with itself).

To summarize our discussion, if a signal $f(t)$ is approximated by another signal $x(t)$ as

$$f(t) \simeq cx(t)$$

then the optimum value of c that minimizes the energy of the error signal in this approximation is given by Eq. (3.11).

Taking our clue from vectors, we say that a signal $f(t)$ contains a component $cx(t)$, where c is given by Eq. (3.11). Note that in vector terminology, $cx(t)$ is the projection of $f(t)$ on $x(t)$. Continuing with the analogy, we say that if the component of a signal $f(t)$ of the form $x(t)$ is zero (that is, $c = 0$), the signals $f(t)$ and $x(t)$ are orthogonal over the interval $[t_1, t_2]$. Therefore, we define the real signals $f(t)$ and $x(t)$ to be orthogonal over the interval $[t_1, t_2]$ if†

$$\int_{t_1}^{t_2} f(t)x(t)\, dt = 0 \tag{3.12}$$

■ **Example 3.1**

For the square signal $f(t)$ shown in Fig. 3.3 find the component in $f(t)$ of the form $\sin t$. In other words, approximate $f(t)$ in terms of $\sin t$

$$f(t) \simeq c\sin t \qquad 0 \le t \le 2\pi$$

so that the energy of the error signal is minimum.

In this case

$$x(t) = \sin t \qquad \text{and} \qquad E_x = \int_0^{2\pi} \sin^2(t)\, dt = \pi$$

From Eq. (3.11), we find

$$c = \frac{1}{\pi} \int_0^{2\pi} f(t)\sin t\, dt = \frac{1}{\pi}\left[\int_0^{\pi} \sin t\, dt + \int_{\pi}^{2\pi} -\sin t\, dt\right] = \frac{4}{\pi} \tag{3.13}$$

---

†For complex signals the definition is modified as in Eq. (3.20).

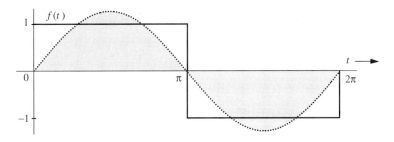

Fig. 3.3 Approximation of square signal in terms of a single sinusoid.

Thus

$$f(t) \simeq \frac{4}{\pi} \sin t \tag{3.14}$$

represents the best approximation of $f(t)$ by the function $\sin t$, which will minimize the error energy. This sinusoidal component of $f(t)$ is shown shaded in Fig. 3.3. By analogy with vectors, we say that the square function $f(t)$ depicted in Fig. 3.3 has a component of signal $\sin t$ and that the magnitude of this component is $4/\pi$. ■

△ **Exercise E3.1**

Show that over an interval $(-\pi \leq t \leq \pi)$, the 'best' approximation of the signal $f(t) = t$ in terms of the function $\sin t$ is $2 \sin t$. Verify that the error signal $e(t) = t - 2 \sin t$ is orthogonal to the signal $\sin t$ over the interval $-\pi \leq t \leq \pi$. Sketch the signals t and $2 \sin t$ over the interval $-\pi \leq t \leq \pi$. ▽

3.1-3 Orthogonality in Complex Signals

So far we have restricted ourselves to real functions of t. To generalize the results to complex functions of t, consider again the problem of approximating a function $f(t)$ by a function $x(t)$ over an interval $(t_1 \leq t \leq t_2)$:

$$f(t) \simeq cx(t) \tag{3.15}$$

where $f(t)$ and $x(t)$ now can be complex functions of t. Recall that the energy E_x of the complex signal $x(t)$ over an interval $[t_1, t_2]$ is

$$E_x = \int_{t_1}^{t_2} |x(t)|^2 \, dt$$

In this case, both the coefficient c and the error

$$e(t) = f(t) - cx(t) \tag{3.16}$$

are complex (in general). For the 'best' approximation, we choose c so that E_e, the energy of the error signal $e(t)$ is minimum. Now,

$$E_e = \int_{t_1}^{t_2} |f(t) - cx(t)|^2 \, dt \tag{3.17}$$

Recall also that

$$|u + v|^2 = (u + v)(u^* + v^*) = |u|^2 + |v|^2 + u^*v + uv^* \tag{3.18}$$

Using this result, we can, after some manipulation, rearrange Eq. (3.17) as

$$E_e = \int_{t_1}^{t_2} |f(t)|^2 \, dt - \left| \frac{1}{\sqrt{E_x}} \int_a^{t_2} f(t)x^*(t) \, dt \right|^2 + \left| c\sqrt{E_x} - \frac{1}{\sqrt{E_x}} \int_{t_1}^{t_2} f(t)x^*(t) \, dt \right|^2$$

Since the first two terms on the right-hand side are independent of c, it is clear that E_e is minimized by choosing c so that the third term on the right-hand side is zero. This yields

$$c = \frac{1}{E_x} \int_{t_1}^{t_2} f(t)x^*(t) \, dt \qquad (3.19)$$

In light of the above result, we need to redefine orthogonality for the complex case as follows: Two complex functions $x_1(t)$ and $x_2(t)$ are orthogonal over an interval $(t_1 \le t \le t_2)$ if

$$\int_{t_1}^{t_2} x_1(t)x_2^*(t) \, dt = 0 \qquad \text{or} \qquad \int_{t_1}^{t_2} x_1^*(t)x_2(t) \, dt = 0 \qquad (3.20)$$

Either equality suffices. This is a general definition of orthogonality, which reduces to Eq. (3.12) when the functions are real.

$\triangle$ **Exercise E3.2**

Show that over an interval $(0 \le t \le 2\pi)$, the 'best' approximation of the square signal $f(t)$ in Fig. 3.3 in terms of the signal e^{jt} is given by $\frac{2}{j\pi} e^{jt}$. Verify that the error signal $e(t) = f(t) - \frac{2}{j\pi} e^{jt}$ is orthogonal to the signal e^{jt}. $\triangledown$

Energy of the Sum of Orthogonal Signals

We know that the square of the length of a sum of two orthogonal vectors is equal to the sum of the squares of the lengths of the two vectors. Thus, if vectors **x** and **y** are orthogonal, and if $\mathbf{z} = \mathbf{x} + \mathbf{y}$, then

$$|\mathbf{z}|^2 = |\mathbf{x}|^2 + |\mathbf{y}|^2$$

We have a similar result for signals. The energy of the sum of two orthogonal signals is equal to the sum of the energies of the two signals. Thus, if signals $x(t)$ and $y(t)$ are orthogonal over an interval $[t_1, t_2]$, and if $z(t) = x(t) + y(t)$, then

$$E_z = E_x + E_y \qquad (3.21)$$

We now prove this result for complex signals of which real signals are a special case. From Eq. (3.18) it follows that

$$\int_{t_1}^{t_2} |x(t) + y(t)|^2 dt = \int_{t_1}^{t_2} |x(t)|^2 dt + \int_{t_1}^{t_2} |y(t)|^2 dt + \int_{t_1}^{t_2} x(t)y^*(t) dt + \int_{t_1}^{t_2} x^*(t)y(t) dt$$

$$= \int_{t_1}^{t_2} |x(t)|^2 dt + \int_{t_1}^{t_2} |y(t)|^2 dt \qquad (3.22)$$

The last result follows from the fact that because of orthogonality, the two integrals of the products $x(t)y^*(t)$ and $x^*(t)y(t)$ are zero [see Eq. (3.20)]. This result can be extended to the sum of any number of mutually orthogonal signals.

3.2 Signal Comparison: Correlation

Section 3.1 has prepared the foundation for signal comparison. Here again, we can benefit by considering the concept of vector comparison. Two vectors $\mathbf{f}$ and $\mathbf{x}$ are similar if $\mathbf{f}$ has a large component along $\mathbf{x}$. In other words, if c in Eq. (3.6) is large, the vectors $\mathbf{f}$ and $\mathbf{x}$ are similar. We could consider c as a quantitative measure of similarity between $\mathbf{f}$ and $\mathbf{x}$. Such a measure, however, would be defective. The degree of similarity between $\mathbf{f}$ and $\mathbf{x}$ should be independent of the lengths of $\mathbf{f}$ and $\mathbf{x}$. If we double the length of $\mathbf{f}$, for example, the degree of similarity between $\mathbf{f}$ and $\mathbf{x}$ should not change. From Eq. (3.6), however, we see that doubling $\mathbf{f}$ doubles the value of c (whereas doubling $\mathbf{x}$ halves the value of c). Our measure is clearly faulty. Similarity between two vectors is indicated by the angle θ between the vectors. The smaller the θ, the greater the similarity, and vice versa. The degree of similarity can therefore be conveniently measured by $\cos \theta$. The larger the $\cos \theta$, the greater is the similarity between the two vectors. Thus, a suitable measure would be $c_n = \cos \theta$, which is given by

$$c_n = \cos \theta = \frac{\mathbf{f} \cdot \mathbf{x}}{|\mathbf{f}| \, |\mathbf{x}|} \tag{3.23}$$

We can readily verify that this measure is independent of the lengths of $\mathbf{f}$ and $\mathbf{x}$. This similarity measure c_n is known as the **correlation coefficient**. Observe that

$$-1 \le c_n \le 1 \tag{3.24}$$

Thus, the magnitude of c_n is never greater than unity. If the two vectors are aligned, the similarity is maximum ($c_n = 1$). Two vectors aligned in opposite directions have the maximum dissimilarity ($c_n = -1$). If the two vectors are orthogonal, the similarity is zero.

We use the same argument in defining a similarity index (the correlation coefficient) for signals. We shall consider the signals over the entire time interval from $-\infty$ to ∞. To make c in Eq. (3.11) independent of energies (sizes) of $f(t)$ and $x(t)$, we must normalize c by normalizing the two signals to have unit energies. Thus, the appropriate similarity index c_n analogous to Eq. (3.23) is given by

$$c_n = \frac{1}{\sqrt{E_f E_x}} \int_{-\infty}^{\infty} f(t)x(t) \, dt \tag{3.25}$$

Observe that multiplying either $f(t)$ or $x(t)$ by any constant has no effect on this index. It is independent of the size (energies) of $f(t)$ and $x(t)$. Using the Schwarz inequality,† we can show that the magnitude of c_n is never greater than 1

$$-1 \le c_n \le 1 \tag{3.26}$$

---

†Schwarz inequality states that for two real energy signals $f(t)$ and $x(t)$

$$\left(\int_{-\infty}^{\infty} f(t)x(t) \, dt \right)^2 \le E_f E_x \tag{3.25n}$$

with equality if and only if $x(t) = Kf(t)$, where K is an arbitrary constant. There is also similar inequality for complex signals.

The Best Friends, Worst Enemies, and Complete Strangers

We can readily verify that if $f(t) = Kx(t)$, then $c_n = 1$ when K is any positive constant, and $c_n = -1$ when K is any negative constant. Also $c_n = 0$ if $f(t)$ and $x(t)$ are orthogonal. Thus, the maximum similarity [when $f(t) = Kx(t)$] is indicated by $c_n = 1$, the maximum dissimilarity [when $f(t) = -Kx(t)$] is indicated by $c_n = -1$. When the two signals are orthogonal, the similarity is zero. Qualitatively speaking, we may view orthogonal signals as unrelated signals. Note that maximum dissimilarity is different from unrelatedness qualitatively. For example, we have the best friends ($c_n = 1$), the worst enemies ($c_n = -1$), and complete strangers, who do not care whether we exist or not ($c_n = 0$). The enemies are not strangers, but in many ways, people who think like us, only in opposite ways.

We can readily extend this discussion to complex signal comparison. We generalize the definition of c_n to include complex signals as

$$c_n = \frac{1}{\sqrt{E_f E_x}} \int_{-\infty}^{\infty} f(t)x^*(t)\, dt \tag{3.27}$$

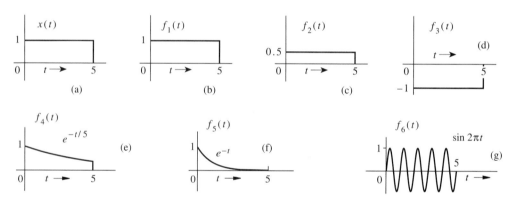

Fig. 3.4 Signals for Example 3.2.

■ **Example 3.2**

Find the correlation coefficient c_n between the pulse $x(t)$ and the pulses $f_i(t)$, $i = 1, 2, 3, 4, 5,$ and 6 illustrated in Fig. 3.4.

We shall compute c_n using Eq. (3.25) for each of the 6 cases. Let us first compute the energies of all the signals.

$$E_x = \int_0^5 x^2(t)\, dt = \int_0^5 dt = 5 \tag{3.28}$$

In the same way we find $E_{f_1} = 5$, $E_{f_2} = 1.25$, and $E_{f_3} = 5$. Also, to determine E_{f_4} and E_{f_5}, we determine the energy E of $e^{-at}u(t)$ over the interval $t = 0$ to T:

$$E = \int_0^T \left(e^{-at}\right)^2 dt = \int_0^T e^{-2at}\, dt = \frac{1}{2a}\left(1 - e^{-2aT}\right)$$

For $f_4(t)$, $a = 1/5$ and $T = 5$. Therefore $E_{f_4} = 2.1617$. For $f_5(t)$, $a = 1$ and $T = \infty$. Therefore $E_{f_5} = 0.5$. The energy of E_{f_6} is given by

$$E_{f_6} = \int_0^5 \sin^2 2\pi t \, dt = 2.5$$

Using Eq. (3.25), the correlation coefficients for the six cases are found as

(1) $\dfrac{1}{\sqrt{(5)(5)}} \displaystyle\int_0^5 dt = 1$ **(2)** $\dfrac{1}{\sqrt{(1.25)(5)}} \displaystyle\int_0^5 (0.5)\, dt = 1$

(3) $\dfrac{1}{\sqrt{(5)(5)}} \displaystyle\int_0^5 (-1)\, dt = -1$ **(4)** $\dfrac{1}{\sqrt{(2.1617)(5)}} \displaystyle\int_0^5 e^{-t/5}\, dt = 0.961$

(5) $\dfrac{1}{\sqrt{(0.5)(5)}} \displaystyle\int_0^5 e^{-t}\, dt = 0.628$ **(6)** $\dfrac{1}{\sqrt{(2.5)(5)}} \displaystyle\int_0^5 \sin 2\pi t \, dt = 0$ ■

Comments on the results: Because $f_1(t) = x(t)$, the two signals have the maximum possible similarity and $c_n = 1$. However, the signal $f_2(t)$ also shows maximum possible similarity with $c_n = 1$. The reason is that we have defined c_n to measure the similarity of the waveshapes; it is independent of the amplitude (strength) of the signals compared. The signal $f_2(t)$ is identical to $x(t)$ in shape; only the amplitude (strength) is different. Hence $c_n = 1$. The signal $f_3(t)$, on the other hand, has the maximum possible dissimilarity with $x(t)$ because it is equal to $-x(t)$. For $f_4(t)$, $c_n = 0.961$, implying a high degree of similarity with $x(t)$. This is reasonable because $f_4(t)$ is very similar to $x(t)$ over the duration of $x(t)$ (for $0 \le t \le 5$). Just by inspection, we notice that the variations or changes in both $x(t)$ and $f_4(t)$ are at similar rates. Such is not the case with $f_5(t)$, where we notice that variations in $f_5(t)$ are generally at a higher rate than those in $x(t)$. There is still considerable similarity; both signals always remain positive, and show no oscillations. Both signals have zero or negligible strength beyond $t = 5$. Thus, $f_5(t)$ is similar to $x(t)$, but not as similar as $f_4(t)$. This is why $c_n = 0.628$ for $f_5(t)$. The signal $f_6(t)$ is orthogonal to $x(t)$, so that $c_n = 0$. This fact appears to indicate that the dissimilarity in this case is not as strong as that of $f_3(t)$ for which $c_n = -1$. This conclusion may seem odd because $f_3(t)$ appears more similar to $x(t)$ than does $f_6(t)$. The dissimilarity between $x(t)$ and $f_3(t)$ is of the nature of antipathy (the worst enemy); in a way they are very similar, but in opposite ways. On the other hand, the dissimilarity of $x(t)$ with $f_6(t)$ stems from the fact that they are almost of different species or from different planets; it is of the nature of being strangers to each other. Hence its dissimilarity with $x(t)$ rates lower than that with $f_3(t)$.

△ **Exercise E3.3**

Show that c_n of signals $f_2(t)$ and $f_3(t)$ in Fig. 3.4 is -1, that of $f_2(t)$ and $f_4(t)$ is 0.961, and that of $f_3(t)$ and $f_6(t)$ is zero. ▽

3.2-1 Application to Signal Detection

Correlation between two signals is an extremely important concept which measures the degree of similarity (agreement or alignment) between the two signals. This concept is widely used for signal processing applications in radar, sonar, digital communication, electronic warfare and many others.

We explain this concept by an example of radar where a signal pulse is transmitted in order to detect a suspected target. If a target is present, the pulse will be reflected by it. If a target is not present, there will be no reflected pulse, just noise. The presence or absence of the reflected pulse confirms the presence or absence of a target. The crucial problem in this procedure is to detect the heavily attenuated, reflected pulse (of known waveform) buried in the unwanted noise signal. In this situation, correlation of the received pulse with the transmitted pulse can be of great help. A similar situation exists in digital communication where we are required to detect the presence of one of the two known waveforms in the presence of noise.

We now explain qualitatively how signal detection using correlation technique is accomplished. Consider the case of binary communication, where two known waveforms are received in a random sequence. Each time we receive a pulse, our task is to determine which of the two (known) waveforms is received. To make the detection easier, we must make the two pulses as dissimilar as possible. Therefore, we should select the pulse that is the negative of the other pulse. This choice gives the highest dissimilarity ($c_n = -1$). This scheme is sometimes called the *antipodal* scheme. We can also use orthogonal pulses which result in $c_n = 0$. In practice, both these options are used, although antipodal is the best in terms of distinguishability between the two pulses.

Now let us consider the antipodal scheme in which the two pulses are $p(t)$ and $-p(t)$. The correlation coefficient c_n of these pulses is -1. Assume that there is no noise or any other imperfections in the transmission. The receiver consists of a correlator which computes the correlation between $p(t)$ and the received pulse. If the correlation is 1, we decide that $p(t)$ is received, and if the correlation is -1, we decide that $-p(t)$ is received. Because of the maximum possible dissimilarity between the two pulses, detection is easier. The situation is almost like that in a fairy tale, where everybody lives happily ever after. In practice, however, several imperfections occur. There is always an unwanted signal (noise) superimposed on the received pulses. Moreover, during transmission, pulses get distorted and dispersed (spread out) in time. Consequently, a received pulse is corrupted by overlapping tails from other pulses. This changes the shape of received pulses, and the correlation coefficient is no more ± 1, but has a smaller magnitude, thus reducing the distinguishability of pulses. We use a *threshold detector*, which decides that if the correlation is positive ($c_n > 0$), the received pulse is $p(t)$, and if the correlation is negative ($c_n < 0$), the received pulse is $-p(t)$.

Suppose, for example, that $p(t)$ has been transmitted. In the ideal case, correlation of this pulse at the receiver would be 1, the maximum possible. Now because of the noise and other imperfections, the correlation is going to be less than 1. In some extreme situation, the noise and overlapping from other pulses can make this pulse so dissimilar to $p(t)$ that the correlation can be a negative amount. In this case, the threshold detector decides that $-p(t)$ has been received, thus causing a detection error. In the same way, if $-p(t)$ is transmitted, the channel noise, pulse distortion, and the overlapping from other pulses could result in a positive correlation, causing a detection error. Our task is to make sure that the transmitted pulses have sufficient energy to keep the relative damage to the pulse caused by noise within a limit and the error probability below acceptable bounds. In the ideal case, the margin provided by the correlation c_n for distinguishing the two pulses is 2 (from 1 to -1 and vice versa). The noise and other imperfections reduce this margin. That is why it is important to start with as large a margin as possible. For this reason the antipodal scheme has the best performance in terms of guarding

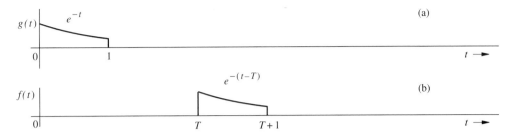

Fig. 3.5   Physical explanation of the correlation function.

against the channel noise and other imperfections. However, as mentioned earlier, because of some other reasons, different schemes such as an orthogonal scheme, where $c_n = 0$, are also used even when they provide a smaller margin (from 0 to 1 and vice versa) in distinguishing the pulses.

Some aspects of pulse dispersion were discussed in Secs. 2.7-5 and 2.7-6. In chapter 4, we shall discuss pulse distortion during transmission. Calculation of error probability in the presence of noise and other imperfections are beyond the scope of this work. For a detailed treatment of this subject, the reader may refer to the author's book on communication systems.[1]

3.2-2   Correlation Functions

Consider an application of correlation to signal detection in a radar, where a signal pulse is transmitted in order to detect a suspected target. If a target is present, the pulse will be reflected by it. In contrast, if the target is not present, there will be no reflected pulse, just noise. By detecting the presence or absence of the reflected pulse we confirm the presence or absence of the target. By measuring the time delay between the transmitted and received (reflected) pulse we determine the distance of the target. Let the transmitted and the reflected pulses be denoted by $g(t)$ and $f(t)$, respectively, as shown in Fig. 3.5. If we were to use Eq. (3.25) directly to measure the correlation coefficient c_n, we would obtain

$$c_n = \frac{1}{\sqrt{E_f E_g}} \int_{-\infty}^{\infty} f(t)g(t)\, dt = 0 \qquad (3.29)$$

Thus, the correlation is zero because the pulses are disjoint (nonoverlapping in time). The integral (3.29) will yield zero value even when the pulses are identical but with relative time shift. To avoid this difficulty, we compare the received pulse $f(t)$ with a delayed pulse $g(t)$ for various values of delay. If for some value of delay parameter there is a strong correlation, we not only detect the presence of the pulse but we also detect the relative time shift of $f(t)$ with respect to $g(t)$. For this reason, instead of using the integral on the right-hand, we use the modified integral $\psi_{fg}(t)$, the **crosscorrelation** function of two real signals $f(t)$ and $g(t)$ defined by†

---

†For complex signals we define

$$\psi_{fg}(t) \equiv \int_{-\infty}^{\infty} f^*(\tau)g(\tau - t)\, d\tau$$

$$\psi_{fg}(t) \equiv \int_{-\infty}^{\infty} f(\tau)g(\tau - t)\,d\tau \tag{3.30}$$

Here τ is a dummy variable, and the pulse $g(\tau - t)$ is the pulse $g(\tau)$ delayed by t seconds with respect to the $f(\tau)$ pulse. Therefore, $\psi_{fg}(t)$ is an indication of similarity (correlation) of the f pulse with g pulse delayed by t seconds. Thus, $\psi_{fg}(t)$ measures the similarity of pulses even if they are disjoint. In the case of signals in Fig. 3.5, $\psi_{fg}(t)$ will show significant correlation around $t = T$. This observation allows us not only to detect the presence of the target, but also to calculate its distance.

Convolution and Correlation

We now examine the close connection between the convolution and the correlation of $f(t)$ and $g(t)$ [given by Eq. (3.30)]. Note that $g(\tau - t)$ is the $g(\tau)$ time-shifted by t. Thus, $\psi_{fg}(t)$ is equal to the area under the product of the f pulse and g pulse time-shifted by t (without time-inversion). In convolution also we follow the same procedure, except that the g pulse is time-inverted before it is shifted by t. This observation suggests that $\psi_{fg}(t)$ is equal to $f(t) * g(-t)$ [the convolution of $f(t)$ with time-inverted $g(t)$], that is,

$$\psi_{fg}(t) = f(t) * g(-t) \tag{3.31}$$

This can be formally proved as follows. Letting $g(-t) = w(t)$,

$$f(t) * g(-t) = f(t) * w(t) = \int_{-\infty}^{\infty} f(\tau)w(t - \tau)\,d\tau = \int_{-\infty}^{\infty} f(\tau)g(\tau - t)\,d\tau = \psi_{fg}(t)$$

To reiterate, $\psi_{fg}(t)$ is equal to the area under the product of the f pulse and g pulse time-shifted by t (without time-inversion), and is given by the convolution of $f(t)$ with $g(-t)$.

△ **Exercise E3.4**

Show that $\psi_{fg}(t)$, the correlation function of $f(t)$ and $g(t)$ in Fig. 2.11 is given by $c(t)$ in Fig. 2.12. ▽

△ **Exercise E3.5**

Show that $\psi_{fg}(t)$, the correlation function of $f(t)$ and $g(t)$ in Fig. 2.12 is given by $c(t)$ in Fig. 2.11. ▽

Autocorrelation Function

Correlation of a signal with itself is called the **autocorrelation**. The autocorrelation function $\psi_f(t)$ of a signal $f(t)$ is defined as

$$\psi_f(t) \equiv \int_{-\infty}^{\infty} f(\tau)f(\tau - t)\,d\tau \tag{3.32}$$

In Chapter 4, we shall show that the autocorrelation function provides valuable spectral information about the signal.

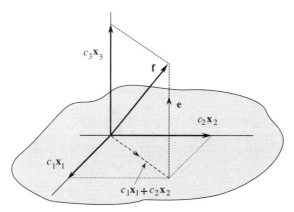

Fig. 3.6   Representation of a vector in three-dimensional space.

3.3   Signal representation by Orthogonal Signal Set

In this section we show a way of representing a signal as a sum of orthogonal signals. Here again we can benefit from the insight gained from a similar problem in vectors. We know that a vector can be represented as a sum of orthogonal vectors, which form the coordinate system of a vector space. The problem in signals is analogous, and the results for signals are parallel to those for vectors. So, let us review the case of vector representation.

3.3-1   Orthogonal Vector Space

Let us investigate a three-dimensional Cartesian vector space described by three mutually orthogonal vectors $\mathbf{x}_1$, $\mathbf{x}_2$, and $\mathbf{x}_3$, as illustrated in Fig. 3.6. First, we shall seek to approximate a three-dimensional vector $\mathbf{f}$ in terms of two mutually orthogonal vectors $\mathbf{x}_1$ and $\mathbf{x}_2$:

$$\mathbf{f} \simeq c_1\mathbf{x}_1 + c_2\mathbf{x}_2$$

The error $\mathbf{e}$ in this approximation is

$$\mathbf{e} = \mathbf{f} - (c_1\mathbf{x}_1 + c_2\mathbf{x}_2)$$

or

$$\mathbf{f} = c_1\mathbf{x}_1 + c_2\mathbf{x}_2 + \mathbf{e}$$

As in the earlier geometrical argument, we see from Fig 3.6 that the length of $\mathbf{e}$ is minimum when $\mathbf{e}$ is perpendicular to the $\mathbf{x}_1$-$\mathbf{x}_2$ plane, and $c_1\mathbf{x}_1$ and $c_2\mathbf{x}_2$ are the projections (components) of $\mathbf{f}$ on $\mathbf{x}_1$ and $\mathbf{x}_2$, respectively. Therefore, the constants c_1 and c_2 are given by formula (3.6). Observe that the error vector is orthogonal to both the vectors $\mathbf{x}_1$ and $\mathbf{x}_2$.

Now, let us determine the 'best' approximation to $\mathbf{f}$ in terms of all three mutually orthogonal vectors $\mathbf{x}_1, \mathbf{x}_2$, and $\mathbf{x}_3$:

$$\mathbf{f} \simeq c_1\mathbf{x}_1 + c_2\mathbf{x}_2 + c_3\mathbf{x}_3 \qquad (3.33)$$

Figure 3.6 shows that a unique choice of c_1, c_2, and c_3 exists, for which (3.33) is no longer an approximation but an equality

$$\mathbf{f} = c_1\mathbf{x}_1 + c_2\mathbf{x}_2 + c_3\mathbf{x}_3 \tag{3.34}$$

In this case, $c_1\mathbf{x}_1, c_2\mathbf{x}_2$, and $c_3\mathbf{x}_3$ are the projections (components) of $\mathbf{f}$ on $\mathbf{x}_1, \mathbf{x}_2$, and $\mathbf{x}_3$, respectively; that is,

$$c_i = \frac{\mathbf{f} \cdot \mathbf{x}_i}{\mathbf{x}_i \cdot \mathbf{x}_i} \tag{3.35a}$$

$$= \frac{1}{|\mathbf{x}_i|^2} \mathbf{f} \cdot \mathbf{x}_i \qquad i = 1, 2, 3 \tag{3.35b}$$

Note that the error in the approximation is zero when $\mathbf{f}$ is approximated in terms of three mutually orthogonal vectors: $\mathbf{x}_1, \mathbf{x}_2$, and $\mathbf{x}_3$. The reason is that $\mathbf{f}$ is a three-dimensional vector, and the vectors $\mathbf{x}_1, \mathbf{x}_2$, and $\mathbf{x}_3$ represent a *complete set* of orthogonal vectors in three-dimensional space. Completeness here means that it is impossible to find another vector $\mathbf{x}_4$ in this space, which is orthogonal to all the three vectors $\mathbf{x}_1, \mathbf{x}_2$, and $\mathbf{x}_3$. Any vector in this space can then be represented (with zero error) in terms of these three vectors. Such vectors are known as **basis** vectors. If a set of vectors $\{\mathbf{x}_i\}$ is not complete, the error in the approximation will generally not be zero. Thus, in the three-dimensional case discussed above, it is generally not possible to represent a vector $\mathbf{f}$ in terms of only two basis vectors without an error.

The choice of basis vectors is not unique. In fact, a set of basis vectors corresponds to a particular choice of coordinate system. Thus, a 3-dimensional vector $\mathbf{f}$ may be represented in many different ways, depending on the coordinate system used.

3.3-2 Orthogonal Signal Space

We start with real signals first, and then extend the discussion to complex signals. We proceed with our signal approximation problem using clues and insights developed for vector approximation. As before, we define orthogonality of a real signal set $x_1(t), x_2(t), \cdots, x_N(t)$ over interval $[t_1, t_2]$ as

$$\int_{t_1}^{t_2} x_m(t)x_n(t)\, dt = \begin{cases} 0 & m \neq n \\ E_n & m = n \end{cases} \tag{3.36}$$

If the energies $E_n = 1$ for all n, then the set is *normalized* and is called an **orthonormal set**. An orthogonal set can always be normalized by dividing $x_n(t)$ by $\sqrt{E_n}$ for all n.

Now, consider approximating a signal $f(t)$ over the interval $[t_1, t_2]$ by a set of N real, mutually orthogonal signals $x_1(t), x_2(t), \ldots, x_N(t)$ as

$$f(t) \simeq c_1x_1(t) + c_2x_2(t) + \cdots + c_Nx_N(t) \tag{3.37a}$$

$$= \sum_{n=1}^{N} c_nx_n(t) \tag{3.37b}$$

The error $e(t)$ in the approximation (3.37)

$$e(t) = f(t) - \sum_{n=1}^{N} c_n x_n(t) \tag{3.38}$$

We show in Appendix 3A that E_e, the energy of the error signal $e(t)$, is minimized if we choose

$$c_n = \frac{\displaystyle\int_{t_1}^{t_2} f(t) x_n(t)\, dt}{\displaystyle\int_{t_1}^{t_2} x_n{}^2(t)\, dt} \tag{3.39a}$$

$$= \frac{1}{E_n} \int_{t_1}^{t_2} f(t) x_n(t)\, dt \qquad n = 1, 2, \ldots, N \tag{3.39b}$$

For this choice of the coefficients c_n, it is shown in Appendix 3A that the error signal energy E_e is given by

$$E_e = \int_{t_1}^{t_2} f^2(t)\, dt - \sum_{n=1}^{N} c_n{}^2 E_n \tag{3.40}$$

Observe that the error energy E_e generally decreases as N, the number of terms, is increased because the term $c_k{}^2 E_k$ is nonnegative. Hence, it is possible that the error energy $\to 0$ as $N \to \infty$. When this happens, the orthogonal signal set is said to be **complete**. In this case, Eq. (3.37b) is no more an approximation but an equality

$$f(t) = c_1 x_1(t) + c_2 x_2(t) + \cdots + c_n x_n(t) + \cdots$$

$$= \sum_{n=1}^{\infty} c_n x_n(t) \qquad t_1 \le t \le t_2 \tag{3.41}$$

where the coefficients c_n are given by Eq. (3.39). Because the error signal energy approaches zero, it follows that the energy of $f(t)$ is now equal to the sum of the energies of its orthogonal components $c_1 x_1(t)$, $c_2 x_2(t)$, $c_3 x_3(t)$, $\cdots$.

The series on the right-hand side of Eq. (3.41) is called the **generalized Fourier series** of $f(t)$ with respect to the set $\{x_n(t)\}$. When the set $\{x_n(t)\}$ is such that the error energy $E_e \to 0$ as $N \to \infty$ for every member of some particular class, we say that the set $\{x_n(t)\}$ is complete on $[t_1, t_2]$ for that class of $f(t)$, and the set $\{x_n(t)\}$ is called a set of **basis functions** or **basis signals**. Unless otherwise mentioned, in future we shall consider only the class of energy signals.

Thus, when the set $\{x_n(t)\}$ is complete, we have the equality (3.41). One subtle point that must be understood clearly is the meaning of equality in Eq. (3.41). *The equality here is not an equality in the ordinary sense, but in the sense that the error energy, that is, the energy of the difference between the two sides of Eq. (3.41), approaches zero.* If the equality exists in the ordinary sense, the error energy is always zero, but the converse is not necessarily true. The error energy can approach zero even though $e(t)$, the difference between the two sides, is nonzero at some isolated instants. The reason is that even if $e(t)$ is nonzero at such instants, the area under $e^2(t)$ is still zero; thus the Fourier series on the right-hand side of

Eq. (3.41) may differ from $f(t)$ at a finite number of points. In fact, when $f(t)$ has a jump discontinuity at $t = t_0$, the corresponding Fourier series at t_0 converges to the mean of $f(t_0{}^+)$ and $f(t_0{}^-)$.

In Eq. (3.41), the energy of the left-hand side is E_f, and the energy of the right-hand side is the sum of the energies of all the orthogonal components.† Thus

$$\int_{t_1}^{t_2} f^2(t)\,dt = c_1{}^2 E_1 + c_2{}^2 E_2 + \cdots$$

$$= \sum_{n=1}^{\infty} c_n{}^2 E_n \qquad (3.42)$$

This equation goes under the name of **Parseval's theorem**. Recall that the signal energy (area under the squared value of a signal) is analogous to the square of the length of a vector in the vector-signal analogy. In vector space we know that the square of the length of a vector is equal to the sum of the squares of the lengths of its orthogonal components. The above equation (3.42) is the statement of this fact as it applies to signals.

Generalization to Complex Signals

The above results can be generalized to complex signals as follows: A set of functions $x_1(t)$, $x_2(t)$, ..., $x_N(t)$ is mutually orthogonal over the interval $[t_1, t_2]$ if

$$\int_{t_1}^{t_2} x_m(t) x_n^*(t)\,dt = \begin{cases} 0 & m \neq n \\ E_n & m = n \end{cases} \qquad (3.43)$$

If this set is complete for a certain class of functions, then a function $f(t)$ in this class can be expressed as

$$f(t) = c_1 x_1(t) + c_2 x_2(t) + \cdots + c_i x_i(t) + \cdots \qquad (3.44)$$

where

$$c_n = \frac{1}{E_n} \int_{t_1}^{t_2} f(t) x_n^*(t)\,dt \qquad (3.45)$$

Equation (3.39) [or Eq. (3.45)] shows one interesting property of the coefficients of c_1, c_2, ..., c_N; the optimum value of any coefficient in the approximation (3.37) is independent of the number of terms used in the approximation. For example, if we have used only one term ($N = 1$) or two terms ($N = 2$) or any number of terms, the optimum value of the coefficient c_1 would be the same [as given by Eq. (3.39)]. The advantage of this approximation of a signal $f(t)$ by a set of mutually orthogonal signals is that we can continue to add terms to the approximation without disturbing the previous terms. This property of **finality** of the values of the coefficients is very

---

†Note that the energy of a signal $cx(t)$ is $c^2 E_x$.

important from a practical point of view.‡

Some Examples of Generalized Fourier Series

Signals are vectors in every sense. Like a vector, a signal can be represents as a sum of its components in a variety of ways. Just as vector coordinate systems are formed by mutually orthogonal vectors (rectangular, cylindrical, spherical), we also have signal coordinate systems (basis signals) formed by a variety of sets of mutually orthogonal signals. There exist a large number of orthogonal signal sets which can be used as basis signals for generalized Fourier series. Some well-known signal sets are trigonometric (sinusoid) functions, exponential functions, Walsh functions, Bessel functions, Legendre polynomials, Laguerre functions, Jacobi polynomials, Hermite polynomials, and Chebyshev polynomials. The functions that concern us most in this book are the trigonometric and the exponential sets discussed in the rest of the chapter.

A Historical Note: Baron Jean-Baptiste-Joseph Fourier (1768-1830)

The Fourier series and integral is a most beautiful and fruitful development, which serves as an indispensable instrument in the treatment of many problems in mathematics, science, and engineering. Maxwell was so taken by the beauty of the Fourier series that he called it a great mathematical poem. In electrical engineering, it is central to the areas of communication, signal processing and several other fields, including antennas, but it was not received enthusiastically by the scientific world when it was presented. In fact, Fourier could not get his results published as a paper.

Fourier, a tailor's son, was orphaned at age 8 and educated at a local military college (run by Benedictine monks), where he excelled in mathematics. The Benedictines prevailed upon the young genius to choose the priesthood as his vocation, but the revolution broke out before he could take his vows. Fourier joined the people's party. But in its early days, the French Revolution, like most revolutions of its kind, liquidated a large segment of the intelligentsia, including prominent scientists such as Lavosier. This persecution caused many intellectuals to leave France to save themselves from a rapidly rising tide of barbarism. Fourier, who was an early enthusiast of the Revolution, narrowly escaped the guillotine twice. It was to the everlasting credit of Napoleon that he stopped the persecution of the intelligentsia and founded new schools to replenish their ranks. The 26-year old Fourier was appointed chair of mathematics at the newly created school École Normale in 1794.[2]

---

‡Contrast this situation with the polynomial approximation of $f(t)$. Suppose we wish to approximate $f(t)$ by a polynomial in t such that the polynomial is equal to $f(t)$ at two points t_1 and t_2. This can be done by choosing a first order polynomial $a_0 + a_1 t$ with

$$f(t_1) = a_0 + a_1 t_1 \quad \text{and} \quad f(t_2) = a_0 + a_1 t_2$$

Solution of these equations yields the desired values of a_0 and a_1. For a three-point approximation, we must choose the polynomial $a_0 + a_1 t + a_2 t^2$ with

$$f(t_i) = a_0 + a_1 t_i + a_2 t_i^2 \quad i = 1, 2, \text{ and } 3$$

The approximation improves with larger number of points (higher-order polynomial), but the coefficients a_0, a_1, a_2, $\cdots$ do not have the finality property. As we increase the number of terms in the polynomial, we need to recalculate the coefficients.

Joseph Fourier (left) and **Napoleon** (right).

Napoleon was the first modern ruler with a scientific education, and he was one of the rare persons who was equally comfortable with soldiers and scientist. The age of Napoleon was one of the most fruitful in the history of science. Napoleon liked to sign himself as a 'member of *institut de France*'(a fraternity of scientists), and he once expressed to Laplace his regret that "force of circumstances has led me so far from the career of a scientist."[3] Many great figures in science and mathematics, including Fourier and Laplace, were honored and promoted by Napoleon. In 1798, he took a group of scientists, artists, and scholars—Fourier among them—on his Egyptian expedition, with the promise of an exciting and historic union of adventure and research. Fourier proved to be a capable administrator of the newly formed Institut d'Egypte, which, incidentally, was responsible for the discovery of the Rosetta Stone. The inscription on this stone in two languages and three scripts (hieroglyphic, demotic, and Greek) enabled Thomas Young and Jean-Francois Champollion, a protege of Fourier, to invent a method of translating hieroglyphic writings of ancient Egypt–the only significant result of Napoleon's Egyptian expedition.

Back in France in 1801, Fourier briefly served in his former position as professor of mathematics at the École Polytechnique in Paris. In 1802 Napoleon appointed him the prefect of Isère (with its headquarters in Grenoble), a position in which Fourier served with distinction. Fourier was created Baron of the Empire by Napoleon in 1809. Later, when Napoleon was exiled to Elba, his route was to take him through Grenoble. Fourier had the route changed to avoid meeting Napoleon, which would have displeased Fourier's new master, the Bourbon King Louis XVIII. Within a year, Napoleon escaped from Elba. On his way home, at Grenoble, Fourier was brought before him in chains. Napoleon scolded Fourier for his ungrateful behavior but reappointed him the prefect of Rhone at Lyons. Within four months Napoleon was defeated at Waterloo, and was exiled to St. Helena, never to return.

Fourier once again was in disgrace as a Bonapartist, and had to pawn his effects to keep himself alive. But through the intercession of a former student, who was now a prefect of Paris, he was appointed director of the statistical bureau of the Seine, a position that allowed him ample time for scholarly pursuits. Later, in 1827, he was elected to the powerful position of perpetual secretary of the Paris Academy of Science, a section of the institute.[4]

While serving as the prefect of Grenoble, Fourier carried on his elaborate investigation of propagation of heat in solid bodies, which led him to the Fourier series and the Fourier integral. On 21 December 1807, he announced these results in a prize paper on the theory of heat. Fourier claimed that an arbitrary function (continuous or with discontinuities) defined in a finite interval by an arbitrarily capricious graph can always be expressed as a sum of sinusoids (Fourier series). The judges, who included the great French mathematicians Laplace, Lagrange, Monge, and LaCroix admitted the novelty and importance of Fourier's work, but criticized it for lack of mathematical rigor and generality. Lagrange thought it incredible that a sum of sines and cosines could add up to anything but an infinitely differentiable function. Moreover, one of the properties of an infinitely differentiable function is that if we know its behavior over an arbitrarily small interval, we can determine its behavior over the entire range (the Taylor-Maclaurin series). Such a function is far from an arbitrary or a capriciously drawn graph.[5] Fourier thought the criticism unjustified but was unable to prove his claim because the tools required for operations with infinite series were not available at the time. However, posterity has proved Fourier to be closer to the truth than his critics. This is the classic conflict between pure mathematicians and physicists or engineers, as we shall see again in the life of Oliver Heaviside (p. 381). In 1829 Dirichlet proved Fourier's claim concerning capriciously drawn functions with a few restrictions (Dirichlet conditions).

Although three of the four judges were in favor of publication, this paper was rejected because of vehement opposition by Lagrange. Fifteen years later, after several attempts and disappointments, Fourier published the results in expanded form as a text, *Theorie analytique de la chaleur*, which is now a classic.

3.4   Trigonometric Fourier Series

Consider a signal set:

$$\{1,\ \cos \omega_0 t,\ \cos 2\omega_0 t,\ \ldots,\ \cos n\omega_0 t,\ \ldots;$$
$$\sin \omega_0 t,\ \sin 2\omega_0 t,\ \ldots,\ \sin n\omega_0 t,\ \ldots\} \tag{3.46}$$

A sinusoid of frequency $n\omega_0$ is called the n**th harmonic** of the sinusoid of frequency ω_0 when n is an integer. In this set the sinusoid of frequency ω_0, called the **fundamental**, serves as an anchor of which all the remaining terms are harmonics. Note that the constant term 1 is the 0th harmonic in this set because $\cos(0 \times \omega_0 t) = 1$. In Appendix 3B we show that this set is orthogonal over any interval of duration $T_0 = 2\pi/\omega_0$, which is the period of the fundamental. Specifically, we have shown that

$$\int_{T_0} \cos n\omega_0 t \cos m\omega_0 t\, dt = \begin{cases} 0 & n \neq m \\ \dfrac{T_0}{2} & m = n \neq 0 \end{cases} \tag{3.47a}$$

$$\int_{T_0} \sin n\omega_0 t \, \sin m\omega_0 t \, dt = \begin{cases} 0 & n \neq m \\ \frac{T_0}{2} & n = m \neq 0 \end{cases} \qquad (3.47b)$$

and

$$\int_{T_0} \sin n\omega_0 t \, \cos m\omega_0 t \, dt = 0 \quad \text{for all } n \text{ and } m \qquad (3.47c)$$

The notation $\int_{T_0}$ means the integral over an interval from $t = t_1$ to $t_1 + T_0$ for any value of t_1. These equations show that the set (3.46) is orthogonal over any contiguous interval of duration T_0. This is the **trigonometric set**, which can be shown to be a complete set.[6, 7] Therefore, we can express a signal $f(t)$ by a trigonometric Fourier series over any interval of duration T_0 seconds as

$$f(t) = a_0 + a_1 \cos \omega_0 t + a_2 \cos 2\omega_0 t + \cdots$$

$$+ \, b_1 \sin \omega_0 t + b_2 \sin 2\omega_0 t + \cdots \qquad t_1 \leq t \leq t_1 + T_0 \qquad (3.48a)$$

or

$$f(t) = a_0 + \sum_{n=1}^{\infty} a_n \cos n\omega_0 t + b_n \sin n\omega_0 t \qquad t_1 \leq t \leq t_1 + T_0 \qquad (3.48b)$$

where

$$\omega_0 = \frac{2\pi}{T_0} \qquad (3.49)$$

Using Eq. (3.39), we can determine the Fourier coefficients a_0, a_n, and b_n. Thus

$$a_n = \frac{\displaystyle\int_{t_1}^{t_1+T_0} f(t) \cos n\omega_0 t \, dt}{\displaystyle\int_{t_1}^{t_1+T_0} \cos^2 n\omega_0 t \, dt} \qquad (3.50)$$

The integral in the denominator of Eq. (3.50) as seen from Eq. (3.47a) (with $m = n$) is $T_0/2$ when $n \neq 0$. Moreover, for $n = 0$, the denominator is T_0. Hence

$$a_0 = \frac{1}{T_0} \int_{t_1}^{t_1+T_0} f(t) \, dt \qquad (3.51a)$$

and

$$a_n = \frac{2}{T_0} \int_{t_1}^{t_1+T_0} f(t) \cos n\omega_0 t \, dt \qquad n = 1, 2, 3, \ldots \qquad (3.51b)$$

Arguing the same way, we obtain

$$b_n = \frac{2}{T_0} \int_{t_1}^{t_1+T_0} f(t) \sin n\omega_0 t \, dt \qquad n = 1, 2, 3, \ldots \qquad (3.51c)$$

Compact Trigonometric Fourier Series

The trigonometric Fourier series in Eq. (3.48) contains sine and cosine terms of the same frequency. We can combine the two terms to obtain a single sinusoid of the same frequency using the trigonometric identity

$$a_n \cos n\omega_0 t + b_n \sin n\omega_0 t = C_n \cos (n\omega_0 t + \theta_n) \qquad (3.52)$$

where

$$C_n = \sqrt{a_n{}^2 + b_n{}^2} \qquad (3.53a)$$

$$\theta_n = \tan^{-1}\left(\frac{-b_n}{a_n}\right) \qquad (3.53b)$$

For consistency, we denote the dc term a_0 by C_0, that is

$$C_0 = a_0 \qquad (3.53c)$$

Using the identity (3.52), the trigonometric Fourier series in Eq. (3.48) can be expressed in the **compact form** of the trigonometric Fourier series as

$$f(t) = C_0 + \sum_{n=1}^{\infty} C_n \cos\left(n\omega_0 t + \theta_n\right) \qquad t_1 \le t \le t_1 + T_0 \qquad (3.54)$$

where the coefficients C_n and θ_n are computed from a_n and b_n using Eqs. (3.53).

Equation 3.51a shows that a_0 (or C_0) is the average value of $f(t)$ (averaged over one period). This value can often be determined by inspection of $f(t)$.

■ **Example 3.3**

Find the compact trigonometric Fourier series for the exponential $e^{-t/2}$ depicted in Fig. 3.7a over the shaded interval $0 \le t \le \pi$.

Because we are required to represent $f(t)$ by the trigonometric Fourier series over the interval $0 \le t \le \pi$ only, $T_0 = \pi$, and the fundamental frequency is

$$\omega_0 = \frac{2\pi}{T_0} = 2$$

Therefore

$$f(t) = a_0 + \sum_{n=1}^{\infty} a_n \cos 2nt + b_n \sin 2nt \qquad 0 \le t \le \pi$$

where [from Eq. (3.51a)]

$$a_0 = \frac{1}{\pi} \int_0^{\pi} e^{-t/2} \, dt = 0.504$$

$$a_n = \frac{2}{\pi} \int_0^{\pi} e^{-t/2} \cos 2nt \, dt = 0.504 \left(\frac{2}{1 + 16n^2}\right)$$

and

$$b_n = \frac{2}{\pi} \int_0^{\pi} e^{-t/2} \sin 2nt \, dt = 0.504 \left(\frac{8n}{1 + 16n^2}\right)$$

Therefore

$$f(t) = 0.504 \left[1 + \sum_{n=1}^{\infty} \frac{2}{1 + 16n^2} \left(\cos 2nt + 4n \sin 2nt\right)\right] \qquad 0 \le t \le \pi$$

To find the compact Fourier series, we compute its coefficients using Eq. (3.53) as

$$C_0 = a_0 = 0.504$$

$$C_n = \sqrt{a_n^2 + b_n^2} = 0.504 \sqrt{\frac{4}{(1+16n^2)^2} + \frac{64n^2}{(1+16n^2)^2}} = 0.504\left(\frac{2}{\sqrt{1+16n^2}}\right)$$

$$\theta_n = \tan^{-1}\left(\frac{-b_n}{a_n}\right) = \tan^{-1}(-4n) = -\tan^{-1} 4n \qquad (3.55)$$

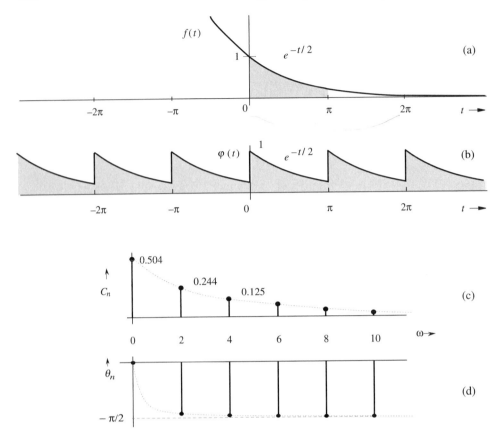

Fig. 3.7 A periodic signal and its Fourier spectra.

The values of C_n and θ_n for the dc and the first seven harmonics are computed from the above equation and displayed in Table 3.1. Using these numerical values, we can express $f(t)$ in the compact trigonometric Fourier series as

$$f(t) = 0.504 + 0.504 \sum_{n=1}^{\infty} \frac{2}{\sqrt{1 + 16n^2}} \cos\left(2nt - \tan^{-1} 4n\right) \qquad 0 \le t \le \pi \qquad (3.56a)$$

$$= 0.504 + 0.244 \cos\left(2t - 75.96°\right) + 0.125 \cos\left(4t - 82.87°\right)$$

$$+ 0.084 \cos\left(6t - 85.24°\right) + 0.063 \cos\left(8t - 86.42°\right) + \cdots \qquad 0 \le t \le \pi \ (3.56b)$$

Remember that the right-hand side represents $e^{-t/2}$ over the interval 0 to π only. Outside this interval, the two sides need not be equal.

Table 3.1

n	0	1	2	3	4	5	6	7
C_n	0.504	0.244	0.125	0.084	0.063	0.0504	0.042	0.036
θ_n	0	−75.96	−82.87	−85.24	−86.42	−87.14	−87.61	−87.95

Periodicity of the Trigonometric Fourier Series

We have shown how an arbitrary signal $f(t)$ may be expressed as a trigonometric Fourier series over any interval of T_0 seconds. In Example 3.3, for instance, we represented $e^{-t/2}$ over the interval from 0 to $\pi/2$ only. The Fourier series found in Eq. (3.56) is equal to $e^{-t/2}$ over this interval alone. Outside this interval the series is not necessarily equal to $e^{-t/2}$. It would be interesting to find out what happens to the Fourier series outside this interval. We now show that the trigonometric Fourier series is a periodic function of period T_0 (the period of the fundamental). Let us denote the trigonometric Fourier series on the right-hand side of Eq. (3.54) by $\varphi(t)$. Therefore

$$\varphi(t) = C_0 + \sum_{n=1}^{\infty} C_n \cos{(n\omega_0 t + \theta_n)} \qquad \text{for all } t$$

and

$$\varphi(t + T_0) = C_0 + \sum_{n=1}^{\infty} C_n \cos{[n\omega_0(t + T_0) + \theta_n]}$$

$$= C_0 + \sum_{n=1}^{\infty} C_n \cos{[(n\omega_0 t + 2n\pi) + \theta_n]}$$

$$= C_0 + \sum_{n=1}^{\infty} C_n \cos{(n\omega_0 t + \theta_n)}$$

$$= \varphi(t) \qquad \text{for all } t \tag{3.57}$$

This result shows that the trigonometric Fourier series is a periodic function of period T_0 (the period of its fundamental). For instance, $\varphi(t)$, the Fourier series on the right-hand side of Eq. (3.56), is a periodic function in which the segment of $f(t)$ in Fig. 3.7a over the interval $(0 \le t \le \pi)$ repeats periodically every π seconds, as illustrated in Fig. 3.7b.[†] Thus, when we represent a signal $f(t)$ by the trigonometric Fourier series over a certain interval of duration T_0, the function $f(t)$ and its Fourier series $\varphi(t)$ need be equal only over that interval of T_0 seconds. Outside this interval, the Fourier series repeats periodically with period T_0.

Now, if the function $f(t)$ were itself to be periodic with period T_0, then a Fourier series representing $f(t)$ over an interval T_0 will also represent $f(t)$ for all t (not just over the interval T_0). Another interesting fact, as seen in Fig. 1.7, is that a periodic signal $f(t)$ can be generated by a periodic repetition of any of its segment of duration T_0. Therefore, the trigonometric Fourier series representing a segment of $f(t)$ of duration T_0 starting at any instant represents $f(t)$ for all t. Therefore, it follows that in computing the coefficients a_0, a_n and b_n, we may use any value for t_1 in Eqs. (3.51). In other words, we may perform this integration over any interval of T_0. Thus the Fourier coefficients of a series representing a periodic signal $f(t)$ (for all t) can be expressed as

$$a_0 = \frac{1}{T_0} \int_{T_0} f(t)\, dt \tag{3.58a}$$

[†]In reality, the series convergence at the points of discontinuity shows about 9% overshoot (Gibbs phenomenon[6]) as discussed later.

$$a_n = \frac{2}{T_0} \int_{T_0} f(t) \cos n\omega_0 t \, dt \qquad n = 1, 2, 3, \dots \qquad (3.58\text{b})$$

and

$$b_n = \frac{2}{T_0} \int_{T_0} f(t) \sin n\omega_0 t \, dt \qquad n = 1, 2, 3, \dots \qquad (3.58\text{c})$$

where $\int_{T_0}$ means that the integration is performed over any interval of T_0 seconds.

The Fourier Spectrum

The compact trigonometric Fourier series in Eq. (3.54) indicates that a periodic signal $f(t)$ can be expressed as a sum of sinusoids of frequencies 0 (dc), ω_0, $2\omega_0$, $\cdots$, $n\omega_0$, $\cdots$, whose amplitudes are C_0, C_1, C_2, ..., C_n, $\cdots$, and whose phases are 0, θ_1, θ_2, $\cdots$, θ_n, $\cdots$, respectively. We can readily plot amplitude C_n vs. ω (**amplitude spectrum**) and θ_n vs. ω (**phase spectrum**). These two plots together are the **frequency spectra** of $f(t)$.

Figures 3.7c and 3.7d show the amplitude and phase spectra for the periodic signal $\varphi(t)$ in Fig. 3.7b. These spectra tell us at a glance the frequency composition of $\varphi(t)$; that is, the amplitudes and phases of various sinusoidal components of $\varphi(t)$. Knowing the frequency spectra, we can reconstruct or synthesize $\varphi(t)$, as shown on the right-hand side of Eq. (3.56). Therefore, the frequency spectra in Figs. 3.7c and 3.7d provide an alternative description—**the frequency-domain description** of $\varphi(t)$. *The time-domain description of $\varphi(t)$ is depicted in Fig. 3.7b. A signal, therefore, has a dual identity: the time-domain identity $\varphi(t)$ and the frequency-domain identity (Fourier spectra). The two identities complement each other; taken together, they provide a better understanding of a signal.*

Series Convergence at Jump Discontinuities

An interesting aspect of a Fourier series is that whenever there is a jump discontinuity in $f(t)$, the series at the point of discontinuity converges to an average of the left-hand and right-hand limits of $f(t)$ at the instant of discontinuity.[†] In Fig. 3.7b, for instance, $\varphi(t)$ is discontinuous at $t = 0$ with $\varphi(0^+) = 1$ and $\varphi(0^-) = e^{-\pi/2} = 0.208$. The corresponding Fourier series converges to a value $(1+0.208)/2 = 0.604$ at $t = 0$. This conclusion is easily verified from Eq. 3.56b by setting $t = 0$.

Existence of the Fourier Series: Dirichlet Conditions

There are two basic conditions for the existence of the Fourier series

1. For the series to exist, the coefficients a_0, a_n, and b_n in Eq. (3.51) must be finite. From Eqs. (3.51a), (3.51b), and (3.51c), it follows that the existence of these coefficients is guaranteed if $f(t)$ is absolutely integrable over one period; that is,

$$\int_{T_0} |f(t)| \, dt < \infty \qquad (3.59)$$

[†]This behavior of the Fourier series is dictated by its error energy minimization property, discussed in Sec. 3.3.

This condition is known as the **weak Dirichlet condition**. If a function $f(t)$ satisfies the weak Dirichlet condition, the existence of a Fourier series is guaranteed, but the series may not converge at every point. For example, if a function $f(t)$ is infinite at some point, then obviously the series representing the function will be nonconvergent at that point. Similarly, if a function has an infinite number of maxima and minima in one period, then the function contains an appreciable amount of components of frequencies approaching infinity. Consequently, the coefficients in the series at higher frequencies do not decay rapidly, so that the series will not converge rapidly or uniformly. Thus, for a convergent Fourier series, in addition to condition (3.59), we require that

2. The function $f(t)$ have only a finite number of maxima and minima in one period, and only a finite number of finite discontinuities in one period. These two conditions are known as the **strong Dirichlet conditions**. We note here that any periodic waveform that can be generated in a laboratory satisfies strong Dirichlet conditions, and hence possesses a convergent Fourier series. Thus, a physical possibility of a periodic waveform is a valid and sufficient condition for the existence of a convergent series.

■ **Example 3.4**

Find the compact trigonometric Fourier series for the periodic square wave $f(t)$ illustrated in Fig. 3.8a, and sketch its amplitude and phase spectra.

Here, the period $T_0 = 2\pi$ and $\omega_0 = 2\pi/T_0 = 1$. Therefore

$$f(t) = a_0 + \sum_{n=1}^{\infty} a_n \cos nt + b_n \sin nt$$

where+

$$a_0 = \frac{1}{T_0} \int_{T_0} f(t)\, dt$$

In the above equation we may integrate $f(t)$ over any interval of duration $T_0 = 2\pi$. Figure 3.8a shows that the best choice for a region of integration is from $-\pi$ to π. Because $f(t) = 1$ only over $(-\frac{\pi}{2}, \frac{\pi}{2})$ and $f(t) = 0$ over the remaining segment,

$$a_0 = \frac{1}{2\pi} \int_{-\pi/2}^{\pi/2} dt = \frac{1}{2} \tag{3.60a}$$

We could have easily deduced that a_0, the average value of $f(t)$, is $\frac{1}{2}$ merely by inspection of $f(t)$ in Fig. 3.8a. Also,

$$a_n = \frac{1}{\pi} \int_{-\pi/2}^{\pi/2} \cos nt\, dt = \frac{2}{n\pi} \sin\left(\frac{n\pi}{2}\right)$$

$$= \begin{cases} 0 & n \text{ even} \\ \frac{2}{\pi n} & n = 1,\, 5,\, 9,\, 13,\, \cdots \\ -\frac{2}{\pi n} & n = 3,\, 7,\, 11,\, 15,\, \cdots \end{cases} \tag{3.60b}$$

$$b_n = \frac{1}{\pi} \int_{-\pi/2}^{\pi/2} \sin nt\, dt = 0 \tag{3.60c}$$

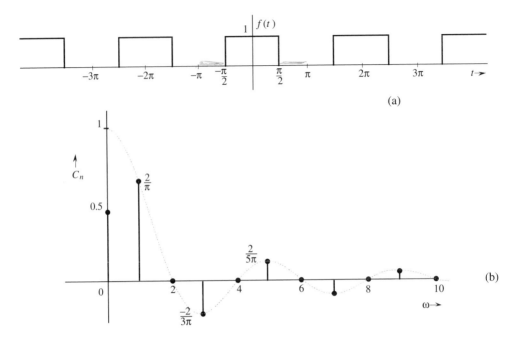

Fig. 3.8 A square pulse periodic signal and its Fourier spectra.

Therefore

$$f(t) = \frac{1}{2} + \frac{2}{\pi}\left(\cos t - \frac{1}{3}\cos 3t + \frac{1}{5}\cos 5t - \frac{1}{7}\cos 7t + \cdots\right) \qquad (3.61)$$

Observe that $b_n = 0$ and all the sine terms are zero. Only the cosine terms appear in the trigonometric series. The series is therefore already in the compact form except that the amplitudes of alternating harmonics are negative. Now, by definition, amplitudes C_n are positive [see Eq. (3.53a)]. The negative sign can be accommodated by a phase of π radians as seen from the trigonometric identity†

$$-\cos x = \cos(x - \pi)$$

Using this fact, we can express the series in (3.61) as

$$f(t) = \frac{1}{2} + \frac{2}{\pi}\left[\cos t + \frac{1}{3}\cos(3t - \pi) + \frac{1}{5}\cos 5t + \frac{1}{7}\cos(7t - \pi) + \frac{1}{9}\cos 9t + \cdots\right]$$

This is precisely the Fourier series in the compact trigonometric form. The amplitudes are

$$C_0 = \tfrac{1}{2}$$

$$C_n = \begin{cases} 0 & n \text{ even} \\ \frac{2}{\pi n} & n \text{ odd} \end{cases}$$

$$\theta_n = \begin{cases} 0 & \text{for all } n \neq 3, 7, 11, 15, \cdots \\ -\pi & n = 3, 7, 11, 15, \cdots \end{cases}$$

---

†Because $\cos(x \pm \pi) = -\cos x$, we could have chosen the phase π or $-\pi$. In fact, $\cos(x \pm N\pi) = -\cos x$ for any odd integral value of N. Therefore, the phase can be chosen as $\pm N\pi$ where N is any convenient odd integer.

Using the above values we could plot amplitude and phase spectra. However, to simplify our task in this special case, we will allow amplitude C_n to take on negative values so that we do not need a phase of $-\pi$ to account for the sign. In other words, phases of all components are zero, so we can discard the phase spectrum and manage with only the amplitude spectrum, as shown in Fig. 3.8b. Observe that this simpler procedure involves no loss of information and that the amplitude spectrum in Fig. 3.8b has the complete information about the Fourier series in (3.61). *Therefore, whenever all sine terms vanish* $(b_n = 0)$, *it is convenient to allow* C_n *to take on negative values.* This procedure permits the spectral information to be conveyed by a single spectrum—the amplitude spectrum. Because C_n can be positive as well as negative, the spectrum is called the *amplitude spectrum* rather than the *magnitude spectrum.* ■

■ **Example 3.5**

Find the compact trigonometric Fourier series for the triangular periodic signal $f(t)$ illustrated in Fig. 3.9a, and sketch the amplitude and phase spectra for $f(t)$.

In this case the period $T_0 = 2$. Hence

$$\omega_0 = \frac{2\pi}{2} = \pi$$

and

$$f(t) = a_0 + \sum_{n=1}^{\infty} a_n \cos n\pi t + b_n \sin n\pi t$$

where

$$f(t) = \begin{cases} 2At & |t| \leq \frac{1}{2} \\ 2A(1 - t) & \frac{1}{2} < t \leq \frac{3}{2} \end{cases}$$

Here it will be advantageous to choose the interval of integration from $-\frac{1}{2}$ to $\frac{3}{2}$ rather than 0 to 2.

A glance at Fig. 3.9a shows that the average value (dc) of $f(t)$ is zero, so that $a_0 = 0$. Also

$$a_n = \frac{2}{2} \int_{-1/2}^{3/2} f(t) \cos n\pi t \, dt$$

$$= \int_{-1/2}^{1/2} 2At \cos n\pi t \, dt + \int_{1/2}^{3/2} 2A(1 - t) \cos n\pi t \, dt$$

The detailed evaluation of the above integrals shows that both have a value of zero. Therefore

$$a_n = 0 \tag{3.62a}$$

$$b_n = \int_{-1/2}^{1/2} 2At \sin n\pi t \, dt + \int_{1/2}^{3/2} 2A(1 - t) \sin n\pi t \, dt$$

The detailed evaluation of these integrals yields

$$b_n = \frac{8A}{n^2 \pi^2} \sin \left(\frac{n\pi}{2} \right) = \begin{cases} 0 & n \text{ even} \\ \frac{8A}{n^2 \pi^2} & n = 1, 5, 9, 13, \cdots \\ -\frac{8A}{n^2 \pi^2} & n = 3, 7, 11, 15, \cdots \end{cases} \tag{3.62b}$$

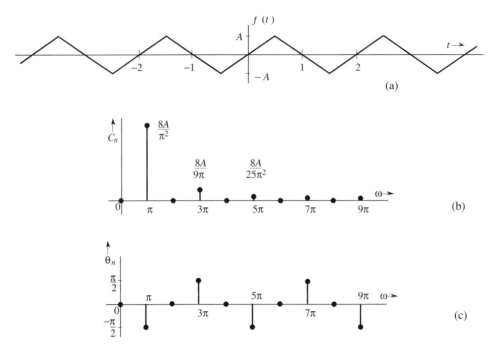

Fig. 3.9 A triangular periodic signal and its Fourier spectra.

Therefore

$$f(t) = \frac{8A}{\pi^2} \left[\sin \pi t - \frac{1}{9} \sin 3\pi t + \frac{1}{25} \sin 5\pi t - \frac{1}{49} \sin 7\pi t + \cdots \right] \qquad (3.63)$$

In order to plot Fourier spectra, the series must be converted into compact trigonometric form as in Eq. (3.54). In this case, sine terms are readily converted into cosine terms with a suitable phase shift. For example,

$$\pm \sin kt = \cos (kt \mp 90°)$$

Using this identity, Eq. (3.63) can be expressed as

$$f(t) = \frac{8A}{\pi^2} \left[\cos (\pi t - 90°) + \frac{1}{9} \cos (3\pi t + 90°) + \frac{1}{25} \cos (5\pi t - 90°) \right.$$

$$\left. + \frac{1}{49} \cos (7\pi t + 90°) + \cdots \right] \qquad (3.64)$$

In this series all the even harmonics are missing. The phases of odd harmonics alternate from $-90°$ to $90°$. Figure 3.9 shows amplitude and phase spectra for $f(t)$. ∎

3.4-1 The Effect of Symmetry

The Fourier series for the periodic signal in Fig. 3.7b (Example 3.3) consists of sine and cosine terms, but the series for the signal $f(t)$ in Fig. 3.8a (Example 3.4) consists of cosine terms only, and the series for the signal $f(t)$ in Fig. 3.9a (Example 3.5) consists of sine terms only. This observation is no accident. We can show that the Fourier series of any even periodic function $f(t)$ consists of cosine terms only and the series for any odd periodic function $f(t)$ consists of sine terms only. Moreover, because of symmetry (even or odd), the information of one period of $f(t)$ is implicit in only half the period, as seen in Figs. 3.8a and 3.9a. In these cases, knowing the signal over a half period and what kind of symmetry (even or odd) is present, we can determine the signal waveform over a complete period. For this reason, the Fourier coefficients in these cases can be computed by integrating over only half the period rather than a complete period. To prove this result, recall that

$$a_0 = \frac{1}{T_0} \int_{-T_0/2}^{T_0/2} f(t)\, dt \tag{3.65a}$$

$$a_n = \frac{2}{T_0} \int_{-T_0/2}^{T_0/2} f(t) \cos n\omega_0 t\, dt \tag{3.65b}$$

$$b_n = \frac{2}{T_0} \int_{-T_0/2}^{T_0/2} f(t) \sin n\omega_0 t\, dt \tag{3.65c}$$

Recall also that $\cos n\omega_0 t$ is an even function and $\sin n\omega_0 t$ is an odd function of t. If $f(t)$ is an even function of t, then $f(t) \cos n\omega_0 t$ is also an even function and $f(t) \sin n\omega_0 t$ is an odd function of t (see Sec. 1.5-1). Therefore, use of Eqs. (1.33a) and (1.33b) yields

$$a_0 = \frac{2}{T_0} \int_0^{T_0/2} f(t)\, dt \tag{3.66a}$$

$$a_n = \frac{4}{T_0} \int_0^{T_0/2} f(t) \cos n\omega_0 t\, dt \tag{3.66b}$$

$$b_n = 0 \tag{3.66c}$$

Similarly, if $f(t)$ is an odd function of t, then $f(t) \cos n\omega_0 t$ is an odd function of t and $f(t) \sin n\omega_0 t$ is an even function of t. Therefore

$$a_0 = a_n = 0 \tag{3.67a}$$

$$b_n = \frac{4}{T_0} \int_0^{T_0/2} f(t) \sin n\omega_0 t\, dt \tag{3.67b}$$

Observe that, because of symmetry, the integration required to compute the coefficients need be performed over only half the period.

  If a periodic signal $f(t)$, shifted by half the period, remains unchanged except for a sign—that is, if

$$f\left(t - \tfrac{T_0}{2}\right) = -f(t)$$

the signal is said to have a **half-wave** symmetry. We can demonstrate that in a signal with a half-wave symmetry, all the even-numbered harmonics vanish (see Prob. 3.4-7). The signal in Fig. 3.9a is a clear example of such a symmetry. This half-wave symmetry is also present in the signal in Fig. 3.8a but in a subtle form. The half-wave symmetry becomes obvious, however, when we subtract the dc component 0.5 from this signal. Note that this signal has a dc component 0.5 and only odd harmonics.

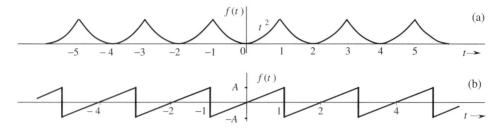

Fig. 3.10   Periodic signals for Exercise E3.6.

△   **Exercise E3.6**
    Find the compact trigonometric Fourier series for the periodic signals depicted in Figs. 3.10a and 3.10b. Sketch their amplitude and phase spectra. Allow C_n to take on negative values if $b_n = 0$ so that the phase spectrum can be eliminated. Hint: Use Eqs. (3.66) and (3.67) for symmetry. Answers:

(a) $f(t) = \frac{1}{3} - \frac{4}{\pi^2} \left(\cos \pi t - \frac{1}{4} \cos 2\pi t + \frac{1}{9} \cos 3\pi t - \frac{1}{16} \cos 4\pi t + \cdots \right)$

$\qquad = \frac{1}{3} + \frac{4}{\pi^2} \left[\cos(\pi t - \pi) + \frac{1}{4} \cos 2\pi t + \frac{1}{9} \cos(3\pi t - \pi) + \cdots \right]$

(b) $f(t) = \frac{2A}{\pi} \left[\sin \pi t - \frac{1}{2} \sin 2\pi t + \frac{1}{3} \sin 3\pi t - \frac{1}{4} \sin 4\pi t + \cdots \right]$

$\qquad = \frac{2A}{\pi} \left[\cos(\pi t - 90°) + \frac{1}{2} \cos(2\pi t + 90°) + \frac{1}{3} \cos(3\pi t - 90°) + \cdots \right]$ ▽

3.4-2   Determining the Fundamental Frequency and Period

   We have seen that every periodic signal can be expressed as a sum of sinusoids of a fundamental frequency ω_0 and its harmonics. One may ask whether a sum of sinusoids of **any** frequencies represents a periodic signal. If so, how does one determine the period? Consider the following three functions:

$$f_1(t) = 2 + 7 \cos \left(\tfrac{1}{2} t + \theta_1 \right) + 3 \cos \left(\tfrac{2}{3} t + \theta_2 \right) + 5 \cos \left(\tfrac{7}{6} t + \theta_3 \right)$$

$$f_2(t) = 2 \cos (2t + \theta_1) + 5 \sin (\pi t + \theta_2)$$

$$f_3(t) = 3 \sin \left(3\sqrt{2}\, t + \theta \right) + 7 \cos \left(6\sqrt{2}\, t + \phi \right)$$

   Recall that every frequency in a periodic signal is an integral multiple of the fundamental frequency ω_0. Therefore, the ratio of any two frequencies is of the form m/n where m and n are integers. This means that the ratio of any two frequencies is a rational number. When the ratio of two frequencies is a rational number, they are said to be **harmonically** related.
   The largest positive number of which all the frequencies are integral multiples is the fundamental frequency. The frequencies in the spectrum of $f_1(t)$ are $\frac{1}{2}$, $\frac{2}{3}$, and $\frac{7}{6}$ (we do not consider dc). The ratios of the successive frequencies are $\frac{3}{4}$ and $\frac{4}{7}$, respectively. Because both these numbers are rational, all the three frequencies in the spectrum are harmonically related and the signal $f_1(t)$ is periodic. The largest

number of which $\frac{1}{2}$, $\frac{2}{3}$, and $\frac{7}{6}$ are integral multiples is $\frac{1}{6}$.† Moreover, $3(\frac{1}{6}) = \frac{1}{2}$, $4(\frac{1}{6}) = \frac{2}{3}$, and $7(\frac{1}{6}) = \frac{7}{6}$. Therefore the fundamental frequency is $\frac{1}{6}$. The three frequencies in the spectrum are the third, fourth, and seventh harmonics. Observe that the fundamental frequency component is absent in this Fourier series.

The signal $f_2(t)$ is not periodic because the ratio of two frequencies in the spectrum is $2/\pi$, which is not a rational number. The signal $f_3(t)$ is periodic because the ratio of frequencies $3\sqrt{2}$ and $6\sqrt{2}$ is $1/2$, a rational number. The greatest common divisor of $3\sqrt{2}$ and $6\sqrt{2}$ is $3\sqrt{2}$. Therefore the fundamental frequency $\omega_0 = 3\sqrt{2}$, and the period

$$T_0 = \frac{2\pi}{(3\sqrt{2})} = \frac{\sqrt{2}}{3}\pi$$

△ **Exercise E3.7**

Determine whether the signal

$$f(t) = \cos\left(\tfrac{2}{3}t + 30°\right) + \sin\left(\tfrac{4}{5}t + 45°\right)$$

is periodic. If it is periodic, find the fundamental frequency and the period. What harmonics are present in $f(t)$?

Answer: Periodic with $\omega_0 = \frac{2}{15}$ and period $T_0 = 15\pi$. The fifth and sixth harmonics.   ▽

3.4-3 The Role of Amplitude and Phase Spectra in Wave Shaping

The trigonometric Fourier series of a signal $f(t)$ shows explicitly the sinusoidal components of $f(t)$. We can synthesize $f(t)$ by adding the sinusoids in the spectrum of $f(t)$. To synthesize the square-pulse periodic signal $f(t)$ of Fig. 3.8a, we add successive harmonics in its spectrum step by step and observe the similarity of the resulting signal to $f(t)$. The Fourier series for this function as found in Example 3.4 is

$$f(t) = \frac{1}{2} + \frac{2}{\pi}\left(\cos t - \frac{1}{3}\cos 3t + \frac{1}{5}\cos 5t - \frac{1}{7}\cos 7t + \cdots\right)$$

We start the synthesis with only the first term in the series ($n = 0$), a constant $\frac{1}{2}$ (dc); this is a gross approximation of the square wave, as shown in Fig. 3.11a. In the next step we add the dc ($n = 0$) and the first harmonic (fundamental), which results in a signal shown in Fig. 3.11b. Observe that the synthesized signal somewhat resembles $f(t)$. It is a smoothed-out version of $f(t)$. The sharp corners in $f(t)$ are not reproduced in this signal because sharp corners indicate rapid changes and their reproduction requires rapidly varying (that is, higher frequency) components, which are excluded. Figure 3.11c shows the sum of dc, first, and third harmonics (even harmonics are absent). As we increase the number of harmonics progressively, as illustrated in Figs. 3.11d (sum up to the fifth harmonic) and 3.11e (sum up to

---

†The largest number of which $\frac{a_1}{b_1}$, $\frac{a_2}{b_2}$, $\cdots$, $\frac{a_m}{b_m}$ are integral multiples is the ratio of the GCF (greatest common factor) of the numerators set $(a_1, a_2, \cdots, a_m)$ to the LCM (least common multiple) of the denominator set $(b_1, b_2, \cdots, b_m)$. For instance, for the set $(\frac{2}{3}, \frac{6}{7}, 2)$, the GCF of the numerator set $(2, 6, 2)$ is 2; the LCM of the denominator set $(3, 7, 1)$ is 21. Therefore, $\frac{2}{21}$ is the largest number of which $\frac{2}{3}, \frac{6}{7}$, and 2 are integral multiples.

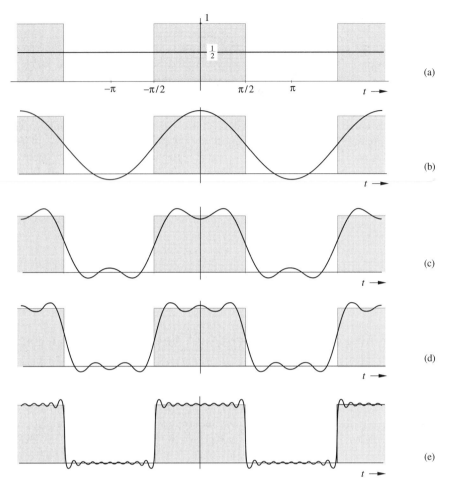

Fig. 3.11 Synthesis of a square pulse periodic signal by successive addition of its harmonics.

the nineteenth harmonic), the edges of the pulses become sharper and the signal resembles $f(t)$ more closely.

Asymptotic Rate of Amplitude Spectrum Decay

The amplitude spectrum indicates the amounts (amplitudes) of various frequency components of $f(t)$. If $f(t)$ is a smooth function, its variations are less rapid. Synthesis of such a function requires predominantly lower-frequency sinusoids and relatively small amounts of rapidly varying (higher frequency) sinusoids. The amplitude spectrum of such a function would decay swiftly with frequency. To synthesize such a function we require fewer terms in the Fourier series for a good approximation. In contrast, a signal with sharp changes, such as jump discontinuities, contains rapid variations and its synthesis requires a relatively large amount of high-frequency components. The amplitude spectrum of such a signal would decay slowly with frequency, and to synthesize such a function, we require many terms in its Fourier series for a good approximation. The square wave $f(t)$ is a discontinuous function with jump discontinuities, and therefore its amplitude spectrum decays rather slowly, as $1/n$ [see Eq. (3.61)]. On the other hand, the triangular pulse periodic signal in Fig. 3.9a is smoother because it is a continuous function (no jump discontinuities). Its spectrum decays rapidly with frequency as $1/n^2$ [see Eq. (3.63)].

We can show[8] that if the first $k - 1$ derivatives of a periodic signal $f(t)$ are continuous and the kth derivative is discontinuous, then its amplitude spectrum C_n decays with frequency at least as rapidly as $1/n^{k+1}$. This result provides a simple and useful means for predicting the asymptotic rate of convergence of the Fourier series. In the case of the square-wave signal (Fig. 3.8a), the zero-th derivative of the signal (the signal itself) is discontinuous, so that $k = 0$. For the triangular periodic signal in Fig. 3.9a, the first derivative is discontinuous; that is, $k = 1$. For this reason, the spectra of these signals decay as $1/n$ and $1/n^2$, respectively.

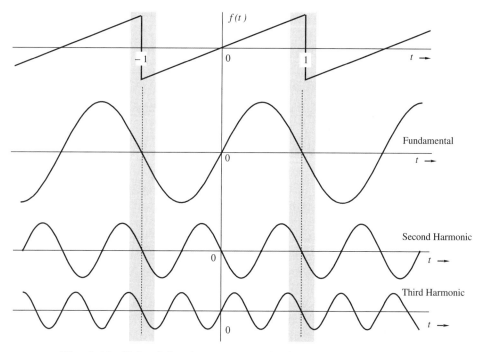

Fig. 3.12 Role of the phase spectrum in shaping a periodic signal.

The Role of the Phase Spectrum

The relationship between the amplitude spectrum and the waveform $f(t)$ is reasonably clear. The relationship between the phase spectrum and the periodic signal waveform is not so direct, however. Yet the phase spectrum plays an equally important role in waveshaping. We can explain this role by considering a signal $f(t)$ that has rapid changes such as jump discontinuities. To synthesize an instantaneous change at a jump discontinuity, the phases of the various sinusoidal components in its spectrum must be such that all (or most) of the harmonic amplitudes will have one sign before the discontinuity and the opposite sign after the discontinuity. This will result in a sharp change in $f(t)$ at the point of discontinuity. We can verify this assertion in any waveform with jump discontinuity. Consider, for example, the sawtooth waveform in Fig. 3.10b. This waveform has a discontinuity at $t = 1$. The Fourier series for this waveform as given in Exercise E3.6b is

$$f(t) = \frac{2A}{\pi} \left[\cos{(\pi t - 90°)} + \tfrac{1}{2} \cos{(2\pi t + 90°)} + \tfrac{1}{3} \cos{(3\pi t - 90°)} \right.$$
$$\left. + \tfrac{1}{4} \cos{(4\pi t + 90°)} + \cdots \right] \tag{3.68}$$

Figure 3.12 shows the first three components of this series. The phases of all the (infinite) components are such that their amplitudes are positive just before $t = 1$ and turn negative just after $t = 1$, the point of discontinuity. The same behavior is also observed at $t = -1$, where similar discontinuity occurs. This sign change in all the harmonics adds up to produce the jump discontinuity. In achieving a sharp change in the waveform, the role of the phase spectrum is crucial. If we try to reconstruct this signal while ignoring the phase spectrum, the result will be a smeared and spread-out waveform. In general the phase spectrum is just as crucial in determining the waveform as is the amplitude spectrum. *The synthesis of any signal $f(t)$ is achieved by using a proper combination of amplitudes and phases of various sinusoids. This unique combination is the Fourier spectrum of $f(t)$.*

Fourier Synthesis of Discontinuous Functions: The Gibbs Phenomenon

Fig. 3.11 shows the square function $f(t)$ and its approximation by a truncated trigonometric Fourier series that includes only the first n harmonics for $n=1, 3, 5$, and 19. The plot of the truncated series approximates closely the function $f(t)$ as n increases, and we expect that the series will converge exactly to $f(t)$ as $n \to \infty$. This is because, as shown in Sec. 3.3, the energy of the difference between $f(t)$ and its Fourier series over one period (the error energy) $\to 0$ as $n \to \infty$. Yet the curious fact, as seen from Fig. 3.11, is that even for large n, the truncated series exhibits an oscillatory behavior and an overshoot approaching a value of about 9% in the vicinity of the discontinuity at the first peak of oscillation. Regardless of the value of n, the overshoot remains at about 9%. This strange behavior appears to contradict the mathematical result derived in Sec. 3.3-2 that the error energy $\to 0$ as $n \to \infty$. In fact, this apparent contradiction puzzled many people at the turn of the century. Josiah Willard Gibbs gave a mathematical explanation of this behavior (now called called the **Gibbs phenomenon**). We can reconcile the two conflicting notions by observing from Fig. 3.11 that the frequency of oscillation of the synthesized signal is n, so the width of the spike with 9% overshoot is approximately $1/2n$. As we increase n, the number of terms in the series, the frequency of oscillation increases and the spike width $1/2n$ diminishes. As $n \to \infty$, the error energy $\to 0$ because the error consists mostly of the spikes, whose widths $\to 0$. Therefore, as $n \to \infty$, the corresponding Fourier series differs from $f(t)$ by about 9% at the immediate left and right of the points of discontinuity, and yet the error energy $\to 0$.

When we use only the first n terms in the Fourier series to synthesize a signal, we are abruptly terminating the series, giving a unit weight to the first n harmonics and zero weight to all the remaining harmonics beyond n. This abrupt termination of the series causes the Gibbs phenomenon in synthesis of discontinuous functions. More discussion on the Gibbs phenomenon, its ramifications, and cure appear in section 4.9.

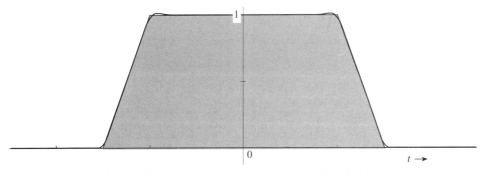

Fig. 3.13 Fourier Synthesis of a continuous signal using first 19 harmonics.

Gibbs phenomenon is present only when there is a jump discontinuity in $f(t)$. When a continuous function $f(t)$ is synthesized using the first n terms of the Fourier series, the synthesized function approaches $f(t)$ for all t as $n \to \infty$. No Gibbs phenomenon appears. Note the absence of the Gibbs phenomenon in Fig. 3.13, where a continuous signal is synthesized using first 19 harmonics. Compare the similar situation for a discontinuous signal in Fig. 3.11.

△ **Exercise E3.8**
By inspection of signals in Figs. 3.7b, 3.10a, and 3.10b, determine the asymptotic rate of decay of their amplitude spectra.
Answer: $1/n, 1/n^2$, and $1/n$, respectively. ▽

A Historical Note on the Gibbs Phenomenon

Normally speaking, troublesome functions with strange behavior are invented by mathematicians, although we rarely see such oddities in practice. In the case of the Gibbs phenomenon, however, the tables were turned. A rather puzzling behavior was observed in such a mundane object as a mechanical wave synthesizer, and then well-known mathematicians of the day were dispatched on the scent of it to discover its hide-out.

Albert Michelson (of Michelson-Morley fame) was an intense, practical man who developed ingenious physical instruments of extraordinary precision, mostly in the field of optics. In 1898 he developed an instrument (the harmonic analyzer) which could compute the first 80 coefficients of the Fourier series of a signal $f(t)$ specified by any graphical description. The harmonic analyzer could also be used as a harmonic synthesizer, which could plot a function $f(t)$ generated by summing the first 80 harmonics (Fourier components) of arbitrary amplitudes and phases. This instrument therefore had the ability of self-checking its operation by analyzing a signal $f(t)$ and then adding the resulting 80 components to see whether the sum yielded a close approximation of $f(t)$.

Michelson found that the instrument checked very well with most of signals analyzed. However, when he tried a discontinuous function, such as a square wave,† a curious behavior was observed. The sum of 80 components showed oscillatory behavior (ringing) with an overshoot of 9% in the vicinity of the points of discontinuity. Moreover, this behavior was a constant feature regardless of the number of terms

---

†Actually it was a periodic sawtooth signal

Albert Michelson (left) and **Willard J. Gibbs** (right).

added. Larger number of terms made the oscillations proportionately faster, but regardless of the number of terms added the overshoot remained 9%. This puzzling behavior caused Michelson to suspect some mechanical defect in his synthesizer. He wrote about his observation in a letter to *Nature* (December 1898). Josiah Willard Gibbs, an eminent mathematical physicist (inventor of vector analysis), and a professor at Yale, investigated and clarified this behavior for a sawtooth periodic signal in a letter to Nature.[9] Later, in 1906, Bôcher generalized the result for any function with discontinuity.[10] It was Bôcher who gave the name *Gibbs phenomenon* to this behavior. Gibbs showed that the peculiar behavior in the synthesis of a square wave was inherent in the behavior of the Fourier series because of nonuniform convergence at the points of discontinuity.

   This, however, is not the end of the story. Both Bôcher and Gibbs were under the impression that this property had remained undiscovered until Gibbs's letter in 1899. It is now known that the Gibbs phenomenon had been observed in 1848 by Wilbraham of Trinity College, Cambridge, who clearly saw the behavior of the sum of the Fourier series components in the periodic sawtooth signal investigated by Gibbs.[11] Apparently, this work was not known to most people, including Gibbs and Bôcher.

3.5 Exponential Fourier Series

   It is shown in Appendix 3C that the set of exponentials $e^{jn\omega_0 t}$ $(n = 0, \pm 1, \pm 2, \ldots)$ is orthogonal over any interval of duration $T_0 = 2\pi/\omega_0$, that is,

$$\int_{T_0} e^{jm\omega_0 t}(e^{jn\omega_0 t})^* \, dt = \int_{T_0} e^{j(m-n)\omega_0 t} \, dt = \begin{cases} 0 & m \neq n \\ T_0 & m = n \end{cases} \qquad (3.69)$$

Moreover, this set is a complete set.[6,7] From Eqs. (3.44) and (3.45), it follows that a signal $f(t)$ can be expressed over an interval of duration T_0 seconds as an exponential Fourier series

$$f(t) = \sum_{n=-\infty}^{\infty} D_n e^{jn\omega_0 t} \tag{3.70}$$

where [see Eq. (3.45)]

$$D_n = \frac{1}{T_0} \int_{T_0} f(t) e^{-jn\omega_0 t}\, dt \tag{3.71}$$

The exponential Fourier series is basically another form of the trigonometric Fourier series. Each sinusoid of frequency ω can be expressed as a sum of two exponentials $e^{j\omega t}$ and $e^{-j\omega t}$. This results in the exponential Fourier series consisting of components of the form $e^{jn\omega_0 t}$ with n varying from $-\infty$ to ∞. The exponential Fourier series in Eq. (3.70) is periodic with period T_0.

In order to see its close connection with the trigonometric series, we shall rederive the exponential Fourier series from the trigonometric Fourier series. A sinusoid in the trigonometric series can be expressed as a sum of two exponentials using Euler's formula:

$$C_n \cos(n\omega_0 t + \theta_n) = \frac{C_n}{2}\left[e^{j(n\omega_0 t + \theta_n)} + e^{-j(n\omega_0 t + \theta_n)} \right]$$

$$= \underbrace{\left(\frac{C_n}{2} e^{j\theta_n} \right)}_{D_n} e^{jn\omega_0 t} + \underbrace{\left(\frac{C_n}{2} e^{-j\theta_n} \right)}_{D_{-n}} e^{-jn\omega_0 t}$$

$$= D_n e^{jn\omega_0 t} + D_{-n} e^{-jn\omega_0 t} \tag{3.72}$$

The compact trigonometric Fourier series of a periodic signal $f(t)$ is given by

$$f(t) = C_0 + \sum_{n=1}^{\infty} C_n \cos(n\omega_0 t + \theta_n)$$

Use of Eq. (3.72) in the above equation (and letting $C_0 = D_0$) yields

$$f(t) = D_0 + \sum_{n=1}^{\infty} D_n e^{jn\omega_0 t} + D_{-n} e^{-jn\omega_0 t}$$

$$= \sum_{n=-\infty}^{\infty} D_n e^{jn\omega_0 t}$$

which is precisely Eq. (3.70) derived earlier. Observe the compactness of expressions (3.70) and (3.71) and compare them to expressions corresponding to the trigonometric Fourier series. These two equations clearly demonstrate the principle virtue of the exponential Fourier series. First, the form of the series is more compact. Second, the mathematical expression for deriving the coefficients of the series is also compact. The exponential series is far more convenient to handle than the trigonometric one. In the system analysis also, the exponential form proves more convenient than the trigonometric form. For these reasons we shall use exponential (rather than trigonometric) representation of signals in the rest of the book.

The connection between the trigonometric and exponential series coefficients is clear in Eq. (3.72):

$$D_n = \tfrac{1}{2}C_n e^{j\theta_n}$$

$$D_{-n} = \tfrac{1}{2}C_n e^{-j\theta_n} \tag{3.73}$$

The connection between the trigonometric and the exponential Series also becomes clear when we substitute $e^{-j\omega t} = \cos \omega t - j \sin \omega t$ in Eq. (3.71) to obtain

$$D_n = \frac{1}{2}(a_n - jb_n) \tag{3.74}$$

■ **Example 3.6**

Find the exponential Fourier series for the signal in Fig. 3.7b (Example 3.3). In this case $T_0 = \pi$, $\omega_0 = 2\pi/T_0 = 2$, and

$$\varphi(t) = \sum_{n=-\infty}^{\infty} D_n e^{j2nt}$$

where

$$D_n = \frac{1}{T_0} \int_{T_0} \varphi(t) e^{-j2nt}\, dt$$

$$= \frac{1}{\pi} \int_0^\pi e^{-t/2}\, e^{-j2nt}\, dt$$

$$= \frac{1}{\pi} \int_0^\pi e^{-(\frac{1}{2}+j2n)t}\, dt$$

$$= \left. \frac{-1}{\pi\left(\frac{1}{2}+j2n\right)} e^{-(\frac{1}{2}+j2n)t}\right|_0^\pi$$

$$= \frac{0.504}{1+j4n} \tag{3.75}$$

and

$$\varphi(t) = 0.504 \sum_{n=-\infty}^{\infty} \frac{1}{1+j4n} e^{j2nt} \tag{3.76a}$$

$$= 0.504\left[1 + \frac{1}{1+j4}e^{j2t} + \frac{1}{1+j8}e^{j4t} + \frac{1}{1+j12}e^{j6t} + \cdots\right.$$

$$\left. + \frac{1}{1-j4}e^{-j2t} + \frac{1}{1-j8}e^{-j4t} + \frac{1}{1-j12}e^{-j6t} + \cdots\right] \tag{3.76b}$$

Observe that the coefficients D_n are complex. Moreover, D_n and D_{-n} are conjugates as expected [see Eq. (3.73)].   ■

3.5-1 Exponential Fourier Spectra

In exponential spectra, we plot coefficients D_n as a function of ω. But since D_n is complex in general, we need two plots: the real and the imaginary parts of D_n, or the magnitude and the angle of D_n. We prefer the latter because of its close connection to the amplitudes and phases of corresponding components of the trigonometric Fourier series. We therefore plot $|D_n|$ vs. ω and $\angle D_n$ vs. ω. This requires that the coefficients D_n be expressed in polar form as $|D_n|e^{j\angle D_n}$.

Comparison of Eqs. (3.51a) and (3.71) (for $n = 0$) shows that

$$D_0 = a_0 = C_0 \tag{3.77a}$$

Equation (3.73) shows that, for real $f(t)$, the twin coefficients D_n and D_{-n} are conjugates, and

$$|D_n| = |D_{-n}| = \frac{1}{2}C_n \qquad n \neq 0 \tag{3.77b}$$

$$\angle D_n = \theta_n \qquad \text{and} \qquad \angle D_{-n} = -\theta_n \tag{3.77c}$$

Thus,

$$D_n = |D_n|e^{j\theta_n} \qquad \text{and} \qquad D_{-n} = |D_n|e^{-j\theta_n} \tag{3.77d}$$

where $|D_n|$ are the magnitudes and $\angle D_n$ are the angles of various exponential components. From Eqs. (3.77) it follows that the magnitude spectrum ($|D_n|$ vs. ω) is an even function of ω and the angle spectrum ($\angle D_n$ vs. ω) is an odd function of ω when $f(t)$ is a real signal.

For the series in Example 3.6 [Eq. (3.76b)], for instance,

$$D_0 = 0.504$$

$$D_1 = \frac{0.504}{1+j4} = 0.122e^{-j75.96°} \implies |D_1| = 0.122, \ \angle D_1 = -75.96°$$

$$D_{-1} = \frac{0.504}{1-j4} = 0.122e^{j75.96°} \implies |D_{-1}| = 0.122, \ \angle D_{-1} = 75.96°$$

and

$$D_2 = \frac{0.504}{1+j8} = 0.0625e^{-j82.87°} \implies |D_2| = 0.0625, \ \angle D_2 = -82.87°$$

$$D_{-2} = \frac{0.504}{1-j8} = 0.0625e^{j82.87°} \implies |D_{-2}| = 0.0625, \ \angle D_{-2} = 82.87°$$

and so on. Note that D_n and D_{-n} are conjugates, as expected [see Eqs. (3.77)].

Figure 3.14 shows the frequency spectra (amplitude and angle) of the exponential Fourier series for the periodic signal $\varphi(t)$ in Fig. 3.7b.

We notice some interesting features of these spectra. First, the spectra exist for positive as well as negative values of ω (the frequency). Second, the amplitude spectrum is an even function of ω and the angle spectrum is an odd function of ω. Finally, we see a close connection between these spectra and the spectra of the corresponding trigonometric Fourier series for $\varphi(t)$ (Figs. 3.7c and d).

What is a Negative Frequency?

The existence of the spectrum at negative frequencies is somewhat disturbing because, by definition, the frequency (number of repetitions per second) is a positive

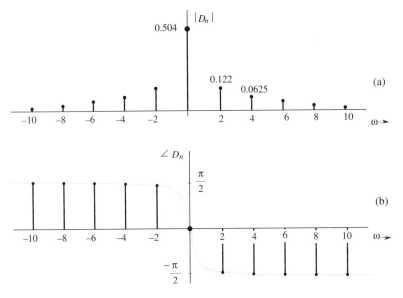

Fig. 3.14 Exponential Fourier spectra for the signal in Fig. 3.7a.

quantity. How do we interpret a negative frequency? Using a trigonometric identity, we can express a sinusoid of a negative frequency $-\omega_0$ as

$$\cos\left(-\omega_0 t + \theta\right) = \cos\left(\omega_0 t - \theta\right)$$

This equation clearly shows that the frequency of a sinusoid $\cos\left(\omega_0 t + \theta\right)$ is $|\omega_0|$, which is a positive quantity. The same conclusion is reached by observing that

$$e^{\pm j\omega_0 t} = \cos\omega_0 t \pm j\sin\omega_0 t$$

Thus, the frequency of exponentials $e^{\pm j\omega_0 t}$ is indeed $|\omega_0|$. How do we then interpret the spectral plots for negative values of ω? A healthier way of looking at the situation is to say that *exponential spectra are a graphical representation of coefficients D_n as a function of ω. Existence of the spectrum at $\omega = -n\omega_0$ is merely an indication of the fact that an exponential component $e^{-jn\omega_0 t}$ exists in the series*. We know that [see Eq. (3.72)] a sinusoid of frequency $n\omega_0$ can be expressed in terms of a pair of exponentials $e^{jn\omega_0 t}$ and $e^{-jn\omega_0 t}$.

Equation (3.77) shows the close connection between the trigonometric spectra (C_n and θ_n) with exponential spectra ($|D_n|$ and $\angle D_n$). The dc components D_0 and C_0 are identical in both spectra. Moreover, the exponential amplitude spectrum $|D_n|$ is half of the trigonometric amplitude spectrum C_n for $n \geq 1$. The exponential angle spectrum $\angle D_n$ is identical to the trigonometric phase spectrum θ_n for $n \geq 0$. We can therefore produce the exponential spectra merely by inspection of trigonometric spectra, and vice versa. The following example demonstrates this feature.

■ **Example 3.7**

The trigonometric Fourier spectra of a certain periodic signal $f(t)$ are shown in Fig. 3.15a. By inspecting these spectra, sketch the corresponding exponential Fourier spectra and verify your results analytically.

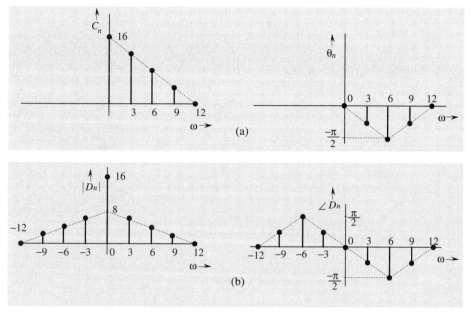

Fig. 3.15 Spectra for Example 3.7.

The trigonometric spectral components exist at frequencies $0, 3, 6$, and 9. The exponential spectral components exist at $0, 3, 6, 9$ and $-3, -6, -9$. Consider first the amplitude spectrum. The dc component remains unchanged; that is, $D_0 = C_0 = 16$. Now $|D_n|$ is an even function of ω and $|D_n| = |D_{-n}| = C_n/2$. Thus, all the remaining spectrum $|D_n|$ for positive n is half the trigonometric amplitude spectrum C_n, and the spectrum $|D_n|$ for negative n is a reflection about the vertical axis of the spectrum for positive n, as shown in Fig. 3.15b.

The angle spectrum $\angle D_n = \theta_n$ for positive n and is $-\theta_n$ for negative n, as depicted in Fig. 3.15b. We shall now verify that both sets of spectra represent the same signal.

Signal $f(t)$ whose trigonometric spectra are shown in Fig. 3.15a, has four spectral components of frequencies $0, 3, 6$, and 9. The dc component is 16. The amplitude and the phase of the component of frequency 3 are 12 and $-\frac{\pi}{4}$, respectively. Therefore, this component can be expressed as $12 \cos{(3t - \frac{\pi}{4})}$. Proceeding in this manner, we can write the Fourier series for $f(t)$ as

$$f(t) = 16 + 12 \cos\left(3t - \tfrac{\pi}{4}\right) + 8 \cos\left(6t - \tfrac{\pi}{2}\right) + 4 \cos\left(9t - \tfrac{\pi}{4}\right)$$

Consider now the exponential spectra in Fig. 3.15b. They contain components of frequencies 0 (dc), ± 3, ± 6, and ± 9. The dc component is $D_0 = 16$. The component e^{j3t} (frequency 3) has magnitude 6 and angle $-\frac{\pi}{4}$. Therefore, this component strength is $6e^{-j\frac{\pi}{4}}$, and it can be expressed as $(6e^{-j\frac{\pi}{4}})e^{j3t}$. Similarly, the component of frequency -3 is $(6e^{j\frac{\pi}{4}})e^{-j3t}$. Proceeding in this manner, $\hat{f}(t)$, the signal corresponding to the spectra in Fig. 3.15b, is

$$\hat{f}(t) = 16 + [6e^{-j\frac{\pi}{4}}e^{j3t} + 6e^{j\frac{\pi}{4}}e^{-j3t}] + [4e^{-j\frac{\pi}{2}}e^{j6t} + 4e^{j\frac{\pi}{2}}e^{-j6t}] + [2e^{-j\frac{\pi}{4}}e^{j9t} + 2e^{j\frac{\pi}{4}}e^{-j9t}]$$

$$= 16 + 6\left[e^{j(3t-\frac{\pi}{4})} + e^{-j(3t-\frac{\pi}{4})}\right] + 4\left[e^{j(6t-\frac{\pi}{2})} + e^{-j(6t-\frac{\pi}{2})}\right] + 2\left[e^{j(9t-\frac{\pi}{4})} + e^{-j(9t-\frac{\pi}{4})}\right]$$

$$= 16 + 12 \cos\left(3t - \tfrac{\pi}{4}\right) + 8 \cos\left(6t - \tfrac{\pi}{2}\right) + 4 \cos\left(9t - \tfrac{\pi}{4}\right)$$

Clearly both sets of spectra represent the same periodic signal. ∎

Bandwidth of a Signal

The difference between the highest and the lowest frequencies of the spectral components of a signal is the **bandwidth** of the signal. The bandwidth of the signal whose exponential spectra are shown in Fig. 3.15b is 9 (in radians). The highest and lowest frequencies are 9 and 0 respectively. Note that the component of frequency 12 has zero amplitude and is nonexistent. Moreover, the lowest frequency is 0, not -9. Recall that the frequencies (in the conventional sense) of the spectral components at $\omega = -3, -6$, and -9 in reality are 3, 6, and 9.† The bandwidth can be more readily seen from the trigonometric spectra in Fig. 3.15a.

■ **Example 3.8**

Find the exponential Fourier series and sketch the corresponding spectra for the impulse train $\delta_{T_0}(t)$ depicted in Fig. 3.16a. From this result sketch the trigonometric spectrum and write the trigonometric Fourier series for $\delta_{T_0}(t)$.

The exponential Fourier series is given by

$$\delta_{T_0}(t) = \sum_{n=-\infty}^{\infty} D_n e^{jn\omega_0 t} \qquad \omega_0 = \frac{2\pi}{T_0} \tag{3.78}$$

where

$$D_n = \frac{1}{T_0} \int_{T_0} \delta_{T_0}(t) e^{-jn\omega_0 t} \, dt$$

Choosing the interval of integration $(\frac{-T_0}{2}, \frac{T_0}{2})$ and recognizing that over this interval $\delta_{T_0}(t) = \delta(t)$,

$$D_n = \frac{1}{T_0} \int_{-T_0/2}^{T_0/2} \delta(t) e^{-jn\omega_0 t} \, dt$$

In this integral the impulse is located at $t = 0$. From the sampling property (1.24a), the integral on the right-hand side is the value of $e^{-jn\omega_0 t}$ at $t = 0$ (where the impulse is located). Therefore

$$D_n = \frac{1}{T_0} \tag{3.79}$$

Substitution of this value in Eq. (3.78) yields the desired exponential Fourier series

$$\delta_{T_0}(t) = \frac{1}{T_0} \sum_{n=-\infty}^{\infty} e^{jn\omega_0 t} \qquad \omega_0 = \frac{2\pi}{T_0} \tag{3.80}$$

Equation (3.79) shows that the exponential spectrum is uniform ($D_n = 1/T_0$) for all the frequencies, as shown in Fig. 3.16b. The spectrum, being real, requires only the amplitude plot. All phases are zero.

To sketch the trigonometric spectrum, we use Eq. (3.77) to obtain

$$C_0 = D_0 = \frac{1}{T_0}$$

$$C_n = 2|D_n| = \frac{2}{T_0} \qquad n = 1, 2, 3, \cdots$$

$$\theta_n = 0$$

---

†Some authors *do* define bandwidth as the difference between the highest and the lowest (negative) frequency in the exponential spectrum. The bandwidth according to this definition is twice that defined here. In reality, this definition defines not the signal bandwidth but the *spectral width* (width of the exponential spectrum of the signal).

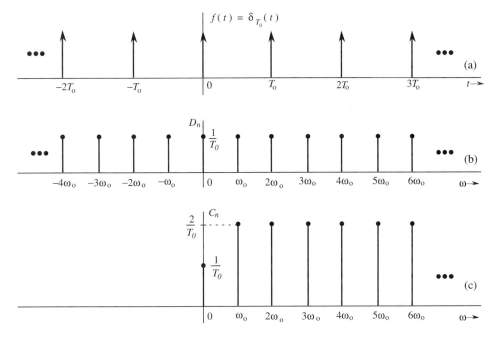

Fig. 3.16 Impulse train and its Fourier spectra.

Figure 3.16c shows the trigonometric Fourier spectrum. From this spectrum we can express $\delta_{T_0}(t)$ as

$$\delta_{T_0}(t) = \frac{1}{T_0}\left[1 + 2(\cos \omega_0 t + \cos 2\omega_0 t + \cos 3\omega_0 t + \cdots)\right] \qquad \omega_0 = \frac{2\pi}{T_0} \qquad (3.81)$$

∎

Effect of Symmetry in Exponential Fourier Series

When $f(t)$ has an even symmetry, $b_n = 0$, and from Eq. (3.74), $D_n = a_n/2$, which is real (positive or negative). Hence, $\angle D_n$ can only be 0 or $\pm\pi$. Moreover, we may compute $D_n = a_n/2$ using Eq. (3.66b), which requires integration over a half period only. Similarly, when $f(t)$ has an odd symmetry, $a_n = 0$, and $D_n = -jb_n/2$ is imaginary (positive or negative). Hence, $\angle D_n$ can only be 0 or $\pm\pi/2$. Moreover, we may compute $D_n = -jb_n/2$ using Eq. (3.67b), which requires integration over a half period only. Note, however, that in the exponential case, we are using the symmetry property indirectly by finding the trigonometric coefficients. We cannot apply it directly to find D_n from eq. (3.71).

△ **Exercise E3.9**

The exponential Fourier spectra of a certain periodic signal $f(t)$ are shown in Fig. 3.17. Determine and sketch the trigonometric Fourier spectra of $f(t)$ by inspection of Fig. 3.17. Now write the (compact) trigonometric Fourier series for $f(t)$.

Answer: $\qquad f(t) = 4 + 6\cos\left(3t - \frac{\pi}{6}\right) + 2\cos\left(6t - \frac{\pi}{4}\right) + 4\cos\left(9t - \frac{\pi}{2}\right) \qquad \triangledown$

△ **Exercise E3.10**

Find the exponential Fourier series and sketch the corresponding Fourier spectrum D_n vs. ω for the full-wave rectified sine wave depicted in Fig. 3.18.

Answer:

$$f(t) = \frac{2}{\pi}\sum_{n=-\infty}^{\infty}\frac{1}{1 - 4n^2}e^{j2nt} \qquad \triangledown$$

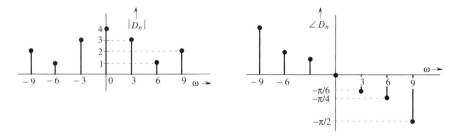

Fig. 3.17 Fourier spectra for the signal in Exercise E3.9.

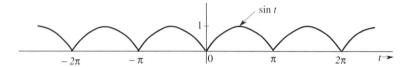

Fig. 3.18 A full-wave rectified sine wave in Exercise E3.10.

3.5-2 Parseval's Theorem

The trigonometric Fourier series of a periodic signal $f(t)$ is given by

$$f(t) = C_0 + \sum_{n=1}^{\infty} C_n \cos{(n\omega_0 t + \theta_n)}$$

Every term on the right-hand side of this equation is a power signal. Moreover, all the Fourier components on the right-hand side are orthogonal over one period. Hence, the power of $f(t)$ is equal to the power of the sum of all the sinusoidal components on the right-hand side. We already demonstrated in Example 1.2 that the power of the sum of sinusoids is equal to the sum of the powers of all the sinusoids. Moreover, the power of a sinusoid of amplitude C_n is $C_n{}^2/2$ regardless of the values of its frequency phase, and the power of a dc term C_0 is $C_0{}^2$. Thus, the power of $f(t)$ is given by

$$P_f = C_0{}^2 + \frac{1}{2} \sum_{n=1}^{\infty} C_n{}^2 \tag{3.82}$$

This result, which is an alternate form of Eq. (3.42), is the **Parseval's theorem** (for Fourier series). It states that the power of a periodic signal is equal to the sum of the powers of its Fourier components.

We can apply the same argument to the exponential Fourier series, which is also made up of the orthogonal components. Hence, the power of a periodic signal $f(t)$ can be expressed as a sum of the powers of its exponential components. In Eq. (1.5d), we showed that the power of an exponential $De^{j\omega_0 t}$ is $|D^2|$. Using this result we can express the power of a periodic signal $f(t)$ in terms of its exponential Fourier series coefficients as

$$P_f = \sum_{n=-\infty}^{\infty} |D_n|^2 \tag{3.83a}$$

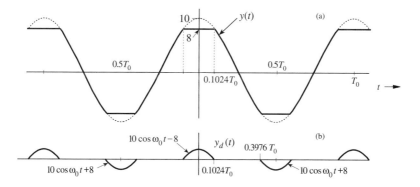

Fig. 3.19 (a) A clipped sinusoid $\cos \omega_0 t$ (b) the distortion component $f_d(t)$ of the signal in (a).

For a real $f(t)$, $|D_{-n}| = |D_n|$. Therefore

$$P_f = D_0{}^2 + 2 \sum_{n=1}^{\infty} |D_n|^2 \tag{3.83b}$$

■ **Example 3.9**

The input signal to an audio amplifier of gain 100 is given by $x(t) = 0.1 \cos \omega_0 t$. Hence, the output is a sinusoid $10 \cos \omega_0 t$. However, the amplifier, being nonlinear at higher amplitude levels, clips all amplitudes beyond ± 8 volts as shown in Fig. 3.19a. We shall determine the harmonic distortion incurred in this operation.

The output $y(t)$ is the clipped signal in Fig. 3.19a. The distortion signal $y_d(t)$, shown in Fig. 3.19b, is the difference between the undistorted sinusoid $10 \cos \omega_0 t$ and the output signal $y(t)$. The signal $y_d(t)$, whose period is T_0 [the same as that of $y(t)$] can be described over the first cycle as

$$y_d(t) = \begin{cases} 10 \cos \omega_0 t - 8 & |t| \le 0.1024 T_0 \\ 10 \cos \omega_0 t + 8 & \frac{T_0}{2} - 0.1024 T_0 \le |t| \le \frac{T_0}{2} + 0.1024 T_0 \\ 0 & \text{everywhere else} \end{cases}$$

Observe that $y_d(t)$ is an even function of t and its mean value is zero. Moreover, $b_n = 0$, and $C_n = a_n$. Hence, its Fourier series can be expressed as

$$y_d(t) = \sum_{n=1}^{\infty} C_n \cos n\omega_0 t$$

As usual, we can compute the coefficients C_n (which is equal to a_n) by integrating $y_d(t) \cos n\omega_0 t$ over one cycle (and then dividing by $2/T_0$). Because $y_d(t)$ has even symmetry, we can find a_n by integrating the expression over half cycle only using Eq. (3.66b). The straightforward evaluation of the appropriate integral yields†

---

†In addition, $y_d(t)$ exhibits half-wave symmetry (see Prob. 3.4-7), where the second half cycle is the negative of the first half cycle. Because of this property, all the even harmonics vanish, and the odd harmonics can be computed by integrating the appropriate expressions over the first half cycle only (from $-T_0/4$ to $T_0/4$) and doubling the resulting values. Moreover, because of even symmetry, we can integrate the appropriate expressions over 0 to $T_0/4$ (instead of from $-T_0/4$ to $T_0/4$) and double the resulting values. In essence, this allows us to compute C_n by integrating the

$$C_n = \begin{cases} \frac{20}{\pi} \left[\frac{\sin[0.6435(n+1)]}{n+1} + \frac{\sin[0.6435(n-1)]}{n-1} \right] - \frac{32}{\pi} \left[\frac{\sin(0.6435n)}{n} \right] & n \text{ odd} \\ 0 & n \text{ even} \end{cases}$$

Computing the coefficients C_1, C_2, C_3, $\cdots$ from this expression, we can write

$$y_d(t) = 1.04 \cos \omega_0 t + 0.733 \cos 3\omega_0 t + 0.311 \cos 5\omega_0 t + \cdots$$

Computing Harmonic Distortion

In this case, we can compute the amount of distortion in the output signal by computing the power of the distortion component $y_d(t)$. Because $y_d(t)$ is an even function of t and because the power in the first half cycle is identical to the power in the second half cycle, we can compute the power by averaging the energy over a quarter cycle. Thus

$$P_{y_d} = \frac{1}{T_0} \int_{-T_0/2}^{T_0/2} y_d^2(t)\, dt$$

$$= \frac{1}{T_0/4} \int_0^{T_0/4} y_d^2(t)\, dt$$

$$= \frac{4}{T_0} \int_0^{0.1024 T_0} (10 \cos \omega_0 t - 8)^2\, dt$$

$$0.865$$

The power of the desired signal $10 \cos \omega_0 t$ is $(10)^2/2 = 50$. Hence, the total harmonic distortion is

$$D_{\text{tot}} = \frac{0.865}{50} \times 100 = 1.73\%$$

The powers of the first harmonic and the third harmonic components of $y_d(t)$ are $(1.04)^2/2 = 0.5408$ and $(0.726)^2/2 = 0.2635$, respectively. Hence, the first harmonic and the third harmonic distortions are

$$D_1 = \frac{0.5408}{50} \times 100 = 1.08\% \text{ and } D_3 = \frac{0.2635}{50} \times 100 = 0.527\% \quad \blacksquare$$

3.6   Numerical Computation of D_n

We can compute D_n numerically using the DFT (the discrete Fourier transform discussed in Sec. 5.2), which uses the samples of a periodic signal $f(t)$ over one period. The sampling interval is T seconds. Hence, there are $N_0 = T_0/T$ number of samples in one period T_0. To find the relationship between D_n and the samples of $f(t)$, consider Eq. (3.71)

---

expression over the quarter cycle only and then quadrupling the resulting values. Thus

$$C_n = a_n = \frac{8}{T_0} \int_0^{0.1024 T_0} [10 \cos \omega_0 t - 8] \cos n\omega_0 t\, dt$$

$$D_n = \frac{1}{T_0} \int_{T_0} f(t)e^{-jn\omega_0 t}\, dt$$

$$= \lim_{T \to 0} \frac{1}{T_0} \sum_{k=0}^{N_0-1} f(kT)e^{-jn\omega_0 kT}\, T$$

$$= \lim_{T \to 0} \frac{1}{N_0} \sum_{k=0}^{N_0-1} f(kT)e^{-jn\Omega_0 k} \tag{3.84}$$

where $f(kT)$ is the kth sample of $f(t)$ and

$$N_0 = \frac{T_0}{T}, \qquad \Omega_0 = \omega_0 T = \frac{2\pi}{N_0} \tag{3.85}$$

In practice, it is impossible to make $T \to 0$ in computing the right-hand side of Eq. (3.84). We can make T small, but not zero, which will cause the data to increase without limit. Thus, we shall ignore the limit on T in Eq. (3.84) with the implicit understanding that T is reasonably small. Nonzero T will result in some computational error, which is inevitable in any numerical evaluation of an integral. The error resulting from nonzero T is called the **aliasing error**, which is discussed in more details in Chapter 5. Thus, we can express Eq. (3.84) as

$$D_n = \frac{1}{N_0} \sum_{k=0}^{N_0-1} f(kT)e^{-jn\Omega_0 k} \tag{3.86a}$$

Now, from Eq. (3.85), $\Omega_0 N_0 = 2\pi$. Hence, $e^{jn\Omega_0(k+N_0)} = e^{jn\Omega_0 k}$ and from Eq. (3.86a), it follows that

$$D_{n+N_0} = D_n \tag{3.86b}$$

Clearly, the Fourier spectrum D_n repeats periodically with period N_0, which causes spectral overlap due to repeating cycles. To understand the nature of this overlap, see Fig. 5.12f. Generally, D_n decays with n and the spectral overlap caused by the periodic repetition will have negligible effect if we use large enough N_0 (the period). The first cycle intersects the second cycle at $n = N_0/2$. Hence, overlapping will be negligible if D_n is very small for $n \geq N_0/2$. Because of the periodicity of D_n, we need to evaluate only N_0 values of D_n over the range $n = -N_0/2$ to $N_0/2 - 1$. However, the DFT (or FFT) computes D_n for $n = 0$ to $N_0 - 1$. The periodicity property $D_{n+N_0} = D_n$ means, beyond $n = N_0/2$, the coefficients represent the values for negative n. For instance, when $N_0 = 32$, $D_{17} = D_{-15}$, $D_{18} = D_{-14}$, $\cdots$, $D_{31} = D_{-1}$. The cycle repeats again from $n = 32$ on.

We can use the efficient FFT (the **fast Fourier transform** discussed in Sec. 5.3) to compute the right-hand side of the above equation. We shall use MATLAB to implement the FFT algorithm. For this purpose, we need samples of $f(t)$ over one period starting at $t = 0$. In this algorithm, it is also preferable (although not necessary) that N_0 be a power of 2, that is $N_0 = 2^m$, where m is an integer.

⊙ **Computer Example C3.1**
    Compute and plot the trigonometric and exponential Fourier spectra for the periodic signal in Fig. 3.7b (Example 3.3).

The samples of $f(t)$ start at $t = 0$ and the last (N_0-th) sample is at $t = T_0 - T$ (the last sample is not at $t = T_0$ because the sample at $t = 0$ is identical to the sample at $t = T_0$, and the next cycle begins at $t = T_0$). At the points of discontinuity, the sample value is taken as the average of the values of the function on two sides of the discontinuity. Thus, in the present case, the first sample (at $t = 0$) is not 1, but $(e^{-\pi/2} + 1)/2 = 0.604$. To determine N_0, we require that D_n for $n \geq N_0/2$ to be negligible. Because $f(t)$ has a jump discontinuity, D_n decays rather slowly as $1/n$. Hence, choice of $N_0 = 200$ is acceptable because the $(N_0/2)$-nd (100th) harmonic is about 0.01 (about 1%) of the fundamental. However, we also require N_0 to be power of 2. Hence, we shall take $N_0 = 256 = 2^8$.

We write and save a MATLAB file (or program) c31.m to compute and plot the exponential Fourier coefficients.

```
% (c31.m)
%M is the number of coefficients to be computed
T0=pi;N0=256;T=T0/N0;M=10;
t=0:T:T*(N0-1); t=t';
f=exp(-t/2);f(1)=0.604;
% fft(f) is the FFT [the sum on the right-hand side of Eq. (3.86)]
Dn=fft(f)/N0
[Dnangle,Dnmag]=cart2pol(real(Dn),imag(Dn));
k=0:length(Dn)-1;k=k';
subplot(211),stem(k,Dnmag)
subplot(212), stem(k,Dnangle)
```

To compute trigonometric Fourier series coefficients, we recall the program c31.m along with commands to convert D_n into C_n and θ_n.

```
c31;clg
C0=Dnmag(1); Cn=2*Dnmag(2:M);
Amplitudes=[C0;Cn]
Angles=Dnangle(1:M);
Angles=Angles*(180/pi);
disp('Amplitudes Angles')
[Amplitudes Angles]
% To Plot the Fourier coefficients
k=0:length(Amplitudes)-1; k=k';
subplot(211),stem(k,Amplitudes)
subplot(212), stem(k,Angles)
```

```
ans =

   Amplitudes Angles
   0.5043         0
   0.2446   -75.9622
   0.1251   -82.8719
   0.0837   -85.2317
   0.0629   -86.4175
   0.0503   -87.1299
   0.0419   -87.6048
   0.0359   -87.9437
   0.0314   -88.1977
   0.0279   -88.3949    ⊙
```

3.7 LTIC System Response to periodic Inputs

A periodic signal can be expressed as a sum of everlasting exponentials (or sinusoids). We also know how to find the response of an LTIC system to an everlasting exponential. From this information we can readily determine the response of an LTIC system to periodic inputs. A periodic signal $f(t)$ with period T_0 can be expressed as an exponential Fourier series

$$f(t) = \sum_{n=-\infty}^{\infty} D_n e^{jn\omega_0 t} \qquad \omega_0 = \frac{2\pi}{T_0}$$

In Sec. 2.4-3, we showed that the response of an LTIC system with transfer function $H(s)$ to an everlasting exponential input e^{st} is also an everlasting exponential $H(s)e^{st}$. Therefore, the system response to everlasting exponential $e^{j\omega t}$ is an everlasting exponential $H(j\omega)e^{j\omega t}$. This input-output pair can be displayed as[‡]

$$\underbrace{e^{j\omega t}}_{\text{input}} \Longrightarrow \underbrace{H(j\omega)e^{j\omega t}}_{\text{output}}$$

Therefore, from linearity property

$$\underbrace{\sum_{n=-\infty}^{\infty} D_n e^{jn\omega_0 t}}_{\text{input } f(t)} \Longrightarrow \underbrace{\sum_{n=-\infty}^{\infty} D_n H(jn\omega_0)e^{jn\omega_0 t}}_{\text{response } y(t)} \qquad (3.87)$$

The response $y(t)$ is obtained in the form of an exponential Fourier series, and is therefore a periodic signal of the same period as that of the input.

We shall demonstrate the utility of these results by the following example.

■ Example 3.9

A full-wave rectifier (Fig. 3.20a) is used to obtain a dc signal from a sinusoid $\sin t$. The rectified signal $f(t)$, depicted in Fig. 3.18, is applied to the input of a a low-pass RC filter, which suppresses the time-varying component and yields a dc component with some residual ripple. Find the filter output $y(t)$. Find also the dc output and the rms value of the ripple voltage.

First, we shall find the Fourier series for the rectified signal $f(t)$, whose period is $T_0 = \pi$. Consequently, $\omega_0 = 2$, and

$$f(t) = \sum_{n=-\infty}^{\infty} D_n e^{j2nt}$$

where

$$D_n = \frac{1}{\pi} \int_0^{\pi} \sin t \; e^{-j2nt} \, dt = \frac{2}{\pi(1 - 4n^2)} \qquad (3.88)$$

[‡]This result applies only to the asymptotically stable systems. This is because when $s = j\omega$, the integral on the right-hand side of Eq. (2.48) does not converge for unstable systems. Moreover, for marginally stable systems also, that integral does not converge in the ordinary sense, and $H(j\omega)$ cannot be obtained by replacing s in $H(s)$ with $j\omega$.

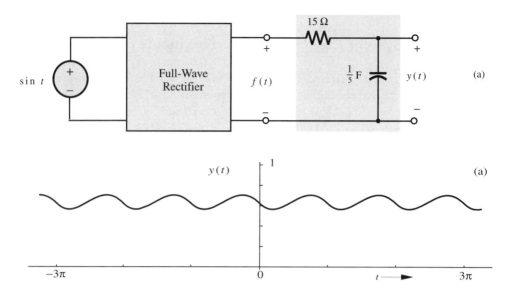

Fig. 3.20 A full-wave rectifier with a lowpass filter and its output.

Therefore

$$f(t) = \sum_{-\infty}^{\infty} \frac{2}{\pi(1 - 4n^2)} e^{j2nt}$$

Next, we find the transfer function of the RC filter in Fig. 3.20a. This filter is identical to the RC circuit in Example 1.11 (Fig. 1.32) for which the differential equation relating the output (capacitor voltage) to the input $f(t)$ was found to be [Eq. (1.60)]

$$(3D + 1)y(t) = f(t)$$

The transfer function $H(s)$ for this system is found from Eq. (2.50) as

$$H(s) = \frac{1}{3s + 1}$$

and

$$H(j\omega) = \frac{1}{3j\omega + 1} \tag{3.89}$$

From Eq. (3.87), the filter output $y(t)$ can be expressed as (with $\omega_0 = 2$)

$$y(t) = \sum_{n=-\infty}^{\infty} D_n H(jn\omega_0)e^{jn\omega_0 t} = \sum_{n=-\infty}^{\infty} D_n H(j2n)e^{j2nt}$$

Substituting D_n and $H(j2n)$ from Eqs. (3.88) and (3.89) in the above equation, we obtain

$$y(t) = \sum_{n=-\infty}^{\infty} \frac{2}{\pi(1 - 4n^2)(j6n + 1)} e^{j2nt} \tag{3.90}$$

Note that the output $y(t)$ is also a periodic signal given by the exponential Fourier series on the right-hand side. The output is numerically computed from the above equation and plotted in Fig. 3.20b.

The output Fourier series coefficient corresponding to $n = 0$ is the dc component of the output, given by $2/\pi$. The remaining terms in the Fourier series constitute the unwanted component called the ripple. We can determine the rms value of the ripple voltage by finding the power of the ripple component using Eq. (3.83). The power of the ripple is the power of all the components except the dc $(n = 0)$. Note that $\hat{D}_n$, the exponential Fourier coefficient for the output $y(t)$ is

$$\hat{D}_n = \frac{2}{\pi(1 - 4n^2)(j6n + 1)}$$

Therefore, from Eq. (3.83b), we have

$$P_{ripple} = 2\sum_{n=1}^{\infty}|D_n|^2 = 2\sum_{n=1}^{\infty}\left|\frac{2}{\pi(1 - 4n^2)(j6n + 1)}\right|^2 = \frac{8}{\pi^2}\sum_{n=1}^{\infty}\frac{1}{(1 - 4n^2)^2(36n^2 + 1)}$$

Numerical computation of the right-hand side yields

$$P_{ripple} = 0.0025, \quad \text{and the ripple rms value} = \sqrt{P_{ripple}} = 0.05$$

This shows that the rms ripple voltage is 5% of the amplitude of the input sinusoid. ∎

Why Use Exponentials?

The exponential Fourier series is just another way of representing trigonometric Fourier series (or vice versa). The two forms carry identical information—no more, no less. The reasons for preferring the exponential form have already been mentioned: This form is more compact, and the expression for deriving the exponential coefficients is also more compact, as compared to those in the trigonometric series. Furthermore, the system response to exponential signals is also simpler (more compact) than the system response to sinusoids. In addition, the mathematical manipulation and handling of exponential form proves much easier than the trigonometric form in the area of signals as well as systems. For these reasons, in our future discussion we shall use the exponential form exclusively.

A minor disadvantage of the exponential form is that it cannot be visualized as easily as sinusoids. For intuitive and qualitative understanding, the sinusoids have the edge over exponentials. Fortunately, this difficulty can be overcome readily because of close connection between exponential and Fourier spectra. For the purpose of mathematical analysis we shall continue to use exponential signals and spectra, but to understand the physical situation intuitively or qualitatively we shall speak in terms of sinusoids and trigonometric spectra. Thus, in future discussions, although all mathematical manipulation will be in terms of exponential spectra, we shall speak of exponential and sinusoids interchangeably when discussing intuitive and qualitative insights and the understanding of physical situations. This is an important point; readers should make an extra effort to familiarize themselves with the two forms of spectra, their relationships, and their convertibility.

Dual Personality of a Signal

The discussion so far shows that a periodic signal has a dual personality—the time domain and frequency domain. It can be described by its waveform or by its Fourier spectra. The time- and frequency-domain descriptions provide complementary insights into a signal. For in-depth perspective, we need to understand both of these identities. It is important for the reader to learn to think of a signal from both of these perspectives. In the next chapter, we shall see that aperiodic signals also have this dual personality. Moreover, we shall show that even LTI systems have this dual personality, which offers complementary insights into the system behavior.

Limitations of the Fourier Series Method of Analysis

We have developed here a method of representing a periodic signal as a weighted sum of everlasting exponentials whose frequencies lie along the $j\omega$-axis in the s-plane. This representation (Fourier series) is valuable in many applications. However, as a tool for analyzing linear systems, it has serious limitations and consequently has limited utility for the following reasons:

1. The Fourier series can be used only for periodic inputs. All practical inputs are aperiodic (remember that a periodic signal starts at $t = -\infty$).

2. This technique can be applied readily to asymptotically stable systems. It cannot handle so easily unstable or even marginally stable systems (see the footnote on p. 219).

The first limitation can be overcome by representing aperiodic signals in terms of everlasting exponentials. This representation can be achieved through the Fourier integral, which may be considered as an extension of the Fourier series. We shall therefore use the Fourier series as a stepping stone to the Fourier integral developed in the next chapter. The second limitation can be overcome by using exponentials e^{st} where s is not restricted to the imaginary axis, but is free to take on complex values. This generalization leads to the Laplace integral, discussed in Chapter 6 (the Laplace transform).

3.8 Appendix

Appendix 3A: Derivation of Eq. (3.39)

The error $e(t)$ in the approximation (3.37)

$$e(t) = f(t) - \sum_{n=1}^{N} c_n x_n(t) \tag{3.91}$$

The error energy E_e is

$$E_e = \int_{t_1}^{t_2} e^2(t)\, dt = \int_{t_1}^{t_2} \left[f(t) - \sum_{n=1}^{N} c_n x_n(t) \right]^2 dt \tag{3.92}$$

Since E_e is a function of N parameters $c_1, c_2, \ldots, c_N$, to minimize E_e, a necessary condition is

$$\frac{\partial E_e}{\partial c_i} = \frac{\partial}{\partial c_i} \int_{t_1}^{t_2} \left[f(t) - \sum_{n=1}^{N} c_n x_n(t) \right]^2 dt = 0 \qquad i = 1, 2, \ldots, N \tag{3.93}$$

When we expand the integrand, we find that all the cross-product terms arising from the orthogonal signals are zero by virtue of orthogonality; that is, all terms of the form $\int x_m(t)x_n(t)\,dt$ with $m \neq n$ vanish. Similarly, the derivative with respect to c_i of all terms not containing c_i is zero. For each i, this observation leaves only two nonzero terms in Eq. (3.93):

$$\frac{\partial}{\partial c_i} \int_{t_1}^{t_2} \left[-2c_i f(t)x_i(t) + c_i^2 x_i^2(t) \right] dt = 0$$

or

$$-2 \int_{t_1}^{t_2} f(t)x_i(t)\,dt + 2c_i \int_{t_1}^{t_2} x_i^2(t)\,dt = 0 \qquad i = 1, 2, \ldots, n$$

Therefore

$$c_i = \frac{\displaystyle\int_{t_1}^{t_2} f(t)x_i(t)\,dt}{\displaystyle\int_{t_1}^{t_2} x_i^2(t)\,dt} = \frac{1}{E_i} \int_{t_1}^{t_2} f(t)x_i(t)\,dt \qquad i = 1, 2, \ldots, N \qquad (3.94)$$

Derivation of Eq. (3.40)

$$E_e = \int_{t_1}^{t_2} \left[f(t) - \sum_{n=1}^{N} c_n x_n(t) \right]^2 dt$$

$$= \int_{t_1}^{t_2} f^2(t)\,dt + \sum_{n=1}^{N} c_n^2 \int_{t_1}^{t_2} x_n^2(t)\,dt - 2 \sum_{n=1}^{N} c_n \int_{t_1}^{t_2} f(t)x_n(t)\,dt$$

Substitution of Eqs. (3.36) and (3.94) in this equation yields

$$E_e = \int_{t_1}^{t_2} f^2(t)\,dt + \sum_{n=1}^{N} c_n^2 E_n - 2 \sum_{n=1}^{N} c_n^2 E_n$$

$$= \int_{t_1}^{t_2} f^2(t)\,dt - \sum_{n=1}^{N} c_n^2 E_n \qquad (3.95)$$

As $N \to \infty$ for a complete orthogonal set, $E_e \to 0$, and the energy of $f(t)$ is equal to the sum of energies of all the orthogonal components $c_1 x_1(t)$, $c_2 x_2(t)$, $c_3 x_3(t)$, $\cdots$.

Appendix 3B: Orthogonality of the Trigonometric Signal Set

Consider an integral I defined by

$$I = \int_{T_0} \cos n\omega_0 t \, \cos m\omega_0 t \, dt \qquad (3.96a)$$

where $\int_{T_0}$ stands for integration over any contiguous interval of T_0 seconds. By using a trigonometric identity (see Sec. B.7-6), Eq. (3.96a) can be expressed as

$$I = \frac{1}{2} \left[\int_{T_0} \cos(n+m)\omega_0 t \, dt + \int_{T_0} \cos(n-m)\omega_0 t \, dt \right] \qquad (3.96b)$$

Since $\cos \omega_0 t$ executes one complete cycle during any interval of T_0 duration, $\cos (n+m)\omega_0 t$ executes $(n+m)$ complete cycles during any interval of duration T_0. Therefore, the first integral in Eq. (3.96b), which represents the area under $(n+m)$ complete cycles of a sinusoid, equals zero. The same argument shows that the second integral in Eq. (3.96b) is also zero, except when $n = m$. Hence I in Eq. (3.96) is zero for all $n \neq m$. When $n = m$, the first integral in Eq. (3.96b) is still zero, but the second integral yields

$$I = \frac{1}{2} \int_{T_0} dt = \frac{T_0}{2}$$

Thus

$$\int_{T_0} \cos n\omega_0 t \cos m\omega_0 t \, dt = \begin{cases} 0 & n \neq m \\ \frac{T_0}{2} & m = n \neq 0 \end{cases} \tag{3.97a}$$

Using similar arguments, we can show that

$$\int_{T_0} \sin n\omega_0 t \sin m\omega_0 t \, dt = \begin{cases} 0 & n \neq m \\ \frac{T_0}{2} & n = m \neq 0 \end{cases} \tag{3.97b}$$

and

$$\int_{T_0} \sin n\omega_0 t \cos m\omega_0 t \, dt = 0 \quad \text{for all } n \text{ and } m \tag{3.97c}$$

Appendix 3C: Orthogonality of the Exponential Signal Set

The set of exponentials $e^{jn\omega_0 t}$ $(n = 0, \pm1, \pm2, \ldots)$ is orthogonal over any interval of duration T_0, that is,

$$\int_{T_0} e^{jm\omega_0 t} (e^{jn\omega_0 t})^* \, dt = \int_{T_0} e^{j(m-n)\omega_0 t} \, dt = \begin{cases} 0 & m \neq n \\ T_0 & m = n \end{cases} \tag{3.98}$$

Let the integral on the left-hand side of Eq. (3.98) be I.

$$I = \int_{T_0} e^{jm\omega_0 t} (e^{jn\omega_0 t})^* \, dt = \int_{T_0} e^{j(m-n)\omega_0 t} \, dt \tag{3.99}$$

The case $m = n$ is trivial. In this case the integrand is unity, and $I = T_0$. When $m \neq n$

$$I = \frac{1}{j(m-n)\omega_0} e^{j(m-n)\omega_0 t} \Big|_{t_1}^{t_1+T_0} = \frac{1}{j(m-n)\omega_0} e^{j(m-n)\omega_0 t_1} \left[e^{j(m-n)\omega_0 T_0} - 1 \right] = 0$$

The last result follows from the fact that $\omega_0 T_0 = 2\pi$, and $e^{j2\pi k} = 1$ for all integral values of k.

3.9 Summary

This chapter discusses the foundations of signal representation in terms of its components. There is a perfect analogy between vectors and signals; the analogy is so strong that the term 'analogy' understates the reality. Signals are not just *like* vectors. Signals *are* vectors. The inner or scalar product of two (real) signals is the area under the product of the two signals. If this inner or scalar product is zero, the signals are said to be orthogonal. A signal $f(t)$ has a component $cx(t)$, where c is the inner product of $f(t)$ and $x(t)$ divided by E_x, the energy of $x(t)$.

A good measure of similarity of two signals $f(t)$ and $x(t)$ is the correlation co-efficient c_n, which is equal to the inner product of $f(t)$ and $x(t)$ divided by $\sqrt{E_f E_x}$. It can be shown that $-1 \leq c_n \leq 1$. The maximum similarity ($c_n = 1$) occurs only when the two signals have the same waveform within a (positive) multiplicative constant, that is, when $f(t) = Kx(t)$. The maximum dissimilarity ($c_n = -1$) oc-curs only when $f(t) = -Kx(t)$. Zero similarity ($c_n = 0$) occurs when the signals are orthogonal. In binary communication, where we are required to distinguish between the two known waveforms in the presence of noise and distortion, select-ing the two waveforms with maximum dissimilarity ($c_n = -1$) provides maximum distinguishability.

Just as a vector can be represented by the sum of its orthogonal components in a complete orthogonal vector space, a signal can also be represented by the sum of its orthogonal components in a complete orthogonal signal space. Such a representation is known as the *generalized Fourier series* representation. A vector can be represented by its orthogonal components in many different ways, depending on the coordinate system used. Similarly a signal can be represented in terms of different orthogonal signal sets of which the trigonometric and the exponential signal sets are two examples. We have shown that the trigonometric and exponential Fourier series are periodic with period equal to that of the fundamental in the set. In this chapter we have shown how a periodic signal can be represented as a sum of (everlasting) sinusoids or exponentials. If the frequency of a periodic signal is ω_0, then it can be expressed as a sum of the sinusoid of frequency ω_0 and its harmonics (trigonometric Fourier series). We can reconstruct the periodic signal from a knowledge of the amplitudes and phases of these sinusoidal components (amplitude and phase spectra).

If a periodic signal has an even symmetry, the Fourier series contains only cosine terms. In contrast, if the signal has an odd symmetry, the Fourier series contains only sine terms. If a periodic signal has a jump discontinuity, the signal is not smooth and requires significant high frequency components to synthesize jumps. Consequently, its amplitude spectrum decays slowly with frequency as $1/n$. If the signal has no jump discontinuities, but its first derivative has a jump discontinuity, the signal is smoother, and its amplitude spectrum decays faster as $1/n^2$. If neither the signal nor its first derivative has jump discontinuities, but the second derivative has discontinuities, then the signal is even more smooth, and its amplitude spectrum decays still faster as $1/n^3$, and so on.

A sinusoid can be expressed in terms of exponentials. Therefore, the Fourier se-ries of a periodic signal can also be expressed as a sum of exponentials (the exponen-tial Fourier series). The exponential form of the Fourier series and the expressions for the series coefficients are more compact than those of the trigonometric Fourier series. Also, the response of LTIC systems to an exponential input is simpler than that for a sinusoidal input. Moreover, the exponential form of representation lends itself better to mathematical manipulations than does the trigonometric form. For these reasons, the exponential form of the series is preferred in modern practice in the areas of signals and systems.

The plots of amplitudes and angles of various exponential components of the Fourier series as functions of the frequency are the exponential Fourier spectra (amplitude and angle spectra) of the signal. Because a sinusoid $\cos \omega_0 t$ can be represented as a sum of two exponentials, $e^{j\omega_0 t}$ and $e^{-j\omega_0 t}$, the frequencies in the exponential spectra range from $-\infty$ to ∞. By definition, the frequency of a signal is always a positive quantity. Presence of a spectral component of a negative frequency $-n\omega_0$ merely indicates that the Fourier series contains terms of the form $e^{-jn\omega_0 t}$. The spectra of the trigonometric and exponential Fourier series are closely related, and one can be found by the inspection of the other.

The Fourier series coefficients C_n or D_n may be computed numerically using the discrete Fourier transform (DFT), which can be implemented by an efficient FFT (fast Fourier transform) algorithm. This method uses N_0 uniform samples of $f(t)$ over one period starting at $t = 0$.

In Sec. 3.7, we discuss a method of finding the response of an LTIC system to a periodic input signal. The periodic input is expressed as an exponential Fourier series, which consists of everlasting exponentials of the form $e^{jn\omega_0 t}$. We also know that response of an LTIC system to an everlasting exponential $e^{jn\omega_0 t}$ is $H(jn\omega_0)e^{jn\omega_0 t}$. The system response is the sum of the system's responses to all the exponential components in the Fourier series for the input. The response is, therefore, also an exponential Fourier series. Thus, the response is also a periodic signal of the same period as that of the input.

References

1. Lathi, B.P., *Modern Digital And Analog Communication Systems*, 2nd Ed., Holt, Rinehart and Winston, New York, 1989.

2. Bell, E. T., *Men of Mathematics*, Simon and Schuster, New York, 1937.

3. Durant, Will, and Ariel, *The Age of Napoleon*, History of Civilization, Part XI, Simon and Schuster, New York, 1975.

4. Calinger, R., 4th ed., *Classics of Mathematics*, Moore Publishing Co., Oak Park, Il., 1982.

5. Lanczos, C., *Discourse on Fourier Series*, Oliver Boyd Ltd., London, 1966.

6. Walker, P.L., *The Theory of Fourier Series and Integrals*, Wiley-Interscience, New York, 1986.

7. Churchill, R. V., and J. W. Brown, *Fourier Series and Boundary Value Problems*, 3rd Ed., McGraw-Hill, New York, 1978.

8. Guillemin, E. A., *Theory of Linear Physical Systems*, Wiley, New York, 1963.

9. Gibbs, W. J., *Nature*, vol. 59, p. 606, April 1899.

10. Bôcher, M., Annals of Mathematics, (2), vol. 7, 1906.

11. Carslaw, H. S., Bulletin, American Mathematical Society, vol. 31, pp. 420-424, Oct. 1925.

Problems

3.1-1 Derive Equation (3.6) in an alternate way by observing that $\mathbf{e} = (\mathbf{f} - c\mathbf{x})$ and

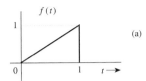
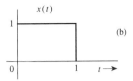

Fig. P3.1-2

$|\mathbf{e}|^2 = (\mathbf{f}-c\mathbf{x}) \cdot (\mathbf{f}-c\mathbf{x}) = |\mathbf{f}|^2 + c^2|\mathbf{x}|^2 - 2c\mathbf{f}\cdot\mathbf{x}$. Hint: Find the value of c to minimize $|\mathbf{e}|^2$.

3.1-2 **(a)** For the signals $f(t)$ and $x(t)$ depicted in Fig. P3.1-2, find the component of the form $x(t)$ contained in $f(t)$. In other words find the optimum value of c in the approximation $f(t) \approx cx(t)$ so that the error signal energy is minimum.
(b) Find the error signal $e(t)$ and its energy E_e. Show that the error signal is orthogonal to $x(t)$, and that $E_f = c^2 E_x + E_e$. Can you explain this result in terms of vectors.

3.1-3 For the signals $f(t)$ and $x(t)$ shown in Fig. P3.1-2, find the component of the form $f(t)$ contained in $x(t)$. In other words, find the optimum value of c in the approximation $x(t) \approx cf(t)$ so that the error signal energy is minimum. What is the error signal energy?

3.1-4 Repeat Prob. 3.1-2 if $x(t)$ is a sinusoid pulse shown in Fig. P3.1-4.

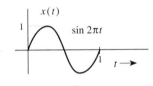

Fig. P3.1-4

3.1-5 If $x(t)$ and $y(t)$ are orthogonal, then show that the energy of the signal $x(t) + y(t)$ is identical to the energy of the signal $x(t) - y(t)$ and is given by $E_x + E_y$. Explain this result using vector concepts. In general, show that for orthogonal signals $x(t)$ and $y(t)$ and for any pair of arbitrary constants c_1 and c_2, the energies of $c_1 x(t) + c_2 y(t)$ and $c_1 x(t) - c_2 y(t)$ are identical, given by $c_1^2 E_x + c_2^2 E_y$.

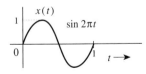
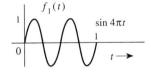
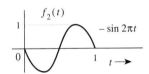

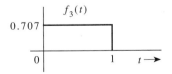
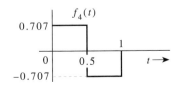

Fig. P3.2-1

3.2-1   Find the correlation coefficient c_n of signal $x(t)$ and each of the four pulses $f_1(t)$, $f_2(t)$, $f_3(t)$, and $f_4(t)$ depicted in Fig. P3.2-1. Which pair of pulses would you select for a binary communication in order to provide maximum margin against noise along the transmission path?

3.3-1   Let $x_1(t)$ and $x_2(t)$ be two signals orthonormal (that is, with unit energies) over an interval from $t = t_1$ to t_2. Consider a signal $f(t)$ where

$$f(t) = c_1 x_1(t) + c_2 x_2(t)   t_1 \le t \le t_2$$

This signal can be represented by a two-dimensional vector $\mathbf{f}(c_1, c_2)$.
(a) Determine the vector representation of the following six signals in the two-dimensional vector space:

    (i) $f_1(t) = 2x_1(t) - x_2(t)$ **(iv)** $f_4(t) = x_1(t) + 2x_2(t)$

    (ii) $f_2(t) = -x_1(t) + 2x_2(t)$ **(v)** $f_5(t) = 2x_1(t) + x_2(t)$

    (iii) $f_3(t) = -x_2(t)$ **(vi)** $f_6(t) = 3x_1(t)$

(b) Point out pairs of mutually orthogonal vectors among these six vectors. Verify that the pairs of signals corresponding to these orthogonal vectors are also orthogonal.

3.4-1   **(a)** Sketch the signal $f(t) = t^2$ for all t and find the trigonometric Fourier series $\varphi(t)$ to represent $f(t)$ over the interval $(-1, 1)$. Sketch $\varphi(t)$ for all values of t.

3.4-2   **(a)** Sketch the signal $f(t) = t$ for all t and find the trigonometric Fourier series $\varphi(t)$ to represent $f(t)$ over the interval $(-\pi, \pi)$. Sketch $\varphi(t)$ for all values of t.

3.4-3   For each of the periodic signals shown in Fig. P3.4-3, find the compact trigonometric Fourier series and sketch the amplitude and phase spectra. If either the sine or cosine terms are absent in the Fourier series, explain why.

3.4-4   **(a)** Find the trigonometric Fourier series for $x(t)$ shown in Fig. P3.4-4.
(b) The signal $x(t)$ is the time-inverted signal $\varphi(t)$ in Fig. 3.7b. Thus, $x(t) = \varphi(-t)$. Hence, the Fourier series for $x(t)$ can be obtained by replacing t with $-t$ in the Fourier series [Eq. (3.56)] for $\varphi(t)$. Verify that the Fourier series thus obtained is identical to that found in part **(a)**.
(c) Show that, in general, time-inversion of a periodic signal does not affect the amplitude spectrum, and the phase spectrum is also unchanged except for the change of sign.

3.4-5   **(a)** Find the trigonometric Fourier series for the periodic signal $x(t)$ depicted in Fig. P3.4-5.
(b) The signal $x(t)$ can be obtained by time-compressing the signal $\varphi(t)$ in Fig. 3.7b by a factor 2. Thus, $x(t) = \varphi(2t)$. Hence, the Fourier series for $x(t)$ can be obtained by replacing t with $2t$ in the Fourier series [Eq. (3.56)] for $\varphi(t)$. Verify that the Fourier series thus obtained is identical to that found in part **(a)**.
(c) Show that, in general, time-compression of a periodic signal by a factor a expands the Fourier spectra by the same factor a. In other words C_0, C_n, and θ_n remain unchanged, but the fundamental frequency is increased by the factor a, thus expanding the spectrum. Similarly time-expansion of a periodic signal by a factor a compresses its Fourier spectra by the factor a.

3.4-6   **(a)** Find the trigonometric Fourier series for the periodic signal $g(t)$ in Fig. P3.4-6. Take advantage of the symmetry.

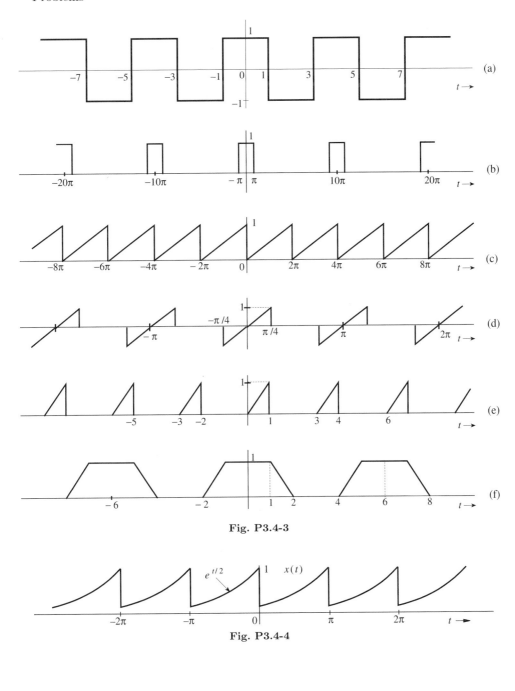

Fig. P3.4-3

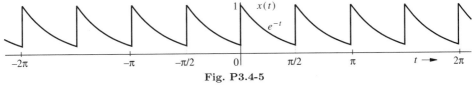

Fig. P3.4-4

Fig. P3.4-5

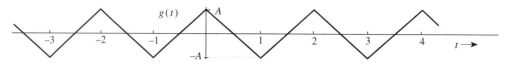

Fig. P3.4-6

(b) Observe that $g(t)$ is identical to $f(t)$ in Fig. 3.9a left-shifted by 0.5 second. Therefore, $g(t) = f(t+0.5)$, and the Fourier series for $g(t)$ can be found by replacing t with $t+0.5$ in Eq. (3.63) [the Fourier series for $f(t)$]. Verify that the Fourier series thus obtained is identical to that found in part **(a)**.

(c) Show that, in general, a time shift of T seconds of a periodic signal does not affect the amplitude spectrum. However, the phase of the nth harmonic is decreased (increased) by $n\omega_0 T$ for a delay (advance) of T seconds.

3.4-7 If the two halves of one period of a periodic signal are of identical shape except that the one is the negative of the other, the periodic signal is said to have a **half-wave symmetry**. If a periodic signal $f(t)$ with a period T_0 satisfies the half-wave symmetry condition, then

$$f\left(t - \frac{T_0}{2}\right) = -f(t)$$

In this case, show that all the even-numbered harmonics vanish, and that the odd-numbered harmonic coefficients are given by

$$a_n = \frac{4}{T_0} \int_0^{T_0/2} f(t) \cos n\omega_0 t \, dt \quad \text{and} \quad b_n = \frac{4}{T_0} \int_0^{T_0/2} f(t) \sin n\omega_0 t \, dt$$

Using these results, find the Fourier series for the periodic signals in Fig. P3.4-7.

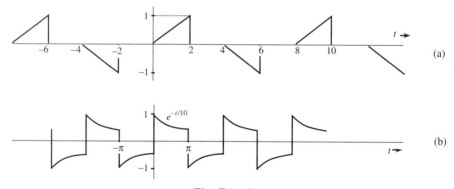

Fig. P3.4-7.

Fig. P3.4-8.

3.4-8   Over a finite interval, a signal can be represented by more than one trigonometric (or exponential) Fourier series. For instance, if we wish to represent $f(t) = t$ over an interval $0 \le t \le 1$ by a Fourier series with fundamental frequency $\omega_0 = 2$, we can draw a pulse $f(t) = t$ over the interval $0 \le t \le 1$ and repeat the pulse every π seconds so that $T_0 = \pi$ and $\omega_0 = 2$ (Fig. P3.4-8a). If we want the fundamental frequency ω_0 to be 4, we repeat the pulse every $\pi/2$ seconds. If we want the series to contain only cosine terms with $\omega_0 = 2$, we construct a pulse $f(t) = |t|$ over $-1 \le t \le 1$, and repeat it every π seconds (Fig. P3.4-8b). The resulting signal is an even function with period π. Hence, its Fourier series will have only cosine terms with $\omega_0 = 2$. The resulting Fourier series represents $f(t) = t$ over $0 \le t \le 1$ as desired. We do not care what it represents outside this interval.

Represent $f(t) = t$ over $0 \le t \le 1$ by a Fourier series that has

(a) $\omega_0 = \frac{\pi}{2}$ and contains all harmonics, but cosine terms only.
(b) $\omega_0 = 2$ and contains all harmonics, but sine terms only.
(c) $\omega_0 = \frac{\pi}{2}$ and contains all harmonics, which are neither exclusively sine nor cosine
(d) $\omega_0 = 1$ and contains only odd harmonics and cosine terms.
(e) $\omega_0 = \frac{\pi}{2}$ and contains only odd harmonics and sine terms.
(f) $\omega_0 = 1$ and contains only odd harmonics, which are neither exclusively sine nor cosine.

Hint: For parts d, e, and f, you need to use half wave symmetry discussed in Prob. 3.4-7. Cosine terms imply possible dc component.

3.4-9   State, with reasons, whether the following signals are periodic or aperiodic. For periodic signals, find the period and state which of the harmonics are present in the series.

(a) $3\sin t + 2\sin 3t$ **(f)** $\sin \frac{5t}{2} + 3\cos \frac{6t}{5} + 3\sin \left(\frac{t}{7} + 30°\right)$

(b) $2 + 5\sin 4t + 4\cos 7t$ **(g)** $\sin 3t + \cos \frac{15}{4}t$

(c) $2\sin 3t + 7\cos \pi t$ **(h)** $(3\sin 2t + \sin 5t)^2$

(d) $7\cos \pi t + 5\sin 2\pi t$ **(i)** $(5\sin 2t)^3$

(e) $3\cos \sqrt{2}t + 5\cos 2t$

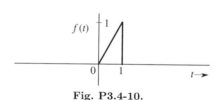

$f(t)$

Fig. P3.4-10.

3.4-10   Find the trigonometric Fourier series for $f(t)$ shown in Fig. P3.4-10 over the interval $[0, 1]$. Use $\omega_0 = 2\pi$. Sketch the Fourier series $\varphi(t)$ for all t. Compute the energy of the error signal $e(t)$ if the number of terms in the Fourier series are N for $N = 1, 2, 3$ and 4.

Hint: Use Eq. (3.40) to compute error energy.

3.4-11   Walsh functions, which can take on only two amplitude values, form a complete set of orthonormal functions and are of great practical importance in digital applications because they can be easily generated by logic circuitry and because multiplication with these functions can be implemented by simply using a polarity reversing switch.

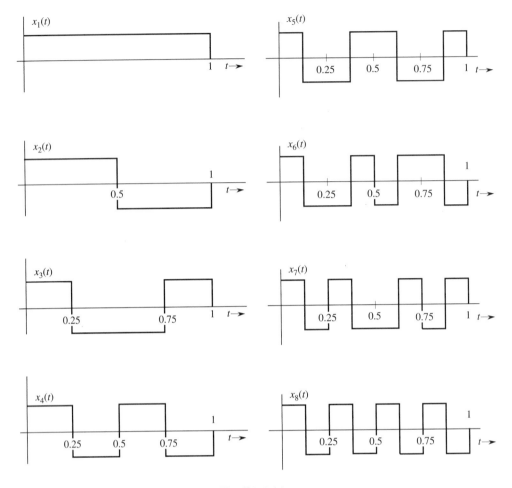

Fig. P3.4-11.

Figure P3.4-11 shows the first eight functions in this set. Represent $f(t)$ in Fig. P3.4-11 over the interval $[0, 1]$ using a Walsh Fourier series using these 8 basis functions. Compute the energy of $e(t)$, the error in the approximation using the first N non-zero terms in the series for $N = 1, 2, 3$ and 4. How does the Walsh series compare with the trigonometric series in Prob. 3.4-10 from the viewpoint of the error energy for a given N?

3.4-12 A set of Legendre polynomials $P_n(t), (n = 0, 1, 2, 3, \cdots)$ forms a complete set of orthogonal functions over the interval $-1 < t < 1$. These polynomials are defined by

$$P_n(t) = \frac{1}{n!\, 2^n} \frac{d^n}{dt^n} (t^2 - 1)^n \qquad n = 0, 1, 2, \cdots$$

Thus

$$P_0(t) = 1, \qquad\qquad P_1(t) = t$$
$$P_2(t) = \frac{1}{2}(3t^2 - 1), \quad P_3(t) = \frac{1}{2}(5t^3 - 3t) \qquad \text{etc.}$$

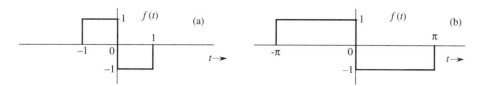

Fig. P3.4-12.

Legendre polynomials are orthogonal. Reader may verify that

$$\int_{-1}^{1} P_m(t)P_n(t)\, dt = \begin{cases} \frac{2}{2m+1} & m = n \\ 0 & m \neq n \end{cases}$$

(a) Represent $f(t)$ in Fig. P3.4-12a by the Legendre Fourier series over the interval $-1 < t < 1$. Compute only first two nonzero coefficients of the series. Compute the energy of $e(t)$, the error for the one and two (non-zero) term approximations.

(b) Represent $f(t)$ in Fig. P3.4-12b using the Legendre Fourier series. Compute the coefficients of the series for the first two non-zero terms.

Hint: Although the series representation is valid only over $-1 < t < 1$, it can be extended to any region by suitable time scaling.

3.5-1 For each of the periodic signals in Fig. P3.4-3, find the exponential Fourier series and sketch the corresponding spectra.

3.5-2 The trigonometric Fourier series of a certain periodic signal is given by

$$f(t) = 3 + \sqrt{3}\cos 2t + \sin 2t + \sin 3t - \tfrac{1}{2}\cos\left(5t + \tfrac{\pi}{3}\right)$$

(a) Sketch the trigonometric Fourier spectra.

(b) By inspection of the spectra in part **a**, sketch the exponential Fourier series spectra.

(c) By inspection of the spectra in part **b**, write the exponential Fourier series for $f(t)$.

Hint: To express the Fourier series in compact form, combine the sine and cosine terms of the same frequency. Moreover, all terms must appear in the cosine form with positive amplitudes. This can always be done by suitably adjusting the phase.

3.5-3 The exponential Fourier series of a certain periodic function is given as

$$f(t) = (2 + j2)e^{-j3t} + j2e^{-jt} + 3 - j2e^{jt} + (2 - j2)e^{j3t}$$

(a) Sketch the exponential Fourier spectra.

(b) By inspection of the spectra in part **a**, sketch the trigonometric Fourier spectra for $f(t)$.

(c) Find the compact trigonometric Fourier series from these spectra.

(d) Find the signal bandwidth.

3.5-4 If a periodic signal $f(t)$ is expressed as an exponential Fourier series

$$f(t) = \sum_{n=-\infty}^{\infty} D_n e^{jn\omega_0 t}$$

(a) Show that the exponential Fourier series for $\hat{f}(t) = f(t - T)$ is given by

$$\hat{f}(t) = \sum_{n=-\infty}^{\infty} \hat{D}_n e^{jn\omega_0 t}$$

in which

$$|\hat{D}_n| = |D_n| \quad \text{and} \quad \angle \hat{D}_n = \angle D_n - n\omega_0 T$$

This result shows that time shifting of a periodic signal by T seconds merely changes the phase spectrum by $n\omega_0 T$. The amplitude spectrum is unchanged.

(b) Show that the exponential Fourier series for $\tilde{f}(t) = f(at)$ is given by

$$\tilde{f}(t) = \sum_{n=-\infty}^{\infty} D_n e^{jn(a\omega_0)t}$$

This result shows that time compression of a periodic signal by a factor a expands its Fourier spectra by the same factor a. Similarly, time expansion of a periodic signal by a factor a compresses its Fourier spectra by the factor a.

3.5-5 **(a)** The Fourier series for the periodic signal in Fig. 3.10a is given in Exercise E3.6. Verify Parseval's theorem for this series, given that

$$\sum_{n=1}^{\infty} \frac{1}{n^4} = \frac{\pi^4}{90}$$

(b) If $f(t)$ is approximated by the first N terms in this series, find N so that the power of the error signal is less than 1% of P_f.

3.5-6 **(a)** The Fourier series for the periodic signal in Fig. 3.10b is given in Exercise E3.6. Verify Parseval's theorem for this series, given that

$$\sum_{n=1}^{\infty} \frac{1}{n^2} = \frac{\pi^2}{6}$$

(b) If $f(t)$ is approximated by the first N terms in this series, find N so that the power of the error signal is less than 10% of P_f.

3.5-7 The signal $f(t)$ in Fig. 3.18 is approximated by the first $2N + 1$ terms (from $n = -N$ to N) in its exponential Fourier series given in Exercise E3.10. Determine the value of N if this $(2N + 1)$-term Fourier series power is to be no less than 99.75% of the power of $f(t)$.

3.6-1 Find the response of an LTIC system with transfer function

$$H(s) = \frac{s}{s^2 + 2s + 3}$$

to the periodic input shown in Fig. 3.7b.

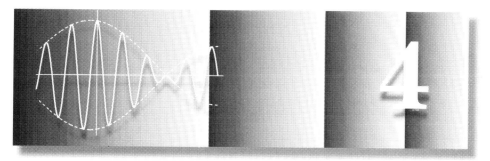

Continuous-Time Signal Analysis: The Fourier Transform

In Chapter 3, we succeeded in representing periodic signals as a sum of (everlasting) sinusoids or exponentials. In this chapter we extend this spectral representation to aperiodic signals.

4.1 Aperiodic Signal Representation by Fourier Integral

Applying a limiting process, we now show that an aperiodic signal can be expressed as a continuous sum (integral) of everlasting exponentials. To represent an aperiodic signal $f(t)$ such as the one depicted in Fig. 4.1a by everlasting exponential signals, let us construct a new periodic signal $f_{T_0}(t)$ formed by repeating the signal $f(t)$ at intervals of T_0 seconds, as illustrated in Fig. 4.1b. The period T_0 is made long enough to avoid overlap between the repeating pulses. The periodic signal $f_{T_0}(t)$ can be represented by an exponential Fourier series. If we let $T_0 \to \infty$, the pulses in the periodic signal repeat after an infinite interval and, therefore

$$\lim_{T_0 \to \infty} f_{T_0}(t) = f(t)$$

Thus, the Fourier series representing $f_{T_0}(t)$ will also represent $f(t)$ in the limit $T_0 \to \infty$. The exponential Fourier series for $f_{T_0}(t)$ is given by

$$f_{T_0}(t) = \sum_{n=-\infty}^{\infty} D_n e^{jn\omega_0 t} \qquad (4.1)$$

where

$$D_n = \frac{1}{T_0} \int_{-T_0/2}^{T_0/2} f_{T_0}(t) e^{-jn\omega_0 t}\, dt \qquad (4.2a)$$

and

$$\omega_0 = \frac{2\pi}{T_0} \qquad (4.2b)$$

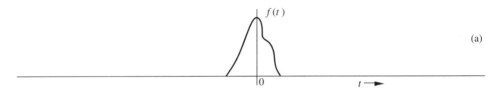

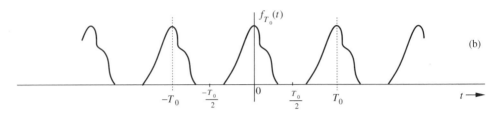

Fig. 4.1   Construction of a periodic signal by periodic extension of $f(t)$.

Observe that integrating $f_{T_0}(t)$ over $(-\frac{T_0}{2}, \frac{T_0}{2})$ is the same as integrating $f(t)$ over $(-\infty, \infty)$. Therefore, Eq. (4.2a) can be expressed as

$$D_n = \frac{1}{T_0} \int_{-\infty}^{\infty} f(t) e^{-jn\omega_0 t}\, dt \qquad (4.2c)$$

It is interesting to see how the nature of the spectrum changes as T_0 increases. To understand this fascinating behavior, let us define $F(\omega)$, a continuous function of ω, as

$$F(\omega) = \int_{-\infty}^{\infty} f(t) e^{-j\omega t}\, dt \qquad (4.3)$$

A glance at Eqs. (4.2c) and (4.3) shows that

$$D_n = \frac{1}{T_0} F(n\omega_0) \qquad (4.4)$$

This means that the Fourier coefficients D_n are $(1/T_0)$ times the samples of $F(\omega)$ uniformly spaced at intervals of ω_0, as depicted in Fig. 4.2a.† Therefore, $(1/T_0)F(\omega)$ is the envelope for the coefficients D_n. We now let $T_0 \to \infty$ by doubling T_0 repeatedly. Doubling T_0 halves the fundamental frequency ω_0 [Eq. (4.2b)], so that there are now twice as many components (samples) in the spectrum. However, by doubling T_0, the envelope $(1/T_0)\, F(\omega)$ is halved, as shown in Fig. 4.2b. If we continue this process of doubling T_0 repeatedly, the spectrum progressively becomes denser while its magnitude becomes smaller. Note, however, that the relative shape of the envelope remains the same [proportional to $F(\omega)$ in Eq. (4.3)]. In the limit as $T_0 \to \infty$, $\omega_0 \to 0$ and $D_n \to 0$. This result means the spectrum is so dense that the spectral components are spaced at zero (infinitesimal) intervals. At the same time, the amplitude of each component is zero (infinitesimal). We have *nothing of everything, yet we have something!* This paradox sounds like *Alice in Wonderland*,

---

†For the sake of simplicity, we assume D_n, and therefore $F(\omega)$, in Fig. 4.2, to be real. The argument, however, is also valid for complex D_n [or $F(\omega)$].

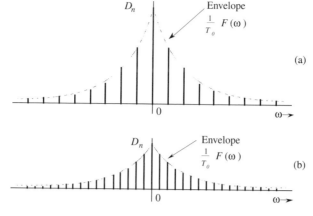

Fig. 4.2   Change in the Fourier spectrum when the period T_0 in Fig. 4.1 is doubled.

but as we shall see, these are the classic characteristics of a very familiar phenomenon.†

Substitution of Eq. (4.4) in Eq. (4.1) yields

$$f_{T_0}(t) = \sum_{n=-\infty}^{\infty} \frac{F(n\omega_0)}{T_0} e^{jn\omega_0 t} \tag{4.5}$$

As $T_0 \to \infty$, ω_0 becomes infinitesimal ($\omega_0 \to 0$). Hence, we shall replace ω_0 by a more appropriate notation, $\Delta\omega$. In terms of this new notation, Eq. (4.2b) becomes

$$\Delta\omega = \frac{2\pi}{T_0}$$

and Eq. (4.5) becomes

$$f_{T_0}(t) = \sum_{n=-\infty}^{\infty} \left[\frac{F(n\Delta\omega)\Delta\omega}{2\pi} \right] e^{(jn\Delta\omega)t} \tag{4.6a}$$

Equation (4.6a) shows that $f_{T_0}(t)$ can be expressed as a sum of everlasting exponentials of frequencies $0, \pm\Delta\omega, \pm 2\Delta\omega, \pm 3\Delta\omega, \ldots$ (the Fourier series). The amount of the component of frequency $n\Delta\omega$ is $[F(n\Delta\omega)\Delta\omega]/2\pi$. In the limit as $T_0 \to \infty$, $\Delta\omega \to 0$ and $f_{T_0}(t) \to f(t)$. Therefore

$$f(t) = \lim_{T_0 \to \infty} f_{T_0}(t) = \lim_{\Delta\omega \to 0} \frac{1}{2\pi} \sum_{n=-\infty}^{\infty} F(n\Delta\omega)e^{(jn\Delta\omega)t}\Delta\omega \tag{4.6b}$$

The sum on the right-hand side of Eq. (4.6b) can be viewed as the area under the function $F(\omega)e^{j\omega t}$, as illustrated in Fig. 4.3. Therefore

$$f(t) = \frac{1}{2\pi} \int_{-\infty}^{\infty} F(\omega)e^{j\omega t}d\omega \tag{4.7}$$

The integral on the right hand side is called the **Fourier integral**. We have now succeeded in representing an aperiodic signal $f(t)$ by a Fourier integral (rather

---

†If nothing else, the reader now has an irrefutable proof of the proposition that 0% ownership of everything is better than 100% ownership of nothing.

than a Fourier series).† This integral is basically a Fourier series (in the limit) with fundamental frequency $\Delta\omega \rightarrow 0$, as seen from Eq. (4.6). The amount of the exponential $e^{jn\Delta\omega t}$ is $F(n\Delta\omega)\Delta\omega/2\pi$. Thus, the function $F(\omega)$ given by Eq. (4.3) acts as a spectral function.

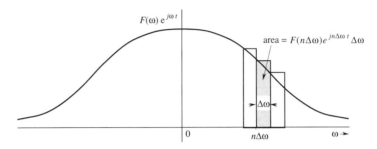

Fig. 4.3   The Fourier series becomes the Fourier integral in the limit as $T_0 \rightarrow \infty$.

We call $F(\omega)$ the **direct** Fourier transform of $f(t)$, and $f(t)$ the **inverse** Fourier transform of $F(\omega)$. The same information is conveyed by the statement that $f(t)$ and $F(\omega)$ are a Fourier transform pair. Symbolically, this statement is expressed as

$$F(\omega) = \mathcal{F}[f(t)] \qquad \text{and} \qquad f(t) = \mathcal{F}^{-1}[F(\omega)]$$

or

$$f(t) \Longleftrightarrow F(\omega)$$

To recapitulate,

$$F(\omega) = \int_{-\infty}^{\infty} f(t)e^{-j\omega t}\,dt \tag{4.8a}$$

and

$$f(t) = \frac{1}{2\pi} \int_{-\infty}^{\infty} F(\omega)e^{j\omega t}\,d\omega \tag{4.8b}$$

It is helpful to keep in mind that the Fourier integral in Eq. (4.8b) is of the nature of a Fourier series with fundamental frequency $\Delta\omega$ approaching zero [Eq. (4.6b)]. Therefore, most of the discussion and properties of Fourier series apply to the Fourier transform as well. We can plot the spectrum $F(\omega)$ as a function of ω. Since $F(\omega)$ is complex, we have both amplitude and angle (or phase) spectra

$$F(\omega) = |F(\omega)|e^{j\angle F(\omega)} \tag{4.9}$$

in which $|F(\omega)|$ is the amplitude and $\angle F(\omega)$ is the angle (or phase) of $F(\omega)$. According to Eq. (4.8a),

$$F(-\omega) = \int_{-\infty}^{\infty} f(t)e^{j\omega t}\,dt$$

From this equation and Eq. (4.8a), it follows that if $f(t)$ is a real function of t, then $F(\omega)$ and $F(-\omega)$ are complex conjugates. Therefore

---

†This derivation should not be considered as a rigorous proof of Eq. (4.7). The situation is not as simple as we have made it appear.[1]

$$|F(-\omega)| = |F(\omega)| \tag{4.10a}$$

$$\angle F(-\omega) = -\angle F(\omega) \tag{4.10b}$$

Thus, for real $f(t)$, the amplitude spectrum $|F(\omega)|$ is an even function, and the phase spectrum $\angle F(\omega)$ is an odd function of ω. This property the (**conjugate symmetry property**) is valid only for real $f(t)$. These results were derived earlier for the Fourier spectrum of a periodic signal [Eq. (3.77)] and should come as no surprise. *The transform $F(\omega)$ is the frequency-domain specification of $f(t)$.*

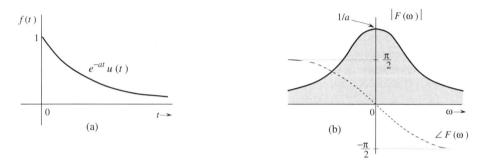

Fig. 4.4 $e^{-at}u(t)$ and its Fourier spectra.

■ **Example 4.1**

Find the Fourier transform of $e^{-at}u(t)$.

By definition [Eq. (4.8a)],

$$F(\omega) = \int_{-\infty}^{\infty} e^{-at}u(t)e^{-j\omega t}\,dt = \int_0^{\infty} e^{-(a+j\omega)t}\,dt = \frac{-1}{a+j\omega}e^{-(a+j\omega)t}\bigg|_0^{\infty}$$

But $|e^{-j\omega t}| = 1$. Therefore, as $t \to \infty$, $e^{-(a+j\omega)t} = e^{-at}e^{-j\omega t} = 0$ if $a > 0$. Therefore

$$F(\omega) = \frac{1}{a+j\omega} \qquad a > 0 \tag{4.11a}$$

Expressing $a + j\omega$ in the polar form as $\sqrt{a^2+\omega^2}\,e^{j\tan^{-1}(\frac{\omega}{a})}$, we obtain

$$F(\omega) = \frac{1}{\sqrt{a^2+\omega^2}}e^{-j\tan^{-1}(\frac{\omega}{a})} \tag{4.11b}$$

Therefore

$$|F(\omega)| = \frac{1}{\sqrt{a^2+\omega^2}} \qquad \text{and} \qquad \angle F(\omega) = -\tan^{-1}\left(\frac{\omega}{a}\right) \tag{4.12}$$

The amplitude spectrum $|F(\omega)|$ and the phase spectrum $\angle F(\omega)$ are depicted in Fig. 4.4b. Observe that $|F(\omega)|$ is an even function of ω, and $\angle F(\omega)$ is an odd function of ω, as expected. ■

Existence of the Fourier Transform

In Example 4.1 we observed that when $a < 0$, the Fourier integral for $e^{-at}u(t)$ does not converge. Hence, the Fourier transform for $e^{-at}u(t)$ does not exist if $a < 0$ (growing exponential). Clearly, not all signals are Fourier-transformable. The existence of the Fourier transform is assured for any $f(t)$ satisfying the Dirichlet conditions mentioned on p. 194-195. The first of these conditions is†

$$\int_{-\infty}^{\infty} |f(t)| \, dt \; < \; \infty \qquad (4.13)$$

Because $|e^{-j\omega t}| = 1$, from Eq. (4.8a), we obtain

$$|F(\omega)| \leq \int_{-\infty}^{\infty} |f(t)| \, dt$$

This inequality shows that the existence of the Fourier transform is assured if condition (4.13) is satisfied. Otherwise, there is no guarantee. We have seen in Example 4.1 that the Fourier transform does not exist for an exponentially growing signal (which violates this condition). Although this condition is sufficient, it is not necessary for the existence of the Fourier transform of a signal. For example, the signal $(\sin at)/t$ violates condition (4.13), but does have a Fourier transform. Any signal that can be generated in practice satisfies the Dirichlet conditions and therefore has a Fourier transform. Thus, the physical existence of a signal is a sufficient condition for the existence of its transform.

Linearity of the Fourier Transform

The Fourier transform is linear; that is, if

$$f_1(t) \Longleftrightarrow F_1(\omega) \qquad \text{and} \qquad f_2(t) \Longleftrightarrow F_2(\omega)$$

then

$$a_1 f_1(t) + a_2 f_2(t) \Longleftrightarrow a_1 F_1(\omega) + a_2 F_2(\omega) \qquad (4.14)$$

The proof is trivial and follows directly from Eq. (4.8a). This result can be extended to any finite number of terms.

4.1-1 Physical Appreciation of the Fourier Transform

In understanding any aspect of the Fourier transform, we should remember that Fourier representation is a way of expressing a signal in terms of everlasting sinusoids (or exponentials). The Fourier spectrum of a signal indicates the relative amplitudes and phases of the sinusoids that are required to synthesize that signal. A periodic signal Fourier spectrum has finite amplitudes and exists at discrete frequencies (ω_0 and its multiples). Such a spectrum is easy to visualize, but the spectrum of an aperiodic signal is not easy to visualize because it has a continuous spectrum. The continuous spectrum concept can be appreciated by considering an analogous, more tangible phenomenon. One familiar example of a continuous distribution is the loading of a beam. Consider a beam loaded with weights $D_1, D_2, D_3, \ldots, D_n$ units at the uniformly spaced points $x_1, x_2, \ldots, x_n$, as shown in Fig. 4.5a. The total load W_T on the beam is given by the sum of these loads at each of the n points:

---

†The remaining Dirichlet conditions are as follows: in any finite interval, $f(t)$ may have only a finite number of maxima and minima and a finite number of finite discontinuities. When these conditions are satisfied, the Fourier integral on the right-hand side of Eq. (4.8b) converges to $f(t)$ at all points where $f(t)$ is continuous and converges to the average of the right-hand and left-hand limits of $f(t)$ at points where $f(t)$ is discontinuous.

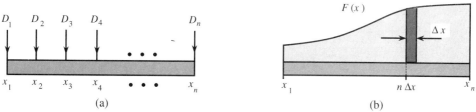

Fig. 4.5   Analogy for Fourier transform.

$$W_T = \sum_{i=1}^{n} D_i$$

Consider now the case of a continuously loaded beam, as depicted in Fig. 4.5b. In this case, although there appears to be a load at every point, the load at any one point is zero. This does not mean that there is no load on the beam. A meaningful measure of load in this situation is not the load at a point, but rather the loading density per unit length at that point. Let $F(x)$ be the loading density per unit length of beam. It then follows that the load over a beam length Δx ($\Delta x \to 0$), at some point x, is $F(x)\Delta x$. To find the total load on the beam, we divide the beam into segments of interval Δx ($\Delta x \to 0$). The load over the nth such segment of length Δx is $F(n\Delta x)\Delta x$. The total load W_T is given by

$$W_T = \lim_{\Delta x \to 0} \sum_{x_1}^{x_n} F(n\Delta x)\,\Delta x$$

$$= \int_{x_1}^{x_n} F(x)\,dx$$

In the case of discrete loading (Fig. 4.5a), the load exists only at the n discrete points. At other points there is no load. On the other hand, in the continuously loaded case, the load exists at every point, but at any specific point x, the load is zero. The load over a small interval Δx, however, is $[F(n\Delta x)]\,\Delta x$ (Fig. 4.5b). Thus, even though the load at a point x is zero, the relative load at that point is $F(x)$.

An exactly analogous situation exists in the case of a signal spectrum. When $f(t)$ is periodic, the spectrum is discrete, and $f(t)$ can be expressed as a sum of discrete exponentials with finite amplitudes:

$$f(t) = \sum_{n} D_n e^{jn\omega_0 t}$$

For an aperiodic signal, the spectrum becomes continuous; that is, the spectrum exists for every value of ω, but the amplitude of each component in the spectrum is zero. The meaningful measure here is not the amplitude of a component of some frequency but the spectral density per unit bandwidth. From Eq. (4.6b) it is clear that $f(t)$ is synthesized by adding exponentials of the form $e^{jn\Delta\omega t}$, in which the contribution by any one exponential component is zero. But the contribution by exponentials in an infinitesimal band $\Delta\omega$ located at $\omega = n\Delta\omega$ is $\frac{1}{2\pi}F(n\Delta\omega)\Delta\omega$, and the addition of all these components yields $f(t)$ in the integral form:

$$f(t) = \lim_{\Delta\omega \to 0} \frac{1}{2\pi} \sum_{n=-\infty}^{\infty} F(n\Delta\omega)e^{(jn\Delta\omega)t}\Delta\omega = \frac{1}{2\pi} \int_{-\infty}^{\infty} F(\omega)e^{j\omega t}\,d\omega \qquad (4.15)$$

The contribution by components within a band $d\omega$ is $\frac{1}{2\pi}F(\omega)\,d\omega = F(\omega)\,d\mathcal{F}$, where $d\mathcal{F}$ is the bandwidth in hertz. Clearly, $F(\omega)$ is the **spectral density** per unit bandwidth (in hertz). It also follows that even if the amplitude of any one component is zero, the relative amount of a component of frequency ω is $F(\omega)$. Although $F(\omega)$ is a spectral density, in practice it is customarily called the **spectrum** of $f(t)$ rather than the spectral density of $f(t)$. Deferring to this convention, we shall call $F(\omega)$ the Fourier spectrum (or Fourier transform) of $f(t)$.

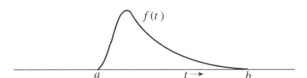

Fig. 4.6   Marvel of the Fourier transform.

A Marvelous Balancing Act

An important point to remember here is that $f(t)$ is represented (or synthesized) by exponentials or sinusoids that are everlasting (not causal). Such conceptualization leads to a rather fascinating picture when we try to visualize the synthesis of a timelimited pulse signal $f(t)$ (Fig. 4.6) by the sinusoidal components in its Fourier spectrum. The signal $f(t)$ exists only over an interval (a, b) and is zero outside this interval. The spectrum of $f(t)$ contains an infinite number of exponentials (or sinusoids) which start at $t = -\infty$ and continue forever. The amplitudes and phases of these components are such that they add up exactly to $f(t)$ over the finite interval (a, b) and add up to zero everywhere outside this interval. Juggling with amplitudes and phases of an infinite number of components to achieve such a perfect and delicate balance boggles the human imagination. Yet the Fourier transform accomplishes it routinely, without much thinking on our part. Indeed, we become so involved in mathematical manipulations that we fail to notice this marvel.

A Word About Notation

In Chapter 2 [Eq. (2.48)], we defined the system transfer function $H(s)$ as

$$H(s) = \int_{-\infty}^{\infty} h(t)e^{-st}\,dt \qquad (4.16)$$

Setting $s = j\omega$ in this equation yields

$$H(j\omega) = \int_{-\infty}^{\infty} h(t)e^{-j\omega t}\,dt \qquad (4.17)$$

The right-hand side is the Fourier transform of $h(t)$, and according to our notation introduced in Eq. (4.3) this is $H(\omega)$, whereas the same entity is denoted by $H(j\omega)$

in the notation introduced in Chapter 2. Thus, to be consistent with the previous notation, we should have denoted the Fourier transform by $F(j\omega)$ rather than $F(\omega)$ in Eq. (4.3). In fact, the notation $F(j\omega)$ for the Fourier transform is often used in the literature. It is, however, a bit clumsy and does not lend itself so easily to manipulation as the notation $F(\omega)$. For this reason we shall continue with the dual notation, while remembering that both $F(\omega)$ and $F(j\omega)$ represent the same entity. This fact is particularly important when we discuss the Laplace transform and filtering in the future, and should be kept in mind throughout the rest of the book. In the same way, we must remember that $H(\omega)$ and $H(j\omega)$ represent the same entity.

4.1-2 LTIC System Response Using the Fourier Transform

We wanted to represent a signal $f(t)$ as a sum of (everlasting) exponentials so that we could find a system response to $f(t)$ as a sum of the system's responses to the exponential components of $f(t)$. Consider an asymptotically stable LTIC system with transfer function $H(s)$. The response of this system to everlasting exponential $e^{j\omega t}$ is $H(\omega)e^{j\omega t}$. Such an input-output pair will be denoted by the directed arrow representation as

$$e^{j\omega t} \implies H(\omega)e^{j\omega t}$$

Therefore

$$e^{j(n\triangle\omega)t} \implies H(n\triangle\omega)e^{j(n\triangle\omega)t}$$

'and

$$\left[\frac{F(n\triangle\omega)\triangle\omega}{2\pi}\right]e^{(jn\triangle\omega)t} \implies \left[\frac{F(n\triangle\omega)H(n\triangle\omega)\triangle\omega}{2\pi}\right]e^{j(n\triangle\omega)t}$$

Using the linearity property

$$\underbrace{\lim_{\triangle\omega\to0}\sum_{n=-\infty}^{\infty}\left[\frac{F(n\triangle\omega)\triangle\omega}{2\pi}\right]e^{(jn\triangle\omega)t}}_{\text{input } f(t)} \implies \underbrace{\lim_{\triangle\omega\to0}\sum_{n=-\infty}^{\infty}\left[\frac{F(n\triangle\omega)H(n\triangle\omega)\triangle\omega}{2\pi}\right]e^{j(n\triangle\omega)t}}_{\text{output } y(t)}$$

The left-hand side is the input $f(t)$ [see Eqs. (4.6a) and (4.6b)], and the right-hand side is the response $y(t)$. Thus

$$y(t) = \frac{1}{2\pi}\lim_{\triangle\omega\to0}\sum_{n=-\infty}^{\infty}F(n\triangle\omega)H(n\triangle\omega)e^{j(n\triangle\omega)t}\triangle\omega$$

$$= \frac{1}{2\pi}\int_{-\infty}^{\infty}F(\omega)H(\omega)e^{j\omega t}\,d\omega$$

$$= \frac{1}{2\pi}\int_{-\infty}^{\infty}Y(\omega)e^{j\omega t}\,d\omega \qquad (4.18)$$

where $Y(\omega)$, the Fourier transform of $y(t)$, is given by†

---

†The relationship (4.19) applies only to the asymptotically stable systems. The reason is that when $s = j\omega$, the integral on the right-hand side of Eq. (2.48) does not converge for unstable systems. Moreover, even for marginally stable systems, that integral does not converges in the ordinary sense, and $H(j\omega)$ [or $H(\omega)$] cannot be obtained by replacing s in $H(s)$ with $j\omega$. As shown in Eq. (4.44b), Eq. (4.19) can be applied to marginally stable systems provided $H(\omega)$ is interpreted as the Fourier transform of $h(t)$ rather than as $H(s)$ with s replaced by $j\omega$.

$$Y(\omega) = F(\omega)H(\omega) \tag{4.19}$$

In conclusion, we showed that for an LTIC system with transfer function $H(s)$, if the input and the output are $f(t)$ and $y(t)$, respectively, and if

$$f(t) \iff F(\omega) \qquad y(t) \iff Y(\omega)$$

then for asymptotically stable systems

$$Y(\omega) = F(\omega)H(\omega)$$

We shall derive this result again later in a more formal way.

The procedure of the frequency-domain method is identical to that of the time-domain method. In the time-domain case we express the input $f(t)$ as a sum of its impulse components; in the frequency-domain case, the input is expressed as a sum of everlasting exponentials (or sinusoids). In the former case, the response $y(t)$ obtained by summing the system's responses to impulse components results in the convolution integral; in the latter case, the response obtained by summing the system's response to everlasting exponential components results in the Fourier integral. These ideas can be expressed mathematically as follows:

1 For the time-domain case

$$\delta(t) \implies h(t) \qquad \text{the impulse response of the system is } h(t)$$

$$f(t) = \int_{-\infty}^{\infty} f(x)\delta(t - x)\, dx \qquad \text{expresses } f(t) \text{ as a sum of impulse components}$$

and

$$y(t) = \int_{-\infty}^{\infty} f(x)h(t - x)\, dx \qquad \text{expresses } y(t) \text{ as a sum of responses to impulse components}$$

2   For the frequency-domain case

$$e^{j\omega t} \implies H(\omega)e^{j\omega t} \qquad \text{the system response to } e^{j\omega t} \text{ is } H(\omega)e^{j\omega t}$$

$$f(t) = \frac{1}{2\pi} \int_{-\infty}^{\infty} F(\omega)e^{j\omega t}\, d\omega \quad \text{shows } f(t) \text{ as a sum of everlasting exponential components}$$

and

$$y(t) = \frac{1}{2\pi} \int_{-\infty}^{\infty} F(\omega)H(\omega)e^{j\omega t}\, d\omega \qquad y(t) \text{ is a sum of responses to exponential components}$$

The frequency-domain view sees a system in terms of its frequency response (system response to various sinusoidal components). It views a signal as a sum of various sinusoidal components. Transmission of a signal through a (linear) system is viewed as transmission of various sinusoidal components of the signal through the system.

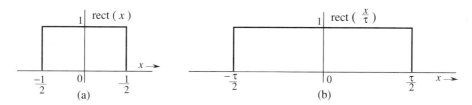

Fig. 4.7   A gate pulse.

4.2   Transforms of Some Useful Functions

For convenience, we now introduce a compact notation for some useful functions such as gate, triangle, and interpolation functions.

Unit Gate Function

We define a unit gate function rect (x) as a gate pulse of unit height and unit width, centered at the origin, as illustrated in Fig. 4.7a:†

$$\text{rect}\,(x) = \begin{cases} 0 & |x| > \frac{1}{2} \\ \frac{1}{2} & |x| = \frac{1}{2} \\ 1 & |x| < \frac{1}{2} \end{cases} \tag{4.20}$$

The gate pulse in Fig. 4.7b is the unit gate pulse rect (x) expanded by a factor τ and therefore can be expressed as rect $(\frac{x}{\tau})$ (see Sec. 1.3-2). Observe that τ, the denominator of the argument of rect$(\frac{x}{\tau})$, indicates the width of the pulse.

Unit Triangle Function

We define a unit triangle function $\Delta(x)$ as a triangular pulse of unit height and unit width, centered at the origin, as shown in Fig. 4.8a

$$\Delta(x) = \begin{cases} 0 & |x| \geq \frac{1}{2} \\ 1 - 2|x| & |x| < \frac{1}{2} \end{cases} \tag{4.21}$$

The pulse in Fig. 4.8b is $\Delta(\frac{x}{\tau})$. Observe that here, as for the gate pulse, the denominator τ of the argument of $\Delta(\frac{x}{\tau})$ indicates the pulse width.

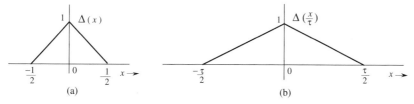

Fig. 4.8   A triangle pulse.

---

†At $x = 0$, we require rect $(x) = 0.5$, because the inverse Fourier transform of a discontinuous signal converges to the mean of its two values at the discontinuity.

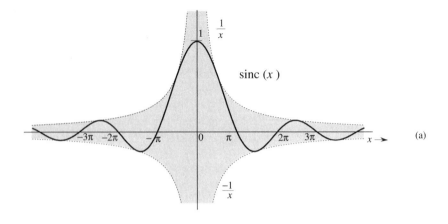

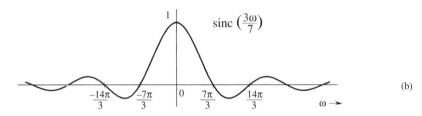

Fig. 4.9 A sinc pulse.

Interpolation Function sinc (x)

The function $\sin x / x$ is the "sine over argument" function denoted by $\text{sinc}\,(x)$.†
This function plays an important role in signal processing. It is also known as the
filtering or interpolating function. We define

$$\text{sinc}\,(x) = \frac{\sin x}{x} \tag{4.22}$$

Inspection of Eq. (4.22) shows that

1. $\text{sinc}\,(x)$ is an even function of x.

2. $\text{sinc}\,(x) = 0$ when $\sin x = 0$ except at $x = 0$, where it appears indeterminate.
    This means that $\text{sinc}\,x = 0$ for $x = \pm\pi, \pm2\pi, \pm3\pi, \ldots$

3. Using L'Hôpital's rule, we find $\text{sinc}\,(0) = 1$.

4. $\text{sinc}\,(x)$ is the product of an oscillating signal $\sin x$ (of period 2π) and a mono-
    tonically decreasing function $1/x$. Therefore, $\text{sinc}\,(x)$ exhibits sinusoidal os-
    cillations of period 2π, with amplitude decreasing continuously as $1/x$.

Figure 4.9a shows $\text{sinc}\,(x)$. Observe that $\text{sinc}\,(x) = 0$ for values of x that are
positive and negative integral multiples of π. Figure 4.9b shows $\text{sinc}\,\left(\frac{3\omega}{7}\right)$. The

---

†$\text{sinc}\,(x)$ is also denoted by $\text{Sa}\,(x)$ in the literature. Some authors define $\text{sinc}\,(x)$ as

$$\text{sinc}\,(x) = \frac{\sin \pi x}{\pi x}$$

argument $\frac{3\omega}{7} = \pi$ when $\omega = \frac{7\pi}{3}$. Therefore, the first zero of this function occurs at $\omega = \frac{7\pi}{3}$.

△ **Exercise E4.1**

Sketch: **(a)** $\mathrm{rect}\left(\frac{x}{8}\right)$ **(b)** $\Delta\left(\frac{\omega}{10}\right)$ **(c)** $\mathrm{sinc}\left(\frac{3\pi\omega}{2}\right)$ **(d)** $\mathrm{sinc}\,(t)\mathrm{rect}\left(\frac{t}{4\pi}\right)$. ▽

■ **Example 4.2**

Find the Fourier transform of $f(t) = \mathrm{rect}\left(\frac{t}{\tau}\right)$ (Fig. 4.10a).

$$F(\omega) = \int_{-\infty}^{\infty} \mathrm{rect}\left(\frac{t}{\tau}\right) e^{-j\omega t}\,dt$$

Since $\mathrm{rect}\left(\frac{t}{\tau}\right) = 1$ for $|t| < \frac{\tau}{2}$, and since it is zero for $|t| > \frac{\tau}{2}$,

$$F(\omega) = \int_{-\tau/2}^{\tau/2} e^{-j\omega t}\,dt$$

$$= -\frac{1}{j\omega}(e^{-j\omega\tau/2} - e^{j\omega\tau/2}) = \frac{2\sin\left(\frac{\omega\tau}{2}\right)}{\omega}$$

$$= \tau\frac{\sin\left(\frac{\omega\tau}{2}\right)}{\left(\frac{\omega\tau}{2}\right)} = \tau\,\mathrm{sinc}\left(\frac{\omega\tau}{2}\right)$$

Therefore

$$\mathrm{rect}\left(\frac{t}{\tau}\right) \Longleftrightarrow \tau\,\mathrm{sinc}\left(\frac{\omega\tau}{2}\right) \tag{4.23}$$

Recall that $\mathrm{sinc}\,(x) = 0$ when $x = \pm n\pi$. Hence, $\mathrm{sinc}\left(\frac{\omega\tau}{2}\right) = 0$ when $\frac{\omega\tau}{2} = \pm n\pi$; that is, when $\omega = \pm\frac{2n\pi}{\tau}, (n = 1, 2, 3, \ldots)$, as depicted in Fig. 4.10b. The Fourier transform $F(\omega)$ shown in Fig. 4.10b exhibits positive and negative values. A negative amplitude can be considered as a positive amplitude with a phase of $-\pi$ or π. We use this observation to plot the amplitude spectrum $|F(\omega)| = |\mathrm{sinc}\left(\frac{\omega\tau}{2}\right)|$ (Fig. 4.10c) and the phase spectrum $\angle F(\omega)$ (Fig. 4.10d). The phase spectrum, which is required to be an odd function of ω, may be drawn in several other ways because a negative sign can be accounted for by a phase of $\pm n\pi$, where n is any odd integer. All such representations are equivalent. ■

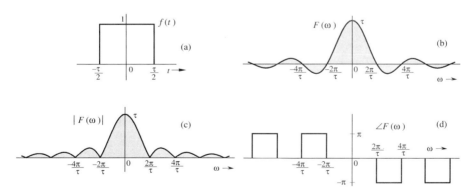

Fig. 4.10 A gate pulse $f(t)$, its Fourier spectrum $F(\omega)$, amplitude spectrum $|F(\omega)|$, and phase spectrum $\angle F(\omega)$.

Fig. 4.11   Unit impulse and its Fourier spectrum.

Bandwidth of rect $\left(\frac{t}{\tau}\right)$

The spectrum $F(\omega)$ in Fig. 4.10 peaks at $\omega = 0$ and decays at higher frequencies. Therefore, rect $\left(\frac{t}{\tau}\right)$ is a lowpass signal with most of the signal energy in lower frequency components. Strictly speaking, because the spectrum extends from 0 to ∞, the bandwidth is ∞. However, much of the spectrum is concentrated within the first lobe (from $\omega = 0$ to $\omega = \frac{2\pi}{\tau}$). Therefore, a rough estimate of the bandwidth of a rectangular pulse of width τ seconds is $\frac{2\pi}{\tau}$ rad/s, or $\frac{1}{\tau}$ Hz.† Note the reciprocal relationship of the pulse width with its bandwidth. We shall observe later that this result is true in general.

■ **Example 4.3**

Find the Fourier transform of the unit impulse $\delta(t)$.

Using the sampling property of the impulse [Eq. (1.24)], we obtain

$$\mathcal{F}[\delta(t)] = \int_{-\infty}^{\infty} \delta(t)e^{-j\omega t}dt = 1 \tag{4.24a}$$

or

$$\delta(t) \Longleftrightarrow 1 \tag{4.24b}$$

Figure 4.11 shows $\delta(t)$ and its spectrum.   ■

■ **Example 4.4**

Find the inverse Fourier transform of $\delta(\omega)$.

On the basis of Eq. (4.8b) and the sampling property of the impulse function,

$$\mathcal{F}^{-1}[\delta(\omega)] = \frac{1}{2\pi}\int_{-\infty}^{\infty} \delta(\omega)e^{j\omega t}\,d\omega = \frac{1}{2\pi}$$

Therefore

$$\frac{1}{2\pi} \Longleftrightarrow \delta(\omega) \tag{4.25a}$$

or

$$1 \Longleftrightarrow 2\pi\delta(\omega) \tag{4.25b}$$

This result shows that the spectrum of a constant signal $f(t) = 1$ is an impulse $2\pi\delta(\omega)$, as illustrated in Fig. 4.12.

The result [Eq. (4.25b)] also could have been anticipated on qualitative grounds. Recall that the Fourier transform of $f(t)$ is a spectral representation of $f(t)$ in terms of everlasting exponential components of the form $e^{j\omega t}$. Now, to represent a constant signal

———————————

†To compute bandwidth, we must consider the spectrum only for positive values of ω. See discussion on p. 212.

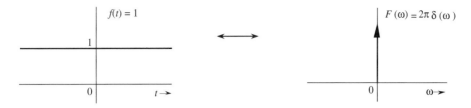

Fig. 4.12 A constant (dc) signal and its Fourier spectrum.

$f(t) = 1$, we need a single everlasting exponential $e^{j\omega t}$ with $\omega = 0$.‡ This results in a spectrum at a single frequency $\omega = 0$. Another way of looking at the situation is that $f(t) = 1$ is a dc signal which has a single frequency $\omega = 0$ (dc). ■

If an impulse at $\omega = 0$ is a spectrum of a dc signal, what does an impulse at $\omega = \omega_0$ represent? We shall answer this question in the next example.

■ **Example 4.5**

Find the inverse Fourier transform of $\delta(\omega - \omega_0)$.

Using the sampling property of the impulse function, we obtain

$$\mathcal{F}^{-1}[\delta(\omega - \omega_0)] = \frac{1}{2\pi} \int_{-\infty}^{\infty} \delta(\omega - \omega_0)e^{j\omega t}\,d\omega = \frac{1}{2\pi}e^{j\omega_0 t}$$

Therefore

$$\frac{1}{2\pi}e^{j\omega_0 t} \Longleftrightarrow \delta(\omega - \omega_0)$$

or

$$e^{j\omega_0 t} \Longleftrightarrow 2\pi\delta(\omega - \omega_0) \tag{4.26a}$$

This result shows that the spectrum of an everlasting exponential $e^{j\omega_0 t}$ is a single impulse at $\omega = \omega_0$. We reach the same conclusion by qualitative reasoning. To represent the everlasting exponential $e^{j\omega_0 t}$, we need a single everlasting exponential $e^{j\omega t}$ with $\omega = \omega_0$. Therefore, the spectrum consists of a single component at frequency $\omega = \omega_0$.

From Eq. (4.26a) it follows that

$$e^{-j\omega_0 t} \Longleftrightarrow 2\pi\delta(\omega + \omega_0) \tag{4.26b}$$

■

■ **Example 4.6**

Find the Fourier transforms of the everlasting sinusoid $\cos \omega_0 t$.

Recall the Euler formula

$$\cos \omega_0 t = \frac{1}{2}(e^{j\omega_0 t} + e^{-j\omega_0 t})$$

---

‡The constant multiplier 2π in the spectrum $[F(\omega) = 2\pi\delta(\omega)]$ may be a bit puzzling. Since $1 = e^{j\omega t}$ with $\omega = 0$, it appears that the Fourier transform of $f(t) = 1$ should be an impulse of strength unity rather than 2π. Recall, however, that in the Fourier transform $f(t)$ is synthesized not by exponentials of amplitude $F(n\Delta\omega)\Delta\omega$, but of amplitude $1/2\pi$ times $F(n\Delta\omega)\Delta\omega$, as seen from Eq. (4.6b). Had we used variable $\mathcal{F}$ (hertz) instead of ω, the spectrum would have been a unit impulse.

Fig. 4.13 A cosine signal and its Fourier spectrum.

Adding Eqs. (4.26a) and (4.26b), and using the above formula, we obtain

$$\cos \omega_0 t \iff \pi[\delta(\omega + \omega_0) + \delta(\omega - \omega_0)] \tag{4.27}$$

The spectrum of $\cos \omega_0 t$ consists of two impulses at ω_0 and $-\omega_0$, as shown in Fig. 4.13. The result also follows from qualitative reasoning. An everlasting sinusoid $\cos \omega_0 t$ can be synthesized by two everlasting exponentials, $e^{j\omega_0 t}$ and $e^{-j\omega_0 t}$. Therefore the Fourier spectrum consists of only two components of frequencies ω_0 and $-\omega_0$. ■

■ **Example 4.7**
Find the Fourier transform of the unit step function $u(t)$.

Trying to find the Fourier transform of $u(t)$ by direct integration leads to an indeterminate result, because

$$U(\omega) = \int_{-\infty}^{\infty} u(t)e^{-j\omega t}\, dt = \int_{0}^{\infty} e^{-j\omega t}\, dt = \frac{-1}{j\omega} e^{-j\omega t}\Big|_{0}^{\infty}$$

Observe that the upper limit of $e^{-j\omega t}$ as $t \to \infty$ yields an indeterminate answer. So we approach this problem by considering $u(t)$ as a decaying exponential $e^{-at}u(t)$ in the limit as $a \to 0$ (Fig. 4.14a). Thus

$$u(t) = \lim_{a \to 0} e^{-at}u(t)$$

and

$$U(\omega) = \lim_{a \to 0} \mathcal{F}\{e^{-at}u(t)\} = \lim_{a \to 0} \frac{1}{a + j\omega} \tag{4.28a}$$

Expressing the right-hand side in terms of its real and imaginary parts yields

$$U(\omega) = \lim_{a \to 0} \left[\frac{a}{a^2 + \omega^2} - j\frac{\omega}{a^2 + \omega^2} \right] \tag{4.28b}$$

$$= \lim_{a \to 0} \left[\frac{a}{a^2 + \omega^2} \right] + \frac{1}{j\omega}$$

The function $a/(a^2 + \omega^2)$ has interesting properties. First, the area under this function (Fig. 4.14b) is π regardless of the value of a

$$\int_{-\infty}^{\infty} \frac{a}{a^2 + \omega^2}\, d\omega = \tan^{-1}\frac{\omega}{a}\Big|_{-\infty}^{\infty} = \pi$$

Second, when $a \to 0$, this function approaches zero for all $\omega \neq 0$, and all its area (π) is concentrated at a single point $\omega = 0$. Clearly, as $a \to 0$, this function approaches an impulse of strength π.† Thus

$$U(\omega) = \pi\delta(\omega) + \frac{1}{j\omega} \tag{4.29}$$

---

†The second term on the right-hand side of Eq. (4.28b), being an odd function of ω, has zero area regardless of the value of a. As $a \to 0$, the second term approaches $1/j\omega$.

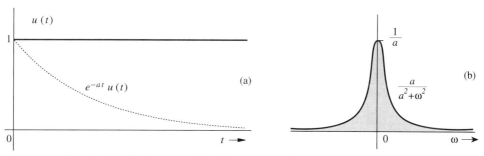

Fig. 4.14 Derivation of the Fourier transform of the step function.

Note that $u(t)$ is not a "true" dc signal because it is not constant over the interval $-\infty$ to ∞. To synthesize a "true" dc, we require only one everlasting exponential with $\omega = 0$ (impulse at $\omega = 0$). The signal $u(t)$ has a jump discontinuity at $t = 0$. It is impossible to synthesize such a signal with a single everlasting exponential $e^{j\omega t}$. To synthesize this signal from everlasting exponentials, we need, in addition to an impulse at $\omega = 0$, all frequency components, as indicated by the term $1/j\omega$ in Eq. (4.29). ■

△ **Exercise E4.2**

Show that the Fourier transform of the sign function $\mathrm{sgn}(t)$ depicted in Fig. 4.15a is $2/j\omega$. Hint: Note that $\mathrm{sgn}(t)$ shifted vertically by 1 yields $2u(t)$ ▽.

△ **Exercise E4.3**

Show that the inverse Fourier transform of $F(\omega)$ illustrated in Fig. 4.15b is $f(t) = \frac{\omega_0}{\pi}\,\mathrm{sinc}\,(\omega_0 t)$. Sketch $f(t)$. ▽.

△ **Exercise E4.4**

Show that $\cos(\omega_0 t + \theta) \iff \pi[\delta(\omega + \omega_0)e^{-j\theta} + \delta(\omega - \omega_0)e^{j\theta}]$.

Hint: $\cos(\omega_0 t + \theta) = \frac{1}{2}[e^{j(\omega_0 t+\theta)} + e^{-j(\omega_0 t+\theta)}]$ ▽

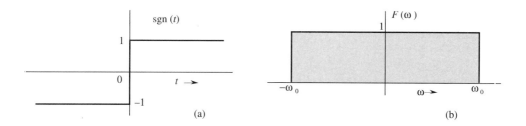

Fig. 4.15

4.3 Some properties of the Fourier Transform

We now study some of the important properties of the Fourier transform and their implications as well as applications. Before embarking on this study, we shall explain an important and pervasive aspect of the Fourier transform: the time-frequency duality.

Table 4.1

A Short Table of Fourier Transforms

	$f(t)$	$F(\omega)$	
1	$e^{-at}u(t)$	$\dfrac{1}{a + j\omega}$	$a > 0$
2	$e^{at}u(-t)$	$\dfrac{1}{a - j\omega}$	$a > 0$
3	$e^{-a|t|}$	$\dfrac{2a}{a^2 + \omega^2}$	$a > 0$
4	$te^{-at}u(t)$	$\dfrac{1}{(a + j\omega)^2}$	$a > 0$
5	$t^n e^{-at}u(t)$	$\dfrac{n!}{(a + j\omega)^{n+1}}$	$a > 0$
6	$\delta(t)$	1	
7	1	$2\pi\delta(\omega)$	
8	$e^{j\omega_0 t}$	$2\pi\delta(\omega - \omega_0)$	
9	$\cos\omega_0 t$	$\pi[\delta(\omega - \omega_0) + \delta(\omega + \omega_0)]$	
10	$\sin\omega_0 t$	$j\pi[\delta(\omega + \omega_0) - \delta(\omega - \omega_0)]$	
11	$u(t)$	$\pi\delta(\omega) + \dfrac{1}{j\omega}$	
12	$\operatorname{sgn} t$	$\dfrac{2}{j\omega}$	
13	$\cos\omega_0 t\, u(t)$	$\dfrac{\pi}{2}[\delta(\omega - \omega_0) + \delta(\omega + \omega_0)] + \dfrac{j\omega}{\omega_0^2 - \omega^2}$	
14	$\sin\omega_0 t\, u(t)$	$\dfrac{\pi}{2j}[\delta(\omega - \omega_0) - \delta(\omega + \omega_0)] + \dfrac{\omega_0}{\omega_0^2 - \omega^2}$	
15	$e^{-at}\sin\omega_0 t\, u(t)$	$\dfrac{\omega_0}{(a+j\omega)^2 + \omega_0^2}$	$a > 0$
16	$e^{-at}\cos\omega_0 t\, u(t)$	$\dfrac{a+j\omega}{(a+j\omega)^2 + \omega_0^2}$	$a > 0$
17	$\operatorname{rect}\left(\dfrac{t}{\tau}\right)$	$\tau\operatorname{sinc}\left(\dfrac{\omega\tau}{2}\right)$	
18	$\dfrac{W}{\pi}\operatorname{sinc}(Wt)$	$\operatorname{rect}\left(\dfrac{\omega}{2W}\right)$	
19	$\Delta\left(\dfrac{t}{\tau}\right)$	$\dfrac{\tau}{2}\operatorname{sinc}^2\left(\dfrac{\omega\tau}{4}\right)$	
20	$\dfrac{W}{2\pi}\operatorname{sinc}^2\left(\dfrac{Wt}{2}\right)$	$\Delta\left(\dfrac{\omega}{2W}\right)$	
21	$\displaystyle\sum_{n=-\infty}^{\infty}\delta(t - nT)$	$\omega_0\displaystyle\sum_{n=-\infty}^{\infty}\delta(\omega - n\omega_0)$	$\omega_0 = \dfrac{2\pi}{T}$
22	$e^{-t^2/2\sigma^2}$	$\sigma\sqrt{2\pi}\,e^{-\sigma^2\omega^2/2}$	

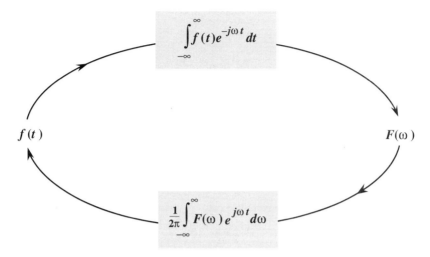

Fig. 4.16 A near symmetry between the direct and the inverse Fourier transforms.

4.3-1 Symmetry of Direct and Inverse Transform Operations: Time-Frequency Duality

Equations (4.8) show an interesting fact: the direct and the inverse transform operations are remarkably similar. These operations, required to go from $f(t)$ to $F(\omega)$ and then from $F(\omega)$ to $f(t)$, are depicted graphically in Fig. 4.16. There are only two minor differences in these operations: the factor 2π appears only in the inverse operator, and the exponential indices in the two operations have opposite signs. Otherwise the two operations are symmetrical.† This observation has far-reaching consequences in the study of the Fourier transform. It is the basis of the so-called duality of time and frequency. *The duality principle may be compared with a photograph and its negative. A photograph can be obtained from its negative, and by using an identical procedure, the negative can be obtained from the photograph.* For any result or relationship between $f(t)$ and $F(\omega)$, there exists a dual result or relationship, obtained by interchanging the roles of $f(t)$ and $F(\omega)$ in the original result (along with some minor modifications arising because of the factor 2π and a sign change). For example, the time-shifting property, to be proved later, states that if $f(t) \Longleftrightarrow F(\omega)$, then

$$f(t - t_0) \Longleftrightarrow F(\omega)e^{-j\omega t_0} \tag{4.30a}$$

The dual of this property (the frequency-shifting property) states that

---

†Of the two differences, the former can be eliminated by change of variable from ω to $\mathcal{F}$ (in hertz). In this case

$$\omega = 2\pi\mathcal{F} \qquad \text{and} \qquad d\omega = 2\pi\, d\mathcal{F}$$

Therefore, the direct and the inverse transforms are given by

$$F(2\pi\mathcal{F}) = \int_{-\infty}^{\infty} f(t)e^{-j2\pi\mathcal{F}t}dt \qquad \text{and} \qquad f(t) = \int_{-\infty}^{\infty} F(2\pi\mathcal{F})e^{j2\pi\mathcal{F}t}d\mathcal{F}$$

This leaves only one significant difference, that of sign change in the exponential index. Otherwise the two operations are symmetrical.

$$f(t)e^{j\omega_0 t} \Longleftrightarrow F(\omega - \omega_0) \qquad (4.30b)$$

Observe the role reversal of time and frequency in these two equations (with the minor difference of the sign change in the exponential index). The value of this principle lies in the fact that *whenever we derive any result, we can be sure that it has a dual.* This possibility can give valuable insights about many unsuspected properties or results in signal processing.

The properties of the Fourier transform are useful not only in deriving the direct and inverse transforms of many functions, but also in obtaining several valuable results in signal processing. The reader should not fail to observe the ever-present duality in this discussion. We begin with the symmetry property, which is one of the consequences of the duality principle discussed.

4.3-2 Symmetry Property

This property states that if

$$f(t) \Longleftrightarrow F(\omega)$$

then

$$F(t) \Longleftrightarrow 2\pi f(-\omega) \qquad (4.31)$$

Proof: According to Eq. (4.8b)

$$f(t) = \frac{1}{2\pi} \int_{-\infty}^{\infty} F(x)e^{jxt}\, dx$$

Hence

$$2\pi f(-t) = \int_{-\infty}^{\infty} F(x)e^{-jxt}\, dx$$

Changing t to ω yields Eq. (4.31).

■ **Example 4.8**

In this example we apply the symmetry property [Eq. (4.31)] to the pair in Fig. 4.17a. From Eq. (4.23) we have

$$\underbrace{\operatorname{rect}\left(\frac{t}{\tau}\right)}_{f(t)} \Longleftrightarrow \underbrace{\tau \operatorname{sinc}\left(\frac{\omega\tau}{2}\right)}_{F(\omega)} \qquad (4.32)$$

Also, $F(t)$ is the same as $F(\omega)$ with ω replaced by t, and $f(-\omega)$ is the same as $f(t)$ with t replaced by $-\omega$. Therefore, the symmetry property (4.31) yields

$$\underbrace{\tau \operatorname{sinc}\left(\frac{\tau t}{2}\right)}_{F(t)} \Longleftrightarrow \underbrace{2\pi \operatorname{rect}\left(\frac{-\omega}{\tau}\right)}_{2\pi f(-\omega)} = 2\pi \operatorname{rect}\left(\frac{\omega}{\tau}\right) \qquad (4.33)$$

In Eq. (4.33) we used the fact that $\operatorname{rect}(-x) = \operatorname{rect}(x)$ because rect is an even function. Figure 4.17b shows this pair graphically. Observe the interchange of the roles of t and ω

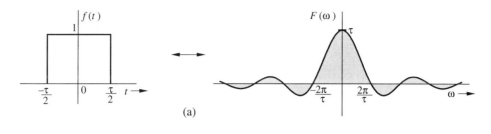

(a)

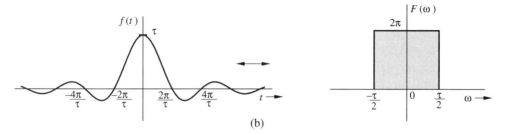

(b)

Fig. 4.17 Symmetry property of the Fourier transform.

(with the minor adjustment of the factor 2π). This result appears as pair 18 in Table 4.1 (with $\tau/2 = W$).

As an interesting exercise, the reader should generate the dual of every pair in Table 4.1 by applying the symmetry property.   ■

△ **Exercise E4.5**

Apply symmetry property to pairs 1, 3, and 9 (Table 4.1) to show that

(a) $\frac{1}{jt+a} \Longleftrightarrow 2\pi e^{a\omega} u(-\omega)$ (b) $\frac{2a}{t^2+a^2} \Longleftrightarrow 2\pi e^{-a|\omega|}$
(c) $\delta(t+t_0) + \delta(t-t_0) \Longleftrightarrow 2\cos t_0\omega$ ▽

4.3-3 The Scaling Property

If

$$f(t) \Longleftrightarrow F(\omega)$$

then, for any real constant a,

$$f(at) \Longleftrightarrow \frac{1}{|a|}F\left(\frac{\omega}{a}\right) \tag{4.34}$$

Proof: For a positive real constant a,

$$\mathcal{F}[f(at)] = \int_{-\infty}^{\infty} f(at)e^{-j\omega t}\,dt = \frac{1}{a}\int_{-\infty}^{\infty} f(x)e^{(-j\omega/a)x}\,dx = \frac{1}{a}F\left(\frac{\omega}{a}\right)$$

Similarly, we can demonstrate that if $a < 0$,

$$f(at) \Longleftrightarrow \frac{-1}{a}F\left(\frac{\omega}{a}\right)$$

Hence follows Eq. (4.34).

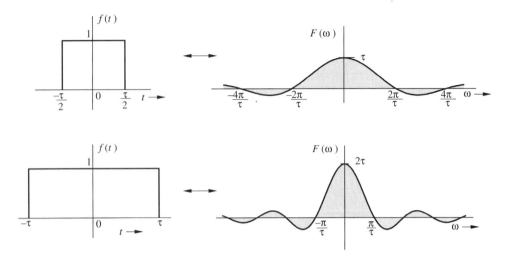

Fig. 4.18   Scaling property of the Fourier transform.

Significance of the Scaling Property

The function $f(at)$ represents the function $f(t)$ compressed in time by a factor a (see Sec. 1.3-2). Similarly, a function $F(\frac{\omega}{a})$ represents the function $F(\omega)$ expanded in frequency by the same factor a. *The scaling property states that time compression of a signal results in its spectral expansion, and time expansion of the signal results in its spectral compression.* Intuitively, compression in time by a factor a means that the signal is varying rapidly by the same factor. To synthesize such a signal, the frequencies of its sinusoidal components must be increased by the factor a, implying that its frequency spectrum is expanded by the factor a. Similarly, a signal expanded in time varies more slowly; hence the frequencies of its components are lowered, implying that its frequency spectrum is compressed. For instance, the signal $\cos 2\omega_0 t$ is the same as the signal $\cos \omega_0 t$ time-compressed by a factor of 2. Clearly, the spectrum of the former (impulse at $\pm 2\omega_0$) is an expanded version of the spectrum of the latter (impulse at $\pm \omega_0$). The effect of this scaling is demonstrated in Fig. 4.18.

Reciprocity of Signal Duration and Its Bandwidth

The scaling property implies that if $f(t)$ is wider, its spectrum is narrower, and vice versa. Doubling the signal duration halves its bandwidth and vice versa. This suggests that the bandwidth of a signal is inversely proportional to the signal duration or width (in seconds). We have already verified this fact for the gate pulse, where we found that the bandwidth of a gate pulse of width τ seconds is $\frac{1}{\tau}$ Hz. More discussion of this interesting topic appears in the literature.[2]

By letting $a = -1$ in Eq. (4.34), we obtain the **time and frequency inversion property**

$$f(-t) \iff F(-\omega) \tag{4.35}$$

■ **Example 4.9**

Find the Fourier transforms of $e^{at}u(-t)$ and $e^{-a|t|}$.

Application of Eq. (4.35) to pair 1 (Table 4.1) yields

$$e^{at}u(-t) \Longleftrightarrow \frac{1}{a - j\omega}$$

Also

$$e^{-a|t|} = e^{-at}u(t) + e^{at}u(-t)$$

Therefore

$$e^{-a|t|} \Longleftrightarrow \frac{1}{a + j\omega} + \frac{1}{a - j\omega} = \frac{2a}{a^2 + \omega^2} \qquad (4.36)$$

The signal $e^{-a|t|}$ and its spectrum are illustrated in Fig. 4.19.   ■

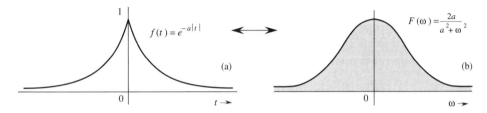

Fig. 4.19 $e^{-a|t|}$ and its Fourier spectrum.

4.3-4   The Time-Shifting Property

If

$$f(t) \Longleftrightarrow F(\omega)$$

then

$$f(t - t_0) \Longleftrightarrow F(\omega)e^{-j\omega t_0} \qquad (4.37a)$$

Proof: By definition,

$$\mathcal{F}[f(t - t_0)] = \int_{-\infty}^{\infty} f(t - t_0)e^{-j\omega t}\, dt$$

Letting $t - t_0 = x$, we have

$$\mathcal{F}[f(t - t_0)] = \int_{-\infty}^{\infty} f(x)e^{-j\omega(x + t_0)}\, dx$$

$$= e^{-j\omega t_0} \int_{-\infty}^{\infty} f(x)e^{-j\omega x}\, dx = F(\omega)e^{-j\omega t_0} \qquad (4.37b)$$

This result shows that *delaying a signal by t_0 seconds does not change its amplitude spectrum. The phase spectrum, however, is changed by $-\omega t_0$.*

Physical Explanation of the Linear Phase

Time delay in a signal causes a linear phase shift in its spectrum. This result can also be derived by heuristic reasoning. Imagine $f(t)$ being synthesized by its Fourier components, which are sinusoids of certain amplitudes and phases. The delayed signal $f(t - t_0)$ can be synthesized by the same sinusoidal components,

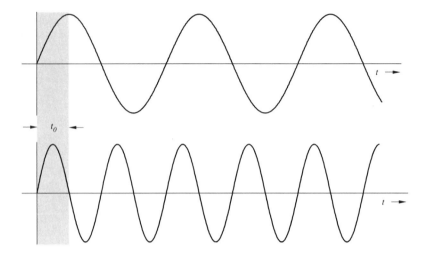

Fig. 4.20 Physical explanation of the time-shifting property.

each delayed by t_0 seconds. The amplitudes of the components remain unchanged. Therefore, the amplitude spectrum of $f(t-t_0)$ is identical to that of $f(t)$. The time delay of t_0 in each sinusoid, however, does change the phase of each component. Now, a sinusoid $\cos \omega t$ delayed by t_0 is given by

$$\cos \omega (t - t_0) = \cos (\omega t - \omega t_0)$$

Therefore a time delay t_0 in a sinusoid of frequency ω manifests as a phase delay of ωt_0. This is a linear function of ω, meaning that higher-frequency components must undergo proportionately higher phase shifts to achieve the same time delay. This effect is depicted in Fig. 4.20 with two sinusoids, the frequency of the lower sinusoid being twice that of the upper. The same time delay t_0 amounts to a phase shift of $\pi/2$ in the upper sinusoid and a phase shift of π in the lower sinusoid. This verifies the fact that *to achieve the same time delay, higher-frequency sinusoids must undergo proportionately higher phase shifts*. The principle of linear phase shift is very important and we shall encounter it again in distortionless signal transmission and filtering applications.

■ **Example 4.10**

Find the Fourier transform of $e^{-a|t-t_0|}$.

This function, shown in Fig. 4.21a, is a time-shifted version of $e^{-a|t|}$ (depicted in Fig. 4.19a). From Eqs. (4.36) and (4.37) we have

$$e^{-a|t-t_0|} \Longleftrightarrow \frac{2a}{a^2 + \omega^2} e^{-j\omega t_0} \tag{4.38}$$

The spectrum of $e^{-a|t-t_0|}$ (Fig. 4.21b) is the same as that of $e^{-a|t|}$ (Fig. 4.19b), except for an added phase shift of $-\omega t_0$.

Observe that the time delay t_0 causes a linear phase spectrum $-\omega t_0$. This example clearly demonstrates the effect of time shift. ■

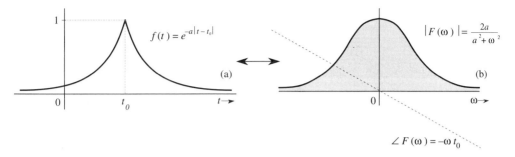

Fig. 4.21 Effect of time-shifting on the Fourier spectrum of a signal.

■ Example 4.11

Find the Fourier transform of the gate pulse $f(t)$ illustrated in Fig. 4.22a.

The pulse $f(t)$ is the gate pulse rect $\left(\frac{t}{\tau}\right)$ in Fig. 4.10a delayed by $\tau/2$ seconds. Hence, according to Eq. (4.37a), its Fourier transform is the Fourier transform of rect $\left(\frac{t}{\tau}\right)$ multiplied by $e^{-j\omega\frac{\tau}{2}}$. Therefore

$$F(\omega) = \tau \operatorname{sinc}\left(\frac{\omega\tau}{2}\right) e^{-j\omega\frac{\tau}{2}}$$

The amplitude spectrum $|F(\omega)|$ (depicted in Fig. 4.22b) of this pulse is the same as that indicated in Fig. 4.10c. But the phase spectrum has an added linear term $-\omega\tau/2$. Hence, the phase spectrum of $f(t)$ is identical to that in Fig. 4.10b plus a linear term $-\omega\tau/2$, as indicated in Fig. 4.22c. ■

△ Exercise E4.6

Using pair 18 and the time-shifting property, show that the Fourier transform of $\operatorname{sinc}[\omega_0(t - T)]$ is $\frac{\pi}{\omega_0} \operatorname{rect}\left(\frac{\omega}{2\omega_0}\right) e^{-j\omega T}$. Sketch the amplitude and phase spectra of the Fourier transform. ▽

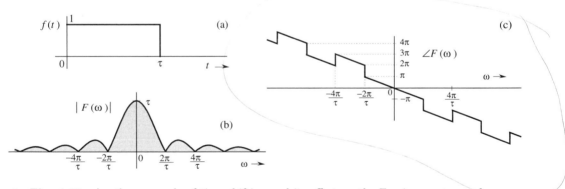

Fig. 4.22 Another example of time-shifting and its effect on the Fourier spectrum of a signal.

4.3-5 The Frequency-Shifting Property

If

$$f(t) \Longleftrightarrow F(\omega)$$

then

$$f(t)e^{j\omega_0 t} \Longleftrightarrow F(\omega - \omega_0) \tag{4.39}$$

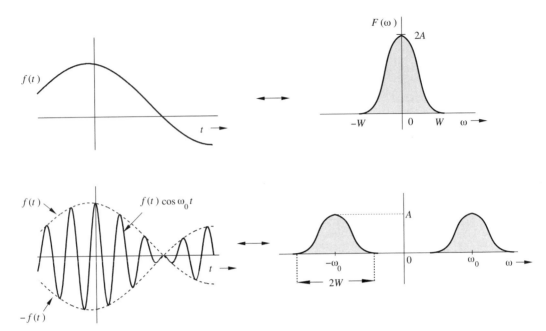

Fig. 4.23   Amplitude modulation of a signal causes spectral shifting.

Proof: By definition,

$$\mathcal{F}[f(t)e^{j\omega_0 t}] = \int_{-\infty}^{\infty} f(t)e^{j\omega_0 t}e^{-j\omega t}dt = \int_{-\infty}^{\infty} f(t)e^{-j(\omega-\omega_0)t}dt = F(\omega - \omega_0)$$

According to this property, the multiplication of a signal by a factor $e^{j\omega_0 t}$ shifts the spectrum of that signal by $\omega = \omega_0$. Note the duality between the time-shifting and the frequency-shifting properties.

Changing ω_0 to $-\omega_0$ in Eq. (4.39) yields

$$f(t)e^{-j\omega_0 t} \iff F(\omega + \omega_0) \tag{4.40}$$

Because $e^{j\omega_0 t}$ is not a real function that can be generated, frequency shifting in practice is achieved by multiplying $f(t)$ by a sinusoid. This assertion follows from the fact that

$$f(t)\cos \omega_0 t = \tfrac{1}{2}[f(t)e^{j\omega_0 t} + f(t)e^{-j\omega_0 t}]$$

From Eqs. (4.39) and (4.40), it follows that

$$f(t)\cos \omega_0 t \iff \tfrac{1}{2}[F(\omega - \omega_0) + F(\omega + \omega_0)] \tag{4.41}$$

This result shows that the multiplication of a signal $f(t)$ by a sinusoid of frequency ω_0 shifts the spectrum $F(\omega)$ by $\pm\omega_0$, as depicted in Fig. 4.23.

Multiplication of a sinusoid $\cos \omega_0 t$ by $f(t)$ amounts to modulating the sinusoid amplitude. This type of modulation is known as **amplitude modulation**. The sinusoid $\cos \omega_0 t$ is called the **carrier**, the signal $f(t)$ is the **modulating signal**, and the signal $f(t)\cos \omega_0 t$ is the **modulated signal**. Further discussion of modulation

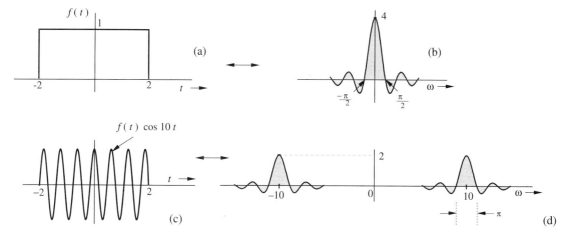

Fig. 4.24   An example of spectral shifting by amplitude modulation.

and demodulation appears in Secs. 4.7 and 4.8.

To sketch a signal $f(t) \cos \omega_0 t$, we observe that

$$f(t) \cos \omega_0 t = \begin{cases} f(t) & \text{when } \cos \omega_0 t = 1 \\ -f(t) & \text{when } \cos \omega_0 t = -1 \end{cases}$$

Therefore, $f(t) \cos \omega_0 t$ touches $f(t)$ when the sinusoid $\cos \omega_0 t$ is at its positive peaks and touches $-f(t)$ when $\cos \omega_0 t$ is at its negative peaks. This means that $f(t)$ and $-f(t)$ act as envelopes for the signal $f(t) \cos \omega_0 t$ (see Fig. 4.23). The signal $-f(t)$ is a mirror image of $f(t)$ about the horizontal axis. Figure 4.23 shows the signal $f(t)$, $f(t) \cos \omega_0 t$ and their spectra.

■ Example 4.12

Find and sketch the Fourier transform of the modulated signal $f(t) \cos 10t$ in which $f(t)$ is a gate pulse rect $(\frac{t}{4})$ as illustrated in Figure 4.24a.

From pair 17 (Table 4.1) we find rect $(\frac{t}{4}) \Longleftrightarrow 4 \operatorname{sinc}(2\omega)$, which is depicted in Fig. 4.24b. From Eq. (4.41) it follows that

$$f(t) \cos 10t \Longleftrightarrow \tfrac{1}{2}[F(\omega + 10) + F(\omega - 10)]$$

In this case, $F(\omega) = 4 \operatorname{sinc}(2\omega)$. Therefore

$$f(t) \cos 10t \Longleftrightarrow 2 \operatorname{sinc}[2(\omega + 10)] + 2 \operatorname{sinc}[2(\omega - 10)]$$

The spectrum of $f(t) \cos 10t$ is obtained by shifting $F(\omega)$ in Fig. 4.24b to the left by 10 and also to the right by 10, and then multiplying it by one-half, as depicted in Fig. 4.24d. ■

△ Exercise E4.7

Sketch signal $e^{-|t|} \cos 10t$. Find the Fourier transform of this signal and sketch its spectrum.

Answer: $F(\omega) = \frac{1}{(\omega - 10)^2 + 1} + \frac{1}{(\omega + 10)^2 + 1}$. The spectrum is that in Fig. 4.19b (with $a = 1$), shifted to ± 10 and multiplied by one-half.   ▽

Application to Modulation

Modulation is used to shift signal spectra. Some of the situations where spectrum shifting is necessary are presented next.

1. If several signals, each occupying the same frequency band, are transmitted simultaneously over the same transmission medium, they will all interfere; it will be impossible to separate or retrieve them at a receiver. For example, if all radio stations decide to broadcast audio signals simultaneously, the receiver will not be able to separate them. This problem is solved by using modulation, whereby each radio station is assigned a distinct carrier frequency. Each station transmits a modulated signal. This procedure shifts the signal spectrum to its allocated band, which is not occupied by any other station. A radio receiver can pick up any station by tuning to the band of the desired station. The receiver must now demodulate the received signal (undo the effect of modulation). Demodulation therefore consists of another spectral shift required to restore the signal to its original band. Note that both modulation and demodulation implement spectral shifting; consequently, demodulation operation is similar to modulation (see Sec. 4.7).

   This method of transmitting several signals simultaneously over a channel by sharing its frequency band is known as **frequency-division multiplexing (FDM)**.

2. For effective radiation of power over a radio link, the antenna size must be of the order of the wavelength of the signal to be radiated. Audio signal frequencies are so low (wavelengths are so large) that impracticably large antennas will be required for radiation. Here, shifting the spectrum to a higher frequency (a smaller wavelength) by modulation solves the problem.

4.3-6 Convolution

The time convolution property and its dual, the frequency convolution property, state that if

$$f_1(t) \Longleftrightarrow F_1(\omega) \qquad \text{and} \qquad f_2(t) \Longleftrightarrow F_2(\omega)$$

then (**time convolution**)

$$f_1(t) * f_2(t) \Longleftrightarrow F_1(\omega)F_2(\omega) \tag{4.42}$$

and (**frequency convolution**)

$$f_1(t)f_2(t) \Longleftrightarrow \frac{1}{2\pi}F_1(\omega) * F_2(\omega) \tag{4.43}$$

Proof: By definition

$$\mathcal{F}|f_1(t) * f_2(t)| = \int_{-\infty}^{\infty} e^{-j\omega t} \left[\int_{-\infty}^{\infty} f_1(\tau)f_2(t - \tau)d\tau \right] dt$$

$$= \int_{-\infty}^{\infty} f_1(\tau) \left[\int_{-\infty}^{\infty} e^{-j\omega t} f_2(t - \tau)dt \right] d\tau$$

The inner integral is the Fourier transform of $f_2(t - \tau)$, given by [time-shifting property in Eq. (4.37)] $F_2(\omega)e^{-j\omega\tau}$. Hence

$$\mathcal{F}[f_1(t)*f_2(t)] = \int_{-\infty}^{\infty} f_1(\tau)e^{-j\omega\tau} F_2(\omega)d\tau = F_2(\omega) \int_{-\infty}^{\infty} f_1(\tau)e^{-j\omega\tau}d\tau = F_1(\omega)F_2(\omega)$$

We have demonstrated in Eq. (2.48) that the transfer function $H(\omega)$ is the Fourier transform of the unit impulse response $h(t)$. Thus

$$h(t) \Longleftrightarrow H(\omega) \tag{4.44a}$$

Application of the time convolution property to $y(t) = f(t) * h(t)$ yields (assuming both $f(t)$ and $h(t)$ are Fourier transformable)

$$Y(\omega) = F(\omega)H(\omega) \tag{4.44b}$$

This is precisely the result proved earlier in Eq. (4.19).†
The frequency convolution property (4.43) can be proved in exactly the same way by reversing the roles of $f(t)$ and $F(\omega)$.

■ **Example 4.13**
Using the time convolution property, show that if

$$f(t) \Longleftrightarrow F(\omega)$$

then

$$\int_{-\infty}^{t} f(\tau)d\tau \Longleftrightarrow \frac{F(\omega)}{j\omega} + \pi F(0)\delta(\omega) \tag{4.45}$$

Because

$$u(t - \tau) = \begin{cases} 1 & \tau \le t \\ 0 & \tau > t \end{cases}$$

it follows that

$$f(t) * u(t) = \int_{-\infty}^{\infty} f(\tau)u(t - \tau)d\tau = \int_{-\infty}^{t} f(\tau)\,d\tau$$

---

†In Eq. (4.44b), $h(t) \Longleftrightarrow H(\omega)$. To understand finer points of Eq. (4.44b), see footnote on p. 243.

Now, from the time convolution property [Eq. 4.42], it follows that

$$f(t) * u(t) = \int_{-\infty}^{t} f(\tau)\,d\tau \Longleftrightarrow F(\omega)\left[\frac{1}{j\omega} + \pi\delta(\omega)\right]$$

$$= \frac{F(\omega)}{j\omega} + \pi F(0)\delta(\omega)$$

In deriving the last result, we used Eq. (1.23a)   ■

△ **Exercise E4.8**

Using the time convolution property, show that $f(t) * \delta(t) = f(t)$ ▽

△ **Exercise E4.9**

Using the time convolution property, show that

$$e^{-at}u(t) * e^{-bt}u(t) = \frac{1}{b-a}\left[e^{-at} - e^{-bt}\right]u(t)$$

Hint: Use property (4.42) to find the Fourier transform of $e^{-at}u(t) * e^{-bt}u(t)$. Then use partial fraction expansion to find its inverse Fourier transform.   ▽

4.3-7 Time Differentiation and Time Integration

If

$$f(t) \Longleftrightarrow F(\omega)$$

then **(time differentiation)**†

$$\frac{df}{dt} \Longleftrightarrow j\omega F(\omega) \tag{4.46}$$

and **(time integration)**

$$\int_{-\infty}^{t} f(\tau)d\tau \Longleftrightarrow \frac{F(\omega)}{j\omega} + \pi F(0)\delta(\omega) \tag{4.47}$$

Proof: Differentiation of both sides of Eq. (4.8b) yields

$$\frac{df}{dt} = \frac{1}{2\pi}\int_{-\infty}^{\infty} j\omega F(\omega)e^{j\omega t}\,d\omega$$

This result shows that

$$\frac{df}{dt} \Longleftrightarrow j\omega F(\omega)$$

---

†Valid only if the transform of df/dt exists.

Table 4.2

Fourier Transform Operations

Operation	$f(t)$	$F(\omega)$
Addition	$f_1(t) + f_2(t)$	$F_1(\omega) + F_2(\omega)$
Scalar multiplication	$k f(t)$	$k F(\omega)$
Symmetry	$F(t)$	$2\pi f(-\omega)$
Scaling (a real)	$f(at)$	$\dfrac{1}{\lvert a \rvert} F\left(\dfrac{\omega}{a}\right)$
Time shift	$f(t - t_0)$	$F(\omega) e^{-j\omega t_0}$
Frequency shift (ω_0 real)	$f(t) e^{j\omega_0 t}$	$F(\omega - \omega_0)$
Time convolution	$f_1(t) * f_2(t)$	$F_1(\omega) F_2(\omega)$
Frequency convolution	$f_1(t) f_2(t)$	$\dfrac{1}{2\pi} F_1(\omega) * F_2(\omega)$
Time differentiation	$\dfrac{d^n f}{dt^n}$	$(j\omega)^n F(\omega)$
Time integration	$\displaystyle \int_{-\infty}^{t} f(x)\, dx$	$\dfrac{F(\omega)}{j\omega} + \pi F(0) \delta(\omega)$

Repeated application of this property yields

$$\frac{d^n f}{dt^n} \Longleftrightarrow (j\omega)^n F(\omega) \tag{4.48}$$

The time-integration property [Eq. (4.47)] has already been proved in Example 4.13.

■ **Example 4.14**

Using the time-differentiation property, find the Fourier transform of the triangle pulse $\Delta(\frac{t}{\tau})$ illustrated in Fig. 4.25a.

To find the Fourier transform of this pulse we differentiate the pulse successively, as illustrated in Fig. 4.25b and c. Because df/dt is constant everywhere, its derivative, $d^2 f/dt^2$, is zero everywhere. But df/dt has jump discontinuities with a positive jump of $2/\tau$ at $t = \pm\frac{\tau}{2}$, and a negative jump of $4/\tau$ at $t = 0$. Recall that the derivative of a signal at a jump discontinuity is an impulse at that point of strength equal to the amount of jump. Hence, $d^2 f/dt^2$, the derivative of df/dt, consists of a sequence of impulses, as depicted in Fig. 4.25c; that is,

$$\frac{d^2 f}{dt^2} = \frac{2}{\tau}[\delta(t + \tfrac{\tau}{2}) - 2\delta(t) + \delta(t - \tfrac{\tau}{2})] \tag{4.49}$$

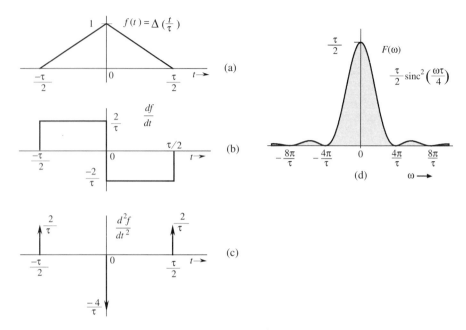

Fig. 4.25   Finding the Fourier transform of a piecewise-linear signal using the time-differentiation property.

From the time-differentiation property (4.48)

$$\frac{d^2 f}{dt^2} \Longleftrightarrow (j\omega)^2 F(\omega) = -\omega^2 F(\omega) \tag{4.50a}$$

Also, from the time-shifting property (4.37)

$$\delta(t - t_0) \Longleftrightarrow e^{-j\omega t_0} \tag{4.50b}$$

Taking the Fourier transform of Eq. (4.49) and using the results in Eqs. (4.50), we obtain

$$-\omega^2 F(\omega) = \tfrac{2}{\tau}[e^{j\frac{\omega\tau}{2}} - 2 + e^{-j\frac{\omega\tau}{2}}] = \tfrac{4}{\tau}\left(\cos \tfrac{\omega\tau}{2} - 1\right) = -\tfrac{8}{\tau} \sin^2 \left(\tfrac{\omega\tau}{4}\right)$$

and

$$F(\omega) = \frac{8}{\omega^2\tau} \sin^2\left(\frac{\omega\tau}{4}\right) = \frac{\tau}{2}\left[\frac{\sin\left(\frac{\omega\tau}{4}\right)}{\frac{\omega\tau}{4}}\right]^2 = \frac{\tau}{2} \operatorname{sinc}^2\left(\frac{\omega\tau}{4}\right) \tag{4.51}$$

The spectrum $F(\omega)$ is depicted in Fig. 4.25d. This procedure of finding the Fourier transform can be applied to any function $f(t)$ made up of straight-line segments with $f(t) \rightarrow 0$ as $|t| \rightarrow \infty$. The second derivative of such a signal yields a sequence of impulses whose Fourier transform can be found by inspection. This example suggests a numerical method of finding the Fourier transform of an arbitrary signal $f(t)$ by approximating the signal by straight-line segments. ■

△ **Exercise E4.10**

  Find the Fourier transform of rect $\left(\frac{t}{\tau}\right)$, using the time-differentiation property. ▽

4.4 Signal Transmission through LTIC Systems

If $f(t)$ and $y(t)$ are the input and output of an LTIC system with transfer function $H(\omega)$, then, as demonstrated in Eq. (4.44b)

$$Y(\omega) = H(\omega)F(\omega) \tag{4.52}$$

This result applies only to asymptotically (and marginally) stable systems because of the reasons discussed in the footnote of p. 243. Moreover, $f(t)$ has to be Fourier transformable. Consequently, exponentially growing inputs cannot be handled by this method.

In Chapter 6, we shall see that the Laplace transform, which is a generalized Fourier transform, is more versatile and capable of analyzing all kinds of LTIC systems whether stable, unstable, or marginally stable. Laplace transform can also handle exponentially growing inputs. Compared to the Laplace transform, the Fourier transform in system analysis is clumsier. Hence, the Laplace transform is preferable to the Fourier transform in LTIC system analysis, and we shall not belabor the application of the Fourier transform to LTIC system analysis. We consider just one example here.

■ **Example 4.15**

Find the zero-state response of a stable LTIC system with transfer function†

$$H(s) = \frac{1}{s+2} \tag{4.53}$$

and the input $f(t) = e^{-t}u(t)$. In this case,

$$F(\omega) = \frac{1}{j\omega + 1} \quad \text{and} \quad H(\omega) = H(s)|_{s=j\omega} = \frac{1}{j\omega + 2}$$

Therefore

$$Y(\omega) = H(\omega)F(\omega)$$

$$= \frac{1}{(j\omega + 2)(j\omega + 1)}$$

Expanding the right-hand side in partial fractions (Sec. B.5)

$$Y(\omega) = \frac{1}{j\omega + 1} - \frac{1}{j\omega + 2} \tag{4.54}$$

and

$$y(t) = (e^{-t} - e^{-2t})u(t) \quad ■$$

△ **Exercise E4.11**

For the system in Example 4.15, show that the zero-input response to the input $e^t u(-t)$ is $y(t) = \frac{1}{3}[e^t u(-t) + e^{-2t}u(t)]$.

Hint: Use pair 2 (Table 4.1) to find the Fourier transform of $e^t u(-t)$. ▽

---

†Stability implies that the region of convergence of $H(s)$ includes the $j\omega$ axis.

4.4-1 Signal Distortion during Transmission

For a system with transfer function $H(\omega)$, if $F(\omega)$ and $Y(\omega)$ are the spectra of the input and the output signals, respectively, then

$$Y(\omega) = F(\omega)H(\omega) \tag{4.55}$$

The transmission of the input signal $f(t)$ through the system changes it into the output signal $y(t)$. Equation (4.55) shows the nature of this change or modification. Here $F(\omega)$ and $Y(\omega)$ are the spectra of the input and the output, respectively. Therefore, $H(\omega)$ is the spectral response of the system. The output spectrum is obtained by the input spectrum multiplied by the spectral response of the system. Equation (4.55), which clearly brings out the spectral shaping (or modification) of the signal by the system, can be expressed in polar form as

$$|Y(\omega)|e^{j\angle Y(\omega)} = |F(\omega)||H(\omega)|e^{j[\angle F(\omega)+\angle H(j\omega)]}$$

Therefore

$$|Y(\omega)| = |F(\omega)|\,|H(\omega)| \tag{4.56a}$$

$$\angle Y(\omega) = \angle F(\omega) + \angle H(\omega) \tag{4.56b}$$

During transmission, the input signal amplitude spectrum $|F(\omega)|$ is changed to $|F(\omega)||H(\omega)|$. Similarly, the input signal phase spectrum $\angle F(\omega)$ is changed to $\angle F(\omega) + \angle H(\omega)$. An input signal spectral component of frequency ω is modified in amplitude by a factor $|H(\omega)|$ and is shifted in phase by an angle $\angle H(\omega)$. Clearly, $|H(\omega)|$ is the amplitude response, and $\angle H(\omega)$ is the phase response of the system. The plots of $|H(\omega)|$ and $\angle H(\omega)$ as functions of ω show at a glance how the system modifies the amplitudes and phases of various sinusoidal inputs. For this reason, $H(\omega)$ is called the **frequency response** of the system. During transmission through the system, some frequency components may be boosted in amplitude, while others may be attenuated. The relative phases of the various components also change. In general, the output waveform will be different from the input waveform.

Distortionless Transmission

In several applications, such as signal amplification or message signal transmission over a communication channel, we require that the output waveform be a replica of the input waveform. In such cases we need to minimize the distortion caused by the amplifier or the communication channel. It is, therefore, of practical interest to determine the characteristics of a system that allows a signal to pass without distortion (**distortionless transmission**).

Transmission is said to be distortionless if the input and the output have identical waveshapes within a multiplicative constant. A delayed output that retains the input waveform is also considered distortionless. Thus, in distortionless transmission, the input $f(t)$ and the output $y(t)$ satisfy the condition

$$y(t) = kf(t - t_d) \tag{4.57}$$

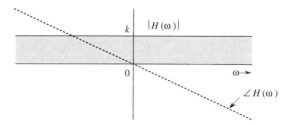

Fig. 4.26 LTIC system Frequency response for distortionless transmission.

The Fourier transform of this equation yields

$$Y(\omega) = kF(\omega)e^{-j\omega t_d}$$

But

$$Y(\omega) = F(\omega)H(\omega)$$

Therefore

$$H(\omega) = k\,e^{-j\omega t_d}$$

This is the transfer function required for distortionless transmission. From this equation it follows that

$$|H(\omega)| = k \qquad\qquad (4.58a)$$

$$\angle H(\omega) = -\omega t_d \qquad\qquad (4.58b)$$

This result shows that for distortionless transmission, the amplitude response $|H(\omega)|$ must be a constant, and the phase response $\angle H(\omega)$ must be a linear function of ω with slope $-t_d$, where t_d is the delay of the output with respect to input (Fig. 4.26).

Intuitive Explanation of the Distortionless Transmission Conditions

It is instructive to derive the conditions for distortionless transmission heuristically. Once again, imagine $f(t)$ to be composed of various sinusoids (its spectral components), which are being passed through a distortionless system. For the distortionless case, the output signal is the input signal multiplied by k and delayed by t_d. To synthesize such a signal, we need exactly the same components as those of $f(t)$, with each component multiplied by k and delayed by t_d. Thus, the system transfer function $H(\omega)$ should be such that each sinusoidal component suffers the same attenuation k and each component undergoes the same time delay of t_d seconds. The first condition requires that

$$|H(\omega)| = k$$

We have seen in our discussion on p. 258 that to achieve the same time delay t_d for every frequency component requires a linear phase delay ωt_d (Fig. 4.20). Therefore

$$\angle H(\omega) = -\omega t_d$$

This equation shows that the time delay resulting from signal transmission through a system is the negative of the slope of the system phase response $\angle H(\omega)$; that is,

$$t_d(\omega) = -\frac{d}{d\omega}\angle H(\omega) \qquad\qquad (4.59)$$

If the slope of $\angle H(\omega)$ is constant (that is, if $\angle H(\omega)$ is linear with ω), all the components are delayed by the same time interval t_d. But if the slope is not constant, the time delay t_d varies with frequency. This variation means that different frequency components undergo different amounts of time delay, and consequently the output waveform will not be a replica of the input waveform. A good way to judge phase linearity is to plot t_d as a function of frequency. For a distortionless system, t_d should be constant over the band of interest.

It is often thought (erroneously) that flatness of amplitude response $|H(\omega)|$ alone can guarantee signal quality. However, a system may have a flat amplitude response and yet distort a signal beyond recognition if the phase response is not linear (t_d not constant).

The Nature of Distortion in Audio and Video Signals

Generally speaking, a human ear can readily perceive amplitude distortion, although it is relatively insensitive to phase distortion. For the phase distortion to become noticeable, the variation in delay [variation in the slope of $\angle H(\omega)$] should be comparable to the signal duration (or the physically perceptible duration, in case the signal itself is long). In the case of audio signals, each spoken syllable can be considered as an individual signal. The average duration of a spoken syllable is of a magnitude of the order of 0.01 to 0.1 seconds. The audio systems may have nonlinear phases, yet no noticeable signal distortion results because in practical audio systems, maximum variation in the slope of $\angle H(\omega)$ is only a small fraction of a millisecond. This is the real truth underlying the statement that "the human ear is relatively insensitive to phase distortion."[3] As a result, the manufacturers of audio equipment make available only $|H(\omega)|$, the amplitude response characteristic of their systems.

For video signals, in contrast, the situation is exactly the opposite. The human eye is sensitive to phase distortion but is relatively insensitive to amplitude distortion. The amplitude distortion in television signals manifests itself as a partial destruction of the relative half-tone values of the resulting picture, which is not readily apparent to the human eye. The phase distortion (nonlinear phase), on the other hand, causes different time delays in different picture elements. The result is a smeared picture, which is readily perceived by the human eye. Phase distortion is also very important in digital communication systems because the nonlinear phase characteristic of a channel causes pulse dispersion (spreading out), which in turn causes pulses to interfere with neighboring pulses. This interference can cause an error in the pulse amplitude at the receiver: a binary **1** may read as **0**, and vice versa.

4.5   Ideal and practical filters

Ideal filters allow distortionless transmission of a certain band of frequencies and suppress all the remaining frequencies. The ideal lowpass filter (Fig. 4.27), for example, allows all components below $\omega = W$ rad/s to pass without distortion and suppresses all components above $\omega = W$. Figure 4.28 illustrates ideal highpass and bandpass filter characteristics.

The ideal lowpass filter in Fig. 4.27a has a linear phase of slope $-t_d$, which results in a time delay of t_d seconds for all its input components of frequencies

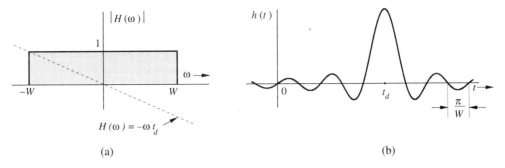

Fig. 4.27   Ideal lowpass filter: its frequency response and impulse response.

below W rad/s. Therefore, if the input is a signal $f(t)$ bandlimited to W rad/s, the output $y(t)$ is $f(t)$ delayed by t_d; that is,

$$y(t) = f(t - t_d)$$

The signal $f(t)$ is transmitted by this system without distortion, but with time delay t_d. For this filter $|H(\omega)| = \text{rect}\left(\frac{\omega}{2W}\right)$ and $\angle H(\omega) = e^{-j\omega t_d}$, so that

$$H(\omega) = \text{rect}\left(\frac{\omega}{2W}\right) e^{-j\omega t_d} \tag{4.60a}$$

The unit impulse response $h(t)$ of this filter is obtained from pair 18 (Table 4.1) and the time-shifting property

$$h(t) = \mathcal{F}^{-1}\left[\text{rect}\left(\frac{\omega}{2W}\right) e^{-j\omega t_d}\right]$$

$$= \frac{W}{\pi} \, \text{sinc}\,[W(t - t_d)] \tag{4.60b}$$

Recall that $h(t)$ is the system response to impulse input $\delta(t)$, which is applied at $t = 0$. Figure 4.27b shows a curious fact: the response $h(t)$ begins even before the input is applied (at $t = 0$). Clearly, the filter is noncausal and therefore physically unrealizable. Similarly, one can show that other ideal filters (such as the ideal highpass or the ideal bandpass filters depicted in Fig. 4.28) are also physically unrealizable.

For a physically realizable system, $h(t)$ must be causal; that is,

$$h(t) = 0 \qquad \text{for } t < 0$$

In the frequency domain, this condition is equivalent to the well-known **Paley-Wiener criterion**, which states that the necessary and sufficient condition for the amplitude response $|H(\omega)|$ to be realizable is

$$\int_{-\infty}^{\infty} \frac{|\ln|H(\omega)||}{1 + \omega^2} d\omega < \infty \tag{4.61}$$

If $H(\omega)$ does not satisfy this condition, it is unrealizable. Note that if $|H(\omega)| = 0$ over any finite band, $|\ln|H(\omega)|| = \infty$ over that band, and the condition (4.61) is violated. If, however, $H(\omega) = 0$ at a single frequency (or a set of discrete frequencies),

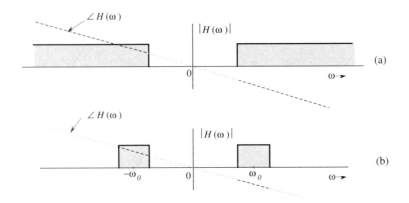

Fig. 4.28 Ideal highpass and bandpass filter frequency response.

Fig. 4.29 Approximate realization of an ideal lowpass filter by truncation of its impulse response.

the integral in Eq. (4.61) may still be finite even though the integrand is infinite. Therefore, for a physically realizable system, $H(\omega)$ may be zero at some discrete frequencies, but it cannot be zero over any finite band. According to this criterion, ideal filter characteristics (Figs. 4.27 and 4.28) are clearly unrealizable.†

The impulse response $h(t)$ in Fig. 4.27 is not realizable. One practical approach to filter design is to cut off the tail of $h(t)$ for $t < 0$. The resulting causal impulse response $\widehat{h}(t)$, where

$$\widehat{h}(t) = h(t)u(t)$$

is physically realizable because it is causal (Fig. 4.29). If t_d is sufficiently large, $\widehat{h}(t)$ will be a close approximation of $h(t)$, and the resulting filter $\widehat{H}(\omega)$ will be a good approximation of an ideal filter. This close realization of the ideal filter is achieved because of the increased value of time-delay t_d. This observation means that the price of close realization is higher delay in the output; this situation is common in noncausal systems. Of course, theoretically, a delay $t_d = \infty$ is needed to realize the ideal characteristics. But a glance at Fig. 4.27b shows that a delay t_d of three or four times $\frac{\pi}{W}$ will make $\widehat{h}(t)$ a reasonably close version of $h(t-t_d)$. For instance, an audio

---

†$|H(\omega)|$ is assumed to be square-integrable, that is,

$$\int_{-\infty}^{\infty} |H(\omega)|^2 \, d\omega$$

is finite. Note that the Paley-Wiener criterion is a criterion for the realizability of the amplitude response $|H(\omega)|$.

filter is required to handle frequencies of up to 20 kHz ($W = 40,000\pi$). In this case, a t_d of about 10^{-4} (0.1 ms) would be a reasonable choice. The truncation operation (cutting the tail of $h(t)$ to make it causal), however, creates some unsuspected problems. We discuss these problems and their cure in Sec. 4.9.

In practice, we can realize a variety of filter characteristics to approach ideal characteristics. Practical (realizable) filter characteristics are gradual, without jump discontinuities in amplitude response. We shall study such filter families (Butterworth and Chebyshev) in Secs. 7.4 and 7.5. Figure 7.17 illustrates the amplitude response of lowpass Butterworth filters.

△ **Exercise E4.12**

Show that a filter with Gaussian transfer function $H(\omega) = e^{-\alpha\omega^2}$ is unrealizable. Demonstrate this fact in two ways: first by showing that its impulse response is noncausal, and then by showing that $|H(\omega)|$ violates the Paley-Wiener criterion.

Hint: Use pair 22 in Table 4.1   ▽

Thinking in Time- and Frequency-Domains: A Two Dimensional View of Signals and Systems

Both signals and systems have dual personalities; the time domain and the frequency domain. For a deeper perspective, we should examine and understand both these identities because they offer complementary insights. An exponential signal, for instance, can be specified by its time domain description such as $e^{-2t}u(t)$ or by its Fourier transform (its frequency domain description) $\frac{1}{j\omega+2}$. The time-domain description depicts the waveform of a signal. The frequency-domain description portrays its spectral composition (relative amplitudes of its sinusoidal (or exponential) components and their phases). For the signal e^{-2t}, for instance, the time-domain description portrays the exponentially decaying signal with a time constant 0.5. The frequency-domain description characterizes it as a lowpass signal, which can be synthesized by sinusoids with amplitudes decaying with frequency roughly as $1/\omega$.

An LTIC system can also be described or specified in the time domain by its impulse response $h(t)$ or in the frequency domain by its transfer function $H(\omega)$. In Sec. 2.7, we studied intuitive insights in the system behavior offered by the impulse response, which consists of characteristic modes of the system. By purely qualitative reasoning, we saw that the system responds well to signals that are similar to the characteristic modes and responds poorly to signals which are very different from those modes. We also saw that the shape of the impulse response $h(t)$ determines the system time constant (speed of response), and pulse dispersion (spreading), which, in turn, determines the rate of pulse transmission.

The transfer function $H(\omega)$ specifies the frequency response; that is, the system response to exponential or sinusoidal input of various frequencies. This is precisely the filtering characteristic of the system.

Experienced electrical engineers instinctively think in both domains (the time and frequency) whenever possible. When they look at a signal, they consider, its waveform, the signal width (duration), and the rate at which the waveform decays. This is basically a time-domain perspective. They also think of the signal in terms of its frequency spectrum–that is, in terms of its sinusoidal components and their relative amplitudes and phases. This is the frequency-domain perspective.

When they think of a system, they think of its impulse response $h(t)$. The width of $h(t)$ indicates the time constant (response time); that is, how fast the system is capable of responding to an input, and how much dispersion (spreading) it will cause. This is the time-domain perspective. From the frequency-domain perspective, these engineers view a system as a filter, which selectively transmits certain frequency components and suppresses the others [frequency response $H(\omega)$]. Knowing the input signal spectrum and the frequency response of the system, they create a mental image of the output signal spectrum. This concept is precisely expressed by $Y(\omega) = F(\omega)H(\omega)$.

We can analyze LTI systems by time-domain techniques or by frequency-domain techniques. Then why learn both? The reason is that the two domains offer complementary insights into system behavior. Some aspects are easily grasped in one domain; other aspects may be easier to see in the other domain. Both the time-domain and the frequency-domain methods are as essential for the study of signals and systems as two eyes are essential to a human being for correct visual perception of reality. A person can see with either eye, but for proper perception of three dimensional-reality, both eyes are essential.

It is important to keep the two domains separate, and not to mix the entities in the two domains. If we are using the frequency domain to determine the system response, we must deal with all signals in terms of their spectra (Fourier transforms) and all systems in terms of their transfer functions. For example, to determine the system response $y(t)$ to an input $f(t)$, we must first convert the input signal into its frequency domain description $F(\omega)$. The system description also must be in the frequency-domain; that is, the transfer function $H(\omega)$. The output signal spectrum $Y(\omega) = F(\omega)H(\omega)$. Thus, the result (output) is also in the frequency domain. To determine the final answer $y(t)$, we must take the inverse transform of $Y(\omega)$.

4.6 Signal Energy

The signal energy E_f of a signal $f(t)$ was defined in Chapter 1 as

$$E_f = \int_{-\infty}^{\infty} |f(t)|^2 \, dt \tag{4.62}$$

Signal energy can be related to the signal spectrum $F(\omega)$ by substituting Eq. (4.8b) in the above equation:

$$E_f = \int_{-\infty}^{\infty} f(t)f^*(t) \, dt = \int_{-\infty}^{\infty} f(t) \left[\frac{1}{2\pi} \int_{-\infty}^{\infty} F^*(\omega)e^{-j\omega t} \, d\omega \right] dt$$

Here we used the fact that $f^*(t)$, being the conjugate of $f(t)$, can be expressed as the conjugate of the right-hand side of Eq. (4.8b). Now, interchanging the order of integration yields

$$E_f = \frac{1}{2\pi} \int_{-\infty}^{\infty} F^*(\omega) \left[\int_{-\infty}^{\infty} f(t)e^{-j\omega t} \, dt \right] d\omega$$

$$= \frac{1}{2\pi} \int_{-\infty}^{\infty} F(\omega)F^*(\omega) \, d\omega$$

$$= \frac{1}{2\pi} \int_{-\infty}^{\infty} |F(\omega)|^2 \, d\omega \tag{4.63}$$

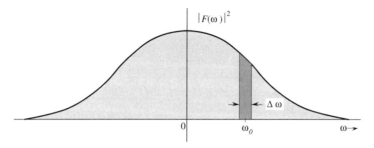

Fig. 4.30   Interpretation of Energy spectral density of a signal.

Consequently,

$$E_f = \int_{-\infty}^{\infty} |f(t)|^2 \, dt = \frac{1}{2\pi} \int_{-\infty}^{\infty} |F(\omega)|^2 \, d\omega \qquad (4.64)$$

This is the statement of the well-known **Parseval's theorem** (for Fourier transform). A similar result was obtained in Eqs. (3.42)and (3.82) for a periodic signal and its Fourier series. This result allows us to determine the signal energy from either the time-domain specification $f(t)$ or the frequency-domain specification $F(\omega)$ of the same signal.

Equation (4.63) can be interpreted to mean that the energy of a signal $f(t)$ results from energies contributed by all the spectral components of the signal $f(t)$. The total signal energy is the area under $|F(\omega)^2|$ (divided by 2π). If we consider a small band $\Delta\omega$ ($\Delta\omega \to 0$), as illustrated in Fig. 4.30, the energy ΔE_f of the spectral components in this band is the area of $|F(\omega)|^2$ under this band (divided by 2π):

$$\Delta E_f = \frac{1}{2\pi} |F(\omega)|^2 \, \Delta\omega = |F(\omega)|^2 \, \Delta\mathcal{F} \qquad \frac{\Delta\omega}{2\pi} = \Delta\mathcal{F} \text{ Hz} \qquad (4.65)$$

Therefore, the energy contributed by the components in this band of $\Delta\mathcal{F}$ (in hertz) is $|F(\omega)|^2\Delta\mathcal{F}$. The total signal energy is the sum of energies of all such bands and is indicated by the area under $|F(\omega)|^2$ as in Eq. (4.63). Therefore, $|F(\omega)|^2$ is the **energy spectral density** (per unit bandwidth in hertz).

For real signals, $F(\omega)$ and $F(-\omega)$ are conjugates, and $|F(\omega)|^2$ is an even function of ω because

$$|F(\omega)|^2 = F(\omega)F^*(\omega) = F(\omega)F(-\omega)$$

Therefore, Eq. (4.63) can be expressed as†

$$E_f = \frac{1}{\pi} \int_0^{\infty} |F(\omega)|^2 \, d\omega \qquad (4.66)$$

The signal energy E_f, which results from contributions from all the frequency components from $\omega = 0$ to ∞, is given by ($1/\pi$ times the area under $|F(\omega)|^2$ from $\omega = 0$ to ∞. It follows that the energy contributed by spectral components of frequencies between ω_1 and ω_2 is

$$\triangle E_f = \frac{1}{\pi} \int_{\omega_1}^{\omega_2} |F(\omega)|^2 \, d\omega \qquad (4.67)$$

---

†In Eq. (4.66) it is assumed that $F(\omega)$ does not contain an impulse at $\omega = 0$. If such an impulse exists, it should be integrated separately with a multiplying factor of $1/2\pi$ rather than $1/\pi$.

■ **Example 4.16**
    Find the energy of signal $f(t) = e^{-at}u(t)$. Determine the frequency W (rad/s) so that the energy contributed by the spectral components of all the frequencies below W is 95% of the signal energy E_f.
    We have

$$E_f = \int_{-\infty}^{\infty} f^2(t)dt = \int_{0}^{\infty} e^{-2at}dt = \frac{1}{2a}$$

We can verify this result by Parseval's theorem. For this signal

$$F(\omega) = \frac{1}{j\omega + a}$$

and

$$E_f = \frac{1}{\pi}\int_{0}^{\infty}|F(\omega)|^2 d\omega = \frac{1}{\pi}\int_{0}^{\infty}\frac{1}{\omega^2 + a^2}d\omega = \frac{1}{\pi a}\tan^{-1}\frac{\omega}{a}\Big|_{0}^{\infty} = \frac{1}{2a}$$

    The band $\omega = 0$ to $\omega = W$ contains 95% of the signal energy, that is, $0.95/2a$. Therefore, from Eq. (4.67) with $\omega_1 = 0$ and $\omega_2 = W$, we obtain

$$\frac{0.95}{2a} = \frac{1}{\pi}\int_{0}^{W}\frac{d\omega}{\omega^2 + a^2} = \frac{1}{\pi a}\tan^{-1}\frac{\omega}{a}\Big|_{0}^{W} = \frac{1}{\pi a}\tan^{-1}\frac{W}{a}$$

or

$$\frac{0.95\pi}{2} = \tan^{-1}\frac{W}{a} \Longrightarrow W = 12.706a \text{ rad/s}$$

This result indicates that the spectral components of $f(t)$ in the band from 0 (dc) to 12.706a rad/s (2.02a Hz) contribute 95% of the total signal energy; all the remaining spectral components (in the band from 12.706a rad/s to ∞) contribute only 5% of the signal energy. ■

△ **Exercise E4.13**
    Use Parseval's theorem to show that the energy of the signal

$$f(t) = \frac{2a}{t^2 + a^2}$$

is $\frac{2\pi}{a}$. Hint: Find $F(\omega)$ using pair 3 and the symmetry property. ▽

The Essential Bandwidth of a Signal

    Spectra of most of the signals extend to infinity. However, because the energy of any practical signal is finite, the signal spectrum must approach 0 as $\omega \to \infty$. Most of the signal energy is contained within a certain band of B Hz, and the energy contributed by the components beyond B Hz is negligible. We can therefore suppress the signal spectrum beyond B Hz with little effect on the signal shape and energy. The bandwidth B is called the **essential bandwidth** of the signal. The criterion for selecting B depends on the error tolerance in a particular application. We may, for example, select B to be that band which contains 95% of the signal energy.[†] This figure may be higher or lower than 95%, depending on the precision needed. Using such a criterion, we can determine the essential bandwidth of a signal. The essential bandwidth B for the signal $e^{-at}u(t)$, using 95% energy criterion, was determined in Example 4.16 to be 2.02a Hz.

[†]For lowpass signals, the essential bandwidth may also be defined as a frequency at which the value of the amplitude spectrum is a small fraction (about 1%) of its peak value. In Example 4.16, for instance, the peak value, which occurs at $\omega = 0$, is $1/a$.

Suppression of all the spectral components of $f(t)$ beyond the essential bandwidth results in a signal $\hat{f}(t)$, which is a close approximation of $f(t)$. If we use the 95% criterion for the essential bandwidth, the energy of the error (the difference) $f(t) - \hat{f}(t)$ is 5% of E_f.

Energy Spectral Density From Autocorrelation Function

Correlation of a function $f(t)$ with itself is its **autocorrelation function** $\psi_f(t)$, which, for a real $f(t)$, is given by [see Eq. (3.32)]

$$\psi_f(t) = \int_{-\infty}^{\infty} f(x)f(x-t)\, dx \qquad (4.68a)$$

Also, from Eq. (3.31) with $g(t) = f(t)$, it follows that

$$\psi_f(t) = f(t) * f(-t) \qquad (4.68b)$$

From Eq. (4.68b) it is clear that

$$\psi_f(-t) = f(-t) * f(t) = \psi_f(t)$$

Therefore, for real $f(t)$, autocorrelation function $\psi_f(t)$ is an even function of t. The Fourier transform of Eq. (4.68b) yields

$$\psi_f(t) \iff |F(\omega)|^2 \qquad (4.69)$$

Therefore, the Fourier transform of the autocorrelation function is its energy spectral density $|F(\omega)|^2$. It is clear that $\psi_f(t)$ provides the spectral information of $f(t)$ directly.

The direct link of the autocorrelation function to the spectral information can be explained intuitively as follows. The autocorrelation function $\psi_f(t)$ is the correlation of a signal with itself delayed by t seconds. A signal $f(t)$ correlates perfectly with itself with zero delay. But as the delay increases, the similarity decreases. Thus, the autocorrelation function $\psi_f(t)$ is a nonincreasing function of t. If $f(t)$ is a slowly varying signal (low frequency signal), it changes slowly with t. Consequently such a signal will show considerable similarity or correlation with itself even for relatively large delay. The autocorrelation function $\psi_f(t)$ decays slowly with t and has a larger width. On the other hand, for a rapidly varying signal, the signal similarity will decrease rapidly with delay t and $\psi_f(t)$ has a smaller width. Thus, the shape of $\psi_f(t)$ has a direct link to spectral information of $f(t)$.

4.7   Application to Communications: Amplitude Modulation

Modulation causes a spectral shift in a signal and is used to gain certain advantages mentioned in Sec. 4.3-5. Broadly speaking, there are two classes of modulation: amplitude (linear) modulation and angle (nonlinear) modulation, which are the subject of the next two sections. In this section, we shall discuss some practical forms of amplitude modulation.

4.7-1   Double Sideband, Suppressed Carrier (DSB-SC) Modulation

In amplitude modulation, the amplitude A of the carrier $A\cos(\omega_c t + \theta_c)$ is varied in some manner with the **baseband** (message)† signal $m(t)$ (known as the

---

†The term baseband is used to designate the band of frequencies of the signal delivered by the source or the input transducer.

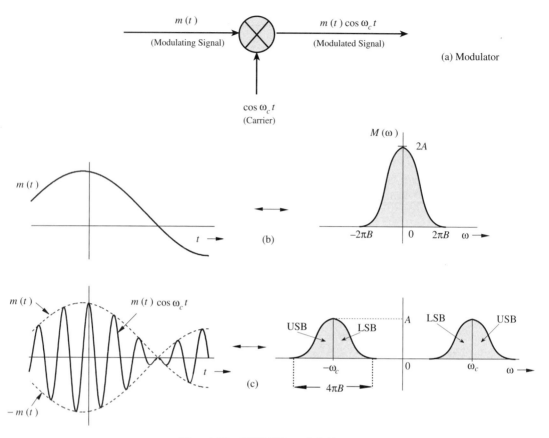

Fig. 4.31 DSB-SC modulation.

modulating signal). The frequency ω_c and the phase θ_c are constant. We can assume $\theta_c = 0$ without a loss of generality. If the carrier amplitude A is made directly proportional to the modulating signal $m(t)$, the modulated signal is $m(t)\cos\omega_c t$ (Fig. 4.31a). As was indicated earlier [Eq. (4.41)], this type of modulation simply shifts the spectrum of $m(t)$ to the carrier frequency (Fig. 4.31c). Thus, if

$$m(t) \iff M(\omega)$$

then

$$m(t)\cos\omega_c t \iff \frac{1}{2}[M(\omega + \omega_c) + M(\omega - \omega_c)] \qquad (4.70)$$

Recall that $M(\omega - \omega_c)$ is $M(\omega)$ shifted to the right by ω_c and $M(\omega + \omega_c)$ is $M(\omega)$ shifted to the left by ω_c. Thus, the process of modulation shifts the spectrum of the modulating signal to the left and the right by ω_c. Note also that if the bandwidth of $m(t)$ is B Hz, then, as indicated in Fig. 4.31c, the bandwidth of the modulated signal is $2B$ Hz. We also observe that the modulated signal spectrum centered at ω_c is composed of two parts: a portion that lies above ω_c, known as the **upper sideband (USB)**, and a portion that lies below ω_c, known as the **lower sideband (LSB)**. Similarly, the spectrum centered at $-\omega_c$ has upper and lower sidebands. This form of modulation is called **double sideband (DSB)** modulation for obvious reason.

The relationship of B to ω_c is of interest. Figure 4.31c shows that $\omega_c \geq 2\pi B$ in order to avoid the overlap of the spectra centered at $\pm\omega_c$. If $\omega_c < 2\pi B$, the spectra overlap and the information of $m(t)$ is lost in the process of modulation, a loss which makes it impossible to get back $m(t)$ from the modulated signal $m(t)\cos \omega_c t$.†

■ **Example 4.17**

For a baseband signal $m(t) = \cos \omega_m t$, find the DSB signal, and sketch its spectrum. Identify the upper and lower sidebands.

We shall work this problem in the frequency-domain as well as the time-domain in order to clarify the basic concepts of DSB-SC modulation. In the frequency-domain approach, we work with the signal spectra. The spectrum of the baseband signal $m(t) = \cos \omega_m t$ is given by

$$M(\omega) = \pi[\delta(\omega - \omega_m) + \delta(\omega + \omega_m)]$$

The spectrum consists of two impulses located at $\pm\omega_m$, as depicted in Fig. 4.32a. The DSB-SC (modulated) spectrum, as indicated by Eq. (4.70), is the baseband spectrum in Fig. 4.32a shifted to the right and the left by ω_c (times one-half), as depicted in Fig. 4.32b. This spectrum consists of impulses at $\pm(\omega_c - \omega_m)$ and $\pm(\omega_c + \omega_m)$. The spectrum beyond ω_c is the upper sideband (USB), and the one below ω_c is the lower sideband (LSB). Observe that the DSB-SC spectrum does not have the component of the carrier frequency ω_c. This is why it is called **double sideband-suppressed carrier (DSB-SC)** .

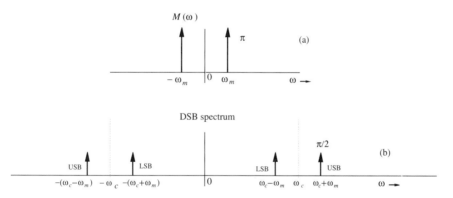

Fig. 4.32 An example of DSB-SC modulation.

In the time-domain approach, we work directly with signals in the time-domain. For the baseband signal $m(t) = \cos \omega_m t$, the DSB-SC signal $\varphi_{\text{DSB-SC}}(t)$ is

$$\varphi_{\text{DSB-SC}}(t) = m(t)\cos \omega_c t$$
$$= \cos \omega_m t \cos \omega_c t$$
$$\frac{1}{2}[\cos(\omega_c + \omega_m)t + \cos(\omega_c - \omega_m)t] \qquad (4.71)$$

---

†Practical factors may impose additional restrictions on ω_c. For instance, in broadcast applications, a radiating antenna can radiate only a narrowband without distortion. This restriction implies that to avoid distortion caused by the radiating antenna, $\omega_c/2\pi B \gg 1$. The broadcast band AM radio, for instance, with $B = 5$ kHz and the band of 550 to 1600 kHz for carrier frequency give a ratio of $\omega_c/2\pi B$ roughly in the range of 100 to 300.

This result shows that when the baseband (message) signal is a single sinusoid of frequency ω_m, the modulated signal consists of two sinusoids: the component of frequency $\omega_c + \omega_m$ (the upper sideband), and the component of frequency $\omega_c - \omega_m$ (the lower sideband). Figure 4.32b illustrates precisely the spectrum of $\varphi_{\text{DSB-SC}}(t)$. Thus, each component of frequency ω_m in the modulating signal results into two components of frequencies $\omega_c + \omega_m$ and $\omega_c - \omega_m$ in the modulated signal. This being a DSB-SC (suppressed carrier) modulation, there is no component of the carrier frequency ω_c on the right-hand side of the above equation as expected.† ■

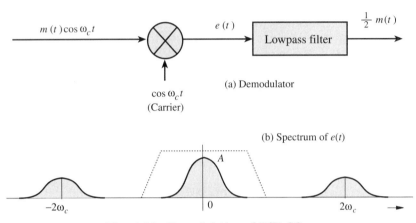

Fig. 4.33   Demodulation of DSB-SC.

Demodulation of DSB-SC Signals

The DSB-SC modulation translates or shifts the frequency spectrum to the left and the right by ω_c (that is, at $+\omega_c$ and $-\omega_c$), as seen from Eq. (4.70). To recover the original signal $m(t)$ from the modulated signal, we must retranslate the spectrum to its original position. The process of recovering the signal from the modulated signal (retranslating the spectrum to its original position) is referred to as **demodulation**, or **detection**. Observe that if the modulated signal spectrum in Fig. 4.31c is shifted to the left and to the right by ω_c (and halved), we obtain the spectrum illustrated in Fig. 4.33b, which contains the desired baseband spectrum in addition to an unwanted spectrum at $\pm 2\omega_c$. The latter can be suppressed by a lowpass filter. Thus, demodulation, which is almost identical to modulation, consists of multiplication of the incoming modulated signal $m(t) \cos \omega_c t$ by a carrier $\cos \omega_c t$ followed by a lowpass filter, as depicted in Fig. 4.33a. We can verify this conclusion directly in the time-domain by observing that the signal $e(t)$ in Fig. 4.33a is

$$e(t) = m(t) \cos^2 \omega_c t$$
$$= \frac{1}{2}[m(t) + m(t) \cos 2\omega_c t] \tag{4.72a}$$

---

†The term suppressed carrier does not necessarily mean absence of the spectrum at the carrier frequency. The term "suppressed carrier" merely implies that there is no discrete component of the carrier frequency. Since no discrete component exists, the spectrum of DSB-SC does not have impulses at $\pm\omega_c$, a fact which further implies that the modulated signal $m(t) \cos \omega_c t$ does not contain a term of the form $k \cos \omega_c t$ (assuming that $m(t)$ has a zero mean value).

Therefore, the Fourier transform of the signal $e(t)$ is

$$E(\omega) = \frac{1}{2}M(\omega) + \frac{1}{4}[M(\omega + 2\omega_c) + M(\omega - 2\omega_c)] \qquad (4.72b)$$

Hence, $e(t)$ consists of two components $\frac{1}{2}m(t)$ and $\frac{1}{2}m(t)\cos 2\omega_c t$, with their spectra, as illustrated in Fig. 4.33b. The spectrum of the second component, being a modulated signal with carrier frequency $2\omega_c$, is centered at $\pm 2\omega_c$. Hence, this component is suppressed by the lowpass filter in Fig. 4.33a. The desired component $\frac{1}{2}M(\omega)$, being a lowpass spectrum (centered at $\omega = 0$), passes through the filter unharmed, resulting in the output $\frac{1}{2}m(t)$.

A possible form of lowpass filter characteristics is depicted (dotted) in Fig. 4.33b. This method of recovering the baseband signal is called **synchronous detection**, or **coherent detection**, where we use a carrier of exactly the same frequency (and phase) as the carrier used for modulation. Thus, for demodulation, we need to generate a local carrier at the receiver in frequency and phase coherence (synchronism) with the carrier used at the modulator. We shall demonstrate in Example 4.18 that both, the phase and frequency synchronism, are extremely critical.

■ **Example 4.18**

Discuss the effect of lack of frequency and phase coherence (synchronism) between the carriers at the modulator (transmitter) and the demodulator (receiver) in DSB-SC.

Let the modulator carrier be $\cos \omega_c t$ (Fig. 4.31a). For the demodulator in Fig. 4.33a, we shall consider two cases: (1) the first case with carrier $\cos(\omega_c t + \theta)$ (phase error of θ) and (2) the second case with carrier $\cos(\omega_c + \Delta\omega)t$ (frequency error $\Delta\omega$).

(a) With the demodulator carrier $\cos(\omega_c t + \theta)$ (instead of $\cos \omega_c t$) in Fig. 4.33a, the multiplier output is $e(t) = m(t)\cos \omega_c t\cos(\omega_c t + \theta)$ instead of $m(t)\cos^2 \omega_c t$. Using the trigonometric identity, we obtain

$$\begin{aligned}
e(t) &= m(t)\cos \omega_c t\,\cos(\omega_c t + \theta) \\
&= \frac{1}{2}m(t)[\cos\theta + \cos(2\omega_c t + \theta)]
\end{aligned}$$

The spectrum of the component $\frac{1}{2}m(t)\cos(2\omega_c t + \theta)$ is centered at $\pm 2\omega_c$. Consequently, it will be filtered out by the lowpass filter at the output. The component $\frac{1}{2}m(t)\cos\theta$ is the signal $m(t)$ multiplied by a constant $\frac{1}{2}\cos\theta$. The spectrum of this component is centered at $\omega = 0$ (lowpass spectrum), and will pass through the lowpass filter at the output, yielding the output $\frac{1}{2}m(t)\cos\theta$.

If θ is constant, the phase asynchronism merely yields an attenuated output (by a factor $\cos\theta$). Unfortunately, in practice, θ is often the phase difference between the carriers generated by two distant generators, and varies randomly with time. This variation would result in an output whose gain varies randomly with time.

(b) In the case of frequency error, the demodulator carrier is $\cos(\omega_c + \Delta\omega)t$. This situation is very similar to the phase error case in (a) with θ replaced by $(\Delta\omega)t$. Following the analysis in part (a), we can express the demodulator product $e(t)$ as

$$\begin{aligned}
e(t) &= m(t)\cos \omega_c t\,\cos(\omega_c + \Delta\omega)t \\
&= \frac{1}{2}m(t)[\cos(\Delta\omega)t + \cos(2\omega_c + \Delta\omega)t]
\end{aligned}$$

The spectrum of the component $\frac{1}{2}m(t)\cos{(2\omega_c + \Delta\omega)t}$ is centered at $\pm(2\omega_c + \Delta\omega)$. Consequently, this component will be filtered out by the lowpass filter at the output. The component $\frac{1}{2}m(t)\cos{(\Delta\omega)t}$ is the signal $m(t)$ multiplied by a low frequency carrier of frequency $\Delta\omega$. The spectrum of this component is centered at $\pm\Delta\omega$. In practice, the frequency error $(\Delta\omega)$ is usually very small. Hence, the signal $\frac{1}{2}m(t)\cos{(\Delta\omega)t}$ (whose spectrum is centered at $\pm\Delta\omega$) is a lowpass signal and passes through the lowpass filter at the output, resulting in the output $\frac{1}{2}m(t)\cos{(\Delta\omega)t}$. The output is the desired signal $m(t)$ multiplied by a very low frequency sinusoid $\cos{(\Delta\omega)t}$. Clearly, the output in this case is not merely an attenuated replica of the desired signal $m(t)$, but represents $m(t)$ multiplied by a time-varying gain $\cos{(\Delta\omega)t}$. If, for instance, the transmitter and the receiver carrier frequencies differ just by 1 Hz, the output will be the desired signal $m(t)$ multiplied by a time-varying signal whose gain goes from the maximum to 0 every half second. This is like some restless child fiddling with the volume control knob of a receiver, going from maximum volume to zero volume every half second. This kind of distortion (called the **beat effect**) is beyond repair. ▪

4.7-2   Amplitude Modulation (AM)

For the suppressed carrier scheme just discussed, a receiver must generate a carrier in frequency and phase synchronism with the carrier at the transmitter that may be located hundreds or thousands of miles away. This situation calls for a sophisticated receiver, which could be quite costly. The other alternative is for the transmitter to transmit a carrier $A \cos \omega_c t$ [along with the modulated signal $m(t) \cos \omega_c t$] so that there is no need to generate a carrier at the receiver. In this case the transmitter needs to transmit much larger power, a rather expensive procedure. In point-to-point communications, where there is one transmitter for each receiver, substantial complexity in the receiver system can be justified, provided there is a large enough saving in expensive high-power transmitting equipment. On the other hand, for a broadcast system with a multitude of receivers for each transmitter, it is more economical to have one expensive high-power transmitter and simpler, less expensive receivers. The second option (transmitting a carrier along with the modulated signal) is the obvious choice in this case. This is the so-called AM (amplitude modulation), in which the transmitted signal $\varphi_{\rm AM}(t)$ is given by

$$\varphi_{\rm AM}(t) = A \cos \omega_c t + m(t) \cos \omega_c t \qquad (4.73a)$$

$$= [A + m(t)] \cos \omega_c t \qquad (4.73b)$$

Recall that the DSB-SC signal is $m(t) \cos \omega_c t$. From Eq. (4.73b) it follows that the AM signal is identical to the DSB-SC signal with $A + m(t)$ as the modulating signal [instead of $m(t)$]. Therefore, to sketch $\varphi_{\rm AM}(t)$, we sketch $A + m(t)$ and $-[A + m(t)]$ and fill in between with the sinusoid of the carrier frequency. Two cases are considered in Fig. 4.34. In the first case, A is large enough so that $A + m(t) \geq 0$ (is nonnegative) for all values of t. In the second case, A is not large enough to satisfy this condition. In the first case, the envelope (Fig. 4.34d) has the same shape as $m(t)$ (although riding on a dc of magnitude A). In the second case the envelope shape is not $m(t)$, for some parts get rectified (Fig. 4.34e). Thus, we can detect the desired signal $m(t)$ by detecting the envelope in the first case. In the second case, such a detection is not possible. We shall see that the envelope detection is

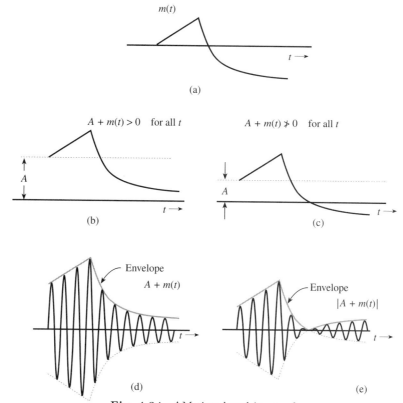

Fig. 4.34   AM signal and its envelope.

an extremely simple and inexpensive operation, which does not require generation of a local carrier for the demodulation. But as seen above the envelope of AM has the information about $m(t)$ only if the AM signal $[A + m(t)] \cos \omega_c t$ satisfies the condition $A + m(t) > 0$ for all t. Thus, the condition for envelope detection of an AM signal is

$$A + m(t) \geq 0 \qquad \text{for all } t \qquad\qquad (4.74)$$

If m_p is the peak amplitude (positive or negative) of $m(t)$ (see Fig. 4.34), then $m(t) \geq -m_p$. Hence, the condition (4.74) is equivalent to†

$$A \geq m_p \qquad\qquad (4.75)$$

Thus the minimum carrier amplitude required for the viability of envelope detection is m_p. This point is clearly illustrated in Fig. 4.34.

We define the modulation index μ as

$$\mu = \frac{m_p}{A} \qquad\qquad (4.76)$$

---

†In case the negative and the positive peak amplitudes are not identical, m_p in condition (4.75) is the absolute negative peak amplitude.

where A is the carrier amplitude. Note that m_p is a constant of the signal $m(t)$. Because $A \geq m_p$ and because there is no upper bound on A, it follows that

$$0 \leq \mu \leq 1 \qquad (4.77)$$

as the required condition for the viability of demodulation of AM by an envelope detector.

When $A < m_p$, Eq. (4.76) shows that $\mu > 1$ (overmodulation). In this case, the option of envelope detection is no longer viable. We then need to use synchronous demodulation. Note that synchronous demodulation can be used for any value of μ (see Prob. 4.7-4). The envelope detector, which is considerably simpler and less expensive than the synchronous detector, can be used only for $\mu \leq 1$.

■ **Example 4.19**

Sketch $\varphi_{\mathrm{AM}}(t)$ for modulation indices of $\mu = 0.5$ (50% modulation) and $\mu = 1$ (100% modulation), when $m(t) = B \cos \omega_m t$. This case is referred to as **tone modulation** because the modulating signal is a pure sinusoid (or tone).

In this case, $m_p = B$ and the modulation index according to Eq. (4.76) is

$$\mu = \frac{B}{A}$$

Hence, $B = \mu A$ and

$$m(t) = B \cos \omega_m t = \mu A \cos \omega_m t$$

Therefore

$$\varphi_{\mathrm{AM}}(t) = [A + m(t)] \cos \omega_c t = A[1 + \mu \cos \omega_m t] \cos \omega_c t \qquad (4.78)$$

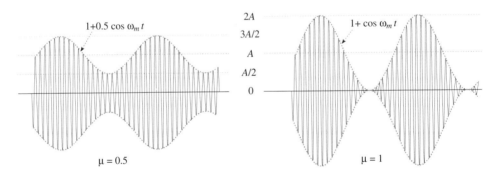

Fig. 4.35   Tone-modulated AM (a) $\mu = 0.5$. (b) $\mu = 1$.

Figures 4.35a and b show the modulated signals corresponding to $\mu = 0.5$ and $\mu = 1$, respectively.   ■

Demodulation of AM: The Envelope Detector

The AM signal can be demodulated coherently by a locally generated carrier (see Prob. 4.7-4). However, coherent, or synchronous, demodulation of AM (with $\mu \leq 1$) will defeat the very purpose of AM and, hence, is rarely used in practice.

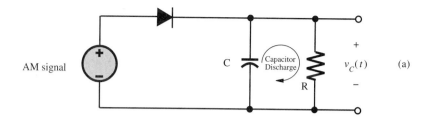

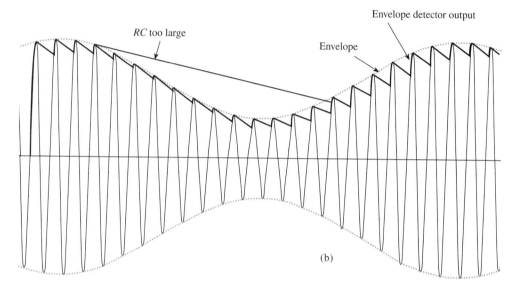

Fig. 4.36   Envelope detector.

We shall consider here one of the noncoherent methods of AM demodulation, the **envelope detection**.†

In an envelope detector, the output of the detector follows the envelope of the (modulated) input signal. The circuit illustrated in Fig. 4.36a functions as an envelope detector. During the positive cycle of the input signal, the diode conducts and the capacitor C charges up to the peak voltage of the input signal (Fig. 4.36b). As the input signal falls below this peak value, the diode is cut off, because the capacitor voltage (which is very nearly the peak voltage) is greater than the input signal voltage, a circumstance causing the diode to open. The capacitor now discharges through the resistor R at a slow rate (with a time constant RC). During the next positive cycle, the same drama repeats. When the input signal becomes greater than the capacitor voltage, the diode conducts again. The capacitor again charges to the peak value of this (new) cycle. As the input voltage falls below the new peak value, the diode cuts off again and the capacitor discharges slowly during the cutoff period, a process that changes the capacitor voltage very slightly.

---

†There are also other methods of noncoherent detection. The rectifier detector consists of a rectifier followed by a lowpass filter. This method is also simple and almost as inexpensive as the envelope detector[4]. The nonlinear detector, although simple and inexpensive, results in a distorted output.

In this manner, during each positive cycle, the capacitor charges up to the peak voltage of the input signal and then decays slowly until the next positive cycle. Thus, the output voltage $v_C(t)$ follows the envelope of the input. The capacitor discharge between positive peaks, however, causes a ripple signal of frequency ω_c in the output. This ripple can be reduced by increasing the time constant RC so that the capacitor discharges very little between the positive peaks ($RC \leq 1/\omega_c$). Making RC too large, however, would make it impossible for the capacitor voltage to follow the envelope (see Fig. 4.36b). Thus, RC should be large compared to $1/\omega_c$ but should be small compared to $1/2\pi B$, where B is the highest frequency in $m(t)$. Incidentally, these two conditions also require that $\omega_c \gg 2\pi B$, a condition necessary for a well-defined envelope.

The envelope-detector output $v_C(t)$ is $A + m(t)$ plus a ripple of frequency ω_c. The dc term A can be blocked out by a capacitor or a simple RC highpass filter. The ripple may be reduced further by another (lowpass) RC filter. In the case of audio signals, the speakers cannot respond to the high frequency ripple, and therefore, they act as lowpass filters themselves.

4.7-3   Single Sideband Modulation (SSB)

Figures 4.37a and 4.37b show the baseband spectrum $M(\omega)$, and the spectrum of the DSB-SC modulated signal $m(t) \cos \omega_c t$. The DSB spectrum in Fig. 4.37b has two sidebands: the upper sideband (USB) and the lower sideband (LSB), both containing complete information of $M(\omega)$ [see Eq. (4.10)]. Clearly, it is redundant to transmit both sidebands, a process which requires twice the bandwidth of the baseband signal. A scheme where only one sideband is transmitted is known as **single sideband (SSB) transmission**, which requires only one-half the bandwidth of the DSB signal. Thus, we transmit only the upper sidebands (Figures 4.37c) or only the lower sidebands (Fig. 4.37d).

An SSB signal can be coherently (synchronously) demodulated. For example, multiplication of a USB signal (Fig. 4.37c) by $\cos \omega_c t$ shifts its spectrum to the left and to the right by ω_c, yielding the spectrum in Fig. 4.37e. Lowpass filtering of this signal yields the desired baseband signal. The case is similar with LSB signal. Hence, demodulation of SSB signals is identical to that of DSB-SC signals, and the synchronous demodulator in Fig. 4.33a can demodulate SSB signals. Note that we are talking of SSB signals without an additional carrier. Hence, they are suppressed carrier signals (SSB-SC).

■ **Example 4.20**
Find the USB (the upper sideband) and LSB (the lower sideband) signals when $m(t) = \cos \omega_m t$. Sketch their spectra, and show that these SSB signals can be demodulated using the synchronous demodulator in Fig. 4.33a.
The DSB-SC signal for this case is

$$\varphi_{\text{DSB-SC}}(t) = m(t) \cos \omega_c t$$
$$= \cos \omega_m t \cos \omega_c t$$
$$\frac{1}{2}[\cos (\omega_c - \omega_m)t + \cos (\omega_c + \omega_m)t] \tag{4.79}$$

As pointed out in Example 4.17, the terms $\frac{1}{2} \cos (\omega_c + \omega_m)t$ and $\frac{1}{2} \cos (\omega_c - \omega_m)t$ represent the upper and lower sidebands, respectively. Figure 4.38a and b show the spectra

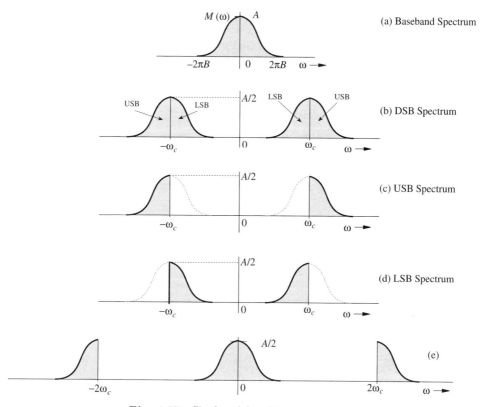

Fig. 4.37 Single sideband transmission.

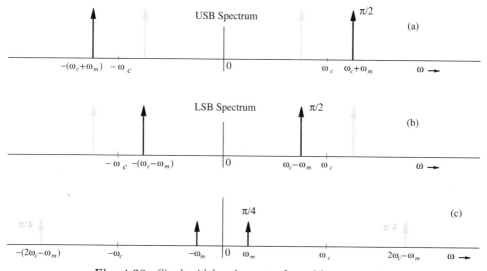

Fig. 4.38 Single sideband spectra for $m(t) = \cos \omega_m t$.

of the upper and lower sidebands. Observe that these spectra can be obtained from the DSB-SC spectrum in Fig. 4.32b by suppressing the undesired sidebands using a proper filter. For instance, the USB signal in Fig. 4.38a can be obtained by passing the DSB-SC signal (Fig. 4.32b) through a highpass filter of cutoff frequency ω_c. Similarly, the LSB signal in Fig. 4.38b can be obtained by passing the DSB-SC signal through a lowpass filter of cutoff frequency ω_c.

If we apply the LSB signal $\frac{1}{2}\cos(\omega_c - \omega_m)t$ to the synchronous demodulator in fig. 4.33a, the multiplier output is

$$e(t) = \frac{1}{2}\cos(\omega_c - \omega_m)t \cos \omega_c t$$
$$= \frac{1}{4}[\cos \omega_m t + \cos(2\omega_c - \omega_m)t]$$

The term $\frac{1}{4}\cos(2\omega_c - \omega_m)t$ is suppressed by the lowpass filter, a fact which results in the desired output $\frac{1}{4}\cos \omega_m t$ (which is $m(t)/4$). The spectrum of this term is $\pi[\delta(\omega + \omega_0) + \delta(\omega - \omega_0)]/4$, as depicted in Fig. 4.38c. In the same way we can show that the USB signal can be demodulated by the synchronous demodulator.

In frequency-domain, demodulation (multiplication by $\cos \omega_c t$ amounts to shifting the LSB spectrum (Fig. 4.38b) to the left and the right by ω_c (times one-half) and then suppressing the high frequency, as illustrated in Fig. 4.38c. The resulting spectrum represents the desired signal $\frac{1}{4}m(t)$. ■

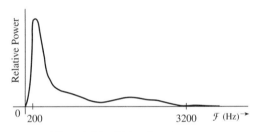

Fig. 4.39   Voice Spectrum.

Generation of SSB Signals

Two methods are commonly used to generate SSB signals. The first method, the **selective-filtering method** uses sharp cutoff filters to eliminate the undesired sideband, and the second method uses phase-shifting networks[4] to achieve the same goal.† We shall consider here only the first method.

The selective-filtering Method is the most commonly used method of generating SSB signals. In this method, a DSB-SC signal is passed through a sharp cutoff filter to eliminate the undesired sideband.

To obtain the USB, the filter should pass all components above ω_c unattenuated and completely suppress all components below ω_c. Such an operation requires an ideal filter, which is unrealizable. It can, however, be realized closely if there is some separation between the passband and the stopband. Fortunately, the voice signal provides this condition, because its spectrum shows little power content at the origin (Fig. 4.39). Moreover, articulation tests show that for speech signals, frequency components below 300 Hz are not important. In other words, we may suppress all speech components below 300 Hz without affecting the intelligibility

---

†Yet another method, known as Weaver's method, is also used to generate SSB signals.

appreciably.‡ Thus, filtering of the unwanted sideband becomes relatively easy for speech signals because we have a 600 Hz transition region around the cutoff frequency ω_c. For signals, which have considerable power at low frequencies (around $\omega = 0$), SSB techniques cause considerable distortion. Such is the case with video signals. Consequently, for video signals, instead of SSB, we use another technique, the **vestigial sideband (VSB)**, which is a compromise between SSB and DSB. It inherits the advantages of SSB and DSB but avoids their disadvantages. VSB signals are relatively easy to generate, and their bandwidth is only slightly (typically 25%) greater than that of the SSB signals. In VSB signals, instead of rejecting one sideband completely (as in SSB), we accept a gradual cutoff off of one sideband[4].

4.8 Angle Modulation

A sinusoid is characterized by its amplitude and angle (which includes its frequency and phase). In amplitude modulated signals, the information content of the baseband (message) signal $m(t)$ appears in the amplitude variations of the carrier. In **angle modulation** discussed in this section, the information content of $m(t)$ is carried by the angle of the carrier. Angle modulation also goes by the name **exponential modulation**.

The generalized angle modulated (or exponentially modulated) carrier can be described as

$$\varphi_{\mathrm{EM}}(t) = A \, \cos\left[\omega_c t + k\psi(t)\right] \tag{4.80}$$

where k is an arbitrary constant and $\psi(t)$, which is a measure of $m(t)$, is obtained by an invertible linear operation on $m(t)$. In other words, $\psi(t)$ is the output of some linear system with a suitable transfer function $H(s)$ when the input is $m(t)$, as depicted in Fig. 4.40.† If $h(t)$ is the unit impulse response of this system; that is, if $h(t) \Longleftrightarrow H(s)$, then

Fig. 4.40 Generation of angle modulated signal.

$$\psi(t) = \int_{-\infty}^{t} m(\alpha)h(t-\alpha)\, d\alpha \tag{4.81}$$

By selecting suitable $h(t)$, we can obtain a variety of subclasses of angle modulation. For instance, if we select $h(t) = u(t)$, the resulting form is the well-known **frequency modulation (FM)**. In contrast, use of $h(t) = \delta(t)$ leads to **phase modulation (PM)**. These are but two of the infinite possibilities. Although in

---

‡Similarly, suppression of speech-signal components above 3500 Hz causes no appreciable change in intelligibility.

†Because $H(s)$ is required to be invertible, we can obtain $m(t)$ by passing $\psi(t)$ through the inverse linear system, which has the transfer function $1/H(s)$.

digital communication, use of phase and frequency modulation is common, the so-called broadcast FM is not FM in the classical sense, but is a generalized angle modulation because of the inclusion of the **preemphasis** filter used to improve its noise suppressing abilities. It is called FM for historical reason in the sense that angle modulation was first conceived and introduced in the form of frequency modulation. The broadcast FM, although originating as true FM in the laboratories, was modified in broadcasting for better performance. Yet, the term FM continued to be used to describe this scheme.

In amplitude modulation, the carrier frequency is constant, but the amplitude changes with $m(t)$. In contrast, in angle modulation, the carrier amplitude is always constant, but the carrier frequency varies continuously with the message $m(t)$. By definition, a sinusoidal signal is expected to have a constant frequency; hence, the variation of frequency with time appears to be contradictory to the conventional definition of a sinusoidal signal frequency. Therefore, we must generalize the notion of a sinusoid so as to make allowance for variation of frequency with time. This generalization leads us to a new concept of **instantaneous frequency**.

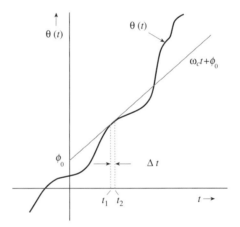

Fig. 4.41   Concept of instantaneous frequency.

4.8-1   The Concept of Instantaneous Frequency

As seen above, the carrier frequency is changing continuously every instant in FM. Prima facie, this does not make much sense because to define a frequency, we must have a sinusoidal signal at least over one cycle with the same frequency. We cannot imagine a sinusoid whose frequency is different at every instant. This problem reminds us of our first encounter with the concept of **instantaneous velocity** in our beginning mechanics course. Until that time, we were used to thinking of velocity as being constant over a time interval and were incapable of even imagining that velocity could vary at each instant. But after some mental struggle, the idea gradually sinks in. We never forget, however, the wonder and amazement that was caused by the idea when it was first introduced. A similar experience awaits the reader with the concept of instantaneous frequency.

Let us consider a generalized sinusoidal signal $\varphi(t)$ given by

$$\varphi(t) = A \cos \theta(t) \tag{4.82}$$

where $\theta(t)$, the **generalized angle**, is a function of t. Figure 4.41 illustrates a hypothetical case of $\theta(t)$. The generalized angle for a conventional sinusoid $A \cos(\omega_c t + \phi_o)$ is $\omega_c t + \phi_o$. This plot, a straight line with a slope ω_c and intercept ϕ_o, is also illustrated in Fig. 4.41. The plot of $\theta(t)$ for the hypothetical case happens to be tangential to the angle $(\omega_c t + \phi_o)$ at some instant t. The crucial point is that over a small interval $\Delta t \to 0$, the signal $\varphi(t) = A \cos \theta(t)$ and the sinusoid $A \cos(\omega_c t + \phi_o)$ are identical; that is,

$$\varphi(t) = A \cos(\omega_c t + \phi_o) \qquad t_1 < t < t_2$$

We are certainly justified in saying that over this small interval Δt, the frequency of $\varphi(t)$, is ω_c. Because $(\omega_c t + \phi_o)$ is tangential to $\theta(t)$, the frequency of $\varphi(t)$ is the slope of its angle $\theta(t)$ over this small interval. We can generalize this concept at every instant and say that the instantaneous frequency ω_i at any instant t is the slope of $\theta(t)$ at t. Thus, for $\varphi(t)$ in Eq. (4.82), the instantaneous frequency $\omega_i(t)$ is given by

$$\omega_i(t) = \frac{d\theta}{dt} \tag{4.83a}$$

$$\theta(t) = \int_{-\infty}^{t} \omega_i(\alpha) \, d\alpha \tag{4.83b}$$

For a conventional sinusoid $A \cos(\omega_c t + \phi_o)$, we have $\theta(t) = \omega_c t + \phi_o$, and $\omega_i(t) = d\theta(t)/dt = \omega_c$, a constant, as desired. Clearly, the generalized definition of instantaneous frequency does not conflict with our old notion of frequency.

Now we can see the possibility of transmitting the information of $m(t)$ by varying the angle θ of a carrier. Two simple possibilities are **phase modulation (PM)** and **frequency modulation (FM)**. In PM, the angle $\theta(t)$ is varied linearly with $m(t)$:

$$\theta(t) = \omega_c t + k_p m(t) \tag{4.84a}$$

where k_p is a constant and ω_c is the carrier frequency. The resulting PM wave is

$$\varphi_{\text{PM}}(t) = A \cos[\omega_c t + k_p m(t)] \tag{4.84b}$$

The instantaneous frequency $\omega_i(t)$ in this case is given by

$$\omega_i(t) = \frac{d\theta}{dt} = \omega_c + k_p \dot{m}(t) \tag{4.84c}$$

Hence in phase modulation, the instantaneous frequency ω_i varies linearly with the derivative of the modulating signal. If the instantaneous frequency ω_i is varied linearly with the modulating signal, we have frequency modulation. Thus, in FM, the instantaneous frequency ω_i is

$$\omega_i(t) = \omega_c + k_f m(t) \tag{4.85a}$$

where k_f is a constant. From Eq. (4.83b), we find the angle $\theta(t)$ as

$$\theta(t) = \int_{-\infty}^{t} [\omega_c + k_f m(\alpha)]\, d\alpha$$

$$= \omega_c t + k_f \int_{-\infty}^{t} m(\alpha)\, d\alpha \tag{4.85b}$$

Here we have assumed the constant term in $\theta(t)$ to be zero without loss of generality. Thus, the FM wave is

$$\varphi_{\mathrm{FM}}(t) = A \cos\left[\omega_c t + k_f \int_{-\infty}^{t} m(\alpha)\, d\alpha\right] \tag{4.85c}$$

Observe that both PM and FM are special cases of the exponentially modulated signal $\varphi_{\mathrm{EM}}(t)$ in Eq. (4.80). If $h(t) = \delta(t)$ in Eq. (4.81), then use of the sampling property of the impulse in Eq. (4.81) yields $\psi(t) = m(t)$, and Eq. (4.80) reduces to PM in Eq. (4.84b). Similarly, if $h(t) = u(t)$, then the fact that $u(t - \alpha) = 1$ over $-\infty < \alpha \le t$ yields $\int m(\alpha) h(t - \alpha)\, d\alpha = \int m(\alpha)\, d\alpha$, and Eq. (4.80) reduces to FM in Eq. (4.85c).

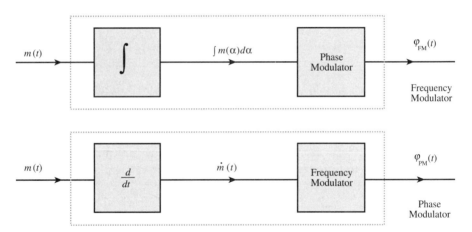

Fig. 4.42   Phase and frequency modulation are inseparable.

All In The Family

Equations (4.84b) and (4.85c) indicate that PM and FM are not only very similar but are inseparable. Replacing $m(t)$ in Eq. (4.84b) with $\int^{t} m(\alpha)\, d\alpha$ changes PM into FM. Thus, a signal that is an FM wave corresponding to $m(t)$ is also the PM wave corresponding to $\int^{t} m(\alpha)\, d\alpha$ (Fig. 4.42a). Similarly, a PM wave corresponding to $m(t)$ is the FM wave corresponding to $\dot{m}(t)$ (Fig. 4.42b).

We conclude that just by looking at an angle-modulated carrier, we cannot tell whether it is FM or PM. In fact, it is meaningless to enquire if a certain angle modulated wave is FM or PM. An analogous situation would be to ask a person (who is married, with children), whether he is a father or a son. The person would be puzzled because he is both, a father (of his child) and a son (of his father).

We have seen that PM and FM are not different kind of modulation, but two special cases of generalized angle modulation. Such a view is very fruitful because it shows the convertibility of one type of angle modulation (such as PM) to another (such as FM). This convertibility is quite clear in Fig. 4.42. For instance, we show later that the bandwidth of FM is approximately $2k_f m_p$, where m_p is the peak amplitude of $m(t)$. We can derive the equivalent result for PM by referring to Fig. 4.42b, which shows that PM is actually the FM when the modulating signal is $\dot{m}(t)$. Clearly, the bandwidth of PM is approximately $2k_p m_p{}'$, where $m_p{}'$ is the peak amplitude of $\dot{m}(t)$. This argument shows that if we analyze one type of angle modulation (such as FM), we could readily extend those results to any other kind. Historically, the angle modulation concept began with FM. Hence, it is customary to analyze FM and then modify those results for other forms, such as PM. But this does not imply that FM is superior to other kinds of angle modulation. On the contrary, PM is superior to FM for most analog signals such as audio and video. Actually, the optimum performance is realized neither by PM nor FM, but by some other form, depending on the nature of the baseband (message) signal.

This discussion also shows that we need not discuss methods of generation and demodulation of each type of modulation. Figure 4.42 clearly indicates that the PM can be generated by an FM generator, and the FM can be generated by a PM generator. One of the methods of generating FM in practice (the Armstrong indirect-FM system) actually integrates $m(t)$ and uses it to phase-modulate a carrier. Similar remarks apply to demodulation of FM and PM.

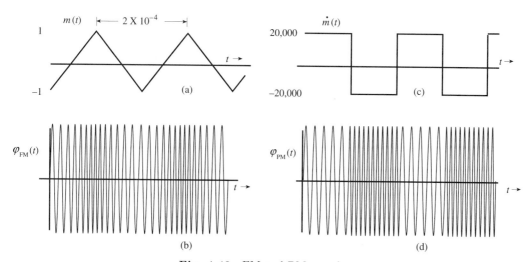

Fig. 4.43   FM and PM waveforms.

■ **Example 4.21**

Sketch FM and PM waves for the modulating signal $m(t)$ depicted in Fig. 4.43a. The constants k_f and k_p are $2\pi(10^5)$ and 10π, respectively, and the carrier frequency $\mathcal{F}_c$ is 100 MHz.

For FM [see Eq. (4.85a)]

$\omega_i = \omega_c + k_f m(t)$. Dividing throughout by 2π, we obtain the equation in terms of the variable $\mathcal{F}$ (frequency in Hz). The instantaneous frequency $\mathcal{F}_i$ is

$$\mathcal{F}_i = \mathcal{F}_c + \frac{k_f}{2\pi} m(t) = 10^8 + 10^5 m(t)$$

$$(\mathcal{F}_i)_{\min} = 10^8 - 10^5 \, |[m(t)]_{\min}| = 99.9 \text{ MHz}$$

$$(\mathcal{F}_i)_{\max} = 10^8 + 10^5 [m(t)]_{\max} = 100.1 \text{ MHz}$$

Because $m(t)$ increases and decreases linearly with time, the instantaneous frequency increases linearly from 99.9 to 100.1 MHz over a half-cycle and decreases linearly from 100.1 to 99.9 MHz over the remaining half-cycle of the modulating signal (Fig. 4.43b).

For PM

PM for $m(t)$ is FM for $\dot{m}(t)$. This assertion also follows from Eq. (4.84c) or Fig. 4.42b.

$$\mathcal{F}_i = \mathcal{F}_c + \frac{k_p}{2\pi} \dot{m}(t) = 10^8 + 5\,\dot{m}(t)$$

$$(\mathcal{F}_i)_{\min} = 10^8 - 5\,|[\dot{m}(t)]_{\min}| = 10^8 - 10^5 = 99.9 \text{ MHz}$$

$$(\mathcal{F}_i)_{\max} = 10^8 + 5\,[\dot{m}(t)]_{\max} = 100.1 \text{ MHz}$$

Because $\dot{m}(t)$ switches back and forth from a value of $-20{,}000$ to $20{,}000$, the carrier frequency switches back and forth from 99.9 to 100.1 MHz every half-cycle of $\dot{m}(t)$, as illustrated in Fig. 4.43d.

This indirect method of sketching PM (using $\dot{m}(t)$ to frequency-modulate a carrier) works as long as $m(t)$ is a continuous signal. If $m(t)$ is discontinuous, $\dot{m}(t)$ contains impulses, and this method is not so convenient. In such a case, a direct approach should be used. This is demonstrated in the next example. ■

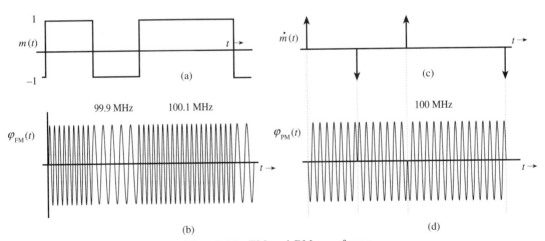

Fig. 4.44   FM and PM waveforms.

■ **Example 4.22**

Sketch FM and PM waves for the digital modulating signal $m(t)$ depicted in Fig. 4.44a. The constants k_f and k_p are $2\pi(10^5)$ and $\pi/2$, respectively, and $\mathcal{F}_c = 100$ MHz.

For FM

$$\mathcal{F}_i = \mathcal{F}_c + \frac{k_f}{2\pi} m(t) = 10^8 + 10^5 m(t)$$

Because $m(t)$ switches back and forth from 1 to -1 and vice versa, the FM wave frequency switches back and forth from 99.9 MHz to 100.1 MHz and vice versa, as shown in Fig. 4.44b. This scheme of a carrier frequency modulation by a digital signal is known as **frequency-shift keying (FSK)**, because the information digits are transmitted by shifting the carrier frequency.

For PM

$$\mathcal{F}_i = \mathcal{F}_c + \frac{k_f}{2\pi} \dot{m}(t) = 10^8 + \frac{1}{4}\dot{m}(t)$$

The derivative $\dot{m}(t) = 0$ everywhere except for impulses of strength ± 2 at the points of discontinuities of $m(t)$ (Fig. 4.44c). This fact means the carrier frequency is $\mathcal{F}_c = 100$ MHz everywhere except at the points of discontinuities, where it changes momentarily by infinite amount. It is not immediately apparent how an instantaneous frequency can be changed by an infinite amount and then changed back to the original frequency in zero time. Let us consider the direct approach.

$$\varphi_{\text{PM}}(t) = A \cos\left[\omega_c t + k_p m(t)\right]$$
$$= A \cos\left[\omega_c t + \frac{\pi}{2} m(t)\right]$$

$$= \begin{cases} A \sin \omega_c t & \text{when } m(t) = -1 \\ -A \sin \omega_c t & \text{when } m(t) = 1 \end{cases}$$

This PM wave, illustrated in Fig. 4.44c, has the same frequency $\mathcal{F}_c = 100$ MHz everywhere. However, there are phase discontinuities of π radians at the instants where impulses of $\dot{m}(t)$ are located. At these instants, the carrier phase shifts by π instantaneously. A finite phase shift in zero time implies infinite instantaneous frequency ($d\theta/dt = \infty$) at these instants. This conclusion agrees with our observation about $\dot{m}(t)$.

This scheme of carrier phase modulation by a digital signal is called **phase-shifting keying (PSK)**, because information digits are transmitted by shifting the carrier phase. Note that PSK may also be viewed as a DSB-SC modulation by $m(t)$.

The amount of phase discontinuity in $\varphi_{\text{PM}}(t)$ at the instant where $m(t)$ is discontinuous is $k_p m_d$, where m_d is the amount of discontinuity in $m(t)$ at that instant. In the present example, the amplitude of $m(t)$ changes by 2 (from -1 to 1) at the discontinuity. Hence, the phase discontinuity in $\varphi_{\text{PM}}(t)$ is $k_p m_d = \frac{\pi}{2}(2) = \pi$ radians, which confirms our earlier result.

When $m(t)$ is a digital signal (as in Fig. 4.44a), $\varphi_{\text{PM}}(t)$ shows a phase discontinuity where $m(t)$ has a jump discontinuity. In such a case the phase deviation $k_p m(t)$ must be restricted to a range $(-\pi, \pi)$ in order to avoid ambiguity in demodulation. For example, if k_p were $3\pi/2$ in the present example, then

$$\varphi_{\text{PM}}(t) = A \cos\left[\omega_c t + \frac{3\pi}{2} m(t)\right]$$

In this case $\varphi_{\text{PM}}(t) = A \sin \omega_c t$ when $m(t) = 1$ or $-1/3$. This will certainly cause ambiguity at the receiver when $A \sin \omega_c t$ is received. Such ambiguity never arises if $k_p m(t)$ is restricted to the range $(-\pi, \pi)$.

The ambiguity arises only when $m(t)$ has jump discontinuities. In such a case, the phase of $\varphi_{\text{PM}}(t)$ changes instantaneously. Because a phase $\varphi_o + 2n\pi$ is indistinguishable from the phase φ_o, ambiguities will be inherent in the demodulator unless the phase variations are limited to the range $(-\pi, \pi)$. For this reason k_p should be small enough to restrict the phase change $k_p m(t)$ to the range $(-\pi, \pi)$.

No such restriction on k_p is required if $m(t)$ is continuous. In this case the phase change is not instantaneous, but gradual over a time, and a phase $\varphi_o + 2n\pi$ will exhibit n additional carrier cycles over the case of phase of only φ_o. This conclusion can also be verified from Example 4.21, where the maximum phase change $\Delta\varphi = 10\pi$.

Because a bandlimited signal cannot have jump discontinuities, we can say that when $m(t)$ is bandlimited, k_p has no restrictions.   ■

4.8-2   Bandwidth of Angle-Modulated Signals

Unlike amplitude modulation, there is no simple relationship between the baseband signal waveform and the corresponding angle modulated waveform. The same is true of their spectra. Because of nonlinear nature of angle modulation, derivation of $\Phi_{EM}(\omega)$, the frequency spectrum of the modulated signal is extremely complicated and can be obtained only for few special cases. Generally, the bandwidth of an angle modulated signal is infinite even when the baseband signal bandwidth is finite. However, most of the signal power (or energy) resides in a finite band. We shall now try to estimate this essential bandwidth of an angle modulated signal.

Let us start with the angle modulated signal in Eq. (4.80), and consider first the case of small k $(k \to 0)$.

$$
\begin{aligned}
\varphi_{EM}(t) &= A \, \cos\left[\omega_c t + k\psi(t)\right] \\
&= A \, \cos\omega_c t \cos\left[k\psi(t)\right] - A \, \sin\omega_c t \sin\left[k\psi(t)\right] \\
&\approx A \, \cos\omega_c t - Ak\psi(t) \sin\omega_c t \qquad k \to 0
\end{aligned}
\tag{4.86}
$$

Comparison of the right-hand side expression with $\varphi_{AM}(t)$ in Eq. (4.73a) shows that the two expressions are very similar. The first term is the carrier, and the second term, representing the sidebands, has the same form as the DSB-SC signal corresponding to the baseband signal $Ak\psi(t)$. The only difference is that the carrier is sine instead of cosine. This is just a matter of carrier phase difference of $\pi/2$. Hence, the bandwidth of the angle modulated signal is the same as that of AM signal corresponding to the baseband signal $\psi(t)$. If $m(t)$ is bandlimited to B Hz, then the bandwidth of $\psi(t)$ is also B Hz.† Hence, the bandwidth of $\varphi_{EM}(t)$ is $2B$ Hz, the same as that of AM. But this true only when $k \to 0$. Let us now consider the general case.

In angle modulation, the carrier frequency is varied from its quiescent value ω_c. Let the maximum deviation of the carrier frequency be $\Delta\omega$. In other words, the carrier frequency varies in the range from $\omega_c - \Delta\omega$ to $\omega_c + \Delta\omega$. Because the carrier frequency always remains in this band of width $2\Delta\omega$ radians/s, could we say that the resulting spectrum also remains within this band and the bandwidth of the angle modulated signal is $2\Delta\omega$? This assertion implies that if a sinusoid takes an instantaneous frequency ω_x, the resulting spectrum is concentrated only at ω_x. This is true only if the carrier has infinite duration. For a finite duration sinusoid of

---

†Because $\psi(t)$ is the output of a linear system when the input is $m(t)$, the bandwidth of $\psi(t)$ cannot be greater than B Hz. Moreover, the filter is invertible. Hence, the filter bandwidth cannot be less than B Hz, or some of the components of $m(t)$ will be lost. So, the bandwidth of $\psi(t)$ cannot be less than B Hz. Hence, it is exactly equal to B Hz.

frequency ω_x, the spectrum is not concentrated at ω_x, but spreads out on both sides of ω_x, as can be seen from Fig. 4.24d in Example 4.12. In a typical angle modulated signal, the carrier frequency is directly proportional to $m(t)$, which changes with t. Hence, the instantaneous frequency will also change with t continuously. Such continuous shift in frequency will cause the spectral spread beyond the band $2\Delta\omega$. Clearly, the bandwidth of the angle modulated signal is somewhat larger than $2\Delta\omega$ rads/s. How much larger? This missing link can be found by looking at the results derived earlier for the case of $k \to 0$. Let us first determine $\Delta\omega$.

From Eq. (4.80), it follows that

$$\omega_i(t) = \omega_c + k\dot{\psi}(t) \tag{4.87}$$

if the peak amplitude of $\dot{\psi}(t)$ is denoted by ψ'_p, then the carrier frequency varies in the range from $\omega_c - k\psi'_p$ to $\omega_c + k\psi'_p$† Therefore

$$\Delta\omega = k\psi'_p \tag{4.88a}$$

The carrier frequency deviation $\Delta\mathcal{F}$ in Hz is

$$\Delta\mathcal{F} = \frac{\Delta\omega}{2\pi} = \frac{k}{2\pi}\psi'_p \tag{4.88b}$$

As demonstrated earlier, because of spectral spreading, the angle modulated signal bandwidth is somewhat larger than $2\Delta\mathcal{F}$. Let the actual bandwidth B_{EM} in Hz be

$$B_{\text{EM}} = 2\Delta\mathcal{F} + X$$
$$= \frac{k}{\pi}\psi'_p + X \tag{4.89}$$

where X is unknown. To determine X, recall that for the case $k \to 0$, we found the bandwidth to be $2B$. But as Eq. (4.89) indicates, this bandwidth is X when $k \to 0$. Therefore, $X = 2B$, and

$$B_{\text{EM}} = 2(\Delta\mathcal{F} + B) \text{ Hz} \tag{4.90}$$

A more rigorous derivation of this result appears in reference 4. Note that when $k \to 0$, $\Delta\mathcal{F} \to 0$ and $\Delta\mathcal{F} \ll B$. On the other hand when k is very large, $\Delta\mathcal{F} \gg B$. The former case is known as the **narrowband** angle modulation and the latter is known as the **wideband** angle modulation.

Recall that for FM, $\dot{\psi}(t) = m(t)$, and $\psi'_p = m_p$, where m_p is the peak amplitude of $m(t)$. Similarly, for PM, $\psi(t) = m(t)$. Hence, $\psi'_p = m'_p$, where m'_p is the peak amplitude of $\dot{m}(t)$. Thus

$$(\Delta\mathcal{F})_{\text{FM}} = \frac{k_f}{\pi}m_p \quad \text{and} \quad (\Delta\mathcal{F})_{\text{PM}} = \frac{k_p}{\pi}m'_p \tag{4.91}$$

We observe an interesting fact in angle modulation. The bandwidth of the modulated signal is adjustable by choosing suitable value of $\Delta\mathcal{F}$ or the constant k (k_f in FM or k_p in PM). Amplitude modulation lacks this feature. The bandwidth of each AM scheme is fixed. It is a general principle in communication theory

---

†This assertion implies an assumption $\dot{\psi}(t)|_{max} = \left|\dot{\psi}(t)|_{min}\right|$

that widening a signal bandwidth makes the signal more immune to noise during transmission. Thus, widening the transmission bandwidth makes angle modulated signals can be made more immune to noise. Moreover, this very property allows us to reduce the signal power required to achieve the same quality of transmission. Thus, angle modulation allows us to exchange signal power for bandwidth.

Also, because of its constant amplitude, angle modulation has a major advantage over amplitude modulation. This feature makes angle modulation less susceptible to nonlinear distortion. We shall see in the following section (Sec. 4.8-3) that no distortion results when we pass an angle modulated signal through a nonlinear device whose output $y(t)$ and the input $x(t)$ are related by $y(t) = x^2(t)$ [in general $y(t) = \sum a_n x^n(t)$]. Such a nonlinearity can be disastrous in amplitude modulated systems. This is the primary reason why angle modulation is used in microwave relay systems, where nonlinear operation of amplifiers and other devices has thus far been unavoidable at the required high power levels. In addition, the constant amplitude of FM gives it a kind of immunity against rapid fading. The effect of amplitude variations caused by rapid fading can be eliminated by using automatic gain control and bandpass limiting[4]. Angle modulation is also less vulnerable than amplitude modulation to small interference from adjacent channels. But the price for all these advantages is paid in terms of increased bandwidth. We can demonstrate that for the same bandwidth, the pulse code modulation (PCM), discussed in Chapter 5, is superior to angle modulation[4].

4.8-3 Generation and Demodulation of Angle Modulated Signals.

In Eq. (4.86), we see that a narrowband angle (or exponential) modulated signal (NBEM) consists of a carrier term and a DSB-SC term whose carrier has a $\pi/2$ phase shift. Hence, we can readily generate this signal using the procedure discussed in Sec. 4.7. Wideband modulation (WBEM) can be obtained from NBEM by passing the NBEM signal through a nonlinear device. Consider, for example, a nonlinear device whose input $x(t)$ and the output $y(t)$ are related by $y(t) = x^2(t)$. If the input is an angle modulated signal $\cos[\omega_c t + k\psi(t)]$, then the output $y(t)$ is given by

$$y(t) = \cos^2[\omega_c t + k\psi(t)]$$
$$= \frac{1}{2} + \frac{1}{2}\cos[2\omega_c t + 2k\psi(t)]$$

If we pass this signal through a bandpass filter centered at $2\omega_c$, the output is

$$z(t) = \frac{1}{2}\cos[2\omega_c t + 2k\psi(t)]$$

Observe that the second order of nonlinearity has doubled the carrier frequency as well as the effective value of k without causing any distortion. In a similar way, we can show that an nth-order of nonlinearity increases n-fold the carrier frequency as well as the effective value of k. This fact allows us the convert the NBEM into WBEM. This is the indirect method of generating angle modulated signal.

We can also generate angle modulated signal by a direct method, which uses a **voltage controlled oscillator (VCO)**. The output of a VCO is a constant amplitude sinusoid, whose instantaneous frequency is directly proportional to an input voltage $m(t)$. Clearly, a VCO is an FM generator. As demonstrated earlier, FM generator, with minor modification, can be used to generate any other form of angle modulation.

Demodulation

We shall discuss here demodulation of FM waves. As explained earlier, FM demodulator, with some minor modification, can be used for demodulation of any other form of angle modulation. Because the instantaneous frequency of FM wave is proportional to the baseband signal $m(t)$, an FM demodulator is a device whose output is proportional to frequency of the input signal. Thus, the gain $H(\omega)$ of an ideal FM demodulator is of the form $c_1\omega + c_2$. An ideal differentiator has this property. If the input to an ideal differentiator is an angle modulated signal $x(t) = \cos[\omega_c t + k\psi(t)]$, the output $y(t)$ is given by

$$y(t) = \frac{dx(t)}{dt} = -[\omega_c + k\dot{\psi}(t)]\sin[\omega_c t + k\psi(t)]$$
$$= [\omega_c + k\dot{\psi}(t)]\sin[\omega_c t + k\psi(t) + \pi]$$

The output is also an angle modulated signal, whose envelope is $\omega_c + k\dot{\psi}(t)$. Hence, an ideal differentiator followed by an envelope detector will result in the output $\omega_c + k\dot{\psi}(t)$. After blocking the dc, we obtain the desired output $k\dot{\psi}(t)$. Recall that for FM, $\psi(t) = \int^t m(\alpha)d\alpha$. Hence, $\dot{\psi}(t) = m(t)$. Another device that can be used as an FM demodulator is a tuned circuit, whose resonant frequency is selected either above or below the carrier frequency of the FM signal to be demodulated. The frequency response of a tuned circuit (below the resonant frequency) is approximately linear with the input frequency (at least over a small band). This scheme suffers from the fact that the slope of $H(\omega)$ of a tuned circuit is linear only over a small band, and therefore causes considerable distortion in the output. This fault can partially be corrected by a balanced discriminator that uses two resonant circuits, one tuned above and the other tuned below ω_c.

These days, a **phase-locked loop (PLL)**, whose performance is superior to any of the methods discussed here (especially in the large noise environment) has become very popular as a demodulator of angle modulated signals because of its reasonable cost. More discussion about modulation and demodulation of angle modulated signals appears in reference 4.

A Historical Note

In the twenties, broadcasting was in its infancy. However, there was a constant search for techniques that would reduce noise (static). Now, since the noise power is proportional to the modulated signal bandwidth (sidebands), attempts were focused on finding a modulation scheme to reduce the bandwidth. It was rumored that a new method had been discovered for eliminating sidebands (no sidebands, no bandwidth!). The concept of FM, where the carrier frequency would be varied in proportion to the message $m(t)$, appeared quite intriguing. The carrier frequency $\omega(t)$ would be varied with time so that $\omega(t) = \omega_c + km(t)$, where k is an arbitrary constant. The carrier frequency will remain within the band from $\omega_c - km_p$ to

$\omega_c + km_p$. The spectrum, centered at ω_c, would have a bandwidth $2km_p$, which is controlled by the arbitrary constant k. By using an arbitrarily small k, we could make the information bandwidth arbitrarily small. This was a passport to communication heaven. Unfortunately, the experimental results showed that something was seriously wrong somewhere. The FM bandwidth was found to be always greater than (at best equal to) the AM bandwidth. In some cases, its bandwidth was several times that of AM.

Careful analysis by Carson showed that the FM bandwidth could never be smaller than that of AM; at best equal to that of AM. Unfortunately, Carson did not recognize the compensating advantage of FM in its ability to suppress noise. Without any justification, he states, "Thus, FM introduces inherent distortion and has no compensating advantages whatsoever."[5] In his later paper he says: "In fact, as more and more schemes are analyzed and tested, and as the essential nature of the problem is clearly perceivable, we are unavoidably forced to the conclusion that static, like the poor, will always be with us."[6] This opinion of one of the ablest mathematicians of the day in the communication industry set back the development of FM. The noise-suppressing advantage of FM was later proved by Major Edwin H. Armstrong[7], a brilliant engineer whose contributions to the field of radio systems are comparable to those of Hertz and Marconi. It was largely the work of Armstrong that was responsible for rekindling the interest in FM. Lamentably, Armstrong, who became despondent over the lengthy, most acrimonious, and expensive court battles with some titans of the communication industry over his patent rights, committed suicide in 1954 by walking out of a window 13 stories above the street.

4.8-4 Frequency-Division Multiplexing

Signal multiplexing allows transmission of several signals on the same channel. In Chapter 5, we shall discuss time-division multiplexing (TDM), where several signals time-share the same channel, such as a cable or an optical fiber. In frequency-division multiplexing (FDM), the use of modulation, as illustrated in Fig. 4.45, makes several signals share the band of the same channel. Each signal is modulated by a different carrier frequency. The various carriers are adequately separated to avoid overlap (or interference) between the spectra of various modulated signals. These carriers are referred to as **subcarriers**. Each signal may use a different kind of modulation (for example, DSB-SC, AM, SSB-SC, VSB-SC, or even FM or PM). The modulated-signal spectra may be separated by a small guard band to avoid interference and facilitate signal separation at the receiver.

When all of the modulated spectra are added, we have a composite signal that may be considered as a baseband signal. Sometimes, this composite baseband signal may be used to further modulate a high-frequency (radio frequency, or RF) carrier for the purpose of transmission.

At the receiver, the incoming signal is first demodulated by the RF carrier to retrieve the composite baseband, which is then bandpass filtered to separate each modulated signal. Then each modulated signal is individually demodulated by an appropriate subcarrier to obtain all the basic baseband signals.

4.9   Data Truncation: Window Functions

We often need to truncate data in diverse situations from numerical computa-

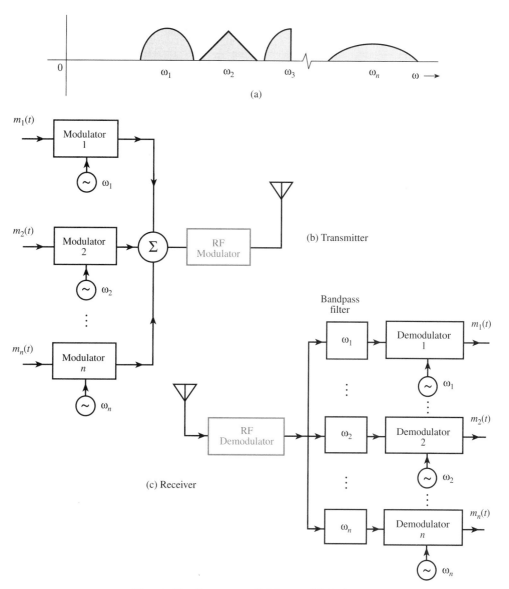

Fig. 4.45   Frequency division multiplexing.

tions to filter design. For example, if we need to compute numerically the Fourier transform of some signal, say $e^{-t}u(t)$, on a computer, we will have to truncate the signal $e^{-t}u(t)$ beyond a sufficiently large value of t (typically five time constants and above). The reason is that in numerical computations, we have to deal with data of finite duration. Similarly, the impulse response $h(t)$ of an ideal lowpass filter is noncausal, and approaches zero asymptotically as $|t| \to \infty$. For a practical design, we may want to truncate $h(t)$ beyond a sufficiently large value of $|t|$ to make $h(t)$ causal. In signal sampling, to eliminate aliasing, we need to truncate the signal spectrum beyond the half sampling frequency $\omega_s/2$, using an anti-aliasing filter.

Again, we may want to synthesize a periodic signal by adding the first n harmonics and truncating all the higher harmonics. These examples show that data truncation can occur in both time and frequency domain. On the surface, truncation appears to be a simple problem of cutting off the data at a point where it is deemed to be sufficiently small. Unfortunately, this is not the case. Simple truncation can cause some unsuspected problems.

Window Functions

Truncation operation may be regarded as multiplying a signal of a large width by a window function of a smaller (finite) width. Simple truncation amounts to using a **rectangular window** $w_R(t)$ (Fig. 4.48a) in which we assign unit weight to all the data within the window width ($|t| < \frac{T}{2}$), and assign zero weight to all the data lying outside the window ($|t| > \frac{T}{2}$). It is also possible to use a window in which the weight assigned to the data within the window may not be constant. In a **triangular window** $w_T(t)$, for example, the weight assigned to data decreases linearly over the window width (Fig. 4.48b).

Consider a signal $f(t)$ and a window function $w(t)$. If $f(t) \Longleftrightarrow F(\omega)$ and $w(t) \Longleftrightarrow W(\omega)$, and if the windowed function $f_w(t) \Longleftrightarrow F_w(\omega)$, then

$$f_w(t) = f(t)w(t) \qquad \text{and} \qquad F_w(\omega) = \frac{1}{2\pi}F(\omega) * W(\omega)$$

According to the width property of convolution, it follows that the width of $F_w(\omega)$ equals the sum of the widths of $F(\omega)$ and $W(\omega)$. Thus, truncation of a signal increases its bandwidth by the amount of bandwidth of $w(t)$. Clearly, the truncation of a signal causes its spectrum to spread (or smear) by the amount of the bandwidth of $w(t)$. Recall that the signal bandwidth is inversely proportional to the signal duration (width). Hence, the wider the window, the smaller is its bandwidth, and the smaller is the **spectral spreading**. This result is predictable because a wider window means we are accepting more data (closer approximation), which should cause smaller distortion (smaller spectral spreading). Smaller window width (poorer approximation) causes more spectral spreading (more distortion). There are also other effects produced by the fact that $W(\omega)$ is really not strictly bandlimited, and its spectrum $\rightarrow 0$ only asymptotically. This causes the spectrum of $F_w(\omega) \rightarrow 0$ asymptotically also at the same rate as that of $W(\omega)$, even though the $F(\omega)$ may be strictly bandlimited. Thus, windowing causes the spectrum of $F(\omega)$ to leak in the band where it is supposed to be zero. This effect is called **leakage**. These twin effects, the spectral spreading and the leakage, will now be clarified by an example.

For an example, let us take $f(t) = \cos \omega_0 t$ and a rectangular window $w_R(t) = \text{rect}(\frac{t}{T})$, illustrated in Fig. 4.46b. The reason for selecting a sinusoid for $f(t)$ is that its spectrum consists of spectral lines of zero width (Fig. 4.46a). This choice will make the effect of spectral spreading and leakage clearly visible. The spectrum of the truncated signal $f_w(t)$ is the convolution of the two impulses of $F(\omega)$ with the sinc spectrum of the window function. Because the convolution of any function with an impulse is the function itself (shifted at the location of the impulse), the resulting spectrum of the truncated signal is ($1/2\pi$ times) the two sinc pulses at $\pm\omega_0$, as depicted in Fig. 4.46c. Comparison of spectra $F(\omega)$ and $F_w(\omega)$ reveals the effects of truncation. These are:

1 The spectral lines of $F(\omega)$ have zero width. But the truncated signal is spread out by $4\pi/T$ about each spectral line. The amount of spread is equal to the

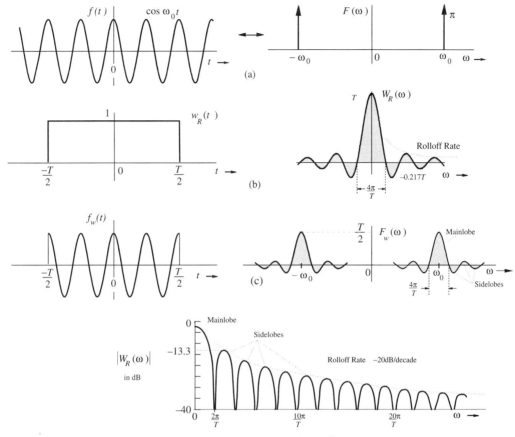

Fig. 4.46   Windowing and its effects.

width of the mainlobe of the window spectrum. One effect of this **spectral spreading** (or smearing) is that if $f(t)$ has two spectral components of frequencies differing by less than $4\pi/T$ rad/s ($2/T$ Hz), they will be indistinguishable in the truncated signal. The result is loss of spectral resolution. We would like the spectral spreading (mainlobe width) to be as small as possible.

2 In addition to the mainlobe spreading, the truncated signal also has sidelobes, which decay slowly with frequency. The spectrum of $f(t)$ is zero everywhere except at $\pm\omega_0$. On the other hand, the truncated signal spectrum $F_w(\omega)$ is zero nowhere because of sidelobes. These sidelobes decay asymptotically as $1/\omega$. Thus, the truncation causes spectral **leakage** in the band where the spectrum of the signal $f(t)$ is zero. The peak **sidelobe** magnitude is 0.217 times the mainlobe magnitude (13.3 dB below the peak mainlobe magnitude). Also, the sidelobes decay at a rate $1/\omega$, which is -6 dB/octave (or -20 dB/decade). This is the **rolloff rate** of sidelobes. We want smaller sidelobes with a faster rate of decay (high rolloff rate). Figure 4.46d shows $|W_R(\omega)|$ (in dB) as a function of ω. This plot clearly shows the mainlobe and sidelobe features, with the first sidelobe amplitude -13.3 dB below the mainlobe amplitude, and the sidelobes decaying at a rate of -6 dB/octave (or -20 dB per decade).

So far, we have discussed the effect of signal truncation (truncation in time domain) on the signal spectrum. Because of time-frequency duality, the effect of spectral truncation (truncation in frequency domain) on the signal shape is similar.

Remedies for Side Effects of Truncation

For better results, we must try to minimize the truncation's twin side effects, the spectral spreading (mainlobe width) and leakage (sidelobe). Let us consider each of these ills.

1 The spectral spread (mainlobe width) of the truncated signal is equal to the bandwidth of the window function $w(t)$. We know that the signal bandwidth is inversely proportional to the signal width (duration). Hence, to reduce the spectral spread (mainlobe width), we need to increase the window width.

2 To improve the leakage behavior, we must search for the cause of the slow decay of sidelobes. In Chapter 3, we saw that the Fourier spectrum decays as $1/\omega$ for a signal with jump discontinuity, and decays as $1/\omega^2$ for a continuous signal whose first derivative is discontinuous, and so on.† Smoothness of a signal is measured by the number of continuous derivatives it possesses. The smoother the signal, the faster the decay of its spectrum. Thus, we can achieve a given leakage behavior by selecting a suitably smooth window.

3 For a given window width, the remedies for the two effects are incompatible. If we try to improve one, the other deteriorates. For instance, among all the windows of a given width, the rectangular window has the smallest spectral spread (mainlobe width), but has high level sidelobes, which decay slowly. A tapered (smooth) window of the same width has smaller and faster decaying sidelobes, but it has a wider mainlobe.‡ But we can compensate for the increased mainlobe width by widening the window. Thus, we can remedy both the side effects of truncation by selecting a suitably smooth window of sufficient width.

There are several well-known tapered-window functions, such as Bartlett (triangular), Hanning (von Hann), Hamming, Blackman, and Kaiser, which truncate the data gradually. These windows offer different tradeoffs with respect to spectral spread (mainlobe width), the peak sidelobe magnitude, and the leakage rolloff rate as indicated in Table 4.3.[8, 9] Observe that all windows are symmetrical about the origin (even functions of t). Because of this feature, $W(\omega)$ is a real function of ω; that is, $\angle W(\omega)$ is either 0 or π. Hence, the phase function of the truncated signal has a minimal amount of distortion.

Figure 4.47 shows two well-known tapered-window functions, the von Hann (or Hanning) window $w_{\mathrm{HAN}}(x)$ and the Hamming window $w_{\mathrm{HAM}}(x)$. We have intentionally used the independent variable x because windowing can be performed in time domain as well as in frequency domain; so x could be t or ω, depending on the application.

---

†This result was demonstrated for periodic signals. However, it applies to aperiodic signals also. This is because we showed in Chapter 4 that if $f_{T_0}(t)$ is a periodic signal formed by periodic extension of an aperiodic signal $f(t)$, then the spectrum of $f_{T_0}(t)$ is ($1/T_0$ times) the samples of $F(\omega)$. Thus, what is true of the decay rate of the spectrum of $f_{T_0}(t)$ is also true of the rate of decay of $F(\omega)$.

‡A tapered window yields a higher mainlobe width because the effective width of a tapered window is smaller than that of the rectangular window (see Sec. 2.7-2 [Eq. (2.67)] for the definition of effective width). Therefore, from the reciprocity of the signal width and its bandwidth, it follows that the rectangular window mainlobe width is smaller than that of a tapered window.

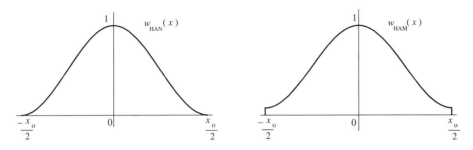

Fig. 4.47 Hanning and Hamming windows.

There are hundreds of windows, each with differing characteristics. But the choice depends on a particular application. The rectangular window has the narrowest mainlobe. The Bartlett (triangle) window (also called the Fejer or Cesaro) is inferior in all respects to the Hanning window. For this reason it is rarely used in practice. Hanning is preferred over Hamming in spectral analysis because it has faster sidelobe decay. For filtering applications, on the other hand, the Hamming window is the choice because it has the smallest sidelobe magnitude for a given mainlobe width. The Hamming window is the most widely used, general purpose window. The Kaiser window, which uses $I_0(\alpha)$, the Bessel function of the order 0, is more versatile and adjustable. Selecting a proper value of α $(0 \leq \alpha \leq 10)$ allows the designer to tailor the window to suit a particular application. The parameter α controls the mainlobe and sidelobe trade-off. When $\alpha = 0$, the Kaiser window is the rectangular window. For $\alpha = 5.4414$, it is the Hamming window, and when $\alpha = 8.885$, it is the Blackman window. As α increases, the mainlobe width increases and the sidelobe level decreases.

Table 4.3

Some Window Functions and Their Characteristics

Window $w(t)$	Mainlobe Width	Rolloff Rate dB/oct	Peak Sidelobe Level in dB
1 Rectangular: $\mathrm{rect}(\frac{t}{T})$	$\frac{4\pi}{T}$	-6	-13.3
2 Bartlett: $\Delta(\frac{t}{2T})$	$\frac{8\pi}{T}$	-12	-26.5
3 Hanning: $0.5\left[1 + \cos\left(\frac{2\pi t}{T}\right)\right]$	$\frac{8\pi}{T}$	-18	-31.5
4 Hamming: $0.54 + 0.46\cos\left(\frac{2\pi t}{T}\right)$	$\frac{8\pi}{T}$	-6	-42.7
5 Blackman: $0.42 + 0.5\cos\left(\frac{2\pi t}{T}\right) + 0.08\cos\left(\frac{4\pi t}{T}\right)$	$\frac{12\pi}{T}$	-18	-58.1
6 Kaiser: $\dfrac{I_0\left[\alpha\sqrt{1-4(\frac{t}{T})^2}\right]}{I_0(\alpha)}$ $1 \leq \alpha \leq 10$	$\frac{11.2\pi}{T}$	-6	-59.9 $(\alpha = 8.168)$

4.9-1 Filter Design Using Windows

We shall design an ideal lowpass filter of bandwidth W rad/s. For this filter, the impulse response $h(t) = \frac{W}{\pi}\text{sinc}\,(Wt)$ (Fig. 4.48c) is noncausal and, therefore, unrealizable. Truncation of $h(t)$ by a suitable window (Fig. 4.48a) makes it realizable, although the resulting filter is now an approximation to the desired ideal filter.† We shall use a rectangular window $w_R(t)$ and a triangular (Bartlett) window $w_T(t)$ to truncate $h(t)$, and then examine the resulting filters. The truncated impulse responses $h_R(t)$ and $h_T(t)$ for the two cases are depicted in Fig. (4.48d).

$$h_R(t) = h(t)w_R(t) \qquad \text{and} \qquad h_T(t) = h(t)w_T(t)$$

Hence, the windowed filter transfer function is the convolution of $H(\omega)$ with the Fourier transform of the window, as illustrated in Fig. 4.48e and f. We make the following observations.

1. The windowed filter spectra show **spectral spreading** at the edges, and instead of a sudden switch there is a gradual transition from the passband to the stopband of the filter. The transition band is smaller ($2\pi/T$ rad/s) for the rectangular case compared to the triangular case ($4\pi/T$ rad/s).

2. Although $H(\omega)$ is bandlimited, the windowed filters are not. But the stopband behavior of the triangular case is superior to that of the rectangular case. For the rectangular window, the leakage in the stopband decreases slowly (as $1/\omega$) compared to that of the triangular window (as $1/\omega^2$). Moreover, the rectangular case has a higher peak sidelobe amplitude compared to that of the triangular window.

4.10 Summary

In Chapter 3 we represented periodic signals as a sum of (everlasting) sinusoids or exponentials (Fourier series). In this chapter we extended this result to aperiodic signals, which are represented by the Fourier integral (instead of the Fourier series). An aperiodic signal $f(t)$ may be regarded as a periodic signal with period $T_0 \rightarrow \infty$, so that the Fourier integral is basically a Fourier series with a fundamental frequency approaching zero. Therefore, for aperiodic signals, the Fourier spectra are continuous. This continuity means that a signal is represented as a sum of sinusoids (or exponentials) of all frequencies over a continuous frequency interval. The Fourier transform $F(\omega)$, therefore, is the spectral density (per unit bandwidth in Hz).

An ever-present aspect of the Fourier transform is the duality between time and frequency, which also implies duality between the signal $f(t)$ and its transform $F(\omega)$. This duality arises because of near-symmetrical equations for direct and inverse Fourier transforms. The duality principle has far-reaching consequences and yields many valuable insights into signal analysis.

The scaling property of the Fourier transform leads to the conclusion that the signal bandwidth is inversely proportional to signal duration (signal width). Time

---

†In addition to truncation, we also need to delay the truncated function by $\frac{T}{2}$ in order to render it causal. However, the time delay only adds a linear phase to the spectrum without changing the amplitude spectrum. For this reason, we shall ignore the delay in order to simplify our discussion.

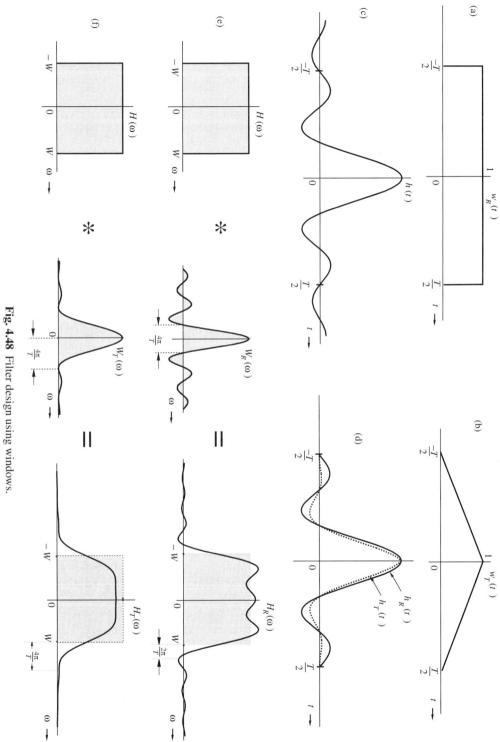

Fig. 4.48 Filter design using windows.

shifting of a signal does not change its amplitude spectrum, but adds a linear phase spectrum. Multiplication of a signal by an exponential $e^{j\omega_0 t}$ results in shifting the spectrum to the right by ω_0. In practice, spectral shifting is achieved by multiplying a signal with a sinusoid such as $\cos \omega_0 t$ (rather than the exponential $e^{j\omega_0 t}$). This process is known as amplitude modulation. Multiplication of two signals results in convolution of their spectra, whereas convolution of two signals results in multiplication of their spectra.

For an LTIC system with the transfer function $H(\omega)$, the input and output spectra $F(\omega)$ and $Y(\omega)$ are related by the equation $Y(\omega) = F(\omega)H(\omega)$. This is valid only for asymptotically stable systems. For distortionless transmission of a signal through an LTIC system, the amplitude response $|H(\omega)|$ of the system must be constant, and the phase response $\angle H(\omega)$ should be a linear function of ω over a band of interest. Ideal filters, which allow distortionless transmission of a certain band of frequencies and suppress all the remaining frequencies, are physically unrealizable (noncausal). In fact, it is impossible to build a physical system with zero gain $[H(\omega) = 0]$ over a finite band of frequencies. Such systems (which include ideal filters) can be realized only with infinite time delay in the response.

The energy of a signal $f(t)$ is equal to $1/2\pi$ times the area under $|F(\omega)^2|$ (Parseval's theorem). The energy contributed by spectral components within a band $\Delta \mathcal{F}$ (in Hz) is given by $|F(\omega)|^2 \Delta \mathcal{F}$. Therefore, $|F(\omega)|^2$ is the energy spectral density per unit bandwidth (in Hz). The energy spectral density $|F(\omega)|^2$ of a signal $f(t)$ is the Fourier transform of the autocorrelation function $\psi_f(t)$ of the signal $f(t)$. Thus, a signal autocorrelation function has a direct link to its spectral information.

The process of modulation shifts the signal spectrum to different frequencies. Modulation is used for many reasons: to transmit several messages simultaneously over the same channel to utilize channel's high bandwidth, to effectively radiate power over a radio link, to shift signal spectrum at higher frequencies to overcome the difficulties associated with signal processing at lower frequencies, to effect the exchange of transmission bandwidth and transmission power required to transmit data at a certain rate. Broadly speaking there are two types of modulation; amplitude and angle modulation. Each of these two classes has several subclasses. Amplitude modulation bandwidth is generally fixed. The bandwidth in angle modulation, however, is controllable. The higher the bandwidth, the more immune is the scheme to noise.

In practice, we often need to truncate data. Truncating data is like viewing it through a window, which permits a view of only certain portions of the data and hides (suppresses) the remainder. Abrupt truncation of data amounts to a rectangular window, which assigns a unit weight to data seen from the window and assigns zero weight to the remaining data. Tapered windows, on the other hand, reduce the weight gradually from 1 to 0. Data truncation can cause some unsuspected problems. For example, in computation of the Fourier transform, windowing (data truncation) causes spectral spreading (spectral smearing) that is characteristic of the window function used. A rectangular window results in the least spreading, but it does so at the cost of a high and oscillatory spectral leakage outside the signal band which decays slowly as $1/\omega$. Compared to a rectangular window, tapered windows in general have larger spectral spreading (smearing), but the spectral leakage is smaller and decays faster with frequency. If we try to reduce spectral leakage by using a smoother window, the spectral spreading increases. Fortunately, the spectral spreading can be reduced by increasing the window width. Therefore, we can achieve a given combination of spectral spread (transition bandwidth) and leakage characteristics by choosing a suitable tapered window function of a sufficiently longer width T.

References

1. Churchill, R.V., and J.W. Brown, *Fourier Series and Boundary Value Problems, 3d ed.*, McGraw-Hill, New York, 1978.

2. Bracewell, R.N., *Fourier Transform and Its Applications, revised 2nd Ed.*, McGraw-Hill, New York, 1986.

3. Guillemin, E.A., *Theory of Linear Physical Systems*, Wiley, New York, 1963.

4. Lathi, B.P., *Modern Digital and Analog Communication Systems, 3rd ed.*, Oxford University Press, New York, 1998.

5. J. Carson, "Notes on Theory of Modulation" *Proc. IRE*, vol 10, February 1922, pp. 57-64.

6. J. Carson, "The Reduction of Atmospheric Disturbances" *Proc. IRE*, vol 16, July 1928, pp. 966-975.

7. Armstrong E.H., "A Method of Reducing Disturbances in Radio Signaling by a System of Frequency Modulation", *Proc. IRE*, vol. 24, May 1936, pp. 689-740.

8. Hamming, R.W., *Digital Filters, 2nd Ed.*, Prentice-Hall, Englewood Cliffs, N.J. 1983.

9. Harris, F.J., "On the Use of Windows for Harmonic Analysis with the Discrete Fourier Transform", *Proc. IEEE*, vol. 66, No. 1, January 1978, pp. 51-83.

Problems

4.1-1 Show that if $f(t)$ is an even function of t, then

$$F(\omega) = 2 \int_0^\infty f(t) \cos \omega t \, dt$$

and if $f(t)$ is an odd function of t, then

$$F(\omega) = -2j \int_0^\infty f(t) \sin \omega t \, dt$$

Hence, prove that if $f(t)$ is a real and even function of t, then $F(\omega)$ is a real and even function of ω. In addition, if $f(t)$ is a real and odd function of t, then $F(\omega)$ is an imaginary and odd function of ω.

4.1-2 Show that for a real $f(t)$, Eq. (4.8b) can be expressed as

$$f(t) = \frac{1}{\pi} \int_0^\infty |F(\omega)| \cos \left[\omega t + \angle F(\omega)\right] d\omega$$

This is the trigonometric form of the Fourier integral. Compare this with the compact trigonometric Fourier series.

4.1-3 A signal $f(t)$ can be expressed as the sum of even and odd components (see Sec. 1.5-2):

$$f(t) = f_e(t) + f_o(t)$$

Fig. P4.1-4

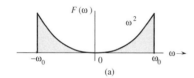

Fig. P4.1-5

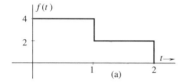

Fig. P4.1-6

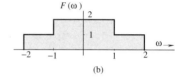

Fig. P4.1-7

(a) If $f(t) \Longleftrightarrow F(\omega)$, show that for real $f(t)$,

$$f_e(t) \Longleftrightarrow \operatorname{Re}[F(\omega)] \qquad \text{and} \qquad f_o(t) \Longleftrightarrow j \operatorname{Im}[F(\omega)]$$

(b) Verify these results by finding the Fourier transforms of the even and odd components of the following signals: **(i)** $u(t)$ **(ii)** $e^{-at}u(t)$.

4.1-4 From definition (4.8a), find the Fourier transforms of the signals $f(t)$ in Fig. P4.1-4.

4.1-5 From definition (4.8a), find the Fourier transforms of the signals depicted in Fig. P4.1-5.

4.1-6 Using Eq. (4.8b), find the inverse Fourier transforms of the spectra in Fig. P4.1-6

4.1-7 Using Eq. (4.8b), find the inverse Fourier transforms of the spectra in Fig. P4.1-7.

4.2-1 Sketch the following functions:
   (a) $\operatorname{rect}\left(\frac{t}{2}\right)$ **(b)** $\Delta\left(\frac{3\omega}{100}\right)$ **(c)** $\operatorname{rect}\left(\frac{t-10}{8}\right)$ **(d)** $\operatorname{sinc}\left(\frac{\pi\omega}{5}\right)$ **(e)** $\operatorname{sinc}\left(\frac{\omega-10\pi}{5}\right)$
   (f) $\operatorname{sinc}\left(\frac{t}{5}\right)\operatorname{rect}\left(\frac{t}{10\pi}\right)$. Hint: $f\left(\frac{x-a}{b}\right)$ is $f\left(\frac{x}{b}\right)$ right-shifted by a.

4.2-2 From definition (4.8a), show that the Fourier transform of $\operatorname{rect}(t-5)$ is $\operatorname{sinc}\left(\frac{\omega}{2}\right)e^{-j5\omega}$. Sketch the resulting amplitude and phase spectra.

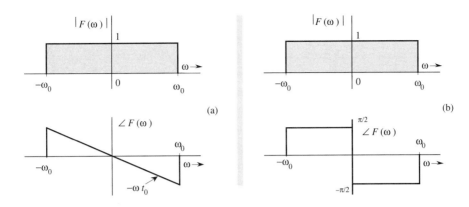

Fig. P4.2-4

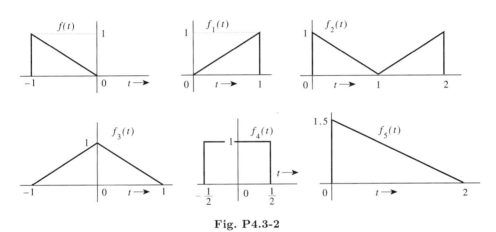

Fig. P4.3-2

4.2-3 From definition (4.8b), show that the inverse Fourier transform of $\text{rect}\left(\frac{\omega-10}{2\pi}\right)$ is $\text{sinc}\,(\pi t)\,e^{j10t}$.

4.2-4 Find the inverse Fourier transform of $F(\omega)$ for the spectra illustrated in Figs. P4.2-4a and b.

Hint: $F(\omega) = |F(\omega)|e^{j\angle F(\omega)}$. This problem illustrates how different phase spectra (both with the same amplitude spectrum) represent entirely different signals.

4.3-1 Apply the symmetry property to the appropriate pair in Table 4.1 to show that
(a) $\frac{1}{2}[\delta(t) + \frac{j}{\pi t}] \Longleftrightarrow u(\omega)$ (b) $\delta(t+T) + \delta(t-T) \Longleftrightarrow 2\cos T\omega$
(c) $\delta(t+T) - \delta(t-T) \Longleftrightarrow 2j\sin T\omega$.

4.3-2 The Fourier transform of the triangular pulse $f(t)$ in Fig. P4.3-2a is expressed as

$$F(\omega) = \frac{1}{\omega^2}(e^{j\omega} - j\omega e^{j\omega} - 1)$$

Using this information, and the time-shifting and time-scaling properties, find the Fourier transforms of the signals $f_i(t)$ ($i = 1, 2, 3, 4, 5$) shown in Fig. P4.3-2.

Hint: See Sec. 1.3 for explanation of various signal operations. Pulses $f_i(t)$ ($i = 2, 3, 4$ can be expressed as a combination of $f(t)$ and $f_1(t)$ with suitable time shift (which may be positive or negative).

4.3-3 Using only the time-shifting property and Table 4.1, find the Fourier transforms of the signals depicted in Fig. P4.3-3.

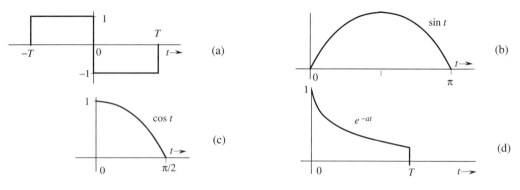

Fig. P4.3-3

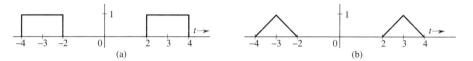

Fig. P4.3-4

Fig. P4.3-5

Hint: Signals in Figs. b, c, and d can be expressed in the form $f(t)[u(t) - u(t - a)]$.

4.3-4 Using the time-shifting property, show that if $f(t) \iff F(\omega)$, then

$$f(t + T) + f(t - T) \iff 2F(\omega) \cos T\omega$$

This is the dual of Eq. (4.41). Using this result and pairs 17 and 19 in Table 4.1, find the Fourier transforms of the signals shown in Fig. P4.3-4.

4.3-5 Prove the following results, which are duals of each other:

$$f(t) \sin \omega_0 t \iff \tfrac{1}{2j}[F(\omega - \omega_0) - F(\omega + \omega_0)]$$
$$\tfrac{1}{2j}[f(t + T) - f(t - T)] \iff F(\omega) \sin T\omega$$

Using the latter result and Table 4.1, find the Fourier transform of the signal in Fig. P4.3-5.

4.3-6 The signals in Fig. P4.3-6 are modulated signals with carrier $\cos 10t$. Find the Fourier transforms of these signals using the appropriate properties of the Fourier transform and Table 4.1. Sketch the amplitude and phase spectra for parts **(a)** and **(b)**.

4.3-7 Using the frequency-shifting property and Table 4.1, find the inverse Fourier transform of the spectra depicted in Fig. P4.3-7.

4.3-8 Using the time convolution property, prove pairs 2, 4, 13 and 14 in Table 2.1 (assume $\lambda < 0$ in pair 2, λ_1 and $\lambda_2 < 0$ in pair 4, $\lambda_1 < 0$ and $\lambda_2 > 0$ in pair 13, and λ_1 and $\lambda_2 > 0$ in pair 14). Hint: You will need partial fraction expansion. For pair 2, you need to apply the result in Eq. (1.23).

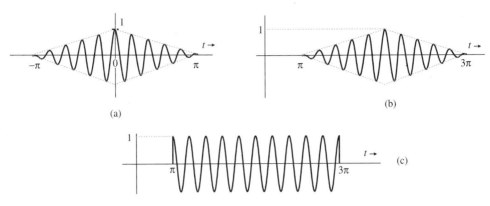

Fig. P4.3-6

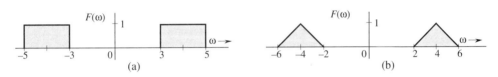

Fig. P4.3-7

4.3-9 A signal $f(t)$ is bandlimited to B Hz. Show that the signal $f^n(t)$ is bandlimited to nB Hz. Hint: Start with $n = 2$. Use frequency convolution property and the width property of convolution.

4.3-10 Find the Fourier transform of the signal in Fig. P4.3-3a by three different methods:
(a) By direct integration using the definition (4.8a).
(b) Using only pair 17 Table 4.1 and the time-shifting property.
(c) Using the time-differentiation and time-shifting properties, along with the fact that $\delta(t) \Longleftrightarrow 1$.
Hint: $1 - \cos 2x = 2\sin^2 x$.

4.3-11 (a) Prove the frequency differentiation property (dual of the time differentiation):

$$-jtf(t) \Longleftrightarrow \frac{d}{d\omega}F(\omega)$$

(b) Using this property and pair 1 (Table 4.1), determine the Fourier transform of $te^{-at}u(t)$.

4.4-1 For an LTIC system with transfer function

$$H(s) = \frac{1}{s+1}$$

find the (zero-state) response if the input $f(t)$ is (a) $e^{-2t}u(t)$ (b) $e^{-t}u(t)$
(c) $e^t u(-t)$ (d) $u(t)$
Hint: For part (d), you need to apply the result in Eq. (1.23).

4.4-2 A stable LTIC system is specified by the transfer function

$$H(\omega) = \frac{-1}{j\omega - 2}$$

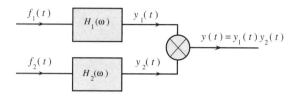

Fig. P4.4-3

Find the impulse response of this system and show that this is a noncausal system. Find the (zero-state) response of this system if the input $f(t)$ is
(a) $e^{-t}u(t)$ **(b)**$e^t u(-t)$.

4.4-3 Signals $f_1(t) = 10^4 \text{rect}\,(10^4 t)$ and $f_2(t) = \delta(t)$ are applied at the inputs of the ideal lowpass filters $H_1(\omega) = \text{rect}\,(\frac{\omega}{40,000\pi})$ and $H_2(\omega) = \text{rect}\,(\frac{\omega}{20,000\pi})$ (Fig. P4.4-3). The outputs $y_1(t)$ and $y_2(t)$ of these filters are multiplied to obtain the signal $y(t) = y_1(t)y_2(t)$.
(a) Sketch $F_1(\omega)$ and $F_2(\omega)$.
(b) Sketch $H_1(\omega)$ and $H_2(\omega)$.
(c) Sketch $Y_1(\omega)$ and $Y_2(\omega)$.
(d) Find the bandwidths of $y_1(t), y_2(t)$, and $y(t)$.
Hint for part **(d)**: Use the convolution property and the width property of convolution to determine the bandwidth of $y_1(t)y_2(t)$.

4.4-4 A lowpass system time constant is often defined as the width of its unit impulse response $h(t)$ (see Sec. 2.7-2). An input pulse $p(t)$ to this system acts like an impulse of strength equal to the area of $p(t)$ if the width of $p(t)$ is much smaller than the system time constant. Assume $p(t)$ to be a lowpass pulse, that is its spectrum is concentrated at low frequencies. Verify this behavior by considering a system whose unit impulse response is $h(t) = \text{rect}\,(\frac{t}{10^{-3}})$. The input pulse is a triangle pulse $p(t) = \Delta(\frac{t}{10^{-6}})$. The area under this pulse is $A = 0.5 \times 10^{-6}$. Show that the system response to this pulse is very nearly the system response to the input $A\delta(t)$.

4.4-5 A lowpass system time constant is often defined as the width of its unit impulse response $h(t)$ (see Sec. 2.7-2). An input pulse $p(t)$ to this system passes practically without distortion, if the width of $p(t)$ is much greater than the system time constant. Assume $p(t)$ to be a lowpass pulse, that is its spectrum is concentrated at low frequencies. Verify this behavior by considering a system whose unit impulse response is $h(t) = \text{rect}\,(\frac{t}{10^{-3}})$. The input pulse is a triangle pulse $p(t) = \Delta(t)$. Show that the system output to this pulse is very nearly $k\,p(t)$, where k is the system gain to a dc signal, that is, $k = H(0)$.

4.4-6 A causal signal $h(t)$ has a Fourier transform $H(\omega)$. If $R(\omega)$ and $X(\omega)$ are the real and the imaginary parts of $H(\omega)$, that is , $H(\omega) = R(\omega) + jX(\omega)$, then show that

$$R(\omega) = \frac{1}{\pi}\int_{-\infty}^{\infty} \frac{X(\omega)}{\omega - y} \quad \text{and} \quad X(\omega) = -\frac{1}{\pi}\int_{-\infty}^{\infty} \frac{R(\omega)}{\omega - y}$$

assuming that $h(t)$ has no impulse at the origin. This pair of integrals defines the **Hilbert transform**.
Hint: Let $h_e(t)$ and $h_o(t)$ be the even and odd components of $h(t)$. Use results in Prob. 4.1-3. See Fig. 1.24 for the relationship between $h_e(t)$ and $h_o(t)$. Recall that $\text{sgn}(t) \Longleftrightarrow 2/j\omega$. Use convolution property.
This problem states one of the important properties of causal systems: that the real and imaginary parts of the transfer function of a causal system are related. If one

specifies the real part, the imaginary part cannot be specified independently. The imaginary part is predetermined by the real part, and vice versa. This result also leads to the conclusion that the magnitude and angle of $H(\omega)$ are related provided all the poles and zeros of $H(\omega)$ lie in the LHP.

4.5-1 Consider a filter with the transfer function

$$H(\omega) = e^{-(k\omega^2 + j\omega t_0)}$$

Show that this filter is physically unrealizable by using the time-domain criterion [noncausal $h(t)$] and the frequency-domain (Paley-Wiener) criterion. Can this filter be made approximately realizable by choosing a sufficiently large t_0? Use your own (reasonable) criterion of approximate realizability to determine t_0.

Hint: Use pair 22 in Table 4.1.

4.5-2 Show that a filter with transfer function

$$H(\omega) = \frac{2(10^5)}{\omega^2 + 10^{10}} e^{-j\omega t_0}$$

is unrealizable. Can this filter be made approximately realizable by choosing a sufficiently large t_0? Use your own (reasonable) criterion of approximate realizability to determine t_0.

Hint: Show that the impulse response is noncausal.

4.5-3 Determine if the filters with the following transfer functions are physically realizable. If they are not realizable, can they be realized exactly or approximately by allowing a finite time delay in the response?

$$H(\omega) = \textbf{(a)}\ 10^{-6} \operatorname{sinc}(10^{-6}\omega) \quad \textbf{(b)}\ 10^{-4}\Delta\left(\frac{\omega}{40,000\pi}\right) \quad \textbf{(c)}\ 2\pi\,\delta(\omega)$$

4.6-1 Show that the energy of a Gaussian pulse

$$f(t) = \frac{1}{\sigma\sqrt{2\pi}} e^{-\frac{t^2}{2\sigma^2}}$$

is $\frac{1}{2\sigma\sqrt{\pi}}$. Verify this result by deriving the energy E_f from $F(\omega)$ using Parseval's theorem.

Hint: See pair 22 in Table 4.1. Use the fact that

$$\int_{-\infty}^{\infty} e^{-x^2/2}\,dx = \sqrt{2\pi}$$

4.6-2 Show that

$$\int_{-\infty}^{\infty} \operatorname{sinc}^2(kx)\,dx = \frac{\pi}{k}$$

Hint: Recognize that the integral is the energy of $f(t) = \operatorname{sinc}(kt)$. Find this energy by using Parseval's theorem.

4.6-3 A lowpass signal $f(t)$ is applied to a squaring device. The squarer output $f^2(t)$ is applied to a lowpass filter of bandwidth $\Delta\mathcal{F}$ Hz (Fig. P4.6-3). Show that if $\Delta\mathcal{F}$ is very small ($\Delta\mathcal{F} \to 0$), then the filter output is a dc signal $y(t) \approx 2E_f\Delta\mathcal{F}$.

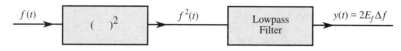

Fig. P4.6-3

Hint: If $f^2(t) \Longleftrightarrow A(\omega)$, then show that $Y(\omega) \approx [4\pi A(0)\Delta\mathcal{F}]\delta(\omega)$ if $\Delta\mathcal{F} \to 0$. Now, show that $A(0) = E_f$.

4.6-4 Generalize Parseval's theorem to show that for real, Fourier transformable signals $f_1(t)$ and $f_2(t)$

$$\int_{-\infty}^{\infty} f_1(t)f_2(t)\, dt = \frac{1}{2\pi}\int_{-\infty}^{\infty} F_1(-\omega)F_2(\omega)\, d\omega = \frac{1}{2\pi}\int_{-\infty}^{\infty} F_1(\omega)F_2(-\omega)\, d\omega$$

4.6-5 For the signal

$$f(t) = \frac{2a}{t^2 + a^2}$$

determine the essential bandwidth B Hz of $f(t)$ such that the energy contained in the spectral components of $f(t)$ of frequencies below B Hz is 99% of the signal energy E_f. Hint: See Exercise E4.5b.

4.7-1 For each of the following 3 baseband signals **(i)** $m(t) = \cos 1000t$ **(ii)** $m(t) = 2\cos 1000t + \cos 2000t$ **(iii)** $m(t) = \cos 1000t \cos 3000t$

(a) Sketch the spectrum of $m(t)$.
(b) Sketch the spectrum of the DSB-SC signal $m(t)\cos 10{,}000t$.
(c) Identify the upper sideband (USB) and the lower sideband (LSB) spectra.
(d) Identify the frequencies in the baseband, and the corresponding frequencies in the DSB-SC, USB and LSB spectra. Explain the nature of frequency shifting in each case.

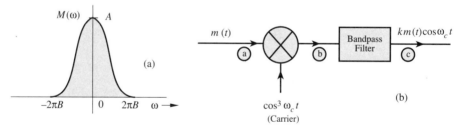

Fig. P4.7-2

4.7-2 You are asked to design a DSB-SC modulator to generate a modulated signal $km(t)\cos \omega_c t$, where $m(t)$ is a signal bandlimited to B Hz (Fig. P4.7-2a). Figure P4.7-2b shows a DSB-SC modulator available in the stock room. The bandpass filter is tuned to ω_c. The carrier generator available generates not $\cos \omega_c t$, but $\cos^3 \omega_c t$.
(a) Explain whether you would be able to generate the desired signal using only this equipment. If so, what is the value of k?
(b) Determine the signal spectra at points b and c, and indicate the frequency bands occupied by these spectra.
(c) What is the minimum usable value of ω_c?
(d) Would this scheme work if the carrier generator output were $\cos^2 \omega_c t$? Explain.
(e) Would this scheme work if the carrier generator output were $\cos^n \omega_c t$ for any integer $n \geq 2$?

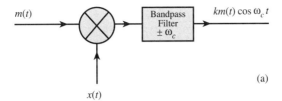

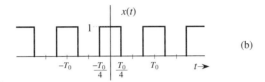

Fig. P4.7-3

4.7-3 In practice, the analog multiplication operation is difficult and expensive. For this reason, in amplitude modulators, it is necessary to find some alternative to multiplication of $m(t)$ with cos $\omega_c t$. Fortunately, for this purpose, we can replace multiplication with switching operation. A similar observation applies to demodulators. In the scheme depicted in Fig. P4.7-3a, the period of the rectangular periodic pulse $x(t)$ shown in Fig. P4.7-3b is $T_0 = 2\pi/\omega_c$. The bandpass filter is centered at $\pm\omega_c$. Note that multiplication by a square periodic pulse $x(t)$ in Fig. P4.7-3b amounts to periodic on-off switching of $m(t)$. This is a relatively simple and inexpensive operation.

Show that this scheme can generate amplitude modulated signal k cos $\omega_c t$. Determine the value of k. Show that the same scheme can also be used for demodulation provided the bandpass filter in Fig. P4.7-3a is replaced by a lowpass (or baseband) filter.

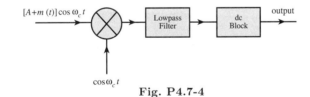

Fig. P4.7-4

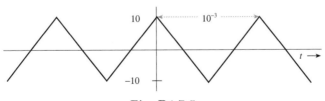

Fig. P4.7-5

4.7-4 Figure P4.7-4 presents a scheme for coherent (synchronous) demodulation. Show that this scheme can demodulate the AM signal $[A + m(t)]$ cos $\omega_c t$ regardless of the value of A.

4.7-5 Sketch the AM signal $[A+m(t)]$ cos $\omega_c t$ for the periodic triangle signal $m(t)$ illustrated in Fig. P4.7-5 corresponding to the modulation index: **(a)** $\mu = 0.5$, **(b)** $\mu = 1$, **(c)** $\mu = 2$, and **(d)** $\mu = \infty$. How do you interpret the case $\mu = \infty$?

4.7-6 For each of the following three baseband signals (a) $m(t) = \cos 100t$ (b) $m(t) = \cos 100t + 2\cos 300t$ (c) $m(t) = \cos 100t \cos 500t$

(i) Sketch the spectrum of $m(t)$.
(ii) Find and sketch the spectrum of the DSB-SC signal $2m(t)\cos 1000t$.
(iii) From the spectrum obtained in (ii), suppress the LSB spectrum to obtain the USB spectrum.
(iv) Knowing the USB spectrum in (ii), write the expression $\varphi_{\mathrm{USB}}(t)$ for the USB signal.
(v) Repeat (iii) and (iv) to obtain the LSB signal $\varphi_{\mathrm{LSB}}(t)$.

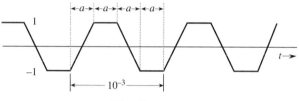

Fig. P4.8-1

4.8-1 Sketch $\varphi_{\mathrm{FM}}(t)$ and $\varphi_{\mathrm{PM}}(t)$ for the modulating signal $m(t)$ depicted in Fig. P4.8-1, given $\omega_c = 2\pi \times 10^7$, $k_f = 2\pi \times 10^5$, and $k_p = 50\pi$.

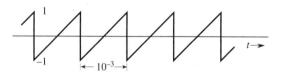

Fig. P4.8-2

4.8-2 A baseband signal $m(t)$ is a periodic saw tooth signal shown in Fig. P4.8-2. Sketch $\varphi_{\mathrm{FM}}(t)$ and $\varphi_{\mathrm{PM}}(t)$ for this $m(t)$ if $\omega_c = 2\pi \times 10^6$, $k_f = 20,000\pi$, and $k_p = \pi/2$. Explain why it is necessary to use $k_p < \pi$ in this case.

4.8-3 For a modulating signal

$$m(t) = 2\cos 100t + 18\cos 2000\pi t$$

Determine the bandwidths of the corresponding $\varphi_{\mathrm{FM}}(t)$ and $\varphi_{\mathrm{PM}}(t)$ if $k_f = 1000\pi$ and $k_p = 1$.

4.8-4 An angle-modulated signal is described by the equation

$$\varphi_{\mathrm{EM}}(t) = 10\cos(\omega_c t + 0.1\sin 2000\pi t)$$

(a) Find the frequency deviation $\Delta\mathcal{F}$ (b) Estimate the bandwidth of $\varphi_{\mathrm{EM}}(t)$.

4.8-5 Repeat Prob. 4.8-4 if

$$\varphi_{\mathrm{EM}}(t) = 5\cos(\omega_c t + 20\sin 1000\pi t + 10\sin 2000\pi t)$$

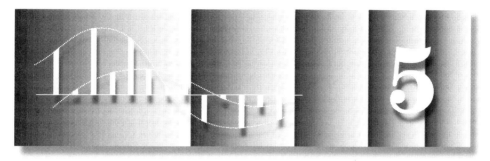

Sampling

A continuous-time signal can be processed by processing its samples through a discrete-time system. For this purpose, it is important to maintain the signal sampling rate sufficiently high so that the original signal can be reconstructed from these samples without error (or with an error within a given tolerance). The necessary quantitative framework for this purpose is provided by the sampling theorem derived in the following section.

5.1   The Sampling Theorem

We now show that a real signal whose spectrum is bandlimited to B Hz $[F(\omega) = 0$ for $|\omega| > 2\pi B]$ can be reconstructed exactly (without any error) from its samples taken uniformly at a rate $\mathcal{F}_s > 2B$ samples per second. In other words, the minimum sampling frequency is $\mathcal{F}_s = 2B$ Hz.†

To prove the sampling theorem, consider a signal $f(t)$ (Fig. 5.1a) whose spectrum is bandlimited to B Hz (Fig. 5.1b).‡ For convenience, spectra are shown as functions of ω as well as of $\mathcal{F}$ (Hz). Sampling $f(t)$ at a rate of $\mathcal{F}_s$ Hz ($\mathcal{F}_s$ samples per second) can be accomplished by multiplying $f(t)$ by an impulse train $\delta_T(t)$(Fig. 5.1c), consisting of unit impulses repeating periodically every T seconds, where $T = 1/\mathcal{F}_s$. The result is the sampled signal $\overline{f}(t)$ resented in Fig. 5.1d. The sampled signal consists of impulses spaced every T seconds (the sampling interval). The nth impulse, located at $t = nT$, has a strength $f(nT)$, the value of $f(t)$ at $t = nT$.

$$\overline{f}(t) = f(t)\delta_T(t) = \sum_n f(nT)\delta(t - nT) \tag{5.1}$$

---

†The theorem stated here (and proved subsequently) applies to lowpass signals. A bandpass signal whose spectrum exists over a frequency band $\mathcal{F}_c - \frac{B}{2} < |\mathcal{F}| < \mathcal{F}_c + \frac{B}{2}$ has a bandwidth of B Hz. Such a signal is uniquely determined by $2B$ samples per second. In general, the sampling scheme is a bit more complex in this case. It uses two interlaced sampling trains, each at a rate of B samples per second (known as second-order sampling). See, for example, the references.[1,2]

‡The spectrum $F(\omega)$ in Fig. 5.1b is shown as real, for convenience. However, our arguments are valid for complex $F(\omega)$ as well.

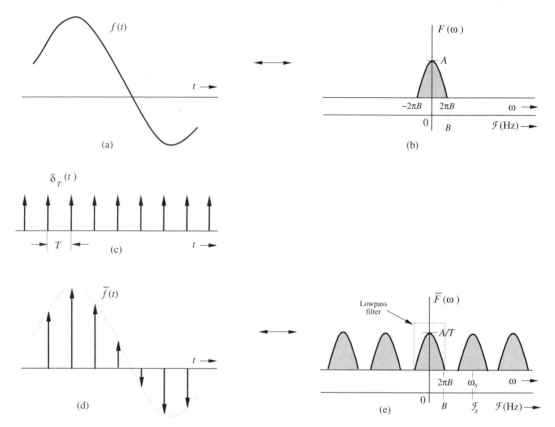

Fig. 5.1   Sampled signal and its Fourier spectrum.

Because the impulse train $\delta_T(t)$ is a periodic signal of period T, it can be expressed as a trigonometric Fourier series already obtained in Example 3.8 [Eq. (3.81)]

$$\delta_T(t) = \frac{1}{T}[1 + 2\cos \omega_s t + 2\cos 2\omega_s t + 2\cos 3\omega_s t + \cdots] \qquad \omega_s = \frac{2\pi}{T} = 2\pi\mathcal{F}_s \quad (5.2)$$

Therefore

$$\overline{f}(t) = f(t)\delta_T(t) = \frac{1}{T}[f(t) + 2f(t)\cos \omega_s t + 2f(t)\cos 2\omega_s t + 2f(t)\cos 3\omega_s t + \cdots] \quad (5.3)$$

To find $\overline{F}(\omega)$, the Fourier transform of $\overline{f}(t)$, we take the Fourier transform of the right-hand side of Eq. (5.3), term by term. The transform of the first term in the brackets is $F(\omega)$. The transform of the second term $2f(t)\cos \omega_s t$ is $F(\omega - \omega_s) + F(\omega + \omega_s)$ [see Eq. (4.41)]. This represents spectrum $F(\omega)$ shifted to ω_s and $-\omega_s$. Similarly, the transform of the third term $2f(t)\cos 2\omega_s t$ is $F(\omega - 2\omega_s) + F(\omega + 2\omega_s)$, which represents the spectrum $F(\omega)$ shifted to $2\omega_s$ and $-2\omega_s$, and so on to infinity. This result means that the spectrum $\overline{F}(\omega)$ consists of $F(\omega)$ repeating periodically with period $\omega_s = \frac{2\pi}{T}$ rad/s, or $\mathcal{F}_s = \frac{1}{T}$ Hz, as depicted in Fig. 5.1e. There is also a constant multiplier $1/T$ in Eq. (5.3). Therefore

$$\overline{F}(\omega) = \frac{1}{T} \sum_{n=-\infty}^{\infty} F(\omega - n\omega_s) \qquad\qquad (5.4)$$

If we are to reconstruct $f(t)$ from $\overline{f}(t)$, we should be able to recover $F(\omega)$ from $\overline{F}(\omega)$. This recovery is possible if there is no overlap between successive cycles of $\overline{F}(\omega)$. Figure 5.1e indicates that this requires

$$\mathcal{F}_s \geq 2B \tag{5.5}$$

Also, the sampling interval $T = 1/\mathcal{F}_s$. Therefore

$$T \leq \frac{1}{2B} \tag{5.6}$$

Thus, as long as the sampling frequency $\mathcal{F}_s$ is greater than twice the signal bandwidth B (in hertz), $\overline{F}(\omega)$ will consist of nonoverlapping repetitions of $F(\omega)$. In such a case, Fig. 5.1e indicates that $f(t)$ can be recovered from its samples $\overline{f}(t)$ by passing the sampled signal $\overline{f}(t)$ through an ideal lowpass filter of bandwidth B Hz. The minimum sampling rate $\mathcal{F}_s = 2B$ required to recover $f(t)$ from its samples $\overline{f}(t)$ is called the **Nyquist rate** for $f(t)$, and the corresponding sampling interval $T = 1/2B$ is called the **Nyquist interval** for $f(t)$.†

■ **Example 5.1**

In this example, we examine the effects of sampling a signal at the Nyquist rate, below the Nyquist rate (undersampling) and above the Nyquist rate (oversampling). Consider a signal $f(t) = \text{sinc}^2(5\pi t)$ (Fig. 5.2a) whose spectrum is $F(\omega) = 0.2\,\Delta(\frac{\omega}{20\pi})$ (Fig. 5.2b). The bandwidth of this signal is 5 Hz (10π rad/s). Consequently, the Nyquist rate is 10 Hz; that is, we must sample the signal at a rate no less than 10 samples/s. The Nyquist interval is $T = 1/2B = 0.1$ second.

Recall that the sampled signal spectrum consists of $(1/T)F(\omega) = \frac{0.2}{T}\,\Delta(\frac{\omega}{20\pi})$ repeating periodically with a period equal to the sampling frequency $\mathcal{F}_s$ Hz. We present this information in the following Table for three sampling rates: $\mathcal{F}_s = 5$ Hz (undersampling), 10 Hz (Nyquist rate), and 20 Hz (oversampling).

sampling frequency $\mathcal{F}_s$	sampling interval T	$\frac{1}{T}F(\omega)$	comments
5 Hz	0.2	$\Delta\left(\frac{\omega}{20\pi}\right)$	Undersampling
10 Hz	0.1	$2\Delta\left(\frac{\omega}{20\pi}\right)$	Nyquist Rate
20 Hz	0.05	$4\Delta\left(\frac{\omega}{20\pi}\right)$	Oversampling

In the first case (undersampling), the sampling rate is 5 Hz (5 samples/sec.), and the spectrum $\frac{1}{T}F(\omega)$ repeats every 5 Hz (10π rad/sec.). The successive spectra overlap, as depicted in Fig. 5.2d, and the spectrum $F(\omega)$ is not recoverable from $\overline{F}(\omega)$; that is, $f(t)$ cannot be reconstructed from its samples $\overline{f}(t)$ in Fig. 5.2c. In the second case, we use the Nyquist sampling rate of 10 Hz (Fig. 5.2e). The spectrum $\overline{F}(\omega)$ consists of back-to-back, nonoverlapping repetitions of $\frac{1}{T}F(\omega)$ repeating every 10 Hz. Hence, $F(\omega)$ can be recovered from $\overline{F}(\omega)$ using an ideal lowpass filter of bandwidth 5 Hz (Fig. 5.2f). Finally, in the last case of oversampling (sampling rate 20 Hz), the spectrum $\overline{F}(\omega)$ consists of nonoverlapping repetitions of $\frac{1}{T}F(\omega)$ (repeating every 20 Hz) with empty band between successive cycles

---

†We have proved that for errorfree recovery of a signal of bandwidth B Hz, the sampling rate $\mathcal{F}_s \geq 2B$. However, in a special case, where $F(\omega)$ contains an impulse at the highest frequency B, the sampling rate must be $\mathcal{F}_s > 2B$ Hz. Such is the case when $f(t) = \sin 2\pi Bt$. This signal is bandlimited to B Hz, but all of its samples are zero when taken at a rate $\mathcal{F}_s = 2B$ (starting at $t = 0$), and $f(t)$ cannot be recovered from its Nyquist samples.

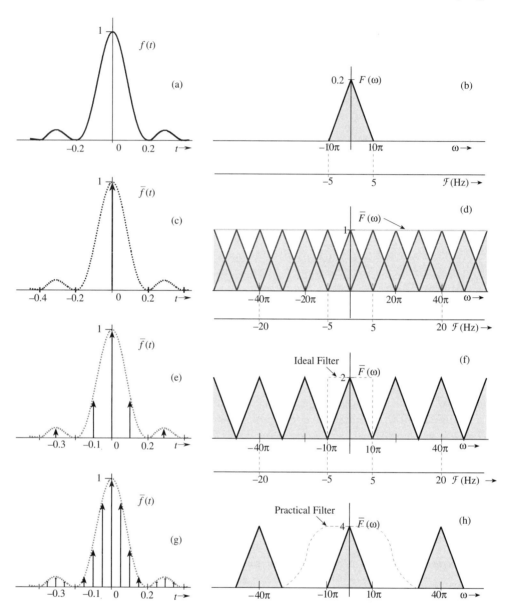

Fig. 5.2   Effect of undersampling and oversampling.

(Fig. 5.2h). Hence, $F(\omega)$ can be recovered from $\overline{F}(\omega)$ using an ideal lowpass filter or even a practical lowpass filter (shown dotted in Fig. 5.2h).† ∎

△ **Exercise E5.1**

Find the Nyquist rate and the Nyquist interval for the signals (a) sinc $(100\pi t)$ and
(b) sinc $(100\pi t)+$ sinc $(50\pi t)$.

---

†The filter should have a constant gain between 0 to 5 Hz, and zero gain beyond 10 Hz. In practice, the gain beyond 10 Hz can be made negligibly small, but not zero.

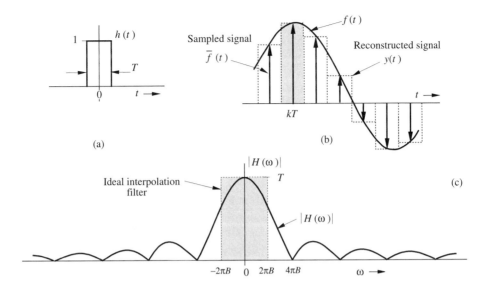

Fig. 5.3 Simple interpolation using a zero-order hold circuit.

Answer: The Nyquist interval is 0.01 second and the Nyquist sampling rate is 100 Hz for both the signals. ▽

5.1-1 Signal Reconstruction: The Interpolation Formula

The process of reconstructing a continuous-time signal $f(t)$ from its samples is also known as **interpolation**. In Sec. 5.1, we saw that a signal $f(t)$ bandlimited to B Hz can be reconstructed (interpolated) exactly from its samples. This reconstruction is accomplished by passing the sampled signal through an ideal lowpass filter of bandwidth B Hz. As seen from Eq. (5.3) [or Fig. 5.1e], the sampled signal contains a component $\frac{1}{T}f(t)$ and to recover $f(t)$ (or $F(\omega)$), the sampled signal must be passed through an ideal lowpass filter of bandwidth B Hz and gain T. Thus, the reconstruction (or interpolating) filter transfer function is

$$H(\omega) = T \, \text{rect}\left(\frac{\omega}{4\pi B}\right) \tag{5.7}$$

The interpolation process here is expressed in the frequency-domain as a filtering operation. Now, we shall examine this process from a different viewpoint, that of the time-domain.

To begin with, let us consider a very simple interpolating filter whose impulse response is $\text{rect}\left(\frac{t}{T}\right)$, depicted in Fig. 5.3a. This is a gate pulse centered at the origin, having unit height, and width T (the sampling interval). We shall find the output of this filter when the input is the sampled signal $\overline{f}(t)$. Each sample in $\overline{f}(t)$, being an impulse, produces at the output a gate pulse of height equal to the strength of the sample. For instance, the kth sample is an impulse of strength $f(kT)$ located at $t - kT$, and can be expressed as $f(kT)\delta(t - kT)$. When this impulse passes through the filter, it produces at the output a gate pulse of height $f(kT)$, centered at $t = kT$ (shown shaded in Fig. 5.3b). Each sample in $\overline{f}(t)$ will generate a corresponding gate pulse resulting in the filter output that is a staircase approximation of $f(t)$, shown dotted in Fig. 5.3b. This filter thus gives a crude form of interpolation.

The transfer function of this filter $H(\omega)$ is the Fourier transform of the impulse response rect $(\frac{t}{T})$. Assuming the Nyquist sampling rate; that is, $T = 1/2B$

$$h(t) = \text{rect}\left(\frac{t}{T}\right) = \text{rect}\,(2Bt)$$

and

$$H(\omega) = T\,\text{sinc}\left(\frac{\omega T}{2}\right) = \frac{1}{2B}\,\text{sinc}\left(\frac{\omega}{4B}\right) \tag{5.8}$$

The amplitude response $|H(\omega)|$ for this filter, illustrated in Fig. 5.3c, explains the reason for the crudeness of this interpolation. This filter, also known as the **zero-order hold** filter, is a poor form of the ideal lowpass filter (shown shaded in Fig. 5.3c) required for exact interpolation.†

We can improve on the zero-order hold filter by using the **first-order hold** filter, which results in a linear interpolation instead of the staircase interpolation. The linear interpolator, whose impulse response is a triangle pulse $\Delta(\frac{t}{2T})$, results in an interpolation in which successive sample tops are connected by straight line segments (see Prob. 5.1-5).

The ideal interpolation filter transfer function obtained in Eq. (5.7) is illustrated in Fig. 5.4a. The impulse response of this filter, the inverse Fourier transform of $H(\omega)$ is

$$h(t) = 2BT\,\text{sinc}\,(2\pi Bt) \tag{5.9a}$$

Assuming the Nyquist sampling rate; that is, $2BT = 1$, then

$$h(t) = \text{sinc}\,(2\pi Bt) \tag{5.9b}$$

This $h(t)$ is depicted in Fig. 5.4b. Observe the very interesting fact that $h(t) = 0$ at all Nyquist sampling instants $(t = \pm\frac{n}{2B})$ except at $t = 0$. When the sampled signal $\overline{f}(t)$ is applied at the input of this filter, the output is $f(t)$. Each sample in $\overline{f}(t)$, being an impulse, generates a sinc pulse of height equal to the strength of the sample, as illustrated in Fig. 5.4c. The process is identical to that depicted in Fig. 5.3b, except that $h(t)$ is a sinc pulse instead of a gate pulse. Addition of the sinc pulses generated by all the samples results in $f(t)$. The kth sample of the input $\overline{f}(t)$ is the impulse $f(kT)\delta(t - kT)$; the filter output of this impulse is $f(kT)h(t - kT)$. Hence, the filter output to $\overline{f}(t)$, which is $f(t)$, can now be expressed as a sum

$$f(t) = \sum_k f(kT)h(t - kT)$$

$$= \sum_k f(kT)\,\text{sinc}\,[2\pi B(t - kT)] \tag{5.10a}$$

$$= \sum_k f(kT)\,\text{sinc}\,(2\pi Bt - k\pi) \tag{5.10b}$$

---

†Figure 5.3a shows that the impulse response of this filter is noncausal, and this filter is not realizable. In practice, we make it realizable by delaying the impulse response by $T/2$. This merely delays the output of the filter by $T/2$.

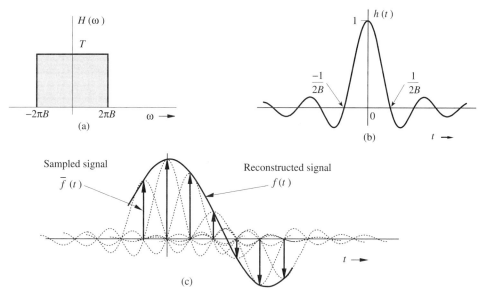

Fig. 5.4   Ideal interpolation.

Equation (5.10) is the **interpolation formula**, which yields values of $f(t)$ between samples as a weighted sum of all the sample values.

■ **Example 5.2**

Find a signal $f(t)$ that is bandlimited to B Hz, and whose samples are

$$f(0) = 1 \quad \text{and} \quad f(\pm T) = f(\pm 2T) = f(\pm 3T) = \cdots = 0$$

where the sampling interval T is the Nyquist interval for $f(t)$; that is, $T = 1/2B$.

We use the interpolation formula (5.10b) to construct $f(t)$ from its samples. Since all but one of the Nyquist samples are zero, only one term (corresponding to $k = 0$) in the summation on the right-hand side of Eq. (5.10b) survives. Thus

$$f(t) = \text{sinc}\,(2\pi Bt)$$

This signal is illustrated in Fig. 5.4b. Observe that this is the only signal that has a bandwidth B Hz and the sample values $f(0) = 1$ and $f(nT) = 0 \,(n \neq 0)$. No other signal satisfies these conditions.   ■

5.1-2   Practical Difficulties in Signal Reconstruction

If a signal is sampled at the Nyquist rate $\mathcal{F}_s = 2B$ Hz, the spectrum $\overline{F}(\omega)$ consists of repetitions of $F(\omega)$ without any gap between successive cycles, as depicted in Fig. 5.5a. To recover $f(t)$ from $\overline{f}(t)$, we need to pass the sampled signal $\overline{f}(t)$ through an ideal lowpass filter, shown dotted in Fig. 5.5a. As seen in Sec. 4.5, such a filter is unrealizable; it can be closely approximated only with infinite time delay in the response. In other words, we can recover the signal $f(t)$ from its samples with infinite time delay. A practical solution to this problem is to sample the signal at a rate higher than the Nyquist rate ($\mathcal{F}_s > 2B$ or $\omega_s > 4\pi B$). The result is $\overline{F}(\omega)$, consisting of repetitions of $F(\omega)$ with a finite band gap between successive

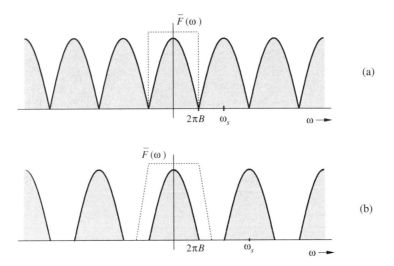

(a)

(b)

Fig. 5.5 Spectra of a signal sampled at (a) the Nyquist rate (b) above the Nyquist rate.

cycles, as illustrated in Fig. 5.5b. We can now recover $F(\omega)$ from $\overline{F}(\omega)$ using a lowpass filter with a gradual cutoff characteristic, shown dotted in Fig. 5.5b. But even in this case, the filter gain must be zero beyond the first cycle of $F(\omega)$ (see Fig. 5.5b). According to the Paley-Wiener criterion, it is impossible to realize even this filter. The only advantage in this case is that the required filter can be closely approximated with a smaller time delay. This fact indicates that it is impossible in practice to recover a bandlimited signal $f(t)$ exactly from its samples, even if the sampling rate is higher than the Nyquist rate. However, as the sampling rate increases, the recovered signal approaches the desired signal more closely.

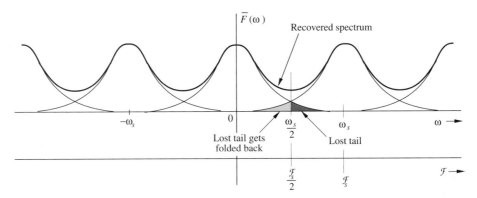

Fig. 5.6 Aliasing effect.

The Treachery of Aliasing

There is another fundamental practical difficulty in reconstructing a signal from its samples. The sampling theorem was proved on the assumption that the signal $f(t)$ is bandlimited. **All practical signals are timelimited**; that is, they are of

finite duration or width. We can demonstrate (see Prob. 5.1-10) that a signal cannot
be timelimited and bandlimited simultaneously. If a signal is timelimited, it cannot
be bandlimited and vice versa (but it can be simultaneously nontimelimited and
nonbandlimited). Clearly, all practical signals, which are necessarily timelimited,
are nonbandlimited; they have infinite bandwidth, and the spectrum $\overline{F}(\omega)$ consists
of overlapping cycles of $F(\omega)$ repeating every $\mathcal{F}_s$ Hz (the sampling frequency), as
illustrated in Fig. 5.6. Because of infinite bandwidth in this case, the spectral overlap
is a constant feature, regardless of the sampling rate. Because of the overlapping
tails, $\overline{F}(\omega)$ no longer has complete information about $F(\omega)$, and it is no longer
possible, even theoretically, to recover $f(t)$ from the sampled signal $\overline{f}(t)$. If the
sampled signal is passed through an ideal lowpass filter, the output is not $F(\omega)$ but
a version of $F(\omega)$ distorted as a result of two separate causes:

1. The loss of the tail of $F(\omega)$ beyond $|\mathcal{F}| > \mathcal{F}_s/2$ Hz;

2. The reappearance of this tail inverted or folded onto the spectrum. Note that
   the spectra cross at frequency $\mathcal{F}_s/2 = 1/2T$ Hz. This frequency is called the
   folding frequency. The spectrum, therefore, folds onto itself at the folding
   frequency. For instance, a component of frequency $\frac{\mathcal{F}_s}{2} + \mathcal{F}_x$ shows up as or
   "impersonates" a component of lower frequency $\frac{\mathcal{F}_s}{2} - \mathcal{F}_x$ in the reconstructed
   signal. Thus, the components of frequencies above $\mathcal{F}_s/2$ reappear as components of frequencies below $\mathcal{F}_s/2$. This tail inversion, known as **spectral
   folding** or **aliasing**, is shown shaded in Fig. 5.6. In this process of aliasing,
   not only are we losing all the components of frequencies above $\mathcal{F}_s/2$ Hz, but
   these very components reappear (aliased) as lower frequency components. This
   reappearance destroys the integrity of the lower frequency components also, as
   depicted in Fig. 5.6.

Aliasing problem is analogous to that of an army with a platoon that has secretly
defected to the enemy side. The platoon is, however, ostensibly loyal to the army.
The army is in double jeopardy. First, the army has lost this platoon as a fighting
force. In addition, during actual fighting, the army will have to contend with the
sabotage by the defectors, and will have to find another loyal platoon to neutralize
the defectors. Thus, the army has lost two platoons in nonproductive activity.

A Solution: The Antialiasing Filter

If you were the commander of the betrayed army, the solution to the problem
would be obvious. As soon as the commander gets wind of the defection, he would
incapacitate, by whatever means, the defecting platoon *before the fighting begins.*
This way he loses only one (the defecting) platoon. This is a partial solution to the
double jeopardy of betrayal, a solution that partly rectifies the problem and reduces
the losses to half.

We follow exactly the same procedure. The potential defectors are all the fre-
quency components beyond $\frac{\mathcal{F}_s}{2} = \frac{1}{2T}$ Hz. We should eliminate (suppress) these
components from $f(t)$ *before sampling* $f(t)$. This way we lose only the components
beyond the folding frequency $\frac{\mathcal{F}_s}{2}$ Hz; these components now cannot reappear to cor-
rupt the components with frequencies below the folding frequency. This suppression
of higher frequencies can be accomplished by an ideal lowpass filter of bandwidth
$\mathcal{F}_s/2$ Hz. This filter is called the **antialiasing filter**. Note that the antialiasing
operation must be performed *before the signal is sampled.*

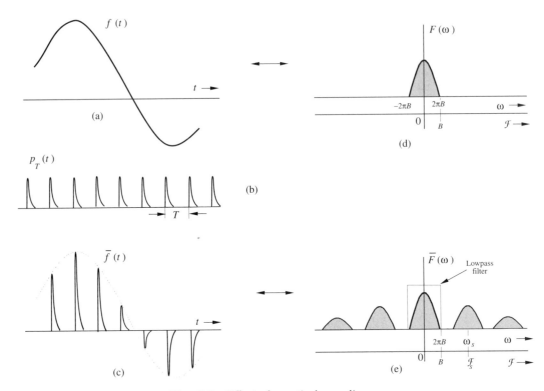

Fig. 5.7   Effect of practical sampling.

The antialiasing filter, being an ideal filter, is unrealizable. In practice, we use a steep cutoff filter, which leaves a sharply attenuated residual spectrum beyond the folding frequency $\mathcal{F}_s/2$.

Practical Sampling

In proving the sampling theorem, we assumed ideal samples obtained by multiplying a signal $f(t)$ by an impulse train which is physically nonexistent. In practice, we multiply a signal $f(t)$ by a train of pulses of finite width, depicted in Fig. 5.7b. The sampled signal is illustrated in Fig. 5.7c. We wonder whether it is possible to recover or reconstruct $f(t)$ from the sampled signal $\overline{f}(t)$ in Fig. 5.7c. Surprisingly, the answer is affirmative, provided that the sampling rate is not below the Nyquist rate. The signal $f(t)$ can be recovered by lowpass filtering $\overline{f}(t)$ as if it were sampled by impulse train.

The plausibility of this result becomes apparent when we consider the fact that reconstruction of $f(t)$ requires the knowledge of the Nyquist sample values. This information is available or built in the sampled signal $\overline{f}(t)$ in Fig. 5.7c because the kth sampled pulse strength is $f(kT)$. To prove the result analytically, we observe that the sampling pulse train $p_T(t)$ depicted in Fig. 5.7b, being a periodic signal, can be expressed as a trigonometric Fourier series

$$p_T(t) = C_0 + \sum_{n=1}^{\infty} C_n \cos\left(n\omega_s t + \theta_n\right) \qquad \omega_s = \frac{2\pi}{T}$$

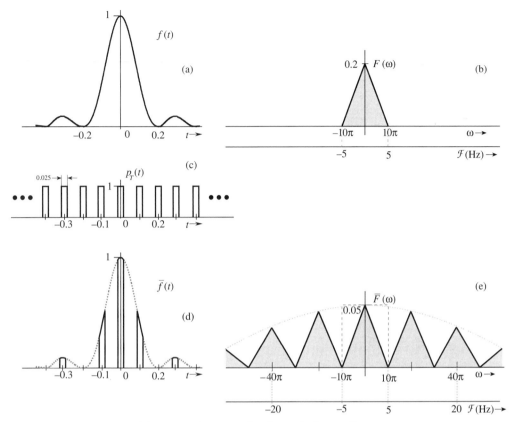

Fig. 5.8   An example of practical sampling.

and

$$\overline{f}(t) = f(t)p_T(t) = f(t)\left[C_0 + \sum_{n=1}^{\infty} C_n \cos\left(n\omega_s t + \theta_n\right)\right]$$

$$= C_0 f(t) + \sum_{n=1}^{\infty} C_n f(t) \cos\left(n\omega_s t + \theta_n\right) \qquad (5.11)$$

The sampled signal $\overline{f}(t)$ consists of $C_0 f(t)$, $C_1 f(t) \cos\left(\omega_s t + \theta_1\right)$, $C_2 f(t) \cos\left(2\omega_s t + \theta_2\right)$, ... Note that the first term $C_0 f(t)$ is the desired signal and all the other terms are modulated signals with spectra centered at $\pm\omega_s, \pm2\omega_s, \pm3\omega_s, \ldots$, as illustrated in Fig. 5.7e. Clearly the signal $f(t)$ can be recovered by lowpass filtering of $\overline{f}(t)$, provided that $\omega_s > 4\pi B$ (or $\mathcal{F}_s > 2B$).

■ **Example 5.3**

   To demonstrate practical sampling, consider a signal $f(t) = \text{sinc}^2(5\pi t)$ sampled by a rectangular pulse sequence $p_T(t)$ illustrated in Fig. 5.8c. The period of $p_T(t)$ is 0.1 second, so that the fundamental frequency (which is the sampling frequency) is 10 Hz. Hence, $\omega_s = 20\pi$. The Fourier series for $p_T(t)$ can be expressed as

$$p_T(t) = C_0 + \sum_{n=1}^{\infty} C_n \cos n\omega_s t$$

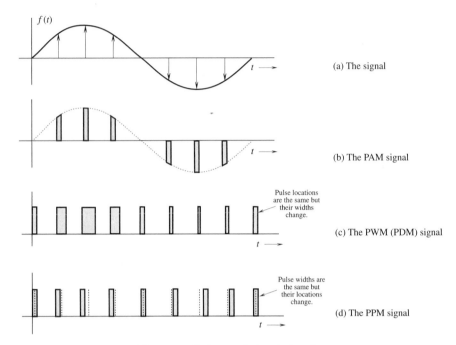

Fig. 5.9 Pulse modulated signals.

Use of Eqs. (3.66) yields $C_0 = \frac{1}{4}$ and $C_n = \frac{2}{n\pi} \sin\left(\frac{n\pi}{4}\right)$; that is,

$$C_0 = \frac{1}{4}, \quad C_1 = \frac{\sqrt{2}}{\pi}, \quad C_2 = \frac{1}{\pi}, \quad C_3 = \frac{\sqrt{2}}{3\pi}, \quad C_4 = 0, \quad C_5 = -\frac{\sqrt{2}}{5\pi}, \ldots$$

Consequently

$$\overline{f}(t) = f(t)p_T(t) = \frac{1}{4}f(t) + \frac{\sqrt{2}}{\pi}f(t)\cos 20\pi t + \frac{1}{\pi}f(t)\cos 40\pi t + \frac{\sqrt{2}}{3\pi}f(t)\cos 60\pi t + \cdots$$

and

$$\overline{F}(\omega) = \frac{1}{4}F(\omega) + \frac{1}{\pi\sqrt{2}}[F(\omega - 20\pi) + F(\omega + 20\pi)] + \frac{1}{2\pi}[F(\omega - 40\pi) + F(\omega + 40\pi)]$$

$$+ \frac{1}{3\pi\sqrt{2}}[F(\omega - 60\pi) + F(\omega + 60\pi)] + \cdots$$

In the present case $F(\omega) = 0.2\,\Delta\left(\frac{\omega}{20\pi}\right)$. The spectrum $\overline{F}(\omega)$ is depicted in Fig. 5.8e. Observe that the spectrum consists of $F(\omega)$ repeating periodically at the interval of 20π rad/s (10 Hz). Hence, there is no overlap between cycles, and $F(\omega)$ can be recovered by using an ideal lowpass filter of bandwidth 5 Hz. An ideal lowpass filter of unit gain (and bandwidth 5 Hz) will allow the first term on the right-side of the above equation to pass fully and suppress all the other terms. Hence, the output $y(t)$ is

$$y(t) = \frac{1}{4}f(t) \quad \blacksquare$$

5.1-3 Some Applications of the Sampling Theorem

The sampling theorem is very important in signal analysis, processing, and transmission because it allows us to replace a continuous-time signal by a discrete sequence of numbers. Processing a continuous-time signal is therefore equivalent to processing a discrete sequence of numbers. Such processing leads us directly into the area of digital filtering. In the field of communication, the transmission of a continuous-time message reduces to the transmission of a sequence of numbers using pulse trains. The continuous-time signal $f(t)$ is sampled, and sample values are used to modify certain parameters of a periodic pulse train. We may vary the amplitudes (Fig. 5.9b), widths (Fig. 5.9c), or positions (Fig. 5.9d) of the pulses in proportion to the sample values of the signal $f(t)$. Accordingly, we have **pulse-amplitude modulation** (PAM), **pulse-width modulation** (PWM), or **pulse position modulation** (PPM). The most important form of pulse modulation today is **pulse code modulation** (PCM), discussed below. In all these cases, instead of transmitting $f(t)$, we transmit the corresponding pulse-modulated signal. At the receiver, we read the information of the pulse-modulated signal and reconstruct the analog signal $f(t)$.

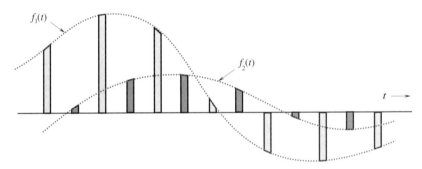

Fig. 5.10 Time-division multiplexing of two signals.

One advantage of using pulse modulation is that it permits the simultaneous transmission of several signals on a time-sharing basis **(time-division multiplexing, or TDM)**. Because a pulse-modulated signal occupies only a part of the channel time, we can transmit several pulse-modulated signals on the same channel by inter-
weaving them. Figure 5.10 shows the TDM of two PAM signals. In this manner, we can multiplex several signals on the same channel by reducing pulse widths.

Another method of transmitting several baseband signals simultaneously is frequency-division multiplexing (FDM), discussed in Sec. 4.8-4. In FDM various signals are multiplexed by sharing the channel bandwidth. The spectrum of each message is shifted to a specific band not occupied by any other signal. The information of various signals is located in nonoverlapping frequency bands of the channel (Fig. 4.45). In a way, TDM and FDM are the duals of each other.

Pulse Code Modulation (PCM)

PCM is the most useful and widely used of all the pulse modulations mentioned. Basically, PCM is a method of converting an analog signal into a digital signal (A/D conversion). An **analog** signal is characterized by the fact that its amplitude can take on any value over a continuous range. Hence, analog signal can take on an infinite number of values. In contrast, a **digital** signal amplitude can take on only a finite number of values. An analog signal can be converted into a digital signal by means of sampling and **quantizing** (rounding off). Sampling an analog signal alone will not yield a digital signal because a sampled analog signal can still take on any value in a continuous range. It is digitized by rounding off its value to one of the closest permissible numbers (or **quantized levels**), as illustrated in Fig. 5.11a. The amplitudes of the analog signal $f(t)$ lie in the range $(-V, V)$. This range is partitioned into L subintervals, each of magnitude $\Delta v = 2V/L$. Next, each sample amplitude is approximated by the midpoint value of the subinterval in which the sample falls (see Fig. 5.11a for $L = 16$). It is clear that each sample is approximated to one of the L numbers. Thus, the signal is digitized with quantized samples taking on any one of the L values. Such a signal is known as an L-**ary digital signal**.

From a practical viewpoint, a binary digital signal (a signal that can take on only two values) is very desirable because of its simplicity, economy, and ease of engineering. We can convert an L-ary signal into a binary signal by using pulse coding. Figure 5.11b shows such a code for the case of $L = 16$. This code, formed by binary representation of the 16 decimal digits from 0 to 15, is known as the **natural binary code (NBC)**. Other possible ways of assigning a binary code exist. Each of the 16 levels to be transmitted is assigned one binary code of four digits. Thus, each sample in this example is encoded by four binary digits. To transmit this binary data, we need to assign a distinct pulse shape to each of the two binary states. One possible way is to assign a negative pulse to a binary **0** and a positive pulse to a binary **1** so that each sample is now transmitted by a group of four binary pulses (pulse code), as depicted in Fig. 5.11b. The resulting signal is a binary PCM signal. The analog signal $f(t)$ is now converted to a (binary) digital signal. A **bi**nary digit is called **bit** for convenience. This contraction of binary digit by bit has become an industry standard abbreviation.

The audio signal bandwidth is about 15 kHz, but subjective tests show that signal articulation (intelligibility) is not affected if all the components above 3400 Hz are suppressed.[3]† Since the objective in telephone communication is intelligibility rather than high fidelity, the components above 3400 Hz are eliminated by a lowpass filter. The resulting signal is then sampled at a rate of 8000 samples per second (8 kHz). This rate is intentionally kept higher than the Nyquist sampling rate of 6.8 kHz to avoid unrealizable filters required for signal reconstruction. Each sample is finally quantized into 256 levels ($L = 256$), which requires a group of eight binary pulses to encode each sample ($2^8 = 256$). Thus, a digitized telephone signal consists of $8 \times 8000 = 64000$ or 64 kbits/s data, requiring 64,000 binary pulses per second for its transmission.

The compact disc (CD) is a recent application of PCM. This is a high-fidelity situation requiring the audio signal bandwidth of 15 kHz. Although the Nyquist sampling rate is only 30 kHz, an actual sampling rate of 44.1 kHz is used for the

---

†Components below 300 Hz may also be suppressed without affecting the articulation.

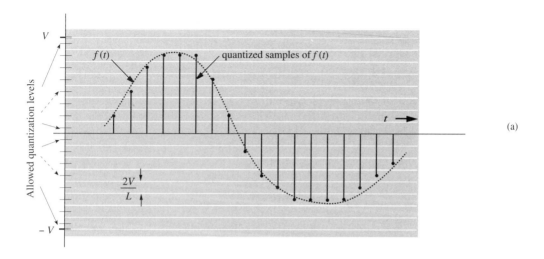

(a)

Digit	Binary equivalent	Pulse code waveform
0	0000	
1	0001	
2	0010	
3	0011	
4	0100	
5	0101	
6	0110	
7	0111	
8	1000	
9	1001	
10	1010	
11	1011	
12	1100	
13	1101	
14	1110	
15	1111	

(b)

Fig. 5.11   Analog-to-digital (A/D) conversion of a signal.

reason mentioned earlier. The signal is quantized into a rather large number of levels ($L = 65,536$) to reduce quantizing error. The binary-coded samples are now recorded on the CD.

Advantages of Digital Signals

Some of the advantages of digital signals over analog signals are listed below:

1. Transmission of digital signals is more rugged than that of analog signals because digital signals can withstand channel noise and distortion much better as long as the noise and the distortion are within limits. The digital (binary) message in Fig. 5.12a is distorted by the channel, as illustrated in Fig. 5.12b. Yet if the distortion remains within a limit, we can recover the data without error because we need only to make a simple binary decision as to whether the received pulse is positive or negative. Figure 5.12c shows the same data with channel distortion and noise. Here again the data can be recovered correctly as long as the distortion and the noise are within limits. Such is not the case with analog messages. Any distortion or noise, no matter how small, will distort the received signal.

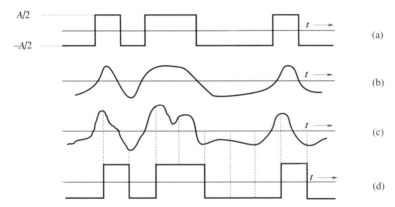

Fig. 5.12   Digital signal: (a) transmitted (b) received distorted signal (without noise) (c) received distorted signal (with noise) (d) regenerated signal at the receiver.

2. The greatest advantage of digital signal transmission over analog transmission, however, is the viability of regenerative repeaters in the former. In an analog transmission system, a message signal, as it travels along the channel (transmission path), grows progressively weaker, whereas the channel noise and the signal distortion, being cumulative, become progressively stronger. Ultimately the signal, overwhelmed by noise and distortion, is mutilated. Amplification is of little help because it enhances the signal and the noise in the same proportion. Consequently, the distance over which an analog message can be transmitted is limited by the transmitted power. If a transmission path is long enough, the channel distortion and noise will accumulate sufficiently to overwhelm even a digital signal. The trick is to set up repeater stations along the transmission path at distances short enough to be able to detect signal pulses before the noise and distortion have a chance to accumulate sufficiently. At each repeater station the pulses are detected, and new, clean pulses are transmitted to the next

repeater station, which, in turn, duplicates the same process. If the noise and distortion remain within limits (which is possible because of the closely spaced repeaters), pulses can be detected correctly.† This way the digital messages can be transmitted over longer distances with greater reliability. In contrast, analog messages cannot be cleaned up periodically, and the transmission is therefore less reliable. The most significant error in PCM comes from quantizing. This error can be reduced as much as desired by increasing the number of quantizing levels, the price of which is paid in an increased bandwidth of the transmission medium (channel).

3. Digital hardware implementation is flexible and permits the use of microprocessors, miniprocessors, digital switching, and large-scale integrated circuits.

4. Digital signals can be coded to yield extremely low error rates and high fidelity, as well as privacy. Also, more sophisticated algorithms can be used to process digital signals.

5. It is easier and more efficient to multiplex several digital signals.

6. Digital signal storage is relatively easy and inexpensive. It also has the ability to search and select information from distant electronic storehouses.

7. Reproduction with digital messages is extremely reliable without deterioration. Analog messages such as photocopies and films, for example, lose quality at each successive stage of reproduction, and have to be transported physically from one distant place to another, often at relatively high cost.

8. The cost of digital hardware continues to halve every two or three years, while performance or capacity doubles over the same time period. And there is no end in sight yet to this breathtaking and relentless exponential progress in digital technology. In recent years we have seen the compact disc—a digital device—bury the analog long-playing record; newspapers transmit photographs in scanned digital form; and more recently the shift in the United States toward a digital standard for high-definition television as opposed to the analog standard embraced by Japan and Europe. In contrast, analog technologies such as paper, video, sound, and film do not decline rapidly in cost. If anything, they become more expensive with time. For these and other reasons, it is only a matter of time before cost/performance curves cross, and digital technologies come to dominate in any given area of communication or storage technologies.

As mentioned earlier, digital signals come from a variety of sources. Some sources such as computers are inherently digital. Some sources are analog, but are converted into digital form by a variety of techniques such as PCM modulation.

△   **Exercise E5.2**
    American Standard Code for Information Interchange (ASCII) has 128 characters which are binary coded. A certain computer generates 100,000 characters/s. Show that
 (a) 7 bits (binary digits) are required to encode each character.
 (b) 700,000 bits/s are required to transmit the computer output.   ▽

A Historical Note
Gottfried Wilhelm Leibnitz (1646-1716) was the first mathematician to work out systematically the binary representation (using **1**'s and **0**'s) for any number.

---

†The error in pulse detection can be made negligible.

He felt a spiritual significance in this discovery, reasoning that **1** representing unity
was clearly a symbol for God, while **0** represented the nothingness. Therefore, if all
numbers can be represented merely by the use of **1** and **0**, surely this is the same
as saying that God created the universe out of nothing!

5.1-4   Dual of the Time-Sampling: The Spectral Sampling Theorem

As in other cases, the sampling theorem has its dual. In Sec. 5.1, we discussed
the time-sampling theorem where we showed that a signal bandlimited to B Hz can
be* reconstructed from the signal samples taken at a rate of $\mathcal{F}_s > 2B$ samples/s.
Note that the signal spectrum exists over the frequency range $-B$ to B Hz. Therefore, $2B$ is the spectral width (not the bandwidth, which is B) of the signal. This
fact means that a signal $f(t)$ can be reconstructed from samples taken at a rate $\mathcal{F}_s$
greater than the spectral width (in Hz) of the signal.

The dual of the time-sampling theorem is the frequency-sampling theorem.
This theorem applies to timelimited signals, which are duals of bandlimited signals.
We now prove that the spectrum $F(\omega)$ of a signal timelimited to τ seconds can be
reconstructed from the samples of $F(\omega)$ taken at a rate $R > \tau$ (the signal width)
samples per Hertz.

Figure 5.13a shows a signal $f(t)$ that is timelimited to τ seconds along with its
Fourier transform $F(\omega)$. Although $F(\omega)$ is complex in general, it is adequate for
our line of reasoning to show $F(\omega)$ as a real function.

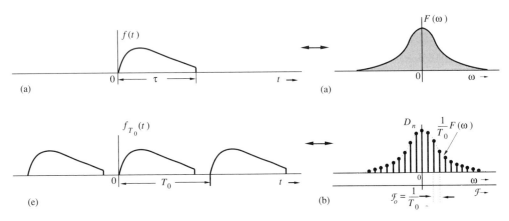

Fig. 5.13   Periodic repetition of a signal amounts to sampling its spectrum.

$$F(\omega) = \int_{-\infty}^{\infty} f(t)e^{-j\omega t}\,dt = \int_{0}^{\tau} f(t)e^{-j\omega t}\,dt \tag{5.12}$$

We now construct $f_{T_0}(t)$, a periodic signal formed by repeating $f(t)$ every T_0
seconds ($T_0 > \tau$), as depicted in Fig. 5.13b. This periodic signal can be expressed
by the exponential Fourier series

$$f_{T_0}(t) = \sum_{n=-\infty}^{\infty} D_n e^{jn\omega_0 t} \qquad \omega_0 = \frac{2\pi}{T_0}$$

where (assuming $\tau < T_0$)

$$D_n = \frac{1}{T_0} \int_0^{T_0} f(t) e^{-jn\omega_0 t}\, dt = \frac{1}{T_0} \int_0^{\tau} f(t) e^{-jn\omega_0 t}\, dt$$

From Eq. (5.12) it follows that

$$D_n = \frac{1}{T_0} F(n\omega_0)$$

This result indicates that the coefficients of the Fourier series for $f_{T_0}(t)$ are $(1/T_0)$ times the sample values of the spectrum $F(\omega)$ taken at intervals of ω_0. This means that the spectrum of the periodic signal $f_{T_0}(t)$ is the sampled spectrum $F(\omega)$, as illustrated in Fig. 5.13b. Now as long as $\tau < T_0$, the successive cycles of $f(t)$ appearing in $f_{T_0}(t)$ do not overlap, so that $f(t)$ can be recovered from $f_{T_0}(t)$. Such recovery implies indirectly that $F(\omega)$ can be reconstructed from its samples. These samples are separated by the fundamental frequency $\mathcal{F}_0 = 1/T_0$ Hz of the periodic signal $f_{T_0}(t)$. The condition for recovery is that $T_0 \geq \tau$; that is†

$$\mathcal{F}_0 \leq \frac{1}{\tau}\ \text{Hz}$$

Therefore, to be able to reconstruct the spectrum $F(\omega)$ from the samples of $F(\omega)$, the samples should be taken at frequency intervals not greater than $\mathcal{F}_0 = 1/\tau$ Hz. If R is the sampling rate (samples/Hz), then

$$R = \frac{1}{\mathcal{F}_0} \geq \tau\ \text{samples/Hz} \qquad (5.13)$$

Spectral Interpolation

The spectrum $F(\omega)$ of a signal $f(t)$ timelimited to τ seconds can be reconstructed from the samples of $F(\omega)$. For this case, using the dual of the approach employed to derive the signal interpolation formula in Eq. (5.10), we obtain the spectral interpolation formula

$$F(\omega) = \sum_n F(n\omega_0)\ \text{sinc}\left(\frac{\omega \tau}{2} - n\pi\right) \qquad \omega_0 = \frac{2\pi}{\tau} \qquad (5.14)$$

■ **Example 5.4**
The spectrum $F(\omega)$ of a unit duration signal $f(t)$ is sampled at the intervals of 1 Hz or 2π rad/s (the Nyquist rate). The samples are:

$$F(0) = 1 \quad \text{and} \quad F(\pm 2\pi n) = 0\ (n = 1,\, 2,\, 3, \ldots)$$

We use the interpolation formula (5.14) to construct $F(\omega)$ from its samples. Since all but one of the Nyquist samples are zero, only one term (corresponding to $n = 0$) in the summation on the right-hand side of Eq. (5.14) survives. Thus, with $F(0) = 1$ and $\tau = 1$, we obtain

$$F(\omega) = \text{sinc}\left(\frac{\omega}{2}\right) \qquad (5.15)$$

For a signal of unit duration this is the only spectrum with the sample values $F(0) = 1$ and $F(2\pi n) = 0\ (n \neq 0)$. No other spectrum satisfies these conditions.   ■

---

†This result assumes that $f(t)$ does not have impulses at $t = 0$ or τ. If impulses do occur, then $\mathcal{F}_0 < 1/\tau$.

5.2 Numerical Computation of the Fourier Transform: The Discrete Fourier Transform (DFT)

Numerical computation of the Fourier transform of $f(t)$ requires sample values of $f(t)$ because a digital computer can work only with discrete data (sequence of numbers). Moreover, a computer can compute $F(\omega)$ only at some discrete values of ω [samples of $F(\omega)$]. We therefore need to relate the samples of $F(\omega)$ to samples of $f(t)$. This task can be accomplished by using the results of the two sampling theorems developed in Sec. 5.1.

We begin with a timelimited signal $f(t)$ (Fig. 5.14a) and its spectrum $F(\omega)$ (Fig. 5.14b). Since $f(t)$ is timelimited, $F(\omega)$ is nonbandlimited. For convenience, we shall show all spectra as functions of the frequency variable $\mathcal{F}$ (in Hertz) rather than ω. According to the sampling theorem, the spectrum $\overline{F}(\omega)$ of the sampled signal $\overline{f}(t)$ consists of $F(\omega)$ repeating every $\mathcal{F}_s$ Hz where $\mathcal{F}_s = 1/T$. This is depicted in Figs. 5.14c and 5.14d.† In the next step, the sampled signal in Fig. 5.14c is repeated periodically every T_0 seconds, as illustrated in Fig. 5.14e. According to the spectral sampling theorem, such an operation results in sampling the spectrum at a rate of T_0 samples per Hz. This sampling rate means that the samples are spaced at $\mathcal{F}_o = 1/T_0$ Hz, as depicted in Fig. 5.14f.

The above discussion shows that, when a signal $f(t)$ is sampled and then periodically repeated, the corresponding spectrum is also sampled and periodically repeated. Our goal is to relate the samples of $f(t)$ to the samples of $F(\omega)$.

Number of Samples

One interesting observation in Figs. 5.14e and 5.14f is that N_0, the number of samples of the signal in Fig. 5.14e in one period T_0 is identical to N_0', the number of samples of the spectrum in Fig. 5.14f in one period $\mathcal{F}_s$. The reason is

$$N_0 = \frac{T_0}{T} \quad \text{and} \quad N_0' = \frac{\mathcal{F}_s}{\mathcal{F}_o} \tag{5.16a}$$

But, because

$$\mathcal{F}_s = \frac{1}{T} \quad \text{and} \quad \mathcal{F}_o = \frac{1}{T_0} \tag{5.16b}$$

$$N_0 = \frac{T_0}{T} = \frac{\mathcal{F}_s}{\mathcal{F}_o} = N_0' \tag{5.16c}$$

Aliasing and Leakage in Numerical Computation

Figure 5.14f shows the presence of aliasing in the samples of the spectrum $F(\omega)$. This aliasing error can be reduced as much as desired by increasing the sampling frequency $\mathcal{F}_s$ (decreasing the sampling interval $T = \frac{1}{\mathcal{F}_s}$). The aliasing can never be eliminated for timelimited $f(t)$, however, because its spectrum $F(\omega)$ is nonbandlimited. Had we started out with a signal having a bandlimited spectrum $F(\omega)$, there would be no aliasing in the spectrum in Fig. 5.14f. Unfortunately such

---

†There is a multiplying constant $1/T$ for the spectrum in Fig. 5.14d [see Eq. (5.4)], but this is irrelevant to our discussion here.

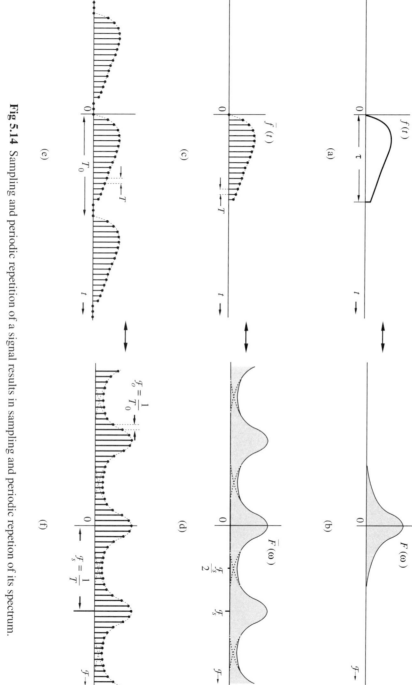

Fig 5.14 Sampling and periodic repetition of a signal results in sampling and periodic repetition of its spectrum.

a signal is nontimelimited and its repetition (in Fig. 5.14e) would result in signal overlapping (aliasing in the time domain). In this case we shall have to contend with errors in signal samples. In other words, in computing the direct or inverse Fourier transform numerically, we can reduce the error as much as we wish, but the error can never be eliminated. This fact is true of numerical computation of direct and inverse Fourier transforms, regardless of the method used. For example, if we determine the Fourier transform by direct integration numerically, using Eq. (4.8a), there will be an error because the interval of integration Δt can never be made zero. Similar remarks apply to numerical computation of the inverse transform. Therefore, we should always keep in mind the nature of this error in our results. In our discussion (Fig. 5.14), we assumed $f(t)$ to be a timelimited signal. If $f(t)$ is not timelimited, we would need to timelimit it because numerical computations can work only with finite data. Further, this data truncation causes error because of spectral spreading (smearing) and leakage, as discussed in Sec. 4.9. The leakage also causes aliasing. Leakage can be reduced by using a tapered window for signal truncation. But this choice increases spectral spreading or smearing. The spectral spreading can be reduced by increasing the window width (i.e. more data), which increases T_0, and reduces $\mathcal{F}_0$ (increases **spectral** or **frequency resolution**).

Picket Fence Effect

The numerical computation method yields only the uniform sample values of $F(\omega)$ [or $f(t)$]. Using this method is like viewing a signal and its spectrum through a "picket-fence." The major peaks of $F(\omega)$ [or $f(t)$] could lie between two samples and may remain hidden, a situation giving a false picture of the reality. Such misleading results can be avoided by using a sufficiently large N_0, the number of samples, which increases resolution. We can also use the spectral interpolation formula [Eq. (5.14)] to determine the values of $F(\omega)$ between samples.

Derivation of the Discrete Fourier transform (DFT)

If $f(kT)$ and $F(r\omega_0)$ are the kth and rth samples of $f(t)$ and $F(\omega)$, respectively, then we define new variables f_k and F_r as

$$f_k = Tf(kT)$$

$$= \frac{T_0}{N_0}f(kT) \tag{5.17a}$$

and

$$F_r = F(r\omega_0) \tag{5.17b}$$

where

$$\omega_0 = 2\pi\mathcal{F}_o = \frac{2\pi}{T_0} \tag{5.17c}$$

We shall now show that f_k and F_r are related by the following equations:†

---

†In Eqs. (5.18a) and (5.18b), the summation is performed from 0 to $N_0 - 1$. It is shown in Eq. (10.12) that the summation may be performed over any successive N_0 values of k or r.

$$F_r = \sum_{k=0}^{N_0-1} f_k e^{-jr\Omega_0 k} \tag{5.18a}$$

$$f_k = \frac{1}{N_0} \sum_{r=0}^{N_0-1} F_r e^{jr\Omega_0 k} \qquad \Omega_0 = \omega_0 T = \frac{2\pi}{N_0} \tag{5.18b}$$

These equations define the direct and the inverse **discrete Fourier transforms (DFT)**, with F_r the direct discrete Fourier transform (DFT) of f_k, and f_k the inverse discrete Fourier transform (IDFT) of F_r. The notation

$$f_k \Longleftrightarrow F_r$$

is also used to indicate that f_k and F_r are a DFT pair. Remember that f_k is T_0/N_0 times the kth sample of $f(t)$ and F_r is the rth sample of $F(\omega)$. Knowing the sample values of $f(t)$, we can compute the sample values of $F(\omega)$—and vice versa—using the DFT. Note, however, that f_k is a function of k ($k = 0, 1, 2, \ldots, N_0 - 1$) rather than of t and that F_r is a function of r ($r = 0, 1, 2, \ldots, N_0 - 1$) rather than of ω. Moreover, both f_k and F_r are periodic sequences of period N_0 (Figs. 5.14e and 5.14f). Such sequences are called N_0-**periodic sequences**. The proof of the DFT relationships in Eqs. (5.18) follows directly from the results of the sampling theorem. The sampled signal $\overline{f}(t)$ (Fig. 5.14c) can be expressed as

$$\overline{f}(t) = \sum_{k=0}^{N_0-1} f(kT)\delta(t - kT) \tag{5.19}$$

Since $\delta(t - kT) \Longleftrightarrow e^{-jk\omega T}$, the Fourier transform of Eq. (5.19) yields

$$\overline{F}(\omega) = \sum_{k=0}^{N_0-1} f(kT)e^{-jk\omega T} \tag{5.20}$$

But from Fig. 5.1e [or Eq. (5.4)], it is clear that over the interval $|\omega| \le \frac{\omega_s}{2}$, $\overline{F}(\omega)$, the Fourier transform of $\overline{f}(t)$ is $\frac{F(\omega)}{T}$, assuming negligible aliasing. Hence

$$F(\omega) = T\overline{F}(\omega) = T\sum_{k=0}^{N_0-1} f(kT)e^{-jk\omega T} \qquad |\omega| \le \frac{\omega_s}{2}$$

and

$$F_r = F(r\omega_0) = T\sum_{k=0}^{N_0-1} f(kT)e^{-jkr\omega_0 T} \tag{5.21}$$

If we let $\omega_0 T = \Omega_0$, then from Eqs. (5.16a) and (5.16b)

$$\Omega_0 = \omega_0 T = 2\pi \mathcal{F}_o T = \frac{2\pi}{N_0} \tag{5.22}$$

Also, from Eq. (5.17a),

$$Tf(kT) = f_k$$

Therefore, Eq. (5.21) becomes

$$F_r = \sum_{k=0}^{N_0-1} f_k e^{-jr\Omega_0 k} \qquad \Omega_0 = \frac{2\pi}{N_0} \tag{5.23}$$

The inverse transform relationship (5.18b) could be derived by using a similar procedure with the roles of t and ω reversed, but here we shall use a more direct proof. To prove the inverse relation in Eq. (5.18b), we multiply both sides of Eq. (5.23) by $e^{jm\Omega_0 r}$ and sum over r as

$$\sum_{r=0}^{N_0-1} F_r e^{jm\Omega_0 r} = \sum_{r=0}^{N_0-1} \left[\sum_{k=0}^{N_0-1} f_k e^{-jr\Omega_0 k} \right] e^{jm\Omega_0 r}$$

By interchanging the order of summation on the right hand side

$$\sum_{r=0}^{N_0-1} F_r e^{jm\Omega_0 r} = \sum_{k=0}^{N_0-1} f_k \left[\sum_{r=0}^{N_0-1} e^{j(m-k)\Omega_0 r} \right]$$

Appendix 5.1 shows that the inner sum on the right hand side is zero for $k \neq m$, and that the sum is N_0 when $k = m$. Therefore the outer sum will have only one nonzero term when $k = m$, and it is $N_0 f_k = N_0 f_m$. Therefore

$$f_m = \frac{1}{N_0} \sum_{r=0}^{N_0-1} F_r e^{jm\Omega_0 r} \qquad \Omega_0 = \frac{2\pi}{N_0} \tag{5.24}$$

Because F_r is N_0-periodic, we need to determine the values of F_r over any one period. It is customary to determine F_r over the range $(0, N_0 - 1)$ rather than over the range $(-\frac{N_0}{2}, \frac{N_0}{2} - 1)$.‡

Choice of T, T_0, and N_0

In DFT computation, we first need to select suitable values for N_0, T, and T_0. For this purpose we should first decide on B, the essential bandwidth (in hertz) of the signal. The sampling frequency $\mathcal{F}_s$ must be at least $2B$; that is,

$$\frac{\mathcal{F}_s}{2} \geq B \tag{5.25a}$$

Moreover, the sampling interval $T = \frac{1}{\mathcal{F}_s}$ [Eq. (5.16b)], and

$$T \leq \frac{1}{2B} \tag{5.25b}$$

Once we pick B, we can choose T according to Eq. (5.25b)). Also,

$$\mathcal{F}_o = \frac{1}{T_0} \tag{5.26}$$

where $\mathcal{F}_o$ is the **frequency resolution** [separation between samples of $F(\omega)$]. Hence, if $\mathcal{F}_o$ is given, we can pick T_0 according to Eq. (5.26). Knowing T_0 and T, we determine N_0 from

---

‡The DFT relationships represent a transform in their own right, and they are exact. If, however, we identify f_k and F_r as the samples of a signal $f(t)$ and of its Fourier transform $F(\omega)$, respectively, then the DFT relationships are approximations because of the aliasing and leakage effects.

$$N_0 = \frac{T_0}{T} \qquad (5.27)$$

Points of Discontinuity

If $f(t)$ has a jump discontinuity at a sampling point, the sample value should be taken as the average of the values on the two sides of the discontinuity because the Fourier representation at a point of discontinuity converges to the average value.

Zero Padding

Recall that observing F_r is like observing the spectrum $F(\omega)$ through a "picket-fence." If the frequency sampling interval $\mathcal{F}_o$ is not sufficiently small, we could miss out on some significant details and obtain a misleading picture. To obtain a higher number of samples, we need to reduce $\mathcal{F}_o$. Because $\mathcal{F}_o = 1/T_0$, a higher number of samples requires us to increase the value of T_0, the period of repetition for $f(t)$. This option increases N_0, the number of samples of $f(t)$, by adding dummy samples with a value of 0. This addition of dummy samples is known as **zero padding**. Thus, zero padding increases the number of samples and may help in getting a better idea of the spectrum $F(\omega)$ from its samples F_r.

Zero Padding Does Not Improve Accuracy or Resolution

One point should be understood clearly: that zero padding merely gives us more samples without improving the accuracy of those sample values. Zero padding will be useful only if the sampling interval T is sufficiently small so that the aliasing error is negligible. If there is a good deal of aliasing to begin with, zero padding will merely give us more samples of the garbage. It can never improve accuracy or frequency resolution in a true sense. The accuracy can be increased only by reducing aliasing, which requires reduction in the signal sampling interval T ($T < 1/2B$, where B is the effective bandwidth of the signal).

■ Example 5.5

A signal $f(t)$ has a duration of 2 ms and an essential bandwidth of 10 kHz. It is desirable to have a frequency resolution of 100 Hz in the DFT ($\mathcal{F}_o = 100$). Determine N_0.

The effective signal duration T_0 is given by

$$T_0 = \frac{1}{\mathcal{F}_o} = \frac{1}{100} = 10 \text{ ms}$$

Since the signal duration is only 2 ms, we need zero padding over 8 ms. Also, $B = 10,000$. Hence, $\mathcal{F}_s = 2B = 20,000$ and $T = 1/\mathcal{F}_s = 50\,\mu s$. Further,

$$N_0 = \frac{\mathcal{F}_s}{\mathcal{F}_o} = \frac{20,000}{100} = 200$$

The **fast Fourier transform** (FFT) algorithm (discussed in Sec. 5.3) is used to compute DFT, where it proves convenient (although not necessary) to select N_0 as a power of 2; that is, $N_0 = 2^n$ (n, integer). Let us choose $N_0 = 256$. Increasing N_0 from 200 to 256 can be used to reduce aliasing error (by reducing T), to improve resolution (by increasing T_0), or a combination of both.

(i) Reducing aliasing error: We maintain the same T_0 so that $\mathcal{F}_o = 100$. Hence

$$\mathcal{F}_s = N_0 \mathcal{F}_o = 256 \times 100 = 25600 \qquad \text{and} \qquad T = \frac{1}{\mathcal{F}_s} = 39\,\mu s$$

Thus, increasing N_0 from 200 to 256 permits us to reduce the sampling interval T from $50\,\mu s$ to $39\,\mu s$ while maintaining the same frequency resolution ($\mathcal{F}_o = 100$).

(ii) **Improving resolution:** Here we maintain the same $T = 50\,\mu s$, which yields

$$T_0 = N_0 T = 256(50 \times 10^{-6}) = 12.8 \text{ ms} \quad \text{and} \quad \mathcal{F}_o = \frac{1}{T_0} = 78.125 \text{ Hz}$$

Thus, increasing N_0 from 200 to 256 can improve the frequency resolution from 100 to 78.125 Hz while maintaining the same aliasing error ($T = 50\,\mu s$).

(iii) **Combination of these two options**

We could choose $T = 45\,\mu s$ and $T_0 = 11.5$ ms so that $\mathcal{F}_o = 86.96$ Hz. ■

■ **Example 5.6**

Use DFT to compute the Fourier transform of $e^{-2t}u(t)$. Plot the resulting Fourier spectra.

We first determine T and T_0. The Fourier transform of $e^{-2t}u(t)$ is $1/(j\omega + 2)$. This lowpass signal is not bandlimited. In Sec. 4.6, we used the energy criterion to compute the essential bandwidth of a signal. Here, we shall present a simpler, but workable alternative to the energy criterion. The essential bandwidth of a signal will be taken as that frequency where $|F(\omega)|$ reduces to 1% of its peak value (see footnote on p. 276). In this case, the peak value occurs at $\omega = 0$, where $|F(0)| = 0.5$. Observe that

$$|F(\omega)| = \frac{1}{\sqrt{\omega^2 + 4}} \approx \frac{1}{\omega} \qquad \omega \gg 2$$

Also, 1% of the peak value is $0.01 \times 0.5 = 0.005$. Hence, the essential bandwidth B is at $\omega = 2\pi B$, where

$$|F(\omega)| \approx \frac{1}{2\pi B} = 0.005 \quad \Rightarrow B = \frac{100}{\pi} \text{ Hz}$$

and from Eq. (5.25b),

$$T \le \frac{1}{2B} = \frac{\pi}{200} = 0.015708$$

Had we used 1% energy criterion to determine the essential bandwidth, following the procedure in Example 4.16, we would have obtained $B = 20.26$ Hz, which is somewhat smaller than the value just obtained by using 1% amplitude criterion.

The second issue is to determine T_0. Because the signal is not timelimited, we have to truncate it at T_0 such that $f(T_0) \ll 1$. A reasonable choice would be $T_0 = 4$ because $f(4) = e^{-8} = 0.000335 \ll 1$. The result is $N_0 = T_0/T = 254.6$, which is not a power of 2. Hence, we choose $T_0 = 4$, and $T = 0.015625 = 1/64$, yielding $N_0 = 256$, which is a power of 2.

Note that there is a great deal of flexibility in determining T and T_0, depending on the accuracy desired and the computational capacity available. We could just as well have chosen $T = 0.03125$, yielding $N_0 = 128$, although this choice would have given a slightly higher aliasing error.

Because the signal has a jump discontinuity at $t = 0$, the first sample (at $t = 0$) is 0.5, the averages of the values on the two sides of the discontinuity. We compute F_r (the DFT) from the samples of $e^{-2t}u(t)$ according to Eq. (5.18a). Note that F_r is the rth sample of $F(\omega)$, and these samples are spaced at $\mathcal{F}_0 = 1/T_0 = 0.25$ Hz. ($\omega_0 = \pi/2$ rad/s.)

Because F_r is N_0-periodic, $F_r = F_{(r+256)}$ so that $F_{256} = F_0$. Hence, we need to plot F_r over the range $r = 0$ to 255 (not 256). Moreover, because of this periodicity, $F_{-r} = F_{(-r+256)}$, and the values of F_r over the range $r = -127$ to -1 are identical to those over the range $r = 129$ to 255. Thus, $F_{-127} = F_{129}$, $F_{-126} = F_{130}$, $\ldots F_{-1} = F_{255}$. In addition, because of the property of conjugate symmetry of the Fourier transform, $F_{-r} = F_r^*$, it follows that $F_{-1} = F_1^*$, $F_{-2} = F_2^*$, $\ldots, F_{-128} = F_{128}^*$. Thus, we need F_r only over the range $r = 0$ to $N_0/2$ (128 in this case).

Figure 5.15 shows the plots of $|F_r|$ and $\angle F_r$ and the exact magnitude and phase spectra (depicted by continuous curves) for comparison. Note the nearly perfect agreement between the two sets of spectra. We have depicted the plot of only the first 28 points rather than all the 128 point because it would have made the figure very crowded, resulting in loss of clarity. The points are at the intervals of $1/T_0 = 1/4$ Hz or $\omega_0 = 1.5708$ rad/s. The 28 samples, therefore, exhibit the plots over the range $\omega = 0$ to $\omega = 28(1.5708) \approx 44$ rad/s or 7 Hz.

In this example, we knew $F(\omega)$ beforehand and hence could make intelligent choices for B (or the sampling frequency $\mathcal{F}_s$). In practice, we generally do not know $F(\omega)$ beforehand. In fact, that is the very thing we are trying to determine. In such a case, we must make an intelligent guess for B or $\mathcal{F}_s$ from circumstantial evidence. We should then continue reducing the value of T and recomputing the transform until the result stabilizes within the desired number of significant digits. The MATLAB program, which implements the DFT using the FFT algorithm, is presented in Example C5.1. ■

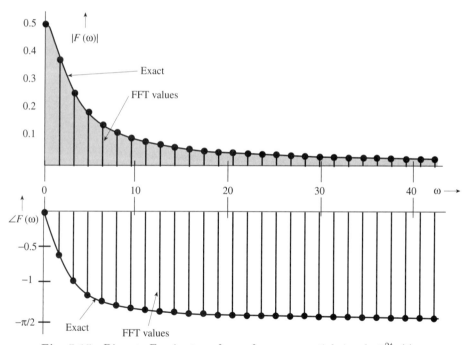

Fig. 5.15 Discrete Fourier transform of an exponential signal $e^{-2t}u(t)$.

⊙ **Computer Example C5.1**

Use DFT (implemented by the FFT, the fast Fourier transform algorithm) to compute the Fourier transform of $e^{-2t}u(t)$.

```
T=0.015625;T0=4;N0=T0/T;
t=0:T:T*(N0-1);t=t';
f=T*exp(-2*t);
```

```
f(1)=T*0.5;
F=fft(f);
[Fp,Fm]=cart2pol(real(F),imag(F));
k=0:N0-1; k=k';
w=2*pi*k/T0;
subplot(211),plot(w(1:128),Fm(1:128))
subplot(212),plot(w(1:128),Fp(1:128))      ⊙
```

■ **Example 5.7**

Use DFT to compute the Fourier transform of $8 \, \text{rect} \, (t)$.

This gate function and its Fourier transform are illustrated in Fig. 5.16a and b. To determine the value of the sampling interval T, we must first decide on the essential bandwidth B. In Fig. 5.16b, we see that $F(\omega)$ decays rather slowly with ω. Hence, the essential bandwidth B is rather large. For instance, at $B = 15.5$ Hz (97.39 rad/s), $F(\omega) = -0.1643$, which is about 2% of the peak at $F(0)$. Hence, the essential bandwidth is well above 16 Hz if we use the 1% criterion for computing the essential bandwidth. However, we shall deliberately take $B = 4$ for two reasons: (1) to show the effect of aliasing and (2) the use of $B > 4$ will give enormous number of samples, which cannot be conveniently displayed on the book size page without losing sight of the essentials. Thus, we shall intentionally accept approximation in order to clarify the concepts of DFT graphically.

The choice of $B = 4$ results in the sampling interval $T = \frac{1}{2B} = \frac{1}{8}$. Looking again at the spectrum in Fig. 5.16b, we see that the choice of the frequency resolution $\mathcal{F}_o = \frac{1}{4}$ Hz is reasonable. Such a choice gives us four samples in each lobe of $F(\omega)$. In this case $T_0 = \frac{1}{\mathcal{F}_o} = 4$ seconds and $N_0 = \frac{T_0}{T} = 32$. The duration of $f(t)$ is only 1 second. We must repeat it every 4 seconds ($T_0 = 4$), as depicted in Fig. 5.16c, and take samples every $\frac{1}{8}$ second. This choice yields 32 samples ($N_0 = 32$). Also,

$$f_k = T f(kT)$$

$$= \frac{1}{8} f(kT)$$

Since $f(t) = 8 \, \text{rect} \, (t)$, the values of f_k are 1, 0, or 0.5 (at the points of discontinuity), as illustrated in Fig. 5.16c. In this figure, f_k is depicted as a function of t as well as k, for convenience.

In the derivation of the DFT, we assumed that $f(t)$ begins at $t = 0$ (Fig. 5.14a), and then took N_0 samples over the interval $(0, T_0)$. In the present case, however, $f(t)$ begins at $-\frac{1}{2}$. This difficulty is easily resolved when we realize that the DFT obtained by this procedure is actually the DFT of f_k repeating periodically every T_0 seconds. Figure 5.16c clearly indicates that repeating the segment of f_k over the interval from -2 to 2 seconds periodically is identical to repeating the segment of f_k over the interval from 0 to 4 seconds. Hence, the DFT of the samples taken from -2 to 2 seconds is the same as that of the samples taken from 0 to 4 seconds. Therefore, regardless of where $f(t)$ starts, we can always take the samples of $f(t)$ and its periodic extension over the interval from 0 to T_0. In the present example, the 32 sample values are

$$f_k = \begin{cases} 1 & 0 \le k \le 3 \quad \text{and} \quad 29 \le k \le 31 \\ 0 & 5 \le k \le 27 \\ 0.5 & k = 4, 28 \end{cases}$$

Observe that the last sample is at $t = 31/8$, not at 4, because the signal repetition starts at $t = 4$, and the sample at $t = 4$ is the same as the sample at $t = 0$. Now, $N_0 = 32$ and $\Omega_0 = 2\pi/32 = \pi/16$. Therefore [see Eq. (5.18a)]

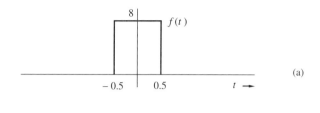

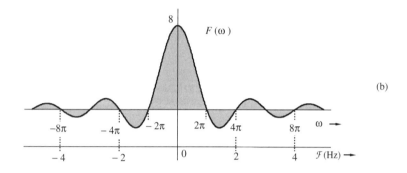

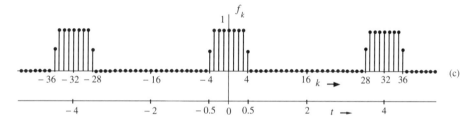

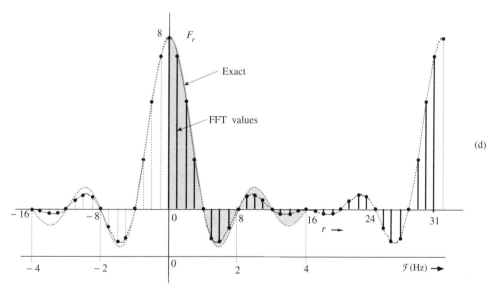

Fig. 5.16   Discrete Fourier Transform of a gate pulse.

$$F_r = \sum_{k=0}^{31} f_k e^{-jr\frac{\pi}{16}k}$$

Values of F_r are computed according to this equation and plotted in Fig. 5.16d.

The samples F_r are separated by $\mathcal{F}_o = \frac{1}{T_0}$ Hz. In this case $T_0 = 4$, so the frequency resolution $\mathcal{F}_o$ is $\frac{1}{4}$ Hz, as desired. The folding frequency $\frac{\mathcal{F}_s}{2} = B = 4$ Hz corresponds to $r = \frac{N_0}{2} = 16$. Because F_r is N_0-periodic ($N_0 = 32$), the values of F_r for $r = -16$ to $n = -1$ are the same as those for $r = 16$ to $n = 31$. For instance, $F_{17} = F_{-15}$, $F_{18} = F_{-14}$, and so on. The DFT gives us the samples of the spectrum $F(\omega)$.

For the sake of comparison, Fig. 5.16d also shows the shaded curve $8\,\text{sinc}\,(\frac{\omega}{2})$, which is the Fourier transform of $8\,\text{rect}\,(t)$. The values of F_r computed from DFT equation show aliasing error, which is clearly seen by comparing the two superimposed plots. The error in F_2 is just about 1.3%. However, the aliasing error increases rapidly with r. For instance, the error in F_6 is about 12%, and the error in F_{10} is 33%. The error in F_{14} is a whopping 72%. The percent error increases rapidly near the folding frequency ($r = 16$) because $f(t)$ has a jump discontinuity, which makes $F(\omega)$ decays slowly as $1/\omega$. Hence, near the folding frequency, the inverted tail (due to aliasing) is very nearly equal to $F(\omega)$ itself. Moreover, the final values are the difference between the exact and the folded values (which are very close to the exact values). Hence, the percent error near the folding frequency ($r = 16$ in this case) is very high, although the absolute error is very small. Clearly, for signals with jump discontinuities, the aliasing error near the folding frequency will always be high (in percentage terms), regardless of the choice of N_0. To ensure a negligible aliasing error at any value r, we must make sure that $N_0 \gg r$. This observation is valid for all signals with jump discontinuities. ∎

⊙ **Computer Example C5.2**

Use DFT (implemented by the FFT algorithm) to compute the Fourier transform of $8\,\text{rect}\,(t)$. Plot the resulting Fourier spectra.

The MATLAB program, which implements this DFT equation using the FFT algorithm, is given next. First we write a MATLAB program to generate 32 samples of f_k, and then we compute the DFT.

```
% (c52.m)
N0=32;k=0:N0-1;
f=[ones(1,4) 0.5 zeros(1,23) 0.5 ones(1,3)];
Fr=fft(f);
subplot(2,1,1),stem(k,f)
xlabel('k');ylabel('fk');
subplot(2,1,2),stem(k,Fr)
xlabel('r');ylabel('Fr');   ⊙
```

5.2-1 Some Properties of DFT

The discrete Fourier transform is basically the Fourier transform of a sampled signal repeated periodically. Hence, the properties derived earlier for the Fourier transform apply to the DFT as well.

1. Linearity

If $f_k \Longleftrightarrow F_r$ and $g_k \Longleftrightarrow G_r$, then

$$a_1 f_k + a_2 g_k \Longleftrightarrow a_1 F_r + a_2 G_r \tag{5.28}$$

The proof is trivial.

2. Conjugate Symmetry

$$F_{N_0-r} = F_r^* \tag{5.29}$$

This follows from the conjugate symmetry property of the Fourier transform ($F_r^* = F_{-r}$) and the periodic property of DFT ($F_{-r} = F_{N_0-r}$). Because of this property, we need compute only half the DFTs for real f_k. The other half are the conjugates.

3. Time Shifting (Circular Shifting)

$$f_{k-n} \Longleftrightarrow F_r e^{-jr\Omega_0 n} \tag{5.30}$$

Proof: Using Eq. (5.18b), we find the inverse DFT of $F_r e^{-jr\Omega_0 n}$ as

$$\frac{1}{N_0} \sum_{r=0}^{N_0-1} F_r e^{-jr\Omega_0 n} e^{jr\Omega_0 k} = \frac{1}{N_0} \sum_{r=0}^{N_0-1} F_r e^{jr\Omega_0(k-n)} = f_{k-n}$$

4. Frequency Shifting

$$f_k e^{jk\Omega_0 m} \Longleftrightarrow F_{r-m} \tag{5.31}$$

Proof: This proof is identical to that of the time shifting property except that we start with Eq. (5.18a).

5. Circular (also called periodic) Convolution

$$f_k \circledast g_k \Longleftrightarrow F_r G_r \tag{5.32a}$$

and

$$f_k g_k \Longleftrightarrow \frac{1}{N_0} F_r \circledast G_r \tag{5.32b}$$

For two N_0-periodic sequences f_k and g_k, the circular convolution is defined by

$$f_k \circledast g_k = \sum_{n=0}^{N_0-1} f_n g_{k-n} = \sum_{n=0}^{N_0-1} g_n f_{k-n} \tag{5.33}$$

To prove (5.32a), we find the DFT of circular convolution $f_k \circledast g_k$ as

$$\sum_{k=0}^{N_0-1} \left(\sum_{n=0}^{N_0-1} f_n g_{k-n} \right) e^{-jr\Omega_0 k} = \sum_{n=0}^{N_0-1} f_n \left(\sum_{k=0}^{N_0-1} g_{k-n} e^{-jr\Omega_0 k} \right)$$

$$= \sum_{n=0}^{N_0-1} f_n (G_r e^{-jr\Omega_0 n}) = F_r G_r$$

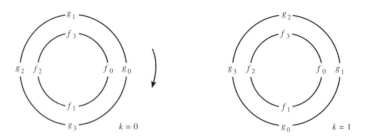

Fig. 5.17 Graphical picture of circular convolution.

Equation (5.32b) can be proved in the same way.

For nonperiodic sequences, the convolution can be visualized in terms of two sequences, where one sequence is fixed and the other sequence is inverted and moved past the fixed sequence, one digit at a time. If the two sequences are N_0-periodic, the same configuration will repeat after N_0 shifts of the sequence. Clearly the convolution $f_k \circledast g_k$ becomes N_0-periodic (circular), and such convolution can be conveniently visualized as illustrated in Fig. 5.17 for the case of $N_0 = 4$. The inner sequence f_k is clockwise and fixed. The outer sequence g_k is inverted so that it becomes counterclockwise. This sequence is now rotated clockwise one unit at a time. We multiply the overlapping numbers and add. For example, the value of $f_k \circledast g_k$ at $k = 0$ (Fig. 5.17) is

$$f_0 g_0 + f_1 g_3 + f_2 g_2 + f_3 g_1$$

and the value of $f_k \circledast g_k$ at $k = 1$ is (Fig. 5.17)

$$f_0 g_1 + f_1 g_0 + f_2 g_3 + f_3 g_2$$

and so on.

Applications of DFT

The DFT is useful not only in the computation of direct and inverse Fourier transforms, but also in other applications such as convolution, correlation, and filtering. Use of the efficient FFT algorithm discussed in Sec. 5.3 makes it particularly appealing.

Linear Convolution

Let $f(t)$ and $g(t)$ be the two signals to be convolved. In general, these signals may have different time durations. To convolve them by using their samples, they must be sampled at the same rate (not below the Nyquist rate of either signal). Let $f_k \, (0 \leq k \leq N_1 - 1)$ and $g_k \, (0 \leq k \leq N_2 - 1)$ be the corresponding discrete sequences representing these samples. Now,

$$c(t) = f(t) * g(t)$$

and if we define three sequences as $f_k = Tf(kT)$, $g_k = Tg(kT)$, and $c_k = Tc(kT)$, then[†]

$$c_k = f_k * g_k$$

[†]We can show that[4] $c_k = \lim_{T \to 0} f_k * g_k$. Because $T \neq 0$ in practice, there will be some error in this equation.

where we define the linear convolution sum of two discrete sequences f_k and g_k as

$$f_k * g_k = \sum_{n=-\infty}^{\infty} f_n g_{k-n}$$

Because of the width property of the convolution, c_k exists for $0 \leq k \leq N_1 + N_2 - 1$. To be able to use the DFT technique of circular convolution, we must make sure that the circular convolution will yield the same result as the linear convolution. In other words, the resulting signal of the circular convolution must have the same length $(N_1 + N_2 - 1)$ as that of the signal resulting from linear convolution. This step can be accomplished by adding $N_2 - 1$ dummy samples of zero value to f_k and $N_1 - 1$ dummy samples of zero value to g_k (zero padding). This procedure changes the length of both f_k and g_k to be $N_1 + N_2 - 1$. The circular convolution now is identical to the linear convolution except that it repeats periodically with period $N_1 + N_2 - 1$. The rigorous proof of this statement is provided in Sec. 10.6-3.

We can use the DFT to find the convolution $f_k * g_k$ in three steps, as follows:

1. Find the DFTs F_r and G_r corresponding to suitably padded f_k and g_k.
2. Multiply F_r by G_r.
3. Find the IDFT of $F_r G_r$. This procedure of convolution, when implemented by the fast Fourier transform algorithm (discussed later), is known as the **fast convolution**.

Filtering

We generally think of filtering in terms of some hardware-oriented solution (namely, building a circuit with RLC components and operational amplifiers). However, filtering also has a software-oriented solution [a computer algorithm that yields the filtered output $y(t)$ for a given input $f(t)$]. This goal can be conveniently accomplished by using the DFT. If $f(t)$ is the signal to be filtered, then F_r, the DFT of f_k, is found. The spectrum F_r is then shaped (filtered) as desired by multiplying F_r by H_r, where H_r are the samples of the filter transfer function $H(\omega)$ $[H_r = H(r\omega_0)]$. Finally, we take the IDFT of $F_r H_r$ to obtain the filtered output y_k $[y_k = Ty(kT)]$. This procedure is demonstrated in the following example.

■ **Example 5.8**

The signal $f(t)$ in Fig. 5.18a is passed through an ideal lowpass filter of transfer function $H(\omega)$ depicted in Fig. 5.18b. Using DFT, find the filter output.

We have already found the 32-point DFT of $f(t)$ (see Fig. 5.16d). Next we multiply F_r by H_r. To find H_r, we recall using $\mathcal{F}_o = \frac{1}{4}$ in computing the 32-point DFT of $f(t)$. Because F_r is 32-periodic, H_r must also be 32-periodic with samples separated by $\frac{1}{4}$ Hz. This fact means that H_r must be repeated every 8 Hz or 16π rad/s (see Fig. 5.18c). The resulting 32 samples of H_r over $(0 \leq \omega \leq 16\pi)$ are as follows:

$$H_r = \begin{cases} 1 & 0 \leq r \leq 7 \quad \text{and} \quad 25 \leq r \leq 31 \\ 0 & 9 \leq r \leq 23 \\ 0.5 & r = 8, 24 \end{cases}$$

We multiply F_r with H_r. The desired output signal samples y_k are found by taking the inverse DFT of $F_r H_r$. The resulting output signal is illustrated in Fig. 5.18d. Table 5.1 gives a printout of f_k, F_r, H_r, Y_r, and y_k. ■

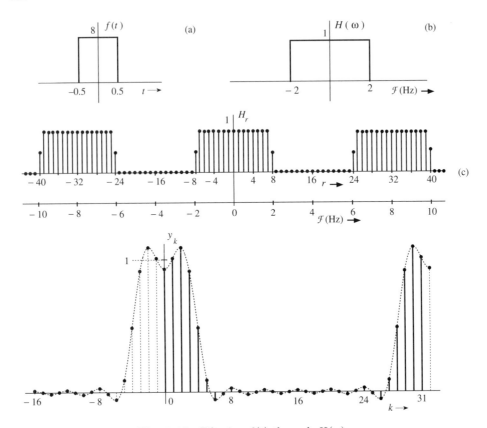

Fig. 5.18   Filtering $f(t)$ through $H(\omega)$.

⊙ **Computer Example C5.3**
Solve Example 5.8 using MATLAB.

The MATLAB program of Example C5.2, where we obtained the 32-point F_r, should be saved as an m-file, e.g., 'c52.m'. We can import F_r in the MATLAB environment by the command 'c52'.

```
c52;
N0=32;k=0:N0-1;
H=[0ones(1,8) 0.5 zeros(1,15) 0.5 ones(1,7)];
Yr=H.*Fr;
yk=ifft(Yr);
stem(k,yk)    ⊙
```

5.3   The Fast Fourier Transform (FFT)

The number of computations required in performing the DFT was dramatically reduced by an algorithm developed by Tukey and Cooley in 1965.[5] This algorithm, known as the **Fast Fourier transform** (FFT), reduces the number of computations from something of the order of N_0^2 to $N_0 \log N_0$. To compute one sample F_r from Eq. (5.18a), we require N_0 complex multiplications and $N_0 - 1$ complex additions. To compute N_0 such values (F_r for $r = 0, 1, \ldots, N_0 - 1$), we require a total of N_0^2

No.	f_k	F_r	H_r	$F_r H_r$	y_k
0	1	8.000	1	8.000	.9285
1	1	7.179	1	7.179	1.009
2	1	5.027	1	5.027	1.090
3	1	2.331	1	2.331	.9123
4	0.5	0.000	1	0.000	.4847
5	0	−1.323	1	−1.323	.08884
6	0	−1.497	1	−1.497	−.05698
7	0	−.8616	1	−.8616	−.01383
8	0	0.000	0.5	0.000	.02933
9	0	.5803	0	0.000	.004837
10	0	.6682	0	0.000	−.01966
11	0	.3778	0	0.000	−.002156
12	0	0.000	0	0.000	.01534
13	0	−.2145	0	0.000	.0009828
14	0	−.1989	0	0.000	−.01338
15	0	−.06964	0	0.000	−.0002876
16	0	0.000	0	0.000	.01280
17	0	−.06964	0	0.000	−.0002876
18	0	−.1989	0	0.000	−.01338
19	0	−.2145	0	0.000	.0009828
20	0	0.000	0	0.000	.01534
21	0	.3778	0	0.000	−.002156
22	0	.6682	0	0.000	−.01966
23	0	.5803	0	0.000	.004837
24	0	0.000	0.5	0.000	.03933
25	0	−.8616	1	−.8616	−.01383
26	0	−1.497	1	−1.497	−.05698
27	0	−1.323	1	−1.323	.08884
28	0.5	0.000	1	0.000	.4847
29	1	2.331	1	2.331	.9123
30	1	5.027	1	5.027	1.090
31	1	7.179	1	7.179	1.009

complex multiplications and $N_0(N_0 - 1)$ complex additions. For a large N_0, these computations can be prohibitively time-consuming, even for a high-speed computer.

Although there are many variations of the original Tukey-Cooley algorithm, these can be grouped into two basic types: **decimation-in-time** and **decimation-in-frequency**. The algorithm is simplified if we choose N_0 to be a power of 2, although such a choice is not essential. For convenience, we define

$$W_{N_0} = e^{-(j2\pi/N_0)} = e^{-j\Omega_0} \tag{5.34}$$

so that

$$F_r = \sum_{k=0}^{N_0-1} f_k W_{N_0}^{kr} \qquad 0 \le r \le N_0 - 1 \tag{5.35a}$$

and

$$f_k = \frac{1}{N_0} \sum_{r=0}^{N_0-1} F_r W_{N_0}^{-kr} \qquad 0 \le k \le N_0 - 1 \tag{5.35b}$$

The Decimation-in-Time Algorithm

Here we divide the N_0-point data sequence f_k into two ($\frac{N_0}{2}$)-point sequences consisting of even- and odd-numbered samples respectively, as follows:

$$\underbrace{f_0, f_2, f_4, \ldots, f_{N_0-2}}_{\text{sequence } g_k}, \underbrace{f_1, f_3, f_5, \ldots, f_{N_0-1}}_{\text{sequence } h_k}$$

Then, from Eq. (5.35a),

$$F_r = \sum_{k=0}^{\frac{N_0}{2}-1} f_{2k} W_{N_0}^{2kr} + \sum_{k=0}^{\frac{N_0}{2}-1} f_{2k+1} W_{N_0}^{(2k+1)r} \tag{5.36}$$

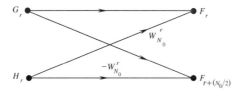

Fig. 5.19 Butterfly.

Also, since

$$W_{\frac{N_0}{2}} = W_{N_0}^2 \tag{5.37}$$

we have

$$F_r = \sum_{k=0}^{\frac{N_0}{2}-1} f_{2k} W_{\frac{N_0}{2}}^{kr} + W_{N_0}^r \sum_{k=0}^{\frac{N_0}{2}-1} f_{2k+1} W_{\frac{N_0}{2}}^{kr}$$

$$= G_r + W_{N_0}^r H_r \qquad 0 \leq r \leq N_0 - 1 \tag{5.38}$$

where G_r and H_r are the $(\frac{N_0}{2})$-point DFTs of the even- and odd-numbered sequences, g_k and h_k, respectively. Also, G_r and H_r, being the $(\frac{N_0}{2})$-point DFTs, are $(\frac{N_0}{2})$-periodic. Hence

$$G_{r+(\frac{N_0}{2})} = G_r$$

$$H_{r+(\frac{N_0}{2})} = H_r \tag{5.39}$$

Moreover,

$$W_{N_0}^{r+(\frac{N_0}{2})} = W_{N_0}^{\frac{N_0}{2}} W_{N_0}^r = e^{-j\pi} W_{N_0}^r = -W_{N_0}^r \tag{5.40}$$

From Eqs. (5.38), (5.39), and (5.40), we obtain

$$F_{r+(\frac{N_0}{2})} = G_r - W_{N_0}^r H_r \tag{5.41}$$

This property can be used to reduce the number of computations. We can compute the first $(\frac{N_0}{2})$ points $(0 \leq n \leq \frac{N_0}{2} - 1)$ of F_r using Eq. (5.38) and compute the last $\frac{N_0}{2}$ points using Eq. (5.41) as

$$F_r = G_r + W_{N_0}^r H_r \qquad 0 \leq r \leq \frac{N_0}{2} - 1 \tag{5.42a}$$

$$F_{r+(\frac{N_0}{2})} = G_r - W_{N_0}^r H_r \qquad 0 \leq r \leq \frac{N_0}{2} - 1 \tag{5.42b}$$

Thus, an N_0-point DFT can be computed by combining the two $(\frac{N_0}{2})$-point DFTs, as in Eq. (5.42). These equations can be represented conveniently by the **signal flow** graph depicted in Fig. 5.19. This structure is known as a **butterfly**. Figure 5.20a shows the implementation of Eqs. (5.39) for the case of $N_0 = 8$.

The next step is to compute the $(\frac{N_0}{2})$-point DFTs G_r and H_r. We repeat the same procedure by dividing g_k and h_k into two $(\frac{N_0}{4})$-point sequences corresponding to the even- and odd-numbered samples. Then we continue this process

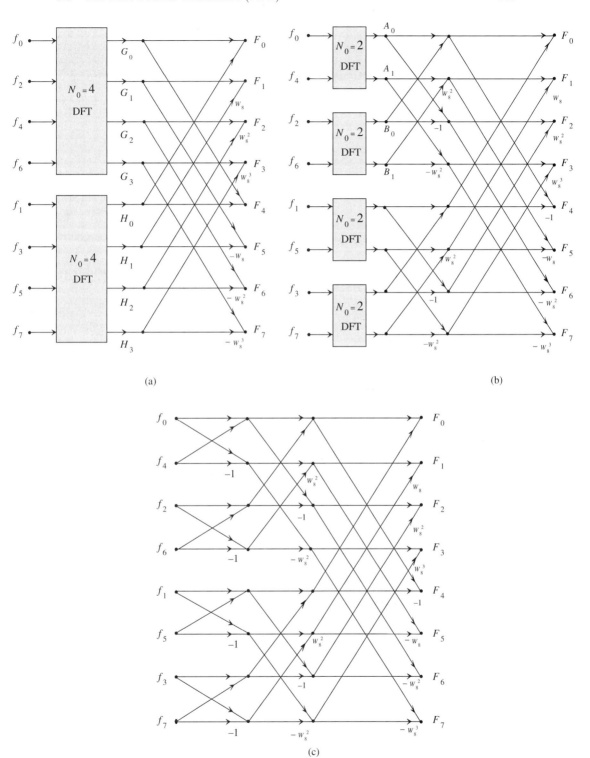

Fig. 5.20 Successive steps in 8-point FFT.

until we reach the one-point DFT. Figures 5.20a, 5.20b, and 5.20c show these steps for the case of $N_0 = 8$. Figure 5.20c shows that the two-point DFTs require no multiplication.

To count the number of computations required in the first step, assume that G_r and H_r are known. Equations (5.42) clearly show that to compute all the N_0 points of the F_r, we require N_0 complex additions and $\frac{N_0}{2}$ complex multiplications† (corresponding to $W_{N_0}^r H_r$).

In the second step, to compute the $(\frac{N_0}{2})$-point DFT G_r from the $(\frac{N_0}{4})$-point DFT, we require $\frac{N_0}{2}$ complex additions and $\frac{N_0}{4}$ complex multiplications. We require an equal number of computations for H_r. Hence, in the second step, there are N_0 complex additions and $\frac{N_0}{2}$ complex multiplications. The number of computations required remains the same in each step. Since a total of $\log_2 N_0$ steps is needed to arrive at a one-point DFT, we require, conservatively, a total of $N_0 \log_2 N_0$ complex additions and $(\frac{N_0}{2}) \log_2 N_0$ complex multiplications, to compute the N_0-point DFT.

The procedure for obtaining IDFT is identical to that used to obtain the DFT except that $W_{N_0} = e^{j(2\pi/N_0)}$ instead of $e^{-j(2\pi/N_0)}$ (in addition to the multiplier $1/N_0$). Another FFT algorithm, the **decimation-in-frequency** algorithm, is similar to the decimation-in-time algorithm. The only difference is that instead of dividing f_k into two sequences of even- and odd-numbered samples, we divide f_k into two sequences formed by the first $\frac{N_0}{2}$ and the last $\frac{N_0}{2}$ digits, proceeding in the same way until a single-point DFT is reached in $\log_2 N_0$ steps. The total number of computations in this algorithm is the same as that in the decimation-in-time algorithm.

5.4 Appendix 5.1

We show that

$$\sum_{k=0}^{N_0-1} e^{jm\Omega_0 k} = \begin{cases} N_0 & m = 0, \pm N_0, \pm 2N_0, \ldots \\ 0 & \text{otherwise} \end{cases} \tag{5.43}$$

Recall that $\Omega_0 N_0 = 2\pi$ and $e^{jm\Omega_0 k} = 1$ for $m = 0, \pm N_0, \pm 2N_0, \ldots$, so that

$$\sum_{k=0}^{N_0-1} e^{jm\Omega_0 k} = \sum_{k=0}^{N_0-1} 1 = N_0 \qquad \text{for} \quad m = 0, \pm N_0, \pm 2N_0, \ldots$$

To compute the sum for other values of m, we note that the sum on the left-hand side of Eq. (5.43) is a geometric progression with common ratio $\alpha = e^{jm\Omega_0}$. Therefore, (see Sec. B.7-4)

$$\sum_{k=0}^{N_0-1} e^{jm\Omega_0 k} = \frac{e^{jm\Omega_0 N_0} - 1}{e^{jm\Omega_0} - 1} = 0 \qquad (e^{jm\Omega_0 N_0} = e^{j2\pi m} = 1)$$

---

†Actually, $\frac{N_0}{2}$ is a conservative figure because some multiplications corresponding to the cases where $W_{N_0}^r = 1, j$, etc., are eliminated.

5.5 Summary

A signal bandlimited to B Hz can be reconstructed exactly from its samples if the sampling rate $\mathcal{F}_s > 2B$ Hz (the sampling theorem). Such a reconstruction, although possible theoretically, poses practical problems such as the need for filters with zero gain over a band (or bands) of frequencies. Such filters are unrealizable, or realizable with infinite time delay. Therefore, in practice, there is always an error in reconstructing a signal from its samples. Moreover, practical signals are not bandlimited, which causes an additional error (aliasing error) in signal reconstruction from its samples. Aliasing error can be eliminated by bandlimiting a signal to its effective bandwidth.

The sampling theorem is very important in signal analysis, processing, and transmission because it allows us to replace a continuous-time signal with a discrete sequence of numbers. Processing a continuous-time signal is therefore equivalent to processing a discrete sequence of numbers. This leads us directly into the area of digital filtering (discrete-time systems). In the field of communication, the transmission of a continuous-time message reduces to the transmission of a sequence of numbers. This opens doors to many new techniques of communicating continuous-time signals by pulse trains.

The dual of the sampling theorem states that for a signal timelimited to τ seconds, its spectrum $F(\omega)$ can be reconstructed from the samples of $F(\omega)$ taken at uniform intervals not greater than $1/\tau$ Hz. In other words, the spectrum should be sampled at a rate not less than τ samples/Hz.

To compute the direct or inverse Fourier transform numerically, we need a relationship between the samples of $f(t)$ and $F(\omega)$. The sampling theorem and its dual provide such a quantitative relationship in the form of a discrete Fourier transform (DFT). The DFT computations are greatly facilitated by a fast Fourier transform (FFT) algorithm, which reduces the number of computations from something of the order of N_0^2 to $N_0 \log N_0$.

References

1. Linden, D.A., "A Discussion of Sampling Theorem," *Proc. IRE*, vol. 47, pp 1219-1226, July 1959.

2. Bracewell, R.N., *The Fourier Transform and Its Applications, 2nd revised ed.*, McGraw-Hill, New York, 1986.

3. Bennett, W.R., *Introduction to Signal Transmission*, McGraw-Hill, New York, 1970.

4. Lathi, B.P., *Linear Systems and Signals*, Berkeley-Cambridge Press, Carmichael, CA, 1992.

5. Cooley, J.W., and J.W., Tukey, "An Algorithm for the Machine Calculation of Complex Fourier Series," *Mathematics of Computation*, Vol. 19, pp. 297-301, April 1965.

Problems

5.1-1 Figure P5.1-1 shows Fourier spectra of signals $f_1(t)$ and $f_2(t)$. Determine the Nyquist sampling rates for signals $f_1(t)$, $f_2(t)$, $f_1^2(t)$, $f_2^3(t)$, and $f_1(t)f_2(t)$.

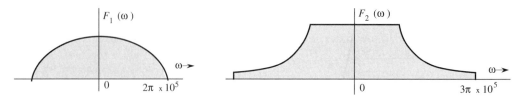

Fig. P5.1-1

5.1-2 Determine the Nyquist sampling rate and the Nyquist sampling interval for the signals **(a)** $\text{sinc}^2 (100\pi t)$ **(b)** $0.01 \text{sinc}^2 (100\pi t)$ **(c)** $\text{sinc} (100\pi t) + 3 \text{sinc}^2 (60\pi t)$ **(d)** $\text{sinc} (50\pi t)\text{sinc} (100\pi t)$.

5.1-3 A signal $f(t) = \text{sinc} (200\pi t)$ is sampled (using uniformly spaced impulses) at a rate of (a) 150 Hz (b) 200 Hz (c) 300 Hz. For each of the three cases (i) sketch the spectrum of the sampled signal, (ii) explain if you can recover the signal $f(t)$ from the sampled signal, (iii) if the sampled signal is passed through an ideal lowpass filter of bandwidth 100 Hz, sketch the spectrum of the output signal.

5.1-4 One realization of a practical zero-order hold circuit is presented in Fig. P5.1-4. (a)Find the unit impulse response of this circuit. Hint: Recall that the impulse response $h(t)$ is the output of the circuit in Fig. P5.1-4 when the input $f(t) = \delta(t)$. (b)Find the transfer function $H(\omega)$, and sketch $|H(\omega)|$. (c)Show that when a sampled signal $\overline{f}(t)$ is applied at the input of this circuit, the output is a staircase approximation of $f(t)$. The sampling interval is T.

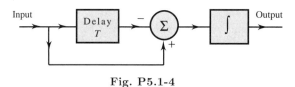

Fig. P5.1-4

5.1-5 **(a)** A first-order hold circuit can also be used to reconstruct a signal $f(t)$ from its samples. The impulse response of this circuit is

$$h(t) = \Delta \left(\frac{t}{2T} \right)$$

where T is the sampling interval. Consider a typical sampled signal $\overline{f}(t)$ and show that this circuit performs the linear interpolation. In other words, the filter output consists of sample tops connected by straight line segments. Follow the procedure discussed in Sec. 5.1-1 (Fig. 5.3b).
(b) Determine the transfer function of this filter, and its amplitude response, and compare it with the ideal filter required for signal reconstruction.
(c) This filter, being noncausal, is unrealizable. By delaying its impulse response, the filter can be made realizable. What is the minimum delay required to make it realizable? How would this delay affect the reconstructed signal and the filter frequency response?
(d) Show that the filter in part **(c)** can be realized by a filter depicted in Fig. P5.1-4 followed by an identical filter in cascade. Hint: show that the impulse response of this circuit is $\Delta(\frac{t}{2T})$ delayed by T seconds.

5.1-6 A signal $f(t) = \text{sinc} (200\pi t)$ is sampled by a periodic pulse train $p_T(t)$ resented in Fig. P5.1-6. Find and sketch the spectrum of the sampled signal. Explain if you

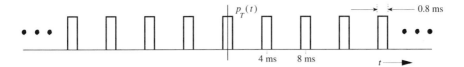

Fig. P5.1-6

will be able to reconstruct $f(t)$ from these samples. If the sampled signal is passed through an ideal lowpass filter of bandwidth 100 Hz and unit gain, find the filter output. What is the filter output if its bandwidth is B Hz, where $100 < B < 150$? What will happen if the bandwidth exceeds 150 Hz?

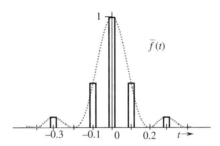

Fig. P5.1-7   Flat top sampling.

5.1-7   In Example 5.3, the sampling of a signal $f(t)$ was accomplished by multiplying the signal by a pulse train $p_T(t)$, resulting in the sampled signal depicted in Fig. 5.8d. This procedure is known as the **natural sampling**. Figure P5.1-7 shows the so called **flat top sampling** of the same signal $f(t) = \text{sinc}^2(5\pi t)$.
(a) Show that the signal $f(t)$ can be recovered from flat top samples if the sampling rate is no less than the Nyquist rate.
(b) Explain how you would recover $f(t)$ from the flat top samples.
(c) Find the expression for the sampled signal spectrum $\overline{F}(\omega)$ and sketch it roughly. Hint: First show that the flat top sampled signal can be generated by passing the signal $f(t)\delta_T(t)$ through a filter whose impulse response is $h(t) = p_T(t)$. For signal recovery from the samples, follow the reverse procedure.

5.1-8   A compact disc (CD) records audio signals digitally by using PCM. Assume the audio signal bandwidth to be 15 kHz.
(a) What is the Nyquist rate?
(b) If the Nyquist samples are quantized into 65536 levels ($L = 65536$) and then binary-coded, what number of binary digits is required to encode a sample.
(c) Determine the number of binary digits/s (bits/s) required to encode the audio signal.
(d) For practical reasons discussed in the text, signals are sampled at a rate well above the Nyquist rate. Practical CDs use 44100 samples/s. If $L = 65536$, determine the number of pulses/s required to encode the signal.

5.1-9   A TV signal (video and audio) has a bandwidth of 4.5 MHz. This signal is sampled, quantized and binary-coded to obtain a PCM (pulse code modulated) signal.
(a) Determine the sampling rate if the signal is to be sampled at a rate 20% above the Nyquist rate.
(b) If the samples are quantized into 1024 levels, what number of binary pulses is required to encode each sample.
(c) Determine the binary pulse rate (bits/s) of the binary coded signal.

5.1-10 Prove that a signal cannot be simultaneously timelimited and bandlimited.

Hint: Show that contrary assumption leads to contradiction. Assume a signal simultaneously timelimited and bandlimited so that $F(\omega) = 0$ for $|\omega| > 2\pi B$. In this case $F(\omega) = F(\omega)\text{rect}\left(\frac{\omega}{4\pi B'}\right)$ for $B' > B$. This fact means that $f(t)$ is equal to $f(t) * 2B'\text{sinc}\,(2\pi B't)$. The latter cannot be timelimited because the sinc function tail extends to infinity.

5.2-1 For a signal $f(t)$ that is timelimited to 10 ms and has an essential bandwidth of 10 kHz, determine N_0, the number of signal samples necessary to compute a power of 2-FFT with a frequency resolution $\mathcal{F}_o$ of at least 50 Hz. Explain if any zero padding is necessary.

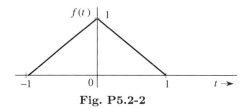

Fig. P5.2-2

5.2-2 To compute the DFT of signal $f(t)$ in Fig. P5.2-2, write the sequence f_k (for $k = 0$ to $N_0 - 1$) if the frequency resolution $\mathcal{F}_o$ must be at least 0.25 Hz. Assume the essential bandwidth (the folding frequency) of $f(t)$ to be at least 3 Hz. Do not compute the DFT; just write the appropriate sequence f_k.

5.2-3 Choose appropriate values for N_0 and T and compute the DFT of the signal $e^{-t}u(t)$. Use two different criteria for determining the effective bandwidth of $e^{-t}u(t)$. Use the bandwidth to be that frequency where the amplitude response drops to 1% of its peak value (at $\omega = 0$). Next, use the 99% energy criterion for determining the bandwidth (see Example 4.16).

5.2-4 Repeat Problem 5.2-3 for the signal

$$f(t) = \frac{2}{t^2 + 1}$$

Hint: $\frac{2}{t^2+1} \Longleftrightarrow 2\pi e^{-|\omega|}$

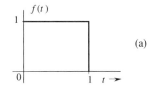
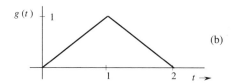

Fig. P5.2-5

5.2-5 For the signals $f(t)$ and $g(t)$ resented in Fig. P5.2-5, write the appropriate sequences f_k and g_k necessary for the computation of the convolution of $f(t)$ and $g(t)$ using DFT. Use $T = \frac{1}{8}$.

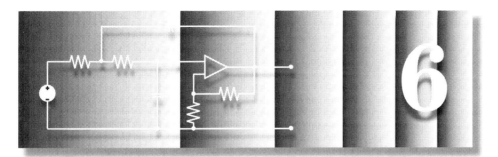

Continuous-Time System Analysis Using the Laplace Transform

The Fourier transform is a tool which allows us to represent a signal $f(t)$ as a continuous sum of exponentials of the form $e^{j\omega t}$, whose frequencies are restricted to the imaginary axis in the complex plane $(s = j\omega)$. As we saw in Chapters 4 and 5, such a representation is quite valuable in the analysis and processing of signals. In the area of system analysis, however, the use of Fourier transform leaves much to be desired. First, the Fourier transform exists only for a restricted class of signals and, therefore, cannot be used for such inputs as growing exponentials. Second, the Fourier transform cannot be used easily to analyze unstable or even marginally stable systems.

6.1 The Laplace Transform

The basic reason for both these difficulties is that for some signals, such as $e^{at}u(t)$ $(a > 0)$, the Fourier transform does not exist because ordinary sinusoids or exponentials of the form $e^{j\omega t}$ (on account of their constant amplitudes) are incapable of synthesizing exponentially growing signals. This problem could be resolved if it were possible to use basis signals of the form e^{st} (instead of $e^{j\omega t}$), where the complex frequency s is not restricted to just the imaginary axis (as in the Fourier transform). This is precisely what is done in the following extended transform known as the **bilateral Laplace transform**, where the frequency variable $s = j\omega$ is generalized to $s = \sigma + j\omega$. Such generalization permits us to use exponentially growing sinusoids to synthesize a signal $f(t)$. Before developing the mathematical operations required for such an extension, we will find it illuminating to have an intuitive understanding of such a generalization.

6.1-1 Intuitive Understanding of the Laplace Transform

If a signal $f(t)$ shown in Fig. 6.1d is not Fourier transformable, we may be able to make it Fourier transformable by multiplying it with a decaying exponential such as $e^{-\sigma t}$.†For example, a signal $e^{2t}u(t)$ can be made Fourier transformable by multiplying it with $e^{-\sigma t}$ with $\sigma > 2$. Let

---

†This assumes that $|f(t)| < M e^{at}$ for some real, positive, and finite numbers M and a. Such signals are called *exponential-order* signals. Signals which do not satisfy this condition may not be Laplace transformable.

as depicted in Fig. 6.1a. The signal $\phi(t)$ is now Fourier transformable and its Fourier components are of the form $e^{j\omega t}$ with frequencies ω varying from $\omega = -\infty$ to ∞. The exponential components $e^{j\omega t}$ and $e^{-j\omega t}$ in the spectrum add to give a sinusoid of frequency ω. The spectrum contains an infinite number of such sinusoids, each having an infinitesimal amplitude. It would be very confusing to draw all these components; hence, in Fig. 6.1b, we show just two typical components. Addition of all such components (infinite in number) results in $\phi(t)$, illustrated in Fig. 6.1a. The exponential spectral components of $\phi(t)$ are of the form $e^{j\omega t}$ with complex frequencies $j\omega$ lying on the imaginary axis from $\omega = -\infty$ to ∞, as shown in Fig. 6.1c.

Figure 6.1a shows a signal $\phi(t) = f(t)e^{-\sigma t}$, Fig. 6.1b shows two of its infinite spectral components, and Fig. 6.1c shows the location in the complex plane of the frequencies of all the spectral components of $\phi(t)$. Now, the desired signal $f(t)$ can be obtained by multiplying $\phi(t)$ with $e^{\sigma t}$. This fact means that we can synthesize $f(t)$ by multiplying each spectral component of $\phi(t)$ with $e^{\sigma t}$ and adding them. But multiplying the spectral components of $\phi(t)$ (sinusoids in Fig. 6.1b) with $e^{\sigma t}$ results in the exponentially growing sinusoids as shown in Fig. 6.1e. Addition of all such exponentially growing sinusoids (infinite in number) results in $f(t)$ in Fig. 6.1d. The spectral components of $\phi(t)$ are of the form $e^{j\omega t}$. Multiplication of these components with $e^{\sigma t}$ results in the spectral components of the form $e^{\sigma t}e^{j\omega t} = e^{(\sigma+j\omega)t}$. Therefore, a component of frequency $j\omega$ in the spectrum of $\phi(t)$ is transformed into a component of frequency $\sigma + j\omega$ in the spectrum of $f(t)$. The location in the complex plane of the frequencies $\sigma + j\omega$ is along a vertical line, depicted in Fig. 6.1f.

It is clear that a signal $f(t)$ can be synthesized with growing everlasting exponentials lying along the path $\sigma + j\omega$, with ω varying from $-\infty$ to ∞. The value of σ is flexible. For example, if $f(t) = e^{2t}u(t)$, then $\phi(t) = f(t)e^{-\sigma t}$ can be made Fourier transformable if we choose $\sigma > 2$. Thus, there are infinite choices for σ. This means the spectrum of $f(t)$ is not unique, and there are infinite possible ways of synthesizing $f(t)$. Nevertheless, σ has a certain minimum value σ_0 for a given $f(t)$ [$\sigma_0 = 2$ for $f(t) = e^{2t}u(t)$]. The region in the complex plane where $\sigma > \sigma_0$ (Fig. 6.1g) is called the **region of convergence** (or **existence**) of the resulting transform of $f(t)$.

All these conclusions reached here by pure heuristic reasoning will now be derived analytically. The frequency $j\omega$ in the Fourier transform will be generalized to $s = \sigma + j\omega$.

6.1-2 Analytical Development of the Bilateral Laplace Transform

Because we are unifying the Fourier and Laplace transforms, we need to use here the unified notation $F(j\omega)$ instead of $F(\omega)$ for the Fourier transform, which is given by

$$F(j\omega) = \int_{-\infty}^{\infty} f(t)e^{-j\omega t}\, dt \tag{6.1}$$

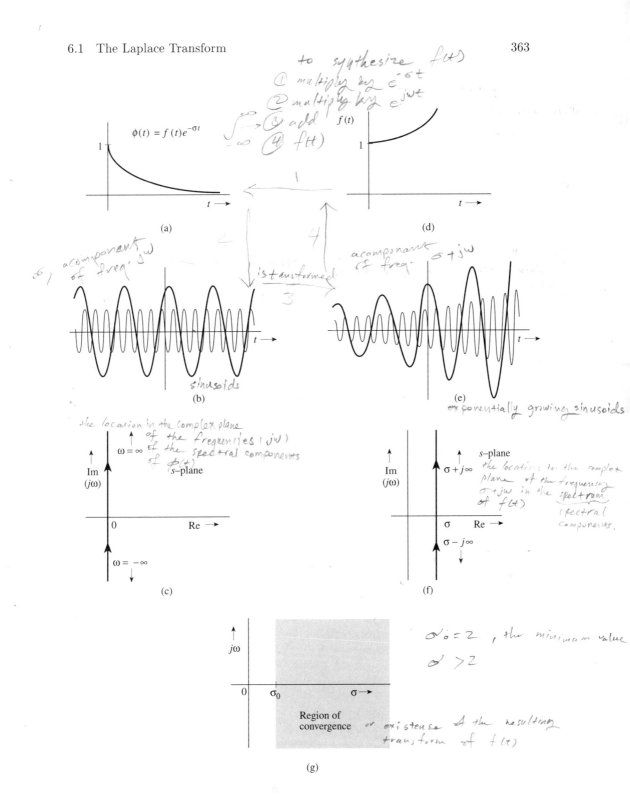

Fig. 6.1   Heuristic explanation of extension of the Fourier transform to Laplace transform.

and

$$2 \quad f(t) = \frac{1}{2\pi} \int_{-\infty}^{\infty} F(j\omega)e^{j\omega t} \, d\omega \tag{6.2}$$

φ(t)   *φ(t)* Fourier transformable.
f(t) is not in (6.1)

Consider now the Fourier transform of $f(t)e^{-\sigma t}$ (σ real)

$$\mathcal{F}[f(t)e^{-\sigma t}] = \int_{-\infty}^{\infty} f(t)e^{-\sigma t}e^{-j\omega t} \, dt \tag{6.3}$$

$$1' \quad F(s + j\omega) = \int_{-\infty}^{\infty} f(t)e^{-(\sigma+j\omega)t} \, dt \tag{6.4}$$

It follows from Eq. (6.1) that the above integral is $F(\sigma + j\omega)$. Thus

$$\mathcal{F}[f(t)e^{-\sigma t}] = F(\sigma + j\omega) \tag{6.5}$$

The inverse Fourier transform of the above equation yields

$$f(t)e^{-\sigma t} = \frac{1}{2\pi} \int_{-\infty}^{\infty} F(\sigma + j\omega)e^{j\omega t} \, d\omega \tag{6.6}$$

Multiplying both sides with $e^{\sigma t}$ yields

$$2' \quad f(t) = \frac{1}{2\pi} \int_{-\infty}^{\infty} F(\sigma + j\omega)e^{(\sigma+j\omega)t} \, d\omega \tag{6.7}$$

The quantity $(\sigma + j\omega)$ is the complex frequency s. We now change the variable in this integral from ω to s. Because $s = \sigma + j\omega$, $d\omega = (1/j) \, ds$. The limits of integration from $\omega = -\infty$ to ∞ now become from $(\sigma - j\infty)$ to $(\sigma + j\infty)$ for the variable s. Recall, however, that for a given $f(t)$, σ has a certain minimum value σ_0, and we can select any value of $\sigma > \sigma_0$. In general, we can let the limits of integration range from $s = c - j\infty$ to $s = c + j\infty$, where $c > \sigma_0$. Thus, Eq. (6.7) becomes

$$2'' \quad f(t) = \frac{1}{2\pi j} \int_{c-j\infty}^{c+j\infty} F(s)e^{st} \, ds \tag{6.8a}$$

$c > \sigma_0$
$c = \sigma \to \sigma_0$

From Eqs. (6.4) and (6.5), we obtain

$$1'' \quad F(s) = \int_{-\infty}^{\infty} f(t)e^{-st} \, dt \tag{6.8b}$$

This pair of equations is known as the **bilateral Laplace transform pair** (or the **two-sided Laplace transform pair**). The bilateral Laplace transform will be denoted symbolically as

$$F(s) = \mathcal{L}[f(t)] \quad \text{and} \quad f(t) = \mathcal{L}^{-1}[F(s)]$$

or simply as

$$f(t) \Longleftrightarrow F(s)$$

Response of LTIC Systems

Equation (6.8a) expresses $f(t)$ as a weighted sum of exponentials of the form e^{st}. This point becomes clear when we write the integral in Eq. (6.8a) as a sum

input →
$$f(t) = \frac{1}{2\pi j} \int_{c-j\infty}^{c+j\infty} F(s)e^{st}\, ds$$

$$\equiv \lim_{\Delta s \to 0} \sum_{n=-\infty}^{\infty} \left[\frac{F(n\Delta s)\Delta s}{2\pi j} \right] e^{(n\Delta s)t} \tag{6.9}$$

Clearly, the Laplace transform expresses $f(t)$ as a sum of everlasting exponentials of the form $e^{(n\Delta s)t}$ along the path $c - j\infty$ to $c + j\infty$ with $c > \sigma_0$. These are exponentially growing (if $c > 0$) or decaying (if $c < 0$) sinusoids. We can readily determine an LTIC system response to the input $f(t)$ by observing that if the system transfer function is $H(s)$, then, as demonstrated in Eq. (2.47), the system response to the (everlasting) exponential $e^{(n\Delta s)t}$ is $H(n\Delta s)e^{(n\Delta s)t}$. From Eq. (6.9), it follows that the system response to the input $f(t)$ is

response
$$y(t) = \lim_{\Delta s \to 0} \sum_{n=-\infty}^{\infty} \left[\frac{F(n\Delta s)H(n\Delta s)\Delta s}{2\pi j} \right] e^{(n\Delta s)t}$$

$$= \frac{1}{2\pi j} \int_{c'-j\infty}^{c'+j\infty} F(s)H(s)e^{st}\, ds \tag{6.10}$$

The path of integration (from $c' - j\infty$ to $c' + j\infty$) in Eq. (6.10) may be different from that in Eq. (6.9) to allow for the convergence of the integral in Eq. (6.10). This subtle point will be discussed later in Sec. 6.8-1. If $y(t) \Longleftrightarrow Y(s)$, then according to Eq. (6.10) it follows that

$$Y(s) = F(s)H(s) \tag{6.11}$$

Here, we have expressed the input $f(t)$ as a sum of exponentials components of the form e^{st}. The system response is then obtained by adding the responses to all these exponential components. The procedure followed here is identical to that in chapter 2 (where the input is expressed as a sum of impulses) or chapter 4 (where the input is expressed as a sum of exponentials of the form $e^{j\omega t}$).

In conclusion, we have shown that for an LTIC system with transfer function $H(s)$, if the input and the output are $f(t)$ and $y(t)$, respectively, and if

$$f(t) \Longleftrightarrow F(s), \qquad y(t) \Longleftrightarrow Y(s)$$

then

$$Y(s) = F(s)H(s) \tag{6.12}$$

We shall derive this result more formally later.

Linearity of the Laplace Transform

The Laplace transform is a linear operator, and the principle of superposition applies; that is, if

$$f_1(t) \Longleftrightarrow F_1(s) \quad \text{and} \quad f_2(t) \Longleftrightarrow F_2(s)$$

then

$$a_1 f_1(t) + a_2 f_2(t) \Longleftrightarrow a_1 F_1(s) + a_2 F_2(s) \tag{6.13}$$

The proof is trivial and follows directly from the definition of the Laplace transform. This result can be extended to any finite number of terms.

The Region of Convergence

Earlier, we discussed the intuitive meaning of the region of convergence (or region of existence) of the Laplace transform $F(s)$. Mathematically, the region of convergence for $F(s)$ is the set of values of s (the region in the complex plane) for which the integral in Eq. (6.8b) defining the direct Laplace transform $F(s)$ converges. This concept will become clear in the following example.

■ **Example 6.1** ~~table~~

For the signal $f(t) = e^{-at}u(t)$, find the Laplace transform $F(s)$ and its region of convergence.

By definition   *from table · pair #5*

$$F(s) = \int_{-\infty}^{\infty} e^{-at}u(t)e^{-st}\, dt$$

Because $u(t) = 0$ for $t < 0$ and $u(t) = 1$ for $t \geq 0$,

$$F(s) = \int_{0}^{\infty} e^{-at}e^{-st}\, dt = \int_{0}^{\infty} e^{-(s+a)t}\, dt = -\left.\frac{1}{s+a}e^{-(s+a)t}\right|_{0}^{\infty} \tag{6.14}$$

Note that s is complex and as $t \to \infty$, the term $e^{-(s+a)t}$ does not necessarily vanish. Here we recall that for a complex number $z = \alpha + j\beta$,

$$e^{-zt} = e^{-(\alpha+j\beta)t} = e^{-\alpha t}e^{-j\beta t}$$

Now $|e^{-j\beta t}| = 1$ regardless of the value of βt. Therefore, as $t \to \infty$, $e^{-zt} \to 0$ only if $\alpha > 0$, and $e^{-zt} \to \infty$ if $\alpha < 0$. Thus

$$\lim_{t\to\infty} e^{-zt} = \begin{cases} 0 & \text{Re } z > 0 \\ \infty & \text{Re } z < 0 \end{cases} \tag{6.15}$$

Clearly

$$\lim_{t\to\infty} e^{-(s+a)t} = \begin{cases} 0 & \text{Re}\,(s+a) > 0 \\ \infty & \text{Re}\,(s+a) < 0 \end{cases}$$

Use of this result in Eq. (6.14) yields

$$F(s) = \frac{1}{s+a} \qquad \text{Re}\,(s+a) > 0 \tag{6.16a}$$

or

$$e^{-at}u(t) \Longleftrightarrow \frac{1}{s+a} \qquad \text{Re }s > -a \tag{6.16b}$$

The region of convergence of $F(s)$ is Re $s > -a$, as shown in the shaded area in Fig. 6.2a. This fact means that the integral defining $F(s)$ in Eq. (6.14) exists only for the values of s in the shaded region in Fig. 6.2a. For other values of s, the integral in Eq. (6.14) does not converge. For this reason the shaded region is called the *region of convergence* (or the *region of existence*) for $F(s)$.

Recall that the Fourier transform of $e^{-at}u(t)$ does not exist for negative values of a. In contrast, the Laplace transform exists for all values of a, and its region of convergence is to the right of the line Re $s = -a$.   ■

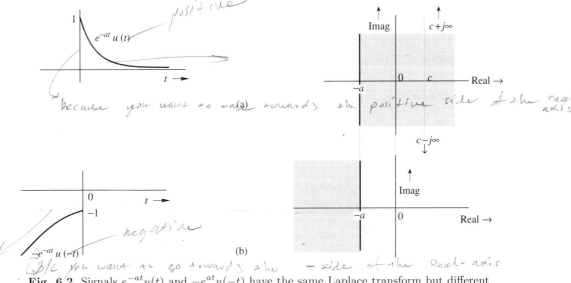

positive

$e^{-at} u(t)$

because you want to ma(a) towards the positive side of the real axis

negative

$-e^{-at} u(-t)$

(b)

(b/c you want to go towards the – side of the Real-axis

Fig. 6.2 Signals $e^{-at}u(t)$ and $-e^{at}u(-t)$ have the same Laplace transform but different regions of convergence.

Role of the Region of Convergence

The region of convergence is required for evaluating the inverse Laplace transform $f(t)$ from $F(s)$ as defined by Eq. (6.8a). The operation of finding the inverse transform requires an integration in the complex plane, which needs some explanation. The path of integration is along $c + j\omega$, with ω varying from $-\infty$ to ∞. Moreover, the path of integration must lie in the region of convergence (or existence) for $F(s)$. For the signal $e^{-at}u(t)$, this is possible if $c > -a$. One possible path of integration is shown (dotted) in Fig. 6.2a. Thus, to obtain $f(t)$ from $F(s)$, the integration in (6.8a) is performed along this path. When we integrate $[1/(s+a)]e^{st}$ along this path, the result is $e^{-at}u(t)$. Such integration in the complex plane requires a background in the theory of functions of complex variables. We can avoid this integration by compiling a table of Laplace transforms (Table 6.1), where the Laplace transform pairs are tabulated for a variety of signals. To find the inverse Laplace transform of say, $1/(s+a)$, instead of using the complex integral (6.8a), we look up the table and find the inverse Laplace transform to be $e^{-at}u(t)$. Although the table given here is rather short, it comprises the functions of most practical interest. A more comprehensive table appears in Doetsch.[1]

The Unilateral Laplace Transform   *causal signals only. $t = 0 \to \infty$*

In order to understand the need for defining a unilateral transform, let us find the Laplace transform of signal $f(t)$ illustrated in Fig. 6.2b

$$f(t) = -e^{-at}u(-t)$$

$(+)$

The Laplace transform of this signal is

$$F(s) = \int_{-\infty}^{\infty} -e^{-at}u(-t)e^{-st}\, dt$$

$s < -a$

$$= \frac{1}{s-\lambda} = \frac{1}{s+a}$$

notice for $f(t) = -e^{-at}u(-t)$ region of convergence $s < -a$

for $f(t) = e^{-at}u(t)$

$s > -a$

Because $u(-t) = 1$ for $t < 0$ and $u(-t) = 0$ for $t > 0$,

$$F(s) = \int_{-\infty}^{0} -e^{-at}e^{-st}\, dt = -\int_{-\infty}^{0} e^{-(s+a)t}\, dt = \frac{1}{s+a} e^{-(s+a)t} \bigg|_{-\infty}^{0}$$

Equation (6.15) shows that

$$\lim_{t \to -\infty} e^{-(s+a)t} = 0 \qquad \text{Re}\,(s+a) < 0$$

Hence

$$F(s) = \frac{1}{s+a} \qquad \text{Re}\, s < -a \tag{6.17}$$

The signal $-e^{-at}u(-t)$ and its region of convergence (Re $s < -a$) are depicted in Fig. 6.2b. Note that the Laplace transforms for the signals $e^{-at}u(t)$ and $-e^{-at}u(-t)$ are identical except for their regions of convergence. Therefore, for a given $F(s)$, there may be more than one inverse transform, depending on the region of convergence. In other words, there is no one-to-one correspondence between $F(s)$ and $f(t)$, unless the region of convergence is specified. This fact increases the complexity in using the Laplace transform. The complexity is the result of trying to handle causal as well as noncausal signals. If we restrict all our signals to the causal type, such an ambiguity does not arise. There is only one inverse transform of $F(s) = 1/(s+a)$, namely, $e^{-at}u(t)$. To find $f(t)$ from $F(s)$, we need not even specify the region of convergence. In summary, if all signals are restricted to the causal type, then, for a given $F(s)$, there is only one inverse transform $f(t)$.†

The unilateral Laplace transform is a special case of the bilateral Laplace transform, where all signals are restricted to being causal; consequently the limits of integration for the integral in Eq. (6.8b) can be taken from 0 to ∞. Therefore, the unilateral Laplace transform $F(s)$ of a signal $f(t)$ is defined as

$$F(s) \equiv \int_{0^-}^{\infty} f(t)e^{-st}\, dt \tag{6.18}$$

We choose 0^- (rather than 0^+ used in some texts) as the lower limit of integration. This convention not only ensures inclusion of an impulse function at $t = 0$, but also allows us to use initial conditions at 0^- (rather than at 0^+) in the solution of differential equations via the Laplace transform. In practice, we are likely to know the initial conditions before the input is applied (at 0^-), not after the input is applied (at 0^+). Other advantages of this convention appear on p. 392.

The unilateral Laplace transform simplifies the system analysis problem considerably, but the price for this simplification is that we cannot analyze noncausal systems or use noncausal inputs. However, in most practical problems this is of little consequence. For this reason, we shall first consider the unilateral Laplace transform and its application to system analysis. (The bilateral Laplace transform is discussed later in Sec. 6.8.)

Observe that basically there is no difference between the unilateral and the bilateral Laplace transform. The unilateral transform is the bilateral transform that deals with a subclass of signals starting at $t = 0$ (causal signals). Therefore, the expression [(Eq. (6.8a)] for the inverse Laplace transform remains unchanged. In practice, the term *Laplace transform* means *the unilateral Laplace transform*.

---

†Actually, $F(s)$ specifies $f(t)$ within a null function $n(t)$ which has the property that the area under $|n(t)|^2$ is zero over any finite interval 0 to t ($t > 0$) (Lerch's theorem). For example, if two functions are identical everywhere except at points of discontinuity, they differ by a null function.

Existence of the Laplace Transform

The variable s in the Laplace transform is complex in general, and it can be expressed as $s = \sigma + j\omega$. By definition

$$F(s) = \int_{0^-}^{\infty} f(t)e^{-st}\, dt$$

$$= \int_{0^-}^{\infty} \left[f(t)e^{-\sigma t}\right] e^{-j\omega t}\, dt$$

Because $|e^{j\omega t}| = 1$, the integral on the right-hand side of this equation converges if

$$\int_{0^-}^{\infty} \left|f(t)e^{-\sigma t}\right|\, dt < \infty \tag{6.19}$$

Hence the existence of the Laplace transform is guaranteed if the integral in (6.19) is finite for some value of σ. Any signal that grows no faster than an exponential signal $Me^{\sigma_0 t}$ for some M and σ_0 satisfies the condition (6.19). Thus, if for some M and σ_0,

$$|f(t)| \le M e^{\sigma_0 t} \tag{6.20}$$

we can choose $\sigma > \sigma_0$ to satisfy (6.19).† The signal e^{t^2}, in contrast, grows at a rate faster than $e^{\sigma_0 t}$, and consequently e^{t^2} is not Laplace transformable. Fortunately such signals (which are not Laplace transformable) are of little consequence from either a practical or a theoretical viewpoint. If σ_0 is the smallest value of σ for which the integral in (6.19) is finite, σ_0 is called the **abscissa** of **convergence** and the region of convergence of $F(s)$ is Re $s > \sigma_0$. The abscissa of convergence for $e^{-at}u(t)$ is $-a$ (the region of convergence is Re $s > -a$).

■ **Example 6.2**
Determine the Laplace transform of (a) $\delta(t)$ (b) $u(t)$ (c) $\cos \omega_0 t\, u(t)$.

(a)

$$\mathcal{L}\left[\delta(t)\right] = \int_{0^-}^{\infty} \delta(t)e^{-st}\, dt$$

Using the sampling property [Eq. (1.24a)], we obtain

$$\mathcal{L}\left[\delta(t)\right] = 1 \qquad \text{for all } s$$

that is

$$\delta(t) \Longleftrightarrow 1 \qquad \text{for all } s \tag{6.21}$$

(b) To find the Laplace transform of $u(t)$, recall that $u(t) = 1$ for $t \ge 0$. Therefore

$$\mathcal{L}\left[u(t)\right] = \int_{0^-}^{\infty} u(t)e^{-st}\, dt = \int_{0^-}^{\infty} e^{-st}\, dt = -\frac{1}{s}e^{-st}\Big|_{0^-}^{\infty}$$

$$\mathcal{L}\left[u(t)\right] = \frac{1}{s} \qquad \text{Re } s > 0 \tag{6.22}$$

---

†Condition (6.20) is sufficient but not necessary for the existence of the Laplace transform. For example $f(t) = 1/\sqrt{t}$ is infinite at $t = 0$ and, (6.20) cannot be satisfied, but the transform of $1/\sqrt{t}$ exists and is given by $\sqrt{\pi/s}$.

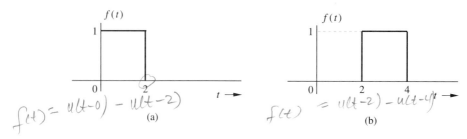

(handwritten below left figure) $f(t) = u(t-0) - u(t-2)$

(handwritten below right figure) $f(t) = u(t-2) - u(t-4)$

Fig. 6.3 Signals for Exercise E6.1.

We also could have obtained this result from (6.16b) by letting $a = 0$.

(c) Because

$$\cos \omega_0 t \, u(t) = \tfrac{1}{2}\left[e^{j\omega_0 t} + e^{-j\omega_0 t}\right] u(t) \tag{6.23}$$

$$\mathcal{L}\left[\cos \omega_0 t \, u(t)\right] = \tfrac{1}{2}\mathcal{L}\left[e^{j\omega_0 t} u(t) + e^{-j\omega_0 t} u(t)\right]$$

From Eq. (6.16), it follows that

$$\mathcal{L}\left[\cos \omega_0 t \, u(t)\right] = \frac{1}{2}\left[\frac{1}{s - j\omega_0} + \frac{1}{s + j\omega_0}\right] \qquad \mathrm{Re}\,(s \pm j\omega) = \mathrm{Re}\,s > 0$$

(handwritten left margin) From table! $\mathcal{L}[\cos \omega_0 t \, u(t)] =$

$$\mathcal{L}\left[\cos \omega_0 t \, u(t)\right] = \frac{s}{s^2 + \omega_0^2} \qquad \boxed{\mathrm{Re}\,s > 0} \tag{6.24}$$

For the unilateral Laplace transform, there is a unique inverse transform of $F(s)$; consequently, there is no need to specify the region of convergence explicitly. For this reason, we shall generally ignore any mention of the region of convergence for unilateral transforms. Recall, also, that in the unilateral Laplace transform it is understood that every signal $f(t)$ is zero for $t < 0$, and it is appropriate to indicate this fact by multiplying the signal by $u(t)$.

△ **Exercise E6.1**

By direct integration, find the Laplace transform $F(s)$ and the region of convergence of $F(s)$ for the signals shown in Fig. 6.3.

Answer: **(a)** $\frac{1}{s}(1 - e^{-2s})$ for all s.   **(b)** $\frac{1}{s}(1 - e^{-2s})e^{-2s}$ for all s.   ▽   *Time shift*

(handwritten) $f(t) = \frac{u(t) - u(t-2)}{= u(t-0)}$ $\frac{1}{s}(e^{-2s} - e^{-4s})$ *property.*

Connection to the Fourier Transform *(handwritten)* $f(t) = u(t-2) - u(t-4)$

The definition of the Laplace transform is identical to that of the Fourier transform with $j\omega$ replaced by s. It is reasonable to expect $F(s)$, the Laplace transform of $f(t)$, to be the same as $F(\omega)$, the Fourier transform of $f(t)$ with $j\omega$ replaced by s. For example, we found the Fourier transform of $e^{-at}u(t)$ to be $1/(j\omega + a)$. Replacing $j\omega$ with s in the Fourier transform results in $1/(s + a)$, which is the Laplace transform as seen from Eq. (6.16b). Unfortunately this procedure is not valid for all $f(t)$. We may use it only if the region of convergence for $F(s)$ includes the imaginary $(j\omega)$ axis. For instance, the Fourier transform of the unit step function is $\pi\delta(\omega) + (1/j\omega)$. The corresponding Laplace transform is $1/s$, and its region of convergence, which is $\mathrm{Re}\,s > 0$, does not include the imaginary axis. In this case the connection between the Fourier and Laplace transforms is not so simple. The reason for this complication is related to the convergence of the Fourier integral, where the path of integration is restricted to the imaginary axis. Because of this restriction, the Fourier integral for the step function does not converge in the ordinary sense as Example 4.7 demonstrates. We had to use a generalized function (impulse) for convergence. The Laplace integral for $u(t)$, in contrast, converges in the ordinary sense, but only for Re $s > 0$, a region forbidden to the Fourier transform. Another interesting fact is that although the Laplace transform is a generalization of the Fourier transform, there are signals (e.g., periodic signals) for which the Laplace transform does not exist, although the Fourier transform exists (but not in the ordinary sense).

6.1-3   Finding the Inverse Transform

Finding the inverse Laplace transform by using the definition (6.8a) requires integration in the complex plane, a subject beyond the scope of this book.[2] For our purpose, we can find the inverse transforms from the transform table 6.1. All we need is to express $F(s)$ as a sum of simpler functions of the form listed in the table. Most of the transforms $F(s)$ of practical interest are **rational functions**; that is, ratios of polynomials in s. Such functions can be expressed as a sum of simpler functions by using partial fraction expansion (see Sec. B.5). Values of s for which $F(s) = 0$ are called the **zeros** of $F(s)$; the values of s for which $F(s) \to \infty$ are called the **poles** of $F(s)$. If $F(s)$ is a rational function of the form $P(s)/Q(s)$, the roots of $P(s)$ are the zeros and the roots of $Q(s)$ are the poles of $F(s)$.

■ **Example 6.3**

Find the inverse Laplace transforms of

(a) $\dfrac{7s - 6}{s^2 - s - 6}$ (b) $\dfrac{2s^2 + 5}{s^2 + 3s + 2}$ (c) $\dfrac{6(s + 34)}{s(s^2 + 10s + 34)}$ (d) $\dfrac{8s + 10}{(s + 1)(s + 2)^3}$

The inverse transform of none of the above functions is directly available in Table 6.1. We need to expand these functions into partial fractions discussed in Sec. B.5.

(a)

$$F(s) = \frac{7s - 6}{(s + 2)(s - 3)}$$

$$= \frac{k_1}{s + 2} + \frac{k_2}{s - 3}$$

To determine k_1, corresponding to the term $(s + 2)$, we cover up (conceal) the term $(s + 2)$ in $F(s)$ and substitute $s = -2$ (the value of s that makes $s + 2 = 0$) in the remaining expression (see Sec. B.5-2)

$$k_1 = \frac{7s - 6}{(s + 2)(s - 3)}\Bigg|_{s = -2} = \frac{-14 - 6}{-2 - 3} = 4$$

Similarly, to determine k_2 corresponding to the term $(s - 3)$, we cover up the term $(s - 3)$ in $F(s)$ and substitute $s = 3$ in the remaining expression

$$k_2 = \frac{7s - 6}{(s + 2)(s - 3)}\Bigg|_{s = 3} = \frac{21 - 6}{3 + 2} = 3$$

Therefore

$$F(s) = \frac{7s - 6}{(s + 2)(s - 3)} = \frac{4}{s + 2} + \frac{3}{s - 3} \tag{6.25a}$$

Table 6.1

A Short Table of (Unilateral) Laplace Transforms

$f(t)$	$F(s)$

1. $\times \delta(t) \;\overset{\circ}{\circ}\; \times \delta(t-0) \xleftarrow{\text{differentiation property}} \longrightarrow \times 1\, e^{-0s}$

2. $\times u(t) \overset{\circ}{\circ} \times u(t-0)$ $\qquad\qquad \times \dfrac{1}{s} e^{-0s}$

2a. $u(t) - u(t-2) \xleftarrow{\text{time shift}}$ $\qquad \dfrac{1}{s}(1 - e^{-2s}) = \dfrac{1}{s} - \dfrac{1}{s} e^{-2s}$

3. $tu(t) \overset{\circ}{\circ} (t-0)\,u(t-0)$ $\qquad \dfrac{1}{s^2} e^{-0s}$

due to freq-shift property

$f(t)\,e^{s_0 t} \Longleftrightarrow F(s-s_0)$

$s_0 = +\lambda$

2b. $u(t-2) - u(t-4) \xleftarrow{\text{time shift}}$ $\qquad \dfrac{1}{s}(e^{-2s} - e^{-4s})$

4. $t^n u(t)$ $\qquad\qquad \dfrac{n!}{s^{n+1}}$

5. $1\, e^{\lambda t} u(t)$ $\qquad\qquad \dfrac{1}{s - \lambda}$

6. $1\, t e^{\lambda t} u(t) \qquad\qquad\longrightarrow \qquad \dfrac{1}{(s-\lambda)^2}$

7. $t^n e^{\lambda t} u(t)$ $\qquad\qquad \dfrac{n!}{(s-\lambda)^{n+1}}$

8a. $\cos bt\, u(t)$ $\qquad\qquad \dfrac{s}{s^2 + b^2}$

due to freq-shift property

$f(t)\,e^{s_0 t} \Longleftrightarrow F(s-s_0)$

$s_0 = -a$

8b. $\sin bt\, u(t)$ $\qquad\qquad \dfrac{b}{s^2 + b^2}$

9a. $e^{-at} \cos bt\, u(t) \qquad\longrightarrow \qquad \dfrac{s + a}{(s+a)^2 + b^2}$

9b. $e^{-at} \sin bt\, u(t)$ $\qquad\qquad \dfrac{b}{(s+a)^2 + b^2}$

10a. $r e^{-at} \cos(bt + \theta)\, u(t)$ $\qquad \dfrac{(r\cos\theta)s + (ar\cos\theta - br\sin\theta)}{s^2 + 2as + (a^2 + b^2)}$

10b. $r e^{-at} \cos(bt + \theta)\, u(t)$ $\qquad \dfrac{0.5 r e^{j\theta}}{s + a - jb} + \dfrac{0.5 r e^{-j\theta}}{s + a + jb}$

10c. $r e^{-at} \cos(bt + \theta)\, u(t)$ $\qquad \dfrac{As + B}{s^2 + 2as + c}$

$r = \sqrt{\dfrac{A^2 c + B^2 - 2ABa}{c - a^2}},\quad \theta = \tan^{-1} \dfrac{Aa - B}{A\sqrt{c - a^2}}$

$b = \sqrt{c - a^2}$

10d. $e^{-at}\left[A\cos bt + \dfrac{B - Aa}{b}\sin bt\right] u(t)$ $\qquad \dfrac{As + B}{s^2 + 2as + c}$

$b = \sqrt{c - a^2}$

Checking the answer

It is easy to make a mistake in partial fraction computations. Fortunately it is simple to check the answer by recognizing that $F(s)$ and its partial fractions must be equal for every value of s if the partial fractions are correct. Let us verify this assertion in Eq. (6.25a) for some convenient value, say $s = 1$. Substitution of $s = 1$ in Eq. (6.25a) yields†

$$-\frac{1}{6} = \frac{4}{3} - \frac{3}{2} = -\frac{1}{6}$$

We can now be sure of our answer with a high margin of confidence. Using Pair 5 (Table 6.1) in Eq. (6.25a), we obtain

$$f(t) = \mathcal{L}^{-1}\left(\frac{4}{s+2} + \frac{3}{s-3}\right) = \left(4e^{-2t} + 3e^{3t}\right)u(t) \qquad (6.25b)$$

(b)

$$F(s) = \frac{2s^2 + 5}{s^2 + 3s + 2} = \frac{2s^2 + 5}{(s+1)(s+2)}$$

Observe that $F(s)$ is an improper function with $\boxed{m = n.}$ In such a case we can express $F(s)$ as a sum of the coefficient b_n (the coefficient of the highest power in the numerator) plus partial fractions corresponding to the poles of $F(s)$ (see Sec. B.5-4). In the present case $b_n = 2$. Therefore

$$F(s) = 2 + \frac{k_1}{s+1} + \frac{k_2}{s+2}$$

where

$$k_1 = \frac{2s^2 + 5}{(s+1)(s+2)}\bigg|_{s=-1} = \frac{2+5}{-1+2} = 7$$

and

$$k_2 = \frac{2s^2 + 5}{(s+1)(s+2)}\bigg|_{s=-2} = \frac{8+5}{-2+1} = -13$$

Therefore

$$F(s) = 2 + \frac{7}{s+1} - \frac{13}{s+2}$$

From Table 6.1, Pairs 1 and 5, we obtain

$$f(t) = 2\delta(t) + \left(7e^{-t} - 13e^{-2t}\right)u(t) \qquad (6.26)$$

(c)

$$F(s) = \frac{6(s+34)}{s(s^2 + 10s + 34)}$$

$$= \frac{6(s+34)}{s(s+5-j3)(s+5+j3)}$$

$$= \frac{k_1}{s} + \frac{k_2}{s+5-j3} + \frac{k_2^*}{s+5+j3}$$

†Because $F(s) = \infty$ at its poles, we should avoid the pole values (-2 and 3 in the present case) for checking. It is possible that the answers may check even if partial fractions are wrong. This situation can occur when two or more errors cancel their effects. But the chances of this problem arising for randomly selected values of s are extremely small.

Note that the coefficients (k_2 and $k_2{}^*$) of the conjugate terms must also be conjugate (see Sec. B.5). Now

$$k_1 = \left.\frac{6(s+34)}{s(s^2 + 10s + 34)}\right|_{s=0} = \frac{6 \times 34}{34} = 6$$

$$k_2 = \left.\frac{6(s+34)}{s(s+5-j3)(s+5+j3)}\right|_{s=-5+j3} = \frac{29+j3}{-3-j5} = -3+j4 \quad \overset{-102+136j}{}$$

Therefore $= \dfrac{6(s+34)}{3[(-5+j3)+5+j3]} = \dfrac{6[(-5+j3)+34]}{(-5+j3)(j6)} = \dfrac{29+j3}{-j5+j^2 3} = \dfrac{29+j3}{-3-j5} = \dfrac{(29+j3)(-3+j5)}{(-3-j5)(-3+j5)} =$

$$k_2{}^* = -3-j4 \qquad \qquad \overline{34}$$

To use Pair 10b (Table 6.1), we need to express k_2 and $k_2{}^*$ in polar form.

$$-3 + j4 = \left(\sqrt{3^2 + 4^2}\right) e^{j\tan^{-1}(4/-3)} = 5e^{j\tan^{-1}(4/-3)}$$

Observe that $\tan^{-1}\left(\frac{4}{-3}\right) \neq \tan^{-1}\left(-\frac{4}{3}\right)$. This fact is evident in Fig. 6.4. Remember that electronic calculators can give answers only for the angles in the first and the fourth quadrant. For this reason it is important to plot the point (e.g. $-3+j4$) in the complex plane, as depicted in Fig. 6.4, and determine the angle. For further discussion of this topic, see Example B.1.

From Fig. 6.4, we observe that

$$k_2 = -3 + j4 = 5e^{j126.9°}$$

so that

$$k_2{}^* = 5e^{-j126.9°}$$

Therefore

$$\boxed{0.5\,r}$$

$$F(s) = \frac{6}{s} + \frac{5e^{j126.9°}}{s+5-j3} + \frac{5e^{-j126.9°}}{s+5+j3} \qquad \qquad 10b$$

From Table 6.1 (Pairs 2 and 10b), we obtain

$$\boxed{r} \qquad f(t) = \left[6 + 10e^{-5t}\cos(3t + 126.9°)\right] u(t) \qquad (6.27)$$

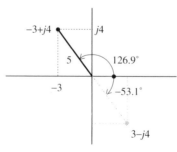

Fig. 6.4 $\tan^{-1}\left(\frac{-4}{3}\right) \neq \tan^{-1}\left(\frac{4}{-3}\right)$.

Alternative Method Using Quadratic Factors

[handwritten: Put in $F(s)$ of 10c]

The above procedure involves considerable manipulation of complex numbers. As indicated by Pair 10c (Table 6.1), the inverse transform of quadratic terms (with complex conjugate poles) can be found directly without having to find first-order partial fractions. We now show the alternative method (also discussed in Sec. B.5-2), which uses this observation. For this purpose we shall express $F(s)$ as

$$F(s) = \frac{6(s+34)}{s(s^2 + 10s + 34)} = \frac{k_1}{s} + \frac{As + B}{s^2 + 10s + 34}$$

We have already determined that $k_1 = 6$ by the (Heaviside) "cover-up" method. Therefore

$$\frac{6(s+34)}{s(s^2 + 10s + 34)} = \frac{6}{s} + \frac{As + B}{s^2 + 10s + 34}$$

Clearing the fractions by multiplying both sides by $s(s^2 + 10s + 34)$ yields

$$6(s+34) = 6\left(s^2 + 10s + 34\right) + s(As + B) = \underline{6s^2} + 60s + 204 + \underline{As^2} + Bs \,.$$

[handwritten: $6s^2 + 6s + 204$]

$$= (6 + A)s^2 + (60 + B)s + 204$$

[handwritten: 0 under (6+A), 6 under (60+B)]

Now, equating the coefficients of s^2 and s on both sides yields

$$0 = (6 + A) \implies A = -6$$

$$6 = 60 + B \implies B = -54$$

and

$$F(s) = \frac{6}{s} + \frac{-6s - 54}{s^2 + 10s + 34} \qquad \text{10c}$$

We now use Pairs 2 and 10c to find the inverse Laplace transform. The parameters for Pair 10c are $A = -6$, $B = -54$, $a = 5$, $c = 34$, and $b = \sqrt{c - a^2} = 3$, and   *[handwritten: 2(5)s]*

$$r = \sqrt{\frac{A^2 c + B^2 - 2ABa}{c - a^2}} = 10, \qquad \theta = \tan^{-1} \frac{Aa - B}{A\sqrt{c - a^2}} = 126.9°$$

Therefore

$$f(t) = [6 + 10e^{-5t} \cos{(3t + 126.9°)}]u(t)$$

which agrees with the previous result.

Short-Cuts

The partial fractions with quadratic terms also can be obtained by using short cuts. We have

$$F(s) = \frac{6(s+34)}{s(s^2 + 10s + 34)} = \frac{6}{s} + \frac{As + B}{s^2 + 10s + 34}$$

We can determine A by eliminating B on the right-hand side. This step can be accomplished by multiplying both sides of the above equation by s and then letting $s \to \infty$. This procedure yields

$$0 = 6 + A \implies A = -6$$

Therefore

$$\frac{6(s+34)}{s(s^2 + 10s + 34)} = \frac{6}{s} + \frac{-6s + B}{s^2 + 10s + 34}$$

To find B, we let s take on any convenient value, say $s = 1$, in this equation to obtain

$$\frac{210}{45} = 6 + \frac{B-6}{45}$$

Multiplying both sides of this equation by 45 yields

$$210 = 270 + B - 6 \implies B = -54$$

a deduction which agrees with the results we found earlier.

(d)

$$F(s) = \frac{8s+10}{(s+1)(s+2)^3}$$

$$= \frac{k_1}{s+1} + \frac{a_0}{(s+2)^3} + \frac{a_1}{(s+2)^2} + \frac{a_2}{s+2}$$

where (see Sec. B.5-2)

$$k_1 = \left. \frac{8s+10}{(s+1)(s+2)^3} \right|_{s=-1} = 2$$

$$a_0 = \left. \frac{8s+10}{(s+1)(s+2)^3} \right|_{s=-2} = 6$$

Quotient rule :
is very lengthy
Continue with the
"Alternative Method"

$$a_1 = \left\{ \frac{d}{ds}\left[\frac{8s+10}{(s+1)(s+2)^3} \right] \right\}_{s=-2} = -2$$

$$a_2 = \frac{1}{2}\left\{ \frac{d^2}{ds^2}\left[\frac{8s+10}{(s+1)(s+2)^3} \right] \right\}_{s=-2} = -2$$

Therefore

$$F(s) = \frac{2}{s+1} + \frac{6}{(s+2)^3} - \frac{2}{(s+2)^2} - \frac{2}{s+2}$$

and

$$f(t) = \left[2e^{-t} + (3t^2 - 2t - 2)e^{-2t}\right]u(t) \tag{6.28}$$

Alternative Method: A Hybrid of Heaviside and Clearing Fractions

In this method, the simpler coefficients k_1 and a_0 are determined by the Heaviside "cover-up" procedure, as discussed earlier. To determine the remaining coefficients, we use the clearing-fraction method (see Sec. B.5-3). Using the values $k_1 = 2$ and $a_0 = 6$ obtained earlier by the Heaviside "cover-up" method, we have

$$\frac{8s+10}{(s+1)(s+2)^3} = \frac{2}{s+1} + \frac{6}{(s+2)^3} + \frac{a_1}{(s+2)^2} + \frac{a_2}{s+2}$$

We now clear fractions by multiplying both sides of the equation by $(s+1)(s+2)^3$. This procedure yields†

$$8s + 10 = 2(s+2)^3 + 6(s+1) + a_1(s+1)(s+2) + a_2(s+1)(s+2)^2$$

$$0s^3 + 0s^2 + 8s + 10 = (2+a_2)s^3 + (12 + a_1 + 5a_2)s^2 + (30 + 3a_1 + 8a_2)s + (22 + 2a_1 + 4a_2)$$

---

†We could have cleared fractions without finding k_1 and a_0. This alternative, however, proves more laborious because it increases the number of unknowns to 4. By predetermining k_1 and a_0, we reduce the unknowns to 2. Moreover, this method provides a convenient check on the solution. This hybrid procedure achieves the best of both methods.

Equating coefficients of s^3 and s^2 on both sides, we obtain

$$0 = (2 + a_2) \implies a_2 = -2$$

$$0 = 12 + a_1 + 5a_2 = 2 + a_1 \implies a_1 = -2$$

We can stop here if we wish, since the two desired coefficients a_1 and a_2 have already been found. However, equating the coefficients of s^1 and s^0 serves as a check on our answers. This step yields

$$8 = 30 + 3a_1 + 8a_2$$

$$10 = 22 + 2a_1 + 4a_2$$

Substitution of $a_1 = a_2 = -2$, obtained earlier, satisfies these equations. This step assures the correctness of our answers.

Another Alternative: A Hybrid of Heaviside and Short-Cuts

In this method, the simpler coefficients k_1 and a_0 are determined by the Heaviside "cover-up" procedure, as discussed earlier. The usual short-cuts are then used to determine the remaining coefficients. Using the values $k_1 = 2$ and $a_0 = 6$, determined earlier by the Heaviside method, we have

$$\frac{8s + 10}{(s+1)(s+2)^3} = \frac{2}{s+1} + \frac{6}{(s+2)^3} + \frac{a_1}{(s+2)^2} + \frac{a_2}{s+2}$$

There are two unknowns, a_1 and a_2. If we multiply both sides by s and then let $s \to \infty$, we eliminate a_1. This procedure yields

$$0 = 2 + a_2 \implies a_2 = -2$$

Therefore

$$\frac{8s + 10}{(s+1)(s+2)^3} = \frac{2}{s+1} + \frac{6}{(s+2)^3} + \frac{a_1}{(s+2)^2} - \frac{2}{s+2}$$

There is now only one unknown, a_1. This value can be determined readily by setting s equal to any convenient value, say $s = 0$. This step yields

$$\frac{10}{8} = 2 + \frac{3}{4} + \frac{a_1}{4} - 1 \implies a_1 = -2 \quad \blacksquare$$

⊙ **Computer Example C6.1**

Find the inverse laplace transform of the following functions using partial fraction expansion method:

$$\text{(a)} \; \frac{2s^2 + 5}{s^2 + 3s + 2} \quad \text{(b)} \frac{2s^2 + 7s + 4}{(s+1)(s+2)^2} \quad \text{(c)} \frac{8s^2 + 21s + 19}{(s+2)(s^2 + s + 7)}$$

(a)
num=[2 0 5]; den=[1 3 2];
[r,p,k]=residue(num,den)

r = -13, 7
p = -2, -1
k = 2

Thus,

$$F(s) = \frac{-13}{s+2} + \frac{7}{s+1} \qquad \text{and} \qquad f(t) = (-13e^{-2t} + 7e^{-t})u(t) + 2\delta(t)$$

(b)
num=[2 7 4]; den=[conv(conv([1 1],[1 2]),[1 2])];
[r, p, k]= residue(num,den);

r = 3, 2, -1
p = -2, -2, -1
k = [ ]

Hence, $F(s) = \dfrac{3}{s+2} + \dfrac{2}{(s+2)^2} - \dfrac{1}{s+1}$ and $f(t) = (3e^{-2t} + 2te^{-2t} - e^{-t})u(t)$

(c)
num=[8 21 19]; den=[conv([0 1 2],[1 1 7])];
[r, p, k]= residue(num,den)
[angle,mag]=cart2pol(real(r),imag(r))

r = 3.5 - 0.4811i, 3.5 + 0.4811i, 1.00
p = -0.5 + 2.5981i, -0.5 - 2.5981i, -2.00
k = [ ]
angle = -0.1366, 0.1366, 0
mag = 3.5329, 3.5329, 1.00

Hence, $F(s) = \dfrac{1}{s+2} + \dfrac{3.5329\, e^{-j0.1366}}{s+0.5 - j2.5981} + \dfrac{3.5329\, e^{j0.1366}}{s+0.5 + j2.5981}$

and

$$f(t) = [\, e^{-2t} + 1.766\, e^{-0.5t} \cos{(2.5981t - 0.1366)}]\, u(t) \quad \odot$$

⊙   **Computer Example C6.2**
   Find (a) the direct Laplace transform of $\sin at + \cos bt$ (b) the inverse Laplace transform of $(as^2)/(s^2 + b^2)$.
   Here we shall use *Symbolic Math Toolbox*, which is a collection of functions for MATLAB used for manipulating and solving symbolic expressions.
   (a)
   f=sym('sin(a*t)+cos(b*t)');
   F=laplace(f)

   F=(a*s^2+b^2*a+s^3+s*a^2)/(s^2+a^2)/(s^2+b^2)

Thus,

$$F(s) = \frac{s^3 + as^2 + a^2s + b^2a}{(s^2 + a^2)(s^2 + b^2)}$$

   (b)
   F=sym('(a*s^2)/(s^2+b^2);
   f=invlaplace(F)

   F=a*dirac(t)-a*b*sin(b*t)

Thus,

$$f(t) = a\delta(t) + ab\sin(bt)u(t) \quad \odot$$

△ **Exercise E6.2**

(i) Show that the Laplace transform of $10e^{-3t}\cos(4t + 53.13°)$ is $\dfrac{6s - 14}{s^2 + 6s + 25}$. Use Pair 10a from Table 6.1.

(ii) Find the inverse Laplace transform of: **(a)** $\dfrac{s + 17}{s^2 + 4s - 5}$ **(b)** $\dfrac{3s - 5}{(s + 1)(s^2 + 2s + 5)}$

(c) $\dfrac{16s + 43}{(s - 2)(s + 3)^2}$

Answers: **(a)** $\left(3e^t - 2e^{-5t}\right)u(t)$ **(b)** $\left[-2e^{-t} + \frac{5}{2}e^{-t}\cos(2t - 36.87°)\right]u(t)$

(c) $\left[3e^{2t} + (t - 3)e^{-3t}\right]u(t)$ ▽

A Historical Note: Marquis Pierre-Simon De Laplace (1749-1827)

The Laplace transform is named after the great French mathematician and astronomer Laplace, who first presented the transform and its applications to differential equations in a paper published in 1779.

Laplace developed the foundations of potential theory and made important contributions to special functions, probability theory, astronomy, and celestial mechanics. In his *Exposition du systeme du Monde* (1796), Laplace formulated a nebular hypothesis of cosmic origin and tried to explain the universe as a pure mechanism. In his *Traite de Mechanique Celeste* (*celestial mechanics*), which completed the work of Newton, Laplace used mathematics and physics to subject the solar system and all heavenly bodies to the laws of motion and the principle of gravitation. Newton had been unable to explain the irregularities of some heavenly bodies; in desperation, he concluded that God himself must intervene now and then to prevent some catastrophes, such as Jupiter eventually falling into the sun (and the moon into the earth) as predicted by Newton's calculations. Laplace proposed to show that these irregularities would correct themselves periodically, and that a little patience—in Jupiter's case, 929 years—would see everything returning automatically to order, and there was no reason why the solar and the stellar systems could not continue to operate by the laws of Newton and Laplace to the end of time.[3]

Laplace presented a copy of *Mechanique Celeste* to Napoleon, who, after reading the book, took Laplace to task for not including God in his scheme: "You have written this huge book on the system of the world without once mentioning the author of the universe." "Sire," Laplace retorted, "I had no need of that hypothesis." Napoleon was not amused, and when he reported this reply to Lagrange, the latter remarked, "Ah, but that is a fine hypothesis. It explains so many things."[4]

Napoleon, following his policy of honoring and promoting scientists, made Laplace the minister of the interior. To Napoleon's dismay, he found the great mathematician-astronomer bringing "the spirit of infinitesimals" into administration, and so had Laplace transferred hastily to the senate.

Oliver Heaviside (1850-1925)

Although Laplace published his transform method to solve differential equations in 1779, the method did not catch on until a century later. It was rediscovered independently in a rather awkward form by an eccentric British engineer, Oliver Heaviside (1850-1925), one of the tragic figures in the history of electrical engineering. Despite his prolific contributions to electrical engineering, he was severely criticized during his lifetime, and was neglected later to the point that hardly a

P.S. de Laplace (left) and Oliver Heaviside (right).

textbook today mentions his name or contributions. With the passage of time, Heaviside becomes more distant, although his studies had a major impact on many aspects of modern electrical engineering. It was Heaviside who made transatlantic communication possible by inventing cable loading, but no one ever mentions him as a pioneer or an innovator in telephony. It was Heaviside who suggested the use of inductive cable loading, but the credit is given to M. Pupin, who actually built the first loading coil. In addition, Heaviside was[5]

- The first to find a solution to the distortionless transmission line;
- The innovator of lowpass filters;
- The first to write Maxwell's equations in modern form;
- The co-discoverer of rate energy transfer by an electromagnetic field;
- An early champion of the now-common phasor analysis;
- An important contributor to the development of vector analysis. In fact, he essentially created the subject independently of Gibbs[6];
- An originator of the use of operational mathematics used to solve linear integro-differential equations, which eventually led to rediscovery of the ignored Laplace transform;
- The first to theorize (along with Kennelly of Harvard) that a conducting layer (the Kennelly-Heaviside layer) of atmosphere exists, which allows radio waves to follow earth's curvature instead of traveling off into space in a straight line;
- The first to posit that an electrical charge would increase in mass as its velocity increases; an anticipation of an aspect of Einstein's special theory of relativity.[7] He also forecast the possibility of superconductivity.

Heaviside was a self-made, self-educated man with only an elementary school education, who eventually became a pragmatically successful mathematical physicist. He began his career as a telegrapher, but increasing deafness forced him to retire at the age of 24. He then devoted himself to the study of electricity. His creative work was disdained by many professional mathematicians because of his lack of formal education and his unorthodox methods.

Heaviside's misfortune was that he was criticized both by mathematicians, who faulted him for lack of rigor, and by men of practice, who faulted him for using too much mathematics and thereby confusing students. Many mathematicians, trying to find solutions to the distortionless transmission line, failed because no rigorous tools were available at the time. Heaviside succeeded because he used mathematics not with rigor, but with insight and intuition. Using his much-maligned operational method, Heaviside successfully attacked problems that the rigid mathematicians could not solve: problems such as the flow of heat in a body of spatially varying conductivity. Heaviside brilliantly used this method in 1895 to demonstrate a fatal flaw in Lord Kelvin's determination of the geological age of the earth by secular cooling; he used the same flow of heat theory as for his cable analysis. Yet the mathematicians of the Royal Society remained unmoved and were not the least impressed by the fact that Heaviside had found the answer to problems no one else could solve. Many mathematicians who examined his work dismissed it with contempt, asserting that his methods were either complete nonsense or a rehash of already-known ideas.[5]

Sir William Preece, the chief engineer of the British Post Office, a savage critic of Heaviside, ridiculed Heaviside's work as too theoretical and, therefore, leading to faulty conclusions. Heaviside's work on transmission lines and loading was dismissed by the British Post Office; this work might have remained hidden, had not Lord Kelvin himself publicly expressed admiration for it.[5]

Heaviside's operational calculus may be formally inaccurate, but in fact it anticipated the current operational methods developed in more recent years.[8] Although his method was not fully understood, it provided correct results. When Heaviside was attacked for the vague meaning of his operational calculus, his pragmatic reply was, "Shall I refuse my dinner because I do not fully understand the process of digestion?"

Heaviside lived as a bachelor hermit, often in near-squalid conditions, and died largely unnoticed, in poverty. His life demonstrates the arrogance and snobbishness of the intellectual establishment, which does not respect creativity unless it is presented in the strict language of the establishment.

6.2   Some Properties of the Laplace Transform

Because it is a generalized form of the Fourier transform, we expect the Laplace transform to have properties similar to those of the Fourier transform. However, we are discussing here mainly the properties of the unilateral Laplace transform, and they differ somewhat from those of the Fourier transform (which is a bilateral transform).

Properties of the Laplace transform are useful not only in the derivation of the Laplace transform of functions but also in the solutions of linear integro-differential equations. A glance at Eqs. (6.8a) and (6.8b) shows that, as in the case of the Fourier transform, there is a certain measure of symmetry in going from $f(t)$ to $F(s)$, and vice versa. This symmetry or duality is also carried over to the properties of the Laplace transform. This fact will be evident in the following development.

1. Time Shifting

This property states that if

$$f(t) \iff F(s)$$

then for $t_0 \geq 0$

$$f(t - t_0) \iff F(s)e^{-st_0} \tag{6.29a}$$

Observe that $f(t)$ starts at $t = 0$, and, therefore, $f(t - t_0)$ starts at $t = t_0$. This fact is implicit, but is not explicitly indicated in Eq. (6.29a). This often leads to inadvertent errors. To avoid such a pitfall, we should restate the property as follows: If

$$f(t)u(t) \iff F(s)$$

then

$$f(t - t_0)u(t - t_0) \iff F(s)e^{-st_0} \qquad\qquad t_0 \geq 0 \tag{6.29b}$$
delay   delay

Proof:

$$\mathcal{L}\left[f(t - t_0)u(t - t_0)\right] = \int_0^\infty f(t - t_0)u(t - t_0)e^{-st}\,dt$$

Setting $t - t_0 = x$, we obtain .

$$\mathcal{L}\left[f(t - t_0)u(t - t_0)\right] = \int_{-t_0}^\infty f(x)u(x)e^{-s(x+t_0)}\,dx$$

Because $u(x) = 0$ for $x < 0$ and $u(x) = 1$ for $x \geq 0$, the limits of integration are from 0 to ∞. Thus

$$\mathcal{L}\left[f(t - t_0)u(t - t_0)\right] = \int_0^\infty f(x)e^{-s(x+t_0)}\,dx$$

$$= e^{-st_0}\int_0^\infty f(x)e^{-sx}\,dx$$

$$= F(s)e^{-st_0}$$

Note that $f(t - t_0)u(t - t_0)$ is the signal $f(t)u(t)$ delayed by t_0 seconds. The time-shifting property states that *delaying a signal by t_0 seconds amounts to multiplying its transform e^{-st_0}*.

This property of the unilateral Laplace transform holds only for positive t_0 because if t_0 were negative, the signal $f(t - t_0)u(t - t_0)$ would not be causal.

We can readily verify this property in Exercise E6.1. If the signal in Fig. 6.3a is $f(t)u(t)$, then the signal in Fig. 6.3b is $f(t - 2)u(t - 2)$. The Laplace transform for the pulse in Fig. 6.3a is $\frac{1}{s}(1 - e^{-2s})$. Therefore, the Laplace transform for the pulse in Fig. 6.3b is $\frac{1}{s}(1 - e^{-2s})e^{-2s}$.

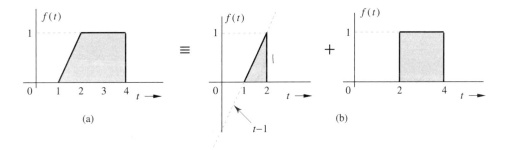

Fig. 6.5 Finding a mathematical description of a function $f(t)$ in Fig. 6.5a.

The time-shifting property proves very convenient in finding the Laplace transform of functions with different descriptions over different intervals, as the following example demonstrates.

■ **Example 6.4**

Find the Laplace transform of $f(t)$ depicted in Fig. 6.5a.

Describing mathematically a function such as the one in Fig. 6.5a is discussed in Sec. 1.4. The function $f(t)$ in Fig. 6.5a can be described as a sum of two components shown in Fig. 6.5b. The equation for the first component is $t - 1$ over $1 \leq t \leq 2$, so that this part of $f(t)$ can be described by $(t - 1)[u(t - 1) - u(t - 2)]$. The second component can be described by $u(t - 2) - u(t - 4)$. Therefore

$$f(t) = (t - 1)\,[u(t - 1) - u(t - 2)] + [u(t - 2) - u(t - 4)]$$

$$= (t - 1)u(t - 1) - (t - 1)u(t - 2) + u(t - 2) - u(t - 4) \qquad (6.30a)$$

The first term on the right-hand side is the signal $tu(t)$ delayed by 1 second. Also, the third and fourth terms are the signal $u(t)$ delayed by 2 and 4 seconds respectively. The second term, however, cannot be interpreted as a delayed version of any entry in Table 6.1. For this reason, we rearrange it as

$$(t - 1)u(t - 2) = (t - 2 + 1)u(t - 2) = (t - 2)u(t - 2) + u(t - 2)$$

We have now expressed the second term in the desired form as $tu(t)$ delayed by 2 seconds plus $u(t)$ delayed by 2 seconds. With this result, Eq. (6.30a) can be expressed as

$$f(t) = (t - 1)u(t - 1) - (t - 2)u(t - 2) - u(t - 4) \qquad (6.30b)$$

Application of the time-shifting property to $tu(t) \Longleftrightarrow 1/s^2$ yields

$$(t - 1)u(t - 1) \Longleftrightarrow \frac{1}{s^2}e^{-s} \quad \text{and} \quad (t - 2)u(t - 2) \Longleftrightarrow \frac{1}{s^2}e^{-2s}$$

Also

$$u(t) \Longleftrightarrow \frac{1}{s} \quad \text{and} \quad u(t - 4) \Longleftrightarrow \frac{1}{s}e^{-4s} \qquad (6.31)$$

Therefore

$$F(s) = \frac{1}{s^2}e^{-s} - \frac{1}{s^2}e^{-2s} - \frac{1}{s}e^{-4s} \qquad (6.32)$$

■

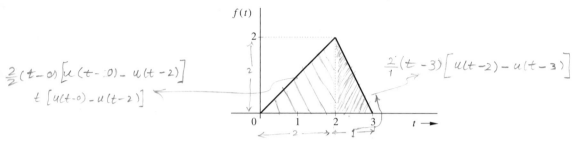

Fig. 6.6 The signal for Exercise E6.3.

(handwritten left margin:)
$$\frac{2}{2}(t-o)\left[u(t-:0)-u(t-2)\right]$$
$$t\left[u(t-o)-u(t-2)\right]$$

(handwritten right of figure:)
$$\frac{2}{1}(t-3)\left[u(t-2)-u(t-3)\right]$$

■ Example 6.5

Find the inverse Laplace transform of *(handwritten: same as 6.3 example.)*

$$F(s) = \frac{s+3+5e^{-2s}}{(s+1)(s+2)}$$

Observe the exponential term e^{-2s} in the numerator of $F(s)$, indicating time-delay. In such a case we should separate $F(s)$ into terms with and without delay factor, as

$$F(s) = \underbrace{\frac{s+3}{(s+1)(s+2)}}_{F_1(s)} + \underbrace{\frac{5e^{-2s}}{(s+1)(s+2)}}_{F_2(s)e^{-2s}}$$

where

$$F_1(s) = \frac{s+3}{(s+1)(s+2)} = \frac{2}{s+1} - \frac{1}{s+2}$$

$$F_2(s) = \frac{5}{(s+1)(s+2)} = \frac{5}{s+1} - \frac{5}{s+2}$$

(handwritten:)
$$= \frac{k_1}{s+1} + \frac{k_2}{s+2}$$
$$k_1 = \frac{s+3}{(s+1)(s+2)}\Big|_{s=-1} = 2$$
$$k_2 = \frac{s+3}{(s+1)(s+2)}\Big|_{s=-2} = -1$$
--- the same

Therefore

$$f_1(t) = \left(2e^{-t} - e^{-2t}\right)u(t)$$

$$f_2(t) = 5\left(e^{-t} - e^{-2t}\right)u(t)$$

Also, because

$$F(s) = F_1(s) + F_2(s)e^{-2s}$$

$$f(t) = f_1(t) + f_2(t-2)$$
$$= \left(2e^{-t} - e^{-2t}\right)u(t) + 5\left[e^{-(t-2)} - e^{-2(t-2)}\right]u(t-2) \quad ■$$

(handwritten: Example 6.4) △ **Exercise E6.3** *(handwritten:)* $= t\left[u(t)-u(t-2)\right] - 2(t-3)\left[u(t-2)-u(t-3)\right] = tu(t) - tu(t-2) - 2(t-3)u(t-2) + 2(t-3)u(t-3)$
Find the Laplace transform of the signal illustrated in Fig. 6.6.

(handwritten:)
$\left[-t-2t+6\right]u(t-2)$
$(-3t+6)\;u(t-2)$

Answer: $\dfrac{1}{s^2}\left(1 - 3e^{-2s} + 2e^{-3s}\right)$ ▽

(handwritten:)
$f(t) = tu(t) - 3(t+2)u(t-2) + 2(t-3)u(t-3)$
$F(s) = \frac{1}{s^2} - 3\frac{1}{s^2}e^{-2s} + 2\frac{1}{s^2}e^{-3s}$

(handwritten: Example 6.5) △ **Exercise E6.4**
Find the inverse Laplace transform of

$$F(s) = \frac{3e^{-2s}}{(s-1)(s+2)}$$

Answer: $\left[e^{t-2} - e^{-2(t-2)}\right]u(t-2)$ ▽

(handwritten:)
$F(s) = \frac{3}{(s-1)(s+2)} = \frac{k_1}{s-1} + \frac{k_2}{s+2}$
$k_1 = \frac{3}{(s-1)(s+2)}\Big|_{s=1} = \frac{3}{3} = 1$
$k_2 = \frac{3}{(s-1)(s+2)}\Big|_{s=-2} = \frac{3}{-3} = -1$
$F(s) = \frac{1}{s-1} - \frac{1}{s+2}$
$f(t) = \left[e^t - e^{-2t}\right]u(t)$
$f(t-2) = \left[e^{(t-2)} - e^{-2(t-2)}\right]u(t-2)$

2. Frequency Shifting

This property states that if

$$f(t) \Longleftrightarrow F(s)$$

then

$$f(t)e^{s_0 t} \Longleftrightarrow F(s - s_0) \tag{6.33}$$

Observe the symmetry (or duality) between this property and the time-shifting property (6.29a). $: f(t)\, u(t) \Longleftrightarrow F(s)$

Proof:

$$\mathcal{L}\left[f(t)e^{s_0 t}\right] = \int_{0^-}^{\infty} f(t)e^{s_0 t}e^{-st}\, dt = \int_{0^-}^{\infty} f(t)e^{-(s-s_0)t}\, dt = F(s - s_0)$$

■ **Example 6.6** $P. 372$

Derive Pair 9a in Table 6.1 from Pair 8a and the frequency-shifting property. The Pair 8a is

$$\cos bt\, u(t) \Longleftrightarrow \frac{s}{s^2 + b^2}$$

From the frequency-shifting property [Eq. (6.33)] with $s_0 = -a$ we obtain

$$e^{-at}\cos bt\, u(t) \Longleftrightarrow \frac{s + a}{(s + a)^2 + b^2} \quad ■$$

△ **Exercise E6.5**

Derive Pair 6 in Table 6.1 from Pair 3 and the frequency-shifting property.   ▽

We are now ready to consider the two of the most important properties of the Laplace transform: the time-differentiation and time-integration properties.

3. The Time-Differentiation Property†

This property states that if

$$f(t) \Longleftrightarrow F(s)$$

then

$$\frac{df}{dt} \Longleftrightarrow sF(s) - f(0^-) \tag{6.34a}$$

Repeated application of this property yields

$$\frac{d^2 f}{dt^2} \Longleftrightarrow s^2 F(s) - sf(0^-) - \dot{f}(0^-) \tag{6.34b}$$

---

†The dual of the time-differentiation property is the frequency-differentiation property, which states that

$$tf(t) \Longleftrightarrow -\frac{d}{ds}F(s)$$

$$\frac{d^n f}{dt^n} \iff s^n F(s) - s^{n-1} f(0^-) - s^{n-2} \dot{f}(0^-) - \cdots - f^{(n-1)}(0^-) \qquad (6.34c)$$

where $f^{(r)}(0^-)$ is $d^r f/dt^r$ at $t = 0^-$.

   Proof:

$$\mathcal{L}\left[\frac{df}{dt}\right] = \int_{0^-}^{\infty} \frac{df}{dt} e^{-st}\, dt$$

Integrating by parts, we obtain

$$\mathcal{L}\left[\frac{df}{dt}\right] = f(t)e^{-st}\Big|_{0^-}^{\infty} + s \int_{0^-}^{\infty} f(t)e^{-st}\, dt$$

For the Laplace integral to converge (that is for $F(s)$ to exist), it is necessary that $f(t)e^{-st} \to 0$ as $t \to \infty$ for the values of s in the region of convergence for $F(s)$. Thus,

$$\mathcal{L}\left[\frac{df}{dt}\right] = -f(0^-) + sF(s)$$

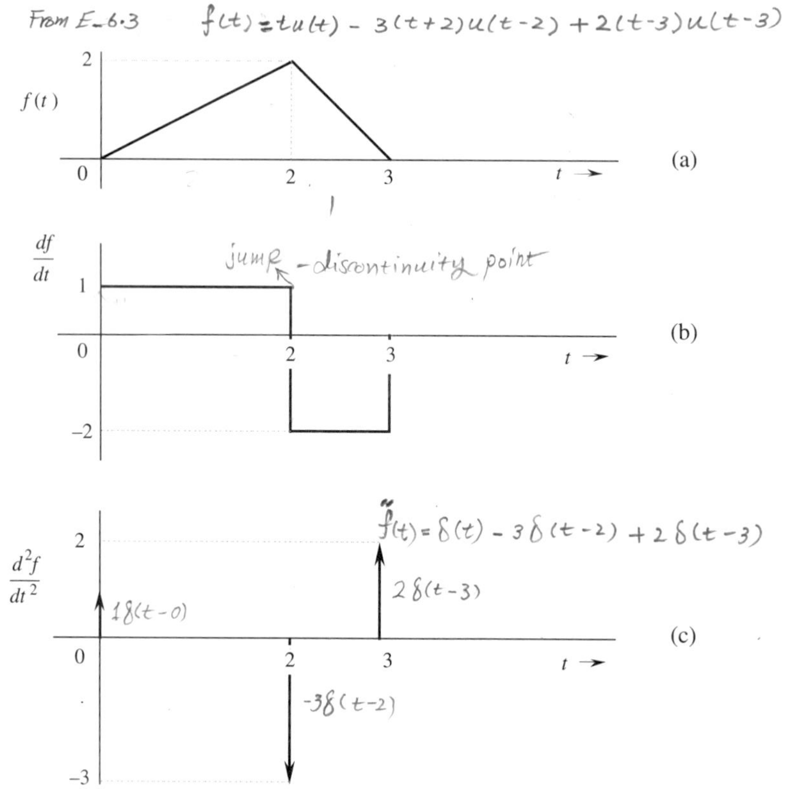

Fig. 6.7 Finding the Laplace transform of a piecewise-linear function using the time-differentiation property.

■ **Example 6.7** ⌣

Find the Laplace transform of the signal $f(t)$ in Fig. 6.7a using Table 6.1 and the time-differentiation and time-shifting properties of the Laplace transform.

Figures 6.7b and 6.8c show the first two derivatives of $f(t)$. Recall that the derivative at a point of jump discontinuity is an impulse of strength equal to the amount of jump [see Eq. (1.27)]. Therefore, obtain $\frac{d^2 f}{dt}$, then use the diffrenetiation property.

$$\frac{d^2 f}{dt^2} = \delta(t) - 3\delta(t-2) + 2\delta(t-3)$$

The Laplace transform of this equation yields

$$\mathcal{L}\left(\frac{d^2 f}{dt^2}\right) = \mathcal{L}\left[\delta(t) - 3\delta(t-2) + 2\delta(t-3)\right]$$

Using the time-differentiation property (6.34b), the time-shifting property (6.29a) and the facts that $f(0^-) = \dot{f}(0^-) = 0$, and $\delta(t) \Longleftrightarrow 1$, we obtain

$$s^2 F(s) - 0 - 0 = 1 - 3e^{-2s} + 2e^{-3s}$$

Therefore

$$F(s) = \frac{1}{s^2}\left(1 - 3e^{-2s} + 2e^{-3s}\right)$$

same answer as E.6.3 which used the time shift property.

which confirms the earlier result in Exercise E6.3. ■

4. The Time-Integration Property†

This property states that if

$$f(t) \Longleftrightarrow F(s)$$

then

$$\int_{0^-}^{t} f(\tau)\, d\tau \Longleftrightarrow \frac{F(s)}{s} \tag{6.35}$$

and

$$\int_{-\infty}^{t} f(\tau)\, d\tau \Longleftrightarrow \frac{F(s)}{s} + \frac{\int_{-\infty}^{0^-} f(\tau)\, d\tau}{s} \tag{6.36}$$

Proof: We define

$$g(t) \equiv \int_{0^-}^{t} f(\tau)\, d\tau$$

so that

$$\frac{d}{dt} g(t) = f(t) \qquad \text{and} \qquad g(0^-) = 0$$

Now, if

$$g(t) \Longleftrightarrow G(s)$$

---

†The dual of the time-integration property is the frequency-integration property, which states that

$$\frac{f(t)}{t} \Longleftrightarrow \int_{s}^{\infty} F(z)\, dz$$

then

$$F(s) = \mathcal{L}\left[\frac{d}{dt}g(t)\right] = sG(s) - g(0^-) = sG(s)$$

Therefore

$$G(s) = \frac{F(s)}{s}$$

or

$$\int_{0^-}^{t} f(\tau)\,d\tau \iff \frac{F(s)}{s}$$

To prove Eq. (6.36), observe that

$$\int_{-\infty}^{t} f(\tau)\,d\tau = \int_{-\infty}^{0^-} f(\tau)\,d\tau + \int_{0^-}^{t} f(\tau)\,d\tau$$

Note that the first term on the right-hand side is a constant. Taking the Laplace transform of the above equation and using Eq. (6.35), we obtain

$$\int_{-\infty}^{t} f(\tau)\,d\tau \iff \frac{\int_{-\infty}^{0^-} f(\tau)\,d\tau}{s} + \frac{F(s)}{s}$$

Scaling

This property states that if

$$f(t) \iff F(s)$$

then for $a > 0$

$$f(at) \iff \frac{1}{a}F\left(\frac{s}{a}\right) \tag{6.37}$$

The proof is identical to that of the scaling property of the Fourier transform derived in Chapter 4 [Eq. (4.34)]. Note that a is restricted to only positive values because if $f(t)$ is causal, then $f(at)$ is anticausal (exists only for $t < 0$) for negative a, and anticausal signals are not permitted in the (unilateral) Laplace transform.

Recall that $f(at)$ is the signal $f(t)$ time-compressed by the factor a, and $F\left(\frac{s}{a}\right)$ is $F(s)$ expanded along the s-scale by the same factor a (see Sec. 1.3-2). The scaling property states that *time-compression of a signal by a factor a causes expansion of its Laplace transform in s-scale by the same factor. Similarly, time-expansion $f(t)$ causes compression of $F(s)$ in s-scale by the same factor.*

5.   Time Convolution and Frequency Convolution

This property states that if

$$f_1(t) \iff F_1(s) \quad \text{and} \quad f_2(t) \iff F_2(s)$$

then (**time convolution property**)

$$f_1(t) * f_2(t) \iff F_1(s)F_2(s) \tag{6.38}$$

and (**frequency convolution property**)

$$f_1(t)f_2(t) \iff \frac{1}{2\pi j}[F_1(s) * F_2(s)] \tag{6.39}$$

Table 6.2

The Laplace Transform Properties

	Operation	$f(t)$	$F(s)$
1	Addition	$f_1(t) + f_2(t)$	$F_1(s) + F_2(s)$
2	Scalar multiplication	$kf(t)$	$kF(s)$
3	Time differentiation	$\dfrac{df}{dt}$	$sF(s) - f(0^-)$
		$\dfrac{d^2 f}{dt^2}$	$s^2 F(s) - sf(0^-) - \dot{f}(0^-)$
		$\dfrac{d^3 f}{dt^3}$	$s^3 F(s) - s^2 f(0^-) - s\dot{f}(0^-) - \ddot{f}(0^-)$
4	Time integration	$\displaystyle\int_{0^-}^{t} f(\tau)\,d\tau$	$\dfrac{1}{s}F(s)$
		$\displaystyle\int_{-\infty}^{t} f(\tau)\,d\tau$	$\dfrac{1}{s}F(s) + \dfrac{1}{s}\displaystyle\int_{-\infty}^{0^-} f(t)\,dt$
5	Time shift	$f(t-t_0)u(t-t_0)$	$F(s)e^{-st_0}\quad t_0 \geq 0$
6	Frequency shift	$f(t)e^{s_0 t}$	$F(s - s_0)$
7 8	Frequency differentiation	$-tf(t)$	$\dfrac{dF(s)}{ds}$
9	Frequency integration	$\dfrac{f(t)}{t}$	$\displaystyle\int_{s}^{\infty} F(z)\,dz$
10	Scaling	$f(at),\ a \geq 0$	$\dfrac{1}{a}F\left(\dfrac{s}{a}\right)$
11	Time convolution	$y(t)=f(t)*h(t)$ $f_1(t)*f_2(t)$	$Y(s)=F(s)H(s)$ LTIC I/o relationship, with Time Conv. $F_1(s)F_2(s)$
12	Frequency convolution	$f_1(t)f_2(t)$	$\dfrac{1}{2\pi j}F_1(s)*F_2(s)$
13	Initial value	$f(0^+)$	$\lim_{s\to\infty} sF(s)\quad (n > m)$
14	Final value	$f(\infty)$	$\lim_{s\to0} sF(s)$ (poles of $sF(s)$ in LHP)

Observe the symmetry (or duality) between the two properties. Proofs of these properties are similar to the proofs in Chapter 4 for the Fourier transform.

Equation (2.48) indicates that $H(s)$, the transfer function of an LTIC system, is the Laplace transform of the system's impulse response $h(t)$; that is,

$$h(t) \iff H(s) \tag{6.40}$$

We can apply the time convolution property to the LTIC input-output relationship $y(t) = f(t) * h(t)$ to obtain

$$Y(s) = F(s)H(s) \tag{6.41}$$

This exact result appeared earlier in Eq. (6.11).

■ **Example 6.8**   table 6.1 #5
                    table 6.2 #11

Using the time convolution property of the Laplace transform, determine $c(t) = e^{at}u(t) * e^{bt}u(t)$.

Fro Eq. (6.38), it follows that

$$C(s) = \frac{1}{(s-a)(s-b)} = \frac{1}{a-b}\left[\frac{1}{s-a} - \frac{1}{s-b}\right]$$

The inverse transform of the above equation yields

$$c(t) = \frac{1}{a-b}(e^{at} - e^{bt})u(t) \quad ■$$

Handwritten annotations:

$C(t) = \dfrac{k_1}{s-a} + \dfrac{k_2}{s-b}$

$k_1 = \dfrac{1}{(s-a)(s-b)}\Big|_{s=a} = \dfrac{1}{a-b}$

$k_2 = \dfrac{1}{(s-a)(s-b)}\Big|_{s=b} = \dfrac{1}{b-a} = \dfrac{-1}{a-b}$

$C(t) = \dfrac{1}{(a-b)(1-A)} - \dfrac{1}{(a-b)(s-b)}$

6.3   Solution of Differential and Integro-Differential Equations

The time-differentiation property of the Laplace transform has set the stage for solving linear differential (or integro-differential) equations with constant coefficients. Because $d^k y/dt^k \iff s^k Y(s)$, the Laplace transform of a differential equation is an algebraic equation which can be readily solved for $Y(s)$. Next we take the inverse Laplace transform of $Y(s)$ to find the desired solution $y(t)$. The following examples demonstrate the Laplace transform procedure for solving linear differential equations with constant coefficients.

■ **Example 6.9**

Solve the second-order linear differential equation

$$(D^2 + 5D + 6)\, y(t) = (D+1)f(t) \tag{6.42a}$$

if the initial conditions are $y(0^-) = 2$, $\dot{y}(0^-) = 1$, and the input $f(t) = e^{-4t}u(t)$.

The equation is

$$\frac{d^2 y}{dt^2} + 5\frac{dy}{dt} + 6y(t) = \frac{df}{dt} + f(t) \tag{6.42b}$$

Let

$$y(t) \iff Y(s)$$

Then from Eq. (6.34)

$$\frac{dy}{dt} \iff sY(s) - y(0^-) = sY(s) - 2$$

and

$$\frac{d^2y}{dt^2} \iff s^2Y(s) - sy(0^-) - \dot{y}(0^-) = s^2Y(s) - 2s - 1$$

Moreover, for $f(t) = e^{-4t}u(t)$,

$$f(t) \iff F(s) = \frac{1}{s+4}, \quad \text{and} \quad \frac{df}{dt} \iff sF(s) - f(0^-) = \frac{s}{s+4} - 0 = \frac{s}{s+4}$$

Taking the Laplace transform of Eq. (6.42b), we obtain

$$\left[s^2Y(s) - 2s - 1\right] + 5\left[sY(s) - 2\right] + 6Y(s) = \frac{s}{s+4} + \frac{1}{s+4} \tag{6.43a}$$

Collecting all the terms of $Y(s)$ and the remaining terms separately on the left-hand side, we obtain

$$\left(s^2 + 5s + 6\right)Y(s) - (2s + 11) = \frac{s+1}{s+4} \tag{6.43b}$$

Therefore

$$\left(s^2 + 5s + 6\right)Y(s) = \underbrace{(2s + 11)}_{\substack{\text{initial} \\ \text{condition} \\ \text{term}}} + \underbrace{\frac{s+1}{s+4}}_{\substack{\text{input} \\ \text{term}}} = \frac{2s^2 + 20s + 45}{s+4}$$

and

$$Y(s) = \frac{2s^2 + 20s + 45}{(s^2 + 5s + 6)(s+4)}$$

$$s = \frac{-5 \mp 1}{2} \Rightarrow s = \frac{-6}{2} = -3 \Rightarrow s+3$$
$$s = \frac{-4}{2}, \quad s - 2 \Rightarrow s+2$$

$$= \frac{2s^2 + 20s + 45}{(s+2)(s+3)(s+4)} = \frac{k_1}{(s+2)} + \frac{k_2}{(s+3)} + \frac{k_3}{(s+4)}$$

Expanding the right-hand side into partial fractions yields

$$k_1 = \frac{2(-2)^2 + 20(-2) + 45}{(s+2)(-2+3)(-2+4)}\bigg|_{s=-2} = \frac{13}{2}$$

$$k_2 = \frac{2(-3)^2 + 20(-3) + 45}{(-3+2)(-3+3)(-3+4)}\bigg|_{s=-3} = \frac{3}{-1}$$

$$Y(s) = \frac{13/2}{s+2} - \frac{3}{s+3} - \frac{3/2}{s+4}$$

$$k_3 = \frac{2(-4)^2 + 20(-4) + 45}{(-4+2)(-4+3)(-4+4)}\bigg|_{s=-4} = \frac{-3}{2}$$

The inverse Laplace transform of the above equation yields

$$y(t) = \left(\frac{13}{2}e^{-2t} - 3e^{-3t} - \frac{3}{2}e^{-4t}\right)u(t) \tag{6.44}$$

∎

   This example demonstrates the ease with which the Laplace transform can solve linear differential equations with constant coefficients. The method is general and can solve a linear differential equation with constant coefficients of any order.

Zero-Input and Zero-State Components of Response

   The Laplace transform method gives the total response, which includes zero-input and zero-state components. It is possible to separate the two components if we so desire. The initial condition terms in the response give rise to the zero-input response. For instance, in Example 6.9, the terms arising due to initial conditions $y(0^-) = 2$ and $\dot{y}(0^-) = 1$ in Eq. (6.43a) generate the zero-input response. These initial condition terms are $-(2s + 11)$, as seen in Eq. (6.43b). The terms on the right-hand side are exclusively due to the input. Equation (6.43b) is reproduced below with the proper labeling of the terms.

$$\left(s^2 + 5s + 6\right) Y\left(s\right) - \left(2s + 11\right) = \frac{s+1}{s+4}$$

so that

$$\left(s^2 + 5s + 6\right) Y\left(s\right) = \underbrace{\left(2s + 11\right)}_{\text{initial condition terms}} + \underbrace{\frac{s+1}{s+4}}_{\text{input terms}}$$

Therefore

$$Y\left(s\right) = \underbrace{\frac{2s+11}{s^2+5s+6}}_{\text{zero-input component}} + \underbrace{\frac{s+1}{(s+4)(s^2+5s+6)}}_{\text{zero-state component}}$$

$$= \left[\frac{7}{s+2} - \frac{5}{s+3}\right] + \left[\frac{-1/2}{s+2} + \frac{2}{s+3} - \frac{3/2}{s+4}\right]$$

Taking the inverse transform of this equation yields

$$y(t) = \underbrace{\left(7e^{-2t} - 5e^{-3t}\right) u(t)}_{\text{zero-input response}} + \underbrace{\left(-\tfrac{1}{2}e^{-2t} + 2e^{-3t} - \tfrac{3}{2}e^{-4t}\right) u(t)}_{\text{zero-state response}}$$

Comments on Initial Conditions at 0^- and at 0^+

The initial conditions in Example 6.9 are $y(0^-) = 2$ and $\dot{y}(0^-) = 1$. If we let $t = 0$ in the total response in Eq. (6.44), we find $y(0) = 2$ and $\dot{y}(0) = 2$, which is at odds with the given initial conditions. Why? The reason is that the initial conditions are given at $t = 0^-$ (just before the input is applied), when only the zero-input response is present. The zero-state response is the result of the input $f(t)$ applied at $t = 0$. Hence, this component does not exist at $t = 0^-$. Consequently, the initial conditions at $t = 0^-$ are satisfied by the zero-input response, not by the total response. We can readily verify in this example that the zero-input response does indeed satisfy the given initial conditions at $t = 0^-$. It is the total response that satisfies the initial conditions at $t = 0^+$, which are generally different from the initial conditions at 0^-.

There also exists a $\mathcal{L}_+$ version of the Laplace transform, which uses the initial conditions at $t = 0^+$ rather than at 0^- (as in our present $\mathcal{L}_-$ version). The $\mathcal{L}_+$ version, which was in vogue till the early sixties, is identical to the $\mathcal{L}_-$ version except the limits of Laplace integral [Eq. (6.18)] are from 0^+ to ∞. Hence, by definition, the origin $t = 0$ is excluded from the domain. This version, still used in some math books, has some serious difficulties. For instance, the Laplace transform of $\delta(t)$ is zero because $\delta(t) = 0$ for $t \geq 0^+$. Moreover, this approach is practically useless in the theoretical study of linear systems because the response obtained by this method cannot be separated into zero-input and zero-state components. As we know, the zero-state component represents the system response as an explicit function of the input, and without knowing this component, it is not possible to assess the effect of the input on the system response. The $\mathcal{L}_+$ version can separate the response in terms of the natural and the forced components, which are not as interesting as the zero-input and the zero-state components. Note that we can always determine the natural and the forced components from the zero-input and the zero-state components (see Eq. (2.51b), but the converse is not true. Because of these and some other problems, electrical engineers (wisely) started discarding the $\mathcal{L}_+$ version in the early sixties.

It is interesting to note the time domain duals of these two Laplace versions. The classical method is the dual of the $\mathcal{L}_+$ method, and the convolution (zero-input/zero-state) method is the dual of the $\mathcal{L}_-$ method. The first pair uses the initial conditions at 0^+, and the second pair uses those at $t = 0^-$. The first pair (the classical method and the $\mathcal{L}_+$ version) is useless in the theoretical study of linear system analysis. It was no coincidence that the $\mathcal{L}_-$ version was adopted immediately after the introduction of the state-space analysis (which uses zero-input/zero-state separation of the input) to the electrical engineering community.

△ **Exercise E6.6**
   Solve

$$\frac{d^2y}{dt^2} + 4\frac{dy}{dt} + 3y(t) = 2\frac{df}{dt} + f(t)$$

for the input $f(t) = u(t)$. The initial conditions are $y(0^-) = 1$ and $\dot{y}(0^-) = 2$.
Answer: $y(t) = \frac{1}{3}(1 + 9e^{-t} - 7e^{-3t})u(t)$ ▽

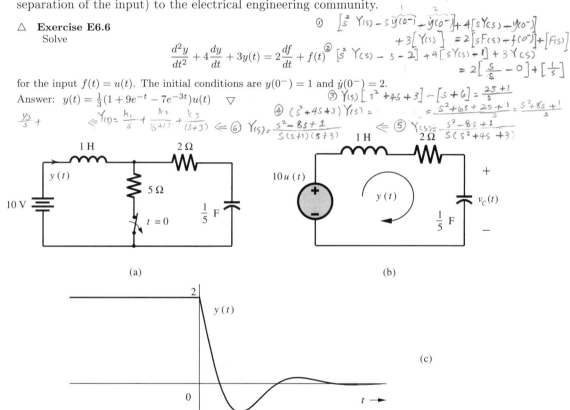

Fig. 6.8 Analysis of a network with a switching action (Example 6.10).

■ **Example 6.10**
   In the circuit of Fig. 6.8a, the switch is in the closed position for a long time before $t = 0$, when it is opened instantaneously. Find the inductor current $y(t)$ for $t \geq 0$.

   When the switch is in the closed position (for a long time), the inductor current is 2 amps and the capacitor voltage is 10 volts. When the switch is opened, the circuit is equivalent to that depicted in Fig. 6.8b, with the initial inductor current $y(0^-) = 2$ and the initial capacitor voltage $v_C(0^-) = 10$. The input voltage is 10 volts, starting at $t = 0$, and, therefore, can be represented by $10u(t)$.

The loop equation of the circuit in Fig. 6.8b is

$$(1)\;\frac{dy}{dt} + 2y(t) + 5\int_{-\infty}^{t} y(\tau)\,d\tau = 10u(t) \tag{6.45}$$

If

$$y(t) \Longleftrightarrow Y(s) \tag{6.46a}$$

then

$$\frac{dy}{dt} \Longleftrightarrow sY(s) - y(0^-) = sY(s) - 2 \tag{6.46b}$$

and [see Eq. (6.36)]

$$\int_{-\infty}^{t} y(\tau)\,d\tau \Longleftrightarrow \frac{Y(s)}{s} + \frac{\int_{-\infty}^{0^-} y(\tau)\,d\tau}{s} \tag{6.46c}$$

Because $y(t)$ is the capacitor current, the integral $\int_{-\infty}^{0^-} y(\tau)\,d\tau$ is $q_C(0^-)$, the capacitor charge at $t = 0^-$, which is given by C times the capacitor voltage at $t = 0^-$. Therefore

$$\int_{-\infty}^{0^-} y(\tau)\,d\tau = q_C(0^-) = Cv_C(0^-) = \frac{1}{5}(10) = 2 \tag{6.47}$$

From Eq. (6.46c) it follows that

$$\int_{-\infty}^{t} y(\tau)\,d\tau \Longleftrightarrow \frac{Y(s)}{s} + \frac{2}{s} \tag{6.48}$$

Taking the Laplace transform of Eq. (6.45) and using Eqs. (6.46a), (6.46b), and (6.48), we obtain

$$sY(s) - 2 + 2Y(s) + \frac{5Y(s)}{s} + \frac{10}{s} = \frac{10}{s}$$

or

$$\left[s + 2 + \frac{5}{s}\right] Y(s) = 2$$

and

$$Y(s) = \frac{2s}{s^2 + 2s + 5}$$

To find the inverse Laplace transform of $Y(s)$, we use Pair 10c (Table 6.1) with values $A = 2, B = 0, a = 1$, and $c = 5$. This yields

$$r = \sqrt{\tfrac{20}{4}} = \sqrt{5},\;\; b = \sqrt{c - a^2} = 2 \;\text{ and }\; \theta = \tan^{-1}(\tfrac{2}{4}) = 26.6°$$

Therefore

$$y(t) = \sqrt{5}\,e^{-t}\cos{(2t + 26.6°)}u(t)$$

This response is shown in Fig. 6.8c. ■

6.3-1   Zero-State Response: The Transfer Function of an LTIC System

Consider an nth-order LTIC system specified by the equation

$$Q(D)y(t) = P(D)f(t)$$

or

$$(D^n + a_{n-1}D^{n-1} + \cdots + a_1D + a_0)y(t) =$$
$$(b_n D^n + b_{n-1}D^{n-1} + \cdots + b_1 D + b_0)f(t) \qquad (6.49)$$

We shall now find the general expression for the zero-state response of an LTIC system. Zero-state response $y(t)$, by definition, is the system response to an input when the system is initially relaxed (in zero state). Therefore, $y(t)$ satisfies the system equation (6.49) with zero initial conditions

$$y(0^-) = \dot{y}(0^-) = \ddot{y}(0^-) = \cdots = y^{(n-1)}(0^-) = 0$$

Moreover, the input $f(t)$ is causal, so that

$$f(0^-) = \dot{f}(0^-) = \ddot{f}(0^-) = \cdots = f^{(n-1)}(0^-) = 0$$

Let

$$y(t) \Longleftrightarrow Y(s) \quad \text{and} \quad f(t) \Longleftrightarrow F(s)$$

Because of zero initial conditions

$$D^r y(t) = \frac{d^r}{dt^r}y(t) \Longleftrightarrow s^r Y(s)$$

$$D^k f(t) = \frac{d^k}{dt^k}f(t) \Longleftrightarrow s^k F(s)$$

Therefore, the Laplace transform of Eq. (6.49) yields

$$\left(s^n + a_{n-1}s^{n-1} + \cdots + a_1 s + a_0\right)Y(s) = \left(b_n s^n + b_{n-1}s^{n-1} + \cdots + b_1 s + b_0\right)F(s)$$

or

$$Y(s) = \frac{b_n s^n + b_{n-1}s^{n-1} + \cdots + b_1 s + b_0}{s^n + a_{n-1}s^{n-1} + \cdots + a_1 s + a_0}F(s) \qquad (6.50a)$$

$$= \frac{P(s)}{Q(s)}F(s) \qquad (6.50b)$$

But we have shown in Eq. (6.41) that $Y(s) = H(s)F(s)$. Consequently,

$$H(s) = \frac{P(s)}{Q(s)} \qquad (6.51)$$

This equation was derived earlier in time-domain [see Eq. (2.50)]. Thus,

$$Y(s) = H(s)F(s) \qquad (6.52)$$

We have already derived this result twice using other approaches. We now have an alternate interpretation (or an alternate definition) for the transfer function $H(s)$.

It is the ratio of $Y(s)$ to $F(s)$ when all the initial conditions are zero (when the system is in zero state). Thus

$$H(s) \equiv \frac{Y(s)}{F(s)} = \frac{\mathcal{L}[\text{zero-state response}]}{\mathcal{L}[\text{input}]} \tag{6.53}$$

We have shown that $Y(s)$, the Laplace transform of the zero-state response $y(t)$, is the product of $F(s)$ and $H(s)$, where $F(s)$ is the Laplace transform of the input $f(t)$ and $H(s)$ is the system transfer function [relating the particular output $y(t)$ to the input $f(t)$]. Observe that the denominator of $H(s)$ is $Q(s)$, the characteristic polynomial of the system. Therefore, *the poles of $H(s)$ are the characteristic roots of the system.* Consequently, the system stability criterion can be stated in terms of the poles of the transfer function of a system, as follows:

1. An LTIC system is asymptotically stable if and only if all the poles of its transfer function $H(s)$ are in the LHP. The poles may be repeated or unrepeated.

2. An LTIC system is unstable if and only if either one or both of the following conditions exist: (i) at least one pole of $H(s)$ is in the RHP; (ii) there are repeated poles of $H(s)$ on the imaginary axis.

3. An LTIC system is marginally stable if and only if there are no poles of $H(s)$ in the RHP, and there are some unrepeated poles on the imaginary axis.

We can now represent the transformed version of the system, as depicted in Fig. 6.9. The input $F(s)$ is the Laplace transform of $f(t)$, and the output $Y(s)$ is the Laplace transform of (the zero-input response) $y(t)$. The system is described by the transfer function $H(s)$. The output $Y(s)$ is the product $F(s)H(s)$.

. The result $Y(s) = H(s)F(s)$ greatly facilitates derivation of the system response to a given input. We shall demonstrate this assertion this by an example.

■ **Example 6.11**

Find the response $y(t)$ of an LTIC system described by the equation

$$\frac{d^2 y}{dt^2} + 5\frac{dy}{dt} + 6y(t) = \frac{df}{dt} + f(t)$$

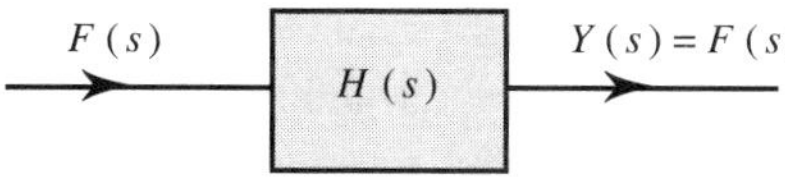

$$F(s) \longrightarrow \boxed{H(s)} \longrightarrow Y(s) = F(s)H(s)$$

Fig. 6.9 The transformed representation of an LTIC system.

if the input $f(t) = 3e^{-5t}u(t)$ and all the initial conditions are zero; that is, the system is in zero-state.

The system equation is

$$\underbrace{\left(D^2 + 5D + 6\right)}_{Q(D)} y(t) = \underbrace{(D+1)}_{P(D)} f(t)$$

Therefore

$$\frac{y(t)}{f(t)} = H(s) = \frac{P(s)}{Q(s)} = \frac{s+1}{s^2 + 5s + 6}$$

Also

$$F(s) = \mathcal{L}\left[3e^{-5t}u(t)\right] = \frac{3}{s+5}$$

and

$$Y(s) = F(s)H(s) = \frac{3(s+1)}{(s+5)(s^2 + 5s + 6)}$$

$$= \frac{3(s+1)}{(s+5)(s+2)(s+3)} = \frac{k_1}{(s+5)} + \frac{k_2}{(s+2)} + \frac{k_3}{(s+3)} \cdots$$

$$= \frac{-2}{s+5} - \frac{1}{s+2} + \frac{3}{s+3}$$

The inverse Laplace transform of this equation is

$$y(t) = \left(-2e^{-5t} - e^{-2t} + 3e^{-3t}\right) u(t) \quad \blacksquare$$

■ **Example 6.12**

Show that the transfer function of

(a) an ideal delay of T seconds is e^{-sT};

(b) an ideal differentiator is s;

(c) an ideal integrator is $1/s$.

(a) Ideal Delay

For an ideal delay of T seconds, the input $f(t)$ and output $y(t)$ are related by

$$y(t) = f(t - T)$$

or

$$Y(s) = F(s)e^{-sT} \qquad \text{[see Eq. (6.29a)]}$$

Therefore

$$H(s) = \frac{Y(s)}{F(s)} = e^{-sT} \tag{6.54}$$

(b) Ideal Differentiator

For an ideal differentiator, the input $f(t)$ and the output $y(t)$ are related by

$$y(t) = \frac{df}{dt}$$

The Laplace transform of this equation yields

$$Y(s) = sF(s) \qquad [f(0^-) = 0 \text{ for a causal signal}]$$

and

$$H(s) = \frac{Y(s)}{F(s)} = s \tag{6.55}$$

(c) Ideal Integrator

For an ideal integrator with zero initial state, that is $y(0^-) = 0$,

$$y(t) = \int_0^t f(\tau)\, d\tau$$

and

$$Y(s) = \frac{1}{s} F(s)$$

Therefore

$$H(s) = \frac{1}{s} \tag{6.56}$$

△ **Exercise E6.7**

For an LTIC system with transfer function

$$H(s) = \frac{s+5}{s^2 + 4s + 3}$$

(a) Describe the differential equation relating the input $f(t)$ and output $y(t)$.

(b) Find the system response $y(t)$ to the input $f(t) = e^{-2t}u(t)$ if the system is initially in zero state.

Answers: (a) $\dfrac{d^2y}{dt^2} + 4\dfrac{dy}{dt} + 3y(t) = \dfrac{df}{dt} + 5f(t)$ (b) $y(t) = \left(2e^{-t} - 3e^{-2t} + e^{-3t}\right)u(t)$ ▽

6.4 Analysis of Electrical Networks: The Transformed Network

Example 6.10 shows how electrical networks may be analyzed by writing the integro-differential equation(s) of the system and then solving these equations by the Laplace transform. We now show that it is also possible to analyze electrical networks directly without having to write the integro-differential equations. This procedure is considerably simpler because it permits us to treat an electrical network as if it were a resistive network. For this purpose, we need to represent a network in the "frequency domain" where all the voltages and currents are represented by their Laplace transforms.

For the sake of simplicity, let us first discuss the case with zero initial conditions. If $v(t)$ and $i(t)$ are the voltage across and the current through an inductor of L henries, then

$$v(t) = L\frac{di}{dt}$$

The Laplace transform of this equation (assuming zero initial current) is

$$V(s) = LsI(s)$$

Similarly, for a capacitor of C farads, the voltage-current relationship is $i(t) = C\frac{dv}{dt}$ and its Laplace transform, assuming zero initial capacitor voltage, yields $I(s) = CsV(s)$; that is,

$$V(s) = \frac{1}{Cs}I(s)$$

For a resistor of R ohms, the voltage-current relationship is $v(t) = Ri(t)$, and its Laplace transform is

$$V(s) = RI(s)$$

Thus, in the "frequency domain," the voltage-current relationships of an inductor and a capacitor are algebraic; these elements behave like resistors of "resistance" Ls and $1/Cs$, respectively. The generalized "resistance" of an element is called its **impedance** and is given by the ratio $V(s)/I(s)$ for the element (under zero initial conditions). The impedances of a resistor of R ohms, an inductor of L henries, and a capacitance of C farads are R, Ls, and $1/Cs$, respectively.

Also, the interconnection constraints (Kirchhoff's laws) remain valid for voltages and currents in the frequency domain. To demonstrate this point, let $v_j(t)$ ($j=1, 2, \ldots, k$) be the voltages across k elements in a loop and let $i_j(t)$ ($j = 1, 2, \ldots, m$) be the j currents entering a node. Then

$$\sum_{j=1}^{k} v_j(t) = 0 \quad \text{and} \quad \sum_{j=1}^{m} i_j(t) = 0 \tag{6.57}$$

Now if

$$v_j(t) \Longleftrightarrow V_j(s) \quad \text{and} \quad i_j(t) \Longleftrightarrow I_j(s)$$

then

$$\sum_{j=1}^{k} V_j(s) = 0 \quad \text{and} \quad \sum_{j=1}^{m} I_j(s) = 0 \tag{6.58}$$

This result shows that if we represent all the voltages and currents in an electrical network by their Laplace transforms, we can treat the network as if it consisted of the "resistances" R, Ls and $1/Cs$ corresponding to a resistor R, an inductor L, and a capacitor C, respectively. The system equations (loop or node) are now algebraic. Moreover, the simplification techniques that have been developed for resistive circuits—equivalent series and parallel impedances, voltage and current divider rules, Thévenin and Norton theorems—can be applied to general electrical networks. The following examples demonstrate these concepts.

■ **Example 6.13**
Find the loop current $i(t)$ in the circuit shown in Fig. 6.10a if all the initial conditions are zero.

In the first step, we represent the circuit in the frequency domain, as illustrated in Fig. 6.10b. All the voltages and currents are represented by their Laplace transforms. The voltage $10u(t)$ is represented by $10/s$ and the (unknown) current $i(t)$ is represented by its Laplace transform $I(s)$. All the circuit elements are represented by their respective impedances. The inductor of 1 henry is represented by s, the capacitor of $1/2$ farad is represented by $2/s$, and the resistor of 3 ohms is represented by 3. We now consider the frequency-domain representation of voltages and currents. The voltage across any element

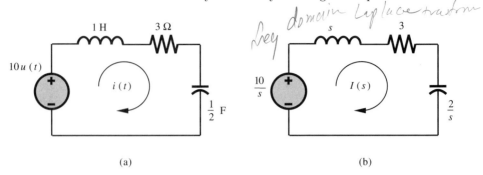

freq domain Laplace transform (handwritten)

Fig. 6.10 A circuit and its transformed version (Example 6.13).

is $I(s)$ times its impedance. Therefore, the total voltage drop in the loop is $I(s)$ times the total loop impedance, and it must be equal to $V(s)$, (transform of) the input voltage. The total impedance in the loop is

$$Z(s) = s + 3 + \frac{2}{s} = \frac{s^2 + 3s + 2}{s}$$

The input "voltage" is $V(s) = 10/s$. Therefore, the "loop current" $I(s)$ is

$$I(s) = \frac{V(s)}{Z(s)} = \frac{\frac{10}{s}}{\frac{s^2+3s+2}{s}} = \frac{10}{s^2 + 3s + 2} = \frac{10}{(s+1)(s+2)} = \frac{10}{s+1} - \frac{10}{s+2}$$

The inverse transform of this equation yields the desired result:

$$i(t) = 10(e^{-t} - e^{-2t})u(t) \qquad \blacksquare$$

Initial Condition Generators

The above discussion, where we assumed zero initial conditions, can be readily extended to the case of nonzero initial conditions because the initial condition in a capacitor or an inductor can be represented by an equivalent source. We now show that a capacitor C with an initial voltage $v(0)$ (Fig. 6.11a) can be represented in the frequency domain by an uncharged capacitor of impedance $1/Cs$ in series with a voltage source of value $v(0)/s$ (Fig. 6.11b) or as the same uncharged capacitor in parallel with a current source of value $Cv(0)$ (Fig. 6.11c). Similarly, an inductor L with an initial current $i(0)$ (Fig. 6.11d) can be represented in the frequency domain by an inductor of impedance Ls in series with a voltage source of value $Li(0)$ (Fig. 6.11e) or by the same inductor in parallel with a current source of value $i(0)/s$ (Fig. 6.11f). To prove this point, consider the terminal relationship of the capacitor in Fig. 6.11a

$$i(t) = C\frac{dv}{dt}$$

The Laplace transform of this equation yields

$$I(s) = C[sV(s) - v(0)]$$

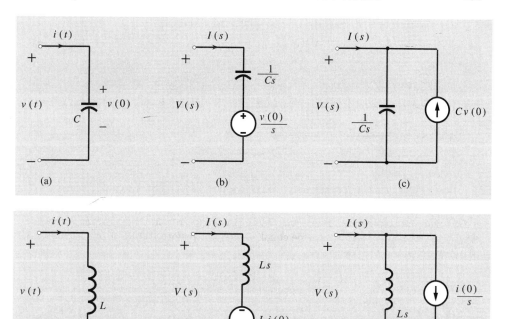

Fig. 6.11 Initial condition generators for a capacitor and an inductor.

This equation can be rearranged as

$$V(s) = \frac{1}{Cs}I(s) + \frac{v(0)}{s} \qquad (6.59a)$$

Observe that $V(s)$ is the voltage (in the frequency domain) across the charged capacitor and $I(s)/Cs$ is the voltage across the same capacitor without any charge. Therefore, the above equation shows that the charged capacitor can be represented by the same capacitor (uncharged) in series with a voltage source of value $v(0)/s$, as depicted in Fig. 6.11b. Equation (6.59a) can also be rearranged as

$$V(s) = \frac{1}{Cs}\left[I(s) + Cv(0)\right] \qquad (6.59b)$$

This equation shows that the charged capacitor voltage $V(s)$ is equal to the uncharged capacitor voltage caused by a current $I(s) + Cv(0)$. This result is reflected precisely in Fig. 6.11c, where the current through the uncharged capacitor is $I(s) + Cv(0)$.†

---

†In the time domain, a charged capacitor C with initial voltage $v(0)$ can be represented as the same capacitor uncharged in series with a voltage source $v(0)u(t)$, or in parallel with a current source $Cv(0)\delta(t)$. Similarly, an inductor L with initial current $i(0)$ can be represented by the same inductor with zero initial current in series with a voltage source $Li(0)\delta(t)$ or with a parallel current source $i(0)u(t)$.

For the inductor in Fig. 6.11d, the terminal equation is

$$v(t) = L\frac{di}{dt}$$

and

$$V(s) = L[sI(s) - i(0)]$$

$$= LsI(s) - Li(0) \tag{6.60a}$$

$$= Ls\left[I(s) - \frac{i(0)}{s}\right] \tag{6.60b}$$

We can verify that Fig. 6.11e satisfies Eq. (6.60a) and that Fig. 6.11f satisfies Eq. (6.60b).

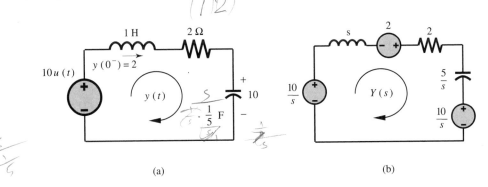

(a)   (b)

Fig. 6.12 A circuit and its transformed version with initial condition generators (Example 6.12).

Let us rework Example 6.10 using these ideas. Figure 6.12a shows the circuit in Fig. 6.8b with the initial conditions $y(0) = 2$ and $v_C(0) = 10$. Figure 6.12b shows the frequency-domain representation (transformed circuit) of the circuit in Fig. 6.12a. The resistor is represented by its impedance 2; the inductor with initial current of 2 amps is represented according to the arrangement in Fig. 6.11e with a series voltage source $Ly(0) = 2$. The capacitor with initial voltage of 10 volts is represented according to the arrangement in Fig. 6.11b with a series voltage source $v(0)/s = 10/s$. Note that the impedance of the inductor is s and that of the capacitor is $5/s$. The input of $10u(t)$ is represented by its Laplace transform $10/s$.

The total voltage in the loop is $\frac{10}{s} + 2 - \frac{10}{s} = 2$, and the loop impedance is $(s + 2 + \frac{5}{s})$. Therefore

$$Y(s) = \frac{2}{s + 2 + \frac{5}{s}}$$

$$= \frac{2s}{s^2 + 2s + 5}$$

which confirms our earlier result in Example 6.10.

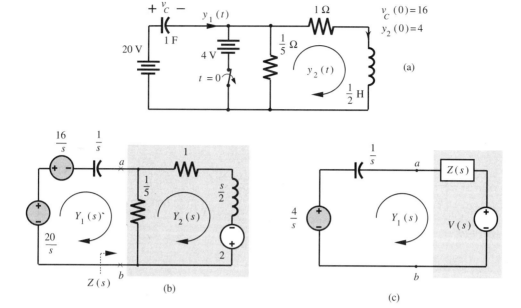

Fig. 6.13 Solving Example 6.14 using initial condition generators and Thèvenin equivalent representation.

■ Example 6.14

The switch in the circuit of Fig. 6.13a is in the closed position for a long time before $t = 0$, when it is opened instantaneously. Find the currents $y_1(t)$ and $y_2(t)$ for $t \geq 0$.

Inspection of this circuit shows that when the switch is closed and the steady-state conditions are reached, the capacitor voltage $v_C = 16$ volts, and the inductor current $y_2 = 4$ amps. Therefore, when the switch is opened (at $t = 0$), the initial conditions are $v_C(0^-) = 16$ and $y_2(0^-) = 4$. Figure 6.13b shows the transformed version of the circuit in Fig. 6.13a. We have used equivalent sources to account for the initial conditions. The initial capacitor voltage of 16 volts is represented by a series voltage of $16/s$ and the initial inductor current of 4 amps is represented by a source of value $Ly_2(0) = 2$.

From Fig. 6.13b, the loop equations can be written directly in the frequency domain as

$$\frac{Y_1(s)}{s} + \frac{1}{5}[Y_1(s) - Y_2(s)] = \frac{4}{s}$$

$$-\frac{1}{5}Y_1(s) + \frac{6}{5}Y_2(s) + \frac{s}{2}Y_2(s) = 2$$

$$\begin{bmatrix} \frac{1}{s} + \frac{1}{5} & -\frac{1}{5} \\ -\frac{1}{5} & \frac{6}{5} + \frac{s}{2} \end{bmatrix} \begin{bmatrix} Y_1(s) \\ Y_2(s) \end{bmatrix} = \begin{bmatrix} \frac{4}{s} \\ 2 \end{bmatrix}$$

Application of Cramer's rule to this equation yields

$$Y_1(s) = \frac{24(s+2)}{s^2 + 7s + 12}$$

$$= \frac{24(s+2)}{(s+3)(s+4)} = \frac{-24}{s+3} + \frac{48}{s+4}$$

and

$$y_1(t) = \left(-24e^{-3t} + 48e^{-4t}\right)u(t)$$

Similarly, we obtain

$$Y_2(s) = \frac{4(s+7)}{s^2 + 7s + 12}$$

$$= \frac{16}{s+3} - \frac{12}{s+4}$$

and

$$y_2(t) = \left(16e^{-3t} - 12e^{-4t}\right)u(t)$$

We also could have computed $Y_1(s)$ and $Y_2(s)$ using Thévenin's theorem by replacing the circuit to the right of the capacitor (right of terminals ab) with its Thévenin equivalent, as shown in Fig. 6.13c. Figure 6.13b shows that the Thévenin impedance $Z(s)$ and the Thévenin source $V(s)$ are:

$$Z(s) = \frac{\frac{1}{5}\left(\frac{s}{2}+1\right)}{\frac{1}{5} + \frac{s}{2} + 1} = \frac{s+2}{5s+12}$$

$$V(s) = \frac{-\frac{1}{5}}{\frac{1}{5} + \frac{s}{2} + 1}\cdot 2 = \frac{-4}{5s+12}$$

According to Fig. 6.13c, the current $Y_1(s)$ is given by

$$Y_1(s) = \frac{\frac{4}{s} - V(s)}{\frac{1}{s} + Z(s)}$$

$$= \frac{24(s+2)}{s^2 + 7s + 12}$$

a conclusion which confirms the earlier result. We may determine $Y_2(s)$ in a similar manner. ∎

■ **Example 6.15**

The switch in the circuit in Fig. 6.14a is at position a for a long time before $t = 0$, when it is moved instantaneously to position b. Determine the current $y_1(t)$ and the output voltage $v_0(t)$ for $t \geq 0$.

Just before switching, the values of the loop currents are 2 and 1, respectively, that is; $y_1(0^-) = 2$ and $y_2(0^-) = 1$.

The equivalent circuits for two types of inductive couplings are illustrated in Figs. 6.14b and 6.14c. For our situation, the circuit in Fig. 6.14c applies. Fig. 6.14d shows the transformed version of the circuit in Fig. 6.14a after switching. Note that the inductors $L_1 + M$, $L_2 + M$, and $-M$ are 3, 4, and -1 henries with impedances $3s$, $4s$, and $-s$ respectively. The initial condition voltages in the three branches are $(L_1 + M)y_1(0) = 6$, $(L_2 + M)y_2(0) = 4$, and $-M[y_1(0) - y_2(0)] = -1$, respectively. The two loop equations of the circuit are†

---

†The time domain equations (loop equations) are

$$L_1 \frac{dy_1}{dt} + (R_1 + R_2)y_1(t) - R_2 y_2(t) + M \frac{dy_2}{dt} = 10u(t)$$

$$M \frac{dy_1}{dt} - R_2 y_1(t) + L_2 \frac{dy_2}{dt} + (R_2 + R_3)y_2(t) = 0$$

The Laplace transform of these equations yields Eq. (6.61).

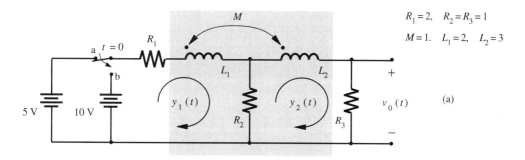

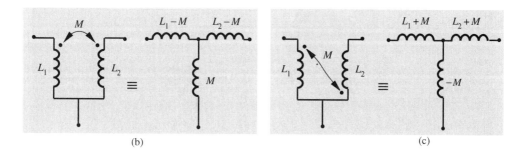

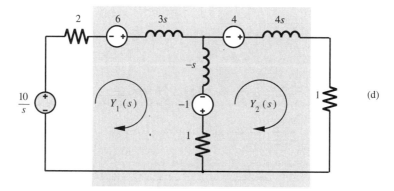

Fig. 6.14 The Laplace transform analysis of a coupled inductive network (Example 6.15) by the transformed circuit method.

$$(2s + 3)Y_1(s) + (s - 1)Y_2(s) = \frac{10}{s} + 5$$

$$(s - 1)Y_1(s) + (3s + 2)Y_2(s) = 5 \qquad (6.61)$$

or

$$\begin{bmatrix} 2s + 3 & s - 1 \\ s - 1 & 3s + 2 \end{bmatrix} \begin{bmatrix} Y_1(s) \\ Y_2(s) \end{bmatrix} \begin{bmatrix} \frac{5s+10}{s} \\ 5 \end{bmatrix}$$

and

$$Y_1(s) = \frac{2s^2 + 9s + 4}{s(s^2 + 3s + 1)}$$

$$= \frac{4}{s} - \frac{1}{s + 0.382} - \frac{1}{s + 2.618}$$

Therefore

$$y_1(t) = \left(4 - e^{-0.382t} - e^{-2.618t}\right) u(t)$$

Similarly

$$Y_2(s) = \frac{s^2 + 2s + 2}{s(s^2 + 3s + 1)}$$

$$= \frac{2}{s} - \frac{1.618}{s + 0.382} + \frac{0.618}{s + 2.618}$$

and

$$y_2(t) = \left(2 - 1.618e^{-0.382t} + 0.618e^{-2.618t}\right) u(t)$$

The output voltage

$$v_0(t) = y_2(t) = \left(2 - 1.618e^{-0.382t} + 0.618e^{-2.618t}\right) u(t) \quad \blacksquare$$

△ **Exercise E6.8**
    For the RLC circuit in Fig. 6.15, the input is switched on at $t = 0$. The initial conditions
are $y(0^-) = 2$ amps and $v_C(0) = 50$ volts. Find the loop current $y(t)$ and the capacitor voltage
$v_C(t)$ for $t \geq 0$.
Answer:

$$y(t) = 10\sqrt{2}e^{-t} \cos{(2t + 81.8°)}u(t), \quad v_C(t) = [24 + 31.62e^{-t} \cos{(2t - 34.7°)}]u(t) \ \triangledown$$

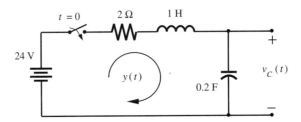

Fig. 6.15 Circuit for Exercise E6.8.

6.4-1 Analysis of Active Circuits

Although we have considered examples of only passive networks so far, the
circuit analysis procedure using the Laplace transform is also applicable to active
circuits. All that is needed is to replace the active elements with their mathematical
models (or equivalent circuits) and proceed as before.

The operational amplifier (depicted by the triangular symbol in Fig. 6.16a)
is a well-known element in modern electronic circuits. The terminals with the
positive and the negative signs correspond to noninverting and inverting terminals,

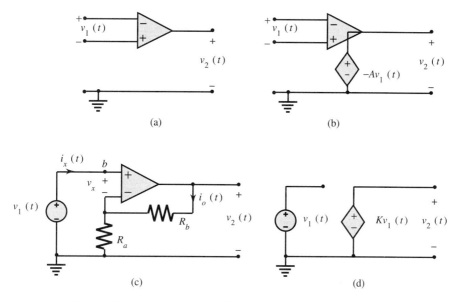

Fig. 6.16 Operational amplifier and its equivalent circuit.

respectively. This means that the polarity of the output voltage v_2 is the same as that of the input voltage at the terminal marked by the positive sign (noninverting). The opposite is true for the inverting terminal, marked by the negative sign.

Figure 6.16b shows the model (equivalent circuit) of the operational amplifier (op amp) in Fig. 6.16a. A typical op amp has a very large gain. The output voltage $v_2 = -Av_1$, where A is typically 10^5 to 10^6. The input impedance is very high (typically $10^6\,\Omega$ for BJT to $10^{12}\,\Omega$ for Bi-FET), and the output impedance is very low (50 to $100\,\Omega$). For most applications, we are justified in assuming the gain A and the input impedance to be infinite and the output impedance to be zero. For this reason we see an ideal voltage source at the output.

Consider now the operational amplifier with resistors R_a and R_b connected, as shown in Fig. 6.16c. This configuration is known as the **noninverting amplifier**. Observe that the input polarities in this configuration are inverted when compared to those in Fig. 6.16a. We now show that the output voltage v_2 and the input voltage v_1 in this case are related by

$$v_2 = Kv_1 \qquad K = 1 + \frac{R_b}{R_a} \qquad (6.62)$$

First, we recognize that because the input impedance and the gain of the operational amplifier approach infinity, the input current i_x and the input voltage v_x in Fig. 6.16c must approach zero. The dependent source in this case is Av_x instead of $-Av_x$ because of the input polarity inversion. The dependent source Av_x (see Fig. 6.16b) at the output will generate current i_o, as illustrated in Fig. 6.16c. Now

$$v_2 = (R_b + R_a)i_o$$

also

$$v_1 = v_x + R_a i_o$$

$$= R_a i_o$$

Therefore

$$\frac{v_2}{v_1} = \frac{R_b + R_a}{R_a} = 1 + \frac{R_b}{R_a} = K$$

or

$$v_2(t) = K v_1(t)$$

The equivalent circuit of the noninverting amplifier is depicted in Fig. 6.16d.

■ **Example 6.16**

The circuit in Fig. 6.17a is called the **Sallen-Key** circuit, which is frequently used in filter design. Find the transfer function $H(s)$ relating the output voltage $v_o(t)$ to the input voltage $v_i(t)$.

We are required to find

$$H(s) = \frac{V_o(s)}{V_i(s)}$$

assuming all initial conditions to be zero.

Figure 6.17b shows the transformed version of the circuit in Fig. 6.17a. The noninverting amplifier is replaced by its equivalent circuit. All the voltages are replaced by their Laplace transforms and all the circuit elements are shown by their impedances. All the initial conditions are assumed to be zero, as required for determining $H(s)$.

We shall use node analysis to derive the result. There are two unknown node voltages, $V_a(s)$ and $V_b(s)$, requiring two node equations.

At node a, $I_{R_1}(s)$, the current in R_1 (leaving the node a), is $[V_a(s) - V_i(s)]/R_1$. Similarly, $I_{R_2}(s)$, the current in R_2 (leaving the node a), is $[V_a(s) - V_b(s)]/R_2$ and $I_{c_1}(s)$, the current in capacitor C_1 (leaving the node a), is $[V_a(s) - V_o(s)]C_1 s = [V_a(s) - K V_b(s)]C_1 s$.

The sum of all the three currents is zero. Therefore

$$\frac{V_a(s) - V_i(s)}{R_1} + \frac{V_a(s) - V_b(s)}{R_2} + [V_a(s) - K V_b(s)] C_1 s = 0$$

or

$$\left(\frac{1}{R_1} + \frac{1}{R_2} + C_1 s \right) V_a(s) - \left(\frac{1}{R_2} + K C_1 s \right) V_b(s) = \frac{1}{R_1} V_i(s) \qquad (6.63a)$$

Similarly, the node equation at node b yields

$$\frac{V_b(s) - V_a(s)}{R_2} + C_2 s V_b(s) = 0$$

or

$$-\frac{1}{R_2} V_a(s) + \left(\frac{1}{R_2} + C_2 s \right) V_b(s) = 0 \qquad (6.63b)$$

The two node equations (6.63a) and (6.63b) in two unknown node voltages $V_a(s)$ and $V_b(s)$ can be expressed in matrix form as

$$\begin{bmatrix} G_1 + G_2 + C_1 s & -(G_2 + K C_1 s) \\ -G_2 & (G_2 + C_2 s) \end{bmatrix} \begin{bmatrix} V_a(s) \\ V_b(s) \end{bmatrix} = \begin{bmatrix} G_1 V_i(s) \\ 0 \end{bmatrix}. \qquad (6.64)$$

where

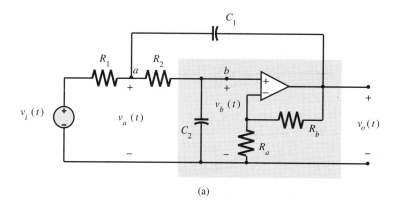

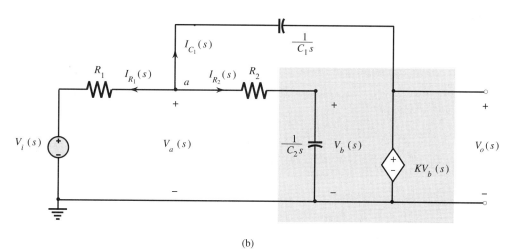

(b)

Fig. 6.17 Sallen-Key circuit and its equivalent.

$$G_1 = \frac{1}{R_1} \quad \text{and} \quad G_2 = \frac{1}{R_2}$$

Application of Cramer's rule to Eq. (6.64) yields

$$\frac{V_b(s)}{V_i(s)} = \frac{G_1 G_2}{C_1 C_2 s^2 + [G_1 G_2 + G_2 C_2 + G_2 C_1(1-K)]s + G_1 G_2}$$

$$= \frac{\omega_0{}^2}{s^2 + 2\alpha s + \omega_0{}^2}$$

where

$$K = 1 + \frac{R_b}{R_a} \quad \text{and} \quad \omega_0{}^2 = \frac{G_1 G_2}{C_1 C_2} = \frac{1}{R_1 R_2 C_1 C_2} \qquad (6.65a)$$

$$2\alpha = \frac{G_1 C_2 + G_2 C_2 + G_2 C_1(1-K)}{C_1 C_2} = \frac{1}{R_1 C_1} + \frac{1}{R_1 C_2} + \frac{1}{R_2 C_2}(1-K) \qquad (6.65b)$$

Now

$$V_o(s) = KV_b(s)$$

Therefore

$$H(s) = \frac{V_o(s)}{V_i(s)} = K\frac{V_b(s)}{V_i(s)} = \frac{K\omega_0{}^2}{s^2 + 2\alpha s + \omega_0{}^2} \tag{6.66}$$

∎

6.4-2   Initial and Final Values

In certain applications, it is desirable to know the values of $f(t)$ as $t \to 0$ and $t \to \infty$ [initial and final values of $f(t)$] from the knowledge of its Laplace transform $F(s)$. Initial and final value theorems provide such information.

The initial value theorem states that if $f(t)$ and its derivative df/dt are both Laplace transformable, then

$$f(0^+) = \lim_{s\to\infty} sF(s) \tag{6.67}$$

provided that the limit on the right-hand side of Eq. (6.67) exists.

The final value theorem states that if both $f(t)$ and df/dt are Laplace transformable, then

$$\lim_{t\to\infty} f(t) = \lim_{s\to 0} sF(s) \tag{6.68}$$

provided that $sF(s)$ has no poles in the RHP or on the imaginary axis. To prove these theorems, we use Eq. (6.34a)

$$sF(s) - f(0^-) = \int_{0^-}^{\infty} \frac{df}{dt}e^{-st}\,dt$$

$$= \int_{0^-}^{0^+} \frac{df}{dt}e^{-st}\,dt + \int_{0^+}^{\infty} \frac{df}{dt}e^{-st}\,dt$$

$$= f(t)\Big|_{0^-}^{0^+} + \int_{0^+}^{\infty} \frac{df}{dt}e^{-st}\,dt$$

$$= f(0^+) - f(0^-) + \int_{0^+}^{\infty} \frac{df}{dt}e^{-st}\,dt$$

Therefore

$$sF(s) = f(0^+) + \int_{0^+}^{\infty} \frac{df}{dt}e^{-st}\,dt$$

and

$$\lim_{s\to\infty} sF(s) = f(0^+) + \lim_{s\to\infty} \int_{0^+}^{\infty} \frac{df}{dt}e^{-st}\,dt$$

$$= f(0^+) + \int_{0^+}^{\infty} \frac{df}{dt}\left(\lim_{s\to\infty} e^{-st}\right)dt$$

$$= f(0^+)$$

Comment: The initial value theorem should be applied only if $F(s)$ is strictly proper ($m < n$), because for $m \geq n$, $\lim_{s \to \infty} sF(s)$ does not exist, and the theorem does not apply.

To prove the final value theorem, we let $s \to 0$ in Eq. (6.34a) to obtain

$$\lim_{s \to 0} \left[sF(s) - f(0^-) \right] = \lim_{s \to 0} \int_{0^-}^{\infty} \frac{df}{dt} e^{-st} \, dt = \int_{0^-}^{\infty} \frac{df}{dt} \, dt$$

$$= f(t) \Big|_{0^-}^{\infty} = \lim_{t \to \infty} f(t) - f(0^-)$$

a deduction which leads to the desired result (6.68)

Comment: The final value theorem applies only if the poles of $sF(s)$ are in the LHP. If there is a pole on the imaginary axis, then $\lim_{s \to 0} sF(s)$ does not exist. If there is a pole in the RHP, $\lim_{t \to \infty} f(t)$ does not exist.

■ **Example 6.17**

Determine the initial and final values of $y(t)$ if its Laplace transform $Y(s)$ is given by

$$Y(s) = \frac{10(2s + 3)}{s(s^2 + 2s + 5)}$$

Equations (6.67) and (6.68) yield

$$y(0^+) = \lim_{s \to \infty} sY(s) = \lim_{s \to \infty} \frac{10(2s + 3)}{(s^2 + 2s + 5)} = 0$$

$$y(\infty) = \lim_{s \to 0} sY(s) = \lim_{s \to 0} \frac{10(2s + 3)}{(s^2 + 2s + 5)} = 6 \quad ■$$

6.5   Block Diagrams

Large systems may consist of an enormous number of components or elements. Analyzing such systems all at once could be next to impossible. Anyone who has seen the circuit diagram of a radio or a TV receiver can appreciate this fact. In such cases, it is convenient to represent a system by suitably interconnected subsystems, each of which can be readily analyzed. Each subsystem can be characterized in terms of its input-output relationships. A linear (sub)system can be characterized by its transfer function $H(s)$. Figure 6.18a shows a block diagram of a system with a transfer function $H(s)$ and its input and output represented by their frequency domain descriptions $F(s)$ and $Y(s)$ respectively.

Subsystems may be interconnected by using three elementary types of interconnections (Figs. 6.18b, 6.18c, 6.18d): cascade, parallel, and feedback. When two transfer functions appear in cascade, as depicted in Fig. 6.18b, the transfer function of the overall system is the product of the two transfer functions. This conclusion follows from the fact that in Fig. 6.18b

$$\frac{Y(s)}{F(s)} = \frac{W(s)}{F(s)} \frac{Y(s)}{W(s)} = H_1(s)H_2(s)$$

This result can be extended to any number of transfer functions in cascade.

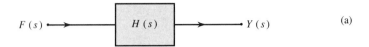

(a)

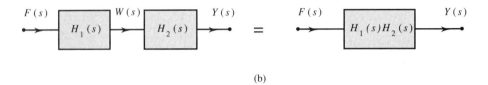

(b)

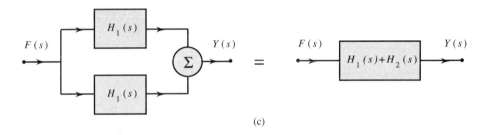

(c)

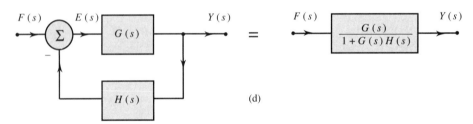

(d)

Fig. 6.18 Elementary connections of blocks and their equivalents.

Similarly, when two transfer functions, $H_1(s)$ and $H_2(s)$, appear in parallel, as illustrated in Fig. 6.18c, the overall transfer function is given by $H_1(s) + H_2(s)$, the sum of the two transfer functions. The proof is trivial. This result can be extended to any number of systems in parallel.

When the output is fed back to the input, as shown in Fig. 6.18d, the overall transfer function $Y(s)/F(s)$ can be computed as follows. The inputs to the summer are $F(s)$ and $-H(s)Y(s)$. Therefore, $E(s)$, the output of the summer, is

$$E(s) = F(s) - H(s)Y(s)$$

But

$$Y(s) = G(s)E(s)$$

$$= G(s)[F(s) - H(s)Y(s)]$$

Therefore

$$Y(s)\,[1 + G(s)H(s)] = G(s)F(s)$$

so that

$$\frac{Y(s)}{F(s)} = \frac{G(s)}{1 + G(s)H(s)} \tag{6.69}$$

Therefore, the feedback loop can be replaced by a single block with the transfer function shown in Eq. (6.69) (see Fig. 6.18d).

In deriving these equations, we implicitly assume that when the output of one subsystem is connected to the input of another subsystem, the latter does not load the former. For example, the transfer function $H_1(s)$ in Fig. 6.18b is computed by assuming that the second subsystem $H_2(s)$ was not connected. This is the same as assuming that $H_2(s)$ does not load $H_1(s)$. In other words, the input-output relationship of $H_1(s)$ will remain unchanged regardless of whether $H_2(s)$ is connected or not. Many modern circuits use op amps with high input impedances, so this assumption is justified. When such an assumption is not valid, $H_1(s)$ must be computed under operating conditions (that is, when $H_2(s)$ is connected).

The MATLAB example C6.3 allows us to determine the transfer function of the feedback system in Fig. 6.18d [Eq. (6.69)], when the transfer functions $G(s)$ and $H(s)$ are given.

⊙ **Computer Example C6.3**

Find the transfer function of the feedback system of Fig. 6.18d when $G(s) = \frac{K}{s(s+8)}$, $H(s) = 1$, and $K = 7$, 16, and 80.

For the sake of generalization, we shall split $G(s)$ into two terms $G_1(s) = K$ and $G_2(s) = \frac{1}{s(s+8)}$. Such generalization allows us to use this program when $G(s)$ is made up of two subsystems in cascade.

```
% (c62.m)
G1num=[0 0 K];G1den=[0 0 1];
G2num=[0 0 1];G2den=[1 8 0];
Hnum=[0 0 1];Hden=[0 0 1];
[Gnum,Gden]=series(G1num,G1den,G2num,G2den);
[ClGnum,ClGden]=feedback(Gnum,Gden,Hnum,Hden);
printsys(ClGnum,ClGden)
```

The feedback transfer function, when $K = 7$, 16, and 80, can be obtained by the following MATLAB commands:

```
K=7;c62
num/den =

           7
    -------------
    s^2 + 8 s + 7

K=16,c62
num/den =

          16
    -------------
    s^2 + 8 s + 16

K=80;c62
num/den =

          80
    -------------
    s^2 + 8 s + 80
```
⊙

6.6   System realization

We now develop a systematic method for realization (or simulation) of an arbitrary nth-order transfer function. Since realization is basically a synthesis problem, there is no unique way of realizing a system. A given transfer function can be realized in many different ways. We present here three different ways of realization: canonical, cascade and parallel realization. The second form of canonical realization is discussed in Appendix 6.1 at the end of this chapter. A transfer function $H(s)$ can be realized by using integrators or differentiators along with summers and multipliers. For practical reasons we avoid the use of differentiators. A differentiator accentuates high-frequency signals, which, by their nature, have large rates of change (large derivative). Signals, in practice, are always corrupted by noise, which happens to be a broad-band signal; that is, the noise contains components of frequencies ranging from low to very high. In processing desired signals by a differentiator, the high-frequency components of noise are amplified disproportionately. Such amplified noise may swamp the desired signal. The integrator, in contrast, tends to suppress a high frequency signal by smoothing it out. In addition, practical differentiators built with op amp circuits tend to be unstable. For these reasons we avoid differentiators in practical realizations.

Consider an LTIC system with a transfer function

$$H(s) = \frac{b_m s^m + b_{m-1} s^{m-1} + \cdots + b_1 s + b_0}{s^n + a_{n-1} s^{n-1} + \cdots + a_1 s + a_0}$$

For large s $(s \to \infty)$

$$H(s) \simeq b_m s^{m-n}$$

Therefore, for $m > n$, the system acts as an $(m - n)$th-order differentiator [see Eq. (6.55)]. For this reason, we restrict $m \leq n$ for practical systems. With this restriction, the most general case is $m = n$ with the transfer function

$$H(s) = \frac{b_n s^n + b_{n-1} s^{n-1} + \cdots + b_1 s + b_0}{s^n + a_{n-1} s^{n-1} + \cdots + a_1 s + a_0} \qquad (6.70)$$

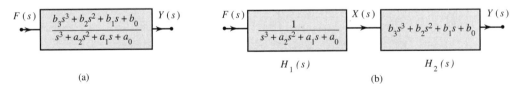

(a)   (b)

Fig. 6.19   Realization of a transfer function in two steps.

6.6-1   Canonical or Direct-Form Realization

Rather than realizing the general nth-order system described by Eq. (6.70), we begin with a specific case of the following third-order system and then extend the results to the nth-order case

$$H(s) = \frac{b_3 s^3 + b_2 s^2 + b_1 s + b_0}{s^3 + a_2 s^2 + a_1 s + a_0} \qquad (6.71)$$

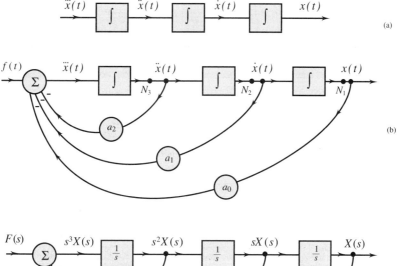

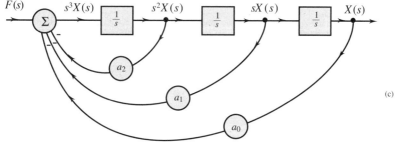

Fig. 6.20 Realization of $H_1(s) = \frac{1}{s^3+a_2s^2+a_1s+a_0}$

For convenience of realization, we shall express this transfer function as a cascade of two transfer functions $H_1(s)$ and $H_2(s)$, as depicted in Fig. 6.19.

$$H(s) = \underbrace{\left(\frac{1}{s^3 + a_2s^2 + a_1s + a_0}\right)}_{H_1(s)} \underbrace{\left(b_3s^3 + b_2s^2 + b_1s + b_0\right)}_{H_2(s)} \tag{6.72}$$

The output of $H_1(s)$ is denoted by $X(s)$, as illustrated in Fig. 6.19b. Therefore

$$Y(s) = \left(b_3s^3 + b_2s^2 + b_1s + b_0\right) X(s) \tag{6.73}$$

and

$$X(s) = \frac{1}{s^3 + a_2s^2 + a_1s + a_0}F(s) \tag{6.74}$$

We shall first realize $H_1(s)$. Equation (6.74) enables us to write the differential equation relating $x(t)$ to $f(t)$ as

$$(D^3 + a_2D^2 + a_1D + a_0)x(t) = f(t) \tag{6.75a}$$

or

$$\ddot{x}(t) + a_2\ddot{x}(t) + a_1\dot{x}(t) + a_0x(t) = f(t) \tag{6.75b}$$

a deduction which yields

$$\ddot{x}(t) = -a_2\ddot{x}(t) - a_1\dot{x}(t) - a_0x(t) + f(t) \tag{6.75c}$$

Our task is to realize a system whose input $f(t)$ and output $x(t)$ satisfy Eq. (6.75c). Let us assume that $\ddot{x}$ is available. Successive integration of $\dddot{x}$ yields $\ddot{x}$, $\dot{x}$, and x (Fig. 6.20a). We can now generate $\dddot{x}$ from $f(t)$, x, $\dot{x}$, and $\ddot{x}$ according to Eq. (6.75c), using the feedback connections, as shown in Fig. 6.20b.† The reader can verify that the system in Fig. 6.20b indeed satisfies Eq. (6.75c). Clearly, the transfer function of this system (Fig. 6.20b) is $H_1(s)$ in Eq. (6.72). The signals $x(t)$, $\dot{x}(t)$, and $\ddot{x}(t)$ are available at points N_1, N_2, and N_3, respectively. If the initial conditions $x(0)$, $\dot{x}(t)$, and $\ddot{x}(0)$ are nonzero, they should be added at points N_1, N_2, and N_3, respectively.‡ Figure 6.20c shows the system of Fig. 6.20b in frequency domain. Each integrator is represented by the transfer function $1/s$ [see Eq. (6.56)]. Signals $f(t)$, $x(t)$, $\dot{x}(t)$, $\ddot{x}(t)$, and $\dddot{x}(t)$ are represented by their Laplace transforms $F(s)$, $X(s)$, $sX(s)$, $s^2X(s)$, and $s^3X(s)$, respectively.

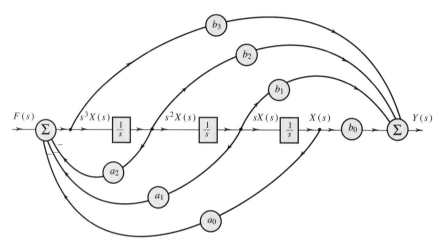

Fig. 6.21 Realization of $H(s) = \dfrac{b_3 s^3 + b_2 s^2 + b_1 s + b_0}{s^3 + a_2 s^2 + a_1 s + a_0}$

Figure 6.20c is a realization of the transfer function $H_1(s)$ in Eq. (6.72). To realize the transfer function $H(s)$, we need to augment $H_1(s)$ so that the final output $Y(s)$ is generated from $X(s)$ according to Eq. (6.73):

$$Y(s) = (b_3 s^3 + b_2 s^2 + b_1 s + b_0)X(s)$$

$$= b_3 s^3 X(s) + b_2 s^2 X(s) + b_1 s X(s) + b_0 X(s)$$

Signals $X(s)$, $sX(s)$, $s^2X(s)$, and $s^3X(s)$ are available at various points in Fig. 6.20c so that $Y(s)$ can be generated by using feedforward connections to the output summer, as depicted in Fig. 6.21. Therefore, the realization in Fig. 6.21 has the desired transfer function $H(s)$ in Eq. (6.71).

---

†It may seem odd that we first assumed the existence of $\dddot{x}$ and generated $\ddot{x}, \dot{x}, x$ by its successive integration, and then in turn generated $\dddot{x}$ from $\ddot{x}, \dot{x}$, and x. This procedure poses a dilemma similar to "Which came first, the chicken or egg?" The problem here is satisfactorily resolved by writing the expression for $\dddot{x}$ at the output of the summer in Fig. 6.20b and verifying that this expression is indeed the same as Eq. (6.75c).

‡For example, the initial condition $x(0)$ can be incorporated into our realization by adding a constant signal of value $x(0)$ at point N_1. This follows from the fact that $\int_0^t \dot{x}(\tau)\,d\tau = x(t) - x(0)$, and $x(t) = x(0) + \int_0^t \dot{x}(\tau)\,d\tau$. Similarly, the initial condition $\dot{x}(0)$ should be added at point N_2, and so on.

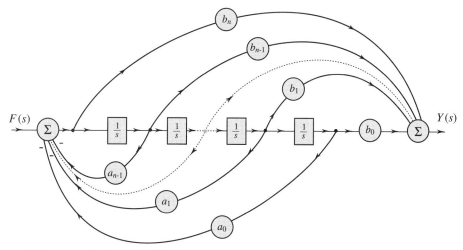

Fig. 6.22 Realization of the nth-order transfer function $H(s)$ in Eq. (6.70).

Generalization of this realization for the nth-order transfer function in Eq. (6.70) is shown in Fig. 6.22. This is one of the two **canonical** realizations (also known as the **controller canonical** or **direct-form** realization). The second canonical realization (**observer** canonical realization) is discussed in Appendix 6.1 at the end of this chapter. Observe that n integrators are required for a realization of an nth-order transfer function. The canonical (direct-form) realization procedure of an nth-order transfer function is systematic and straightforward, as illustrated in Fig. 6.22. After examining this figure, we can summarize the procedure as follows:

1. Draw an input summer followed by n integrators in cascade.

2. Draw the n feedback connections from the output of each of the n integrators to the input summer. The n feedback coefficients are $a_0, a_1, a_2, \cdots, a_{n-1}$, respectively, and the feedback connections have negative signs (for subtraction) at the input summer.

3. Draw the $n + 1$ feedforward connections to the output summer from the outputs of all the n integrators and the input summer. The $n + 1$ feedforward coefficients are $b_0, b_1, b_2, \cdots, b_n$, respectively, and the feedforward connections have positive signs (for addition) at the output summer. Observe that the connections a_k and b_k start from the same point. Thus, the connections a_0 and b_0 start at the same point, and so do connections a_1 and b_1, and so on.

Note that a_n is assumed to be unity and does not appear explicitly anywhere in the realization. If $a_n \neq 1$, then $H(s)$ should be normalized by dividing both its numerator and its denominator by a_n.

■ **Example 6.18**

Find the canonical realization of the following transfer functions.

(a) $\dfrac{5}{s+2}$ (b) $\dfrac{s+5}{s+7}$ (c) $\dfrac{s}{s+7}$ (d) $\dfrac{4s+28}{s^2+6s+5}$

All four of these transfer functions are special cases of $H(s)$ in Eq. (6.70).

(a) In this case, the transfer function is of the first order ($n = 1$); therefore, we need only one integrator for its realization. The feedback and feedforward coefficients are

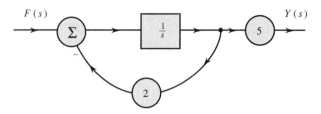

$F(s)$ $Y(s)$

Fig. 6.23 Realization of $\frac{5}{s+2}$

$$a_0 = 2 \quad \text{and} \quad b_0 = 5 \quad b_1 = 0$$

The realization is depicted in Fig. 6.23. Because $n = 1$, there is a single feedback connection from the output of the integrator to the input summer with coefficient $a_0 = 2$. For $n = 1$, there are $n + 1 = 2$ feedforward connections in general. However, in this case, $b_1 = 0$, and there is only one feedforward connection with coefficient $b_0 = 5$ from the output of the integrator to the output summer. Observe that because there is only one signal to be summed at the output summer, a summer is not needed. For this reason, the output summer is omitted in Fig. 6.23.

(b)
$$H(s) = \frac{s+5}{s+7}$$

The realization appears in Fig. 6.24. Here $H(s)$ is a first order transfer function with $a_0 = 7$ and $b_0 = 5$, $b_1 = 1$. There is a single feedback connection (with coefficient 7) from the integrator output to the input summer. There are two feedforward connections from the outputs of the integrator and the input summer to the output summer.†

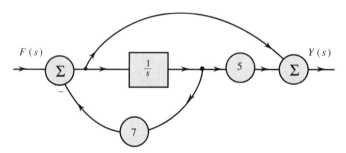

$F(s)$ $Y(s)$

Fig. 6.24 Realization of $\frac{s+5}{s+7}$

(c)
$$H(s) = \frac{s}{s+7}$$

This first-order transfer function is similar to that in **(b)**, except that $b_0 = 0$. Therefore, the realization is similar to that in Fig. 6.24 with the feedforward connection from the output of the integrator missing, as depicted in Fig. 6.25a. Also, because there is only

---

†When $m = n$ (as in this case), $H(s)$ can also be realized in another way by recognizing that
$$H(s) = 1 - \frac{2}{s+7}$$

We now realize $H(s)$ as a parallel combination of two transfer functions, as indicated by the above equation.

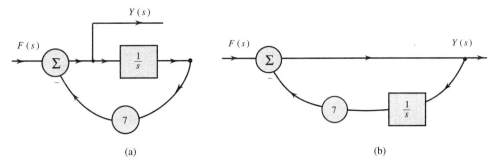

Fig. 6.25 Realization of $\frac{s}{s+7}$

one signal to be summed at the output summer, the output summer is omitted. The realization in Fig. 6.25a is redrawn in a more convenient form, as illustrated in Fig. 6.25b.

(d)

$$H(s) = \frac{4s + 28}{s^2 + 6s + 5}$$

This is a second-order system with $b_0 = 28$, $b_1 = 4$, $b_2 = 0$, $a_0 = 5$, $a_1 = 6$.

Figure 6.26 shows a realization with two feedback connections and two feedforward connections. ■

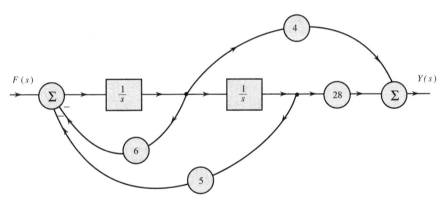

Fig. 6.26 Realization of $\frac{4s+28}{s^2+6s+5}$

△ **Exercise E6.9**
   Realize the transfer function

$$H(s) = \frac{2s}{s^2 + 6s + 25} \qquad \triangledown$$

6.6-2 Cascade and Parallel Realizations

An nth-order transfer function $H(s)$ can be expressed as a product or a sum of n first-order transfer functions. Accordingly, we can also realize $H(s)$ as a cascade (series) or parallel form of these n first-order transfer functions. Consider, for instance, the transfer function in part **(d)** of the last example:

$$H(s) = \frac{4s + 28}{s^2 + 6s + 5}$$

We can express $H(s)$ as

$$H(s) = \frac{4s + 28}{(s + 1)(s + 5)} = \underbrace{\left(\frac{4s + 28}{s + 1}\right)}_{H_1(s)} \underbrace{\left(\frac{1}{s + 5}\right)}_{H_2(s)} \tag{6.76a}$$

We can also express $H(s)$ as a sum of partial fractions as

$$H(s) = \frac{4s + 28}{(s + 1)(s + 5)} = \underbrace{\frac{6}{s + 1}}_{H_3(s)} - \underbrace{\frac{2}{s + 5}}_{H_4(s)} \tag{6.76b}$$

Equation (6.76) gives us the option of realizing $H(s)$ as a cascade of $H_1(s)$ and $H_2(s)$, as shown in Fig. 6.27a, or a parallel of $H_3(s)$ and $H_4(s)$, as depicted in Fig. 6.27b. Each of the first-order transfer functions in Figs. 6.27a or 6.27b can be realized by using a single integrator, as discussed in Example 6.18.

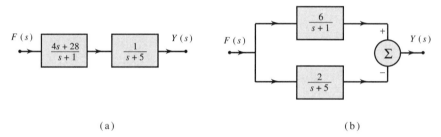

(a)   (b)

Fig. 6.27 Realization of $\frac{4s+28}{(s+1)(s+5)}$: (a) cascade form (b) parallel form.

We have presented here three forms of realization (canonical, cascade, and parallel). The second canonical realization is developed in Appendix 6.1. However, this discussion by no means exhausts all the possibilities. Moreover, in the cascade form, there are different ways of grouping the factors in the numerator and the denominator of $H(s)$. Accordingly, several cascade forms are possible.

From a practical viewpoint, parallel and cascade forms are preferable because parallel and certain cascade forms are numerically less sensitive than canonical forms to small parameter variations in the system. Qualitatively, this difference can be explained by the fact that in a canonical realization all the coefficients interact with each other, and a change in any coefficient will be magnified through its repeated influence from feedback and feedforward connections. In a parallel realization, in contrast, the change in a coefficient will affect only a localized segment; the case with a cascade realization is similar.

In the above examples of cascade and parallel realization, we have separated $H(s)$ into first-order factors. For $H(s)$ of higher orders, we could group $H(s)$ into factors, not all of which are necessarily of the first order. For example, if $H(s)$ is a third-order transfer function, we could realize this function as a cascade (or a parallel) combination of a first-order and a second-order factor.

Realization of Complex Conjugate Poles

The complex poles in $H(s)$ should be realized as a second-order (quadratic) factor because we cannot implement multiplication by complex numbers. Consider, for example,

$$H(s) = \frac{10s + 50}{(s + 3)(s^2 + 4s + 13)}$$

$$= \frac{10s + 50}{(s + 3)(s + 2 - j3)(s + 2 + j3)}$$

$$= \frac{2}{s + 3} - \frac{1 + j2}{s + 2 - j3} - \frac{1 - j2}{s + 2 + j3}$$

We cannot realize first-order transfer functions individually with the poles $-2 \pm j3$ because they require multiplication by complex numbers in the feedback and the feedforward paths. Therefore, we need to combine the conjugate poles and realize them as a second-order transfer function.† In the present case, we can express $H(s)$ as

$$H(s) = \left(\frac{10}{s + 3} \right) \left(\frac{s + 5}{s^2 + 4s + 13} \right) \tag{6.77a}$$

$$= \frac{2}{s + 3} - \frac{2s - 8}{s^2 + 4s + 13} \tag{6.77b}$$

Now we can realize $H(s)$ in cascade form using Eq. (6.77a) or in parallel form using Eq. (6.77b).

Realization of Repeated Poles

When repeated poles occur, the procedure for canonical and cascade realization is exactly the same as above. In parallel realization, however, the procedure requires a special precaution, as explained in Example 6.19 below.

■ **Example 6.19**

Determine the parallel realization of

$$H(s) = \frac{7s^2 + 37s + 51}{(s + 2)(s + 3)^2}$$

$$= \frac{5}{s + 2} + \frac{2}{s + 3} - \frac{3}{(s + 3)^2}$$

This third-order transfer function should require no more than three integrators. But if we try to realize each of the three partial fractions separately, we require four integrators because one of the terms is second-order. This difficulty can be avoided by observing that the terms $1/(s + 3)$ and $1/(s + 3)^2$ can be realized with a cascade of two subsystems, each having a transfer function $1/(s + 3)$, as shown in Fig. 6.28. Each of the three first-order transfer functions in Fig. 6.28 may now be realized as in Fig. 6.23. ■

△ **Exercise E6.10**

Find a canonical, a cascade, and a parallel realization of

$$H(s) = \frac{s + 3}{s^2 + 7s + 10} = \left(\frac{s + 3}{s + 2} \right) \left(\frac{1}{s + 5} \right) \qquad \triangledown$$

---

†It is possible to realize complex, conjugate poles indirectly by using a cascade of two first-order transfer functions. A transfer function with poles $-a \pm jb$ can be realized by using a cascade of two identical first-order transfer functions, each having a pole at $-a$. (See Prob. 6.6-7.)

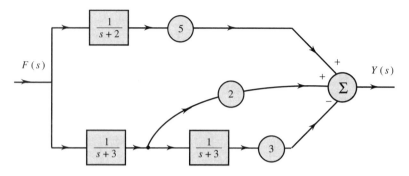

Fig. 6.28 Parallel realization of $\frac{7s^2+37s+51}{(s+2)(s+3)^2}$

6.6-3 System Realization Using Operational Amplifiers

In this section, we discuss practical implementation of the realizations described in the previous subsection. Earlier we saw that the basic elements required for the synthesis of an LTIC system (or a given transfer function) are (scalar) multipliers, integrators, and summers (or adders). All these elements can be realized by operational amplifier (op amp) circuits, as explained below.

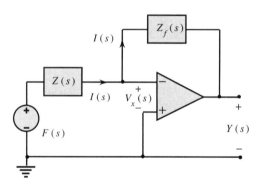

Fig. 6.29 A basic inverting configuration op amp circuit.

Operational Amplifier Circuits

Figure 6.29 shows an op amp circuit in the frequency domain (the transformed circuit). Because the input impedance of the op amp is infinite (very high), all of the current $I(s)$ flows in the feedback path, as illustrated in Fig. 6.29. Moreover $V_x(s)$, the voltage at the input of the op amp, is zero (very small) because of the infinite (very large) gain of the op amp. Therefore, for all practical purposes,

$$Y(s) = -I(s)Z_f(s)$$

Moreover, because $v_x \approx 0$,

$$I(s) = \frac{F(s)}{Z(s)}$$

Substitution of the second equation in the first yields

$$Y(s) = -\frac{Z_f(s)}{Z(s)}F(s)$$

Therefore, the op amp circuit in Fig. 6.29 has the transfer function

$$H(s) = -\frac{Z_f(s)}{Z(s)} \qquad (6.78)$$

By properly choosing $Z(s)$ and $Z_f(s)$, we can obtain a variety of transfer functions, as the following development shows.

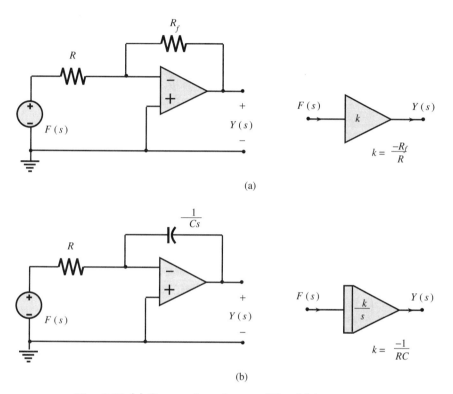

(a)

(b)

Fig. 6.30 (a) Op amp inverting amplifier (b) integrator.

The Scalar Multiplier

If we use a resistor R_f in the feedback and a resistor R at the input (Fig. 6.30a), then $Z_f(s) = R_f$, $Z(s) = R$, and

$$H(s) = -\frac{R_f}{R} \qquad (6.79a)$$

The system acts as a scalar multiplier (or an amplifier) with a negative gain $\frac{R_f}{R}$. A positive gain can be obtained by using two such multipliers in cascade or by using a single noninverting amplifier, as depicted in Fig. 6.16c. Figure 6.30a also shows the compact symbol used in circuit diagrams for a scalar multiplier.

The Integrator

If we use a capacitor C in the feedback and a resistor R at the input (Fig. 6.30b), then $Z_f(s) = 1/Cs$, $Z(s) = R$, and

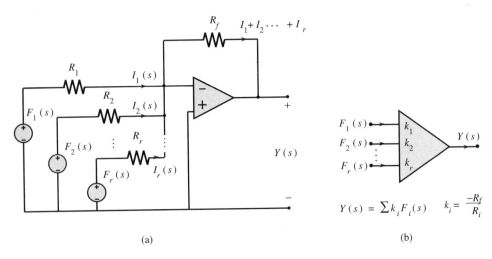

Fig. 6.31 Op amp summing and amplifying circuit.

$$H(s) = \left(-\frac{1}{RC}\right)\frac{1}{s} \tag{6.79b}$$

The system acts as an ideal integrator with a gain $-1/RC$. Figure 6.30b also shows the compact symbol used in circuit diagrams for an integrator.

The Summer

Consider now the circuit in Fig. 6.31a with r inputs $F_1(s)$, $F_2(s)$, ..., $F_r(s)$. As usual, the input voltage $V_x(s) \simeq 0$ because the gain of op amp $\to \infty$. Moreover, the current going into the op amp is very small ($\simeq 0$) because the input impedance $\to \infty$. Therefore, the total current in the feedback resistor R_f is $I_1(s) + I_2(s) + \cdots + I_r(s)$. Moreover, because $V_x(s) = 0$,

$$I_j(s) = \frac{F_j(s)}{R_j} \qquad j = 1, 2, \ldots, r$$

Also

$$\begin{aligned}
Y(s) &= -R_f \left[I_1(s) + I_2(s) + \cdots + I_r(s) \right] \\
&= -\left[\frac{R_f}{R_1} F_1(s) + \frac{R_f}{R_2} F_2(s) + \cdots + \frac{R_f}{R_r} F_r(s) \right] \\
&= k_1 F_1(s) + k_2 F_2(s) + \cdots + k_r F_r(s)
\end{aligned} \tag{6.80}$$

where

$$k_i = \frac{-R_f}{R_i}$$

Clearly, the circuit in Fig. 6.31 serves a summer and an amplifier with any desired gain for each of the input signals. Figure 6.31b shows the compact symbol used in circuit diagrams for a summer with r inputs.

■ **Example 6.20**

Using op amp circuits, realize the canonical form of the transfer function

$$H(s) = \frac{2s + 5}{s^2 + 4s + 10}$$

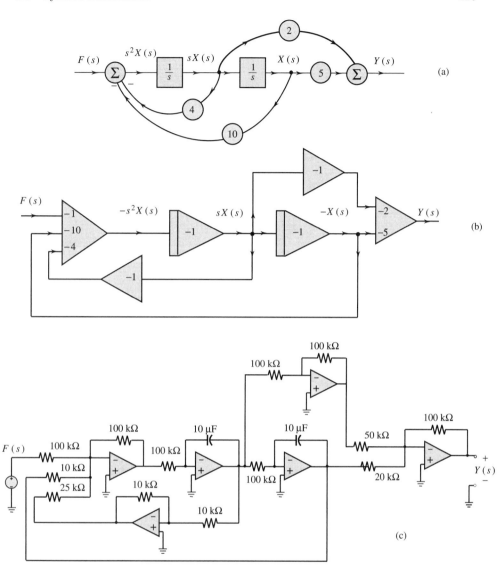

Fig. 6.32 Op amp realization of a second-order transfer function $\frac{2s+5}{s^2+4s+10}$

The basic canonical realization is shown in Fig. 6.32a. Signals at various points are also indicated in the realization. Op amp elements (multipliers, integrators, and summers) change the polarity of the output signals. To incorporate this fact, we modify the canonical realization in Fig. 6.32a to that depicted in Fig. 6.32b. In Fig. 6.32a, the successive outputs of the summer and the integrators are $s^2X(s)$, $sX(s)$, and $X(s)$ respectively. Because of polarity reversals in op amp circuits, these outputs are $-s^2X(s)$, $sX(s)$, and $-X(s)$ respectively in Fig. 6.32b. This polarity reversal requires corresponding modifications in the signs of feedback and feedforward gains. According to Fig. 6.32a

$$s^2X(s) = F(s) - 4sX(s) - 10X(s)$$

Therefore

$$-s^2X(s) = -F(s) + 4sX(s) + 10X(s)$$

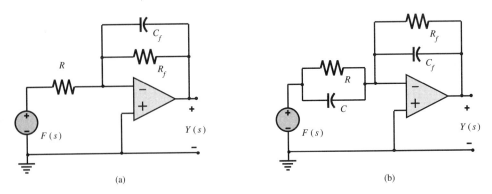

Fig. 6.33 Op amp circuits for exercise E6.11.

Because the summer gains are always negative (see Fig. 6.31b), we rewrite the above equation as

$$-s^2 X(s) = -1[F(s)] - 4[-sX(s)] - 10[-X(s)]$$

Figure 6.32b shows the implementation of this equation. The hardware realization appears in Fig. 6.32c. Both integrators have a unity gain which requires $RC = 1$. We have used $R=100$ kΩ and $C = 10\mu$F. The gain of 10 in the outer feedback path is obtained in the summer by choosing the feedback resistor of the summer to be 100 kΩ and an input resistor of 10 kΩ. Similarly, the gain of 4 in the inner feedback path is obtained by using the corresponding input resistor of 25 kΩ. The gains of 2 and 5, required in the feedforward connections, are obtained by using a feedback resistor of 100 kΩ and input resistors of 50 kΩ and 20 kΩ respectively.

The op amp realization in Fig. 6.32 is not necessarily the one that uses the fewest op amps. It is possible to avoid the two inverting op amps (with gain -1) in Fig. 6.32 by adding signal $sX(s)$ to the input and output summers directly, using the noninverting amplifier configuration in Fig. 6.16. There are also circuits (such as Sallen-Key) which realize a first- or second-order transfer function using only one op amp. ∎

△ **Exercise E6.11**
  Show that the transfer functions of the op amp circuits in Figs. 6.33a and 6.33b are $H_1(s)$ and $H_2(s)$, respectively, where

$$H_1(s) = \frac{-R_f}{R}\left(\frac{a}{s+a}\right) \qquad a = \frac{1}{R_f C_f}$$

$$H_2(s) = -\frac{C}{C_f}\left(\frac{s+b}{s+a}\right) \qquad a = \frac{1}{R_f C_f} \quad b = \frac{1}{RC} \quad \triangledown$$

6.7   Application to Feedback and Controls

Generally, systems are designed to produce a desired output $y(t)$ for a given input $f(t)$. Using the given performance criteria, we can design a system shown in Fig. 6.34a. Ideally, such an open-loop system should yield the desired output. In practice, however, the system characteristics change with time, as a result of aging or replacement of some components, or because of changes in the environment in which the system is operating. Hence, for a given input, the output of the system will also change with time. This condition is clearly undesirable in precision systems.

A possible solution to this problem is to apply an input that is not a predetermined function of time, but which will change to counteract the effects of changing system characteristics and the environment. In short, we must provide a correction at the system input to account for the undesired changes mentioned above. But these changes are generally unpredictable, and it is difficult to program appropriate corrections to the input. However, the difference between the actual output and the desired output clearly indicates the suitable correction to be applied to the system input. Hence, we could make the input $f(t)$ proportional to the desired output, and feed back the actual output $y(t)$ to the input for comparison. The difference acts as the corrected input to the system. The input to the system is therefore continuously adjusted to obtain the desired response. Such systems (Fig. 6.34b) are called **feedback** or **closed-loop** systems for obvious reasons. We observe thousands of examples of feedback systems around us in everyday life. Most social, economical, educational, and political processes are, in fact, feedback processes. The human body itself is a fine example of a feedback system; almost all of our actions are the product of feedback mechanism. The human sensors such as eyes, ears, nose, tongue, and touch are continuously monitoring the state of our system. This information is fed back to the brain, which acts as a controller to apply corrected inputs through our motor mechanisms and thus accomplish the desired objective.

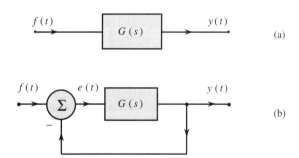

Fig. 6.34 Open-loop and closed-loop (feedback) systems.

Consider the process of driving an automobile. Our senses are continuously feeding the information to the brain. Eyes report seeing a child on the road. The brain will immediately apply input to the arms to steer the car away from the child. There is a red light ahead. Again the brain will apply the input to the legs to brake the car. Suddenly, ears report an ambulance siren; the brain will apply corrective input to steer the car off the street temporarily. A foul smell is reported by the nose. The brain will again take measures to speed away the car from the spot. In this example, the child, the red light, the ambulance siren, and the foul smell are unpredictable changes in the environment (fed back to the input by our senses). Despite these unpredictable changes in the environment, the process of reaching the destination is completed because of feedback. The foregoing is an example of multiple variable feedback.

A feedback system can address the problems arising because of unwanted disturbances such as random-noise signals in electronic systems, a gust of wind affecting a tracking antenna, a meteorite hitting a spacecraft, and the rolling motion of antiaircraft gun platforms mounted on ships or moving tanks. Feedback may also be used to reduce nonlinearities in a system, or control its rise time (or bandwidth). Feedback is used to achieve, with a given system, the desired objective within a given tolerance, despite partial ignorance of the system and the environment. A feedback system, thus, has an ability for supervision and self-correction in the face of changes in the system parameters, and external disturbances (change in the environment). Consider the feedback amplifier in Fig. 6.35. Let the forward amplifier gain $G = 10,000$. One hundredth of the output is fed back to the input ($H = 0.01$). The gain T of the feedback amplifier is obtained by [see Eq. ((6.69)]

$$T = \frac{G}{1 + GH} = \frac{10,000}{1 + 100} = 99.01$$

Suppose that because of aging or replacement of some transistors, the gain G of the forward amplifier changes from 10,000 to 20,000. The new gain of the feedback amplifier is given by

$$T = \frac{G}{1 + GH} = \frac{20,000}{1 + 200} = 99.5$$

Observe that 100% variation in the forward gain G causes only 0.5% variation in the feedback amplifier gain T.

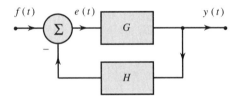

Fig. 6.35 Effects of positive and negative feedback.

Now, consider what happens when we add (instead of subtract) the signal fed back to the input. Such addition means the sign on the feedback connection is $+$ instead of $-$ (which is same as changing the sign of H in Fig. 6.35). Consequently

$$T = \frac{G}{1 - GH}$$

If we let $G = 10,000$ as before and $H = 0.9 \times 10^{-4}$, then

$$T = \frac{10,000}{1 - 0.9(10^4)(10^{-4})} = 100,000$$

Suppose that because of aging or replacement of some transistors, the gain of the forward amplifier changes to 11,000. The new gain of the feedback amplifier is

$$T = \frac{11,000}{1 - 0.9(11,000)(10^{-4})} = 1,100,000$$

Observe that in this case, mere 10% increase in the forward gain G caused 1000% increase in the gain T (from 100,000 to 1,100,000). Clearly, the amplifier is very sensitive to parameter variations. This behavior is exactly opposite of what was observed earlier, when the signal fed back was subtracted from the input.

What is the difference between the two situations? Crudely speaking, the former case is called the **negative feedback** and the latter is the **positive feedback**. We shall later see that, generally, feedback system cannot be described in such black and white terms. The positive feedback increases gain, but tends to make the system more sensitive to parameter variations. It could also lead to instability. In the above case, when $G = 111,111$, $GH = 1$, and $T = \infty$. The cause of instability is that the signal fed back in this case is exactly equal to the input signal itself because $GH = 1$. Hence, once a signal is applied, no matter how small and how short in duration, it comes back to reinforce the input undiminished, which further passes to the output, and is fed back again and again and again. In essence, the signal perpetuates itself forever. This perpetuation, even when the input ceases to exist, is precisely the symptom of instability.

6.7-1   Analysis of a Simple Control System

Figure 6.36a represents an automatic position control system, which can be used to control the angular position of a heavy object (e.g., a tracking antenna, an anti-aircraft gun mount, or the position of a ship). The input θ_i is the desired angular position of the object, which can be set at any given value. The actual angular position θ_o of the object (the output) is measured by a potentiometer whose wiper is mounted on the output shaft. The difference between the output θ_o and the input θ_i is amplified; the amplified output, which is proportional to $\theta_o - \theta_i$, is applied to the motor input. If $\theta_o - \theta_i = 0$ (the output being equal to the desired angle), there is no input to the motor, and the motor stops. But if $\theta_o \neq \theta_i$, there will be a nonzero input to the motor, which will turn the shaft until $\theta_o = \theta_i$. It is evident that by setting the input potentiometer at a desired position in this system, we can control the angular position of a heavy remote object.

The block diagram of this system is shown in Fig. 6.36b. The amplifier gain is K, where K is adjustable. Let the motor (with load) transfer function that relates the output angle θ_o to the motor input voltage be $G(s)$ [see Eq. (1.65)]. This feedback arrangement is identical to that in Fig. 6.18d with $H(s) = 1$. Hence, $T(s)$, the (closed-loop) system transfer function relating the output θ_o to the input θ_i, is

$$T(s) = \frac{KG(s)}{1 + KG(s)}$$

From this equation, we shall investigate the behavior of the automatic position control system in Fig. 6.36a for a step and a ramp input.

Step Input

If we desire to change the angular position of the object instantaneously, we need to apply a step input. We may then want to know how long the system takes to position itself at the desired angle, whether it reaches the desired angle, and whether it reaches the desired position smoothly (monotonically) or oscillates about the final position. If the system oscillates, we may want to know how long it takes for the oscillations to settle down. All these questions can be readily answered by finding the output $\theta_o(t)$ when the input $\theta_i(t) = u(t)$. A step input implies the instantaneous change in the angle. This input would be one of the most difficult

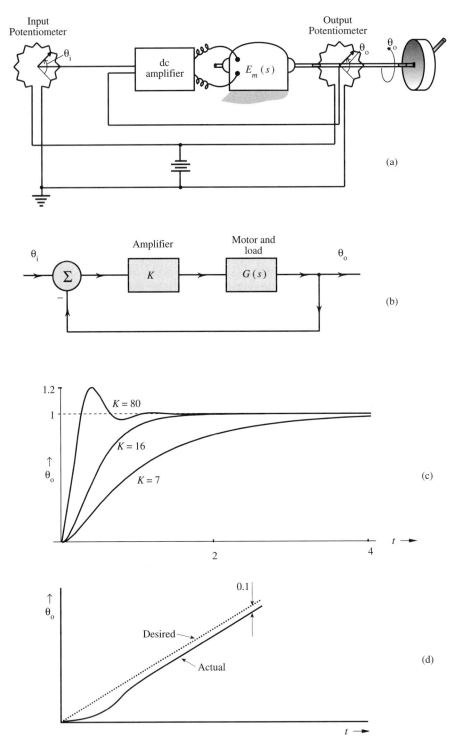

Fig. 6.36 (a) An automatic position control system (b) its block diagram (c) the unit step response (d) the unit ramp response.

to follow; if the system can perform well for this input, it is likely to give a good account of itself under most other expected situations. This is the reason why we test control systems for a step input.

For the step input $\theta_i(t) = u(t)$, $\Theta_i(s) = 1/s$ and

$$\Theta_o(s) = \frac{1}{s}T(s) = \frac{KG(s)}{s[1 + KG(s)]}$$

Let the motor (with load) transfer function relating the load angle $\theta_o(t)$ to the motor input voltage be $G(s) = \frac{1}{s(s+8)}$ [see Eq. (1.65)]. This yields

$$\Theta_o(s) = \frac{\frac{K}{s(s+8)}}{s\left[1 + \frac{K}{s(s+8)}\right]} = \frac{K}{s(s^2 + 8s + K)}$$

Let us investigate the system behavior for three different values of gain K.

1. $K = 7$

$$\Theta_o(s) = \frac{7}{s(s^2 + 8s + 7)} = \frac{7}{s(s+1)(s+7)}$$

$$= \frac{1}{s} - \frac{\frac{7}{6}}{s+1} + \frac{\frac{1}{6}}{s+7}$$

and

$$\theta_o(t) = \left(1 - \tfrac{7}{6}e^{-t} + \tfrac{1}{6}e^{-7t}\right)u(t)$$

This response, illustrated in Fig. 6.36c, appears rather sluggish. To speed up the response let us increase the gain to, say, 80.

2. $K = 80$

$$\Theta_o(s) = \frac{80}{s(s^2 + 8s + 80)} = \frac{80}{s(s+4-j8)(s+4+j8)}$$

$$= \frac{1}{s} + \frac{\frac{\sqrt{5}}{4}e^{j153°}}{s+4-j8} + \frac{\frac{\sqrt{5}}{4}e^{-j153°}}{s+4+j8}$$

and

$$\theta_o(t) = \left[1 + \tfrac{\sqrt{5}}{2}e^{-4t}\cos\left(8t + 153°\right)\right]u(t)$$

This response, depicted in Fig. 6.36c, is certainly faster than in the earlier case ($K = 7$), but unfortunately the improvement is achieved at the cost of ringing (oscillations) with high overshoot. In the present case the **percent overshoot** PO is 21%. The response reaches its peak value at **peak time** $t_p = 0.393$ seconds. The **rise time**, defined as the time required for the response to rise form 10% to 90% of its steady-state value, indicates the speed of response.† In the present case $t_r = 0.175$ seconds. The steady-state value of the response is unity so that the **steady-state error** is zero. Theoretically it takes infinite time for the response to

---

†**Delay time** t_d, defined as the time required for the response to reach 50% of its steady-state value, is another indication of speed. For the present case, $t_d = 0.141$ seconds

reach the desired value of unity. In practice, however, we may consider the response to have settled to the final value if it closely approaches the final value. A widely accepted measure of closeness is within 2% of the final value. The time required for the response to reach and stay within 2% of the final value is called the settling time t_s.† In Fig. 6.36c, we find $t_s \approx 1$ second. A good system has a small overshoot, a small value of t_r and t_s and a small steady-state error.

A large overshoot, as in the present case, may be unacceptable in many applications. Let us try to determine K (the gain) which yields fastest response without oscillations. Complex characteristic roots lead to oscillations; to avoid oscillations, the characteristic roots should be real. In the present case the characteristic polynomial is $s^2 + 8s + K$. For $K > 16$, the characteristic roots are complex; for $K < 16$, the roots are real. The fastest response without oscillations is obtained by choosing $K = 16$. We now consider this case.

3. $K = 16$

$$\Theta_o(s) = \frac{16}{s(s^2 + 8s + 16)} = \frac{16}{s(s+4)^2}$$

$$= \frac{1}{s} - \frac{1}{s+4} - \frac{4}{(s+4)^2}$$

and

$$\theta_o(t) = \left[1 - (4t+1)e^{-4t}\right]u(t)$$

This response also appears in Fig. 6.36c. The system with $K > 16$ is said to be **underdamped** (oscillatory response), whereas the system with $K < 16$ is said to be **overdamped**. For $K = 16$, the system is said to be **critically damped**.

There is a trade-off between undesirable overshoot and rise time. Reducing overshoots leads to higher rise time (sluggish system). In practice, a small overshoot may be acceptable, which is still faster than the critical damping. Note that percent overshoot PO and peak time t_p are meaningless for the overdamped or critically damped cases. In addition to adjusting gain K, we may need to augment the system with some type of compensator if the specifications on overshoot and the speed of response are too stringent.

Ramp Input

If the antiaircraft gun in Fig. 6.36a is tracking an enemy plane moving with a uniform velocity, the gun-position angle must increase linearly with t. Hence, the input in this case is a ramp; that is, $\theta_i(t) = tu(t)$. Let us find the response of the system to this input when $K = 80$. In this case, $\Theta_i(s) = \frac{1}{s^2}$, and

$$\Theta_o(s) = \frac{80}{s^2(s^2 + 8s + 80)} = -\frac{0.1}{s} + \frac{1}{s^2} + \frac{0.1(s-2)}{s^2 + 8s + 80}$$

Use of Table 6.1 yields

$$y(t) = \left[-0.1 + t + \frac{1}{8}e^{-8t}\cos(8t + 36.87°)\right]u(t)$$

---

†Typical percentage values used are 2% to 5% for t_s.

This response, sketched in Fig. 6.36d, shows that there is a steady-state error $e_r = 0.1$ radian. In many cases such a small steady-state error may be tolerable. If, however, a zero steady-state error to a ramp input is required, this system in its present form is unsatisfactory. We must add some form of compensator to the system.

⊙ **Computer Example C6.4**

Find the step response of the feedback system in Fig. 6.36b with $G(s) = \frac{1}{s(s+8)}$ if $K = 7$, 16, and 80. Find the unit ramp response of this system for the case $K = 80$.

In computer example C6.3, we have obtained the transfer functions of this feedback system. Here, we shall redo this part again in another way to illustrate the use of 'conv' command when the denominator of $G(s)$ is made up of two factors $D_1(s)$ and $D_2(s)$. The command 'conv' multiplies $D_1(s)$ with $D_2(s)$ and gives the coefficients of the product $D_1(s)D_2(s)$. To find the step and ramp response for a given value of K, we first create an m-file c64a.m as

```
% (c64a.m)
Gnum=[0 0 K];Gden=conv([0 1 0],[0 1 8]);
Hnum=[0 0 1];Hden=[0 0 1];
[NumTF,DenTF]=feedback(Gnum,Gden,Hnum,Hden);
step(NumTF,DenTF)
```

To plot the step response (as in Fig. 6.36c), we create another file c64b.m as:

```
K=7;c64a; hold on,
K=16;c64a; K=80;c64a
```

The unit ramp response of this system is the same as the unit step response of a system with transfer function $T(s)/s$. Hence, we can use the file c64a to find the ramp response. To plot the ramp response, for $K = 80$, we create c64c.m file as follows:

```
K=80; c64a;
NumTFr=NumTF; DenTFr=conv([0 1 0],DenTF);
printsys(NumTFr,DenTFr);
step(NumTFr,DenTFr)   ⊙
```

Design Specifications

The above discussion has given the reader some idea of the various specifications a control system might require. In general we may be required to design a control system to meet some or all of the following specifications:

1. **Transient Response**

   (a) Specified overshoot to step input.

   (b) Specified rise time t_r and /or delay time t_d.

   (c) Specified settling time t_s.

2. **Steady-State Error**

   Specified steady-state error to certain expected inputs such as step, ramp, or parabolic inputs. In control systems, the transient response is generally specified for the step input. The reason is that the step input represents a sudden jump discontinuity. Hence if a system has an acceptable transient response for step input, it is likely to have acceptable transient response for most of the practical inputs. Steady-state errors, however, must be specified for typical inputs of the system. For the given system one must determine what kind of input (step, ramp, etc) are likely to occur, and then specify acceptable steady-state requirements for these inputs.

3. **Sensitivity**
   The system should satisfy a specified sensitivity specifications to some system parameter variations, or to certain disturbances. Sensitivity analysis will not be considered here.

6.7-2 Analysis of a Second-Order System

The transient response depends upon the location of poles and zeros of the transfer function $T(s)$. For a general case, however, there is no quick way of predicting transient response parameters (PO, t_r, t_s) from the knowledge of poles and zeros of $T(s)$. However, for a second-order system with no zeros, there is a direct relationship between the pole locations and the transient response. In such a case, the pole locations can be immediately determined from the knowledge of the transient parameter specifications. As we shall see, the study of the second-order system can be used to study many higher-order systems. For this reason we shall now study the behavior of a second-order system in detail.

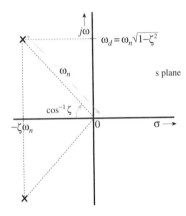

Fig. 6.37 Pole location of a second-order system.

Let us consider a second-order transfer function $T(s)$, given by

$$T(s) = \frac{\omega_n^2}{s^2 + 2\zeta\omega_n s + \omega_n^2} \tag{6.81}$$

The poles of $T(s)$ are $-\zeta\omega_n \pm j\omega_n\sqrt{1-\zeta^2}$, as depicted in Fig. 6.37. These are complex when the **damping ratio** $\zeta < 1$ (underdamped case), and are real for $\zeta \geq 1$. $\zeta = 1$ represents the critically damped and $\zeta > 1$, the overdamped case. Smaller ζ means smaller damping, leading to higher overshoot and faster response. For the unit step input $F(s) = 1/s$, and

$$Y(s) = \frac{\omega_n^2}{s(s^2 + 2\zeta\omega_n s + \omega_n^2)} = \frac{1}{s} - \frac{s + 2\zeta\omega_n}{s^2 + 2\zeta\omega_n s + \omega_n^2}$$

Use of Table 6.1 (pairs 1 and 10c) yields

$$y(t) = \left[1 - \frac{1}{\sqrt{1-\zeta^2}}e^{-\zeta\omega_n t}\sin(\omega_n\sqrt{1-\zeta^2}t + \cos^{-1}\zeta)\right]u(t) \tag{6.82}$$

The nature of this response for the underdamped case ($\zeta < 1$) is illustrated in Fig. 6.38. The response decays exponentially as $e^{-\zeta\omega_n t}$. Hence, the time constant of response is $1/\zeta\omega_n$. It takes four time constants for an exponential to decay to slightly less than 2% of its initial value. Hence, the settling time t_s is given by

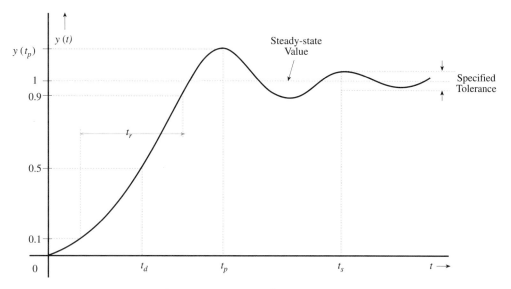

Fig. 6.38 Unit step response of a second-order system.

$$t_s = \frac{4}{\zeta\omega_n} \tag{6.83}$$

To determine the PO (percent overshoot), we find t_p, the peak time where the peak overshoot occurs. At $t = t_p$, $dy/dt = 0$. The straightforward solution of this equation yields

$$t_p = \frac{\pi}{\omega_n\sqrt{1-\zeta^2}}$$

Moreover, PO is given by

$$\text{PO} = \frac{y(t_p) - 1}{1} \times 100\% = 100\, e^{-\zeta\pi/\sqrt{1-\zeta^2}} \tag{6.84}$$

Figure 6.39 shows PO as a function of ζ. For a second-order system, the PO is directly related to the damping ratio ζ.

We may now proceed to determine similar expressions for t_r and t_d. Unfortunately, these turn out to be transcendental equations. However, these equations can be solved on a computer. Results of these computation appear in Fig. 6.39. Although precise expressions for t_r and t_d cannot be found, useful approximations are[9]

$$t_r \approx \frac{1 - 0.4167\zeta + 2.917\zeta^2}{\omega_n} \qquad t_d \approx \frac{1.1 + 0.125\zeta + 0.469\zeta^2}{\omega_n} \tag{6.85}$$

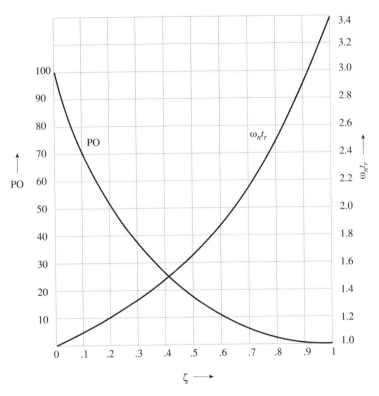

Fig. 6.39 (a) Plots of PO, t_s and t_r as functions of damping ratio ζ.

Clearly, for a second-order system the pole locations determine the transient behavior of the system. We observe that the closer the poles to the $j\omega$-axis, the smaller the value of ζ, and the larger the PO. Generally speaking, poles should not be too close to the $j\omega$-axis, a positioning which leaves too little safety margin for stability and makes the system sensitive to parameter variations. Hence it is generally desirable to have a larger value for ζ (small PO). For a fast response it is desirable to have small values for t_r, t_s, t_p and t_d.

This discussion shows that for a second-order system in Eq. 6.81, all the transient parameters (PO, t_r, t_s, t_p and t_d) are related to the pole location of $T(s)$. From the point of view of the system design, it will be convenient to draw the contours representing different values of transient parameters in the s-plane. For instance, each radial line drawn from the origin in the s-plane represents a constant ζ line (see Fig. 6.37). Since the PO is directly related to ζ (Fig. 6.39), each radial line also represents a line of constant PO, as depicted in Fig. 6.40. Similarly, each vertical line represents constant $\zeta\omega_n$ (see Fig. 6.37). Because $t_s = 4/\zeta\omega_n$, the lines representing constant settling time are vertical lines, as shown in Fig. 6.40. This figure also shows the contours for constant t_r. These contours allows us to determine by inspection, the important transient characteristics (PO, t_r, t_s) of a second-order system from the knowledge of its pole locations. Moreover, if we are required to synthesize a second-order system to meet some given transient specifications, we can find the desired $T(s)$ with the help of this figure.

As an example consider the position control system in Fig. 6.36a. Let the

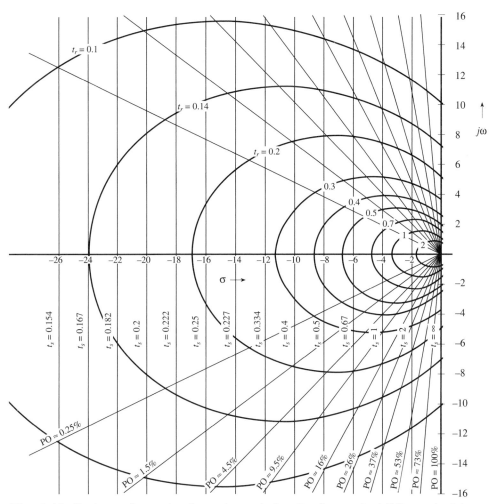

Fig. 6.40 Contours of second-order system pole location for constant PO, constant t_s, and constant t_r in s plane.

transient specifications for this system be given as

$$\text{PO} \le 16\%, \qquad t_r \le 0.5 \text{ seconds} \qquad t_s \le 2 \text{ seconds} \tag{6.86}$$

We delineate appropriate contours in Fig. 6.40 to meet the above specifications. The shaded region defined by these contours in Fig. 6.41 meets all the three requirements. Hence $T(s)$ must be chosen so that both of its poles lie in the shaded region. The transfer function $T(s)$ for the closed-loop system in Fig. 6.36a is given by

$$T(s) = \frac{KG(s)}{1 + KG(s)} = \frac{K}{s^2 + 8s + K} \tag{6.87a}$$

This shows that locations of the poles of $T(s)$ can be adjusted by changing the gain K. We must choose the gain K so that the poles lie in the shaded region in Fig. 6.41. The poles of $T(s)$ are the roots of the characteristic equation

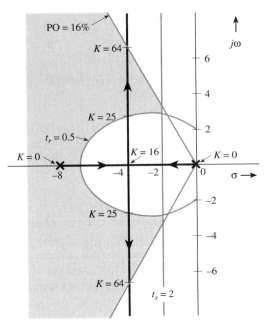

Fig. 6.41 Designing a second-order system to meet a given transient specifications.

$$s^2 + 8s + K = 0 \qquad (6.87\text{b})$$

Hence the poles are

$$s_{1,2} = -4 \pm \sqrt{16 - K} \qquad (6.87\text{c})$$

The poles s_1 and s_2 (the roots of the characteristic equation) move along a certain path in the s-plane as we vary K from 0 to ∞. When $K = 0$, the poles are -8, 0. For $K < 16$, the poles are real and both poles move towards a value -4 as K varies from 0 to 16 (overdamping). For $K = 16$, both poles coincide at -4 (critical damping). For $K > 16$, the poles become complex with values $-4 \pm j\sqrt{K - 16}$ (underdamping) Since the real part of the poles is -4 for all $K > 16$, the path of the poles is vertical as illustrated in Fig. 6.41. One pole moves up and the other (its conjugate) moves down along the vertical line passing through -4. We can label the values of K for several points along these paths, as depicted in Fig. 6.41. Each of these paths represents a locus of a pole of $T(s)$ or a locus of a root of the characteristic equation of $T(s)$ as K is varied from 0 to ∞. For this reason this set of paths is called the **root locus**. The root locus gives us the information as to how the poles of the closed-loop transfer function $T(s)$ move as the gain K is varied from 0 to ∞. In our design problem, we must choose a value of K such that the poles of $T(s)$ lie in the shaded region shown in Fig. 6.41. This figure shows that the system will meet the given specifications [Eq. (6.86)] for $25 \le K \le 64$. For $K = 64$, for instance we have

$$\text{PO} = 16\%, \qquad t_r = 0.2 \text{ seconds}, \qquad t_s = \frac{4}{4} = 1 \text{ seconds} \qquad (6.88)$$

Higher-order Systems

Our discussion, so far, has been limited to second-order $T(s)$ only. If $T(s)$ has additional poles which are far away to the left of $j\omega$-axis, they have only a negligible effect on the transient behavior of the system. The reason is that the time constants of such poles are considerably smaller when compared to the time constant of the complex conjugate poles near the $j\omega$-axis. Consequently, the exponentials arising because of poles far away from the $j\omega$-axis die quickly compared to those arising because of poles located near the $j\omega$-axis. In addition, the coefficients of the former terms are much smaller than unity. Hence, they are also very small to begin with and decay rapidly. The poles near the $j\omega$-axis are called the **dominant poles**. A criterion commonly used is that any pole which is five times as far from the $j\omega$-axis as the dominant poles contributes negligibly to the step response, and the transient behavior of a higher-order system is often reduced to that of a second-order system. In addition, a closely placed pole-zero pair (called **dipole**), contributes negligibly to the transient behavior. For this reason, many of the pole-zero configurations in practice reduce to two or three poles with one or two zeros. Workers in the field have prepared charts for transient behavior of these systems for several such pole-zero combinations, which may be used to design most of the higher-order systems.

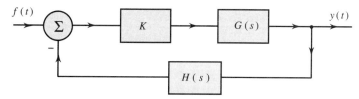

Fig. 6.42 A feedback system with a variable gain K.

6.7-3   Root Locus

The example in Sec. 6.7-2 gives a good idea of the utility of root locus in design of control systems. Surprisingly, root locus can be sketched quickly using certain basic rules provided by W.R. Evans in 1948.[†] With the ready availability of computers, root locus can be produced easily. Nevertheless, understanding these rules can be a great help in developing the intuition needed for design. We shall present here the rules, but omit the proofs of some of them.

We begin with a feedback system depicted in Fig. 6.42, which is identical to Fig. 6.18d, except for the explicit representation of a variable gain K. The system in Fig. 6.36a is a special case with $H(s) = 1$. For the system in Fig. 6.42,

$$T(s) = \frac{KG(s)}{1 + KG(s)H(s)} \tag{6.89a}$$

The characteristic equation of this system is[‡]

[†]This procedure was developed as early as 1868 in Maxwell's paper "On Governors".
[‡]This characteristic equation is also valid when the gain K is in the feedback path [lumped with $H(s)$] rather than in the forward path. The equation applies as long as the gain K is in the loop at any point. Hence, the root locus rules discussed here apply to all such cases.

$$1 + KG(s)H(s) = 0 \qquad\qquad (6.89b)$$

We shall consider the paths of the roots of $1 + KG(s)H(s) = 0$ as K varies from 0 to ∞. When the loop is opened, the transfer function is $KG(s)H(s)$. For this reason, we refer to $KG(s)H(s)$ as the **open-loop transfer function**. The rules for sketching the root locus are as follows.

1. Root loci begin ($K = 0$) at the open-loop poles and terminate on the open-loop zeros ($K = \infty$). This fact means that the number of loci is exactly n, the order of the open-loop transfer function. Let $G(s)H(s) = N(s)/D(s)$, where $N(s)$ and $D(s)$ are polynomials of powers m and n, respectively. Hence, $1 + KG(s)H(s) = 0$ implies $D(s) + KN(s) = 0$. Therefore, $D(s) = 0$ when $K = 0$. In this case, the roots are poles of $G(s)H(s)$; that is, the open-loop poles. Similarly, when $K \to \infty$, $D(s) + KN(s) = 0$ implies $N(s) = 0$. Hence, the roots are the open-loop zeros. For the system in Fig. 6.36a, the open-loop transfer function is $K/s(s + 8)$. The open loop poles are 0 and -8 and the zeros (where $K/s(s + 8) = 0$ are both ∞. We can verify from Fig. 6.41 that the root loci do begin at 0 and -8 and terminate at ∞.

2. A real axis segment is a part of the root locus if the sum of the real axis poles and zeros of $G(s)H(s)$ that lie to the right of the segment is odd. Moreover, the root loci are symmetric about real axis.

   We can readily verify in Fig. 6.41 that the real axis segment to the right of -8 has only one pole (and no zeros). Hence, this segment is a part of the root locus.

3. The $n - m$ root loci terminate at ∞ at angles $k\pi/(n - m)$ for $k = 1, 3, 5, \ldots$.

   Note that, according to rule 1, m loci terminate on the open loop zeros, and the remaining $n - m$ loci terminate at ∞ according to this rule. In Fig. 6.41, we verify that $n - m = 2$ loci terminate at ∞ at angles $k\pi/2$ for $k = 1$ and 3.

   Now we shall make an interesting observation. If a transfer function $G(s)$ has m (finite) zeros and n poles, then $\lim_{s\to\infty} G(s) = s^m/s^n = 1/s^{n-m}$. Hence, $G(s)$ has $n - m$ zeros at ∞. This fact shows that although $G(s)$ has only m finite zeros, there are additional $n - m$ zeros at ∞. According to rule 1, m loci terminate on m finite zeros, and according to this rule the remaining $n - m$ loci terminate at ∞, which are also zeros of $G(s)$. This result means all loci begin on open loop poles and terminate on open loop zeros.

4. The centroid of the asymptotes (point where the asymptotes converge) of the $(n - m)$ loci that terminate at ∞ is

$$\sigma = \frac{(p_1 + p_2 + \cdots + p_n) - (z_1 + z_2 + \cdots + z_m)}{(n - m)}$$

   where $p_1, p_2, \ldots, p_n$ are the poles and $z_1, z_2, \ldots, z_m$ are the zeros, respectively, of the open-loop transfer function.
   Figure 6.41 verifies that the centroid of the loci is $[(-8 + 0) - 0]/2 = -4$.

5. There are additional rules, which allow us to compute the points where the loci intersect and where they cross the $j\omega$ axis to enter in the right-half plane. These rules allow us to draw a quick and rough sketch of the root loci. But

the ready availability of computers and programs makes it much easier to draw actual loci. The first four rules are still very helpful for a quick sketching of the root loci.

Understanding these rules can be helpful in design of control systems as demonstrated later. They are an aid in determining what modifications should be made (or what kind of compensator to add) to the open loop transfer function in order to meet given design specifications.

■ **Example 6.21**

Using the four rules of the root loci, sketch the root locus for a system with open loop transfer function

$$KG(s)H(s) = \frac{K}{s(s+2)(s+4)}$$

1. Rule 1: For this $G(s)H(s)$, $n = 3$. Hence, there are three root loci, which begin at the poles of $G(s)H(s)$; that is, at 0, -2 and -4.
2. Rule 2: There are odd numbers of poles to the right of the real axis segment $s < -4$, and $-2 < s < 0$. Hence, these segments are the part of root locus. In other words, the entire real axis in the left-half plane, except the segment between -2 and -4, is a part of the root locus.
3. Rule 3: $n - m = 3$. Hence, (all) the three loci terminate at ∞ along asymptotes at angles $k\pi/3$ for $k = 1$, 3 and 5. Thus, the asymptote angles are 60°, 120° and 180°.
4. Rule 4: The centroid (where all the three asymptotes converge) is $(0 - 2 - 4)/3 = -2$. We draw three asymptotes starting at -2 at angles 60°, 120° and 180°, as shown in Fig. 6.43. This information suffices to give an idea about the root locus. The actual root loci are also shown in Fig. 6.43. Two of the asymptotes cross over to the RHP, behavior which shows that for some range of K, the system becomes unstable. ■

⊙ **Computer Example C6.5**
Solve Example 6.21 using MATLAB.
The MATLAB commands to find the root locus for this case are:

```
num=[0 0 0 1];
den=conv(conv([1 0],[1 2]),[1 4]);
rlocus(num,den),grid   ⊙
```

6.7-4 Steady-State Errors

Steady-state specifications impose additional constraints on the closed-loop transfer function $T(s)$. The steady-state error is the difference between the desired output [reference $f(t)$] and the actual output $y(t)$. Thus $e(t) = f(t) - y(t)$, and

$$\begin{aligned} E(s) &= F(s) - Y(s) \\ &= F(s)\left[1 - \frac{Y(s)}{F(s)}\right] \\ &= F(s)[1 - T(s)] \end{aligned} \tag{6.90}$$

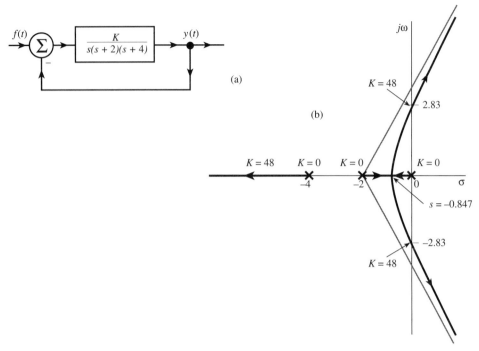

Fig. 6.43 A third-order feedback system and its root locus.

The steady-state error e_{ss} is the value of $e(t)$ as $t \to \infty$. This value can be readily obtained from the final-value theorem [Eq. (6.68)]:

$$e_{\mathrm{ss}} = \lim_{s \to 0} sE(s) = \lim_{s \to 0} sF(s)[1 - T(s)] \tag{6.91}$$

1. For the unit step input, the steady-state error e_s is given by

$$e_s = \lim_{s \to 0} [1 - T(s)] = 1 - T(0) \tag{6.92}$$

If $T(0) = 1$, the steady-state error to unit-step input is zero.

2. For a unit ramp input, $F(s) = 1/s^2$ and e_r, the steady-state error, is given by

$$e_r = \lim_{s \to 0} \frac{1 - T(s)}{s} \tag{6.93}$$

If $T(0) \neq 1$, $e_r = \infty$. Hence for a finite steady-state error to ramp input, a necessary condition is $T(0) = 1$, implying zero steady-state error to step input. Assuming $T(0) = 1$ and applying L'Hopital's rule to Eq. (6.93), we have

$$e_r = \lim_{s \to 0} [-\dot{T}(s)] = -\dot{T}(0) \tag{6.94}$$

3. Using a similar argument, we can show that for a unit parabolic input $t = (t^2/2)u(t)$, and $F(s) = 1/s^3$ and e_p, the steady state error is

$$e_p = -\frac{\ddot{T}(0)}{2} \tag{6.95}$$

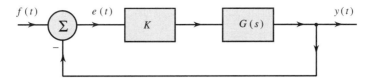

Fig. 6.44 A unity feedback system with variable gain K.

assuming $T(0) = 1$ and $\dot{T}(0) = 0$.

Many systems in practice have unity feedback, as depicted in Fig. 6.44. In such a case, the steady-state error analysis is greatly simplified. Let us define **positional error constant** K_p, **velocity error constant** K_v, and **acceleration error constant** K_a as

$$K_p = \lim_{s \to 0}[KG(s)], \qquad K_v = \lim_{s \to 0} s[KG(s)], \qquad K_a = \lim_{s \to 0} s^2[KG(s)] \qquad (6.96)$$

Because $T(s) = KG(s)/1 + KG(s)$, from Eq. (6.90), we obtain

$$E(s) = \frac{1}{1 + KG(s)}F(s)$$

The steady-state errors are given by

$$e_s = \lim_{s \to 0} s\frac{1/s}{1 + KG(s)} = \frac{1}{1 + \lim_{s \to 0}[KG(s)]} = \frac{1}{1 + K_p} \qquad (6.97a)$$

$$e_r = \lim_{s \to 0} s\frac{1/s^2}{1 + KG(s)} = \frac{1}{\lim_{s \to 0} s[KG(s)]} = \frac{1}{K_v} \qquad (6.97b)$$

$$e_p = \lim_{s \to 0} s\frac{1/s^3}{1 + KG(s)} = \frac{1}{\lim_{s \to 0} s^2[KG(s)]} = \frac{1}{K_a} \qquad (6.97c)$$

For the system in Fig. 6.36a,

$$G(s) = \frac{1}{s(s + 8)}$$

Hence, from Eq. (6.96)

$$K_p = \infty, \qquad K_v = \frac{K}{8}, \qquad K_a = 0 \qquad (6.98)$$

Substitution of these values in Eq. (6.97) yields

$$e_s = 0, \qquad e_r = \frac{8}{K}, \qquad e_p = \infty$$

A system where $G(s)$ has one pole at the origin (as the present case) is designated as **type 1** system. Such a system can track position of an object with zero error ($e_s = 0$), and yields a constant error in tracking an object moving with constant velocity ($e_r =$ a constant). But the type 1 system is not suitable for tracking a constant acceleration object.

If $G(s)$ has no poles at the origin, then K_p is finite and $K_v = K_a = 0$. Thus, for

$$G(s) = \frac{(s+2)}{(s+1)(s+10)}$$

$K_p = K/5$ and $K_v = K_a = 0$. Hence, $e_s = 5/(5 + K)$ and $e_r = e_p = \infty$. Such systems are designated as **type 0** systems. These systems have finite e_s, but infinite e_r and e_p. These systems may be acceptable for step inputs (position control), but not for ramp or parabolic inputs (tracking velocity or acceleration).

If $G(s)$ has two poles at the origin, the system is designated as **type 2** system. In this case $K_p = K_v = \infty$, and $K_a = $ finite. Hence, $e_s = e_r = 0$ and e_p is finite.

In general, if $G(s)$ has q poles at the origin, it is a type q system. Clearly, for a unity feedback system, increasing the number of poles at the origin in $G(s)$ improves the steady-state performance. However, this procedure increases n and reduces the magnitude of σ, the centroid of the root locus asymptotes. This shifts the root locus towards the $j\omega$-axis with consequent deterioration in the transient performance and the system stability.

It should be remembered that the results in Eqs. (6.96) and (6.97) apply only to unity feedback systems (Fig. 6.44). Steady-state error specifications in this case are translated in terms of constraints on the open-loop transfer function $KG(s)$. In contrast, the results in Eqs. (6.92) through (6.95) apply to unity as well as nonunity feedback systems, and are more general. Steady-state-error specifications in this case are translated in terms of constraints on the closed-loop transfer function $T(s)$.

The unity feedback system in Fig. 6.36a is type 1 system. We have designed this system earlier to meet the following transient specifications:

$$\text{PO} = 16\%, \qquad t_r \le 0.5, \qquad t_s \le 2 \tag{6.99}$$

Let us further specify that the system meet the following steady-state specifications:

$$e_s = 0, \qquad e_r \le 0.15$$

For this case, we already found $e_s = 0$, $e_r = 8/K$ and $e_p = \infty$ [see Eq. (6.98)]. But we require $e_r \le 0.15$. Therefore

$$\frac{8}{K} \le 0.15 \quad \Rightarrow K \ge 53.34 \tag{6.100}$$

Turning to Fig. 6.41, we note that the poles of $T(s)$ lie in the acceptable region to meet transient specifications ($25 < K < 64$). Equation (6.100) shows that we must use $K \ge 53.34$ to meet steady-state performance. Therefore to meet both the transient and the steady-state specifications we must set the gain in the range $53.34 < K < 64$. The smallest steady-state error for a ramp input is obtained for $K = 64$. For this case,

$$e_r = \frac{8}{K} = \frac{8}{64} = 0.125$$

Thus if the system is to meet the transient performance in Eq. (6.99), the minimum $e_r = 0.125$. We can do no better. In case we are required to have $e_r < 0.125$

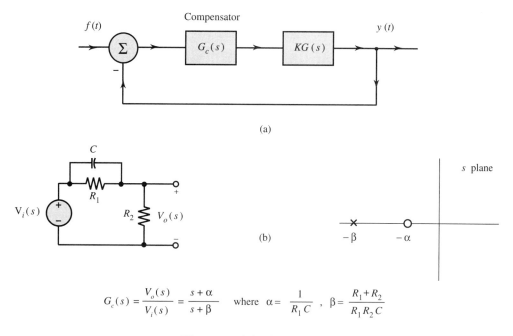

(a)

(b)

$$G_c(s) = \frac{V_o(s)}{V_i(s)} = \frac{s+\alpha}{s+\beta} \quad \text{where} \quad \alpha = \frac{1}{R_1 C} \;, \quad \beta = \frac{R_1+R_2}{R_1 R_2 C}$$

Fig. 6.45 A lead compensator.

while maintaining the same transient performance, we will have to use some kind of compensation.

6.7-5 Compensation

The synthesis problem for the position control system in Fig. 6.36a is a very simple example where the transient and steady-state specifications could be met by simple adjustment of gain K. In many cases, it may be impossible to meet both sets of specification (transient and steady state) by simple adjustment of the gain K. We may be able to satisfy one set of specifications or the other but not both. Consider again the system in Fig. 6.36a, with the following specifications:

$$\text{PO} = 16\% \qquad t_r \le 0.5 \qquad t_s \le 2 \qquad e_s = 0 \qquad e_r \le 0.05$$

To meet the steady-state specification we must have

$$\frac{8}{K} \le 0.05 \quad \Rightarrow K \ge 160$$

But Fig. 6.41 indicates that for $K > 64$, the poles of $T(s)$ move out of the region acceptable for transient performance. Clearly, we can meet either the transient or the steady-state specification but not both. In such case we must add some kind of compensation, which will modify the root locus to meet all the specifications. A little familiarity with root-locus techniques gives the insight and judgment needed to choose a proper compensator transfer function. Figure 6.41 indicates that shifting the root locus to the left will accomplish the desired performance. We can place a compensator of transfer function $G_c(s)$ in series with $G(s)$ (Fig. 6.45a) and select

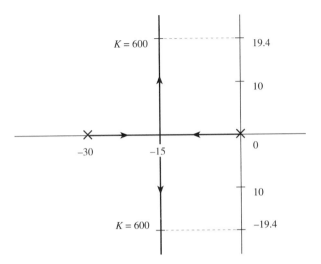

Fig. 6.46 Root locus of the system in Fig. 6.36a after lead compensation.

the poles and zeros of $G_c(s)$ in such a way as to shift the centroid of the root locus to the left. If $G_c(s)$ has a single pole and a single zero, then choosing the pole farther to the left of the zero would shift the centroid to the left according to the fourth rule of the root locus. Thus, we use

$$G_c(s) = \frac{s + \alpha}{s + \beta} \qquad \beta > \alpha$$

Such a compensator is called the **lead compensator**, which is readily realized using a simple RC circuit shown in Fig. 6.45b. We have a wide choice for values of α and β. To simplify design, let us choose $\alpha = 8$ and $\beta = 30$. This choice yields

$$G_c(s)[K\,G(s)] = \left(\frac{s + 8}{s + 30}\right) \frac{K}{s(s+8)} = \frac{K}{s(s+30)}$$

To simplify our discussion we deliberately chose $\alpha = 8$ to cancel the pole of $G(s)$. In practice, we do not necessarily have to cancel the pole of $G(s)$. In the present situation, $\sigma = (-30 + 0)/2 = 15$, and the new root locus appears in Fig. 6.46. Observe that the situation has improved considerably by shifting the centroid from -4 to -15. If we select $K = 600$, we have

$$T(s) = \frac{\frac{600}{s(s+30)}}{1 + \frac{600}{s(s+30)}} = \frac{600}{s^2 + 30s + 600}$$

In this case $\omega_n = \sqrt{600} = 24.5$. Also $\zeta\omega_n = 15$. Hence, $\zeta = 15/24.5 = 0.61$ and $t_s = 4/\zeta\omega_n = 4/15 = 0.266$. From Eq. (6.84), we find PO=8.9%. Moreover, from Fig. 6.39 [or Eq. (6.85)], for $\zeta = 0.61$, we find $\omega_n t_r = 1.83$ so that $t_r = 1.83/24.5 = 0.0747$. We also have

$$K_p = \infty \qquad K_v = \frac{K}{30} = \frac{600}{30} = 20 \quad \Rightarrow e_s = 0 \qquad e_r = \frac{1}{20} = 0.05$$

The system meets all the specifications and more.

We shall now discuss a compensation which is primarily used to improve the steady-state performance. For unity feedback systems, the steady-state performance of a system is improved by placing an integrator in the forward path of $G(s)$. This procedure increases the system type, thus increasing K_p, K_v, K_a, etc. The compensator in this case is $G_c(s) = 1/s$.

In this scheme the compensator is an ideal integrator. Hence, this scheme is known as **integral control**. Design of an ideal integrator necessitates elaborate and expensive equipment. Hence, such a compensator is used where cost considerations are not very important. For example, integrating gyroscopes are used for this purpose in aircraft. In most cases a **lag compensator** (described below) which closely approximates the a behavior of an integrator is used. A lag compensator transfer function is given by

$$G_c(s) = \frac{s + \alpha}{s + \beta} \qquad \alpha > \beta$$

and

$$G_c(0) = \frac{\alpha}{\beta}$$

A lag compensator can be readily realized using a simple RC circuit depicted in Fig. 6.47. For a unity feedback system, addition of a compensator $G_c(s)$ causes all the error constants K_p, K_v, K_a, etc., to be multiplied by $G_c(0)$. Thus, the lag compensator increases K_p, K_v, $K_{a,}$, etc. by a factor (α/β), thereby reducing steady-state errors.

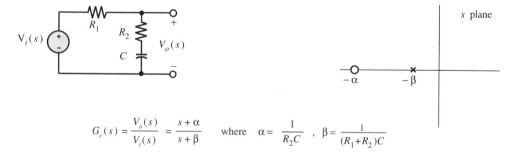

$$G_c(s) = \frac{V_o(s)}{V_i(s)} = \frac{s + \alpha}{s + \beta} \qquad \text{where} \quad \alpha = \frac{1}{R_2 C} \quad , \quad \beta = \frac{1}{(R_1 + R_2)C}$$

Fig. 6.47 A lag compensator.

A lag compensator improves the steady-state performance, but in general degrades the transient performance. Since $\alpha > \beta$, the magnitude of σ, the root locus centroid, is reduced. This reduction causes the root locus to shift toward the $j\omega$-axis, with the consequent deterioration of the transient performance. This side effect of a lag compensator can be made negligible by choosing α and β such that $\alpha - \beta$ is very small, but the ratio α/β is high. Such a pair of pole and zero act like a dipole and has only a negligible effect on the transient behavior of the system. The root locus also changes very little. We can realize such a dipole by placing both the pole and the zero of $G_c(s)$ close to the origin (α and $\beta \to 0$). For instance, if we select $\alpha = 0.1$ and $\beta = 0.01$, the centroid will be shifted by only a negligible amount $(\alpha - \beta)/(n - m) = 0.09/(n - m)$. However, since $\alpha/\beta = 10$, all the error constants are increased by a factor 10. Thus, we can have our cake and eat it too!

We can improve the transient and the steady-state performance simultaneously by using a combination of lead and lag networks.

6.7-6 Stability Considerations

In practice, we rarely use positive feedback because, as explained earlier, such systems are prone to instability and are very sensitive to changes in the system parameters or environment. Would negative feedback make a system stable and less sensitive to unwanted changes? Not necessarily! The reason is that if a feedback were truly negative, the system would be stable. But a system that has negative feedback at one frequency may have a positive feedback at some other frequency because of phase shift in the transmission path. In other words, a feedback system, generally, cannot be described in black and white terms such as having positive or negative feedback. Let us clarify this statement by an example.

Consider the case $G(s)H(s) = 1/s(s+2)(s+4)$. The root locus of this system appears in Fig. 6.43. This system shows negative feedback at lower frequencies. But because of phase shift at higher frequencies, the feedback becomes positive. Consider the loop gain $G(s)H(s)$ at a frequency $\omega = 2.83$ (at $s = j2.83$).

$$G(j\omega)H(j\omega) = \frac{1}{j\omega(j\omega + 2)(j\omega + 4)}$$

At $\omega = 2.83$

$$G(j2.83)H(j2.83) = \frac{1}{j2.83(j2.83 + 2)(j2.83 + 4)}$$

$$= \frac{1}{48}e^{-j180°} = -\frac{1}{48}$$

Recall that the overall gain (transfer function) $T(s)$ is

$$T(s) = \frac{KG(s)}{1 + KG(s)H(s)}$$

At frequency $s = j2.83$ ($\omega = 2.83$), the gain is

$$T(j2.83) = \frac{KG(j2.83)}{1 - \frac{K}{48}}$$

As long as K remains below 48, the system is stable, but for $K = 48$, the system gain goes to ∞, and the system becomes unstable. The feedback, which was negative below $\omega = 2.83$ (because the phase shift has not reached $-180°$), becomes positive. If there is enough gain ($K = 48$) at this frequency, the signal fed back is equal to the input signal, and the signal perpetuates itself for ever. In other words, the signal starts generating (oscillating) at this frequency, which is precisely the instability. Note that the system remains unstable for all values of $K > 48$. This is clear from the root locus in Fig. 6.43, which shows that the two branches cross over to the RHP for $K > 48$. The crossing point is $s = j2.83$.

This discussion shows that the same system, which has negative feedback at lower frequency may have positive feedback at higher frequency. For this reason, feedback system are quite prone to instability, and the designer has to pay a great deal of attention to this aspect. Root locus does indicate the region of stability.

6.8   The Bilateral Laplace Transform

Situations involving noncausal signals and/or systems cannot be handled by the (unilateral) Laplace transform discussed so far. These cases can be analyzed by the **bilateral** (or **two-sided**) Laplace transform defined by

$$F(s) = \int_{-\infty}^{\infty} f(t)e^{-st}\, dt \qquad (6.101\text{a})$$

and $f(t)$ can be obtained from $F(s)$ by the inverse transformation

$$f(t) = \frac{1}{2\pi j} \int_{c-j\infty}^{c+j\infty} F(s)e^{st}\, ds \qquad (6.101\text{b})$$

Observe that the unilateral Laplace transform discussed so far is a special case of the bilateral Laplace transform, where the signals are restricted to the causal type. Basically, the two transforms are the same. For this reason we use the same notation for the bilateral Laplace transform.

Earlier we showed that the Laplace transforms of $e^{-at}u(t)$ and of $-e^{at}u(-t)$ are identical. The only difference involves their regions of convergence. The region of convergence for the former is $\operatorname{Re} s > -a$; that for the latter is $\operatorname{Re} s < -a$, as illustrated in Fig. 6.2. Clearly, the inverse Laplace transform of $F(s)$ is not unique unless the region of convergence is specified. If we restrict all our signals to the causal type, however, this ambiguity does not arise. The inverse transform of $1/(s + a)$ is $e^{-at}u(t)$. Thus, in the unilateral Laplace transform, we can ignore the region of convergence in determining the inverse transform of $F(s)$.

We now show that any bilateral transform can be expressed in terms of two unilateral transforms. It is, therefore, possible to evaluate bilateral transforms from a table of unilateral transforms.

Consider the function $f(t)$ appearing in Fig. 6.48a. We separate $f(t)$ into two components, $f_1(t)$ and $f_2(t)$, representing the positive time (**causal**) component and the negative time (**anticausal**) component of $f(t)$ respectively (Figs. 6.48b and 6.48c):

$$f_1(t) = f(t)u(t) \qquad (6.102\text{a})$$

$$f_2(t) = f(t)u(-t) \qquad (6.102\text{b})$$

The bilateral Laplace transform of $f(t)$ is given by

$$
\begin{aligned}
F(s) &= \int_{-\infty}^{\infty} f(t)e^{-st}\, dt \\[1mm]
&= \int_{-\infty}^{0} f_2(t)e^{-st}\, dt + \int_{0}^{\infty} f_1(t)e^{-st}\, dt \\[1mm]
&= F_2(s) + F_1(s) \qquad (6.103)
\end{aligned}
$$

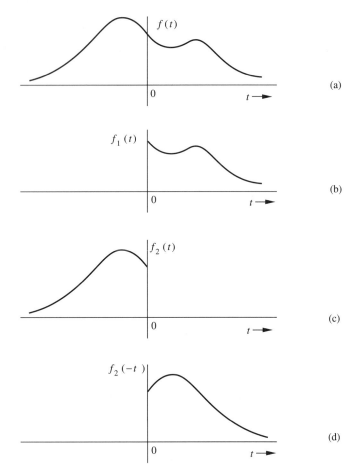

Fig. 6.48 Expressing a signal as a sum of causal and anticausal components.

where $F_1(s)$ is the Laplace transform of the causal component $f_1(t)$ and $F_2(s)$ is the Laplace transform of the anticausal component $f_2(t)$. But $F_2(s)$ is given by

$$F_2(s) = \int_{-\infty}^{0} f_2(t)e^{-st}\, dt$$

$$= \int_{0}^{\infty} f_2(-t)e^{st}\, dt$$

Therefore

$$F_2(-s) = \int_{0}^{\infty} f_2(-t)e^{-st}\, dt \qquad (6.104)$$

It is clear that $F_2(-s)$ is the Laplace transform of $f_2(-t)$, which is causal (Fig. 6.48d), so $F_2(-s)$ can be found from the unilateral transform table. Changing the sign of s in $F_2(-s)$ yields $F_2(s)$.

To summarize, the bilateral transform $F(s)$ in Eq. (6.103) can be computed from the unilateral transforms in two steps:

1) Split $f(t)$ into its causal and anticausal components, $f_1(t)$ and $f_2(t)$, respectively.

2) The signals $f_1(t)$ and $f_2(-t)$ are both causal. Take the (unilateral) Laplace transform of $f_1(t)$ and add to it the (unilateral) Laplace transform of $f_2(-t)$, with s replaced by $-s$. This procedure gives the (bilateral) Laplace transform of $f(t)$.

Since $f_1(t)$ and $f_2(-t)$ are both causal, $F_1(s)$ and $F_2(-s)$ are both unilateral Laplace transforms. Let σ_{c1} and σ_{c2} be the abscissas of convergence of $F_1(s)$ and $F_2(-s)$, respectively. This statement implies that $F_1(s)$ exists for all s with $\mathrm{Re}\,s > \sigma_{c1}$, $F_2(-s)$ exists for all s with $\mathrm{Re}\,s > \sigma_{c2}$, and $F_2(s)$ exists for all s with $\mathrm{Re}\,s < -\sigma_{c2}$. Because $F(s) = F_1(s) + F_2(s)$, $F(s)$ exists for all s such that

$$\sigma_{c1} < \mathrm{Re}\,s < -\sigma_{c2} \qquad (6.105)$$

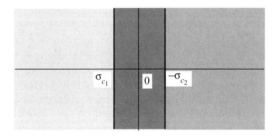

Region of convergence for causal component of $f(t)$.

Region of convergence for anticausal component of $f(t)$.

Region (strip) of convergence for the entire $f(t)$.

Fig. 6.49.

The regions of convergence (or existence) of $F_1(s)$, $F_2(s)$, and $F(s)$ are shown in Fig. 6.49. Because $F(s)$ is finite for all values of s lying in the strip of convergence ($\sigma_{c1} < \mathrm{Re}\,s < -\sigma_{c2}$), poles of $F(s)$ must lie outside this strip. The poles of $F(s)$ arising from the causal component $f_1(t)$ lie to the left of the **strip (region) of convergence**, and those arising from its anticausal component $f_2(t)$ lie to its right (see Fig. 6.49). This fact is of crucial importance in finding the inverse bilateral transform.

As an example, consider

$$f(t) = e^{bt}u(-t) + e^{at}u(t) \qquad (6.106)$$

We already know the Laplace transform of the causal component

$$e^{at}u(t) \Longleftrightarrow \frac{1}{s-a} \qquad \mathrm{Re}\,s > a \qquad (6.107)$$

For the anticausal component, $f_2(t) = e^{bt}u(-t)$, we have

$$f_2(-t) = e^{-bt}u(t) \iff \frac{1}{s+b} \qquad \text{Re } s > -b$$

so that

$$F_2(s) = \frac{1}{-s+b} = \frac{-1}{s-b} \qquad \text{Re } s < b$$

Therefore

$$e^{bt}u(-t) \iff \frac{-1}{s-b} \qquad \text{Re } s < b \tag{6.108}$$

and the Laplace transform of $f(t)$ in Eq. (6.106) is

$$F(s) = -\frac{1}{s-b} + \frac{1}{s-a} \qquad \text{Re } s > a \quad \text{and} \quad \text{Re } s < b$$

$$= \frac{a-b}{(s-b)(s-a)} \qquad a < \text{Re } s < b \tag{6.109}$$

Figure 6.50 shows $f(t)$ and the region of convergence of $F(s)$ for various values of a and b. Equation (6.109) indicates that the region of convergence of $F(s)$ does not exist if $a > b$, which is precisely the case in Fig. 6.50g. Observe that the poles of $F(s)$ are outside (on the edges) of the region of convergence. The poles of $F(s)$ because of the anticausal component of $f(t)$ lie to the right of the region of convergence, and those due to the causal component of $f(t)$ lie to its left.

■ **Example 6.22**
  Find the inverse Laplace transform of

$$F(s) = \frac{-3}{(s+2)(s-1)}$$

if the region of convergence is **(a)** $-2 < \text{Re } s < 1$ **(b)** $\text{Re } s > 1$ **(c)** $\text{Re } s < -2$.

(a) $\qquad\qquad F(s) = \frac{1}{s+2} - \frac{1}{s-1}$

Now, $F(s)$ has poles at -2 and 1. The strip of convergence is $-2 < \text{Re } s < 1$. The pole at -2, being to the left of the strip of convergence, corresponds to the causal signal. The pole at 1, being to the right of the strip of convergence, corresponds to the anticausal signal. Equations (6.107) and (6.108) yield

$$f(t) = e^{-2t}u(t) + e^{t}u(-t)$$

  (b) Both poles lie to the left of the region of convergence, so both poles correspond to causal signals. Therefore

$$f(t) = (e^{-2t} - e^{t})u(t)$$

  (c) Both poles lie to the right of the region of convergence, so both poles correspond to anticausal signals, and

$$f(t) = (-e^{-2t} + e^{t})u(-t)$$

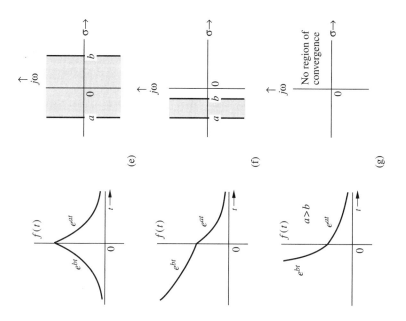

Figure 6.50 Some signals and their regions of conversions.

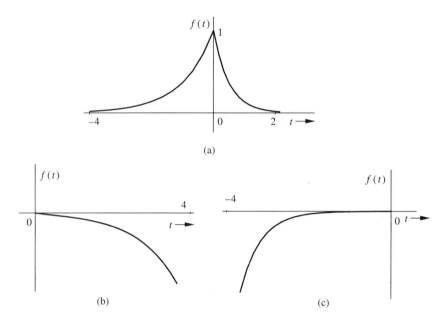

Fig. 6.51 Three possible inverse transforms of $\frac{-3}{(s+2)(s-1)}$.

Figure 6.51 shows the three inverse transforms corresponding to the same $F(s)$ but with different regions of convergence. ■

6.8-1 Linear System Analysis Using the Bilateral Transform

Since the bilateral Laplace transform can handle noncausal signals, we can analyze noncausal LTIC systems using the bilateral Laplace transform. We have shown that the (zero-state) output $y(t)$ is given by

$$y(t) = \mathcal{L}^{-1}\left[F(s)H(s)\right] \tag{6.110}$$

This expression is valid only if $F(s)H(s)$ exists. The region of convergence of $F(s)H(s)$ is the region where both $F(s)$ and $H(s)$ exist. In other words, the region of convergence of $F(s)H(s)$ is the region common to the regions of convergence of both $F(s)$ and $H(s)$. These ideas are clarified in the following examples.

■ **Example 6.23**

Find the current $y(t)$ for the RC circuit in Fig. 6.52a if the voltage $f(t)$ is

$$f(t) = e^t u(t) + e^{2t} u(-t)$$

The transfer function $H(s)$ of the circuit is given by

$$H(s) = \frac{s}{s+1}$$

Because $h(t)$ is a causal function, the region of convergence of $H(s)$ is $\operatorname{Re} s > -1$. Next, the bilateral Laplace transform of $f(t)$ is given by

$$F(s) = \frac{1}{s-1} - \frac{1}{s-2} = \frac{-1}{(s-1)(s-2)} \qquad 1 < \operatorname{Re} s < 2$$

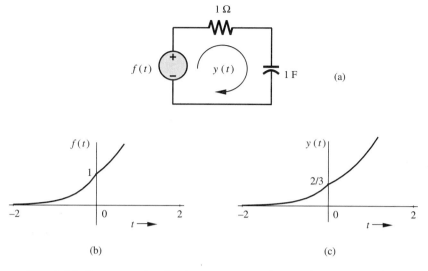

Fig. 6.52 Response of a circuit to a noncausal input (Example 6.23).

The response $y(t)$ is the inverse transform of $F(s)H(s)$

$$y(t) = \mathcal{L}^{-1}\left[\frac{-s}{(s+1)(s-1)(s-2)}\right]$$

$$= \mathcal{L}^{-1}\left[\frac{1}{6}\frac{1}{s+1} + \frac{1}{2}\frac{1}{s-1} - \frac{2}{3}\frac{1}{s-2}\right]$$

The region of convergence of $F(s)H(s)$ is that region of convergence common to both $F(s)$ and $H(s)$. This is $1 < \text{Re } s < 2$. The poles $s = \pm 1$ lie to the left of the region of convergence and, therefore, correspond to causal signals; the pole $s = 2$ lies to the right of the region of convergence and thus represents an anticausal signal. Hence

$$y(t) = \tfrac{1}{6}e^{-t}u(t) + \tfrac{1}{2}e^{t}u(t) + \tfrac{2}{3}e^{2t}u(-t)$$

Figure 6.52c shows $y(t)$. Note that in this example, if

$$f(t) = e^{-4t}u(t) + e^{-2t}u(-t)$$

then the region of convergence of $F(s)$ is $-4 < \text{Re } s < -2$. Here no region of convergence exists for $F(s)H(s)$, and the response $y(t)$ goes to infinity. ∎

■ **Example 6.24**

Find the response $y(t)$ of a noncausal system with the transfer function

$$H(s) = \frac{-1}{s-1} \qquad \text{Re } s < 1$$

to the input $f(t) = e^{-2t}u(t)$.

We have

$$F(s) = \frac{1}{s+2} \qquad \text{Re } s > -2$$

and

$$Y(s) = F(s)H(s) = \frac{-1}{(s-1)(s+2)}$$

The region of convergence of $F(s)H(s)$ is, therefore, the region $-2 < \operatorname{Re} s < 1$. By partial fraction expansion

$$Y(s) = \frac{-1/3}{s-1} + \frac{1/3}{s+2} \qquad -2 < \operatorname{Re} s < 1$$

and

$$y(t) = \tfrac{1}{3}\left[e^t u(-t) + e^{-2t} u(t)\right]$$

Note that the pole of $H(s)$ lies in the RHP at 1. Yet the system is not unstable. The pole(s) in the RHP may indicate instability or noncausality, depending on its location with respect to the region of convergence of $H(s)$. For example, if $H(s) = -1/(s-1)$ with $\operatorname{Re} s > 1$, the system is causal and unstable, with $h(t) = -e^t u(t)$. In contrast, if $H(s) = -1/(s-1)$ with $\operatorname{Re} s < 1$, the system is noncausal and stable, with $h(t) = e^t u(-t)$.
∎

■ **Example 6.25**

Find the response $y(t)$ of a system with the transfer function

$$H(s) = \frac{1}{s+5} \qquad \operatorname{Re} s > -5$$

and the input

$$f(t) = e^{-t} u(t) + e^{-2t} u(-t)$$

The input $f(t)$ is of the type depicted in Fig. 6.50g, and the region of convergence for $F(s)$ does not exist. In this case, we must determine separately the system response to each of the two input components, $f_1(t) = e^{-t} u(t)$ and $f_2(t) = e^{-2t} u(-t)$.

$$F_1(s) = \frac{1}{s+1} \qquad \operatorname{Re} s > -1$$

$$F_2(s) = \frac{-1}{s+2} \qquad \operatorname{Re} s < -2$$

If $y_1(t)$ and $y_2(t)$ are the system responses to $f_1(t)$ and $f_2(t)$, respectively, then

$$Y_1(s) = \frac{1}{(s+1)(s+5)} \qquad \operatorname{Re} s > -1$$

$$= \frac{1/4}{s+1} - \frac{1/4}{s+5}$$

so that

$$y_1(t) = \tfrac{1}{4}\left(e^{-t} - e^{-5t}\right) u(t)$$

and

$$Y_2(s) = \frac{-1}{(s+2)(s+5)} \qquad -5 < \operatorname{Re} s < -2$$

$$= \frac{-1/3}{s+2} + \frac{1/3}{s+5}$$

so that

$$y_2(t) = \tfrac{1}{3}\left[e^{-2t}u(-t) + e^{-5t}u(t)\right]$$

Therefore

$$y(t) = y_1(t) + y_2(t)$$

$$= \tfrac{1}{3}e^{-2t}u(-t) + \left(\tfrac{1}{4}e^{-t} + \tfrac{1}{12}e^{-5t}\right)u(t) \quad \blacksquare$$

6.9   Appendix 6.1: Second Canonical Realization

An nth-order transfer function can also be realized by a second canonical (**observer canonical**) form. As in the case of the first canonical, we begin with a realization of a third-order transfer function in Eq. (6.71)

$$H(s) = \frac{Y(s)}{F(s)} = \frac{b_3 s^3 + b_2 s^2 + b_1 s + b_0}{s^3 + a_2 s^2 + a_1 s + a_0} \tag{6.111}$$

Therefore

$$(s^3 + a_2 s^2 + a_1 s + a_0)Y(s) = (b_3 s^3 + b_2 s^2 + b_1 s + b_0)F(s)$$

Transporting all but the first term on the right-hand side yields

$$s^3 Y(s) = b_3 s^3 F(s) + [-a_2 Y(s) + b_2 F(s)]s^2 + [-a_1 Y(s) + b_1 F(s)]s$$
$$+ [-a_0 Y(s) + b_0 F(s)]$$

Dividing throughout by s^3 yields

$$Y(s) = b_3 F(s) + \frac{1}{s}[-a_2 Y(s) + b_2 F(s)] + \frac{1}{s^2}[-a_1 Y(s) + b_1 F(s)]$$
$$+ \frac{1}{s^3}[-a_0 Y(s) + b_0 F(s)] \tag{6.112}$$

Therefore, $Y(s)$ can be generated by adding four signals appearing on the right-hand side of Eq. (6.112). We shall build $Y(s)$ step by step, adding one component at a time. Figure 6.53a shows only the first component; that is, $b_3 F(s)$. Figure 6.53b shows $Y(s)$ formed by the first two components, $b_3 F(s)$ and $\frac{1}{s}[-a_2 Y(s) + b_2 F(s)]$. Observe that the term $a_2 Y(s)$ is obtained from $Y(s)$ itself. We add $-a_2 Y(s)$ to $b_2 F(s)$ and pass it through an integrator to obtain $\frac{1}{s}[-a_2 Y(s) + b_2 F(s)]$. Figure 6.53c shows $Y(s)$ built up from the first three components. Finally, Fig. 6.53d shows $Y(s)$ built up from all the components. This is the final form, which represents an alternative realization of $H(s)$ in Eq. (6.111).

This realization can be readily generalized for an nth-order transfer function in Eq. (6.70) using n integrators.

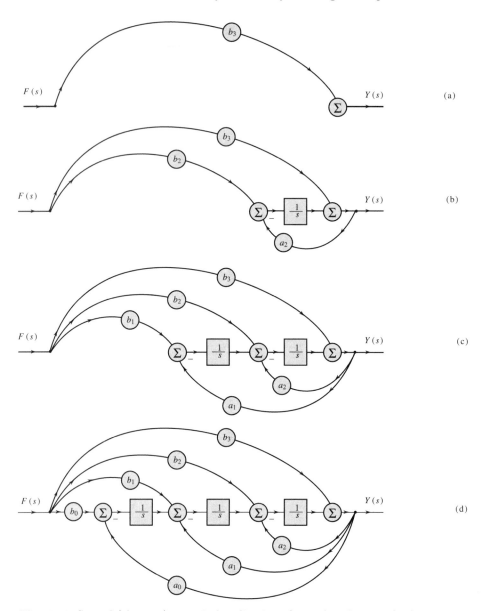

Fig. 6.53 Second (observer) canonical realization of an nth-order transfer function.

6.10 Summary

The Fourier transform cannot be used directly in analysis of unstable, or even marginally stable systems. Moreover, the inputs must also be restricted to Fourier transformable signals, which leaves out exponentially growing signals. Both these limitations are the result of the fact that the spectral components used in the Fourier transform to synthesize a signal $f(t)$ are ordinary sinusoids or exponentials of the form $e^{j\omega t}$, whose frequencies are restricted to the $j\omega$ axis in the complex plane.

These spectral components are incapable of synthesizing exponentially growing signals. We extend the Fourier transform by generalizing the frequency variable from $s = j\omega$ to $s = \sigma + j\omega$. The resulting transform is the Laplace transform, which can analyze all types of LTIC systems. The Laplace transform can also handle exponentially growing signals.

The system response to an everlasting exponential e^{st} is also an everlasting exponential $H(s)e^{st}$, where $H(s)$ is the system transfer function. We can view the Laplace transform as a tool by which a signal is expressed as a sum of everlasting exponentials e^{st}. The relative amount of a component e^{st} is $F(s)$. Therefore $F(s)$, the Laplace transform of $f(t)$, represents the spectrum of exponential components of $f(t)$. Moreover, $H(s)$ is the system response (or gain) to a spectral component e^{st}, and the output signal spectrum is the input spectrum $F(s)$ times the spectral response (gain) $H(s)$ $[Y(s) = F(s)H(s)]$.

The Laplace transform changes integro-differential equations of LTIC systems into algebraic equations. Therefore, solving these integro-differential equations reduces to solving algebraic equations. The Laplace transform method cannot generally be used for time-varying parameter systems or for nonlinear systems.

The transfer function of a system may also be defined as a ratio of the Laplace transform of the output to the Laplace transform of the input when all initial conditions are zero (system in zero state). If $F(s)$ is the Laplace transform of the input $f(t)$ and $Y(s)$ is the Laplace transform of the corresponding output $y(t)$ (when all initial conditions are zero), then $Y(s) = F(s)H(s)$, where $H(s)$ is the system transfer function. The system transfer function $H(s)$ is the Laplace transform of the system impulse response $h(t)$. Like the impulse response $h(t)$, the transfer function $H(s)$ is also an external description of the system.

Electrical circuit analysis can also be carried out by using a transformed circuit method, in which all signals (voltages and currents) are represented by their Laplace transforms, all elements by their impedances (or admittances), and initial conditions by their equivalent sources (initial condition generators). In this method, a network can be analyzed as if it were a resistive circuit.

Large systems can be considered as suitably interconnected subsystems represented by blocks. Each subsystem, being a smaller system, can be readily analyzed and represented by its input-output relationship, such as its transfer function. Analysis of large systems can be carried out with the knowledge of input-output relationships of its subsystems and the nature of interconnection of various subsystems.

LTIC systems can be realized by scalar multipliers, summers, and integrators. A given transfer function can be synthesized in many different ways. Canonical, cascade, and parallel forms of realization are discussed. In practice, all the building blocks (scalar multipliers, summers, and integrators) can be obtained from operational amplifiers.

Feedback systems are closed-loop systems mainly used to counteract the effects of unpredictable variations in the system parameters, the load, and the environment. Such systems are designed for a specified speed and the steady-state error. For speed, the useful transient parameters are rise time, peak time, and settling time. Percent overshoot indicates how smoothly the system output rises to its final value. We can relate the steady-state error to system parameters. In many cases, adjusting the amplifier gain K can result in the desired performance. If the requirements cannot be satisfied by mere adjustment of gain, then some form of compensation must be used. The loci of the characteristic roots of the system are called the root locus, which proves extremely convenient in designing a feedback system.

Most of the input signals and practical systems are causal. Consequently, we are required most of the time to deal with causal signals. When all signals are restricted to the causal type, the Laplace transform analysis is greatly simplified; the region of convergence of a signal becomes irrelevant to the analysis process. This special case of the Laplace transform (which is restricted to causal signals) is called the unilateral Laplace transform. Much of the chapter deals with this variety of Laplace transform. Section 6.8 discusses the general Laplace transform (the bilateral Laplace transform), which can handle causal and noncausal signals and systems. In the bilateral transform, the inverse transform of $F(s)$ is not unique but depends on the region of convergence of $F(s)$. Thus the region of convergence plays a very crucial role in the bilateral Laplace transform.

References

1. Doetsch, G., *Introduction to the Theory and Applications of the Laplace Transformation with a Table of Laplace Transformations*, Springer Verlag, New York, 1974.

2. LePage, W.R., *Complex Variables and the Laplace Transforms for Engineers*, McGraw-Hill, New York, 1961.

3. Durant, Will, and Ariel, *The Age of Napoleon, The Story of Civilization Series*, Part XI, Simon and Schuster, New York, 1975.

4. Bell, E.T., *Men of Mathematics*, Simon and Schuster, New York, 1937.

5. Nahin, P.J., "Oliver Heaviside: Genius and Curmudgeon," IEEE Spectrum, vol. 20, pp. 63-69, July 1983.

6. Berkey, D., *Calculus*, 2nd ed., Saunder's College Publishing, Philadelphia, Pa. 1988.

7. Encyclopaedia Britannica, *Micropaedia IV*, 15th ed., Chicago, IL. 1982.

8. Churchill, R.V., *Operational Mathematics*, 2nd ed, McGraw-Hill, New York, 1958.

9. Yang, J.S. and Levine, W.S. Chapter 10 in *The Control Handbook*, CRC Press, 1996.

Problems

6.1-1 By direct integration [Eq. (6.8b)] find the Laplace transforms and the region of convergence of the following functions:

(a) $u(t) - u(t-1)$ (e) $\cos \omega_1 t \cos \omega_2 t \, u(t)$

(b) $te^{-t}u(t)$ (f) $\cosh (at) \, u(t)$

(c) $t \cos \omega_0 t \, u(t)$ (g) $\sinh (at) \, u(t)$

(d) $(e^{2t} - 2e^{-t})u(t)$ (h) $e^{-2t} \cos (5t + \theta) \, u(t)$

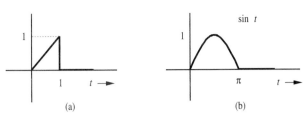 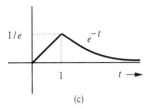

Fig. P6.1-2

6.1-2 By direct integration find the Laplace transforms of the signals shown in Fig. P6.1-2.

6.1-3 Find the inverse (unilateral) Laplace transforms of the following functions:

(a) $\dfrac{2s+5}{s^2+5s+6}$

(f) $\dfrac{s+2}{s(s+1)^2}$

(b) $\dfrac{3s+5}{s^2+4s+13}$

(g) $\dfrac{1}{(s+1)(s+2)^4}$

(c) $\dfrac{(s+1)^2}{s^2-s-6}$

(h) $\dfrac{s+1}{s(s+2)^2(s^2+4s+5)}$

(d) $\dfrac{5}{s^2(s+2)}$

(i) $\dfrac{s^3}{(s+1)^2(s^2+2s+5)}$

(e) $\dfrac{2s+1}{(s+1)(s^2+2s+2)}$

6.2-1 Find the Laplace transforms of the following functions using only Table 6.1 and the time-shifting property (if needed) of the unilateral Laplace transform:

(a) $u(t)-u(t-1)$ (e) $te^{-t}u(t-\tau)$

(b) $e^{-(t-\tau)}u(t-\tau)$ (f) $\sin[\omega_0(t-\tau)]\,u(t-\tau)$

(c) $e^{-(t-\tau)}u(t)$ (g) $\sin[\omega_0(t-\tau)]\,u(t)$

(d) $e^{-t}u(t-\tau)$ (h) $\sin\omega_0 t\,u(t-\tau)$

6.2-2 Using only Table 6.1 and the time-shifting property, determine the Laplace transform of the signals in Fig. P6.1-2.

Hint: See Sec. 1.4 for discussion of expressing such signals analytically.

6.2-3 Find the inverse Laplace transforms of the following functions:

(a) $\dfrac{(2s+5)e^{-2s}}{s^2+5s+6}$

(c) $\dfrac{e^{-(s-1)}+3}{s^2-2s+5}$

(b) $\dfrac{se^{-3s}+2}{s^2+2s+2}$

(d) $\dfrac{e^{-s}+e^{-2s}+1}{s^2+3s+2}$

6.2-4 The Laplace transform of a causal periodic signal can be determined from the knowledge of the Laplace transform of its first cycle (period).

(a) If the Laplace transform of $f(t)$ in Fig. P6.2-4a is $F(s)$, then show that $G(s)$, the Laplace transform of $g(t)$ [Fig. P6.2-4b], is

$$G(s)=\frac{F(s)}{1-e^{-sT_0}}\qquad \mathrm{Re}\,s>0$$

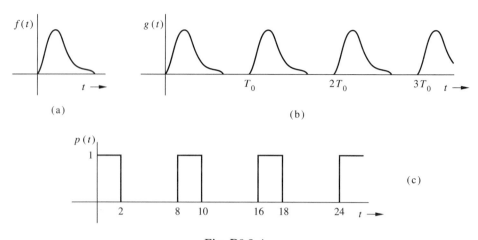

Fig. P6.2-4

(b) Using this result, find the Laplace transform of the signal $p(t)$ illustrated in Fig. P6.2-4c.

Hint: $1 + x + x^2 + x^3 + \cdots = \frac{1}{1-x}$ for $|x| < 1$.

6.2-5 Starting only with the fact that $\delta(t) \iff 1$, build Pairs 2 through 10b in Table 6.1, using various properties of the Laplace transform. Hint: $u(t)$ is the integral of $\delta(t)$, $tu(t)$ is integral of $u(t)$ [or second integral of $\delta(t)$], and so on.

6.2-6 **(a)** Find the Laplace transform of the pulses in Fig. 6.3 in the text by using only the time-differentiation property, time-shifting property, and the fact that $\delta(t) \iff 1$.
(b) In Example 6.7, the Laplace transform of $f(t)$ is found by finding the Laplace transform of $d^2 f/dt^2$. Find the Laplace transform of $f(t)$ in that example by finding the Laplace transform of df/dt.
Hint for part **(b)**: df/dt can be expressed as a sum of step functions (delayed by various amounts) whose transforms can be determined readily.

6.3-1 Using the Laplace transform, solve the following differential equations:

(a) $\left(D^2 + 3D + 2\right) y(t) = Df(t)$ if $y(0^-) = \dot{y}(0^-) = 0$ and $f(t) = u(t)$

(b) $\left(D^2 + 4D + 4\right) y(t) = (D + 1)f(t)$ if $y(0^-) = 2$, $\dot{y}(0^-) = 1$ and $f(t) = e^{-t}u(t)$

(c) $\left(D^2 + 6D + 25\right) y(t) = (D + 2)f(t)$ if $y(0^-) = \dot{y}(0^-) = 1$ and $f(t) = 25u(t)$

6.3-2 Solve the differential equations in Prob. 6.3-1 using the Laplace transform. In each case determine the zero-input and zero-state components of the solution.

6.3-3 Solve the following simultaneous differential equations using the Laplace transform, assuming all initial conditions to be zero and the input $f(t) = u(t)$:

$$\textbf{(a)}(D + 3)y_1(t) - 2y_2(t) = f(t)$$
$$- 2y_1(t) + (2D + 4)y_2(t) = 0$$
$$\textbf{(b)}(D + 2)y_1(t) - (D + 1)y_2(t) = 0$$
$$- (D + 1)y_1(t) + (2D + 1)y_2(t) = f(t)$$

Determine the transfer functions relating outputs $y_1(t)$ and $y_2(t)$ to the input $f(t)$.

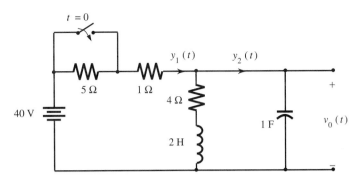

Fig. P6.3-4

6.3-4 For the circuit in Fig. P6.3-4, the switch is in open position for a long time before $t = 0$, when it is closed instantaneously.

(a) Write loop equations (in time domain) for $t \geq 0$.

(b) Solve for $y_1(t)$ and $y_2(t)$ by taking the the Laplace transform of loop equations found in part (a).

6.3-5 For each of the systems described by the following differential equations, find the system transfer function:

(a) $\dfrac{d^2 y}{dt^2} + 11\dfrac{dy}{dt} + 24y(t) = 5\dfrac{df}{dt} + 3f(t)$

(b) $\dfrac{d^3 y}{dt^3} + 6\dfrac{d^2 y}{dt^2} - 11\dfrac{dy}{dt} + 6y(t) = 3\dfrac{d^2 f}{dt^2} + 7\dfrac{df}{dt} + 5f(t)$

(c) $\dfrac{d^4 y}{dt^4} + 4\dfrac{dy}{dt} = 3\dfrac{df}{dt} + 2f(t)$

6.3-6 For each of the systems specified by the following transfer functions, find the differential equation relating the output $y(t)$ to the input $f(t)$:

(a) $H(s) = \dfrac{s+5}{s^2 + 3s + 8}$ (b) $H(s) = \dfrac{s^2 + 3s + 5}{s^3 + 8s^2 + 5s + 7}$

(c) $H(s) = \dfrac{5s^2 + 7s + 2}{s^2 - 2s + 5}$

6.3-7 For a system with transfer function

$$H(s) = \frac{s+5}{s^2 + 5s + 6}$$

(a) Find the (zero-state) response if the input $f(t)$ is:

(i) $e^{-3t}u(t)$ (ii) $e^{-4t}u(t)$ (iii) $e^{-4(t-5)}u(t-5)$ (iv) $e^{-4(t-5)}u(t)$ (v) $e^{-4t}u(t-5)$.

(b) For this system write the differential equation relating the output $y(t)$ to the input $f(t)$.

6.3-8 Repeat Prob. 6.3-7 if

$$H(s) = \frac{2s+3}{s^2 + 2s + 5}$$

and the input $f(t)$ is: (a) $10u(t)$ (b) $u(t-5)$.

6.3-9 Repeat Prob. 6.3-7 if

$$H(s) = \frac{s}{s^2 + 9}$$

and the input $f(t) = (1 - e^{-t})u(t)$.

6.3-10 For an LTIC system with zero initial conditions (system initially in zero state), if an input $f(t)$ produces an output $y(t)$, then show that:
(a) the input df/dt produces an output dy/dt, and
(b) the input $\int_0^t f(\tau)\,d\tau$ produces an output $\int_0^t y(\tau)\,d\tau$. Hence, show that the unit step response of a system is an integral of the impulse response; that is, $\int_0^t h(\tau)\,d\tau$.

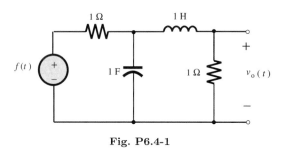

Fig. P6.4-1

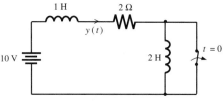

Fig. P6.4-2

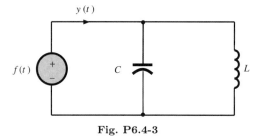

Fig. P6.4-3

6.4-1 Find the zero-state response $v_0(t)$ of the network in Fig. P6.4-1 if the input voltage $f(t) = te^{-t}u(t)$. Find the transfer function relating the output $v_0(t)$ to the input $f(t)$. From the transfer function, write the differential equation relating $v_0(t)$ to $f(t)$.

6.4-2 The switch in the circuit of Fig. P6.4-2 is closed for a long time and then opened instantaneously at $t = 0$. Find and sketch the current $y(t)$.

6.4-3 Find the current $y(t)$ for the parallel resonant circuit in Fig. P6.4-3 if the input is
(a) $f(t) = A \cos \omega_0 t\, u(t)$
(b) $f(t) = A \sin \omega_0 t\, u(t)$
$$\omega_0{}^2 = \frac{1}{LC}$$

Assume all initial conditions to be zero.

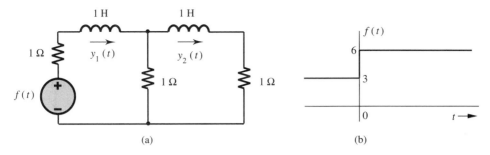

Fig. P6.4-4

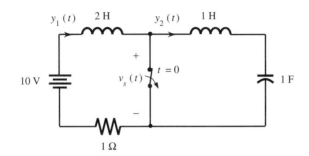

Fig. P6.4-5

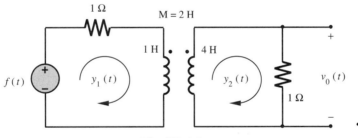

Fig. P6.4-6

6.4-4 Find the loop currents $y_1(t)$ and $y_2(t)$ for $t \geq 0$ in the circuit of Fig. P6.4-4a for the input $f(t)$ in Fig. P6.4-4b.

6.4-5 For the network in Fig. P6.4-5 the switch is in a closed position for a long time before $t = 0$, when it is opened instantaneously. Find $y_1(t)$ and $v_s(t)$ for $t \geq 0$.

6.4-6 Find the output voltage $v_0(t)$ for $t \geq 0$ for the circuit in Fig. P6.4-6, if the input $f(t) = 100u(t)$. The system is in zero state initially.

6.4-7 Find the output voltage $v_0(t)$ for the network in Fig. P6.4-7 if the initial conditions are $i_L(0) = 1$ A and $v_C(0) = 3$ V. (Hint: Use the parallel form of initial condition generators.)

6.4-8 For the network in Fig. P6.4-8, the switch is in position a for a long time and then is moved to position b instantaneously at $t = 0$. Determine the current $y(t)$ for $t > 0$. Hint: Use the Thévenin equivalent.

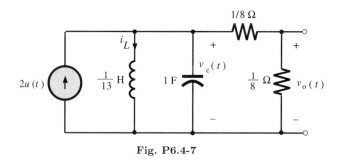

Fig. P6.4-7

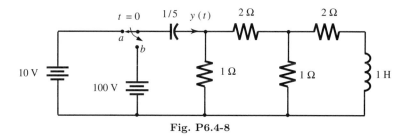

Fig. P6.4-8

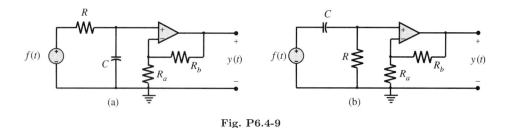

Fig. P6.4-9

6.4-9 Show that the transfer function which relates the output voltage $y(t)$ to the input voltage $f(t)$ for the op amp circuit in Fig. P6.4-9a is given by

$$H(s) = \frac{Ka}{s+a} \quad \text{where} \quad K = 1 + \frac{R_b}{R_a} \quad \text{and} \quad a = \frac{1}{RC}$$

and that the transfer function for the circuit in Fig. P6.4-9b is given by

$$H(s) = \frac{Ks}{s+a}$$

6.4-10 For a second-order op amp circuit in Fig. P6.4-10, show that the transfer function $H(s)$ relating the output voltage $v_0(t)$ to the input voltage $f(t)$ given by

$$H(s) = \frac{-s}{s^2 + 8s + 12}$$

6.4-11 Using the initial and final value theorems, find the initial and final value of the zero-state response of a system with the transfer function

$$H(s) = \frac{6s^2 + 3s + 10}{2s^2 + 6s + 5}$$

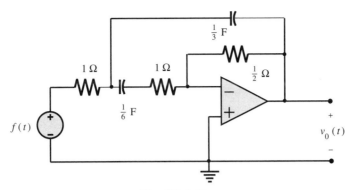

Fig. P6.4-10

and the input **(a)** $u(t)$ **(b)** $e^{-t}u(t)$.

6.5-1 Figure P6.5-1a shows two resistive ladder segments. The transfer function of each segment (ratio of output to input voltage) is $\frac{1}{2}$. Figure P6.5-1b shows these two segments connected in cascade.

(a) Is the transfer function (ratio of output to input voltage) of this cascaded network $\left(\frac{1}{2}\right)\left(\frac{1}{2}\right) = \frac{1}{4}$?

(b) If your answer is affirmative, verify the answer by direct computation of the transfer function. Does this computation confirm the earlier value $\frac{1}{4}$? If not, why?

(c) Repeat the problem with $R_3 = R_4 = 200$. Does this result suggest the answer to the problem in part **(b)**?

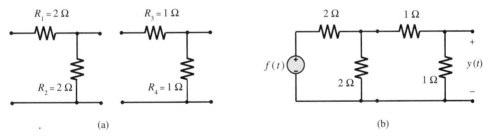

         (a)   (b)

Fig. P6.5-1

6.5-2 In Fig. 6.18, $h_1(t)$ and $h_2(t)$ are the impulse responses of systems with transfer functions $H_1(s)$ and $H_2(s)$, respectively. Determine the impulse response of the cascade, and parallel combination of $H_1(s)$ and $H_2(s)$ shown in Fig. 6.18b and c.

6.6-1 Realize

$$H(s) = \frac{s(s+2)}{(s+1)(s+3)(s+4)}$$

by canonical, series, and parallel forms.

6.6-2 Repeat Problem 6.6-1 if

(a) $H(s) = \dfrac{3s(s+2)}{(s+1)(s^2+2s+2)}$

(b) $H(s) = \dfrac{2s-4}{(s+2)(s^2+4)}$

6.6-3 Repeat Problem 6.6-1 if

$$H(s) = \frac{2s+3}{5s(s+2)^2(s+3)}$$

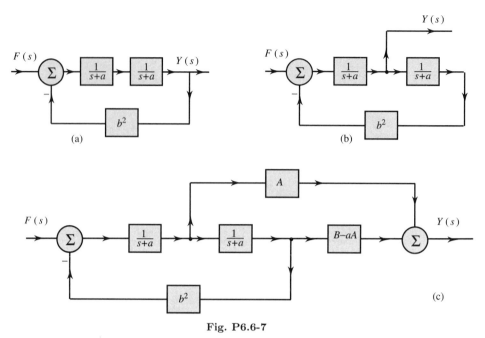

Fig. P6.6-7

Hint: Normalize the highest-power coefficient in the denominator to unity.

6.6-4   Repeat Problem 6.6-1 if

$$H(s) = \frac{s(s+1)(s+2)}{(s+5)(s+6)(s+8)}$$

Hint: Here $m = n = 3$.

6.6-5   Repeat Problem 6.6-1 if

$$H(s) = \frac{s^3}{(s+1)^2(s+2)(s+3)}$$

6.6-6   Repeat Problem 6.6-1 if

$$H(s) = \frac{s^3}{(s+1)(s^2+4s+13)}$$

6.6-7   In this problem we show how a pair of complex conjugate poles may be realized using a cascade of two first-order transfer functions. Show that the transfer functions of the block diagrams in Figs. P6.6-7a and b are

(a) $H(s) = \dfrac{1}{(s+a)^2 + b^2} = \dfrac{1}{s^2 + 2as + (a^2 + b^2)}$

(b) $H(s) = \dfrac{s+a}{(s+a)^2 + b^2} = \dfrac{s+a}{s^2 + 2as + (a^2 + b^2)}$

Hence, show that the transfer function of the block diagram in Fig. P6.6-7c is

(c) $H(s) = \dfrac{As + B}{(s+a)^2 + b^2} = \dfrac{As + B}{s^2 + 2as + (a^2 + b^2)}$

6.6-8   Show op amp realization of the following transfer functions:

(i) $\dfrac{-10}{s+5}$ (ii) $\dfrac{10}{s+5}$ (iii) $\dfrac{s+2}{s+5}$

6.6-9 Show two different op amp circuit realizations of the transfer function

$$H(s) = \frac{s+2}{s+5} = 1 - \frac{3}{s+5}$$

6.6-10 Show op amp canonical realization of the transfer function

$$H(s) = \frac{3s+7}{s^2 + 4s + 10}$$

6.6-11 Show op amp canonical realization of the transfer function

$$H(s) = \frac{s^2 + 5s + 2}{s^2 + 4s + 13}$$

6.7-1 Determine the rise time t_r, the settling time t_s, the PO, and the steady-state errors e_s, e_r, and e_p for each of the following systems, whose transfer functions are:

(a) $\dfrac{9}{s^2 + 3s + 9}$ (b) $\dfrac{4}{s^2 + 3s + 4}$ (c) $\dfrac{95}{s^2 + 10s + 100}$

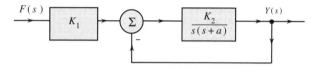

Fig. P6.7-2

6.7-2 For a position control system depicted in Fig. P6.7-2, the unit step response shows the peak time $t_p = \pi/4$, the PO = 9%, and the steady-state value of the output for the unit step input is $y_{ss} = 2$. Determine K_1, K_2 and a.

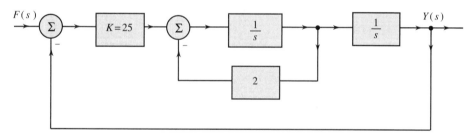

Fig. P6.7-3

6.7-3 For a position control system illustrated in Fig. P6.7-3 the following specifications are imposed: $t_r \leq 0.3$, $t_s \leq 1$, PO $\leq 30\%$ and $e_s = 0$. Which of these specifications cannot be met by the system for any value of K? Which specifications can be met by simple adjustment of K?

6.7-4 Open loop transfer functions of four closed-loop systems are given below. In each case, give a rough sketch of the root locus.

(a) $\dfrac{K(s+1)}{s(s+3)(s+5)}$ (c) $\dfrac{K(s+5)}{s(s+3)}$

(b) $\dfrac{K(s+1)}{s(s+3)(s+5)(s+7)}$ (d) $\dfrac{K(s+1)}{s(s+4)(s^2 + 2s + 2)}$

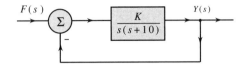

Fig. P6.7-5

6.7-5 For a unity feedback system shown in Fig. P6.7-5, We are required to meet the specifications PO $\leq$ 16%, $t_r \leq 0.2$, $t_s \leq 0.5$, $e_s = 0$ and $e_r \leq 0.06$. Is it possible to meet these specifications just by adjusting K? If not, suggest a suitable form of compensator and find the resulting PO, t_s, t_r, e_s and e_r.

6.8-1 Find the region of convergence, if it exists, of the (bilateral) Laplace transform of the following signals:

    (a) $e^{tu(t)}$ **(b)** $e^{-tu(t)}$ **(c)** $\dfrac{1}{1+t^2}$ **(d)** $\dfrac{1}{1+e^t}$ **(e)** e^{-kt^2}

6.8-2 Find the (bilateral) Laplace transform and the corresponding region of convergence for the following signals:

    (a) $e^{-|t|}$ **(b)** $e^{-|t|}\cos t$ **(c)** $e^{t}u(t) + e^{2t}u(-t)$

    (d) $e^{-tu(t)}$ **(e)** $e^{tu(-t)}$ **(f)** $\cos \omega_0 t\, u(t) + e^{t}\, u(-t)$

6.8-3 Find the inverse (bilateral) Laplace transforms of the following functions:

    (a) $\dfrac{2s+5}{(s+2)(s+3)}$ $-3 < \sigma < -2$

    (b) $\dfrac{2s-5}{(s-2)(s-3)}$ $2 < \sigma < 3$

    (c) $\dfrac{2s+3}{(s+1)(s+2)}$ $\sigma > -1$

    (d) $\dfrac{2s+3}{(s+1)(s+2)}$ $\sigma < -2$

    (e) $\dfrac{3s^2 - 2s - 17}{(s+1)(s+3)(s-5)}$ $-1 < \sigma < 5$

6.8-4 Find

$$\mathcal{L}^{-1}\left[\frac{2s^2 - 2s - 6}{(s+1)(s-1)(s+2)}\right]$$

    if the region of convergence is **(a)** Re $s > 1$ **(b)** Re $s < -2$
    (c) $-1 <$ Re $s < 1$ **(d)** $-2 <$ Re $s < -1$

6.8-5 For a causal LTIC system having a transfer function $H(s) = \frac{1}{s+1}$ find the output $y(t)$ if the input $f(t)$ is given by

    (a) $e^{-|t|/2}$ **(d)** $e^{2t}u(t) + e^{t}u(-t)$

    (b) $e^{t}u(t) + e^{2t}u(-t)$ **(e)** $e^{-t/4}u(t) + e^{-t/2}u(-t)$

    (c) $e^{-t/2}u(t) + e^{-t/4}u(-t)$ **(f)** $e^{-3t}u(t) + e^{-2t}u(-t)$

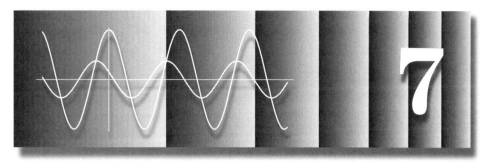

Frequency Response and
Analog Filters

Filtering is an important area of signal processing. We have already discussed ideal filters in Chapter 4. In this chapter, we shall discuss the practical filter characteristics and their design. Filtering characteristics of a filter are indicated by its response to sinusoids of various frequencies varying from 0 to ∞. Such characteristics are called the frequency response of the filter. Let us start with determining the frequency response of an LTIC system.

Recall that for $h(t)$, we use the notation $H(\omega)$ for its Fourier transform and $H(s)$ for its Laplace transform. Also, when the system is causal and asymptotically stable, all the poles of $H(s)$ lie in the LHP. Hence, the region of convergence for $H(s)$ includes the $j\omega$ axis, and we can obtain the Fourier transform $H(\omega)$ by substituting $s = j\omega$ in the corresponding Laplace transform $H(s)$ (see P. 370). Therefore, $H(j\omega)$ and $H(\omega)$ represent the same entity when the system is asymptotically stable. In this and later chapters, we shall often find it convenient to use the notation $H(j\omega)$ instead of $H(\omega)$.

7.1 Frequency Response of an LTIC System

In this section we find the system response to sinusoidal inputs. In Sec. 2.4-3 we showed that an LTIC system response to an everlasting exponential input $f(t) = e^{st}$ is also an everlasting exponential $H(s)e^{st}$. As before, we represent an input-output pair using an arrow directed from the input to the output as

$$e^{st} \Longrightarrow H(s)e^{st} \qquad (7.1)$$

Setting $s = \pm j\omega$ in this relationship yields

$$e^{j\omega t} \Longrightarrow H(j\omega)e^{j\omega t} \qquad (7.2a)$$

$$e^{-j\omega t} \Longrightarrow H(-j\omega)e^{-j\omega t} \qquad (7.2b)$$

Addition of these two relationships yields

$$2\cos \omega t \Longrightarrow H(j\omega)e^{j\omega t} + H(-j\omega)e^{-j\omega t} = 2\mathrm{Re}\left[H(j\omega)e^{j\omega t}\right] \qquad (7.3)$$

471

We can express $H(j\omega)$ in the polar form as

$$H(j\omega) = |H(j\omega)|\, e^{j\angle H(j\omega)} \qquad (7.4)$$

With this result, the relationship (7.3) becomes

$$\cos \omega t \Longrightarrow |H(j\omega)| \cos\left[\omega t + \angle H(j\omega)\right]$$

In other words, the system response $y(t)$ to a sinusoidal input $\cos \omega t$ is given by

$$y(t) = |H(j\omega)| \cos\left[\omega t + \angle H(j\omega)\right] \qquad (7.5a)$$

Using a similar argument, we can show that the system response to a sinusoid $\cos(\omega t + \theta)$ is

$$y(t) = |H(j\omega)| \cos\left[\omega t + \theta + \angle H(j\omega)\right] \qquad (7.5b)$$

This result, where we have let $s = j\omega$, is valid only for asymptotically stable systems because the relationship (7.1) applies only for the values of s lying in the region of convergence for $H(s)$. For the case of unstable and marginally stable systems, this region does not include the imaginary axis $s = j\omega$ (see also the footnote on p. 243).

Equation (7.5) shows that for a sinusoidal input of radian frequency ω, the system response is also a sinusoid of the same frequency ω. *The amplitude of the output sinusoid is $|H(j\omega)|$ times the input amplitude, and the phase of the output sinusoid is shifted by $\angle H(j\omega)$ with respect to the input phase* (see Fig. 7.1). For instance, if a certain system has $|H(j10)| = 3$ and $\angle H(j10) = -30°$, then the system amplifies a sinusoid of frequency $\omega = 10$ by a factor of 3 and delays its phase by $30°$. The system response to an input $5\cos(10t + 50°)$ is $3 \times 5\cos(10t + 50° - 30°) = 15\cos(10t + 20°)$.

Clearly $|H(j\omega)|$ is the system **gain**, and a plot of $|H(j\omega)|$ versus ω shows the system gain as a function of frequency ω. This function is more commonly known as the **amplitude response**. Similarly, $\angle H(j\omega)$ is the **phase response** and a plot of $\angle H(j\omega)$ versus ω shows how the system modifies or changes the phase of the input sinusoid. These two plots together, as functions of ω, are called the **frequency response of the system**. Observe that $H(j\omega)$ has the information of $|H(j\omega)|$ and $\angle H(j\omega)$. For this reason, $H(j\omega)$ is also called the **frequency response of the system**. The frequency response plots show at a glance how a system responds to sinusoids of various frequencies. Thus, the frequency response of a system represents its filtering characteristic.

■ **Example 7.1**
Find the frequency response (amplitude and phase response) of a system whose transfer function is

$$H(s) = \frac{s + 0.1}{s + 5}$$

Also, find the system response $y(t)$ if the input $f(t)$ is **(a)** $\cos 2t$ **(b)** $\cos(10t - 50°)$.

In this case

$$H(j\omega) = \frac{j\omega + 0.1}{j\omega + 5}$$

Recall that the magnitude of a complex number is equal to the square root of the sum of the squares of its real and imaginary parts. Hence

$$|H(j\omega)| = \frac{\sqrt{\omega^2 + 0.01}}{\sqrt{\omega^2 + 25}} \quad \text{and} \quad \angle H(j\omega) = \tan^{-1}\left(\frac{\omega}{0.1}\right) - \tan^{-1}\left(\frac{\omega}{5}\right)$$

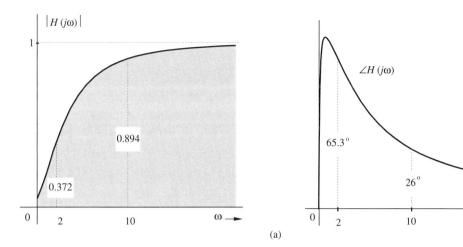

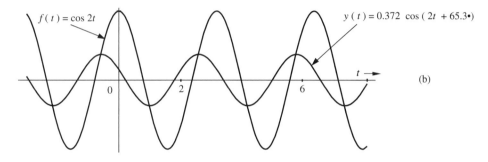

Fig. 7.1 Frequency response of an LTIC system in Example 7.1.

Both the amplitude and the phase response are depicted in Fig. 7.1a as functions of ω. These plots furnish the complete information about the frequency response of the system to sinusoidal inputs.

(a) For the input $f(t) = \cos 2t$, $\omega = 2$, and

$$|H(j2)| = \frac{\sqrt{(2)^2 + 0.01}}{\sqrt{(2)^2 + 25}} = 0.372$$

$$\angle H(j2) = \tan^{-1}\left(\frac{2}{0.1}\right) - \tan^{-1}\left(\frac{2}{5}\right) = 87.1° - 21.8° = 65.3°$$

We also could have read these values directly from the frequency response plots in Fig. 7.1a corresponding to $\omega = 2$. This result means that for a sinusoidal input with frequency $\omega = 2$, the amplitude gain of the system is 0.372, and the phase shift is 65.3°. In other words, the output amplitude is 0.372 times the input amplitude, and the phase of the output is shifted with respect to that of the input by 65.3°. Therefore, the system response to an input $\cos 2t$ is

$$y(t) = 0.372 \cos(2t + 65.3°)$$

The input $\cos 2t$ and the corresponding system response $0.372 \cos(2t + 65.34°)$ are illustrated in Fig. 7.1b.

(b) For the input $\cos{(10t-50°)}$, instead of computing the values $|H(j\omega)|$ and $\angle H(j\omega)$ as in part **(a)**, we shall read them directly from the frequency response plots in Fig. 7.1a corresponding to $\omega = 10$. These are:

$$|H(j10)| = 0.894 \qquad \text{and} \qquad \angle H(j10) = 26°$$

Therefore, for a sinusoidal input of frequency $\omega = 10$, the output sinusoid amplitude is 0.894 times the input amplitude and the output sinusoid is shifted with respect to the input sinusoid by $26°$. Therefore, $y(t)$, the system response to an input $\cos{(10t - 50°)}$, is

$$y(t) = 0.894\cos{(10t - 50° + 26°)} = 0.894\cos{(10t - 24°)}$$

If the input were $\sin{(10t - 50°)}$, the response would be $0.894\sin{(10t - 50° + 26°)} = 0.894\sin{(10t - 24°)}$.

The frequency response plots in Fig. 7.1a show that the system has highpass filtering characteristics; it responds well to sinusoids of higher frequencies (ω well above 5), and suppresses sinusoids of lower frequencies (ω well below 5). ∎

⊙ **Computer Example C7.1**
Plot the frequency response of the transfer functions $H(s) = \frac{s+5}{s^2+3s+2}$.

```
num=[1 5];
den=[1 3 2];
w=.1:.01:100;
axis([log10(.1) log10(100) -50 50])
[mag,phase,w]=bode(num,den,w);
subplot(211),semilogx(w,20*log10(mag))
subplot(212),semilogx(w,phase)    ⊙
```

∎ **Example 7.2**
Find and sketch the frequency response (amplitude and phase response) for
(a) an ideal delay of T seconds;
(b) an ideal differentiator;
(c) an ideal integrator.

(a) Ideal delay of T seconds: The transfer function of an ideal delay is [see Eq. (6.54)]

$$H(s) = e^{-sT}$$

Therefore

$$H(j\omega) = e^{-j\omega T}$$

Consequently

$$|H(j\omega)| = 1 \qquad \text{and} \qquad \angle H(j\omega) = -\omega T \qquad\qquad (7.6)$$

This amplitude and phase response is shown in Fig. 7.2a. The amplitude response is constant (unity) for all frequencies. The phase shift increases linearly with frequency with a slope of $-T$. This result can be explained physically by recognizing that if a sinusoid $\cos{\omega t}$ is passed through an ideal delay of T seconds, the output is $\cos{\omega(t - T)}$. The output sinusoid amplitude is the same as that of the input for all values of ω. Therefore, the amplitude response (gain) is unity for all frequencies. Moreover, the output $\cos{\omega(t - T)} = \cos{(\omega t - \omega T)}$ has a phase shift $-\omega T$ with respect to the input $\cos{\omega t}$. Therefore, the phase response is linearly proportional to the frequency ω with a slope $-T$.

(b) An ideal differentiator: The transfer function of an ideal differentiator is [see Eq. (6.55)]

$$H(s) = s$$

Therefore

$$H(j\omega) = j\omega = \omega e^{j\pi/2}$$

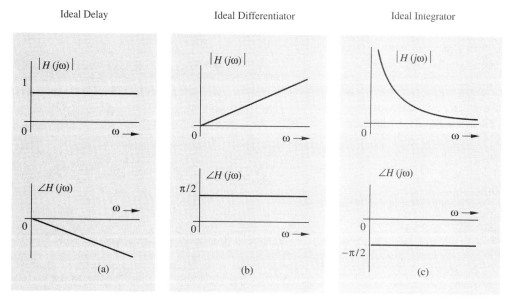

Fig. 7.2 Frequency response of an ideal (a) delay (b) differentiator (c) integrator.

Consequently

$$|H(j\omega)| = \omega \qquad \text{and} \qquad \angle H(j\omega) = \tfrac{\pi}{2} \tag{7.7}$$

This amplitude and phase response is depicted in Fig. 7.2b. The amplitude response increases linearly with frequency, and phase response is constant ($\pi/2$) for all frequencies. This result can be explained physically by recognizing that if a sinusoid $\cos \omega t$ is passed through an ideal differentiator, the output is $-\omega \sin \omega t = \omega \cos{(\omega t + \tfrac{\pi}{2})}$. Therefore, the output sinusoid amplitude is ω times the input amplitude; that is, the amplitude response (gain) increases linearly with frequency ω. Moreover, the output sinusoid undergoes a phase shift $\tfrac{\pi}{2}$ with respect to the input $\cos \omega t$. Therefore, the phase response is constant ($\pi/2$) with frequency.

In an ideal differentiator, the amplitude response (gain) is proportional to frequency $[|H(j\omega)| = \omega]$, so that the higher-frequency components are enhanced (see Fig. 7.2b). All practical signals are contaminated with noise, which, by its nature, is a broad-band (rapidly varying) signal containing components of very high frequencies. A differentiator can increase the noise disproportionately to the point of drowning out the desired signal. This is the reason why ideal differentiators are avoided in practice.

(c) An ideal integrator: The transfer function of an ideal integrator is [see Eq. (6.56)]

$$H(s) = \frac{1}{s}$$

Therefore

$$H(j\omega) = \frac{1}{j\omega} = \frac{-j}{\omega} = \frac{1}{\omega}e^{-j\pi/2}$$

Consequently

$$|H(j\omega)| = \frac{1}{\omega} \qquad \text{and} \qquad \angle H(j\omega) = -\frac{\pi}{2} \tag{7.8}$$

This amplitude and phase response is illustrated in Fig. 7.2c. The amplitude response is inversely proportional to frequency, and the phase shift is constant ($-\pi/2$) with frequency.

This result can be explained physically by recognizing that if a sinusoid $\cos \omega t$ is passed through an ideal integrator, the output is $\frac{1}{\omega} \sin \omega t = \frac{1}{\omega} \cos \left(\omega t - \frac{\pi}{2}\right)$. Therefore, the amplitude response is inversely proportional to ω, and the phase response is constant $(-\pi/2)$ with frequency.

Because its gain is $1/\omega$, the ideal integrator suppresses higher-frequency components but enhances lower-frequency components with $\omega < 1$. Consequently, noise signals (if they do not contain an appreciable amount of very low frequency components) are suppressed (smoothed out) by an integrator.   ■

△ **Exercise E7.1**
     Find the response of an LTIC system specified by

$$\frac{d^2 y}{dt^2} + 3\frac{dy}{dt} + 2y(t) = \frac{df}{dt} + 5f(t)$$

if the input is a sinusoid $20 \sin (3t + 35°)$

Answer: $10.23 \sin (3t - 61.91°)$ ▽

7.1-1 Steady-State Response to Causal Sinusoidal Inputs

So far we discussed the LTIC system response to everlasting sinusoidal inputs (starting at $t = -\infty$). In practice, we are more interested in causal sinusoidal inputs (sinusoids starting at $t = 0$). Consider the input $e^{j\omega t} u(t)$, which starts at $t = 0$ rather than at $t = -\infty$. In this case $F(s) = 1/(s + j\omega)$. Moreover, according to Eq. (6.51) $H(s) = P(s)/Q(s)$, where $Q(s)$ is the characteristic polynomial given by $Q(s) = (s - \lambda_1)(s - \lambda_2) \cdots (s - \lambda_n)$. Hence,

$$Y(s) = F(s)H(s) = \frac{P(s)}{(s - \lambda_1)(s - \lambda_2) \cdots (s - \lambda_n)(s - j\omega)}$$

In the partial fraction expansion of the right-hand side, let the coefficients corresponding to the n terms $(s - \lambda_1)$, $(s - \lambda_2) \cdots (s - \lambda_n)$ be $k_1, k_2, \cdots k_n$. The coefficient corresponding to the last term $(s - j\omega)$ is $P(s)/Q(s)|_{s=j\omega} = H(j\omega)$. Hence,

$$Y(s) = \sum_{i=1}^{n} \frac{k_i}{s - \lambda_i} + \frac{H(j\omega)}{s - j\omega}$$

and

$$y(t) = \underbrace{\sum_{i=1}^{n} k_i e^{\lambda_i t} u(t)}_{\text{transient component } y_{tr}(t)} + \underbrace{H(j\omega)e^{j\omega t} u(t)}_{\text{steady-state component } y_{ss}(t)} \qquad (7.9)$$

For an asymptotically stable system, the characteristic mode terms $e^{\lambda_i t}$ decay with time, and, therefore, constitute the so-called **transient** component of the response. The last term $H(j\omega)e^{j\omega t}$ persists forever, and is the **steady-state** component of the response given by

$$y_{ss}(t) = H(j\omega)e^{j\omega t} u(t)$$

From the argument that led to Eq. (7.5a), it follows that for a causal sinusoidal input $\cos \omega t$, the steady-state response $y_{ss}(t)$ is given by

$$y_{ss}(t) = |H(j\omega)| \cos [\omega t + \angle H(j\omega)]u(t) \qquad (7.10)$$

In summary, $|H(j\omega)| \cos [\omega t + \angle H(j\omega)]$ is the total response to everlasting sinusoid $\cos \omega t$, and is also the steady-state response to the same input applied at $t = 0$.

7.2 Bode Plots

Sketching frequency response plots is considerably facilitated by the use of logarithmic scales. The amplitude and phase response plots as a function of ω on a logarithmic scale are known as the **Bode plots**. By using the asymptotic behavior of the amplitude and the phase response, we can sketch these plots with remarkable ease, even for higher-order transfer functions.

Let us consider a system with the transfer function

$$H(s) = \frac{K(s + a_1)(s + a_2)}{s(s + b_1)(s^2 + b_2s + b_3)} \tag{7.11a}$$

where the second-order factor $(s^2 + b_2s + b_3)$ is assumed to have complex conjugate roots. We shall rearrange Eq. (7.11a) in the form

$$H(s) = \frac{Ka_1a_2}{b_1b_3} \frac{\left(\dfrac{s}{a_1} + 1\right)\left(\dfrac{s}{a_2} + 1\right)}{s\left(\dfrac{s}{b_1} + 1\right)\left(\dfrac{s^2}{b_3} + \dfrac{b_2}{b_3}s + 1\right)} \tag{7.11b}$$

and

$$H(j\omega) = \frac{Ka_1a_2}{b_1b_3} \frac{\left(1 + \dfrac{j\omega}{a_1}\right)\left(1 + \dfrac{j\omega}{a_2}\right)}{j\omega\left(1 + \dfrac{j\omega}{b_1}\right)\left[1 + j\dfrac{b_2\omega}{b_3} + \dfrac{(j\omega)^2}{b_3}\right]} \tag{7.11c}$$

This equation shows that $H(j\omega)$ is a complex function of ω. The amplitude response $|H(j\omega)|$ and the phase response $\angle H(j\omega)$ are given by

$$|H(j\omega)| = \frac{Ka_1a_2}{b_1b_3} \frac{\left|1 + \dfrac{j\omega}{a_1}\right|\left|1 + \dfrac{j\omega}{a_2}\right|}{|j\omega|\left|1 + \dfrac{j\omega}{b_1}\right|\left|1 + j\dfrac{b_2\omega}{b_3} + \dfrac{(j\omega)^2}{b_3}\right|} \tag{7.12a}$$

and

$$\angle H(j\omega) = \angle\left(1 + \frac{j\omega}{a_1}\right) + \angle\left(1 + \frac{j\omega}{a_2}\right) - \angle j\omega$$

$$- \angle\left(1 + \frac{j\omega}{b_1}\right) - \angle\left[1 + \frac{jb_2\omega}{b_3} + \frac{(j\omega)^2}{b_3}\right] \tag{7.12b}$$

From Eq. (7.12b) we see that the phase function consists of the addition of only three kinds of terms: (i) the phase of $j\omega$, which is 90° for all values of ω, (ii) the phase for the first-order term of the form $1 + \dfrac{j\omega}{a}$, and (iii) the phase of the second-order term

$$\left[1 + \frac{jb_2\omega}{b_3} + \frac{(j\omega)^2}{b_3}\right]$$

We can plot these three basic phase functions for ω in the range 0 to ∞ and then, using these plots, we can construct the phase function of any transfer by properly adding these basic responses. Note that if a particular term is in the numerator, its phase is added, but if the term is in the denominator, its phase is subtracted. This makes it easy to plot the phase function $\angle H(j\omega)$ as a function of ω. Computation of $|H(j\omega)|$, unlike that of the phase function, however, involves the multiplication and division of various terms. This is a formidable task, especially when we have to plot this function for the entire range of ω (0 to ∞).

We know that a log operation converts multiplication and division to addition and subtraction. So, instead of plotting $|H(j\omega)|$, why not plot $\log|H(j\omega)|$ to simplify our task? We can take advantage of the fact that logarithmic units are desirable in several applications, where the variables considered have a very large range of variation. This is particularly true in frequency response plots, where we may have to plot frequency response over a frequency range starting from a very low frequency near 0 to a very high frequency in the range of 10^{10} or higher. A linear plot for such a large range will bury much of the useful information. Also, the amplitude response may have a very large dynamic range from a low of 10^{-6} to a high of 10^6. A linear plot would be unsuitable for such a situation. Therefore, logarithmic plots not only simplify our task of plotting, but, fortunately, they are also desirable in this situation. The logarithmic unit is the **decibel** and is equal to twenty times the logarithm of the quantity (log to the base 10). Therefore, $20\log_{10}|H(j\omega)|$ is simply the log amplitude in decibels (dB). Thus, instead of plotting $|H(j\omega)|$, we shall plot $20\log_{10}|H(j\omega)|$ as a function of ω. These plots (log amplitude and phase) are called **Bode plots**. For the transfer function in Eq. (7.12a), the **log amplitude** is

$$20\log|H(j\omega)| = 20\log\frac{Ka_1a_2}{b_1b_3} + 20\log\left|1 + \frac{j\omega}{a_1}\right| + 20\log\left|1 + \frac{j\omega}{a_2}\right| - 20\log|j\omega|$$

$$- 20\log\left|1 + \frac{j\omega}{b_1}\right| - 20\log\left|1 + \frac{jb_2\omega}{b_3} + \frac{(j\omega)^2}{b_3}\right| \qquad (7.13)$$

The term $20\log(Ka_1a_2/b_1b_3)$ is a constant. We observe that the log amplitude is a sum of four basic terms corresponding to (i) a constant, (ii) a pole or zero at the origin ($20\log|j\omega|$), (iii) a first-order pole or zero ($20\log|1 + j\omega/a|$), and (iv) complex conjugate poles or zeros ($20\log|1 + j\omega b_2/b_3 + (j\omega)^2/b_3|$). We can sketch these four basic terms as functions of ω and use them to construct the log-amplitude plot of any desired transfer function. Let us discuss each of the terms.

1. Constant ka_1a_2/b_1b_3

The log amplitude of this term is also a constant, $20\log(Ka_1a_2/b_1b_3)$. The phase contribution from this term is zero.

2. Pole (or Zero) at the origin

Log Magnitude

Such a pole gives rise to the term $-20\log|j\omega|$, which can be expressed as

$$-20\log|j\omega| = -20\log\omega$$

This function can be plotted as a function of ω. However, we can effect further simplification by using the logarithmic scale for the variable ω itself. Let us define a new variable u such that

$$u = \log \omega \qquad (7.14)$$

Hence

$$-20 \log \omega = -20u \qquad (7.15a)$$

The log-amplitude function $-20u$ is plotted as a function of u in Fig. 7.3a. This is a straight line with a slope of -20. It crosses the u-axis at $u=0$. The ω-scale ($u = \log \omega$) also appears in Fig. 7.3a. Semilog graphs can be conveniently used for plotting, and we can directly plot ω on semilog paper. A ratio of 10 is a **decade** and a ratio of 2 is known as an **octave**. Furthermore, a decade along the ω-scale is equivalent to 1 unit along the u-scale. We can also show that a ratio of 2 (an octave) along the ω-scale equals to 0.3010 (which is $\log_{10} 2$) along the u-scale.†

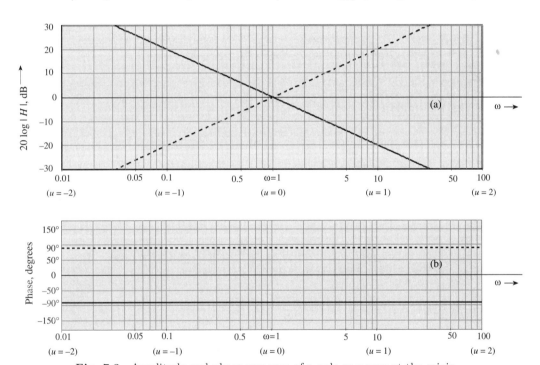

Fig. 7.3   Amplitude and phase response of a pole or a zero at the origin.

---

†This point can be shown as follows: Let ω_1 and ω_2 along the ω-scale correspond to u_1 and u_2 along the u-scale. Then $\log \omega_1 = u_1$, and $\log \omega_2 = u_2$. Then

$$u_2 - u_1 = \log_{10} \omega_2 - \log_{10} \omega_1 = \log_{10}(\omega_2/\omega_1)$$

Thus, if

$$(\omega_2/\omega_1) = 10 \quad \text{(which is a decade)}$$

then

$$u_2 - u_1 = \log_{10} 10 = 1$$

and if

$$(\omega_2/\omega_1) = 2 \quad \text{(which is an octave)}$$

then

$$u_2 - u_1 = \log_{10} 2 = 0.3010$$

Note that equal increments in u are equivalent to equal ratios on the ω-scale. Thus, one unit along the u-scale is the same as one decade along the ω-scale. This means that the amplitude plot has a slope of -20 dB/decade or $-20(0.3010) = -6.02$ dB/octave (commonly stated as 6 dB/octave). Moreover, the amplitude plot crosses the ω-axis at $\omega = 1$, since $u = \log_{10}\omega = 0$ when $\omega = 1$.

For the case of a zero at the origin, the log-amplitude term is $20\log\omega$. This is a straight line passing through $\omega = 1$ and having a slope of 20 dB/decade (or 6 dB/octave). This plot is a mirror image about the ω-axis of the plot for a pole at the origin and is shown dotted in Fig. 7.3a.

Phase

The phase function corresponding to the pole at the origin is $-\angle j\omega$ [see Eq. (7.12b)]. Thus

$$\angle H(j\omega) = -\angle j\omega = -90°   \tag{7.15b}$$

The phase is constant $(-90°)$ for all values of ω, as depicted in Fig. 7.3b. For a zero at the origin, the phase is $\angle j\omega = 90°$. This is a mirror image of the phase plot for a pole at the origin and is shown dotted in Fig. 7.3b.

3. First-Order Pole (or Zero)

The Log Magnitude

The log amplitude because of a first-order pole at $-a$ is $-20\log\left|1 + \dfrac{j\omega}{a}\right|$. Let us investigate the asymptotic behavior of this function for extreme values of ω ($\omega \ll a$ and $\omega \gg a$).

(a) For $\omega \ll a$,

$$-20\log\left|1 + \frac{j\omega}{a}\right| \approx -20\log 1 = 0   \tag{7.16}$$

Hence, the log-amplitude function $\to 0$ asymptotically for $\omega \ll a$ (Fig. 7.4a).

(b) For the other extreme case, where $\omega \gg a$,

$$-20\log\left|1 + \frac{j\omega}{a}\right| \approx -20\log\left(\frac{\omega}{a}\right)   \tag{7.17a}$$

$$= -20\log\omega + 20\log a   \tag{7.17b}$$

$$= -20u + 20\log a$$

This represents a straight line (when plotted as a function of u, the log of ω) with a slope of -20 dB/decade (or -6 dB/octave). When $\omega = a$, the log amplitude is zero [Eq. (7.17b)]. Hence, this line crosses the ω-axis at $\omega = a$, as illustrated in Fig. 7.4a. Note that the asymptotes in (a) and (b) meet at $\omega = a$.

The exact log amplitude for this pole is

$$-20\log\left|1 + \frac{j\omega}{a}\right| = -20\log\left(1 + \frac{\omega^2}{a^2}\right)^{\frac{1}{2}}$$

$$= -10\log\left(1 + \frac{\omega^2}{a^2}\right)   \tag{7.18}$$

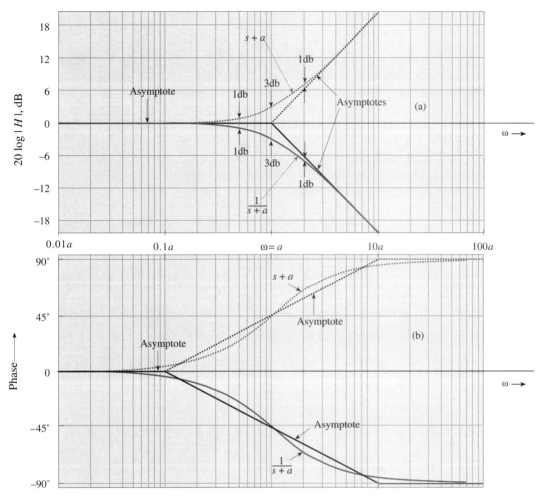

Fig. 7.4 Amplitude and phase response of a first-order pole or zero.

This exact log magnitude function also appears in Fig. 7.4a. Observe that the actual plot and the asymptotic plots are very close. A maximum error of 3 dB occurs at $\omega = a$. This frequency is known as the **corner frequency** or **break frequency**. The error everywhere else is less than 3 dB. A plot of the error as a function of ω is shown in Fig. 7.5a. This figure shows that the error at one octave above or below the corner frequency is 1 dB and the error at two octaves above or below the corner frequency is 0.3 dB. The actual plot can be obtained by adding the error to the asymptotic plot.

The amplitude response for a zero at $-a$ (shown dotted in Fig. 7.4a) is identical to that of the pole at $-a$ with a sign change, and therefore is the mirror image (about the 0-dB line) of the amplitude plot for a pole at $-a$.

Phase

The phase for the first-order pole at $-a$ is

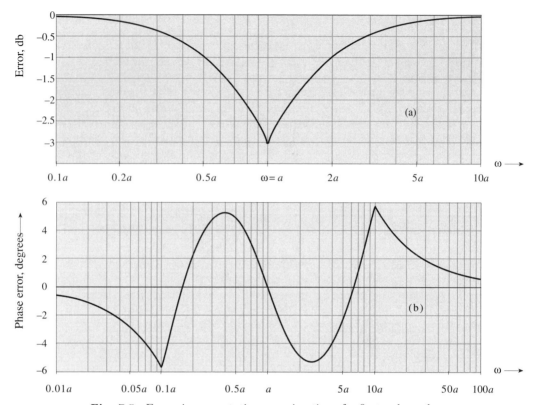

Fig. 7.5 Errors in asymptotic approximation of a first-order pole.

$$\angle H\left(j\omega\right) = -\angle\left(1 + \frac{j\omega}{a}\right) = -\tan^{-1}\left(\frac{\omega}{a}\right)$$

Let us investigate the asymptotic behavior of this function. For $\omega \ll a$,

$$-\tan^{-1}\left(\frac{\omega}{a}\right) \approx 0$$

and, for $\omega \gg a$,

$$-\tan^{-1}\left(\frac{\omega}{a}\right) \approx -90°$$

The actual plot along with the asymptotes is depicted in Fig. 7.4b. In this case, we use a three-line segment asymptotic plot for greater accuracy. The asymptotes are (i) a phase angle of $0°$ for $\omega \le a/10$, (ii) a phase angle of $-90°$ for $\omega \ge 10a$, and a straight line with a slope $-45°$/decade connecting these two asymptotes (from $\omega = a/10$ to $10a$) crossing the ω axis at $\omega = a/10$. It can be seen from Fig. 7.4b that the asymptotes are very close to the curve and the maximum error is $5.7°$. A plot of the error as a function of ω is illustrated in Fig. 7.5b. The actual plot can be obtained by adding the error to the asymptotic plot. The phase function for a pole at $-a$ is shown in Fig. 7.4b. The phase for a zero at $-a$ (shown dotted in Fig.

7.4b) is identical to that of the pole at $-a$ with a sign change, and therefore is the mirror image (about the $0°$ line) of the phase plot for a pole at $-a$.

4. Second-Order Pole (or Zero)

Let us consider the second-order pole in Eq. (7.11a). The denominator term is $s^2 + b_2 s + b_3$. We shall introduce the often-used standard form $s^2 + 2\zeta\omega_n s + \omega_n^2$ instead of $s^2 + b_2 s + b_3$. With this form, the log amplitude function for the second-order term in Eq. (7.13) becomes

$$-20\log\left|1 + 2j\zeta\frac{\omega}{\omega_n} + \left(\frac{j\omega}{\omega_n}\right)^2\right| \tag{7.19a}$$

and the phase function is

$$-\angle\left[1 + 2j\zeta\frac{\omega}{\omega_n} + \left(\frac{j\omega}{\omega_n}\right)^2\right] \tag{7.19b}$$

The Log Magnitude

The log amplitude is given by

$$\log\text{ amplitude} = -20\log\left|1 + 2j\zeta\left(\frac{\omega}{\omega_n}\right) + \left(\frac{j\omega}{\omega_n}\right)^2\right| \tag{7.20}$$

For $\omega \ll \omega_n$, the log amplitude becomes

$$\log\text{ amplitude} \approx -20\log 1 = 0 \tag{7.21}$$

For $\omega \gg \omega_n$, the log amplitude is

$$\log\text{ amplitude} \approx -20\log\left|\left(-\frac{\omega}{\omega_n}\right)^2\right| = -40\log\left(\frac{\omega}{\omega_n}\right) \tag{7.22a}$$

$$= -40\log\omega - 40\log\omega_n \tag{7.22b}$$

$$= -40u - 40\log\omega_n \tag{7.22c}$$

The two asymptotes are (i) zero for $\omega < \omega_n$, and (ii) $-40u - 40\log\omega_n$ for $\omega > \omega_n$. The second asymptote is a straight line with a slope of -40 dB/decade (or -12dB/octave) when plotted against the log ω scale. It begins at $\omega = \omega_n$ [see Eq. (7.22b)]. The asymptotes are depicted in Fig. 7.6a. The exact log amplitude is given by [see Eq. (7.20)]

$$\log\text{ amplitude} = -20\log\left\{\left[1 - \left(\frac{\omega}{\omega_n}\right)^2\right]^2 + 4\zeta^2\left(\frac{\omega}{\omega_n}\right)^2\right\}^{\frac{1}{2}} \tag{7.23}$$

Clearly, the log amplitude in this case involves a parameter ζ. For each value of ζ, we have a different plot. For complex conjugate poles,† $\zeta < 1$. Hence, we must sketch a family of curves for a number of values of ζ in the range 0 to 1. This is illustrated in Fig. 7.6a. The error between the actual plot and the asymptotes

---

†For $\zeta \geq 1$, the two poles in the second-order factor are no longer complex but real, and each of these two real poles can be dealt with as separate first-order factors.

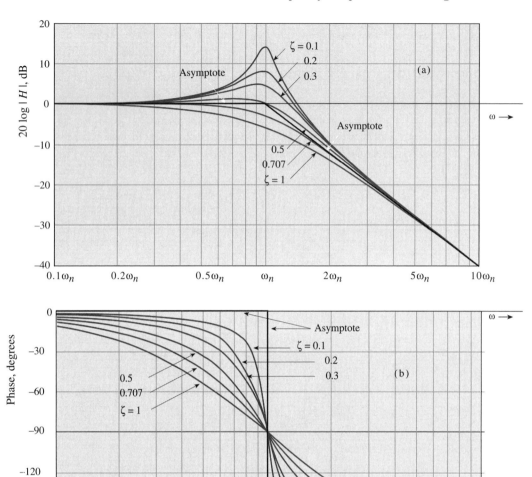

Fig. 7.6   Amplitude and phase response of a second-order pole.

is shown in Fig. 7.7. The actual plot can be obtained by adding the error to the asymptotic plot.

For second-order zeros (complex conjugate zeros), the plots are mirror images (about the 0-dB line) of the plots depicted in Fig. 7.6a. Note the resonance phenomenon of the complex conjugate poles. This phenomenon is barely noticeable for $\zeta > 0.707$ but becomes pronounced as $\zeta \to 0$.

Phase

The phase function for second-order poles, as apparent in Eq. (7.19b), is

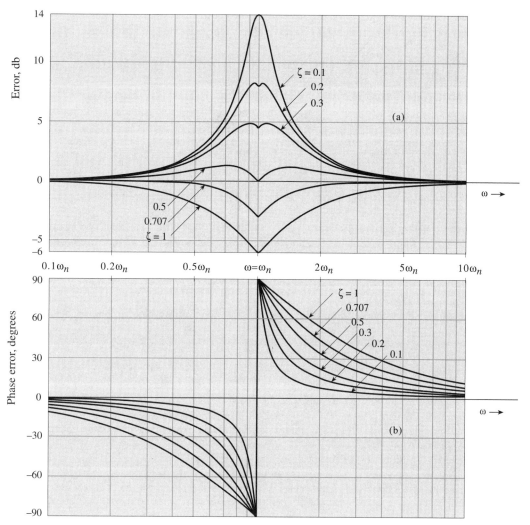

Fig. 7.7   Errors in the asymptotic approximation of a second-order pole.

$$\angle H(j\omega) = -\tan^{-1}\left[2\zeta\left(\frac{\omega}{\omega_n}\right)\bigg/1-\left(\frac{\omega}{\omega_n}\right)^2\right] \tag{7.24}$$

For $\omega \ll \omega_n$,

$$\angle H(j\omega) \approx 0$$

For $\omega \gg \omega_n$,

$$\angle H(j\omega) \simeq -180°$$

Hence, the phase $\to -180°$ as $\omega \to \infty$. As in the case of amplitude, we also have a family of phase plots for various values of ζ, as illustrated in Fig. 7.6b. A convenient asymptote for the phase of complex conjugate poles is a step function that is $0°$ for $\omega < \omega_n$ and $-180°$ for $\omega > \omega_n$. An error plot for such an asymptote is shown in Fig. 7.7 for various values of ζ. The exact phase is the asymptotic value plus the error.

For complex conjugate zeros, the amplitude and phase plots are mirror images of those for complex conjugate poles. We shall demonstrate the application of these techniques with two examples.

■ **Example 7.3**
Sketch the Bode plots for the transfer function

$$H(s) = \frac{20s(s + 100)}{(s + 2)(s + 10)}$$

First, we write the transfer function in normalized form

$$H(s) = \frac{20 \times 100}{2 \times 10} \frac{s\left(1 + \frac{s}{100}\right)}{\left(1 + \frac{s}{2}\right)\left(1 + \frac{s}{10}\right)} = 100 \frac{s\left(1 + \frac{s}{100}\right)}{\left(1 + \frac{s}{2}\right)\left(1 + \frac{s}{10}\right)} \tag{7.25}$$

Here, the constant term is 100; that is, 40 dB ($20 \log 100 = 40$). This term can be added to the plot by simply relabeling the horizontal axis (from where the asymptotes begin) as the 40 dB line (see Fig. 7.8a). Such a step implies shifting the horizontal axis upward by 40 dB. This is precisely what is desired.

In addition, we have two first-order poles at -2 and -10, one zero at the origin, and one zero at -100.

Step 1. For each of these terms, we draw an asymptotic plot as follows (see Fig. 7.8a):

(i) For the zero at the origin draw a straight line with a slope of 20 dB/decade passing through $\omega = 1$.

(ii) For the pole at -2, draw a straight line with a slope of -20 dB/decade (for $\omega > 2$) beginning at the corner frequency $\omega = 2$.

(iii) For the pole at -10, draw a straight line with a slope of -20 dB/decade beginning at the corner frequency $\omega = 10$.

(iv) For the zero at -100, draw a straight line with a slope of 20 dB/decade beginning at the corner frequency $\omega = 100$.

Step 2. Add all the asymptotes, as depicted in Fig. 7.8a.

Step 3. Apply the following corrections (see Fig. 7.5a):

(i) The correction at $\omega = 1$ because of the corner frequency at $\omega = 2$ is -1 dB. The correction at $\omega = 1$ because of the corner frequencies at $\omega = 10$ and $\omega = 100$ are quite small (see Fig. 7.5a) and may be ignored. Hence, the net correction at $\omega = 1$ is -1 dB.

(ii) The correction at $\omega = 2$ because of the corner frequency at $\omega = 2$ is -3 dB, and the correction because of the corner frequency at $\omega = 10$ is -0.17 dB. The correction because of the corner frequency $\omega = 100$ can be safely ignored. Hence the net correction at $\omega = 2$ is -3.17 dB.

(iii) The correction at $\omega = 10$ because of the corner frequency at $\omega = 10$ is -3 dB, and the correction because of the corner frequency at $\omega = 2$ is -0.17 dB. The correction because of $\omega = 100$ is 0.04 dB and may be ignored. Hence the net correction at $\omega = 10$ is -3.17 dB.

(iv) The correction at $\omega = 100$ because of the corner frequency at $\omega = 100$ is 3 dB, and the corrections because of the other corner frequencies may be ignored.

(v) The corrections at $\omega = 4$ and $\omega = 5$ (because of corner frequencies at $\omega = 2$ and 10) are found to be about -1.75 dB each.

With these corrections, the resulting amplitude plot is illustrated in Fig. 7.8a.

Phase Plots

We draw the asymptotes corresponding to each of the four factors:

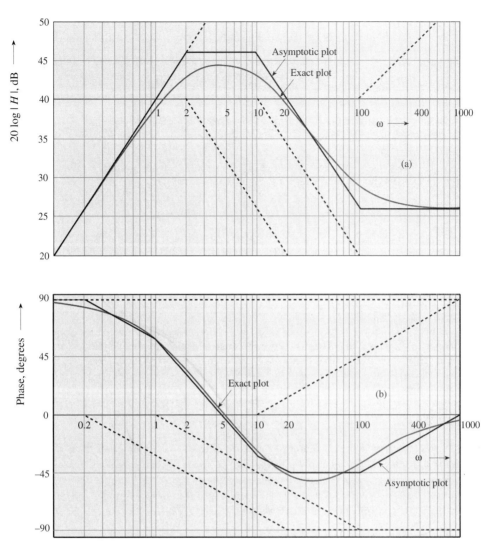

Fig. 7.8   Amplitude and phase response of the second-order system in Example 7.3.

(i) The zero at the origin causes a 90° phase shift.

(ii) The pole at $s = -2$ gives rise to a $-45°$/decade asymptote starting at $\omega = 0.2$, which goes up to $\omega = 20$. For $\omega < 0.2$, the asymptote is $0°$, and, for $\omega > 20$, the asymptotic value is $-90°$.

(iii) The pole at $s = -10$ has an asymptote with a zero value for $-\infty < \omega < 1$ and a slope of $-45°$/decade beginning at $\omega = 1$ and going up to $\omega = 100$. The asymptotic value for $\omega > 100$ is $-90°$.

(iv) The zero at $s = -100$ gives rise to an asymptote with a $45°$/decade slope, beginning at $\omega = 10$ and going up to $\omega = 1000$. For $\omega < 10$, the asymptotic value is $0°$, and, for $\omega > 1000$, the asymptotic value is $90°$. All the asymptotes are added, as shown in Fig. 7.8b. The appropriate corrections are applied from Fig. 7.5b, and the exact phase plot is depicted in Fig. 7.8b.   ∎

◼ **Example 7.4**

Sketch the amplitude and phase response (Bode plots) for the transfer function

$$H(s) = \frac{10(s + 100)}{s^2 + 2s + 100} = 10 \frac{\left(1 + \frac{s}{100}\right)}{1 + \frac{s}{50} + \frac{s^2}{100}} \tag{7.26}$$

Here, the constant term is 10; that is, 20 dB (20 log 10 = 20). To add this term, we simply label the horizontal axis (from where the asymptotes begin) as the 20 dB line as before(see Fig. 7.9a).

In addition, we have a real zero at $s = -100$ and a pair of complex conjugate poles. When we express the second-order factor in standard form, *

$$s^2 + 2s + 100 = s^2 + 2\zeta\omega_n s + \omega_n^2$$

we have

$$\omega_n = 10 \qquad \text{and} \qquad \zeta = 0.1$$

Step 1. Draw an asymptote of -40 dB/decade (-12 dB/octave) starting at $\omega = 10$ for the complex conjugate poles, and draw another asymptote of 20 dB/decade, starting at $\omega = 100$ for the (real) zero.

Step 2. Add both asymptotes.

Step 3. Apply the correction at $\omega = 100$, where the correction because of the corner frequency $\omega = 100$ is 3 dB. The correction because of the corner frequency $\omega = 10$ may be ignored. Next, apply the correction at $\omega = 10$, where the correction because of the corner frequency $\omega = 10$ is 13.90 dB (see Fig. 7.7a for $\zeta = 0.1$). We may find corrections at a few more points. The resulting plot is illustrated in Fig. 7.9a.

Phase Plot

The asymptote for the complex conjugate poles is a step function with a jump of $-90°$ at $\omega = 10$, and the asymptote for the zero at $s = -100$ is a straight line with a slope of $45°$/decade, leveling off at $\omega = 10$ and $\omega = 100$ to $0°$ and $90°$, respectively. The two asymptotes add to give the sawtooth shown in Fig. 7.9b. We now apply the corrections from Fig. 7.7b and Fig. 7.5b to obtain the exact plot.   ◼

⊙ **Computer Example C7.2**

Solve Examples 7.3 and 7.4 using M-file functions in MATLAB.

Frequency response may be plotted using several functions. For the purpose of Bode plots, however, the most suitable file is bode.m, which can be used as demonstrated here.

```
% For Example 7.3
num=[20 2000 0];den=[1 12 20];
bode(num,den)
% For Example 7.4
num=[0 10 1000];den=[1 2 100];
bode(num,den)  ⊙
```

Poles and Zeros in RHP

In our discussion so far, we have assumed the poles and zeros of the transfer function to be in the LHP. What if some of the poles and/or zeros of $H(s)$ lie in the RHP? If there is a pole in the RHP, the integral in Eq. (2.49) with $s = j\omega$ does not

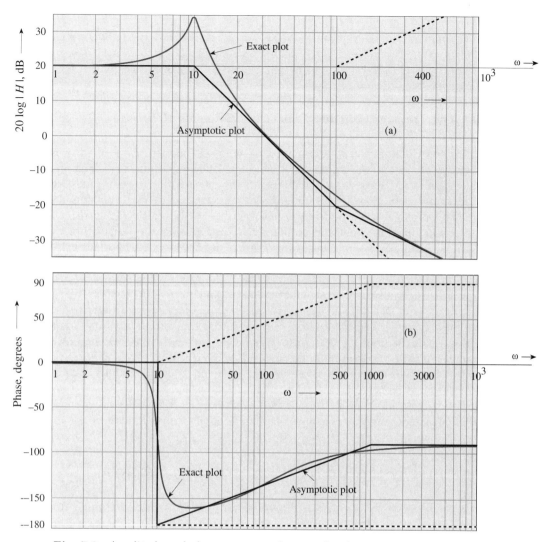

Fig. 7.9 Amplitude and phase response of a second-order system in Example 7.4.

converge, and $H(j\omega)$ is meaningless. Anyway, such systems (unstable) are useless for any signal processing application. For this reason, we shall consider only the case of the RHP zero. We can readily show that the amplitude function of a RHP zero at a is identical to that of a LHP zero at $-a$. The reason is that

$$\left|1 + \frac{j\omega}{a}\right| = \left|1 - \frac{j\omega}{a}\right| = \left(1 + \frac{\omega^2}{a^2}\right)^{\frac{1}{2}}$$

Therefore, the log amplitude plots remain unchanged whether the zero(s) are in the LHP or the RHP. However, the phase of the RHP zero at $s = a$ is

$$\angle(j\omega - a) = \angle - (a - j\omega) = \pi + \tan^{-1}\left(\frac{-\omega}{a}\right) = \pi - \tan^{-1}\left(\frac{\omega}{a}\right)$$

whereas the phase of the LHP pole at $s = -a$ is $-\tan^{-1}(\omega/a)$.

The complex conjugate zeros in RHP give rise to a term $s^2 - 2\zeta\omega_n s + \omega_n^2$, which is identical to the term $s^2 + 2\zeta\omega_n s + \omega_n^2$ with a sign change in ζ. Hence, from Eqs. (7.23) and (7.24) it follows that the amplitudes are identical, but the phases are of opposite signs for the two terms.

7.2-1   The Transfer Function From the Frequency Response

In the previous section we were given the transfer function of a system. From a knowledge of the transfer function, we developed techniques for determining the system response to sinusoidal inputs. We can also reverse the procedure to determine the transfer function of a system if the system response to a sinusoidal input is known. This problem has significant practical utility. If we are given a system in a black box with only the input and output terminals available, the transfer function has to be determined by experimental measurements at the input and output terminals. The frequency response to sinusoidal inputs is one of the possibilities that is very attractive because of the simple nature of the measurements involved. One only needs to apply a sinusoidal signal at the input and observe the output. We find the amplitude gain $|H(j\omega)|$ and the output phase shift $\angle H(j\omega)$ (with respect to the input sinusoid) for various values of ω over the entire range from 0 to ∞. This information yields the frequency response plots (Bode plots) when plotted against $\log \omega$. From these plots we determine the appropriate asymptotes by taking advantage of the fact that the slopes of all asymptotes must be multiples of ± 20 dB/decade if the transfer function is a rational function (function which is not necessarily a ratio of two polynomials in s). From the asymptotes, the corner frequencies are obtained. Corner frequencies determine the poles and zeros of the transfer function.

7.3   Control System Design Using Frequency Response

Figure 7.10a shows a basic closed-loop system, whose open-loop transfer function (transfer function when the loop is opened) is $KG(s)H(s)$. The closed-loop transfer function is [see Eq. (6.69)

$$T(s) = \frac{KG(s)}{1 + KG(s)H(s)}$$

The time-domain method of control system design discussed in Sec. 6.7 works only when the transfer function of the plant (the system to be controlled) is known and is a rational function (ratio of two polynomials in s). The input-output description of practical systems is often unknown and is more likely to be nonrational. A system containing an ideal time delay (dead time) is an example of a nonrational system. In such cases, we can determine the frequency response of the open-loop system empirically and use this data to design the (closed-loop) system. In this section we shall discuss feedback system design procedure based on frequency response description of a system. However, the frequency response design method is not as convenient as the time-domain design method from the viewpoint of the transient and the steady-state error specifications. Consequently, the time-domain method in

Sec. 6.7 and the frequency response method should be considered as complementary rather than as alternatives or as rivals.

Frequency response information can be presented in a various forms of which Bode plot is one. The same information can be presented by the **Nyquist plot** also known as the **polar plot** or by the **Nichols plot** also known as log-magnitude versus angle plot. Here, we shall discuss the techniques using Bode and Nyquist plots only. Figure 7.10b shows Bode plots for the open-loop transfer function $K/s(s+2)(s+4)$ when $K = 24$. The same information is presented in polar form in the corresponding Nyquist plot in Fig. 7.10c. For example, at $\omega = 1$, $|H(j\omega)| = 2.6$ and $\angle H(j\omega) = -130.6°$. We plot a point at a distance 2.6 units and at an angle $-130.6°$ from the horizontal axis. This point is labeled as $\omega = 1$ for identification (see Fig. 7.10c). We plot such points for several frequencies in the range from $\omega = 0$ to ∞ and draw a smooth curve through them to obtain the Nyquist plot. The same information is presented in Cartesian form in Nichols plot. For instance, at $\omega = 1$, the log magnitude is $20 \log 2.6 = 8.3$ dB, and the phase at $\omega = 1$ is $-130.6°$. We plot a point at coordinates $x = 8.3$, $y = -130.6°$ and label this point as $\omega = 1$ for identification. We do this for several values of ω from 0 to ∞ and draw a curve through these points to obtain Nichols plot. Using Bode or Nyquist (or Nichols) plots of the open-loop transfer function, we can readily investigate stability aspect of the corresponding closed-loop system as discussed below.

7.3-1   Relative Stability: Gain and Phase margins

For the system in Fig. 7.10a, the characteristic equation is $1 + KG(s)H(s) = 0$ and the characteristic roots are the roots of $KG(s)H(s) = -1$. The system becomes unstable when the root loci cross over to RHP. The crossing occurs on the imaginary axis where $s = j\omega$ (see Fig. 6.43). Hence, at the verge of instability (marginal stability)

$$KG(j\omega)H(j\omega) = -1 = 1\,e^{\pm j\pi}$$

Thus, at the verge of instability, the magnitude and angle of the open loop gain $KG(j\omega)H(j\omega)$ are

$$|KG(j\omega)H(j\omega)| = 1, \quad \text{and} \quad \angle G(j\omega)H(j\omega) = \pm\pi$$

Thus, on the verge of instability, the open-loop transfer function has unity gain and phase of $\pm\pi$. In order to understand the significance of these conditions, let us consider the system in Fig. 7.10a, with open-loop transfer function $K/s(s+2)(s+4)$. The Bode plot for this transfer function (for $K = 24$) is depicted in Fig. 7.10b. The root locus for this system is illustrated in Fig. 6.43. The loci cross over to RHP for $K > 48$. For $K < 48$, the system is stable. Let us consider the case $K = 24$. Figure 7.10b shows Bode plots when $K = 24$. Let the frequency where the angle plot crosses $-180°$ be ω_p (the **phase crossover frequency**). Observe that at ω_p, the gain is 0.5 or -6 dB. This shows that the gain K will have to double (to value 48) to have unity gain, which is the verge of instability. For this reason we say that the system has a gain margin $\alpha_M = 6$ dB. On the other hand, if ω_g is the frequency where the gain is unity or 0 dB, (the **gain crossover frequency**), then at this frequency, the open-loop phase is $-157.5°$. The phase will have to decrease from

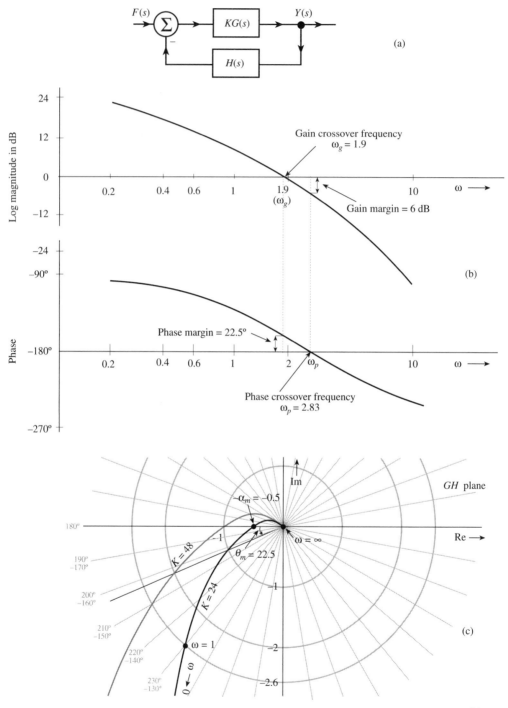

Fig. 7.10 Gain and Phase margins of a system with open-loop transfer function $\frac{24}{s(s+2)(s+4)}$.

this value to $-180°$ before the system becomes unstable. Thus, the system has a phase margin of $\theta_M = 22.5°$. Clearly, the gain and phase margins are measures of relative stability of the system.

Figure 7.10c shows that the Nyquist plot of $KG(s)H(s)$ crosses the real axis at -0.5 for $K = 24$. If we double K to a value 48, the magnitude of every point doubles (but the phase is unchanged). This step expands the Nyquist plot by a factor 2. Hence, for $K = 48$, the Nyquist plot lies on the real axis at -1; that is, $KG(j\omega)H(j\omega) = -1$, and the system becomes unstable. For $K > 48$, the plot crosses and goes beyond the point -1. Thus, the critical point -1 lies inside the curve; that is, the curve encircles the critical point -1. When the Nyquist plot of an open-loop transfer function encircles the critical point -1, the corresponding closed-loop system becomes unstable. This statement, roughly speaking, is the well-known **Nyquist criterion** in a simplified form.† For the Nyquist plot in Fig. 7.10b (for $K = 24$), the gain will have to double before the system becomes unstable. Thus, the gain margin is 2 (6 dB) in this case. In general, if the Nyquist plot crosses the negative real axis at $-\alpha_m$, then the gain margin is $1/\alpha_m$. Similarly, if $-\pi + \theta_m$ is the angle at which the Nyquist plot crosses the unit circle, the phase margin is θ_m. In the present case, $\theta_m = 22.5°$.

In order to protect a system from instability because of variations in system parameters (or in the environment), the system should be designed with reasonable gain and phase margins. Small margins indicate that the poles of the closed-loop system are in the LHP, but too close to the $j\omega$ axis. The transient response of such systems will have a large overshoot (PO). On the other hand, very large (positive) gain and phase margins may indicate a sluggish system. Generally, a gain margin higher than 6 dB and a phase margin of about $30°$ to $60°$ are considered desirable. Design specifications for transient performance are often given in terms of gain and phase margins.

7.3-2   Transient Performance in Terms of Frequency Response

For a second-order system in Eq. (6.81), we saw the dependence of the transient response (PO, t_r, t_d and t_s) on the dominant pole location. Using this knowledge, we developed in Sec. 6.7 a procedure for designing a control system for a specified transient performance. In order to develop such a procedure from the knowledge of system's frequency response (rather than its transfer function), we must know the relationship between the frequency response and the transient response of the system in Eq. (6.81). Figure 7.11 shows the frequency response of a second-order system in Eq. (6.81). The peak frequency response M_p (the maximum value of the amplitude response), which occurs at frequency ω_p, indicates relative stability of the system. Higher peak response generally indicates smaller ζ (see Fig. 7.6a), which implies poles closer to the imaginary axis, and less relative stability. Higher M_p also means higher PO (the step response overshoot). Generally acceptable values of M_p in practice range from 1.1 to 1.5. The 3-dB bandwidth ω_b of the frequency response indicates the speed of the system. We can show that ω_b and t_r

---

†The Nyquist criterion states as follows: A closed curve C_s in the s plane enclosing m zeros and n poles of an open-loop transfer function $W(s)$ maps into a closed curve C_w in the W plane encircling the origin of the W plane $m - n$ times, in the same direction as that of C_s. If $n - m$ is negative, then the encirclement is in the opposite direction.

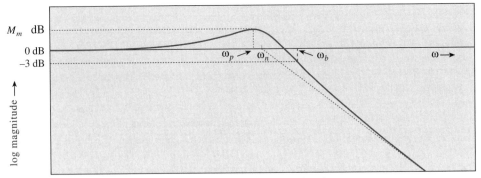

Fig. 7.11 Frequency response of a second-order system.

are inversely proportional. Hence, higher ω_b indicates smaller t_r (faster response). For the second-oder system in Eq. (6.81), we have

$$T(j\omega) = \frac{\omega_n^2}{(j\omega)^2 + 2j\zeta\omega_n\omega + \omega_n^2}$$

To find M_p, we let $d|T(j\omega)|/d\omega = 0$. From the solution of this equation, we find

$$M_p = \frac{1}{2\zeta\sqrt{1-\zeta^2}} \qquad \zeta \leq 0.707$$

$$\omega_p = \omega_n\sqrt{1-\zeta^2} \qquad \zeta \leq 0.707$$

$$\omega_b = \omega_n\left[(1-2\zeta^2) + \sqrt{4\zeta^4 - 4\zeta^2 + 2}\right]^{1/2} \qquad (7.27)$$

These equations show that we can determine ζ and ω_n from M_p and ω_p. Knowledge of ζ and ω_n allow us to determine the transient parameters, such as, PO, t_r and t_s as seen from Eqs. (6.83), (6.84) and (6.85). Conversely, if we are given certain transient specifications PO, t_r, and t_s, we can determine the required M_p and ω_p. Thus, the problem now reduces to designing a system, which has a certain M_p and ω_p for the closed-loop frequency response. In practice, we know the open-loop system frequency response. So the ultimate problem reduces to relating the frequency response of the closed-loop system to that of the the open-loop system. To do this, we shall consider the case of unity feedback system, where the feedback transfer function is $H(s) = 1$.† The closed-loop transfer function in this case is

$$T(s) = \frac{KG(s)}{1 + KG(s)}$$

and

$$T(j\omega) = \frac{KG(j\omega)}{1 + KG(j\omega)}$$

Let

$$T(j\omega) = Me^{j\alpha(\omega)}, \quad \text{and} \quad KG(j\omega) = x(\omega) + jy(\omega)$$

---

†The results for the unity feedback can be extended to nonunity feedback systems.

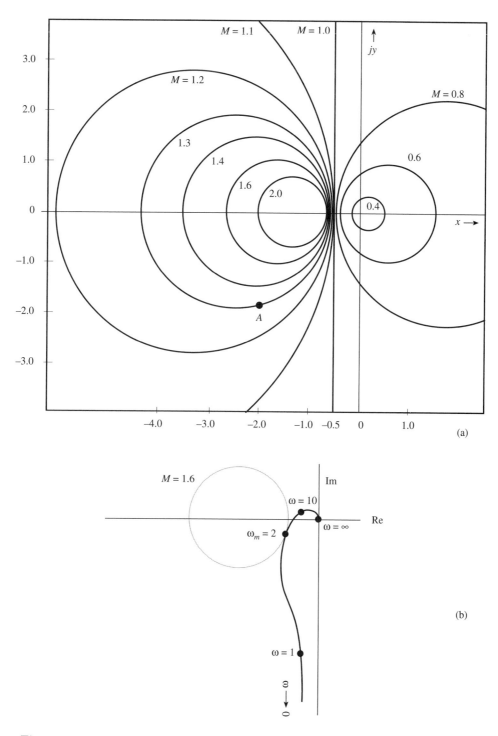

Fig. 7.12 Relationship between the open-loop and the closed-loop frequency response.

Consequently

$$Me^{j\alpha(\omega)} = \frac{x + jy}{1 + x + jy}$$

Straightforward manipulation of this equation yields

$$\left(x + \frac{M^2}{M^2 - 1}\right)^2 + y^2 = \frac{M^2}{(M^2 - 1)^2}$$

This is an equation of a circle centered at $[-\frac{M^2}{M^2-1} \quad 0]$ and of radius $\frac{M}{M^2-1}$ in the $KG(j\omega)$ plane. Figure 7.12a shows circles for various values of M. Because M is the closed-loop system amplitude response, these circles are the contours of constant amplitude response of the closed-loop system. For example, the point $A = -2 - j1.85$ lies on the circle $M = 1.3$. This means, at a frequency where the open-loop transfer function is $G(j\omega) = -2 - j1.85$, the corresponding closed-loop transfer function amplitude response is 1.3.†

To obtain the closed-loop frequency response, we superimpose on these contours the Nyquist plot of the open-loop transfer function $KG(j\omega)$. For each point of $KG(j\omega)$, we can determine the corresponding value of M, the closed-loop amplitude response. From similar contours for constant α (the closed-loop phase response), we can determine the closed loop phase response. Thus, the complete closed-loop frequency response can be obtained from this plot. We are primarily interested in finding M_p, the peak value of M and ω_p, the frequency where it occurs. Figure 7.12b indicates how these values may be determined. The circle to which the Nyquist plot is tangent corresponds to M_p, and the frequency at which the Nyquist plot is tangential to this circle is ω_p. For the system, whose Nyquist plot appears in Fig. 7.12b, $M_p = 1.6$ and $\omega_p = 2$. From these values, we can estimate ζ and ω_n, and determine the transient parameters PO, t_r and t_s.

In designing systems, we first determine M_p and ω_p required to meet the given transient specifications from Eqs. (7.27). The Nyquist plot in conjunction with M circles suggests how these values of M_p and ω_p may be realized. In many cases, a mere change in gain K of the open-loop transfer function will suffice. Increasing K expands the Nyquist plot and changes the values M_p and ω_p correspondingly. If this is not enough, we should consider some form of compensation such as lag and/or lead networks. Using a computer, one can quickly observe the effect of particular form of compensation on M_p and ω_p.

7.4 Filter Design by Placement of Poles and Zeros OF H(s)

In this section we explore the strong dependence of frequency response on the location of poles and zeros of $H(s)$. This dependence points to a simple intuitive procedure to filter design. A systematic filter design procedure to meet given specifications is discussed later in Secs. 7.5, 7.6, and 7.7.

7.4-1 Dependence of Frequency Response on poles and Zeros of H(s)

Frequency response of a system is basically the information about the filtering capability of the system. We now examine the close connection that exists between

---

†We can find similar contours for constant α (the closed-loop phase response).

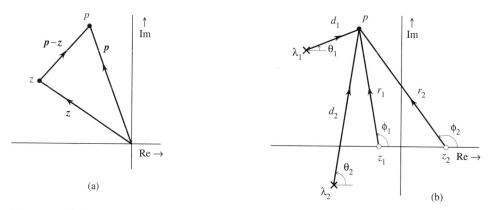

Fig. 7.13 (a)vector representation of complex numbers (b) vector representation of factors of $H(s)$.

the pole-zero locations of a system transfer function and its frequency response (or filtering characteristics). A system transfer function can be expressed as

$$H(s) = \frac{P(s)}{Q(s)} = b_n \frac{(s - z_1)(s - z_2) \cdots (s - z_n)}{(s - \lambda_1)(s - \lambda_2) \cdots (s - \lambda_n)} \tag{7.28a}$$

where z_1, z_2, ..., z_n are the zeros of $H(s)$ and the characteristic roots λ_1, λ_2, ..., λ_n are the poles of $H(s)$. Now the value of the transfer function $H(s)$ at some frequency $s = p$ is

$$H(s)|_{s=p} = b_n \frac{(p - z_1)(p - z_2) \cdots (p - z_n)}{(p - \lambda_1)(p - \lambda_2) \cdots (p - \lambda_n)} \tag{7.28b}$$

This equation consists of factors of the form $p - z_i$ and $p - \lambda_i$. The factor $p - z$ is a complex number represented by a vector drawn from point z to the point p in the complex plane, as illustrated in Fig. 7.13a. The length of this line segment is $|p - z|$, the magnitude of $p - z$. The angle of this directed line segment (with horizontal axis) is $\angle(p - z)$. To compute $H(s)$ at $s = p$, we draw line segments from all poles and zeros of $H(s)$ to the point p, as shown in Fig. 7.13b. The vector connecting a zero z_i to the point p is $p - z_i$. Let the length of this vector be r_i, and let its angle with the horizontal axis be ϕ_i. Then $p - z_i = r_i e^{j\phi_i}$. Similarly, the vector connecting a pole λ_i to the point p is $p - \lambda_i = d_i e^{j\theta_i}$, where d_i and θ_i are the length and the angle (with the horizontal axis), respectively, of the vector $p - \lambda_i$. Now from Eq. (7.28b) it follows that

$$H(s)|_{s=p} = b_n \frac{(r_1 e^{j\phi_1})(r_2 e^{j\phi_2}) \cdots (r_n e^{j\phi_n})}{(d_1 e^{j\theta_1})(d_2 e^{j\theta_2}) \cdots (d_n e^{j\theta_1})}$$

$$= b_n \frac{r_1 r_2 \cdots r_n}{d_1 d_2 \cdots d_n} e^{j[(\phi_1 + \phi_2 + \cdots + \phi_n) - (\theta_1 + \theta_2 + \cdots + \theta_n)]}$$

Therefore

$$|H(s)|_{s=p} = b_n \frac{r_1 r_2 \cdots r_n}{d_1 d_2 \cdots d_n}$$

$$= b_n \frac{\text{product of the distances of zeros to } p}{\text{product of the distances of poles to } p} \tag{7.29a}$$

and

$$\angle H(s)|_{s=p} = (\phi_1 + \phi_2 + \cdots + \phi_n) - (\theta_1 + \theta_2 + \cdots + \theta_n)$$

$$= \text{sum of zero angles to } p - \text{sum of pole angles to } p \quad (7.29b)$$

Using this procedure, we can determine $H(s)$ for any value of s. To compute the frequency response $H(j\omega)$, we use $s = j\omega$ (a point on the imaginary axis), connect all poles and zeros to the point $j\omega$, and determine $|H(j\omega)|$ and $\angle H(j\omega)$ from Eqs. (7.29). We repeat this procedure for all values of ω from 0 to ∞ to obtain the frequency response.

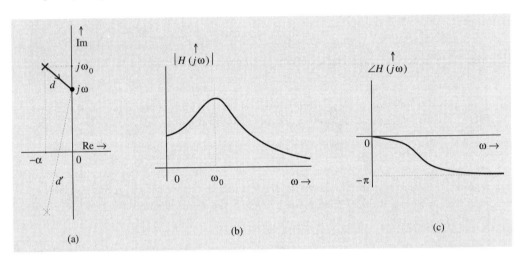

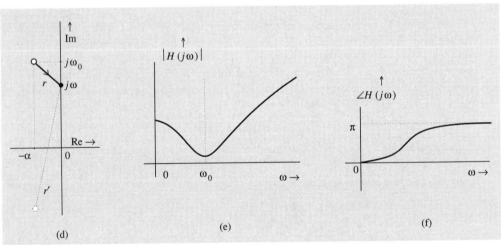

Fig. 7.14 The role of poles and zeros in determining the frequency response of an LTIC system.

Gain Enhancement by a Pole

To understand the effect of poles and zeros on the frequency response, consider a hypothetical case of a single pole $-\alpha + j\omega_0$, as depicted in Fig. 7.14a. To find the amplitude response $|H(j\omega)|$ for a certain value of ω, we connect the pole to the point $j\omega$ (Fig. 7.14a). If the length of this line is d, then $|H(j\omega)|$ is proportional to $1/d$.

$$|H(j\omega)| = \frac{K}{d} \tag{7.30}$$

where the exact value of constant K is not important at this point. As ω increases from zero up, d decreases progressively until ω reaches the value ω_0. As ω increases beyond ω_0, d increases progressively. Therefore, according to Eq. (7.30), the amplitude response $|H(j\omega)|$ increases from $\omega = 0$ until $\omega = \omega_0$, and it decreases continuously as ω increases beyond ω_0, as illustrated in Fig. 7.14b. Therefore, a pole at $-\alpha + j\omega_0$ results in a frequency-selective behavior that enhances the gain at the frequency ω_0 (resonance). Moreover, as the pole moves closer to the imaginary axis (as α is reduced), this enhancement (resonance) becomes more pronounced. This is because α, the distance between the pole and $j\omega_0$ (d corresponding to $j\omega_0$), becomes smaller, which increases the gain K/d. In the extreme case, when $\alpha = 0$ (pole on the imaginary axis), the gain at ω_0 goes to infinity. Repeated poles further enhance the frequency-selective effect. To summarize, we can enhance a gain at a frequency ω_0 by placing a pole opposite the point $j\omega_0$. The closer the pole is to $j\omega_0$, the higher is the gain at ω_0, and the gain variation is more rapid (more frequency-selective) in the vicinity of frequency ω_0. Note that a pole must be placed in the LHP for stability.

Here we have considered the effect of a single complex pole on the system gain. For a real system, a complex pole $-\alpha + j\omega_0$ must accompany its conjugate $-\alpha - j\omega_0$. We can readily show that the presence of the conjugate pole does not appreciably change the frequency-selective behavior in the vicinity of ω_0. This is because the gain in this case is K/dd', where d' is the distance of a point $j\omega$ from the conjugate pole $-\alpha - j\omega_0$. Because the conjugate pole is far from $j\omega_0$, there is no dramatic change in the length d' as ω varies in the vicinity of ω_0. There is a gradual increase in the value of d' as ω increases, which leaves the frequency-selective behavior as it was originally, with only minor changes.

Gain Suppression by a Zero

Using the same argument, we observe that zeros at $-\alpha \pm j\omega_0$ (Fig. 7.14d) will have exactly the opposite effect of suppressing the gain in the vicinity of ω_0, as shown in Fig. 7.14e). A zero on the imaginary axis at $j\omega_0$ will totally suppress the gain (zero gain) at frequency ω_0. Repeated zeros will further enhance the effect. Also, a closely-placed pair of a pole and a zero (dipole) tend to cancel out each other's influence on the frequency response. Clearly, a proper placement of poles and zeros can yield a variety of frequency-selective behavior. Using these observations, we can design lowpass, highpass, bandpass, and bandstop (or notch) filters.

Phase response can also be computed graphically. In Fig. 7.14a, angles formed by the complex conjugate poles $-\alpha \pm j\omega_0$ at $\omega = 0$ (the origin) are equal and opposite. As ω increases from 0 up, the angle θ_1 because of the pole $-\alpha + j\omega_0$, which has a negative value at $\omega = 0$, is reduced in magnitude; the angle θ_2 because of the pole $-\alpha - j\omega_0$, which has a positive value at $\omega = 0$, increases in magnitude. As a result, $\theta_1 + \theta_2$, the sum of the two angles, increases continuously, approaching a value π as $\omega \to \infty$. The resulting phase response $\angle H(j\omega) = -(\theta_1 + \theta_2)$ is illustrated in Fig. 7.14c. Similar arguments apply to zeros at $-\alpha \pm j\omega_0$. The resulting phase response $\angle H(j\omega) = (\phi_1 + \phi_2)$ is depicted in Fig. 7.14f.

We now focus on simple filters, using the intuitive insights gained in this discussion. The discussion is essentially qualitative.

7.4-2 Lowpass Filters

A typical lowpass filter has a maximum gain at $\omega = 0$. Therefore, we need to place a pole (or poles) on the real axis opposite the origin $(j\omega = 0)$, as shown in Fig. 7.15a. The transfer function of this system is

$$H(s) = \frac{\omega_c}{s + \omega_c}$$

We have chosen the numerator of $H(s)$ to be ω_c in order to normalize the dc gain $H(0)$ to unity. If d is the distance from the pole $-\omega_c$ to a point $j\omega$ (Fig. 7.15a), then

$$|H(j\omega)| = \frac{\omega_c}{d}$$

with $H(0) = 1$. As ω increases, d increases and $|H(j\omega)|$ decreases monotonically with ω, as illustrated in Fig. 7.15d by label $n = 1$. This is clearly a lowpass filter with gain enhanced in the vicinity of $\omega = 0$.

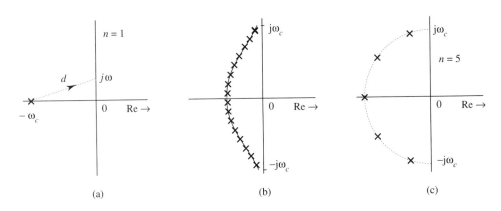

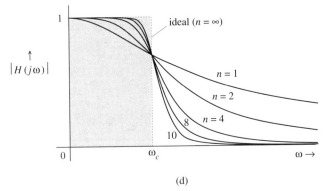

Fig. 7.15 Pole-zero configuration and the amplitude response of a lowpass (Butterworth) filter.

Wall-to-wall Poles

An ideal lowpass filter characteristic (shaded) in Fig. 7.15d, has a constant gain of unity up to frequency ω_c. Then the gain drops suddenly to 0 for $\omega > \omega_c$. To achieve the ideal lowpass characteristic, we need enhanced gain over the entire frequency band from 0 to ω_c. We know that to enhance a gain at any frequency ω, we need to place a pole opposite ω. To achieve an enhanced gain for all frequencies over the band (0 to ω_c), we need to place a pole opposite every frequency in this band. In other words, we need a **continuous wall of poles** facing the imaginary axis opposite the frequency band 0 to ω_c (and from 0 to $-\omega_c$ for conjugate poles), as depicted in Fig. 7.15b. At this point, the optimum shape of this wall is not obvious because our arguments are qualitative and intuitive. Yet, it is certain that to have enhanced gain (constant gain) at every frequency over this range, we need an infinite number of poles on this wall. We can show that for a maximally flat† response over the frequency range (0 to ω_c), the wall is a semicircle with an infinite number of poles uniformly distributed along the wall.[1] In practice, we compromise by using a finite number (n) of poles with less-than-ideal characteristics. Figure 7.15c shows the pole configuration for a fifth-order ($n = 5$) filter. The amplitude response for various values of n are illustrated in Fig. 7.15d. As $n \to \infty$, the filter response approaches the ideal. This family of filters is known as the **Butterworth** filters. There are also other families. In **Chebyshev** filters, the wall shape is a semiellipse rather than a semicircle. The characteristics of a Chebyshev filter are inferior to those of Butterworth over the passband $(0, \omega_c)$, where the characteristics show a rippling effect instead of the maximally flat response of Butterworth. But in the stopband $(\omega > \omega_c)$, Chebyshev behavior is superior in the sense that Chebyshev filter gain drops faster than that of the Butterworth.

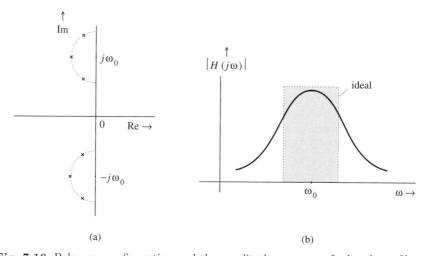

(a)   (b)

Fig. 7.16 Pole-zero configuration and the amplitude response of a bandpass filter.

7.4-3   Bandpass Filters

The shaded characteristic in Fig. 7.16b shows the ideal bandpass filter gain. In the bandpass filter, the gain is enhanced over the entire passband. Our earlier

---

†Maximally flat amplitude response means the first $2n - 1$ derivatives of $|H(j\omega)|$ are zero at $\omega = 0$.

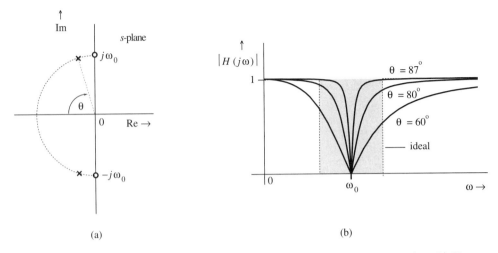

Fig. 7.17 Pole-zero configuration and the amplitude response of a bandstop (notch) filter.

discussion indicates that this can be realized by a wall of poles opposite the imaginary axis in front of the passband centered at ω_0. (There is also a wall of conjugate poles opposite $-\omega_0$.) Ideally, an infinite number of poles is required. In practice, we compromise by using a finite number of poles and accepting less-than-ideal characteristics (Fig. 7.16).

7.4-4 Notch (Bandstop) Filters

An ideal notch filter amplitude response (shown shaded in Fig. 7.17b) is a complement of the amplitude response of an ideal bandpass filter. Its gain is zero over a small band centered at some frequency ω_0 and is unity over the remaining frequencies. Realization of such a characteristic requires an infinite number of poles and zeros. Let us consider a practical second-order notch filter to obtain zero gain at a frequency $\omega = \omega_0$. For this purpose we must have zeros at $\pm j\omega_0$. The requirement of unity gain at $\omega = \infty$ requires the number of poles to be equal to the number of zeros ($m = n$). This ensures that for very large values of ω, the product of the distances of poles from ω will be equal to the product of the distances of zeros from ω. Moreover, unity gain at $\omega = 0$ requires a pole and the corresponding zero to be equidistant from the origin. For example, if we use two (complex conjugate) zeros, we must have two poles; the distance from the origin of the poles and of the zeros should be the same . This requirement can be met by placing the two conjugate poles on the semicircle of radius ω_0, as depicted in Fig. 7.17a. The poles can be anywhere on the semicircle to satisfy the equidistance condition. Let the two conjugate poles be at angles $\pm\theta$ with respect to the negative real axis. Recall that a pole and a zero in the vicinity tend to cancel out each other's influences. Therefore, placing poles closer to zeros (selecting θ closer to $\pi/2$) results in a rapid recovery of the gain from value 0 to 1 as we move away from ω_0 in either direction. Figure 7.17b shows the gain $|H(j\omega)|$ for three different values of θ.

■ **Example 7.5**

Design a second-order notch filter to suppress 60 Hz hum in a radio receiver.

We use the poles and zeros in Fig. 7.17a with $\omega_0 = 120\pi$. The zeros are at $s = \pm j\omega_0$. The two poles are at $-\omega_0 \cos\theta \pm j\omega_0 \sin\theta$. The filter transfer function is (with $\omega_0 = 120\pi$)

$$H(s) = \frac{(s - j\omega_0)(s + j\omega_0)}{(s + \omega_0 \cos\theta + j\omega_0 \sin\theta)(s + \omega_0 \cos\theta - j\omega_0 \sin\theta)}$$

$$= \frac{s^2 + \omega_0^2}{s^2 + (2\omega_0 \cos\theta)s + \omega_0^2} = \frac{s^2 + 142122.3}{s^2 + (753.98 \cos\theta)s + 142122.3}$$

and

$$|H(j\omega)| = \frac{-\omega^2 + 142122.3}{\sqrt{(-\omega^2 + 142122.3)^2 + (753.98\omega \cos\theta)^2}}$$

The closer the poles are to the zeros (closer the θ to $\frac{\pi}{2}$), the faster the gain recovery from 0 to 1 on either side of $\omega_0 = 120\pi$. Figure 7.17b shows the amplitude response for three different values of θ. This example is a case of very simple design. To achieve zero gain over a band, we need an infinite number of poles as well as of zeros. ∎

⊙ **Computer Example C7.3**
Plot the amplitude response of the transfer function

$$H(s) = \frac{s^2 + \omega_0^2}{s^2 + (2\omega_0 \cos\theta)s + \omega_0^2}$$

of a second order notch filter for $\omega_0 = 120\pi$ and $\theta = 60°, 80°$, and $87°$.

```
w0=120*pi;
theta=[60 80 87]*(pi/180);
for m=1:length(theta)
    num=[1 0 w0^2];
    den=[1 2*w0*cos(theta(m)) w0^2];
    w=0:.5:1000; w=w';
    [mag,phase,w]=bode(num,den,w);
    plot(w,mag),hold on,axis([0 1000 0 1.1])
end      ⊙
```

Figures 7.16b and 7.17b show that a notch (stopband) filter frequency response is a complement of the bandpass filter frequency response. If $H_{BP}(s)$ and $H_{BS}(s)$ are the transfer functions of a bandpass and a bandstop filter (both centered at the same frequency), then

$$H_{BS}(s) = 1 - H_{BP}(s)$$

Therefore, a bandstop filter transfer function may also be obtained from the corresponding bandpass filter transfer function. The case of lowpass and highpass filters is similar. If $H_{LP}(s)$ and $H_{HP}(s)$ are the transfer functions of a lowpass and a highpass filter respectively (both with the same cutoff frequency), then

$$H_{HP}(s) = 1 - H_{LP}(s)$$

Therefore, a highpass filter transfer function may also be obtained from the corresponding lowpass filter transfer function.

△ **Exercise E7.2**

   Using the qualitative method of sketching the frequency response, show that the system with the pole-zero configuration in Fig. 7.18a is a highpass filter, and that with the configuration in Fig. 7.18b is a bandpass filter.   ▽

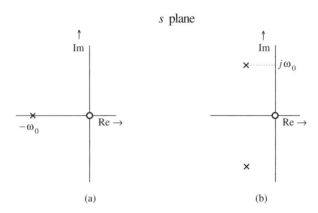

Fig. 7.18 Pole-zero configuration of a highpass filter in Exercise E7.2.

7.4-5 Practical Filters and Their Specifications

   For ideal filters everything is black and white; the gains are either zero or unity over certain bands. As we saw earlier, real life does not permit such a world view. Things have to be gray or shades of gray. In practice, we can realize a variety of filter characteristics which can only approach ideal characteristics.

   An ideal filter has a passband (unity gain) and a stopband (zero gain) with a sudden transition from the passband to the stopband. There is no transition band. For practical (or realizable) filters, on the other hand, the transition from the passband to the stopband (or vice versa) is gradual, and takes place over a finite band of frequencies. Moreover, for realizable filters, the gain cannot be zero over a finite band (Paley-Wiener condition). As a result, there can no true stopband for practical filters. We therefore define a **stopband** to be a band over which the gain is below some small number G_s, as illustrated in Fig. 7.19. Similarly, we define a **passband** to be a band over which the gain is between 1 and some number G_p ($G_p < 1$), as shown in Fig. 7.19. We have selected the passband gain of unity for convenience. It could be any constant. Usually the gains are specified in terms of decibels. This is simply 20 times the log (to base 10) of the gain. Thus

$$\hat{G}(\text{dB}) = 20 \log_{10} G$$

A gain of unity is 0 dB and a gain of $\sqrt{2}$ is 3.01 dB, usually approximated by 3 dB. Sometimes the specification may be in terms of attenuation, which is the negative of the gain in dB. Thus a gain of $1/\sqrt{2}$; that is, 0.707, is -3 dB, but is an attenuation of 3 dB.

   In our design procedure we assume that G_p (**minimum passband gain**) and G_s (**maximum stopband gain**) are specified. Figure 7.19 shows the passband, the stopband, and the transition band for typical lowpass, bandpass, highpass, and

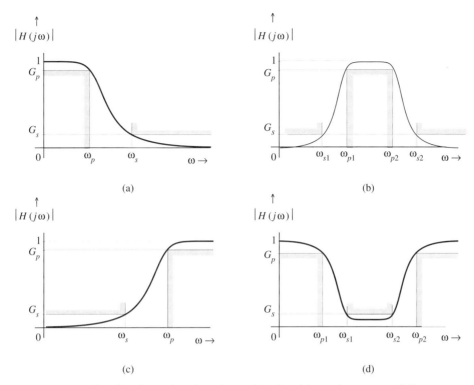

Fig. 7.19   Passband, stopband, and transitionband in various types of filters.

bandstop filters. In this chapter we shall discuss the design procedures for these four types of filters. Fortunately, the highpass, bandpass, and bandstop filters can be obtained from a basic lowpass filter by simple frequency transformations. For example, replacing s with ω_c/s in the lowpass filter transfer function results in a highpass filter. Similarly, other frequency transformations yield the bandpass and bandstop filters. Hence, it is necessary to develop a design procedure only for a basic lowpass filter. Using appropriate transformations, we can then design other types of filters. We shall consider here two well known families of filters: the Butterworth and the Chebyshev filters.

7.5   Butterworth Filters

The amplitude response $|H(j\omega)|$ of an nth order Butterworth lowpass filter is given by

$$|H(j\omega)| = \frac{1}{\sqrt{1 + \left(\dfrac{\omega}{\omega_c}\right)^{2n}}} \tag{7.31}$$

Observe that at $\omega = 0$, the gain $|H(j0)|$ is unity and at $\omega = \omega_c$, the gain $|H(j\omega_c)| = 1/\sqrt{2}$ or -3 dB. The gain drops by a factor $\sqrt{2}$ at $\omega = \omega_c$. Because the power is proportional to the amplitude squared, the power ratio (output power to input

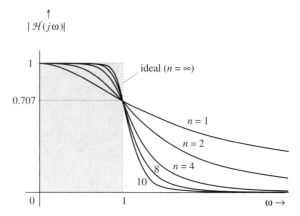

Fig. 7.20 Amplitude response of a normalized lowpass Butterworth filter.

power) drops by a factor 2 at $\omega = \omega_c$. For this reason ω_c is called the **half-power frequency** or the **3 dB-cutoff frequency** (amplitude ratio of $\sqrt{2}$ is 3 dB).

Normalized Filter

In the design procedure it proves most convenient to consider a normalized filter $\mathcal{H}(s)$, whose half-power frequency is 1 rad/s ($\omega_c = 1$). For such a filter, the amplitude characteristic in Eq. (7.31) reduces to

$$|\mathcal{H}(j\omega)| = \frac{1}{\sqrt{1 + \omega^{2n}}} \tag{7.32}$$

We can prepare a table of normalized transfer functions $\mathcal{H}(s)$ which yield the frequency response in Eq. (7.32) for various values of n. Once the normalized transfer function is obtained, we can obtain the desired transfer function $H(s)$ for any value of ω_c by simple frequency scaling, where we replace s by s/ω_c in the normalized $\mathcal{H}(s)$.

The amplitude response $|\mathcal{H}(j\omega)|$ of the normalized lowpass Butterworth filters is depicted in Fig. 7.20 for various values of n. From Fig. 7.20 we observe the following:

1. The Butterworth amplitude response decreases monotonically. Moreover, the first $2n - 1$ derivatives of the amplitude response are zero at $\omega = 0$. For this reason this characteristic is called **maximally flat** at $\omega = 0$. Observe that a constant characteristic (ideal) is maximally flat for all $\omega < 1$. In the Butterworth filter we try to retain this property at least at the origin.†

2. The filter gain is 1 (0 dB) at $\omega = 0$ and 0.707 (−3 dB) at $\omega = 1$ for all n. Therefore, the 3-dB (or half power) bandwidth is 1 rad/s for all n.

3. For large n, the amplitude response approaches the ideal characteristic.

To determine the corresponding transfer function $\mathcal{H}(s)$, recall that $\mathcal{H}(-j\omega)$ is the complex conjugate of $\mathcal{H}(j\omega)$. Therefore

$$\mathcal{H}(j\omega)\mathcal{H}(-j\omega) = |\mathcal{H}(j\omega)|^2 = \frac{1}{1 + \omega^{2n}}$$

---

†Butterworth filter exhibits maximally flat characteristic also at $\omega = \infty$.

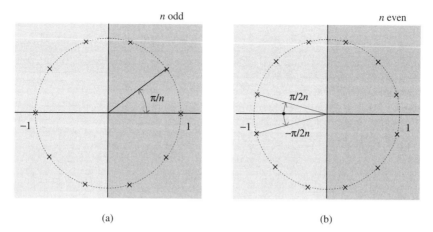

Fig. 7.21 Poles of a normalized even-order lowpass Butterworth Filter transfer function and its conjugate.

Substituting $s = j\omega$ in this equation, we obtain

$$\mathcal{H}(s)\mathcal{H}(-s) = \frac{1}{1 + (s/j)^{2n}}$$

The poles of $\mathcal{H}(s)\mathcal{H}(-s)$ are given by

$$s^{2n} = -(j)^{2n}$$

In this result we use the fact that $-1 = e^{j\pi(2k-1)}$ for integral values of k, and $j = e^{j\pi/2}$ to obtain

$$s^{2n} = e^{j\pi(2k-1+n)} \qquad k \text{ integer}$$

This equation yields the poles of $\mathcal{H}(s)\mathcal{H}(-s)$ as

$$s_k = e^{\frac{j\pi}{2n}(2k+n-1)} \qquad k = 1, 2, 3, \ldots, 2n \qquad (7.33)$$

Observe that all poles have a unit magnitude; that is, they are located on a unit circle in the s-plane separated by angle π/n, as illustrated in Fig. 7.21 for odd and even n. Since $\mathcal{H}(s)$ is stable and causal, its poles must lie in the LHP. The poles of $\mathcal{H}(-s)$ are the mirror images of the poles of $\mathcal{H}(s)$ about the vertical axis. Hence, the poles of $\mathcal{H}(s)$ are those in the LHP and the poles of $\mathcal{H}(-s)$ are those in the RHP in Fig. 7.21. The poles corresponding to $\mathcal{H}(s)$ are obtained by setting $k = 1, 2, 3, \ldots, n$ in Eq. (7.33); that is

$$s_k = e^{\frac{j\pi}{2n}(2k+n-1)}$$
$$= \cos\frac{\pi}{2n}(2k + n - 1) + j\sin\frac{\pi}{2n}(2k + n - 1) \qquad k = 1, 2, 3, \ldots, n \qquad (7.34)$$

and $\mathcal{H}(s)$ is given by

$$\mathcal{H}(s) = \frac{1}{(s - s_1)(s - s_2)\cdots(s - s_n)} \qquad (7.35)$$

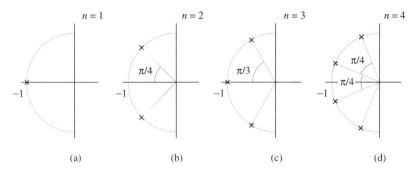

Fig. 7.22   Poles of a normalized lowpass Butterworth filter of various orders.

For instance, from Eq. (7.34), we find the poles of $\mathcal{H}(s)$ for $n = 4$ to be at angles $5\pi/8$, $7\pi/8$, $9\pi/8$, and $11\pi/8$. These lie on the unit circle, as shown in Fig. 7.22, and are given by $-0.3827 \pm j0.9239$, $-0.9239 \pm j0.3827$. Hence, $\mathcal{H}(s)$ can be expressed as

$$\mathcal{H}(s) = \frac{1}{(s + 0.3827 - j0.9239)(s + 0.3827 + j0.9239)(s + 0.9239 - j0.3827)(s + 0.9239 + j0.3827)}$$

$$= \frac{1}{(s^2 + 0.7654s + 1)(s^2 + 1.8478s + 1)}$$

$$= \frac{1}{s^4 + 2.6131s^3 + 3.4142s^2 + 2.6131s + 1}$$

We can proceed in this way to find $\mathcal{H}(s)$ for any value of n. In general

$$\mathcal{H}(s) = \frac{1}{B_n(s)} = \frac{1}{s^n + a_{n-1}s^{n-1} + \cdots + a_1 s + 1} \tag{7.36}$$

where $B_n(s)$ is the Butterworth polynomial of the nth order. Table 7.1 shows the coefficients $a_1, a_2, \ldots, a_{n-2}, a_{n-1}$ for various values of n; Table 7.2 shows $B_n(s)$ in factored form. In these Tables, we read for $n = 4$

$$\mathcal{H}(s) = \frac{1}{s^4 + 2.6131s^3 + 3.4142s^2 + 2.6131s + 1}$$

$$= \frac{1}{(s^2 + 0.7654s + 1)(s^2 + 1.8478s + 1)}$$

a result, which confirms our earlier computations.

We can also find the normalized Butterworth filter transfer function using MATLAB function [z,p,k]=buttap(n) to find poles,zeros and the gain factor of a normalized nth-order Butterworth filter.

⊙ **Computer Example C7.4**

Using MATLAB, find poles, zeros, and the gain factor of a normalized 4th-order Butterworth filter.

[z,p,k]=buttap(4)

MATLAB returns poles, zeros, and the gain factor k, which is unity for all orders.   ⊙

Table 7.1: Coefficients of Butterworth Polynomial $B_n(s) = s^n + a_{n-1}s^{n-1} + \cdots + a_1 s + 1$

n	a_1	a_2	a_3	a_4	a_5	a_6	a_7	a_8	a_9
2	1.41421356								
3	2.00000000	2.00000000							
4	2.61312593	3.41421356	2.61312593						
5	3.23606798	5.23606798	5.23606798	3.23606798					
6	3.86370331	7.46410162	9.14162017	7.46410162	3.86370331				
7	4.49395921	10.09783468	14.59179389	14.59179389	10.09783468	4.49395921			
8	5.12583090	13.13707118	21.84615097	25.68835593	21.84615097	13.13707118	5.12583090		
9	5.75877048	16.58171874	31.16343748	41.98635573	41.98635573	31.16343748	16.58171874	5.75877048	
10	6.39245322	20.43172909	42.80206107	64.88239627	74.23342926	64.88239627	42.80206107	20.43172909	6.39245322

Table 7.2: Butterworth Polynomials in Factorized Form

n	$B_n(s)$
1	$s+1$
2	$s^2 + 1.41421356s + 1$
3	$(s+1)(s^2 + s + 1)$
4	$(s^2 + 0.76536686s + 1)(s^2 + 1.84775907s + 1)$
5	$(s+1)(s^2 + 0.61803399s + 1)(s^2 + 1.93180339s + 1)$
6	$(s^2 + 0.51763809s + 1)(s^2 + 1.41421356s + 1)(s^2 + 1.93185165s + 1)$
7	$(s+1)(s^2 + 0.44504187s + 1)(s^2 + 1.24697960s + 1)(s^2 + 1.80193774s + 1)$
8	$(s^2 + 0.39018064s + 1)(s^2 + 1.11114047s + 1)(s^2 + 1.66293922s + 1)(s^2 + 1.96157056s + 1)$
9	$(s+1)(s^2 + 0.34729636s + 1)(s^2 + s + 1)(s^2 + 1.53208889s + 1)(s^2 + 1.87938524s + 1)$
10	$(s^2 + 0.31286893s + 1)(s^2 + 0.90798100s + 1)(s^2 + 1.41421356s + 1)(s^2 + 1.78201305s + 1)(s^2 + 1.97537668s + 1)$

Frequency Scaling

Although Tables 7.1 and 7.2 are for normalized Butterworth filters with 3 dB bandwidth $\omega_c = 1$, the results can be extended to any value of ω_c by simply replacing s by s/ω_c. This step implies replacing ω by ω/ω_c in Eq. (7.32). For example, the second-order Butterworth filter for $\omega_c = 100$ can be obtained from Table 7.1 by replacing s by $s/100$. This step yields

$$H(s) = \frac{1}{\left(\frac{s}{100}\right)^2 + \sqrt{2}\left(\frac{s}{100}\right) + 1}$$

$$= \frac{1}{s^2 + 100\sqrt{2}s + 10^4} \tag{7.37}$$

The amplitude response $|H(j\omega)|$ of the filter in Eq. (7.37) is identical to that of normalized $|\mathcal{H}(j\omega)|$ in Eq. (7.32), expanded by a factor 100 along the horizontal (ω) axis (frequency scaling).

Determination of n, the Filter Order

If $\hat{G}_x$ is the gain of a lowpass Butterworth filter in dB units at $\omega = \omega_x$, then according to Eq. (7.31)

$$\hat{G}_x = 20\log_{10}|H(j\omega_x)| = -10\log\left[1 + \left(\frac{\omega_x}{\omega_c}\right)^{2n}\right]$$

Substitution of the specifications in Fig. 7.19a (gains $\hat{G}_p$ at ω_p and $\hat{G}_s$ at ω_s) in this equation yields

$$\hat{G}_p = -10\log\left[1 + \left(\frac{\omega_p}{\omega_c}\right)^{2n}\right]$$

$$\hat{G}_s = -10\log\left[1 + \left(\frac{\omega_s}{\omega_c}\right)^{2n}\right]$$

or

$$\left(\frac{\omega_p}{\omega_c}\right)^{2n} = 10^{-\hat{G}_p/10} - 1 \tag{7.38a}$$

$$\left(\frac{\omega_s}{\omega_c}\right)^{2n} = 10^{-\hat{G}_s/10} - 1 \tag{7.38b}$$

Dividing (7.38b) by (7.38a), we obtain

$$\left(\frac{\omega_s}{\omega_p}\right)^{2n} = \left[\frac{10^{-\hat{G}_s/10} - 1}{10^{-\hat{G}_p/10} - 1}\right]$$

and

$$n = \frac{\log\left[\left(10^{-\hat{G}_s/10} - 1\right) / \left(10^{-\hat{G}_p/10} - 1\right)\right]}{2\log(\omega_s/\omega_p)} \tag{7.39}$$

Also from Eq. (7.38a)

$$\omega_c = \frac{\omega_p}{\left[10^{-\hat{G}_p/10} - 1\right]^{1/2n}} \tag{7.40}$$

Alternatively, from Eq. (7.38b)

$$\omega_c = \frac{\omega_s}{\left[10^{-\hat{G}_s/10} - 1\right]^{1/2n}} \tag{7.41}$$

■ **Example 7.6**

Design a Butterworth lowpass filter to meet the specifications (Fig. 7.23):

(i) Passband gain to lie between 1 and $G_p = 0.794\,(\hat{G}_p = -2\,\mathrm{dB})$ for $0 \le \omega < 10$.

(ii) Stopband gain not to exceed $G_s = 0.1\,(\hat{G}_s = -20\,\mathrm{dB})$ for $\omega \ge 20$.

Step 1: Determine n

Here $\omega_p = 10$, $\omega_s = 20$, $\hat{G}_p = -2$ dB, and $\hat{G}_s = -20$ dB. Substituting these values in Eq. (7.39) yields

$$n = 3.701$$

Since n can only be an integer, we choose

$$n = 4$$

Step 2: Determine ω_c

Substitution of $n = 4$, $\omega_p = 10$ in Eq. (7.40) yields

$$\omega_c = 10.693$$

Alternately, substitution of $n = 4$ in Eq. (7.41) yields

$$\omega_c = 11.261$$

Because we selected $n = 4$ rather than 3.701, we obtain two different values of ω_c. Choice of $\omega_c = 10.693$ will satisfy exactly the requirement $G_p = 0.794$ over the passband $(0, 10)$, and will surpass the requirement $G_s = 0.1$ in the stopband $\omega \ge 20$. On the other hand, choice of $\omega_c = 11.261$ will exactly satisfy the requirement on G_s but will oversatisfy the requirement for G_p. Let us choose the former case ($\omega_c = 10.693$).

Step 3: Determine the normalized transfer function $\mathcal{H}(s)$

The normalized fourth-order transfer function $\mathcal{H}(s)$ is found from Table 7.1 as

$$\mathcal{H}(s) = \frac{1}{s^4 + 2.6131s^3 + 3.4142s^2 + 2.6131s + 1}$$

Step 4: Determine the final filter transfer function $H(s)$

The desired transfer function with $\omega_c = 10.693$ is obtained by replacing s with $s/10.693$ in the normalized transfer function $\mathcal{H}(s)$ as

$$H(s) = \frac{1}{\left(\frac{s}{10.693}\right)^4 + 2.6131\left(\frac{s}{10.693}\right)^3 + 3.4142\left(\frac{s}{10.692}\right)^2 + 2.6131\left(\frac{s}{10.693}\right) + 1}$$

$$= \frac{13073.7}{s^4 + 27.942s^3 + 390.4s^2 + 3194.88s + 13073.7}$$

$$= \frac{13073.7}{(s^2 + 8.1844s + 114.34)(s^2 + 19.758s + 114.34)}$$

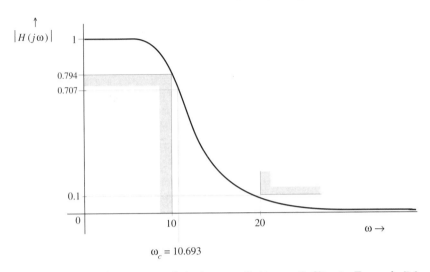

Fig. 7.23 Amplitude response of the lowpass Butterworth filter in Example 7.6.

The amplitude response of this filter is given by Eq. (7.31) with $n = 4$ and $\omega_c = 10.693$

$$|H(j\omega)| = \frac{1}{\sqrt{(\frac{\omega}{10.693})^8 + 1}}$$

Figure 7.23 shows the amplitude response of this filter.

We could also have used the alternate value of $\omega_c = 11.261$. This choice would result in a slightly different design. Either of the two designs satisfies the specifications. ■

⊙ **Computer Example C7.5**
Solve Example 7.6 using MATLAB

```
% Step 1: Determine n
wp=10; ws=20; Gp=-2; Gs=-20;
P1=-Gs/10; P2=-Gp/10; Wsp=ws/wp;
nc=log((10^P1-1)/(10^P2-1))/(2*log(Wsp));
n=ceil(nc);
% Step 2: Determine Wc (option that satisfies passband requirement
% exactly and may oversatisfy stopband requirement).
Wc=wp/(10^P2-1)^(1/(2*n));
% Step 3: Determine the normalized transfer function H(s)
for k=1:n
    A=(2*(k-1)+n+1)/(2*n);
    Sk=cos(A*pi)+j*sin(A*pi);
    s=[s Sk];
end
s=s';
num1=[0 1]; den1=poly([s']);
% Step 4: Determine the final filter transfer function H(s)
num2=[0 Wc^n]; den2=poly(Wc*[s']);
fprintf('Filter Order is n = %i\n',n)
fprintf('Cutoff Frequency of the Filter is Wc = %.4f\n',Wc)
disp('Poles of the transfer function are'),s
```

```
disp('The normalized fourth-order transfer function is')
printsys(abs(num1),abs(den1))
disp('The transfer function with s replaced by s/Wc is')
printsys(abs(num2),abs(den2))
% Step 5: Amplitude response of the filter
w=0:.01:40; w=w';
[mag,phase,w]=bode(num2,den2,w);
plot(w,mag)
Filter Order is n = 4
Cutoff Frequency of the Filter is Wc = 10.6934
Poles of the transfer function are
s = -0.3827 - 0.9239i
-0.9239 - 0.3827i
-0.9239 + 0.3827i
-0.3827 + 0.9239i
The normalized fourth-order transfer function is
```

$$\text{num/den} = \frac{1}{s\hat{\ }4 + 2.613\ s\hat{\ }3 + 3.414\ s\hat{\ }2 + 2.613\ s + 1}$$

The transfer function with s replaced by s/Wc is

$$\text{num/den} = \frac{13,000}{s\hat{\ }4 + 27.94\ s\hat{\ }3 + 390.4\ s\hat{\ }2 + 3195\ s + 1.3\ e{+}004} \odot$$

Using M-files from MATLAB Signal Processing Toolbox

We can also compute the desired filter transfer function using appropriate M-files from the *Signal Processing Toolbox* as shown in the next few examples.

$\odot$ **Computer Example C7.6**

Using M-file functions in MATLAB, design a lowpass Butterworth filter to meet the specifications in Example 7.6.

```
Wp=10;Ws=20;Gp=-2;Gs=-20;
[n,Wc]=buttord(Wp,Ws,-Gp,-Gs,'s');
[num,den]=butter(n,Wc,'s');
```

Here num and den are the coefficients of the numerator and the denominator polynomials of the desired filter. In this example, the matlab answer is $n + 1$ element vectors as num= 0 0 0 0 16081 and den= 1 29 433 3732 16081; that is,

$$H(s) = \frac{16081}{s^4 + 29s^3 + 433s^2 + 3732s + 16081}$$

This is the alternate solution where the passband specifications are exceeded, but the stopband specifications are satisfied exactly. On the other hand, the solution in Example C7.5 exceeds the stopband specifications, but satisfies the passband specifications exactly because of use of Eq. (7.40) [rather than Eq. (7.41)]. To plot amplitude response, we can use the last three functions from Example C7.5. $\odot$

△ **Exercise E7.3**

Determine n, the order of the lowpass Butterworth filter to meet the following specifications:
$\hat{G}_p = -0.5\,\text{dB}$, $\hat{G}_s = -20\,\text{dB}$, $\omega_p = 100$, and $\omega_s = 200$.

Answer: 5. ▽

7.6 Chebyshev Filters

The amplitude response of a normalized Chebyshev lowpass filter is given by

$$|\mathcal{H}(j\omega)| = \frac{1}{\sqrt{1 + \epsilon^2 C_n{}^2(\omega)}} \tag{7.42}$$

where $C_n(\omega)$, the nth-order Chebyshev polynomial, is given by

$$C_n(\omega) = \cos\left(n \cos^{-1}\omega\right) \tag{7.43a}$$

An alternative expression for $C_n(\omega)$ is

$$C_n(\omega) = \cosh\left(n \cosh^{-1}\omega\right) \tag{7.43b}$$

The form (7.43a) is most convenient to compute $C_n(\omega)$ for $|\omega| < 1$ and form (7.43b) is convenient for computing $C_n(\omega)$ for $|\omega| > 1$. We can show[1] that $C_n(\omega)$ is also expressible in polynomial form, as shown† in Table 7.3 for $n = 1$ to 10.

The normalized Chebyshev lowpass amplitude response [Eq. (7.42)] is depicted in Fig. 7.24 for $n = 6$ and $n = 7$. We make the following general observations:

Table 7.3: Chebyshev Polynomials

n	$C_n(\omega)$
0	1
1	ω
2	$2\omega^2 - 1$
3	$4\omega^3 - 3\omega$
4	$8\omega^4 - 8\omega^2 + 1$
5	$16\omega^5 - 20\omega^3 + 5\omega$
6	$32\omega^6 - 48\omega^4 + 18\omega^2 - 1$
7	$64\omega^7 - 112\omega^5 + 56\omega^3 - 7\omega$
8	$128\omega^8 - 256\omega^6 + 160\omega^4 - 32\omega^2 + 1$
9	$256\omega^9 - 576\omega^7 + 432\omega^5 - 120\omega^3 + 9\omega$
10	$512\omega^{10} - 1280\omega^8 + 1120\omega^6 - 400\omega^4 + 50\omega^2 - 1$

---

†The Chebyshev polynomial $C_n(\omega)$ has the property[1]
$$C_n(\omega) = 2\omega C_{n-1}(\omega) - C_{n-2}(\omega) \qquad n > 2$$
Thus, knowing that
$$C_0(\omega) = 1 \quad \text{and} \quad C_1(\omega) = \omega$$
we can construct $C_n(\omega)$ for any value of n. For example,
$$C_2(\omega) = 2\omega C_1(\omega) - C_0(\omega) = 2\omega^2 - 1$$
and so on.

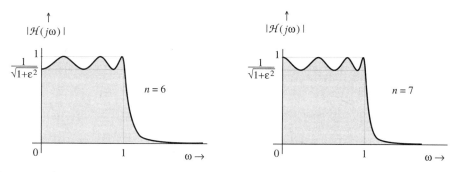

Fig. 7.24 Amplitude response of normalized sixth- and seventh-order lowpass Chebyshev filters.

1. The Chebyshev amplitude response has ripples in the passband and is smooth (monotonic) in the stopband. The passband is $0 \leq \omega \leq 1$, and there is a total of n maxima and minima over the passband $0 \leq \omega \leq 1$.

2. From Table 7.3, we observe that

$$C_n^2(0) = \begin{cases} 0 & n \text{ odd} \\ 1 & n \text{ even} \end{cases} \tag{7.44}$$

Therefore, the dc gain is

$$|\mathcal{H}(0)| = \begin{cases} 1 & n \text{ odd} \\ \frac{1}{\sqrt{1+\epsilon^2}} & n \text{ even} \end{cases} \tag{7.45}$$

3. The parameter ϵ controls the height of ripples. In the passband, r, the ratio of the maximum gain to the minimum gain is

$$r = \sqrt{1+\epsilon^2} \tag{7.46a}$$

This ratio r, specified in decibels, is

$$\hat{r} = 20 \log \sqrt{1+\epsilon^2} = 10 \log_{10}(1+\epsilon^2) \tag{7.46b}$$

so that

$$\epsilon^2 = 10^{\hat{r}/10} - 1 \tag{7.47}$$

Because all the ripples in the passband are of equal height, the Chebyshev polynomials $C_n(\omega)$ are known as **equal-ripple functions**.

4. The ripple is present only over the passband $0 \leq \omega \leq 1$. At $\omega = 1$, the amplitude response is $1/\sqrt{1+\epsilon^2} = 1/r$. For $\omega > 1$, the gain decreases monotonically.

5. For Chebyshev filters, the ripple $\hat{r}$ dB takes the place of $\hat{G}_p$ (the minimum gain in the passband). For example, $\hat{r} \leq 2$ dB specifies that the gain variations of more than 2 dB cannot be tolerated in the passband. In the Butterworth filter $\hat{G}_p = -2$ dB means the same thing.

6. If we reduce the ripple, the passband behavior improves, but it does so at the cost of stopband behavior. As r is decreased (ϵ is reduced), the gain in the stopband increases, and vice-versa. Hence, there is a tradeoff between the allowable passband ripple and the desired attenuation in the stopband. Note

that the extreme case $\epsilon = 0$ yields zero ripple, but the filter now becomes an allpass filter, as seen from Eq. 7.42, by letting $\epsilon = 0$.

7. Finally, the Chebyshev filter has a sharper cutoff (smaller transition band) than the same-order Butterworth filter,† but this is achieved at the expense of inferior passband behavior (rippling).

Determination of n (Filter Order)

For a normalized Chebyshev filter, the gain $\hat{G}$ in dB [see Eq. (7.42)] is

$$\hat{G} = -10 \log \left[1 + \epsilon^2 C_n{}^2(\omega)\right]$$

The gain is $\hat{G}_s$ at ω_s. Therefore

$$\hat{G}_s = -10 \log \left[1 + \epsilon^2 C_n{}^2(\omega_s)\right] \tag{7.48}$$

or

$$\epsilon^2 C_n{}^2(\omega_s) = 10^{-\hat{G}_s/10} - 1$$

Use of Eq. (7.43b) and Eq.(7.47) in the above equation yields

$$\cosh\left[n \cosh^{-1}(\omega_s)\right] = \left[\frac{10^{-\hat{G}_s/10} - 1}{10^{\hat{r}/10} - 1}\right]^{1/2}$$

Hence

$$n = \frac{1}{\cosh^{-1}(\omega_s)} \cosh^{-1} \left[\frac{10^{-\hat{G}_s/10} - 1}{10^{\hat{r}/10} - 1}\right]^{1/2} \tag{7.49a}$$

Note that these equations are for normalized filters, where $\omega_p = 1$. For a general case, we replace ω_s with $\frac{\omega_s}{\omega_p}$ to obtain

$$n = \frac{1}{\cosh^{-1}(\omega_s/\omega_p)} \cosh^{-1} \left[\frac{10^{-\hat{G}_s/10} - 1}{10^{\hat{r}/10} - 1}\right]^{1/2} \tag{7.49b}$$

Pole Locations

We could follow the procedure of the Butterworth filter to obtain the pole locations of the Chebyshev filter. The procedure is straightforward but tedious and does not yield any special insight into our development. The Butterworth filter poles lie on a semicircle. We can show that the poles of an nth-order normalized Chebyshev filter lie on a semiellipse of the major and minor semiaxes $\cosh x$ and $\sinh x$, respectively, where[1]

$$x = \frac{1}{n} \sinh^{-1} \left(\frac{1}{\epsilon}\right) \tag{7.50}$$

The Chebyshev filter poles are

$$s_k = -\sin \left[\frac{(2k-1)\pi}{2n}\right] \sinh x + j \cos \left[\frac{(2k-1)\pi}{2n}\right] \cosh x \quad k = 1, 2, \cdots, n \tag{7.51}$$

---

†We can show[2] that at higher frequencies (in the stopband), the Chebyshev filter gain is smaller than the comparable Butterworth filter gain by about $6(n - 1)$ dB.

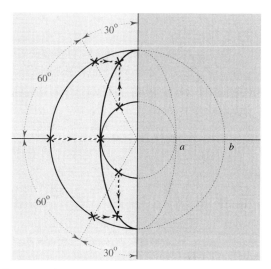

Fig. 7.25 Poles of a normalized third-order lowpass Chebyshev filter transfer function and its conjugate.

The geometrical construction for determining the pole location is depicted in Fig. 7.25 for $n = 3$. A similar procedure applies to any n; it consists of drawing two semicircles of radii $a = \sinh x$ and $b = \cosh x$. We now draw radial lines along the corresponding Butterworth angles and locate the nth-order Butterworth poles (shown by crosses) on the two circles. The location of the kth Chebyshev pole is the intersection of the horizontal projection and the vertical projection from the corresponding kth Butterworth poles on the outer and the inner circle, respectively.

The transfer function $\mathcal{H}(s)$ of the normalized nth-order lowpass Chebyshev filter is

$$\mathcal{H}(s) = \frac{K_n}{C'_n(s)} = \frac{K_n}{s^n + a_{n-1}s^{n-1} + \cdots + a_1 s + a_0} \qquad (7.52)$$

The constant K_n is selected to have proper dc gain, as shown in Eq. (7.45). As a result

$$K_n = \begin{cases} a_0 & n \text{ odd} \\[2mm] \dfrac{a_0}{\sqrt{1+\epsilon^2}} = \dfrac{a_0}{10^{\hat{r}/20}} & n \text{ even} \end{cases} \qquad (7.53)$$

The design procedure is considerably simplified by ready-made tables of the polynomial $C'_n(s)$ in Eq. (7.52) or the pole locations of $\mathcal{H}(s)$. Table 7.4 lists the coefficients $a_0, a_1, a_2, \cdots, a_{n-1}$ of the polynomial $C'_n(s)$ in Eq. (7.52) for $\hat{r} = 0.5$, 1, 2, and 3 dB ripples corresponding to the values of $\epsilon = 0.3493$, 0.5088, 0.7648, and 0.9976, respectively. Table 7.5 lists the poles of various Chebyshev filters for the same values of $\hat{r}$ (and ϵ). Tables listing more extensive values of $\hat{r}$ (or ϵ) can be found in the literature. We can also use MATLAB functions for this purpose.

⊙ **Computer Example C7.7**
    Using MATLAB, find poles, zeros, and the gain factor of a normalized 3rd-order Chebyshev filter with $\hat{r} = 2$ dB.

    [z,p,k]=cheb1ap(3,2)   ⊙

Table 7.4:   Chebyshev Filter Coefficients of the Denominator Polynomial
$C'_n(s) = s^n + a_{n-1}s^{n-1} + a_{n-2}s^{n-2} + \cdots + a_1 s + a_0$

n	a_0	a_1	a_2	a_3	a_4	a_5	a_6
1	2.8627752					0.5 db ripple	
2	1.5162026	1.4256245				$(\hat{r} = 0.5)$	
3	0.7156938	1.5348954	1.2529130				
4	0.3790506	1.0254553	1.7168662	1.1973856			
5	0.1789234	0.7525181	1.3095747	1.9373675	1.1724909		
6	0.0947626	0.4323669	1.1718613	1.5897635	2.1718446	1.1591761	
7	0.0447309	0.2820722	0.7556511	1.6479029	1.8694079	2.4126510	1.1512176

n	a_0	a_1	a_2	a_3	a_4	a_5	a_6
1	1.9652267					1 db ripple	
2	1.1025103	1.0977343				$(\hat{r} = 1)$	
3	0.4913067	1.2384092	0.9883412				
4	0.2756276	0.7426194	1.4539248	0.9528114			
5	0.1228267	0.5805342	0.9743961	1.6888160	0.9368201		
6	0.0689069	0.3070808	0.9393461	1.2021409	1.9308256	0.9282510	
7	0.0307066	0.2136712	0.5486192	1.3575440	1.4287930	2.1760778	0.9231228

n	a_0	a_1	a_2	a_3	a_4	a_5	a_6
1	1.3075603					2 db ripple	
2	0.8230604	0.8038164				$(\hat{r} = 2)$	
3	0.3268901	1.0221903	0.7378216				
4	0.2057651	0.5167981	1.2564819	0.7162150			
5	0.0817225	0.4593491	0.6934770	1.4995433	0.7064606		
6	0.0514413	0.2102706	0.7714618	0.8670149	1.7458587	0.7012257	
7	0.0204228	0.1660920	0.3825056	1.1444390	1.0392203	1.9935272	0.6978929

n	a_0	a_1	a_2	a_3	a_4	a_5	a_6
1	1.0023773					3 db ripple	
2	0.7079478	0.6448996				$(\hat{r} = 3)$	
3	0.2505943	0.9283480	0.5972404				
4	0.1769869	0.4047679	1.1691176	0.5815799			
5	0.0626391	0.4079421	0.5488626	1.4149874	0.5744296		
6	0.0442467	0.1634299	6990977	6906098	1.6628481	0.5706979	
7	0.0156621	0.1461530	0.3000167	1.0518448	0.8314411	1.9115507	0.5684201

Table 7.5:   Chebyshev Filter Pole Locations

n	$\hat{r} = 0.5$	$\hat{r} = 1$	$\hat{r} = 2$	$\hat{r} = 3$
1	-2.8628	-1.9652	-1.3076	-1.0024
2	$-0.7128 \pm j1.0040$	$-0.5489 \pm j0.8951$	$-0.4019 \pm j0.8133$	$-0.3224 \pm j0.7772$
3	-0.6265 $-0.3132 \pm j1.0219$	-0.4942 $-0.2471 \pm j0.9660$	-0.3689 $-0.1845 \pm j0.9231$	-0.2986 $-0.1493 \pm j0.9038$
4	$-0.1754 \pm j1.0163$ $-0.4233 \pm j0.4209$	$-0.1395 \pm j0.9834$ $-0.3369 \pm j0.4073$	$-0.1049 \pm j0.9580$ $-0.2532 \pm j0.3968$	$-0.0852 \pm j0.9465$ $-0.2056 \pm j0.3920$
5	-0.3623 $-0.1120 \pm j1.0116$ $-0.2931 \pm j0.6252$	-0.2895 $-0.0895 \pm j0.9901$ $-0.2342 \pm j0.6119$	-0.2183 $-0.0675 \pm j0.9735$ $-0.1766 \pm j0.6016$	-0.1775 $-0.0549 \pm j0.9659$ $-0.1436 \pm j0.5970$
6	$-0.0777 \pm j1.0085$ $-0.2121 \pm j0.7382$ $-0.2898 \pm j0.2702$	$-0.0622 \pm j0.9934$ $-0.1699 \pm j0.7272$ $-0.2321 \pm j0.2662$	$-0.0470 \pm j0.9817$ $-0.1283 \pm j0.7187$ $-0.1753 \pm j0.2630$	$-0.0382 \pm j0.9764$ $-0.1044 \pm j0.7148$ $-0.1427 \pm j0.2616$
7	-0.2562 $-0.0570 \pm j1.0064$ $-0.1597 \pm j0.8071$ $-0.2308 \pm j0.4479$	-0.2054 $-0.0457 \pm j0.9953$ $-0.1281 \pm j0.7982$ $-0.1851 \pm j0.4429$	-0.1553 $-0.0346 \pm j0.9866$ $-0.0969 \pm j0.7912$ $-0.1400 \pm j0.4391$	-0.1265 $-0.0281 \pm j0.9827$ $-0.0789 \pm j0.7881$ $-0.1140 \pm j0.4373$
8	$-0.0436 \pm j1.0050$ $-0.1242 \pm j0.8520$ $-0.1859 \pm j0.5693$ $-0.2193 \pm j0.1999$	$-0.0350 \pm j0.9965$ $-0.0997 \pm j0.8447$ $-0.1492 \pm j0.5644$ $-0.1760 \pm j0.1982$	$-0.0265 \pm j0.9898$ $-0.0754 \pm j0.8391$ $-0.1129 \pm j0.5607$ $-0.1332 \pm j0.1969$	$-0.0216 \pm j0.9868$ $-0.0614 \pm j0.8365$ $-0.0920 \pm j0.5590$ $-0.1085 \pm j0.1962$
9	-0.1984 $-0.0345 \pm j1.0040$ $-0.0992 \pm j0.8829$ $-0.1520 \pm j0.6553$ $-0.1864 \pm j0.3487$	-0.1593 $-0.0277 \pm j0.9972$ $-0.0797 \pm j0.8769$ $-0.1221 \pm j0.6509$ $-0.1497 \pm j0.3463$	-0.1206 $-0.0209 \pm j0.9919$ $-0.0603 \pm j0.8723$ $-0.0924 \pm j0.6474$ $-0.1134 \pm j0.3445$	-0.0983 $-0.0171 \pm j0.9896$ $-0.0491 \pm j0.8702$ $-0.0753 \pm j0.6459$ $-0.0923 \pm j0.3437$
10	$-0.0279 \pm j1.0033$ $-0.0810 \pm j0.9051$ $-0.1261 \pm j0.7183$ $-0.1589 \pm j0.4612$ $-0.1761 \pm j0.1589$	$-0.0224 \pm j0.9978$ $-0.1013 \pm j0.7143$ $-0.0650 \pm j0.9001$ $-0.1277 \pm j0.4586$ $-0.1415 \pm j0.1580$	$-0.0170 \pm j0.9935$ $-0.0767 \pm j0.7113$ $-0.0493 \pm j0.8962$ $-0.0967 \pm j0.4567$ $-0.1072 \pm j0.1574$	$-0.0138 \pm j0.9915$ $-0.0401 \pm j0.8945$ $-0.0625 \pm j0.7099$ $-0.0788 \pm j0.4558$ $-0.0873 \pm j0.1570$

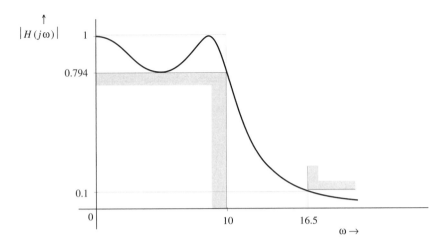

Fig. 7.26   Amplitude response of the lowpass Chebyshev filter in Example 7.7.

■ **Example 7.7**

Design a Chebyshev lowpass filter to satisfy the following criteria (Fig. 7.26):

The ratio $\hat{r} \leq 2$ dB over a passband $0 \leq \omega \leq 10$ ($\omega_p = 10$). The stopband gain $\hat{G}_s \leq -20$ dB for $\omega > 16.5$ ($\omega_s = 16.5$).

Observe that the specifications are the same as those in Example 7.6, except for the transition band. Here the transition band is from 10 to 16.5, whereas in Example 7.6 it is 10 to 20. Despite this stringent requirement, we shall find that Chebyshev requires a lower-order filter than the Butterworth filter found in Example 7.6.

Step 1: Determining n

According to Eq. (7.49b), we have

$$n = \frac{1}{\cosh^{-1}(1.65)} \cosh^{-1} \left[\frac{10^2 - 1}{10^{0.2} - 1} \right]^{1/2} = 2.999$$

Because n must be an integer, we select $n = 3$. Observe that even with more stringent requirements, the Chebyshev filter requires only $n = 3$. The passband behavior of the Butterworth filter, however, is superior (maximally flat at $\omega = 0$) compared to that of the Chebyshev, which has rippled passband characteristics.

Step 2: Determining $\mathcal{H}(s)$

We may use the Table 7.4 to determine $\mathcal{H}(s)$. For $n = 3$ and $\hat{r} = 2$ dB, we read the coefficients of the denominator polynomial of $\mathcal{H}(s)$ as $a_0 = 0.3269$, $a_1 = 1.0222$, and $a_2 = 0.7378$. Also in Eq. (7.53), for odd n, the numerator is given by a constant $K_n = a_0 = 0.3269$. Therefore,

$$\mathcal{H}(s) = \frac{0.3269}{s^3 + 0.7378s^2 + 1.0222s + 0.3269} \tag{7.54}$$

Because there are infinite possible combinations of n and $\hat{r}$, Table 7.4 (or 7.5) can list values of the denominator coefficients for values of $\hat{r}$ in quantum increments only.[1] For the values of n and $\hat{r}$ not listed in the Table, we can compute pole locations from Eq. (7.51). For the sake of demonstration, we now recompute $\mathcal{H}(s)$ using this method. In this case, the value of ϵ is [see Eq. (7.47)]

$$\epsilon = \sqrt{10^{\hat{r}/10} - 1} = \sqrt{10^{0.2} - 1} = 0.7647$$

From Eq. (7.50)

$$x = \frac{1}{n}\sinh^{-1}\frac{1}{\epsilon} = \frac{1}{3}\sinh^{-1}(1.3077) = 0.3610$$

Now from Eq. (7.51), we have $s_1 = -0.1844 + j0.9231$, $s_2 = -0.3689$, and $s_3 = -0.1844 - j0.9231$. Therefore

$$\mathcal{H}(s) = \frac{K_n}{(s + 0.3689)(s + 0.1844 + j0.9231)(s + 0.1844 - j0.9231)}$$

$$= \frac{K_n}{s^3 + 0.7378s^2 + 1.0222s + 0.3269} = \frac{0.3269}{s^3 + 0.7378s^2 + 1.0222s + 0.3269}$$

which confirms the earlier result.

Step 3: Determining $H(s)$

Recall that $\omega_p = 1$ for the normalized transfer function. For $\omega_p = 10$, the desired transfer function $H(s)$ can be obtained from the normalized transfer function $\mathcal{H}(s)$ by replacing s with $s/\omega_p = s/10$. Therefore

$$H(s) = \frac{0.3269}{(\frac{s}{10} + 0.3689)(\frac{s}{10} + 0.1844 + j0.9231)(\frac{s}{10} + 0.1844 - j0.9231)}$$

$$= \frac{326.9}{s^3 + 7.378s^2 + 102.22s + 326.9}$$

In the present case, $\hat{r} = 2$ dB means that [see Eq. (7.47)]

$$\epsilon^2 = 10^{0.2} - 1 = 0.5849$$

The frequency response is [see Eq. (7.42) and Table 7.3]

$$|\mathcal{H}(j\omega)| = \frac{1}{\sqrt{1 + 0.5849(4\omega^3 - 3\omega)^2}}$$

This is the normalized filter amplitude response. The actual filter response $|H(j\omega)|$ is obtained by replacing ω with $\frac{\omega}{\omega_p}$; that is, with $\frac{\omega}{10}$ in $\mathcal{H}(j\omega)$

$$|H(j\omega)| = \frac{1}{\sqrt{1 + 0.5849\left[4\left(\frac{\omega}{10}\right)^3 - 3\left(\frac{\omega}{10}\right)^2\right]}}$$

$$= \frac{10^3}{\sqrt{9.3584\omega^6 - 1403.76\omega^4 + 52640\omega^2 + 10^6}}$$

Observe that despite more stringent specifications than those in Example 7.6, the Chebyshev filter requires $n = 3$ compared to the Butterworth filter in Example 7.6, which requires $n = 4$. Figure 7.26 shows the amplitude response. ∎

⊙ **Computer Example C7.8**

Design a lowpass Chebyshev filter for the specifications in Example 7.7 using functions from *Signal Processing Toolbox* in MATLAB.

```
Wp=10;Ws=16.5;r=2;Gs=-20;
[n,Wp]=cheb1ord(Wp,Ws,r,-Gs,'s');
[num,den]=cheby1(n,r,Wp,'s');
```

MATLAB returns $n = 3$ and num= 0 0 0 326.8901, den= 1 7.3782 102.219 326.8901; that is,

$$H(s) = \frac{326.8901}{s^3 + 7.3782s^2 + 102.219s + 326.8901}$$

a result, which agrees with the solution in Example 7.7. To plot amplitude response, we can use the last three functions from Example C7.5. $\odot$

△ **Exercise E7.4**
   Determine n, the order of the lowpass Butterworth filter to meet the following specifications: $\hat{G}_p = -0.5\,\text{dB}$, $\hat{G}_s = -20\,\text{dB}$, $\omega_p = 100$, and $\omega_s = 200$.
Answer: 5. ▽

△ **Exercise E7.5**
   Determine n (the order) and the transfer function of a Chebyshev filter to meet the following specifications: $\hat{r} = 2$ dB, $\hat{G}_s = -20$ dB, $\omega_p = 10$ rad/s, and $\omega_s = 28$ rad/s.
Answer: $n = 2$

$$H(s) = \frac{50.5823}{(s + 4.0191 + j6.8937)(s + 4.0191 - j6.8937)} = \frac{50.5823}{s^2 + 8.0381s + 63.6768}$$

Hint: In this case $K_2 = \frac{a_0}{\sqrt{1+\epsilon^2}}$ ▽

7.6-1 Inverse Chebyshev Filters

The passband behavior of the Chebyshev filters exhibits ripples and the stopband is smooth. Generally, passband behavior is more important and we would prefer that the passband have smooth response. However, ripples can be tolerated in the stopband as long as they meet a given specification. The **inverse Chebyshev** filter does exactly that. Both, the Butterworth and the Chebyshev filters, have finite poles and no finite zeros. The inverse Chebyshev has finite zeros and poles. It exhibits maximally flat passband response and equal-ripple stopband response.

The inverse Chebyshev response can be obtained from the Chebyshev in two steps as follows: Let $\mathcal{H}_C(\omega)$ be the Chebyshev amplitude response given in Eq. (7.42). In the first step, we subtract $|\mathcal{H}_C(\omega)|^2$ from 1 to obtain a highpass filter characteristic where the stopband (from 0 to 1) has ripples and the passband (from 1 to ∞) is smooth. In the second step, we interchange the stopband and passband by frequency transformation where ω is replaced by $1/\omega$. This step inverts the passband from the range 1 to ∞ to the range 0 to 1, and the stopband is now from 1 to ∞. Moreover, the passband is now smooth and the stopband has ripples. This is precisely the inverse Chebyshev amplitude response $|\mathcal{H}(\omega)|$ given by

$$|\mathcal{H}(\omega)|^2 = 1 - |\mathcal{H}_C(1/\omega)|^2 = \frac{\epsilon^2 C_n^2(1/\omega)}{1 + \epsilon^2 C_n^2(1/\omega)}$$

where $C_n(\omega)$ are the nth-order Chebyshev polynomials listed in Table 7.3.

The inverse Chebyshev filters are preferable to the Chebyshev filters in many ways. For example, the passband behavior, especially for small ω, is better for the inverse Chebyshev than for the Chebyshev or even for the Butterworth filter of the same order. The inverse Chebyshev also has the smallest transition band of the three filters. Moreover, the phase function (or time-delay) characteristic of the inverse Chebyshev filter is better than that of the Chebyshev filter.[2] Both the Chebyshev and inverse Chebyshev filter require the same order n to meet a given set of specifications.[1] But the inverse Chebyshev realization requires more elements and thus is less economical than the Chebyshev filter. But the inverse Chebyshev does require fewer elements than a comparable performance Butterworth filter. Rather than give complete development of inverse Chebyshev filters, we shall solve a problem using MATLAB functions from the *Signal Processing Toolbox*.

⊙ **Computer Example C7.9**
Design a lowpass inverse Chebyshev filter for the specifications in Example 7.7, using functions from *Signal Processing Toolbox* in MATLAB.

Wp=10;Ws=16.5;Gp=-2;Gs=-20;
[n,Ws]=cheb2ord(Wp,Ws,-Gp,-Gs,'s');
[num,den]=cheby2(n,-Gs,Ws,'s')

MATLAB returns $n = 3$ and num= 0 5 0 1805.9, den= 1 23.2 256.4 1805.9; that is,

$$H(s) = \frac{5s^2 + 1805.9}{s^3 + 23.2s^2 + 256.4s + 1805.9}$$

To plot amplitude response, we can use the last three functions from Example C7.5.   ⊙

7.6-2 Elliptic Filters

Recall our discussion in Sec. 7.4 that placing a zero on the imaginary axis (at $s = j\omega$) causes the gain ($|H(j\omega)|$) to go to zero (infinite attenuation). We can realize a sharper cutoff characteristic by placing a zero (or zeros) near $\omega = \omega_s$. Butterworth and Chebyshev filters do not make use of zeros in $\mathcal{H}(s)$. But an elliptic filter does. This is the reason for the superiority of the elliptic filter.

A Chebyshev filter has a smaller transition band compared to that of a Butterworth filter because a Chebyshev filter allows rippling in the passband (or stopband). If we allow ripple in both the passband and the stopband, we can achieve further reduction in the transition band. Such is the case with elliptic (or Cauer) filters, whose normalized amplitude response is given by

$$|\mathcal{H}(j\omega)| = \frac{1}{\sqrt{1 + \epsilon^2 R_n^{\,2}(\omega)}}$$

where $R_n(\omega)$ is the nth-order Chebyshev rational function determined from the specific ripple characteristic. The parameter ϵ controls the ripple. The gain at ω_p ($\omega_p = 1$ for the normalized case) is $\frac{1}{\sqrt{1+\epsilon^2}}$.

The elliptic filter is the most efficient if we can tolerate ripples in both the passband and the stopband. For a given transition band, it provides the largest ratio of the passband gain to stopband gain, or for a given ratio of passband to stopband gain, it requires the smallest transition band. In compensation, however, we must accept ripples in both the passband and the stopband. In addition, because of zeros in the numerator of $\mathcal{H}(s)$, the elliptic filter response decays at a slower rate at frequencies higher than ω_s. For instance, the amplitude response of a third-order elliptic filter decays at a rate of only -6 dB/octave at very high frequencies. This is because the filter has two zero and three poles. The two zeros increase the amplitude response at a rate of 12 dB/octave, and the three poles reduce the amplitude response at a rate of -18 dB/octave, thus giving a net decay rate of -6 dB/octave. For the Butterworth and Chebyshev filters, there are no zeros in $\mathcal{H}(s)$.

Therefore, their amplitude response decays at a rate of -18 dB/octave. However, the rate of decay of the amplitude response is seldom important as long as we meet our specification of a given G_s at ω_s.

Calculation of pole-zero locations of elliptic filters is much more complicated than that in Butterworth or even Chebyshev filters. Fortunately, this task is greatly simplified by computer programs and extensive ready-made design Tables available in the literature.[3] The MATLAB function [z,p,k]=ellipap(n,-Gp,-Gs) in *Signal Processing Toolbox* determines poles, zeros, and the gain factor of a normalized analog elliptic lowpass filter of order n with a minimum passband gain Gp dB, and maximum stopband gain Gs dB. The normalized passband edge is 1 rad/s.

⊙ **Computer Example C7.10**

Design the lowpass elliptic filter for the specifications in Example 7.7 using functions from *Signal Processing Toolbox* in MATLAB.

> **Wp=10;Ws=16.5;Gp=-2;Gs=-20;**
> **[n,Wp]=ellipord(Wp,Ws,-Gp,-Gs,'s');**
> **[num,den]=ellip(n,-Gp,-Gs,Wp,'s')**

MATLAB returns $n = 3$ and num= 0 2.7881 0 481.1626, den= 1 7.261 106.9991 481.1626; that is,

$$H(s) = \frac{2.7881s^2 + 481.1626}{s^3 + 7.261s^2 + 106.9991s + 481.1626}$$

To plot amplitude response, we can use the last three functions from Example C7.5.   ⊙

7.7 Frequency Transformations

Earlier we saw how a lowpass filter transfer function of arbitrary specifications can be obtained from a normalized lowpass filter using frequency scaling. Using certain frequency transformations, we can obtain transfer functions of highpass, bandpass, and bandstop filters from a basic lowpass filter (the prototype filter) design. For example, a highpass filter transfer function can be obtained from the prototype lowpass filter transfer function by replacing s with ω_p/s. Similar transformations allow us to design bandpass and bandstop filters from appropriate lowpass prototype filters.

The prototype filter may be of any kind, such as Butterworth, Chebyshev, elliptic, and so on. We first design a suitable prototype lowpass filter $\mathcal{H}_p(s)$. In the next step, we replace s with a proper transformation $T(s)$ to obtain the desired highpass, bandpass, or bandstop filter.

7.7-1 Highpass Filters

Figure 7.27a shows an amplitude response of a typical highpass filter. The appropriate lowpass prototype response required for the design of a highpass filter in Fig. 7.27a is depicted in Fig. 7.27b. We must first determine this prototype filter transfer function $\mathcal{H}_p(s)$ with the passband $0 \leq \omega \leq 1$ and the stopband $\omega \geq \omega_p/\omega_s$. The desired transfer function of the highpass filter to satisfy specifications in Fig. 7.27a is then obtained by replacing s with $T(s)$ in $\mathcal{H}_p(s)$, where

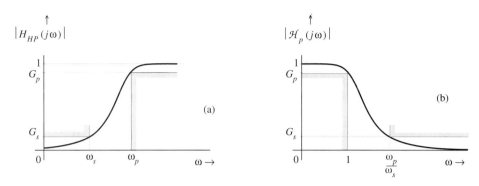

Fig. 7.27   Frequency transformation for highpass filters.

$$T(s) = \frac{\omega_p}{s} \tag{7.55}$$

■ **Example 7.8**

Design a Chebyshev highpass filter with the amplitude response specifications illustrated in Fig. 7.28a with $\omega_s = 100$, $\omega_p = 165$, $G_s = 0.1\,(-20\,\text{dB})$, and $G_p = 0.794\,(-2\,\text{dB})$.

Step 1: Determine the prototype lowpass filter

The prototype lowpass filter has $\hat{\omega}_p = 1$ and $\hat{\omega}_s = 165/100 = 1.65$. This means the prototype filter in Fig. 7.27b has a passband $0 \le \omega \le 1$ and a stopband $\omega \ge 1.65$, as shown in Fig. 7.28b. Also, $G_p = 0.794\,(-2\,\text{dB})$ and $G_s = 0.1\,(-20\,\text{dB})$. We already designed a Chebyshev filter with these specifications in Example 7.7. The transfer function of this filter is [Eq. (7.54)]

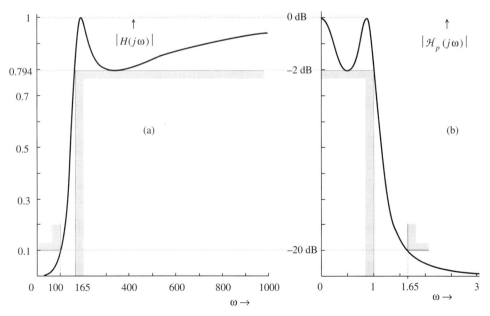

Fig. 7.28   Highpass Filter Design for Example 7.8.

$$\mathcal{H}_p(s) = \frac{0.3269}{s^3 + 0.7378s^2 + 1.0222s + 0.3269}$$

The amplitude response of this prototype filter is depicted in Fig. 7.28b.

Step 2: Substitute s with $T(s)$ in $\mathcal{H}_p(s)$

The desired highpass filter transfer function $H(s)$ is obtained from $\mathcal{H}_p(s)$ by replacing s with $T(s) = \omega_p/s = 165/s$. Therefore

$$H(s) = \frac{0.3269}{\left(\frac{165}{s}\right)^3 + 0.7378\left(\frac{165}{s}\right)^2 + 1.0222\left(\frac{165}{s}\right) + 0.3269}$$

$$= \frac{s^3}{s^3 + 515.94s^2 + 61445.75s + 13742005}$$

The amplitude response $|H(j\omega)|$ for this filter is illustrated in Fig. 7.28a.   ■

⊙ **Computer Example C7.11**

Design the highpass filter for the specifications in Example 7.8 using functions from *Signal Processing Toolbox* in MATLAB. We shall give here MATLAB functions for all types of filters.

```
Ws=100;Wp=165;Gp=-2;Gs=-20;
% Butterworth
[n,Wn]=buttord(Wp,Ws,-Gp,-Gs,'s')
[num,den]=butter(n,Wn,'high','s')
% Chebyshev
[n,Wn]=cheb1ord(Wp,Ws,-Gp,-Gs,'s')
[num,den]=cheby1(n,-Gp,Wn,'high','s')
% Inverse Chebyshev
[n,Wn]=cheb2ord(Wp,Ws,-Gp,-Gs,'s')
[num,den]=cheby2(n,-Gs,Wn,'high','s')
% Elliptic
[n,Wn]=ellipord(Wp,Ws,-Gp,-Gs,'s')
[num,den]=ellip(n,-Gp,-Gs,Wn,'high','s')
```

To plot amplitude response, we can use the last three functions in Example C7.5.   ⊙

7.7-2 Bandpass Filters

Figure 7.29a shows an amplitude response of a typical bandpass filter. To design such a filter, we first find $\mathcal{H}_p(s)$, the transfer function of a prototype lowpass filter, to meet the specifications in Fig. 7.29b, where ω_s is given by the smaller of

$$\frac{\omega_{p_1}\omega_{p_2} - \omega_{s_1}{}^2}{\omega_{s_1}(\omega_{p_2} - \omega_{p_1})} \quad \text{or} \quad \frac{\omega_{s_2}{}^2 - \omega_{p_1}\omega_{p_2}}{\omega_{s_2}(\omega_{p_2} - \omega_{p_1})} \tag{7.56}$$

Now, the desired transfer function of the bandpass filter to satisfy the specifications in Fig. 7.29a is obtained from $\mathcal{H}_p(s)$ by replacing s with $T(s)$, where

$$T(s) = \frac{s^2 + \omega_{p_1}\omega_{p_2}}{(\omega_{p_2} - \omega_{p_1})s} \tag{7.57}$$

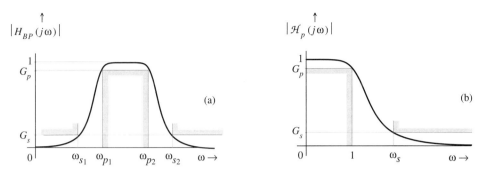

Fig. 7.29 Frequency transformation for bandpass filters.

■ **Example 7.9**

Design a Chebyshev bandpass filter with the amplitude response specifications shown in Fig. 7.30a with $\omega_{p_1} = 1000$, $\omega_{p_2} = 2000$, $\omega_{s_1} = 450$, $\omega_{s_2} = 4000$, $G_s = 0.1\,(-20\,\mathrm{dB})$, and $G_p = 0.891\,(-1\,\mathrm{dB})$. Observe that for Chebyshev filter, $G_p = -1$ dB is equivalent to $\hat{r} = 1$ dB.

The solution is executed in two steps: in the first step, we determine the lowpass prototype filter transfer function $\mathcal{H}_p(s)$. In the second step, the desired bandpass filter transfer function is obtained from $\mathcal{H}_p(s)$ by substituting s with $T(s)$, the lowpass to bandpass transformation in Eq. (7.57).

Step 1: Find $\mathcal{H}_p(s)$, the lowpass prototype filter transfer function.

This is done in 3 substeps as follows:

Step 1.1: Find ω_s for the prototype filter.

The frequency ω_s is found [using Eq. (7.56)], to be the smaller of

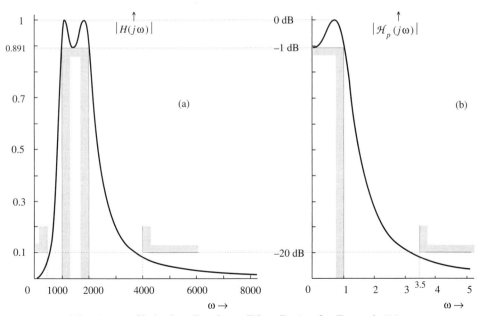

Fig. 7.30 Chebyshev Bandpass Filter Design for Example 7.9.

$$\frac{(1000)(2000) - (450)^2}{450(2000 - 1000)} = 3.99 \quad \text{and} \quad \frac{(4000)^2 - (1000)(2000)}{4000(2000 - 1000)} = 3.5$$

which is 3.5.

Step 1.2: Determine n

We now need to design a prototype lowpass filter in Fig. 7.29b with $\hat{G}_p = -1$ dB, $\hat{G}_s = -20$ dB, $\omega_p = 1$, and $\omega_s = 3.5$, as illustrated in Fig. 7.30b. The Chebyshev filter order n required to meet these specifications is obtained from Eq. (7.49b) (or Eq. (7.49a) because, in this case, $\omega_p = 1$), as

$$n = \frac{1}{\cosh^{-1}(3.5)} \cosh^{-1} \left[\frac{10^2 - 1}{10^{0.1} - 1} \right]^{\frac{1}{2}} = 1.904$$

a result, which is rounded up to $n = 2$.

Step 1.3: Determine the prototype filter transfer function $\mathcal{H}_p(s)$

We can obtain the transfer function of the second-order Chebyshev filter by computing its poles for $n = 2$ and $\hat{r} = 1$ ($\epsilon = 0.5088$) using Eq. (7.51). However, since Table 7.4 lists the denominator polynomial for $\hat{r} = 1$ and $n = 2$, we need not perform the computations and may use the ready-made transfer function directly as

$$\mathcal{H}_p(s) = \frac{0.9826}{s^2 + 1.0977s + 1.1025} \tag{7.58}$$

Here we used Eq. (7.53) to find the numerator $K_n = \frac{a_0}{\sqrt{1+\epsilon^2}} = \frac{1.1025}{\sqrt{1.2589}} = 0.9826$. The amplitude response of this prototype filter is depicted in Fig. 7.30b.

Step 2: Find the desired bandpass filter transfer function $H(s)$ using the lowpass to bandpass transformation.

Finally, the desired bandpass filter transfer function $H(s)$ is obtained from $\mathcal{H}_p(s)$ by replacing s with $T(s)$, where [see Eq. (7.57)]

$$T(s) = \frac{s^2 + 2(10)^6}{1000s}$$

Replacing s with $T(s)$ in the right-hand side of Eq. (7.58) yields the final bandpass transfer function

$$H(s) = \frac{9.826(10)^5 s^2}{s^4 + 1097.7s^3 + 5.1025(10)^6 s^2 + 2.195(10)^9 s + 4(10)^{12}}$$

The amplitude response $|H(j\omega)|$ of this filter is shown in Fig. 7.30a. ∎

We may use a similar procedure for the Butterworth filter. Compared to Chebyshev design, Butterworth filter design involves two additional steps. First, we need to compute the cutoff frequency ω_c of the prototype filter. For a Chebyshev filter, the critical frequency happens to the frequency where the gain is G_p. This frequency is $\omega = 1$ in the prototype filter. For Butterworth, on the other hand, the critical frequency is the half power (or 3 dB-cutoff) frequency ω_c, which is not necessarily the frequency where the gain is G_p. To find the transfer function of the Butterworth prototype filter, it is essential to know ω_c. Once we know ω_c, the prototype filter transfer function is obtained by replacing s with s/ω_c in the normalized transfer function $\mathcal{H}(s)$. This step is also unnecessary in the Chebyshev filter design. We shall demonstrate the procedure for the Butterworth filter design by an example below.

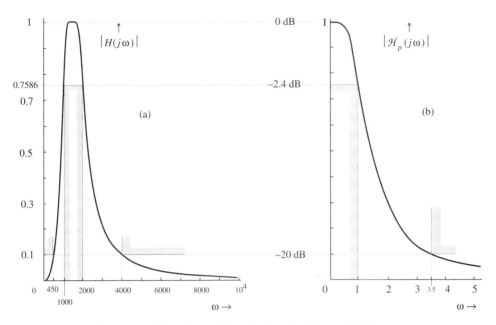

Fig. 7.31 Butterworth Bandpass Filter Design for Example 7.10.

■ **Example 7.10**

Design a Butterworth bandpass filter with the amplitude response specifications illustrated in Fig. 7.31a with $\omega_{p_1} = 1000$, $\omega_{p_2} = 2000$, $\omega_{s_1} = 450$, $\omega_{s_2} = 4000$, $G_p = 0.7586\,(-2.4\,\text{dB})$, and $G_s = 0.1\,(-20\,\text{dB})$.

As in the previous example, the solution is executed in two steps: in the first step, we determine the lowpass prototype filter transfer function $\mathcal{H}_p(s)$. In the second step, the desired bandpass filter transfer function is obtained from $\mathcal{H}_p(s)$ by substituting s with $T(s)$, the lowpass to bandpass transformation in Eq. (7.57).

Step 1: Find $\mathcal{H}_p(s)$, the lowpass prototype filter transfer function.
This goal is accomplished in 5 substeps used in the design of the lowpass Butterworth filter (see Example 7.6):

Step 1.1: Find ω_s for the prototype filter.
For the prototype lowpass filter transfer function $\mathcal{H}_p(s)$ with the amplitude response shown in Fig. 7.31b, the frequency ω_s is found [using Eq. (7.56)] to be the smaller of

$$\frac{(1000)(2000) - (450)^2}{450(2000 - 1000)} = 3.99 \quad \text{and} \quad \frac{(4000)^2 - (1000)(2000)}{4000(2000 - 1000)} = 3.5$$

which is 3.5, as depicted in Fig. 7.31b.

Step 1.2: Determine n
For a prototype lowpass filter in Fig. 7.29b, $\hat{G}_p = -2.4$ dB, $\hat{G}_s = -20$ dB, $\omega_p = 1$, and $\omega_s = 3.5$. Hence, according to Eq. (7.39), the Butterworth filter order n required to meet these specifications is

$$n = \frac{1}{2 \log 3.5} \log\left[\frac{10^2 - 1}{10^{0.24} - 1}\right] = 1.955$$

which is rounded up to $n = 2$.

Step 1.3: Determine ω_c

In this step (which is not necessary for the Chebyshev design), we determine the 3 dB cutoff frequency ω_c for the prototype filter. Use of Eq. (7.41) yields

$$\omega_c = \frac{3.5}{(10^2 - 1)^{1/4}} = 1.10958$$

Step 1.4: Determine the normalized transfer function $\mathcal{H}(s)$

The normalized second-order lowpass Butterworth transfer function from Table 7.1 is

$$\mathcal{H}(s) = \frac{1}{s^2 + \sqrt{2}s + 1}$$

This is the transfer function of a normalized filter (meaning that $\omega_c = 1$).

Step 1.5: Determine the prototype filter transfer function $\mathcal{H}_p(s)$

The prototype filter transfer function $\mathcal{H}_p(s)$ is obtained by substituting s with $s/\omega_c = s/1.10958$ in the normalized transfer function $\mathcal{H}(s)$ found in step 1.4 as

$$\mathcal{H}_p(s) = \frac{(1.10958)^2}{s^2 + \sqrt{2}(1.10958)s + (1.10958)^2} = \frac{1.231}{s^2 + 1.5692s + 1.2312} \tag{7.59}$$

The amplitude response of this prototype filter is illustrated in Fig. 7.31b.

Step 2: Find the desired bandpass filter transfer function $H(s)$ using the lowpass to bandpass transformation.

Finally the desired bandpass filter transfer function $H(s)$ is obtained from $\mathcal{H}_p(s)$ by replacing s with $T(s)$, where [see Eq. (7.57)]

$$T(s) = \frac{s^2 + 2(10)^6}{1000s}$$

Replacing s with $T(s)$ in the right-hand side of Eq. (7.59) yields the final bandpass transfer function

$$H(s) = \frac{1.2312(10)^6 \, s^2}{s^4 + 1569 \, s^3 + 5.2312(10)^6 \, s^2 + 3.1384(10)^9 s + 4(10)^{12}}$$

The amplitude response $|H(j\omega)|$ of this filter is shown in Fig. 7.31a. ◼

⊙ **Computer Example C7.12**

Design a bandpass filter for the specifications in Example 7.10 using functions from *Signal Processing Toolbox* in MATLAB. We shall give here MATLAB functions for the four types of filters.

For bandpass filters, we use the same functions as those used for lowpass filter in Examples C7.6, C7.8-C7.10, with one difference: Wp and Ws are 2 element vectors as Wp=[Wp1 Wp2], Ws=[Ws1 Ws2].

```
Wp=[1000 2000];Ws=[450 4000];Gp=-2.4;Gs=-20;
% Butterworth
[n,Wn]=buttord(Wp,Ws,-Gp,-Gs,'s')
[num,den]=butter(n,Wn,'s')
% Chebyshev
[n,Wn]=cheb1ord(Wp,Ws,-Gp,-Gs,'s');
[num,den]=cheby1(N,-Gp,Wn,'s')
```

```
% Inverse Chebyshev
[n,Ws]=cheb2ord(Wp,Ws,-Gp,-Gs,'s');
[num,den]=cheby2(n,-Gs,Ws,'s')
% Elliptic filter
[n,Wn]=ellipord(Wp,Ws,-Gp,-Gs,'s');
[num,den]=ellip(n,-Gp,-Gs,Wn,'s')   ⊙
```

To plot amplitude response, we can use the last three functions from Example C7.5.

7.7-3 Bandstop Filters

Figure 7.32a shows an amplitude response of a typical bandstop filter. To design such a filter, we first find $\mathcal{H}_p(s)$, the transfer function of a prototype lowpass filter, to meet the specifications in Fig. 7.32b, where ω_s is given by the smaller of

$$\frac{(\omega_{p_2} - \omega_{p_1})\omega_{s_1}}{\omega_{p_1}\omega_{p_2} - \omega_{s_1}^2} \quad \text{or} \quad \frac{(\omega_{p_2} - \omega_{p_1})\omega_{s_2}}{\omega_{s_2}^2 - \omega_{p_1}\omega_{p_2}} \tag{7.60}$$

The desired transfer function of the bandstop filter to satisfy the specifications in Fig. 7.32a is obtained from $\mathcal{H}_p(s)$ by replacing s with $T(s)$, where

$$T(s) = \frac{(\omega_{p_2} - \omega_{p_1})s}{s^2 + \omega_{p_1}\omega_{p_2}} \tag{7.61}$$

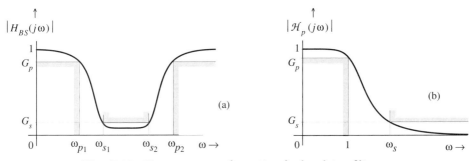

Fig. 7.32 Frequency transformation for bandstop filters.

■ **Example 7.11**
Design a Butterworth bandstop filter with the specifications depicted in Fig. 7.33a with $\omega_{p_1} = 60$, $\omega_{p_2} = 260$, $\omega_{s_1} = 100$, $\omega_{s_2} = 150$, $G_p = 0.776\,(-2.2\,\text{dB})$, and $G_s = 0.1\,(-20\,\text{dB})$.

In the first step we shall determine the prototype lowpass filter transfer function $\mathcal{H}_p(s)$, and in the second step we use the lowpass to bandstop transformation in Eq. (7.61) to obtain the desired bandstop filter transfer function $H(s)$.

Step 1: Find $\mathcal{H}_p(s)$, the lowpass prototype filter transfer function.
This goal is accomplished in 5 substeps used in the design of the lowpass Butterworth filter (see Example 7.6):

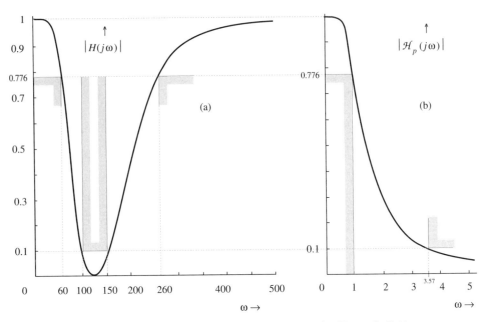

Fig. 7.33 Butterworth Bandstop Filter Design for Example 7.11.

Step 1.1: Find ω_s for the prototype filter.

For the prototype lowpass filter transfer function $\mathcal{H}_p(s)$ with the specifications illustrated in Fig. 7.32b, the frequency ω_s is found [using Eq. (7.60)] to be the smaller of

$$\frac{(100)(260 - 60)}{(260)(60) - 100^2} = 3.57 \quad \text{and} \quad \frac{150(260 - 60)}{150^2 - (260)(60)} = 4.347$$

which is 3.57, as shown in Fig. 7.33b.

Step 1.2: Determine n

For the prototype lowpass filter in Fig. 7.33b, $\hat{G}_p = -2.2$ dB, $\hat{G}_s = -20$ dB, $\omega_p = 1$, and $\omega_s = 3.57$. According to Eq. (7.39), the Butterworth filter order n required to meet these specifications is

$$n = \frac{1}{2 \log(3.57)} \left[\frac{10^2 - 1}{10^{0.22} - 1} \right] = 1.9689$$

We round up the value of n to 2.

Step 1.3: Determine ω_c

The half power frequency ω_c for the prototype Butterworth filter, using Eq. (7.40) with $\omega_p = 1$, is

$$\omega_c = \frac{\omega_p}{(10^{-\hat{G}_p/10} - 1)^{\frac{1}{4}}} = \frac{1}{(10^{0.22} - 1)^{\frac{1}{4}}} = 1.1096$$

Step 1.4: Determine the normalized transfer function

The transfer function of the second-order normalized Butterworth filter from the Table 7.1 is

$$\mathcal{H}(s) = \frac{1}{s^2 + \sqrt{2}s + 1} \tag{7.62}$$

Step 1.5: Determine the prototype filter transfer function $\mathcal{H}_p(s)$

The prototype filter transfer function $\mathcal{H}_p(s)$ is obtained by substituting s with $s/\omega_c = s/1.1096$ in the normalized transfer function $\mathcal{H}(s)$ in step 1.4. This move yields

$$\mathcal{H}_p(s) = \frac{1}{(\frac{s}{\omega_c})^2 + \sqrt{2}\frac{s}{\omega_c} + 1} = \frac{1.2312}{s^2 + 1.5692s + 1.2312} \tag{7.63}$$

The amplitude response of this prototype filter is depicted in Fig. 7.33b.

Step 2: Find the desired bandstop filter transfer function $H(s)$ using the lowpass to bandstop transformation

Finally, the desired transfer function $H(s)$ of the bandpass filter with specifications illustrated in Fig. 7.33a is obtained from $\mathcal{H}_p(s)$ by replacing s with $T(s)$, where [see Eq. (7.61)]

$$T(s) = \frac{200s}{s^2 + 15,600}$$

Replacing s with $T(s)$ in the right-hand side of Eq. (7.63) yields the final bandstop transfer function

$$H(s) = \frac{1.2312}{(\frac{200s}{s^2+15,600})^2 + 1.5692(\frac{200s}{s^2+15,600}) + 1.2312}$$

$$= \frac{(s^2 + 15600)^2}{s^4 + 254.9s^3 + 63690.9s^2 + (3.977)10^6 s + (2.433)10^8}$$

The amplitude response $|H(j\omega)|$ is shown in Fig. 7.33a.   ∎

⊙ **Computer Example C7.13**

Design the bandstop filter for the specifications in Example 7.11 using functions from *Signal Processing Toolbox* in MATLAB. We shall give here MATLAB functions for all the four types of filters.

```
Wp=[60 260]; Ws=[100 150]; Gp=-2.2;Gs=-20;
% Butterworth
[n,Wn]=buttord(Wp,Ws,-Gp,-Gs,'s')
[num,den]=butter(n,Wn,'stop','s')
% Chebyshev
[n,Wn]=cheb1ord(Wp,Ws,-Gp,-Gs,'s')
[num,den]=cheby1(n,-Gp,Wn,'stop','s')
% Inverse Chebyshev
[n,Wn]=cheb2ord(Wp,Ws,-Gp,-Gs,'s')
[num,den]=cheby2(n,-Gs,Wn,'stop','s')
% Elliptic
[n,Wn]=ellipord(Wp,Ws,-Gp,-Gs,'s')
[num,den]=ellip(n,-Gp,-Gs,Wn,'stop','s')    ⊙
```

7.8   Filters to Satisfy Distortionless transmission Conditions

The purpose of a filter is to suppress unwanted frequency components and to transmit the desired frequency components without distortion. In Sec. 4.4, we saw that this requires the filter amplitude response to be constant and the phase response to be a linear function of ω over the passband.

The filters discussed so far have stressed the constancy of the amplitude response. The linearity of the phase response has been ignored. As we saw earlier, the human ear is sensitive to amplitude distortion but somewhat insensitive to phase distortion. For this reason filters in audio application are designed primarily for constant amplitude response, and the phase response is only a secondary consideration.

We also saw earlier that the human eye is sensitive to phase distortion and relatively insensitive to amplitude distortion. Therefore, in video applications we cannot ignore phase distortion. In pulse communication, both the amplitude and the phase distortion are important for correct information transmission. Thus, in practice, we also need to design filters primarily for phase linearity in video applications. In pulse communication applications, it is important to have filters with constant amplitude response and a linear phase response. We shall briefly discuss some aspects and approaches to the design of such filters. More discussion appears in the literature.[2]

We showed [see Eq. (4.59)] that the time delay t_d resulting from the signal transmission through a filter is the negative of the slope of the filter phase response $\angle H(j\omega)$; that is,

$$t_d(\omega) = -\frac{d}{d\omega}\angle H(j\omega) \tag{7.64}$$

If the slope of $\angle H(j\omega)$ is constant over the desired band (that is, if $\angle H(j\omega)$ is linear with ω), all the components are delayed by the same time interval t_d. In this case the output is a replica of the input, assuming that all components are attenuated equally; that is, $|H(j\omega)| = $ constant over the passband.

If the slope of the phase response is not constant, t_d, the time delay, varies with frequency. This variation means that different frequency components undergo different amounts of time delay, and consequently the output waveform will not be a replica of the input waveform even if the amplitude response is constant over the passband. A good way of judging phase linearity is to plot t_d as a function of frequency. For a distortionless system, t_d (the negative slope of $\angle H(j\omega)$) should be constant over the band of interest. This is in addition to the requirement of constancy of the amplitude response.

Generally speaking, the two requirements of distortionless transmission conflict. The more we approach the ideal amplitude response, the further we deviate from the ideal phase response. The sharper the cutoff characteristic (smaller the transition band), the more nonlinear is the phase response near the transition band. We can verify this fact from Fig. 7.34, which shows the delay characteristic of the Butterworth and the Chebyshev family of filters. The Chebyshev filter, which has a sharper cutoff than that of the Butterworth, shows considerably more variation in time delay of various frequency components as compared to that of the Butterworth.

For the applications where the phase linearity is also important, there are two possible approaches:

1 If $t_d = $ constant (phase linearity) is the primary requirement, we design a filter for which t_d is maximally flat around $\omega = 0$ and accept the resulting amplitude response, which may not be so flat nor have a sharp cutoff characteristic.

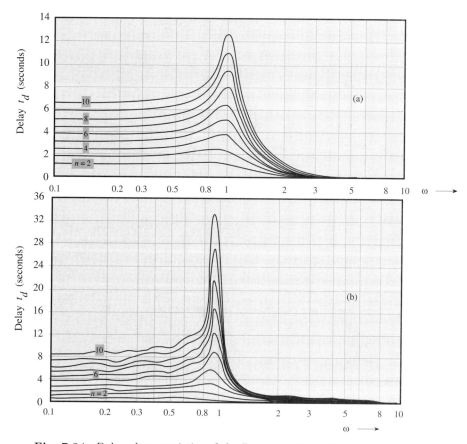

Fig. 7.34 Delay characteristics of the Butterworth and Chebyshev filters.

Contrast this with the Butterworth filter, which is designed to yield the maximally flat amplitude response at $\omega = 0$ without any attention to the phase response. A family of filters which yields a maximally flat t_d goes under the name **Bessel-Thomson** filters, which uses the nth-order Bessel polynomial in the denominator of nth-order $H(s)$.

2 If both amplitude and phase response are important, we start with a filter to satisfy the amplitude response specifications, disregarding the phase response specifications. We cascade this filter with another filter, an **equalizer**, whose amplitude response is flat for all frequencies (**the allpass filter**) and whose t_d characteristic is complementary to that of the main filter in such a way that their composite phase characteristic is approximately linear. The cascade thus has linear phase and the amplitude response of the main filter (as required).

Allpass Filters

An allpass filter has equal number of poles and zeros. All the poles are in the LHP (left half plane) for stability. All the zeros are mirror images of the poles about the imaginary axis. In other words, for every pole at $-a + jb$, there is a zero at $a + jb$. Thus, all the zeros are in the RHP. Any filter with this kind of pole-zero configuration is an allpass filter; that is, its amplitude response is constant for all frequencies. We can verify this assertion by considering a transfer function with a pole at $-a + jb$ and a zero at $a + jb$:

$$H(s) = \frac{s - a - jb}{s + a - jb} \quad \text{and} \quad H(j\omega) = \frac{j\omega - a - jb}{j\omega + a - jb} = \frac{-a + j(\omega - b)}{a + j(\omega - b)}$$

Therefore

$$|H(j\omega)| = \frac{\sqrt{(-a)^2 + (\omega - b)^2}}{\sqrt{a^2 + (\omega - b)^2}} = 1 \tag{7.65}$$

$$\angle H(j\omega) = \tan^{-1}\left[\frac{\omega - b}{-a}\right] - \tan^{-1}\left[\frac{\omega - b}{a}\right]$$

$$= \pi - \tan^{-1}\left[\frac{\omega - b}{a}\right] - \tan^{-1}\left[\frac{\omega - b}{a}\right] = \pi - 2\tan^{-1}\left[\frac{\omega - b}{a}\right] \tag{7.66}$$

Observe that although the amplitude response is unity regardless of pole-zero locations, the phase response depends on the locations of poles (or zeros). By placing poles in proper locations, we can obtain a desirable phase response that is complementary to the phase response of the main filter.

7.9   Summary

The response of an LTIC system with transfer function $H(s)$ to an everlasting sinusoid of frequency ω is also an everlasting sinusoid of the same frequency. The output amplitude is $|H(j\omega)|$ times the input amplitude, and the output sinusoid is shifted in phase with respect to the input sinusoid by $\angle H(j\omega)$ radians. The plot of $|H(j\omega)|$ vs ω indicates the amplitude gain of sinusoids of various frequencies and is called the *amplitude response* of the system. The plot of $\angle H(j\omega)$ vs ω indicates the phase shift of sinusoids of various frequencies and is called the phase response.

Plotting of the frequency response is remarkably simplified by using logarithmic units for amplitude as well as frequency. Such plots are known as the Bode plots. The use of logarithmic units makes it possible to add (rather than multiply) the amplitude response of four basic types of factors that occur in transfer functions: (1) a constant (2) a pole or a zero at the origin (3) a first order pole or a zero, and (4) complex conjugate poles or zeros. For phase plots, we use linear units for phase and logarithmic units for the frequency. The phases corresponding to the three basic types of factors mentioned above add. The asymptotic properties of the amplitude and phase responses allow their plotting with remarkable ease even for transfer functions of high orders.

The frequency response of a system is determined by the locations in the complex plane of poles and zeros of its transfer function. We can design frequency selective filters by proper placement of its transfer function poles and zeros. Placing a pole (a zero) near a frequency $j\omega_0$ in the complex plane enhances (suppresses) the frequency response at the frequency $\omega = \omega_0$. With this concept, a proper combination of poles and zeros at suitable locations can yield desired filter characteristics.

Two families of analog filters are considered: Butterworth and Chebyshev. The Butterworth filter has a maximally flat amplitude response over the passband.

The Chebyshev amplitude response has ripples in the passband. On the other hand, the behavior of the Chebyshev filter in the stopband is superior to that of the Butterworth filter. The design procedure for lowpass filters can be readily applied to highpass, bandpass, and bandstop filters by using appropriate frequency transformations discussed in Sec. 7.7.

Allpass filters have a constant gain but a variable phase with respect to frequency. Therefore, placing an allpass filter in cascade with a system leaves its amplitude response unchanged, but alters its phase response. Thus, an allpass filter can be used to modify the phase response of a system.

References

1. Wai-Kai Chen, *Passive and active Filters*, Wiley, New York, 1986.

2. Van Valkenberg, M.E., *Analog Filter Design*, Holt, Rinehart and Winston, New York, 1982.

3. Christian E., and E. Eisenmann, *Filter Design Tables and Graphs*, Transmission Networks International, Inc., Knightdale, N.C., 1977.

Problems

7.1-1 For an LTIC system described by the transfer function

$$H(s) = \frac{s+2}{s^2 + 5s + 4}$$

find the response to the following everlasting sinusoidal inputs: **(a)** $5\cos(2t + 30°)$ **(b)** $10\sin(2t + 45°)$ **(c)** $10\cos(3t + 40°)$. Observe that these are everlasting sinusoids.

7.1-2 For an LTIC system described by the transfer function

$$H(s) = \frac{s+3}{(s+2)^2}$$

find the steady-state system response to the following inputs:
(a) $10u(t)$ **(b)** $\cos(2t + 60°)u(t)$ **(c)** $\sin(3t - 45°)u(t)$ **(d)** $e^{j3t}u(t)$

7.1-3 For an allpass filter specified by the transfer function

$$H(s) = \frac{-(s - 10)}{s + 10}$$

find the system response to the following (everlasting) inputs: **(a)** $e^{j\omega t}$ **(b)** $\cos(\omega t + \theta)$ **(c)** $\cos t$ **(d)** $\sin 2t$ **(e)** $\cos 10t$ **(f)** $\cos 100t$.
Comment on the filter response.

7.2-1 Sketch Bode plots for the following transfer functions:

(a) $\dfrac{s(s + 100)}{(s + 2)(s + 20)}$ **(b)** $\dfrac{(s + 10)(s + 20)}{s^2(s + 100)}$ **(c)** $\dfrac{(s + 10)(s + 200)}{(s + 20)^2(s + 1000)}$

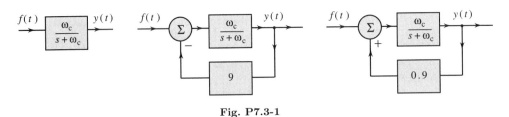

Fig. P7.3-1

7.2-2 Repeat Prob. 7.2-1 if

(a) $\dfrac{s^2}{(s+1)(s^2+4s+16)}$ (b) $\dfrac{s}{(s+1)(s^2+14.14s+100)}$ (c) $\dfrac{(s+10)}{s(s^2+14.14s+100)}$

7.3-1 Feedback can be used to increase (or decrease) the system bandwidth. Consider a system in Fig. P7.3-1a with transfer function $G(s) = \frac{\omega_c}{s+\omega_c}$.
(a) Show that the 3 dB bandwidth of this system is ω_c.
(b) To increase the bandwidth of this system, we use negative feedback with $H(s) = 9$, as depicted in Fig. P7.3-1b. Show that the 3 dB bandwidth of this system is $10\omega_c$.
(c) To decrease the bandwidth of this system, we use positive feedback with $H(s) = -0.9$, as illustrated in Fig. P7.3-1c. Show that the 3 dB bandwidth of this system is $\omega_c/10$.
(d) The system gain at dc times its 3 dB bandwidth is called the **gain-bandwidth product** of a system. Show that this product is the same for all the three systems in Fig. P7.3-1. This result shows that if we increase the bandwidth, the gain decreases and vice versa.

7.4-1 Using the graphical method of Sec 7.4-1, draw a rough sketch of the amplitude and phase response of an LTIC system described by the transfer function

$$H(s) = \frac{s^2 - 2s + 50}{s^2 + 2s + 50} = \frac{(s-1-j7)(s-1+j7)}{(s+1-j7)(s+1+j7)}$$

What kind of filter is this?

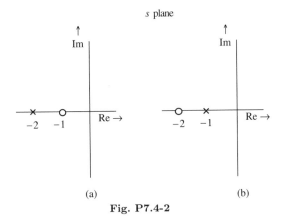

(a)   (b)

Fig. P7.4-2

7.4-2 Using the graphical method of Sec. 7.4-1, draw a rough sketch of the amplitude and phase response of LTIC systems whose pole-zero plots are shown in Fig. P7.4-2.

7.4-3 Design a second-order bandpass filter with center frequency $\omega = 10$. The gain should be zero at $\omega = 0$ and at $\omega = \infty$. Select poles at $-a \pm j10$. Leave your answer in terms of a. Explain the influence of a on the frequency response.

7.5-1 Determine the transfer function $H(s)$ and the amplitude response $H(j\omega)$ for a third-order lowpass Butterworth filter if the 3 dB cutoff frequency $\omega_c = 100$. Find your answer without using Tables 7.1 or 7.2. Verify your answer using either of these Tables.

7.5-2 Determine n, the order of a lowpass Butterworth filter, and the corresponding cutoff frequency ω_c required to satisfy the following lowpass filter specifications. Find both the values of ω_c, the one that oversatisfies the passband specifications, and the one that oversatisfies the stopband specifications.

(a) $\hat{G}_p \geq -0.5$ dB, $\hat{G}_s \leq -20$ dB, $\omega_p = 100$ rad/s, and $\omega_s = 200$ rad/s.
(b) $G_p \geq 0.9885$, $G_s \leq 10^{-3}$, $\omega_p = 1000$ rad/s, and $\omega_s = 2000$ rad/s.
(c) The gain at $3\omega_c$ is required to be no greater than -50 dB.

7.5-3 Find the transfer function $H(s)$ and the amplitude response $H(j\omega)$ for a lowpass Butterworth filter to satisfy the specifications: $\hat{G}_p \geq -3$ dB, $\hat{G}_s \leq -14$ dB, $\omega_p = 100,000$ rad/s, and $\omega_s = 150,000$ rad/s. It is desirable to oversatisfy (if possible) the requirement of $\hat{G}_s$. Determine the $\hat{G}_p$ and $\hat{G}_s$ of your design.

7.6-1 Repeat Prob. 7.5-1 for a Chebyshev filter. Do not use Tables.

7.6-2 Design a lowpass Chebyshev filter to satisfy the specifications: $\hat{G}_p \geq -1$ dB, $\hat{G}_s \leq -22$ dB, $\omega_p = 100$ rad/s, and $\omega_s = 200$ rad/s.

7.6-3 Design a lowpass Chebyshev filter to satisfy the specifications: $\hat{G}_p \geq -2$ dB, $\hat{G}_s \leq -25$ dB, $\omega_p = 10$ rad/s, and $\omega_s = 15$ rad/s.

7.6-4 Design a lowpass Chebyshev filter whose 3 dB cutoff frequency is ω_c, and the gain drops to -50 dB at $3\omega_c$.

7.7-1 Find the transfer function $H(s)$ for a highpass Butterworth filter to satisfy the specifications: $\hat{G}_s \leq -20$ dB, $\hat{G}_p \geq -1$ dB, $\omega_s = 10$, and $\omega_p = 20$.

7.7-2 Find the transfer function $H(s)$ for a highpass Chebyshev filter to satisfy the specifications: $\hat{G}_s \leq -22$ dB, $\hat{G}_p \geq -1$ dB, $\omega_s = 10$, and $\omega_p = 20$

7.7-3 Find the transfer function $H(s)$ for a Butterworth bandpass filter to satisfy the specifications: $\hat{G}_s \leq -17$ dB, $\hat{G}_p \geq -3$ dB, $\omega_{p_1} = 100$ rad/s, $\omega_{p_2} = 250$ rad/s, and $\omega_{s_1} = 40$ rad/s, $\omega_{s_2} = 500$ rad/s.

7.7-4 Find the transfer function $H(s)$ for a Chebyshev bandpass filter to satisfy the specifications: $\hat{G}_s \leq -17$ dB, $\hat{r} \leq 1$ dB, $\omega_{p_1} = 100$ rad/s, $\omega_{p_2} = 250$ rad/s, and $\omega_{s_1} = 40$ rad/s, $\omega_{s_2} = 500$ rad/s.

7.7-5 Find the transfer function $H(s)$ for a Butterworth bandstop filter to satisfy the specifications: $\hat{G}_s \leq -24$ dB, $\hat{G}_p \geq -3$ dB, $\omega_{p_1} = 20$ rad/s, $\omega_{p_2} = 60$ rad/s, and $\omega_{s_1} = 30$ rad/s, $\omega_{s_2} = 38$ rad/s.

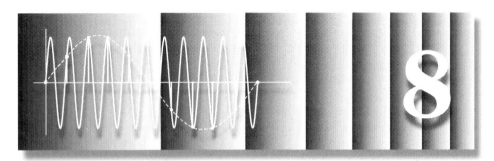

Discrete-Time
Signals and Systems

In this chapter we introduce the basic concepts of discrete-time signals and systems.

8.1 Introduction

Signals specified over a continuous range of t are **continuous-time signals**, denoted by the symbols $f(t), y(t)$, etc. Systems whose inputs and outputs are continuous-time signals are **continuous-time systems**. In contrast, signals defined only at discrete instants of time are **discrete-time signals**. Systems whose inputs and outputs are discrete-time signals are called **discrete-time systems**. A digital computer is a familiar example of this type of system. We consider here uniformly spaced discrete instants such as $\ldots, -2T, -T, 0, T, 2T, 3T, \ldots, kT, \ldots$. Discrete-time signals can therefore be specified as $f(kT)$, $y(kT)$, and so on (k, integer). We further simplify this notation to $f[k], y[k]$, etc., where it is understood that $f[k] = f(kT)$ and that k is an integer. A typical discrete-time signal, depicted in Fig. 8.1, is therefore a sequence of numbers. This signal may be denoted by $f(kT)$ and viewed as a function of time t where signal values are specified at $t = kT$. It may also be denoted by $f[k]$ and viewed as a function of k (k, integer). For instance, a continuous-time exponential $f(t) = e^{-t}$, when sampled every $T = 0.1$ second, results in a discrete-time signal $f(kT)$ given by

$$f(kT) = e^{-kT} = e^{-0.1k}$$

Clearly, this signal is a function of k and may be expressed as $f[k]$. We can plot this signal as a function of t or as a function of k (k, integer). The representation $f[k]$ is more convenient and will be followed throughout this book. A discrete-time signal therefore may be viewed as a sequence of numbers, and a discrete-time system may be seen as processing a sequence of numbers $f[k]$ and yielding as output another sequence of numbers $y[k]$.

540

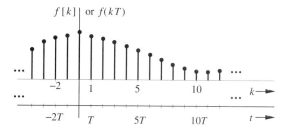

Fig. 8.1 A discrete-time signal.

Discrete-time signals arise naturally in situations which are inherently discrete-time, such as population studies, amortization problems, national income models, and radar tracking. They may also arise as a result of sampling continuous-time signals in sampled data systems, digital filtering, and so on. Digital filtering is a particularly interesting application in which continuous-time signals are processed by discrete-time systems, using appropriate interfaces at the input and output, as illustrated in Fig. 8.2. A continuous-time signal $f(t)$ is first sampled to convert it into a discrete-time signal $f[k]$, which is then processed by a discrete-time system to yield the output $y[k]$. A continuous-time signal $y(t)$ is finally constructed from $y[k]$. We shall use the notations C/D and D/C for continuous-to-discrete-time and discrete-to-continuous-time conversion. Using the interfaces in this manner, we can process a continuous-time signal with an appropriate discrete-time system. As we shall see later in our discussion, discrete-time systems have several advantages over continuous-time systems. For this reason, there is an accelerating trend toward processing continuous-time signals with discrete-time systems.

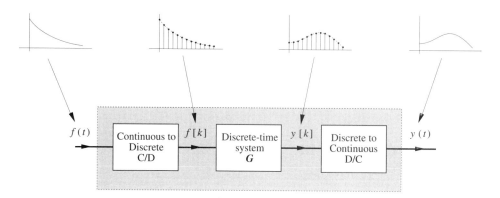

Fig. 8.2 Processing a continuous-time signal by a discrete-time system.

8.2 Some Useful Discrete-Time Signal Models

We now discuss some important discrete-time signal models which are encountered frequently in the study of discrete-time signals and systems.

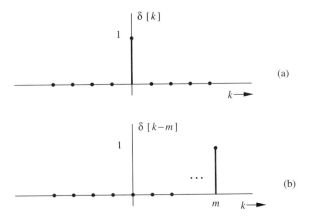

Fig. 8.3 Discrete-time impulse function.

1. Discrete-Time Impulse Function $\delta[k]$

The discrete-time counterpart of the continuous-time impulse function $\delta(t)$ is $\delta[k]$, defined by

$$\delta[k] = \begin{cases} 1 & k = 0 \\ 0 & k \neq 0 \end{cases} \tag{8.1}$$

This function, also called the unit impulse sequence, is shown in Fig. 8.3a. The time-shifted impulse sequence $\delta[k-m]$ is depicted in Fig. 8.3b. Unlike its continuous-time counterpart $\delta(t)$, this is a very simple function without any mystery.

Later, we shall express an arbitrary input $f[k]$ in terms of impulse components. The (zero-state) system response to input $f[k]$ can then be obtained as the sum of system responses to impulse components of $f[k]$.

2. Discrete-Time Unit Step Function $u[k]$

The discrete-time counterpart of the unit step function $u(t)$ is $u[k]$ (Fig. 8.4), defined by

$$u[k] = \begin{cases} 1 & \text{for } k \geq 0 \\ 0 & \text{for } k < 0 \end{cases} \tag{8.2}$$

If we want a signal to start at $k = 0$ (so that it has a zero value for all $k < 0$), we need only multiply the signal with $u[k]$.

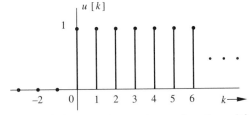

Fig. 8.4 A discrete-time unit step function $u[k]$.

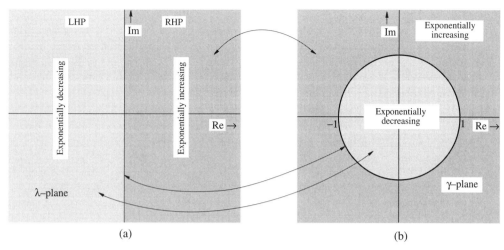

Fig. 8.5   The λ-plane, the γ-plane and their mapping.

3.   Discrete-Time Exponential γ^k

A continuous-time exponential $e^{\lambda t}$ can be expressed in an alternate form as

$$e^{\lambda t} = \gamma^t \qquad (\gamma = e^\lambda \text{ or } \lambda = \ln \gamma) \qquad (8.3a)$$

For example, $e^{-0.3t} = (0.7408)^t$ because $e^{-0.3} = 0.7408$. Conversely, $4^t = e^{1.386t}$ because $\ln 4 = 1.386$, that is, $e^{1.386} = 4$. In the study of continuous-time signals and systems we prefer the form $e^{\lambda t}$ rather than γ^t. The discrete-time exponential can also be expressed in two forms as

$$e^{\lambda k} = \gamma^k \qquad (\gamma = e^\lambda \text{ or } \lambda = \ln \gamma) \qquad (8.3b)$$

For example, $e^{3k} = (e^3)^k = (20.086)^k$. Similarly, $5^k = e^{1.609k}$ because $5 = e^{1.609}$. In the study of discrete-time signals and systems, unlike the continuous-time case, the form γ^k proves more convenient than the form $e^{\lambda k}$. Because of unfamiliarity with exponentials with bases other than e, exponentials of the form γ^k may seem inconvenient and confusing at first. The reader is urged to plot some exponentials to acquire a sense of these functions.

Nature of γ^k: The signal $e^{\lambda k}$ grows exponentially with k if Re $\lambda > 0$ (λ in RHP), and decays exponentially if Re $\lambda < 0$ (λ in LHP). It is constant or oscillates with constant amplitude if Re $\lambda = 0$ (λ on the imaginary axis). Clearly, the location of λ in the complex plane indicates whether the signal $e^{\lambda k}$ grows exponentially, decays exponentially, or oscillates with constant frequency (Fig. 8.5a). A constant signal ($\lambda = 0$) is also an oscillation with zero frequency. We now find a similar criterion for determining the nature of γ^k from the location of γ in the complex plane.

Figure 8.5a shows a complex plane (λ-plane). Consider a signal $e^{j\Omega k}$. In this case, $\lambda = j\Omega$ lies on the imaginary axis (Fig. 8.5a), and therefore is a constant-amplitude oscillating signal. This signal $e^{j\Omega k}$ can be expressed as γ^k, where $\gamma = e^{j\Omega}$.

Because the magnitude of $e^{j\Omega}$ is unity, $|\gamma| = 1$. Hence, when λ lies on the imaginary axis, the corresponding γ lies on a circle of unit radius, centered at the origin (the **unit circle** illustrated in Fig. 8.5b). Therefore, a signal γ^k oscillates with constant amplitude if γ lies on the unit circle. Remember, also, that a constant signal ($\lambda = 0$, $\gamma = 1$) is an oscillating signal with zero frequency. Thus, the imaginary axis in the λ-plane maps into the unit circle in the γ-plane.

Next consider the signal $e^{\lambda k}$, where λ lies in the left-half plane in Fig. 8.5a. This means $\lambda = a + jb$, where a is negative ($a < 0$). In this case, the signal decays exponentially. This signal can be expressed as γ^k, where

$$\gamma = e^\lambda = e^{a+jb} = e^a\, e^{jb}$$

and

$$|\gamma| = |e^a|\,|e^{jb}| = e^a \qquad \text{because } |e^{jb}| = 1$$

Also, a is negative ($a < 0$). Hence, $|\gamma| = e^a < 1$. This result means that the corresponding γ lies inside the unit circle. Therefore, a signal γ^k decays exponentially if γ lies within the unit circle (Fig. 8.5b). If, in the above case we had selected a to be positive, (λ in the right-half plane), then $|\gamma| > 1$, and γ lies outside the unit circle. Therefore, a signal γ^k grows exponentially if γ lies outside the unit circle (Fig. 8.5b).

To summarize, the imaginary axis in the λ-plane maps into the unit circle in the γ-plane. The left-half plane in the λ-plane maps into the inside of the unit circle and the right-half of the λ-plane maps into the outside of the unit circle in the γ-plane, as depicted in Fig. 8.5. This fact means that the signal γ^k grows exponentially with k if γ is outside the unit circle ($|\gamma| > 1$), and decays exponentially if γ is inside the unit circle ($|\gamma| < 1$). The signal is constant or oscillates with constant amplitude if γ is on the unit circle ($|\gamma| = 1$).

Observe that

$$\gamma^{-k} = \left(\frac{1}{\gamma}\right)^k \tag{8.4}$$

Figures 8.6a and 8.6b show plots of $(0.8)^k$, and $(-0.8)^k$, respectively. Figures 8.6c and 8.6d show plots of $(0.5)^k$, and $(1.1)^k$, respectively. These plots verify our earlier conclusions about the location of γ and the nature of signal growth. Observe that a signal $(-\gamma)^k$ alternates sign successively (is positive for even values of k and negative for odd values of k, as depicted in Fig. 8.6b). Also, the exponential $(0.5)^k$ decays faster than $(0.8)^k$. The exponential $(0.5)^k$ can also be expressed as 2^{-k} because $(0.5)^{-1} = 2$ [see Eq. (8.4)].

△ **Exercise E8.1**

Sketch signals (a) $(1)^k$ (b) $(-1)^k$ (c) $(0.5)^k$ (d) $(-0.5)^k$ (e) $(0.5)^{-k}$ (f) 2^{-k} (g) $(-2)^k$. Express these exponentials as γ^k, and plot γ in the complex plane for each case. Verify that γ^k decays exponentially with k if γ lies inside the unit circle, and that γ^k grows with k if γ is outside the unit circle. If γ is on the unit circle, γ^k is constant or oscillates with a constant amplitude.

Hint: $(1)^k = 1$ for all k. However, $(-1)^k = 1$ for even values of k and is -1 for odd values of k. Therefore, $(-1)^k$ switches back and forth from 1 to -1 (oscillates with a constant amplitude). Note also that Eq. (8.4) yields $(0.5)^{-k} = 2^k$ ▽

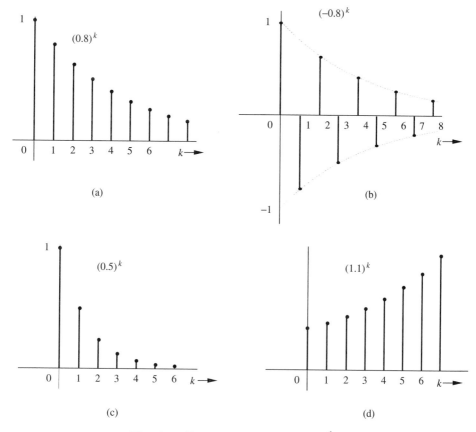

Fig. 8.6 discrete-time exponentials γ^k.

△ **Exercise E8.2**
   (a) Show that (i) $(0.25)^{-k} = 4^k$ (ii) $4^{-k} = (0.25)^k$ (iii) $e^{2t} = (7.389)^t$ (iv) $e^{-2t} = (0.1353)^t = (7.389)^{-t}$ (v) $e^{3k} = (20.086)^k$ (vi) $e^{-1.5k} = (0.2231)^k = (4.4817)^{-k}$
   (b) Show that (i) $2^k = e^{0.693k}$ (ii) $(0.5)^k = e^{-0.693k}$ (iii) $(0.8)^{-k} = e^{0.2231k}$ ▽

⊙ **Computer Example C8.1**
   Sketch the discrete-time signals (a) $(-0.5)^k$ (b) $(2)^{-k}$ (c) $(-2)^k$

   (a) k=0:5; k=k'; fk1=(-0.5).^k; stem(k,fk)
   (b) k=0:5; k=k'; fk=2.^(-k); stem(k,fk)
   (c) k=0:5; k=k';fk=(-2).^k; stem(k,fk3)   ⊙

4. Discrete-Time Exponential $e^{j\Omega k}$

   A general discrete-time exponential $e^{j\Omega k}$ (also called **phasor**) is a complex valued function of k and therefore its graphical description requires two plots (real part and imaginary part or magnitude and angle). To avoid two plots, we shall plot the values of $e^{j\Omega k}$ in the complex plane for various values of k, as illustrated in Fig. 8.7. The function $f[k] = e^{j\Omega k}$ takes on values e^{j0}, $e^{j\Omega}$, $e^{j2\Omega}$, $e^{j3\Omega}$, ... at $k = 0, 1, 2, 3, \ldots$, respectively. For the sake of simplicity we shall ignore the negative values of k for the time being. Note that

Locus of $e^{j\Omega k}$

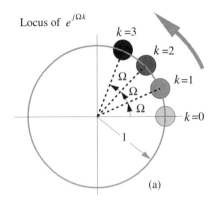

Locus of $e^{-j\Omega k}$

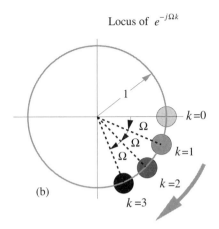

Fig. 8.7   Locus of (a) $e^{j\Omega k}$ (b) $e^{-j\Omega k}$.

$$e^{j\Omega k} = re^{j\theta}, \qquad r = 1, \quad \text{and} \quad \theta = k\Omega$$

This fact shows that the magnitude and angle of $e^{j\Omega k}$ are 1 and $k\Omega$, respectively. Therefore, the points e^{j0}, $e^{j\Omega}$, $e^{j2\Omega}$, $e^{j3\Omega}$, ..., $e^{jk\Omega}$, ... lie on a circle of unit radius (unit circle) at angles 0, Ω, 2Ω, 3Ω, ..., $k\Omega$, ... respectively, as shown in Fig. 8.7a. For each unit increase in k, the function $f[k] = e^{j\Omega k}$ moves along the unit circle counterclockwise by an angle Ω. Therefore, the locus of $e^{j\Omega k}$ may be viewed as a phasor rotating counterclockwise at a uniform speed of Ω radians per unit sample interval. The exponential $e^{-j\Omega k}$, on the other hand, takes on values $e^{j0} = 1$, $e^{-j\Omega}$, $e^{-j2\Omega k}$, $e^{-j3\Omega}$, ... at $k = 0, 1, 2, 3, \ldots$, as depicted in Fig. 8.7b. Therefore, $e^{-j\Omega k}$ may be viewed as a phasor rotating clockwise at a uniform speed of Ω radians per unit sample interval.

Using Euler's formula, we can express an exponential $e^{j\Omega k}$ in terms of sinusoids of the form $\cos{(\Omega k + \theta)}$, and vice versa

$$e^{j\Omega k} = (\cos{\Omega k} + j\sin{\Omega k}) \tag{8.5a}$$

$$e^{-j\Omega k} = (\cos{\Omega k} - j\sin{\Omega k}) \tag{8.5b}$$

These equations show that **the frequency of both** $e^{j\Omega k}$ **and** $e^{-j\Omega k}$ **is** Ω (radians/sample). Therefore, the frequency of $e^{j\Omega k}$ is $|\Omega|$. Because of Eqs. (8.5), exponentials and sinusoids have similar properties and peculiarities. The discrete-time sinusoids will be considered next.

5.   Discrete-Time Sinusoid $\cos{(\Omega k + \theta)}$

A general discrete-time sinusoid can be expressed as $C \cos{(\Omega k + \theta)}$, where C is the *amplitude*, Ω is the *frequency* (in radians per sample), and θ is the *phase* (in radians). Figure 8.8 shows a discrete-time sinusoid $\cos{(\frac{\pi}{12}k + \frac{\pi}{4})}$.

Here we make one basic observation. Because $\cos{(-x)} = \cos{(x)}$,

$$\cos{(-\Omega k + \theta)} = \cos{(\Omega k - \theta)} \tag{8.6}$$

This shows that both $\cos{(\Omega k + \theta)}$ and $\cos{(-\Omega k + \theta)}$ have the same frequency (Ω). Therefore, **the frequency of** $\cos{(\Omega k + \theta)}$ **is** $|\Omega|$.

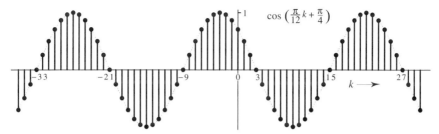

Fig. 8.8   A discrete-time sinusoid $\cos(\frac{\pi}{12}k + \frac{\pi}{4})$.

⊙ **Computer Example C8.2**
Sketch the discrete-time sinusoid $\cos\left(\frac{\pi}{12}k + \frac{\pi}{4}\right)$

```
k=-36:30; k=k';
fk=cos(k*pi/12+pi/4);
stem(k,fk)        ⊙
```

Sampled Continuous-Time Sinusoid Yields a Discrete-Time Sinusoid

A continuous-time sinusoid $\cos \omega t$ sampled every T seconds yields a discrete-time sequence whose kth element (at $t = kT$) is $\cos \omega kT$. Thus, the sampled signal $f[k]$ is given by

$$f[k] = \cos \omega kT$$
$$= \cos \Omega k \qquad \text{where } \Omega = \omega T \qquad (8.7)$$

Clearly, a continuous-time sinusoid $\cos \omega t$ sampled every T seconds yields a discrete-time sinusoid $\cos \Omega k$, where $\Omega = \omega T$. Superficially, it may appear that a discrete-time sinusoid is a continuous-time sinusoid's cousin in a striped suit. As we shall see, however, some of the properties of discrete-time sinusoids are very different from those of continuous-time sinusoids. In the continuous-time case, the period of a sinusoid can take on any value; integral, fractional, or even irrational. The discrete-time signal, in contrast, is specified only at integral values of k. Therefore, the period must be an integer (in terms of k) or an integral multiple of T (in terms of variable t).

Some Peculiarities of Discrete-Time Sinusoids

There are two unexpected properties of discrete-time sinusoids which distinguish them from their continuous-time relatives.

1. A continuous-time sinusoid is always periodic regardless of the value of its frequency ω. But a discrete-time sinusoid $\cos \Omega k$ is periodic only if Ω is 2π times some rational number ($\frac{\Omega}{2\pi}$ is a rational number).
2. A continuous-time sinusoid $\cos \omega t$ has a unique waveform for each value of ω. In contrast, a sinusoid $\cos \Omega k$ does not have a unique waveform for each value of Ω. In fact, discrete-time sinusoids with frequencies separated by multiples of 2π are identical. Thus, a sinusoid $\cos \Omega k = \cos (\Omega + 2\pi)k = \cos (\Omega + 4\pi)k = \cdots$. We now examine each of these peculiarities.

1   Not All Discrete-Time Sinusoids Are Periodic

A discrete-time signal $f[k]$ is said to be N_0-periodic if

$$f[k] = f[k + N_0] \qquad (8.8)$$

for some positive integer N_0. The smallest value of N_0 that satisfies Eq. (8.8) is the **period** of $f[k]$. Figure 8.9 shows an example of a periodic signal of period 6. Observe that each period contains 6 samples (or values). If we consider the first cycle to start at $k = 0$, the last sample (or value) in this cycle is at $k = N_0 - 1 = 5$ (not at $k = N_0 = 6$). Note also that, by definition, a periodic signal must begin at $k = -\infty$ (everlasting signal) for the reasons discussed in Sec. 1.2-4.

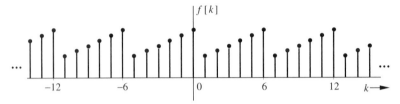

Fig. 8.9   Discrete-time periodic signal.

If a signal $\cos \Omega k$ is N_0-periodic, then

$$\cos \Omega k = \cos \Omega(k + N_0)$$
$$= \cos (\Omega k + \Omega N_0)$$

This result is possible only if ΩN_0 is an integral multiple of 2π; that is,

$$\Omega N_0 = 2\pi m \qquad m \text{ integer}$$

or

$$\frac{\Omega}{2\pi} = \frac{m}{N_0} \tag{8.9a}$$

Because both m and N_0 are integers, Eq. (8.9a) implies that the sinusoid $\cos \Omega k$ is periodic only if $\frac{\Omega}{2\pi}$ is a rational number. In this case the period N_0 is given by [Eq. (8.9a)]

$$N_0 = m \left(\frac{2\pi}{\Omega} \right) \tag{8.9b}$$

To compute N_0, we must choose the smallest value of m that will make $m\left(\frac{2\pi}{\Omega}\right)$ an integer. For example, if $\Omega = \frac{4\pi}{17}$, then the smallest value of m that will make $m\frac{2\pi}{\Omega} = m\frac{17}{2}$ an integer is 2. Therefore

$$N_0 = m\frac{2\pi}{\Omega} = 2\frac{17}{2} = 17$$

Using a similar argument, we can show that this discussion also applies to a discrete-time exponential $e^{j\Omega k}$. Thus, a discrete-time exponential $e^{j\Omega k}$ is periodic only if $\frac{\Omega}{2\pi}$ is a rational number.†

Physical Explanation of the Periodicity Relationship

Qualitatively, this result can be explained by recognizing that a discrete-time sinusoid $\cos \Omega k$ can be obtained by sampling a continuous-time sinusoid $\cos \Omega t$ at unit time interval $T = 1$; that is, $\cos \Omega t$ sampled at $t = 0, 1, 2, 3, \ldots$. This fact

---

†We can also demonstrate this point by observing that if $e^{j\Omega k}$ is N_0-periodic, then

$$e^{j\Omega k} = e^{j\Omega(k+N_0)} = e^{j\Omega k}e^{j\Omega N_0}$$

This result is possible only if $\Omega N_0 = 2\pi m$ (m, an integer). This conclusion leads to Eq. (8.9b).

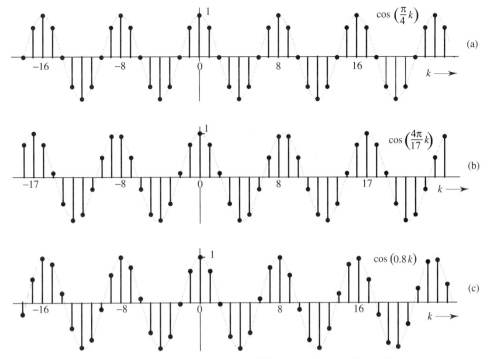

Fig. 8.10 Physical explanation of the periodicity relationship.

means $\cos \Omega t$ is the envelope of $\cos \Omega k$. Since the period of $\cos \Omega t$ is $2\pi/\Omega$, there are $2\pi/\Omega$ number of samples (elements) of $\cos \Omega k$ in one cycle of its envelope. This number may or may not be an integer.

Figure 8.10 shows three sinusoids $\cos\left(\frac{\pi}{4}k\right)$, $\cos\left(\frac{4\pi}{17}k\right)$, and $\cos\left(0.8k\right)$. Figure 8.10a shows $\cos\left(\frac{\pi}{4}k\right)$, for which there are exactly 8 samples in each cycle of its envelope $\left(\left(\frac{2\pi}{\Omega} = 8\right)\right.$. Thus, $\cos\left(\frac{\pi}{4}k\right)$ repeats every cycle of its envelope. Clearly, $\cos\left(4k/\pi\right)$ is periodic with period 8. On the other hand, Fig. 8.10b, which shows $\cos\left(\frac{4\pi}{17}k\right)$, has an average of $\frac{2\pi}{\Omega} = 8.5$ samples (not an integral number) in one cycle of its envelope. Therefore, the second cycle of the envelope will not be identical to the first cycle. But there are 17 samples (an integral number) in 2 cycles of its envelope. Hence, the pattern becomes repetitive every 2 cycles of its envelope. Therefore, $\cos\left(\frac{4\pi}{17}k\right)$ is also repetitive but its period is 17 samples (two cycles of its envelope). This observation indicates that a signal $\cos \Omega k$ is periodic only if we can fit an integral number (N_0) of samples in m integral number of cycles of its envelope so that the pattern becomes repetitive every m cycles of its envelope. Because the period of the envelope is $\frac{2\pi}{\Omega}$, we conclude that

$$N_0 = m\left(\frac{2\pi}{\Omega}\right)$$

which is precisely the condition of periodicity in Eq. (8.9b). If $\frac{\Omega}{2\pi}$ is irrational, it is impossible to fit an integral number (N_0) of samples in an integral number (m) of cycles of its envelope, and the pattern can never become repetitive. For instance, the sinusoid $\cos\left(0.8k\right)$ in Figure 8.10c has an average of 2.5π samples (an irrational number) per envelope cycle, and the pattern can never be made repetitive over any integral number (m) of cycles of its envelope; so $\cos\left(0.8k\right)$ is not periodic.

△ **Exercise E8.3**
State with reasons if the following sinusoids are periodic. If periodic, find the period.
(i) $\cos\left(\frac{3\pi}{7}k\right)$ (ii) $\cos\left(\frac{10}{7}k\right)$ (iii) $\cos\left(\sqrt{\pi}k\right)$
Ans: (i) Periodic: period $N_0 = 14$. (ii) and (iii) Aperiodic: $\Omega/2\pi$ irrational.   ▽

⊙ **Computer Example C8.3**
Sketch and verify if $\cos\left(\frac{3\pi}{7}k\right)$ is periodic.

According to Eq. (8.9b), the smallest value of m that will make $N_0 = m\left(\frac{2\pi}{\Omega}\right) = m\left(\frac{14}{3}\right)$ an integer is 3. Therefore, $N_0 = 14$. This result means $\cos\left(\frac{3\pi}{7}k\right)$ is periodic and its period is 14 samples in three cycles of its envelop. This assertion can be verified by the following MATLAB commands:

```
t=-5*pi:pi/100:5*pi; t=t';
ft=cos(3*pi*t/7);
plot(t,ft,':'), hold on
k=-15:15; k=k';
fk=cos(k*3*pi/7);
stem(k,fk), hold off       ⊙
```

2   Nonuniqueness of Discrete-Time Sinusoid Waveforms

A continuous-time sinusoid $\cos\omega t$ has a unique waveform for every value of ω in the range 0 to ∞. Increasing ω results in a sinusoid of ever increasing frequency. Such is not the case for the discrete-time sinusoid $\cos\Omega k$ because

$$\cos\left(\Omega \pm 2\pi m\right)k = \cos\left(\Omega k \pm 2\pi mk\right)$$

Now, if m is an integer, mk is also an integer, and the above equation reduces to

$$\cos\left(\Omega \pm 2\pi m\right)k = \cos\Omega k \qquad m \quad \text{integer} \qquad (8.10)$$

This result shows that a discrete-time sinusoid of frequency Ω is indistinguishable from a sinusoid of frequency Ω plus or minus an integral multiple of 2π. This statement certainly does not apply to continuous-time sinusoids.

This result means that discrete-time sinusoids of frequencies separated by integral multiples of 2π are identical. The most dramatic consequence of this fact is that a discrete-time sinusoid $\cos\left(\Omega k + \theta\right)$ has a unique waveform only for the values of Ω over a range of 2π. We may select this range to be 0 to 2π, or π to 3π, or even $-\pi$ to π. The important thing is that the range must be of width 2π. A sinusoid of any frequency outside this interval is identical to a sinusoid of frequency within this range of width 2π. We shall select this range $-\pi$ to π and call it **the fundamental range of frequencies**. Thus, a sinusoid of any frequency Ω is identical to some sinusoid of frequency Ω_f in the fundamental range $-\pi$ to π. Consider, for example, sinusoids of frequencies $\Omega = 8.7\pi$ and 9.6π. We can add or subtract any integral multiple of 2π from these frequencies and the sinusoids will still remain unchanged. To reduce these frequencies to the fundamental range $(-\pi$ to $\pi)$, we need to subtract $4 \times 2\pi = 8\pi$ from 8.7π and subtract $5 \times 2\pi = 10\pi$ from 9.6π, to yield frequencies 0.7π and -0.4π, respectively. Thus

$$\cos\left(8.7\pi k + \theta\right) = \cos\left(0.7\pi k + \theta\right)$$

$$\cos\left(9.6\pi k + \theta\right) = \cos\left(-0.4\pi k + \theta\right) \qquad (8.11)$$

This result shows that a sinusoid $\cos(\Omega k + \theta)$ *can always be expressed as* $\cos(\Omega_f k + \theta)$, *where* $-\pi \le \Omega_f < \pi$ *(the fundamental frequency range)*. The reader should get used to the fact that the range of discrete-time frequencies is only 2π. We may select this range to be from $-\pi$ to π or from 0 to 2π, or any other interval of width 2π. It is most convenient to use the range from $-\pi$ to π. At times, however, we shall find it convenient to use the range from 0 to 2π. Thus, in the discrete-time world, frequencies can be considered to lie only in the fundamental frequency range (from $-\pi$ to π, for instance). Sinusoids of frequencies outside the fundamental frequencies do exist technically. But physically, they cannot be distinguished from the sinusoids of frequencies within the fundamental range. Thus, a discrete-time sinusoid of any frequency, no matter how high, is identical to a sinusoid of some frequency within the fundamental range ($-\pi$ to π).

The above results, derived for discrete-time sinusoids, are also applicable to discrete-time exponentials of the form $e^{j\Omega k}$. For example

$$e^{j(\Omega \pm 2\pi m)k} = e^{j\Omega k} e^{\pm j 2\pi m k} = e^{j\Omega k} \qquad m, \text{ integer} \qquad (8.12)$$

Here we have used the fact that $e^{\pm j 2\pi n} = 1$ for all integral values of n. This result means that discrete-time exponentials of frequencies separated by integral multiples of 2π are identical.

Further Reduction in the Frequency Range of Distinguishable Discrete-Time Sinusoids

We shall now show that the range of frequencies that can be distinguished can be further reduced from $(-\pi, \pi)$ to $(0, \pi)$. According to Eq. (8.6), $\cos(-\Omega k + \theta) = \cos(\Omega k - \theta)$. In other words, the frequencies in the range (0 to $-\pi$) can be expressed as frequencies in the range (0 to π) with opposite phase. For example, the second sinusoid in Eq. (8.11) can be expressed as

$$\cos(9.6\pi k + \theta) = \cos(-0.4\pi k + \theta) = \cos(0.4\pi k - \theta) \qquad (8.13)$$

This result shows that a sinusoid of any frequency Ω can always be expressed as a sinusoid of a frequency $|\Omega_f|$, where $|\Omega_f|$ lies in the range 0 to π. Note, however, a possible sign change in the phases of the two sinusoids. In other words, a discrete-time sinusoid of any frequency, no matter how high, is identical in every respect to a sinusoid within the fundamental frequency range, such as $-\pi$ to π. In contrast, a discrete-time sinusoid of any frequency, no matter how high, can be expressed, with a possible sign change in phase, as a sinusoid of frequency in the range $(0, \pi)$; that is, within **half the fundamental frequency range**.

A systematic procedure to reduce the frequency of a sinusoid $\cos(\Omega k + \theta)$ is to express Ω as†

$$\Omega = \Omega_f + 2\pi m \qquad |\Omega_f| \le \pi, \quad \text{and } m \text{ an integer} \qquad (8.14)$$

This procedure is always possible. The reduced frequency of the sinusoid $\cos(\Omega k + \theta)$ is then $|\Omega_f|$.

---

†Equation (8.14) can also be expressed as $\Omega_f = \Omega|_{\text{modulo } 2\pi}$

■ **Example 8.1**

Consider sinusoids of frequencies Ω equal to (a) 0.5π (b) 1.6π (c) 2.5π (d) 5.6π (e) 34.116. Each of these sinusoids is equivalent to a sinusoid of some frequency $|\Omega_f|$ in the range 0 to π. We shall now determine these frequencies. This goal is readily accomplished by expressing the frequency Ω as in Eq. (8.14).

(a) The frequency 0.5π is in the range (0 to π) so that it cannot be reduced further.

(b) The frequency $1.6\pi = 2\pi - 0.4\pi$, and $\Omega_f = -0.4\pi$. Therefore, a sinusoid of frequency 1.6π can be expressed as a sinusoid of frequency $|\Omega_f| = 0.4\pi$.

(c) $2.5\pi = 2\pi + 0.5\pi$, and $\Omega_f = 0.5\pi$. Therefore, a sinusoid of frequency 2.5π can be expressed as a sinusoid of frequency $|\Omega_f| = 0.5\pi$.

(d) $5.6\pi = 3(2\pi) - 0.4\pi$, and $\Omega_f = -0.4\pi$. Therefore, a sinusoid of frequency 5.6π can be expressed as a sinusoid of frequency $|\Omega_f| = 0.4\pi$.

(e) $34.116 = 5(2\pi) + 2.7$, and $\Omega_f = 2.7$. Therefore, a sinusoid of frequency 34.116 can be expressed as a sinusoid of frequency $|\Omega_f| = 2.7$. ■

The fundamental range frequencies can be determined by using a simple graphical artifice as follows: mark all the frequencies on a tape using a linear scale, starting with zero frequency. Now wind this tape continuously around the two poles, one at $|\Omega_f| = 0$ and the other at $|\Omega_f| = \pi$, as illustrated in Fig. 8.11. The reduced value of any frequency marked on the tape is its projection on the horizontal ($|\Omega_f|$) axis. For instance, the reduced frequency corresponding to $\Omega = 1.6\pi$ is 0.4π (the projection of 1.6π on the horizontal Ω_f axis). Similarly, frequencies 2.5π, 5.6π, and 34.116 correspond to frequencies 0.5π, 0.4π, and 2.7 on the $|\Omega_f|$ axis.

△ **Exercise E8.4**

Show that the sinusoids of frequencies $\Omega =$ (a) 2π (b) 3π (c) 5π (d) 3.2π (e) 22.1327 (f) $\pi + 2$ can be expressed as sinusoids of frequencies (a) 0 (b) π (c) π (d) 0.8π (e) 3 (f) $\pi - 2$, respectively. ▽.

△ **Exercise E8.5**

Show that a discrete-time sinusoid of frequency $\pi + x$ can be expressed as a sinusoid with frequency $\pi - x$ ($0 \le x \le \pi$). This fact shows that a sinusoid with frequency above π by amount x has the frequency identical to a sinusoid of frequency below π by the same amount x, and the maximum rate of oscillation occurs at $\Omega = \pi$. As Ω increases beyond π, the rate of oscillation actually decreases. ▽.

⊙ **Computer Example C8.4**

In the fundamental range of frequencies from $-\pi$ to π find a sinusoid that is indistinguishable from the sinusoid $\cos\left(\frac{3\pi}{7}k\right)$. Verify by plotting these two sinusoids that they are indeed identical.

The sinusoid $\cos\left(\frac{3\pi}{7}k\right)$ is identical to the sinusoid $\cos\left(\frac{3\pi}{7} - 2\pi\right)k = \cos\left(-\frac{11\pi}{7}k\right) = \cos\left(\frac{11\pi}{7}k\right)$. We may verify that these two sinusoids are identical.

```
k=-15:15; k=k';
fk1=cos(3*pi*k/7);
fk2=cos(11*pi*k/7);
stem(k,fk1,'x'),hold on,
stem(k,fk2),hold off      ⊙
```

Physical Explanation of Nonuniqueness of Discrete-Time Sinusoids

Nonuniqueness of discrete-time sinusoids is easy to prove mathematically. But why does it happen physically? We now give here two different physical explanations of this intriguing phenomenon.

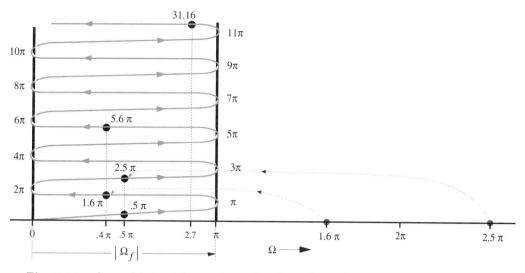

Fig. 8.11 A graphical artifice to determine the reduced frequency of a discrete-time sinusoid.

The First Explanation

Recall that sampling a continuous-time sinusoid $\cos \Omega t$ at unit time intervals ($T = 1$) generates a discrete-time sinusoid $\cos \Omega k$. Thus, by sampling at unit intervals, we generate a discrete-time sinusoid of frequency Ω (rad/sample) from a continuous-time sinusoid of frequency Ω (rad/s). Superficially, it appears that since a continuous-time sinusoid waveform is unique for each value of Ω, the resulting discrete-time sinusoid must also have a unique waveform for each Ω. Recall, however, that there is a unit time interval between samples. If a continuous-time sinusoid executes several cycles during unit time (between successive samples), it will not be visible in its samples. The sinusoid may just as well not have executed those cycles. Another low frequency continuous-time sinusoid could also give the same samples. Figure 8.12 shows how the samples of two very different continuous-time sinusoids of different frequencies generate identical discrete-time sinusoid. This illustration explains why two discrete-time sinusoids whose frequencies Ω are nominally different have the same waveform.

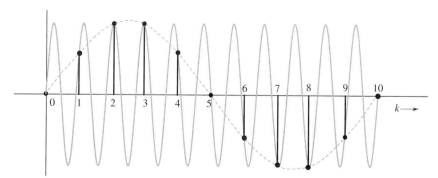

Fig. 8.12 Physical explanation of nonuniqueness of Discrete-time sinusoid waveforms.

Human Eye is a Lowpass Filter

Figure 8.12 also brings out one interesting fact; that a human eye is a lowpass filter. Both the continuous-time sinusoids in Fig. 8.12 have the same set of samples. Yet, when we see the samples, we interpret them as the samples of the lower frequency sinusoid. The eye does not see (or cannot reconstruct) the wiggles of the higher frequency sinusoid between samples because the eye is basically a lowpass filter.

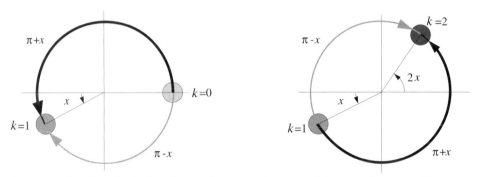

Fig. 8.13   Another physical explanation of nonuniqueness of discrete-time sinusoid waveforms.

The Second Explanation

Here we shall present a quantitative argument using a discrete-time exponential rather than a discrete-time sinusoid. As explained earlier, a discrete-time exponential $e^{j\Omega k}$ can be viewed as a phasor rotating counterclockwise at a uniform angular velocity of Ω rad/sample, as shown in Fig. 8.7a. A similar argument shows that the exponential $e^{-j\Omega k}$ is a phasor rotating clockwise at a uniform angular velocity of Ω radians per sample, as depicted in Fig. 8.7b. The angular velocity of both these rotating phasors is Ω rad. Therefore, as the frequency Ω increases, the angular velocity also increases. This, however, is true only for values of Ω in the range 0 to π. Something very interesting happens when the frequency Ω increases beyond π. Let $\Omega = \pi + x$ where $x < \pi$. Figure 8.13a shows the phasor progressing from $k = 0$ to $k = 1$, and Fig. 8.13b shows the same phasor progressing from $k = 1$ to $k = 2$. Because the phasor rotates at a speed of $\Omega = \pi + x$ radians/sample, the phasor angles at $k = 0$, 1, and 2 are 0, $\pi + x$ and $2\pi + 2x = 2x$, respectively. In both the figures, the phasor is progressing counterclockwise at a velocity of $(\pi + x)$ rad/sample. But we may also interpret this motion as the phasor moving clockwise (shown in gray) at a lower speed of $(\pi - x)$ rad/sample. Either of these interpretations describes the phasor motion correctly. If this motion could be seen by a human eye, which is a lowpass filter, it will automatically interpret the speed as $\pi - x$, the lower of the two speeds. This is the stroboscopic effect observed in movies, where at certain speeds, carriage wheels appear to move backwards.†

---

†A stroboscope is a source of light that flashes periodically on an object, thus generating a sampled image of that object. When a stroboscope flashes on a rotating object, such as a wheel, the wheel appears to rotate at a certain speed. Now increase the actual speed of rotation (while maintaining the same flashing rate). If the speed is increased beyond some critical value, the wheels appear to rotate backwards because of the lowpass filtering effect described above in the text. As we continue to increase the speed further, the backward rotation appears to slow down continuously to zero speed (where the wheels appear stationary), and reverse the direction again. This effect is often observed in movies in scenes with running carriages. A movie reel consists of a sequence of photographs shot at discrete instants, and is basically a sampled signal.

Thus, in a signal $e^{j\Omega k}$, the frequency $\Omega = \pi + x$ appears as frequency $\pi - x$. Therefore, as Ω increases beyond π, the actual frequency decreases, until at $\Omega = 2\pi$ ($x = \pi$), the actual frequency is zero ($\pi - x = 0$). As we increase Ω beyond 2π, the same cycle of events repeats. For instance, $\Omega = 2.5\pi$ is the same as $\Omega = 0.5\pi$.

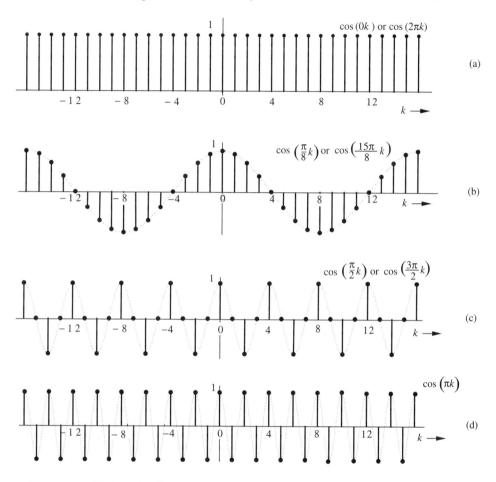

Fig. 8.14 Highest Oscillation Rate in a Discrete-Time sinusoid occurs at $\Omega = \pi$.

Highest Oscillation Rate in a Discrete-Time Sinusoid Occurs at $\Omega = \pi$

This discussion shows that the highest rate of oscillation occurs for the frequency $\Omega = \pi$. The rate of oscillation increases continuously as Ω increases from 0 to π, then decreases as Ω increases from π to 2π. Recall that a frequency $\pi + x$ appears as the frequency $\pi - x$. The frequency $\Omega = 2\pi$ ($x = \pi$) is the same as the frequency $\Omega = 0$ (constant signal). These conclusions can be verified from Fig. 8.14, which shows sinusoids of frequencies $\Omega = $ (a) 0 or 2π (b) $\frac{\pi}{8}$ or $\frac{15\pi}{8}$ (c) $\frac{\pi}{2}$ or $\frac{3\pi}{2}$ (d) π.

6. Exponentially Varying Discrete-Time Sinusoid $\gamma^k \cos(\Omega k + \theta)$

This is a sinusoid $\cos(\Omega k + \theta)$ with an exponentially varying amplitude γ^k. It is obtained by multiplying the sinusoid $\cos(\Omega k + \theta)$ by an exponential γ^k. Figure 8.15

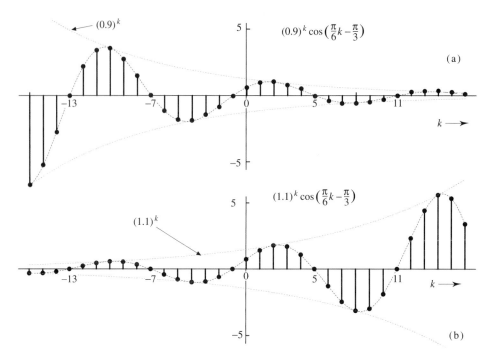

Fig. 8.15 Examples of exponentially varying discrete-time sinusoids.

shows signals $(0.9)^k \cos \left(\frac{\pi}{6} k - \frac{\pi}{3} \right)$, and $(1.1)^k \cos \left(\frac{\pi}{6} k - \frac{\pi}{3} \right)$. Observe that if $|\gamma| < 1$, the amplitude decays, and if $|\gamma| > 1$, the amplitude grows exponentially.

8.2-1   Size of a Discrete-Time Signal

Arguing along the lines similar to those used in continuous-time signals, the size of a discrete-time signal $f[k]$ will be measured by its energy E_f defined by

$$E_f = \sum_{k=-\infty}^{\infty} |f[k]|^2 \tag{8.15}$$

This definition is valid for real or complex $f[k]$. For this measure to be meaningful, the energy of a signal must be finite. A necessary condition for the energy to be finite is that the signal amplitude must $\to 0$ as $|k| \to \infty$. Otherwise the sum in Eq. (8.15) will not converge. If E_f is finite, the signal is called an **energy signal**.

In some cases, for instance, when the amplitude of $f[k]$ does not $\to 0$ as $|k| \to \infty$, then the signal energy is infinite, and a more meaningful measure of the signal in such a case would be the time average of the energy (if it exists), which is the signal power P_f defined by

$$P_f = \lim_{N \to \infty} \frac{1}{2N+1} \sum_{-N}^{N} |f[k]|^2 \tag{8.16}$$

For periodic signals, the time averaging need be performed only over one period in view of the periodic repetition of the signal. If P_f is finite and nonzero, the signal is called a **power signal**. As in the continuous-time case, a discrete-time signal can either be an energy signal or a power signal, but cannot be both at the same time. Some signals are neither energy nor power signals.

△ **Exercise E8.7**

(a) Show that the signal $a^k u[k]$ is an energy signal of energy $\frac{1}{1-|a|^2}$ if $|a| < 1$. It is a power signal of power $P_f = 0.5$ if $|a| = 1$. It is neither an energy signal nor a power signal if $|a| > 1$. ▽

8.3   Sampling Continuous-Time Sinusoid and Aliasing

On the surface, the fact that discrete-time sinusoids of frequencies differing by $2\pi m$ are identical may appear innocuous. But in reality it creates a serious problem for processing continuous-time signals by digital filters. A continuous-time sinusoid $f(t) = \cos \omega t$ sampled every T seconds ($t = kT$) results in a discrete-time sinusoid $f[k] = \cos \omega kT$. Thus, the sampled signal $f[k]$ is given by

$$f[k] = \cos \omega kT$$
$$= \cos \Omega k \qquad \text{where } \Omega = \omega T$$

Recall that the discrete-time sinusoids $\cos \Omega k$ have unique waveforms only for the values of frequencies in the range $\Omega \le \pi$ or $\omega T \le \pi$ (fundamental frequency range). We know that a sinusoid of frequency $\Omega > \pi$ appears as a sinusoid of a lower frequency $\Omega \le \pi$. For a sampled continuous-time sinusoid, this fact means that samples of a sinusoid of frequency $\omega > \pi/T$ appear as samples of a sinusoid of lower frequency $\omega \le \pi/T$. The mechanism of how the samples of continuous-time sinusoids of two (or more) different frequencies can generate the same discrete-time signal is shown in Fig. 8.12. *This phenomenon is known as* **aliasing** *because, through sampling, two entirely different analog sinusoids take on the same "discrete-time" identity.*

Aliasing causes ambiguity in digital signal processing, which makes it impossible to determine the true frequency of the sampled signal. Therefore, aliasing is highly undesirable and should be avoided. To avoid aliasing, the frequencies of the continuous-time sinusoids to be processed should be kept within the range $\omega T \le \pi$ or $\omega \le \pi/T$. Under this condition, the question of ambiguity or aliasing does not arise because any continuous-time sinusoid of frequency in this range has a unique waveform when it is sampled. Therefore, if ω_h is the highest frequency to be processed, then, to avoid aliasing,

$$\omega_h \le \frac{\pi}{T} \qquad \qquad (8.17a)$$

If $\mathcal{F}_h$ is the highest frequency in Hertz, $\mathcal{F}_h = \omega_h/2\pi$, and, according to Eq. (8.17a),

$$\mathcal{F}_h \le \frac{1}{2T} \qquad \qquad (8.17b)$$

or

$$T \le \frac{1}{2\mathcal{F}_h} \qquad \qquad (8.17c)$$

This result shows that discrete-time signal processing places the limit on the highest frequency $\mathcal{F}_h$ that can be processed for a given value of the sampling interval T

according to Eq. (8.17b). But we can process a signal of any frequency (without aliasing) by choosing a sufficiently low value of T according to Eq. (8.17c). The sampling rate or sampling frequency $\mathcal{F}_s$ is the reciprocal of the sampling interval T, and, according to Eq. (8.17c),†

$$\mathcal{F}_s = \frac{1}{T} \geq 2\mathcal{F}_h \tag{8.18}$$

This result, which is a special case of the **sampling theorem** (proved in Chapter 5), states that to process a continuous-time sinusoid by a discrete-time system, the sampling rate must not be less than twice the frequency (in Hz) of the sinusoid. In short, **a sampled sinusoid must have a minimum of two samples per cycle.** For a sampling rate below this minimum value, the output signal will be aliased, which means the signal will be mistaken for a sinusoid of lower frequency.

Equation 8.18 indicates that $\mathcal{F}_h$, the highest frequency that can be processed, is half the sampling frequency $\mathcal{F}_s$. This means the range of frequencies that can be processed without aliasing is from 0 to $\mathcal{F}_s/2$

$$0 \leq \mathcal{F} \leq \frac{\mathcal{F}_s}{2} \tag{8.19}$$

Frequencies greater than $\mathcal{F}_s/2$ (half the sampling frequency) will be aliased and appear as frequencies lower than $\mathcal{F}_s/2$. The aliasing appears as a folding back of frequencies about $\mathcal{F}_s/2$. Hence, this frequency is also known as the folding frequency. The details of this folding are explained more fully in Fig. 5.6.

The folding process is multilayered, as depicted in Fig. 8.11. The spectrum first folds back at the folding frequency, and then again folds forward at the origin, then back again at the folding frequency, and so on. We can find the aliased frequency (the reduced frequency) using an equation similar to Eq. (8.14) applicable to sampled continuous-time sinusoids.

We saw that a continuous-time sinusoid of frequency ω appears as a discrete-time sinusoid of frequency $\Omega = \omega T$. Hence, if ω_f is the reduced (aliased) frequency corresponding to a sinusoid of frequency ω, then, according to Eq. (8.14)

$$\omega T = \omega_f T + 2\pi m \qquad |\omega_f|T \leq \pi, \quad \text{and } m \text{ an integer} \tag{8.20}$$

When we express the radian frequencies in Hertz ($\omega = 2\pi\mathcal{F}$, etc.), and use the fact that the sampling frequency $\mathcal{F}_s = \frac{1}{T}$, Eq. (8.20) becomes

$$\mathcal{F} = \mathcal{F}_f + m\mathcal{F}_s \qquad |\mathcal{F}_f| \leq \frac{\mathcal{F}_s}{2}, \quad \text{and } m \text{ an integer} \tag{8.21}$$

Thus, if a continuous-time sinusoid of frequency $\mathcal{F}$ Hz is sampled at a rate of $\mathcal{F}_s$ Hz (samples/second), the resulting samples would appear as if they had come from a continuous-time sinusoid of a lower (aliased) frequency $|\mathcal{F}_f|$. For instance, if a continuous-time sinusoid of frequency 10 kHz were sampled at a rate of 3 kHz (3000 samples/second), the resulting samples will appear as if they had come from a continuous-time sinusoid of frequency 1 kHz because $10,000 = 1,000 + 3(3000)$. Note, however, if the frequency of a sinusoid is less than the folding frequency $\mathcal{F}_s/2$ (half the sampling frequency), there is no aliasing. Thus, the condition for the absence of aliasing is that the frequency of a sinusoid must be less than half the sampling frequency (the folding frequency).

---

†In some special cases, where the signal spectrum contains an impulse at $\mathcal{F}_h$, the sampling rate $\mathcal{F}_s$ must be greater than $2\mathcal{F}_h$ (see footnote on p.321)

■ **Example 8.2**

Determine the maximum sampling interval T that can be used in a discrete-time oscillator which generates a sinusoid of 50 kHz.

Here the highest frequency $\mathcal{F}_h = 50$ kHz. Therefore, according to Eq. (8.17c)

$$T \le \frac{1}{2\mathcal{F}_h} = 10\,\mu s$$

The sampling interval must not be greater than $10\,\mu s$. The minimum sampling frequency is $\mathcal{F}_s = \frac{1}{T} = 100$ kHz. If we use $T = 10\,\mu s$, the oscillator output will exhibit two samples per cycle. If we require the oscillator output to have 20 samples per cycle, then we must use $T = 1\,\mu s$ (sampling frequency $\mathcal{F}_s = 1$ MHz).   ■

■ **Example 8.3**

A discrete-time amplifier uses a sampling interval $T = 25\,\mu s$. What is the highest frequency of a signal that can be processed with this amplifier without aliasing?

According to Eq. (8.17b)

$$\mathcal{F}_h = \frac{1}{2T} = 20 \text{ kHz}   ■$$

■ **Example 8.4**

A sampler with sampling interval $T = 0.001$ second (1 ms.) samples continuous-time sinusoids of the following frequencies: (a) 400 Hz (b) 1 kHz (c) 1.4 kHz (d) 1.6kHz (e) 3.522 kHz. Determine the aliased frequencies of the resulting sampled signals.

The sampling frequency is $\mathcal{F}_s = 1/T = 1,000$. The folding frequency $\mathcal{F}_s/2 = 500$. Hence, sinusoids below 500 Hz will not be aliased and sinusoids of frequency above 500 Hz will be aliased. Using Eq. (8.21), we find:

(a)   400 Hz is less than 500 Hz (the folding frequency, which is half the sampling frequency $\mathcal{F}_s$). Hence, there is no aliasing.

(b) $1000 = 0 + 1000$ so that $\mathcal{F}_f = 0$ and the aliased frequency ($|\mathcal{F}_f|$) is zero. The sampled signal appears as samples of a dc signal.

(c)$1400 = 400 + 1000$ so that $\mathcal{F}_f = 400$ and the aliased frequency ($|\mathcal{F}_f|$) is 400 Hz. The sampled signal appears as samples of a signal of frequency 400 Hz.

(d) $1600 = -400 + 2(1000)$ so that $\mathcal{F}_f = -400$ and the aliased frequency ($|\mathcal{F}_f|$) is 400 Hz. The sampled signal appears as samples of a signal of frequency 400 Hz.

(e) $3522 = -478 + 4(1000)$ so that $\mathcal{F}_f = -478$ and the aliased frequency ($|\mathcal{F}_f|$) is 478 Hz. The sampled signal appears as samples of a signal of frequency 478 Hz.

Graphically, we can solve this problem using the artifice in Fig. 8.11. The folding frequency is 500 Hz instead of π. In case (a), the frequency 400 Hz is below the folding frequency 500 Hz. Hence, the samples of this sinusoid will not be aliased. For case (b), the frequency 1000 Hz, when folded back at 500 Hz terminates at the origin $\mathcal{F} = 0$. Hence, the aliased frequency is 0. For case (c), the frequency 1400 Hz folds back at 500 Hz, then folds forward at 0, and terminates at 400 Hz. Similarly, for case (d), the frequency 1600 Hz folds back at 500, then folds forward at 0, and folds back again at 500 Hz to terminate at 400 Hz, and so on.   ■

8.4   Useful Signal Operations

Signal operations discussed for continuous-time systems also apply to discrete-time systems with some modification in time scaling. Since the independent variable in our signal description is time, the operations are called time shifting, time inversion (or time reversal), and time scaling. However, this discussion is valid for functions having independent variables other than time (e.g., frequency or distance).

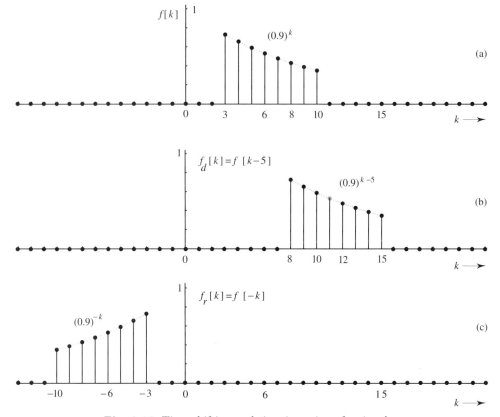

Fig. 8.16 Time-shifting and time inversion of a signal.

8.4-1 Time Shifting

Following the argument used for continuous-time signals, we can show that to time shift a signal $f[k]$ by m units, we replace k with $k - m$. Thus, $f[k - m]$ represents $f[k]$ time shifted by m units. If m is positive, the shift is to the right (delay). If m is negative, the shift is to the left (advance). Thus, $f[k - 2]$ is $f[k]$ delayed (right-shifted) by 2 units, and $f[k + 2]$ is $f[k]$ advanced (left-shifted) by 2 units. The signal $f_d[k]$ in Fig. 8.16b, being the signal in Fig. 8.16a delayed by 5 units, is the same as $f[k]$ with k replaced by $k - 5$. Now, $f[k] = (0.9)^k$ for $3 \le k \le 10$. Therefore, $f_d[k] = (0.9)^{k-5}$ for $3 \le k - 5 \le 10$ or $8 \le k \le 15$, as illustrated in Fig. 8.16b.

8.4-2 Time Inversion (or Reversal)

Following the argument used for continuous-time signals, we can show that to time invert a signal $f[k]$, we replace k with $-k$. This operation rotates the signal about the vertical axis. Figure 8.16c shows $f_r[k]$, which is the time-inverted signal $f[k]$ in Fig. 8.16a. The expression for $f_r[k]$ is the same as that for $f[k]$ with k replaced by $-k$. Because $f[k] = (0.9)^k$ for $3 \le k \le 10$, $f_r[k] = (0.9)^{-k}$ for $3 \le -k \le 10$; that is, $-3 \ge k \ge -10$, as shown in Fig. 8.16c.

8.4-3 Time Scaling

Following the argument used for continuous-time signals, we can show that to time scale a signal $f[k]$ by a factor a, we replace k with ak. However, because the discrete-time argument k can take only integral values, certain restrictions and changes in the procedure are necessary.

Time Compression: Decimation or Downsampling

Consider a signal

$$f_c[k] = f[2k] \tag{8.22}$$

The signal $f_c[k]$ is the signal $f[k]$ compressed by a factor 2. Observe that $f_c[0] = f[0]$, $f_c[1] = f[2]$, $f_c[2] = f[4]$, and so on. This fact shows that $f_c[k]$ is made up of even numbered samples of $f[k]$. The odd numbered samples of $f[k]$ are missing (Fig. 8.17b).† This operation loses part of the data, and that is why such time compression is called **decimation** or **downsampling**. In the continuous-time case, time compression merely speeds up the signal without loss of data. In general, $f[mk]$ (m integer) consists of only every mth sample of $f[k]$.

Time Expansion

Consider a signal

$$f_e[k] = f\left[\frac{k}{2}\right] \tag{8.23}$$

The signal $f_e[k]$ is the signal $f[k]$ expanded by a factor 2. According to Eq. (8.23), $f_e[0] = f[0]$, $f_e[1] = f[1/2]$, $f_e[2] = f[1]$, $f_e[3] = f[3/2]$, $f_e[4] = f[2]$, $f_e[5] = f[5/2]$, $f_e[6] = f[3]$, and so on. Now, $f[k]$ is defined only for integral values of k, and is zero (or undefined) for all fractional values of k. Therefore, for $f_e[k]$, its odd numbered samples $f_e[1]$, $f_e[3]$, $f_e[5]$, ... are all zero (or undefined), as depicted in Fig. 8.17c. In general, a function $f_e[k] = f[k/m]$ (m integer) is defined for $k = 0, \pm m, \pm 2m, \pm 3m, \ldots$, and is zero (or undefined) for all remaining values of k.

Interpolation

In the time-expanded signal in fig. 8.17c, the missing odd numbered samples can be reconstructed from the nonzero valued samples using some suitable interpolation formula. Figure 8.17d shows such an interpolated function $f_i[k]$, where the missing samples are constructed using an ideal lowpass filter interpolation formula (5.10b). In practice, we may use a realizable interpolation, such as a linear interpolation, where $f_i[1]$ is taken as the mean of $f_i[0]$ and $f_i[2]$. Similarly, $f_i[3]$ is taken as the mean of $f_i[2]$ and $f_i[4]$, and so on. This process of time expansion and inserting the missing samples using an interpolation is called **interpolation** or **upsampling**. In this operation, we increase the number of samples.

$\triangle$ **Exercise E8.6**

Show that for a linearly interpolated function $f_i[k] = f[k/2]$, the odd numbered samples interpolated values are $f_i[k] = \frac{1}{2}\left\{f[\frac{k-1}{2}] + f[\frac{k+1}{2}]\right\}$. $\triangledown$

---

†Odd numbered samples of $f[k]$ can be retained (and even numbered samples omitted) by using the transform

$$f_c[k] = f[2k+1]$$

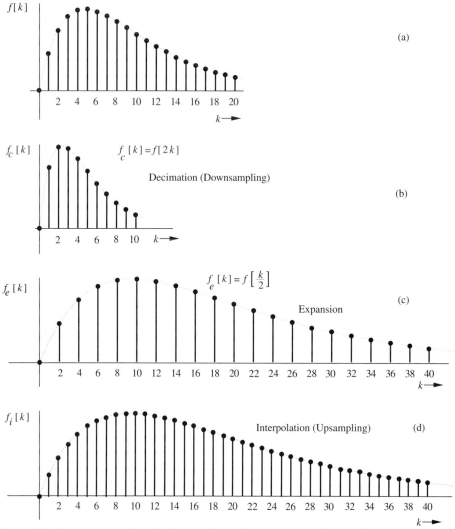

Fig. 8.17   Time compression (decimation) and time expansion (interpolation) of a signal.

8.5   Examples of Discrete-Time Systems

We shall give here three examples of discrete-time systems. In the first two examples, the signals are inherently discrete-time. In the third example, a continuous-time signal is processed by a discrete-time system, as illustrated in Fig. 8.2, by discretizing the signal through sampling.

■ **Example 8.5**
A person makes a deposit (the input) in a bank regularly at an interval of T (say, 1 month). The bank pays a certain interest on the account balance during the period T and mails out a periodic statement of the account balance (the output) to the depositor. Find the equation relating the output $y[k]$ (the balance) to the input $f[k]$ (the deposit).

In this case, the signals are inherently discrete-time. Let

$$f[k] = \text{the deposit made at the } k\text{th discrete instant}$$
$$y[k] = \text{the account balance at the } k\text{th instant computed}$$
$$\text{immediately after the } k\text{th deposit } f[k] \text{ is received}$$
$$r = \text{interest per dollar per period } T$$

The balance $y[k]$ is the sum of (i) the previous balance $y[k-1]$, (ii) the interest on $y[k-1]$ during the period T, and (iii) the deposit $f[k]$

$$
\begin{aligned}
y[k] &= y[k-1] + ry[k-1] + f[k] \\
&= (1+r)y[k-1] + f[k]
\end{aligned}
\tag{8.24}
$$

or

$$y[k] - ay[k-1] = f[k] \qquad\qquad a = 1 + r \tag{8.25a}$$

In this example the deposit $f[k]$ is the input (cause) and the balance $y[k]$ is the output (effect).

We can express Eq. (8.25a) in an alternate form. The choice of index k in Eq. (8.25a) is completely arbitrary, so we can substitute $k+1$ for k to obtain

$$y[k+1] - ay[k] = f[k+1] \tag{8.25b}$$

We also could have obtained Eq. (8.25b) directly by realizing that $y[k+1]$, the balance at the $(k+1)$st instant, is the sum of $y[k]$ plus $ry[k]$ (the interest on $y[k]$) plus the deposit (input) $f[k+1]$ at the $(k+1)$st instant.

For a hardware realization of such a system, we rewrite Eq. (8.25a) as

$$y[k] = ay[k-1] + f[k] \tag{8.25c}$$

Figure 8.18 shows the hardware realization of this equation using a single time delay of T units.† To understand this realization, assume that $y[k]$ is available. Delaying it by T, we generate $y[k-1]$. Next, we generate $y[k]$ from $f[k]$ and $y[k-1]$ according to Eq. (8.25c).

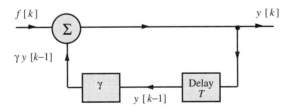

Fig. 8.18 Realization of the savings account system.

A withdrawal is a negative deposit. Therefore, this formulation can handle deposits as well as withdrawals. It also applies to a loan payment problem with the initial value $y[0] = -M$, where M is the amount of the loan. A loan is an initial deposit with a negative value. Alternately, we may treat a loan of M dollars taken at $k = 0$ as an input of $-M$ at $k = 0$ [see Prob. 9.4-9]. ■

---

†The time delay in Fig. 8.18 need not be T. The use of any other value will result in a time-scaled output.

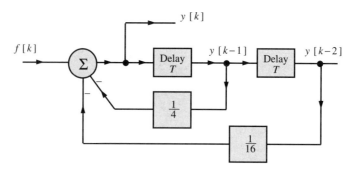

Fig. 8.19 Realization of a second-order discrete-time system in Example 8.6.

■ **Example 8.6**

In the kth semester, $f[k]$ number of students enroll in a course requiring a certain textbook. The publisher sells $y[k]$ new copies of the book in the kth semester. On the average, one quarter of students with books in saleable condition resell their books at the end of the semester, and the book life is three semesters. Write the equation relating $y[k]$, the new books sold by the publisher, to $f[k]$, the number of students enrolled in the kth semester, assuming that every student buys a book.

In the kth semester, the total books $f[k]$ sold to students must be equal to $y[k]$ (new books from the publisher) plus used books from students enrolled in the two previous semesters (because the book life is only three semesters). There are $y[k-1]$ new books sold in the $(k-1)$st semester, and one quarter of these books; that is, $\frac{1}{4}y[k-1]$ will be resold in the kth semester. Also, $y[k-2]$ new books are sold in the $(k-2)$nd semester, and one quarter of these; that is, $\frac{1}{4}y[k-2]$ will be resold in the $(k-1)$st semester. Again a quarter of these; that is, $\frac{1}{16}y[k-2]$ will be resold in the kth semester. Therefore, $f[k]$ must be equal to the sum of $y[k]$, $\frac{1}{4}y[k-1]$, and $\frac{1}{16}y[k-2]$.

$$y[k] + \tfrac{1}{4}y[k-1] + \tfrac{1}{16}y[k-2] = f[k] \tag{8.26a}$$

Equation (8.26a) can also be expressed in an alternative form by realizing that this equation is valid for any value of k. Therefore, replacing k by $k+2$, we obtain

$$y[k+2] + \tfrac{1}{4}y[k+1] + \tfrac{1}{16}y[k] = f[k+2] \tag{8.26b}$$

This is the alternative form of Eq. (8.26a).

For a realization of a system with this input-output equation, we rewrite Eq. (8.26a) as

$$y[k] = -\tfrac{1}{4}y[k-1] - \tfrac{1}{16}y[k-2] + f[k] \tag{8.26c}$$

Figure 8.19 shows a hardware realization of Eq. (8.26c) using two time delays (here the time delay T is a semester). To understand this realization, assume that $y[k]$ is available. Then, by delaying it successively, we generate $y[k-1]$ and $y[k-2]$. Next we generate $y[k]$ from $f[k]$, $y[k-1]$, and $y[k-2]$ according to Eq. (8.26c). ■

Equations (8.25) and (8.26) are examples of difference equations; the former is a first-order and the latter is a second-order difference equation. Difference equations also arise in numerical solution of differential equations.

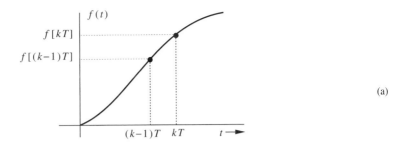

(a)

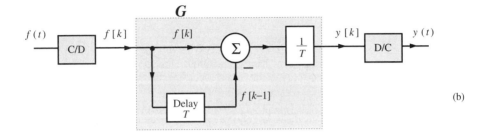

(b)

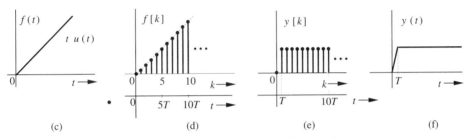

(c)   (d)   (e)   (f)

Fig. 8.20 Digital differentiator and its realization.

■ **Example 8.7: Digital Differentiator**

Design a discrete-time system, like the one in Fig. 8.2, to differentiate continuous-time signals. Determine the sampling interval if this differentiator is used in an audio system where the input signal bandwidth is below 20 kHz.

In this case, the output $y(t)$ is required to be the derivative of the input $f(t)$. The discrete-time processor (system) G processes the samples of $f(t)$ to produce the discrete-time output $y[k]$. Let $f[k]$ and $y[k]$ represent the samples T seconds apart of the signals $f(t)$ and $y(t)$, respectively; that is,

$$f[k] = f(kT) \qquad \text{and} \qquad y[k] = y(kT) \tag{8.27}$$

The signals $f[k]$ and $y[k]$ are the input and the output for the discrete-time system G. Now, we require that

$$y(t) = \frac{df}{dt} \tag{8.28}$$

Therefore, at $t = kT$ (see Fig. 8.20a)

$$y(kT) = \left. \frac{df}{dt} \right|_{t=kT}$$

$$= \lim_{T \to 0} \frac{1}{T} \left[f(kT) - f[(k-1)T] \right]$$

By using the notation in Eq. (8.27), the above equation can be expressed as

$$y[k] = \lim_{T \to 0} \frac{1}{T} \{f[k] - f[k-1]\}$$

This is the input-output relationship for G required to achieve our objective. In practice, the sampling interval T cannot be zero. Assuming T to be sufficiently small, the above equation can be expressed as

$$y[k] \simeq \frac{1}{T} \{f[k] - f[k-1]\} \tag{8.29}$$

The approximation improves as T approaches 0. A discrete-time processor G to realize Eq. (8.29) is shown inside the shaded box in Fig. 8.20b. The system in Fig. 8.20b acts as a differentiator. This example shows how a continuous-time signal can be processed by a discrete-time system.

To determine the sampling interval T, we note that the highest frequency that will appear at the input is 20 kHz; that is, $\mathcal{F}_h = 20,000$. Hence, according to Eq. (8.17c)

$$T \le \frac{1}{40,000} = 25 \ \mu s$$

To gain some insight into this method of signal processing, let us consider the differentiator in Fig. 8.20b with a ramp input $f(t) = t$, depicted in Fig. 8.20c. If the system were to act as a differentiator, then the output $y(t)$ of the system should be the unit step function $u(t)$. Let us investigate how the system performs this particular operation and how well it achieves the objective.

The samples of the input $f(t) = t$ at the interval of T seconds act as the input to the discrete-time system G. These samples, denoted by a compact notation $f[k]$, are, therefore,

$$f[k] = f(t)|_{t=kT} = t|_{t=kT} \qquad t \ge 0$$
$$= kT \qquad k \ge 0$$

Figure 8.20d shows the sampled signal $f[k]$. This signal acts as an input to the discrete-time system G. Figure 8.20b shows that the operation of G consists of subtracting a sample from the previous (delayed) sample and then multiplying the difference with $1/T$. From Fig. 8.20d, it is clear that the difference between the successive samples is a constant $kT - (k-1)T = T$ for all samples, except for the sample at $k = 0$ (because there is no previous sample at $k = 0$). The output of G is $1/T$ times the difference T, which is unity for all values of k, except $k = 0$, where it is zero. Therefore, the output $y[k]$ of G consists of samples of unit values for $k \ge 1$, as illustrated in Fig. 8.20e. The D/C (discrete-time to continuous-time) converter converts these samples into a continuous-time signal $y(t)$, as shown in Fig. 8.20f. Ideally, the output should have been $y(t) = u(t)$. This deviation from the ideal is caused by the fact that we have used a nonzero sampling interval T. As T approaches zero, the output $y(t)$ approaches the desired output $u(t)$. ■

△ **Exercise E8.8**

Design a discrete-time system, such as in Fig. 8.2, to integrate continuous-time signals.

Hint: If $f(t)$ and $y(t)$ are the input and the output of an integrator, then $\frac{dy}{dt} = f(t)$. Approximation (similar to that in Example 8.7) of this equation at $t = kT$ yields $y[k] - y[k-1] = Tf[k]$. Show a realization of this system.   ▽

Practical Realization of Discrete-Time Systems

These examples show that the basic elements required in the realization of discrete-time systems are time delays, scalar multipliers, and adders (summers).

We show in Chapter 11 that this is generally true of discrete-time systems. The discrete-time systems can be realized in two ways:

1. By using digital computers which readily perform the operations of adding, multiplying, and delaying. Minicomputers and microprocessors are well suited for this purpose, especially for signals with frequencies below 100 kHz.

2. By using special-purpose time-delay devices that have been developed in the last two decades. These include monolithic MOS charge-transfer devices (CTD) such as charge-coupled devices (CCD) and bucket brigade devices (BBD), which are implemented on silicon substrate as integrated circuit elements. In addition, there are surface acoustic wave (SAW) devices built on piezoelectric substrates. Systems using these devices are less expensive but are not as reliable or as accurate as the digital systems. Digital systems are preferable for signals below 100 kHz. Systems using CTD are suitable and competitive with those using SAW devices in the frequency range 1 kHz to 20 MHz. At frequencies higher than 20 MHz, SAW devices are preferred and are the only realistic choice for frequencies higher than 50 MHz. Systems using SAW devices with frequencies in the range of 10 MHz to 1 GHz are implemented routinely.[1]

There is a basic difference between continuous-time systems and analog systems. The same is true of discrete-time and digital systems. This is fully explained in Secs. 1.7-6 and 1.7-7.† For historical reasons, digital computers (rather than time-delay elements, such as CCD or SAW devices) were used in the realization of early discrete-time systems. Because of this fact, the terms *digital filters* and *discrete-time systems* are used synonymously in the literature. This distinction is irrelevant in the analysis of discrete-time systems. For this reason, in this book, the term *digital filters* implies *discrete-time systems*, and *analog filters* means *continuous-time systems*. Moreover, the terms C/D (continuous-to-discrete-time) and D/C will be used interchangeably with terms A/D (analog-to-digital) and D/A, respectively.

Advantages of Digital Signal Processing

1. Digital filters have a greater degree of precision and stability. They can be perfectly duplicated without having to worry about component value tolerances as in analog case.
2. Digital filters are more flexible. Their characteristics can be easily altered simply by changing the program.
3. A greater variety of filters can be realized by digital systems.
4. Very low frequency filters, if realized by continuous-time systems, require prohibitively bulky components. Such is not the case with digital filters.
5. Digital signals can be stored easily on magnetic tapes or disks without deterioration of signal quality.
6. More sophisticated signal processing algorithms can be used to process digital signals.
7. Digital filters can be time shared, and therefore can serve a number of inputs simultaneously.

---

†The terms *discrete-time* and *continuous-time* qualify the nature of a signal along the time axis (horizontal axis). The terms *analog* and *digital*, in contrast, qualify the nature of the signal amplitude (vertical axis).

8. Using integrated circuit technology, they can be fabricated in small packages requiring low power consumption.

Some more advantages of using digital signals are listed in Sec. 5.1-3.

8.6 Summary

Signals specified only at discrete instants such as $t = 0, T, 2T, 3T, \ldots, kT$ are discrete-time signals. Basically, it is a sequence of numbers. Such a signal may be viewed as a function of time t, where the signal is defined or specified only at $t = kT$ with k any positive or negative integer. The signal therefore may be denoted as $f(kT)$. Alternately, such a signal may be viewed as a function of k, where k is any positive or negative integer. The latter approach results in a more compact notation such as $f[k]$, which is convenient and easier to manipulate. A system whose inputs and outputs are discrete-time signals is a discrete-time system.

In the study of continuous-time systems, exponentials with the natural base; that is, exponentials of the form $e^{\lambda t}$, where λ is complex in general, are more natural and convenient. In contrast, in the study of discrete-time systems, exponentials with a general base; that is, exponentials of the form γ^k, where γ is complex in general, are more convenient. One form of exponential can be readily converted to the other form by noting that $e^{\lambda k} = \gamma^k$, where $\gamma = e^\lambda$, or $\lambda = \ln \gamma$, and λ as well as γ are complex in general. The exponential γ^k grows exponentially with k if $|\gamma| > 1$ (γ outside the unit circle), and decays exponentially if $|\gamma| < 1$ (γ within the unit circle). If $|\gamma| = 1$; that is, if γ lies on the unit circle, the exponential is either a constant or oscillates with a constant amplitude.

Discrete-time sinusoids have two properties not shared by their continuous-time cousins. First, a discrete-time sinusoid $\cos \Omega k$ is periodic only if $\Omega/2\pi$ is a rational number. Second, discrete-time sinusoids whose frequencies Ω differ by an integral multiple of 2π are identical. Consequently, a discrete-time sinusoid of any frequency Ω is identical to some discrete-time sinusoid whose frequency lies in the interval $-\pi$ to π (called the fundamental frequency range). Further, because $\cos(-\Omega k + \theta) = \cos(\Omega k - \theta)$, a sinusoid of a frequency in the range from $-\pi$ to 0 can be expressed as a sinusoid of frequency in the range 0 to π. Thus, a discrete-time sinusoid of any frequency can be expressed as a sinusoid of frequency in the range 0 to π. Thus, in practice, a discrete-time sinusoid frequency is at most π. The highest rate of oscillation in a discrete-time sinusoid occurs when its frequency is π. In a given time, a sinusoid of frequency other than π will have a fewer number of cycles (or oscillations) than the sinusoid of frequency π. This peculiarity of nonuniqueness of waveforms in discrete-time sinusoids of different frequencies has a far reaching consequences in signal processing by discrete-time systems.

One useful measure of the size of a discrete-time signal is its energy defined by the sum $\sum_k |f[k]|^2$, if it is finite. If the signal energy is infinite, the proper measure is its power, if it exists. The signal power is the time average of its energy (averaged over the entire time interval from $k = -\infty$ to ∞). For periodic signals, the time averaging need be performed only over one period in view of the periodic repetition of the signal. Signal power is also equal to the mean squared value of the signal (averaged over the entire time interval from $k = -\infty$ to ∞).

Sampling a continuous-time sinusoid $\cos(\omega t + \theta)$ at uniform intervals of T seconds results in a discrete-time sinusoid $\cos(\Omega k + \theta)$, where $\Omega = \omega T$. A continuous time sinusoid of frequency $\mathcal{F}$ Hz must be sampled at a rate no less than $2\mathcal{F}$ Hz. Otherwise, the resulting sinusoid is aliased; that is, it appears as a sampled version of a sinusoid of lower frequency.

Discrete-time signals classification is identical to that of continuous-time signals, discussed in chapter 1.

A signal $f[k]$ delayed by m time units (right-shifted) is given by $f[k-m]$. On the other hand, $f[k]$ advanced (left-shifted) by m time units is given by $f[k+m]$. A signal $f[k]$, when time inverted, is given by $f[-k]$. These operations are the same as those for the continuous-time case. The case of time scaling, however, is somewhat different because of the discrete nature of variable k. Unlike the continuous-time case, where time compression results in the same data at a higher speed, time compression in the discrete-time case eliminates part of the data. Consequently, this operation is called *decimation* or *downsampling*. Time expansion operation of discrete-time signals results in time expanding the signal, thus creating zero-valued samples in between. We can reconstruct the zero-valued samples using interpolation from the nonzero samples. The interpolation, thus, creates additional samples in between using the interpolation process. For this reason, this operation is called *interpolation* or *upsampling*.

Discrete-time systems may be used to process discrete-time signals, or to process continuous-time signals using appropriate interfaces at the input and output. At the input, the continuous-time input signal is converted into a discrete-time signal through sampling. The resulting discrete-time signal is now processed by the discrete-time system yielding a discrete-time output. The output interface now converts the discrete-time output into a continuous-time output. Discrete-time systems are characterized by difference equations.

Discrete-time systems can be realized by using scalar multipliers, summers, and time delays. These operations can be readily performed by digital computers. Time delays also can be obtained from charge coupled devices (CCD), bucket brigade devices (BBD), and surface acoustic wave devices (SAW). Several advantages of discrete-time systems over continuous-time systems are discussed in Sec. 8.5. Because of these advantages, discrete-time systems are replacing continuous-time systems in several applications.

References

1.   Milstein, L. B., and P.K. Das, "Surface Acoustic wave Devices," IEEE Communication Society Magazine, vol. 17, No. 5, pp. 25-33, September 1979.

Problems

8.2-1   The following signals are in the form $e^{\lambda k}$. Express them in the form γ^k: **(a)** $e^{-0.5k}$ **(b)** $e^{0.5k}$ **(c)** $e^{-j\pi k}$ **(d)** $e^{j\pi k}$. In each case show the locations of λ and γ in the complex plane. Verify that an exponential is growing if γ lies outside the unit circle (or if λ lies in the RHP), is decaying if γ lies within the unit circle (or if λ lies in the

LHP), and has a constant amplitude if γ lies on the unit circle (or if λ lies on the imaginary axis).

8.2-2 Repeat Prob. 8.2-1 for the exponentials **(a)** $e^{-(1+j\pi)k}$ **(b)** $e^{-(1-j\pi)k}$ **(c)** $e^{(1+j\pi)k}$
(d) $e^{(1-j\pi)k}$ **(e)** $e^{-(1+j\frac{\pi}{3})k}$ **(f)** $e^{(1-j\frac{\pi}{3})k}$.

8.2-3 State with reason(s) if the following signals are periodic: **(a)** $\cos{(0.5\pi k + 0.2)}$ **(b)** $\cos{(\sqrt{2}\pi k + 1.2)}$ **(c)** $\sin{(0.5k + \frac{\pi}{3})}$ **(d)** $e^{j\frac{\pi}{3}k}$. If periodic, determine the period.

8.2-4 Repeat Prob. 8.2-3 for the signals **(a)** $\cos{(0.6\pi k + 0.2)}$ **(b)** $\cos{(0.8\pi k + 1.2)}$ **(c)** $\cos{(0.6\pi k + 0.3)} + 3\sin{(0.5\pi k + 0.4)} + 8\cos{(0.8\pi k - \frac{\pi}{3})}$
Hint: For parts **(a)** and **(b)**, use Eq. (8.9b) to determine the period. For part **(c)**, if N_1, N_2, and N_3 are the periods of the three sinusoids, the widths of m_1, m_2, and m_3 number of cycles (m_1, m_2, m_3 integers) of the three sinusoids must be equal for the signal to be periodic.

8.2-5 Determine the fundamental range frequency Ω_f for the sinusoids of the frequencies $\Omega = $ **(a)** 0.8π **(b)** 1.2π **(c)** 6.9 **(d)** 3.7π **(e)** 22.9π . For each case, determine also the lowest frequency which can be used to describe these sinusoids.

8.2-6 Show that $\cos{(0.6\pi k + \frac{\pi}{6})} + \sqrt{3}\cos{(1.4\pi k + \frac{\pi}{3})} = 2\cos{(0.6\pi k - \frac{\pi}{6})}$.

8.2-7 Express the following exponentials in the form $e^{j(\Omega k + \theta)}$, where $0 \le \Omega < 2\pi$:

  (a) $e^{j(8.2\pi k + \theta)}$ **(b)** $e^{j4\pi k}$ **(c)** $e^{-j1.95k}$ **(d)** $e^{-j10.7\pi k}$.

8.2-8 If $f[k] = (0.8)^k u[k]$, find the energies of $f[k]$, $-f[k]$, and $cf[k]$.

8.2-9 Find the energies of the signals depicted in Figs. P8.2-9.

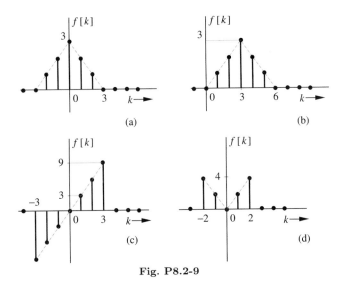

Fig. P8.2-9

8.2-10 Find the powers of the signals $(1)^k$, $(-1)^k$, $u[k]$, $(-1)^k u[k]$, and $\cos{[\frac{\pi}{3}k + \frac{\pi}{6}]}$.

8.2-11 Find the powers of the signals illustrated in Fig. P10.1-4 and P10.1-5 (Chapter 10).

8.2-12 Show that the power of a signal $\mathcal{D}e^{j\frac{2\pi}{N_0}k}$ is $|\mathcal{D}|^2$. Hence, show that the power of a signal

$$f[k] = \sum_{r=0}^{N_0-1} \mathcal{D}_r e^{jr\frac{2\pi}{N_0}k} \qquad \text{is} \qquad P_f = \sum_{r=0}^{N_0-1} |\mathcal{D}_r|^2$$

8.3-1 A discrete-time processor uses a sampling interval $T = 0.5\,\mu s$. What is the highest frequency of a signal that can be processed with this processor without aliasing? If a signal of frequency 2 MHz is sampled by this processor, what would be the (aliased) frequency of the resulting sampled signal?

8.3-2 A sampler with sampling interval is $T = 0.1$ second samples a continuous-time sinusoids $10\cos{(11\pi t + \frac{\pi}{6})}$ and $5\cos{(29\pi t - \frac{\pi}{6})}$. Find the expressions for the resulting discrete-time sinusoids. Hint: Reduce all discrete-time frequencies to the smallest value (in the range 0 to π).

8.3-3 (a) A signal $10\cos 2000\pi t + \sqrt{2}\sin 3000\pi t + 2\cos{(5000\pi t + \frac{\pi}{4})}$ is sampled at a rate of 4000 Hz (4000 samples/second). Find the resulting sampled signal. Does this sampling rate cause any aliasing? Explain.
(b) Determine the maximum sampling interval T that can be used to sample the signal in (a) without aliasing.

8.3-4 A sampler with sampling interval is $T = 10^{-4}$ seconds samples continuous-time sinusoids of the following frequencies: $\mathcal{F} =$ (i) 1500 Hz   (ii) 8,500 Hz   (iii) 10 kHz   (iv) 11.5 kHz   (v) 32 kHz   (vi) 9600 Hz. Determine of what (continuous-time) frequency these samples appear in each case.

8.4-1 For the signal shown in Fig. P8.2-9b, sketch the signals: (a) $f[-k]$ (b) $f[k+6]$ (c) $f[k-6]$ (d) $f[3k]$ (e) $f\left[\frac{k}{3}\right]$ without the interpolation of the missing samples.

8.4-2 Repeat Prob. 8.4-1 for the signal depicted in Fig. P8.2-9c.

8.4-3 Sketch the signals (a) $u[k-2] - u[k-6]$ (b) $k\{u[k] - u[k-7]\}$ (c) $(k-2)\{u[k-2] - u[k-6]\}$ (d) $(-k+8)\{u[k-6] - u[k-9]\}$ (e) $(k-2)\{u[k-2] - u[k-6]\} + (-k+8)\{u[k-6] - u[k-9]\}$

8.4-4 Describe each of the signals in Fig. P8.2-9 by a single expression valid for all k. Hint: Use the expressions of the form in Prob. 8.4-3.

8.4-5 If the energy of a signal $f[k]$ is E_f, then show that the energy of $f[k-m]$ is also E_f.

8-4-6 If the power of a periodic signal $f[k]$ is P_f, find the powers and the rms values of (a) $-f[k]$ (b) $f[-k]$ (c) $f[k-m]$ (d) $cf[k]$. Comment.

8.5-1 A cash register output $y[k]$ represents the total cost of k items rung up by a cashier. The input $f[k]$ is the cost of the kth item.

(a) Write the difference equation relating $y[k]$ to $f[k]$.
(b) Realize this system using a time-delay element.
(c) Redo the problem if there is a 10% sales tax.

8.5-2 Let $p[k]$ be the population of a certain country at the beginning of the kth year. The birth and death rates of the population during any year are 3.3% and 1.3%, respectively. If $i[k]$ is the total number of immigrants entering the country during the kth year, write the difference equation relating $p[k+1]$, $p[k]$, and $i[k]$.
Hint: Assume that the immigrants enter the country throughout the year at a uniform rate, so that their average birth and death rates need to be averaged.

8.5-3 For an integrator, the output $y(t)$ is the area under the input $f(t)$ from $t = 0$ to t. Show that the equation of a digital integrator is

$$y[k] - y[k-1] \approx Tf[k-1]$$

If an input $u[k]$ is applied to such an integrator, show that the output is a ramp $kT\,u[k]$. Hint: Set $k = 0, 1, 2, 3, \ldots$ successively in this equation to find $y[k]$.

In Exercise E8.8, using a slightly different approach, we found the integrator equation to be $y[k] - y[k-1] = Tf[k]$. Show that the response of this integrator to a unit step input $u[k]$ is $kT\,u[k] + T\,u[k]$, which approaches the ramp $kT\,u[k]$ as $T \to 0$.

8.5-4   A moving average is used to detect a trend of a rapidly fluctuating variable such as the stock market average. A variable may fluctuate (up and down) daily, masking its long-term (secular) trend. We can discern the long-term trend by smoothing or averaging the past N values of the variable. For the stock market average, we may consider a five-day moving average $y[k]$ to be the mean of the past five days' market closing values $f[k], f[k-1], \ldots, f[k-4]$.
(a)   Write the difference equation relating $y[k]$ to the input $f[k]$.
(b)   Using time-delay elements, realize the five-day moving average filter.

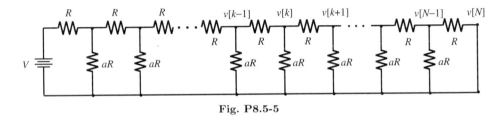

Fig. P8.5-5

8.5-5   The voltage at the kth node of a resistive ladder in Fig. P8.5-5 is $v[k]$ $(k = 0, 1, 2, \ldots, N)$. Show that $v[k]$ satisfies the second-order difference equation

$$v[k+2] - Av[k+1] + v[k] = 0 \qquad A = 2 + \tfrac{1}{a}$$

Hint: Consider the node equation at the kth node with voltage $v[k]$.

Time-Domain Analysis of Discrete-Time Systems

In this chapter we discuss time-domain analysis of LTID (linear time-invariant discrete-time systems). The procedure is parallel to that for continuous-time systems, with minor differences.

9.1 Discrete-Time System equations

Difference Equations

Equations (8.25), (8.26), and (8.29) are examples of difference equations. Equations (8.25) and (8.29) are first-order difference equations, and Eq. (8.26) is a second-order difference equation. All these equations are linear, with constant (not time-varying) coefficients. Before giving a general form of an nth-order difference equation, we recall that a difference equation can be written in two forms; the first form uses delay terms such as $y[k-1]$, $y[k-2]$, $f[k-1]$, $f[k-2]$, ..., etc., and the alternate form uses advance terms such as $y[k+1]$, $y[k+2]$, ..., etc. Both forms are useful. We start here with a general nth-order difference equation, using advance operator form:

$$y[k+n] + a_{n-1}y[k+n-1] + \cdots + a_1y[k+1] + a_0y[k] =$$

$$b_mf[k+m] + b_{m-1}f[k+m-1] + \cdots + b_1f[k+1] + b_0f[k] \qquad (9.1)$$

Observe that the coefficient of $y[k+n]$ can be assumed to be unity without loss of generality. If this coefficient is other than unity, we can divide the equation throughout by the value of the coefficient of $y[k+n]$ to normalize the equation to the form in (9.1).

Causality Condition

The left-hand side of Eq. (9.1) consists of the output at instants $k+n$, $k+n-1$, $k+n-2$, and so on. The right-hand side of Eq. (9.1) consists of the input at instants $k+m$, $k+m-1$, $k+m-2$, and so on. For a causal system the output cannot depend on future input values. This fact shows that when the system equation is in the advance operator form (9.1), the causality requires $m \leq n$. For a general causal case, $m = n$, and Eq. (9.1) can be expressed as

$$y[k+n] + a_{n-1}y[k+n-1] + \cdots + a_1y[k+1] + a_0y[k] =$$
$$b_nf[k+n] + b_{n-1}f[k+n-1] + \cdots + b_1f[k+1] + b_0f[k] \qquad (9.2a)$$

where some of the coefficients on both sides can be zero. However, the coefficient of $y[k+n]$ is normalized to unity. Equation (9.2a) is valid for all values of k. Therefore, the equation is still valid if we replace k by $k - n$ throughout the equation [see Eqs. (8.25a) and (8.25b)]. Such replacement yields the alternative form (the delay operator form) of Eq. (9.2a):

$$y[k] + a_{n-1}y[k-1] + \cdots + a_1y[k-n+1] + a_0y[k-n] =$$
$$b_nf[k] + b_{n-1}f[k-1] + \cdots + b_1f[k-n+1] + b_0f[k-n] \qquad (9.2b)$$

We shall designate Form (9.2a) the **advance operator form**, and Form (9.2b) the **delay operator form**.

9.1-1   Initial Conditions and Iterative Solution of Difference Equations

Equation (9.2b) can be expressed as

$$y[k] = -a_{n-1}y[k-1] - a_{n-2}y[k-2] - \cdots - a_0y[k-n]$$
$$+ b_nf[k] + b_{n-1}f[k-1] + \cdots + b_0f[k-n] \qquad (9.2c)$$

This equation shows that $y[k]$, the output at the kth instant, is computed from $2n+1$ pieces of information. These are the past n values of the output: $y[k-1]$, $y[k-2]$, ..., $y[k-n]$, the past n values of the input: $f[k-1]$, $f[k-2]$, ..., $f[k-n]$, and the present value of the input $f[k]$. If the input is causal, then $f[-1] = f[-2] = \ldots = f[-n] = 0$, and we need only n initial conditions $y[-1], y[-2], \ldots, y[-n]$. This result allows us to compute iteratively or recursively the output $y[0]$, $y[1]$, $y[2]$, $y[3]$, ..., and so on.† For instance, to find $y[0]$ we set $k = 0$ in Eq. (9.2c). The left-hand side is $y[0]$, and the right-hand side contains terms $y[-1]$, $y[-2]$, ..., $y[-n]$, and the inputs $f[0]$, $f[-1]$, $f[-2]$, ..., $f[-n]$. Therefore, to begin with, we must know the n initial conditions $y[-1], y[-2], \ldots, y[-n]$. Knowing these conditions and the input $f[k]$, we can iteratively find the response $y[0]$, $y[1]$, $y[2]$, ..., and so on. The following examples demonstrate this procedure. This method basically reflects the manner in which a computer would solve a difference equation, given the input and initial conditions.

---

†For this reason Eq. (9.2) is called a **recursive difference equation**. However in Eq. (9.2), if $a_0 = a_1 = a_2 = \cdots = a_{n-1} = 0$, then, according to Eq. (9.2c), determination of the present output $y[k]$ does not require the past values $y[k-1], y[k-2], \ldots$, etc. For this reason, when $a_i = 0$, $(i = 0, 1, \ldots, n-1)$, the difference Eq. (9.2) is **nonrecursive**. This classification is important in designing and realizing digital filters. In this chapter, however, this classification is not important. The analysis techniques developed here (and in Chap. 12) apply to general recursive and nonrecursive systems. Observe that a nonrecursive system is a special case of a recursive system with $a_0 = a_1 = \ldots = a_{n-1} = 0$. Design of recursive and nonrecursive systems is discussed in Chapter 12.

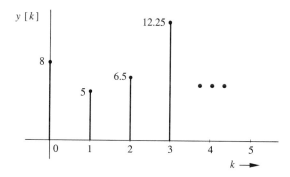

Fig. 9.1 Iterative solution of a difference equation in Example 9.1.

■ **Example 9.1**
Solve iteratively

$$y[k] - 0.5y[k-1] = f[k] \qquad (9.3a)$$

with initial condition $y[-1] = 16$ and causal input $f[k] = k^2$ (starting at $k = 0$). This equation can be expressed as

$$y[k] = 0.5y[k-1] + f[k] \qquad (9.3b)$$

If we set $k = 0$ in this equation, we obtain

$$\begin{aligned} y[0] &= 0.5y[-1] + f[0] \\ &= 0.5(16) + 0 = 8 \end{aligned}$$

Now, setting $k = 1$ in Eq. (9.3b) and using the value $y[0] = 8$ (computed in the first step) and $f[1] = (1)^2 = 1$, we obtain

$$y[1] = 0.5(8) + (1)^2 = 5$$

Next, setting $k = 2$ in Eq. (9.3b) and using the value $y[1] = 5$ (computed in the previous step) and $f[2] = (2)^2$, we obtain

$$y[2] = 0.5(5) + (2)^2 = 6.5$$

Continuing in this way iteratively, we obtain

$$y[3] = 0.5(6.5) + (3)^2 = 12.25$$
$$y[4] = 0.5(12.25) + (4)^2 = 22.125$$

. . . .

The output $y[k]$ is depicted in Fig. 9.1.   ■

We now present one more example of iterative solution—this time for a second-order equation. Iterative method can be applied to a difference equation in delay or advance operator form. In Example 9.1 we considered the former. Let us now apply the iterative method to the advance operator form.

■ **Example 9.2**
Solve iteratively

$$y[k+2] - y[k+1] + 0.24y[k] = f[k+2] - 2f[k+1] \qquad (9.4)$$

with initial conditions $y[-1] = 2$, $y[-2] = 1$ and a causal input $f[k] = k$ (starting at $k = 0$). The system equation can be expressed as

$$y[k+2] = y[k+1] - 0.24y[k] + f[k+2] - 2f[k+1] \qquad (9.5)$$

Setting $k = -2$ and then substituting $y[-1] = 2$, $y[-2] = 1$, $f[0] = f[-1] = 0$ (recall that $f[k] = k$ starting at $k = 0$), we obtain

$$y[0] = 2 - 0.24(1) + 0 - 0 = 1.76$$

Setting $k = -1$ in Eq. (9.5) and then substituting $y[0] = 1.76$, $y[-1] = 2$, $f[1] = 1$, $f[0] = 0$, we obtain

$$y[1] = 1.76 - 0.24(2) + 1 - 0 = 2.28$$

Setting $k = 0$ in Eq. (9.5) and then substituting $y[0] = 1.76$, $y[1] = 2.28$, $f[2] = 2$ and $f[1] = 1$ yields

$$y[2] = 2.28 - 0.24(1.76) + 2 - 2(1) = 1.8576$$

and so on. ■

Note carefully the iterative or recursive nature of the computations. From the n initial conditions (and input) we obtained $y[0]$ first. Then, using this value of $y[0]$ and the previous $n-1$ initial conditions (along with the input), we found $y[1]$. Next, using $y[0]$, $y[1]$ along with the initial conditions and input, we obtained $y[2]$, and so on. This method is general and can be applied to a difference equation of any order. The iterative procedure cannot be applied to continuous-time systems. It is interesting that the hardware realization of Eq. (9.3a) depicted in Fig. 8.17 (with $\gamma = 0.5$) generates the solution precisely in this (iterative) fashion.

△ **Exercise E9.1**
Using the iterative method, find the first three terms of $y[k]$ if
$$y[k+1] - 2y[k] = f[k]$$
The initial condition is $y[-1] = 10$ and the input $f[k] = 2$ starting at $k = 0$.
Answer: $y[0] = 20$, $y[1] = 42$, and $y[2] = 86$.   ▽

⊙ **Computer Example C9.1**
Solve Example 9.2 using MATLAB.

```
%(c91.m)
Y=[1 2]; Y=Y';
k=-2:2; k=k';
f=[0 0 0 ]';
for m=1:length(k)-2
    y=Y(m+1)-0.24*Y(m)+f(m+2)-2*f(m+1);
    Y=[Y;y];
        F=m;f=[f;F];
end
stem(k,Y)
[k Y]
c91

ans =
-2.0000    1.0000
-1.0000    2.0000
      0    1.7600
 1.0000    2.2800
 2.0000    0.8576        ⊙
```

We shall see in the future that the solution of a difference equation obtained in this direct (iterative) way is useful in many situations. Despite the many uses of this method, a closed-form solution of a difference equation is far more useful in study of the system behavior and its dependence on the input and the various system parameters. For this reason we shall develop a systematic procedure to analyze discrete-time systems along lines similar to those used for continuous-time systems.

Operational Notation

In difference equations it is convenient to use operational notation similar to that used in differential equations for the sake of compactness. In continuous-time systems we used the operator D to denote the operation of differentiation. For discrete-time systems we shall use the operator E to denote the operation for advancing the sequence by one time unit. Thus

$$E f[k] \equiv f[k+1]$$

$$E^2 f[k] \equiv f[k+2]$$

$$\dots\dots\dots\dots$$

$$E^n f[k] \equiv f[k+n] \tag{9.6}$$

The first-order difference equation of the savings account problem was found to be [see Eq. (8.25b)]

$$y[k+1] - a y[k] = f[k+1] \tag{9.7}$$

Using the operational notation, we can express this equation as

$$E y[k] - a y[k] = E f[k]$$

or

$$(E - a) y[k] = E f[k] \tag{9.8}$$

The second-order difference equation (8.26b)

$$y[k+2] + \tfrac{1}{4} y[k+1] + \tfrac{1}{16} y[k] = f[k+2]$$

can be expressed in operational notation as

$$\left(E^2 + \tfrac{1}{4} E + \tfrac{1}{16}\right) y[k] = E^2 f[k]$$

A general nth-order difference Eq. (9.2a) can be expressed as

$$(E^n + a_{n-1} E^{n-1} + \cdots + a_1 E + a_0) y[k] =$$
$$(b_n E^n + b_{n-1} E^{n-1} + \cdots + b_1 E + b_0) f[k] \tag{9.9a}$$

or

$$Q[E] y[k] = P[E] f[k] \tag{9.9b}$$

where $Q[E]$ and $P[E]$ are nth-order polynomial operators

$$Q[E] = E^n + a_{n-1}E^{n-1} + \cdots + a_1 E + a_0 \tag{9.10}$$

$$P[E] = b_n E^n + b_{n-1}E^{n-1} + \cdots + b_1 E + b_0 \tag{9.11}$$

Response of Linear Discrete-Time Systems

Following the procedure used for continuous-time systems on p. 105 (footnote), we can show that Eq. (9.9) is a linear equation (with constant coefficients). A system described by such an equation is a linear time-invariant discrete-time (LTID) system. We can verify, as in the case of LTIC systems (see footnote on p. 105), that the general solution of Eq. (9.9) consists of zero-input and zero-state components.

9.2   System response to Internal Conditions: The Zero-Input Response

The zero-input response $y_0[k]$ is the solution of Eq. (9.9) with $f[k] = 0$; that is,

$$Q[E]y_0[k] = 0 \tag{9.12a}$$

or

$$(E^n + a_{n-1}E^{n-1} + \cdots + a_1 E + a_0)y_0[k] = 0 \tag{9.12b}$$

or

$$y_0[k+n] + a_{n-1}y_0[k+n-1] + \cdots + a_1 y_0[k+1] + a_0 y_0[k] = 0 \tag{9.12c}$$

We can solve this equation systematically. But even a cursory examination of this equation points to its solution. This equation states that a linear combination of $y_0[k]$ and advanced $y_0[k]$ is zero *not for some values of* k, *but for all* k. Such situation is possible *if and only if* $y_0[k]$ and advanced $y_0[k]$ have the same form. Only an exponential function γ^k has this property as the following equation indicates.

$$\gamma^{k+m} = \gamma^m \gamma^k \tag{9.13}$$

This equation shows that the delayed γ^k is a constant times γ^k. Therefore, the solution of Eq. (9.12) must be of the form

$$y_0[k] = c\gamma^k \tag{9.14}$$

To determine c and γ, we substitute this solution in Eq. (9.12). Equation (9.14) yields

$$Ey_0[k] = y_0[k+1] = c\gamma^{k+1}$$

$$E^2 y_0[k] = y_0[k+2] = c\gamma^{k+2}$$

$$\cdots\cdots\cdots\cdots\cdots\cdots\cdots\cdots \tag{9.15}$$

$$E^n y_0[k] = y_0[k+n] = c\gamma^{k+n}$$

Substitution of these results in Eq. (9.12b) yields

$$c(\gamma^n + a_{n-1}\gamma^{n-1} + \cdots + a_1\gamma + a_0)\gamma^k = 0 \tag{9.16}$$

For a nontrivial solution of this equation

$$(\gamma^n + a_{n-1}\gamma^{n-1} + \cdots + a_1\gamma + a_0) = 0 \tag{9.17a}$$

or

$$Q[\gamma] = 0 \tag{9.17b}$$

Our solution $c\gamma^k$ [Eq. (9.14)] is correct, provided that γ satisfies Eq. (9.17). Now, $Q[\gamma]$ is an nth-order polynomial and can be expressed in the factorized form (assuming all distinct roots):

$$(\gamma - \gamma_1)(\gamma - \gamma_2)\cdots(\gamma - \gamma_n) = 0 \tag{9.17c}$$

Clearly, γ has n solutions $\gamma_1, \gamma_2, \cdots, \gamma_n$ and, therefore, Eq. (9.12) also has n solutions $c_1\gamma_1^k, c_2\gamma_2^k, \cdots, c_n\gamma_n^k$. In such a case we have shown (see footnote on p. 106) that the general solution is a linear combination of the n solutions. Thus

$$y_0[k] = c_1\gamma_1^k + c_2\gamma_2^k + \cdots + c_n\gamma_n^k \tag{9.18}$$

where $\gamma_1, \gamma_2, \cdots, \gamma_n$ are the roots of Eq. (9.17) and $c_1, c_2, \ldots, c_n$ are arbitrary constants determined from n auxiliary conditions, generally given in the form of initial conditions. The polynomial $Q[\gamma]$ is called the **characteristic polynomial** of the system, and

$$Q[\gamma] = 0 \tag{9.19}$$

is the **characteristic equation** of the system. Moreover, $\gamma_1, \gamma_2, \cdots, \gamma_n$, the roots of the characteristic equation, are called **characteristic roots** or **characteristic values** (also **eigenvalues**) of the system. The exponentials $\gamma_i^k (i = 1, 2, \ldots, n)$ are the **characteristic modes** or **natural modes** of the system. A characteristic mode corresponds to each characteristic root of the system, and the *zero-input response is a linear combination of the characteristic modes of the system.*

Repeated Roots

In the discussion so far we have assumed the system to have n distinct characteristic roots $\gamma_1, \gamma_2, \ldots, \gamma_n$ with corresponding characteristic modes $\gamma_1^k, \gamma_2^k, \ldots, \gamma_n^k$. If two or more roots coincide (repeated roots), the form of characteristic modes is modified. Direct substitution shows that if a root γ repeats r times (root of multiplicity r), the characteristic modes corresponding to this root are $\gamma^k, k\gamma^k, k^2\gamma^k, \ldots, k^{r-1}\gamma^k$. Thus, if the characteristic equation of a system is

$$Q[\gamma] = (\gamma - \gamma_1)^r(\gamma - \gamma_{r+1})(\gamma - \gamma_{r+2})\cdots(\gamma - \gamma_n) \tag{9.20}$$

the zero-input response of the system is

$$y_0[k] = (c_1 + c_2 k + c_3 k^2 + \cdots + c_r k^{r-1})\gamma_1^k + c_{r+1}\gamma_{r+1}^k + c_{r+2}\gamma_{r+2}^k + \cdots + c_n\gamma_n^k \tag{9.21}$$

Complex Roots

As in the case of continuous-time systems, the complex roots of a discrete-time system must occur in pairs of conjugates so that the system equation coefficients are real. Complex roots can be treated exactly as we would treat real roots. However, just as in the case of continuous-time systems, we can eliminate dealing with complex numbers by using the real form of the solution.

First we express the complex conjugate roots γ and γ^* in polar form. If $|\gamma|$ is the magnitude and β is the angle of γ, then

$$\gamma = |\gamma|e^{j\beta} \qquad \text{and} \qquad \gamma^* = |\gamma|e^{-j\beta}$$

The zero-input response is given by

$$y_0[k] = c_1 \gamma^k + c_2(\gamma^*)^k$$
$$= c_1|\gamma|^k e^{j\beta k} + c_2|\gamma|^k e^{-j\beta k}$$

For a real system, c_1 and c_2 must be conjugates so that $y_0[k]$ is a real function of k. Let

$$c_1 = \tfrac{c}{2}e^{j\theta} \qquad \text{and} \qquad c_2 = \tfrac{c}{2}e^{-j\theta}$$

Then

$$y_0[k] = \tfrac{c}{2}|\gamma|^k \left[e^{j(\beta k + \theta)} + e^{-j(\beta k + \theta)} \right]$$

$$= c|\gamma|^k \cos(\beta k + \theta) \tag{9.22}$$

where c and θ are arbitrary constants determined from the auxiliary conditions. This is the solution in real form, which avoids dealing with complex numbers.

■ **Example 9.3**

(a) For an LTID system described by the difference equation

$$y[k+2] - 0.6y[k+1] - 0.16y[k] = 5f[k+2] \tag{9.23a}$$

find the total response if the initial conditions are $y[-1] = 0$ and $y[-2] = \tfrac{25}{4}$, and if the input $f[k] = 4^{-k}u[k]$. In this example we shall determine the zero-input component $y_0[k]$ only. The zero-state component is determined later in Example 9.7.

The system equation in operational notation is

$$(E^2 - 0.6E - 0.16)y[k] = 5E^2 f[k] \tag{9.23b}$$

The characteristic polynomial is

$$\gamma^2 - 0.6\gamma - 0.16 = (\gamma + 0.2)(\gamma - 0.8)$$

The characteristic equation is

$$(\gamma + 0.2)(\gamma - 0.8) = 0 \tag{9.24}$$

The characteristic roots are $\gamma_1 = -0.2$ and $\gamma_2 = 0.8$. The zero-input response is

$$y_0[k] = c_1(-0.2)^k + c_2(0.8)^k \tag{9.25}$$

To determine arbitrary constants c_1 and c_2, we set $k = -1$ and -2 in Eq. (9.25), then substitute $y_0[-1] = 0$ and $y_0[-2] = \tfrac{25}{4}$ to obtain†

$$\left. \begin{array}{l} 0 = -5c_1 + \tfrac{5}{4}c_2 \\[2mm] \tfrac{25}{4} = 25c_1 + \tfrac{25}{16}c_2 \end{array} \right\} \implies \begin{array}{l} c_1 = \tfrac{1}{5} \\[2mm] c_2 = \tfrac{4}{5} \end{array}$$

Therefore

$$y_0[k] = \tfrac{1}{5}(-0.2)^k + \tfrac{4}{5}(0.8)^k \tag{9.26}$$

---

†The initial conditions $y[-1]$ and $y[-2]$ are the conditions given on the total response. But because the input does not start until $k = 0$, the zero-state response is zero for $k < 0$. Hence, at $k = -1$ and -2 the total response consists of only the zero-input component, so that $y[-1] = y_0[-1]$ and $y[-2] = y_0[-2]$.

The reader can verify this solution by computing the first few terms using the iterative method (see Examples 9.1 and 9.2).

(b) A similar procedure may be followed for repeated roots. For instance, for a system specified by the equation

$$(E^2 + 6E + 9)y[k] = (2E^2 + 6E)f[k]$$

determine $y_0[k]$, the zero-input component of the response if the initial conditions are $y_0[-1] = -\frac{1}{3}$ and $y_0[-2] = -\frac{2}{9}$.

The characteristic polynomial is $\gamma^2 + 6\gamma + 9 = (\gamma + 3)^2$, and we have a repeated characteristic root at $\gamma = -3$. The characteristic modes are $(-3)^k$ and $k(-3)^k$. Hence, the zero-input response is

$$y_0[k] = (c_1 + c_2 k)(-3)^k$$

We can determine the arbitrary constants c_1 and c_2 from the initial conditions following the procedure in part (a). It is left as an exercise for the reader to show that $c_1 = 4$ and $c_2 = 3$ so that

$$y_0[k] = (4 + 3k)(-3)^k$$

(c) For the case of complex roots, let us find the zero-input response of an LTID system described by the equation

$$(E^2 - 1.56E + 0.81)y[k] = (E + 3)f[k]$$

when the initial conditions are $y_0[-1] = 2$ and $y_0[-2] = 1$.

The characteristic polynomial is $(\gamma^2 - 1.56\gamma + 0.81) = (\gamma - 0.78 - j0.45)(\gamma - 0.78 + j0.45)$. The characteristic roots are $0.78 \pm j0.45$; that is, $0.9e^{\pm j\frac{\pi}{6}}$. Thus, $|\gamma| = 0.9$ and $\beta = \pi/6$, and the zero-input response, according to Eq. (9.22), is given by

$$y_0[k] = c(0.9)^k \cos\left(\tfrac{\pi}{6}k + \theta\right)$$

To determine the arbitrary constants c and θ, we set $k = -1$ and -2 in this equation and substituting the initial conditions $y_0[-1] = 2$, $y_0[-2] = 1$, we obtain

$$2 = \frac{c}{0.9}\cos\left(-\frac{\pi}{6} + \theta\right) = \frac{c}{0.9}\left[\frac{\sqrt{3}}{2}\cos\theta + \frac{1}{2}\sin\theta\right]$$

$$1 = \frac{c}{(0.9)^2}\cos\left(-\frac{\pi}{3} + \theta\right) = \frac{c}{0.81}\left[\frac{1}{2}\cos\theta + \frac{\sqrt{3}}{2}\sin\theta\right]$$

or

$$\frac{\sqrt{3}}{1.8}c\cos\theta + \frac{1}{1.8}c\sin\theta = 2$$

$$\frac{1}{1.62}c\cos\theta + \frac{\sqrt{3}}{1.62}c\sin\theta = 1$$

These are two simultaneous equations in two unknowns $c\cos\theta$ and $c\sin\theta$. Solution of these equations yields

$$c\cos\theta = 2.308$$

$$c\sin\theta = -0.397$$

Dividing $c\sin\theta$ by $c\cos\theta$ yields

$$\tan \theta = \frac{-0.397}{2.308} = \frac{-0.172}{1}$$

$$\theta = \tan^{-1}(-0.172) = -0.17 \text{ rad.}$$

Substituting $\theta = -0.17$ radian in $c \cos \theta = 2.308$ yields $c = 2.34$ and

$$y_0[k] = 2.34(0.9)^k \cos\left(\tfrac{\pi}{6}k - 0.17\right) \qquad k \geq 0$$

Observe that here we have used radian unit for both β and θ. We also could have used the degree unit, although it is not recommended. The important consideration is to be consistent and to use the same units for both β and θ. ■

△ **Exercise E9.2**
   Find and sketch the zero-input response for the systems described by the following equations:
   (a) $y[k+1] - 0.8y[k] = 3f[k+1]$ (b) $y[k+1] + 0.8y[k] = 3f[k+1]$.
   In each case the initial condition is $y[-1] = 10$. Verify the solutions by computing the first
three terms using the iterative method.
   Answer: **(a)** $8(0.8)^k$ **(b)** $-8(-0.8)^k$.   ▽

△ **Exercise E9.3**
   Find the zero-input response of a system described by the equation

$$y[k] + 0.3y[k-1] - 0.1y[k-2] = f[k] + 2f[k-1]$$

   The initial conditions are $y_0[-1] = 1$ and $y_0[-2] = 33$. Verify the solution by computing the
first three terms iteratively.
   Answer: $y_0[k] = (0.2)^k + 2(-0.5)^k$ ▽

△ **Exercise E9.4**
   Find the zero-input response of a system described by the equation

$$y[k] + 4y[k-2] = 2f[k]$$

   The initial conditions are $y_0[-1] = -\frac{1}{2\sqrt{2}}$ and $y_0[-2] = \frac{1}{4\sqrt{2}}$. Verify the solution by computing
the first three terms iteratively.
   Answer: $y_0[k] = (2)^k \cos\left(\frac{\pi}{2}k - \frac{3\pi}{4}\right)$ ▽

⊙ **Computer Example C9.2**
   Find and sketch the zero-input response for the system described by

$$(E^2 - 1.56E + 0.81)y[k] = (E+3)f[k]$$

using the initial conditions y[-1]=2, and y[-2]=1.

```
y=[1 2]; y=y';
k=-2:16; k=k';
for m=1:length(k)-2
    Y=1.56*y(m+1)-0.81*y(m);
    Yzi=[Yzi;Y];
end
stem(k,Yzi)      ⊙
```

9.3 The Unit Impulse Response h[k]

Consider an nth-order system specified by the equation

$$(E^n + a_{n-1}E^{n-1} + \cdots + a_1 E + a_0)y[k] =$$
$$(b_n E^n + b_{n-1}E^{n-1} + \cdots + b_1 E + b_0)f[k] \qquad (9.27a)$$

or

$$Q[E]y[k] = P[E]f[k] \qquad (9.27b)$$

The unit impulse response $h[k]$ is the solution of this equation for the input $\delta[k]$ with all the initial conditions zero; that is,

$$Q[E]h[k] = P[E]\delta[k] \qquad (9.28)$$

subject to initial conditions

$$h[-1] = h[-2] = \cdots = h[-n] = 0 \qquad (9.29)$$

Equation (9.28) can be solved to determine $h[k]$ iteratively or in a closed form. The following example demonstrates the iterative solution.

■Example 9.4: Iterative Determination of h[k]

Find $h[k]$, the unit impulse response of a system described by the equation

$$y[k] - 0.6y[k-1] - 0.16y[k-2] = 5f[k] \qquad (9.30)$$

To determine the unit impulse response, we must let the input $f[k] = \delta[k]$ and the output $y[k] = h[k]$ in the above equation. The resulting equation is

$$h[k] - 0.6h[k-1] - 0.16h[k-2] = 5\delta[k] \qquad (9.31)$$

subject to zero initial state; that is, $h[-1] = h[-2] = 0$.

Setting $k = 0$ in this equation yields

$$h[0] - 0.6(0) - 0.16(0) = 5(1) \implies h[0] = 5$$

Next, setting $k = 1$ in Eq. (9.31) and using $h[0] = 5$, we obtain

$$h[1] - 0.6(5) - 0.16(0) = 5(0) \implies h[1] = 3 \quad ■$$

Continuing this way, we can determine any number of terms of $h[k]$. Unfortunately, such a solution does not yield a closed-form expression for $h[k]$. Nevertheless, determining a few values of $h[k]$ can be useful in determining the closed-form solution, as the following development shows.

The Closed-Form Solution of h[k]

Recall that $h[k]$ is the system response to input $\delta[k]$, which is zero for $k > 0$. We know that when the input is zero, only the characteristic modes can be sustained by the system. Therefore, $h[k]$ must be made up of characteristic modes for $k > 0$. At $k = 0$, it may have some nonzero value A_0, so that $h[k]$ can be expressed as

$$h[k] = A_0\delta[k] + y_n[k] \qquad (9.32)$$

where, $y_n[k]$ is a linear combination of the characteristic modes. In Appendix 9.1 we show that

$$A_0 = \frac{b_0}{a_0} \tag{9.33}$$

Moreover, because $h[k]$ is causal, we must multiply $y_n[k]$ by $u[k]$. Therefore,‡

$$h[k] = \frac{b_0}{a_0}\delta[k] + y_n[k]u[k] \tag{9.34}$$

The n unknown coefficients in $y_n[k]$ (on the right-hand side) can be determined from a knowledge of n values of $h[k]$. Fortunately, it is a straightforward task to determine values of $h[k]$ iteratively, as demonstrated in Example 9.4. We compute n values $h[0], h[1], h[2], \cdots, h[n-1]$ iteratively. Now, setting $k = 0, 1, 2, \cdots, n-1$ in Eq. (9.34), we can determine the n unknowns in $y_n[k]$. This point will become clear in the following example.

■ **Example 9.5**

Determine the unit impulse response $h[k]$ for a system in Example 9.4 specified by the equation

$$y[k] - 0.6y[k-1] - 0.16y[k-2] = 5f[k]$$

This equation can be expressed in the advance operator form as

$$y[k+2] - 0.6y[k+1] - 0.16y[k] = 5f[k+2] \tag{9.35}$$

or

$$(E^2 - 0.6E - 0.16)y[k] = 5E^2 f[k] \tag{9.36}$$

The characteristic polynomial is

$$\gamma^2 - 0.6\gamma - 0.16 = (\gamma + 0.2)(\gamma - 0.8)$$

The characteristic modes are $(-0.2)^k$ and $(0.8)^k$. Therefore

$$y_n[k] = c_1(-0.2)^k + c_2(0.8)^k \tag{9.37}$$

Also, from Eq. (9.36), we have $a_0 = -0.16$ and $b_0 = 0$. Therefore, according to Eq. (9.34)

$$h[k] = [c_1(-0.2)^k + c_2(0.8)^k]u[k] \tag{9.38}$$

To determine c_1 and c_2, we need to find two values of $h[k]$ iteratively. This step is already taken in Example 9.4, where we determined that $h[0] = 5$ and $h[1] = 3$. Now, setting $k = 0$ and 1 in Eq. (9.38) and using the fact that $h[0] = 5$ and $h[1] = 3$, we obtain

$$\left.\begin{array}{c} 5 = c_1 + c_2 \\ 3 = -0.2c_1 + 0.8c_2 \end{array}\right\} \implies \begin{array}{c} c_1 = 1 \\ c_2 = 4 \end{array}$$

---

‡We showed that $h[k]$ consists of characteristic modes only for $k > 0$. Hence, the characteristic mode terms in $h[k]$ start at $k = 1$. To reflect this behavior, they should be expressed in the form $\gamma_j^k u[k-1]$. But because $u[k-1] = u[k] - \delta[k]$, $A_j\gamma_j^k u[k-1] = A_j\gamma_j^k u[k] - A_j\delta[k]$, and $h[k]$ can be expressed in terms of exponentials $\gamma_j^k u[k]$ (which start at $k = 0$), plus an impulse at $k = 0$.

Therefore

$$h[k] = \left[(-0.2)^k + 4(0.8)^k\right] u[k] \qquad (9.39)$$

∎

⊙ **Computer Example C9.3**
Solve Example 9.5 using Matlab.
There are several ways of finding impulse response by Matlab. In this method, we specify the system and the input by a, b, and f vectors. The response is plotted for 20 points. This program can be used to find the discrete-time system response to any input.

```
N=20;
b=[5 0 0];
a=[1 -0.6 -0.16];
f=[1,zeros(1,N-1)];
y=filter(b,a,f);
k=0:1:N-1;
stem(k,y)
xlabel('k');ylabel('h[k]');   ⊙
```

Comment
Although determination of the impulse response $h[k]$ using the procedure in this section is relatively simple, in Chapter 11 we shall discuss another, much simpler, method of z-transform.
In the present method, when $a_0 = 0$, A_0 cannot be determined from Eq. (9.33). It is shown in Appendix 9.1 that in such a case the impulse response contains an additional impulse at $k = 1$, and $h[k]$ can be expressed as

$$h[k] = A_0 \delta[k] + A_1 \delta[k-1] + y_n[k]u[k]$$

Now $h[k]$ contains $n + 2$ unknowns (A_0, A_1, and n coefficients of $y_n[k]$), which can be determined from $n + 2$ values of $h[k]$ obtained iteratively.

△ **Exercise E9.5**
Find $h[k]$, the unit impulse response of the LTID systems specified by the following equations:
(a) $y[k + 2] - 5y[k + 1] + 6y[k] = 8f[k + 1] - 19f[k]$
(b) $y[k + 2] - 4y[k + 1] + 4y[k] = 2f[k + 2] - 2f[k + 1]$

Answers:
(a) $h[k] = -\frac{19}{6}\delta[k] + [\frac{3}{2}(2)^k + \frac{5}{3}(3)^k]u[k]$
(b) $h[k] = (2 + k)2^k u[k]$ ▽

9.4 System Response to External Input: The Zero-State Response

The zero-state response $y[k]$ is the system response to an input $f[k]$ when the system is in zero state. In this section we shall assume that systems are in zero state unless mentioned otherwise, so that the zero-state response will be the total response of the system. Here we follow the procedure parallel to that used in the continuous-time case by expressing an arbitrary input $f[k]$ as a sum of impulse

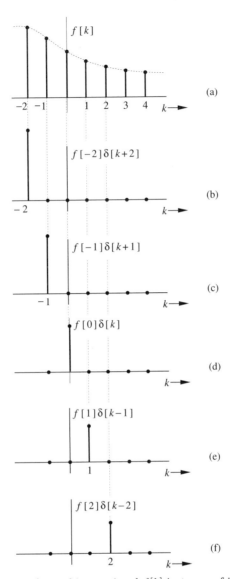

Fig. 9.2 Representation of an arbitrary signal $f[k]$ in terms of impulse components.

components. Figure 9.2 shows how a signal $f[k]$ in Fig. 9.2a can be expressed as a sum of impulse components such as those depicted in Figs. 9.2 b, c, d, e, and f. The component of $f[k]$ at $k = m$ is $f[m]\delta[k - m]$, and $f[k]$ is the sum of all these components summed from $m = -\infty$ to ∞. Therefore

$$f[k] = f[0]\delta[k] + f[1]\delta[k - 1] + f[2]\delta[k - 2] + \cdots$$
$$+ f[-1]\delta[k + 1] + f[-2]\delta[k + 2] + \cdots$$
$$= \sum_{m=-\infty}^{\infty} f[m]\delta[k - m] \qquad (9.40)$$

If we knew the system response to impulse $\delta[k]$, the system response to any arbitrary input could be obtained by summing the system response to various impulse components. Let $h[k]$ be the system response to impulse input $\delta[k]$. We shall use the notation

$$f[k] \Longrightarrow y[k]$$

to indicate the input and the corresponding response of the system. Thus, if

$$\delta[k] \Longrightarrow h[k]$$

then because of time invariance

$$\delta[k - m] \Longrightarrow h[k - m]$$

and because of linearity

$$f[m]\delta[k - m] \Longrightarrow f[m]h[k - m]$$

and again because of linearity

$$\underbrace{\sum_{m=-\infty}^{\infty} f[m]\delta[k - m]}_{f[k]} \Longrightarrow \underbrace{\sum_{m=-\infty}^{\infty} f[m]h[k - m]}_{y[k]}$$

The left-hand side is $f[k]$ [see Eq. (9.40)], and the right-hand side is the system response $y[k]$ to input $f[k]$. Therefore

$$y[k] = \sum_{m=-\infty}^{\infty} f[m]h[k - m] \tag{9.41}$$

The summation on the right-hand side is known as the **convolution sum** of $f[k]$ and $h[k]$, and is represented symbolically by $f[k] * h[k]$

$$f[k] * h[k] = \sum_{m=-\infty}^{\infty} f[m]h[k - m] \tag{9.42}$$

Properties of the Convolution Sum

The structure of the convolution sum is similar to that of the convolution integral. Moreover, the properties of the convolution sum are similar to those of the convolution integral. We shall enumerate these properties here without proof. The proofs are similar to those for the convolution integral and may be derived by the reader.

1. **The Commutative Property**

$$f_1[k] * f_2[k] = f_2[k] * f_1[k] \tag{9.43}$$

2. **The Distributive Property**

$$f_1[k] * (f_2[k] + f_3[k]) = f_1[k] * f_2[k] + f_1[k] * f_3[k] \tag{9.44}$$

3. **The Associative Property**

$$f_1[k] * (f_2[k] * f_3[k]) = (f_1[k] * f_2[k]) * f_3[k] \tag{9.45}$$

4. **The Shifting Property**
If

$$f_1[k] * f_2[k] = c[k]$$

then+

$$f_1[k - m] * f_2[k - n] = c[k - m - n] \tag{9.46}$$

5. **The Convolution with an Impulse**

$$f[k] * \delta[k] = f[k] \tag{9.47}$$

6. **The Width Property**

If $f_1[k]$ and $f_2[k]$ have lengths of m and n elements respectively, then the length of $c[k]$ is $m + n - 1$ elements.† The width property may appear to be violated in some special cases as explained on p. 136.

Causality and Zero-State Response

In deriving Eq. (9.41), we assumed the system to be linear and time-invariant. There were no other restrictions on either the input signal or the system. In practice, almost all of the input signals are causal, and a majority of the systems are also causal. These restrictions further simplify the limits of the sum in Eq. (9.41). If the input $f[k]$ is causal, $f[m] = 0$ for $m < 0$. Similarly, if the system is causal (that is, if $h[k]$ is causal), then $h[x] = 0$ for negative x, so that $h[k - m] = 0$ when $m > k$. Therefore, if $f[k]$ and $h[k]$ are both causal, the product $f[m]h[k - m] = 0$ for $m < 0$ and for $m > k$, and it is nonzero only for the range $0 \leq m \leq k$. Therefore, Eq. (9.41) in this case reduces to

$$y[k] = \sum_{m=0}^{k} f[m]h[k - m] \tag{9.48}$$

We shall evaluate the convolution sum first by an analytical method and later with graphical aid.

■ **Example 9.6**
Determine $c[k] = f[k] * g[k]$ for

$$f[k] = (0.8)^k u[k] \qquad \text{and} \qquad g[k] = (0.3)^k u[k] \tag{9.49}$$

We have

$$c[k] = \sum_{m=-\infty}^{\infty} f[m]g[k - m]$$

Note that

$$f[m] = (0.8)^m u[m] \qquad \text{and} \qquad g[k - m] = (0.3)^{k-m} u[k - m] \tag{9.50}$$

Both $f[k]$ and $g[k]$ are causal. Therefore, [see Eq. (9.48)]

---

†The width of a signal is one less than the number of its elements (length). For instance, the signal in Fig. 9.3h has 6 elements, but has a width of only 5. Thus, if $f_1[k]$ and $f_2[k]$ have widths of W_1 and W_2, respectively, then the width of $c[k]$ is $W_1 + W_2$.

$$c[k] = \sum_{m=0}^{k} f[m]g[k-m]$$

$$= \sum_{m=0}^{k} (0.8)^m u[m] (0.3)^{k-m} u[k-m] \tag{9.51}$$

In the above summation, m lies between 0 and k ($0 \le m \le k$). Therefore, if $k \ge 0$, then both m and $k - m \ge 0$, so that $u[m] = u[k-m] = 1$. If $k < 0$, m is negative because m lies between 0 and k, and $u[m] = 0$. Therefore, Eq. (9.51) becomes

$$c[k] = \sum_{m=0}^{k} (0.8)^m (0.3)^{k-m} \qquad k \ge 0$$

$$= 0 \qquad k < 0$$

and

$$c[k] = (0.3)^k \sum_{m=0}^{k} \left(\frac{0.8}{0.3}\right)^m u[k]$$

This is a geometric progression with common ratio $(0.8/0.3)$. From Sec. B.7-4 we have

$$c[k] = (0.3)^k \frac{(0.8)^{k+1} - (0.3)^{k+1}}{(0.3)^k (0.8 - 0.3)} u[k]$$

$$= 2 \left[(0.8)^{k+1} - (0.3)^{k+1} \right] u[k] \tag{9.52}$$

△ **Exercise E9.6** ■

Show that $(0.8)^k u[k] * u[k] = 5[1 - (0.8)^{k+1}]u[k]$ ▽

Convolution Sum from a Table

Just as in the continuous-time case, we have prepared a table of convolution sums (Table 9.1) from which convolution sums may be determined directly for a variety of signal pairs. For example, the convolution in Example 9.6 can be read directly from this table (Pair 4) as

$$(0.8)^k u[k] * (0.3)^k u[k] = \frac{(0.8)^{k+1} - (0.3)^{k+1}}{0.8 - 0.3} u[k] = 2[(0.8)^{k+1} - (0.3)^{k+1}]u[k]$$

We shall demonstrate the use of the convolution table in the following example.

■ **Example 9.7**

Find the (zero-state) response $y[k]$ of an LTID system described by the equation

$$y[k+2] - 0.6y[k+1] - 0.16y[k] = 5f[k+2]$$

if the input $f[k] = 4^{-k}u[k]$.

The unit impulse response of this system is obtained in Example 9.5.

$$h[k] = [(-0.2)^k + 4(0.8)^k]u[k] \tag{9.53}$$

Therefore

$$y[k] = f[k] * h[k]$$

$$= (4)^{-k} u[k] * \left[(-0.2)^k u[k] + 4(0.8)^k u[k] \right]$$

$$= (4)^{-k} u[k] * (-0.2)^k u[k] + (4)^{-k} u[k] * 4(0.8)^k u[k] \tag{9.54}$$

TABLE 9.1: Convolution Sums

No.	$f_1[k]$	$f_2[k]$	$f_1[k] * f_2[k] = f_2[k] * f_1[k]$	
1	$\delta[k-j]$	$f[k]$	$f[k-j]$	
2	$\gamma^k u[k]$	$u[k]$	$\left[\dfrac{1-\gamma^{k+1}}{1-\gamma}\right]u[k]$	
3	$u[k]$	$u[k]$	$(k+1)u[k]$	
4	$\gamma_1^k u[k]$	$\gamma_2^k u[k]$	$\left[\dfrac{\gamma_1^{k+1}-\gamma_2^{k+1}}{\gamma_1-\gamma_2}\right]u[k]$ $\quad \gamma_1 \neq \gamma_2$	
5	$\gamma_1^k u[k]$	$\gamma_2^k u[-(k+1)]$	$\dfrac{\gamma_1}{\gamma_2-\gamma_1}\gamma_1^k u[k] + \dfrac{\gamma_2}{\gamma_2-\gamma_1}\gamma_2^k u[-(k+1)]$	$\vert\gamma_2\vert > \vert\gamma_1\vert$
6	$k\gamma_1^k u[k]$	$\gamma_2^k u[k]$	$\dfrac{\gamma_1\gamma_2}{(\gamma_1-\gamma_2)^2}\left[\dfrac{\gamma_2^k-\gamma_1^k}{\gamma_2-\gamma_1} + \dfrac{\gamma_1-\gamma_2}{\gamma_2}k\gamma_1^k\right]u[k]$	$\gamma_1 \neq \gamma_2$
7	$ku[k]$	$ku[k]$	$\dfrac{1}{6}k(k-1)(k+1)u[k]$	
8	$\gamma^k u[k]$	$\gamma^k u[k]$	$(k+1)\gamma^k u[k]$	
9	$\gamma^k u[k]$	$ku[k]$	$\left[\dfrac{\gamma(\gamma^k-1)+k(1-\gamma)}{(1-\gamma)^2}\right]u[k]$	
10	$\vert\gamma_1\vert^k\cos(\beta k+\theta)u[k]$	$\gamma_2^k u[k]$	$\dfrac{1}{R}\left[\vert\gamma_1\vert^{k+1}\cos[\beta(k+1)+\theta-\phi]-\gamma_2^{k+1}\cos(\theta-\phi)\right]u[k]$	γ_2 real
			$R=\left[\vert\gamma_1\vert^2+\gamma_2^2-2\vert\gamma_1\vert\gamma_2\cos\beta\right]^{1/2}$	
			$\phi=\tan^{-1}\left[\dfrac{\vert\gamma_1\vert\sin\beta}{\vert\gamma_1\vert\cos\beta-\gamma_2}\right]$	

Note that
$$(4)^{-k}u[k] = \left(\tfrac{1}{4}\right)^k u[k] = (0.25)^k u[k]$$
Therefore
$$y[k] = (0.25)^k u[k] * (-0.2)^k u[k] + 4(0.25)^k u[k] * (0.8)^k u[k]$$

We use Pair 4 (Table 9.1) to find the above convolution sums.

$$y[k] = \left[\frac{(0.25)^{k+1} - (-0.2)^{k+1}}{0.25 - (-0.2)} + 4\frac{(0.25)^{k+1} - (0.8)^{k+1}}{0.25 - 0.8}\right] u[k]$$

$$= \left(2.22\left[(0.25)^{k+1} - (-0.2)^{k+1}\right] - 7.27\left[(0.25)^{k+1} - (0.8)^{k+1}\right]\right) u[k]$$

$$= \left[-5.05(0.25)^{k+1} - 2.22(-0.2)^{k+1} + 7.27(0.8)^{k+1}\right] u[k]$$

Recognizing that
$$\gamma^{k+1} = \gamma(\gamma)^k$$

We can express $y[k]$ as

$$y[k] = \left[-1.26(0.25)^k + 0.444(-0.2)^k + 5.81(0.8)^k\right] u[k]$$

$$= \left[-1.26(4)^{-k} + 0.444(-0.2)^k + 5.81(0.8)^k\right] u[k] \qquad (9.55)$$

∎

△ **Exercise E9.7**
Show that $(0.8)^{k+1}u[k] * u[k] = 4[1 - 0.8(0.8)^k]u[k]$
Use convolution table. Recognize that $(0.8)^{k+1} = 0.8(0.8)^k$ ▽

△ **Exercise E9.8**
Show that $k\,3^{-k}u[k] * (0.2)^k u[k] = \frac{15}{4}[(0.2)^k - (1 - \frac{2}{3}k)3^{-k}]u[k]$
Hint: Use convolution table. Recognize that $3^{-k} = (\tfrac{1}{3})^k$ ▽

△ **Exercise E9.9**
Using convolution table, show that $e^{-k}u[k] * 2^{-k}u[k] = \frac{2}{2-e}[e^{-k} - \frac{e}{2}2^{-k}]u[k]$
Hint: $e^{-k} = (\tfrac{1}{e})^k$ and $2^{-k} = (0.5)^k$ ▽

⊙ **Computer Example C9.4**
Find and sketch the zero-state response for the system described by

$$(E^2 + 6E + 9)y[k] = (2E^2 + 6E)f[k]$$

for the input $f[k] = 4^k u[k]$.

```
k=0:11;
b=[2 6 0];
a=[1 6 9];
f=4.^(k);
y=filter(b,a,f);
stem(k,y)
xlabel('k');ylabel('y[k]');   ⊙
```

9.4-1 Graphical Procedure for the Convolution Sum

The steps in evaluating the convolution sum are parallel to those followed in evaluating the convolution integral. The convolution sum of causal signals $f[k]$ and $g[k]$ is given by

$$c[k] = \sum_{m=0}^{k} f[m]g[k-m]$$

We first plot $f[m]$ and $g[k-m]$ as functions of m (not k), because the summation is over m. Functions $f[m]$ and $g[m]$ are the same as $f[k]$ and $g[k]$, plotted respectively as functions of m (see Fig. 9.3). The convolution operation can be performed as follows:

1. Invert $g[m]$ about the vertical axis ($m = 0$) to obtain $g[-m]$ (Fig. 9.3d). Figure 9.3e shows both $f[m]$ and $g[-m]$.
2. Time shift $g[-m]$ by k units to obtain $g[k-m]$. For $k > 0$, the shift is to the right (delay); for $k < 0$, the shift is to the left (advance). Figures 9.3f and 9.3g show $g[k-m]$ for $k > 0$ and for $k < 0$, respectively.
3. Next we multiply $f[m]$ and $g[k-m]$ and add all the products to obtain $c[k]$. The procedure is repeated for each value of k over the range $-\infty$ to ∞.

We shall demonstrate by an example the graphical procedure for finding the convolution sum. Although both the functions in this example are causal, the procedure is applicable to general case.

■ **Example 9.8**
Find
$$c[k] = f[k] * g[k]$$

where $f[k]$ and $g[k]$ are depicted in Figs. 9.3a and 9.3b, respectively.
We are given
$$f[k] = (0.8)^k \text{ and } g[k] = (0.3)^k$$
Therefore
$$f[m] = (0.8)^m \quad \text{and} \quad g[k-m] = (0.3)^{k-m}$$

Figure 9.3f shows the general situation for $k \geq 0$. The two functions $f[m]$ and $g[k-m]$ overlap over the interval $0 \leq m \leq k$. Therefore

$$c[k] = \sum_{m=0}^{k} f[m]g[k-m]$$

$$= \sum_{m=0}^{k} (0.8)^m (0.3)^{k-m}$$

$$= (0.3)^k \sum_{m=0}^{k} \left(\frac{0.8}{0.3}\right)^m$$

$$= 2\left[(0.8)^{k+1} - (0.3)^{k+1}\right] \qquad k \geq 0 \qquad \text{(see Sec. B.7-4)}$$

For $k < 0$, there is no overlap between $f[m]$ and $g[k-m]$, as shown in Fig. 9.3g so that

$$c[k] = 0 \qquad k < 0$$

and

$$c[k] = 2\left[(0.8)^{k+1} - (0.3)^{k+1}\right] u[k] \tag{9.56}$$

which agrees with the earlier result in Eq. (9.52). ■

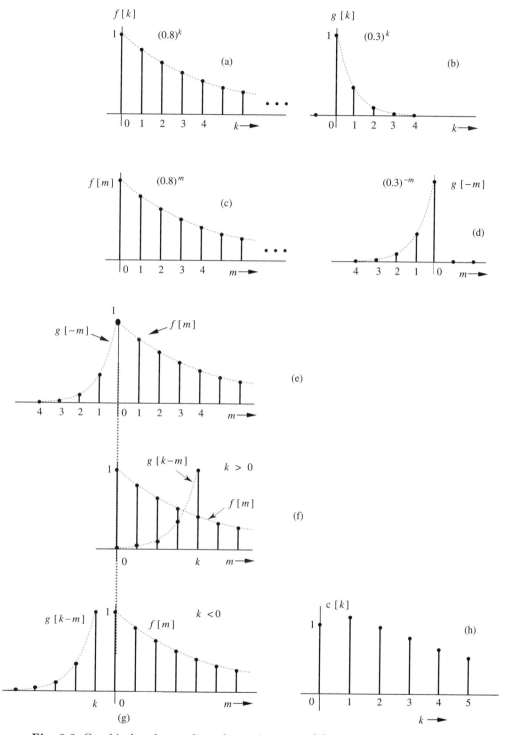

Fig. 9.3 Graphical understanding of convolution of $f[k]$ and $g[k]$ for Example 9.7.

△ **Exercise E9.10**
Find $(0.8)^k u[k] * u[k]$ graphically and sketch the result.
Answer: $5(1 - (0.8)^{k+1})u[k]$ ▽

An Alternative Form of Graphical Procedure: The Sliding Tape Method

This algorithm is convenient when the sequences $f[k]$ and $g[k]$ are short or when they are available only in graphical form. The algorithm is basically the same as the graphical procedure in Fig. 9.3. The only difference is that instead of presenting the data as graphical plots, we display it as a sequence of numbers on tapes. Otherwise the procedure is the same, as will become clear in the following example.

■ **Example 9.9**
Using the sliding tape method, convolve the two sequences $f[k]$ and $g[k]$ depicted in Fig. 9.4a and 9.4b, respectively. In this procedure we write the sequences $f[k]$ and $g[k]$ in the slots of two tapes: f tape and g tape (Fig. 9.4c). Now leave the f tape stationary (to correspond to $f[m]$). The $g[-m]$ tape is obtained by time inverting the $g[m]$ tape about the origin ($k = 0$) so that the slots corresponding to $f[0]$ and $g[0]$ remain aligned (Fig. 9.4d). We now shift the inverted tape by k slots, multiply values on two tapes in adjacent slots, and add all the products to find $c[k]$. Figures 9.4d, e, f, g, h, i, and j show the cases for $k = 0, 1, 2, 3, 4, 5$, and 6, respectively. For the case of $k = 0$, for example (Fig. 9.4d)

$$c[0] = 0 \times 1 = 0$$

For $k = 1$ (Fig. 9.4e)

$$c[1] = (0 \times 1) + (1 \times 1) = 1$$

Similarly,

$$c[2] = (0 \times 1) + (1 \times 1) + (2 \times 1) = 3$$
$$c[3] = (0 \times 1) + (1 \times 1) + (2 \times 1) + (3 \times 1) = 6$$
$$c[4] = (0 \times 1) + (1 \times 1) + (2 \times 1) + (3 \times 1) + (4 \times 1) = 10$$
$$c[5] = (0 \times 1) + (1 \times 1) + (2 \times 1) + (3 \times 1) + (4 \times 1) + (5 \times 1) = 15$$
$$c[6] = (0 \times 1) + (1 \times 1) + (2 \times 1) + (3 \times 1) + (4 \times 1) + (5 \times 1) = 15$$

Figure 9.4j shows that $c[k] = 15$ for $k \geq 5$. Moreover, the two tapes are nonoverlapping for $k < 0$, so that $c[k] = 0$ for $k < 0$. Figure 9.4k shows the plot of $c[k]$.   ■

⊙ **Computer Example C9.5**
Using Matlab, find the convolution of $f[k]$ and $g[k]$ depicted in Fig. 9.5.

```
f=[0 1 2 3 2 1];
g=[1 1 1 1 1 1];
k=0:1:length(f)+length(g)-2;
c=conv(f,g);
stem(k,c)   ⊙
```

An Array Form of Graphical Procedure

The convolution sum can also be obtained from the array formed by sequences $f[k]$ and $g[k]$. This procedure, although convenient from a computational viewpoint, fails to give proper understanding of the convolution mechanism. The procedure is explained in Prob. 9.4-16.

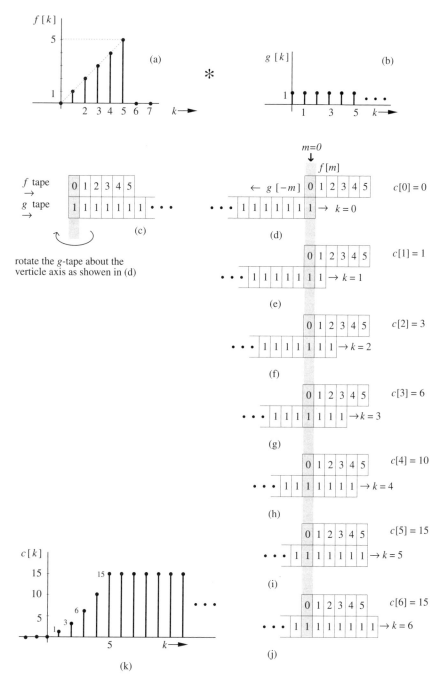

Fig. 9.4 Sliding tape algorithm for discrete-time convolution.

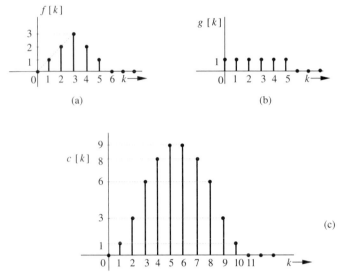

Fig. 9.5 Signals for Exercise E9.11.

△ **Exercise E9.11**

Using the graphical procedure of Example 9.9 (sliding-tape technique), show that $f[k]*g[k] = c[k]$ in Fig. 9.5. Verify the width property of convolution.   ▽

9.4-2   A Very Special Function For LTID Systems: The Everlasting Exponential z^k

In Sec. 2.4-3, we showed that there exists one signal for which the response of an LTIC system is the same as the input within a multiplicative constant. The response of an LTIC system to an everlasting exponential input e^{st} is $H(s)e^{st}$, where $H(s)$ is the system transfer function. We now show that for an LTID system, the same role is played by an everlasting exponential z^k. The system response $y[k]$ in this case is given by

$$y[k] = h[k] * z^k$$

$$= \sum_{m=-\infty}^{\infty} h[m]z^{k-m}$$

$$= z^k \sum_{m=-\infty}^{\infty} h[m]z^{-m}$$

For causal $h[k]$, the limits on the sum on the right-hand side would range from 0 to ∞. In any case, this sum is a function of z. Let us denote it by $H[z]$. Thus,†

$$y[k] = H[z]z^k \tag{9.57a}$$

where

$$H[z] = \sum_{m=-\infty}^{\infty} h[m]z^{-m} \tag{9.57b}$$

---

†This result is valid only for the values of z for which the $\sum_{m=-\infty}^{\infty} h[m]z^{-m}$ exists (or converges).

Note that $H[z]$ is a constant for a given z. Thus, the input and the output are the same (within a multiplicative constant) for the everlasting exponential input z^k.

$H[z]$, which is called the **transfer function** of the system, is a function of the complex variable z. We can define the transfer function $H[z]$ of an LTID system from Eq. (9.57a) as

$$H[z] = \left.\frac{\text{output signal}}{\text{input signal}}\right|_{\text{Input=everlasting exponential } z^k} \tag{9.58}$$

$\triangle$ **Exercise E9.12**

Show that the transfer function of the digital differentiator in Example 8.7 (big shaded block in Fig. 8.20b) is given by $H[z] = \frac{z-1}{Tz}$. $\triangledown$

9.4-3   Total Response

The total response of an LTID system can be expressed as a sum of the zero-input and zero-state components:

$$\text{Total response} = \underbrace{\sum_{j=1}^{n} c_j \gamma_j^k}_{\text{Zero-input component}} + \underbrace{f[k] * h[k]}_{\text{Zero-state component}}$$

In this expression, the zero-input component should be appropriately modified for the case of repeated roots. We have developed procedures to determine these two components. From the system equation we find the characteristic roots and characteristic modes. The zero-input response is a linear combination of the characteristic modes. From the system equation we also determine $h[k]$, the impulse response, as discussed in Sec. 9.3. Knowing $h[k]$ and the input $f[k]$, we find the zero-state response as the convolution of $f[k]$ and $h[k]$. The arbitrary constants $c_1, c_2, \ldots, c_n$ in the zero-input response are determined from the n auxiliary conditions. For the system described by the equation

$$y[k+2] - 0.6y[k+1] - 0.16y[k] = 5f[k+2]$$

with initial conditions $y[-1] = 0, y[-2] = \frac{25}{4}$ and input $f[k] = (4)^{-k}u[k]$, we have determined the two components of the response in Examples 9.3a and 9.7 respectively. From the results in these examples, the total response for $k \geq 0$ is

$$\text{Total response} = \underbrace{0.2(-0.2)^k + 0.8(0.8)^k}_{\text{Zero-input component}} \underbrace{-1.26(4)^{-k} + 0.444(-0.2)^k + 5.81(0.8)^k}_{\text{Zero-state component}}$$

$$\tag{9.59}$$

Natural and Forced Response

The characteristic modes of this system are $(-0.2)^k$ and $(0.8)^k$. The zero-input component is made up of characteristic modes exclusively as expected, but the characteristic modes also appear in the zero-state response. When all the characteristic mode terms in the total response are lumped together, the resulting component is the **natural response**. The remaining part of the total response that is made up of noncharacteristic modes is the **forced response**. For the present case, Eq. (9.59) yields

$$\text{Total response} = \underbrace{0.644(-0.2)^k + 6.61(0.8)^k}_{\text{Natural response}} - \underbrace{1.26(4)^{-k}}_{\text{Forced response}} \qquad k \geq 0 \qquad (9.60)$$

9.5 Classical solution of Linear Difference Equations

As in the case of LTIC systems, we can analyze LTID systems by using the classical method, where the response is obtained as a sum of natural and forced components of the response.

Finding Natural and Forced Response

As explained earlier, the *natural response* of a system consists of all the characteristic mode terms in the response. The remaining noncharacteristic mode terms form the *forced response*. If $y_n[k]$ and $y_\phi[k]$ denote the natural and the forced response respectively, then the total response is given by

$$\text{Total response} = \underbrace{y_n[k]}_{\text{modes}} + \underbrace{y_\phi[k]}_{\text{nonmodes}} \qquad (9.61)$$

Because the total response $y_n[k] + y_\phi[k]$ is a solution of the system equation (9.9), we have

$$Q[E]\,(y_n[k] + y_\phi[k]) = P[E]f[k] \qquad (9.62)$$

But since $y_n[k]$ is made up of characteristic modes,

$$Q[E]y_n[k] = 0 \qquad (9.63)$$

Substitution of this equation in Eq. (9.62) yields

$$Q[E]y_\phi[k] = P[E]f[k] \qquad (9.64)$$

The natural response is a linear combination of characteristic modes. The arbitrary constants (multipliers) are determined from suitable auxiliary conditions usually given as $y[0], y[1], \ldots, y[n-1]$. We now turn our attention to the forced response.

Forced Response

We have shown that the forced response $y_\phi[k]$ satisfies the system Equation (9.64)

$$Q[E]y_\phi[k] = P[E]f[k]$$

By definition, the forced response contains only nonmode terms. To determine the forced response, we shall use a method of undetermined coefficients, the same method used for the continuous-time system. However, rather than retracing all the steps of the continuous-time system, we shall present a table (Table 9.2) listing the inputs and the corresponding forms of forced function with undetermined co-efficients. These coefficients can be determined by substituting $y_\phi[k]$ in Eq. (9.64) and equating the coefficients of similar terms.

Table 9.2

Input $f[k]$		Forced Response $y_\phi[k]$
1.	$r^k \quad r \neq \gamma_i \, (i = 1, 2, \cdots, n)$	cr^k
2.	$r^k \quad r = \gamma_i$	ckr^k
3.	$\cos(\beta k + \theta)$	$c \cos(\beta k + \phi)$
4.	$\left(\displaystyle\sum_{i=0}^{m} a_i k^i \right) r^k$	$\left(\displaystyle\sum_{i=0}^{m} c_i k^i \right) r^k$

Note: By definition, $y_\phi[k]$ cannot have any characteristic mode terms. If any terms shown in the right-hand column for the forced response should also be a characteristic mode of the system, the correct form of the forced response must be modified to $k^i y_\phi[k]$, where i is the smallest integer that will prevent $k^i y_\phi[k]$ from having a characteristic mode term. For example, when the input is r^k, the forced response in the right-hand column is of the form cr^k. But if r^k happens to be a natural mode of the system, the correct form of the forced response is ckr^k (see Pair 2).

■ **Example 9.10**
   Solve

$$(E^2 - 5E + 6)y[k] = (E - 5)f[k] \tag{9.65}$$

if the input $f[k] = (3k + 5)u[k]$ and the auxiliary conditions are $y[0] = 4, y[1] = 13$.
   The characteristic equation is

$$\gamma^2 - 5\gamma + 6 = (\gamma - 2)(\gamma - 3) = 0$$

Therefore, the natural response is

$$y_n[k] = B_1(2)^k + B_2(3)^k \tag{9.66}$$

   To find the form of forced response $y_\phi[k]$, we use Table 9.2, Pair 4 with $r = 1, m = 1$.
This yields

$$y_\phi[k] = c_1 k + c_0$$

Therefore

$$y_\phi[k + 1] = c_1(k + 1) + c_0 = c_1 k + c_1 + c_0$$

$$y_\phi[k + 2] = c_1(k + 2) + c_0 = c_1 k + 2c_1 + c_0$$

Also

$$f[k] = 3k + 5$$

and

$$f[k+1] = 3(k+1) + 5 = 3k + 8$$

Substitution of the above results in Eq. (9.64) yields

$$c_1 k + 2c_1 + c_0 - 5(c_1 k + c_1 + c_0) + 6(c_1 k + c_0) = 3k + 8 - 5(3k + 5)$$

or

$$2c_1 k - 3c_1 + 2c_0 = -12k - 17$$

Comparison of similar terms on the two sides yields

$$\left.\begin{array}{r} 2c_1 = -12 \\ -3c_1 + 2c_0 = -17 \end{array}\right\} \implies \begin{array}{l} c_1 = -6 \\ c_2 = -\frac{35}{2} \end{array}$$

Therefore

$$y_\phi[k] = -6k - \frac{35}{2}$$

The total response is

$$\begin{aligned} y[k] &= y_n[k] + y_\phi[k] \\ &= B_1(2)^k + B_2(3)^k - 6k - \frac{35}{2} \qquad k \geq 0 \end{aligned} \tag{9.67}$$

To determine arbitrary constants B_1 and B_2 we set $k = 0$ and 1 and substitute the initial conditions $y[0] = 4, y[1] = 13$ to obtain

$$\left.\begin{array}{r} 4 = B_1 + B_2 - \frac{35}{2} \\ 13 = 2B_1 + 3B_2 - \frac{47}{2} \end{array}\right\} \implies \begin{array}{l} B_1 = 28 \\ B_2 = \frac{-13}{2} \end{array}$$

Therefore

$$y_n[k] = 28(2)^k - \frac{13}{2}(3)^k \tag{9.68}$$

and

$$y[k] = \underbrace{28(2)^k - \frac{13}{2}(3)^k}_{y_n[k]} \underbrace{- 6k - \frac{35}{2}}_{y_\phi[k]} \tag{9.69}$$

∎

⊙ **Computer Example C9.6**
Solve Example 9.10 using MATLAB.

```
%(c96.m)
Y=[4 13]; Y=Y';
k=0:10; k=k';
f=[5 8];f=f';
for m=1:length(k)-2
    y=5*Y(m+1)-6*Y(m)+f(m+1)-5*f(m);
    Y=[Y; y];
    F=3*(m+1)+5;f=[f;F];
end
stem(k,Y)    ⊙
```

A Comment on Initial Conditions

This method requires auxiliary conditions $y[0], y[1], \ldots, y[n-1]$ for the reasons explained on p. 142. If we are given the initial conditions $y[-1], y[-2], \ldots, y[-n]$, we can derive the conditions $y[0], y[1], \ldots, y[n-1]$ using iterative procedure.

An Exponential Input

As in the case of continuous-time systems, we can show that for a system specified by the equation

$$Q[E]y[k] = P[E]f[k] \tag{9.70}$$

the forced response for the exponential input $f[k] = r^k$ is given by

$$y_\phi[k] = H[r]r^k \qquad r \neq \gamma_i \tag{9.71}$$

where

$$H[r] = \frac{P[r]}{Q[r]} \tag{9.72}$$

The proof follows from the fact that if the input $f[k] = r^k$, then from Table 9.2 (Pair 4), $y_\phi[k] = cr^k$. Therefore

$$E^i f[k] = f[k+i] = r^{k+i} = r^i r^k \qquad \text{and} \qquad P[E]f[k] = P[r]r^k$$
$$E^j y_\phi[k] = y_\phi[k+1] = cr^{k+j} = cr^j r^k \qquad \text{and} \qquad Q[E]y[k] = cQ[r]r^k$$

so that Eq. (9.70) reduces to

$$cQ[r]r^k = P[r]r^k$$

which yields $c = P[r]/Q[r] = H[r]$.

This result is valid only if r is not a characteristic root of the system. If r is a characteristic root, the forced response is ckr^k where c is determined by substituting

$y_\phi[k]$ in the system equation and equating coefficients of similar terms on the two sides. Observe that the exponential r^k includes a wide variety of signals such as a constant C, a sinusoid $\cos(\beta k + \theta)$, and an exponentially growing or decaying sinusoid $|\gamma|^k \cos(\beta k + \theta)$.

1. A Constant Input $f[k] = C$

This is a special case of exponential Cr^k with $r = 1$. Therefore, from Eq. (9.71), we have

$$y_\phi[k] = C\frac{P[1]}{Q[1]} = CH[1] \tag{9.73}$$

2. A Sinusoidal Input

The input $e^{j\beta k}$ is an exponential r^k with $r = e^{j\beta}$. Hence

$$y_\phi[k] = H[e^{j\beta}]e^{j\beta k} = \frac{P[e^{j\beta}]}{Q[e^{j\beta}]}e^{j\beta k}$$

Similarly, for the input $e^{-j\beta k}$

$$y_\phi[k] = H[e^{-j\beta}]e^{-j\beta k}$$

Consequently, if the input $f[k] = \cos \beta k = \frac{1}{2}(e^{j\beta k} + e^{-j\beta k})$,

$$y_\phi[k] = \frac{1}{2}\left\{H[e^{j\beta}]e^{j\beta k} + H[e^{-j\beta}]e^{-j\beta k}\right\}$$

Since the two terms on the right-hand side are conjugates

$$y_\phi[k] = \text{Re}\left\{H[e^{j\beta}]e^{j\beta k}\right\}$$

If

$$H[e^{j\beta}] = |H[e^{j\beta}]|e^{j\angle H[e^{j\beta}]}$$

$$y_\phi[k] = \text{Re}\left\{\left|H[e^{j\beta}]\right|e^{j(\beta k + \angle H[e^{j\beta}])}\right\}$$

$$= |H[e^{j\beta}]|\cos\left(\beta k + \angle H[e^{j\beta}]\right) \qquad (9.74a)$$

Using a similar argument, we can show that for the input

$$f[k] = \cos\left(\beta k + \theta\right)$$

$$y_\phi[k] = |H[e^{j\beta}]|\cos\left(\beta k + \theta + \angle H[e^{j\beta}]\right) \qquad (9.74b)$$

■ **Example 9.11**

For a system specified by the equation

$$(E^2 - 3E + 2)y[k] = (E + 2)f[k]$$

find the forced response for the input $f[k] = (3)^k u[k]$.

In this case

$$H[r] = \frac{P[r]}{Q[r]} = \frac{r + 2}{r^2 - 3r + 2}$$

and the forced response to input $(3)^k u[k]$ is $H[3](3)^k$; that is,

$$y_\phi[k] = \frac{3 + 2}{(3)^2 - 3(3) + 2}(3)^k = \frac{5}{2}(3)^k \qquad k \ge 0 \quad ■$$

■ **Example 9.12**

For an LTID system described by the equation

$$(E^2 - E + 0.16)y[k] = (E + 0.32)f[k]$$

determine the forced response $y_\phi[k]$ if the input is

$$f[k] = \cos\left(2k + \frac{\pi}{3}\right)u[k]$$

Here

$$H[r] = \frac{P[r]}{Q[r]} = \frac{r + 0.32}{r^2 - r + 0.16}$$

For the input $\cos\left(2k + \frac{\pi}{3}\right)u[k]$, the forced response is

$$y_\phi[k] = \left|H[e^{j2}]\right| \cos\left(2k + \frac{\pi}{3} + \angle H[e^{j2}]\right) u[k]$$

where

$$H[e^{j2}] = \frac{e^{j2} + 0.32}{(e^{j2})^2 - e^{j2} + 0.16} = \frac{(-0.416 + j0.909) + 0.32}{(-0.654 - j0.757) - (-0.416 + j0.909) + 0.16}$$

$$= 0.548 e^{j3.294}$$

Therefore

$$\left|H[e^{j2}]\right| = 0.548 \qquad \text{and} \qquad \angle H[e^{j2}] = 3.294$$

so that

$$y_\phi[k] = 0.548 \cos\left(2k + \frac{\pi}{3} + 3.294\right)u[k]$$

$$= 0.548 \cos\left(2k + 4.34\right)u[k] \quad \blacksquare$$

Assessment of the Classical Method

The remarks in Chapter 2 concerning the classical method for solving differential equations (p. 147) also apply to difference equations.

9.6 System Stability

Just as in a continuous-time system, we define a discrete-time system to be **asymptotically stable** if, and only if, the zero-input response approaches zero as $k \to \infty$. If the zero-input response grows without bound as $k \to \infty$, the system is **unstable**. If the zero-input response neither approaches zero nor grows without bound, but remains within a finite limit as $k \to \infty$, the system is **marginally stable**. In the last case, the zero-input response approaches a constant or oscillates with a constant amplitude. Recall that the zero-input response consists of the characteristic modes of the system. The mode corresponding to a characteristic root γ is γ^k. To be more general, let γ be complex so that

$$\gamma = |\gamma|e^{j\beta} \qquad \text{and} \qquad \gamma^k = |\gamma|^k e^{j\beta k}$$

Since the magnitude of $e^{j\beta k}$ is always unity regardless of the value of k, the magnitude of γ^k is $|\gamma|^k$. Therefore

$$\text{if } |\gamma| < 1, \quad \gamma^k \to 0 \qquad \text{as } k \to \infty$$

$$\text{if } |\gamma| > 1, \quad \gamma^k \to \infty \qquad \text{as } k \to \infty$$

$$\text{and if } |\gamma| = 1, \quad |\gamma|^k = 1 \qquad \text{for all } k$$

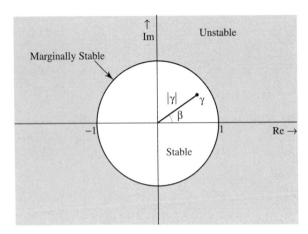

Fig. 9.6 Characteristic roots location and system stability.

It is clear that a system is asymptotically stable if and only if

$$|\gamma_i| < 1 \qquad i = 1, 2, \cdots, n$$

These results can be grasped more effectively in terms of the location of characteristic roots in the complex plane. Figure 9.6 shows a circle of unit radius, centered at the origin in a complex plane. Our discussion clearly shows that if all characteristic roots of the system lie inside this circle (**unit circle**), $|\gamma_i| < 1$ for all i and the system is asymptotically stable. On the other hand, even if one characteristic root lies outside the unit circle, the system is unstable. If none of the characteristic roots lie outside the unit circle, but some simple (unrepeated) roots lie on the circle itself, the system is marginally stable. If two or more characteristic roots coincide on the unit circle (repeated roots), the system is unstable. The reason is that for repeated roots, the zero-input response is of the form $k^{r-1}\gamma^k$, and if $|\gamma| = 1$, then $|k^{r-1}\gamma^k| = k^{r-1} \to \infty$ as $k \to \infty$.† Note, however, that repeated roots inside the unit circle do not cause instability. Figure 9.7 shows the characteristic modes corresponding to characteristic roots at various locations in the complex plane. To summarize:

1. An LTID system is asymptotically stable if and only if all the characteristic roots are inside the unit circle. The roots may be simple or repeated.

2. An LTID system is unstable if and only if either one or both of the following conditions exist: (i) at least one root is outside the unit circle; (ii) there are repeated roots on the unit circle.

3. An LTID system is marginally stable if and only if there are no roots outside the unit circle and there are some unrepeated roots on the unit circle.

---

†If the development of discrete-time systems is parallel to that of continuous-time systems, we wonder why the parallel breaks down here. Why, for instance, aren't LHP and RHP the regions demarcating stability and instability? The reason lies in the form of the characteristic modes. In continuous-time systems we chose the form of characteristic mode as $e^{\lambda_i t}$. In discrete-time systems we choose the form (for computational convenience) to be γ_i^k. Had we chosen this form to be $e^{\lambda_i k}$ where $\gamma_i = e^{\lambda_i}$, then LHP and RHP (for the location of λ_i) again would demarcate stability and instability. The reason is that if $\gamma = e^\lambda$, $|\gamma| = 1$ implies $|e^\lambda| = 1$, and therefore $\lambda = j\omega$. This shows that the unit circle in γ plane maps into the imaginary axis in the λ plane.

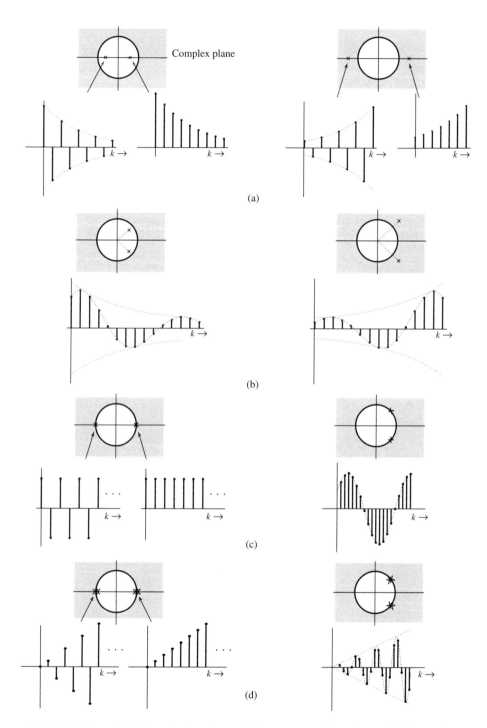

Fig. 9.7 Characteristic roots location and the corresponding characteristic modes.

■ **Example 9.13**

Determine whether the systems specified by the following equations are asymptotically stable, marginally stable, or unstable. In each case plot the characteristic roots in the complex plane.

(a) $y[k+2] + 2.5y[k+1] + y[k] = f[k+1] - 2f[k]$

(b) $y[k] - y[k-1] + 0.21y[k-2] = 2f[k-1] + 3f[k-2]$

(c) $y[k+3] + 2y[k+2] + \frac{3}{2}y[k+1] + \frac{1}{2}y[k] = f[k+1]$

(d) $(E^2 - E + 1)^2 y[k] = (3E + 1)f[k]$

(a) The characteristic polynomial is

$$\gamma^2 + 2.5\gamma + 1 = (\gamma + 0.5)(\gamma + 2)$$

The characteristic roots are -0.5 and -2. Because $|-2| > 1$ (-2 lies outside the unit circle), the system is unstable (Fig. 9.8a).

(b) The characteristic polynomial is

$$\gamma^2 - \gamma + 0.21 = (\gamma - 0.3)(\gamma - 0.7)$$

The characteristic roots are 0.3 and 0.7, both of which lie inside the unit circle. The system is asymptotically stable (Fig. 9.8b).

(c) The characteristic polynomial is

$$\gamma^3 + 2\gamma^2 + \frac{3}{2}\gamma + \frac{1}{2} = (\gamma + 1)(\gamma^2 + \gamma + \frac{1}{2}) = (\gamma + 1)(\gamma + 0.5 - j0.5)(\gamma + 0.5 + j0.5)$$

The characteristic roots are -1, $-0.5 \pm j0.5$ (Fig. 9.8c). One of the characteristic roots is on the unit circle and the remaining two roots are inside the unit circle. The system is marginally stable.

(d) The characteristic polynomial is

$$(\gamma^2 - \gamma + 1)^2 = \left(\gamma - \frac{1}{2} - j\frac{\sqrt{3}}{2}\right)^2 \left(\gamma - \frac{1}{2} + j\frac{\sqrt{3}}{2}\right)^2$$

The characteristic roots are $\frac{1}{2} \pm j\frac{\sqrt{3}}{2} = 1e^{\pm j\frac{\pi}{3}}$ repeated twice, and they lie on the unit circle (Fig. 9.8d). The system is unstable.   ■

△ **Exercise E9.13**

Find and sketch the location in the complex plane of the characteristic roots of the system specified by the following equation:

$$(E + 1)(E^2 + 6E + 25)y[k] = 3Ef[k]$$

Is this a stable, unstable, or marginally stable system? Answer: unstable.   ▽

△ **Exercise E9.14**

Repeat Prob. E9.13 for

$$(E - 1)^2(E + 0.5)y[k] = (E^2 + 2E + 3)f[k]$$

Answer: unstable.   ▽

9.6-1 System Response to Bounded Inputs

As in the case of continuous-time systems, asymptotically stable discrete-time systems have the property that every bounded input produces a bounded output.

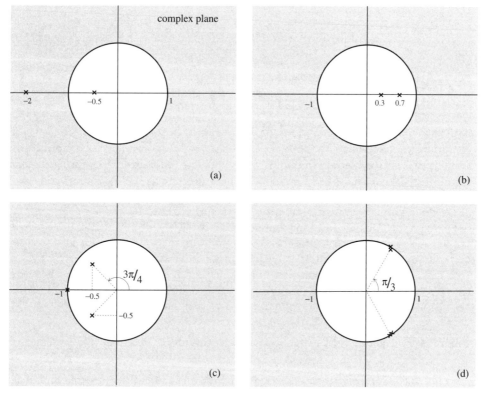

Fig. 9.8 Location of characteristic roots for systems in Example 9.13.

In contrast, for a marginally stable and unstable system it is possible to find a bounded input that produces unbounded output. To prove this result we recall that

$$y[k] = h[k] * f[k]$$

$$= \sum_{m=-\infty}^{\infty} h[m]f[k-m]$$

and

$$|y[k]| = \left| \sum_{m=-\infty}^{\infty} h[m]f[k-m] \right|$$

$$\leq \sum_{m=-\infty}^{\infty} |h[m]| \, |f[k-m]|$$

If $f[k]$ is bounded, then $|f[k-m]| < K_1 < \infty$, and

$$|y[k]| \leq K_1 \sum_{m=-\infty}^{\infty} |h[m]|$$

Clearly the output is bounded if the summation on the right-hand side is bounded; that is, if

$$\sum_{k=-\infty}^{\infty} |h[k]| < K_2 < \infty \tag{9.75}$$

For an asymptotically stable system

$$h[k] = (c_1\gamma_1^k + c_2\gamma_2^k + \cdots + c_n\gamma_n^k)u[k]$$

with the characteristic roots $\gamma_1, \gamma_2, \ldots \gamma_n$ lying within the unit circle. Under these conditions $h[k]$ satisfies Eq. (9.75).† Therefore, the response of an asymptotically stable system to a bounded input is also bounded. Moreover, we can show that for an unstable or marginally stable system, the response is unbounded for some bounded input. As seen in Sec. 2.6-1, these results lead to the formulation of an alternate definition of system stability. A system is said to be stable in the sense of having bounded output for every bounded input (BIBO stability) if and only if its impulse response $h[k]$ satisfies Eq. (9.75). Condition (9.75) is sufficient for a system to produce bounded output for any bounded input. It is relatively easy to show that it is also a necessary condition; that is, for a system that violates (9.75), there exists a bounded input that produces unbounded output (see Prob. 9.6-2). Note that an asymptotically stable system is always BIBO-stable.‡

9.6-2 Intuitive Insights Into System Behavior

The intuitive insights into the behavior of continuous-time systems and their qualitative proofs, discussed in Sec. 2.7, also apply to discrete-time systems. For this reason, we shall merely mention here some of these insights without discussion or proof. The interested reader should read Sec. 2.7 for more explanation.

The system's behavior is strongly influenced by the characteristic roots (or modes) of the system. The system responds strongly to input signals similar to its characteristic modes, and responds poorly to inputs very different from its characteristic modes. In fact, when the input is a characteristic mode of the system, the response goes to infinity provided that the mode is a nondecaying signal. This is the resonance phenomenon. The width of an impulse response $h[k]$ indicates the response time (time required to respond fully to an input) of the system. It is the time constant of the system.† Discrete-time pulses are generally dispersed when

---

†This conclusion follows from the fact that (see Sec. B.7-4)

$$\sum_{k=-\infty}^{\infty} |\gamma_i^k|u[k] = \sum_{k=0}^{\infty} |\gamma_i|^k = \frac{1}{1 - |\gamma_i|} \qquad |\gamma_i| < 1$$

Therefore, if $|\gamma_i| < 1$ for $i = 1, 2, \ldots, n$,

$$\sum_{k=-\infty}^{\infty} |h[k]| \le \sum_{i=1}^{n} \frac{1}{1 - |\gamma_i|} < \infty$$

Although we have assumed all the roots to be distinct in this derivation, it is also valid for repeated roots provided that they lie inside the unit circle.

‡The converse is not true. See footnote on p. 152.

†This part of the discussion applies to systems with impulse response $h[k]$ that is a mostly positive (or mostly negative) pulse.

passed through a discrete-time system. The amount of dispersion (or spreading out) is equal to the system time constant (or width of $h[k]$).

9.7 Appendix 9.1: Determining the Impulse Response

For a discrete-time system specified by Eq. (9.27), we have demonstrated that

$$h[k] = A_0\delta[k] + y_n[k]u[k] \tag{9.76}$$

We now show that

$$A_0 = \frac{b_0}{a_0} \tag{9.77}$$

To prove this point, we substitute Eq. (9.76) in Eq. (9.28) to obtain

$$Q[E]\,(A_0\delta[k] + y_n[k]u[k]) = P[E]\delta[k] \tag{9.78}$$

Because $y_n[k]u[k]$ is a sum of characteristic modes,‡ [see Eq. (9.12a)]

$$Q[E]\,(y_n[k]u[k]) = 0 \quad k \geq 0 \tag{9.79}$$

Equation (9.78) now reduces to

$$A_0 Q[E]\delta[k] = P[E]\delta[k] \qquad k \geq 0$$

or

$$A_0(\delta[k+n] + a_{n-1}\delta[k+n-1] + \cdots + a_1\delta[k+1] + a_0\delta[k]) = b_n\delta[k+n]$$
$$+ b_{n-1}\delta[k+n-1] + \cdots + b_1\delta[k+1] + b_0\delta[k] \quad k \geq 0$$

If we set $k = 0$ in this equation and recognize that $\delta[0] = 1$ and $\delta[m] = 0$ when $m \neq 0$, all but the last terms vanish on both sides, yielding

$$A_0 a_0 = b_0$$

and

$$A_0 = \frac{b_0}{a_0}$$

A Special Case: $a_0 = 0$

In this case $A_0 = \frac{b_0}{a_0}$ becomes indeterminate. The procedure has to be modified slightly. When $a_0 = 0$, $Q[E]$ can be expressed as $E\hat{Q}[E]$, and Eq. (9.28) can be expressed as

$$E\hat{Q}[E]h[k] = P[E]\delta[k]$$

If we recognize that $\frac{1}{E}$ is a delay operator, this equation can be rearranged as

---

‡Note that
$$Q[E]\,(y_n[k]) = 0$$
for all k. But if we restrict the mode terms to be causal, the equation is valid only for $k \geq 0$; that is,
$$Q[E]\,(y_n[k]u[k]) = 0 \qquad k \geq 0$$

$$\hat{Q}[E]h[k] = P[E]\left(\tfrac{1}{E}\delta[k]\right) = P[E]\delta[k-1]$$

In this case the input vanishes not for $k \geq 1$, but for $k \geq 2$. Therefore, the response consists not only of the zero-input term and an impulse $A_0\delta[k]$ (at $k = 0$), but also of an impulse $A_1\delta[k-1]$ (at $k = 1$). Therefore

$$h[k] = A_0\delta[k] + A_1\delta[k-1] + y_n[k]u[k]$$

We can determine the unknowns A_0, A_1, and the n coefficients in $y_n[k]$ from the $n+2$ number of initial values $h[0]$, $h[1]$, $\cdots$, $h[n+1]$, determined as usual from the iterative solution of the equation $Q[E]h[k] = P[E]\delta[k]$. Similarly, if $a_0 = a_1 = 0$, we need to use the form $h[k] = A_0\delta[k] + A_1\delta[k-1] + A_2\delta[k-2] + y_n[k]u[k]$. The $n+3$ unknown constants are determined from the $n+3$ values $h[0]$, $h[1]$, $\cdots$, $h[n+1]$, $h[n+2]$, determined iteratively, and so on.

9.8 Summary

This chapter discusses time-domain analysis of LTID (linear, time-invariant, discrete-time) systems. The analysis is parallel to that of LTIC systems, with some minor differences. Discrete-time systems are described by difference equations. For an nth-order system, n auxiliary conditions must be specified for a unique solution to an input starting at $k = 0$. Characteristic modes are discrete-time exponentials of the form γ^k corresponding to an unrepeated root γ, and the modes are of the form $k^i\gamma^k$ corresponding to a repeated root γ.

The unit impulse function $\delta[k]$ is a sequence of a single number of unit value at $k = 0$. The unit impulse response $h[k]$ of a discrete-time system is a linear combination of its characteristic modes.†

The zero-state response (response due to external input) of a linear system is obtained by breaking the input into impulse components and then adding the system responses to all the impulse components. The sum of the system responses to the impulse components is in the form of a sum, known as the convolution sum, whose structure and properties are similar to the convolution integral. The system response is obtained as the convolution sum of the input $f[k]$ with the system's impulse response $h[k]$. Therefore, the knowledge of the system's impulse response allows us to determine the system response to any arbitrary input.

LTID systems have a very special relationship to the everlasting exponential signal z^k because the response of an LTID system to such an input signal is the same signal within a multiplicative constant. The response of an LTID system to the everlasting exponential input z^k is $H[z]z^k$, where $H[z]$ is the transfer function of the system.

The stability criterion in terms of the location of characteristic roots of the system can be summarized as follows:

1. An LTID system is asymptotically stable if and only if all the characteristic roots are inside the unit circle. The roots may be repeated or unrepeated.

2. An LTID system is unstable if and only if either one or both of the following

---

†There is a possibility of an impulse $\delta[k]$ in addition to characteristic modes.

conditions exist: (i) at least one root is outside the unit circle, (ii) there are repeated roots on the unit circle.

3. An LTID system is marginally stable if and only if there are no roots outside the unit circle, and there are some unrepeated roots on the unit circle.

According to an alternate definition of stability—the bounded-input bounded-output (BIBO) stability—a system is stable if and only if every bounded input produces a bounded output. Otherwise the system is unstable. An asymptotically stable system is always BIBO-stable. The converse is not necessarily true.

Difference equations of LTID systems can also be solved by the classical method, where the response is obtained as a sum of natural and forced responses. These are not the same as the zero-input and zero-state components, although they satisfy the same equations, respectively. Although simple, this method suffers from the fact that it is applicable to a restricted class of input signals, and the system response cannot be expressed as an explicit function of the input. These limitations make it useless in the theoretical study of systems.

Problems

9.1-1 Solve iteratively (first three terms only):

(a) $y[k + 1] - 0.5y[k] = 0$, with $y[-1] = 10$

(b)$y[k + 1] + 2y[k] = f[k + 1]$, with $f[k] = e^{-k}u[k]$ and $y[-1] = 0$

9.1-2 Solve the following equation iteratively (first three terms only):

$$y[k] - 0.6y[k - 1] - 0.16y[k - 2] = 0 \text{ with } y[-1] = -25,\ y[-2] = 0$$

9.1-3 Solve iteratively the second-order difference Eq. (8.26b) in chapter 8 (first three terms only), assuming $y[-1] = y[-2] = 0$ and $f[k] = 100u[k]$.

9.1-4 Solve the following equation iteratively (first three terms only):

$$y[k + 2] + 3y[k + 1] + 2y[k] = f[k + 2] + 3f[k + 1] + 3f[k]$$

with $f[k] = (3)^k u[k]$, $y[-1] = 3$, and $y[-2] = 2$

9.1-5 Repeat Prob. 9.1-4 if

$$y[k] + 2y[k - 1] + y[k - 2] = 2f[k] - f[k - 1]$$

with $f[k] = (3)^{-k}u[k]$, $y[-1] = 2$, and $y[-2] = 3$.

9.2-1 Solve

$$y[k + 2] + 3y[k + 1] + 2y[k] = 0 \text{ if } y[-1] = 0,\ y[-2] = 1$$

9.2-2 Solve

$$y[k + 2] + 2y[k + 1] + y[k] = 0 \text{ if } y[-1] = 1,\ y[-2] = 1$$

9.2-3 Solve

$$y[k + 2] - 2y[k + 1] + 2y[k] = 0 \text{ if } y[-1] = 1,\ y[-2] = 0$$

9.2-4 Find $v[k]$, the voltage at the kth node of the resistive ladder depicted in Fig. P8.5-5 in Chapter 8, if $V = 100$ volts and $a = 2$.

Hint: The difference equation for $v[k]$ is given in Prob. 8.4-5. The auxiliary conditions are $v[0] = 100$ and $v[N] = 0$.

9.3-1 Find the unit impulse response $h[k]$ of a system specified by the equation

$$y[k + 1] + 2y[k] = f[k]$$

9.3-2 Repeat Prob. 9.3-1 if

$$y[k + 1] + 2y[k] = f[k + 1]$$

9.3-3 Repeat Prob. 9.3-1 if

$$(E^2 - 6E + 9)y[k] = Ef[k]$$

9.3-4 Repeat Prob. 9.3-1 if

$$y[k] - 6y[k - 1] + 25y[k - 2] = 2f[k] - 4f[k - 1]$$

9.3-5 For the general nth-order difference Eq. (9.9), if

$$a_0 = a_1 = a_2 = \cdots = a_{n-1} = 0$$

the resulting equation is called a **nonrecursive** difference equation.
(a) Show that the impulse response of a system described by this equation is

$$h[k] = b_n \delta[k] + b_{n-1}\delta[k - 1] + \cdots + b_1\delta[k - n + 1] + b_0\delta[k - n]$$

Hint: Express the system equation in delay-operator form.
(b) Find the impulse response of a nonrecursive LTID system described by the equation

$$y[k] = 3f[k] - 5f[k - 1] - 2f[k - 3]$$

Observe that the impulse response has only a finite (n) number of nonzero elements. For this reason, such systems are called **finite impulse response (FIR)** systems. For a general recursive case [Eq. (9.9)], the impulse response has an infinite number of nonzero elements, and such systems are called **infinite impulse response (IIR)** systems.

9.4-1 Find the (zero-state) response $y[k]$ of an LTID system whose unit impulse response is

$$h[k] = (-2)^k u[k]$$

and the input is $f[k] = e^{-k}u[k]$.

9.4-2 Repeat Prob. 9.4-1 if

$$h[k] = \tfrac{1}{2}[\delta[k] - (-2)^k]u[k]$$

9.4-3 Repeat Prob. 9.4-1 if the input $f[k] = (3)^{k+2}u[k]$, and

$$h[k] = [(2)^k + 3(-5)^k]u[k]$$

9.4-4 Repeat Prob. 9.4-1 if the input $f[k] = (3)^{-k}u[k]$, and

$$h[k] = 3k(2)^k u[k]$$

9.4-5 Repeat Prob. 9.4-1 if the input $f[k] = (2)^k u[k]$, and

$$h[k] = (3)^k \cos\left(\tfrac{\pi}{3}k - 0.5\right)u[k]$$

9.4-6 Find the total response of a system specified by the equation

$$y[k+1] + 2y[k] = f[k+1]$$

if $y[-1] = 10$, and the input $f[k] = e^{-k}u[k]$.

9.4-7 Repeat Prob. 9.4-1 if the impulse response $h[k] = (0.5)^k u[k]$, and the input $f[k]$ is
(a) $2^k u[k]$ (b) $2^{(k-3)}u[k]$ (c) $2^k u[k-2]$.

Hint: You may need to use the shift property (9.46) of the convolution.

9.4-8 In the savings account problem described in Example 8.5 (chapter 8), a person deposits $500 at the beginning of every month, starting at $k = 0$ with the exception at $k = 4$, when instead of depositing $500, she withdraws $1000. Find $y[k]$ if the interest rate is 1% per month ($r = 0.01$).

Hint: Because the deposit starts at $k = 0$, the initial condition $y[-1] = 0$. Withdrawal is a deposit of negative amount.

9.4-9 To pay off a loan of M dollars in N payments using a fixed monthly payment of P dollars, show that

$$P = \frac{r(1+r)^N}{(1+r)^N - 1}M$$

where r is the interest rate per dollar per month.

Hint: This problem can be modeled by Eq. (8.24) with the payments of P dollars starting at $k = 1$. The problem can be approached in two ways: (1) Consider the loan as the initial condition $y_0[0] = -M$, and the input $f[k] = Pu[k-1]$. The loan balance is the sum of the zero-input component (due to the initial condition) and the zero-state component $h[k] * f[k]$. (2) In the second approach the loan is considered as an input $-M$ at $k = 0$; that is, $-M\delta[k]$. The total input is, therefore, $f[k] = -M\delta[k] + Pu[k-1]$, and the loan balance is $h[k] * f[k]$. Because the loan is paid off in N payments, set $y[N] = 0$.

9.4-10 A person receives an automobile loan of $10,000 from a bank at the interest rate of 1.5% per month. His monthly payment is $500, with the first payment due one month after he receives the loan. Compute the number of payments required to pay off the loan. Note that the last payment may not be exactly $500.

Hint: Follow the procedure in Prob. 9.4-9 to determine the balance $y[k]$. To determine N, the number of payments, set $y[N] = 0$. In general, N will not be an integer. The number of payments K is the largest integer $\le N$. The residual payment is $|y[K]|$.

9.4-11 Using the sliding-tape algorithm, show that

(a) $u[k] * u[k] = (k+1)u[k]$

(b) $(u[k] - u[k-m]) * u[k] = (k+1)u[k] - (k-m+1)u[k-m]$

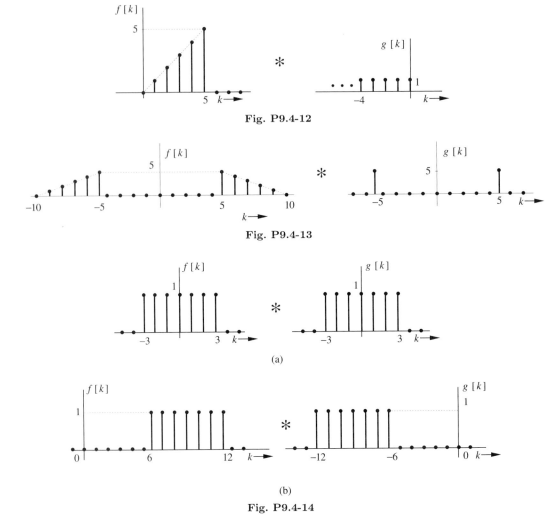

Fig. P9.4-12

Fig. P9.4-13

(a)

(b)

Fig. P9.4-14

9.4-12   Using the sliding-tape algorithm, find $f[k] * g[k]$ for the signals shown in Fig. P9.4-12.

9.4-13   Repeat Prob. 9.4-12 for the signals shown in Fig. P9.4-13.

9.4-14   Repeat Prob. 9.4-12 for the signals depicted in Fig. P9.4-14.

9.4-15   The convolution sum in Eq. (9.48) can be expressed in a matrix form as

$$
\underbrace{\begin{bmatrix} y[0] \\ y[1] \\ \cdots \\ y[k] \end{bmatrix}}_{\mathbf{y}}
\underbrace{\begin{bmatrix} h[0] & 0 & 0 & \cdots & 0 \\ h[1] & h[0] & 0 & \cdots & 0 \\ \cdots\cdots\cdots\cdots\cdots\cdots\cdots\cdots\cdots \\ h[k] & h[k-1] & \cdots & \cdots & h[0] \end{bmatrix}}_{\mathbf{H}}
\underbrace{\begin{bmatrix} f[0] \\ f[1] \\ \cdots \\ f[k] \end{bmatrix}}_{\mathbf{f}}
$$

or

$$\mathbf{y} = \mathbf{H}\mathbf{f}$$

and
$$\mathbf{f} = \mathbf{H}^{-1}\mathbf{y}$$

Knowing $h[k]$ and the output $y[k]$, we can determine the input $f[k]$. This operation is the reverse of the convolution and is known as the **deconvolution**. Moreover, knowing $f[k]$ and $y[k]$, we can determine $h[k]$. This can be done by expressing the above matrix equation as $k + 1$ simultaneous equations in terms of $k + 1$ unknowns $h[0]$, $h[1]$, ... , $h[k]$. These equations can readily be solved iteratively. Thus, we can synthesize a system that yields a certain output $y[k]$ for a given input $f[k]$.

(a) Design a system (that is, determine $h[k]$) that will yield the output sequence (8, 12, 14, 15, 15.5, 15.75, $\cdots$) for the input sequence (1, 1, 1, 1, 1, 1, $\cdots$).

(b) For a system with the impulse response sequence (1, 2, 4, $\cdots$), the output sequence was (1, 7/3, 43/9, $\cdots$). Determine the input sequence.

9.4-16 The sliding-tape method is conceptually quite valuable in understanding the convolution mechanism. Numerical convolution can also be performed from the arrays using the sets $f[0]$, $f[1]$, $f[2]$, ..., and $g[0]$, $g[1]$, $g[2]$, ..., as depicted in Fig. P9.4-16. The ijth element (element in the ith row and jth column) is given by $g[i]f[j]$. We add the elements of the array along its diagonals to produce $c[k] = f[k] * g[k]$. For example, if we sum the elements corresponding to the first diagonal of the array, we obtain $c[0]$. Similarly, if we sum along the second diagonal, we obtain $c[1]$, and so on. Draw the array for the signals $f[k]$ and $g[k]$ in Example 9.9, and find $f[k] * g[k]$.

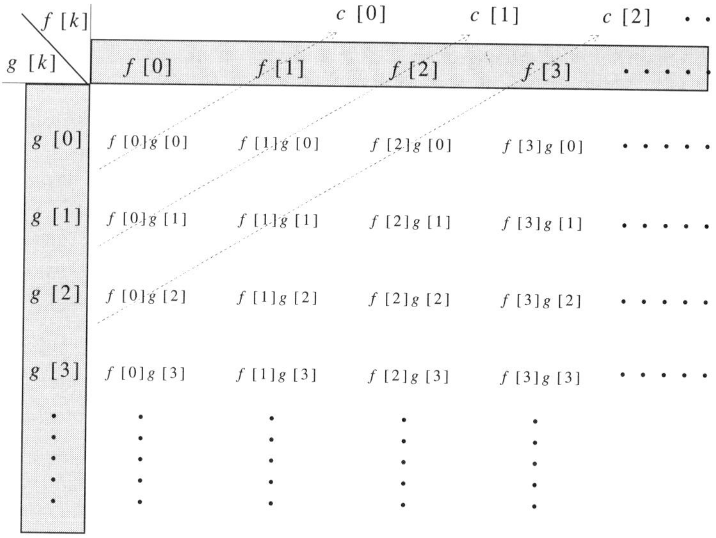

Fig. P9.4-16

9.4-17 Using Eq. (9.58), show that the transfer function of a unit delay is $H[z] = 1/z$.

9.5-1 Using the classical method, solve
$$y[k + 1] + 2y[k] = f[k + 1]$$
with the input $f[k] = e^{-k}u[k]$, and the auxiliary condition $y[0] = 1$.

9.5-2 Using the classical method, solve
$$y[k] + 2y[k - 1] = f[k - 1]$$

with the input $f[k] = e^{-k}u[k]$ and the auxiliary condition $y[-1] = 0$. Hint: You will have to determine the auxiliary condition $y[0]$ using the iterative method.

9.5-3 **(a)** Using the classical method, solve

$$y[k+2] + 3y[k+1] + 2y[k] = f[k+2] + 3f[k+1] + 3f[k]$$

with the input $f[k] = (3)^k$ and the auxiliary conditions $y[0] = 1, y[1] = 3$.
(b) Repeat **(a)** if the auxiliary conditions are $y[-1] = y[-2] = 1$. Hint: Using the iterative method, determine $y[0]$ and $y[1]$.

9.5-4 Using the classical method, solve

$$y[k] + 2y[k-1] + y[k-2] = 2f[k] - f[k-1]$$

with the input $f[k] = 3^{-k}u[k]$ and the auxiliary conditions $y[0] = 2$ and $y[1] = -\frac{13}{3}$.

9.5-5 Using the classical method, solve

$$(E^2 - E + 0.16)y[k] = Ef[k]$$

with the input $f[k] = (0.2)^k u[k]$ and the auxiliary conditions $y[0] = 1, y[1] = 2$. Hint: The input is a natural mode of the system.

9.5-6 Using the classical method, solve

$$(E^2 - E + 0.16)y[k] = Ef[k]$$

with the input $f[k] = \cos\left(\frac{\pi k}{2} + \frac{\pi}{3}\right)u[k]$ and the initial conditions $y[-1] = y[-2] = 0$. Hint: Find $y[0]$ and $y[1]$ iteratively.

9.6-1 Each of the following equations specifies an LTID system. Determine whether these systems are asymptotically stable, unstable, or marginally stable.

(a) $y[k+2] + 0.6y[k+1] - 0.16y[k] = f[k+1] - 2f[k]$

(b) $(E^2 + 1)(E^2 + E + 1)y[k] = Ef[k]$

(c) $(E-1)^2(E + \frac{1}{2})y[k] = (E + 2)f[k]$

(d) $y[k] + 2y[k-1] + 0.96y[k-2] = 2f[k-1] + 3f[k-3]$

(e) $(E^2 - 1)(E^2 + 1)y[k] = f[k]$

9.6-2 In Sec. 9.6 we showed that for BIBO stability in an LTID system, it is sufficient for its impulse response $h[k]$ to satisfy Eq. (9.75). Show that this is also a necessary condition for the system to be BIBO-stable. In other words, show that if Eq. (9.75) is not satisfied, there exists a bounded input that produces unbounded output.
Hint: Assume that a system exists for which $h[k]$ violates Eq. (9.75), yet its output is bounded for every bounded input. Establish contradiction in this statement by considering an input $f[k]$ defined by $f[k_1 - m] = 1$ when $h[m] > 0$ and $f[k_1 - m] = -1$ when $h[m] < 0$, where k_1 is some fixed integer.

9.6-3 Show that a marginally stable system is BIBO-unstable. Verify your result by considering a system with characteristic roots on the unit circle and show that for the input of the form of the natural mode (which is bounded), the response is unbounded.

Fourier Analysis of Discrete-Time Signals

In Chapters 3, 4, and 6, we studied the ways of representing a continuous-time signal as a sum of sinusoids or exponentials. In this chapter we shall discuss similar development for discrete-time signals. Our approach is parallel to that used for continuous-time signals. We first represent a periodic $f[k]$ as a Fourier series formed by a discrete-time exponential (or sinusoid) and its harmonics. Later we extend this representation to an aperiodic signal $f[k]$ by considering $f[k]$ as a limiting case of a periodic signal with the period approaching infinity.

10.1 Periodic Signal Representation by Discrete-Time Fourier Series (DTFS)

A periodic signal of period N_0 is called an N_0-periodic signal. Figure 8.9 shows an example of a periodic signal of period 6. A continuous-time periodic signal of period T_0 can be represented as a trigonometric Fourier series consisting of a sinusoid of the fundamental frequency $\omega_0 = \frac{2\pi}{T_0}$, and all its harmonics (sinusoids of frequencies that are integral multiples of ω_0). The exponential form of the Fourier series consists of exponentials e^{j0t}, $e^{\pm j\omega_0 t}$, $e^{\pm j2\omega_0 t}$, $e^{\pm j3\omega_0 t}$, ... For a parallel development of the discrete time case, recall that the frequency of a sinusoid of period N_0 is $\Omega_0 = 2\pi/N_0$. Hence, an N_0-periodic discrete-time signal $f[k]$ can be represented by a discrete-time Fourier series with fundamental frequency $\Omega_0 = \frac{2\pi}{N_0}$ and its harmonics. As in the continuous-time case, we may use a trigonometric or an exponential form of the Fourier series. Because of its compactness and ease of mathematical manipulations, the exponential form is preferable to the trigonometric. For this reason we shall bypass the trigonometric form and go directly to the exponential form of the discrete-time Fourier series.

The exponential Fourier series consists of the exponentials e^{j0k}, $e^{\pm j\Omega_0 k}$, $e^{\pm j2\Omega_0 k}$, ..., $e^{\pm jn\Omega_0 k}$, ..., and so on. There would be an infinite number of harmonics, except for the property proved in Sec. 8.2: that discrete-time exponentials whose frequencies are separated by 2π (or integral multiples of 2π) are identical because

$$e^{j(\Omega \pm 2\pi)k} = e^{j\Omega k}e^{\pm 2\pi k} = e^{j\Omega k} \qquad (10.1)$$

The consequence of this result is that the rth harmonic is identical to the $(r+N_0)$th harmonic. To demonstrate this, let g_n denote the nth harmonic $e^{jn\Omega_0 k}$. Then

$$g_{r+N_0} = e^{j(r+N_0)\Omega_0 k} = e^{j(r\Omega_0 k + 2\pi k)} = e^{jr\Omega_0 k} = g_r \qquad (10.2)$$

and

$$g_r = g_{r+N_0} = g_{r+2N_0} = \cdots = g_{r+mN_0} \qquad m, \text{ integer} \qquad (10.3)$$

Thus, the first harmonic is identical to the (N_0+1)st harmonic, the second harmonic is identical to the (N_0+2)nd harmonic, and so on. In other words, there are only N_0 independent harmonics, and they range over an interval 2π (because the harmonics are separated by $\Omega_0 = \frac{2\pi}{N_0}$. We may choose these N_0 independent harmonics as $e^{jr\Omega_0 k}$ over $0 \le r \le N_0 - 1$, or over $-1 \le r \le N_0 - 2$, or over $1 \le r \le N_0$, or over any other suitable choice for that matter. Every one of these sets will have the same harmonics, although in different order. Let us take the first choice ($0 \le r \le N_0-1$). This choice corresponds to exponentials $e^{jr\Omega_0 k}$ for $r = 0, 1, 2, \ldots, N_0 - 1$. The Fourier series for an N_0-periodic signal $f[k]$ consists of only these N_0 harmonics, and can be expressed as

$$f[k] = \sum_{r=0}^{N_0-1} \mathcal{D}_r e^{jr\Omega_0 k} \qquad \Omega_0 = \frac{2\pi}{N_0} \qquad (10.4)$$

To compute coefficients $\mathcal{D}_r$ in the Fourier series (10.4), we multiply both sides of (10.4) by $e^{-jm\Omega_0 k}$ and sum over k from $k = 0$ to $(N_0 - 1)$.

$$\sum_{k=0}^{N_0-1} f[k]e^{-jm\Omega_0 k} = \sum_{k=0}^{N_0-1} \sum_{r=0}^{N_0-1} \mathcal{D}_r e^{j(r-m)\Omega_0 k} \qquad (10.5)$$

The right-hand sum, after interchanging the order of summation, results in

$$\sum_{r=0}^{N_0-1} \mathcal{D}_r \left[\sum_{k=0}^{N_0-1} e^{j(r-m)\Omega_0 k} \right] \qquad (10.6)$$

The inner sum, according to Eq. (5.43), is zero for all values of $r \ne m$. It is nonzero with a value N_0 only when $r = m$. This fact means the outside sum has only one term $\mathcal{D}_m N_0$ (corresponding to $r = m$). Therefore, the right-hand side of Eq. (10.5) is equal to $\mathcal{D}_m N_0$, and

$$\sum_{k=0}^{N_0-1} f[k]e^{-jm\Omega_0 k} = \mathcal{D}_m N_0$$

and

$$\mathcal{D}_m = \frac{1}{N_0} \sum_{k=0}^{N_0-1} f[k]e^{-jm\Omega_0 k} \qquad (10.7)$$

We now have a discrete-time Fourier series (DTFS) representation of an N_0-periodic signal $f[k]$ as

$$f[k] = \sum_{r=0}^{N_0-1} \mathcal{D}_r e^{jr\Omega_0 k} \tag{10.8}$$

where

$$\mathcal{D}_r = \frac{1}{N_0} \sum_{k=0}^{N_0-1} f[k] e^{-jr\Omega_0 k} \qquad \Omega_0 = \frac{2\pi}{N_0} \tag{10.9}$$

Observe that DTFS equations (10.8) and (10.9) are identical (within a scaling constant) to the DFT equations (5.18b) and (5.18a).† Therefore, we can compute the DTFS coefficients using the efficient FFT algorithm.

10.1-1 Fourier Spectra of a Periodic Signal $f[k]$

The Fourier series consists of N_0 components

$$\mathcal{D}_0, \; \mathcal{D}_1 e^{j\Omega_0 k}, \; \mathcal{D}_2 e^{j2\Omega_0 k}, \; \ldots, \; \mathcal{D}_{N_0-1} e^{j(N_0-1)\Omega_0 k}$$

The frequencies of these components are $0, \Omega_0, 2\Omega_0, \ldots, (N_0-1)\Omega_0$ where $\Omega_0 = 2\pi/N_0$. The amount of the rth harmonic is $\mathcal{D}_r$. We can plot this amount $\mathcal{D}_r$ (the Fourier coefficient) as a function of Ω. Such a plot, called the **Fourier spectrum** of $f[k]$, gives us, at a glance, the graphical picture of the amounts of various harmonics of $f[k]$.

In general, the Fourier coefficients $\mathcal{D}_r$ are complex, and they can be represented in the polar form as

$$\mathcal{D}_r = |\mathcal{D}_r| e^{j\angle \mathcal{D}_r} \tag{10.10}$$

The plot of $|\mathcal{D}_r|$ vs. Ω is called the amplitude spectrum and that of $\angle \mathcal{D}_r$ vs. Ω is called the angle (or phase) spectrum. These two plots together are the frequency spectra of $f[k]$. Knowing these spectra, we can reconstruct or synthesize $f[k]$ according to Eq. (10.8). Therefore, the Fourier (or frequency) spectra, which are an alternative way of describing a signal $f[k]$, are in every way equivalent (in terms of the information) to the plot of $f[k]$ as a function of k. The Fourier spectra of a signal constitute the **frequency-domain** description of $f[k]$, in contrast to the time-domain description, where $f[k]$ is specified as a function of time (k).

The results are very similar to the representation of a continuous-time periodic signal by an exponential Fourier series except that, generally, the continuous-time signal spectrum bandwidth is infinite, and consists of an infinite number of exponential components (harmonics). The spectrum of the discrete-time periodic signal, in contrast, is bandlimited and has at most N_0 components.

Periodic Extension of Fourier Spectrum

Note that if $\phi[r]$ is an N_0-periodic function of r, then

$$\sum_{r=0}^{N_0-1} \phi[r] = \sum_{r=<N_0>} \phi[r] \tag{10.11}$$

---

†If we let $f[k] = N_0 f_k$ and $\mathcal{D}_r = F_r$, Eqs. (10.8) and (10.9) are identical to Eqs. (5.18b) and (5.18a), respectively.

where $r = <N_0>$ indicates summation over any N_0 consecutive values of r. This follows because the right-hand side of Eq. (10.11) is the sum of all the N_0 consecutive values of $\phi[r]$. Because $\phi[r]$ is periodic, this sum must be the same regardless of where we start the first term. Now $e^{-jr\Omega_0 k}$ is N_0-periodic because

$$e^{-jr\Omega_0(k+N_0)} = e^{-jr\Omega_0 k}e^{-j2\pi r} = e^{-jr\Omega_0 k}$$

Therefore, if $f[k]$ is N_0-periodic, $f[k]e^{-jr\Omega_0 k}$ is also N_0-periodic. Hence, from Eq. (10.9) it follows that $\mathcal{D}_r$ is also N_0-periodic, as is $\mathcal{D}_r e^{jr\Omega_0 k}$. Now, because of property (10.11), we can express Eqs. (10.8) and (10.9) as

$$f[k] = \sum_{r=<N_0>} \mathcal{D}_r e^{jr\Omega_0 k} \tag{10.12}$$

and

$$\mathcal{D}_r = \frac{1}{N_0} \sum_{k=<N_0>} f[k]e^{-jr\Omega_0 k} \tag{10.13}$$

If we plot $\mathcal{D}_r$ for all values of r (rather than only $0 \leq r \leq N_0 - 1$), then the spectrum $\mathcal{D}_r$ is N_0-periodic. Moreover, Eq. (10.12) shows that $f[k]$ can be synthesized by not only the N_0 exponentials corresponding to $0 \leq r \leq N_0 - 1$, but by any successive N_0 exponentials in this spectrum, starting at any value of r (positive or negative). For this reason, it is customary to show the spectrum $\mathcal{D}_r$ for all values of r (not just over the interval $0 \leq r \leq N_0 - 1$). Yet we must remember that to synthesize $f[k]$ from this spectrum, we need to add only N_0 consecutive components.

The spectral components $\mathcal{D}_r$ are separated by the frequency $\Omega_0 = \frac{2\pi}{N_0}$, and there are a total of N_0 components repeating periodically along the Ω axis. Thus, on the frequency scale Ω, $\mathcal{D}_r$ repeats every 2π intervals. Equations (10.12) and (10.13) show that both $f[k]$ and its spectrum $\mathcal{D}_r$ are periodic and both have exactly the same number of components (N_0) over one period. The period of $f[k]$ is N_0 and that of $\mathcal{D}_r$ is 2π radians.

Equation (10.13) shows that $\mathcal{D}_r$ is complex in general, and $\mathcal{D}_{-r}$ is the conjugate of $\mathcal{D}_r$ if $f[k]$ is real. Thus

$$|\mathcal{D}_r| = |\mathcal{D}_{-r}| \quad \text{and} \quad \angle\mathcal{D}_r = -\angle\mathcal{D}_{-r} \tag{10.14}$$

so that the amplitude spectrum $|\mathcal{D}_r|$ is an even function, and $\angle\mathcal{D}_r$ is an odd function of r (or Ω). All these concepts will be clarified by the examples to follow.

■ **Example 10.1**

Find the discrete-time Fourier series (DTFS) for $f[k] = \sin 0.1\pi k$ (Fig. 10.1a). Sketch the amplitude and phase spectra.

In this case the sinusoid $\sin 0.1\pi k$ is periodic because $\frac{\Omega}{2\pi} = \frac{1}{20}$ is a rational number and the period N_0 is [see Eq. (8.9b)]

$$N_0 = m\left(\frac{2\pi}{\Omega}\right) = m\left(\frac{2\pi}{0.1\pi}\right) = 20m$$

The smallest value of m that makes $20m$ an integer is $m = 1$. Therefore, the period $N_0 = 20$, so that $\Omega_0 = \frac{2\pi}{N_0} = 0.1\pi$, and from Eq. (10.12)

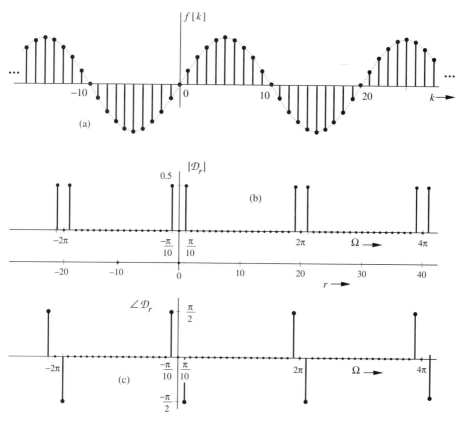

Fig. 10.1 Discrete-time sinusoid $\sin 0.1\pi k$ and its Fourier spectra.

$$f[k] = \sum_{r=<20>} \mathcal{D}_r e^{j0.1\pi rk}$$

where the sum is performed over any 20 consecutive values of r. We shall select the range $-10 \leq r < 10$ (values of r from -10 to 9). This choice corresponds to synthesizing $f[k]$ using the spectral components in the fundamental frequency range ($-\pi \leq \Omega < \pi$). Thus,

$$f[k] = \sum_{r=-10}^{9} \mathcal{D}_r e^{j0.1\pi rk}$$

where, according to Eq. (10.13),

$$\mathcal{D}_r = \frac{1}{20} \sum_{k=-10}^{9} \sin 0.1\pi k\, e^{-j0.1\pi rk}$$

$$= \frac{1}{20} \sum_{k=-10}^{9} \frac{1}{2j} \left(e^{j0.1\pi k} - e^{-j0.1\pi k} \right) e^{-j0.1\pi rk}$$

$$= \frac{1}{40j} \left[\sum_{k=-10}^{9} e^{j0.1\pi k(1-r)} - \sum_{k=-10}^{9} e^{-j0.1\pi k(1+r)} \right]$$

In these sums, r takes on all values between -10 and 9. From Eq. (5.43) it follows that the first sum on the right-hand side is zero for all values of r except $r = 1$, when the sum is equal to $N_0 = 20$. Similarly, the second sum is zero for all values of r except $r = -1$, when it is equal to $N_0 = 20$. Therefore

$$\mathcal{D}_1 = \frac{1}{2j} \quad \text{and} \quad \mathcal{D}_{-1} = -\frac{1}{2j}$$

and all other coefficients are zero. The corresponding Fourier series is given by

$$f[k] = \sin 0.1\pi k = \frac{1}{2j}\left(e^{j0.1\pi k} - e^{-j0.1\pi k}\right) \tag{10.15}$$

Here the fundamental frequency $\Omega_0 = 0.1\pi$, and there are only two nonzero components:

$$\mathcal{D}_1 = \frac{1}{2j} = \frac{1}{2}e^{-j\pi/2}$$

$$\mathcal{D}_{-1} = -\frac{1}{2j} = \frac{1}{2}e^{j\pi/2}$$

Therefore

$$|\mathcal{D}_1| = |\mathcal{D}_{-1}| = \frac{1}{2}$$

$$\angle \mathcal{D}_1 = -\frac{\pi}{2} \quad \text{and} \quad \angle \mathcal{D}_{-1} = \frac{\pi}{2}$$

Figures 10.1b and c shows the sketch of $\mathcal{D}_r$ for the interval $(-10 \le r < 10)$. According to Eq. (10.15a), there are only two components corresponding to $r = 1$ and -1. The remaining 18 coefficients are zero. The rth component $\mathcal{D}_r$ is the amplitude of the frequency $r\Omega_0 = 0.1r\pi$. Therefore, the frequency interval corresponding to $-10 \le r < 10$ is $-\pi \le \Omega < \pi$, as depicted in Figs. 10.1b and c. This spectrum in the interval $-10 \le r < 10$ (or $-\pi \le \Omega < \pi$) is sufficient to specify the frequency-domain description (Fourier series), and we can synthesize $f[k]$ by adding these spectral components. Because of the periodicity property discussed in Sec. 10.1-1, the spectrum $\mathcal{D}_r$ is a periodic function of r with period $N_0 = 20$. For this reason, we repeat the spectrum with period $N_0 = 20$ (or $\Omega = 2\pi$), as illustrated in Figs. 10.1b and c, which are periodic extensions of the spectrum in the range $-10 \le r < 10$. Observe that the amplitude spectrum is an even function and the angle or phase spectrum is an odd function of r (or Ω) as expected.

The result (10.15) is a trigonometric identity, and could have been obtained immediately without the formality of finding the Fourier coefficients. We have intentionally chosen this trivial example to introduce the reader gently to the new concept of the discrete-time Fourier series and its periodic nature. The Fourier series is a way of expressing a periodic signal $f[k]$ in terms of exponentials of the form $e^{jr\Omega_0 k}$ and its harmonics. The result in Eq. (10.15) is merely a statement of the (obvious) fact that $\sin 0.1\pi k$ can be expressed as a sum of two exponentials $e^{j0.1\pi k}$ and $e^{-j0.1\pi k}$.

Because of the periodicity property of the discrete-time exponentials $e^{jr\Omega_0 k}$, the Fourier series components can be selected in any range of length $N_0 = 20$ (or $\Omega = 2\pi$). For example, if we select the frequency range $0 \le \Omega < 2\pi$ (or $0 \le r < 20$), we obtain the Fourier series as

$$f[k] = \sin 0.1\pi k = \frac{1}{2j}\left(e^{j0.1\pi k} - e^{-j1.9\pi k}\right) \tag{10.16}$$

This series is equivalent to that in Eq. (10.15) because, as seen in Sec. 8.2, the two exponentials $e^{j1.9\pi k}$ and $e^{-j0.1\pi k}$ are identical.

We could have selected the spectrum over any other range of width $\Omega = 2\pi$ in Figs. 10.1b and c as a valid discrete-time Fourier series. The reader may satisfy himself by proving that such a spectrum starting anywhere (and of width $\Omega = 2\pi$) is equivalent to the same two components on the right-hand side of Eq. (10.15). ∎

△ **Exercise E10.1**

From the spectrum in Fig. 10.1 write the Fourier series corresponding to the interval $-10 \geq r > -30$ (or $-\pi \geq \Omega > -3\pi$). Show that this Fourier is equivalent to that in Eq. (10.15). ▽

△ **Exercise E10.2**

Find the period and the DTFS for

$$f[k] = 4\cos 0.2\pi k + 6\sin 0.5\pi k$$

over the interval $0 \leq r \leq 19$. Use Eq. (10.9) to compute $\mathcal{D}_r$.
Answer:

$$N_0 = 20, \quad \text{and} \quad f[k] = 2e^{j0.2\pi k} + (3e^{-j\pi/2})e^{j0.5\pi k} + (3e^{j\pi/2})e^{j1.5\pi k} + 2e^{j1.8\pi k} \quad ▽$$

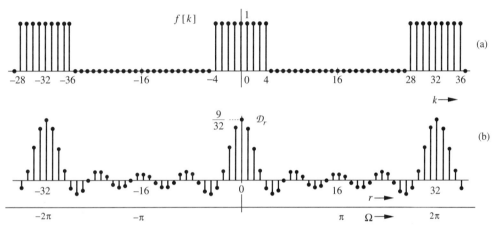

Fig. 10.2 Periodic sampled gate pulse and its Fourier spectrum.

■ **Example 10.2**

Find the discrete-time Fourier series for the periodic sampled gate function shown in Fig. 10.2a.

In this case $N_0 = 32$ and $\Omega_0 = \frac{2\pi}{32} = \frac{\pi}{16}$. Therefore

$$f[k] = \sum_{r=<32>} \mathcal{D}_r e^{jr\frac{\pi}{16}k} \tag{10.17}$$

where

$$\mathcal{D}_r = \frac{1}{32} \sum_{k=<32>} f[k] e^{-jr\frac{\pi}{16}k} \tag{10.18}$$

For our convenience, we shall choose the interval $-16 \leq k \leq 15$ for the summation (10.18), although any other interval of the same width (32 points) would give the same result.

$$\mathcal{D}_r = \frac{1}{32} \sum_{k=-16}^{15} f[k] e^{-jr\frac{\pi}{16}k}$$

Now $f[k] = 1$ for $-4 \leq k \leq 4$ and is zero for all other values of k. Therefore

$$\mathcal{D}_r = \frac{1}{32} \sum_{k=-4}^{4} e^{-jr\frac{\pi}{16}k}$$

This is a geometric progression with a common ratio $e^{-j\frac{\pi}{16}r}$. Therefore,[see Sec. (B.7-4)]

$$\mathcal{D}_r = \frac{1}{32}\left[\frac{e^{-j\frac{5\pi r}{16}} - e^{j\frac{4\pi r}{16}}}{e^{-j\frac{\pi r}{16}} - 1}\right]$$

$$= \left(\frac{1}{32}\right)\frac{e^{-j\frac{0.5\pi r}{16}}\left[e^{-j\frac{4.5\pi r}{16}} - e^{j\frac{4.5\pi r}{16}}\right]}{e^{-j\frac{0.5\pi r}{16}}\left[e^{-j\frac{0.5\pi r}{16}} - e^{j\frac{0.5\pi r}{16}}\right]}$$

$$= \left(\frac{1}{32}\right)\frac{\sin\left(\frac{4.5\pi r}{16}\right)}{\sin\left(\frac{0.5\pi r}{16}\right)}$$

$$= \left(\frac{1}{32}\right)\frac{\sin\left(4.5r\Omega_0\right)}{\sin\left(0.5r\Omega_0\right)} \qquad \Omega_0 = \frac{\pi}{16} \qquad\qquad (10.19)$$

This spectrum (with its periodic extension) is depicted in Fig. 10.2b.† ∎

⊙ **Computer Example C10.1**
Do Example 10.2 using MATLAB.

```
N0=32;k=0:N0-1;
f=[ones(1,5) zeros(1,23) ones(1,4)];
Fr=1/32*fft(f);
r=k;
stem(k,Fr),grid    ⊙
```

10.2   Aperiodic Signal representation by Fourier Integral

In Sec. 10.1 we succeeded in representing periodic signals as a sum of (everlasting) exponentials. In this section we extend this representation to aperiodic signals. The procedure is identical conceptually to that in Chapter 4 used for continuous-time signals.

Applying a limiting process, we now show that aperiodic signals $f[k]$ can be expressed as a continuous sum (integral) of everlasting exponentials. To represent an aperiodic signal $f[k]$ such as the one illustrated in Fig. 10.3a by everlasting exponential signals, let us construct a new periodic signal $f_{N_0}[k]$ formed by repeating the signal $f[k]$ every N_0 units , as shown in Fig. 10.3b. The period N_0 is made long enough to avoid overlap between the repeating cycles ($N_0 \geq 2N + 1$). The periodic signal $f_{N_0}[k]$ can be represented by an exponential Fourier series. If we let $N_0 \to \infty$, the signal $f[k]$ repeats after an infinite interval, and therefore

$$\lim_{N_0\to\infty} f_{N_0}[k] = f[k]$$

---

†In this example we have used the same equations as those for DFT in Example C5.2, with a minor difference. In the present example, the values of $f[k]$ at $k = 4$ and -4 are taken as 1, whereas in Example 5.3 these values are 0.5. This is the reason for the slight difference in spectra in Fig. 10.2b and Fig. 5.16d. Unlike continuous-time signals, discrete-time signals can have no discontinuity.

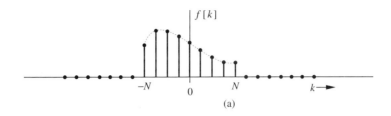

(a)

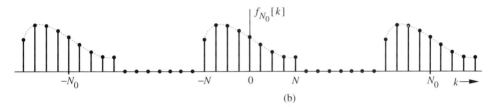

(b)

Fig. 10.3   Generation of a periodic signal by periodic extension of a signal $f[k]$.

Thus, the Fourier series representing $f_{N_0}[k]$ will also represent $f[k]$ in the limit $N_0 \to \infty$. The exponential Fourier series for $f_{N_0}[k]$ is given by

$$f_{N_0}[k] = \sum_{r=<N_0>} \mathcal{D}_r e^{jr\Omega_0 k} \qquad \Omega_0 = \frac{2\pi}{N_0} \qquad (10.20)$$

where

$$\mathcal{D}_r = \frac{1}{N_0} \sum_{k=-\infty}^{\infty} f[k] e^{-jr\Omega_0 k} \qquad (10.21)$$

The limits for the sum on the right-hand side of Eq. (10.21) should be from $-N$ to N. But because $f[k] = 0$ for $|k| > N$, it does not matter if the limits are taken from $-\infty$ to ∞.

It is interesting to see how the nature of the spectrum changes as N_0 increases. To understand this fascinating behavior, let us define $F(\Omega)$, a continuous function of Ω, as

$$F(\Omega) = \sum_{k=-\infty}^{\infty} f[k] e^{-j\Omega k} \qquad (10.22)$$

From this definition and Eq. (10.21), we have

$$\mathcal{D}_r = \frac{1}{N_0} F(r\Omega_0) \qquad (10.23)$$

This result shows that the Fourier coefficients $\mathcal{D}_r$ are $(1/N_0)$ times the samples of $F(\Omega)$ taken every Ω_0 rad/s.† Therefore, $(1/N_0)F(\Omega)$ is the envelope for the coefficients $\mathcal{D}_r$. We now let $N_0 \to \infty$ by doubling N_0 repeatedly. Doubling N_0 halves the fundamental frequency Ω_0 so the spacing between successive spectral components (harmonics) is halved, and there are now twice as many components (samples) in

---

†For the sake of simplicity we assume $\mathcal{D}_r$ and therefore $F(\Omega)$ to be real. The argument, however, is also valid for complex $\mathcal{D}_r$ [or $F(\Omega)$].

the spectrum. At the same time, by doubling N_0, the envelope of the coefficients $\mathcal{D}_r$ is halved, as seen from Eq. (10.23). If we continue this process of doubling N_0 repeatedly, the number of components doubles in each step; the spectrum progressively becomes denser while its magnitude $\mathcal{D}_r$ becomes smaller. Note, however, that the relative shape of the envelope remains the same [proportional to $F(\Omega)$ in Eq. (10.22)]. In the limit, as $N_0 \to \infty$, the fundamental frequency $\Omega_0 \to 0$, and $\mathcal{D}_r \to 0$. The separation between successive harmonics, which is Ω_0, is approaching zero (infinitesimal), and the spectrum becomes so dense that it appears continuous. But as the number of harmonics increases indefinitely, the harmonic amplitudes $\mathcal{D}_r$ become vanishingly small (infinitesimal). We have a strange situation of having **nothing of everything**. This phenomenon is already discussed in Chapter 4, where we showed that these are the classic characteristics of a familiar phenomenon (the density function).

Let us see what happens mathematically as the period $N_0 \to \infty$. According to Eq. (10.22)

$$F(r\Omega_0) = \sum_{k=-\infty}^{\infty} f[k]e^{-jr\Omega_0 k} \tag{10.24}$$

Using Eqs. (10.23) and (10.21), we can express Eq. (10.20) as

$$f_{N_0}[k] = \frac{1}{N_0} \sum_{r=<N_0>} F(r\Omega_0)e^{jr\Omega_0 k} \tag{10.25a}$$

$$= \sum_{r=<N_0>} F(r\Omega_0)e^{jr\Omega_0 k} \left(\frac{\Omega_0}{2\pi}\right) \tag{10.25b}$$

In the limit as $N_0 \to \infty$, $\Omega_0 \to 0$ and $f_{N_0}[k] \to f[k]$. Therefore

$$f[k] = \lim_{\Omega_0 \to 0} \sum_{r=<N_0>} \left[\frac{F(r\Omega_0)\Omega_0}{2\pi}\right] e^{jr\Omega_0 k} \tag{10.26}$$

As $N_0 \to 0$, Ω_0 becomes infinitesimal ($\Omega_0 \to 0$). For this reason it will be appropriate to replace Ω_0 with an infinitesimal notation $\triangle\Omega$:

$$\triangle\Omega = \frac{2\pi}{N_0} \tag{10.27}$$

Equation (10.26) can be expressed as

$$f[k] = \lim_{\triangle\Omega \to 0} \sum_{r=<N_0>} \left[\frac{F(r\triangle\Omega)\triangle\Omega}{2\pi}\right] e^{jr\triangle\Omega k} \tag{10.28}$$

$$= \lim_{\triangle\Omega \to 0} \frac{1}{2\pi} \sum_{r=<N_0>} F(r\triangle\Omega)e^{jr\triangle\Omega k}\triangle\Omega \tag{10.29}$$

The range $r = <N_0>$ implies the interval of N_0 number of harmonics, which is $N_0\triangle\Omega = 2\pi$ according to Eq. (10.27). In the limit, the right-hand side of Eq. (10.29) becomes the integral

$$f[k] = \frac{1}{2\pi} \int_{2\pi} F(\Omega) e^{jk\Omega} \, d\Omega \tag{10.30}$$

where $\int_{2\pi}$ indicates integration over any continuous interval of 2π. The spectrum $F(\Omega)$ is given by [Eq. (10.22)]

$$F(\Omega) = \sum_{k=-\infty}^{\infty} f[k] e^{-j\Omega k} \tag{10.31}$$

The integral on the right-hand side of Eq. (10.30) is called the **Fourier integral**. We have now succeeded in representing an aperiodic signal $f[k]$ by a Fourier integral (rather than a Fourier series). This integral is basically a Fourier series (in the limit) with fundamental frequency $\Delta\Omega \to 0$, as seen in Eq. (10.28). The amount of the exponential $e^{jr\Delta\Omega k}$ is $F(r\Delta\Omega)\Delta\Omega/2\pi$. Thus, the function $F(\Omega)$ given by Eq. (10.31) acts as a spectral function, which indicates the relative amounts of various exponential components of $f[k]$.

We call $F(\Omega)$ the (direct) discrete-time Fourier transform (DTFT) of $f[k]$, and $f[k]$ the inverse discrete-time Fourier transform (IDTFT) of $F(\Omega)$. This can be represented as

$$F(\Omega) = \mathcal{F}\{f[k]\} \qquad \text{and} \qquad f[k] = \mathcal{F}^{-1}\{F(\Omega)\}$$

The same information is conveyed by the statement that $f[k]$ and $F(\Omega)$ are a (discrete-time) Fourier transform pair. Symbolically, this is expressed as

$$f[k] \Longleftrightarrow F(\Omega)$$

The Fourier transform $F(\Omega)$ is the frequency-domain description of $f[k]$.

10.2-1 Nature of Fourier Spectra

We now discuss several important features of the discrete-time Fourier transform and the spectra associated with it.

The Fourier Spectra are Continuous Functions of Ω.
It is helpful to keep in mind that the Fourier integral in Eq. (10.30) is basically a Fourier series with fundamental frequency $\Delta\Omega$ approaching zero [Eq. (10.28)]. Therefore, most of the discussion and properties of Fourier series apply to the Fourier transform as well. The successive harmonics are separated by the fundamental frequency $\Delta\Omega$, which approaches zero. This fact makes the spectra continuous functions of Ω.

The Fourier Spectra are Periodic Functions of Ω with Period 2π
According to Eq. (10.31) it follows that

$$F(\Omega + 2\pi) = \sum_{k=-\infty}^{\infty} f[k] e^{-j(\Omega+2\pi)k} = \sum_{k=-\infty}^{\infty} f[k] e^{-j\Omega k} e^{-j2\pi k} = F(\Omega) \tag{10.32}$$

Clearly, the spectrum $F(\Omega)$ is a continuous and periodic function of Ω with period 2π. We must remember, however, that to synthesize $f[k]$, we need to use the spectrum over a frequency interval of only 2π, starting at any value of Ω [see Eq. (10.30)].

As a matter of convenience, we shall choose this interval to be the fundamental frequency range $(-\pi,\ \pi)$. It is, therefore, not necessary to show discrete-time-signal spectra beyond the fundamental range, although we often do so.

Conjugate Symmetry of $F(\Omega)$ for real $f[k]$

From Eq. (10.31), we obtain

$$F(-\Omega) = \sum_{k=-\infty}^{\infty} f[k]e^{j\Omega k}$$

The right-hand side of this equation is the conjugate of the right-hand side of Eq. (10.31) for real $f[k]$. Therefore, for real $f[k]$, $F(\Omega)$ and $F(-\Omega)$ are conjugates; that is, $F(-\Omega) = F^*(\Omega)$.

Since $F(\Omega)$ is generally complex, we have both amplitude and angle (or phase) spectra

$$F(\Omega) = |F(\Omega)|e^{j\angle F(\Omega)} \qquad\qquad (10.33)$$

Because of conjugate symmetry of $F(\Omega)$, it follows that

$$|F(\Omega)| = |F(-\Omega)| \qquad\qquad (10.34a)$$

$$\angle F(\Omega) = -\angle F(-\Omega) \qquad\qquad (10.34b)$$

Therefore, the amplitude spectrum $|F(\Omega)|$ is an even function of Ω and the phase spectrum $\angle F(\Omega)$ is an odd function of Ω for real $f[k]$.

Linearity of the DTFT

According to Eq. (10.31), it follows that if

$$f_1[k] \Longleftrightarrow F_1(\Omega) \qquad \text{and} \qquad f_2[k] \Longleftrightarrow F_2(\Omega)$$

then

$$a_1 f_1[k] + a_2 f_2[k] \Longleftrightarrow a_1 F_1(\Omega) + a_2 F_2(\Omega) \qquad\qquad (10.35)$$

Existence of the DTFT

Because $|e^{-jk\Omega}| = 1$, from Eq. (10.31), it follows that the existence of $F(\Omega)$ is guaranteed if $f[k]$ is absolutely summable; that is,

$$\sum_{k=-\infty}^{\infty} |f[k]| < \infty \qquad\qquad (10.36)$$

This condition is sufficient but not necessary for the existence of $F(\Omega)$. For instance, the signal $f[k] = \sin k/k$ violates the condition (10.36), but does have DTFT (see Example 10.6).

Physical Appreciation of the Discrete-Time Fourier Transform

In understanding any aspect of the Fourier transform, we should remember that Fourier representation is a way of expressing a signal $f[k]$ as a sum of everlasting

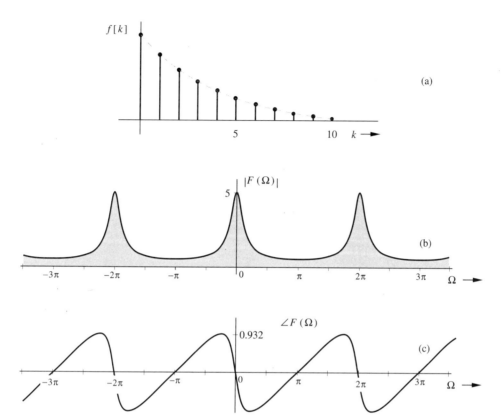

Fig. 10.4 Exponential $\gamma^k u[k]$ and its frequency spectra.

exponentials (or sinusoids). The Fourier spectrum of a signal indicates the relative amplitudes and phases of the exponentials (or sinusoids) required to synthesize $f[k]$. If $f[k]$ is periodic, then its Fourier spectrum has finite amplitudes and exists at discrete frequencies (Ω_0 and its multiples). Such a spectrum is easy to visualize, but the spectrum of an aperiodic signal is not easy to visualize because it is continuous. The physical meaning of the continuous spectrum is fully explained in Sec. 4.1-1.

■ **Example 10.3**

Find the DTFT of $f[k] = \gamma^k u[k]$.

$$F(\Omega) = \sum_{k=0}^{\infty} \gamma^k e^{-j\Omega k}$$

$$= \sum_{k=0}^{\infty} (\gamma e^{-j\Omega})^k$$

This is a geometric progression with a common ratio $\gamma e^{-j\Omega}$. Therefore, [see Sec. (B.7-4)]

$$F(\Omega) = \frac{1}{1 - (\gamma e^{-j\Omega})}$$

provided that $|\gamma e^{-j\Omega}| < 1$. But because $|e^{-j\Omega}| = 1$, this condition implies $|\gamma| < 1$. Therefore

$$F(\Omega) = \frac{1}{1 - \gamma e^{-j\Omega}} \qquad\qquad |\gamma| < 1 \tag{10.37}$$

If $|\gamma| > 1$, $F(\Omega)$ does not converge. This result is in conformity with condition (10.36). From Eq. (10.37)

$$F(\Omega) = \frac{1}{1 - \gamma \cos \Omega + j\gamma \sin \Omega} \tag{10.38}$$

so that

$$|F(\Omega)| = \frac{1}{\sqrt{(1 - \gamma \cos \Omega)^2 + (\gamma \sin \Omega)^2}} \tag{10.39a}$$

$$= \frac{1}{\sqrt{1 + \gamma^2 - 2\gamma \cos \Omega}}$$

$$\angle F(\Omega) = -\tan^{-1}\left[\frac{\gamma \sin \Omega}{1 - \gamma \cos \Omega}\right] \tag{10.39b}$$

Figure 10.4 shows $f[k] = \gamma^k u[k]$ and its spectra for $\gamma = 0.8$. Observe that the frequency spectra are continuous and periodic functions of Ω with the period 2π. As explained earlier, we need to use the spectrum only over the frequency interval of 2π. We often select this interval to be the fundamental frequency range $(-\pi, \pi)$.

The amplitude spectrum $|F(\Omega)|$ is an even function and the phase spectrum $\angle F(\Omega)$ is an odd function of Ω. ■

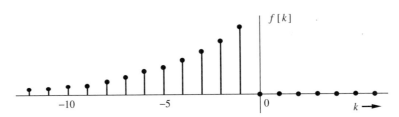

Fig. 10.5   Exponential $\gamma^k u[-(k+1)]$.

■ **Example 10.4**

Find the DTFT of $\gamma^k u[-(k+1)]$ depicted in Fig. 10.5.

$$F(\Omega) = \sum_{k=-\infty}^{\infty} \gamma^k u[-(k+1)] e^{-j\Omega k} = \sum_{k=-1}^{-\infty} (\gamma e^{-j\Omega})^k = \sum_{k=-1}^{-\infty} (\frac{1}{\gamma} e^{j\Omega})^{-k}$$

Setting $k = -m$ yields

$$f[k] = \sum_{m=1}^{\infty} (\frac{1}{\gamma} e^{j\Omega})^m = \frac{1}{\gamma} e^{j\Omega} + \left(\frac{1}{\gamma} e^{j\Omega}\right)^2 + \left(\frac{1}{\gamma} e^{j\Omega}\right)^3 + \cdots$$

This is a geometric series with a common ratio $e^{j\Omega}/\gamma$. Therefore, from Sec. B.7-4,

$$F(\Omega) = \frac{1}{\gamma e^{-j\Omega} - 1} \qquad |\gamma| > 1 \tag{10.40}$$

$$= \frac{1}{(\gamma \cos \Omega - 1) - j\gamma \sin \Omega}$$

Therefore

$$|F(\Omega)| = \frac{1}{\sqrt{1 + \gamma^2 - 2\gamma \cos \Omega}}$$

$$\angle F(\Omega) = \tan^{-1}\left[\frac{\gamma \sin \Omega}{\gamma \cos \Omega - 1}\right] \tag{10.41}$$

The Fourier transform (and the frequency spectra) for this signal is identical to that of $f[k] = \gamma^k u[k]$. Yet there is no ambiguity in determining the IDTFT of $F(\Omega) = \frac{1}{\gamma e^{-j\Omega} - 1}$ because of the restrictions on the value of γ in each case. If $|\gamma| < 1$, then the inverse transform is $f[k] = \gamma^k u[k]$. If $|\gamma| > 1$, it is $f[k] = \gamma^k[-(k+1)]$. ■

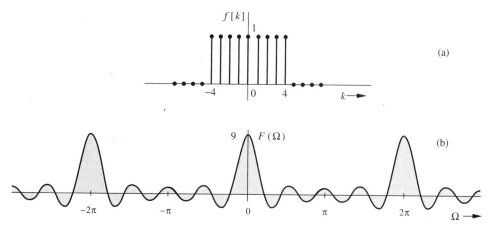

Fig. 10.6   Discrete-time gate pulse and its Fourier spectrum.

■ **Example 10.5**

Find the DTFT of the discrete-time rectangular pulse illustrated in Fig. 10.6a. This pulse is also known as the 9-point rectangular window function.

$$F(\Omega) = \sum_{k=-\infty}^{\infty} f[k] e^{-j\Omega k}$$

$$= \sum_{k=-\frac{M-1}{2}}^{\frac{M-1}{2}} (e^{-j\Omega})^k \qquad M = 9 \tag{10.42}$$

This is a geometric progression with a common ratio $e^{-j\Omega}$ and [see Sec. B.7-4]

$$F(\Omega) = \frac{e^{-j\frac{M+1}{2}\Omega} - e^{j\frac{M-1}{2}\Omega}}{e^{-j\Omega} - 1}$$

$$= \frac{e^{-j\Omega/2}\left(e^{-j\frac{M}{2}\Omega} - e^{j\frac{M}{2}\Omega}\right)}{e^{-j\Omega/2}(e^{-j\Omega/2} - e^{j\Omega/2})}$$

$$= \frac{\sin\left(\frac{M}{2}\Omega\right)}{\sin(0.5\Omega)} \tag{10.43}$$

$$= \frac{\sin(4.5\Omega)}{\sin(0.5\Omega)} \qquad \text{for } M = 9 \tag{10.44}$$

Figure 10.6b shows the spectrum $F(\Omega)$ for $M = 9$.

   Note that the spectrum $\mathcal{D}_r$ in Fig. 10.2b [Eq. (10.19)] is a sampled version of $F(\Omega)$ in Fig. 10.6b [Eq. (10.44)]:

$$\mathcal{D}_r = \frac{1}{32}F(r\Omega_0) \qquad \Omega_0 = \frac{\pi}{16}$$

Therefore, $F(\Omega)$ in Fig. 10.6b is the envelope of $\mathcal{D}_r$ (within a multiplicative constant 32) in Fig. 10.2b. The reason for this behavior is discussed later in Sec. 10.6.   ∎

⊙ **Computer Example C10.2**
Do Example 10.5 using MATLAB.

```
N0=512;
f=[ones(1,5) zeros(1,N0-9) ones(1,4)];
F=fft(f);
r=0:N0-1;
W=r.*2*pi/512;
plot(W,F);
xlabel('W');ylabel('F(W)');grid on;   ⊙
```

■ **Example 10.6**
   Find the inverse DTFT of the rectangular pulse spectrum $F(\Omega) = \text{rect}\left(\frac{\Omega}{2\Omega_c}\right)$ with $\Omega_c = \frac{\pi}{4}$ and repeating at the intervals of 2π, as shown in Fig. 10.7a.
   According to Eq. (10.30)

$$f[k] = \frac{1}{2\pi}\int_{-\pi}^{\pi} F(\Omega)e^{jk\Omega}\,d\Omega$$

$$= \frac{1}{2\pi}\int_{-\Omega_c}^{\Omega_c} e^{jk\Omega}\,d\Omega$$

$$= \frac{1}{j2\pi k}e^{jk\Omega}\Big|_{-\Omega_c}^{\Omega_c}$$

$$= \frac{\sin(\Omega_c k)}{\pi k}$$

$$= \frac{\Omega_c}{\pi}\,\text{sinc}(\Omega_c k) \tag{10.45}$$

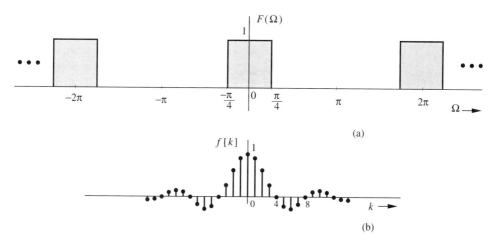

Fig. 10.7 Inverse Discrete-time Fourier transform of a periodic gate spectrum.

The signal $f[k]$ is depicted in Fig. 10.7b (for the case $\Omega_c = \pi/4$). ■

△ **Exercise E10.3**

    Find and sketch the amplitude and phase spectra of the DTFT of the signal $f[k] = \gamma^{|k|}$ with $|\gamma| < 1$.

Answer:

$$F(\Omega) = \frac{1 - \gamma^2}{1 - 2\gamma \cos \Omega + \gamma^2} \qquad \triangledown$$

10.3 Properties of the DTFT

    The *linearity property* [Eq. (10.35)] of DTFT has been already discussed. Other useful properties of the DTFT are as follows:

Time and Frequency Inversion

$$f[-k] \Longleftrightarrow F(-\Omega) \qquad (10.46)$$

From Eq. (10.31), the DTFT of $f[-k]$ is

$$\mathrm{DTFT}\{f[-k]\} = \sum_{k=-\infty}^{\infty} f[-k]e^{-j\Omega k} = \sum_{m=-\infty}^{\infty} f[m]e^{j\Omega m} = F(-\Omega)$$

Multiplication by k: Frequency Differentiation

$$k f[k] \Longleftrightarrow j\frac{dF(\Omega)}{d\Omega} \qquad (10.47)$$

The result follows immediately by differentiating both sides of Eq. (10.31) with respect to Ω.

Time-Shifting Property

If

$$f[k] \Longleftrightarrow F(\Omega)$$

then

$$f[k - k_0] \Longleftrightarrow F(\Omega)e^{-jk_0\Omega} \qquad\qquad k_0 \text{ an integer} \qquad (10.48)$$

This property can be proved by direct substitution in the equation defining the direct transform. From Eq. (10.31) we obtain

$$f[k - k_0] \Longleftrightarrow \sum_{k=-\infty}^{\infty} f[k - k_0]e^{-j\Omega k} = \sum_{m=-\infty}^{\infty} f[m]e^{-j\Omega[m+k_0]}$$

$$= e^{-j\Omega k_0} \sum_{k=-\infty}^{\infty} f[m]e^{-j\Omega m} = e^{-jk_0\Omega}F(\Omega)$$

This result shows that *delaying a signal by k_0 units does not change its amplitude spectrum. The phase spectrum, however, is changed by $-k_0\Omega$.* This added phase is a linear function of Ω with slope $-k_0$.

Physical Explanation of the Linear Phase

Time delay in a signal causes a linear phase shift in its spectrum. The heuristic explanation of this result is exactly parallel to that for continuous-time signals given in Sec. 4.3-4 (see Fig. 4.20).

Frequency-Shifting Property

If

$$f[k] \Longleftrightarrow F(\Omega)$$

then

$$f[k]e^{jk\Omega_s} \Longleftrightarrow F(\Omega - \Omega_s) \qquad\qquad (10.49)$$

This property is the dual of the time-shifting property. To prove this property, we have from Eq. (10.31)

$$f[k]e^{jk\Omega_s} \Longleftrightarrow \sum_{k=-\infty}^{\infty} f[k]e^{jk\Omega_s}e^{-j\Omega k} = \sum_{k=-\infty}^{\infty} f[k]e^{-j[\Omega-\Omega_s]} = F[\Omega - \Omega_s]$$

Time and Frequency Convolution Property

If

$$f_1[k] \Longleftrightarrow F_1(\Omega) \quad \text{and} \quad f_2[k] \Longleftrightarrow F_2(\Omega)$$

then

$$f_1[k] * f_2[k] \Longleftrightarrow F_1(\Omega)F_2(\Omega) \qquad\qquad (10.50a)$$

and

$$f_1[k]f_2[k] \Longleftrightarrow \frac{1}{2\pi}F_1(\Omega) * F_2(\Omega) \qquad\qquad (10.50b)$$

where

$$f_1[k] * f_2[k] = \sum_{m=-\infty}^{\infty} f_1[m]f_2[k-m]$$

and

$$F_1(\Omega) * F_2(\Omega) = \int_{2\pi} F_1(u)F_2(\Omega - u)\, du$$

The time convolution property is proved in Chapter 11 [Eq. (11.18)]. All we have to do is replace z with $e^{j\Omega}$. To prove the frequency-convolution property (10.50b), we have

$$f_1[k]f_2[k] \iff \sum_{k=-\infty}^{\infty} f_1[k]f_2[k]e^{-j\Omega k} = \sum_{k=-\infty}^{\infty} f_2[k]\left[\frac{1}{2\pi}\int_{2\pi} F_1(u)e^{-jku}\, du\right]e^{-j\Omega k}$$

Interchanging the order of summation and integration, we obtain

$$f_1[k]f_2[k] \iff \frac{1}{2\pi}\int_{2\pi} F_1(u)\left[\sum_{k=-\infty}^{\infty} f_2[k]e^{-j(\Omega-u)k}\right] du = \frac{1}{2\pi}\int_{2\pi} F_1(u)F_2(\Omega-u)\, du$$

Parseval's Theorem

If

$$f[k] \iff F(\Omega)$$

then E_f, the energy of $f[k]$, is given by

$$E_f = \sum_{k=-\infty}^{\infty} |f[k]|^2 = \frac{1}{2\pi}\int_{2\pi} |F(\Omega)|^2\, d\Omega \qquad (10.51)$$

In order to prove this property, we have from Eq. (10.31)

$$F^*(-\Omega) = \sum_{k=-\infty}^{\infty} f^*[k]e^{-j\Omega k} \qquad (10.52a)$$

This result shows that

$$f^*[k] \iff F^*(-\Omega) \qquad (10.52b)$$

Now

$$\sum_{k=-\infty}^{\infty} |f[k]|^2 = \sum_{k=-\infty}^{\infty} f^*[k]f[k] = \sum_{k=-\infty}^{\infty} f^*[k]\left[\frac{1}{2\pi}\int_{2\pi} F(\Omega)e^{j\Omega k}\, d\Omega\right]$$

$$= \frac{1}{2\pi}\int_{2\pi} F(\Omega)\left[\sum_{k=-\infty}^{\infty} f^*[k]e^{j\Omega k}\right]$$

$$= \frac{1}{2\pi}\int_{2\pi} F(\Omega)F^*(\Omega)\, d\Omega = \frac{1}{2\pi}\int_{2\pi} |F(\Omega)|^2\, d\Omega$$

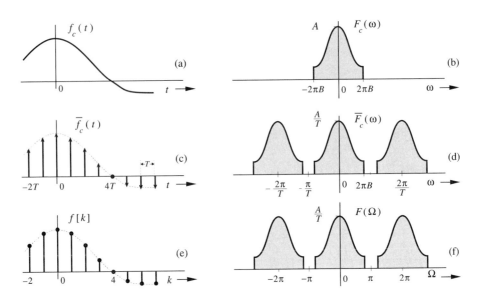

Fig. 10.8   Connection between the DTFT and the Fourier transform.

10.4   DTFT Connection with the Continuous-Time Fourier Transform

Consider a continuous-time signal $f_c(t)$ (Fig. 10.8a) with the Fourier transform $F_c(\omega)$. This signal may or may not be bandlimited. For convenience, we shall assume the signal to be bandlimited to B Hz (Fig. 10.8b). This signal is sampled with a sampling interval T. The sampling rate may or may not be above the Nyquist rate. Again, for convenience, we shall assume that the sampling rate is at least equal to the Nyquist rate; that is, $T \leq 1/2B$. The sampled signal $\overline{f}_c(t)$ (Fig. 10.8c) can be expressed as

$$\overline{f}_c(t) = \sum_{k=-\infty}^{\infty} f_c(kT)\delta(t - kT)$$

The continuous-time Fourier transform of the above equation yields

$$\overline{F}_c(\omega) = \sum_{k=-\infty}^{\infty} f_c(kT)e^{-jkT\omega} \tag{10.53}$$

In Sec. 5.1 (Fig. 5.1e), we have shown that $\overline{F}_c(\omega)$ is $F_c(\omega)/T$ repeating periodically with a period $\omega_s = 2\pi/T$, as illustrated in Fig. 10.8d. Let us construct a discrete-time signal $f[k]$ such that its kth element value is equal to the value of the kth sample of $f_c(t)$, as depicted in Fig. 10.8e; that is,

$$f[k] = f_c(kT)$$

Now, $F(\Omega)$, the DTFT of $f[k]$, is given by

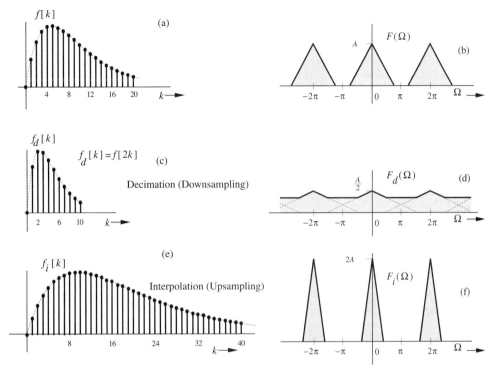

Fig. 10.9   Spectra of the decimated and interpolated signals.

$$F(\Omega) = \sum_{k=-\infty}^{\infty} f[k]e^{-jk\Omega} = \sum_{k=-\infty}^{\infty} f_c(kT)e^{-jk\Omega} \qquad (10.54)$$

Comparison of (10.54) with (10.53) shows that

$$F(\Omega) = \overline{F}_c\left(\frac{\Omega}{T}\right) \qquad (10.55)$$

Thus, $F(\Omega)$ can be obtained from $\overline{F}_c(\omega)$ by replacing ω with Ω/T. Therefore, $F(\Omega)$ is identical to $\overline{F}_c(\omega)$ frequency-scaled by a factor T, as shown in Fig. 10.8f.

In the above derivation, we did not have to use the assumption that $f_c(t)$ is bandlimited or that $f_c(t)$ is sampled at a rate at least equal to its Nyquist rate. If the signal were not bandlimited, the only difference in our discussion would be in the sketches in Fig. 10.8. For instance, $F_c(\omega)$ in Fig. 10.8b would not be bandlimited. This fact would cause overlapping (aliasing) of repeating cycles of $F_c(\omega)$ in Figs. 10.8d and f. A similar thing happens when the signal $f_c(t)$ is sampled below the Nyquist rate.†

10.4-1   DTFT of Decimated and Interpolated Signals

We can use Eq. (10.55) to find the DTFT of decimated and interpolated sig-

---

†In case the signal is not bandlimited and/ot the sampling rate is below its Nyquist rate, the samples $f_c(kT)$ can be interpreted as the Nyquist samples of the inverse Fourier transform of the first cycle (centered at $\omega = 0$) of $T\overline{F}_c(\omega)$.

nals, which are explained in Fig. 8.17. Consider a signal $f[k]$ and its DTFT $F(\Omega)$, as illustrated in Figs. 10.9a and b. Figure 10.9c shows the decimated signal $f[2k]$. If $f[k]$ is considered to be the sample sequence of a continuous-time signal $f_c(t)$, then $f[2k]$ is the sample sequence of $f_c(2t)$, whose Fourier transform is given by $\frac{1}{2}F_c(\omega/2)$ according to Eq. (4.34).† As seen in Eq. (10.55), $F(\Omega)$ is $F_c(\Omega/T)$ repeating periodically with period 2π, and the DTFT of $f[2k]$ is $\frac{1}{2}F_c(\Omega/2T)$ repeating periodically with period 2π. Note that $F(\Omega/2T)$ is $F(\Omega/T)$ time-expanded by factor 2, as shown in Figs. 10.9b and d. If we use mth-order decimation; that is, if we select every mth element in the sequence, the resulting decimated sequence will be $f[mk]$. Using the above argument, it follows that in the fundamental frequency range, the DTFT of $f[mk]$ will be $1/m$ times $F(\Omega/m)$, and it repeats periodically with period 2π. If m is too large, so that the first cycle of $F(\Omega/m)$ goes beyond π, the successive cycles of $F(\Omega/m)$ will overlap, as illustrated in Fig. 10.9d.

Now consider the interpolated signal $f_i[k]$ in Fig. 10.9c. This signal is $f[k/2]$ (obtained by expanding $f[k]$ by factor 2), with the alternate (missing) points filled by interpolated values obtained by ideal lowpass filtering. If the envelope of $f[k]$ is $f_c(t)$, then the envelope of $f_i[k]$ is $f_c(t/2)$, whose Fourier transform is $2F_c(2\omega)$. Thus, in the fundamental frequency range, the spectrum of $f_i[k]$ is 2 times $F(\Omega)$ compressed by factor 2 along the frequency axis, and periodically repeating with period 2π as depicted in Fig. 10.9f. Using a similar argument, we can generalize this result for a time-expanded signal $f[k/m]$ with missing values filled by ideal interpolation. In this case, the spectrum in the fundamental frequency range will be m times $F(m\Omega)$ [$F(\Omega)$ frequency-compressed by factor m] repeating periodically with period 2π.

10.5 Discrete-Time Linear System Analysis by DTFT

Consider a linear time-invariant discrete-time system with the unit impulse response $h[k]$. We shall find the (zero-state) system response $y[k]$ for the input $f[k]$. Because

$$y[k] = f[k] * h[k] \tag{10.56}$$

According to Eq. (10.50a) it follows that

$$Y(\Omega) = F(\Omega)H(\Omega) \tag{10.57}$$

where $F(\Omega)$, $Y(\Omega)$, and $H(\Omega)$ are DTFTs of $f[k]$, $y[k]$, and $h[k]$, respectively; that is,

$$f[k] \Longleftrightarrow F(\Omega), \quad y[k] \Longleftrightarrow Y(\Omega), \quad \text{and} \quad h[k] \Longleftrightarrow H(\Omega)$$

This result is similar to that obtained for continuous-time systems.

The Frequency Response of an LTID System

Equation (9.57a) states that the response to an everlasting exponential input z^k of an LTID system with transfer function $H[z]$ is $H[z]z^k$. If we let $z = e^{j\Omega}$, then

---

†Here, $F_c(\omega)$ should be interpreted as the first cycle (centered at $\omega = 0$) of $T\overline{F}_c(\omega)$

$z^k = e^{j\Omega k}$, and the response to an everlasting exponential input $e^{j\Omega k}$ of an LTID system with transfer function $H[z]$ is $H[e^{j\Omega}]e^{j\Omega k}$. This result can be represented as the input-output pair with a directed arrow notation as usual

$$e^{j\Omega k} \implies H[e^{j\Omega}]e^{j\Omega k}$$

Also, according to Eq. (9.57b) with $z = e^{j\Omega}$, it follows that

$$H[e^{j\Omega}] = \sum_{k=-\infty}^{\infty} h[k]e^{-j\Omega k}$$

Observe that the right-hand side of the above equation is $H(\Omega)$, the DTFT of $h[k]$. Therefore

$$H[e^{j\Omega}] = H(\Omega)$$

and

$$e^{j\Omega k} \implies H(\Omega)e^{j\Omega k}$$

Clearly $H(\Omega)$ is the LTID system frequency response. Following the argument in Sec. 7.1, the amplitude response of the system is $|H(\Omega)|$, and the phase response is $\angle H(\Omega)$. We shall discuss this topic again in greater details in Chapter 12.

Equation (10.57) states that the frequency spectrum of the output signal is the product of the frequency spectrum of the input signal and the frequency response of the system. From Eq. (10.57), we have

$$|Y(\Omega)| = |F(\Omega)|\,|H(\Omega)| \tag{10.58}$$

and

$$\angle Y(\Omega) = \angle F(\Omega) + \angle H(\Omega) \tag{10.59}$$

This result shows that the output amplitude spectrum is the product of the input amplitude spectrum and the amplitude response of the system. The output phase spectrum is the sum of the input phase spectrum and the phase response of the system.

We can also interpret Eq. (10.57) in terms of the frequency-domain viewpoint, which sees a system in terms of its frequency response (system response to various exponential or sinusoidal components). Frequency-domain views a signal as a sum of various exponential or sinusoidal components. Transmission of a signal through a (linear) system is viewed as transmission of various exponential or sinusoidal components of the input signal through the system. This concept can be understood by displaying the input-output relationships by a directed arrow as follows:

$$e^{j\Omega k} \implies H(\Omega)e^{j\Omega k} \qquad \text{the system response to } e^{j\Omega k} \text{ is } H(\Omega)e^{j\Omega k}$$

$$f[k] = \frac{1}{2\pi} \int_{2\pi} F(\Omega)e^{j\Omega k}\, d\Omega \quad \text{shows } f[k] \text{ as a sum of everlasting exponential components}$$

and from Eq. (10.57)

$$y[k] = \frac{1}{2\pi} \int_{2\pi} F(\Omega)H(\Omega)e^{j\Omega k}\, d\Omega \quad y[k] \text{ as a sum of responses to all input components}$$

which is precisely the relationship in Eq. (10.57). Thus, $F(\Omega)$ is the input spectrum and $Y(\Omega)$, the output spectrum (of the exponential components), is $F(\Omega)H(\Omega)$.

■ **Example 10.7**

For a system with unit impulse response $h[k] = (0.5)^k u[k]$, determine the (zero-state) response $y[k]$ for the input $f[k] = (0.8)^k u[k]$.

According to Eq. (10.57)

$$Y(\Omega) = F(\Omega)H(\Omega)$$

From the results in Eq. (10.37), we obtain

$$F(\Omega) = \frac{1}{1 - 0.8e^{-j\Omega}} = \frac{e^{j\Omega}}{e^{j\Omega} - 0.8} \qquad (10.60)$$

Also, $H(\Omega)$ is the DTFT of $(0.5)^k u[k]$, which is obtained from Eq. (10.37) by substituting $\gamma = 0.5$:

$$H(\Omega) = \frac{1}{1 - 0.5e^{-j\Omega}} = \frac{e^{j\Omega}}{e^{j\Omega} - 0.5} \qquad (10.61)$$

Therefore

$$Y(\Omega) = \frac{e^{j2\Omega}}{(e^{j\Omega} - 0.5)(e^{j\Omega} - 0.8)}$$

We can express the right-hand side as a sum of two first-order terms (modified partial fraction expansion as discussed in Sec. B.5-5) as follows:†

$$\frac{Y(\Omega)}{e^{j\Omega}} = \frac{e^{j\Omega}}{(e^{j\Omega} - 0.5)(e^{j\Omega} - 0.8)}$$

$$= \frac{-\frac{5}{3}}{e^{j\Omega} - 0.5} + \frac{\frac{8}{3}}{e^{j\Omega} - 0.8}$$

Consequently,

$$Y(\Omega) = -\left(\frac{5}{3}\right)\frac{e^{j\Omega}}{e^{j\Omega} - 0.5} + \left(\frac{8}{3}\right)\frac{e^{j\Omega}}{e^{j\Omega} - 0.8}$$

$$= -\left(\frac{5}{3}\right)\frac{1}{1 - 0.5e^{-j\Omega}} + \left(\frac{8}{3}\right)\frac{1}{1 - 0.8e^{-j\Omega}}$$

According to Eq. (10.37), the inverse DTFT of this equation is

$$y[k] = \left[-\tfrac{5}{3}(0.5)^k + \tfrac{8}{3}(0.8)^k\right] u[k]$$

This example demonstrates the procedure of determining an LTID system response using DTFT. It is similar to the method of Fourier transform in analysis of LTIC systems. As in the case of Fourier transform, this method can be used only if the system is asymptotically stable and if the input signal is DTF-transformable. We shall not belabor this method further because it is clumsier and more restricted than the z-transform method discussed in the next chapter. In the z-transform, we generalize the frequency variable $j\Omega$ to $\sigma + j\Omega$ so that the resulting exponentials can grow or decay with k. This procedure is

---

†Here $Y(\Omega)$ is a function of variable $e^{j\Omega}$. Hence, $x = e^{j\Omega}$ for the purpose of comparison with the expression in Sec. B.5-5.

similar to what we did in the continuous-time case by generalizing the frequency variable $j\omega$ to $s = \sigma + j\omega$ (from Fourier to Laplace transform) . ■

10.6 Signal processing by DFT and FFT

In this section, we use **DFT** (developed in Chapter 5) as a tool, which allows us to utilize a digital computer for digital signal processing. This signal processing includes spectral analysis of digital signals and LTID system analysis. By spectral analysis, we mean determining the discrete time Fourier series (DTFS) of periodic signals and determining $F(\Omega)$ from $f[k]$ (and vice versa) for aperiodic signals. As a tool for LTID system analysis, DFT can be used as a software oriented solution to digital filtering. DFT can be implemented on a digital computer by an efficient algorithm, the *fast Fourier transform (FFT)* also discussed in Chapter 5. The DFT (using FFT) is truly the workhorse of modern digital signal processing.

10.6-1 Computation of Discrete-Time Fourier Series (DTFS)

The discrete-time Fourier series (DTFS) equations (10.8) and (10.9) are identical to the DFT equations (5.18b) and (5.18a) within a scaling constant N_0. If we let $f[k] = N_0 f_k$ and $\mathcal{D}_r = F_r$ in Eqs. (10.9) and (10.8), we obtain

$$F_r = \sum_{k=0}^{N_0-1} f_k e^{-jr\Omega_0 k}$$

$$f_k = \frac{1}{N_0} \sum_{r=0}^{N_0-1} F_r e^{jr\Omega_0 k} \qquad\qquad \Omega_0 = \frac{2\pi}{N_0}$$

This is precisely the DFT pair in Eqs. (5.18). For instance, to compute the DTFS for the periodic signal in Fig. 10.2a, we use the values of $f_k = f[k]/N_0$ as

$$f_k = \begin{cases} \frac{1}{32} & 0 \le k \le 4 \quad \text{and} \quad 28 \le k \le 31 \\ 0 & 5 \le k \le 27 \end{cases}$$

We use these values in the FFT algorithm discussed in Sec. 5.2-2 to obtain F_r, which is the same as $\mathcal{D}_r$.

10.6-2 Computation of Direct and Inverse DTFT

Spectral analysis of digital signals requires determination of DTFT and IDTFT [determining $F(\Omega)$ from $f[k]$ and vice versa]. This determination could be accomplished by using the DTFT equations [Eqs. (10.30) and (10.31)] directly on a digital computer. However, there are two difficulties in implementation of these equations on a digital computer.

1. Equation (10.31) involves summing an infinite number of terms, which is not possible because it requires infinite computer time.

2. Equation (10.30) requires integration which can only be performed approximately on a computer because a computer approximates an integral by a sum.

The first problem can be surmounted either by restricting the analysis only to a finite length $f[k]$ or by truncating $f[k]$ by a suitable window. The error because

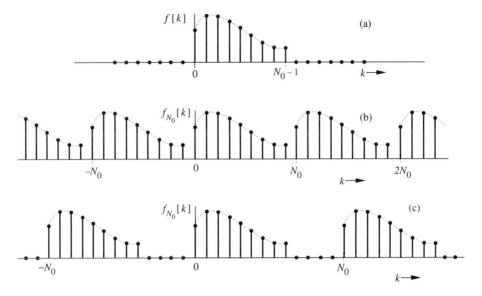

Fig. 10.10   DFT computation of a finite length signal.

of windowing may be reduced by using a wider and a tapered window. We shall see that if $f[k]$ has a finite duration, the samples of $F(\Omega)$ can be computed using a finite sum (rather than an integral). This solves the second problem. Moreover, $f[k]$ is uniquely determined from these samples of $F(\Omega)$.

In order to derive appropriate relationships, consider the signal $f[k]$ starting at $k = 0$, and with a finite length N_0, as shown in Fig. 10.10a. Let us construct a periodic signal $f_{N_0}[k]$ by repeating $f[k]$ periodically at intervals of N_0, as illustrated in Fig. 10.10b. We can represent the periodic signal $f_{N_0}[k]$ by the discrete-time Fourier series (DTFS) as [see Eqs. (10.8) and (10.9)]

$$f_{N_0}[k] = \sum_{r=0}^{N_0-1} \mathcal{D}_r e^{jr\Omega_0 k} \qquad \Omega_0 = \frac{2\pi}{N_0} \qquad (10.62)$$

where

$$\mathcal{D}_r = \frac{1}{N_0} \sum_{k=0}^{N_0-1} f[k] e^{-jr\Omega_0 k} \qquad (10.63)$$

In Eq. (10.63) we used the fact that $f_{N_0}[k] = f[k]$ for $k = 0, 1, 2, \ldots, N_0 - 1$.

By definition, the DTFT of $f[k]$ is

$$F(\Omega) = \sum_{k=0}^{N_0-1} f[k] e^{-j\Omega k} \qquad (10.64)$$

According to Eqs. (10.63) and (10.64), it follows that $N_0 \mathcal{D}_r$ is $F(r\Omega_0)$, the rth sample of $F(\Omega)$. For convenience, we denote this sample by F_r. Thus

$$N_0 \mathcal{D}_r = F(r\Omega_0) = F_r \qquad (10.65)$$

Therefore, $N_0 \mathcal{D}_r$ are the samples of $F(\Omega)$ taken uniformly at the frequency intervals of Ω_0. But because $\Omega_0 = 2\pi/N_0$, there are exactly N_0 number of these samples of $F(\Omega)$ over the fundamental frequency interval of 2π. According to Eqs. (10.62) and (10.65), it follows that

$$f_{N_0}[k] = \frac{1}{N_0} \sum_{r=0}^{N_0-1} F_r e^{jr\Omega_0 k} \tag{10.66}$$

where [from Eqs. (10.63) and (10.65)]

$$F_r = \sum_{k=0}^{N_0-1} f[k] e^{-jr\Omega_0 k} \tag{10.67}$$

Because $f_{N_0}[k] = f[k]$ for $k = 0, 1, 2, \ldots, (N_0 - 1)$, we can express the Eq. (10.66) as

$$f[k] = \frac{1}{N_0} \sum_{r=0}^{N_0-1} F_r e^{jr\Omega_0 k} \qquad k = 0, 1, 2, \ldots, N_0 - 1 \tag{10.68}$$

Moreover, we need to determine F_r, the samples of the DTFT, only over the interval $0 \leq \Omega < 2\pi$. Therefore, Eq. (10.67) can be expressed as

$$F_r = \sum_{k=0}^{N_0-1} f[k] e^{-jr\Omega_0 k} \qquad r = 0, 1, 2, \ldots, N_0 - 1 \tag{10.69}$$

Equations (10.68) and (10.69) are precisely the DFT pair derived in Eqs. (5.18a) and (5.18b). These equations relate F_r [the samples of $F(\Omega)$] to $f[k]$ and vice versa. Here F_r is the DFT of $f[k]$ and $f[k]$ is the IDFT (inverse DFT) of F_r. This relationship is also denoted by the bidirectional arrow notation of a transform as

$$f[k] \Longleftrightarrow F_r$$

To repeat, F_r, the DFT of an N_0-point sequence $f[k]$, is a set of uniform samples of its DTFT $F(\Omega)$ taken at frequency intervals of $\Omega_0 = \frac{2\pi}{N_0}$. The sequence $f[k]$ and its DFT F_r are related by Eqs. (10.68) and (10.69). Observe that there are N_0 elements in $f[k]$. Also, there are exactly N_0 elements in F_r (over the frequency range 2π). The DFT relationships are finite sums and can be readily computed on a digital computer using the efficient fast Fourier transform (FFT) algorithm.

Properties of DFT

We list some of the important properties of DFT proved in Chapter 5. from the preceding discussion, it follows that these properties of DFT also apply to DTFT samples of a finite length $f[k]$.

1. **Linearity:** If $f[k] \Longleftrightarrow F_r$ and $g[k] \Longleftrightarrow G_r$, then

$$a_1 f[k] + a_2 g[k] \Longleftrightarrow a_1 F_r + a_2 G_r \tag{10.70}$$

2. **Conjugate Symmetry:** For real $f[k]$

$$F_{N_0-r} = F_r^*$$ (10.71)

There is a conjugate symmetry about $N_0/2$, which enables us to determine roughly half the values of F_r from the other half of the values, when $f[k]$ is real. For instance, in a 7-point DFT, $F_6 = F_1^*$, $F_5 = F_2^*$ and $F_4 = F_3^*$. In an 8-point DFT, $F_7 = F_1^*$, $F_6 = F_2^*$, $F_5 = F_3^*$, and so on.

3. **Time Shifting (Circular Shifting):**

$$f[k-n] \iff F_r e^{-jr\Omega_0 n}$$ (10.72)

4. **Frequency Shifting:**

$$f[k]e^{jk\Omega_0 m} \iff F_{r-m}$$ (10.73)

5. **Circular (or Periodic) Convolution:**

$$f[k] \circledast g[k] \iff F_r G_r$$ (10.74a)

and

$$f[k]g[k] \iff \frac{1}{N_0}F_r \circledast G_r$$ (10.74b)

where the circular (or periodic) convolution of two N_0-point periodic sequences $f[k]$ and $g[k]$ is defined as

$$f[k] \circledast g[k] = \sum_{n=0}^{N_0-1} f[n]g[k-n] = \sum_{n=0}^{N_0-1} g[n]f[k-n]$$ (10.75)

Caution in Interpreting DFT and IDFT

Equations (10.68) and (10.69) allow us to compute samples of DTFT and IDTFT for a finite length signal on a digital computer. To avoid certain pitfalls, we must understand clearly the nature of functions synthesized by the sums on the right-hand side of these equations. According to Eq. (10.66), it follows that the sum on the right-hand side of Eq. (10.68) is $f_{N_0}[k]$, which is a periodic signal of which $f[k]$ is the first cycle. Similarly, the sum on the right-hand side of Eq. (10.69) is periodic. This is because $F_r = N_0 \mathcal{D}_r$, which is periodic. Therefore, both the DFT equations are periodic. We require only part of these results (over one cycle) to compute the samples of $F(\Omega)$ from $f[k]$ and vice versa. That is why we placed the restriction that k or $r = 0, 1, 2, \ldots, N_0 - 1$ in Eqs. (10.68) and (10.69).

Signal $f[k]$ can start at any value of k.

In deriving the above results, we assumed that the signal $f[k]$ starts at $k = 0$. This restriction, fortunately, is not necessary. We now show that this procedure can be applied to $f[k]$ starting at any instant. Recall that the DFT found by this procedure is actually the DFT of $f_{N_0}[k]$, which is a periodic extension of $f[k]$ with period N_0. In other words, $f_{N_0}[k]$ can be generated from $f[k]$ by placing $f[k]$ and reproduction thereof end to end ad infinitum. Consider now the signal $f[k]$ in Fig. 10.6 in which $f[k]$ begins at $k = -4$. The periodic extension of this signal is depicted in Fig. 10.2a for $N_0 = 32$. A careful glance at this figure shows that this periodic signal can be constructed by any segment of length 32 and repeating it periodically by placing it end to end ad infinitum. We may choose a segment over the range $k = -16$ to 15 or a segment over the range $k = 0$ to 31, or any other segment of length 32. The reader should satisfy himself that the periodic extension of any such segment yields the same periodic signal $f_{N_0}[k]$. Therefore, the DFT corresponding to the periodic signal $f_{N_0}[k]$ is the DFT of any of its segment of N_0 length starting at any point. So the signal $f[k]$ may start at any point. All we need is to construct a periodic signal $f_{N_0}[k]$ which is a periodic extension of $f[k]$, then compute the DFT according to Eq. (10.68) using the sample values in the range $k = 0$ to $N_0 - 1$. For instance, the signal $f[k]$ in Fig. 10.6 does not start at $k = 0$. But we can construct its periodic extension $f_{N_0}[k]$, as shown in Fig. 10.2a, and use the values for $k = 0, 1, 2, \ldots, 31$ in Eq. (10.69) to compute F_r. These values are

$$f[k] = \begin{cases} 1 & 0 \le k \le 4 \quad \text{and} \quad 28 \le k \le 31 \\ 0 & 5 \le k \le 27 \end{cases}$$

Hence, according to Eq. 10.69

$$F_0 = \sum_{k=0}^{N_0-1} f[k] = 9$$

and

$$F_1 = e^{-j\Omega_0} + e^{-j2\Omega_0} + e^{-j3\Omega_0} + e^{-j4\Omega_0} + e^{-j28\Omega_0} + e^{-j29\Omega_0} + e^{-j30\Omega_0} + e^{-j31\Omega_0}$$

Because $\Omega_0 = \frac{\pi}{16}$, we recognize that $e^{-j31\Omega_0} = e^{j\Omega_0}$, $e^{-j30\Omega_0} = e^{j2\Omega_0}$, and so on. Hence

$$F_1 = 1 + 2(\cos \Omega_0 + \cos 2\Omega_0 + \cos 3\Omega_0 + \cos 4\Omega_0) = 7.8865$$

Note that $F_0 = 9$ is the first sample of $F(\Omega)$, $F_1 = F(\frac{\pi}{16})$ is the second sample, and so on. The samples are spaced $\pi/16$ radians apart, giving a total of 32 samples in the fundamental frequency range. The reader may confirm these values from from Eq. (10.44).

■ Example 10.8

Find the DFT of a 3-point signal $f[k]$ illustrated in Fig. 10.11a.

Figure 10.11a shows $f[k]$ (solid line) and $f_{N_0}[k]$ obtained by periodic extension of $f[k]$ (shown by dotted lines). In this case

$$N_0 = 3 \quad \text{and} \quad \Omega_0 = \frac{2\pi}{3}$$

From Eq. (10.69), we obtain

$$F_r = \sum_{k=0}^{2} f[k]e^{-jr\left(\frac{2\pi}{3}\right)k} = 3 + 2e^{-jr\frac{2\pi}{3}} + 3e^{-jr\frac{4\pi}{3}}$$

Therefore

$$F_0 = 3 + 2 + 3 = 8$$

$$F_1 = 3 + 2e^{-j\frac{2\pi}{3}} + 3e^{-j\frac{4\pi}{3}} = 3 + \left(-1 - j\sqrt{3}\right) + \left(-\frac{3}{2} + j\frac{3\sqrt{3}}{2}\right) = \left(\frac{1}{2} + j\frac{\sqrt{3}}{2}\right) = e^{j\frac{\pi}{3}}$$

$$F_2 = 3 + 2e^{-j\frac{4\pi}{3}} + 3e^{-j\frac{8\pi}{3}} = 3 + \left(-1 + j\sqrt{3}\right) + \left(-\frac{3}{2} - j\frac{3\sqrt{3}}{2}\right) = \left(\frac{1}{2} - j\frac{\sqrt{3}}{2}\right) = e^{-j\frac{\pi}{3}}$$

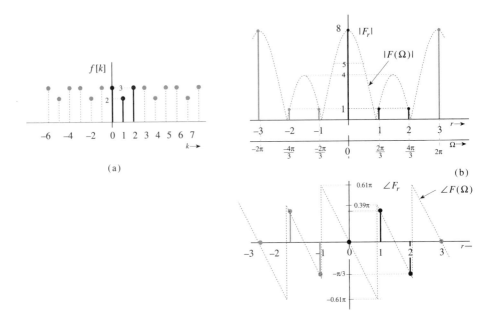

Fig. 10.11   Computation of DFT of a signal $f[k]$.

The magnitudes and angles of F_r are shown in the Table below.

r	0	1	2
$|F_r|$	8	1	1
$\angle F_r$	0	$\frac{\pi}{3}$	$-\frac{\pi}{3}$

Observe that $F_2 = F_1^*$ (the conjugate symmetry property). Because $f[k]$ is a 3-point sequence, F_r (its DFT) is also a 3-point sequence, which repeats periodically. Figure 10.11b shows F_r and $\angle F_r$. Recall that the DFT gives only the sample values of $F(\Omega)$. We want to know if DFT has enough samples to give a reasonably good idea of DTFT. The DTFT for $f[k]$ is given by [Eq. (10.31)]

$$F(\Omega) = \sum_{k=0}^{2} f[k]e^{-j\Omega k} = 3 + 2e^{-j\Omega} + 3e^{-j2\Omega} = e^{-j\Omega}[2 + 3e^{j\Omega} + 3e^{-j\Omega}] = e^{-j\Omega}(2 + 6 \cos \Omega)$$

The amplitude and phase spectra are given by

$$|F(\Omega)| = |2 + 6 \cos \Omega|$$

$$\angle F(\Omega) = \begin{cases} -\Omega & \text{when } (2 + 6 \cos \Omega) > 0 \\ \pi - \Omega & \text{when } (2 + 6 \cos \Omega) < 0 \end{cases}$$

Figure 10.11b shows $|F(\Omega)|$ and $\angle F(\Omega)$ (dotted). Observe that DFT values are exactly equal to DTFT values at the sampling frequencies; there is no approximation. This is always true of DFT of a finite length $f[k]$. However, if $f[k]$ is obtained by truncating or windowing a longer sequence, we shall see that the DFT gives only approximate sample values of the DTFT.

The DFT in this example has too few points to give a reasonable picture of the DTFT. The peak of DTFT appearing between the second and the third sample (between $r = 1$ and 2), for instance, is missed by the DFT. The two valleys of the DTFT are also missed. We definitely need more points in the DFT for an acceptable resolution. This goal is accomplished by **zero padding**, explained below. ∎

Use of Zero Padding

The DFT yields the samples of DTFT at the frequency intervals of $\Omega_0 = 2\pi/N_0$, where Ω_0 is the frequency resolution. Seeing the DTFT through DFT is like viewing $F(\Omega)$ through a *picket fence*. Only the spectral components at the sampled frequencies (which are integral multiples of Ω_0) will be visible. But frequency components lying in between will be hidden behind the picket fence. If DFT has too few points, major peaks and valleys of $F(\Omega)$ existing between the DFT points (sampled frequencies) will not be seen, thus giving an erroneous view of the spectrum $F(\Omega)$. This is precisely the case in Example 10.8. Actually, using the interpolation formula, it is possible to compute any number of values of DTFT from the DFT. But having to use the interpolation formula really defeats the purpose of DFT. We therefore seek to reduce Ω_0 so that the number of samples is increased for a better view of the DTFT.

Because $\Omega_0 = 2\pi/N_0$, we can reduce Ω_0 by increasing N_0, the length of $f[k]$. For a finite length sequence, the only way to increase N_0 is by appending sufficient number of zero valued points to $f[k]$. This procedure of zero padding is depicted in Fig. 10.10c. Recall that N_0 is the period of $f_{N_0}[k]$, which is formed by periodic repetition of $f[k]$. By appending sufficient number of zeros to $f[k]$, as illustrated in Fig. 10.10c, we can increase the period N_0 as much as we wish, thereby increasing the number of points of the DFT. Recall that the number of points of both $f[k]$ as well as F_r are identical (N_0). We shall rework example 10.8 using zero padding to increase the number of samples of DTFT.

∎ **Example 10.9**
Do Example 10.8 by padding three zeros to $f[k]$.
Figure 10.12a shows the zero padded $f[k]$, and the corresponding $f_{N_0}[k]$. Now $f[k]$ is a 6-point sequence. Hence,

$$N_0 = 6, \quad \text{and} \quad \Omega_0 = \frac{2\pi}{6} = \frac{\pi}{3}$$

and from Eq. (10.69), we obtain

$$F_r = \sum_{k=0}^{5} f[k]e^{-jr\left(\frac{\pi}{3}\right)k} = 3 + 2e^{-jr\frac{\pi}{3}} + 3e^{-jr\frac{2\pi}{3}}$$

Therefore

$$F_0 = 3 + 2 + 3 = 8$$

$$F_1 = 3 + 2e^{-j\frac{\pi}{3}} + 3e^{-j\frac{2\pi}{3}} = 3 + \left(1 - j\sqrt{3}\right) + \left(-\frac{3}{2} - j\frac{3\sqrt{3}}{2}\right) = \left(\frac{5}{2} - j\frac{5\sqrt{3}}{2}\right) = 5e^{-j\frac{\pi}{3}}$$

In the same way, we find

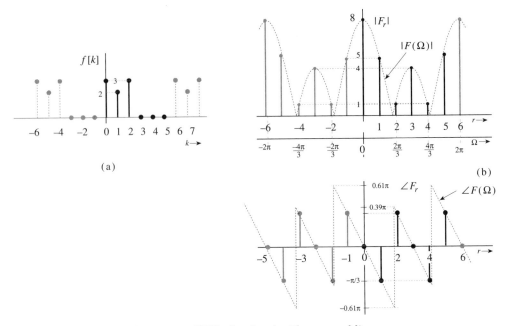

Fig. 10.12   DFT of a signal with zero padding.

$$F_2 = 3 + 2\,e^{-j\frac{2\pi}{3}} + 3\,e^{-j\frac{4\pi}{3}} = e^{j\frac{\pi}{3}}$$

$$F_3 = 3 + 2\,e^{-j\pi} + 3\,e^{-j2\pi} = 4$$

$$F_4 = 3 + 2\,e^{-j\frac{4\pi}{3}} + 3\,e^{-j\frac{8\pi}{3}} = e^{-j\frac{\pi}{3}}$$

$$F_5 = 3 + 2\,e^{-j\frac{5\pi}{3}} + 3\,e^{-j\frac{10\pi}{3}} = 5e^{j\frac{\pi}{3}}$$

The magnitude and angles of F_r are shown in the Table below

r	0	1	2	3	4	5
$|F_r|$	8	5	1	4	1	5
$\angle F_r$	0	$-\frac{\pi}{3}$	$\frac{\pi}{3}$	0	$-\frac{\pi}{3}$	$\frac{\pi}{3}$

Observe that $F_5 = F_1^*$ and $F_4 = F_2^*$ as expected from the conjugate symmetry property. Figure 10.12b shows the plots of $|F_r|$ and $\angle F_r$. Observe that we now have a 6-point DFT, which provides 6 samples of the DTFT spaced at the frequency interval of $\pi/3$ (in contrast to $2\pi/3$ spacing in Example 10.8). The samples corresponding to $r = 0, 2$, and 4 in Example 10.9 are identical to the samples corresponding to $r = 0, 1$, and 2 in Example 10.8. The DFT spectrum in Fig. 10.12b contains all the three samples appearing in Fig. 10.11b plus 3 more samples in between. Clearly, the zero padding allows us a better assessment of the DTFT. But even in this case, the valleys of the $F(\Omega)$ are missed by this (6-point) DFT.

⊙ **Computer Example C10.3**
   Use MATLAB to do Example 10.8 to yield 32 sample values of DTFT.
   Because the signal length is 3, we need to pad 29 zeros to the signal.

```
N0=32;k=0:N-1;
f=[3 2 3 zeros(1,N0-3)];
Fr=fft(f)
r=k;
subplot(3,1,1)
stem(k,f);
xlabel('k');ylabel('f[k]');
subplot(3,1,2)
stem(r,abs(Fr))
xlabel('r');ylabel('Fr');
subplot(3,1,3);
stem(r,angle(Fr))
xlabel('r');ylabel('angle Fr');grid on;   ⊙
```

■ **Example 10.10**

For the signal $f[k]$ in Example 10.8, we need a frequency resolution $\Omega_0 = \pi/6$ for a reasonable view of $F(\Omega)$. Determine the number of zeros needed to be padded, and write the expression for computing the DFT.

For a resolution of $\Omega_0 = \pi/6$, the padded length N_0 of the signal is

$$N_0 = \frac{2\pi}{\Omega_0} = \frac{2\pi}{\pi/6} = 12$$

The padded length required is 12. Therefore we need to pad 9 zeros to $f[k]$. In this case

$$F_r = \sum_{k=0}^{11} f[k]e^{-jr\left(\frac{\pi}{6}\right)k}$$

However, the last 9 samples of $f[k]$ are zero. Therefore, the above equation reduces to

$$F_r = \sum_{k=0}^{2} f[k]e^{-jr\left(\frac{\pi}{6}\right)k} \quad ■$$

△ **Exercise E10.4**

A 3-point signal $f[k]$ is specified by $f[0] = 2$, $f[-1] = f[1] = 1$, and $f[k] = 0$ for all other k. Show that the DFT of this signal is $F_0 = 4$, $F_1 = 1$, and $F_2 = 1$. Find $F(\Omega)$, the DTFT of this signal, and verify that the DFT is equal to the samples of the DTFT at intervals of $\omega_0 = 2\pi/N_0 = 2\pi/3$. ▽

△ **Exercise E10.5**

Show that the 8-point DFT of the signal $f[k]$ in Exercise E10.4 is

$$F_0 = 4, \; F_1 = 3.4142, \; F_2 = 2, \; F_3 = 0.5858, \; F_4 = 0, \; F_5 = 0.5858, \; F_6 = 2, \; F_7 = 3.4142$$

Observe the conjugate symmetry of F_r about $r = N_0/2 = 4$. ▽

Practical Choice of N_0 The value of N_0 is determined by the desired resolution Ω_0. However, there is another consideration in selecting a value of N_0. If we are using the FFT algorithm to compute the DFT, then for efficient computation of

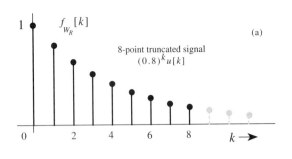

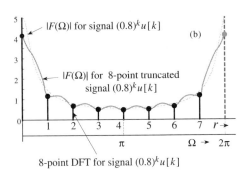

Fig. 10.13 DFT computations for $f[k] = (0.8)^k u[k]$ using an 8-point rectangular window.

DFT, N_0 should be a power of 2. Hence, we often pad sufficient numbers of zeros to ensure this requirement.

Effect of Signal truncation

So far we have considered only finite length sequences. For such sequences, we can readily find the N_0-point DFT, where N_0 is at least equal to the length of the sequence. How do we handle signals of infinite length? It is practically impossible to process infinite length sequences, because they generally require an infinite number of computations. Fortunately, every practical signal $f[k]$ must decay with k (because of a finite energy requirement), and such signal becomes negligibly small beyond $k \geq N_0$ for some suitable value of N_0. For instance, the signal $f[k] = (0.6)^k$ has infinite length. However, $f[k] \leq 0.00028$ at $k \geq 16$. Hence, we may truncate this signal beyond $k = 16$ (or even a little earlier). Straightforward signal truncation in this manner amounts to using a rectangular window with a unit weight for the data in the range $0 \leq k \leq N_0$, and zero weight for the data beyond $k = N_0$. Such a truncation results in Gibbs phenomenon with consequent oscillations in the spectrum of the truncated signal as demonstrated in the following example.

■ **Example 10.11**

The signal $f[k] = (0.8)^k u[k]$ has infinite length. Find the DFT of this signal using an 8-point rectangular window.

The DTFT of this signal obtained in Eq. (10.37) is

$$F(\Omega) = \frac{1}{1 - 0.8e^{-j\Omega}} \quad \text{and} \quad |F(\Omega)| = \frac{1}{\sqrt{1.64 - 1.6\cos\Omega}}$$

The 8-point rectangular window function is

$$w_R[k] = \begin{cases} 1 & 0 \leq k \leq 7 \\ 0 & \text{otherwise} \end{cases}$$

The windowed signal is

$$f_{w_R}[k] = \begin{cases} (0.8)^k & 0 \leq k \leq 7 \\ 0 & \text{otherwise} \end{cases}$$

The DFT of the windowed signal is obtained using the FFT algorithm. Figure 10.13a shows the windowed signal and Fig. 10.13b shows the corresponding 8-point DFT. The dotted curve shows the DTFT amplitude $|F(\Omega)|$ of the complete (untruncated) signal $(0.8)^k u[k]$ for comparison. The unbroken oscillating curve is the plot of the DTFT amplitude for the truncated (8-point) signal. The oscillations are because of the Gibbs phenomenon arising from the rectangular window. Observe the interesting fact that the 8 DFT values computed from the truncated signal are exactly equal to the 8 samples of the DTFT of the complete signal.

10.6-3 Discrete Time Filtering (Convolution) Using DFT

The DFT is useful not only in the computation of direct and inverse Fourier transforms, but also in other applications such as filtering, convolution, and correlation. Use of the efficient FFT algorithm makes DFT particularly appealing. We generally think of filtering in terms of some hardware-oriented solution (using summers, multipliers, and delay elements). However, filtering also has a software-oriented solution [a computer algorithm that yields the filtered output $y(t)$ for a given input $f(t)$]. Such filtering can be conveniently accomplished by using the DFT.

Filtering can be accomplished either in the frequency domain or in the time domain. In the frequency domain approach, for a given input $f[k]$, we are required to find the output $y[k]$ of a filter with a given transfer function $H(\Omega)$. In the time domain approach, for a given input $f[k]$, we are required to find the output $y[k]$ of a filter with a given impulse response $h[k]$. In the time domain, the output is obtained by (linear) convolution of $f[k]$ with $h[k]$. In the frequency domain, the output is obtained as an IDFT of $F_r H_r$ with $F_r = F(r\Omega_0)$ and $H_r = H(r\Omega_0)$. Because the frequency domain method appears as a substep of the time domain method, we shall consider here only the time domain (convolution) method.

We would like to perform linear convolution required in filtering operation using DFT (utilizing FFT algorithm) because of its computational efficiency. However, DFT can be used to evaluate only the circular convolution, not the linear convolution. Fortunately, linear convolution can be made equivalent to circular convolution by suitably padding the two sequences with zeros. This statement, introduced in Chapter 5, will now be proved.

When is Linear Convolution Equivalent to Circular Convolution?

The *circular (or periodic) convolution*, is explained in Sec. 5.2-1. In circular convolution, both sequences to be convolved are N_0-periodic. If $f[k]$ and $g[k]$ are both N_0-periodic, their periodic (or circular) convolution $c[k]$ is defined as

$$c[k] = f[k] \circledast g[k] = \sum_{m=0}^{N_0-1} f[m]g[k-m] \qquad (10.76)$$

Note that the circular convolution differs from the regular (linear) convolution by the fact that the summation is over one period (starting at any point). In the linear convolution, the summation is from $-\infty$ to ∞. The result of a periodic convolution is also an N_0-periodic sequence.

Suppose we wish to convolve two finite length sequences $f[k]$ and $h[k]$ of length N_f and N_h, respectively. Let $y[k]$ be the (linear) convolution of these sequences; that is,

$$y[k] = f[k] * h[k] \tag{10.77}$$

From the width property of the convolution (Sec. 9.4), the length of $y[k]$ is $(N_f + N_h - 1)$. Let the DTFT of the sequences $f[k]$, $h[k]$, and $y[k]$ be $F(\Omega)$, $H(\Omega)$, and $Y(\Omega)$, respectively. Then, from Eq. (10.50a), we have

$$Y(\Omega) = F(\Omega)H(\Omega) \tag{10.78}$$

We know that the DFT of an N_0-point sequence is the set of uniform samples of its DTFT at frequency interval $\Omega_0 = 2\pi/N_0$. Therefore, if Y_r is the N_0-point DFT of $y[k]$, then

$$Y_r = Y(r\Omega_0) \qquad \Omega_0 = \frac{2\pi}{N_0} \tag{10.79}$$

According to Eq. (10.78) it follows that

$$Y_r = F_r H_r \tag{10.80}$$

where F_r and H_r are the rth samples of $F(\Omega)$ and $H(\Omega)$, respectively. For Eq. (10.80) to be valid, F_r and H_r must be compatible for multiplication. In other words, both must be N_0-point sequences if Y_r is an N_0-point sequence. But we know that $f[k]$, $h[k]$, and $y[k]$ are N_f, N_h, and $N_0 = (N_f + N_h - 1)$-point sequences. Hence, we must pad $N_h - 1$ zeros to $f[k]$ and pad $N_f - 1$ zeros to $h[k]$ to ensure that F_r, H_r, and Y_r are all $N_0 = (N_f + N_h - 1)$-point sequences. Once we compute F_r and H_r (after suitably zero-padding $f[k]$ and $h[k]$), we take the IDFT of $F_r H_r$ to obtain $y[k]$. However, we have shown in Chapter 5 [Eq. (5.32a)] that $F_r H_r$ is the DFT of a circular convolution of $f[k]$ and $h[k]$; that is $f[k] \circledast h[k] \iff F_r H_r$. This result means $y[k]$ is equal to the circular convolution of (suitably padded) $f[k]$ and $h[k]$. Thus, $y[k]$ is a periodic sequence whose first period is the linear convolution of (unpadded) $f[k]$ and $h[k]$. To summarize, $y[k]$, which is the linear convolution of $f[k]$ and $h[k]$, is also equal to the circular convolution of suitably padded $f[k]$ and $h[k]$. This is an extremely important result. The system response is given by the linear convolution of $f[k]$ and $h[k]$. But the preceding result allows us to compute this convolution as if it were the circular convolution of (suitably padded) $f[k]$ and $h[k]$. This, in turn, allows us to use DFT to perform the computations.

The procedure to find the (linear) convolution of $f[k]$ and $h[k]$, whose lengths are N_f and N_h, respectively, can be summarized in four steps as follows:

1. Pad $N_h - 1$ zeros to $f[k]$ and $N_f - 1$ zeros to $h[k]$.
2. Find F_r and H_r, the DFTs of the zero-padded sequences $f[k]$ and $h[k]$.
3. Multiply F_r by H_r to obtain Y_r.
4. The desired convolution $y[k]$ is the IDFT of Y_r.

Filtering In the Frequency Domain   If we are given the filter transfer function $H(\Omega)$, we know H_r. We compute F_r, the DFT of $f[k]$. Then follow the steps 3 and 4 to obtain the output $y[k]$.

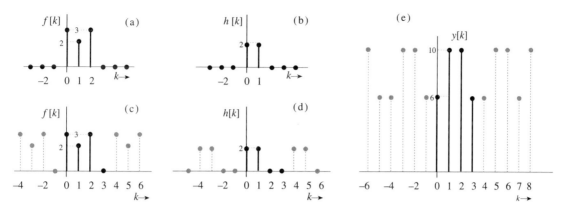

Fig. 10.14 Linear convolution of signals $f[k]$ and $h[k]$ using DFT.

∎ **Example 10.12**

Find $y[k]$, the output of an LTID filter with impulse response $h[k]$ to an input $f[k]$ illustrated in Figs. 10.14a and b.

We obtain the convolution $y[k]$ in four steps listed above.

1. In this case $N_f = 3$ and $N_h = 2$. Therefore, we pad 1 zero to $f[k]$ and 2 zeros to $h[k]$, as depicted in Figs. 10.14c and d, respectively.

2. Now, $N_0 = 4$ and $\Omega_0 = \pi/2$. The DFTs F_r and H_r of the zero-padded sequences $f[k]$ and $h[k]$ are given by

$$F_r = \sum_{k=0}^{2} f[k] e^{-jr\Omega_0 k} = \sum_{k=0}^{2} f[k] e^{-jr\frac{\pi}{2}k}$$

$$= 3 + 2e^{-j\frac{\pi}{2}r} + 3e^{-j\pi r}$$

$$H_r = \sum_{k=0}^{1} h[k] e^{-jr\Omega_0 k} = \sum_{k=0}^{1} h[k] e^{-jr\frac{\pi}{2}k}$$

$$= 2 + 2e^{-j\frac{\pi}{2}r}$$

Substituting $r = 0, 1, 2, 3$, we obtain

$F_0 = 8$	$F_1 = -2j$	$F_2 = 4$	$F_3 = 2j$
$H_0 = 4$	$H_1 = 2\sqrt{2}e^{-j\frac{\pi}{4}}$	$H_2 = 0$	$H_3 = 2\sqrt{2}e^{j\frac{\pi}{4}}$

3. Multiply F_r by H_r to obtain Y_r. This step yields

$$Y_0 = 32 \qquad Y_1 = 4\sqrt{2}e^{-j\frac{3\pi}{4}} \qquad Y_2 = 0 \qquad Y_3 = 4\sqrt{2}e^{j\frac{3\pi}{4}}$$

4. The desired convolution is the IDFT of Y_r, given by

$$y[k] = \frac{1}{4}\sum_{r=0}^{3} Y_r e^{jr\Omega_0 k} = \frac{1}{4}\sum_{r=0}^{3} Y_r e^{jr\frac{\pi}{2}k}$$

$$= \frac{1}{4}\left(Y_0 + Y_1 e^{j\frac{\pi}{2}k} + Y_2 e^{j\pi k} + Y_3 e^{j\frac{3\pi}{2}k}\right)$$

Substitution of $r = 0$, 1, 2, 3 in this equation yields

$$y[0] = 6 \qquad y[1] = 10 \qquad y[2] = 10 \qquad y[3] = 6$$

Figure 10.14e shows $y[k]$ and its periodic extension (dotted). The IDFT yields the periodic signal. We need only the first cycle.   ■

⊙ **Computer Example C10.4**
Use MATLAB to do Example 10.12. Find the answer by direct convolution as well as by using DFT.

In performing the convolution via DFT, we shall use the command 'fft(f,L)', which gives the FFT of a sequence f with sufficient zeros padded to make its length equal to L.

```
f=[3 2 3];
h=[2 2];
L=length(f)+length(h)-1;
k=0:1:L-1;
% Linear convolution: Direct approach
y1=conv(f,h);
subplot(2,1,2)
stem(k,y1)
% Linear Convolution:via DFT
FE=fft(f,L);
HE=fft(h,L);
y2=ifft(FE.*HE);
subplot(2,1,1);
stem(k,y2)   ⊙
```

⊙ **Computer Example C10.5**
Use MATLAB to convolve $(0.8)^k u[k]$ and $(0.5)^k u[k]$. Find the answer by direct convolution and using DFT.

Both the signals have infinite duration. Hence, we must truncate them beyond some suitable value of k, where both functions become negligible. For this purpose $k = 32$ is a reasonable choice.

```
R=32;
m=0:1:R-1;
f=(0.8).^m;
h=(0.5).^m;
L=length(f)+length(h)-1;
k=0:1:L-1;
% Linear convolution: Direct approach
y1=conv(f,h);
subplot(2,1,2)
stem(k,y1)
% Linear Convolution:via DFT
FE=fft(f,L);
HE=fft(h,L);
y2=ifft(FE.*HE);
k=0:1:L-1;
subplot(2,1,1);stem(k,y2)
stem(k,y2)   ⊙
```

△ **Exercise E10.6**

Show that the circular convolution of two 3-point sequences 3, 2, 3, and 1, 2, 3 , (both starting at $k = 0$) is also a 3-point sequence 15, 17, 16. Find the answer by using DFT, and verify your answer using the graphical method explained in Fig. 5.17. Using the sliding tape method (Fig. 9.4), show that the linear convolution of these sequences is a 5-point sequence 3, 8, 16, 12, 9. Indicate how you will obtain the linear convolution of these two sequences using circular convolution. ▽

△ **Exercise E10.7**

The input $f[k]$ of an LTID system is a 4-point sequence 1, 1, 1, 1, and $h[k]$, the impulse response of the system, is a 3-point sequence 1, 2, 1. Both sequences start at $k = 0$. Using DFT, show that the output $y[k]$ is a 6-point sequence 1, 3, 4, 4, 3, 1 starting at $k = 0$. Verify your answer by deriving the output as a linear convolution of the two sequences using the sliding tape method. ▽

Efficacy of DFT in Convolution Computation

Using the sliding tape algorithm discussed in Chapter 9, we can perform the convolution in Example 10.12 with a mere 6 (real) multiplications and 2 additions. The DFT method discussed in this section appears much too laborious, and the use of DFT for convolution may seem questionable. Recall, however, our discussion in Sec. 5.3, which showed that the use of FFT algorithm to compute DFT reduces the number of computations dramatically, especially for large N_0. For small length sequences, the direct convolution method, such as the sliding tape method is faster than the DFT method. But for long sequences, the DFT method using the FFT algorithm is much faster and far more efficient.† The method of finding convolution using the fast Fourier transform (FFT) is known as the *fast convolution.*

Block Filtering or Convolution

In practice, the length of an input signal may be very large, whereas a computer processing such a signal may have a limited memory. To process such long sequences, we can section the input signal into blocks of a length small enough to be processed by a given computer, and then add the outputs resulting from all the input blocks. This procedure can be used because the operation is linear. Such a procedure would be desirable even if the processing computer had an unlimited memory. For longer inputs, we must wait a long time before the input is fed to the computer (before it even starts processing), and then an even longer time for processing the large amount of data. Consequently, there is a long delay in the output. Sectioning the input allows the output to have a smaller processing delay. We shall now discuss two such method of block processing, **overlap and add** method and **overlap and save** method. Either of the methods requires the same number of computations; hence, which method is used is a matter of choice.

Overlap and Add Method

Consider a filtering operation, where the output $y[k]$ is the convolution of the input $f[k]$ and a FIR filter impulse response $h[k]$ of length M. In practice, usually

---

†It can be shown that in convolution of two sequences, each of length N_0, the number of computations required is on the order of N_0^2, whereas for DFT, using FFT algorithm, the number of computations required is only on the order of $N_0 \log_2 N_0$.

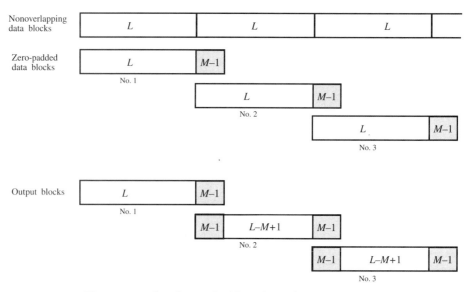

Fig. 10.15   Overlap and add method of block filtering.

M is much smaller than the length of the input. Figure 10.15 illustrates a long input sequence sectioned into nonoverlapping blocks of a manageable length L. The number inside each block indicates its length. Let us assume that $L \gg M$. We now process each block of the input data in sequence. To be able to use the circular convolution for performing the linear convolution, we need to pad $M - 1$ zeros at the end of each data block.‡ Figure 10.15 shows each data block augmented by $M - 1$ zeros. The augmented portion of the block is shown shaded. Observe that the augmented (zero-padded) blocks of the input, each of length $L + M - 1$, now overlap.

The output sequence corresponding to each block also has a length $L + M - 1$ (recall that the length of the circular convolution of two sequences, each of length $L + M - 1$, is also $L + M - 1$). The output sequences, therefore, also overlap, as shown in Fig. 10.15. The total output is given by the sum of all these overlapping output blocks of length $L + M - 1$. The contents of the two successive blocks are added wherever they overlap. This method is known as the *overlap and add* method.

■ **Example 10.13**

Using overlap and add method of block filtering, find the response $y[k]$ of an LTID system, whose impulse response $h[k]$ and the input $f[k]$ are shown in Fig. 10.16.

The output $y[k]$ is a linear convolution of $f[k]$ and $h[k]$. Let us use $L = 3$ for the block convolution. Also $M = 2$. Hence, we need to break the input sequence in blocks of 3 digits and pad each block with $M - 1 = 1$ zero, as depicted in Fig. 10.16. We convolve each of these blocks with $h[k]$ using DFT, as demonstrated in Example 10.11.

In this case, $N_0 = 4$ and $\Omega_0 = \pi/4$. The DFTs F_r and H_r of the zero-padded sequences $f[k]$ and $h[k]$ are given by

$$F_r = \sum_{k=0}^{2} f[k]e^{-jr\frac{\pi}{2}k} \qquad \text{and} \qquad H_r = \sum_{k=0}^{1} h[k]e^{-jr\frac{\pi}{2}k}$$

‡We also pad $h[k]$ with $L - 1$ zeros so that the length of the padded $h[k]$ is $L + M - 1$.

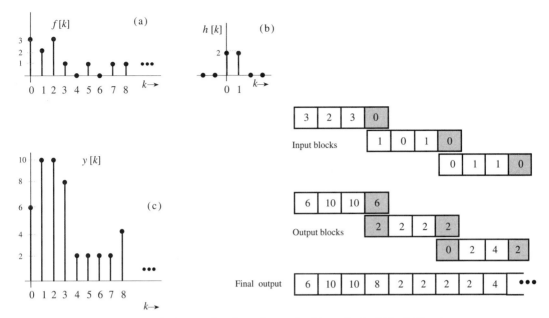

Fig. 10.16   An Example of overlap and add method of block filtering.

Also, $Y_r = F_r H_r$. We compute the values of F_r, H_r, and Y_r using these equations for each block:

For the first block,

$$F_r = 3 + 2e^{-j\frac{\pi}{2}r} + 3e^{-j\pi r}, \qquad H_r = 2 + 2e^{-j\frac{\pi}{2}r} \qquad \text{and} \qquad Y_r = F_r H_r$$

Also

$$y[k] = \frac{1}{4} \sum_{r=0}^{3} Y_r e^{jr\frac{\pi}{2}k} = \frac{1}{4}\left(Y_0 + Y_1 e^{j\frac{\pi}{2}k} + Y_2 e^{j\pi k} + Y_3 e^{j\frac{3\pi}{2}k}\right)$$

Substituting $r = 0, 1, 2, 3$, we obtain

$F_0 = 8$	$F_1 = -2j$	$F_2 = 4$	$F_3 = 2j$
$H_0 = 4$	$H_1 = 2\sqrt{2}e^{-j\frac{\pi}{4}}$	$H_2 = 0$	$H_3 = 2\sqrt{2}e^{j\frac{\pi}{4}}$
$Y_0 = 32$	$Y_1 = 4\sqrt{2}e^{-j\frac{3\pi}{4}}$	$Y_2 = 0$	$Y_3 = 4\sqrt{2}e^{j\frac{3\pi}{4}}$
$y[0] = 6$	$y[1] = 10$	$y[2] = 10$	$y[3] = 6$

Using the same procedure for the second block, we obtain

$$y[0] = 2, \quad y[1] = 2 \quad y[2] = 2, \quad y[3] = 2$$

For the third block, we obtain

$$y[0] = 0, \quad y[1] = 2 \quad y[2] = 4, \quad y[3] = 2$$

Figure 10.16 shows the overlapping input and output blocks, and the convolution sequence obtained by adding the output blocks.

The procedure using DFT given here is much more laborious than the direct convolution by the sliding tape method (Sec. 9.4). The reason is that we did not use FFT algorithm to compute DFT here. In this example, with a rather small N_0, even FFT will

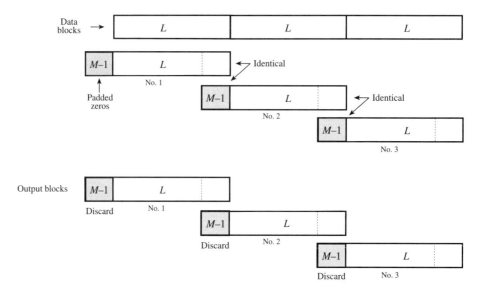

Fig. 10.17   Overlap and Save method of block filtering.

have more number of computations than the direct convolution. In practice, we generally deal with much larger values of N_0, where use of DFT (utilizing FFT algorithm) pays off. It will be informative for the reader to find linear convolution (using the sliding tape method) of one unpadded data block with unpadded $h[k]$. Next, find the circular convolution of the same data block padded with 1 zero and $h[k]$ padded with 2 zeros. Use the graphical method of circular convolution illustrated in Fig. 5.17. Verify that you get the same answer in both cases.   ■

Overlap and Save Method

In the *overlap and save* method also, the input sequence is sectioned into nonoverlapping blocks of a manageable length L. As before, each block is augmented with $M - 1$ data points. But unlike the previous method, this method places augmented points at the beginning of each block. The augmented $M - 1$ data points of a block are the same as the last $M - 1$ points of the previous block so that the last $M - 1$ data points of each block also appear as the first $M - 1$ data points of the succeeding block. The exception is the first block, where the first $M - 1$ data points are taken as zeros, as shown in Fig. 10.17. We now convolve each of these blocks with $h[k]$ (padded by $L - 1$ zeros). As before, DFT is used to perform convolution.

The output sequence corresponding to each block also has a length $L + M - 1$. We discard the first $M - 1$ data points and save the last L data points from each output block, as depicted in Fig. 10.17. The total output is given by combining all the saved blocks in sequence. This method is known as the *overlap and save* method. More details of this method can be found in the literature.[1]

■ **Example 10.14**

Using overlap and save method of block filtering, find the response $y[k]$ of an LTID system, whose impulse response $h[k]$ and the input are illustrated in Fig. 10.16.

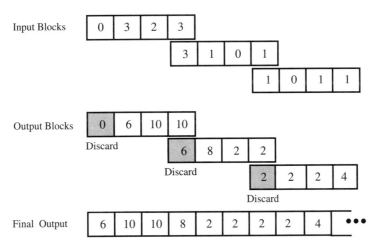

Fig. 10.18   An Example of overlap and save method of block filtering.

We follow the procedure given in Fig. 10.17 to section the input data as shown in Fig. 10.18. Note that the first $M - 1 = 1$ data point of the first block is padded (zero) and the last $M - 1 = 1$ point of each block also appears as the first point of the next block. Each block of length $L - M + 1 = 4$ is now convolved (using DFT) with $h[k]$ (padded with two zeros). The DFT procedure is already explained in Example 10.13. We shall omit the details here. the resulting output blocks are depicted in Fig. 10.18. The first $M - 1 = 1$ point of each output block is discarded. The total output is given by combining all the saved blocks in sequence.   ■

⊙   **Computer Example C10.6**
Use MATLAB to do example 10.13 (overlap and add method).
Here, we use the MATLAB command 'fftfilt(h,f,M)' to perform convolution using overlap and add method with blocks of length M. This m-file is available in *Signal Processing Toolbox*.

```
f=[3 2 3 1 0 1 0 1 1];
h=[2 2];
L=length(f)+length(h)-1;
k=0:1:L-2;
y=fftfilt(h,f,3);
stem(k,real(y));   ⊙
```

△   **Exercise E10.8**
The input $f[k]$ of an LTID system is a sequence 1, 0, -1, 2, ..., and $h[k]$, the impulse response of the system is a 3-point sequence 3, 2, 3. Both sequences start at $k = 0$. Using block convolution with $L = 2$, show that the output is 3, 2, 0, 4, ... Derive your answer using both methods of block filtering.   ▽

10.7   Generalization of the DTFT to the Z-Transform

LTID systems can be analyzed using DTFT. This method, however, has the following limitations:

1. Existence of the DTFT is guaranteed only for absolutely summable signals [see

Eq. (10.36)]. The DTFT does not exist for exponentially growing signals. This means the DTFT method can be applied only for a limited class of inputs.

2. Moreover, this method can be applied only to asymptotically stable systems; it cannot be used for unstable or even marginally stable systems.

These are serious limitations in the study of LTID system analysis. Actually it is the first limitation that is also the cause of the second limitation. Because DTFT is incapable of handling growing signals, it is incapable of handling unstable or marginally stable systems.† Our goal is, therefore, to extend the concept of DTFT so that it can handle exponentially growing signals.

We may wonder what causes this limitation on DTFT so that it is incapable of handling exponentially growing signals. Recall that in DTFT, we are synthesizing an arbitrary signal $f[k]$ using sinusoids or exponentials of the form $e^{j\Omega k}$. These signals are sinusoids with constant amplitudes. They are incapable of synthesizing exponentially growing signals no matter how many such components we add. Our hope, therefore, lies in trying to synthesize $f[k]$ using exponentially growing sinusoids or exponentials. This goal can be accomplished by generalizing the frequency variable $j\Omega$ to $\sigma + j\Omega$; that is, by using exponentials of the form $e^{(\sigma+j\Omega)}$ instead of exponentials $e^{j\Omega}$. The procedure is almost identical to that used in extending the Fourier transform to the Laplace transform in Sec. 6.1. The intuitive argument is identical to that discussed in Sec. 6.1, and the reader may wish to review it to refresh his memory. Here we shall go straight to the analytical development.

As in the case of Fourier to Laplace, it is desirable to use the notation $F(j\Omega)$ instead of $F(\Omega)$ for the DTFT in order to unify the DTFT and the generalized transform (z-transform). Thus,

$$F(j\Omega) = \sum_{k=-\infty}^{\infty} f[k]\, e^{-j\Omega k} \tag{10.81}$$

and

$$f[k] = \frac{1}{2\pi} \int_{-\pi}^{\pi} F(j\Omega)\, e^{j\Omega k}\, d\Omega \tag{10.82}$$

Consider now the DTFT of $f[k]\, e^{-\sigma k}$ (σ real)

$$\text{DTFT}\,[f[k]\, e^{-\sigma k}] = \sum_{k=-\infty}^{\infty} f[k]\, e^{-\sigma k}\, e^{-j\Omega k} \tag{10.83}$$

$$= \sum_{k=-\infty}^{\infty} f[k]\, e^{-(\sigma+j\Omega)k} \tag{10.84}$$

It follows from Eq. (10.81) that the above sum is $F(\sigma + j\Omega)$. Thus

$$\text{DTFT}\,[f[k]\, e^{-\sigma k}] = \sum_{k=-\infty}^{\infty} f[k]\, e^{-(\sigma+j\Omega)k} = F(\sigma + j\Omega) \tag{10.85}$$

---

†Recall that the output of an unstable system grows exponentially. Also, the output of a marginally stable system to characteristic mode input grows with time.

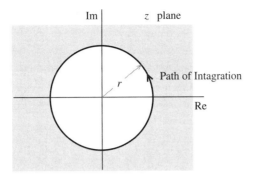

Fig. 10.19 Contour of integration for the z-transform.

Hence, the inverse DTFT of $F(\sigma + j\Omega)$ is $f[k]\, e^{-\sigma k}$. Therefore

$$f[k]\, e^{-\sigma k} = \frac{1}{2\pi} \int_{-\pi}^{\pi} F(\sigma + j\Omega)\, e^{j\Omega k}\, d\Omega \qquad (10.86)$$

Multiplying both sides of the above equation by $e^{\sigma k}$ yields

$$f[k] = \frac{1}{2\pi} \int_{-\pi}^{\pi} F(\sigma + j\Omega)\, e^{(\sigma + j\Omega)k}\, d\Omega \qquad (10.87)$$

Let us define a new variable z as

$$z = e^{\sigma + j\Omega} \qquad \text{so that} \qquad \ln z = \sigma + j\Omega \qquad \text{and} \qquad \frac{1}{z}\, dz = j\, d\Omega \qquad (10.88)$$

Because $z = e^{\sigma + j\Omega}$ is complex, we can express it as $z = r\, e^{j\Omega}$, where $r = e^{\sigma}$. Thus, z lies on a circle of radius r, and as Ω varies from $-\pi$ to π, z circumambulates along this circle, completing exactly one rotation in counterclockwise direction, as illustrated in Fig. 10.19. Changing to variable z in Eq. (10.87) yields

$$f[k] = \frac{1}{2\pi j} \oint F(\ln z)\, z^{k-1}\, dz \qquad (10.89a)$$

and from Eq. (10.85) we obtain

$$F(\ln z) = \sum_{k=-\infty}^{\infty} f[k]\, z^{-k} \qquad (10.89b)$$

where the integral $\oint$ indicates a contour integral around a circle of radius r in counterclockwise direction.

The above two equations are the desired extensions. They are, however, in a clumsy form. For the sake of convenience, we make a notational change by noting that $F(\ln z)$ is a function of z. Let us denote it by a simpler notation $F[z]$. Thus, Eqs. (10.89) become

$$f[k] = \frac{1}{2\pi j} \oint F[z]\, z^{k-1}\, dz \qquad (10.90)$$

and

$$F[z] = \sum_{k=-\infty}^{\infty} f[k]\, z^{-k} \qquad (10.91)$$

This is the (bilateral) z-transform pair. Equation (10.90) expresses $f[k]$ as a continuous sum of exponentials of the form $z^k = e^{(\sigma + j\Omega)k} = r^k\, e^{j\Omega k}$. Thus, by selecting a proper value for r (or σ), we can make the exponential grow (or decay) at any exponential rate we desire.

If we let $\sigma = 0$, we have $z = e^{j\Omega}$ and

$$F[z] = F(\ln z) = F(j\Omega) = F(\Omega) \qquad (10.92)$$

Thus, the familiar DTFT is just a special case of the z-transform $F[z]$ obtained by letting $z = e^{j\Omega}$.

10.8   Summary

This chapter deals with analysis and processing of discrete-time signals. For analysis, our approach is parallel to that used in continuous-time signals. We first represent a periodic $f[k]$ as a Fourier series formed by a discrete-time exponential and its harmonics. Later we extend this representation to an aperiodic signal $f[k]$ by considering $f[k]$ as a limiting case of a periodic signal with period approaching infinity. Periodic signals are represented by discrete-time Fourier series (DTFS); aperiodic signals are represented by the discrete-time Fourier transform (DTFT). The development, although similar to that of continuous-time signals, also reveals some significant differences. The basic difference in the two cases arises because a continuous-time exponential $e^{j\omega t}$ has a unique waveform for every value of ω in the range $-\infty$ to ∞. In contrast, a discrete-time exponential $e^{j\Omega k}$ has a unique waveform only for values of Ω in a continuous interval of 2π. Therefore, if Ω_0 is the fundamental frequency, then at most $\frac{2\pi}{\Omega_0}$ number of exponentials in the Fourier series are independent. Consequently, the discrete-time exponential Fourier series has only $N_0 = \frac{2\pi}{\Omega_0}$ terms.

The discrete-time Fourier transform (DTFT) of an aperiodic signal is a continuous function of Ω and is periodic with period 2π. We can synthesize $F(\Omega)$ from its spectral components in any band of width 2π. Linear time-invariant discrete-time (LTID) systems can be analyzed using DTFT if the input signals are DTF-transformable and if the system is stable. Analysis of unstable (or marginally stable) systems and/or exponentially growing inputs can be performed by z-transform, which is a generalized DTFT. The relationship of DTFT to z-transform is similar to that of the Fourier transform to the Laplace transform. Whereas the z-transform is superior to DTFT for analysis of LTID systems, DTFT is preferable in signal analysis.

If $H(\Omega)$ is the DTFT of the system's impulse response $h[k]$, then $|H(\Omega)|$ is the amplitude response, and $\angle H(\Omega)$ is the phase response of the system. Moreover, if $F(\Omega)$ and $Y(\Omega)$ are the DTFTs of the input $f[k]$ and the corresponding output $y[k]$, then $Y(\Omega) = H(\Omega)F(\Omega)$. Therefore the output spectrum is the product of the input spectrum and the system's frequency response.

The numerical computations in modern digital signal processing can be conveniently performed with the discrete Fourier transform (DFT) introduced in Chapter

5. The DFT computations can be very efficiently executed by using the fast Fourier transform (FFT) algorithm. The DFT is indeed the workhorse of modern digital signal processing. The discrete-time Fourier transform (DTFT) and the inverse discrete-time Fourier transform (IDTFT) can be computed using the DFT. For an N_0-point signal $f[k]$, its DFT yields exactly N_0 samples of $F(\Omega)$ at frequency intervals of $2\pi/N_0$. We can obtain a larger number of samples of $F(\Omega)$ by padding sufficient number of zero valued samples to $f[k]$. The N_0-point DFT of $f[k]$ gives exact values of the DTFT samples if $f[k]$ has a finite length N_0. If the length of $f[k]$ is infinite, we need to truncate $f[k]$ using the appropriate window function.

Because of the convolution property, we can compute convolution of two signals $f[k]$ and $h[k]$ using DFT. For this purpose, we need to pad both the signals by a suitable number of zeros so as to make the linear convolution of the two signals identical to the circular (or periodic) convolution of the padded signals. Large blocks of data may be processed by sectioning the data into smaller blocks and processing such smaller blocks in sequence. Such a procedure requires smaller memory and reduces the processing time.

References

1. Mitra, S.K., *Digital Signal processing: A Computer Based Approach*, McGraw-Hill, New York, 1998.

Problems

10.1-1 Find the discrete-time Fourier series (DTFS) and sketch their spectra $|\mathcal{D}_r|$ and $\angle\mathcal{D}_r$ for $0 \le r \le N_0 - 1$ for the following periodic signal:

$$f[k] = 4\cos 2.4\pi k + 2\sin 3.2\pi k$$

Hint: Reduce frequencies to the fundamental range $(0 \le \Omega \le 2\pi)$. The fundamental frequency Ω_0 is the largest number of which the frequencies appearing in the Fourier series are integral multiples.

10.1-2 Repeat Prob. 10.1-1 if $f[k] = \cos 2.2\pi k \cos 3.3\pi k$.

10.1-3 Repeat Prob. 10.1-1 if $f[k] = 2\cos 3.2\pi(k - 3)$.

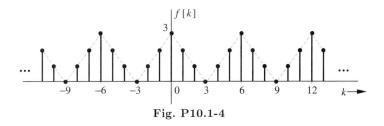

Fig. P10.1-4

10.1-4 Find the discrete-time Fourier series and the corresponding amplitude and phase spectra for the $f[k]$ shown in Fig. P10.1-4.

10.1-5   Repeat Prob. 10.1-4 for the $f[k]$ depicted in Fig. P10.1-5.

10.1-6   Repeat Prob. 10.1-4 for the $f[k]$ illustrated in Fig. P10.1-6.

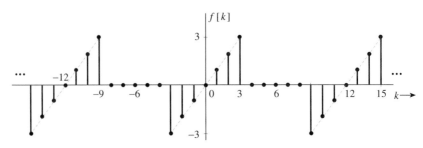

Fig. P10.1-5

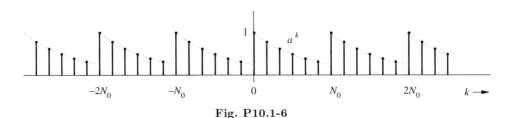

Fig. P10.1-6

10.1-7   A signal $f[k]$ is approximated in terms of another signal $x[k]$ over an interval ($N_1 \leq k \leq N_1$) as

$$f[k] \approx cx[k] \qquad N_1 \leq k \leq N_2$$

(a) Show that for the best approximation that minimizes the energy of the error signal $e[k] = f[k] - cx[k]$ over the same interval

$$c = \frac{1}{E_x} \sum_{k=N_1}^{N_2} f[k]x^*[k]$$

(b) If $c = 0$, the discrete-time signals $f[k]$ and $x[k]$ are said to be orthogonal over the interval ($N_1 \leq k \leq N_2$). Use this observation to define the orthogonality of discrete-time signals.

(c) Show that the set of signals $e^{jr\Omega_0 k}$ for $r = 0, 1, 2, 3, \ldots, N_0 - 1$ is orthogonal over an interval ($0 \leq k \leq N_0 - 1$). Hence, find the exponential Fourier series (DTFS) using the result in part **(a)**.

Hint: Recall that if w is complex, then $|w|^2 = ww^*$.

10.1-8   An N_0-periodic signal $f[k]$ is represented by its DTFS as in Eq. (10.8). Prove Parseval's theorem (for DTFS), which states that

$$\frac{1}{N_0} \sum_{k=<N_0>} |f[k]|^2 = \sum_{r=<N_0>} |\mathcal{D}_r|^2$$

Earlier [Eq. (10.51)], we obtained the Pareseval's theorem for DTFT.

Hint: If w is complex, then $|w|^2 = ww^*$, and use Eq. (5.43).

10.2-1 For the following signals, find the DTFT directly, using the definition in Eq. (10.31).

(a) $f[k] = \delta[k]$ (b) $\delta[k - k_0]$ (c) $a^k u[k-1]$ $|a| < 1$

(d) $f[k] = a^k u[k+1]$ $|a| < 1$.

In each case, sketch the signal and its amplitude spectrum. Sketch phase spectra for parts (a) and (b) only.

10.2-2 Find the DTFT for the signals shown in Fig. P8.2-9 (Chapter 8).

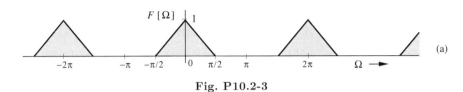

Fig. P10.2-3

10.2-3 Find the inverse DTFT for the spectrum depicted in Fig. P10.2-3.

10.3-1 Using the time-shifting property and the results in Examples 10.3 and 10.5, find the DTFT of (a) $a^k\{u[k] - u[k-10]\}$ (b) $u[k] - u[k-9]$.

10.3-2 Using appropriate properties and the result in Example 10.3, find the DTFT of (a) $(k+1)a^k u[k]$ $(|a| < 1)$ (b) $a^k \cos \Omega_0 k \, u[k]$.

10.4-1 For the spectrum $F(\Omega)$ in Fig. P10.2-3
(a) Find and sketch its IDTFT $f[k]$.
(b) Sketch $f[2k]$, $f[4k]$, and find their DTFTs.
(c) Sketch $f[k/2]$ and fill in the alternate missing samples using ideal interpolation (upsampling by a factor 2). Find the DTFT of the resulting interpolated (upsampled) signal $f_i[k]$.

10.5-1 Using the DTFT method, find the zero-state response $y[k]$ of a causal system with frequency response

$$H(\Omega) = \frac{e^{j\Omega} + 0.32}{e^{j2\Omega} + e^{j\Omega} + 0.16}$$

and the input

$$f[k] = (-0.5)^k u[k]$$

10.5-2 Repeat Prob. 10.5-1 if

$$H(\Omega) = \frac{e^{j\Omega} - 0.5}{(e^{j\Omega} + 0.5)(e^{j\Omega} - 1)}$$

and

$$f[k] = 3^{-(k+1)} u[k]$$

10.5-3 Repeat Prob. 10.5-1 if

$$H(\Omega) = \frac{e^{j\Omega}}{e^{j\Omega} - 0.5}$$

and

$$f[k] = 0.8^k u[k] + 2(2)^k u[-(k+1)]$$

10.6-1 Find the DFT of a 3-point signal $f[k]$ specified by $f[-1] = f[0] = 3$, $f[1] = 2$ and $f[k] = 0$ otherwise. Now determine $F(\Omega)$, the DTFT of $f[k]$, and verify that DFT values are the samples of $F(\Omega)$.

(b) Show that the 3-point DFT of this signal is identical to that of the signal $f[k]$ in Fig. 10.11a. Can you explain why? Does this mean the DTFTs of the two signals are also identical? Determine the DTFTs of the two signals and see if they are identical (for all values of Ω).

(c) Find the 8-point DFT of $f[k]$.

10.6-2   **(a)** Find the 4-point and 8-point DFT of a 4-point signal specified by the sequence 1, 2, 2, 1 starting at $k = 0$.

(b) Find $F(\Omega)$, the DTFT of $f[k]$, and verify the DFT values from $F(\Omega)$.

10.6-3   **(a)** Find the DFT of the signal $f[k] = \delta[k]$. Find also $F(\Omega)$, the DTFT of $\delta[k]$, and verify the DFT values from $F(\Omega)$. Note that this is a 1-point signal ($N_0 = 1$).

(b) Show that the DFT of $f[k] = \delta[k - m]$ is the same as the DFT of $\delta[k]$ for any integral value of m. Explain this behavior.

(c) Repeat part **(a)** for the N_0-point DFT (found by padding $N_0 - 1$ zeros to $\delta[k]$). Explain this DFT from $F(\Omega)$ found in part **(a)**.

10.6-4   **(a)** Find the DFT of the N_0-point signal $f[k] = u[k] - u[k - N_0]$. Find $F(\Omega)$, the DTFT of $f[k]$ and verify that the DFT values are the uniform samples of DTFT at frequency intervals of $\Omega_0 = 2\pi/N_0$.

(b) Is the DTFT found in part **(a)** an adequate frequency-domain description of $f[k]$. If not, what needs to be done to obtain a reasonably adequate DFT?

10.6-5   **(a)** Find the 5-point and 8-point DFT of the signal $f[k]$ illustrated in Fig. P8.2-9d.

(b) Find $F(\Omega)$, the DTFT of $f[k]$, and verify the DFT values from $F(\Omega)$.

Fig. P10.6-6

10.6-6   **(a)** Using the graphical method shown in Fig. 5.17, find the circular convolution of the sequences $f[k]$ and $g[k]$, depicted in Fig. P10.6-6.

(b) Using the sliding tape method (Fig. 9.4), find the linear convolution of the first cycles (over the range $0 \le k \le 3$) of the sequences $f[k]$ and $g[k]$. Is the result same as that found in part **(a)**?

(c) The circular convolution can be made equivalent to the linear convolution by suitably padding (the first cycle of) the sequences $f[k]$ and $g[k]$ with zeros. How many zeros do you need to pad to $f[k]$ and $h[k]$? After suitably padding these sequences, perform the circular convolution using the graphical method illustrated in Fig. 5.17, and show that it is equivalent to the linear convolution found in part **(b)**.

(d) Find the circular convolution of $f[k]$ and $g[k]$ obtained in part **(a)** using DFT.

(e) Find the linear convolution of $f[k]$ and $g[k]$ obtained in part **(b)** using DFT.

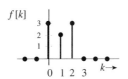
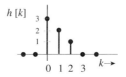

Fig. P10.6-7

10.6-7   Find the output of a system with impulse response $h[k]$ and the input $f[k]$ shown in
Fig. P10.6-7 by the following methods: (i)using linear convolution of $f[k]$ and $g[k]$ by
sliding tape method (ii) using circular convolution of suitably padded $f[k]$ and $g[k]$
using the graphical method, depicted in Fig. 5.17 (iii) using DFT.

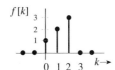
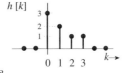

Fig. P10.6-8

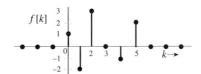

Fig. P10.6-9

10.6-8   **(a)** Repeat Prob. 10.6-7 for the signals $f[k]$ and $h[k]$ illustrated in Fig. P10.6-8.

10.6-9   Using both the methods of block filtering (overlap and add, and overlap and save),
find the output of a filter with impulse response $h[k]$ and the input $f[k]$ shown in
Fig. P10.6-9. Take $L = 3$. Verify that the linear convolution of the input sequence
$f[k] = \{1, -2, 3, 0, -1, 2, \ldots\}$ with $h[k] = \{2, 2\}$ gives the same output as that
found by the block filtering methods.

10.6-10   Find the 16-point IDFT of $F(\Omega)$ in Fig. 10.7a. Do the values of IDFT agree with the
values of $f[k]$ found in Eq. 10.45. If not, why not?

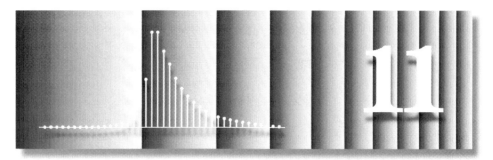

Discrete-Time System Analysis Using the Z-Transform

The counterpart of the Laplace transform for discrete-time systems is the z-transform. The Laplace transform converts integro-differential equations into algebraic equations. In the same way, the z-transforms changes difference equations into algebraic equations, thereby simplifying the analysis of discrete-time systems. The z-transform method of analysis of discrete-time systems parallels the Laplace transform method of analysis of continuous-time systems, with some minor differences. In fact, we shall see that *the z-transform is the Laplace transform in disguise*.

The behavior of discrete-time systems (with some differences) is similar to that of continuous-time systems. The frequency-domain analysis of discrete-time systems is based on the fact (proved in Sec. 9.4-2) that the response of a linear time-invariant discrete-time (LTID) system to an everlasting exponential z^k is also the same exponential (within a multiplicative constant), given by $H[z]z^k$. We then express an input $f[k]$ as a sum of (everlasting) exponentials of the form z^k. The system response to $f[k]$ is then found as a sum of the system's responses to all these exponential components. The tool which allows us to represent an arbitrary input $f[k]$ as a sum of (everlasting) exponentials of the form z^k is the z-transform.

11.1 The Z-Transform

In the last Chapter, we extended the discrete-time Fourier transform to derive the pair of equations defining the z-transform as

$$F[z] \equiv \sum_{k=-\infty}^{\infty} f[k]z^{-k} \tag{11.1}$$

$$f[k] = \frac{1}{2\pi j} \oint F[z]z^{k-1} \, dz \tag{11.2}$$

where the symbol $\oint$ indicates an integration in counterclockwise direction around a closed path in the complex plane (see Fig. 11.1). As in the case of the Laplace transform, we need not worry about this integral at this point because inverse z-transforms of many signals of engineering interest can be found in a z-transform Table. The direct and inverse z-transforms can be expressed symbolically as

$$F[z] = \mathcal{Z}\{f[k]\} \qquad \text{and} \qquad f[k] = \mathcal{Z}^{-1}\{F[z]\}$$

or simply as

$$f[k] \iff F[z]$$

Note that

$$\mathcal{Z}^{-1}[\mathcal{Z}\{f[k]\}] = f[k] \qquad \text{and} \qquad \mathcal{Z}\left[\mathcal{Z}^{-1}\{F[z]\}\right] = F[z]$$

Following the earlier argument, we can find an LTID system response to an input $f[k]$ using the steps as follows:

$$z^k \implies H[z]z^k \qquad \text{the system response to } z^k \text{ is } H[z]z^k$$

$$f[k] = \frac{1}{2\pi j}\oint F[z]z^{k-1}\,dz \quad \text{shows } f[k] \text{ as a sum of everlasting exponential components}$$

and

$$y[k] = \frac{1}{2\pi j}\oint F[z]H[z]z^{k-1}\,dz \quad \text{shows } y[k] \text{ as a sum of responses to exponential components}$$

$$= \frac{1}{2\pi j}\oint Y[z]z^{k-1}\,dz$$

where

$$Y[z] = F[z]H[z]$$

In conclusion, we have shown that for an LTID system with transfer function $H[z]$, if the input and the output are $f[k]$ and $y[k]$, respectively, and if

$$f[k] \iff F[z] \qquad y[k] \iff Y[z]$$

then

$$Y[z] = F[z]H[z]$$

We shall derive this result more formally later.

Linearity of the $\mathcal{Z}$-Transform

Like the Laplace transform, the z-transform is a linear operator. If

$$f_1[k] \iff F_1[z] \quad \text{and} \quad f_2[k] \iff F_2[z]$$

then

$$a_1 f_1[k] + a_2 f_2[k] \iff a_1 F_1[z] + a_2 F_2[z] \tag{11.3}$$

The proof is trivial and follows from the definition of the z-transform. This result can be extended to finite sums.

The Unilateral $\mathcal{Z}$-Transform

For the same reasons discussed in Chapter 6, we first start with a simpler version of the z-transform, the **unilateral** z-**transform**, that is restricted only to the analysis of causal systems with causal inputs (signals starting at $k = 0$). The more general *bilateral* z-*transform* is discussed later in Sec. 11.7. In the unilateral case, the signals are restricted to be causal; that is, they start at $k = 0$. The definition of the unilateral transform is the same as that of the bilateral [Eq. (11.1)] except that the limits of the sum are from 0 to ∞

$$F[z] \equiv \sum_{k=0}^{\infty} f[k]z^{-k} \tag{11.4}$$

where z is complex in general. The expression for the inverse z-transform in Eq. (11.2) remains valid for the unilateral case also.

The Region of Convergence of $\mathbf{F[z]}$

The sum in Eq. (11.1) [or (11.4)] defining the direct z-transform $F[z]$ may not converge (exist) for all values of z. The values of z (the region in the complex plane) for which the sum in Eq. (11.1) converges (or exists) is called the **region of convergence** (or **region of existence**) of $F[z]$. This concept will become clear in the following example.

■ **Example 11.1**
Find the z-transform and the corresponding region of convergence for the signal $\gamma^k u[k]$.
By definition

$$F[z] = \sum_{k=0}^{\infty} \gamma^k u[k] z^{-k}$$

Since $u[k] = 1$ for all $k \geq 0$,

$$F[z] = \sum_{k=0}^{\infty} \left(\frac{\gamma}{z}\right)^k$$

$$= 1 + \left(\frac{\gamma}{z}\right) + \left(\frac{\gamma}{z}\right)^2 + \left(\frac{\gamma}{z}\right)^3 + \cdots + \cdots \tag{11.5}$$

It is helpful to remember the following well-known geometric progression and its sum:

$$1 + x + x^2 + x^3 + \cdots = \frac{1}{1-x} \qquad \text{if} \quad |x| < 1 \tag{11.6}$$

Use of Eq. (11.6) in Eq. (11.5) yields

$$F[z] = \frac{1}{1 - \frac{\gamma}{z}} \qquad \left|\frac{\gamma}{z}\right| < 1$$

$$= \frac{z}{z - \gamma} \qquad |z| > |\gamma| \tag{11.7}$$

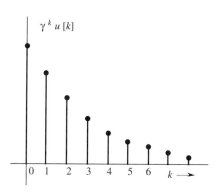

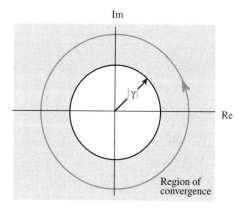

Fig. 11.1 $\gamma^k u[k]$ and the region of convergence of its z-transform.

Observe that $F[z]$ exists only for $|z| > |\gamma|$. For $|z| < |\gamma|$, the sum in Eq. (11.5) does not converge; it goes to infinity. Therefore, the region of convergence (or existence) of $F[z]$ is the shaded region outside the circle of radius $|\gamma|$, centered at the origin, in the z-plane, as depicted in Fig. 11.1b.   ■

The region of convergence is required for evaluating $f[k]$ from $F[z]$, according to Eq. (11.2). The integral in Eq. (11.2) is a contour integral implying integration in a counterclockwise direction along a closed path centered at the origin and satisfying the condition $|z| > |\gamma|$. Thus, any circular path centered at the origin and with a radius greater than $|\gamma|$ (Fig. 11.1b) will suffice. We can show that the integral in Eq. (11.2) along any such path (with a radius greater than $|\gamma|$) yields the same result, namely $f[k]$. Such integration in the complex plane requires a background in the theory of functions of complex variables. We can avoid this integration by compiling a table of z-transforms (Table 11.1), where z-transform pairs are tabulated for a variety of signals. To find the inverse z-transform of say, $z/(z - \gamma)$, instead of using the complex integration in (11.2), we consult the table and find the inverse z-transform of $z/(z - \gamma)$ as $\gamma^k u[k]$. Although the table given here is rather short, it comprises the functions of most practical interest.

The bilateral z-transform is defined by Eq. (11.1) with the limits of the right-hand sum from $-\infty$ to ∞ instead of from 0 to ∞. The situation of the z-transform regarding the uniqueness of the inverse transform is parallel to that of the Laplace transform. For the bilateral case, the inverse z-transform is not unique unless the region of convergence is specified. For the unilateral case, the inverse transform is unique; the region of convergence need not be specified to determine the inverse z-transform. For this reason, we shall ignore the region of convergence in the unilateral z-transform Table 11.1.

Existence of the $\mathcal{Z}$-Transform

By definition

$$F[z] = \sum_{k=0}^{\infty} f[k]z^{-k} = \sum_{k=0}^{\infty} \frac{f[k]}{z^k}$$

The existence of the z-transform is guaranteed if

$$|F[z]| \le \sum_{k=0}^{\infty} \frac{|f[k]|}{|z|^k} < \infty$$

for some $|z|$. Any signal $f[k]$ that grows no faster than an exponential signal $r_0{}^k$, for some r_0, satisfies this condition. Thus, if

$$|f[k]| \le r_0{}^k \qquad \text{for some } r_0 \tag{11.8}$$

then

$$|F[z]| \le \sum_{k=0}^{\infty} \left(\frac{r_0}{|z|}\right)^k = \frac{1}{1 - \frac{r_0}{|z|}} \qquad |z| > r_0$$

Therefore, $F[z]$ exists for $|z| > r_0$. All practical signals satisfy (11.8) and are therefore z-transformable. Some signal models (e.g. γ^{k^2}) which grow faster than the exponential signal $r_0{}^k$ (for any r_0) do not satisfy (11.8) and therefore are not z-transformable. Fortunately, such signals are of little practical or theoretical interest.

■ **Example 11.2**
   Find the z-transforms of **(a)** $\delta[k]$ **(b)** $u[k]$ **(c)** $\cos \beta k \, u[k]$ **(d)** signal shown in Fig. 11.2.
   Recall that by definition

$$F[z] = \sum_{k=0}^{\infty} f[k]z^{-k}$$

$$= f[0] + \frac{f[1]}{z} + \frac{f[2]}{z^2} + \frac{f[3]}{z^3} + \cdots \tag{11.9}$$

(a)   For $f[k] = \delta[k]$, $f[0] = 1$ and $f[2] = f[3] = f[4] = \cdots = 0$. Therefore

$$\delta[k] \Longleftrightarrow 1 \qquad \text{for all } z \tag{11.10}$$

(b)   For $f[k] = u[k]$, $f[0] = f[1] = f[3] = \cdots = 1$. Therefore

$$F[z] = 1 + \frac{1}{z} + \frac{1}{z^2} + \frac{1}{z^3} + \cdots$$

From Eq. (11.6) it follows that

$$F[z] = \frac{1}{1 - \frac{1}{z}} \qquad \left|\frac{1}{z}\right| < 1$$

$$= \frac{z}{z - 1} \qquad |z| > 1$$

Therefore

$$u[k] \Longleftrightarrow \frac{z}{z - 1} \qquad |z| > 1 \tag{11.11}$$

(c)   Recall that $\cos \beta k = \left(e^{j\beta k} + e^{-j\beta k}\right)/2$. Moreover, according to Eq. (11.7),

$$e^{\pm j\beta k}u[k] \Longleftrightarrow \frac{z}{z - e^{\pm j\beta}} \qquad |z| > |e^{\pm j\beta}| = 1$$

Therefore

$$F[z] = \frac{1}{2}\left[\frac{z}{z-e^{j\beta}} + \frac{z}{z-e^{-j\beta}}\right]$$

$$= \frac{z(z-\cos\beta)}{z^2 - 2z\cos\beta + 1} \qquad |z| > 1$$

(d) Here $f[0] = f[1] = f[2] = f[3] = f[4] = 1$ and $f[5] = f[6] = \cdots = 0$. Therefore, according to Eq. (11.9)

$$F[z] = 1 + \frac{1}{z} + \frac{1}{z^2} + \frac{1}{z^3} + \frac{1}{z^4}$$

$$= \frac{z^4 + z^3 + z^2 + z + 1}{z^4}$$

We can also express this result in a closed form by summing the geometric progression on the right-hand side of the above equation, using the formula in Sec. B.7-4. Here the common ratio $r = \frac{1}{z}$, $M = 0$, and $N = 4$, so that

$$F[z] = \frac{(\frac{1}{z})^5 - (\frac{1}{z})^0}{\frac{1}{z} - 1} = \frac{z}{z-1}(1 - z^{-5}) \quad \blacksquare$$

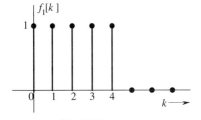

Fig. 11.2

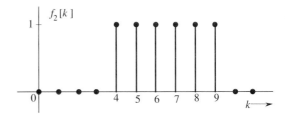

Fig. 11.3

△ **Exercise E11.1**
   (a) Find the z-transform of a signal shown in Fig. 11.3. (b) Using Pair 12a (Table 11.1) find the z-transform of $f[k] = 20.65(\sqrt{2})^k \cos\left(\frac{\pi}{4}k - 1.415\right)u[k]$.

Answers: (a) $F[z] = \dfrac{z^5 + z^4 + +z^3 + z^2 + z + 1}{z^9}$ or $\dfrac{z}{z-1}(z^{-4} - z^{-10})$

(b) $\dfrac{z(3.2z + 17.2)}{z^2 - 2z + 2}$ ▽

11.1-1   Finding the Inverse Transform

As in the Laplace transform, we shall avoid the integration in the complex plane required to find the inverse z-transform [Eq. (11.2)] by using the (unilateral) transform Table. Many of the transforms $F[z]$ of practical interest are rational functions (ratio of polynomials in z). Such functions can be expressed as a sum of simpler functions using partial fraction expansion. This method works because for every transformable $f[k]$ defined for $k \geq 0$, there is a corresponding unique $F[z]$ defined for $|z| > r_0$ (where r_0 is some constant), and vice versa.

Table 11.1: (Unilateral) z-Transform Pairs

	$f[k]$	$F[z]$
1	$\delta[k-j]$	z^{-j}
2	$u[k]$	$\dfrac{z}{z-1}$
3	$ku[k]$	$\dfrac{z}{(z-1)^2}$
4	$k^2 u[k]$	$\dfrac{z(z+1)}{(z-1)^3}$
5	$k^3 u[k]$	$\dfrac{z(z^2+4z+1)}{(z-1)^4}$
6	$\gamma^{k-1} u[k-1]$	$\dfrac{1}{z-\gamma}$
7	$\gamma^k u[k]$	$\dfrac{z}{z-\gamma}$
8	$k\gamma^k u[k]$	$\dfrac{\gamma z}{(z-\gamma)^2}$
9	$k^2 \gamma^k u[k]$	$\dfrac{\gamma z(z+\gamma)}{(z-\gamma)^3}$
10	$\dfrac{k(k-1)(k-2)\cdots(k-m+1)}{\gamma^m m!}\gamma^k u[k]$	$\dfrac{z}{(z-\gamma)^{m+1}}$
11a	$\|\gamma\|^k \cos \beta k\, u[k]$	$\dfrac{z(z-\|\gamma\|\cos\beta)}{z^2 - (2\|\gamma\|\cos\beta)z + \|\gamma\|^2}$
11b	$\|\gamma\|^k \sin \beta k\, u[k]$	$\dfrac{z\|\gamma\|\sin\beta}{z^2 - (2\|\gamma\|\cos\beta)z + \|\gamma\|^2}$
12a	$r\|\gamma\|^k \cos(\beta k + \theta)u[k]$	$\dfrac{rz[z\cos\theta - \|\gamma\|\cos(\beta-\theta)]}{z^2 - (2\|\gamma\|\cos\beta)z + \|\gamma\|^2}$
12b	$r\|\gamma\|^k \cos(\beta k + \theta)u[k] \qquad \gamma = \|\gamma\|e^{j\beta}$	$\dfrac{(0.5re^{j\theta})z}{z-\gamma} + \dfrac{(0.5re^{-j\theta})z}{z-\gamma^*}$
12c	$r\|\gamma\|^k \cos(\beta k + \theta)u[k]$	$\dfrac{z(Az+B)}{z^2 + 2az + \|\gamma\|^2}$
	$r = \sqrt{\dfrac{A^2\|\gamma\|^2 + B^2 - 2AaB}{\|\gamma\|^2 - a^2}}$	
	$\beta = \cos^{-1}\dfrac{-a}{\|\gamma\|},\ \ \theta = \tan^{-1}\dfrac{Aa-B}{A\sqrt{\|\gamma\|^2 - a^2}}$	

■ **Example 11.3**

Find the inverse z-transform of

(a) $\dfrac{8z - 19}{(z - 2)(z - 3)}$ (b) $\dfrac{z(2z^2 - 11z + 12)}{(z - 1)(z - 2)^3}$ (c) $\dfrac{2z(3z + 17)}{(z - 1)(z^2 - 6z + 25)}$

(a) Expanding $F[z]$ into partial fractions yields

$$F[z] = \frac{8z - 19}{(z - 2)(z - 3)} = \frac{3}{z - 2} + \frac{5}{z - 3}$$

From Table 11.1, Pair 6, we obtain

$$f[k] = \left[3(2)^{k-1} + 5(3)^{k-1}\right] u[k - 1] \tag{11.12a}$$

If we expand rational $F[z]$ into partial fractions directly, we shall always obtain an answer that is multiplied by $u[k - 1]$ because of the nature of Pair 6 in Table 11.1. This form is rather awkward as well as inconvenient. We prefer the form that is multiplied by $u[k]$ rather than $u[k - 1]$. A glance at Table 11.1 shows that the z-transform of every signal that is multiplied by $u[k]$ has a factor z in the numerator. This observation suggests that we expand $F[z]$ into *modified partial fractions*, where each term has a factor z in the numerator. This goal can be accomplished by expanding $F[z]/z$ into partial fractions and then multiplying both sides by z. We shall demonstrate this procedure by reworking part **(a)** in Example 11.3. For this case

$$\frac{F[z]}{z} = \frac{8z - 19}{z(z - 2)(z - 3)}$$

$$= \frac{(-19/6)}{z} + \frac{(3/2)}{z - 2} + \frac{(5/3)}{z - 3}$$

Multiplying both sides by z yields

$$F[z] = -\frac{19}{6} + \frac{3}{2}\left(\frac{z}{z - 2}\right) + \frac{5}{3}\left(\frac{z}{z - 3}\right)$$

From Pairs 1 and 7 in Table 11.1, it follows that

$$f[k] = -\tfrac{19}{6}\delta[k] + \left[\tfrac{3}{2}(2)^k + \tfrac{5}{3}(3)^k\right] u[k] \tag{11.12b}$$

The reader can verify that this answer is equivalent to that in Eq. (11.12a) by computing $f[k]$ in both cases for $k = 0, 1, 2, 3, \cdots$, and then comparing the results. The form in Eq. (11.12b) is more convenient than that in Eq. (11.12a). For this reason, we shall always expand $F[z]/z$ rather than $F[z]$ into partial fractions and then multiply both sides by z to obtain modified partial fractions of $F[z]$, which have a factor z in the numerator.

(b) $F[z] = \dfrac{z(2z^2 - 11z + 12)}{(z - 1)(z - 2)^3}$

and

$$\frac{F[z]}{z} = \frac{2z^2 - 11z + 12}{(z - 1)(z - 2)^3}$$

$$= \frac{k}{z - 1} + \frac{a_0}{(z - 2)^3} + \frac{a_1}{(z - 2)^2} + \frac{a_2}{(z - 2)}$$

where

$$k = \left.\frac{2z^2 - 11z + 12}{(z-1)(z-2)^3}\right|_{z=1} = -3$$

$$a_0 = \left.\frac{2z^2 - 11z + 12}{(z-1)(z-2)^3}\right|_{z=2} = -2$$

Therefore

$$\frac{F[z]}{z} = \frac{2z^2 - 11z + 12}{(z-1)(z-2)^3} = \frac{-3}{z-1} - \frac{2}{(z-2)^3} + \frac{a_1}{(z-2)^2} + \frac{a_2}{(z-2)} \tag{11.13}$$

We can determine a_1 and a_2 by clearing fractions or by using the short cuts discussed in Sec. B.5-3. For example, to determine a_2, we multiply both sides of Eq. (11.13) by z and let $z \to \infty$. This yields

$$0 = -3 - 0 + 0 + a_2 \implies a_2 = 3$$

This result leaves only one unknown, a_1, which is readily determined by letting z take any convenient value, say $z = 0$, on both sides of Eq. (11.13). This step yields

$$\frac{12}{8} = 3 + \frac{1}{4} + \frac{a_1}{4} - \frac{3}{2}$$

Multiplying both sides by 8 yields

$$12 = 24 + 2 + 2a_1 - 12 \implies a_1 = -1$$

Therefore

$$\frac{F[z]}{z} = \frac{-3}{z-1} - \frac{2}{(z-2)^3} - \frac{1}{(z-2)^2} + \frac{3}{z-2}$$

and

$$F[z] = -3\frac{z}{z-1} - 2\frac{z}{(z-2)^3} - \frac{z}{(z-2)^2} + 3\frac{z}{z-2}$$

Now the use of Table 11.1, Pairs 7 and 10, yields

$$f[k] = \left[-3 - 2\frac{k(k-1)}{8}(2)^k - \frac{k}{2}(2)^k + 3(2)^k \right] u[k]$$

$$= -[3 + \tfrac{1}{4}(k^2 + k - 12)2^k]u[k]$$

(c) Complex Poles

$$F[z] = \frac{2z(3z + 17)}{(z-1)(z^2 - 6z + 25)} = \frac{2z(3z + 17)}{(z-1)(z - 3 - j4)(z - 3 + j4)}$$

Poles of $F[z]$ are 1, $3 + j4$, and $3 - j4$. Whenever there are complex conjugate poles, the problem can be worked out in two ways. In the first method we expand $F[z]$ into (modified) first-order partial fractions. In the second method, rather than obtaining one factor corresponding to each complex conjugate pole, we obtain quadratic factors corresponding to each pair of complex conjugate poles. This procedure is explained below.

Method of First-Order Factors

$$\frac{F[z]}{z} = \frac{2(3z+17)}{(z-1)(z^2-6z+25)} = \frac{2(3z+17)}{(z-1)(z-3-j4)(z-3+j4)}$$

We find the partial fraction of $F[z]/z$ using the Heaviside "cover-up" method:

$$\frac{F[z]}{z} = \frac{2}{z-1} + \frac{1.6e^{-j2.246}}{z-3-j4} + \frac{1.6e^{j2.246}}{z-3+j4}$$

and

$$F[z] = 2\frac{z}{z-1} + (1.6e^{-j2.246})\frac{z}{z-3-j4} + (1.6e^{j2.246})\frac{z}{z-3+j4}$$

The inverse transform of the first term on the right-hand side is $2u[k]$. The inverse transform of the remaining two terms (complex conjugate poles) can be obtained from Pair 12b (Table 11.1) by identifying $\frac{r}{2} = 1.6$, $\theta = -2.246$ rad., $\gamma = 3+j4 = 5e^{j0.927}$, so that $|\gamma| = 5$, $\beta = 0.927$. Therefore

$$f[k] = \left[2 + 3.2(5)^k \cos(0.927k - 2.246)\right] u[k]$$

Method of Quadratic Factors

$$\frac{F[z]}{z} = \frac{2(3z+17)}{(z-1)(z^2-6z+25)} = \frac{2}{z-1} + \frac{Az+B}{z^2-6z+25}$$

Multiplying both sides by z and letting $z \to \infty$, we find

$$0 = 2 + A \implies A = -2$$

and

$$\frac{2(3z+17)}{(z-1)(z^2-6z+25)} = \frac{2}{z-1} + \frac{-2z+B}{z^2-6z+25}$$

To find B, we let z take any convenient value, say $z = 0$. This step yields

$$\frac{-34}{25} = -2 + \frac{B}{25}$$

Multiplying both sides by 25 yields

$$-34 = -50 + B \implies B = 16$$

Therefore

$$\frac{F[z]}{z} = \frac{2}{z-1} + \frac{-2z+16}{z^2-6z+25}$$

and

$$F[z] = \frac{2z}{z-1} + \frac{z(-2z+16)}{z^2-6z+25}$$

We now use Pair 12c where we identify $A = -2$, $B = 16$, $|\gamma| = 5$, $a = -3$. Therefore

$$r = \sqrt{\frac{100+256-192}{25-9}} = 3.2, \quad \beta = \cos^{-1}(\tfrac{3}{5}) = 0.927 \text{ rad., and}$$

$$\theta = \tan^{-1}(\tfrac{-10}{-8}) = -2.246 \text{ rad., so that}$$

$$f[k] = \left[2 + 3.2(5)^k \cos(0.927k - 2.246)\right] u[k] \quad \blacksquare$$

The procedure for finding partial fractions using MATLAB was demonstrated in chapter 6. The same program can be used in this case, except that we have to find the modified partial fractions here. This goal is readily accomplished by dividing $F[z]$ by z and then taking the partial fractions. We shall demonstrate this procedure with an example.

⊙ **Computer Example C11.1**
Solve Example 11.3a using MATLAB.

```
num=[8 -19]; den=[conv([1 -2],[1 -3]) 0];
[r, p, k]= residue(num,den)
% We could also express den=[1 -5 6 0]

r =
        1.6667
        1.5000
       -3.1667
p =
        3
        2
        0
k =
        []
```

Hence,

$$F[z] = -3.1667 + \frac{1.5z}{z-2} + \frac{1.6667z}{z-3} \quad ⊙$$

△ **Exercise E11.2**
Find the inverse z-transform of the following functions:

(a) $\dfrac{z(2z-1)}{(z-1)(z+0.5)}$ (b) $\dfrac{1}{(z-1)(z+0.5)}$

(c) $\dfrac{9}{(z+2)(z-0.5)^2}$ (d) $\dfrac{5z(z-1)}{z^2 - 1.6z + 0.8}$

Answer: **(a)** $\left[\frac{2}{3} + \frac{4}{3}(-0.5)^k\right] u[k]$ **(b)** $-2\delta[k] + \left[\frac{2}{3} + \frac{4}{3}(-0.5)^k\right] u[k]$

(c) $18\delta[k] - [0.72(-2)^k + 17.28(0.5)^k - 14.4k(0.5)^k]u[k]$

(d) $\frac{5\sqrt{5}}{2}\left(\frac{2}{\sqrt{5}}\right)^k \cos(0.464k + 0.464)u[k]$. Hint: $\sqrt{0.8} = \frac{2}{\sqrt{5}}$. ▽

Inverse Transform by Expansion of F[z] in Power Series of z^{-1}

By definition

$$F[z] = \sum_{k=0}^{\infty} f[k]z^{-k}$$

$$= f[0] + \frac{f[1]}{z} + \frac{f[2]}{z^2} + \frac{f[3]}{z^3} + \cdots$$

$$= f[0]z^0 + f[1]z^{-1} + f[2]z^{-2} + f[3]z^{-3} + \cdots$$

This result is a power series in z^{-1}. Therefore, if we can expand $F[z]$ into a power series in z^{-1}, the coefficients of this power series can be identified as $f[0]$, $f[1]$, $f[2]$,

$f[3]$, $\cdots$, and so on. A rational $F[z]$ can be expanded into a power series of z^{-1} by dividing its numerator by the denominator. Consider, for example,

$$F[z] = \frac{z^2(7z - 2)}{(z - 0.2)(z - 0.5)(z - 1)}$$

$$= \frac{7z^3 - 2z^2}{z^3 - 1.7z^2 + 0.8z - 0.1}$$

To obtain a series expansion in powers of z^{-1}, we divide the numerator by the denominator as follows:

$$
\begin{array}{r}
7 + 9.9z^{-1} + 11.23z^{-2} + 11.87z^{-3} + \cdots \\
z^3 - 1.7z^2 + 0.8z - 0.1 \overline{)\,7z^3 - 2z^2} \\
\underline{7z^3 - 11.9z^2 + 5.60z - 0.7} \\
9.9z^2 - 5.60z + 0.7 \\
\underline{9.9z^2 - 16.83z + 7.92 - 0.99z^{-1}} \\
11.23z - 7.22 + 0.99z^{-1} \\
\underline{11.23z - 19.09 + 8.98z^{-1}} \\
11.87 - 7.99z^{-1}
\end{array}
$$

Thus

$$F[z] = \frac{z^2(7z - 2)}{(z - 0.2)(z - 0.5)(z - 1)} = 7 + 9.9z^{-1} + 11.23z^{-2} + 11.87z^{-3} + \cdots$$

Therefore

$$f[0] = 7, \ f[1] = 9.9, \ f[2] = 11.23, \ f[3] = 11.87, \ \cdots, \text{ and so on.}$$

We give here a simple MATLAB program to find the first N terms of the inverse z-transform.

⊙ **Computer Example C11.2**
   Using MATLAB, find the first 10 values ($f[0]$ through $f[9]$) of the inverse z-transform of $F[z]$ in the above example.

```
num=[7 -2 0 0]; den=[1 -1.7 0.8 -0.1];
f=dimpulse(num, den, 10)
% We could also write den=conv(conv([1 -0.2],[1 -0.5]),[1 -1])

f =
     7.0000
     9.9000
    11.2300
    11.8710
    12.1867
    12.3436
    12.4218
    12.4609
    12.4805
    12.4902     ⊙
```

Although this procedure yields $f[k]$ directly, it does not provide a closed-form solution. For this reason, it is not very useful unless we want to know only the first few terms of the sequence $f[k]$.

△ **Exercise E11.3**

Using long division to find the power series in z^{-1}, show that the inverse z-transform of $z/(z - 0.5)$ is $(0.5)^k u[k]$ or $(2)^{-k}u[k]$.   ▽

Relationship Between h[k] and H[z]

For an LTID system, if $h[k]$ is its unit impulse response, then in Eq. (9.57b) we defined $H[z]$, the system transfer function, as

$$H[z] = \sum_{k=-\infty}^{\infty} h[k]z^{-k}$$

For causal systems, the limits on the sum are from $k = 0$ to ∞. This equation shows that the transfer function $H[z]$ is the z-transform of the impulse response $h[k]$ of an LTID system; that is

$$h[k] \iff H[z] \tag{11.14}$$

This important result relates the impulse response $h[k]$, which is a time-domain specification of a system, to $H[z]$, which is a frequency-domain specification of a system. The result is parallel to that for LTIC systems.

△ **Exercise E11.4**

Redo Exercise E9.5 by taking the inverse z-transform of $H[z]$.   ▽

11.2 Some properties of the Z-Transform

The z-transform properties are useful in the derivation of z-transforms of many functions and also in the solution of linear difference equations with constant coefficients. Here we consider a few important properties of the z-transform.

Right Shift (Delay)

If

$$f[k]u[k] \iff F[z]$$

then

$$f[k - 1]u[k - 1] \iff \frac{1}{z}F[z] \tag{11.15a}$$

and

$$f[k - m]u[k - m] \iff \frac{1}{z^m}F[z] \tag{11.15b}$$

and

$$f[k - 1]u[k] \iff \frac{1}{z}F[z] + f[-1] \tag{11.16a}$$

Repeated application of this property yields

$$f[k - 2]u[k] \Longleftrightarrow \frac{1}{z}\left[\frac{1}{z}F[z] + f[-1]\right] + f[-2]$$

$$= \frac{1}{z^2}F[z] + \frac{1}{z}f[-1] + f[-2] \qquad (11.16b)$$

and

$$f[k - m]u[k] \Longleftrightarrow z^{-m}F[z] + z^{-m}\sum_{k=1}^{m}f[-k]z^{k} \qquad (11.16c)$$

   Proof:

$$\mathcal{Z}\{f[k - m]u[k - m]\} = \sum_{k=0}^{\infty}f[k - m]u[k - m]z^{-k}$$

Recall that $f[k - m]u[k - m] = 0$ for $k < m$, so that the limits on the summation on the right-hand side can be taken from $k = m$ to ∞. Therefore

$$\mathcal{Z}\{f[k - m]u[k - m]\} = \sum_{k=m}^{\infty}f[k - m]z^{-k}$$

$$= \sum_{r=0}^{\infty}f[r]z^{-(r+m)}$$

$$= \frac{1}{z^m}\sum_{r=0}^{\infty}f[r]z^{-r}$$

$$= \frac{1}{z^m}F[z]$$

To prove Eq. (11.16c), we have

$$\mathcal{Z}\{f[k - m]u[k]\} = \sum_{k=0}^{\infty}f[k - m]z^{-k} = \sum_{r=-m}^{\infty}f[r]z^{-(r+m)}$$

$$= z^{-m}\left[\sum_{r=-m}^{-1}f[r]z^{-r} + \sum_{r=0}^{\infty}f[r]z^{-r}\right]$$

$$= z^{-m}\sum_{k=1}^{m}f[-k]z^{k} + z^{-m}F[z]$$

Left Shift (Advance)
   If

$$f[k]u[k] \Longleftrightarrow F[z]$$

then

$$f[k + 1]u[k] \Longleftrightarrow zF[z] - zf[0] \qquad (11.17a)$$

Repeated application of this property yields

$$f[k+2]u[k] \Longleftrightarrow z\left\{z\left(F[z]-zf[0]\right)-f[1]\right\}$$
$$= z^2 F[z] - z^2 f[0] - z f[1] \tag{11.17b}$$

and

$$f[k+m]u[k] \Longleftrightarrow z^m F[z] - z^m \sum_{k=0}^{m-1} f[k]z^{-k} \tag{11.17c}$$

Proof: By definition

$$\mathcal{Z}\left\{f[k+m]u[k]\right\} = \sum_{k=0}^{\infty} f[k+m]z^{-k}$$

$$= \sum_{r=m}^{\infty} f[r]z^{-(r-m)}$$

$$= z^m \sum_{r=m}^{\infty} f[r]z^{-r}$$

$$= z^m \left[\sum_{r=0}^{\infty} f[r]z^{-r} - \sum_{r=0}^{m-1} f[r]z^{-r}\right]$$

$$= z^m F[z] - z^m \sum_{r=0}^{m-1} f[r]z^{-r}$$

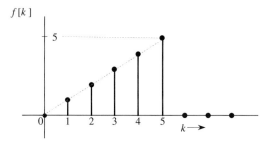

Fig. 11.4   Signal for Example 11.4.

■ **Example 11.4**

Find the z-transform of the signal $f[k]$ depicted in Fig. 11.4.

The signal $f[k]$ can be expressed as a product of k and a gate pulse $u[k] - u[k-6]$. Therefore

$$f[k] = k\left\{u[k] - u[k-6]\right\}$$
$$= ku[k] - ku[k-6]$$

We cannot find the z-transform of $ku[k-6]$ directly by using the right-shift property [Eq. (11.15b)]. So we rearrange it in terms of $(k-6)u[k-6]$ as follows:

$$f[k] = ku[k] - [(k-6)u[k-6] + 6u[k-6]]$$

We can now find the z-transform of the bracketed term by using the right-shift property [Eq. (11.15b)]. Because $u[k] \Longleftrightarrow \frac{z}{z-1}$

$$u[k-6] \Longleftrightarrow \frac{1}{z^6}\frac{z}{z-1} = \frac{1}{z^5(z-1)}$$

Also, because $ku[k] \Longleftrightarrow \frac{z}{(z-1)^2}$

$$(k-6)u[k-6] \Longleftrightarrow \frac{1}{z^6}\frac{z}{(z-1)^2} = \frac{1}{z^5(z-1)^2}$$

Therefore

$$F[z] = \frac{z}{(z-1)^2} - \frac{1}{z^5(z-1)^2} - \frac{6}{z^5(z-1)}$$

$$= \frac{z^6 - 6z + 5}{z^5(z-1)^2} \quad \blacksquare$$

$\triangle$ **Exercise E11.5**

Using only the fact that $u[k] \Longleftrightarrow \frac{z}{z-1}$ and the right-shift property [Eq. (11.15)], find the z-transforms of the signals in Figs. 11.2 and 11.3. The answers are given in Example 11.2d and Exercise E11.1a. $\triangledown$

Convolution

The time convolution property and the frequency convolution property state that if

$$f_1[k] \Longleftrightarrow F_1[z] \quad \text{and} \quad f_2[k] \Longleftrightarrow F_2[z],$$

then (**time convolution**)

$$f_1[k] * f_2[k] \Longleftrightarrow F_1[z]F_2[z] \tag{11.18}$$

and (**frequency convolution**)

$$f_1[k]f_2[k] \Longleftrightarrow \frac{1}{2\pi j}\oint F_1[u]F_2\left[\frac{z}{u}\right]u^{-1}\,du \tag{11.19}$$

Proof: These properties apply to causal as well as noncausal sequences. For this reason, we shall prove them for the more general case of noncausal sequences, where the convolution sum ranges from $-\infty$ to ∞. To prove the time convolution, we have

$$\mathcal{Z}\{f_1[k] * f_2[k]\} = \mathcal{Z}\left[\sum_{m=-\infty}^{\infty} f_1[m]f_2[k-m]\right]$$

$$= \sum_{k=-\infty}^{\infty} z^{-k}\sum_{m=-\infty}^{\infty} f_1[m]f_2[k-m]$$

Interchanging the order of summation,

$$\mathcal{Z}\left[f_1[k] * f_2[k]\right] = \sum_{m=-\infty}^{\infty} f_1[m] \sum_{k=-\infty}^{\infty} f_2[k-m]z^{-k}$$

$$= \sum_{m=-\infty}^{\infty} f_1[m] \sum_{r=-\infty}^{\infty} f_2[r]z^{-(r+m)}$$

$$= \sum_{m=-\infty}^{\infty} f_1[m]z^{-m} \sum_{r=-\infty}^{\infty} f_2[r]z^{-r}$$

$$= F_1[z]F_2[z]$$

To prove the frequency convolution, we start with

$$\mathcal{Z}\left\{f_1[k]f_2[k]\right\} = \sum_{k=-\infty}^{\infty} f_1[k]f_2[k]z^{-k}$$

$$= \frac{1}{2\pi j} \sum_{k=-\infty}^{\infty} f_2[k]z^{-k} \oint F_1[u]u^{k-1}\, du$$

Interchanging the order of summation and integration

$$\mathcal{Z}\left[f_1[k]f_2[k]\right] = \frac{1}{2\pi j} \oint F_1[u] \left[\sum_{k=-\infty}^{\infty} f_2[k] \left(\frac{z}{u}\right)^{-k}\right] u^{-1}\, du$$

$$= \frac{1}{2\pi j} \oint F_1[u]F_2\left[\frac{z}{u}\right] u^{-1}\, du$$

LTID System Response

It is interesting to apply the time convolution property to the LTID input-output equation $y[k] = f[k]*h[k]$. In Eq. (11.14), we have shown that $h[k] \Longleftrightarrow H[z]$. Hence, according to Eq. (11.18), it follows that

$$Y[z] = F[z]H[z] \qquad (11.20)$$

Earlier in the chapter, we derived this important result using informal arguments.

Multiplication by γ^k

If
$$f[k]u[k] \Longleftrightarrow F[z]$$

then
$$\gamma^k f[k]u[k] \Longleftrightarrow F\left[\frac{z}{\gamma}\right] \qquad (11.21)$$

Proof:

$$\mathcal{Z}\{\gamma^k f[k]u[k]\} = \sum_{k=0}^{\infty} \gamma^k f[k]z^{-k} = \sum_{k=0}^{\infty} f[k] \left(\frac{z}{\gamma}\right)^{-k} = F\left[\frac{z}{\gamma}\right]$$

△ **Exercise E11.6**

Using Eq. (11.21), derive Pairs 7 and 8 in Table 11.1 from Pairs 2 and 3, respectively. ▽

Multiplication by k (Scaling in the z-Domain

If

$$f[k]u[k] \Longleftrightarrow F[z]$$

then

$$k f[k]u[k] \Longleftrightarrow -z \frac{d}{dz} F[z] \tag{11.22}$$

Proof:

$$-z \frac{d}{dz} F[z] = -z \frac{d}{dz} \sum_{k=0}^{\infty} f[k]z^{-k} = -z \sum_{k=0}^{\infty} -k f[k]z^{-k-1}$$

$$= \sum_{k=0}^{\infty} k f[k]z^{-k} = \mathcal{Z}\left\{k f[k]u[k]\right\}$$

△ **Exercise E11.7**

Using Eq. (11.22), derive Pairs 3 and 4 in Table 11.1 from Pair 2. Similarly, derive Pairs 8 and 9 from Pair 7. ▽

Initial and Final Value

For a causal $f[k]$,

$$f[0] = \lim_{z \to \infty} F[z] \tag{11.23a}$$

This result follows immediately from Eq. (11.9)

We can also show that if $(z-1)F(z)$ has no poles outside the unit circle, then

$$\lim_{N \to \infty} f(N) = \lim_{z \to 1} (z-1)F(z) \tag{11.23b}$$

11.3 Z-Transform Solution of Linear Difference Equations

The time-shifting (left- or right-shift) property has set the stage for solving linear difference equations with constant coefficients. As in the case of the Laplace transform with differential equations, the z-transform converts difference equations into algebraic equations which are readily solved to find the solution in the z-domain. Taking the inverse z-transform of the z-domain solution yields the desired time-domain solution. The following examples demonstrate the procedure.

■ **Example 11.5**

Solve

$$y[k+2] - 5y[k+1] + 6y[k] = 3f[k+1] + 5f[k] \tag{11.24}$$

if the initial conditions are $y[-1] = \frac{11}{6}$, $y[-2] = \frac{37}{36}$, and the input $f[k] = (2)^{-k}u[k]$.

Table 11.2

$\mathcal{Z}$- Transform Operations

Operation	$f[k]$	$F[z]$
Addition	$f_1[k] + f_2[k]$	$F_1[z] + F_2[z]$
Scalar multiplication	$af[k]$	$aF[z]$
Right-shift	$f[k-m]u[k-m]$	$\dfrac{1}{z^m}F[z]$
	$f[k-m]u[k]$	$\dfrac{1}{z^m}F[z] + \dfrac{1}{z^m}\displaystyle\sum_{k=1}^{m} f[-k]z^k$
	$f[k-1]u[k]$	$\dfrac{1}{z}F[z] + f[-1]$
	$f[k-2]u[k]$	$\dfrac{1}{z^2}F[z] + \dfrac{1}{z}f[-1] + f[-2]$
	$f[k-3]u[k]$	$\dfrac{1}{z^3}F[z] + \dfrac{1}{z^2}f[-1] + \dfrac{1}{z}f[-2] + f[-3]$
Left-shift	$f[k+m]u[k]$	$z^m F[z] - z^m \displaystyle\sum_{k=0}^{m-1} f[k]z^{-k}$
	$f[k+1]u[k]$	$zF[z] - zf[0]$
	$f[k+2]u[k]$	$z^2 F[z] - z^2 f[0] - zf[1]$
	$f[k+3]u[k]$	$z^3 F[z] - z^3 f[0] - z^2 f[1] - zf[2]$
Multiplication by γ^k	$\gamma^k f[k]u[k]$	$F\left[\dfrac{z}{\gamma}\right]$
Multiplication by k	$kf[k]u[k]$	$-z\dfrac{d}{dz}F[z]$
Time Convolution	$f_1[k] * f_2[k]$	$F_1[z]F_2[z]$
Frequency Convolution	$f_1[k]f_2[k]$	$\dfrac{1}{2\pi j}\oint F_1[u]F_2\left[\dfrac{z}{u}\right]u^{-1}\,du$
Initial value	$f[0]$	$\lim_{z\to\infty} F[z]$
Final value	$\lim_{N\to\infty} f[N]$	$\lim_{z\to 1}(z-1)F[z]$ poles of $(z-1)F[z]$ inside the unit circle.

As we shall see, difference equations can be solved by using the right-shift or the left-shift property. Because the difference equation (11.24) is in advance-operator form, the use of the left-shift property in Eqs. (11.17a) and (11.17b) may seem appropriate for its solution. Unfortunately, as seen from Eqs. (11.17a) and (11.17b), these properties require a knowledge of auxiliary conditions $y[0], y[1], \cdots, y[n-1]$ rather than of the initial conditions $y[-1], y[-2], \cdots, y[-n]$, which are generally given. This difficulty can be overcome by expressing the difference equation (11.24) in delay operator form (obtained by replacing k with $k-2$) and then using the right-shift property.† Equation (11.24) in delay operator form is

$$y[k] - 5y[k-1] + 6y[k-2] = 3f[k-1] + 5f[k-2] \qquad (11.25)$$

We now use the right-shift property to take the z-transform of this equation. But before proceeding, we must be clear about the meaning of a term like $y[k-1]$. Does it mean $y[k-1]u[k-1]$ or $y[k-1]u[k]$? The answer becomes clear when we recognize that the use of the unilateral transform implies that we are considering the situation for $k \geq 0$, and that every signal in Eq. (11.25) must be counted from $k = 0$. Therefore, the term $y[k-j]$ means $y[k-j]u[k]$. Remember also that although we are considering the situation for $k \geq 0$, $y[k]$ is present even before $k = 0$ (in the form of initial conditions). Now

$$y[k]u[k] \Longleftrightarrow Y[z]$$

$$y[k-1]u[k] \Longleftrightarrow \frac{1}{z}Y[z] + y[-1] = \frac{1}{z}Y[z] + \frac{11}{6}$$

$$y[k-2]u[k] \Longleftrightarrow \frac{1}{z^2}Y[z] + \frac{1}{z}y[-1] + y[-2] = \frac{1}{z^2}Y[z] + \frac{11}{6z} + \frac{37}{36}$$

Also

$$f[k] = (2)^{-k}u[k] = (2^{-1})^k u[k] = (0.5)^k u[k] \Longleftrightarrow \frac{z}{z-0.5}$$

$$f[k-1]u[k] \Longleftrightarrow \frac{1}{z}F[z] + f[-1] = \frac{1}{z}\frac{z}{z-0.5} + 0 = \frac{1}{z-0.5}$$

$$f[k-2]u[k] \Longleftrightarrow \frac{1}{z^2}F[z] + \frac{1}{z}f[-1] + f[-2] = \frac{1}{z^2}F[z] + 0 + 0 = \frac{1}{z(z-0.5)}$$

Note that for causal input $f[k]$,

$$f[-1] = f[-2] = \cdots = f[-n] = 0$$

Hence

$$f[k-r]u[k] \Longleftrightarrow \frac{1}{z^r}F[z]$$

Taking the z-transform of Eq. (11.25) and substituting the above results, we obtain

$$Y[z] - 5\left[\frac{1}{z}Y[z] + \frac{11}{6}\right] + 6\left[\frac{1}{z^2}Y[z] + \frac{11}{6z} + \frac{37}{36}\right] = \frac{3}{z-0.5} + \frac{5}{z(z-0.5)} \qquad (11.26a)$$

or

$$\left(1 - \frac{5}{z} + \frac{6}{z^2}\right)Y[z] - \left(3 - \frac{11}{z}\right) = \frac{3}{z-0.5} + \frac{5}{z(z-0.5)} \qquad (11.26b)$$

---

†Another approach is to find $y[0], y[1], y[2], \cdots, y[n-1]$ from $y[-1], y[-2], \cdots, y[-n]$ iteratively, as in Sec. 9.1-1, and then apply the left-shift property to Eq. (11.24)

and

$$\left(1 - \frac{5}{z} + \frac{6}{z^2}\right) Y[z] = \left(3 - \frac{11}{z}\right) + \frac{3z + 5}{z(z - 0.5)}$$

$$= \frac{3z^2 - 9.5z + 10.5}{z(z - 0.5)}$$

Multiplication of both sides by z^2 yields

$$\left(z^2 - 5z + 6\right) Y[z] = \frac{z\left(3z^2 - 9.5z + 10.5\right)}{(z - 0.5)}$$

so that

$$Y[z] = \frac{z(3z^2 - 9.5z + 10.5)}{(z - 0.5)(z^2 - 5z + 6)} \tag{11.27}$$

and

$$\frac{Y[z]}{z} = \frac{3z^2 - 9.5z + 10.5}{(z - 0.5)(z - 2)(z - 3)}$$

$$= \frac{(26/15)}{z - 0.5} - \frac{(7/3)}{z - 2} + \frac{(18/5)}{z - 3}$$

Therefore

$$Y[z] = \frac{26}{15}\left(\frac{z}{z - 0.5}\right) - \frac{7}{3}\left(\frac{z}{z - 2}\right) + \frac{18}{5}\left(\frac{z}{z - 3}\right)$$

and

$$y[k] = \left[\frac{26}{15}(0.5)^k - \frac{7}{3}(2)^k + \frac{18}{5}(3)^k\right] u[k] \tag{11.28}$$

∎

This example demonstrates the ease with which linear difference equations with constant coefficients can be solved by z-transform. This method is general; it can be used to solve a single difference equation or a set of simultaneous difference equations of any order as long as the equations are linear with constant coefficients.

Comment

Sometimes auxiliary conditions $y[0], y[1], \cdots, y[n-1]$ (instead of initial conditions $y[-1], y[-2], \cdots, y[-n]$) are given to solve a difference equation. In this case, the equation can be solved by expressing it in the advance operator form and then using the left-shift property (see Exercise E11.9 below).

△ **Exercise E11.8**

Solve the equation below if the initial conditions are $y[-1] = 2$, $y[-2] = 0$, and the input $f[k] = u[k]$:

$$y[k + 2] - \frac{5}{6}y[k + 1] + \frac{1}{6}y[k] = 5f[k + 1] - f[k]$$

Answer: $y[k] = \left[12 - 15(\frac{1}{2})^k + \frac{14}{3}(\frac{1}{3})^k\right] u[k]$ ▽

△ **Exercise E11.9**

Solve the following equation if the auxiliary conditions are $y[0] = 1$, $y[1] = 2$, and the input $f[k] = u[k]$:

$$y[k + 2] + 3y[k + 1] + 2y[k] = f[k + 1] + 3f[k]$$

Answer: $y[k] = \left[\frac{2}{3} + 2(-1)^k - \frac{5}{3}(-2)^k\right] u[k]$ ▽

Zero-Input and Zero-State Components

In Example 11.5 we found the total solution of the difference equation. It is relatively easy to separate the solution into zero-input and zero-state components. All we have to do is to separate the response into terms arising from the input and terms arising from initial conditions. We can separate the response in Eq. (11.26b) as follows:

$$\left(1 - \frac{5}{z} + \frac{6}{z^2}\right) Y[z] - \underbrace{\left(3 - \frac{11}{z}\right)}_{\text{initial condition terms}} = \underbrace{\frac{3}{z - 0.5} + \frac{5}{z(z - 0.5)}}_{\text{terms arising from input}} \tag{11.29}$$

Therefore

$$\left(1 - \frac{5}{z} + \frac{6}{z^2}\right) Y[z] = \underbrace{\left(3 - \frac{11}{z}\right)}_{\text{initial condition terms}} + \underbrace{\frac{(3z + 5)}{z(z - 0.5)}}_{\text{input terms}}$$

Multiplying both sides by z^2 yields

$$\left(z^2 - 5z + 6\right) Y[z] = \underbrace{z(3z - 11)}_{\text{initial condition terms}} + \underbrace{\frac{z(3z + 5)}{z - 0.5}}_{\text{input terms}}$$

and

$$Y[z] = \underbrace{\frac{z(3z - 11)}{z^2 - 5z + 6}}_{\text{zero-input response}} + \underbrace{\frac{z(3z + 5)}{(z - 0.5)(z^2 - 5z + 6)}}_{\text{zero-state response}} \tag{11.30}$$

We expand both terms on the right-hand side into modified partial fractions to yield

$$Y[z] = \underbrace{\left[5\left(\frac{z}{z - 2}\right) - 2\left(\frac{z}{z - 3}\right)\right]}_{\text{zero-input}} + \underbrace{\left[\frac{26}{15}\left(\frac{z}{z - 0.5}\right) - \frac{22}{3}\left(\frac{z}{z - 2}\right) + \frac{28}{5}\left(\frac{z}{z - 3}\right)\right]}_{\text{zero-state}}$$

and

$$y[k] = \left[\underbrace{5(2)^k - 2(3)^k}_{\text{zero-input}} - \underbrace{\frac{22}{3}(2)^k + \frac{28}{5}(3)^k + \frac{26}{15}(0.5)^k}_{\text{zero-state}}\right] u[k]$$

$$= \left[-\frac{7}{3}(2)^k + \frac{18}{5}(3)^k + \frac{26}{15}(0.5)^k\right] u[k]$$

a conclusion, which agrees with the result in Eq. (11.28).

△ **Exercise E11.10**
　Solve
$$y[k + 2] - \frac{5}{6}y[k + 1] + \frac{1}{6}y[k] = 5f[k + 1] - f[k]$$
if the initial conditions are $y[-1] = 2$, $y[-2] = 0$, and the input $f[k] = u[k]$. Separate the response into zero-input and zero-state components.

Answer:

$$y[k] = \left\{ \underbrace{[3(\tfrac{1}{2})^k - \tfrac{4}{3}(\tfrac{1}{3})^k]}_{\text{zero-input}} + \underbrace{[12 - 18(\tfrac{1}{2})^k + 6(\tfrac{1}{3})^k]}_{\text{zero-state}} \right\} u[k]$$

$$= \left[12 - 15(\tfrac{1}{2})^k + \tfrac{14}{3}(\tfrac{1}{3})^k \right] u[k] \quad \triangledown$$

11.3-1   Zero-State Response of LTID Systems: The Transfer Function

Consider an nth-order LTID system specified by the difference equation

$$Q[E]y[k] = P[E]f[k] \tag{11.31a}$$

or

$$(E^n + a_{n-1}E^{n-1} + \cdots + a_1 E + a_0)y[k] =$$
$$(b_n E^n + b_{n-1}E^{n-1} + \cdots + b_1 E + b_0)f[k] \tag{11.31b}$$

or

$$y[k+n] + a_{n-1}y[k+n-1] + \cdots + a_1 y[k+1] + a_0 y[k]$$
$$= b_n f[k+n] + \cdots + b_1 f[k+1] + b_0 f[k] \tag{11.31c}$$

We now derive the general expression for the zero-state response; that is, the system response to input $f[k]$ when all the initial conditions $y[-1] = y[-2] = \cdots = y[-n] = 0$ (zero state). The input $f[k]$ is assumed to be causal so that $f[-1] = f[-2] = \cdots = f[-n] = 0$.

Equation (11.31c) can be expressed in the delay operator form as

$$y[k] + a_{n-1}y[k-1] + \cdots + a_0 y[k-n]$$
$$= b_n f[k] + b_{n-1}f[k-1] + \cdots + b_0 f[k-n] \tag{11.31d}$$

Because $y[-r] = f[-r] = 0$ for $r = 1, 2, \ldots, n$

$$y[k-m]u[k] \Longleftrightarrow \frac{1}{z^m} Y[z]$$

$$f[k-m]u[k] \Longleftrightarrow \frac{1}{z^m} F[z] \qquad m = 1, 2, \ldots, n$$

Now the z-transform of Eq. (11.31d) is given by

$$\left(1 + \frac{a_{n-1}}{z} + \frac{a_{n-2}}{z^2} + \cdots + \frac{a_0}{z^n} \right) Y[z] = \left(b_n + \frac{b_{n-1}}{z} + \frac{b_{n-2}}{z^2} + \cdots + \frac{b_0}{z^n} \right) F[z]$$

Multiplication of both sides by z^n yields

$$(z^n + a_{n-1}z^{n-1} + \cdots + a_1 z + a_0)Y[z]$$
$$= (b_n z^n + b_{n-1}z^{n-1} + \cdots + b_1 z + b_0)F[z]$$

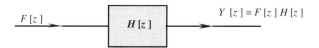

Fig. 11.5   The transformed representation of an LTID system.

Therefore

$$Y[z] = \left(\frac{b_n z^n + b_{n-1} z^{n-1} + \cdots + b_1 z + b_0}{z^n + a_{n-1} z^{n-1} + \cdots + a_1 z + a_0} \right) F[z] \tag{11.32}$$

$$= \frac{P[z]}{Q[z]} F[z] \tag{11.33}$$

We have shown in Eq. (11.20) that $Y[z] = F[z]H[z]$. Hence, it follows that

$$H[z] = \frac{P[z]}{Q[z]} = \frac{b_n z^n + b_{n-1} z^{n-1} + \cdots + b_1 z + b_0}{z^n + a_{n-1} z^{n-1} + \cdots + a_1 z + a_0} \tag{11.34}$$

As in the case of LTIC systems, this result leads to an alternative definition of the LTID system transfer function as the ratio of $Y[z]$ to $F[z]$ (assuming all initial conditions zero).

$$H[z] \equiv \frac{Y[z]}{F[z]} = \frac{\mathcal{Z}[\text{zero-state response}]}{\mathcal{Z}[\text{input}]} \tag{11.35}$$

Because $Y[z]$, the z-transform of the zero-state response $y[k]$, is the product of $F[z]$ and $H[z]$, we can represent an LTID system in the frequency domain by a block diagram, as illustrated in Fig. 11.5. Just as in continuous-time systems, we can represent discrete-time systems in the transformed manner by representing all signals by their z-transforms and all system components (or elements) by their transfer functions.

Observe that the denominator of $H[z]$ is $Q[z]$, the characteristic polynomial of the system. *Therefore the poles of $H[z]$ are the characteristic roots of the system.* Consequently, the system stability criterion can be stated in terms of the poles of the transfer function of an LTID system as follows:

1. An LTID system is asymptotically stable if and only if all the poles of its transfer function $H[z]$ lie inside a unit circle (centered at the origin) in the complex plane. The poles may be repeated or unrepeated.

2. An LTID system is unstable if and only if either one or both of the following conditions exist: (i) at least one pole of $H[z]$ is outside the unit circle; (ii) there are repeated poles of $H[z]$ on the unit circle.

3. An LTID system is marginally stable if and only if there are no poles of $H[z]$ outside the unit circle, and there are some unrepeated poles on the unit circle.

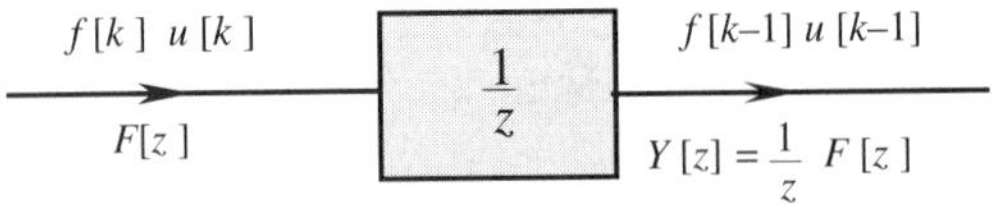

Fig. 11.6   Ideal unit delay and its transfer function.

■ **Example 11.6: The Transfer Function of a Unit Delay**
Show that the transfer function of a unit delay is $1/z$.
If the input to the unit delay is $f[k]u[k]$, then its output (Fig. 11.6) is given by

$$y[k] = f[k-1]u[k-1]$$

The z-transform of this equation yields [see Eq. (11.15a)]

$$Y[z] = \frac{1}{z}F[z]$$

$$= H[z]F[z]$$

It follows that the transfer function of the unit delay is

$$H[z] = \frac{1}{z} \qquad\qquad (11.36)$$

■

■ **Example 11.7**
Find the response $y[k]$ of an LTID system described by the difference equation

$$y[k+2] + y[k+1] + 0.16y[k] = f[k+1] + 0.32f[k]$$

or

$$(E^2 + E + 0.16)y[k] = (E + 0.32)f[k]$$

for the input $f[k] = (-2)^{-k}u[k]$ and with all the initial conditions zero (system in zero state initially).
From the difference equation we find

$$H[z] = \frac{P[z]}{Q[z]} = \frac{z + 0.32}{z^2 + z + 0.16}$$

For the input $f[k] = (-2)^{-k}u[k] = [(-2)^{-1}]^k u(k) = (-0.5)^k u[k]$

$$F[z] = \frac{z}{z + 0.5}$$

and

$$Y[z] = F[z]H[z] = \frac{z(z + 0.32)}{(z^2 + z + 0.16)(z + 0.5)}$$

Therefore

$$\frac{Y[z]}{z} = \frac{(z + 0.32)}{(z^2 + z + 0.16)(z + 0.5)} = \frac{(z + 0.32)}{(z + 0.2)(z + 0.8)(z + 0.5)}$$

$$= \frac{2/3}{z + 0.2} - \frac{8/3}{z + 0.8} + \frac{2}{z + 0.5} \qquad\qquad (11.37)$$

so that

$$Y[z] = \frac{2}{3}\left(\frac{z}{z+0.2}\right) - \frac{8}{3}\left(\frac{z}{z+0.8}\right) + 2\left(\frac{z}{z+0.5}\right) \tag{11.38}$$

and

$$y[k] = \left[\tfrac{2}{3}(-0.2)^k - \tfrac{8}{3}(-0.8)^k + 2(-0.5)^k\right]u[k] \qquad \blacksquare$$

⊙ **Computer Example C11.3**

Solve Example 11.7 using MATLAB. Plot $y[k]$ for $0 \le k \le 10$.

```
k=0:10;
b=[0 1 0.32];
a=[1 1 0.16];
f=(-2).^(-k);
y=filter(b,a,f);
stem(k,y)
xlabel('k');ylabel('y[k]')   ⊙
```

△ **Exercise E11.11**

A discrete-time system is described by the following transfer function:

$$H[z] = \frac{z - 0.5}{(z + 0.5)(z - 1)}$$

(a) Find the system response to input $f[k] = 3^{-(k+1)}u[k]$ if all initial conditions are zero. **(b)** Write the difference equation relating the output $y[k]$ to input $f[k]$ for this system.

Answers: **(a)** $y[k] = \frac{1}{3}\left[\frac{1}{2} - 0.8(-0.5)^k + 0.3\left(\frac{1}{3}\right)^k\right]u[k]$

(b) $y[k+2] - 0.5y[k+1] - 0.5y[k] = f[k+1] - 0.5f[k]$ ▽

11.4 System Realization

We now discuss ways to realize an nth-order discrete-time system described by a transfer function

$$H[z] = \frac{b_n z^n + b_{n-1}z^{n-1} + \cdots + b_1 z + b_0}{z^n + a_{n-1}z^{n-1} + \cdots + a_1 z + a_0} \tag{11.39}$$

This transfer function is identical to the general nth-order continuous-time transfer function $H(s)$ in Eq. (6.70) with s replaced by z. It is reasonable to believe that the realization of $H[z]$ in (11.39) would be identical to that of $H(s)$ with s replaced by z. Fortunately this happens to be the case. In realizations of $H(s)$ the basic element used was an integrator with transfer function $1/s$. In realizations of $H[z]$ the basic element is unit delay with transfer function $1/z$. Therefore, all the realizations of $H(s)$ studied in Sec. 6.6 are also the realizations of $H[z]$ if we replace integrators by unit delays. To demonstrate this point, consider a realization of a third-order transfer function.

$$H[z] = \frac{b_3 z^3 + b_2 z^2 + b_1 z + b_0}{z^3 + a_2 z^2 + a_1 z + a_0} \tag{11.40}$$

Figure 11.7 shows Fig. 6.21 with all the integrators (with transfer function $1/s$) replaced with unit delays (with transfer function $1/z$). We shall now show that this

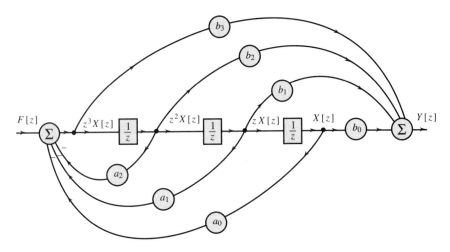

Fig. 11.7   A canonical realization of $H[z]$.

realization indeed represents $H[z]$ in Eq. (11.40). Let the signal at the output of the third delay be $X[z]$. Consequently, signals at the inputs of the second and the first delay are $zX[z]$ and $z^2X[z]$. The first summer output $z^3X[z]$ is equal to the sum of the four inputs to that summer. Therefore

$$z^3X[z] = -a_2z^2X[z] - a_1zX[z] - a_0X[z] + F[z]$$

so that

$$(z^3 + a_2z^2 + a_1z + a_0)X[z] = F[z] \tag{11.41}$$

Moreover, $Y[z]$, the output of the second summer, is equal to the sum of four signals to that summer. Therefore

$$Y[z] = (b_3z^3 + b_2z^2 + b_1z + b_0)X[z] \tag{11.42}$$

From Eqs. (11.41) and (11.42), it follows that

$$\frac{Y[z]}{F[z]} = \frac{b_3z^3 + b_2z^2 + b_1z + b_0}{z^3 + a_2z^2 + a_1z + a_0}$$

This result shows that Fig. 11.7 is indeed a realization of $H[z]$ in Eq. (11.40). Similarly, the cascade and parallel realizations of the continuous-time case are directly applicable to discrete-time systems, with integrators replaced by unit delays. The second canonical realization developed in Appendix 6.1 also applies to discrete-time case with $1/s$ replaced by $1/z$.

■ **Example 11.8**

Realize the following transfer functions, using only the cascade form for part **a** and using only the parallel form for part **b**.

(a) $H[z] = \dfrac{4z + 28}{z^2 + 6z + 5}$ (b) $H[z] = \dfrac{7z^2 + 37z + 51}{(z + 2)(z + 3)^2}$

Identical transfer functions for continuous-time systems are realized in Figs. 6.27 and 6.28. ■

△ **Exercise E11.12**

Give the canonical realization of the following transfer functions. (a) $\dfrac{2}{z + 5}$ (b) $\dfrac{z + 8}{z + 5}$

(c) $\dfrac{z}{z + 5}$ (d) $\dfrac{2z + 3}{z^2 + 6z + 25}$

Answer: See Example 6.18. Replace $1/s$ by $1/z$ and make appropriate changes in coefficients. ▽

11.5   Connection between the Laplace and the Z-Transform

We now show that discrete-time systems also can be analyzed using the Laplace transform. In fact, we shall see that *the z-transform is the Laplace transform in disguise* and that discrete-time systems can be analyzed as if they were continuous-time systems.

So far we have considered the discrete-time signal as a sequence of numbers and not as an electrical signal (voltage or current). Similarly, we have considered a discrete-time system as a mechanism that processes a sequence of numbers (input) to yield another sequence of numbers (output). The system was built by using delays (along with adders and multipliers) that delay sequences of numbers, not electrical signals (voltages or currents). A digital computer is a perfect example: every signal is a sequence of numbers, and the processing involves delaying sequences of numbers (along with addition and multiplication).

Consider a discrete-time system with transfer function $H[z]$ and an input $f[k]$, as shown in Fig. 11.8a. We can think of (or generate, for that matter) a corresponding continuous-time signal $\tilde{f}(t)$ consisting of impulses spaced T seconds apart. Let the kth impulse of strength be $f[k]$ as depicted in Fig. 11.8b. Thus

$$\tilde{f}(t) = \sum_{k=0}^{\infty} f[k]\delta(t - kT) \qquad (11.43)$$

Figure 11.8 shows $f[k]$ and corresponding $\tilde{f}(t)$. Let us now consider a system identical in structure to the discrete-time system with transfer function $H[z]$, except that the delays in $H[z]$ are replaced by elements that delay continuous-time signals (such as voltages or currents). If a continuous-time impulse $\delta(t)$ is applied to such a delay of T seconds, the output will be $\delta(t - T)$. The continuous-time transfer function of such a delay is e^{-sT} [see Eq. (6.54)]. Hence, the delay elements with transfer function $1/z$ in the realization of $H[z]$ will be replaced by the delay elements with transfer function e^{-sT} in the realization of the corresponding $\tilde{H}(s)$. This step is the same as z being replaced by e^{sT}. Therefore, the transfer function of this system is $H[z]$ with z replaced by e^{sT}. Thus $\tilde{H}(s) = H[e^{sT}]$. Now whatever operations are performed by the discrete-time system $H[z]$ on $f[k]$ (Fig. 11.8a) are also performed by the corresponding continuous-time system $H[e^{sT}]$ on the impulse sequence $\tilde{f}(t)$

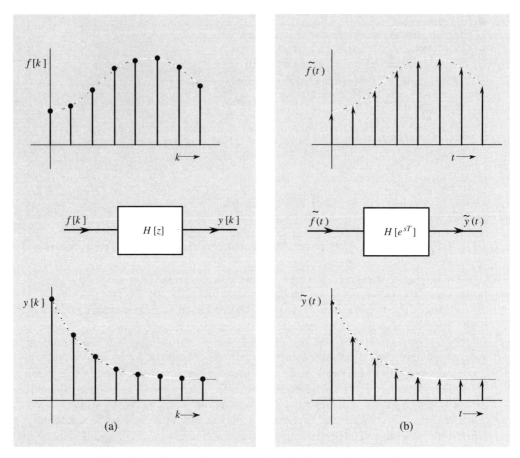

Fig. 11.8   Connection between the Laplace and z-transform.

(Fig. 11.8b). The delaying of a sequence in $H[z]$ would amount to the delaying of an impulse train in $H[e^{sT}]$. The case of adding and multiplying operations is similar. In other words, one-to-one correspondence of the two systems is preserved in every aspect. Therefore, if $y[k]$ is the output of the discrete-time system in Fig. 11.8a, then $\tilde{y}(t)$, the output of the continuous-time system in Fig. 11.8b, would be a sequence of impulses whose kth impulse strength is $y[k]$. Thus

$$\tilde{y}(t) = \sum_{k=0}^{\infty} y[k]\delta(t - kT) \tag{11.44}$$

The system in Fig. 11.8b, being a continuous-time system, can be analyzed by using the Laplace transform. If

$$\tilde{f}(t) \Longleftrightarrow \tilde{F}(s) \quad \text{and} \quad \tilde{y}(t) \Longleftrightarrow \tilde{Y}(s)$$

then

$$\tilde{Y}(s) = H[e^{sT}]\tilde{F}(s) \tag{11.45}$$

Also

$$\tilde{F}(s) = \mathcal{L}\left[\sum_{k=0}^{\infty} f[k]\delta(t - kT)\right]$$

Now, because the Laplace transform of $\delta(t - kT)$ is e^{-skT}

$$\tilde{F}(s) = \sum_{k=0}^{\infty} f[k]e^{-skT} \qquad (11.46)$$

Similarly

$$\tilde{Y}(s) = \sum_{k=0}^{\infty} y[k]e^{-skT} \qquad (11.47)$$

Substitution of Eqs. (11.46)and (11.47) in Eq. (11.45) yields

$$\sum_{k=0}^{\infty} y[k]e^{-skT} = H[e^{sT}]\left[\sum_{k=0}^{\infty} f[k]e^{-skT}\right]$$

By introducing a new variable $z = e^{sT}$, this equation can be expressed as

$$\sum_{k=0}^{\infty} y[k]z^{-k} = H[z]\sum_{k=0}^{\infty} f[k]z^{-k}$$

or

$$Y[z] = H[z]F[z]$$

where

$$F[z] = \sum_{k=0}^{\infty} f[k]z^{-k} \quad \text{and} \quad Y[z] = \sum_{k=0}^{\infty} y[k]z^{-k}$$

It is clear from this discussion that the z-transform can be considered as a Laplace transform with a change of variable $z = e^{sT}$ or $s = (1/T)\ln z$. On the other hand, we may consider the z-transform as an independent transform in its own right. Note that the transformation $z = e^{sT}$ transforms the imaginary axis in the s-plane ($s = j\omega$) into a unit circle in the z-plane ($z = e^{sT} = e^{j\omega T}$, or $|z| = 1$). The LHP and RHP in the s-plane map into the inside and the outside, respectively, of the unit circle in the z-plane.

11.6   Sampled-data (Hybrid) Systems

Sampled-data systems are hybrid systems consisting of discrete-time as well as continuous-time subsystems. Consider, for example, a fire control system. In this case, the problem is to search and track a moving target and fire a projectile for a direct hit. The data obtained from the search and tracking radar is discrete-time data because of a scanning operation, which results in sampling of azimuth, elevation, and the target velocity. This data is now fed to a digital (discrete-time) processor, which performs extensive computations. The computer output is then fed to a continuous-time plant, such as a gun mount, which accordingly positions

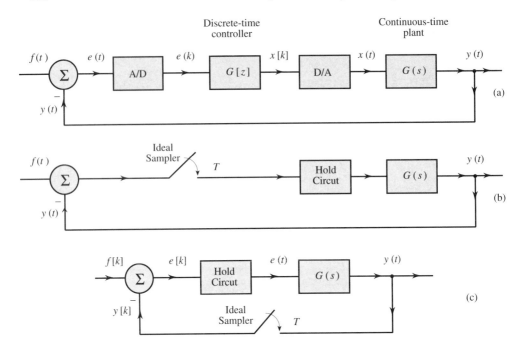

Fig. 11.9   Typical sampled-data systems.

itself at a certain position and fires. Another example is attitude-control problem in a spacecraft, where the information about the actual spacecraft attitude is fed back to a digital processor, which generates corrective input to be applied to the spacecraft, which is a continuous-time system. In automatic periodic quality check in production line, the discrete-data obtained from the periodic check, after some digital processing, generates the corrective input to be applied to a continuous-time plant. In complex control systems, use of digital processor as a controller or a compensator for continuous-time plants is growing rapidly.

In time-sharing systems, where, for economic reasons, certain facilities are shared by several systems, the signals are, by nature, discrete-time or sampled. In regulator type control systems, where an output variable must be maintained at a constant value, the external disturbance and plant parameters variations are usually so slow that continuous monitoring (or feedback) is unnecessary. It is adequate to sample the output periodically and then feed back this discrete-data. In such cases, feedback transducers, data-processing facilities and possibly long and expensive feedback communication facilities can be shared among several control systems.

Figure 11.9 shows some typical sampled-data systems. Figure 11.9a contains a digital processor, whereas in Fig. 11.9b, the sampled signal is directly applied to D/A converter (the hold circuit) without further digital processing. Figure 11.9c shows a practical system, where the input signal itself is a discrete-time signal $f[k]$, and the sampler is in the feedback path. This system is equivalent to that in Fig. 11.9b. How do we analyze such hybrid systems, where continuous-time and discrete-time signals intermingle? An effective strategy in such a situation is to relate the samples of the output to those of the input. But, this procedure yields information about the output only at sampling instants. We can overcome this difficulty by taking the samples at instants in between samples using the **modified z-transform** as explained later.

In sampled-data systems, the discrete-time signals are often obtained as a result of sampling continuous-time signals. These samples are narrow pulses, which may be considered as impulses, provided the pulse width is small compared to the system time constant. Thus, in the following discussion, a discrete-time signal, when it appears in conjunction with a continuous-time system, is a sequence of impulses rather than a sequence of numbers. Hence, a discrete-time signal $f[k]$ can also be considered as continuous-time signal $f(t)$, where

$$f(t) = \sum_k f[k]\delta(t - kT)$$

Observe an interesting fact: in this representation a discrete-time unit impulse $\delta[k]$ is the continuous-time unit impulse $\delta(t)$. Thus, at the input of a discrete-time processor, a discrete-time signal $f[k]$ is just a sequence of numbers. But at the input of a continuous-time system, $f[k]$ is a sequence of impulses. There are appropriate converters at the interface of discrete-time and continuous-time systems to carry out signal conversion to appropriate forms.

To begin with, consider a basic continuous-time system (Fig. 11.10a) with transfer function $H(s)$. The input $f(t)$ is sampled and the sampled signal $f[k]$ is applied to the input of $H(s)$. Although $y(t)$, the output of this system, is continuous, we shall endeavor to find the values of $y(t)$ only at the discrete instants $t = kT$. Such an analysis is relatively simple using the method of z-transform. For this purpose, we consider as if the output is sampled by an hypothetical sampler shown dotted in Fig. 11.10a. Now, we shall relate the input samples $f[k]$ and the output samples $y[k]$. Let $h[k]$ be the unit impulse response relating the output samples to the input samples. In other words, $y[k] = h[k] * f[k]$. Recall also that an unit impulse $\delta[k]$ is $\delta(t)$ when considered in conjunction with a continuous-time system. Hence, $h[k]$, the unit impulse response is the sampled version of the system's unit impulse response $h(t)$. Thus,

$$h[k] = h(kT)$$

where T is the sampling interval. For instance, if $H(s) = \frac{1}{s-\lambda}$, then $h(t) = e^{\lambda t}$ and

$$h[k] = e^{\lambda kT}$$

Therefore, the equivalent discrete-time transfer function $H[z]$ of this system is given by

$$H[z] = \mathcal{Z}\{h[k]\}$$

$$= \mathcal{Z}[e^{\lambda kT}]$$

$$= \frac{z}{z - e^{\lambda T}} \qquad (11.49)$$

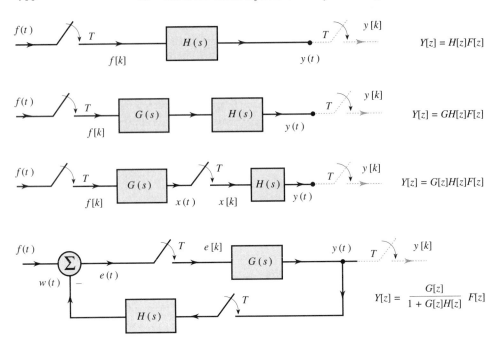

Fig. 11.10   Computing the output in hybrid or sampled-data systems.

Thus, $H[z]$ is the discrete-time transfer function of $H(s) = \frac{1}{s+\lambda}$ that relates $y[k]$ (the output samples) to the discrete-time input $f[k]$.†

If we have two systems with transfer functions $G(s)$ and $H(s)$ in cascade (Fig. 11.10b), the equivalent transfer $T[z] \neq G[z]H[z]$, but is $GH[z]$, where $G[z]$, $H[z]$ and $GH[z]$ correspond to discrete-time transfer functions of $G(s)$, $H(s)$ and $G(s)H(s)$, respectively. For instance, if

$$G(s) = \frac{1}{s+2} \qquad \text{and} \qquad H(s) = \frac{1}{s}$$

Then, according to Eq. (11.49)

$$G[z] = \frac{z}{z - e^{-2T}} \qquad \text{and} \qquad H[z] = \frac{z}{z-1}$$

However, the continuous-time system transfer function is $G(s)H(s)$, where

$$G(s)H(s) = \frac{1}{s(s+2)} = \frac{1}{2}\left[\frac{1}{s} - \frac{1}{s+2}\right]$$

And from Eq. (11.49)

---

†Using this procedure, we have listed $H(s)$ and corresponding $H[z]$ in Table 12.1 in Chapter 12. In this Table, $H[z]$ is multiplied with a scaling factor T, which results in $H[z] = \frac{Tz}{z-e^{\lambda T}}$. For the purpose of the sampled data application, the extra factor T should be ignored throughout in Table 12.1.

$$T[z] = \frac{1}{2}\left(\frac{z}{z-1} - \frac{z}{z - e^{-2T}}\right) \neq G[z]H[z]$$

In this case, we use the notation $GH[z]$ for $T[z]$. Thus, $GH[z] \neq G[z]H[z]$, but is the discrete-time transfer function which corresponds to $G(s)H(s)$.

For the system in Fig. 11.10c,

$$Y[z] = H[z]X[z] = H[z]G[z]F[z] \qquad \text{so that} \qquad T[z] = G[z]H[z]$$

For the system in Fig. 11.10d,

$$E[z] = F[z] - W[z]$$

Moreover,

$$W[z] = H[z]Y[z]$$
$$Y[z] = G[z]E[z]$$
$$= G[z]\left(F[z] - W[z]\right)$$
$$= G[z](F[z] - H[z]Y[z])$$

Hence

$$Y[z] = \frac{G[z]}{1 + G[z]H[z]}F[z]$$

Consequently

$$T[z] = \frac{G[z]}{1 + G[z]H[z]}$$

■ **Example 11.9**

Find the output samples $y[k]$ for the sampled-data system illustrated in Fig. 11.11a when the input is a unit step function $u(t)$, the sampling interval $T = 0.5$ second and

$$G_c[z] = \frac{z}{z-1} \qquad \text{and} \qquad G(s) = \frac{1}{s+4}$$

This system has a discrete-time controller and a continuous-time plant.† To find the transfer function of this system, we observe that

$$Y[z] = G[z]X[z], \qquad X[z] = G_c[z]E[z], \qquad \text{and} \qquad E[z] = F[z] - Y[z]$$

Hence

$$Y[z] = G_c[z]G[z](F[z] - Y[z])$$

and the system transfer function $T[z]$ is

$$T[z] = \frac{Y[z]}{F[z]} = \frac{G_c[z]G[z]}{1 + G_c[z]G[z]}$$

---

†The block diagram in Fig. 11.11a does not show the appropriate converters required at the interface of discrete-time and continuous-time systems; these are implied. Thus, the output of the sampler, which consists of impulse sequence, is converted into sequence of numbers to act as the input to the discrete-time controller. Similarly, the output of a discrete-time controller, which is a sequence of numbers, is converted to a sequence of impulses to act as an input to the plant.

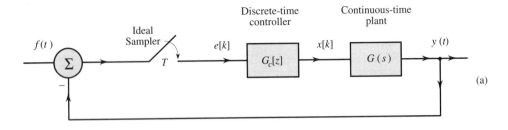

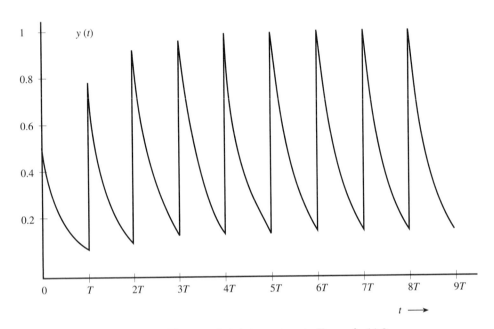

(a)

Fig. 11.11   The sampled-data system in Example 11.9.

For $G(s) = 1/(s+4)$ and $T = 0.5$, we find, from Eq. (11.49), $G[z] = \frac{z}{z-e^{-2}} = \frac{z}{z-0.1353}$. Also $G_c[z] = z/(z-1)$. Substitution of these expressions in $T[z]$ yields

$$T[z] = \frac{z^2}{(z-0.394)(z-0.174)}$$

The output $Y[z]$ is given by $Y[z] = T[z]F[z]$. For the step input $f(t) = u(t)$, the corresponding sampled signal is $u[k]$ so that $F[z] = z/(z-1)$. Hence,

$$Y[z] = T[z]F[z] = \frac{z^3}{(z-1)(z-0.394)(z-0.174)}$$

and

$$\frac{Y[z]}{z} = \frac{z^2}{(z-1)(z-0.394)(z-0.174)}$$

$$= \frac{1}{z-1} - \frac{0.583}{z-0.394} + \frac{0.083}{z-0.174}$$

Hence

$$Y[z] = \frac{z}{z-1} - \frac{0.583z}{z-0.394} + \frac{0.083z}{z-0.174}$$

and

$$y[k] = 1 - 0.583(0.394)^k + 0.083(0.174)^k$$

This response is depicted in Fig. 11.11. ■

11.6-1 Response Between Sampling Instants: The Modified Z-Transform

The above analysis yields the output only at sampling instants. We can readily find the response between successive sampling instants by using the *modified z-transform*. This goal can be accomplished by considering the response at another set of sampling instants $t = (k + \mu)T$, where $0 < \mu < 1$.

Consider the system in Fig. 11.10a with $H(s) = 1/(s-\lambda)$. The impulse response is $h(t) = e^{\lambda t}$ and its samples at instants $t = (k + \mu)T$ are

$$h(t, \mu) = e^{\lambda(k+\mu)T} = e^{\lambda\mu T}\left[e^{\lambda kT}\right] \tag{11.50a}$$

The corresponding z-transfer function is

$$H[z, \mu] = e^{\lambda\mu T}\frac{z}{z - e^{\lambda T}} = \frac{ze^{\lambda\mu T}}{z - e^{\lambda T}} \tag{11.50b}$$

In this manner, We can prepare a table of modified z-transform. When we use $H[z, \mu]$ instead of $H[z]$ in our analysis, we obtain the response at instants $t = (k + \mu)T$. By using different values of μ in the range 0 to T, we can obtain the complete response $y(t)$.

■ **Example 11.10**

Find the output $y(t)$ for all t in Example 11.9.

In Example 11.9, we found the response $y[k]$ only at the sampling instants. To find the output values between sampling instants, we use the modified z-transform. The procedure is the same as before, except that we use modified z-transform corresponding to continuous-time systems and signals. For the system $G(s) = 1/s + 4$ with $T = 0.5$, the modified z-transform [Eq. (11.50b) with $\lambda = -4$, and $T = 0.5$] is

$$H[z, \mu] = e^{\lambda\mu T}\frac{z}{z - e^{-2}} = e^{-2\mu}\frac{z}{z - 0.1353}$$

Moreover to find the modified z-transform corresponding to $f(t) = u(t)$ [$\lambda = 0$ in Eq. (11.50a)], we have $F[z, \mu] = z/(z - 1)$. Substitution of these expressions in those found in Example 11.9, we obtain

$$Y[z, \mu] = e^{-2\mu}\left[\frac{z}{z-1} - \frac{0.583z}{z-0.394} + \frac{0.083z}{z-0.174}\right]$$

From Eqs. (11.50), we obtain the inverse (modified) z-transform of this equation as

$$y[(k+\mu)T] = e^{-2\mu}\left[1 - 0.583(0.394)^k + 0.083(-0.174)^k\right] \qquad 0 < \mu < 1$$

The complete response is also shown in Fig. 11.11. ■

Design of Sampled-Data Systems

As with continuous-time control systems, sampled-data systems are designed to meet certain transient (PO, t_r, t_s, etc.) and steady-state specifications. The design procedure follows along the lines similar to those used for continuous-time systems. We begin with a general second-order system. the relationship between closed-loop pole locations and the corresponding transient parameters PO, t_r, t_s, ... are determined. Hence, for a given transient specifications, an acceptable region in the z-plane where the dominant poles of the closed-loop transfer function $T[z]$ should lie is determined. Next, we sketch the root locus for the system. The rules for sketching the root locus are the same as those for continuous-time systems. If the root locus passes through the acceptable region, the transient specifications can be met by simple adjustment of the gain K. If not, we must use a compensator, which will steer the root locus in the acceptable region.

11.7   The Bilateral Z-Transform

Situations involving noncausal signals or systems cannot be handled by the (unilateral) z-transform discussed so far. Such cases can be analyzed by the **bilateral** (or two-sided) z-transform defined by

$$F[z] \equiv \sum_{k=-\infty}^{\infty} f[k]z^{-k}$$

The inverse z-transform is given by

$$f[k] = \frac{1}{2\pi j} \oint F[z]z^{k-1}\,dz$$

These equations define the bilateral z-transform. The unilateral z-transform discussed so far is a special case, where the input signals are restricted to be causal. Restricting signals in this way results in considerable simplification in the region of convergence. Earlier, we showed that

$$\gamma^k u[k] \iff \frac{z}{z-\gamma} \qquad |z| > |\gamma| \tag{11.51}$$

In contrast, the z-transform of the signal $-\gamma^k u[-(k+1)]$, illustrated in Fig. 11.12a, is

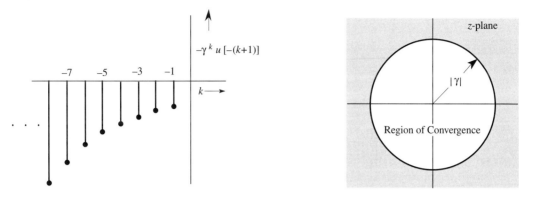

Fig. 11.12 $-\gamma^k u[-(k+1)]$ and the region of convergence of its z-transform.

$$\mathcal{Z}\left\{-\gamma^k u[-(k+1)]\right\} = \sum_{-\infty}^{-1} -\gamma^k z^{-k} = \sum_{-\infty}^{-1} -\left(\frac{\gamma}{z}\right)^k$$

$$= -\left[\frac{z}{\gamma} + \left(\frac{z}{\gamma}\right)^2 + \left(\frac{z}{\gamma}\right)^3 + \cdots\right]$$

$$= 1 - \left[1 + \frac{z}{\gamma} + \left(\frac{z}{\gamma}\right)^2 + \left(\frac{z}{\gamma}\right)^3 + \cdots\right]$$

$$= 1 - \frac{1}{1 - \frac{z}{\gamma}} \qquad \left|\frac{z}{\gamma}\right| < 1$$

$$= \frac{z}{z - \gamma} \qquad |z| < |\gamma|$$

Therefore

$$\mathcal{Z}\left\{-\gamma^k u[-(k+1)]\right\} = \frac{z}{z - \gamma} \qquad |z| < |\gamma| \qquad (11.52)$$

A comparison of Eqs. (11.51) with (11.52) shows that the z-transform of $\gamma^k u[k]$ is identical to that of $-\gamma^k u[-(k+1)]$. The regions of convergence, however, are different. In the former case, $F[z]$ converges for $|z| > |\gamma|$; in the latter, $F[z]$ converges for $|z| < |\gamma|$ (see Fig. 11.12b). Clearly, the inverse transform of $F[z]$ is not unique unless the region of convergence is specified. If we restrict all our signals to be causal, however, this ambiguity does not arise. The inverse transform of $z/(z - \gamma)$ is $\gamma^k u[k]$ even without specifying the region of convergence. Thus, in the unilateral transform, we can ignore the region of convergence in determining the inverse z-transform of $F[z]$.

■ **Example 11.11**
    Determine the z-transform of

$$f[k] = (0.9)^k u[k] + (1.2)^k u[-(k+1)]$$

$$= f_1[k] + f_2[k]$$

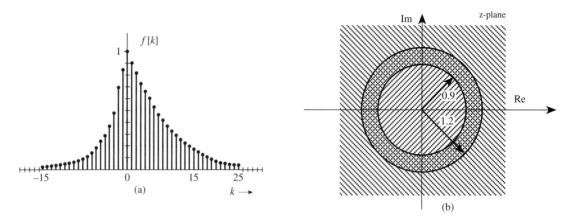

Fig. 11.13   Signal $f[k]$ for Example 11.11.

From the results in Eqs. (11.51) and (11.52), we have

$$F_1[z] = \frac{z}{z - 0.9} \qquad |z| > 0.9$$

$$F_2[z] = \frac{-z}{z - 1.2} \qquad |z| < 1.2$$

The common region where both $F_1[z]$ and $F_2[z]$ converge is $0.9 < |z| < 1.2$ (Fig. 11.13a). Hence

$$F[z] = F_1[z] + F_2[z]$$

$$= \frac{z}{z - 0.9} - \frac{z}{z - 1.2}$$

$$= \frac{-0.3z}{(z - 0.9)(z - 1.2)} \qquad 0.9 < |z| < 1.2 \qquad (11.53)$$

The sequence $f[k]$ and the region of convergence of $F[z]$ are depicted in Fig. 11.13.

■

■ **Example 11.12**
    Find the inverse z-transform of

$$F[z] = \frac{-z(z + 0.4)}{(z - 0.8)(z - 2)}$$

if the region of convergence is **(a)** $|z| > 2$ **(b)** $|z| < 0.8$ **(c)** $0.8 < |z| < 2$.

(a)
$$\frac{F[z]}{z} = \frac{-(z + 0.4)}{(z - 0.8)(z - 2)}$$

$$= \frac{1}{z - 0.8} - \frac{2}{z - 2}$$

and
$$F[z] = \frac{z}{z - 0.8} - 2\frac{z}{z - 2}$$

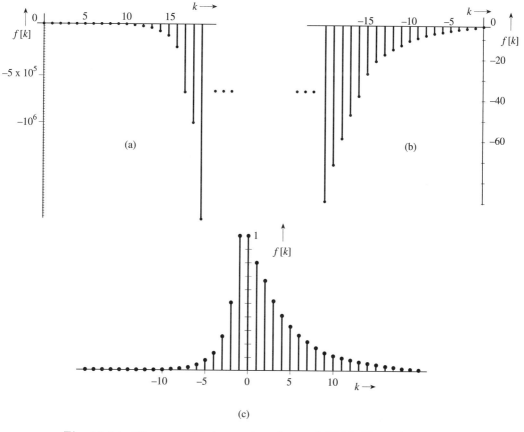

Fig. 11.14   Three possible inverse transforms of $F[z]$ in Example 11.12.

Since the region of convergence is $|z| > 2$, both terms correspond to causal sequences and

$$f[k] = \left[(0.8)^k - 2(2)^k\right] u[k]$$

This sequence appears in Fig. 11.14a.

(b)   In this case, $|z| < 0.8$, which is less than the magnitudes of both poles. Hence, both terms correspond to anticausal sequences, and

$$f[k] = \left[-(0.8)^k + 2(2)^k\right] u\left[-(k+1)\right]$$

This sequence appears in Fig. 11.14b.

(c)   In this case, $0.8 < |z| < 2$; the part of $F[z]$ corresponding to the pole at 0.8 is a causal sequence, and the part corresponding to the pole at 2 is an anticausal sequence:

$$f[k] = (0.8)^k u[k] + 2(2)^k u[-(k+1)]$$

This sequence appears in Fig. 11.14c.   ■

$\triangle$ **Exercise E11.13**
   Find the inverse z-transform of

$$F[z] = \frac{z}{z^2 + \frac{5}{6}z + \frac{1}{6}} \qquad \qquad \frac{1}{2} > |z| > \frac{1}{3}$$

Answer: $6(-\frac{1}{3})^k u[k] + 6(-\frac{1}{2})^k u[-(k+1)]$ $\triangledown$

Inverse Transform by Expansion of F[z] in Power Series of z

   We have

$$F[z] = \sum_k f[k] z^{-k}$$

For an anticausal sequence, which exists only for $k \le -1$, this equation becomes

$$F[z] = f[-1]z + f[-2]z^2 + f[-3]z^3 + \cdots$$

We can find the inverse z-transform of $F[z]$ by dividing the numerator polynomial with the denominator polynomial, both in ascending powers of z, to obtain a polynomial in ascending powers of z. Thus, to find the inverse transform of $z/(z-0.5)$ (when the region of convergence is $|z| < 0.5$), we divide z with $-0.5 + z$ to obtain $-2z - 4z^2 - 8z^3 - \cdots$. Hence, $f[-1] = -2$, $f[-2] = -4$, $f[-3] = -8$ and so on.

11.7-1   Analysis of LTID Systems Using the Bilateral $\mathcal{Z}$-Transform

   Because the bilateral z-transform can handle noncausal signals, we can analyze noncausal linear systems using this transform. The zero-state response $y[k]$ is given by

$$y[k] = \mathcal{Z}^{-1}\{F[z]H[z]\}$$

provided that $F[z]H[z]$ exists. The region of convergence of $F[z]H[z]$ is the region where both $F[z]$ and $H[z]$ exist, a fact which means that the region is common to the convergence of both $F[z]$ and $H[z]$.

■ **Example 11.13**
   For a causal system specified by the transfer function

$$H[z] = \frac{z}{z - 0.5}$$

find the zero-state response to input

$$f[k] = (0.8)^k u[k] + 2(2)^k u[-(k+1)]$$

The z-transform of this signal is found from Example 11.12 (part **c**) as

$$F[z] = \frac{-z(z+0.4)}{(z-0.8)(z-2)} \qquad 0.8 < |z| < 2$$

Therefore

$$Y[z] = F[z]H[z] = \frac{-z^2(z+0.4)}{(z-0.5)(z-0.8)(z-2)}$$

Since the system is causal, the region of convergence of $H[z]$ is $|z| > 0.5$. The region of convergence of $F[z]$ is $0.8 < |z| < 2$. The common region of convergence for $F[z]$ and $H[z]$ is $0.8 < |z| < 2$. Therefore

$$Y[z] = \frac{-z^2(z+0.4)}{(z-0.5)(z-0.8)(z-2)} \qquad 0.8 < |z| < 2$$

Expanding $Y[z]$ into modified partial fractions yields

$$Y[z] = -\frac{z}{z-0.5} + \frac{8}{3}\left(\frac{z}{z-0.8}\right) - \frac{8}{3}\left(\frac{z}{z-2}\right) \qquad 0.8 < |z| < 2$$

The poles at 0.5 and 0.8 are enclosed within the ring of convergence and therefore correspond to the causal part, and the pole at 2 is outside the ring of convergence and corresponds to the anticausal part of $Y[z]$. Therefore

$$y[k] = \left[-(0.5)^k + \tfrac{8}{3}(0.8)^k\right]u[k] + \tfrac{8}{3}(2)^k u[-(k+1)] \qquad \blacksquare$$

■ **Example 11.14**

For the system in Example 11.13 find the zero-state response to input

$$f[k] = \underbrace{(0.8)^k u[k]}_{f_1[k]} + \underbrace{(0.6)^k u[-(k+1)]}_{f_2[k]}$$

The z-transforms of the causal and anticausal components $f_1[k]$ and $f_2[k]$ of the output are

$$F_1[z] = \frac{z}{z-0.8} \qquad |z| > 0.8$$

$$F_2[z] = \frac{-z}{z-0.6} \qquad |z| < 0.6$$

Observe that a common region of convergence for $F_1[z]$ and $F_2[z]$ does not exist. Therefore $F[z]$ does not exist. In such a case we take advantage of the superposition principle and find $y_1[k]$ and $y_2[k]$, the system responses to $f_1[k]$ and $f_2[k]$, separately. The desired response $y[k]$ is the sum of $y_1[k]$ and $y_2[k]$. Now

$$H[z] = \frac{z}{z-0.5} \qquad |z| > 0.5$$

$$Y_1[z] = F_1[z]H[z] = \frac{z^2}{(z-0.5)(z-0.8)} \qquad |z| > 0.8$$

$$Y_2[z] = F_2[z]H[z] = \frac{-z^2}{(z-0.5)(z-0.6)} \qquad 0.5 < |z| < 0.6$$

Expanding $Y_1[z]$ and $Y_2[z]$ into modified partial fractions yields

$$Y_1[z] = -\frac{5}{3}\left(\frac{z}{z-0.5}\right) + \frac{8}{3}\left(\frac{z}{z-0.8}\right) \qquad |z| > 0.8$$

$$Y_2[z] = 5\left(\frac{z}{z-0.5}\right) - 6\left(\frac{z}{z-0.6}\right) \qquad 0.5 < |z| < 0.6$$

Therefore

$$y_1[k] = \left[-\tfrac{5}{3}(0.5)^k + \tfrac{8}{3}(0.8)^k\right]u[k]$$

$$y_2[k] = 5(0.5)^k u[k] + 6(0.6)^k u[-(k+1)]$$

and

$$y[k] = y_1[k] + y_2[k]$$
$$= \left[\tfrac{10}{3}(0.5)^k + \tfrac{8}{3}(0.8)^k \right] u[k] + 6(0.6)^k u[-(k+1)] \quad \blacksquare$$

△   **Exercise E11.14**

For a causal system in Example 11.13, find the zero-state response to input

$$f[k] = \left(\tfrac{1}{4} \right)^k u[k] + 5(3)^k u[-(k+1)]$$

Answer: $\left[-(\tfrac{1}{4})^k + 3(\tfrac{1}{2})^k \right] u[k] + 6(3)^k u[-(k+1)]$ ▽

11.8 Summary

In this chapter we discuss the analysis of linear, time-invariant, discrete-time (LTID) systems by z-transform. The z-transform is an extension of the DTFT with the frequency variable $j\Omega$ generalized to $\sigma + j\Omega$. Such an extension allows us to synthesize discrete-time signals by using exponentially growing (discrete-time) sinusoids. The relationship of the z-transform to the DTFT is identical to that of the Laplace transform to the Fourier. Because of the generalization of the frequency variable, we can analyze all kinds of LTID systems and also handle exponentially growing inputs.

The z-transform changes the difference equations of LTID systems into algebraic equations. Therefore, solving these difference equations reduces to solving algebraic equations.

The transfer function $H[z]$ of an LTID system is equal to the ratio of the z-transform of the output to the z-transform of the input when all initial conditions are zero. Therefore, if $F[z]$ is the z-transform of the input $f[k]$ and $Y[z]$ is the z-transform of the corresponding output $y[k]$ (when all initial conditions are zero), then $Y[z] = H[z]F[z]$. For a system specified by the difference equation $Q[E]y[k] = P[E]f[k]$, the transfer function $H[z] = P[z]/Q[z]$. Moreover, $H[z]$ is the z-transform of the system impulse response $h[k]$. We also showed in Chapter 9 that the system response to an everlasting exponential z^k is $H[z]z^k$.

LTID systems can be realized by scalar multipliers, summers, and time delays. A given transfer function can be synthesized in many different ways. Canonical, cascade and parallel forms of realization are discussed. The realization procedure is identical to that for continuous-time systems.

In Sec. 11.5, we showed that discrete-time systems can be analyzed by the Laplace transform as if they were continuous-time systems. In fact, we showed that the z-transform is the Laplace transform with a change in variable.

In practice, we often have to deal with hybrid systems consisting of discrete-time and continuous-time subsystems. Feedback hybrid systems are also called sampled-data systems. In such systems, we can relate the samples of the output to those of the input. However, the output is generally a continuous-time signal. The output values during the successive sampling intervals can be found by using the modified z-transform.

The majority of the input signals and practical systems are causal. Consequently, we are required to deal with causal signals most of the time. When all signals are restricted to the causal type, the z-transform analysis is greatly simplified; the region of convergence of a signal becomes irrelevant to the analysis process. This special case of z-transform (which is restricted to causal signals) is called the unilateral z-transform. Much of the chapter deals with this transform. Section 11.7 discusses the general variety of the z-transform (bilateral z-transform), which can handle causal and noncausal signals and systems. In the bilateral transform, the inverse transform of $F[z]$ is not unique, but depends on the region of convergence of $F[z]$. Thus, the region of convergence plays a crucial role in the bilateral z-transform.

Problems

11.1-1 Using the definition of the z-transform, show that

$$\textbf{(a)}\ \gamma^{k-1}u[k-1] \iff \frac{1}{z-\gamma} \qquad \textbf{(c)}\ \frac{\gamma^k}{k!}u[k] \iff e^{\gamma/z}$$

$$\textbf{(b)}\ u[k-m] \iff \frac{z}{z^m(z-1)} \qquad \textbf{(d)}\ \frac{(\ln a)^k}{k!}u[k] \iff a^{1/z}$$

11.1-2 Using only the z-transform Table 11.1, show that

$\textbf{(a)}\ 2^{k+1}u[k-1] + e^{k-1}u[k] \iff \frac{4}{z-2} + \frac{z}{e(z-e)}$

$\textbf{(b)}\ k\gamma^k u[k-1] \iff \frac{\gamma z}{(z-\gamma)^2}$

Hint: Express $u[k-1]$ in terms of $u[k]$.

$\textbf{(c)}\ \left[2^{-k}\cos\left(\frac{\pi}{3}k\right)\right]u[k-1] \iff \frac{0.25(z-1)}{z^2-0.5z+0.25}$

Hint: See the hint for part **b**.

$\textbf{(d)}\ k(k-1)(k-2)2^{k-3}u[k-m] \iff \frac{6z}{(z-2)^4}$ for m=0, 1, 2, or 3.

Hint: Examine what happens to the function if $u[k-m]$ is replaced by $u[k]$.

11.1-3 Find the inverse z-transform of

$$\textbf{(a)}\ \frac{z(z-4)}{z^2-5z+6} \qquad\qquad \textbf{(g)}\ \frac{z(z-2)}{z^2-z+1}$$

$$\textbf{(b)}\ \frac{z-4}{z^2-5z+6} \qquad\qquad \textbf{(h)}\ \frac{2z^2-0.3z+0.25}{z^2+0.6z+0.25}$$

$$\textbf{(c)}\ \frac{(e^{-2}-2)z}{(z-e^{-2})(z-2)} \qquad\qquad \textbf{(i)}\ \frac{2z(3z-23)}{(z-1)(z^2-6z+25)}$$

$$\textbf{(d)}\ \frac{z(2z+3)}{(z-1)(z^2-5z+6)} \qquad\qquad \textbf{(j)}\ \frac{z(3.83z+11.34)}{(z-2)(z^2-5z+25)}$$

$$\textbf{(e)}\ \frac{z(-5z+22)}{(z+1)(z-2)^2} \qquad\qquad \textbf{(k)}\ \frac{z^2(-2z^2+8z-7)}{(z-1)(z-2)^3}$$

$$\textbf{(f)}\ \frac{z(1.4z+0.08)}{(z-0.2)(z-0.8)^2}$$

11.1-4 Find the first three terms of $f[k]$ if

$$F[z] = \frac{2z^3+13z^2+z}{z^3+7z^2+2z+1}$$

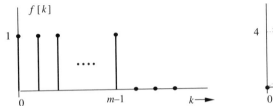

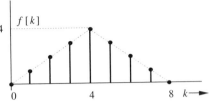

Fig. P11.2-1   Fig. P11.2-2

Find your answer by expanding $F[z]$ as a power series in z^{-1}.

11.1-5 By expanding

$$F[z] = \frac{\gamma z}{(z - \gamma)^2}$$

as a power series in z^{-1}, show that $f[k] = k\gamma^k u[k]$.

11.2-1 For a discrete-time signal shown in Fig. P11.2-1 show that

$$F[z] = \frac{1 - z^{-m}}{1 - z^{-1}}$$

11.2-2 Find the z-transform of the signal illustrated in Fig. P11.2-2. Solve this problem in two ways, as in Examples 11.2d and 11.4. Verify that the two answers are equivalent.

11.2-3 Using only the fact that $\gamma^k u[k] \iff \frac{z}{z - \gamma}$ and properties of the z-transform, find the z-transform of

(a) $k^2 \gamma^k u[k]$ (c) $a^k \left[u[k] - u[k - m] \right]$

(b) $k^3 u[k]$ (d) $ke^{-2k} u[k - m]$

11.2-4 Using only Pair 1 in Table 11.1 and appropriate properties of the z-transform, derive iteratively pairs 2 through 9. In other words, first derive Pair 2. Then use Pair 2 (and Pair 1, if needed) to derive Pair 3, and so on. However, pair 6 should be derived after pair 7.

11.3-1 Solve Prob. 9.4-9 by the z-transform method.

11.3-2 Solve

$$y[k + 1] + 2y[k] = f[k + 1]$$

with $y[0] = 1$ and $f[k] = e^{-(k-1)} u[k]$

11.3-3 Find the output $y[k]$ of an LTID system specified by the equation

$$2y[k + 2] - 3y[k + 1] + y[k] = 4f[k + 2] - 3f[k + 1]$$

if the initial conditions are $y[-1] = 0$, $y[-2] = 1$, and the input $f[k] = (4)^{-k} u[k]$.

11.3-4 Solve Prob. 11.3-3 if instead of initial conditions $y[-1], y[-2]$ you are given the auxiliary conditions $y[0] = \frac{3}{2}$ and $y[1] = \frac{35}{4}$.

11.3-5 Solve

$$4y[k + 2] + 4y[k + 1] + y[k] = f[k + 1]$$

with $y[-1] = 0$, $y[-2] = 1$, and $f[k] = u[k]$.

11.3-6 Solve

$$y[k + 2] - 3y[k + 1] + 2y[k] = f[k + 1]$$

if $y[-1] = 2$, $y[-2] = 3$, and $f[k] = (3)^k u[k]$.

11.3-7 Solve

$$y[k + 2] - 2y[k + 1] + 2y[k] = f[k]$$

with $y[-1] = 1$, $y[-2] = 0$, and $f[k] = u[k]$.

11.3-8 Solve

$$y[k] + 2y[k - 1] + 2y[k - 2] = f[k - 1] + 2f[k - 2]$$

with $y[0] = 0$, $y[1] = 1$, and $f[k] = e^k u[k]$.

11.3-9 (a) Find the zero-state response of an LTID system with transfer function

$$H[z] = \frac{z}{(z + 0.2)(z - 0.8)}$$

and the input $f[k] = e^{(k+1)} u[k]$.

(b) Write the difference equation relating the output $y[k]$ to input $f[k]$.

11.3-10 Repeat Prob. 11.3-9 if $f[k] = u[k]$ and

$$H[z] = \frac{2z + 3}{(z - 2)(z - 3)}$$

11.3-11 Repeat Prob. 11.3-9 if

$$H[z] = \frac{6(5z - 1)}{6z^2 - 5z + 1}$$

and the input $f[k]$ is (a) $(4)^{-k} u[k]$ (b) $(4)^{-(k-2)} u[k - 2]$ (c) $(4)^{-(k-2)} u[k]$
(d) $(4)^{-k} u[k - 2]$.

11.3-12 Repeat Prob. 11.3-9 if $f[k] = u[k]$ and

$$H[z] = \frac{2z - 1}{z^2 - 1.6z + 0.8}$$

11.3-13 Find the transfer functions corresponding to each of the systems specified by difference equations in Probs. 11.3-2, 11.3-3, 11.3-5, and 11.3-8.

11.3-14 Find $h[k]$, the unit impulse response of the systems described by the following equations:

(a) $y[k] + 3y[k - 1] + 2y[k - 2] = f[k] + 3f[k - 1] + 3f[k - 2]$
(b) $y[k + 2] + 2y[k + 1] + y[k] = 2f[k + 2] - f[k + 1]$
(c) $y[k] - y[k - 1] + 0.5y[k - 2] = f[k] + 2f[k - 1]$

11.3-15 Find $h[k]$, the unit impulse response of the systems in Probs. 11.3-9, 11.3-10, and 11.3-12.

11.4-1 Show a canonical, a cascade and a parallel realization of the following transfer functions:

(a) $H[z] = \dfrac{z(3z - 1.8)}{z^2 - z + 0.16}$

(b) $H[z] = \dfrac{5z + 2.2}{z^2 + z + 0.16}$

(c) $H[z] = \dfrac{3.8z - 1.1}{(z - 0.2)(z^2 - 0.6z + 0.25)}$

11.4-2 Give cascade and parallel realizations of the following transfer functions:

(a) $\dfrac{z(1.6z - 1.8)}{(z - 0.2)(z^2 + z + 0.5)}$ (b) $\dfrac{z(2z^2 + 1.3z + 0.96)}{(z + 0.5)(z - 0.4)^2}$

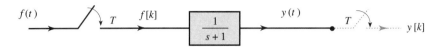

Fig. P11.6-1.

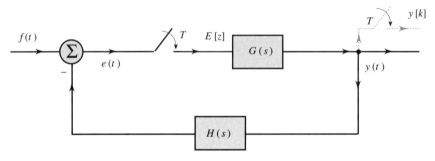

Fig. P11.6-2.

11.4-3   Realize a system whose transfer function is

$$H[z] = \frac{2z^4 + z^3 + 0.8z^2 + 2z + 8}{z^4}$$

11.4-4   Realize a system whose transfer function is given by

$$H[z] = \sum_{k=0}^{6} kz^{-k}$$

11.6-1   Determine $y[k]$, the output samples for the system depicted in Fig. P11.6-1 if the input $f(t) = e^{-2t}u(t)$.

11.6-2   For the sampled-data system in Fig. P11.6-2, show that the z-transfer function is

$$T[z] = \frac{G[z]}{1 + GH[z]}$$

where the transfer function $GH[z]$ corresponds to $G(s)H(s)$ in Table 12.1.

11.6-3   For the sampled-data system in Fig. P11.6-3, show that the output $Y[z]$ is given by

$$Y[z] = \frac{FG[z]}{1 + GH[z]}$$

Note that $FG[z]$ corresponds to the entry for $F(s)G(s)$ in Table 12.1. It is not the same as $F[z]G[z]$. In this case, it is not possible to express the output as $Y[z] = T[z]F[z]$. Consequently, the z-transfer function of such systems does not exist, and their analysis is little more complicated.

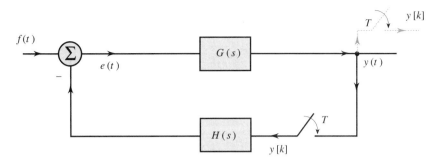

Fig. P11.6-3.

11.7-1 Find the z-transform (if it exists) and the corresponding region of convergence for each of the following signals:

(a) $(0.8)^k u[k] + 2^k u[-(k+1)]$

(b) $2^k u[k] - 3^k u[-(k+1)]$

(c) $(0.8)^k u[k] + (0.9)^k u[-(k+1)]$

(d) $\left[(0.8)^k + 3(0.4)^k\right] u[-(k+1)]$

(e) $\left[(0.8)^k + 3(0.4)^k\right] u[k]$

(f) $(0.8)^k u[k] + 3(0.4)^k u[-(k+1)]$

11.7-2 Find the inverse z-transform of

$$F[z] = \frac{(e^{-2} - 2)z}{(z - e^{-2})(z - 2)}$$

when the region of convergence is

(a) $|z| > 2$ (b) $e^{-2} < |z| < 2$ (c) $|z| < e^{-2}$.

11.7-3 Determine the zero-state response of a system having a transfer function

$$H[z] = \frac{z}{(z + 0.2)(z - 0.8)} \qquad\qquad |z| > 0.8$$

and an input $f[k]$ given by

(a) $f[k] = e^k u[k]$

(b) $f[k] = 2^k u[-(k+1)]$

(c) $f[k] = e^k u[k] + 2^k u[-(k+1)]$

11.7-4 For the system in Problem 11.7-3, determine the zero-state response if the input

$$f[k] = 2^k u[k] + u[-(k+1)]$$

11.7-5 For the system in Problem 11.7-3, determine the zero-state response if the input

$$f[k] = e^{-2k}[-(k+1)]$$

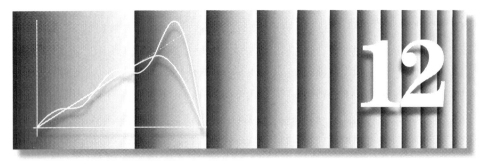

Frequency Response and Digital Filters

Filtering characteristics of a system are specified by its frequency response. For this reason it is important to study frequency response of discrete-time systems, which is very similar to the frequency response of continuous-time systems with some significant differences.

12.1 Frequency Response of Discrete-Time Systems

For (asymptotically stable) continuous-time systems we showed that the system response to an input $e^{j\omega t}$ is $H(j\omega)e^{j\omega t}$, and that the response to an input $\cos \omega t$ is $|H(j\omega)| \cos [\omega t + \angle H(j\omega)]$. Similar results hold for discrete-time systems. We now show that for an (asymptotically stable) LTID system, the system response to an input $e^{j\Omega k}$ is $H[e^{j\Omega}]e^{j\Omega k}$ and the response to an input $\cos \Omega k$ is $|H[e^{j\Omega}]| \cos (\Omega k + \angle H[e^{j\Omega}])$.

The proof is similar to the one used in continuous-time systems. In Sec. 9.4-2 we showed that an LTID system response to an (everlasting) exponential z^k is also an (everlasting) exponential $H[z]z^k$. It is helpful to represent this relationship by a directed arrow notation as

$$z^k \implies H[z]z^k \tag{12.1}$$

Setting $z = e^{\pm j\Omega}$ in this relationship yields

$$e^{j\Omega k} \implies H[e^{j\Omega}]e^{j\Omega k} \tag{12.2a}$$
$$e^{-j\Omega k} \implies H[e^{-j\Omega}]e^{-j\Omega k} \tag{12.2b}$$

Addition of these two equations yields

$$2\cos \Omega k \implies H[e^{j\Omega}]e^{j\Omega k} + H[e^{-j\Omega}]e^{-j\Omega k} = 2\mathrm{Re}\left(H[e^{j\Omega}]e^{j\Omega k}\right) \tag{12.3}$$

Expressing $H[e^{j\Omega}]$ in the polar form

$$H[e^{j\Omega}] = |H[e^{j\Omega}]|e^{j\angle H[e^{j\Omega}]} \tag{12.4}$$

Eq. (12.3) can be expressed as

716

$$\cos \Omega k \implies |H[e^{j\Omega}]| \cos \left(\Omega k + \angle H[e^{j\Omega}] \right) \tag{12.5}$$

In other words, the system response $y[k]$ to a sinusoidal input $\cos \Omega k$ is given by

$$y[k] = |H[e^{j\Omega}]| \cos \left(\Omega k + \angle H[e^{j\Omega}] \right) \tag{12.6a}$$

Following the same argument, the system response to a sinusoid $\cos (\Omega k + \theta)$ is

$$y[k] = |H[e^{j\Omega}]| \cos \left(\Omega k + \theta + \angle H[e^{j\Omega}] \right) \tag{12.6b}$$

This result applies only to asymptotically stable systems because Eq. (12.1) is valid only for values of z lying in the region of convergence of $H[z]$. For $z = e^{j\Omega}$, z lies on the unit circle ($|z| = 1$). The region of convergence for unstable and marginally stable systems does not include the unit circle.

This important result shows that the response of an asymptotically stable LTID system to a discrete-time sinusoidal input of frequency Ω is also a discrete-time sinusoid of the same frequency. *The amplitude of the output sinusoid is $|H[e^{j\Omega}]|$ times the input amplitude, and the phase of the output sinusoid is shifted by $\angle H[e^{j\Omega}]$ with respect to the input phase.* Clearly $|H[e^{j\Omega}]|$ is the amplitude gain, and a plot of $|H[e^{j\Omega}]|$ versus Ω is the amplitude response of the discrete-time system. Similarly, $\angle H[e^{j\Omega}]$ is the phase response of the system, and a plot of $\angle H[e^{j\Omega}]$ vs Ω shows how the system modifies or shifts the phase of the input sinusoid. Note that $H[e^{j\Omega}]$ incorporates the information of both amplitude and phase response and therefore is called the **frequency response** of the system.

These results, although parallel to those for continuous-time systems, differ from them in one significant aspect. In the continuous-time case, the frequency response is $H(j\omega)$. A parallel result for the discrete-time case would lead to frequency response $H[j\Omega]$. Instead, we found the frequency response to be $H[e^{j\Omega}]$. This deviation causes some interesting differences between the behavior of continuous-time and discrete-time systems.

Steady-State Response to Causal Sinusoidal Input

As in the case of continuous-time systems, we can show that the response of an LTID system to a causal sinusoidal input $\cos \Omega k \, u[k]$ is $y[k]$ in Eq. (12.6a), plus a natural component consisting of the characteristic modes (see Prob. 12.1-4). For a stable system, all the modes decay exponentially, and only the sinusoidal component in Eq. (12.6a) persists. For this reason, this component is called the sinusoidal *steady-state* response of the system. Thus, $y_{ss}[k]$, the steady-state response of a system to a causal sinusoidal input $\cos \Omega k \, u[k]$, is

$$y_{ss}[k] = |H[e^{j\Omega}]| \cos \left(\Omega k + \angle H[e^{j\Omega}] \right)$$

System Response to Sampled Continuous-Time Sinusoids

So far we have considered the system response of a discrete-time system to a discrete-time sinusoid $\cos \Omega k$ (or exponential $e^{j\Omega k}$). In practice, the input may be a sampled continuous-time sinusoid $\cos \omega t$ (or an exponential $e^{j\omega t}$). When a sinusoid $\cos \omega t$ is sampled with sampling interval T, the resulting signal is a discrete-time sinusoid $\cos \omega kT$. Therefore, all the results developed in this section apply if we substitute ωT for Ω:

$$\Omega = \omega T$$

■ Example 12.1

For a system specified by the equation

$$y[k+1] - 0.8y[k] = f[k+1]$$

find the system response to the input **(a)** $1^k = 1$ **(b)** $\cos\left[\frac{\pi}{6}k - 0.2\right]$
(c) a sampled sinusoid $\cos 1500t$ with sampling interval $T = 0.001$.

The system equation can be expressed as

$$(E - 0.8)y[k] = Ef[k]$$

Therefore, the transfer function of the system is

$$H[z] = \frac{z}{z - 0.8} = \frac{1}{1 - 0.8z^{-1}}$$

The frequency response is

$$H[e^{j\Omega}] = \frac{1}{1 - 0.8e^{-j\Omega}} \tag{12.7}$$

$$= \frac{1}{1 - 0.8(\cos \Omega - j\sin \Omega)}$$

$$= \frac{1}{(1 - 0.8\cos \Omega) + j0.8\sin \Omega}$$

Therefore

$$\left|H[e^{j\Omega}]\right| = \frac{1}{\sqrt{(1 - 0.8\cos \Omega)^2 + (0.8\sin \Omega)^2}}$$

$$= \frac{1}{\sqrt{1.64 - 1.6\cos \Omega}} \tag{12.8a}$$

and

$$\angle H[e^{j\Omega}] = -\tan^{-1}\left[\frac{0.8\sin \Omega}{1 - 0.8\cos \Omega}\right] \tag{12.8b}$$

The amplitude response $|H[e^{j\Omega}]|$ can also be obtained by observing that $|H|^2 = HH^*$. Therefore

$$\left|H[e^{j\Omega}]\right|^2 = H[e^{j\Omega}]H^*[e^{j\Omega}]$$

$$= H[e^{j\Omega}]H[e^{-j\Omega}] \tag{12.9}$$

From Eq. (12.7) it follows that

$$\left|H[e^{j\Omega}]\right|^2 = \left(\frac{1}{1 - 0.8e^{-j\Omega}}\right)\left(\frac{1}{1 - 0.8e^{j\Omega}}\right)$$

$$= \frac{1}{1.64 - 1.6\cos \Omega}$$

which yields the result found earlier in Eq. (12.8a).

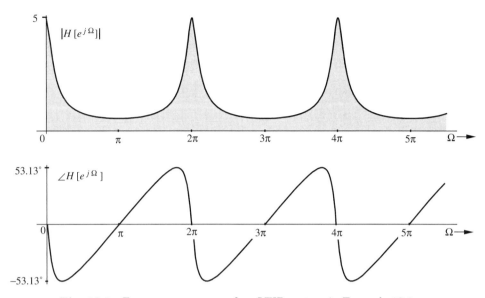

Fig. 12.1 Frequency response of an LTID system in Example 12.1.

Figure 12.1 shows plots of amplitude and phase response as functions of Ω. We now compute the amplitude and the phase response for the various inputs:

(a) $f[k] = 1^k = 1$

Since $1^k = (e^{j\Omega})^k$ with $\Omega = 0$, the amplitude response is $H[e^{j0}]$. From Eq. (12.8a) we obtain

$$H[e^{j0}] = \frac{1}{\sqrt{1.64 - 1.6\cos(0)}} = \frac{1}{\sqrt{0.04}} = 5 = 5\angle 0$$

Therefore

$$|H[e^{j0}]| = 5 \quad \text{and} \quad \angle H[e^{j0}] = 0$$

These values also can be read directly from Figs. 12.1a and 12.1b, respectively, corresponding to $\Omega = 0$. Therefore, the system response to input 1 is

$$y[k] = 5(1^k) = 5 \tag{12.10}$$

(b) $f[k] = \cos\left[\frac{\pi}{6}k - 0.2\right]$

Here $\Omega = \frac{\pi}{6}$. According to Eqs. (12.8)

$$\left|H[e^{j\pi/6}]\right| = \frac{1}{\sqrt{1.64 - 1.6\cos\frac{\pi}{6}}} = 1.983$$

$$\angle H[e^{j\pi/6}] = -\tan^{-1}\left[\frac{0.8\sin\frac{\pi}{6}}{1 - 0.8\cos\frac{\pi}{6}}\right] = -0.916 \text{ rad.}$$

These values also can be read directly from Figs. 12.1a and 12.1b, respectively, corresponding to $\Omega = \frac{\pi}{6}$. Therefore

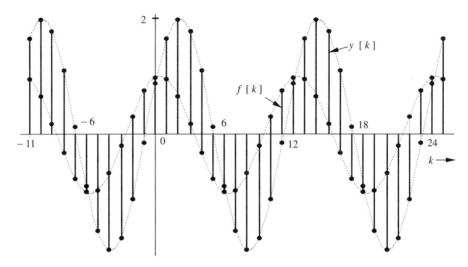

Fig. 12.2 Sinusoidal input and the corresponding output of an LTID system in Example 12.1.

$$y[k] = 1.983 \cos \left(\tfrac{\pi}{6} k - 0.2 - 0.916 \right)$$

$$= 1.983 \cos \left(\tfrac{\pi}{6} k - 1.116 \right) \tag{12.11}$$

Figure 12.2 shows the input $f[k]$ and the corresponding system response.

(c) A sinusoid cos $1500t$ sampled every T seconds ($t = kT$) results in a discrete-time sinusoid

$$f[k] = \cos 1500kT \tag{12.12}$$

For $T = 0.001$, the input is

$$f[k] = \cos (1.5k) \tag{12.13}$$

In this case, $\Omega = 1.5$. According to Eqs. (12.8a) and (12.8b)

$$\left| H[e^{j1.5}] \right| = \frac{1}{\sqrt{1.64 - 1.6 \cos (1.5)}} = 0.809 \tag{12.14}$$

$$\angle H[e^{j1.5}] = - \tan^{-1} \left[\frac{0.8 \sin (1.5)}{1 - 0.8 \cos (1.5)} \right] = -0.702 \text{ rad} \tag{12.15}$$

These values also could be read directly from Fig. 12.1 corresponding to $\Omega = 1.5$. Therefore

$$y[k] = 0.809 \cos (1.5k - 0.702) \quad \blacksquare$$

⊙ **Computer Example C12.1**
Using MATLAB, find the frequency response of the system in Example 12.1.

```
num=[1 0]; den=[1 -0.8];
W=-pi:pi/100:pi;
H=freqz(num,den,W);
mag=abs(H);
phase=180/pi*unwrap(angle(H));
subplot(2,1,1);
plot(W,mag);grid;
subplot(2,1,2);
plot(W,phase);grid   ⊙
```

Comment

Figure 12.1 shows plots of amplitude and phase response as functions of Ω. These plots as well as Eqs. (12.8) indicate that the frequency response of a discrete-time system is a continuous (rather than discrete) function of frequency Ω. There is no contradiction here. This behavior is merely an indication of the fact that the frequency variable Ω is continuous (takes on all possible values) and therefore the system response exists at every value of Ω.

△ **Exercise E12.1**

For a system specified by the equation

$$y[k+1] - 0.5y[k] = f[k]$$

find the amplitude and phase response. Find the system response to sinusoidal input $\cos\left(1000t - \frac{\pi}{3}\right)$ sampled every $T = 0.5$ ms.
Answer:

$$\left|H[e^{j\Omega}]\right| = \frac{1}{\sqrt{1.25 - \cos\Omega}} \qquad \angle H[e^{j\Omega}] = -\tan^{-1}\left[\frac{\sin\Omega}{\cos\Omega - 0.5}\right]$$

$$y[k] = 1.639\cos\left(0.5k - \frac{\pi}{3} - 0.904\right) = 1.639\cos\left(0.5k - 1.951\right) \quad \triangledown$$

△ **Exercise E12.2**

Show that for an ideal delay ($H[z] = 1/z$), the amplitude response $|H[e^{j\Omega}]| = 1$, and the phase response $\angle H[e^{j\Omega}] = -\Omega$. Thus, a pure time-delay does not affect the amplitude gain of sinusoidal input, but it causes a phase shift (delay) of Ω radians in a discrete sinusoid of frequency Ω. Thus, in the case of an ideal delay, we see that the phase shift at the output is proportional to the frequency of the input sinusoid (linear phase shift). $\triangledown$

The Periodic Nature of the Frequency Response

Figure 12.1 shows that for the system in Example 12.1, the frequency response $H[e^{j\Omega}]$ is a periodic function of Ω with period 2π. This fact is not a coincidence. Unlike the frequency response of a continuous-time system, the frequency response of every LTID system is a periodic function of Ω with period 2π. This fact follows from the very structure of the frequency response $H[e^{j\Omega k}]$. Its argument $e^{j\Omega k}$ is a periodic function of Ω with period 2π. This fact will automatically render $H[e^{j\Omega k}]$ periodic. There is a physical reason for this periodicity and the periodicity of $H[e^{j\Omega k}]$ should not come as a surprise. We know that discrete-time sinusoids separated by values of Ω in integral multiples of 2π are identical. Therefore, the system response to such sinusoids (or exponentials) is also identical. Thus for discrete-time systems, we need to plot the frequency response only over the frequency range from $-\pi$ to π (or from 0 to 2π). In a real sense, discrete-time sinusoids of frequencies outside the fundamental range of frequencies do not exist (although they exist in a technical sense).

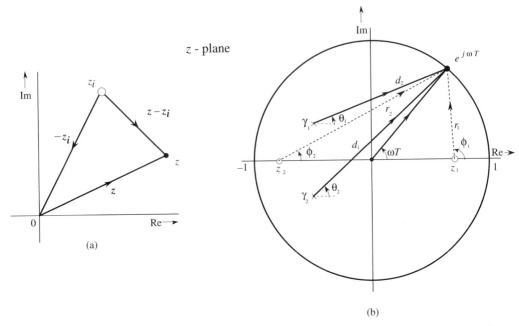

Fig. 12.3   (a) vector representation of complex numbers (b) vector representation of factors of $H[z]$.

12.2 Frequency Response From Pole-Zero Location

The frequency response (amplitude and phase response) of a system are determined by pole-zero locations of the transfer function $H[z]$. Just as in continuous-time systems, it is possible to determine quickly the amplitude and the phase response and to obtain physical insight into the filter characteristics of a discrete-time system by using a graphical technique. Consider the transfer function

$$H[z] = b_n \frac{(z - z_1)(z - z_2)\cdots(z - z_n)}{(z - \gamma_1)(z - \gamma_2)\cdots(z - \gamma_n)} \tag{12.16}$$

We can compute $H[z]$ graphically using the concepts discussed in Sec. 7.3. The directed line segment from z_i to z in the complex plane (Fig. 12.3a) represents the complex number $z - z_i$. The length of this segment is $|z - z_i|$ and its angle with the horizontal axis is $\angle(z - z_i)$.

In filtering applications, the inputs are often the sampled continuous-time sinusoids. Earlier, we showed that a sampled continuous-time sinusoid $\cos \omega t$ appears as a discrete-time sinusoid $\cos \Omega k$ ($\Omega = \omega T$). The appropriate function for computing the frequency response in such a situation, therefore, is $H[e^{j\omega T}]$ ($\Omega = \omega T$). To compute the frequency response $H[e^{j\omega T}]$ we evaluate $H[z]$ at $z = e^{j\omega T}$. But for $z = e^{j\omega T}$, $|z| = 1$ and $\angle z = \omega T$ so that $z = e^{j\omega T}$ represents a point on the unit circle at an angle ωT with the horizontal. We now connect all zeros (z_1, z_2, ..., z_n) and all poles (γ_1, γ_2, ..., γ_n) to the point $e^{j\omega T}$, as indicated in Fig. 12.3b. Let r_1, r_2, ..., r_n be the lengths and ϕ_1, ϕ_2, ..., ϕ_n be the angles, respectively, of the straight lines connecting z_1, z_2, ..., z_n to the point $e^{j\omega T}$. Similarly, let d_1, d_2, ..., d_n be the lengths and θ_1, θ_2, ..., θ_n be the angles, respectively, of the lines connecting γ_1, γ_2, ..., γ_n to $e^{j\omega T}$. Then

$$H[e^{j\omega T}] = H[z]|_{z=e^{j\omega T}} = b_n \frac{(r_1 e^{j\phi_1})(r_2 e^{j\phi_2}) \cdots (r_n e^{j\phi_n})}{(d_1 e^{j\theta_1})(d_2 e^{j\theta_2}) \cdots (d_n e^{j\theta_n})} \tag{12.17}$$

$$= b_n \frac{r_1 r_2 \cdots r_n}{d_1 d_2 \cdots d_n} e^{j[(\phi_1 + \phi_2 + \cdots + \phi_n) - (\theta_1 + \theta_2 + \cdots + \theta_n)]} \tag{12.18}$$

Therefore

$$|H[e^{j\omega T}]| = b_n \frac{r_1 r_2 \cdots r_n}{d_1 d_2 \cdots d_n}$$

$$= b_n \frac{\text{product of the distances of zeros to } e^{j\omega T}}{\text{product of distances of poles to } e^{j\omega T}} \tag{12.19a}$$

and

$$\angle H[e^{j\omega T}] = (\phi_1 + \phi_2 + \cdots \phi_n) - (\theta_1 + \theta_2 + \cdots + \theta_n)$$

$$= \text{sum of zero angles to } e^{j\omega T} - \text{sum of pole angles to } e^{j\omega T} \tag{12.19b}$$

In this manner, we can compute the frequency response $H[e^{j\omega T}]$ for any value of ω by selecting the point on the unit circle at an angle ωT corresponding to that value of ω. This point is $e^{j\omega T}$. In summary, to compute the frequency response $H[e^{j\omega T}]$, we connect all poles and zeros to this point and determine $|H[e^{j\omega T}]|$ and $\angle H[e^{j\omega T}]$ using the above equations. We repeat this procedure for all values of ωT from 0 to π to obtain the frequency response.

Controlling Gain by Placement of Poles and Zeros

The nature of the influence of pole and zero locations on the frequency response is similar to that observed in continuous-time systems with a minor difference. In place of the imaginary axis of the continuous-time systems, we have a unit circle in the discrete-time case. The nearer the pole (or zero) is to a point $e^{j\omega T}$ (on the unit circle) representing some frequency ω, the more influence that pole (or zero) wields on the amplitude response at that frequency because the length of the vector joining that pole (or zero) to the point $e^{j\omega T}$ is small. The proximity of a pole (or zero) has similar effect on the phase response. From Eq. (12.19a), it is clear that to enhance the amplitude response at a frequency ω we should place a pole as close as possible to the point $e^{j\omega T}$ (on the unit circle) representing that frequency ω. Similarly, to suppress the amplitude response at a frequency ω, we should place a zero as close as possible to the point $e^{j\omega T}$ on the unit circle. Placing repeated poles or zeros will further enhance their influence.

Total suppression of signal transmission at any frequency can be achieved by placing a zero on the unit circle at a point corresponding to that frequency. This is the principle of the notch (bandstop) filter.

Placing a pole or a zero at the origin does not influence the amplitude response because the length of the vector connecting the origin to any point on the unit circle is unity. However, a pole (a zero) at the origin generates angle $-\omega T$ (ωT) in

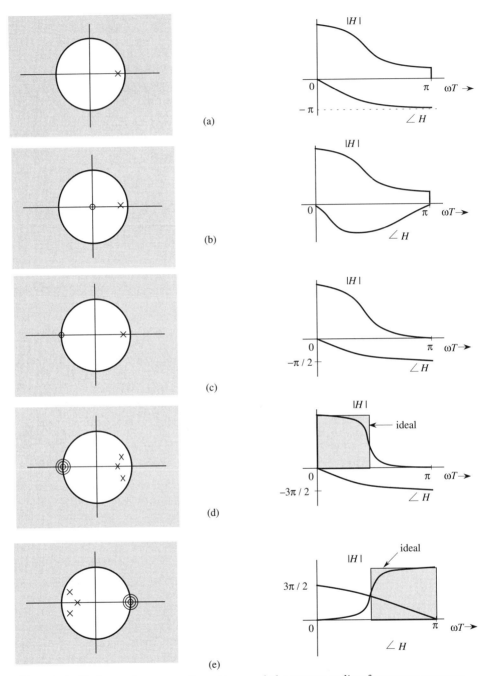

Fig. 12.4 Various pole-zero configurations and the corresponding frequency response.

$\angle H[e^{j\omega T}]$. The phase spectrum $-\omega T$ is a linear function of frequency and therefore represents a pure time-delay of T seconds (see Eq. (10.48) or Exercise E12.2). Therefore, a pole (a zero) at the origin causes a time delay (time advance) of T seconds in the response. There is no change in the amplitude response.

For a stable system, all the poles must be located inside the unit circle. The zeros may lie anywhere. Also, for a physically realizable system, $H[z]$ must be a proper fraction, that is, $n \geq m$. If, to achieve a certain amplitude response, we require $m > n$, we can still make the system realizable by placing a sufficient number of poles at the origin. This will not change the amplitude response but it will increase the time delay of the response.

In general, a pole at a point has the opposite effect of a zero at that point. Placing a zero closer to a pole tends to cancel the effect of that pole on the frequency response.

Lowpass Filters

A lowpass filter has a maximum gain at $\omega = 0$, which corresponds to point $e^{j0T} = 1$ on the unit circle. Clearly, placing a pole inside the unit circle near the point $z = 1$ (Fig. 12.4a) would result in a lowpass response. The corresponding amplitude and phase response appears in Fig. 12.4a. For smaller values of ω, the point $e^{j\omega T}$ (a point on the unit circle at an angle ωT) is closer to the pole, and consequently the gain is higher. As ω increases, the distance of the point $e^{j\omega T}$ from the pole increases. Consequently the gain decreases, resulting in a lowpass characteristic. Placing a zero at the origin does not change the amplitude response but it does modify the phase response, as illustrated in Fig. 12.4b. Placing a zero at $z = -1$, however, changes both the amplitude and phase response (Fig. 12.4c). The point $z = -1$ corresponds to frequency $\omega = \pi/T$ ($z = e^{j\omega T} = e^{j\pi} = -1$). Consequently, the amplitude response now becomes more attenuated at higher frequencies, with a zero gain at $\omega T = \pi$. We can approach ideal lowpass characteristics by using more poles staggered near $z = 1$ (but within the unit circle). Figure 12.4d shows a third-order lowpass filter with three poles near $z = 1$ and a third-order zero at $z = -1$, with corresponding amplitude and phase response. For an ideal lowpass filter we need an enhanced gain at every frequency in the band $(0, \omega_c)$. This can be achieved by placing a continuous wall of poles (requiring an infinite number of poles) opposite this band.

Highpass Filters

A highpass filter has a small gain at lower frequencies and a high gain at higher frequencies. Such a characteristic can be realized by placing a pole or poles near $z = -1$ because we want the gain at $\omega T = \pi$ to be the highest. Placing a zero at $z = 1$ further enhances suppression of gain at lower frequencies. Figure 12.4e shows a possible pole-zero configuration of the third-order highpass filter with corresponding amplitude and phase response.

■ **Example 12.2: Bandpass Filter**

Using trial-and-error, design a tuned (bandpass) filter with zero transmission at 0 Hz and also at 500 Hz. The resonant frequency is required to be 125 Hz. The highest frequency to be processed is $\mathcal{F}_h = 500$ Hz.

Because $\mathcal{F}_h = 500$, we require $T \leq \frac{1}{1000}$ [see Eq. (8.17)]. Let us select $T = 10^{-3}$. Since the amplitude response is zero at $\omega = 0$ and $\omega = 1000\pi$, we need to place zeros at $e^{j\omega T}$ corresponding to $\omega = 0$ and $\omega = 1000\pi$. For $\omega = 0, z = e^{j\omega T} = 1$; for $\omega = 1000\pi$ (with $T = 10^{-3}$), $e^{j\omega T} = -1$. Hence, there must be zeros at $z = \pm 1$. Moreover, we need enhanced frequency response at $\omega = 250\pi$. This frequency (with $\omega T = \pi/4$) corresponds to $z = e^{j\omega T} = e^{j\pi/4}$. Therefore, to enhance the frequency response at this frequency, we

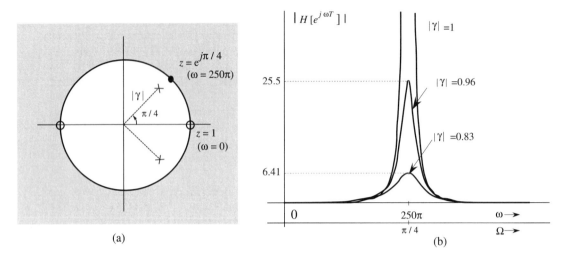

(a)

(b)

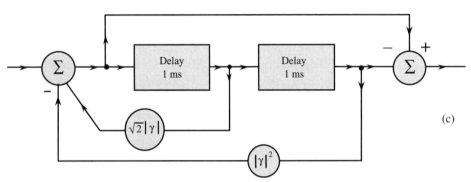

(c)

Fig. 12.5   Designing a bandpass filter in Example 12.2.

place a pole in its vicinity (near $e^{j\pi/4}$). Because this is a complex pole we also need its conjugate near $e^{-j\pi/4}$, as indicated in Fig. 12.5a. Let us choose these poles λ_1 and λ_2 as

$$\lambda_1 = |\gamma|e^{j\pi/4} \quad \text{and} \quad \lambda_2 = |\gamma|e^{-j\pi/4}$$

where $|\gamma| < 1$ for stability. The closer the value of $|\gamma|$ is to the unit circle, the more sharply peaked is the response around $\omega = 250\pi$. We also have a zeros at ± 1. Hence

$$H[z] = K\frac{(z-1)(z+1)}{(z-|\gamma|e^{j\pi/4})(z-|\gamma|e^{-j\pi/4})} = K\frac{z^2-1}{z^2-\sqrt{2}|\gamma|z+|\gamma|^2} \tag{12.20}$$

For convenience we shall choose $K = 1$. The amplitude response is given by

$$\left|H[e^{j\omega T}]\right| = \frac{|e^{j2\omega T}-1|}{|e^{j\omega T}-|\gamma|e^{j\pi/4}|\,|e^{j\omega T}-|\gamma|e^{-j\pi/4}|}$$

Now, using Eq.(12.9), we obtain

$$\left|H[e^{j\omega T}]\right|^2 = \frac{2(1-\cos 2\omega T)}{\left[1+|\gamma|^2-2|\gamma|\cos\left(\omega T-\frac{\pi}{4}\right)\right]\left[1+|\gamma|^2-2|\gamma|\cos\left(\omega T+\frac{\pi}{4}\right)\right]} \tag{12.21}$$

Figure 12.5b shows the amplitude response for values of $|\gamma| = 0.83, 0.96$, and 1. As expected, the gain is zero at $\omega = 0$ and at 500 Hz ($\omega = 1000\pi$). The gain peaks at about 125 Hz ($\omega = 250\pi$). The resonance (peaking) becomes pronounced as $|\gamma|$ approaches 1. Fig. 12.5c shows a canonical realization of this filter [see Eq. (12.20)].   ■

⊙ **Computer Example C12.2**
   Using MATLAB, find the frequency response of the bandpass filter in Example 12.2 for $|\gamma| = 0.96$.

```
gamma=0.96;
num=[1 0 -1]; den=[1 -sqrt(2)*gamma gamma^2];
W=-pi:pi/100:pi;
H=freqz(num,den,W);
mag=abs(H);
phase=180/pi*unwrap(angle(H));
subplot(2,1,1)
plot(W,mag)
subplot(2,1,2)
plot(W,phase)   ⊙
```

■ **Example 12.3:   Notch (Bandstop) Filter**
   Design a second-order notch filter to have zero transmission at 250 Hz and a sharp recovery of gain to unity on both sides of 250 Hz. The highest significant frequency to be processed is $\mathcal{F}_h = 500$ Hz.
   In this case, $T \leq 1/2\mathcal{F}_h = 10^{-3}$. Let us choose $T = 10^{-3}$. For the frequency 250 Hz, $\omega T = 2\pi(250)T = \pi/2$. Thus, the frequency 250 Hz is represented by a point $e^{j\omega T} = e^{j\pi/2} = j$ on the unit circle, as depicted in Fig. 12.6a. Since we need zero transmission at this frequency, we must place a zero at $z = e^{j\pi/2} = j$ and its conjugate at $z = e^{-j\pi/2} = -j$. We also require a sharp recovery of gain on both sides of frequency 250 Hz. To accomplish this goal, we place two poles close to the two zeros in order to cancel out the effect of the zeros as we move away from the point j (corresponding to frequency 250 Hz). For this reason, let us use poles at $\pm ja$ with $a < 1$ for stability. The closer the poles are to the zeros (the closer the a to 1), the faster is the gain recovery on either side of 250 Hz. The resulting transfer function is

$$H[z] = K\frac{(z-j)(z+j)}{(z-ja)(z+ja)} = K\frac{z^2+1}{z^2+a^2}$$

The dc gain (gain at $\omega = 0$, or $z = 1$) of this filter is

$$H[1] = K\frac{2}{1+a^2}$$

Because we require a dc gain of unity, we must select $K = \frac{1+a^2}{2}$. The transfer function is therefore

$$H[z] = \frac{(1+a^2)(z^2+1)}{2(z^2+a^2)} \tag{12.22}$$

and according to Eq. (12.9)

$$|H[e^{j\omega T}]|^2 = \frac{(1+a^2)^2}{4}\frac{(e^{j2\omega T}+1)(e^{-j2\omega T}+1)}{(e^{j2\omega T}+a^2)(e^{-j2\omega T}+a^2)} = \frac{(1+a^2)^2(1+\cos 2\omega T)}{2(1+a^4+2a^2\cos 2\omega T)}$$

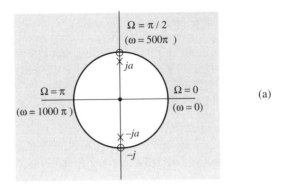

(a)

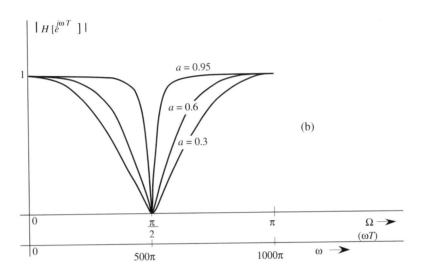

(b)

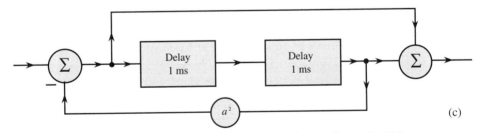

(c)

Fig. 12.6   Designing a notch (bandstop) filter in Example 12.3.

Figure 12.6b shows $|H[e^{j\omega T}]|$ for values of $a = 0.3$, 0.6, and 0.95. Figure 12.6c shows a realization of this filter.   ■

△ **Exercise E12.3**

Using the graphical argument, show that a filter with transfer function

$$H[z] = \frac{z - 0.9}{z}$$

acts as a highpass filter. Make a rough sketch of the amplitude response.   ▽

12.3 Digital Filters

Digital filters and some of their advantages were discussed in Sec. 8.5. Digital filters can be classified as either **recursive** or **nonrecursive.**

Recursive Filters

The terms *recursive* and *nonrecursive* are easily understood in terms of a specific example. Consider a third-order system with the transfer function

$$H[z] = \frac{b_3 z^3 + b_2 z^2 + b_1 z + b_0}{z^3 + a_2 z^2 + a_1 z + a_0} \qquad (12.23)$$

The input $f[k]$ and the corresponding output $y[k]$ of this system are related by the difference equation.

$$y[k] + a_2 y[k-1] + a_1 y[k-2] + a_0 y[k-3] =$$
$$b_3 f[k] + b_2 f[k-1] + b_1 f[k-2] + b_0 f[k-3] \qquad (12.24a)$$

or

$$y[k] = \underbrace{-a_2 y[k-1] - a_1 y[k-2] - a_0 y[k-3]}_{\text{output terms}}$$
$$+ \underbrace{b_3 f[k] + b_2 f[k-1] + b_1 f[k-2] + b_0 f[k-3]}_{\text{input terms}} \qquad (12.24b)$$

Here $y[k]$, the output at instant k, is determined by the input values $f[k]$, $f[k-1]$, $f[k-2]$, and $f[k-3]$ as well as by the past output values $y[k-1]$, $y[k-2]$, and $y[k-3]$. The output is therefore determined iteratively or recursively from the its past values. To compute the present output of this third-order system, we use the past three values of the output. In general, for an nth-order system we use the past n values of the output. Such a system is called a **recursive system**.

An interesting feature of a recursive system is that once an output exists, it tends to propagate itself forever because of its recursive nature. This is also seen from the canonical realization of $H[z]$ in Fig. 12.7a. Once an input (any input) is applied, the feedback connections loop the output continuously back into the system, and the output continues forever. This propagation of the output in perpetuity occurs because of the nonzero values of coefficients a_0, a_1, a_2, ..., a_{n-1}. These coefficients [appearing in the denominator of $H[z]$ in Eq. (12.23)] are the **recursive coefficients**. The coefficients b_0, b_1, b_2, ..., b_n (appearing in the numerator of $H[z]$) are the **nonrecursive coefficients**. If an input $\delta[k]$ is applied at the input of a recursive filter, the response $h[k]$ will continue forever up to $k = \infty$. For this reason, recursive filters are also known as **infinite impulse response (IIR)** filters.

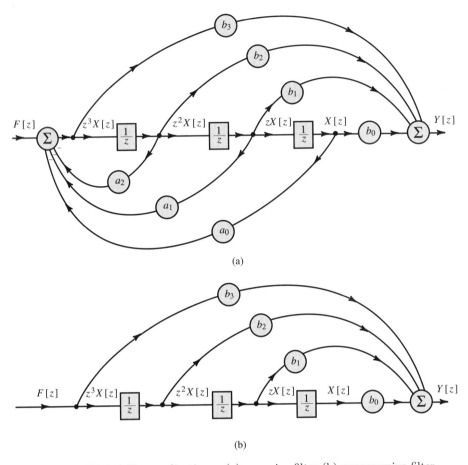

(a)

(b)

Fig. 12.7 Digital filter realization: (a) recursive filter (b) nonrecursive filter.

Nonrecursive Filters

If the recursive coefficients a_0, a_1, and a_2 are zero, $H[z]$ in Eq. (12.23) reduces to

$$H[z] = \frac{b_3 z^3 + b_2 z^2 b_1 z + b_0}{z^3} \tag{12.25a}$$

$$= b_3 + \frac{b_2}{z} + \frac{b_1}{z^2} + \frac{b_0}{z^3} \tag{12.25b}$$

The difference equation corresponding to this system now reduces to

$$y[k] = b_3 f[k] + b_2 f[k-1] + b_1 f[k-2] + b_0 f[k-3] \tag{12.26}$$

Note that $y[k]$ is now computed from the present and the three past values of the input $f[k]$. Such filters are called **nonrecursive filters**. Figure 12.7b shows a canonical realization of $H[z]$, which is identical to the realization in Fig. 12.7a, with all the feedback connections eliminated. If we apply an input $f[k] = \delta[k]$ to this system, the input will pass through the system and will be completely out of the system by $k = 4$. There are no feedback connections to perpetuate the output. Therefore, $h[k]$ will be nonzero only for $k = 0$, 1, 2, and 3. For an nth-order nonrecursive filter, $h[k]$ is zero for $k > n$. Therefore, the duration of $h[k]$ is finite for a nonrecursive filter. For this reason, nonrecursive filters are also known as **finite impulse response** (FIR) filters. Nonrecursive filters are a special case of recursive filters, in which all the recursive coefficients a_0, a_1, a_2, ..., a_{n-1} are zero.

An nth-order nonrecursive filter transfer function is given by ($a_0 = a_1 = a_2 = \cdots = a_{n-1} = 0$).

$$H[z] = \frac{b_n z^n + b_{n-1} z^{n-1} + \cdots + b_1 z + b_0}{z^n} \tag{12.27a}$$

$$= b_n + \frac{b_{n-1}}{z} + \frac{b_{n-2}}{z^2} + \cdots + \frac{b_1}{z^{n-1}} + \frac{b_0}{z^n} \tag{12.27b}$$

The inverse z-transform of this equation yields

$$h[k] = b_n \delta[k] + b_{n-1} \delta[k-1] + \cdots + b_1 \delta[k-n+1] + b_0 \delta[k-n] \tag{12.28}$$

Observe that $h[k] = 0$ for $k > n$.

Because nonrecursive filters are a special case of recursive filters, we expect the performance of recursive filters to be superior. This expectation is true in the sense that a given amplitude response can be achieved by a recursive filter of an order smaller than that required for the corresponding nonrecursive filter. However, nonrecursive filters have the advantage of having linear phase characteristics. Recursive filters can realize linear phase only approximately.

12.4   Filter Design Criteria

A digital filter processes discrete-time signals to yield a discrete-time output. Digital filters can also process analog signals by converting them into discrete-time signals. If the input is a continuous-time signal $f(t)$, it is converted into a discrete-time signal $f[k] = f(kT)$ by a C/D (continuous-time to discrete-time) converter. The signal $f[k]$ is now processed by a "digital" (meaning discrete-time) system with transfer function $H[z]$. The output $y[k]$ of $H[z]$ is then converted into an "analog" (meaning continuous-time) signal $y(t)$. The system in Fig. 12.8a, therefore, acts as a continuous-time (or "analog") system. Our objective is to determine the "digital" (discrete-time) processor $H[z]$ that will make the system in Fig. 12.8a equivalent to a desired "analog" (continuous-time) system with transfer function $H_a(s)$, shown in Fig. 12.8b.

We may strive to make the two systems behave similarly in the time-domain or in the frequency-domain. Accordingly, we have two different design procedures. Let us now determine the equivalence criterion of the two systems in the time-domain and in the frequency-domain.

12.4-1   The Time-Domain Equivalence Criterion

By time-domain equivalence we mean that for the same input $f(t)$, the output $y(t)$ of the system in Fig. 12.8a is equal to the output $y(t)$ of the system in Fig.

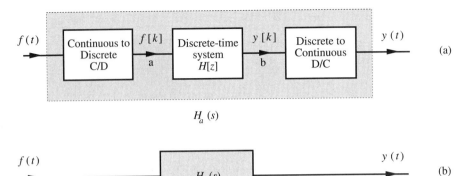

Fig. 12.8 Analog filter realization with a digital filter.

12.8b. Therefore $y(kT)$, the samples of the output in Fig. 12.8b, are identical to $y[k]$, the output of $H[z]$ in Fig. 12.8a.

The output $y(t)$ of the system in Fig. 12.8b is†

$$y(t) = \int_{-\infty}^{\infty} f(\tau)h_a(t-\tau)\,d\tau$$

An integral is a sum in the limit. Therefore, the above equation can be expressed as

$$y(t) = \lim_{\Delta\tau\to 0} \sum_{m=-\infty}^{\infty} f(m\Delta\tau)h_a(t-m\Delta\tau)\Delta\tau \qquad (12.29a)$$

For our purpose it is convenient to use the notation T for $\Delta\tau$ in the above equation. Such a change of notation yields

$$y(t) = \lim_{T\to 0} T \sum_{m=-\infty}^{\infty} f(mT)h_a(t-mT) \qquad (12.29b)$$

The response at the kth sampling instant is $y(kT)$ obtained by setting $t = kT$ in the above equation:

$$y(kT) = \lim_{T\to 0} T \sum_{m=-\infty}^{\infty} f(mT)h_a[(k-m)T] \qquad (12.29c)$$

In Fig. 12.8a, the input to $H[z]$ is $f(kT) = f[k]$. If $h[k]$ is the unit impulse response of $H[z]$, then $y[k]$, the output of $H[z]$, is given by

$$y[k] = \sum_{m=-\infty}^{\infty} f[m]h[k-m] \qquad (12.30)$$

---

†For the sake of generality, we are assuming a noncausal system. The argument and the results are also valid for causal systems.

If the two systems are to be equivalent, $y(kT)$ in Eq. (12.29c) must be equal to $y[k]$ in Eq. (12.30). Therefore

$$h[k] = \lim_{T \to 0} T h_a(kT) \qquad (12.31)$$

This is the time-domain criterion for equivalence of the two systems. according to this criterion, $h[k]$, the unit impulse response of $H[z]$ in Fig. 12.8a, must be equal to T times the samples of $h_a(t)$, the unit impulse response of the system in Fig. 12.8b, assuming that $T \to 0$. This is known as the **impulse invariance criterion** of filter design.†

12.4-2   The Frequency-Domain Equivalence Criterion

In Sec. 2.4-3 [Eq. (2.47)], we proved that for an analog system with transfer function $H_a(s)$, the system response $y(t)$ to the everlasting exponential input $f(t) = e^{st}$ is also an everlasting exponential

$$y(t) = H_a(s)e^{st} \qquad (12.32)$$

Similarly, in Eq. (9.57a), we showed that for a discrete-time system with transfer function $H[z]$, the system response $y[k]$ to an everlasting exponential input $f[k] = z^k$ is also an everlasting exponential $H[z]z^k$:

$$y[k] = H[z]z^k \qquad (12.33)$$

If the systems in Figs. 12.8a and 12.8b are equivalent, then the response of both systems to an everlasting exponential input $f(t) = e^{st}$ should be the same. A continuous-time signal $f(t) = e^{st}$ sampled every T seconds results in a discrete-time signal

$$f[k] = e^{skT}$$
$$= z^k \qquad \text{with } z = e^{sT}$$

This discrete-time exponential z^k is applied at the input of $H[z]$ in Fig. 12.8a, whose response is

$$y[k] = H[z]z^k \Big|_{z=e^{sT}}$$
$$= H[e^{sT}]e^{skT} \qquad (12.34)$$

Now, for the system in Fig. 12.8b, $y(kT)$, the kth sample of the output $y(t)$ in Eq. (12.32), is

$$y(kT) = H_a(s)e^{skT} \qquad (12.35)$$

If the two systems are to be equivalent, a necessary condition is that $y[k]$ in Eq. (12.34) must be equal to $y(kT)$ in Eq. (12.35). This condition means that

$$H[e^{sT}] = H_a(s) \qquad (12.36)$$

This is the frequency-domain criterion for equivalence of the two systems. It should be remembered, however, that with this criterion we are ensuring only that the

---

†Because T is a constant, some authors ignore the factor T, which yields alternate criterion $h[k] = h_a(kT)$. Ignoring T merely scales the amplitude response of the resulting filter.

digital filter's response matches exactly that of the desired analog filter at sampling instants. If we want the two responses to match at every value of t, we must have $T \to 0$. Therefore

$$\lim_{T \to 0} H[e^{sT}] = H_a(s) \tag{12.37}$$

A Practical Difficulty

Both of these criteria for filter design require the condition $T \to 0$ for realizing a digital filter equivalent to a given analog filter. However, this condition is impossible in practice because it necessitates an infinite sampling rate, resulting in an infinite data rate. In practice, we must choose a small but nonzero T to achieve a compromise between the two conflicting requirements, namely closeness of approximation and system cost.

This approximation, however, does not mean that the system in Fig. 12.8a is inferior to that in Fig. 12.8b, because often $H_a(s)$ itself is an approximation to what we are seeking. For example, in lowpass filter design we strive to design a system with ideal lowpass characteristics. Failing that, however, we settle for some approximation such as Butterworth lowpass transfer functions. In fact, it is entirely possible that $H[z]$, which is an approximation to $H_a(s)$, may be a better approximation to the desired characteristics than is $H_a(s)$ itself.

12.5   Recursive Filter Design by the Time-Domain Criterion: The Impulse Invariance Method

The time-domain design criterion for the equivalence of the systems in Figs. 12.8a and 12.8b is [see Eq. (12.31)]

$$h[k] = \lim_{T \to 0} T h_a(kT) \tag{12.38}$$

where $h[k]$ is the unit impulse response of $H[z]$, $h_a(t)$ is the unit impulse response of $H_a(s)$, and T is the sampling interval in Fig. 12.8a.

As indicated earlier, it is impractical to let $T \to 0$. In practice, T is chosen to be small but nonzero. We have already discussed the effect of aliasing and the consequent distortion in the frequency response caused by nonzero T. Assuming that we have selected a suitable value of T, we can ignore the condition $T \to 0$, and Eq. (12.38) can be expressed as

$$h[k] = T h_a(kT) \tag{12.39}$$

The z-transform of this equation yields

$$H[z] = T \mathcal{Z}\Big(h_a(kT)\Big) \tag{12.40}$$

This result yields the desired transfer function $H[z]$.

Let us consider a first-order transfer function

$$H_a(s) = \frac{c}{s - \lambda} \tag{12.41a}$$

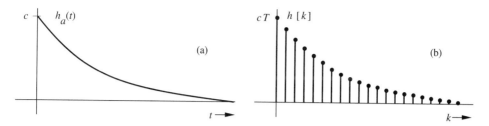

Fig. 12.9 Procedure for the impulse invariance method of filter design.

The impulse response $h(t)$ of this filter is the inverse Laplace transform of $H_a(s)$, which in this case is

$$h_a(t) = ce^{\lambda t}u(t) \tag{12.41b}$$

The corresponding digital filter unit sample response $h[k]$ is given by Eq. (12.39)

$$h[k] = Th_a(kT) = Tce^{k\lambda T} \tag{12.42}$$

Figures 12.9a and b show $h_a(t)$ and $h[k]$. According to Eq. (12.40), $H[z]$ is T times the z-transform of $h[k]$. Thus,

$$H[z] = \frac{Tcz}{z - e^{\lambda T}} \tag{12.43}$$

The procedure of finding $H[z]$ can be systematized for any nth-order system. First we express an nth-order analog transfer function $H_a(s)$ as a sum of partial fractions as

$$H_a(s) = \sum_{i=1}^{n} \frac{c_i}{s - \lambda_i} \tag{12.44}$$

Then the corresponding $H[z]$ is given by

$$H[z] = T \sum_{i=1}^{n} \frac{c_i z}{z - e^{\lambda_i T}} \tag{12.45}$$

This transfer function can be readily realized as a parallel combination of the n first-order systems if all the n poles of $H_a(s)$ are real. The complex conjugate poles, if any, must be realized as a single second-order term. Table 12.1 lists several pairs of $H_a(s)$ and their corresponding $H[z]$.

Choosing the Sampling Interval T

If $\mathcal{F}_h$ is the highest frequency to be processed, then the sampling interval must be no greater than $1/2\mathcal{F}_h$ in order to avoid signal aliasing. However, in the impulse invariance method, there is yet another consideration, which must also be taken into account. Consider a hypothetical frequency response $H_a(j\omega)$ (Fig. 12.10a) that we wish to realize using a digital filter, as illustrated in Fig. 12.8a. Let us assume that we have an equivalent digital filter that meets the time-domain equivalence criterion in Eq. (12.37); that is ,

TABLE 12.1

	$H_a(s)$	$h_a(t)$	$h[k]$	$H[z]$
1	K	$K\delta(t)$	$TK\delta[k]$	TK
2	$\dfrac{1}{s}$	$u(t)$	$Tu[k]$	$\dfrac{Tz}{z-1}$
3	$\dfrac{1}{s^2}$	t	kT^2	$\dfrac{T^2 z}{(z-1)^2}$
4	$\dfrac{1}{s^3}$	$\dfrac{t^2}{2}$	$\dfrac{k^2 T^3}{2}$	$\dfrac{T^3 z(z+1)}{2(z-1)^3}$
5	$\dfrac{1}{s-\lambda}$	$e^{\lambda t}$	$Te^{\lambda kT}$	$\dfrac{Tz}{z-e^{\lambda T}}$
6	$\dfrac{1}{(s-\lambda)^2}$	$te^{\lambda t}$	$kT^2 e^{\lambda kT}$	$\dfrac{T^2 z e^{\lambda T}}{(z-e^{\lambda T})^2}$
7	$\dfrac{As+B}{s^2+2as+c}$	$Tre^{-at}\cos(bt+\theta)$	$Tre^{-akT}\cos(bkT+\theta)$	$\dfrac{Trz\left[z\cos\theta - e^{-aT}\cos(bT-\theta)\right]}{z^2 - (2e^{-aT}\cos bT)z + e^{-2aT}}$

$$r=\sqrt{\frac{A^2 c+B^2-2ABa}{c-a^2}}$$

$$b=\sqrt{c-a^2}$$

$$\theta=\tan^{-1}\left(\frac{Aa-B}{A\sqrt{c-a^2}}\right)$$

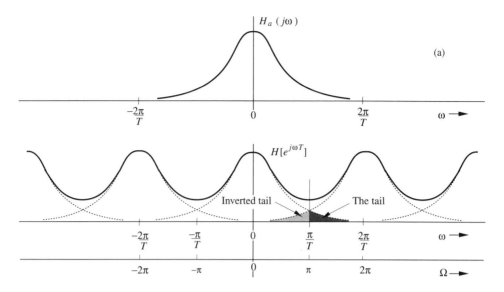

Fig. 12.10 Aliasing in digital filters, and a choice of the sampling interval T.

$$h[k] = \lim_{T \to 0} T h_a(kT)$$

In Chapter 5 (Fig. 5.6), we showed that the Fourier transform of the samples of $h_a(t)$ consists of periodic repetition of $H_a(j\omega)$ with period equal to the sampling frequency $\omega_s = 2\pi/T = 2\pi \mathcal{F}_s$.† Also $H_a(j\omega)$ is not generally bandlimited. Hence, aliasing among various repeating cycles cannot be prevented, as depicted in Fig. 12.10b. The resulting spectrum will be different from the desired spectrum, especially at higher frequencies. If $H_a(j\omega)$ were to be bandlimited; that is, if $H_a(j\omega) = 0$ for $|\omega| > \omega_0$, then the overlap could be avoided if we select the period $2\pi/T > 2\omega_0$. However, according to the Paley-Wiener criterion [Eq. (4.61)], every practical system frequency response is nonbandlimited, and the cycle overlap is inevitable. However, for frequencies beyond some ω_0, if $|H_a(j\omega)|$ is a negligible fraction, say 1%, of $H_a(j\omega)|_{\max}$, then we can consider[1] $H_a(j\omega)$ to be essentially bandlimited to ω_0, and we can select

$$T = \frac{\pi}{\omega_0} \tag{12.46}$$

■ **Example 12.4**

   Design a digital filter to realize the first-order analog lowpass Butterworth filter with the transfer function

$$H_a(s) = \frac{\omega_c}{s + \omega_c} \qquad \omega_c = 10^5 \tag{12.47}$$

---

†How can we apply the discussion in Chapter 5, which applies to impulse samples of continuous-time signals, to discrete-time signals? Recall our discussion in Sec. 10.4 (Fig. 10.8), where we showed that the spectrum of discrete-time signal is just a scaled version of the spectrum of the impulse samples of the corresponding continuous-time signal.

For this filter, we find the corresponding $H[z]$ according to Eq. (12.43) (or pair 5 in Table 12.1) as

$$H[z] = \frac{\omega_c T z}{z - e^{-\omega_c T}} \tag{12.48}$$

Next, we select the value of T according to Eq. (12.46). We find the essential filter bandwidth ω_0, where the filter gain is 1% of the maximum filter gain. Here we use

$$|H_a(j\omega)| = \left| \frac{\omega_c}{\sqrt{\omega^2 + \omega_c^2}} \right|$$

In this case $|H_a(j\omega)|_{\max} = 1$. Hence, the essential bandwidth ω_0 is that frequency where the $|H_a(j\omega_0)| = 0.01$. Observe that

$$|H_a(j\omega)| \approx \frac{\omega_c}{\omega} \qquad \omega \gg \omega_c$$

Hence,

$$|H_a(j\omega_0)| \approx \frac{\omega_c}{\omega_0} = 0.01 \quad \Longrightarrow \quad \omega_0 = 100\omega_c = 10^7$$

Thus, for good results, we should select $T = \frac{\pi}{\omega_0} = 10^{-7}\pi$. However, for the sake of demonstrating aliasing effect of the overlapping cycles, we shall deliberately select a lower value of ω_0 (higher T). Let us select

$$T = \left(\frac{\pi}{10\omega_c} \right) = 10^{-6}\pi \quad \text{and} \quad \omega_c T = \frac{\pi}{10} \tag{12.49}$$

Substitution of this value in Eq. (12.48) yields

$$H[z] = \frac{0.3142 z}{z - 0.7304} \tag{12.50}$$

A canonical realization of this filter is shown in Fig. 12.11a by following the procedure in Sec. 11.4 (see Example 6.18c, Fig. 6.25b). Note the recursive nature of the filter. To find the frequency response of this digital filter, we rewrite $H[z]$ as

$$H[z] = \frac{0.3142}{1 - 0.7304 z^{-1}}$$

Therefore

$$H[e^{j\omega T}] = \frac{0.3142}{1 - 0.7304 e^{-j\omega T}} = \frac{0.3142}{(1 - 0.7304 \cos \omega T) + j0.7304 \sin \omega T}$$

Consequently

$$\left| H[e^{j\omega T}] \right| = \frac{0.3142}{\sqrt{(1 - 0.7304 \cos \omega T)^2 + (0.7304 \sin \omega T)^2}} = \frac{0.3142}{\sqrt{1.533 - 1.4608 \cos \omega T}}$$

$$\angle H[e^{j\omega T}] = -\tan^{-1} \left(\frac{0.7304 \sin \omega T}{1 - 0.7304 \cos \omega T} \right)$$

Substituting $T = \pi \times 10^{-6}$ in the above equations, we obtain

$$\left| H[e^{j\omega T}] \right| = \frac{0.1\pi}{\sqrt{1.533 - 1.4608 \cos (\pi \times 10^{-6}\omega)}} \tag{12.51a}$$

$$\angle H[e^{j\omega T}] = -\tan^{-1} \left[\frac{0.7304 \sin (\pi \times 10^{-6}\omega)}{1 - 0.7304 \cos (\pi \times 10^{-6}\omega)} \right] \tag{12.51b}$$

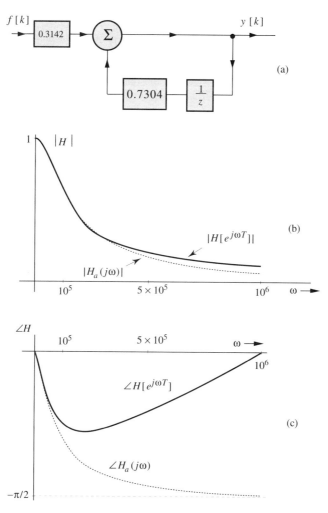

Fig. 12.11 An example of filter design by the impulse invariance method: (a) filter realization (b) amplitude response (c) phase response.

Also, according to Eq. (12.47) (with $\omega_c = 10^5$)

$$H_a(j\omega) = \frac{\omega_c}{j\omega + \omega_c} = \frac{10^5}{j\omega + 10^5} \tag{12.52}$$

Therefore

$$|H_a(j\omega)| = \frac{10^5}{\sqrt{\omega^2 + 10^{10}}} \quad \text{and} \quad \angle H_a(j\omega) = -\tan^{-1}\frac{\omega}{10^5}$$

Figures 12.11b and 12.11c show the amplitude and phase response of the analog and the (normalized)† digital filter over the frequency range $0 \le \omega \le \pi/T = 10^6$. Observe that

---

†The frequency response $H[e^{j\omega T}]$ is higher than the desired frequency response $H_a(j\omega)$ because of aliasing. We can partially correct this difference by multiplying $H[z]$ with a normalizing constant, forcing the resulting $H[e^{j\omega T}]$ to be equal to $H_a(j\omega)$ at $\omega = 0$. The normalizing constant K is defined as the ratio of $H_a(j0)$ to $H[e^{j0}] = H[1]$, which in this case is $1/1.653 = 0.858$. The normalized amplitude response in Fig. 12.11b is that of $0.858\left(\frac{0.1\pi z}{z - 0.7304}\right)$.

the behavior of the analog and the digital filter is very close over the range $\omega \le \omega_c = 10^5$. However, for higher frequencies, there is considerable aliasing, especially in the phase spectrum. ∎

⊙ **Computer Example C12.3**
Using MATLAB, find the impulse invariance digital filter to realize the first-order analog Butterworth filter in Example 12.4.

The analog filter transfer function is $10^5/(s + 10^5)$ and the sampling interval $T = 10^{-6}\pi$. A suitable MATLAB function to solve this problem is 'impinvar'. The input data are the coefficients of the numerator and the denominator polynomials of $H_a(s)$ [entered as $(n + 1)$-element vectors num and den] and the sampling interval T. MATLAB returns b and a, the numerator and the denominator polynomial coefficients of the desired digital filter $H[z]$.

In designing impulse invariance filter, we use the criterion $h[k] = Th_a(kT)$ in Eq. (12.39), whereas most books, including MATLAB, use the criterion $h[k] = h_a(kT)$. Hence, out answer will be T times the answer returned by MATLAB. To correct this discrepancy, we multiply num by T.

    T=pi/10^6; num=T*[0 10^5]; den=[1 10^5];
    Fs=1/T;
    [b,a]=impinvar(num,den,Fs)

MATLAB returns b=0.3142 and a=1 -0.7304. Therefore

$$H[z] = \frac{0.3142z}{z - 0.7304}$$

a conclusion which agrees with our result in Eq. (12.50). To plot the amplitude and the phase response, we can use the last 8 functions in Example C12.1.   ⊙

△ **Exercise E12.4**
Design a digital filter to realize an analog transfer function

$$H_a(s) = \frac{20}{s + 20}$$

Answer: $H[z] = \frac{20Tz}{z - e^{-20T}}$ with $T = \frac{\pi}{2000}$ ▽

Limitations of the Impulse Invariance Method

The impulse invariance method is handicapped by aliasing. Consequently this method can be used to design filters where $H_a(j\omega)$ becomes negligible beyond some frequency B Hz. This condition restricts the procedure to lowpass and bandpass filters. The impulse invariance method cannot be used for highpass or bandstop filters. Moreover, to reduce aliasing effects the sampling rate has to be very high, which makes its implementation costly. In general, the frequency-domain method discussed in the next section is superior to this method.

12.6 Recursive Filter Design by the Frequency-Domain Criterion: The Bilinear Transformation Method

The bilinear transformation method discussed in this section is preferable to the impulse invariance method in filtering problems where the gains are constant over certain bands (piecewise constant amplitude response). This condition exists in lowpass, bandpass, highpass, and bandstop filters. Moreover, this method requires a lower sampling rate compared to the impulse invariance method because of the absence of aliasing. In addition, the filter rolloff characteristics are sharper with this method compared to those obtained using the impulse invariance method. The absence of aliasing is the result of one-to-one mapping from s plane to z plane inherent in this method.

The frequency-domain design criterion is [see Eq. (12.37)]

$$\lim_{T \to 0} H[e^{sT}] = H_a(s) \tag{12.53}$$

Let us consider the following power series for the hyperbolic tangent (see Sec. B.7-3)

$$\tanh\left(\frac{sT}{2}\right) \equiv \frac{e^{sT/2} - e^{-sT/2}}{e^{sT/2} + e^{-sT/2}} = \left[\frac{sT}{2} - \frac{1}{3}\left(\frac{sT}{2}\right)^3 + \frac{2}{15}\left(\frac{sT}{2}\right)^5 + \cdots\right] \tag{12.54}$$

For small T ($T \to 0$), we can ignore the higher-order terms in the infinite series on the right-hand side to yield

$$\lim_{T \to 0} \left(\frac{e^{sT/2} - e^{-sT/2}}{e^{sT/2} + e^{-sT/2}}\right) = \frac{sT}{2}$$

Therefore, as $T \to 0$

$$s = \left(\frac{2}{T}\right) \frac{e^{sT/2} - e^{-sT/2}}{e^{sT/2} + e^{-sT/2}}$$

$$= \left(\frac{2}{T}\right) \frac{e^{sT} - 1}{e^{sT} + 1}$$

Equation (12.53) now can be expressed as

$$H[e^{sT}] = H_a\left(\frac{2}{T} \frac{e^{sT} - 1}{e^{sT} + 1}\right) \tag{12.55a}$$

From this result, it follows that

$$H[z] = H_a\left(\frac{2}{T} \frac{z - 1}{z + 1}\right) \tag{12.55b}$$

$$= H_a(s)|_{s = \frac{2}{T}\frac{z-1}{z+1}} \tag{12.55c}$$

Therefore, we can obtain $H[z]$ from $H_a(s)$ by using the transformation[†]

$$s = \left(\frac{2}{T}\right)\frac{z - 1}{z + 1} \tag{12.57}$$

This transformation is known as the **bilinear transformation**.

[†]There exist other transformations, which can be used to derive $H[z]$ from $H_a(s)$. We start with the power series

$$e^{-sT} = 1 - sT + \tfrac{1}{2}(sT)^2 - \tfrac{1}{3}(sT)^3 + \cdots$$

In the limit as $T \to 0$, all but the first two terms on the right-hand side can be ignored. This yields

$$s = \tfrac{1}{T}(1 - e^{-sT})$$

This results in a transformation

Choice of T in Bilinear Transformation Method

Because of the absence of aliasing in the bilinear transformation method, the value of the sampling interval T can be much smaller compared to the impulse invariance method. By absence of aliasing we mean only the kind of aliasing observed in impulse invariance method (Fig. 12.10b). The signal aliasing, which limits the highest usable frequency, is still present. Thus if the highest frequency to be processed is $\mathcal{F}_h$ Hz, then to avoid signal aliasing, we must use [see Eq. (8.17c)]

$$T \leq \frac{1}{2\mathcal{F}_h} \tag{12.58}$$

■ **Example 12.5**

Using the bilinear transformation, synthesize

$$H_a(s) = \frac{\omega_c}{s + \omega_c} \qquad \omega_c = 10^5$$

From Eq. (12.57), we obtain

$$H[z] = \frac{\omega_c}{\left(\dfrac{2}{T}\dfrac{z-1}{z+1}\right) + \omega_c}$$

$$= \frac{\omega_c T(z+1)}{(2+\omega_c T)z - (2-\omega_c T)}$$

We should use Eq. (12.58) to select a suitable value for T. However, to facilitate comparison with the impulse invariance method, we choose here the same value for T as that in Example 12.4: $T = \frac{\pi}{10\omega_c}$. The substitution of $\omega_c T = \pi/10$ in the above equation yields

$$H[z] = 0.1357 \left(\frac{z+1}{z - 0.7284} \right)$$

Hence

$$H[e^{j\omega T}] = \frac{0.1357(e^{j\omega T} + 1)}{e^{j\omega T} - 0.7284}$$

$$= \frac{0.1357(\cos \omega T + 1 + j \sin \omega T)}{\cos \omega T - 0.7284 + j \sin \omega T} \qquad T = \pi \times 10^{-6}$$

From this we obtain

$$\left| H[e^{j\omega T}] \right| = \left[\frac{0.024(1 + \cos \omega T)}{1 - 0.9518 \cos \omega T} \right]^{1/2} \qquad T = \pi \times 10^{-6} \tag{12.59a}$$

and

$$s = \frac{1}{T}\left(1 - \frac{1}{z}\right) = \frac{z-1}{Tz} \tag{12.56a}$$

Similarly, starting with the power series for e^{sT} yields the transformation

$$s = \frac{1}{T}(z - 1) \tag{12.56b}$$

These are strikingly simple transformations, which work reasonably well for lowpass and bandpass filters with low resonant frequencies. They cannot be used for highpass and bandstop filters, however, and they are inferior to bilinear transformation. The transformation in Eq. (12.56b) also has a stability problem.

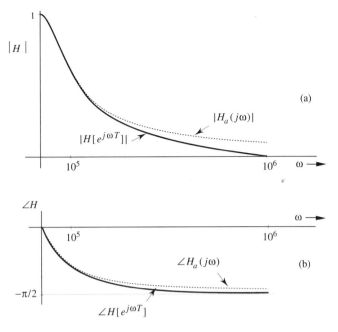

Fig. 12.12 Bilinear transformation method of design: (a) amplitude response (b) phase response.

$$\angle H[e^{j\omega T}] = \tan^{-1}\frac{\sin\,\omega T}{1+\cos\,\omega T} - \tan^{-1}\frac{\sin\,\omega T}{\cos\,\omega T - 0.7285} \qquad (12.59b)$$

Figure 12.12 shows $|H|$ and $\angle H$ as computed from Eqs. (12.59). Compare these with the filter characteristics obtained from the impulse invariant method (Fig. 12.11). ■

⊙ **Computer Example C12.4**

Using MATLAB, find the bilinear transformed digital filter to realize the first-order analog Butterworth filter in Example 12.5.

The analog filter transfer function is $10^5/(s+10^5)$ and the sampling interval $T = 10^{-6}\pi$. Hence, the sampling frequency Fs$= 10^6/\pi$. A suitable MATLAB function to solve this problem is 'bilinear'. The input data are the coefficients of the numerator and the denominator polynomials of $H_a(s)$ [entered as $(n+1)$-element vectors num and den] and the sampling frequency Fs Hz. MATLAB returns b and a, the numerator and the denominator polynomial coefficients of the desired digital filter $H[z]$.

Fs=10^6/pi; num=[0 10^5];den=[1 10^5];
[b,a]=bilinear(num,den,Fs)

MATLAB returns b=0.1358 0.1358 and a=1 -0.7285. Therefore

$$H[z] = \frac{0.1358(z+1)}{z - 0.7285}$$

which agrees with the answer found in Example 12.5. To plot the amplitude and the phase response, we can use the last 8 functions in Example C12.1.   ⊙

Frequency Prewarping Inherent in Bilinear Transformation

Figure 12.12 shows that $|H[e^{j\omega T}]| \simeq |H_a(j\omega)|$ for small ω. For large values of ω, the error increases. Moreover, $|H[e^{j\omega T}]| = 0$ at $\omega = \pi/T$. In fact, it appears as if the entire frequency band (0 to ∞) in $|H_a(j\omega)|$ is compressed within the range $(0, \frac{\pi}{T})$ in $H[e^{j\omega T}]$. Such warping of the frequency scale is peculiar to this transformation. To understand this behavior, consider Eq. (12.55a) with $s = j\omega$

$$H[e^{j\omega T}] = H_a\left(\frac{2}{T}\,\frac{e^{j\omega T}-1}{e^{j\omega T}+1}\right) = H_a\left(\frac{2}{T}\,\frac{e^{\frac{j\omega T}{2}}-e^{\frac{-j\omega T}{2}}}{e^{\frac{j\omega T}{2}}+e^{\frac{-j\omega T}{2}}}\right) = H_a\left(j\tfrac{2}{T}\tan\tfrac{\omega T}{2}\right)$$

Therefore, response of the resulting digital filter at some frequency ω_d is

$$H[e^{j\omega_d T}] = H_a\left(j\tfrac{2}{T}\tan\tfrac{\omega_d T}{2}\right)$$
$$= H_a(j\omega_a) \qquad\qquad (12.60)$$

where

$$\omega_a = \frac{2}{T}\tan\frac{\omega_d T}{2} \qquad\qquad (12.61a)$$

Thus, in the resulting digital filter, the behavior of the desired response $H_a(j\omega)$ at some frequency ω_a appears not at ω_a but at frequency ω_d, where [from Eq. (12.61a)]

$$\omega_d = \frac{2}{T}\tan^{-1}\frac{\omega_a T}{2} \qquad\qquad (12.61b)$$

Figure 12.13a shows the plot of ω_d as a function of ω_a. For small ω_a, the curve in Fig. 12.13a is practically linear, so $\omega_d \simeq \omega_a$. At higher values of ω_a, there is considerable diversion in the values of ω_a and ω_d. Thus, the digital filter imitates the desired analog filter at low frequencies, but at higher frequencies there is considerable distortion. Using this method, if we are trying to synthesize a filter to realize $H_a(j\omega)$ depicted in Fig. 12.13b, the resulting digital filter frequency response will be, as illustrated in Fig. 12.13c. The analog filter behavior in the entire range of ω_a from 0 to ∞ is compressed in the digital filter in the range of ω_d from 0 to π/T. This is as if a promising 20 year old man, who, after learning that he has only a year to live, tries to crowd his last year with every possible adventure, passion, and sensation that a normal human being would have experienced in an entire lifetime. This compression and frequency warping effect is the peculiarity of the bilinear transformation.

There are two ways of overcoming frequency warping. The first is to reduce T (increase the sampling rate) so that the signal bandwidth is kept well below $\frac{\pi}{T}$ and $\omega_a \simeq \omega_d$ over the desired frequency band. This step is easy to execute, but it requires a higher sampling rate (lower T) than necessary. The second approach, known as **prewarping**, solves the problem without unduly reducing T.

12.6-1   Bilinear Transformation Method with Prewarping

In prewarping, we start not with the desired $H_a(j\omega)$ but with a prewarped $H_a(j\omega)$ in such a way that the warping because of bilinear transformation will compensate for the prewarping exactly. The idea here is to begin with a distorted analog filter (prewarping) so that the distortion caused by bilinear transformation will be canceled by the built-in (prewarping) distortion. The idea is similar to

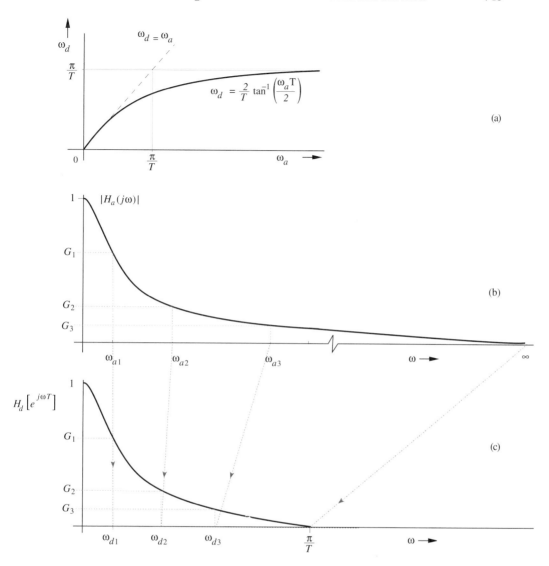

Fig. 12.13 Frequency warping in bilinear transformation: (a) mapping relationship of analog and digital frequencies (b) analog response (c) corresponding digital response.

the one used in prestressed concrete, in which a concrete beam is precompressed initially. When loaded, the beam experiences tension, which is canceled by the built-in compression.

Usually the prewarping is done at certain critical frequencies rather than over the entire band. The final filter behavior is exactly equal to the desired behavior at these selected frequencies. Such a filter is adequate for most filtering problems if we choose the critical frequencies properly.

If we require a filter to have gains $g_1, g_2, \ldots, g_m$ at frequencies (critical frequencies) $\omega_1, \omega_2, \ldots, \omega_m$ respectively, then we must start with an analog filter $H'(j\omega)$ which has gains $g_1, g_2, \ldots, g_m$ at frequencies $\omega_1', \omega_2', \ldots, \omega_m'$ respectively, where

$$\omega_i{}' = \frac{2}{T} \tan \frac{\omega_i T}{2} \qquad\qquad i = 1, 2, \cdots, m \qquad\qquad (12.62)$$

This results in prewarped filter $H'(j\omega)$. Application of the bilinear transformation (12.57) to this filter yields the desired digital filter which has gains g_1, g_2, $\ldots$, g_m at frequencies ω_1, ω_2, $\ldots$, ω_m respectively. This is because, according to Eq. (12.61a), the behavior of the analog filter at a frequency $\omega_i{}'$ appears in the digital filter at frequency

$$\frac{2}{T} \tan^{-1}\left(\frac{\omega_i{}'T}{2}\right) = \frac{2}{T} \tan^{-1}\left[\tan\left(\frac{\omega_i T}{2}\right)\right] = \omega_i$$

We clarify these ideas with an example of a lowpass Butterworth filter.

■ **Example 12.6**

Design a lowpass filter with the following specifications: The gain of unity at $\omega = 0$, and the gain is to be no less than -2 dB ($G_p = 0.785$) over the passband $0 \le \omega \le 10$. The gain is to be no greater than -11 dB ($G_s = 0.2818$) over the stopband $\omega \ge 15$. The highest frequency to be processed is $\omega_h = 35$ rad/s, which yields $T \le \pi/35$. Let us use $T = \pi/35$.

The specifications for a Butterworth filter for this design are $\omega_p = 8$, $\omega_s = 15$, $\hat{G}_p = -2$ dB, and $\hat{G}_s = -11$ dB. In the first step, we prewarp the critical frequencies ω_p and ω_s according to Eq. (12.62):

$$\omega_p{}' = \tfrac{2}{T} \tan \tfrac{\omega_p T}{2} = \tfrac{70}{\pi} \tan\left(\tfrac{4\pi}{35}\right) = 8.3623$$

$$\omega_s{}' = \tfrac{2}{T} \tan \tfrac{\omega_s T}{2} = \tfrac{70}{\pi} \tan\left(\tfrac{15\pi}{70}\right) = 17.7696$$

In the second step, we design a Butterworth filter with critical frequencies $\omega_p{}' = 8.3623$ and $\omega_s{}' = 17.7696$ with $\hat{G}_p = -2$ dB and $\hat{G}_s = -11$ dB. The value of n (order of the filter) is obtained from Eq. (7.39):

$$n = \frac{\log[(10^{1.1} - 1)/(10^{0.2} - 1)]}{2\log(17.7696/8.3623)} = 1.9405$$

We round up the value of n to 2. There are two possible values of ω'_c. We shall choose the one given by equation (7.41), which satisfies the stopband specifications exactly, but oversatisfies that in the passband. This choice yields the 3-dB cutoff frequency $\omega_c{}'$ as

$$\omega_c{}' = \frac{17.7696}{(10^{1.1} - 1)^{\frac{1}{4}}} = 9.6308$$

From Table 7.1, we find the prewarped filter transfer function $H_a(s)$ for $n = 2$ and $\omega_c{}' = 9.6308$ as

$$H_a(s) = \frac{1}{\left(\frac{s}{9.6308}\right)^2 + \sqrt{2}\left(\frac{s}{9.6308}\right) + 1} = \frac{92.7529}{s^2 + 13.62s + 92.7529}$$

Finally, we obtain $H[z]$ from $H_a(s)$, using the bilinear transformation

$$s = \left(\frac{2}{T}\right)\frac{z-1}{z+1} = \left(\frac{70}{\pi}\right)\frac{z-1}{z+1}$$

This substitution yields

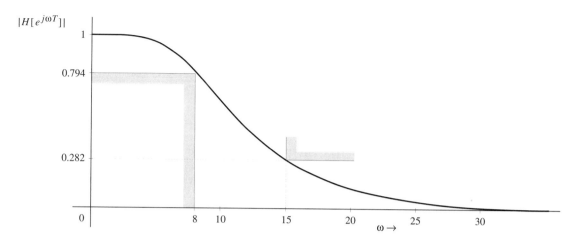

Fig. 12.14   Amplitude response of the filter in Example 12.6.

$$H[z] = H_a(s)\big|_{s=\left(\frac{70}{\pi}\right)\frac{z-1}{z+1}}$$

$$= \frac{92.7529}{s^2 + 13.62s + 92.7529}\bigg|_{s=\left(\frac{70}{\pi}\right)\frac{z-1}{z+1}}$$

$$= \frac{0.1039(z+1)^2}{z^2 - 0.9045z + 0.3201}$$

The frequency response of this filter is given by

$$H\left[e^{j\omega T}\right] = \frac{0.1039(e^{j\omega T}+1)^2}{e^{j2\omega T} - 0.9045e^{j\omega T} + 0.3201}$$

The amplitude response $|H[e^{j\omega T}]|$, with $T = 70/\pi$, is depicted in Fig. 12.14.   ■

Summary of the Bilinear Transformation Method with Prewarping

In the bilinear transformation method with prewarping, all the critical frequencies ω_i are transformed (prewarped) by the equation

$$\omega_i' = \frac{2}{T} \tan \frac{\omega_i T}{2} \qquad i = 1, 2, \cdots, m \qquad (12.63a)$$

The prewarped cutoff frequency ω_c', determined by using prewarped critical frequencies, is used to find the prewarped analog filter transfer function $H_a(s)$. Finally, we replace s with $\frac{2}{T}\frac{z-1}{z+1}$ in $H_a(s)$ to obtain the desired digital filter transfer function $H[z]$

$$H[z] = H_a(s)\big|_{s=\frac{2}{T}\frac{z-1}{z+1}} \qquad (12.63b)$$

A Simplified Procedure

The above procedure can be simplified by observing that the scaling factor $\frac{2}{T}$ is irrelevant in this manipulation and can be ignored. Instead of using Eqs. (12.63a) and (12.63b), we can use the simplified equations

$$\omega_i' = \tan \frac{\omega_i T}{2} \qquad\qquad i = 1, 2, \cdots, m \qquad\qquad (12.64a)$$

and

$$s = \frac{z - 1}{z + 1} \qquad\qquad (12.64b)$$

This simplification works because the factor $2/T$ in Eq. (12.63a) is a frequency scaling factor, and ignoring it in Eq. (12.64a) results in the pretransformed filter that is scaled by a factor $2/T$ in the frequency scale. This scaling is undone by using Eq. (12.64b) instead of Eq. (12.63b).

To demonstrate the procedure, we shall redo Example 12.6 using this simplification. In the first step, we prewarp the critical frequencies ω_p and ω_s according to Eq. (12.64a):

$$\omega_p' = \tan \tfrac{\omega_p T}{2} = \tan\left(\tfrac{4\pi}{35}\right) = 0.3753$$

$$\omega_s' = \tan \tfrac{\omega_s T}{2} = \tan\left(\tfrac{15\pi}{70}\right) = 0.7975$$

In the second step, we design a Butterworth filter with critical frequencies $\omega_p' = 0.3753$ and $\omega_s' = 0.7975$ with $\hat{G}_p = -2$ dB and $\hat{G}_s = -11$ dB. The value of n (order of the filter) is found from Eq. (7.39):

$$n = \frac{\log[(10^{1.1} - 1)/(10^{0.2} - 1)]}{2\log(0.7975/0.3753)} = 1.9405$$

We round up the value of n to 2. Also from Eq. (7.41), we find the 3-dB cutoff frequency ω_c' as

$$\omega_c' = \frac{0.7975}{(10^{1.1} - 1)^{\frac{1}{4}}} = 0.4322$$

From Table 7.1, we find the prewarped filter transfer function $H_a(s)$ for $n = 2$ and $\omega_c' = 0.4322$ as

$$H_a(s) = \frac{1}{\left(\frac{s}{0.4322}\right)^2 + \sqrt{2}\left(\frac{s}{0.4322}\right) + 1} = \frac{0.1868}{s^2 + 0.6112s + 0.1868}$$

Finally, we obtain $H[z]$ from $H_a(s)$, using the simplified bilinear transformation in Eq. (12.64b):

$$s = \frac{z - 1}{z + 1}$$

Therefore

$$H[z] = H_a(s)\big|_{s=\frac{z-1}{z+1}} = \frac{0.1868}{s^2 + 0.6112s + 0.1868}\bigg|_{s=\frac{z-1}{z+1}} = \frac{0.1039(z + 1)^2}{z^2 + 0.904z + 0.3201}$$

which is identical to the result obtained earlier.

⊙ **Computer Example C12.5**

Design a lowpass digital filter to meet the specifications in Example 12.6, using bilinear transformation with prewarping.

We shall give here MATLAB functions to design the four types of approximations: Butterworth, Chebyshev, inverse Chebyshev, and elliptic. The input data asks for frequencies so normalized that the sampling radian frequency is 2. This requirement means the sampling radian frequency, which is $2\pi/T$, must be normalized to 2. Therefore, all the radian frequencies can be normalized by multiplying each of them by T/π. In the present case, $T = \pi/35$ so that $T/\pi = 1/35$. Thus, to normalize any radian frequency, we just divide it by 35. The normalized Wp and Ws are 8/35 and 15/35, respectively.

Wp=8/35;Ws=15/35;Gp=-2;Gs=-11;T=pi/35;

```
% Butterworth
[n,Wn]=buttord(Wp,Ws,-Gp,-Gs);
[b,a]=butter(n,Wn)

% Chebyshev
[n,Wn]=cheb1ord(Wp,Ws,-Gp,-Gs);
[b,a]=cheby1(n,-Gp,Wn)

% Inverse Chebyshev
[n,Wn]=cheb2ord(Wp,Ws,-Gp,-Gs);
[b,a]=cheby2(n,-Gs,Wn)

% Elliptic
[n,Wn]=ellipord(Wp,Ws,-Gp,-Gs);
[b,a]=ellip(n,-Gp,-Gs,Wn)

% Plotting Amplitude and Phase Response
W=0:.001:pi;W=W';
H=freqz(b,a,W);
w=W/T;
mag=abs(H);
phase=180/pi*unwrap(angle(H));
subplot(2,1,1);
plot(w,mag);grid;
subplot(2,1,2);
plot(w,phase);grid
```

MATLAB returns b=0.1039 0.2078 0.1039 and a=1 -0.9045 0.3201 for Butterworth option. Therefore

$$H[z] = \frac{0.1039(z+1)^2}{z^2 - 0.9045z + 0.3201}$$

a result, which agrees with the answer found in Example 12.7.   ⊙

△ **Exercise E12.5**

Design a first-order lowpass Butterworth filter using the prewarping method so that the analog and digital gains are identical at $\omega = 0$ and at the 3-dB cutoff frequency ω_c. Use $T = \pi/4\omega_c$.
Answer:

$$H[z] = \frac{0.8284(z+1)}{2.8284z - 1.1716} = \frac{0.2929(z+1)}{z - 0.4142}$$

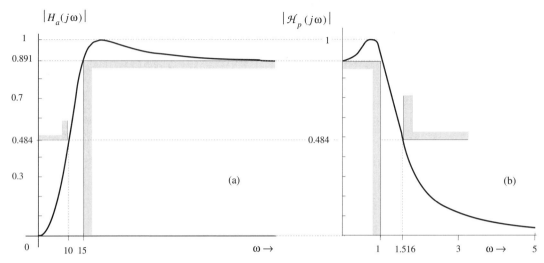

Fig. 12.15   Chebyshev highpass filter design using bilinear transformation with prewarping method.

$\triangledown$

■ Example 12.7: Highpass Filter Design

Design a 1-dB ripple Chebyshev highpass filter with the following specifications (depicted by the brick walls in Fig. 12.15a): The stopband gain $\hat{G}_s \leq -6.3$ dB ($G_s \leq 0.484$) over the stopband $0 \leq \omega \leq 10$ ($\omega_s = 10$). The ripple $\hat{r} \leq 1$ dB ($G_p \geq 0.891$) over a passband $\omega \geq 15$ ($\omega_p = 15$). The highest frequency to be processed is $\omega_h = 80$ radians/s.

In order to select a suitable value of T, we use Eq. (12.58) to avoid signal aliasing:

$$T \leq \frac{\pi}{\omega_h} = \frac{\pi}{80} \implies \text{Let us choose} \quad T = \frac{\pi}{100}$$

The critical frequencies are $\omega_s = 10$ and $\omega_p = 15$. The prewarped critical frequencies, according to Eq. (12.64a), are

$$\omega_s' = \tan \frac{\omega_s T}{2} = \tan \left(\frac{\pi}{20}\right) = 0.1584$$

$$\omega_p' = \tan \frac{\omega_p T}{2} = \tan \left(\frac{3\pi}{40}\right) = 0.24$$

In the second step, we design a prewarped Chebyshev highpass filter with critical frequencies $\omega_s' = 0.1584$ and $\omega_p' = 0.24$ with $\hat{r} = -1$ dB and $\hat{G}_s = -6.3$ dB (Fig. 12.15b). Following the procedure in Sec. 7.6-1, we first design a prototype lowpass filter with specifications, as indicated in Fig. 12.15b. Observe that the critical frequencies of the prototype filter are 1 (passband) and $\omega_p'/\omega_s' = 1.515$ (stopband) as explained in Sec. 7.6-1

The value of n needed to satisfy these specifications is given by [see Eq. (7.49b)]

$$n = \frac{1}{\cosh^{-1}\left(\frac{1.515}{1}\right)} \cosh^{-1} \left[\frac{(10^{0.63} - 1)}{10^{0.1} - 1}\right]^{1/2} = 1.988$$

We round up the value of n to 2. From Table 7.4 (Chebyshev filter with $\hat{r} = 1$ and $n = 2$) we obtain the following prototype transfer function

$$\mathcal{H}_p(s) = \frac{0.982614}{s^2 + 1.0978s + 1.1025} \qquad (12.65)$$

Next, to obtain the desired highpass transfer function, we replace s with ω_p'/s in the above prototype transfer function [see Eq. (7.55)]. To obtain the desired digital transfer function $H[z]$, we then replace s with $\frac{z-1}{z+1}$ [the bilinear transformation in Eq. (12.64b)]. This two-step operation may be combined in a single-step transformation as

$$s \Longrightarrow \frac{\omega_p'}{\left(\frac{z-1}{z+1}\right)} = \frac{\omega_p'(z+1)}{(z-1)} \qquad (12.66)$$

In this case $\omega_p' = 0.24$ so that we replace s with $\frac{0.24(z+1)}{z-1}$ in the prototype transfer function in Eq. (12.65) to obtain the desired digital transfer function

$$H[z] = \frac{0.6902(z-1)^2}{z^2 - 1.4678z + 0.6298}$$

The continuous curve in Fig. 12.15a shows the amplitude response of this filter. ∎

⊙ **Computer Example C12.6**

Design a highpass digital filter to meet the specifications in Example 12.7, using bilinear transformation with prewarping.

As before, we shall give here MATLAB functions to design the four basic types of approximations. The input data asks for frequencies so normalized that the sampling radian frequency is 2. As explained in Example C12.5, all the radian frequencies can be normalized by multiplying each of them by T/π. In the present case, $T = \pi/100$ so that $T/\pi = 1/100$. Thus, to normalize any radian frequency, we just divide it by 100. The normalized Wp and Ws are $15/100$ and $10/100$, respectively.

Wp=0.15;Ws=0.1;-Gp=1;-Gs=6.3;T=pi/100;

```
% Butterworth
[n,Wn]=Buttord(Wp,Ws,-Gp,-Gs);
[b,a]=butter(n,Wn,'high');

% Chebyshev
[n,Wn]=cheb1ord(Wp,Ws,-Gp,-Gs);
[b,a]=cheby1(n,-Gp,Wn,'high');

% Inverse Chebyshev
[n,Wn]=cheb2ord(Wp,Ws,-Gp,-Gs);
[b,a]=cheby2(n,-Gs,Wn,'high')

% Elliptic
[n,Wn]=ellipord(Wp,Ws,-Gp,-Gs);
[b,a]=ellip(n,-Gp,-Gs,Wn,'high')
```

MATLAB returns b=0.6902 -1.3804 0.6902 and a=1 -1.4678 0.6298 for Chebyshev option. Therefore

$$H[z] = \frac{0.6902(z-1)^2}{z^2 - 1.4678z + 0.6298}$$

which agrees with the answer found in Example 12.6. To plot the amplitude and the phase response, we can use the last 9 functions in Example C12.5. ⊙

Bandpass and Bandstop Filters

For bandpass and bandstop filters, we follow a similar procedure. All the critical frequencies are first prewarped using the simplified form in Eq. (12.64a). Next, we determine a prototype lowpass filter, which is then converted to the desired analog filter using appropriate transformations discussed in Sec. 7.6. Finally, we use the bilinear transformation in Eq. (12.64b) to obtain the desired digital filter. As in the case of the highpass filter (discussed in Example 12.7), we can combine the two transformations into a single transformation. For the bandpass filter, we first use the transformation

$$T(s) = \frac{s^2 + \omega_{p_1}{}'\omega_{p_2}{}'}{(\omega_{p_2}{}' - \omega_{p_1}{}')s}$$

and then use the bilinear transformation in Eq. (12.64b). Thus, in the first step we replace s in the prototype transfer function $\mathcal{H}_p(s)$ with $T(s)$ (the frequency transformation). In the second step we replace s with $\frac{z-1}{z+1}$ (the simplified bilinear transformation). Thus, the final transformation is equivalent to replacing s with $T_{bp}[z]$ in the prototype filter transfer function $\mathcal{H}_p(s)$, where

$$
\begin{aligned}
T_{bp}[z] = T(s)|_{s=\frac{z-1}{z+1}} &= \frac{s^2 + \omega_{p_1}{}'\omega_{p_2}{}'}{(\omega_{p_2}{}' - \omega_{p_1}{}')s}\Big|_{s=\frac{z-1}{z+1}} \\
&= \frac{(z-1)^2 + \omega_{p_1}{}'\omega_{p_2}{}'(z+1)^2}{(\omega_{p_2}{}' - \omega_{p_1}{}')(z^2 - 1)} \\
&= \frac{(\omega_{p_1}{}'\omega_{p_2}{}' + 1)z^2 + 2(\omega_{p_1}{}'\omega_{p_2}{}' - 1)z + (\omega_{p_1}{}'\omega_{p_2}{}' + 1)}{(\omega_{p_2}{}' - \omega_{p_1}{}')(z^2 - 1)}
\end{aligned}
$$

Using the same argument, we can show that for the bandstop filter, the desired digital filter transfer function $H[z]$ can be obtained from the corresponding bandstop prototype filter $\mathcal{H}_p(s)$ by replacing s with $T_{bs}[z]$, which is the reciprocal of $T_{bp}[z]$. Both these transformations can be expressed in a more compact form as

$$T_{bp}[z] = \frac{z^2 + 2az + 1}{b(z^2 - 1)} \tag{12.67a}$$

$$T_{bs}[z] = \frac{b(z^2 - 1)}{z^2 + 2az + 1} \tag{12.67b}$$

where

$$a = \frac{\omega_{p_1}{}'\omega_{p_2}{}' - 1}{\omega_{p_1}{}'\omega_{p_2}{}' + 1} \qquad \text{and} \qquad b = \frac{\omega_{p_2}{}' - \omega_{p_1}{}'}{\omega_{p_1}{}'\omega_{p_2}{}' + 1} \tag{12.68}$$

Thus, a digital filter transfer function $H[z]$ can be obtained from $\mathcal{H}_p(s)$ by replacing s with $T_{bp}[z]$ for the bandpass filter, and replacing s with $T_{bs}[z]$ for the bandstop filter.

■ **Example 12.8**

Design a digital Butterworth bandpass filter with amplitude response specifications illustrated by the brick walls in Fig. 12.16a with $\omega_{p_1} = 1000$, $\omega_{p_2} = 2000$, $\omega_{s_1} = 450$, $\omega_{s_2} = 4000$, $G_p = 0.7852\,(-2.1\,\mathrm{dB})$, and $G_s = 0.1\,(-20\,\mathrm{dB})$. Take $T = \pi/10,000$.

The solution is executed in 3 steps: In the first step, we determine the prewarped critical frequencies. In the second step, the lowpass prototype filter transfer function $\mathcal{H}_p(s)$ is found from the prewarped critical frequencies. Finally, the desired $H[z]$ is found from $\mathcal{H}_p(s)$ using the lowpass analog to bandpass digital transformation by replacing s in $\mathcal{H}_p(s)$ with $T_{bp}[z]$.

Step 1: Find prewarped critical frequencies

The prewarped frequencies $\omega_{p_1}{}'$, $\omega_{p_2}{}'$, $\omega_{s_1}{}'$, and $\omega_{s_2}{}'$ corresponding to the four critical frequencies ω_{p_1}, ω_{p_2}, ω_{s_1}, and ω_{s_2} using Eq. (12.64a):

$$\omega_{s_1}{}' = \tan\frac{\omega_{s_1}T}{2}$$

$$= \tan\left(\frac{450\pi}{20,000}\right) = 0.0708$$

$$\omega_{p_1}{}' = \tan\left(\frac{1000\pi}{20,000}\right) = 0.1584$$

$$\omega_{p_2}{}' = \tan\left(\frac{2000\pi}{20,000}\right) = 0.3249$$

$$\omega_{s_2}{}' = \tan\left(\frac{4000\pi}{20,000}\right) = 0.7265$$

Step 2: Find $\mathcal{H}_p(s)$, the prewarped lowpass prototype analog filter

This procedure with 5 substeps is identical to step 1 in the design of an analog bandpass filter discussed in Example 7.10 (Sec. 7.6-2). The 5 substeps are:

Step 2.1: Find $\omega_s{}'$ for the prototype filter. For the prototype lowpass filter transfer function $\mathcal{H}_p(s)$ with amplitude response, as depicted in Fig. 12.16b. The frequency $\omega_s{}'$ is found [using Eq. (7.56)] to be the smaller of

$$\frac{(0.1584)(0.3249) - (0.0708)^2}{0.0708(0.3249 - 0.1584)} = 3.939 \quad \text{and} \quad \frac{(0.7265)^2 - (0.1584)(0.3249)}{0.7265(0.3249 - 0.1584)} = 3.937$$

which is 3.937. We now have a prototype lowpass filter in Fig. 12.16b with $\hat{G}_p = -2.1$ dB, $\hat{G}_s = -20$ dB, $\omega_p{}' = 1$, and $\omega_s{}' = 3.937$.

Step 2.2: Determine the filter order n

The order of the Butterworth filter from Eq. (7.39) is

$$n = \frac{1}{2\log 3.937}\log\left[\frac{10^2 - 1}{10^{0.21} - 1}\right] = 1.8498$$

which is rounded up to $n = 2$.

Step 2.3: Determine the cutoff frequency $\omega_c{}'$ of the prototype filter

In this step (which is not necessary for the Chebyshev design), we determine the 3-dB cutoff frequency $\omega_c{}'$ for the prototype filter using any one of the Eqs. (7.40) or (7.41).

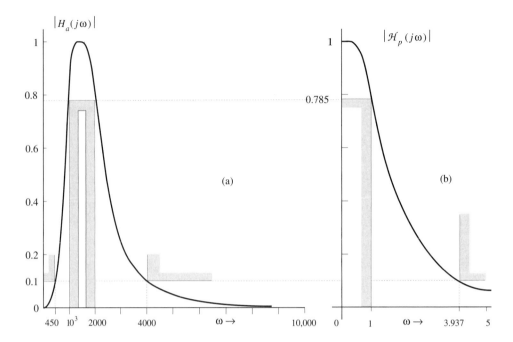

Fig. 12.16   Butterworth bandpass filter design using the bilinear transformation with prewarping.

Each equation gives a different answer, in general. However either answer will satisfy the specifications. Let us select Eq. (7.41), which yields

$$\omega_c{'} = \frac{\omega_s'}{(10^2 - 1)^{1/4}} = 1.248$$

Step 2.4: Find the normalized filter transfer function
The normalized second-order Butterworth filter transfer function (from Table 7.1) is

$$\mathcal{H}(s) = \frac{1}{s^2 + \sqrt{2}s + 1}$$

Step 2.5: Find the prototype filter transfer function $\mathcal{H}_p(s)$
The prototype filter transfer function $\mathcal{H}_p(s)$ is obtained by substituting s with $s/\omega_c{'} = s/1.248$ in the normalized transfer function $\mathcal{H}(s)$ found in step 4. This substitution yields

$$\mathcal{H}_p(s) = \frac{(1.248)^2}{s^2 + \sqrt{2}(1.248)s + (1.248)^2} = \frac{1.5575}{s^2 + 1.7649s + 1.5575}$$

Step 3: Find $H[z]$ by using the bilinear transformation
Finally, the desired transfer function $H[z]$ of the bandpass filter is obtained from $\mathcal{H}_p(s)$ by replacing s with $T_{bp}[z]$ from Eqs. (12.67) and (12.68). From Eq. (12.68), we obtain

$$a = \frac{\omega_{p_1}{'}\omega_{p_2}{'} - 1}{\omega_{p_1}{'}\omega_{p_2}{'} + 1} = \frac{-0.9485}{1.0515} = -0.9021 \quad \text{and} \quad b = \frac{\omega_{p_2}{'} - \omega_{p_1}{'}}{\omega_{p_1}{'}\omega_{p_2}{'} + 1} = \frac{0.1665}{1.0515} = 0.1583$$

Substitution of these values in Eq. (12.67) yields

$$T_{bp}[z] = \frac{z^2 + 2az + 1}{b(z^2 - 1)} = \frac{6.317(z^2 - 1.8042z + 1)}{z^2 - 1}$$

The desired bandpass filter transfer function $H[z]$ is obtained from $\mathcal{H}_p(s)$ by substituting s with $T_{bp}[z]$:

$$H[z] = \frac{0.02964(z^2 - 1)^2}{z^4 - 3.119z^3 + 3.926z^2 - 2.354z + 0.576}$$

The amplitude response $|H[e^{j\omega T}]|$ of this filter is illustrated in Fig. 12.16a. ■

⊙ **Computer Example C12.7**
Design a bandpass digital filter to meet the specifications in Example 12.8, using bilinear transformation with prewarping.

As before, we shall give here MATLAB functions to design the four basic types of approximations. In this case, Wp and Ws are 2-element vectors: Wp=[1000 2000] and Ws=[450 4000]. The input data asks for frequencies so normalized that the sampling radian frequency is 2. As explained in Example C12.5, all the radian frequencies can be normalized by multiplying each of them by T/π. In the present case, $T = \pi/10^4$ so that $T/\pi = 1/10^4$. Thus, to normalize any radian frequency, we just divide it by 10^4. The normalized Wp and Ws are [0.1 0.2] and [0.045 0.4], respectively.

Wp=[0.1 0.2];Ws=[0.045 0.4];Gp=-2.1;Gs=-20;

% Butterworth
[n,Wn]=buttord(Wp,Ws,-Gp,-Gs);
[b,a]=butter(n,Wn)

% Chebyshev
[n,Wn]=cheb1ord(Wp,Ws,-Gp,-Gs)
[b,a]=cheby1(n,-Gp,Wn)

% Inverse Chebyshev
[n,Wn]=cheb2ord(Wp,Ws,-Gp,-Gs)
[b,a]=cheby2(n,-Gs,Wn)

% Elliptic
[n,Wn]=ellipord(Wp,Ws,-Gp,-Gs)
[b,a]=ellip(n,-Gp,-Gs,Wn)

MATLAB gives b=0.0296 0 -0.0593 0 0.0296 and a=1 -3.119 3.9259 -2.3539 0.576 for Butterworth option. Therefore

$$H[z] = \frac{0.0296(z^2 - 1)^2}{z^4 - 3.119z^3 + 3.9259z^2 - 2.3539z + 0.5760}$$

a result, which agrees with the answer found in Example 12.8. To plot the amplitude and the phase response, we can use the last 9 functions in Example C12.5.   ⊙

⊙ **Computer Example C12.8**
Using bilinear transformation with prewarping, design a bandstop digital filter to meet the following specifications: Wp=[450 4000], Ws=[1000 2000], $\hat{G}_p = -2.1$ dB , and $\hat{G}_s = -20$ dB. Use $T = \pi/10^4$.

As before, we give here MATLAB functions to design the four types of approximations. The input data asks for frequencies so normalized that the sampling radian frequency is 2. As explained in Example C12.5, all the radian frequencies can be normalized
by multiplying each of them by T/π. In the present case, $T = \pi/10^4$ so that $T/\pi = 1/10^4$.
Thus, to normalize any radian frequency, we just divide it by 10^4.

Ws=[0.1 0.2];Wp=[0.045 0.4];Gp=-2.1;Gs=-20;

```
% Butterworth
[n,Wn]=buttord(Wp,Ws,-Gp,-Gs)
[b,a]=butter(n,Wn,'stop')

% Chebyshev
[n,Wn]=cheb1ord(Wp,Ws,-Gp,-Gs)
[b,a]=cheby1(n,-Gp,Wn,'stop')

% Inverse Chebyshev
[n,Wn]=cheb2ord(Wp,Ws,-Gp,-Gs);
[b,a]=cheby2(n,-Gs,Wn,'stop')

% Elliptic
[n,Wn]=ellipord(Wp,Ws,-Gp,-Gs);
[b,a]=ellip(n,-Gp,-Gs,Wn,'stop')
```

MATLAB returns b=0.3762 -1.3575 1.9711 -1.3575 0.3762 and a=1 -2.2523 2.0563
-1.2053 0.4197 for Chebyshev option. Therefore

$$H[z] = \frac{0.3762(z^4 - 3.6084z^3 + 5.2395z^2 - 3.6084z + 1)}{z^4 - 2.2523z^3 + 2.0563z^2 - 1.2053z + 0.4197}$$

To plot the amplitude and the phase response, we can use the last 9 functions in Example
C12.5. $\odot$

12.7 Nonrecursive Filters

The recursive filters are very sensitive to coefficient accuracy. Inaccuracies in
their implementation, especially too short a word length, may change their behavior
drastically and even make them unstable. Moreover, the recursive filter designs are
well established only for amplitude responses that are piecewise constant, such as
lowpass, bandpass, highpass, and bandstop filters. In contrast, a nonrecursive filter
can be designed to have an arbitrarily shaped frequency response. In addition,
nonrecursive filters can be designed to have a linear phase response. On the other
hand, if a recursive filter can be found to do the job of a nonrecursive filter, the
recursive filter is of lower order; that is, it is faster (with less processing delay) and
requires less memory. If processing delay is not critical, the nonrecursive filter is the
obvious choice. They also have an important place in non-audio applications, where
a linear phase response is important. We shall review the concept of nonrecursive
systems briefly.

As discussed in Sec. 12.3, nonrecursive filters may be viewed as recursive filters,
where all the feedback or recursive coefficients are zero; that is , when

$$a_0 = a_1 = a_2 = \cdots = a_{n-1} = 0$$

Consequently, the transfer function of the resulting nth-order nonrecursive filter is

$$H[z] = \frac{b_n z^n + b_{n-1} z^{n-1} + \cdots + b_1 z + b_0}{z^n} \tag{12.69a}$$

$$= b_n + \frac{b_{n-1}}{z} + \frac{b_{n-2}}{z^2} + \cdots + \frac{b_1}{z^{n-1}} + \frac{b_0}{z^n} \tag{12.69b}$$

Now, by definition, $H[z]$ is the z-transform of $h[k]$:

$$H[z] = \sum_{k=0}^{\infty} h[k] z^{-k}$$

$$= h[0] + \frac{h[1]}{z} + \frac{h[2]}{z^2} + \cdots + \frac{h[n]}{z^n} + \frac{h[n+1]}{z^{n+1}} + \cdots \tag{12.70}$$

Comparison of this equation with Eq. (12.69b) shows that $h[k] = 0$ for $k > n$, and Eq. (12.70) becomes

$$H[z] = h[0] + \frac{h[1]}{z} + \frac{h[2]}{z^2} + \cdots + \frac{h[n]}{z^n} \tag{12.71a}$$

$$= \frac{h[0]z^n + h[1]z^{n-1} + \cdots + h[n-1]z + h[n]}{z^n} \tag{12.71b}$$

where

$$h[k] = \begin{cases} b_{n-k} & 0 \le k \le n \\ 0 & k > n \end{cases} \tag{12.72}$$

The impulse response $h[k]$ has a finite width of $(n+1)$ elements. Hence, these filters are finite impulse response (FIR) filters. We shall use the terms nonrecursive and FIR interchangeably. Similarly, the terms recursive and IIR (infinite impulse response) will be used interchangeably in our future discussion.

The impulse response $h[k]$ can be expressed as

$$h[k] = h[0]\delta[k] + h[1]\delta[k-1] + \cdots + h[n]\delta[k-n] \tag{12.73}$$

The frequency response of this filter is obtained from Eq. (12.71a) as

$$H[e^{j\omega T}] = h[0] + h[1]e^{-j\omega T} + \cdots + h[n]e^{-jn\omega T} \tag{12.74a}$$

$$= \sum_{k=0}^{n} h[k]e^{-jk\omega T} \tag{12.74b}$$

Filter Realization

The nonrecursive (FIR) filter in Eq. (12.69a) is a special case of a general filter with all feedback (or recursive) coefficients zero. Therefore, the realization of this filter is the same as that of the nth-order recursive filter with all the feedback connections omitted. Figure 12.7b shows a canonical realization of this filter. It is easy to verify from this figure that for the input $\delta[k]$, the output is $h[k]$ given in Eq. (12.73).

The filter in Fig. 12.7b is a tapped delay line with successive taps at unit delay (T seconds). Such a filter is known as a **transversal** filter. Tapped analog delays are integrated circuits, which are available commercially. In these circuits the time delay is implemented by using charge transfer devices, which sample the input signal every T seconds (unit delay) and transfer the successive values of the samples to m storage cells. The stored signal at the kth tap is the input signal delayed by k time units (kT seconds). The sampling interval can be varied electronically over a wide range. Time delay can also be obtained by using shift registers.

12.7-1 Symmetry Conditions for Linear Phase Response

Consider an nth-order finite impulse response (FIR) filter described by the transfer function $H[z]$ [Eq. (12.69) or (12.71)] and the corresponding impulse response $h[k]$ [Eq. (12.73)]. We now show that if $h[k]$ is either symmetric (Fig. 12.17a) or antisymmetric (Fig. 12.17b) about its center point, the filter phase response is a linear function of ω. We consider a case where n is even. To avoid too much abstractness, we choose some convenient value for n, say $n = 4$, to demonstrate our point. It will then be easier to understand the generalization to the nth-order case.

For $n = 4$, the impulse response in Eq. (12.73) reduces to

$$h[k] = h[0]\delta[k] + h[1]\delta[k-1] + h[2]\delta[k-2] + h[3]\delta[k-3] + h[4]\delta[k-4]$$

The transfer function $H[z]$ in Eq. (12.71b) reduces to

$$H[z] = h[0] + \frac{h[1]}{z} + \frac{h[2]}{z^2} + \frac{h[3]}{z^3} + \frac{h[4]}{z^4} \tag{12.75a}$$

$$= z^{-2}\left(h[0]z^2 + h[1]z + h[2] + h[3]z^{-1} + h[4]z^{-2} \right) \tag{12.75b}$$

Therefore, the frequency response is

$$H[e^{j\omega T}] = e^{-j2\omega T}\left(h[0]e^{j2\omega T} + h[1]e^{j\omega T} + h[2] + h[3]e^{-j\omega T} + h[4]e^{-j2\omega T} \right) \tag{12.76}$$

If $h[k]$ is symmetric about its center point ($k = 2$ in this case), then

$$h[0] = h[4], \qquad h[1] = h[3]$$

and the frequency response reduces to

$$H[e^{j\omega T}] = e^{-j2\omega T}\left(h[0]\left[e^{j2\omega T} + e^{-j2\omega T}\right] + h[2] + h[1]\left[e^{j\omega T} + e^{-j\omega T}\right] \right)$$

$$= e^{-j2\omega T}\left(h[2] + 2h[1]\cos\omega T + 2h[0]\cos 2\omega T \right) \tag{12.77}$$

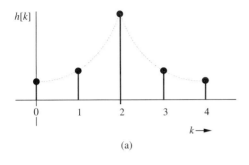

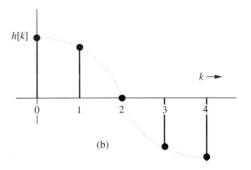

Fig. 12.17 Symmetry conditions for linear phase frequency response in nonrecursive filters.

The quantity inside the parenthesis is real; it may be positive over some bands of frequencies and negative over other bands. This quantity represents the amplitude response $|H[e^{j\omega T}]|$.† The phase response is given by

$$\angle H[e^{j\omega T}] = -2\omega T$$

The phase response is a linear function of ω. The time delay is the negative of the slope of $\angle H[e^{j\omega T}]$ with respect to ω, which is $2T$ seconds in this case [see Eq. (4.59)].

If $h[k]$ is antisymmetric about its center point, then the antisymmetry about the center point requires that $h[k] = 0$ at the center point‡ [see Fig. 12.17b]. Thus, in this case

$$h[0] = -h[4], \qquad h[1] = -h[3], \qquad h[2] = 0$$

and the frequency response reduces to

$$H[e^{j\omega T}] = e^{-j2\omega T}\left(h[0](e^{j2\omega T} - e^{-j2\omega T}) + h[1](e^{j\omega T} - e^{-j\omega T})\right)$$

$$= 2je^{-j2\omega T}\left(h[1]\sin\omega T + h[0]\sin 2\omega T\right)$$

$$= 2e^{j(\frac{\pi}{2} - 2\omega T)}\left(h[1]\sin\omega T + h[0]\sin 2\omega T\right)$$

Thus, the phase response in this case is

$$\angle H[e^{j\omega T}] = \frac{\pi}{2} - 2\omega T$$

---

†Strictly speaking, $|H[e^{j\omega T}]|$ cannot be negative. Recall, however, that the only restriction on amplitude is that it cannot be complex. It has to be real; it can be positive or negative. We should have used some other notation such as $A(\omega)$ to denote the amplitude response. But this would create too many related functions causing possible confusion. Another alternative is to incorporate the negative sign of the amplitude in the phase response, which will be increased (or decreased) by π over the band where the amplitude response is negative. This alternative will still maintain the phase linearity.
‡Antisymmetry property requires that $h[k] = -h[-k]$ at the center point also. This condition is possible only if $h[k] = 0$ at this point.

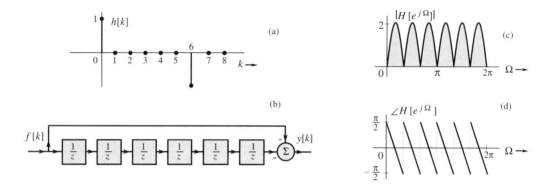

Fig. 12.18   Comb filter: Its impulse and frequency response.

The phase response in this case is also a linear function of ω. The system has the time delay (the negative slope of $\angle H[e^{j\omega T}]$ with respect to ω) of $2T$ seconds (2 units), the same as in the symmetric case. The only difference is that the phase response has a constant term $\pi/2$.

We can obtain similar results for odd values of n (see Prob. 12.7-1). This result can be generalized for an nth-order case to show that the phase response is linear, and the time delay is $\frac{nT}{2}$ seconds (or $\frac{n}{2}$ units).

■ **Example 12.9: Comb Filter**

Determine the transfer function and the frequency response of a sixth-order comb filter whose impulse response is given by

$$h[k] = \delta[k] - \delta[k-6]$$

This impulse response is illustrated in Fig. 12.18a. Its canonical realization is depicted in Fig. 12.18b. Observe that $h[k]$ is antisymmetric about $k = 3$. Also

$$H[z] = \sum_{-\infty}^{\infty} f[k] z^{-k} = 1 - z^{-6} = \frac{z^6 - 1}{z^6} \tag{12.78}$$

The frequency response is given by

$$H[e^{j\Omega}] = 1 - e^{-j6\Omega}$$
$$= e^{-j3\Omega}(e^{j3\Omega} - e^{-j3\Omega})$$
$$= 2j e^{-j3\Omega} \sin 3\Omega$$
$$= 2 e^{j\left(\frac{\pi}{2} - 3\Omega\right)} \sin 3\Omega$$

The term $\sin 3\Omega$ can be positive as well as negative. Therefore

$$|H[e^{j\Omega}]| = 2|\sin 3\Omega|$$

and $\angle H[e^{j\Omega}]$ is as indicated in Fig. 12.8d. The amplitude response, illustrated in Fig. 12.18c, is shaped like a comb with periodic nulls. The filter can be realized by the structure in Fig. 12.7b. Since $h[k] = b_{6-k}$ [see Eq. (12.72)], $b_0 = -1$, $b_1 = b_2 = b_3 = b_4 = b_5 = 0$, and $b_6 = 1$.

Using the same argument, the reader can show that an nth-order comb filter transfer function is

$$H[z] = \frac{z^n - 1}{z^n} \tag{12.79}$$

and

$$H[e^{j\Omega}] = 2e^{j\left(\frac{\pi}{2} - \frac{n\Omega}{2}\right)} \sin \frac{n\Omega}{2} \quad \blacksquare$$

12.8 Nonrecursive Filter Design

As in the case of recursive (IIR) filters, nonrecursive filters can be designed by using the time-domain and the frequency-domain equivalence criteria. In the time-domain equivalence criterion, the digital filter impulse response is made identical to the samples of the desired (analog) filter impulse response. In the frequency-domain equivalence criterion, the digital filter frequency response samples at uniform frequency intervals are matched to the desired analog filter frequency response samples. This method is also known as the **frequency sampling** or the **spectral sampling** method.

12.8-1 Time-Domain Equivalence Method of FIR Filter Design

The time-domain equivalence method (also known as the **Fourier series** method) of design of FIR filters is identical to that for IIR filters discussed in Sec. 12.5, except that FIR filter impulse response must be of finite duration. Therefore, the desired impulse response must be truncated to have finite duration. Truncating the impulse response abruptly will result in oscillatory frequency response because of the Gibbs phenomenon discussed in Sec. 3.4-3. In some filtering applications the oscillatory frequency response (which decays slowly as $1/\omega$) in the stopband may not be acceptable. By using a tapered window function for truncation of $h[k]$, the oscillatory behavior can be reduced or even eliminated at the cost of increasing the transition band as discussed in Sec. 4.9. Note that the impulse response of an nth-order FIR filter has $n + 1$ samples. Hence, for truncating $h[k]$ for an n-th order filter, we must use an N_0-point window, where $N_0 = n + 1$. Several window functions and their tradeoffs appear in Table 12.2.

Design Procedure

Much of the discussion so far has been rather general. We shall now give some concrete examples of such filter design. Because we want the reader to be focussed on the procedure, we shall intentionally choose a small value for n (the filter order) to avoid getting distracted by a jungle of data. The procedure, however, is general and it can be applied to any value of n.

The steps in the time-domain equivalence design method are:

1. Determine the filter impulse response $h[k]$
In the first step, we find the impulse response $h[k]$ of the desired filter. According to the time-domain equivalence criterion in Eq. (12.31),

$$h[k] = Th_a(kT) \tag{12.80}$$

TABLE 12.2

Some Window Function and Their Characteristics

Window $w[k]$	Mainlobe Width	Rolloff Rate dB/octave	Peak Sidelobe Level in dB
1 Rectangular: $-M \leq k \leq M$ $M = \frac{N_0-1}{2}$ $\mathrm{rect}\left(\frac{k}{N_0-1}\right)$	$\frac{4\pi}{N_0}$	-6	-13.3
2 Bartlett: $\Delta\left(\frac{k}{N_0-1}\right)$	$\frac{8\pi}{N_0}$	-12	-26.5
3 Hanning: $0.5\left[1 + \cos\left(\frac{2\pi k}{N_0-1}\right)\right]$	$\frac{8\pi}{N_0}$	-18	-31.5
4 Hamming: $0.54 + 0.46\cos\left(\frac{2\pi k}{N_0-1}\right)$	$\frac{8\pi}{N_0}$	-6	-42.7
5 Blackman: $0.42 + 0.5\cos\left(\frac{2\pi k}{N_0-1}\right) + 0.08\left(\frac{4\pi k}{N_0-1}\right)$	$\frac{12\pi}{N_0}$	-18	-58.1
6 Kaiser: $\dfrac{I_0\left[\alpha\sqrt{1-4\left(\frac{k}{N_0-1}\right)^2}\right]}{I_0(\alpha)}$ $1 \leq \alpha \leq 10$	$\frac{11.2\pi}{N_0}$	-6	-59.9 $(\alpha = 8.168)$

where $h_a(t)$ is the impulse response of the analog filter $H_a(s)$. Therefore, $h_a(t)$ is the inverse Laplace transform of $H_a(s)$ or the inverse Fourier transform of $H_a(j\omega)$. Thus,

$$h_a(t) = \frac{1}{2\pi}\int_{-\pi/T}^{\pi/T} H_a(j\omega)e^{j\omega t}\,d\omega \qquad (12.81a)$$

Recall that a digital filter frequency response is periodic with the first period in the frequency range $-\frac{\pi}{T} \leq \omega < \frac{\pi}{T}$. Hence, the best we could hope is to realize the equivalence of $H_a(j\omega)$ within this range. For this reason, the limits of integration are taken from $-\pi/T$ to π/T. Therefore, according to Eq. (12.80)

$$h[k] = Th_a(kT) = \frac{T}{2\pi}\int_{-\pi/T}^{\pi/T} H_a(j\omega)e^{j\omega kT}\,d\omega \qquad (12.81b)$$

2. Windowing

For linear phase filters, we generally start with zero phase filters for which $H_a(j\omega)$ is either real or imaginary. The impulse response $h_a(t)$ is either an even or odd function of t (see Prob. 4.1-1). In either case, $h_a(t)$ is centered at $t = 0$ and has infinite duration in general. But $h[k]$ must have only a finite duration and it must start at $k = 0$ (causal) for filter realizability. Consequently, the $h[k]$ found in step 1 needs to be truncated using an N_0-point window and then delayed by $\frac{N_0-1}{2}$ to make it causal. This delay produces the desired linear-phase frequency response.

Straight truncation of data amounts to using a rectangular window, which has a unit weight over the window width, and zero weight for all other k. We saw that although such a window gives the smallest transition band, it results in a slowly decaying oscillatory frequency response in the stopband. This behavior can be corrected by using a tapered window of a suitable width.

*

3. Filter Frequency Response and Realization

Knowing $h[0]$, $h[1]$, $h[2]$, $\cdots$, $h[n]$, we determine $H[z]$ using Eq. (12.71) and the frequency response $H[e^{j\omega T}]$ using Eq. (12.74). We now realize the truncated $h[k]$ using the structure in Fig. 12.7b.

Optimality of the Procedure

The procedure outlined here using a rectangular window function is the optimum in the sense that the energy of the error (difference) between the desired frequency response $H_a(j\omega)$ and the realized frequency response $H[e^{j\omega T}]$ is the minimum for a given N_0. This conclusion follows from the fact that the resulting filter frequency response $H[e^{j\omega T}]$ is given by

$$H[e^{j\omega T}] = \sum_k h[k]e^{-j\omega kT}$$

This frequency response is an approximation of the desired frequency response $H_a(j\omega)$ because of the truncation of $h[k]$. Thus,

$$H_a(j\omega) \approx \sum_k h[k]e^{-j\omega kT}$$

How do we select $h[k]$ for the best approximation in the sense of minimizing the energy of the error $H_a(j\omega) - H[e^{j\omega T}]$? We have already solved this problem in Sec. 3.3-2. The above equation shows that the right-hand side is the finite term exponential Fourier series for $H_a(j\omega)$ with period $2\pi/T$. As seen from Eq. (12.81b), $h[k]$ are the Fourier coefficients. We also know that a finite Fourier series is the optimum (in the sense of minimizing the error energy) for a given N_0 according to the finality property of the Fourier coefficients discussed in Sec. 3.3-2.† Clearly, this choice of $h[k]$ is optimum in the sense of the minimum mean squared error. For the obvious reason, this method is also known as the Fourier series method.

■ Example 12.10

Design an ideal lowpass filter for audio band with cutoff frequency 20 kHz. Use a sixth-order nonrecursive filter using rectangular and Hamming windows. The highest frequency to be processed is $\mathcal{F}_h = 40$ kHz.

In this case $n = 6$ and $N_0 = n + 1 = 7$. First we shall choose a suitable value for T. According to Eq. (12.58)

$$T \leq \frac{\pi}{\omega_h} = \frac{1}{2\mathcal{F}_h} = 12.5 \times 10^{-6}$$

---

†Note that this finite term Fourier series corresponds to the rectangular window function. For windows other than rectangular, the optimality does not hold.

Recall that a continuous-time sinusoid of frequency ω, during digital processing appears as a discrete-time sinusoid of frequency $\Omega = \omega T$. The cutoff frequency $\omega_c = 2\pi(20,000) = 40,000\pi$ appears as a discrete-time sinusoid of frequency

$$\Omega_c = \omega_c T = 40,000\pi(12.5 \times 10^{-6}) = \frac{\pi}{2}, \text{ and } \omega_c = \frac{\pi}{2T}$$

The desired (zero phase) filter frequency response is depicted in Fig. 12.19a on both ω and Ω scales. Recall that the digital frequency range is from $-\pi$ to π only. We wish to design an ideal lowpass filter of cutoff frequency $\omega_c = \frac{\pi}{2T}$ rad/s. The frequency response has a period of 2π on Ω scale, and $2\pi/T$ on ω scale. Rather than substitute $T = 12.5 \times 10^{-6}$, it is convenient to leave T as an unknown in our computations and substitute the value only in the end. Thus, we shall use the radian cutoff frequency $\omega_c = \pi/2T$.

The impulse response of the desired ideal lowpass (zero phase) filter in Fig. 12.19a is (Table 4.1, Pair 18)

$$h_a(t) = \frac{1}{2T} \operatorname{sinc}\left(\frac{\pi t}{2T}\right) \tag{12.82}$$

and according to the impulse invariance criterion [Eq. (12.39)]

$$h[k] = Th_a(kT) = \frac{1}{2} \operatorname{sinc}\left(\frac{\pi k T}{2T}\right) = \frac{1}{2} \operatorname{sinc}\left(\frac{\pi k}{2}\right) \tag{12.83}$$

Figure 12.19b shows $h[k]$. To make this filter realizable, we need to truncate it using a suitable N_0-point window, then delay the truncated $h[k]$ by $\frac{N_0-1}{2}$ units. In the present example, $N_0 = 7$. Figure 12.19c shows the impulse response truncated by a 7-point rectangular window and Fig. 12.19d shows the truncated $h[k]$ delayed by $\frac{N_0-1}{2} = 3$ units.

Note that the noncausal filter in this case is made realizable at the cost of a delay of $t = 3T$ seconds. This constant delay of $\frac{nT}{2}$ is what produces a linear phase characteristic. The rectangular windowed, causal filter impulse response $h_R[k]$ is the truncated $h[k]$ in Fig. 12.19d delayed by $3T$.

$$h_R[k] = h[k-3] = \frac{1}{2} \operatorname{sinc}\left[\frac{\pi(k-3)}{2}\right] \qquad 0 \le k \le 6 \tag{12.84}$$

The values of the coefficient $h_R[k]$ are shown in Table 12.3. Also

$$H[z] = \sum_{k=0}^{6} h[k] z^{-k}$$

$$= -\frac{1}{3\pi} + \frac{1}{\pi} z^{-2} + \frac{1}{2} z^{-3} + \frac{1}{\pi} z^{-4} - \frac{1}{3\pi} z^{-6}$$

$$= z^{-3}\left(-\frac{1}{3\pi} z^3 + \frac{1}{\pi} z + \frac{1}{2} + + \frac{1}{\pi} z^{-1} - \frac{1}{3\pi} z^{-3}\right) \tag{12.85}$$

Hence, the frequency response $H_R[e^{j\omega T}]$ is

$$H_R[e^{j\omega T}] = \sum_{k=0}^{6} h_R[k] e^{-j\omega k T}$$

$$= e^{-j3\omega T}\left[\frac{1}{2} + \frac{1}{\pi}\left(e^{j\omega T} + e^{-j\omega T}\right) - \frac{1}{3\pi}\left(e^{j3\omega T} + e^{-j3\omega T}\right)\right]$$

$$= e^{-j3\omega T}\left[\frac{1}{2} + \frac{2}{\pi}\cos\omega T - \frac{2}{3\pi}\cos 3\omega T\right]$$

$$= e^{-j3\omega T}\left[\frac{1}{2} + \frac{2}{\pi}\cos\left(\frac{\omega}{80,000}\right) - \frac{2}{3\pi}\cos\left(\frac{3\omega}{80,000}\right)\right] \tag{12.86}$$

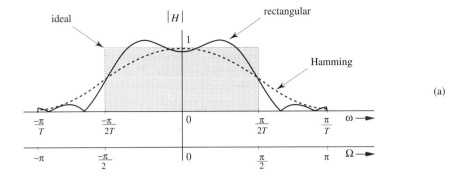

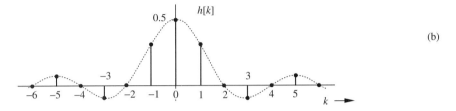

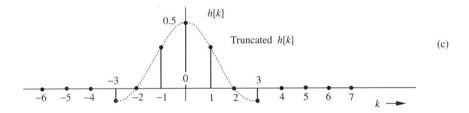

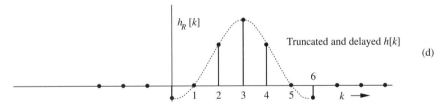

Fig. 12.19 Nonrecursive method of lowpass filter design.

The term $e^{-j3\omega T}$ is a linear phase representing the delay of $3T$ seconds. The magnitude of the bracketed term, depicted in Fig. 12.19a by a solid curve, exhibits oscillatory behavior which decays rather slowly over the stopband. Although increasing n (the system order) improves the frequency response, its oscillatory nature persists (Gibbs phenomenon).† In some filtering applications, the oscillatory characteristic (which decays slowly as $1/\omega$) in the stopband may not be acceptable. By using a tapered window func-

---

†Eq. (12.86) is identical to the first three terms in Eq. (3.61) except that the former is a function of ω and the latter is a function of t. Clearly, $H[e^{j\omega T}]$ is a truncated Fourier series for the gate function. As we increase n, $H[e^{j\omega T}]$ converges to the gate function in the manner depicted in Fig. 3.11. Regardless of the value of n, however, $H[e^{j\omega T}]$ exhibits oscillatory behavior because of the Gibbs phenomenon.

TABLE 12.3

	Rectangular Window		Hamming Window	
k	$h_R[k]$	$w_H[k]$	$h_H[k]$	
0	$-1/3\pi$	0.08	-0.00848	
1	0	0.31	0	
2	$1/\pi$	0.77	0.245	
3	$1/2$	1	0.5	
4	$1/\pi$	0.77	0.245	
5	0	0.31	0	
6	$-1/3\pi$	0.08	-0.00848	

tion such as a Hamming window, the oscillatory behavior can be eliminated at the cost of increasing the transition band (from passband to stopband). The Hamming window function is given by

$$w_H[k] = \begin{cases} 0.54 + 0.46\cos\left(\frac{2\pi k}{N_0 - 1}\right) & -\frac{N_0 - 1}{2} \le |k| \le \frac{N_0 - 1}{2} \\ 0 & \text{otherwise} \end{cases} \tag{12.87}$$

In our case $N_0 = n + 1 = 7$. Hence,

$$w_H[k] = \begin{cases} 0.54 + 0.46\cos\left(\frac{\pi k}{3}\right) & -3 \le |k| \le 3 \\ 0 & \text{otherwise} \end{cases}$$

Table 12.3 also shows the (delayed) Hamming window coefficients $w_H[k]$ and the corresponding impulse response $h_H[k] = h[k]w_H[k]$. The frequency response of the Hamming window filter is

$$H_H[e^{j\omega T}] = e^{-j3\omega T}\left[\tfrac{1}{2} + 0.245\left(e^{j\omega T} + e^{-j\omega T}\right) - 0.00848\left(e^{j3\omega T} + e^{-j3\omega T}\right)\right]$$

$$= e^{-j3\omega T}\left[\tfrac{1}{2} + 0.49\cos\omega T - 0.01696\cos 3\omega T\right]$$

With the coefficients $h_R[k]$ (or $h_H[k]$) in Table 12.3, the desired filter can be realized by using six delay elements, as depicted in Fig. 12.7.

According to Eqs. (12.86), we have

$$\left|H_R[e^{j\omega T}]\right| = \left|\tfrac{1}{2} + \tfrac{2}{\pi}\cos\omega T - \tfrac{2}{3\pi}\cos 3\omega T\right| \tag{12.88a}$$

and

$$\angle H_R[e^{j\omega T}] = \begin{cases} -3\omega T & \text{when} \quad \tfrac{1}{2} + \tfrac{2}{\pi}\cos\omega T - \tfrac{2}{3\pi}\cos 3\omega T \ge 0 \\ \pi - 3\omega T & \text{when} \quad \tfrac{1}{2} + \tfrac{2}{\pi}\cos\omega T - \tfrac{2}{3\pi}\cos 3\omega T < 0 \end{cases} \tag{12.88b}$$

For the Hamming filter

$$|H_H[e^{j\omega T}]| = \tfrac{1}{2} + 0.49\cos\omega T - 0.01696\cos 3\omega T \qquad \angle H_H[e^{j\omega T}] = -3\omega T \qquad (12.88c)$$

In either filter, the phase response is a linear function of ω with slope $-3T$, indicating time delay of $3T$. Note that both $h_R[k]$ and $h_h[k]$ are symmetric about $k = 3$. ■

⊙ **Computer Example C12.9**

Using MATLAB, find the frequency response of the lowpass filter in Example 12.10 for 98th-order filter. Plot the frequency response for rectangular and Hamming window filters.

```
N0=99;
m=(N0-1)/2;
k=0:N0-1;
h1=(1/2)*sinc((k-m)/2);
num1=h1;
den1=[1, zeros(1,N0-1)];
W=-pi:pi/100:pi;
H1=freqz(num1,den1,W);
mag1=abs(H1);
phase1=180/pi*unwrap(angle(H1));
for i=1:N0
k=i-1;
h2(i)=(1/2)*sinc((k-m)/2)*(0.54+0.46*cos(pi*(k-m)/m));
end
num2=h2;
den2=[1, zeros(1,N0-1)];
W=-pi:pi/100:pi;
H2=freqz(num2,den2,W);
mag2=abs(H2);
phase2=180/pi*unwrap(angle(H2));
subplot(2,1,1);
plot(W,mag1,W,mag2);grid;
subplot(2,1,2);
plot(W,phase1,W,phase2);grid    ⊙
```

△ **Exercise E12.6**

If we were to use $n = 8$ filter in Example 12.12, show that the filter transfer function for the bartlett (triangular) window is

$$H[z] = \frac{-0.02653z^6 + 0.2387z^4 + 0.5z^3 + 0.2387z^2 + 0.02653}{z^6}$$

Observe that in this case the filter order is reduced by 2 because the two end-points have a zero value for the Bartlett window. ▽

■ **Example 12.11**

Design a differentiator using a tenth-order nonrecursive filter.

In this case $n = 10$ and $N_0 = 11$. The transfer function of an ideal differentiator is $H_a(s) = s$. Therefore,

$$H_a(j\omega) = j\omega \qquad\qquad |\omega| \le \frac{\pi}{T}$$

and according to Eq. (12.81b)†

$$h[k] = \frac{T}{2\pi} \int_{-\pi/T}^{\pi/T} j\omega e^{j\omega kT} \, d\omega$$

$$= \frac{\cos k\pi}{kT} \tag{12.89}$$

This is a noncausal filter. We truncate it using an 11-point window and then delay it by $\frac{N_0 - 1}{2} = 5$ units to make it realizable. The desired impulse response using a rectangular window is

$$h_R[k] = h[k-5] \qquad 0 \le k \le 10$$

Table 12.4 shows this impulse response under the rectangular window. The frequency response is

$$H_R[e^{j\omega T}] = \sum_{k=0}^{10} h_R[k] e^{-jkT\omega}$$

Substitution of values of $h_R[k]$ from Table 12.4 in this equation yields

$$H_R[e^{j\omega T}] = \frac{1}{T} e^{-j5\omega T} \left[(e^{j\omega T} - e^{-j\omega T}) - \frac{1}{2}(e^{j2\omega T} - e^{-j2\omega T}) + \frac{1}{3}(e^{j3\omega T} - e^{-j3\omega T}) \right.$$

$$\left. - \frac{1}{4}(e^{j4\omega T} - e^{-j4\omega T}) + \frac{1}{5}(e^{j5\omega T} - e^{-j5\omega T}) \right]$$

$$= \frac{j}{T} e^{-j5\omega T} \left[2\sin \omega T - \sin 2\omega T + \frac{2}{3}\sin 3\omega T - \frac{1}{2}\sin 4\omega T + \frac{2}{5}\sin 5\omega T \right]$$

$$= \frac{1}{T} e^{j(\frac{\pi}{2} - 5\omega T)} \left[2\sin \omega T - \sin 2\omega T + \frac{2}{3}\sin 3\omega T - \frac{1}{2}\sin 4\omega T + \frac{2}{5}\sin 5\omega T \right]$$

Hence

$$\left| H_R[e^{j\omega T}] \right| = \frac{1}{T} \left| 2\sin \omega T - \sin 2\omega T + \frac{2}{3}\sin 3\omega T - \frac{1}{2}\sin 4\omega T + \frac{2}{5}\sin 5\omega T \right| \tag{12.90a}$$

and

$$\angle H_R[e^{j\omega T}) = \begin{cases} \frac{\pi}{2} - 5\omega T & \omega > 0 \\ -\frac{\pi}{2} + 5\omega T & \omega < 0 \end{cases} \tag{12.90b}$$

The terms $\pm 5\omega T$ represent the time delay of $5T$ seconds. Note that because $H_a(j\omega) = j\omega$,

$$|H_a(j\omega)| = |\omega|$$

$$\angle H_a(j\omega) = \begin{cases} \pi/2 & \omega > 0 \\ -\pi/2 & \omega < 0 \end{cases} \tag{12.91}$$

---

†Note that according to Eq. (12.81a)

$$h_a(0) = \frac{1}{2\pi} \int_{-\pi/2T}^{\pi/2T} j\omega \, d\omega = 0$$

Therefore, $h[0] = 0$.

Therefore, the phase characteristic of $H_R[e^{j\omega T}]$ is identical to the desired (ideal) differentiator with a delay of $5T$ seconds. Figure 12.20 shows the amplitude response $|H|$ of the realized filter and the $|H_a|$ of the ideal (desired) differentiator. The amplitude response $|H_R[e^{j\omega T}]|$, which is oscillatory, can be improved by using a tapered window function to truncate $h_a(t)$ gradually. The Hamming window function $\omega_H[k]$ [Eq. (12.87)] for $n = 10$ is given by

$$w_H[k] = \begin{cases} 0.54 + 0.46\cos\left(\frac{\pi k}{5}\right) & |k| \le 5 \\ 0 & \text{otherwise} \end{cases}$$

Table 12.4 shows the (delayed) Hamming coefficients $w_H[k]$ and the corresponding filter impulse response $h_H[k] = h[k]w_H[k]$. Note that $h_R[k]$ and $h_H[k]$ are both antisymmetric about $k = 5$. Using the Hamming coefficients from Table 12.4, $H_H[e^{j\omega T}]$, the windowed filter frequency response, is given by

$$H_H[e^{j\omega T}] = \frac{1}{T}e^{j\left(\frac{\pi}{2}-5\omega T\right)}\left[1.8242\sin\omega T - 0.682\sin 2\omega T + 0.2652\sin 3\omega T\right.$$

$$\left.-0.0839\sin 4\omega T + 0.032\sin 5\omega T\right]$$

Figure 12.20 shows the amplitude response of the ideal, the rectangular-windowed, and the Hamming-windowed filter characteristics. The phase response of both the realized filters are identical, but with a delay of $5T$ seconds. The amplitude response of the windowed (Hamming) filter is practically the same over a large band (two thirds of the band) as that for the ideal filter, and hence is preferable to the unwindowed (or rectangular-windowed) filter.

TABLE 12.4

		Rectangular Window	Hamming Window	
k	$h_R[k]$	$w_H[k]$	$h_H[k]$	
0	$1/5T$	0.08	$0.016/T$	
1	$-1/4T$	0.1678	$-0.04195/T$	
2	$1/3T$	0.3978	$0.1326/T$	
3	$-1/2T$	0.6821	$-0.3410/T$	
4	$1/T$	0.9121	$0.9121/T$	
5	0	1	0	
6	$-1/T$	0.9121	$-0.9121/T$	
7	$1/2T$	0.6821	$0.3410/T$	
8	$-1/3T$	0.3978	$-0.1326/T$	
9	$1/4T$	0.1678	$0.04195/T$	
10	$-1/5T$	0.08	$-0.016/T$	

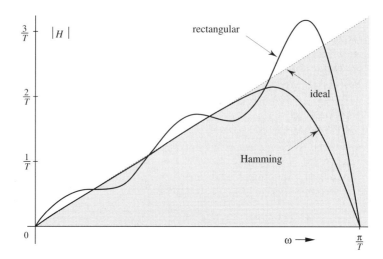

Fig. 12.20   An ideal differentiator design by a tenth-order nonrecursive filter using rectangular and Hamming windows.

The amplitude response of the Hamming filter is practically ideal up to frequency $\omega \approx (2/3)(\pi/T) = 2\pi/3T$. If we wish to design a digital differentiator for audio application, for instance, where the highest frequency is, say 20 kHz, we should select this frequency to be less than $2\pi/3T$. Thus,

$$2\pi \times 20,000 \leq \frac{2\pi}{3T} \implies T \leq 16.67 \ \mu\text{s}$$

Thus, a choice of $T \leq 16.67 \ \mu$s would result in a desired differentiator. This example (also Fig. 12.20) shows that a Hamming window shrinks the passband. This is generally true of tapered windows. To compensate for this shrinkage, we start with a passband somewhat larger (typically 25% larger) than the design passband. In the present case, for instance, selecting $T = 16.67 \ \mu$s would make the passband $\omega_h = \pi/T \approx 2\pi(30,000)$, which is 50% higher than the design passband of 20,000 Hz.   ■

⊙ **Computer Example C12.10**
Using MATLAB, find the frequency response of the 48th-order digital differentiator in Example 12.10. Plot the frequency response for rectangular and Hamming window differentiators.

```
N0=49;
m=(N0-1)/2;
k=0:N0-1;
T=1;
h1=cos((k-m)*pi)./((k-m)*T);h1(25)=0;
num1=h1;
den1=[1, zeros(1,N0-1)];
W=-pi:pi/100:pi;
H1=freqz(num1,den1,W);
mag1=abs(H1);
phase1=180/pi*unwrap(angle(H1));
h2=cos((k-m)*pi)./((k-m)*T).*(0.54+0.46*cos(pi*(k-m)./m));h2(25)=0;
```

```
num2=h2;
den2=den1;
H2=freqz(num2,den2,W);
mag2=abs(H2);
phase2=180/pi*unwrap(angle(H2));
subplot(2,1,1);
plot(W,mag1,W,mag2);
subplot(2,1,2);
plot(W,phase2,W,phase1);        ⊙
```

△ **Exercise E12.7**

Design a sixth-order nonrecursive filter to realize $H_a(j\omega) = \triangle(\omega T/\pi)$

Answer:

$$h[k] = \begin{cases} \frac{1}{4}\text{sinc}^2\left(\frac{\pi k}{4}\right) & |k| \le 3 \\ 0 & |k| > 3 \end{cases} \qquad \triangledown$$

12.8-2   Nonrecursive Filter Design by the Frequency-Domain Criterion: The Frequency Sampling Method

The frequency-domain criterion is [see Eq. (12.36)]

$$H_a(s) = H\left[e^{sT}\right]$$

In this case we shall realize this equality for real frequencies; that is, for $s = j\omega$:

$$H_a(j\omega) = H\left[e^{j\omega T}\right] \tag{12.92}$$

For an nth-order filter, there are only $N_0 = n + 1$ elements in $h[k]$, and we can hope to force the two frequency spectra in Eq. (12.92) to be equal only at N_0 points. Because the spectral width is $\frac{2\pi}{T}$, we choose these frequencies $\omega_0 = \frac{2\pi}{N_0 T}$ rad/s apart; that is ,

$$\omega_0 = \frac{2\pi}{N_0 T} \tag{12.93}$$

We require that

$$H_a(jr\omega_0) = H\left[e^{jr\omega_0 T}\right] \qquad r = 0, 1, 2, , \cdots, N_0 - 1 \tag{12.94}$$

Note that these are the samples of the periodic extension of $H_a(j\omega)$ (or $H[e^{j\omega T}]$). Because we force the frequency response of the filter to be equal to the desired frequency response at N_0 equidistant frequencies in the spectrum, this method is known as the **frequency sampling** or the **spectral sampling** method.

In order to find the filter transfer function, we first determine the filter impulse response $h[k]$. Thus, our problem is to determine the filter impulse response from the knowledge of the N_0 uniform samples of the periodic extension of the filter frequency response $H[e^{j\omega T}]$. But $H[e^{j\omega T}]$ is the DTFT of $h[k]$ [see Eq. (10.92)]. Hence, as shown in Sec. 10.6-2 [Eqs. (10.68) and (10.69)], $h[k]$ and $H[e^{jr\omega_0 T}]$ (the N_0 uniform samples of $H[e^{j\omega T}]$) are the DFT pair with $\Omega_0 = \omega_0 T$. Hence, the desired $h[k]$ is the IDFT of $H[e^{j\omega T}]$, given by

$$h[k] = \frac{1}{N_0} \sum_{r=0}^{N_0-1} H[e^{jr\omega_0 T}] e^{jrk\omega_0 T}$$

$$= \frac{1}{N_0} \sum_{r=0}^{N_0-1} H[e^{jr\omega_0 T}] e^{j\frac{2\pi rk}{N_0}} \qquad k = 0,\ 1,\ 2,\ ,\ \cdots, N_0-1 \quad (12.95)$$

Note that $H[e^{jr\omega_0 T}] = H_a(jr\omega_0)$ are known [Eq. (12.94)]. We can use IFFT to compute the N_0 values of $h[k]$. From these values of $h[k]$, we can determine the filter transfer function $H[z]$ as

$$H[z] = \sum_{k=0}^{N_0-1} h[k] z^{-k} \tag{12.96}$$

Linear Phase (Constant Delay) Filters

We desire that $H_a(j\omega) = H[e^{j\omega T}]$. The filter featured in Eqs. (12.95) and (12.96) satisfies this condition only at the N_0 values of ω. Between samples, the frequency response, especially the phase response, could deviate considerably. If we want a linear phase characteristic, the procedure is slightly modified. First, we start with a zero phase (or a constant phase $\pm\frac{\pi}{2}$) response. For such a frequency response $h[k]$ is an even (or odd) function of k (see Prob. 4.1-1). In either case, $h[k]$ is centered at $k = 0$. To realize this noncausal filter, we need to delay $h[k]$ by $(N_0 - 1)/2$ units. Such a delay amounts to multiplying $H[e^{j\omega T}]$ with $e^{-j\frac{N_0-1}{2}\omega T}$. Thus, the delay of $h[k]$ does not alter the filter amplitude response, but the phase response changes by $-(N_0 - 1)\omega T/2$, which is a linear function of ω. Hence, we are assured that the filter is realizable (causal) and has a linear phase. Thus, if we wish to realize a frequency response $H[e^{j\omega T}]$, we begin with $H[e^{j\omega T}]e^{-j\frac{N_0-1}{2}\omega T}$ and find the IDFT of its N_0 samples. The resulting IDFT at $k = 0, 1, 2, 3, \cdots, N_0 - 1$ is the desired impulse response, which is causal, and the corresponding phase response is linear.

Note that $\omega_0 T = \frac{2\pi}{N_0}$, and the N_0 uniform samples of $H[e^{j\omega T}]e^{-j\frac{N_0-1}{2}\omega T}$ are

$$H_r = H[e^{jr\omega_0 T}] e^{-jr\frac{N_0-1}{2}\omega_0 T}$$

$$= H\left[e^{j\frac{2\pi r}{N_0 T}}\right] e^{-jr\pi\frac{N_0-1}{N_0}} \tag{12.97}$$

Recall that the N_0 samples H_r are the uniform samples of the periodic extension of $H[e^{j\omega T}]e^{-j\frac{N_0-1}{2}\omega T}$. Hence, Eq. (12.97) applies to samples of the frequency range from $0 \le \omega \le \frac{\pi}{T}$. The remaining samples are obtained by using the conjugate symmetry property $H_r = H^*_{N_0-r}$. The desired $h[k]$ is the IDFT of H_r; that is,

$$h[k] = \frac{1}{N_0} \sum_{r=0}^{N_0-1} H_r e^{j\frac{2\pi rk}{N_0}} \qquad k = 0,\ 1,\ 2,\ \cdots, N_0-1 \tag{12.98}$$

and

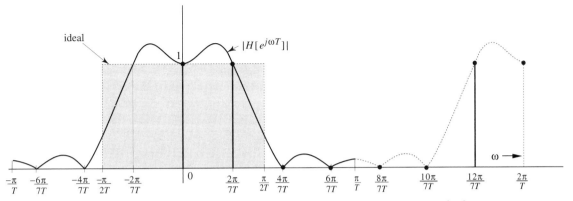

Fig. 12.21 Lowpass filter design using the frequency sampling method.

$$H[z] = \sum_{k=0}^{N_0-1} h[k] z^{-k} \tag{12.99}$$

This procedure will now be explained by an example.

■ Example 12.12

Using the frequency sampling method, design a sixth-order nonrecursive ideal lowpass filter of cutoff frequency $\frac{\pi}{2T}$ rad/s.

The frequency response $H[e^{j\omega T}]$ of an ideal lowpass filter is shown (shaded) in Fig. 12.21. In this case

$$N_0 = 7, \quad \frac{N_0 - 1}{N_0} = \frac{6}{7}, \quad \omega_0 = \frac{2\pi}{N_0 T} = \frac{2\pi}{7T}$$

The seven samples H_r in Eq. (12.97) are

$$H_r = H\left[e^{j\frac{2\pi r}{7T}}\right] e^{-j\frac{6\pi r}{7}}$$

Thus

$$H_0 = 1, \ H_1 = e^{-j\frac{6\pi}{7}}, \ H_2 = H_3 = 0$$

The remaining three samples should be determined using the conjugate property of DFT, $H_r = H_{N_0-r}^*$, that is $H_r = H_{7-r}^*$. Thus

$$H_4 = H_3^* = 0, \ H_5 = H_2^* = 0, \ H_6 = H_1^* = e^{j\frac{6\pi}{7}}$$

The desired $h[k]$ is the IDFT of H_r given by [Eq. (12.98)]

$$h[k] = \sum_{r=0}^{6} H_r e^{j\frac{2\pi r k}{7}} \qquad k = 0, 1, 2, 3, \cdots, 6$$

We may compute this IDFT by using the IFFT algorithm or by straightforward substitution of values of H_r in the above equation as

$$h[0] = \frac{1}{7} \sum_{r=0}^{6} H_r = \frac{1}{7}[1 + e^{-j6\pi/7} + e^{j6\pi/7}] = \frac{1}{7}\left(1 + 2\cos\frac{6\pi}{7}\right) = -0.1146$$

$$h[1] = \frac{1}{7} \sum_{r=0}^{6} H_r e^{j\frac{2\pi r}{7}} = \frac{1}{7}[1 + e^{-j4\pi/7} + e^{j4\pi/7}] = \frac{1}{7}\left(1 + 2\cos\frac{4\pi}{7}\right) = 0.0792$$

$$h[2] = \frac{1}{7} \sum_{r=0}^{6} H_r e^{jr\frac{4\pi r}{7}} = \frac{1}{7}[1 + e^{-j2\pi/7} + e^{j2\pi/7}] = \frac{1}{7}\left(1 + 2\cos\frac{2\pi}{7}\right) = 0.3209$$

$$h[3] = \frac{1}{7} \sum_{r=0}^{6} H_r e^{j\frac{6\pi r}{7}} = \frac{1}{7}[1 + 1 + 1] = 0.4285$$

Similarly, we can show that

$$h[4] = 0.3209, \quad h[5] = 0.0792, \quad \text{and} \quad h[6] = -0.01146$$

Observe that $h[k]$ is symmetrical about its center point $k = 3$ as expected. Compare these values with those found by the impulse invariance method in Table 12.3 for a rectangular window. Although the two sets of values are different, they are comparable. What is the difference in the two filters? The impulse invariance filter optimizes the design with respect to all frequencies. It minimizes the mean squared value of the difference between the desired and the realized frequency response. The frequency sampling method, in contrast, realizes a filter whose frequency response matches exactly to the desired frequency response at N_0 uniformly spaced frequencies. The mean squared error in this design will generally be higher than that in the impulse invariance method.

The filter transfer function is

$$H[z] = -0.1146 + \frac{0.0792}{z} + \frac{0.3209}{z^2} + \frac{0.4285}{z^3} + \frac{0.3209}{z^4} + \frac{0.0792}{z^5} - \frac{0.1146}{z^6}$$

$$= \frac{-0.1146z^6 + 0.0792z^5 + 0.3209z^4 + 0.4285z^3 + 0.3209z^2 + 0.0792z - 0.1146}{z^6}$$

and

$$H[e^{j\omega T}] = -0.1146 + 0.0792e^{-j\omega T} + 0.3209e^{-j2\omega T} + 0.4285e^{-j3\omega T}$$

$$+ 0.3209e^{-j4\omega T} + 0.0792e^{-j5\omega T} - 0.1146e^{-j6\omega T}$$

$$= e^{-j3\omega T}[0.4285 + 0.6418\cos\omega T + 0.1584\cos 2\omega T - 0.2292\cos 3\omega T]$$

The magnitude of this response (Fig. 12.21) shows that the realized filter values match exactly the desired response at the N_0 sample points. The time delay adds a linear phase $-3T\omega$ to the filter characteristic.   ■

⊙ **Computer Example C12.11**

Using MATLAB, find $h[k]$ and the corresponding $H[e^{j\omega T}]$ for the frequency sampling filter in Example 12.12.

```
N0=7;
H=[1 1 0 0 0 0 1];
for i=1:N0
```

```
     r=i-1;
     Hr(i)=H(i)*exp(-j*r*pi*(N0-1)/N0);
end
k=0:6
hk=ifft(Hr);
subplot(2,1,1);
stem(k,hk);
xlabel('k');ylabel('h[k]');
M=512
hE=[hk zeros(1,M-7)]
HE=fft(hE);
subplot(2,1,2);
r=0:M-1;
W=r.*2*pi/512
plot(W,abs(HE));
xlabel('W');ylabel('F(W)');grid;       ⊙
```

An Alternate Method Using Frequency Sampling Filters

We now show an alternative approach to the frequency sampling method, which uses an N_0-order comb filter in cascade with a parallel bank of $N_0 - 1$ first-order filters. This structure forms a frequency sampling filter. We start with Eqs. (6.55). The transfer function $H[z]$ of the filter is first obtained by taking the z-transform of $h[k]$ in Eq. (12.98):

$$H[z] = \sum_{k=0}^{N_0-1} h[k] z^{-k}$$

$$\frac{1}{N_0} \sum_{k=0}^{N_0-1} \left[\sum_{r=0}^{N_0-1} H_r e^{j\frac{2\pi kr}{N_0}} \right] z^{-k}$$

$$= \frac{1}{N_o} \sum_{r=0}^{N_0-1} H_r \sum_{k=0}^{N_0-1} (e^{j\frac{2\pi r}{N_0}} z^{-1})^k$$

The second sum on the right-hand side is a geometric series, and using the result in B.7-4, we have

$$H[z] = \frac{1}{N_0} \sum_{r=0}^{N_0-1} H_r \left[\frac{1 - z^{-N_0} e^{j2\pi r}}{1 - e^{j\frac{2\pi r}{N_0}} z^{-1}} \right]$$

$$= \frac{1}{N_0} \sum_{r=0}^{N_0-1} H_r \left[\frac{z^{-N_0}(z^{N_0} - 1)}{z^{-1}(z - e^{j\frac{2\pi r}{N_0}})} \right]$$

Hence,

$$H[z] = \underbrace{\frac{z^{N_0} - 1}{N_0 z^{N_0}}}_{H_1[z]} \underbrace{\sum_{r=0}^{N_0-1} \frac{z H_r}{z - e^{j\frac{2\pi r}{N_0}}}}_{H_2[z]} \qquad (12.100)$$

Observe that we do not need to perform IDFT (or IFFT) computations to obtain the desired filter transfer function. All we need is the values of the frequency samples H_r, which are given. Equation (12.100) shows that the desired filter is realized as a cascade of two filters with transfer functions $H_1[z]$ and $H_2[z]$. Also, $H_1[z]$ is the transfer function of an N_0-order comb filter (see Example 12.9). The second filter with transfer function $H_2[z]$ is a parallel combination of $n + 1$ first-order filters, whose poles lie on the unit circle at $e^{j\frac{2\pi r}{N_0}}$ ($r = 0, 1, 2, \cdots, N_0 - 1$). For the lowpass or bandpass filters many coefficients H_r appearing in $H_2[z]$ are zero. Recall that in Example 12.12 (lowpass filter) four out of seven coefficients are zero. Thus, in practice the final filter is usually much simpler than it appears in Eq. (12.100). As a result, this method may require a fewer number of computations (multiplications and additions) compared to those in the filter obtained by the direct method (using IDFT).

The poles of the frequency sampling filter are complex in general because they lie on the unit circle. Therefore, we must combine the conjugate poles to form quadratic transfer functions to realize them. All these points will be clarified by designing the filter in Example 12.12 by this method.

■ **Example 12.13**

Redo Example 12.12 using the method of frequency sampling filter.

In this case $n = 6$, $N_0 = n + 1 = 7$. Also

$$H_2 = H_3 = H_4 = H_5 = 0$$

and

$$H_0 = 1, \; H_1 = e^{-j6\pi/7}, \; H_6 = e^{j6\pi/7}$$

Substituting these values in the transfer function $H[z]$ of the desired filter in Eq. (12.100), we obtain

$$H[z] = \underbrace{\frac{z^7 - 1}{7z^7}}_{H_1[z]} \underbrace{\left[\frac{z}{z - 1} + \frac{ze^{-j6\pi/7}}{z - e^{j2\pi/7}} + \frac{ze^{j6\pi/7}}{z - e^{-j2\pi/7}} \right]}_{H_2[z]}$$

We combine the last two terms on the right-hand side corresponding to complex conjugate poles to obtain

$$H[z] = \frac{z^7 - 1}{7z^7} \left[\frac{z}{z - 1} + \frac{z(2z \cos \frac{6\pi}{7} - 2 \cos \frac{8\pi}{7})}{z^2 - (2 \cos \frac{2\pi}{7} z + 1)} \right]$$

$$= \underbrace{\frac{z^7 - 1}{7z^7}}_{H_1[z]} \underbrace{\left[\frac{z}{z - 1} - \frac{1.802z(z - 1)}{z^2 - 1.247z + 1} \right]}_{H_2[z]}$$

We can realize this filter by placing the comb filter $H_1[z]$ in cascade with $H_2[z]$, which consists of a first-order and a second-order filter in parallel. ■

Pole-Zero Cancellation in Frequency Sampling Filters

In the method of frequency sampling filters, we make an intriguing observation that the required nonrecursive (FIR) filter is realized by a cascade of $H_1[z]$ and

$H_2[z]$. However, $H_2[z]$ is recursive (IIR). This strange fact should alert us to the possibility of something interesting going on in this filter. For a nonrecursive filter there can be no poles other than those at the origin [see Eq. (12.69a)]. In the frequency sampling filter [Eq. (12.100)], in contrast, $H_2[z]$ has N_0 poles at $e^{jr\Omega_0}$ ($r = 0, 1, 2, \cdots, n$). All these poles lie on the unit circle at equally spaced points. These poles simply *cannot be* in a nonrecursive filter. They must somehow get canceled along the way somewhere. This is precisely what happens. The zeros of $H_1[z]$ are exactly where the poles of $H_2[z]$ are because

$$z^{N_0} - 1 = (z - e^{j0\frac{2\pi}{N_0}})(z - e^{j\frac{2\pi}{N_0}})(z - e^{j2\frac{2\pi}{N_0}}) \cdots (z - e^{jn\frac{2\pi}{N_0}})$$

Thus, the poles of $H_2[z]$ are canceled by the zeros of $H_1[z]$, rendering the final filter nonrecursive.

Pole-zero cancellation in this filter is a potential cause for mischief because such a perfect cancellation assumes exact realization of both $H_1[z]$ and $H_2[z]$. Such a realization requires the use of infinite precision arithmetic, which is a practical impossibility because of quantization effects. Imperfect cancellation of poles and zeros means there will still be poles on the unit circle, and the filter will not have a finite impulse response. More serious, however, is the fact that the resulting system will be marginally stable. Such a system provides no damping of the round-off noise that is introduced in the computations. In fact, such noise tends to increase with time, and may render the filter useless.

We can partially mitigate this problem by moving both the poles (of $H_2[z]$) and zeros (of $H_1[z]$) to a circle of radius $r = 1 - \epsilon$, where ϵ is a small positive number $\rightarrow 0$. This artifice will make the overall system asymptotically stable.

Spectral Sampling with Windowing

The frequency sampling method can be modified to take advantage of windowing. We first design a frequency sampling filter using a value N_0' that is much higher than the design value N_0. The result is a filter that matches with the desired frequency response at a very large number (N_0') of points. Then we use a suitable N_0-point window to truncate the N_0'-point impulse response. This procedure yields the final design of a desired order.

12.9 Summary

The response of an LTID system with transfer function $H[z]$ to an everlasting sinusoid of frequency Ω is also an everlasting sinusoid of the same frequency. The output amplitude is $|H[e^{j\Omega}]|$ times the input amplitude, and the output sinusoid is shifted in phase with respect to the input sinusoid by $\angle H[e^{j\Omega}]$ radians. The plot of $|H[e^{j\Omega}]|$ vs Ω indicates the amplitude gain of sinusoids of various frequencies and is called the *amplitude response* of the system. The plot of $\angle H[e^{j\Omega}]$ vs Ω indicates the phase shift of sinusoids of various frequencies and is called the phase response. The frequency response of an LTID system is a periodic function of Ω with period 2π. This periodicity is the result of the fact that discrete-time sinusoids with frequencies differing by an integral multiple of 2π are identical.

Frequency response of a system is determined by locations in the complex plane of poles and zeros of its transfer function. We can design frequency selective filters by proper placement of its transfer function poles and zeros. Placing a pole (or a zero) near the point $e^{j\Omega_0}$ in the complex plane enhances (or suppresses) the frequency response at the frequency $\Omega = \Omega_0$. Using this concept, a proper combination of poles and zeros at suitable locations can yield desired filter characteristics.

Digital filters are classified into recursive and nonrecursive filters. The duration of the impulse response of a recursive filter is infinite; that of the nonrecursive filter is finite. For this reason, recursive filters are also called infinite impulse response (IIR) filters, and nonrecursive filters are called finite impulse response (FIR) filters.

Digital filters can process analog signals using A/D and D/A converters. Procedures for designing a digital filter that behaves like a given analog filter are discussed. A digital filter can simulate the behavior of a given analog filter either in time-domain or in frequency-domain. This situation leads to two different design procedures, one using a time-domain equivalence criterion and the other a frequency-domain equivalence criterion.

For recursive or IIR filters, the time-domain equivalence criterion yields the impulse invariance method, and the frequency-domain equivalence criterion yields the bilinear transformation method. The impulse invariance method is handicapped by the aliasing problem, and cannot be used for highpass and bandstop filters. The bilinear transformation method, which is generally superior to the impulse invariance method, suffers from the frequency scale warping effect. However, this effect can be neutralized by prewarping.

For nonrecursive or FIR filters, the time-domain equivalence criterion leads to the method of windowing (Fourier series method), and the frequency-domain equivalence criterion leads to the method of frequency sampling. Because nonrecursive filters are a special case of recursive filters, we expect the performance of recursive filters to be superior. This statement is true in the sense that a given amplitude response can be achieved by a recursive filter of an order smaller than that required for the corresponding nonrecursive filter. However, nonrecursive filters have the advantage that it can realize an arbitrarily shaped amplitude response and a linear phase characteristic. Recursive filters are good for a piecewise constant amplitude response and they can realize linear phase only approximately. To realize a linear phase characteristic in nonrecursive filters, the impulse response $h[k]$ must be either symmetric or antisymmetric about its center point.

References

1.   Mitra, S.K., *Digital Signal processing: A Computer-Based Approach*, McGraw-Hill, New York, 1998.

Problems

12.1-1 Find the amplitude and phase response of the digital filters depicted in Fig. P12.1-1.

12.1-2 Find the amplitude and the phase response of the filters shown in Fig. P12.1-2. Hint: Express $H[e^{j\Omega}]$ as $e^{-j2.5\Omega}H_a[e^{j\Omega}]$.

12.1-3 For an LTID system specified by the equation

$$y[k+1] - 0.5y[k] = f[k+1] + 0.8f[k]$$

(a) Find the amplitude and the phase response.

(b) Find the system response $y[k]$ for the input $f[k] = \cos\left(0.5k - \frac{\pi}{3}\right)$.

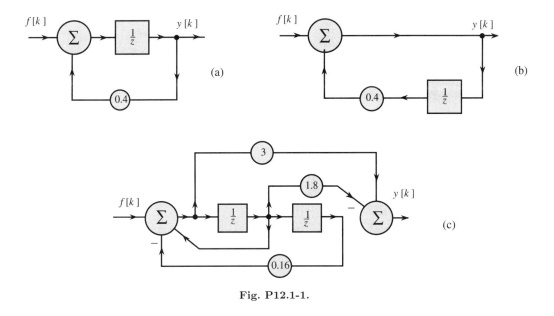

Fig. P12.1-1.

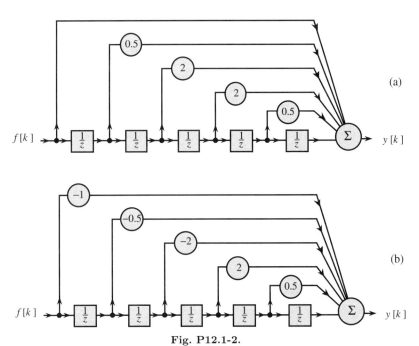

Fig. P12.1-2.

12.1-4 For an asymptotically stable LTID system, show that the steady-state response to input $e^{j\Omega k}u[k]$ is $H[e^{j\Omega}]e^{j\Omega k}u[k]$. The steady-state response is that part of the response which does not decay with time and persists forever.

Hint: Follow the procedure parallel to that used for continuous-time systems in Sec. 7.1-1.

12.1-5 (a) A digital filter has the sampling interval $T = 50\,\mu s$. Determine the highest frequency that can be processed by this filter without aliasing.

(b) If the highest frequency to be processed is 50 kHz, determine the minimum value of the sampling frequency $\mathcal{F}_s$, and the maximum value of the sampling interval T that can be used.

12.2-1 Pole-zero configurations of certain filters are shown in Fig. P12.2-1. Sketch roughly the amplitude response of these filters.

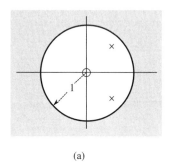
(a)

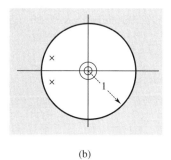
(b)

Fig. P12.2-1.

12.2-2 **(a)** Realize a digital filter whose transfer function is given by

$$H[z] = K \frac{z+1}{z-a}$$

(b) Sketch the amplitude response of this filter, assuming $|a| < 1$.

(c) The amplitude response of this lowpass filter is maximum at $\Omega = 0$. The 3-dB bandwidth is the frequency where the amplitude response drops to 0.707 (or $1/\sqrt{2}$) times its maximum value. Determine the 3-dB bandwidth of this filter when $a = 0.2$.

12.2-3 Design a digital notch filter to reject frequency 5000 Hz completely, and to have a sharp recovery on either side of 5000 Hz to a gain of unity. The highest frequency to be processed is 20 kHz ($\mathcal{F}_h = 20,000$). Hint: See Example 12.3. The zeros should be at $e^{\pm j\omega T}$ for ω corresponding to 5000 Hz, and the poles are at $ae^{\pm j\omega T}$ with $a < 1$. Leave your answer in terms of a. Realize this filter using the canonical form. Find the amplitude response of the filter.

12.2-4 Show that a first-order LTID system with a pole at $z = r$ and a zero at $z = \frac{1}{r}$ ($r \le 1$) is an allpass filter. In other words, show that the amplitude response $|H[e^{j\Omega}]|$ of a system with the transfer function

$$H[z] = \frac{z - \frac{1}{r}}{z - r} \qquad r \le 1$$

is constant with frequency. This is a first-order allpass filter.

Hint: Show that the ratio of the distances of any point on the unit circle from the zero (at $z = \frac{1}{r}$) and the pole (at $z = r$) is a constant $\frac{1}{r}$.

Generalize this result to show that an LTID system with two poles at $z = re^{\pm j\theta}$ and two zeros at $z = \frac{1}{r}e^{\pm j\theta}$ ($r \le 1$) is an allpass filter. In other words, show that the amplitude response of a system with the transfer function

$$H[z] = \frac{(z - \frac{1}{r}e^{j\theta})(z - \frac{1}{r}e^{-j\theta})}{(z - re^{j\theta})(z - re^{-j\theta})} = \frac{z^2 - (\frac{2}{r}\cos\theta)z + \frac{1}{r^2}}{z^2 - (2r\cos\theta)z + r^2} \qquad r \le 1$$

is constant with frequency.

12.4-1 **(a)** A lowpass digital filter with a sampling interval $T = 50\,\mu s$ has a cutoff frequency 10 kHz. If the value of T in this filter is changed to $25\,\mu s$, determine the new cutoff frequency of the filter. Repeat the problem if T is changed to $100\,\mu s$.

12.5-1 **(a)** Using the impulse invariance criterion, design a digital filter to realize an analog filter with transfer function

$$H_a(s) = \frac{7s + 20}{2(s^2 + 7s + 10)}$$

Assume the filter to be bandlimited to the frequency where the gain drops to about 1% of its maximum value.
(b) Show a canonical and a parallel realization of the filter.

12.5-2 Using the impulse invariance criterion, design a digital filter to realize the second-order analog Butterworth filter with transfer function

$$H_a(s) = \frac{1}{s^2 + \sqrt{2}s + 1}$$

Assume the filter to be bandlimited to the frequency where the gain drops to 1% of its maximum value.

12.5-3 Design a digital integrator using the impulse invariance method. Find and sketch the amplitude response, and compare it with that of the ideal integrator. If this integrator is used primarily for integrating audio signals (whose bandwidth is 20 kHz), determine a suitable value for T.

12.5-4 An oscillator by definition is a source (no input) which generates a sinusoid of a certain frequency ω_0. Therefore an oscillator is a system whose zero-input response is a sinusoid of the desired frequency. Find the transfer function of a digital oscillator to oscillate at 10 kHz, by two methods:
(a) Choose $H[z]$ directly so that its zero-input response is a discrete-time sinusoid of frequency $\Omega = \omega T$ corresponding to 10 kHz.
(b) Choose $H_a(s)$ whose zero-input response is an analog sinusoid of 10 kHz. Now use the impulse invariance method to determine $H[z]$.
In both methods select T so that there are 10 samples in each cycle of the sinusoid.
(c) Show a canonical realization of the oscillator.

12.5-5 A variant of the impulse invariance method is the **step invariance** method of digital filter synthesis. In this method, for a given $H_a(s)$, we design $H[z]$ in Fig. 12.8a such that $y(kT)$ in Fig. 12.8b is identical to $y[k]$ in Fig. 12.8a when $f(t) = u(t)$.
(a) Show that in general

$$H[z] = \frac{z-1}{z} \mathcal{Z}\left[\left(\mathcal{L}^{-1}\frac{H_a(s)}{s}\right)_{t=kT}\right]$$

(b) Using this method, design $H[z]$ when

$$H_a(s) = \frac{\omega_c}{s + \omega_c}$$

(c) Synthesize a discrete-time integrator using the step invariant method and compare its amplitude response with that of the ideal integrator.

12.5-6 Synthesize a discrete-time differentiator and integrator, using the ramp invariance method. In this method, for a given $H_a(s)$, we design $H[z]$ such that $y(kT)$ in Fig. 12.8b is identical to $y[k]$ in Fig. 12.8a when $f(t) = tu(t)$.

12.5-7 In an impulse invariance design, show that if $H_a(s)$ is a transfer function of a stable system, the corresponding $H[z]$ is also a transfer function of a stable system.

A pole of $H_a(s)$ at $s = s_i$ is transformed into a pole of $H[z]$ at $z = z_i = e^{s_i t}$. Show that if Re $s_i < 0$, then $|z_i| < 1$.

12.6-1   (a) Design a digital differentiator, using the bilinear transformation.
(b) Show a realization of this filter.
(c) Find and sketch the amplitude response of this filter and compare it with that of the ideal differentiator.
(d) If this filter is used primarily for processing audio signals (voice and music) up to 20 kHz, determine a suitable value for T.

12.6-2   Repeat Prob. 12.6-1 for a digital integrator.

12.6-3   Design a digital lowpass Butterworth filter using the bilinear transformation with prewarping to satisfy the following specifications: $\hat{G}_p = -2$ dB, $\hat{G}_s = -11$ dB, $\omega_p = 100\pi$ rad/s, and $\omega_s = 200\pi$ rad/s. The highest significant frequency is 250 Hz. It is desirable to oversatisfy (if possible) the requirement of $\hat{G}_s$.

12.6-4   Repeat Prob. 12.6-3 for a chebyshev filter.

12.6-5   Design a digital highpass Butterworth filter using the bilinear transformation with prewarping to satisfy the following specifications: $\hat{G}_p = -2$ dB, $\hat{G}_s = -10$ dB, $\omega_p = 150\pi$ rad/s, and $\omega_s = 100\pi$ rad/s. The highest significant frequency is 200 Hz.

12.6-6   Repeat Prob. 12.6-5 for a chebyshev filter.

12.6-7   Design a digital bandpass Butterworth filter using the bilinear transformation with prewarping to satisfy the following specifications: $\hat{G}_p = -2$ dB, $\hat{G}_s = -12$ dB, $\omega_{p_1} = 120$ rad/s, $\omega_{p_2} = 300$ rad/s, and $\omega_{s_1} = 45$ rad/s, $\omega_{s_1} = 450$. The highest significant frequency is 500 Hz.

12.6-8   Repeat Prob. 12.6-7 for a chebyshev filter.

12.6-9   Design a digital bandstop Chebyshev filter using the bilinear transformation with prewarping to satisfy the following specifications: $\hat{G}_p = -1$ dB, $\hat{G}_s = -22$ dB, $\omega_{p_1} = 40$ rad/s, $\omega_{p_2} = 195$ rad/s, and $\omega_{s_1} = 80$ rad/s, $\omega_{s_1} = 120$. The highest significant frequency is 200 Hz.

12.6-10   The bilinear transformation is actually a one-to-one mapping relationship between the s-plane and the z-plane. Show that the transformation

$$s = K \frac{z - 1}{z + 1}$$

maps the $j\omega$-axis in the s-plane into a unit circle in the z-plane. The LHP and the RHP of the s-plane map into the interior and the exterior, respectively, of the unit circle in the z-plane. Show that if $H_a(s)$ represents a stable system, then the corresponding $H[z]$ also represents a stable system.
Hint: $s = \sigma + j\omega$ and $z = (K + s)/(K + s)$.

12.7-1   Show that the frequency response of a third-order ($n = 3$) FIR filter with the impulse response $h[k]$ that is symmetric about its center point is given by

$$H[e^{j\omega T}] = 2e^{-j1.5\omega T}\left(h[1] \cos \frac{\omega T}{2} + h[0] \cos \frac{3\omega T}{2} \right)$$

and for the antisymmetric case

$$H[e^{j\omega T}] = 2e^{j(\frac{\pi}{2} - 1.5\omega T)}\left(h[1] \sin \frac{\omega T}{2} + h[0] \sin \frac{3\omega T}{2} \right)$$

12.7-2   Another form of comb filter is given by the transfer function

$$H[z] = \frac{z^n + 1}{z^n}$$

(a) Find the impulse response of this filter for $n = 6$.
(b) Find canonical realization of this filter.
(c) Find and sketch the amplitude and phase response of this filter.

12.8-1 **(a)** Design a nonrecursive (FIR) filter with $n = 14$ to approximate an ideal lowpass filter with a cutoff frequency at 20 kHz. Use the sampling frequency $\mathcal{F}_s = 200$ kHz.
(b) Modify the design in (a), using the Hamming window.

12.8-2 Using the impulse invariance method, design a tenth-order ($n = 10$) nonrecursive (FIR) filter to approximate the ideal bandpass characteristic depicted in Fig. P12.8-2. Use $T = 2 \times 10^{-3}$.

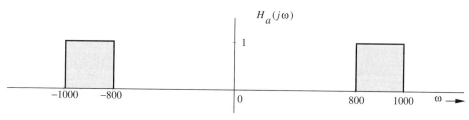

Fig. P12.8-2

12.8-3 Determine $|H[e^{j\omega T}]|$, using the von Hann window for a fifth-order nonrecursive (FIR) filter in Example 12.10 (to approximate an ideal lowpass filter) and for the tenth-order nonrecursive filter in Example 12.11 (to approximate an ideal differentiator).

12.8-4 An ideal $\pi/2$ phase shifter (also known as the Hilbert transformer) is given by

$$H_a(j\omega) = \begin{cases} -j & 0 \le \omega \le \pi/T \\ j & 0 \ge \omega \ge -\pi/T \end{cases}$$

(a) Sketch $|H_a(j\omega)|$ and $\angle H_a(j\omega)$.
(b) Using the impulse invariance method, design a fourteenth-order ($N = 14$) nonrecursive (FIR) filter to approximate $H_a(j\omega)$. Determine the resulting $|H[e^{j\omega T}]|$ and $\angle H[e^{j\omega T}]$, using the rectangular and the Hamming windows.

12.8-5 Design a tenth-order digital differentiator in Example 12.11 using the frequency sampling method. Compare the values of this design with those found in Example 12.11 using the impulse invariance method.

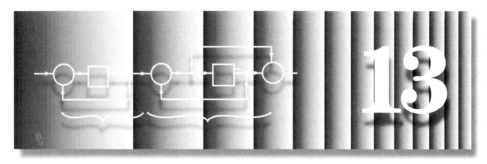

State-Space Analysis

So far we have been describing systems in terms of equations relating certain output to an input (the input-output relationship). This type of description is an *external description* of a system (system viewed from the input and output terminals). As noted in Chapter 1, such a description may be inadequate in some cases, and we need a systematic way of finding system's *internal description*. State space analysis of systems meets this need. In this method, we first select a set of key variables, called the **state variables**, in the system. The state variables have the property that every possible signal or variable in the system at any instant t can be expressed in terms of the state variables and the input(s) at that instant t. If we know all the state variables as a function of t, we can determine every possible signal or variable in the system at any instant with a relatively simple relationship. The system description in this method consists of two parts:

1 Finding the equation(s) relating the state variables to the input(s) (**the state equation**).

2 Finding the output variables in terms of the state variables (**the output equation**).

The analysis procedure, therefore, consists of solving the state equation first, and then solving the output equation. The state space description is capable of determining every possible system variable (or output) from the knowledge of the input and the initial state (conditions) of the system. For this reason it is an *internal description* of the system.

By its nature, the state variable analysis is eminently suited for multiple-input, multiple-output (MIMO) systems. In addition, the state-space techniques are useful for several other reasons, including the following:

1. Time-varying parameter systems and nonlinear systems can be characterized effectively with state-space descriptions.

2. State equations lend themselves readily to accurate simulation on analog or digital computers.

3. For second-order systems ($n = 2$), a graphical method called **phase-plane analysis** can be used on state equations, whether they are linear or nonlinear.

784

4. State equations can yield a great deal of information about a system even when they are not solved explicitly.

This chapter requires some understanding of matrix algebra. Section B.6 is a self-contained treatment of matrix algebra, which should be more than adequate for the purposes of this chapter.

13.1 Introduction

From the discussion in Chapter 1, we know that to determine a system's response(s) at any instant t, we need to know the system's inputs during its entire past, from $-\infty$ to t. If the inputs are known only for $t > t_0$, we can still determine the system output(s) for any $t > t_0$, provided we know certain initial conditions in the system at $t = t_0$. These initial conditions collectively are called the **initial state** of the system (at $t = t_0$).

The state of a system at any instant t_0 is the smallest set of numbers $x_1(t_0)$, $x_2(t_0)$, ... , $x_n(t_0)$ which is sufficient to determine the behavior of the system for all time $t > t_0$ when the input(s) to the system is known for $t > t_0$. The variables x_1, x_2, ... , x_n are known as **state variables.**

The initial conditions of a system can be specified in many different ways. Consequently, the system state can also be specified in many different ways. This means that state variables are not unique. The concept of a system state is very important. We know that an output $y(t)$ at any instant $t > t_0$ can be determined from the initial state $\{x(t_0)\}$ and a knowledge of the input $f(t)$ during the interval (t_0, t). Therefore, the output $y(t_0)$ (at $t = t_0$) is determined from the initial state $\{x(t_0)\}$ and the input $f(t)$ during the interval (t_0, t_0). The latter is $f(t_0)$. Hence, the output at any instant is determined completely from a knowledge of the system state and the input at that instant. This result is also valid for multiple-input, multiple-output (MIMO) systems, where every possible system output at any instant t is determined completely from a knowledge of the system state and the input(s) at the instant t. These ideas should become clear from the following example of an RLC circuit.

■ **Example 13.1**
Find a state-space description of the RLC circuit shown in Fig. 13.1. Verify that all possible system outputs at some instant t can be determined from a knowledge of system state and the input at that instant t.

It is known that inductor currents and capacitor voltages in an RLC circuit can be used as one possible choice of state variables. For this reason, we shall choose x_1 (the capacitor voltage) and x_2 (the inductor current) as our state variables.

The node equation at the intermediate node is

$$i_3 = i_1 - i_2 - x_2$$

but $i_3 = 0.2\dot{x}_1$, $i_1 = 2(f - x_1)$, $i_2 = 3x_1$. Hence

$$0.2\dot{x}_1 = 2(f - x_1) - 3x_1 - x_2$$

or

$$\dot{x}_1 = -25x_1 - 5x_2 + 10f$$

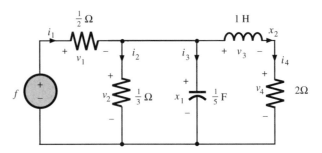

Fig. 13.1 *RLC* network for Example 13.1.

This is the first state equation. To obtain the second state equation, we sum the voltages in the extreme right loop formed by C, L, and the $2\,\Omega$ resistor so that they are equal to zero:

$$-x_1 + \dot{x}_2 + 2x_2 = 0$$

or

$$\dot{x}_2 = x_1 - 2x_2$$

Thus, the two state equations are

$$\dot{x}_1 = -25x_1 - 5x_2 + 10f \qquad (13.1a)$$

$$\dot{x}_2 = x_1 - 2x_2 \qquad (13.1b)$$

Every possible output can now be expressed as a linear combination of x_1, x_2, and f. From Fig. 13.1, we have

$$v_1 = f - x_1$$

$$i_1 = 2(f - x_1)$$

$$v_2 = x_1$$

$$i_2 = 3x_1$$

$$i_3 = i_1 - i_2 - x_2 = 2(f - x_1) - 3x_1 - x_2 = -5x_1 - x_2 + 2f$$

$$i_4 = x_2$$

$$v_4 = 2i_4 = 2x_2$$

$$v_3 = x_1 - v_4 = x_1 - 2x_2 \qquad (13.2)$$

This set of equations is known as the **output equation** of the system. It is clear from this set that every possible output at some instant t can be determined from a knowledge of $x_1(t)$, $x_2(t)$, and $f(t)$, the system state and the input at the instant t. Once we solve the state equations (13.1) to obtain $x_1(t)$ and $x_2(t)$, we can determine every possible output for any given input $f(t)$. ■

   If we already have a system equation in the form of an nth-order differential equation, we can convert it into a state equation as follows. Consider the system equation

$$\frac{d^n y}{dt^n} + a_{n-1}\frac{d^{n-1}y}{dt^{n-1}} + \cdots + a_1\frac{dy}{dt} + a_0 y = f(t) \qquad (13.3)$$

One possible set of initial conditions is $y(0)$, $\dot{y}(0)$, ... , $y^{(n-1)}(0)$. Let us define y, $\dot{y}$, $\ddot{y}$, ... , $y^{(n-1)}$ as the state variables and, for convenience, let us rename the n state variables as x_1, x_2, ... , x_n:

$$x_1 = y$$
$$x_2 = \dot{y}$$
$$x_3 = \ddot{y}$$
$$\vdots$$
$$x_n = y^{(n-1)} \tag{13.4}$$

According to Eq. (13.4), we have

$$\dot{x}_1 = x_2$$
$$\dot{x}_2 = x_3$$
$$\vdots$$
$$\dot{x}_{n-1} = x_n$$

and, according to Eq. (13.3),

$$\dot{x}_n = -a_{n-1}x_n - a_{n-2}x_{n-1} - \cdots - a_1 x_2 - a_0 x_1 + f \tag{13.5a}$$

These n simultaneous first-order differential equations are the state equations of the system. The output equation is

$$y = x_1 \tag{13.5b}$$

For continuous-time systems, the state equations are n simultaneous first-order differential equations in n state variables x_1, x_2, ... , x_n of the form

$$\dot{x}_i = g_i(x_1, x_2, \ldots, x_n, f_1, f_2, \ldots, f_j) \qquad i = 1, 2, \ldots, n$$

where f_1, f_2, ... , f_n are the j system inputs. For a linear system, these equations reduce to a simpler linear form

$$\dot{x}_k = a_{k1}x_1 + a_{k2}x_2 + \cdots + a_{kn}x_n + b_{k1}f_1 + b_{k2}f_2 + \cdots + b_{kj}f_j \qquad k = 1, 2, \ldots, n \tag{13.6a}$$

and the output equations are of the form

$$y_m = c_{m1}x_1 + c_{m2}x_2 + \cdots + c_{mn}x_n + d_{m1}f_1 + d_{m2}f_2 + \cdots + d_{mj}f_j \qquad m = 1, 2, \ldots, k \tag{13.6b}$$

The set of Equations (13.6a) and (13.6b) is called a **dynamical** equation. When it is used to describe a system, it is called the **dynamical-equation description** or **state-variable** description of the system. The n simultaneous first-order state equations are also known as the **normal-form** equations.

These equations can be written more conveniently in matrix form:

$$
\begin{bmatrix} \dot{x}_1 \\ \dot{x}_2 \\ \cdots \\ \dot{x}_n \end{bmatrix} = \begin{bmatrix} a_{11} & a_{12} & \cdots & a_{1n} \\ a_{21} & a_{22} & \cdots & a_{2n} \\ \cdots \cdots \cdots \cdots \cdots \cdots \\ a_{n1} & a_{n2} & \cdots & a_{nn} \end{bmatrix} \begin{bmatrix} x_1 \\ x_2 \\ \cdots \\ x_n \end{bmatrix} + \begin{bmatrix} b_{11} & b_{12} & \cdots & b_{1j} \\ b_{21} & b_{22} & \cdots & b_{2j} \\ \cdots \cdots \cdots \cdots \cdots \cdots \\ b_{n1} & b_{n2} & \cdots & b_{nj} \end{bmatrix} \begin{bmatrix} f_1 \\ f_2 \\ \cdots \\ f_j \end{bmatrix} \tag{13.7a}
$$

$$
\underbrace{}_{\dot{\mathbf{x}}} \qquad \underbrace{}_{\mathbf{A}} \qquad \underbrace{}_{\mathbf{x}} \qquad \underbrace{}_{\mathbf{B}} \qquad \underbrace{}_{\mathbf{f}}
$$

and

$$
\begin{bmatrix} y_1 \\ y_2 \\ \cdots \\ y_k \end{bmatrix} = \begin{bmatrix} c_{11} & c_{12} & \cdots & c_{1n} \\ c_{21} & c_{22} & \cdots & c_{2n} \\ \cdots \cdots \cdots \cdots \cdots \cdots \\ c_{k1} & c_{k2} & \cdots & c_{kn} \end{bmatrix} \begin{bmatrix} x_1 \\ x_2 \\ \cdots \\ x_n \end{bmatrix} + \begin{bmatrix} d_{11} & d_{12} & \cdots & d_{1j} \\ d_{21} & d_{22} & \cdots & d_{2j} \\ \cdots \cdots \cdots \cdots \cdots \cdots \\ d_{k1} & d_{k2} & \cdots & d_{kj} \end{bmatrix} \begin{bmatrix} f_1 \\ f_2 \\ \cdots \\ f_j \end{bmatrix} \tag{13.7b}
$$

$$
\underbrace{}_{\mathbf{y}} \qquad \underbrace{}_{\mathbf{C}} \qquad \underbrace{}_{\mathbf{x}} \qquad \underbrace{}_{\mathbf{D}} \qquad \underbrace{}_{\mathbf{f}}
$$

or

$$
\dot{\mathbf{x}} = \mathbf{A}\mathbf{x} + \mathbf{B}\mathbf{f} \tag{13.8a}
$$

$$
\mathbf{y} = \mathbf{C}\mathbf{x} + \mathbf{D}\mathbf{f} \tag{13.8b}
$$

Equation (13.8a) is the state equation and Eq. (13.8b) is the output equation. The vectors $\mathbf{x}$, $\mathbf{y}$, and $\mathbf{f}$ are the state vector, the output vector, and the input vector, respectively.

For discrete-time systems, the state equations are n simultaneous first-order difference equations. Discrete-time systems are discussed in Sec. 13.6.

13.2   A Systematic Procedure for Determining State Equations

We shall discuss here a systematic procedure for determining the state-space description of linear time-invariant systems. In particular, we shall consider two types of systems: (1) RLC networks and (2) systems specified by block diagrams or nth-order transfer functions.

13.2-1   Electrical Circuits

The method used in Example 13.1 proves effective in most of the simple cases. The steps are as follows:

1. Choose all independent capacitor voltages and inductor currents to be the state variables.

2. Choose a set of loop currents; express the state variables and their first derivatives in terms of these loop currents.

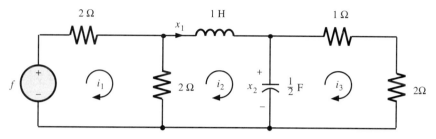

Fig. 13.2   *RLC* network for Example 13.2.

3. Write the loop equations and eliminate all variables other than state variables (and their first derivatives) from the equations derived in Steps 2 and 3.

■ **Example 13.2**

Write the state equations for the network shown in Fig. 13.2.

Step 1. There is one inductor and one capacitor in the network. Therefore, we shall choose the inductor current x_1 and the capacitor voltage x_2 as the state variables.

Step 2. The relationship between the loop currents and the state variables can be written by inspection:

$$x_1 = i_2 \tag{13.9a}$$

$$\tfrac{1}{2}\dot{x}_2 = i_2 - i_3 \tag{13.9b}$$

Step 3. The loop equations are

$$4i_1 - 2i_2 = f \tag{13.10a}$$

$$2(i_2 - i_1) + \dot{x}_1 + x_2 = 0 \tag{13.10b}$$

$$-x_2 + 3i_3 = 0 \tag{13.10c}$$

Now we eliminate i_1, i_2, and i_3 from Eqs. (13.9) and (13.10) as follows. From Eq. (13.10b), we have

$$\dot{x}_1 = 2(i_1 - i_2) - x_2$$

We can eliminate i_1 and i_2 from this equation by using Eqs. (13.9a) and (13.10a) to obtain

$$\dot{x}_1 = -x_1 - x_2 + \tfrac{1}{2}f$$

The substitution of Eqs. (13.9a) and (13.10c) in Eq. (13.9b) yields

$$\dot{x}_2 = 2x_1 - \tfrac{2}{3}x_2$$

These are the desired state equations. We can express them in matrix form as

$$\begin{bmatrix} \dot{x}_1 \\ \dot{x}_2 \end{bmatrix} = \begin{bmatrix} -1 & -1 \\ 2 & -\tfrac{2}{3} \end{bmatrix} \begin{bmatrix} x_1 \\ x_2 \end{bmatrix} + \begin{bmatrix} \tfrac{1}{2} \\ 0 \end{bmatrix} f \tag{13.11}$$

The derivation of state equations from loop equations is facilitated considerably by choosing loops in such a way that only one loop current passes through each of the inductors or capacitors. ■

An Alternative procedure

We can also determine the state equations by the following procedure:

1. Choose all independent capacitor voltages and inductor currents to be the state variables.

2. Replace each capacitor by a fictitious voltage source equal to the capacitor voltage, and replace each inductor by a fictitious current source equal to the inductor current. This step will transform the RLC network into a network consisting only of resistors, current sources, and voltage sources.

3. Find the current through each capacitor and equate it to $C\dot{x}_i$, where x_i is the capacitor voltage. Similarly, find the voltage across each inductor and equate it to $L\dot{x}_j$, where x_j is the inductor current.

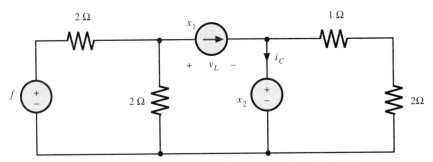

Fig. 13.3   Equivalent circuit of the network in Fig. 13.2.

■ **Example 13.3**

Using the above procedure, write the state equations for the network in Fig. 13.2.

In the network in Fig. 13.2, we replace the inductor by a current source of current x_1 and the capacitor by a voltage source of voltage x_2, as shown in Fig. 13.3. The resulting network consists of five resistors, two voltage sources, and one current source. We can determine the voltage v_L across the inductor and the current i_c through the capacitor by using the principle of superposition. This step can be accomplished by inspection. For example, v_L has three components arising from three sources. To compute the component due to f, we assume that $x_1 = 0$ (open circuit) and $x_2 = 0$ (short circuit). Under these conditions, all of the network to the right of the $2\,\Omega$ resistor is opened, and the component of v_L due to f is the voltage across the $2\,\Omega$ resistor. This voltage is clearly $\frac{1}{2}f$. Similarly, to find the component of v_L due to x_1, we short f and x_2. The source x_1 sees an equivalent resistor of 1Ω across it, and hence $v_L = -x_1$. Continuing the process, we find that the component of v_L due to x_2 is $-x_2$. Hence

$$v_L = \dot{x}_1 = \frac{1}{2}f - x_1 - x_2 \qquad\qquad (13.12a)$$

Using the same procedure, we find

$$i_c = \frac{1}{2}\dot{x}_2 = x_1 - \frac{1}{3}x_2 \qquad\qquad (13.12b)$$

These equations are identical to the state equations (13.11) obtained earlier.† ∎

13.2-2   State Equations From Transfer Function

It is relatively easy to determine the state equations of a system specified by its transfer function. Consider, for example, a first-order system with the transfer function

$$H(s) = \frac{1}{s+a} \tag{13.13}$$

The system realization appears in Fig. 13.4. The integrator output serves as a natural state variable since, in practical realization, initial conditions are placed on the integrator output. From Fig. 13.4, we have

$$\dot{x} = -ax + f \tag{13.14a}$$

$$y = x \tag{13.14b}$$

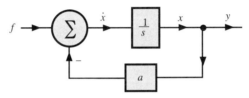

Fig. 13.4

In Sec. 6.6 we saw that a given transfer function can be realized in several ways. Consequently, we should be able to obtain different state-space descriptions of the same system by using different realizations. This assertion will be clarified by the following example.

∎ **Example 13.4**

Determine the state-space description of a system specified by the transfer function

$$H(s) = \frac{2s+10}{s^3 + 8s^2 + 19s + 12} \tag{13.15a}$$

$$= \left(\frac{2}{s+1}\right)\left(\frac{s+5}{s+3}\right)\left(\frac{1}{s+4}\right) \tag{13.15b}$$

$$= \frac{\frac{4}{3}}{s+1} - \frac{2}{s+3} + \frac{\frac{2}{3}}{s+4} \tag{13.15c}$$

---

†This procedure requires modification if the system contains all-capacitor voltage source tie sets or all-inductor current source cut sets. In the case of all-capacitor voltage source tie sets, all capacitor voltages cannot be independent. One capacitor voltage can be expressed in terms of the remaining capacitor voltages and the voltage source(s) in that tie set. Consequently, one of the capacitor voltages should not be used as a state variable, and that capacitor should not be replaced by a voltage source. Similarly, in all-inductor current source tie sets, one inductor should not be replaced by a current source. If there are all-capacitor tie sets or all-inductor cut sets only, no further complications occur. In all-capacitor-voltage source tie sets and/or all-inductor-current source cut sets, we have additional difficulties in that the terms involving derivatives of the input may occur. This problem can be solved by redefining the state variables. The final state variables will not be capacitor voltages and inductor currents.

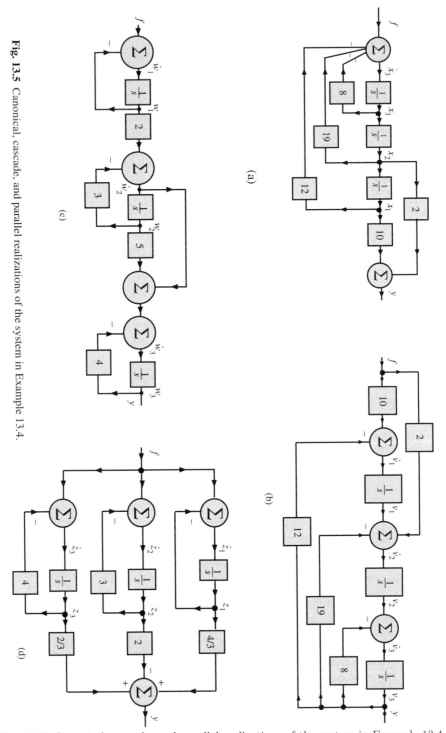

(a)

(b)

(c)

(d)

Fig. 13.5 Canonical, cascade, and parallel realizations of the system in Example 13.4.

Fig. 13.5 Canonical, cascade, and parallel realizations of the system in Example 13.4.

Using the procedure developed in Sec. 6.6, we shall realize $H(s)$ in Eq. (13.15) with four different realizations: (i) the controller canonical form [Eq. (13.15a)], (ii) the observer canonical form [Eq. (13.15a)], (iii) cascade realization [Eq. (13.15b)] and (iv) parallel realization [Eq. (13.15c)]. These realizations are depicted in Figs. 13.5a, 13.5b, 13.5c, and 13.5d, respectively. As mentioned earlier, the output of each integrator serves as a natural state variable.

1. Canonical Forms

Here we shall realize the system using the first (controller) canonical form discussed in Sec. 6.6-1. If we choose the state variables to be the three integrator outputs x_1, x_2, and x_3, then, according to Fig. 13.5a,

$$\dot{x}_1 = x_2$$

$$\dot{x}_2 = x_3 \qquad (13.16a)$$

$$\dot{x}_3 = -12x_1 - 19x_2 - 8x_3 + f$$

Also, the output y is given by

$$y = 10x_1 + 2x_2 \qquad (13.16b)$$

Equations (13.16a) are the state equations, and Eq. (13.16b) is the output equation. In matrix form we have

$$\begin{bmatrix} \dot{x}_1 \\ \dot{x}_2 \\ \dot{x}_3 \end{bmatrix} = \underbrace{\begin{bmatrix} 0 & 1 & 0 \\ 0 & 0 & 1 \\ -12 & -19 & -8 \end{bmatrix}}_{\mathbf{A}} \begin{bmatrix} x_1 \\ x_2 \\ x_3 \end{bmatrix} + \underbrace{\begin{bmatrix} 0 \\ 0 \\ 1 \end{bmatrix}}_{\mathbf{B}} f \qquad (13.17a)$$

and

$$y = \underbrace{\begin{bmatrix} 10 & 2 & 0 \end{bmatrix}}_{\mathbf{C}} \begin{bmatrix} x_1 \\ x_2 \\ x_3 \end{bmatrix} \qquad (13.17b)$$

We can also realize $H(s)$ by using the second (observer) canonical form (discussed in Appendix 6.1), as shown in Fig. 13.5b. If we label the output of the three integrators from left to right as the state variables v_1, v_2, and v_3, then, according to Fig. 13.5b,

$$\dot{v}_1 = -12v_3 + 10f$$

$$\dot{v}_2 = v_1 - 19v_3 + 2f \qquad (13.18a)$$

$$\dot{v}_3 = v_2 - 8v_3$$

and the output y is given by

$$y = v_3 \qquad (13.18b)$$

Hence

$$\begin{bmatrix} \dot{v}_1 \\ \dot{v}_2 \\ \dot{v}_3 \end{bmatrix} = \underbrace{\begin{bmatrix} 0 & 0 & -12 \\ 1 & 0 & -19 \\ 0 & 1 & -8 \end{bmatrix}}_{\hat{\mathbf{A}}} \begin{bmatrix} v_1 \\ v_2 \\ v_3 \end{bmatrix} + \underbrace{\begin{bmatrix} 10 \\ 2 \\ 0 \end{bmatrix}}_{\hat{\mathbf{B}}} f \qquad (13.19a)$$

and

$$y = \underbrace{\begin{bmatrix} 0 & 0 & 1 \end{bmatrix}}_{\check{\mathbf{C}}} \begin{bmatrix} v_1 \\ v_2 \\ v_3 \end{bmatrix} \qquad (13.19b)$$

Observe closely the relationship between the state-space descriptions of $H(s)$ that use the controller canonical form [Eqs. (13.17)] and those using the observer canonical form [Eqs. (13.19)]. The $\mathbf{A}$ matrices in these two cases are the transpose of one another; also, the $\mathbf{B}$ of one is the transpose of $\mathbf{C}$ in the other, and vice versa. Hence

$$(\mathbf{A})^T = \hat{\mathbf{A}}$$

$$(\mathbf{B})^T = \hat{\mathbf{C}} \qquad (13.20)$$

$$(\mathbf{C})^T = \hat{\mathbf{B}}$$

This is no coincidence. This duality relation is generally true.[1]

2. Series Realization

The three integrator outputs w_1, w_2, and w_3 in Fig. 13.5c are the state variables. The state equations are

$$\dot{w}_1 = -w_1 + f \qquad (13.21a)$$

$$\dot{w}_2 = 2w_1 - 3w_2 \qquad (13.21b)$$

$$\dot{w}_3 = 5w_2 + \dot{w}_2 - 4w_3 \qquad (13.21c)$$

and the output equation is

$$y = w_3$$

The elimination of $\dot{w}_2$ from Eq. (13.21c) by using Eq. (13.21b) converts these equations into the desired state form

$$\begin{bmatrix} \dot{w}_1 \\ \dot{w}_2 \\ \dot{w}_3 \end{bmatrix} = \begin{bmatrix} -1 & 0 & 0 \\ 2 & -3 & 0 \\ 2 & 2 & -4 \end{bmatrix} \begin{bmatrix} w_1 \\ w_2 \\ w_3 \end{bmatrix} + \begin{bmatrix} 1 \\ 0 \\ 0 \end{bmatrix} f \qquad (13.22a)$$

and

$$y = \begin{bmatrix} 0 & 0 & 1 \end{bmatrix} \begin{bmatrix} w_1 \\ w_2 \\ w_3 \end{bmatrix} \qquad (13.22b)$$

3. Parallel Realization (Diagonal Representation)

The three integrator outputs z_1, z_2, and z_3 in Fig. 13.5d are the state variables. The state equations are

$$\dot{z}_1 = -z_1 + f$$

$$\dot{z}_2 = -3z_2 + f$$

$$\dot{z}_3 = -4z_3 + f \qquad (13.23a)$$

and the output equation is

$$y = \tfrac{4}{3}z_1 - 2z_2 + \tfrac{2}{3}z_3 \tag{13.23b}$$

Therefore, the equations in the matrix form are

$$\begin{bmatrix} \dot{z}_1 \\ \dot{z}_2 \\ \dot{z}_n \end{bmatrix} = \begin{bmatrix} -1 & 0 & 0 \\ 0 & -3 & 0 \\ 0 & 0 & -4 \end{bmatrix} \begin{bmatrix} z_1 \\ z_2 \\ z_3 \end{bmatrix} + \begin{bmatrix} 1 \\ 1 \\ 1 \end{bmatrix} f \tag{13.24a}$$

$$y = \begin{bmatrix} \tfrac{4}{3} & -2 & \tfrac{2}{3} \end{bmatrix} \begin{bmatrix} z_1 \\ z_2 \\ z_3 \end{bmatrix} \tag{13.24b}$$

■

⊙ **Computer Example C13.1**
Solve Example 13.4 using MATLAB.
Caution: The convention of MATLAB for labeling state variables $x_1, x_2, \ldots, x_n$ in a block diagram, such as shown in Fig. 13.5a, is reversed. What we label x_1 is x_n, and x_2 is x_{n-1}, and so on.

num=[2 10]; den=[1 8 19 12];
[A,B,C,D]=tf2ss(num,den)
% In order to find the transfer function from A, B, C, and D, use
[num, den]=ss2tf(A,B,C,D)
printsys(num,den) ⊙

A General Case

It is clear that a system has several state-space descriptions. Notable among these are the canonical-form variables and the diagonalized variables (in the parallel realization). State equations in these forms can be written immediately by inspection of the transfer function. Consider the general nth-order transfer function

$$H(s) = \frac{b_m s^m + b_{m-1} s^{m-1} + \cdots + b_1 s + b_0}{s^n + a_{n-1} s^{n-1} + \cdots + a_1 s + a_0} \tag{13.25a}$$

$$= \frac{b_m s^m + b_{m-1} s^{m-1} + \cdots + b_1 s + b_0}{(s - \lambda_1)(s - \lambda_2) \cdots (s - \lambda_n)}$$

$$= \frac{k_1}{s - \lambda_1} + \frac{k_2}{s - \lambda_2} + \cdots + \frac{k_n}{s - \lambda_n} \tag{13.25b}$$

Figures 13.6a and 13.6b show the realizations of $H(s)$, using the controller canonical form [Eq. (13.25a)] and the parallel form [Eq. (13.25b)], respectively.

The n integrator outputs $x_1, x_2, \ldots, x_n$ in Fig. 13.6a are the state variables. It is clear that

$$\dot{x}_1 = x_2$$
$$\dot{x}_2 = x_3$$
$$\cdots\cdots\cdots \tag{13.26a}$$
$$\dot{x}_{n-1} = x_n$$
$$\dot{x}_n = -a_{n-1}x_n - a_{n-2}x_{n-1} - \cdots - a_1 x_2 - a_0 x_1 + f$$

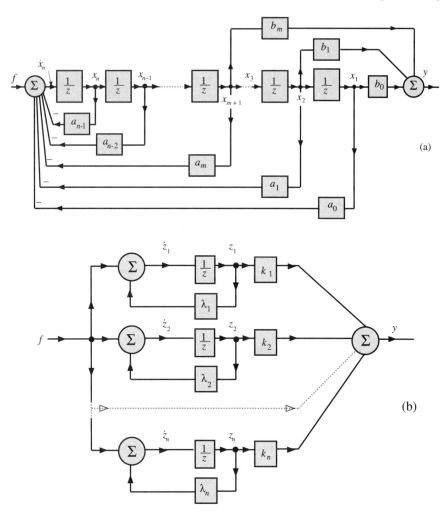

Fig. 13.6   Controller canonical and parallel realizations for an nth order LTIC system.

and output y is

$$y = b_0 x_1 + b_1 x_2 + \cdots + b_m x_{m+1} \qquad (13.26b)$$

or

$$\begin{bmatrix} \dot{x}_1 \\ \dot{x}_2 \\ \cdots \\ \dot{x}_{n-1} \\ \dot{x}_n \end{bmatrix} = \begin{bmatrix} 0 & 1 & 0 & \cdots & 0 & 0 \\ 0 & 0 & 1 & \cdots & 0 & 0 \\ \hdotsfor{6} \\ 0 & 0 & 0 & \cdots & 0 & 1 \\ -a_0 & -a_1 & -a_2 & \cdots & -a_{n-2} & -a_{n-1} \end{bmatrix} \begin{bmatrix} x_1 \\ x_2 \\ \cdots \\ x_{n-1} \\ x_n \end{bmatrix} + \begin{bmatrix} 0 \\ 0 \\ \cdots \\ 0 \\ 1 \end{bmatrix} f$$

$$\tag{13.27a}$$

and

$$y = \begin{bmatrix} b_0 & b_1 & \cdots & b_m & 0 & \cdots & 0 \end{bmatrix} \begin{bmatrix} x_1 \\ x_2 \\ \vdots \\ x_n \end{bmatrix} \qquad (13.27b)$$

Observe that these equations (state equations and output equation) can be written immediately by inspection of $H(s)$.

The n integrator outputs $z_1, z_2, \ldots, z_n$ in Fig. 13.6b are the state variables. It is clear that

$$\dot{z}_1 = \lambda_1 z_1 + f$$

$$\dot{z}_2 = \lambda_2 z_2 + f$$

$$\cdots \cdots \cdots \cdots \qquad (13.28a)$$

$$\dot{z}_n = \lambda_n z_n + f$$

and

$$y = k_1 z_1 + k_2 z_2 + \cdots + k_n z_n \qquad (13.28b)$$

or

$$\begin{bmatrix} \dot{z}_1 \\ \dot{z}_2 \\ \cdots \\ \dot{z}_{n-1} \\ \dot{z}_n \end{bmatrix} = \begin{bmatrix} \lambda_1 & 0 & \cdots & 0 & 0 \\ 0 & \lambda_2 & \cdots & 0 & 0 \\ \cdots & \cdots & \cdots & \cdots & \cdots \\ 0 & 0 & \cdots & \lambda_{n-1} & 0 \\ 0 & 0 & \cdots & 0 & \lambda_n \end{bmatrix} \begin{bmatrix} z_1 \\ z_2 \\ \cdots \\ z_{n-1} \\ z_n \end{bmatrix} + \begin{bmatrix} 1 \\ 1 \\ \cdots \\ 1 \\ 1 \end{bmatrix} f \qquad (13.29a)$$

and

$$y = \begin{bmatrix} k_1 & k_2 & \cdots & k_{n-1} & k_n \end{bmatrix} \begin{bmatrix} z_1 \\ z_2 \\ \vdots \\ z_{n-1} \\ z_n \end{bmatrix} \qquad (13.29b)$$

The state equation (13.29a) and the output equation (13.29b) can be written immediately by inspection of the transfer function $H(s)$ in Eq. (13.25b). Observe that the diagonalized form of the state matrix [Eq. (13.29a)] has the transfer function poles as its diagonal elements. The presence of repeated poles in $H(s)$ will modify the procedure slightly. The handling of these cases is discussed in Sec. 6.6.

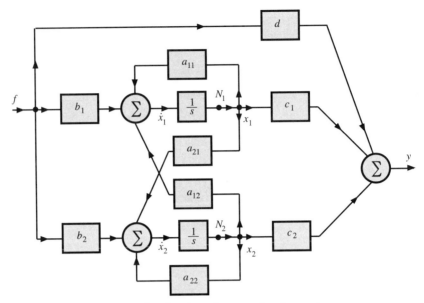

Fig. 13.7 Realization of a second-order system.

It is clear from the above discussion that a state-space description is not unique. For any realization of $H(s)$ using integrators, scalar multipliers, and adders, a corresponding state-space description exists. Since there are many possible realizations of $H(s)$, there are many possible state-space descriptions.

Realization

Consider a second-order system with a single input f, a single output y, and two state variables, x_1 and x_2. The system equations are

$$\dot{x}_1 = a_{11}x_1 + a_{12}x_2 + b_1 f$$

$$\dot{x}_2 = a_{21}x_1 + a_{22}x_2 + b_2 f \qquad (13.30a)$$

and

$$y = c_1 x_1 + c_2 x_2 + df \qquad (13.30b)$$

Figure 13.7 shows the block diagram of the realized system. The initial conditions $x_1(0)$ and $x_2(0)$ should be applied at N_1 and N_2. This procedure can be easily extended to general multiple-input, multiple-output systems with n state variables.

13.3 Solution of State Equations

The state equations of a linear system are n simultaneous linear differential equations of the first order. We studied the techniques of solving linear differential equations in Chapters 2 and 6. The same techniques can be applied to state equations without any modification. However, it is more convenient to carry out the solution in the framework of matrix notation.

These equations can be solved in both the time domain and frequency domain (Laplace transform). The latter requires fewer new concepts and is therefore easier to deal with than the time-domain solution. For this reason, we shall first consider the Laplace transform solution.

13.3-1   Laplace Transform Solution of State Equations

The kth state equation [Eq. (13.6a)] is of the form

$$\dot{x}_k = a_{k1}x_1 + a_{k2}x_2 + \cdots + a_{kn}x_n + b_{k1}f_1 + b_{k2}f_2 + \cdots + b_{kj}f_j \qquad (13.31a)$$

We shall take the Laplace transform of this equation. Let

$$x_k(t) \Longleftrightarrow X_k(s)$$

so that

$$\dot{x}_k(t) \Longleftrightarrow sX_k(s) - x_k(0)$$

Also, let

$$f_i(t) \Longleftrightarrow F_i(s)$$

The Laplace transform of Eq. (13.31a) yields

$$sX_k(s) - x_k(0) = a_{k1}X_1(s) + a_{k2}X_2(s) + \cdots + a_{kn}X_n(s) + b_{k1}F_1(s)$$
$$+ b_{k2}F_2(s) + \cdots + b_{kj}F_j(s) \qquad (13.31b)$$

Taking the Laplace transforms of all n state equations, we obtain

$$s\underbrace{\begin{bmatrix} X_1(s) \\ X_2(s) \\ \cdots \\ X_n(s) \end{bmatrix}}_{\mathbf{X}(s)} - \underbrace{\begin{bmatrix} x_1(0) \\ x_2(0) \\ \cdots \\ x_n(0) \end{bmatrix}}_{\mathbf{x}(0)} = \underbrace{\begin{bmatrix} a_{11} & a_{12} & \cdots & a_{1n} \\ a_{21} & a_{22} & \cdots & a_{2n} \\ \cdots\cdots\cdots\cdots\cdots\cdots \\ a_{n1} & a_{n2} & \cdots & a_{nn} \end{bmatrix}}_{\mathbf{A}} \underbrace{\begin{bmatrix} X_1(s) \\ X_2(s) \\ \cdots \\ X_n(s) \end{bmatrix}}_{\mathbf{X}(s)}$$

$$+ \underbrace{\begin{bmatrix} b_{11} & b_{12} & \cdots & b_{1j} \\ b_{21} & b_{22} & \cdots & b_{2j} \\ \cdots\cdots\cdots\cdots\cdots\cdots \\ b_{n1} & b_{n2} & \cdots & b_{nj} \end{bmatrix}}_{\mathbf{B}} \underbrace{\begin{bmatrix} F_1(s) \\ F_2(s) \\ \cdots \\ F_j(s) \end{bmatrix}}_{\mathbf{F}(s)} \qquad (13.32a)$$

Defining the vectors, as indicated above, we have

$$s\mathbf{X}(s) - \mathbf{x}(0) = \mathbf{A}\mathbf{X}(s) + \mathbf{B}\mathbf{F}(s)$$

or

$$s\mathbf{X}(s) - \mathbf{A}\mathbf{X}(s) = \mathbf{x}(0) + \mathbf{B}\mathbf{F}(s)$$

and

$$(s\mathbf{I} - \mathbf{A})\mathbf{X}(s) = \mathbf{x}(0) + \mathbf{B}\mathbf{F}(s) \qquad (13.32b)$$

where $\mathbf{I}$ is the $n \times n$ identity matrix. From Eq. 13.32b, we have

$$\mathbf{X}(s) = (s\mathbf{I} - \mathbf{A})^{-1}[\mathbf{x}(0) + \mathbf{B}\mathbf{F}(s)] \qquad (13.33a)$$

$$= \mathbf{\Phi}(s)[\mathbf{x}(0) + \mathbf{B}\mathbf{F}(s)] \qquad (13.33b)$$

where

$$\mathbf{\Phi}(s) = (s\mathbf{I} - \mathbf{A})^{-1} \qquad (13.34)$$

Thus, from Eq. (13.33b),

$$\mathbf{X}(s) = \mathbf{\Phi}(s)\mathbf{x}(0) + \mathbf{\Phi}(s)\mathbf{B}\mathbf{F}(s) \qquad (13.35a)$$

and

$$\mathbf{x}(t) = \underbrace{\mathcal{L}^{-1}[\mathbf{\Phi}(s)]\mathbf{x}(0)}_{\text{zero-input component}} + \underbrace{\mathcal{L}^{-1}[\mathbf{\Phi}(s)\mathbf{B}\mathbf{F}(s)]}_{\text{zero-state component}} \qquad (13.35b)$$

Equation (13.35b) gives the desired solution. Observe the two components of the solution. The first component yields $\mathbf{x}(t)$ when the input $f(t) = 0$. Hence the first component is the zero-input component. In a similar manner, we see that the second component is the zero-state component.

■ Example 13.5

Find the state vector $\mathbf{x}(t)$ for the system whose state equation is given by

$$\dot{\mathbf{x}} = \mathbf{A}\mathbf{x} + \mathbf{B}\mathbf{f}$$

where

$$\mathbf{A} = \begin{bmatrix} -12 & \frac{2}{3} \\ -36 & -1 \end{bmatrix} \qquad \mathbf{B} = \begin{bmatrix} \frac{1}{3} \\ 1 \end{bmatrix} \qquad \mathbf{f}(t) = u(t)$$

and the initial conditions are $x_1(0) = 2$, $x_2(0) = 1$.

From Eq. (13.33b), we have

$$\mathbf{X}(s) = \mathbf{\Phi}(s)[\mathbf{x}(0) + \mathbf{B}\mathbf{F}(s)]$$

Let us first find $\mathbf{\Phi}(s)$. We have

$$(s\mathbf{I} - \mathbf{A}) = s\begin{bmatrix} 1 & 0 \\ 0 & 1 \end{bmatrix} - \begin{bmatrix} -12 & \frac{2}{3} \\ -36 & -1 \end{bmatrix} = \begin{bmatrix} s + 12 & -\frac{2}{3} \\ 36 & s + 1 \end{bmatrix}$$

and

$$\mathbf{\Phi}(s) = (s\mathbf{I} - \mathbf{A})^{-1} = \begin{bmatrix} \frac{s+1}{(s+4)(s+9)} & \frac{2/3}{(s+4)(s+9)} \\ \frac{-36}{(s+4)(s+9)} & \frac{s+12}{(s+4)(s+9)} \end{bmatrix} \qquad (13.36a)$$

Now, $\mathbf{x}(0)$ is given as

$$\mathbf{x}(0) = \begin{bmatrix} 2 \\ 1 \end{bmatrix}$$

Also, $F(s) = \frac{1}{s}$, and

$$\mathbf{B}\mathbf{F}(s) = \begin{bmatrix} \frac{1}{3} \\ 1 \end{bmatrix} \frac{1}{s} = \begin{bmatrix} \frac{1}{3s} \\ \frac{1}{s} \end{bmatrix}$$

Therefore

$$\mathbf{x}(0) + \mathbf{B}F(s) = \begin{bmatrix} 2 + \frac{1}{3s} \\ 1 + \frac{1}{s} \end{bmatrix} = \begin{bmatrix} \frac{6s+1}{3s} \\ \frac{s+1}{s} \end{bmatrix}$$

and

$$\mathbf{X}(s) = \mathbf{\Phi}(s)[\mathbf{x}(0) + \mathbf{B}F(s)]$$

$$= \begin{bmatrix} \frac{s+1}{(s+4)(s+9)} & \frac{\frac{2}{3}}{(s+4)(s+9)} \\ \frac{-36}{(s+4)(s+9)} & \frac{s+12}{(s+4)(s+9)} \end{bmatrix} \begin{bmatrix} \frac{6s+1}{3s} \\ \frac{s+1}{s} \end{bmatrix}$$

$$= \begin{bmatrix} \frac{2s^2+3s+1}{s(s+4)(s+9)} \\ \frac{s-59}{(s+4)(s+9)} \end{bmatrix}$$

$$= \begin{bmatrix} \frac{\frac{1}{36}}{s} - \frac{\frac{21}{20}}{s+4} + \frac{\frac{136}{45}}{s+9} \\ -\frac{\frac{63}{5}}{s+4} + \frac{\frac{68}{5}}{s+9} \end{bmatrix}$$

The inverse Laplace transform of this equation yields

$$\begin{bmatrix} x_1(t) \\ x_2(t) \end{bmatrix} = \begin{bmatrix} \left(\frac{1}{36} - \frac{21}{20}e^{-4t} + \frac{136}{45}e^{-9t} \right)u(t) \\ \left(-\frac{63}{5}e^{-4t} + \frac{68}{5}e^{-9t} \right)u(t) \end{bmatrix} \tag{13.36b}$$

∎

⊙ **Computer Example C13.2**
Solve Example 13.5 using MATLAB.
Caution: See caution in Example C13.1.

```
A=[-12 2/3;-36 -1]; B=[1/3; 1];
C=[0 0]; D=0;
x0=[2;1];
t=0:.01:3; t=t';
f=ones(length(t),1);
[y,x]=lsim(A,B,C,D,f,t,x0);
plot(t,x)    ⊙
```

The Output

The output equation is given by

$$\mathbf{y} = \mathbf{C}\mathbf{x} + \mathbf{D}\mathbf{f}$$

and

$$\mathbf{Y}(s) = \mathbf{C}\mathbf{X}(s) + \mathbf{D}\mathbf{F}(s)$$

The substitution of Eq. (13.33b) into this equation yields

$$\mathbf{Y}(s) = \mathbf{C}\{\mathbf{\Phi}(s)[\mathbf{x}(0) + \mathbf{BF}(s)]\} + \mathbf{DF}(s)$$

$$= \underbrace{\mathbf{C\Phi}(s)\mathbf{x}(0)}_{\text{zero-input response}} + \underbrace{[\mathbf{C\Phi}(s)\mathbf{B} + \mathbf{D}]\mathbf{F}(s)}_{\text{zero-state response}} \qquad (13.37)$$

The zero-state response (that is, the response $\mathbf{Y}(s)$ when $\mathbf{x}(0){=}\mathbf{0}$), is given by

$$\mathbf{Y}(s) = [\mathbf{C\Phi}(s)\mathbf{B} + \mathbf{D}]\mathbf{F}(s) \qquad (13.38a)$$

Note that the transfer function of a system is defined under the zero-state condition [see Eq. (6.53)]. The matrix $\mathbf{C\Phi}(s)\mathbf{B} + \mathbf{D}$ is the **transfer function matrix** $\mathbf{H}(s)$ of the system, which relates the responses $y_1, y_2, \ldots, y_k$ to the inputs $f_1, f_2, \ldots, f_j$:

$$\mathbf{H}(s) = \mathbf{C\Phi}(s)\mathbf{B} + \mathbf{D} \qquad (13.38b)$$

and the zero-state response is

$$\mathbf{Y}(s) = \mathbf{H}(s)\mathbf{F}(s) \qquad (13.39)$$

The matrix $\mathbf{H}(s)$ is a $k \times j$ matrix (k is the number of outputs and j is the number of inputs). The ijth element $H_{ij}(s)$ of $H(s)$ is the transfer function that relates the output $y_i(t)$ to the input $f_j(t)$.

■ Example 13.6

Let us consider a system with a state equation

$$\begin{bmatrix} \dot{x}_1 \\ \dot{x}_2 \end{bmatrix} = \begin{bmatrix} 0 & 1 \\ -2 & -3 \end{bmatrix} \begin{bmatrix} x_1 \\ x_2 \end{bmatrix} + \begin{bmatrix} 1 & 0 \\ 1 & 1 \end{bmatrix} \begin{bmatrix} f_1 \\ f_2 \end{bmatrix} \qquad (13.40a)$$

and an output equation

$$\begin{bmatrix} y_1 \\ y_2 \\ y_3 \end{bmatrix} = \begin{bmatrix} 1 & 0 \\ 1 & 1 \\ 0 & 2 \end{bmatrix} \begin{bmatrix} x_1 \\ x_2 \end{bmatrix} + \begin{bmatrix} 0 & 0 \\ 1 & 0 \\ 0 & 1 \end{bmatrix} \begin{bmatrix} f_1 \\ f_2 \end{bmatrix} \qquad (13.40b)$$

In this case,

$$\mathbf{A} = \begin{bmatrix} 0 & 1 \\ -2 & -3 \end{bmatrix} \quad \mathbf{B} = \begin{bmatrix} 1 & 0 \\ 1 & 1 \end{bmatrix} \quad \mathbf{C} = \begin{bmatrix} 1 & 0 \\ 1 & 1 \\ 0 & 2 \end{bmatrix} \quad \mathbf{D} = \begin{bmatrix} 0 & 0 \\ 1 & 0 \\ 0 & 1 \end{bmatrix} \qquad (13.40c)$$

and

$$\mathbf{\Phi}(s) = (s\mathbf{I} - \mathbf{A})^{-1} = \begin{bmatrix} s & -1 \\ 2 & s+3 \end{bmatrix}^{-1} = \begin{bmatrix} \frac{s+3}{(s+1)(s+2)} & \frac{1}{(s+1)(s+2)} \\ \frac{-2}{(s+1)(s+2)} & \frac{s}{(s+1)(s+2)} \end{bmatrix} \qquad (13.41)$$

Hence, the transfer function matrix $\mathbf{H}(s)$ is given by

$$\mathbf{H}(s) = \mathbf{C}\mathbf{\Phi}(s)\mathbf{B} + \mathbf{D}$$

$$= \begin{bmatrix} 1 & 0 \\ 1 & 1 \\ 0 & 2 \end{bmatrix} \begin{bmatrix} \frac{s+3}{(s+1)(s+2)} & \frac{1}{(s+1)(s+2)} \\ \frac{-2}{(s+1)(s+2)} & \frac{s}{(s+1)(s+2)} \end{bmatrix} \begin{bmatrix} 1 & 0 \\ 1 & 1 \end{bmatrix} + \begin{bmatrix} 0 & 0 \\ 1 & 0 \\ 0 & 1 \end{bmatrix}$$

$$= \begin{bmatrix} \frac{s+4}{(s+1)(s+2)} & \frac{1}{(s+1)(s+2)} \\ \frac{s+4}{s+2} & \frac{1}{s+2} \\ \frac{2(s-2)}{(s+1)(s+2)} & \frac{s^2+5s+2}{(s+1)(s+2)} \end{bmatrix} \qquad\qquad (13.42)$$

and the zero-state response is

$$\mathbf{Y}(s) = \mathbf{H}(s)\mathbf{F}(s)$$

Remember that the ijth element of the transfer function matrix in Eq. (13.42) represents the transfer function that relates the output $y_i(t)$ to the input $f_j(t)$. Thus, the transfer function that relates the output y_3 to the input f_2 is $H_{32}(s)$, where

$$H_{32}(s) = \frac{s^2 + 5s + 2}{(s+1)(s+2)} \qquad \blacksquare$$

⊙ **Computer Example C13.3**
Solve Example 13.6 using MATLAB.
Caution: The common factor $(s+1)$ in two of the transfer functions in Eq. (13.42) are canceled. The MATLAB answer gives transfer function with common factor.

    A=[0 1;-2 -3]; B=[1 0;1 1];
    C=[1 0;1 1;0 2]; D=[0 0;1 0;0 1];
    [num1,den1]=ss2tf(A,B,C,D,1)
    [num2,den2]=ss2tf(A,B,C,D,2)   ⊙

Characteristic Roots (Eigenvalues) of a Matrix

It is interesting to observe that the denominator of every transfer function in Eq. (13.42) is $(s+1)(s+2)$ with the exception of $H_{21}(s)$ and $H_{22}(s)$, where the cancellation of the factor $(s+1)$ occurs. This fact is no coincidence. We see that the denominator of every element of $\mathbf{\Phi}(s)$ is $|s\mathbf{I} - \mathbf{A}|$ because $\mathbf{\Phi}(s) = (s\mathbf{I} - \mathbf{A})^{-1}$, and the inverse of a matrix has its determinant in the denominator. Since $\mathbf{C}$, $\mathbf{B}$, and $\mathbf{D}$ are matrices with constant elements, we see from Eq. (13.38b) that the denominator of $\mathbf{\Phi}(s)$ will also be the denominator of $\mathbf{H}(s)$. Hence, the denominator of every element of $\mathbf{H}(s)$ is $|s\mathbf{I} - \mathbf{A}|$, except for the possible cancellation of the common factors mentioned earlier. In other words, the poles of all transfer functions of the system are also the zeros of the polynomial $|s\mathbf{I} - \mathbf{A}|$. *Therefore, the zeros of the polynomial $|s\mathbf{I} - \mathbf{A}|$ are the characteristic roots of the system.* Hence, the characteristic roots of the system are the roots of the equation

$$|s\mathbf{I} - \mathbf{A}| = 0 \qquad\qquad (13.43a)$$

Since $|s\mathbf{I} - \mathbf{A}|$ is an nth-order polynomial in s with n zeros $\lambda_1, \lambda_2, \ldots, \lambda_n$, we can write Eq. (13.43a) as

$$|s\mathbf{I} - \mathbf{A}| = s^n + a_{n-1}s^{n+1} + \cdots + a_1s + a_0$$

$$= (s - \lambda_1)(s - \lambda_2)\cdots(s - \lambda_n) = 0 \qquad (13.43\text{b})$$

For the system in Example 13.6,

$$|s\mathbf{I} - \mathbf{A}| = \begin{vmatrix} s & 0 \\ 0 & s \end{vmatrix} - \begin{vmatrix} 0 & 1 \\ -2 & -3 \end{vmatrix}$$

$$= \begin{vmatrix} s & -1 \\ 2 & s+3 \end{vmatrix}$$

$$= s^2 + 3s + 2 \qquad (13.44\text{a})$$

$$= (s + 1)(s + 2) \qquad (13.44\text{b})$$

Hence

$$\lambda_1 = -1 \quad \text{and} \quad \lambda_2 = -2$$

Equation (13.43) is known as the **characteristic equation of the matrix A**, and $\lambda_1, \lambda_2, \ldots, \lambda_n$ are the characteristic roots of $\mathbf{A}$. The term **eigenvalue**, meaning "characteristic value" in German, is also commonly used in the literature. Thus, we have shown that the characteristic roots of a system are the eigenvalues (characteristic values) of the matrix $\mathbf{A}$.

At this point, the reader will recall that if $\lambda_1, \lambda_2, \ldots, \lambda_n$ are the poles of the transfer function, then the zero-input response is of the form

$$y_0(t) = c_1 e^{\lambda_1 t} + c_2 e^{\lambda_2 t} + \cdots + c_n e^{\lambda_n t} \qquad (13.45)$$

This fact is also obvious from Eq. (13.38). The denominator of every element of the zero-input response matrix $\mathbf{C\Phi}(s)\mathbf{x}(0)$ is $|s\mathbf{I} - \mathbf{A}| = (s - \lambda_1)(s - \lambda_2)\cdots(s - \lambda_n)$. Therefore, the partial-fraction expansion and the subsequent inverse Laplace transform will yield a zero-input component of the form in Eq. (13.45).

13.3-2 Time-Domain Solution of State Equations

The state equation is

$$\dot{\mathbf{x}} = \mathbf{Ax} + \mathbf{Bf} \qquad (13.46)$$

We now show that the solution of the vector differential Equation (13.46) is

$$\mathbf{x}(t) = e^{\mathbf{A}t}\mathbf{x}(0) + \int_0^t e^{\mathbf{A}(t-\tau)}\mathbf{Bf}(\tau)\,d\tau \qquad (13.47)$$

Before proceeding further, we must define the exponential of the matrix appearing in Eq. (13.47). An exponential of a matrix is defined by an infinite series identical to that used in defining an exponential of a scalar. We shall define

$$e^{\mathbf{A}t} = \mathbf{I} + \mathbf{A}t + \frac{\mathbf{A}^2 t^2}{2!} + \frac{\mathbf{A}^3 t^3}{3!} + \cdots + \frac{\mathbf{A}^n t^n}{n!} + \cdots + \qquad (13.48\text{a})$$

$$= \sum_{k=0}^{\infty} \frac{\mathbf{A}^k t^k}{k!} \qquad (13.48\text{b})$$

Thus, if

$$\mathbf{A} = \begin{bmatrix} 0 & 1 \\ 2 & 1 \end{bmatrix}$$

then

$$\mathbf{A}t = \begin{bmatrix} 0 & 1 \\ 2 & 1 \end{bmatrix} t = \begin{bmatrix} 0 & t \\ 2t & t \end{bmatrix} \tag{13.49}$$

and

$$\frac{\mathbf{A}^2 t^2}{2!} = \begin{bmatrix} 0 & 1 \\ 2 & 1 \end{bmatrix} \begin{bmatrix} 0 & 1 \\ 2 & 1 \end{bmatrix} \frac{t^2}{2} = \begin{bmatrix} 2 & 1 \\ 2 & 3 \end{bmatrix} \frac{t^2}{2} = \begin{bmatrix} t^2 & \frac{t^2}{2} \\ t^2 & \frac{3t^2}{2} \end{bmatrix} \tag{13.50}$$

and so on.

We can show that the infinite series in Eq. (13.48) is absolutely and uniformly convergent for all values of t. Consequently, it can be differentiated or integrated term by term. Thus, to find $(d/dt)e^{\mathbf{A}t}$, we differentiate the series on the right-hand side of Eq. (13.48a) term by term:

$$\frac{d}{dt}e^{\mathbf{A}t} = \mathbf{A} + \mathbf{A}^2 t + \frac{\mathbf{A}^3 t^2}{2!} + \frac{\mathbf{A}^4 t^3}{3!} + \cdots \tag{13.51a}$$

$$= \mathbf{A}\left[\mathbf{I} + \mathbf{A}t + \frac{\mathbf{A}^2 t^2}{2!} + \frac{\mathbf{A}^3 t^3}{3!} + \cdots\right]$$

$$= \mathbf{A}e^{\mathbf{A}t} \tag{13.51b}$$

Note that the infinite series on the right-hand side of Eq. (13.51a) also may be expressed as

$$\frac{d}{dt}e^{\mathbf{A}t} = \left[\mathbf{I} + \mathbf{A}t + \frac{\mathbf{A}^2 t^2}{2!} + \frac{\mathbf{A}^3 t^3}{3!} + \cdots + \cdots\right]\mathbf{A}$$

$$= e^{\mathbf{A}t}\mathbf{A}$$

Hence

$$\frac{d}{dt}e^{\mathbf{A}t} = \mathbf{A}e^{\mathbf{A}t} = e^{\mathbf{A}t}\mathbf{A} \tag{13.52}$$

Also note that from the definition (13.48a), it follows that

$$e^{\mathbf{0}} = \mathbf{I} \tag{13.53a}$$

where

$$\mathbf{I} = \begin{bmatrix} 1 & 0 \\ 0 & 1 \end{bmatrix}$$

If we premultiply or postmultiply the infinite series for $e^{\mathbf{A}t}$ [Eq. (13.48a)] by an infinite series for $e^{-\mathbf{A}t}$, we find that

$$(e^{-\mathbf{A}t})(e^{\mathbf{A}t}) = (e^{\mathbf{A}t})(e^{-\mathbf{A}t}) = \mathbf{I} \tag{13.53b}$$

In Sec. B.6-3, we show that

$$\frac{d}{dt}(\mathbf{PQ}) = \frac{d\mathbf{P}}{dt}\mathbf{Q} + \mathbf{P}\frac{d\mathbf{Q}}{dt}$$

Using this relationship, we observe that

$$\frac{d}{dt}[e^{-\mathbf{A}t}\mathbf{x}] = \left(\frac{d}{dt}e^{-\mathbf{A}t}\right)\mathbf{x} + e^{-\mathbf{A}t}\dot{\mathbf{x}}$$

$$= -e^{-\mathbf{A}t}\mathbf{A}\mathbf{x} + e^{-\mathbf{A}t}\dot{\mathbf{x}} \tag{13.54}$$

We now premultiply both sides of Eq. (13.46) by $e^{-\mathbf{A}t}$ to yield

$$e^{-\mathbf{A}t}\dot{\mathbf{x}} = e^{-\mathbf{A}t}\mathbf{A}\mathbf{x} + e^{-\mathbf{A}t}\mathbf{Bf} \tag{13.55a}$$

or

$$e^{-\mathbf{A}t}\dot{\mathbf{x}} - e^{-\mathbf{A}t}\mathbf{A}\mathbf{x} = e^{-\mathbf{A}t}\mathbf{Bf} \tag{13.55b}$$

A glance at Eq. (13.54) shows that the left-hand side of Eq. (13.55b) is $\frac{d}{dt}[e^{-\mathbf{A}t}]$. Hence

$$\frac{d}{dt}[e^{-\mathbf{A}t}] = e^{-\mathbf{A}t}\mathbf{Bf}$$

The integration of both sides of this equation from 0 to t yields

$$e^{-\mathbf{A}t}\mathbf{x}\Big|_0^t = \int_0^t e^{-\mathbf{A}\tau}\mathbf{Bf}(\tau)\,d\tau \tag{13.56a}$$

or

$$e^{-\mathbf{A}t}\mathbf{x}(t) - \mathbf{x}(0) = \int_0^t e^{-\mathbf{A}\tau}\mathbf{Bf}(\tau)\,d\tau \tag{13.56b}$$

Hence

$$e^{-\mathbf{A}t}\mathbf{x} = \mathbf{x}(0) + \int_0^t e^{-\mathbf{A}\tau}\mathbf{Bf}(\tau)\,d\tau \tag{13.56c}$$

Premultiplying Eq. (13.56c) by $e^{\mathbf{A}t}$ and using Eq. (13.53b), we have

$$\mathbf{x}(t) = \underbrace{e^{\mathbf{A}t}\mathbf{x}(0)}_{\text{zero-input component}} + \underbrace{\int_0^t e^{\mathbf{A}(t-\tau)}\mathbf{Bf}(\tau)\,d\tau}_{\text{zero-state component}} \tag{13.57a}$$

This is the desired solution. The first term on the right-hand side represents $x(t)$ when the input $f(t) = 0$. Hence it is the zero-input component. The second term, by a similar argument, is seen to be the zero-state component.

The results of Eq. (13.57a) can be expressed more conveniently in terms of the matrix convolution. We can define the convolution of two matrices in a manner similar to the multiplication of two matrices, except that the multiplication of two elements is replaced by their convolution. For example,

$$\begin{bmatrix} f_1 & f_2 \\ f_3 & f_4 \end{bmatrix} * \begin{bmatrix} g_1 & g_2 \\ g_3 & g_4 \end{bmatrix} = \begin{bmatrix} (f_1 * g_1 + f_2 * g_3) & (f_1 * g_2 + f_2 * g_4) \\ (f_3 * g_1 + f_4 * g_3) & (f_3 * g_2 + f_4 * g_4) \end{bmatrix}$$

Using this definition of matrix convolution, we can express Eq. (13.57a) as

$$\mathbf{x}(t) = e^{\mathbf{A}t}\mathbf{x}(0) + e^{\mathbf{A}t} * \mathbf{B}\mathbf{f}(t) \tag{13.57b}$$

Note that the limits of the convolution integral [Eq. (13.57a)] are from 0 to t. Hence, all the elements of $e^{\mathbf{A}t}$ in the convolution term of Eq. (13.57b) are implicitly assumed to be multiplied by $u(t)$.

The result of Eq. (13.57) can be easily generalized for any initial value of t. It is left as an exercise for the reader to show that the solution of the state equation can be expressed as

$$\mathbf{x}(t) = e^{\mathbf{A}(t-t_0)}\mathbf{x}(t_0) + \int_{t_0}^{t} e^{\mathbf{A}(t-\tau)}\mathbf{B}\mathbf{f}(\tau)\,d\tau \tag{13.58}$$

Determining $e^{\mathbf{A}t}$

The exponential $e^{\mathbf{A}t}$ required in Eq. (13.57) can be computed from the definition in Eq. (13.51a). Unfortunately, this is an infinite series, and its computation can be quite laborious. Moreover, we may not be able to recognize the closed-form expression for the answer. There are several efficient methods of determining $e^{\mathbf{A}t}$ in closed form. It is shown in Sec. B.6-5 that for an $n \times n$ matrix $\mathbf{A}$,

$$e^{\mathbf{A}t} = \beta_0\mathbf{I} + \beta_1\mathbf{A} + \beta_2\mathbf{A}^2 + \cdots + \beta_{n-1}\mathbf{A}^{n-1} \tag{13.59a}$$

where

$$
\begin{bmatrix} \beta_0 \\ \beta_1 \\ \cdots \\ \beta_{n-1} \end{bmatrix}
=
\begin{bmatrix}
1 & \lambda_1 & \lambda_1^2 & \cdots & \lambda_1^{n-1} \\
1 & \lambda_2 & \lambda_2^2 & \cdots & \lambda_2^{n-1} \\
\cdots\cdots\cdots\cdots\cdots\cdots\cdots \\
1 & \lambda_n & \lambda_n^2 & \cdots & \lambda_n^{n-1}
\end{bmatrix}^{-1}
\begin{bmatrix} e^{\lambda_1 t} \\ e^{\lambda_2 t} \\ \cdots \\ e^{\lambda_n t} \end{bmatrix}
$$

and $\lambda_1, \lambda_2, \ldots, \lambda_n$ are the n characteristic values (eigenvalues) of $\mathbf{A}$.

We can also determine $e^{\mathbf{A}t}$ by comparing Eqs. (13.57a) and (13.35b). It is clear that

$$e^{\mathbf{A}t} = \mathcal{L}^{-1}[\mathbf{\Phi}(s)] \tag{13.59b}$$

$$= \mathcal{L}^{-1}[(s\mathbf{I} - \mathbf{A})^{-1}] \tag{13.59c}$$

Thus, $e^{\mathbf{A}t}$ and $\mathbf{\Phi}(s)$ are a Laplace transform pair. To be consistent with Laplace transform notation, $e^{\mathbf{A}t}$ is often denoted by $\boldsymbol{\phi}(t)$, **the state transition matrix** (STM):

$$e^{\mathbf{A}t} = \boldsymbol{\phi}(t)$$

■ **Example 13.7**

Find the solution to the problem in Example 13.5 using the time-domain method. For this case, the characteristic roots are given by

$$|s\mathbf{I} - \mathbf{A}| = \begin{vmatrix} s+12 & -\frac{2}{3} \\ 36 & s+1 \end{vmatrix} = s^2 + 13s + 36 = (s+4)(s+9) = 0$$

The roots are $\lambda_1 = -4$ and $\lambda_2 = -9$, so

$$\begin{bmatrix} \beta_0 \\ \beta_1 \end{bmatrix} = \begin{bmatrix} 1 & -4 \\ 1 & -9 \end{bmatrix}^{-1} \begin{bmatrix} e^{-4t} \\ e^{-9t} \end{bmatrix} = \frac{1}{5} \begin{bmatrix} 9e^{-4t} - 4e^{-9t} \\ e^{-4t} - e^{-9t} \end{bmatrix}$$

and

$$e^{\mathbf{A}t} = \beta_0 \mathbf{I} + \beta_1 \mathbf{A}$$

$$= \left(\frac{9}{5}e^{-4t} - \frac{4}{5}e^{-9t} \right) \begin{bmatrix} 1 & 0 \\ 0 & 1 \end{bmatrix} + \left(\frac{1}{5}e^{-4t} - \frac{1}{5}e^{-9t} \right) \begin{bmatrix} -12 & \frac{2}{3} \\ -36 & -1 \end{bmatrix}$$

$$= \begin{bmatrix} \left(\frac{-3}{5}e^{-4t} + \frac{8}{5}e^{-9t} \right) & \frac{2}{15}(e^{-4t} - e^{-9t}) \\ \frac{36}{5}(-e^{-4t} + e^{-9t}) & \left(\frac{8}{5}e^{-4t} - \frac{3}{5}e^{-9t} \right) \end{bmatrix} \tag{13.60}$$

The zero-input component is given by [see Eq. (13.57a)]

$$e^{\mathbf{A}t}\mathbf{x}(0) = \begin{bmatrix} \left(-\frac{3}{5}e^{-4t} + \frac{8}{5}e^{-9t} \right) & \frac{2}{15}(e^{-4t} - e^{-9t}) \\ \frac{36}{5}(-e^{-4t} + e^{-9t}) & \left(\frac{8}{5}e^{-4t} - \frac{3}{5}e^{-9t} \right) \end{bmatrix} \begin{bmatrix} 2 \\ 1 \end{bmatrix}$$

$$= \begin{bmatrix} \left(\frac{-16}{15}e^{-4t} + \frac{46}{15}e^{-9t} \right) u(t) \\ \left(\frac{-64}{5}e^{-4t} + \frac{69}{5}e^{-9t} \right) u(t) \end{bmatrix} \tag{13.61a}$$

Note the presence of $u(t)$ in Eq. (13.61a), indicating that the response begins at $t = 0$.
The zero-state component is $e^{\mathbf{A}t} * \mathbf{B}f$ [see Eq. (13.57b)], where

$$\mathbf{B}f = \begin{bmatrix} \frac{1}{3} \\ 1 \end{bmatrix} u(t) = \begin{bmatrix} \frac{1}{3}u(t) \\ u(t) \end{bmatrix}$$

and

$$e^{\mathbf{A}t} * \mathbf{B}f(t) = \begin{bmatrix} (\frac{-3}{5}e^{-4t} + \frac{8}{5}e^{-9t})u(t) & \frac{2}{15}(e^{-4t} - e^{-9t})u(t) \\ \frac{36}{5}(-e^{-4t} + e^{-9t}u(t)) & (\frac{8}{5}e^{-4t} - \frac{3}{5}e^{-9t})u(t) \end{bmatrix} * \begin{bmatrix} \frac{1}{3}u(t) \\ u(t) \end{bmatrix}$$

Note again the presence of the term $u(t)$ in every element of $e^{\mathbf{A}t}$. This is the case because
the limits of the convolution integral run from 0 to t [Eq. (13.56)]. Thus

$$e^{\mathbf{A}t} * \mathbf{B}f(t) = \begin{bmatrix} \left(-\frac{3}{5}e^{-4t} + \frac{8}{5}e^{-9t} \right) u(t) * \frac{1}{3}u(t) & \frac{2}{15}(e^{-4t} - e^{-9t})u(t) * u(t) \\ \frac{36}{5}(-e^{-4t} + e^{-9t})u(t) * \frac{1}{3}u(t) & \left(\frac{8}{5}e^{-4t} - \frac{3}{5}e^{-9t} \right) u(t) * u(t) \end{bmatrix}$$

$$= \begin{bmatrix} -\frac{1}{15}e^{-4t}u(t) * u(t) + \frac{2}{5}e^{-9t}u(t) * u(t) \\ -\frac{4}{5}e^{-4t}u(t) * u(t) + \frac{9}{5}e^{-9t}u(t) * u(t) \end{bmatrix}$$

Substitution for the above convolution integrals from the convolution table (Table 2.1) yields

$$e^{\mathbf{A}t} * \mathbf{Bf}(t) = \begin{bmatrix} -\frac{1}{60}(1 - e^{-4t})u(t) + \frac{2}{45}(1 - e^{-9t})u(t) \\[2mm] -\frac{1}{5}(1 - e^{-4t})u(t) + \frac{1}{5}(1 - e^{-9t})u(t) \end{bmatrix}$$

$$= \begin{bmatrix} (\frac{1}{36} + \frac{1}{60}e^{-4t} - \frac{2}{45}e^{-9t})u(t) \\[2mm] \frac{1}{5}(e^{-4t} - e^{-9t})u(t) \end{bmatrix} \tag{13.61b}$$

The sum of the two components [Eq. (13.61a) and Eq. (13.61b)] now gives the desired solution for $\mathbf{x}(t)$:

$$\mathbf{x}(t) = \begin{bmatrix} x_1(t) \\ x_2(t) \end{bmatrix} = \begin{bmatrix} \left(\frac{1}{36} - \frac{21}{20}e^{-4t} + \frac{136}{45}e^{-9t}\right)u(t) \\[2mm] \left(\frac{-63}{5}e^{-4t} + \frac{68}{5}e^{-9t}\right)u(t) \end{bmatrix} \tag{13.61c}$$

This result is consistent with the solution obtained by using the frequency-domain method [see Eq. (13.36b)]. Once the state variables x_1 and x_2 are found for $t \geq 0$, all the remaining variables are determined from the output equation. ■

The Output

The output equation is given by

$$\mathbf{y}(t) = \mathbf{Cx}(t) + \mathbf{Df}(t)$$

The substitution of the solution for $\mathbf{x}$ [Eq. (13.57)] in this equation yields

$$\mathbf{y}(t) = \mathbf{C}[e^{\mathbf{A}t}\mathbf{x}(0) + e^{\mathbf{A}t} * \mathbf{Bf}(t)] + \mathbf{Df}(t) \tag{13.62a}$$

Since the elements of $\mathbf{B}$ are constants,

$$e^{\mathbf{A}t} * \mathbf{Bf}(t) = e^{\mathbf{A}t}\mathbf{B} * \mathbf{f}(t)$$

With this result, Eq. (13.62a) becomes

$$\mathbf{y}(t) = \mathbf{C}[e^{\mathbf{A}t}\mathbf{x}(0) + e^{\mathbf{A}t}\mathbf{B} * \mathbf{f}(t)] + \mathbf{Df}(t) \tag{13.62b}$$

Now recall that the convolution of $f(t)$ with the unit impulse $\delta(t)$ yields $f(t)$. Let us define a $j \times j$ diagonal matrix $\boldsymbol{\delta}(t)$ such that all its diagonal terms are unit impulse functions. It is then obvious that

$$\boldsymbol{\delta}(t) * \mathbf{f}(t) = \mathbf{f}(t)$$

and Eq. (13.62b) can be expressed as

$$\mathbf{y}(t) = \mathbf{C}[e^{\mathbf{A}t}\mathbf{x}(0) + e^{\mathbf{A}t}\mathbf{B} * \mathbf{f}(t)] + \mathbf{D}\boldsymbol{\delta}(t) * \mathbf{f}(t) \tag{13.63a}$$

$$= \mathbf{C}e^{\mathbf{A}t}\mathbf{x}(0) + [\mathbf{C}e^{\mathbf{A}t}\mathbf{B} + \mathbf{D}\boldsymbol{\delta}(t)] * \mathbf{f}(t) \tag{13.63b}$$

With the notation $\boldsymbol{\phi}(t)$ for $e^{\mathbf{A}t}$, Eq. (13.63b) may be expressed as

$$\mathbf{y}(t) = \underbrace{\mathbf{C}\boldsymbol{\phi}(t)\mathbf{x}(0)}_{\text{zero-input response}} + \underbrace{[\mathbf{C}\boldsymbol{\phi}(t)\mathbf{B} + \mathbf{D}\boldsymbol{\delta}(t)] * \mathbf{f}(t)}_{\text{zero-state response}} \qquad (13.63c)$$

The zero-state response; that is, the response when $\mathbf{x}(0) = \mathbf{0}$, is

$$\mathbf{y}(t) = [\mathbf{C}\boldsymbol{\phi}(t)\mathbf{B} + \mathbf{D}\boldsymbol{\delta}(t)] * \mathbf{f}(t) \qquad (13.64a)$$

$$= \mathbf{h}(t) * \mathbf{f}(t) \qquad (13.64b)$$

where

$$\mathbf{h}(t) = \mathbf{C}\boldsymbol{\phi}(t)\mathbf{B} + \mathbf{D}\boldsymbol{\delta}(t) \qquad (13.65)$$

The matrix $\mathbf{h}(t)$ is a $k \times j$ matrix known as the **impulse response matrix**. The reason for this designation is obvious. The ijth element of $\mathbf{h}(t)$ is $h_{ij}(t)$, which represents the zero-state response y_i when the input $f_j(t) = \delta(t)$ and when all other inputs (and all the initial conditions) are zero. It can also be seen from Eq. (13.39) and (13.64b) that

$$\mathcal{L}[\mathbf{h}(t)] = \mathbf{H}(s)$$

■ **Example 13.8**

For the system described by Eqs. (13.40a) and (13.40b), determine $e^{\mathbf{A}t}$ using Eq. (13.59b):

$$\boldsymbol{\phi}(t) = e^{\mathbf{A}t} = \mathcal{L}^{-1}\boldsymbol{\Phi}(s)$$

This problem was solved earlier with frequency-domain techniques. From Eq. (13.41), we have

$$\boldsymbol{\phi}(t) = \mathcal{L}^{-1}\begin{bmatrix} \frac{s+3}{(s+1)(s+2)} & \frac{1}{(s+1)(s+2)} \\ \frac{-2}{(s+1)(s+2)} & \frac{s}{(s+1)(s+2)} \end{bmatrix}$$

$$= \mathcal{L}^{-1}\begin{bmatrix} \frac{2}{s+1} - \frac{1}{s+2} & \frac{1}{s+1} - \frac{1}{s+2} \\ \frac{-2}{s+1} + \frac{2}{s+2} & \frac{-1}{s+1} + \frac{2}{s+2} \end{bmatrix}$$

$$= \begin{bmatrix} 2e^{-t} - e^{-2t} & e^{-t} - e^{-2t} \\ -2e^{-t} + 2e^{-2t} & -e^{-t} + 2e^{-2t} \end{bmatrix}$$

The same result is obtained in Sec. B.6-5 by using Eq. (13.59a) [see Eq. (B.84)].

Also, $\boldsymbol{\delta}(t)$ is a diagonal $j \times j$ or 2×2 matrix:

$$\boldsymbol{\delta}(t) = \begin{bmatrix} \delta(t) & 0 \\ 0 & \delta(t) \end{bmatrix}$$

Substituting the matrices $\boldsymbol{\phi}(t)$, $\boldsymbol{\delta}(t)$, $\mathbf{C}$, $\mathbf{D}$, and $\mathbf{B}$ [Eq. (13.40c)] into Eq. (13.65), we have

$$\mathbf{h}(t) = \begin{bmatrix} 1 & 0 \\ 1 & 1 \\ 0 & 2 \end{bmatrix}\begin{bmatrix} 2e^{-t} - e^{-2t} & e^{-t} - e^{-2t} \\ -2e^{-t} + 2e^{-2t} & -e^{-+2} + e^{-2t} \end{bmatrix}\begin{bmatrix} 1 & 0 \\ 1 & 1 \end{bmatrix} + \begin{bmatrix} 0 & 0 \\ 1 & 0 \\ 0 & 1 \end{bmatrix}\begin{bmatrix} \delta(t) & 0 \\ 0 & \delta(t) \end{bmatrix}$$

$$= \begin{bmatrix} 3e^{-t} - 2e^{-2t} & e^{-t} - e^{-2t} \\ \delta(t) + 2e^{-2t} & e^{-2t} \\ -6e^{-t} + 8e^{-2t} & \delta(t) - 2e^{-2t} + 4e^{-2t} \end{bmatrix} \qquad (13.66)$$

The reader can verify that the transfer-function matrix $\mathbf{H}(s)$ in Eq. (13.42) is the Laplace transform of the unit-impulse response matrix $\mathbf{h}(t)$ in Eq. (13.66). ■

13.4   Linear Transformation of State Vectors

In Sec. 13.1 we saw that the state of a system can be specified in several ways. The sets of all possible state variables must be related—in other words, if we are given one set of state variables, we should be able to relate it to any other set. We are particularly interested in a linear type of relationship. Let $x_1, x_2, \ldots, x_n$ and $w_1, w_2, \ldots, w_n$ be two different sets of state variables specifying the same system. Let these sets be related by linear equations as

$$w_1 = p_{11}x_1 + p_{12}x_2 + \cdots + p_{1n}x_n$$

$$w_2 = p_{21}x_1 + p_{22}x_2 + \cdots + p_{2n}x_n$$

$$\ldots \ldots \ldots \ldots \ldots \ldots \ldots \ldots \ldots \ldots \ldots \ldots \ldots$$

(13.67a)

$$w_n = p_{n1}x_1 + p_{n2}x_2 + \cdots + p_{nn}x_n$$

or

$$\underbrace{\begin{bmatrix} w_1 \\ w_2 \\ \ldots \\ w_n \end{bmatrix}}_{\mathbf{w}} = \underbrace{\begin{bmatrix} p_{11} & p_{12} & \cdots & p_{1n} \\ p_{21} & p_{22} & \cdots & p_{2n} \\ \ldots \ldots \ldots \ldots \ldots \\ p_{n1} & p_{n2} & \cdots & p_{nn} \end{bmatrix}}_{\mathbf{P}} \underbrace{\begin{bmatrix} x_1 \\ x_2 \\ \ldots \\ x_n \end{bmatrix}}_{\mathbf{x}}$$

(13.67b)

Defining the vector $\mathbf{w}$ and matrix $\mathbf{P}$, as shown above, we can write Eq. (13.67b) as

$$\mathbf{w} = \mathbf{Px}$$

(13.67c)

and

$$\mathbf{x} = \mathbf{P}^{-1}\mathbf{w}$$

(13.67d)

Thus, the state vector $\mathbf{x}$ is transformed into another state vector $\mathbf{w}$ through the linear transformation in Eq. (13.67c).

If we know $\mathbf{w}$, we can determine $\mathbf{x}$ from Eq. (13.67d), provided that $\mathbf{P}^{-1}$ exists. This is equivalent to saying that $\mathbf{P}$ is a nonsingular matrix† ($|\mathbf{P}| \neq 0$). Thus, if $\mathbf{P}$ is a nonsingular matrix, the vector $\mathbf{w}$ defined by Eq. (13.67c) is also a state vector. Consider the state equation of a system

$$\dot{\mathbf{x}} = \mathbf{Ax} + \mathbf{Bf}$$

(13.68a)

If

$$\mathbf{w} = \mathbf{Px}$$

(13.68b)

then

$$\mathbf{x} = \mathbf{P}^{-1}\mathbf{w}$$

---

†This condition is equivalent to saying that all n equations in Eq. (13.67a) are linearly independent; that is, none of the n equations can be expressed as a linear combination of the remaining equations.

and

$$\dot{\mathbf{x}} = \mathbf{P}^{-1}\dot{\mathbf{w}}$$

Hence the state equation (13.68a) now becomes

$$\mathbf{P}^{-1}\dot{\mathbf{w}} = \mathbf{A}\mathbf{P}^{-1}\mathbf{w} + \mathbf{B}\mathbf{f}$$

or

$$\dot{\mathbf{w}} = \mathbf{P}\mathbf{A}\mathbf{P}^{-1}\mathbf{w} + \mathbf{P}\mathbf{B}\mathbf{f} \qquad (13.68\text{c})$$

$$= \hat{\mathbf{A}}\mathbf{w} + \hat{\mathbf{B}}\mathbf{f} \qquad (13.68\text{d})$$

where

$$\hat{\mathbf{A}} = \mathbf{P}\mathbf{A}\mathbf{P}^{-1} \qquad (13.69\text{a})$$

and

$$\hat{\mathbf{B}} = \mathbf{P}\mathbf{B} \qquad (13.69\text{b})$$

Equation (13.68d) is a state equation for the same system, but now it is expressed in terms of the state vector $\mathbf{w}$.

The output equation is also modified. Let the original output equation be

$$\mathbf{y} = \mathbf{C}\mathbf{x} + \mathbf{D}\mathbf{f}$$

In terms of the new state variable $\mathbf{w}$, this equation becomes

$$\mathbf{y} = \mathbf{C}(\mathbf{P}^{-1}\mathbf{w}) + \mathbf{D}\mathbf{f}$$

$$= \hat{\mathbf{C}}\mathbf{w} + \mathbf{D}\mathbf{f}$$

where

$$\hat{\mathbf{C}} = \mathbf{C}\mathbf{P}^{-1} \qquad (13.69\text{c})$$

■ **Example 13.9**

The state equations of a certain system are given by

$$\begin{bmatrix} \dot{x}_1 \\ \dot{x}_2 \end{bmatrix} = \begin{bmatrix} 0 & 1 \\ -2 & -3 \end{bmatrix} \begin{bmatrix} x_1 \\ x_2 \end{bmatrix} + \begin{bmatrix} 1 \\ 2 \end{bmatrix} f(t) \qquad (13.70\text{a})$$

Find the state equations for this system when the new state variables w_1 and w_2 are

$$w_1 = x_1 + x_2$$

$$w_2 = x_1 - x_2$$

or

$$\begin{bmatrix} w_1 \\ w_2 \end{bmatrix} = \begin{bmatrix} 1 & 1 \\ 1 & -1 \end{bmatrix} \begin{bmatrix} x_1 \\ x_2 \end{bmatrix} \qquad (13.70\text{b})$$

According to Eq. (13.70b), the state equation for the state variable $\mathbf{w}$ is given by

$$\dot{\mathbf{w}} = \hat{\mathbf{A}}\mathbf{w} + \hat{\mathbf{B}}\mathbf{f}$$

where [see Eq. (13.69)]

$$\hat{\mathbf{A}} = \mathbf{P}\mathbf{A}\mathbf{P}^{-1} = \begin{bmatrix} 1 & 1 \\ 1 & -1 \end{bmatrix} \begin{bmatrix} 0 & 1 \\ -2 & -3 \end{bmatrix} \begin{bmatrix} 1 & 1 \\ 1 & -1 \end{bmatrix}^{-1}$$

$$= \begin{bmatrix} 1 & 1 \\ 1 & -1 \end{bmatrix} \begin{bmatrix} 0 & 1 \\ -2 & -3 \end{bmatrix} \begin{bmatrix} \frac{1}{2} & \frac{1}{2} \\ \frac{1}{2} & -\frac{1}{2} \end{bmatrix}$$

$$= \begin{bmatrix} -2 & 0 \\ 3 & -1 \end{bmatrix}$$

and

$$\hat{\mathbf{B}} = \mathbf{P}\mathbf{B} = \begin{bmatrix} 1 & 1 \\ 1 & -1 \end{bmatrix} \begin{bmatrix} 1 \\ 2 \end{bmatrix} = \begin{bmatrix} 3 \\ -1 \end{bmatrix}$$

Therefore

$$\begin{bmatrix} \dot{w}_1 \\ \dot{w}_2 \end{bmatrix} = \begin{bmatrix} -2 & 0 \\ 3 & -1 \end{bmatrix} \begin{bmatrix} w_1 \\ w_2 \end{bmatrix} + \begin{bmatrix} 3 \\ -1 \end{bmatrix} f(t)$$

This is the desired state equation for the state vector $\mathbf{w}$. The solution of this equation requires a knowledge of the initial state $\mathbf{w}(0)$. This can be obtained from the given initial state $\mathbf{x}(0)$ by using Eq. (13.70b). ■

⊙ **Computer Example C13.4**
Solve Example 13.9 using MATLAB.

```
A=[0 1;-2 -3]; B=[1; 2];
P=[1 1;1 -1];
Ahat=P*A*inv(P)
Bhat=P*B    ⊙
```

Invariance of Eigenvalues

We have seen (Sec. 13.3) that the poles of all possible transfer functions of a system are the eigenvalues of the matrix $\mathbf{A}$. If we transform a state vector from $\mathbf{x}$ to $\mathbf{w}$, the variables $w_1, w_2, \ldots, w_n$ are linear combinations of $x_1, x_2, \ldots, x_n$ and therefore may be considered as outputs. Hence, the poles of the transfer functions relating $w_1, w_2, \ldots, w_n$ to the various inputs must also be the eigenvalues of matrix $\mathbf{A}$. On the other hand, the system is also specified by Eq. (13.68d). This means that the poles of the transfer functions must be the eigenvalues of $\hat{\mathbf{A}}$. Therefore, the eigenvalues of matrix $\mathbf{A}$ remain unchanged for the linear transformation of variables represented by Eq. (13.67), and the eigenvalues of matrix $\mathbf{A}$ and matrix $\hat{\mathbf{A}}(\hat{\mathbf{A}} = \mathbf{P}\mathbf{A}\mathbf{P}^{-1})$ are identical, implying that the characteristic equations of $\mathbf{A}$ and $\hat{\mathbf{A}}$ are also identical. This result also can be proved alternately as follows.
Consider the matrix $\mathbf{P}(s\mathbf{I} - \mathbf{A})\mathbf{P}^{-1}$. We have

$$\mathbf{P}(s\mathbf{I} - \mathbf{A})\mathbf{P}^{-1} = \mathbf{P}s\mathbf{I}\mathbf{P}^{-1} - \mathbf{P}\mathbf{A}\mathbf{P}^{-1} = s\mathbf{P}\mathbf{I}\mathbf{P}^{-1} - \hat{\mathbf{A}} = s\mathbf{I} - \hat{\mathbf{A}}$$

Taking the determinants of both sides, we obtain

$$|\mathbf{P}||s\mathbf{I} - \mathbf{A}||\mathbf{P}^{-1}| = |s\mathbf{I} - \hat{\mathbf{A}}|$$

The determinants $|\mathbf{P}|$ and $|\mathbf{P}^{-1}|$ are reciprocals of each other. Hence

$$|s\mathbf{I} - \mathbf{A}| = |s\mathbf{I} - \hat{\mathbf{A}}| \qquad\qquad (13.71)$$

This is the desired result. We have shown that the characteristic equations of $\mathbf{A}$ and $\hat{\mathbf{A}}$ are identical. Hence the eigenvalues of $\mathbf{A}$ and $\hat{\mathbf{A}}$ are identical.

In Example 13.9, matrix $\mathbf{A}$ is given as

$$\mathbf{A} = \begin{bmatrix} 0 & 1 \\ -2 & -3 \end{bmatrix}$$

The characteristic equation is

$$|s\mathbf{I} - \mathbf{A}| = \begin{vmatrix} s & -1 \\ 2 & s+3 \end{vmatrix} = s^2 + 3s + 2 = 0$$

Also

$$\hat{\mathbf{A}} = \begin{bmatrix} -2 & 0 \\ 3 & -1 \end{bmatrix}$$

and

$$|s\mathbf{I} - \hat{\mathbf{A}}| = \begin{bmatrix} s+2 & 0 \\ -3 & s+1 \end{bmatrix} = s^2 + 3s + 2 = 0$$

This result verifies that the characteristic equations of $\mathbf{A}$ and $\hat{\mathbf{A}}$ are identical.

13.4-1 Diagonalization of Matrix A

For several reasons, it is desirable to make matrix $\mathbf{A}$ diagonal. If $\mathbf{A}$ is not diagonal, we can transform the state variables such that the resulting matrix $\hat{\mathbf{A}}$ is diagonal.† One can show that for any diagonal matrix $\mathbf{A}$, the diagonal elements of this matrix must necessarily be $\lambda_1, \lambda_2, \ldots, \lambda_n$ (the eigenvalues) of the matrix. Consider the diagonal matrix $\mathbf{A}$:

$$\mathbf{A} = \begin{bmatrix} a_1 & 0 & 0 & 0 \\ 0 & a_2 & 0 & 0 \\ \multicolumn{4}{c}{\dots\dots\dots\dots\dots} \\ 0 & 0 & 0 & a_n \end{bmatrix}$$

The characteristic equation is given by

---

†In this discussion we assume distinct eigenvalues. If the eigenvalues are not distinct, we can reduce the matrix to a modified diagonalized (Jordan) form.

$$|s\mathbf{I} - \mathbf{A}| = \begin{bmatrix} (s - a_1) & 0 & 0 & \cdots & 0 \\ 0 & (s - a_2) & 0 & \cdots & 0 \\ \multicolumn{5}{c}{\dotfill} \\ 0 & 0 & 0 & \cdots & (s - a_n) \end{bmatrix} = 0$$

or

$$(s - a_1)(s - a_2) \cdots (s - a_n) = 0$$

Hence, the eigenvalues of $\mathbf{A}$ are $a_1, a_2, \ldots, a_n$. The nonzero (diagonal) elements of a diagonal matrix are therefore its eigenvalues $\lambda_1, \lambda_2, \ldots, \lambda_n$. We shall denote the diagonal matrix by a special symbol, Λ:

$$\Lambda = \begin{bmatrix} \lambda_1 & 0 & 0 & \cdots & 0 \\ 0 & \lambda_2 & 0 & \cdots & 0 \\ \multicolumn{5}{c}{\dotfill} \\ 0 & 0 & 0 & \cdots & \lambda_n \end{bmatrix} \tag{13.72}$$

Let us now consider the transformation of the state vector A such that the resulting matrix $\hat{\mathbf{A}}$ is a diagonal matrix Λ.

   Consider the system

$$\dot{\mathbf{x}} = \mathbf{A}\mathbf{x} + \mathbf{B}\mathbf{f}$$

We shall assume that $\lambda_1, \lambda_2, \ldots, \lambda_n$, the eigenvalues of $\mathbf{A}$, are distinct (no repeated roots). Let us transform the state vector $\mathbf{x}$ into the new state vector $\mathbf{z}$, using the transformation

$$\mathbf{z} = \mathbf{P}\mathbf{x} \tag{13.73a}$$

Then, after the development of Eq. (13.68c), we have

$$\dot{\mathbf{z}} = \mathbf{P}\mathbf{A}\mathbf{P}^{-1}\mathbf{z} + \mathbf{P}\mathbf{B}\mathbf{f} \tag{13.73b}$$

We desire the transformation to be such that $\mathbf{P}\mathbf{A}\mathbf{P}^{-1}$ is a diagonal matrix Λ given by Eq. (13.72), or

$$\dot{\mathbf{z}} = \Lambda\mathbf{z} + \hat{\mathbf{B}}\mathbf{f} \tag{13.73c}$$

Hence

$$\Lambda = \mathbf{P}\mathbf{A}\mathbf{P}^{-1} \tag{13.74a}$$

or

$$\Lambda\mathbf{P} = \mathbf{P}\mathbf{A} \tag{13.74b}$$

We know Λ and $\mathbf{A}$. Equation (13.74b) therefore can be solved to determine $\mathbf{P}$.

■ **Example 13.10**
   Find the diagonalized form of the state equation for the system in Example 13.9.
   In this case,

$$\mathbf{A} = \begin{bmatrix} 0 & 1 \\ -2 & -3 \end{bmatrix}$$

We found $\lambda_1 = -1$ and $\lambda_2 = -2$. Hence

$$\Lambda = \begin{bmatrix} -1 & 0 \\ 0 & -2 \end{bmatrix}$$

and Eq. (13.74b) becomes

$$\begin{bmatrix} -1 & 0 \\ 0 & -2 \end{bmatrix} \begin{bmatrix} p_{11} & p_{12} \\ p_{21} & p_{22} \end{bmatrix} = \begin{bmatrix} p_{11} & p_{12} \\ p_{21} & p_{22} \end{bmatrix} \begin{bmatrix} 0 & 1 \\ -2 & -3 \end{bmatrix}$$

Equating the four elements on two sides, we obtain

$$-p_{11} = -2p_{12} \tag{13.75a}$$

$$-p_{12} = p_{11} - 3p_{12} \tag{13.75b}$$

$$-2p_{21} = -2p_{22} \tag{13.75c}$$

$$-2p_{22} = p_{21} - 3p_{22} \tag{13.75d}$$

The reader will immediately recognize that Eqs. (13.75a) and (13.75b) are identical. Similarly, Eqs. (13.75c) and (13.75d) are identical. Hence two equations may be discarded, leaving us with only two equations [Eqs. (13.75a) and (13.75c)] and four unknowns. This observation means there is no unique solution. There is, in fact, an infinite number of solutions. We can assign any value to p_{11} and p_{21} to yield one possible solution.† If $p_{11} = k_1$ and $p_{21} = k_2$, then from Eqs. (13.75a) and (13.75c) we have $p_{12} = k_1/2$ and $p_{22} = k_2$:

$$\mathbf{P} = \begin{bmatrix} k_1 & \frac{k_1}{2} \\ k_2 & k_2 \end{bmatrix} \tag{13.75e}$$

We may assign any values to k_1 and k_2. For convenience, let $k_1 = 2$ and $k_2 = 1$. This substitution yields

$$\mathbf{P} = \begin{bmatrix} 2 & 1 \\ 1 & 1 \end{bmatrix} \tag{13.75f}$$

The transformed variables [Eq. (13.73a)] are

$$\begin{bmatrix} z_1 \\ z_2 \end{bmatrix} = \begin{bmatrix} 2 & 1 \\ 1 & 1 \end{bmatrix} \begin{bmatrix} x_1 \\ x_2 \end{bmatrix} = \begin{bmatrix} 2x_1 + x_2 \\ x_1 + x_2 \end{bmatrix} \tag{13.76}$$

Thus, the new state variables z_1 and z_2 are related to x_1 and x_2 by Eq. (13.76). The system equation with $\mathbf{z}$ as the state vector is given by [see Eq. (13.73c)]

$$\dot{\mathbf{z}} = \Lambda \mathbf{z} + \hat{\mathbf{B}} \mathbf{f}$$

---

†If, however, we want the state equations in diagonalized form, as in Eq. (13.29a), where all the elements of $\hat{\mathbf{B}}$ matrix are unity, there is a unique solution. The reason is that the equation $\hat{\mathbf{B}} = \mathbf{PB}$, where all the elements of $\hat{\mathbf{B}}$ are unity, imposes additional constraints. In the present example, this condition will yield $p_{11} = \frac{1}{2}$, $p_{12} = \frac{1}{4}$, $p_{21} = \frac{1}{3}$, and $p_{22} = \frac{1}{3}$. The relationship between $\mathbf{z}$ and $\mathbf{x}$ is then

$$z_1 = \tfrac{1}{2}x_1 + \tfrac{1}{4}x_2 \quad \text{and} \quad z_2 = \tfrac{1}{3}x_1 + \tfrac{1}{3}x_2$$

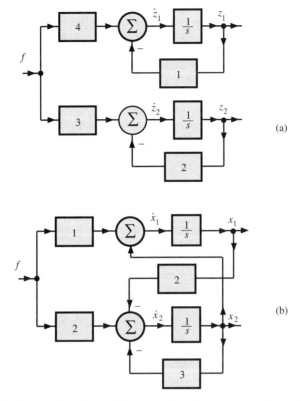

Fig. 13.8   Two realizations of the second-order system in Example 13.10.

where

$$\hat{\mathbf{B}} = \mathbf{PB} = \begin{bmatrix} 2 & 1 \\ 1 & 1 \end{bmatrix} \begin{bmatrix} 1 \\ 2 \end{bmatrix} = \begin{bmatrix} 4 \\ 3 \end{bmatrix}$$

Hence

$$\begin{bmatrix} \dot{z}_1 \\ \dot{z}_2 \end{bmatrix} = \begin{bmatrix} -1 & 0 \\ 0 & -2 \end{bmatrix} \begin{bmatrix} z_1 \\ z_2 \end{bmatrix} + \begin{bmatrix} 4 \\ 3 \end{bmatrix} f \qquad (13.77a)$$

or

$$\dot{z}_1 = -z_1 + 4f$$

$$\dot{z}_2 = -2z_2 + 3f \qquad (13.77b)$$

Note the distinctive nature of these state equations. Each state equation involves only one variable and therefore can be solved by itself. A general state equation has the derivative of one state variable equal to a linear combination of all state variables. Such is not the case with the diagonalized matrix Λ. Each state variable z_i is chosen so that it is uncoupled from the rest of the variables; hence a system with n eigenvalues is split into n decoupled systems, each with an equation of the form

$$\dot{z}_i = \lambda_i z_i + (\text{input terms})$$

This fact also can be readily seen from Fig. 13.8a, which is a realization of the system represented by Eq. (13.77). In contrast, consider the original state equations [see Eq. 13.70a)]

$$\dot{x}_1 = x_2 + f(t)$$

$$\dot{x}_2 = -2x_1 - 3x_2 + 2f(t)$$

A realization for these equations is shown in Fig. 13.8b. It can be seen from Fig. 13.8a that the states z_1 and z_2 are decoupled, whereas the states x_1 and x_2 (Fig. 13.8b) are coupled. It should be remembered that Figs. 13.8a and 13.8b are simulations of the same system.† ■

⊙ **Computer Example C13.5**
Solve Example 13.10 using MATLAB.
Caution: The answer for $\hat{\mathbf{B}}$ is not unique.

A=[0 1;-2 -3]; B=[1; 2];
[V, L]=eig(A);
P=inv(V);
Lambda=P*A*inv(P);
Bhat=P*B   ⊙

13.5   Controllability and Observability

Consider a diagonalized state-space description of a system

$$\dot{\mathbf{z}} = \Lambda\mathbf{z} + \hat{\mathbf{B}}\mathbf{f} \tag{13.78a}$$

and

$$\mathbf{Y} = \hat{\mathbf{C}}\mathbf{z} + \mathbf{D}\mathbf{f} \tag{13.78b}$$

We shall assume that all n eigenvalues λ_1, λ_2, ... , λ_n are distinct. The state equations (13.78a) are of the form

$$\dot{z}_m = \lambda_m z_m + \hat{b}_{m1}f_1 + \hat{b}_{m2}f_2 + \cdots + \hat{b}_{mj}f_j \qquad m = 1, 2, \ldots n$$

If $\hat{b}_{m1}$, $\hat{b}_{m2}$, ... , $\hat{b}_{mj}$ (the mth row in matrix $\hat{\mathbf{B}}$) are all zero, then

$$\dot{z}_m = \lambda_m z_m$$

and the variable z_m is uncontrollable because z_m is not connected to any of the inputs. Moreover, z_m is decoupled from all the remaining $(n-1)$ state variables because of the diagonalized nature of the variables. Hence, there is no direct or indirect coupling of z_m with any of the inputs, and the system is uncontrollable. In contrast, if at least one element in the mth row of $\hat{\mathbf{B}}$ is nonzero, z_m is coupled to at least one input and is therefore controllable. *Thus, a system with a diagonalized*

---

†Here we only have a simulated state equation; the outputs are not shown. The outputs are linear combinations of state variables (and inputs). Hence, the output equation can be easily incorporated into these diagrams (see Fig. 13.7).

state [Eqs. (13.78)] is completely controllable if and only if the matrix $\hat{\mathbf{B}}$ has no row of zero elements.

The outputs [Eq.(13.78b)] are of the form

$$y_i = \hat{c}_{i1}z_1 + \hat{c}_{i2}z_2 + \cdots + \hat{c}_{in}z_n + \sum_{m=1}^{j} d_{im}f_m$$

If $\hat{c}_{im} = 0$, then the state z_m will not appear in the expression for y_i. Since all the states are decoupled because of the diagonalized nature of the equations, the state z_m cannot be observed directly or indirectly (through other states) at the output y_i. Hence the mth mode $e^{\lambda_m t}$ will not be observed at the output y_i. If $\hat{c}_{1m}$, $\hat{c}_{2m}$, $\ldots$, $\hat{c}_{km}$ (the mth column in matrix $\hat{\mathbf{C}}$) are all zero, the state z_m will not be observable at any of the k outputs, and the state z_m is unobservable. In contrast, if at least one element in the mth column of $\hat{\mathbf{C}}$ is nonzero, z_m is observable at least at one output. *Thus, a system with diagonalized equations of the form in Eqs. (13.78) is completely observable if and only if the matrix $\hat{\mathbf{C}}$ has no column of zero elements.* In the above discussion, we assumed distinct eigenvalues; for repeated eigenvalues, the modified criteria can be found in the literature.[1,2]

If the state-space description is not in diagonalized form, it may be converted into diagonalized form using the procedure in Example 13.10. It is also possible to test for controllability and observability even if the state-space description is in undiagonalized form.[1,2]†

■ **Example 13.11**

Investigate the controllability and observability of the systems in Figs. 13.9a and 13.9b.

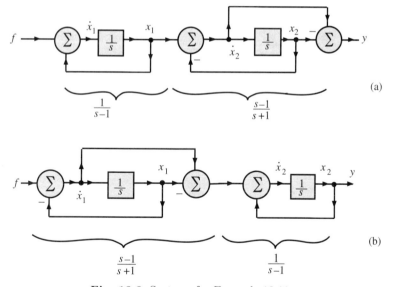

Fig. 13.9 Systems for Example 13.11.

---

†We can show that a system is completely controllable if and only if the $n \times nj$ composite matrix $[\mathbf{B}, \mathbf{AB}, \mathbf{A}^2\mathbf{B}, \ldots, \mathbf{A}^{n-1}\mathbf{B}]$ has a rank n. Similarly, a system is completely observable if and only if the $n \times nk$ composite matrix $[\mathbf{C}', \mathbf{A}'\mathbf{C}', \mathbf{A}'^2\mathbf{C}', \ldots, \mathbf{A}'^{n-1}\mathbf{C}']$ has a rank n.

In both cases, the state variables are identified as the two integrator outputs, x_1 and x_2. The state equations for the system in Fig. 13.9a are

$$\dot{x}_1 = x_1 + f$$
$$\dot{x}_2 = x_1 - x_2 \tag{13.79}$$

and

$$y = x_1 - 2x_2$$

Hence

$$\mathbf{A} = \begin{bmatrix} 1 & 0 \\ 1 & -1 \end{bmatrix}, \quad \mathbf{B} = \begin{bmatrix} 1 \\ 0 \end{bmatrix}, \quad \mathbf{C} = [1 \quad -2], \quad \mathbf{D} = 0$$

$$|s\mathbf{I} - \mathbf{A}| = \begin{vmatrix} s - 1 & 0 \\ -1 & s + 1 \end{vmatrix} = (s - 1)(s + 1)$$

Therefore

$$\lambda_1 = 1 \quad \text{and} \quad \lambda_2 = -1$$

and

$$\Lambda = \begin{bmatrix} 1 & 0 \\ 0 & -1 \end{bmatrix} \tag{13.80}$$

We shall now use the procedure in Sec. 13.4-1 to diagonalize this system. According to Eq. (13.74b), we have

$$\begin{bmatrix} 1 & 0 \\ 0 & -1 \end{bmatrix} \begin{bmatrix} p_{11} & p_{12} \\ p_{21} & p_{22} \end{bmatrix} = \begin{bmatrix} p_{11} & p_{12} \\ p_{21} & p_{22} \end{bmatrix} \begin{bmatrix} 1 & 0 \\ 1 & -1 \end{bmatrix}$$

The solution of this equation yields

$$p_{12} = 0 \quad \text{and} \quad -2p_{21} = p_{22}$$

Choosing $p_{11} = 1$ and $p_{21} = 1$, we have

$$p = \begin{bmatrix} 1 & 0 \\ 1 & -2 \end{bmatrix}$$

and

$$\hat{\mathbf{B}} = \mathbf{PB} = \begin{bmatrix} 1 & 0 \\ 1 & -2 \end{bmatrix} \begin{bmatrix} 1 \\ 0 \end{bmatrix} = \begin{bmatrix} 1 \\ 1 \end{bmatrix} \tag{13.81a}$$

All the rows of $\hat{\mathbf{B}}$ are nonzero. Hence the system is controllable. Also,

$$\mathbf{Y} = \mathbf{Cx}$$
$$= \mathbf{CP}^{-1}\mathbf{z}$$
$$= \hat{\mathbf{C}}\mathbf{z} \tag{13.81b}$$

and

$$\hat{\mathbf{C}} = \mathbf{CP}^{-1} = [1 \quad -2] \begin{bmatrix} 1 & 0 \\ 1 & -2 \end{bmatrix}^{-1} = [1 \quad -2] \begin{bmatrix} 1 & 0 \\ \frac{1}{2} & -\frac{1}{2} \end{bmatrix} = [0 \quad 1] \tag{13.81c}$$

The first column of $\hat{\mathbf{C}}$ is zero. Hence the mode z_1 (corresponding to $\lambda_1 = 1$) is unobservable. The system is therefore controllable but not observable. We come to the same conclusion by realizing the system with the state variables z_1 and z_2, whose state equations are

$$\dot{\mathbf{z}} = \Lambda\mathbf{z} + \hat{\mathbf{B}}f$$

$$y = \hat{\mathbf{C}}\mathbf{z}$$

According to Eqs. (13.80) and (13.81), we have

$$\dot{z}_1 = z_1 + f$$

$$\dot{z}_2 = -z_2 + f$$

and

$$y = z_2$$

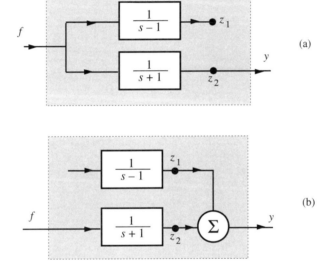

Fig. 13.10   Equivalent of the systems in Fig. 13.9.

Figure 13.10a shows a realization of these equations. It is clear that each of the two modes is controllable, but the first mode (corresponding to $\lambda = 1$) is not observable at the output.

The state equations for the system in Fig. 13.9b are

$$\dot{x}_1 = -x_1 + f$$

$$\dot{x}_2 = -2x_1 + x_2 + f \qquad (13.82)$$

and

$$y = x_2$$

Hence

$$\mathbf{A} = \begin{bmatrix} -1 & 0 \\ -2 & 1 \end{bmatrix}, \quad \mathbf{B} = \begin{bmatrix} 1 \\ 1 \end{bmatrix}, \quad \mathbf{C} = \begin{bmatrix} 0 & 1 \end{bmatrix}, \quad \mathbf{D} = 0$$

$$|s\mathbf{I} - \mathbf{A}| = \begin{vmatrix} s+1 & 0 \\ -1 & s-1 \end{vmatrix} = (s+1)(s-1)$$

so that $\lambda_1 = -1$, $\lambda_2 = 1$, and

$$\Lambda = \begin{bmatrix} -1 & 0 \\ 0 & 1 \end{bmatrix} \tag{13.83}$$

Diagonalizing the matrix, we have

$$\begin{bmatrix} 1 & 0 \\ 0 & -1 \end{bmatrix} \begin{bmatrix} p_{11} & p_{12} \\ p_{21} & p_{22} \end{bmatrix} = \begin{bmatrix} p_{11} & p_{12} \\ p_{21} & p_{22} \end{bmatrix} \begin{bmatrix} -1 & 0 \\ -2 & 1 \end{bmatrix}$$

The solution of this equation yields $p_{11} = -p_{12}$ and $p_{22} = 0$. Choosing $p_{11} = -1$ and $p_{21} = 1$, we obtain

$$\mathbf{P} = \begin{bmatrix} -1 & 1 \\ 1 & 0 \end{bmatrix}$$

and

$$\hat{\mathbf{B}} = \mathbf{PB} = \begin{bmatrix} -1 & 1 \\ 1 & 0 \end{bmatrix} \begin{bmatrix} 1 \\ 1 \end{bmatrix} = \begin{bmatrix} 0 \\ 1 \end{bmatrix} \tag{13.84a}$$

$$\hat{\mathbf{C}} = \mathbf{CP}^{-1} = \begin{bmatrix} 0 & 1 \end{bmatrix} \begin{bmatrix} 0 & 1 \\ 1 & 1 \end{bmatrix} = \begin{bmatrix} 1 & 1 \end{bmatrix} \tag{13.84b}$$

The first row of $\hat{\mathbf{B}}$ is zero. Hence the second mode (corresponding to $\lambda_1 = 1$) is not controllable. However, since none of the columns of $\hat{\mathbf{C}}$ vanish, both modes are observable at the output. Hence the system is observable but not controllable.

We reach to the same conclusion by realizing the system with the state variables z_1 and z_2. The two state equations are

$$\dot{\mathbf{z}} = \Lambda \mathbf{z} + \hat{\mathbf{B}} f$$

$$y = \hat{\mathbf{C}} \mathbf{z}$$

From Eqs. (13.83) and (13.84), we have

$$\dot{z}_1 = z_1$$

$$\dot{z}_2 = -z_2 + f$$

and thus

$$y = z_1 + z_2 \tag{13.85}$$

Figure 13.10b shows a realization of these equations. Clearly, each of the two modes is observable at the output, but the mode corresponding to $\lambda_1 = 1$ is not controllable. ∎

⊙ **Computer Example C13.6**
Solve Example 13.11 using MATLAB.

A=[1 0;1 -1]; B=[1; 0]; C=[1 -2];
[V, L]=eig(A);
P=inv(V);

Ahat=P*A*inv(P);
Bhat=P*B
Chat=C*inv(P) $\odot$

13.5-1 Inadequacy of the Transfer Function Description of a System

Example 13.11 demonstrates the inadequacy of the transfer function to describe an LTI system in general. The systems in Figs. 13.9a and 13.9b both have the same transfer function

$$H(s) = \frac{1}{s+1}$$

Yet the two systems are very different. Their true nature is revealed in Figs. 13.10a and 13.10b, respectively. Both the systems are unstable, but their transfer function $H(s) = \frac{1}{s+1}$ does not give any hint of it. The system in Fig. 13.9a appears stable from the external terminals, but it is internally unstable. The system in Fig. 13.9b, on the other hand, will show instability at the external terminals, but its transfer function $H(s) = \frac{1}{s+1}$ is silent about it. The system in Fig. 13.9a is controllable but not observable, whereas the system in Fig. 13.9b is observable but not controllable.

The transfer function description of a system looks at a system only from the input and output terminals. Consequently, the transfer description can specify only the part of the system which is coupled to the input and the output terminals. Figures 13.10a and 13.10b show that in both cases only a part of the system that has a transfer function $H(s) = \frac{1}{s+1}$ is coupled to the input and the output terminals. This is the reason why both systems have the same transfer function $H(s) = \frac{1}{s+1}$.

The state variable description (Eqs. 13.79 and 13.82), on the other hand, contains all the information about these systems to describe them completely. The reason is that the state variable description is an internal description, not the external description obtained from the system behavior at external terminals.

Mathematically, the reason the transfer function fails to describe these systems completely is the fact that their transfer function has a common factor $s - 1$ in the numerator and denominator; this common factor is canceled out with a consequent loss of the information about these systems. Such a situation occurs when a system is uncontrollable and/or unobservable. If a system is both controllable and observable (which is the case with most of the practical systems) the transfer function describes the system completely. In such a case the internal and external descriptions are equivalent.

13.6 State-Space Analysis of Discrete-Time Systems

We have shown that an nth-order differential equation can be expressed in terms of n first-order differential equations. In the following analogous procedure, we show that an nth-order difference equation can be expressed in terms of n first-order difference equations.

Consider the z-transfer function

$$H[z] = \frac{b_m z^m + b_{m-1} z^{m-1} + \cdots + b_1 z + b_0}{z^n + a_{n-1} z^{n-1} + \cdots + a_1 z + a_0} \tag{13.86a}$$

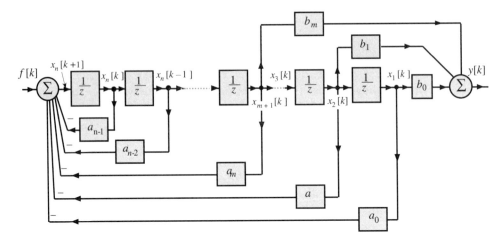

Fig. 13.11   Controller canonical realization of an nth-order discrete-time system.

The input $f[k]$ and the output $y[k]$ of this system are related by the difference equation

$$(E^n + a_{n-1}E^{n-1} + \cdots + a_1 E + a_0)y[k] =$$
$$(b_m E^m + b_{m-1}E^{m-1} + \cdots + b_1 E + b_0)f[k] \quad (13.86b)$$

The controller canonical realization of this equation is illustrated in Fig. 13.11. Signals appearing at the outputs of n delay elements are denoted by $x_1[k]$, $x_2[k]$, $\ldots$, $x_n[k]$. The input of the first delay is $x_n[k+1]$. We can now write n equations, one at the input of each delay:

$$x_1[k + 1] = x_2[k]$$
$$x_2[k + 1] = x_3[k]$$
$$\dotsb \quad (13.87)$$
$$x_{n-1}[k + 1] = x_n[k]$$
$$x_n[k + 1] = -a_0 x_1[k] - a_1 x_2[k] - \cdots - a_{n-1} x_n[k] + f[k]$$

and

$$y(k) = b_0 x_1(k) + b_1 x_2(k) + \cdots + b_m x_{m+1}(k) \quad (13.88)$$

Equations (13.87) are n first-order difference equations in n variables $x_1(k)$, $x_2(k)$, $\ldots$, $x_n(k)$. These variables should immediately be recognized as state variables, since the specification of the initial values of these variables in Fig. 13.11 will uniquely determine the response $y[k]$ for a given $f[k]$. Thus, Eqs. (13.87) represent the state equations, and Eq. (13.88) is the output equation. In matrix form we can write these equations as

$$
\underbrace{\begin{bmatrix} x_1[k+1] \\ x_2[k+1] \\ \cdots \\ x_{n-1}[k+1] \\ x_n[k+1] \end{bmatrix}}_{\mathbf{x}[k+1]} = \underbrace{\begin{bmatrix} 0 & 1 & 0 & \cdots & 0 & 0 \\ 0 & 0 & 1 & \cdots & 0 & 0 \\ \cdot & \cdot & \cdot & \cdots & \cdot & \cdot \\ 0 & 0 & 0 & \cdots & 0 & 1 \\ -a_0 & -a_1 & -a_2 & \cdots & -a_{n-2} & -a_{n-1} \end{bmatrix}}_{\mathbf{A}} \underbrace{\begin{bmatrix} x_1[k] \\ x_2[k] \\ \cdots \\ x_{n-1}[k] \\ x_n[k] \end{bmatrix}}_{\mathbf{x}[k]} + \underbrace{\begin{bmatrix} 0 \\ 0 \\ \cdots \\ 0 \\ 1 \end{bmatrix}}_{\mathbf{B}} f[k]
$$

$$(13.89a)$$

and

$$
\mathbf{y}[k] = \underbrace{\begin{bmatrix} b_0 & b_1 & \cdots & b_m \end{bmatrix}}_{\mathbf{C}} \begin{bmatrix} x_1[k] \\ x_2[k] \\ \vdots \\ x_{m+1}[k] \end{bmatrix}
$$

$$(13.89b)$$

In general,

$$\mathbf{x}[k+1] = \mathbf{A}\mathbf{x}[k] + \mathbf{B}f[k] \tag{13.90a}$$

$$\mathbf{y}[k] = \mathbf{C}\mathbf{x}[k] + \mathbf{D}f[k] \tag{13.90b}$$

Here we have represented a discrete-time system with state equations in controller canonical form. There are several other possible representations, as discussed in Sec. 13.2. We may, for example, realize the system by using a series, parallel, or observer canonical form. In all cases, the output of each delay element qualifies as a state variable. We then write the equation at the input of each delay element. The n equations thus obtained are the n state equations.

13.6-1   Solution in State-Space

Consider the state equation

$$\mathbf{x}[k+1] = \mathbf{A}\mathbf{x}[k] + \mathbf{B}f[k] \tag{13.91}$$

From this equation it follows that

$$\mathbf{x}[k] = \mathbf{A}\mathbf{x}[k-1] + \mathbf{B}f[k-1] \tag{13.92a}$$

and

$$\mathbf{x}[k-1] = \mathbf{A}\mathbf{x}[k-2] + \mathbf{B}f[k-2] \tag{13.92b}$$

$$\mathbf{x}[k-2] = \mathbf{A}\mathbf{x}[k-3] + \mathbf{B}f[k-3] \tag{13.92c}$$

$$\cdots\cdots\cdots\cdots\cdots\cdots\cdots\cdots\cdots\cdots$$

$$\mathbf{x}[1] = \mathbf{A}\mathbf{x}[0] + \mathbf{B}f[0]$$

Substituting Eq. (13.92b) in Eq. (13.92a), we obtain

$$\mathbf{x}[k] = \mathbf{A}^2\mathbf{x}[k-2] + \mathbf{A}\mathbf{B}\mathbf{f}[k-2] + \mathbf{B}\mathbf{f}[k-1]$$

Substituting Eq. (13.92c) in this equation, we obtain

$$\mathbf{x}[k] = \mathbf{A}^3\mathbf{x}[k-3] + \mathbf{A}^2\mathbf{B}\mathbf{f}[k-3] + \mathbf{A}\mathbf{B}\mathbf{f}[k-2] + \mathbf{B}\mathbf{f}[k-1]$$

Continuing in this way, we obtain

$$\mathbf{x}[k] = \mathbf{A}^k\mathbf{x}[0] + \mathbf{A}^{k-1}\mathbf{B}\mathbf{f}[0] + \mathbf{A}^{k-2}\mathbf{B}\mathbf{f}[1] + \cdots + \mathbf{B}\mathbf{f}[k-1]$$

$$= \mathbf{A}^k\mathbf{x}[0] + \sum_{j=0}^{k-1} \mathbf{A}^{k-1-j}\mathbf{B}\mathbf{f}[j] \tag{13.93a}$$

The upper limit on the summation in Eq. (13.93a) is nonnegative. Hence $k \geq 1$, and the summation is recognized as the convolution sum

$$\mathbf{A}^{k-1}u[k-1] * \mathbf{B}\mathbf{f}[k]$$

Hence

$$\mathbf{x}[k] = \underbrace{\mathbf{A}^k\mathbf{x}[0]}_{\text{zero-input}} + \underbrace{\mathbf{A}^{k-1}u[k-1] * \mathbf{B}\mathbf{f}[k]}_{\text{zero-state}} \tag{13.93b}$$

and

$$\mathbf{y}[k] = \mathbf{C}\mathbf{x} + \mathbf{D}\mathbf{f}$$

$$= \mathbf{C}\mathbf{A}^k\mathbf{x}[0] + \sum_{j=0}^{k-1} \mathbf{C}\mathbf{A}^{k-1-j}\mathbf{B}\mathbf{f}[j] + \mathbf{D}\mathbf{f} \tag{13.94a}$$

$$= \mathbf{C}\mathbf{A}^k\mathbf{x}[0] + \mathbf{C}\mathbf{A}^{k-1}u[k-1] * \mathbf{B}\mathbf{f}[k] + \mathbf{D}\mathbf{f} \tag{13.94b}$$

In Sec. B.6-5, we showed that

$$\mathbf{A}^k = \beta_0\mathbf{I} + \beta_1\mathbf{A} + \beta_2\mathbf{A}^2 + \cdots + \beta_{n-1}\mathbf{A}^{n-1} \tag{13.95a}$$

where (assuming n distinct eigenvalues of $\mathbf{A}$)

$$\begin{bmatrix} \beta_0 \\ \beta_1 \\ \cdots \\ \beta_{n-1} \end{bmatrix} = \begin{bmatrix} 1 & \lambda_1 & \lambda_1^2 & \cdots & \lambda_1^{n-1} \\ 1 & \lambda_2 & \lambda_2^2 & \cdots & \lambda_2^{n-1} \\ \cdots\cdots\cdots\cdots\cdots \\ 1 & \lambda_n & \lambda_n^2 & \cdots & \lambda_n^{n-1} \end{bmatrix}^{-1} \begin{bmatrix} \lambda_1^k \\ \lambda_2^k \\ \cdots \\ \lambda_n^k \end{bmatrix} \tag{13.95b}$$

and $\lambda_1, \lambda_2, \ldots, \lambda_n$ are the n eigenvalues of $\mathbf{A}$.

We can also determine $\mathbf{A}^k$ from the z-transform formula, which will be derived later in Eq. (13.102):

$$\mathbf{A}^k = \mathcal{Z}^{-1}[(\mathbf{I} - z^{-1}\mathbf{A})^{-1}] \tag{13.95c}$$

■ **Example 13.12**

Give a state-space description of the system in Fig. 13.12. Find the output $y[k]$ if the input $f[k] = u[k]$ and the initial conditions are $x_1[0] = 2$ and $x_2[0] = 3$.

The state equations are [see Eq. (13.89)]

$$\begin{bmatrix} x_1[k+1] \\ x_2[k+1] \end{bmatrix} = \begin{bmatrix} 0 & 1 \\ -\frac{1}{6} & \frac{5}{6} \end{bmatrix} \begin{bmatrix} x_1[k] \\ x_2[k] \end{bmatrix} + \begin{bmatrix} 0 \\ 1 \end{bmatrix} f \tag{13.96a}$$

and

$$y[k] = [\,-1 \quad 5\,] \begin{bmatrix} x_1[k] \\ x_2[k] \end{bmatrix} \tag{13.96b}$$

To find the solution [Eq. (13.94)], we must first determine $\mathbf{A}^k$. The characteristic equation of $\mathbf{A}$ is

$$|\lambda \mathbf{I} - \mathbf{A}| = \begin{vmatrix} \lambda & -1 \\ \frac{1}{6} & \lambda - \frac{5}{6} \end{vmatrix} = \lambda^2 - \frac{5}{6}\lambda + \frac{1}{6} = \left(\lambda - \frac{1}{3}\right)\left(\lambda - \frac{1}{2}\right) = 0$$

Hence, $\lambda_1 = \frac{1}{3}$ and $\lambda_2 = \frac{1}{2}$ are the eigenvalues of $\mathbf{A}$ and [see Eq. (13.95)]

$$\mathbf{A}^k = \beta_0 \mathbf{I} + \beta_1 \mathbf{A}$$

where [see Eq. (B.95b)]

$$\begin{bmatrix} \beta_0 \\ \beta_1 \end{bmatrix} = \begin{bmatrix} 1 & \frac{1}{3} \\ 1 & \frac{1}{2} \end{bmatrix}^{-1} \begin{bmatrix} (\frac{1}{3})^k \\ (\frac{1}{2})^k \end{bmatrix} = \begin{bmatrix} 3 & -2 \\ -6 & 6 \end{bmatrix} \begin{bmatrix} (3)^{-k} \\ (2)^{-k} \end{bmatrix} = \begin{bmatrix} 3(3)^{-k} - 2(2)^{-k} \\ -6(3)^{-k} + 6(2)^{-k} \end{bmatrix}$$

and

$$\mathbf{A}^k = [\,3(3)^{-k} - 2(2)^{-k}\,] \begin{bmatrix} 1 & 0 \\ 0 & 1 \end{bmatrix} + [\,-6(3)^{-k} + 6(2)^{-k}\,] \begin{bmatrix} 0 & 1 \\ -\frac{1}{6} & \frac{5}{6} \end{bmatrix}$$

$$= \begin{bmatrix} 3(3)^{-k} - 2(2)^{-k} & -6(3)^{-k} + 6(2)^{-k} \\ (3)^{-k} - (2)^{-k} & -2(3)^{-k} + 3(2)^{-k} \end{bmatrix} \tag{13.97}$$

We can now determine the state vector $\mathbf{x}[k]$ from Eq. (13.93b). Since we are interested in the output $y[k]$, we shall use Eq. (13.94b) directly. Note that

$$\mathbf{CA}^k = [\,-1 \quad 5\,]\,\mathbf{A}^k = [\,2(3)^{-k} - 3(2)^{-k} \quad -4(3)^{-k} + 9(2)^{-k}\,] \tag{13.98}$$

and the zero-input response is $\mathbf{CA}^k \mathbf{x}[0]$, with

$$\mathbf{x}[0] = \begin{bmatrix} 2 \\ 3 \end{bmatrix}$$

Hence, the zero-input response is

$$\mathbf{CA}^k \mathbf{x}[0] = -8(3)^{-k} + 21(2)^{-k} \tag{13.99a}$$

The zero-state component is given by the convolution sum of $\mathbf{CA}^{k-1} u[k-1]$ and $\mathbf{B}f[k]$. Using the shifting property of the convolution sum [Eq. (9.46)], we can obtain the zero-state component by finding the convolution sum of $\mathbf{CA}^k u[k]$ and $\mathbf{B}f[k]$ and then replacing

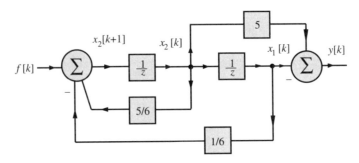

Fig. 13.12   System for Example 13.12.

k with $k-1$ in the result. We use this procedure because the convolution sums are listed in Table 9.1 for functions of the type $f[k]u[k]$ rather than $f[k]u[k-1]$.

$$\mathbf{CA}^k u[k] * \mathbf{B}f[k] = [\, 2(3)^{-k} - 3(2)^{-k} \quad -4(3)^{-k} + 9(2)^{-k}\,] * \begin{bmatrix} 0 \\ u[k] \end{bmatrix}$$

$$= -4(3)^{-k} * u[k] + 9(2)^{-k} * u[k]$$

Using Table 9.1 (Pair 2a), we obtain

$$\mathbf{CA}^k u[k] * \mathbf{B}f[k] = -4\left[\frac{1 - 3^{-(k+1)}}{1 - \frac{1}{3}}\right] u[k] + 9\left[\frac{1 - 2^{-(k+1)}}{1 - \frac{1}{2}}\right] u[k]$$

$$= [12 + 6(3^{-(k+1)}) - 18(2^{-(k+1)})]u[k]$$

Now the desired (zero-state) response is obtained by replacing k by $k-1$. Hence

$$\mathbf{CA}^k u[k] * \mathbf{B}f[k-1] = [12 + 6(3)^{-k} - 18(2)^{-k}]u[k-1] \tag{13.99b}$$

It follows that

$$y[k] = [-8(3)^{-k} + 21(2)^{-k}u[k] + [12 + 6(3)^{-k} - 18(2)^{-k}]u[k-1] \tag{13.100a}$$

This is the desired answer. We can simplify this answer by observing that $12 + 6(3)^{-k} - 18(2)^{-k} = 0$ for $k = 0$. Hence, $u[k-1]$ may be replaced by $u[k]$ in Eq. (13.99b), and

$$y[k] = [12 - 2(3)^{-k} + 3(2)^{-k}]u[k] \tag{13.100b}$$

■

⊙ **Computer Example C13.7**
Solve Example 13.12 using MATLAB.

```
A=[0 1;-1/6 5/6]; B=[0; 1]; C=[-1 5]; D=0;
x0=[2;3];
k=0:25;
u=ones(1,26);
[y,x]=dlsim(A,B,C,D,u,x0);
stem(k,y)     ⊙
```

⊙ **Computer Example C13.8**
Using MATLAB find the zero-state response of the system in Example 13.12.

A=[0 1;-1/6 5/6]; B=[0; 1]; C=[-1 5]; D=0;
[num,den]=ss2tf(A,B,C,D);
k=0:25;
u=ones(1:length(k));
y=filter(num,den,u);
stem(k,y)   ⊙

13.6-2   The $\mathcal{Z}$-Transform Solution

The z-transform of Eq. (13.91) is given by

$$z\mathbf{X}[z] - z\mathbf{x}[0] = \mathbf{A}\mathbf{X}[z] + \mathbf{B}\mathbf{F}[z]$$

Therefore

$$(z\mathbf{I} - \mathbf{A})\mathbf{X}[z]) = z\mathbf{x}[0] + \mathbf{B}\mathbf{F}[z]$$

and

$$\mathbf{X}[z] = (z\mathbf{I} - \mathbf{A})^{-1}z\mathbf{x}[0] + (z\mathbf{I} - \mathbf{A})^{-1}\mathbf{B}\mathbf{F}[z]$$
$$= (\mathbf{I} - z^{-1}\mathbf{A})^{-1}\mathbf{x}[0] + (z\mathbf{I} - \mathbf{A})^{-1}\mathbf{B}\mathbf{F}[z] \qquad (13.101a)$$

Hence

$$\mathbf{x}[k] = \underbrace{\mathcal{Z}^{-1}[(\mathbf{I} - z^{-1}\mathbf{A})^{-1}]\mathbf{x}[0]}_{\text{zero-input component}} + \underbrace{\mathcal{Z}^{-1}[(z\mathbf{I} - \mathbf{A})^{-1}\mathbf{B}\mathbf{F}[z]]}_{\text{zero-state component}} \qquad (13.101b)$$

A comparison of Eq. (13.101b) with Eq. (13.93b) shows that

$$\mathbf{A}^k = \mathcal{Z}^{-1}[(\mathbf{I} - z^{-1}\mathbf{A})^{-1}] \qquad (13.102)$$

The output equation is given by

$$\mathbf{Y}[z] = \mathbf{C}\mathbf{X}[z] + \mathbf{D}\mathbf{F}[z]$$
$$= \mathbf{C}[(\mathbf{I} - z^{-1}\mathbf{A})^{-1}\mathbf{x}[0] + (z\mathbf{I} - \mathbf{A})^{-1}\mathbf{B}\mathbf{F}[z]] + \mathbf{D}\mathbf{F}[z]$$
$$= \mathbf{C}(\mathbf{I} - z^{-1}\mathbf{A})^{-1}\mathbf{x}[0] + [\mathbf{C}(z\mathbf{I} - \mathbf{A})^{-1}\mathbf{B} + \mathbf{D}]\mathbf{F}[z] \qquad (13.103a)$$
$$= \underbrace{\mathbf{C}(\mathbf{I} - z^{-1}\mathbf{A})^{-1}\mathbf{x}[0]}_{\text{zero-input response}} + \underbrace{\mathbf{H}[z]\mathbf{F}[z]}_{\text{zero-state response}}$$

where

$$\mathbf{H}[z] = \mathbf{C}(z\mathbf{I} - \mathbf{A}^{-1})\mathbf{B} + \mathbf{D} \qquad (13.103b)$$

Note that $\mathbf{H}[z]$ is the transfer function matrix of the system, and $H_{ij}[z]$, the ijth element of $\mathbf{H}[z]$, is the transfer function relating the output $y_i(k)$ to the input $f_j(k)$. If we define $\mathbf{h}[k]$ as

$$\mathbf{h}[k] = \mathcal{Z}^{-1}[\mathbf{H}[z]]$$

then $\mathbf{h}[k]$ represents the unit impulse function response matrix of the system. Thus, $h_{ij}[k]$, the ijth element of $\mathbf{h}(k)$, represents the zero-state response $y_i(k)$ when the input $f_j(k) = \delta[k]$ and all other inputs are zero.

■ **Example 13.13**

Using the z-transform, find the response $y[k]$ for the system in Example 13.12. According to Eq. (13.103a)

$$\mathbf{Y}[z] = \begin{bmatrix} -1 & 5 \end{bmatrix} \begin{bmatrix} 1 & -\frac{1}{z} \\ \frac{1}{6z} & 1-\frac{5}{6z} \end{bmatrix}^{-1} \begin{bmatrix} 2 \\ 3 \end{bmatrix} + \begin{bmatrix} -1 & 5 \end{bmatrix} \begin{bmatrix} z & -1 \\ \frac{1}{6} & z-\frac{5}{6} \end{bmatrix}^{-1} \begin{bmatrix} 0 \\ \frac{z}{z-1} \end{bmatrix}$$

$$= \begin{bmatrix} -1 & 5 \end{bmatrix} \begin{bmatrix} \frac{z(6z-5)}{6z^2-5z+1} & \frac{6z}{6z^2-5z+1} \\ \frac{-z}{6z^2-5z+1} & \frac{6z^2}{6z^2-5z+1} \end{bmatrix} \begin{bmatrix} 2 \\ 3 \end{bmatrix} + \begin{bmatrix} -1 & 5 \end{bmatrix} \begin{bmatrix} \frac{z}{(z-1)(z^2-\frac{5}{6}z+\frac{1}{6})} \\ \frac{z^2}{(z-1)(z^2-\frac{5}{6}z+\frac{1}{6})} \end{bmatrix}$$

$$= \frac{13z^2-3z}{z^2-\frac{5}{6}z+\frac{1}{6}} + \frac{(5z-1)z}{(z-1)(z^2-\frac{5}{6}z+\frac{1}{6})}$$

$$= \frac{-8z}{z-\frac{1}{3}} + \frac{21z}{z-\frac{1}{2}} + \frac{12z}{z-1} + \frac{12z}{z-1} + \frac{6z}{z-\frac{1}{3}} - \frac{18z}{z-\frac{1}{2}}$$

Therefore

$$y[k] = \underbrace{[-8(3)^{-k} + 21(2)^{-k}}_{\text{zero-input response}} + \underbrace{12 + 6(3)^{-k} - 18(2)^{-k}}_{\text{zero-state response}}]u[k] \qquad ■$$

Linear Transformation, Controllability, and Observability

The procedure for linear transformation is parallel to that in the continuous-time case (Sec. 13.4). If $\mathbf{w}$ is the transformed-state vector given by

$$\mathbf{w} = \mathbf{P}\mathbf{x}$$

then

$$\mathbf{w}[k+1] = \mathbf{P}\mathbf{A}\mathbf{P}^{-1}\mathbf{w}[k] + \mathbf{P}\mathbf{B}\mathbf{f}$$

and

$$\mathbf{y}[k] = (\mathbf{C}\mathbf{P}^{-1})\mathbf{w} + \mathbf{D}\mathbf{f}$$

Controllability and observability may be investigated by diagonalizing the matrix.

13.7 Summary

An nth-order system can be described in terms of n key variables—the state variables of the system. The state variables are not unique, but can be selected in a variety of ways. Every possible system output can be expressed as a linear combination of the state variables and the inputs. Therefore the state variables describe the entire system, not merely the relationship between certain input(s) and output(s). For this reason, the state variable description is an internal description of the system. Such a description is therefore the most general system description, and it contains the information of the external descriptions, such as the impulse response and the transfer function. State-variable description can also be extended to time-varying parameter systems and nonlinear systems. An external system description may not describe a system completely.

The state equations of a system can be written directly from the knowledge of the system structure, from the system equations, or from the block diagram representation of the system. State equations consist of a set of n first-order differential equations and can be solved by time-domain or frequency-domain (transform) methods. Because a set of state variables is not unique, we can have a variety of state-space descriptions of the same system. It is possible to transform one given set of state variables into another by a linear transformation. Using such a transformation, we can see clearly which of the system states are controllable and which are observable.

References

1. Kailath, Thomas, *Linear Systems*, Prentice-Hall, Englewood Cliffs, N.J., 1980.

2. Zadeh, L., and C. Desoer, *Linear System Theory*, McGraw-Hill, New York, 1963.

Problems

13.1-1 Convert each of the following second-order differential equations into a set of two first-order differential equations (state equations). State which of the sets represent nonlinear equations.

(a) $\ddot{y} + 10\dot{y} + 2y = f$
(b) $\ddot{y} + 2e^{y}\dot{y} + \log y = f$
(c) $\ddot{y} + \phi_1(y)\dot{y} + \phi_2(y)y = f$

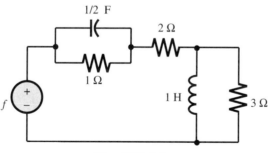

Fig. P13.2-1

13.2-1 Write the state equations for the RLC network in Fig. P13.2-1.

13.2-2 Write the state and output equations for the network in Fig. P13.2-2.

13.2-3 Write the state and output equations for the network in Fig. P13.2-3.

13.2-4 Write the state and output equations for the electrical network in Fig. P13.2-4.

13.2-5 Write the state and output equations for the network in Fig. P13.2-5.

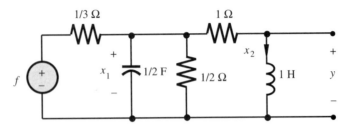

Fig. P13.2-2

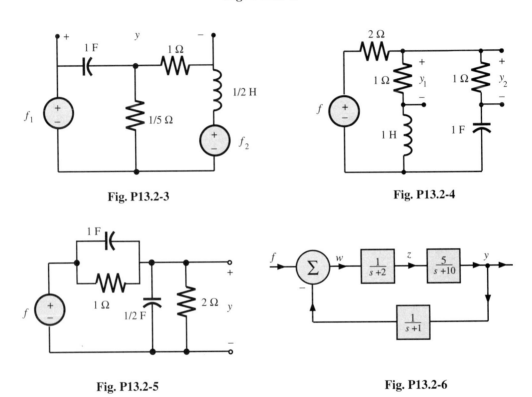

Fig. P13.2-3   Fig. P13.2-4

Fig. P13.2-5   Fig. P13.2-6

13.2-6 Write the state and output equations of the system shown in Fig. P13.2-6.

13.2-7 Write the state and output equations of the system shown in Fig. P13.2-7.

13.2-8 Write the different sets of state equations (two canonical, series, and parallel forms) and the output equation for a system having a transfer function

$$H(s) = \frac{3s + 10}{s^2 + 7s + 12}$$

13.2-9 Repeat Problem 13.2-8 if

(a) $H(s) = \dfrac{4s}{(s+1)(s+2)^2}$ (b) $H(s) = \dfrac{s^3 + 7s^2 + 12s}{(s+1)^3(s+2)}$

13.3-1 Find the state vector $\mathbf{x}(t)$ using the Laplace transform method if

$$\dot{\mathbf{x}} = \mathbf{A}\mathbf{x} + \mathbf{B}\mathbf{f}$$

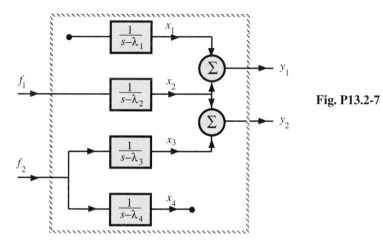

Fig. P13.2-7

where

$$\mathbf{A} = \begin{bmatrix} 0 & 2 \\ -1 & -3 \end{bmatrix} \quad \text{and} \quad \mathbf{B} = \begin{bmatrix} 0 \\ 1 \end{bmatrix} \quad \mathbf{x}(0) = \begin{bmatrix} 2 \\ 1 \end{bmatrix} \quad \text{and} \quad f(t) = 0$$

13.3-2 Repeat Prob. 13.3-1 if

$$\mathbf{A} = \begin{bmatrix} -5 & -6 \\ 1 & 0 \end{bmatrix} \quad \mathbf{B} = \begin{bmatrix} 1 \\ 0 \end{bmatrix} \quad \mathbf{x}(0) = \begin{bmatrix} 5 \\ 4 \end{bmatrix} \quad \text{and} \quad f(t) = \sin 100t$$

13.3-3 Repeat Prob. 13.3-1 if

$$\mathbf{A} = \begin{bmatrix} -2 & 0 \\ 1 & -1 \end{bmatrix} \quad \mathbf{B} = \begin{bmatrix} 1 \\ 0 \end{bmatrix} \quad \mathbf{x}(0) = \begin{bmatrix} 0 \\ -1 \end{bmatrix} \quad \text{and} \quad f(t) = u(t)$$

13.3-4 Repeat Prob. 13.3-1 if

$$\mathbf{A} = \begin{bmatrix} -1 & 1 \\ 0 & -2 \end{bmatrix} \quad \mathbf{B} = \begin{bmatrix} 1 & 1 \\ 0 & 1 \end{bmatrix} \quad \mathbf{x}(0) = \begin{bmatrix} 1 \\ 2 \end{bmatrix} \quad \mathbf{f} = \begin{bmatrix} u(t) \\ \delta(t) \end{bmatrix}$$

13.3-5 Using the Laplace transform method, find the response y if

$$\dot{\mathbf{x}} = \mathbf{A}\mathbf{x} + \mathbf{B}f(t)$$

$$y = \mathbf{C}\mathbf{x} + \mathbf{D}f(t)$$

where

$$\mathbf{A} = \begin{bmatrix} -3 & 1 \\ -2 & 0 \end{bmatrix} \quad \mathbf{B} = \begin{bmatrix} 1 \\ 0 \end{bmatrix} \quad \mathbf{C} = [0 \quad 1] \quad \mathbf{D} = 0$$

and

$$f(t) = u(t) \qquad \mathbf{x}(0) = \begin{bmatrix} 2 \\ 0 \end{bmatrix}$$

13.3-6 Repeat Prob. 13.3-5 if

$$\mathbf{A} = \begin{bmatrix} -1 & 1 \\ -1 & -1 \end{bmatrix} \quad \mathbf{B} = \begin{bmatrix} 0 \\ 1 \end{bmatrix} \quad \mathbf{C} = \begin{bmatrix} 1 & 1 \end{bmatrix} \quad \mathbf{D} = 1$$

$$f(t) = u(t) \qquad \mathbf{x}(0) = \begin{bmatrix} 2 \\ 1 \end{bmatrix}$$

13.3-7 The transfer function $H(s)$ in Prob. 13.2-8 is realized as a cascade of $H_1(s)$ followed by $H_2(s)$, where

$$H_1(s) = \frac{1}{s+3} \quad \text{and} \quad H_2(s) = \frac{3s+10}{s+4}$$

Let the outputs of these subsystems be state variables x_1 and x_2, respectively. Write the state equations and the output equation for this system and verify that $\mathbf{H}(s) = \mathbf{C}\boldsymbol{\phi}(s)\mathbf{B} + \mathbf{D}$.

13.3-8 Find the transfer-function matrix $\mathbf{H}(s)$ for the system in Prob. 13.3-5.

13.3-9 Find the transfer-function matrix $\mathbf{H}(s)$ for the system in Prob. 13.3-6.

13.3-10 Find the transfer-function matrix $\mathbf{H}(s)$ for the system

$$\dot{\mathbf{x}} = \mathbf{A}\mathbf{x} + \mathbf{B}\mathbf{f}$$

$$y = \mathbf{C}\mathbf{x} + \mathbf{D}\mathbf{f}$$

where

$$\mathbf{A} = \begin{bmatrix} 0 & 1 \\ -1 & -2 \end{bmatrix} \quad \mathbf{B} = \begin{bmatrix} 0 & 1 \\ 1 & 0 \end{bmatrix} \quad \mathbf{f} = \begin{bmatrix} f_1(t) \\ f_2(t) \end{bmatrix}$$

$$\mathbf{C} = \begin{bmatrix} 1 & 2 \\ 4 & 1 \\ 1 & 1 \end{bmatrix} \quad \mathbf{D} = \begin{bmatrix} 0 & 0 \\ 0 & 0 \\ 1 & 0 \end{bmatrix}$$

13.3-11 Repeat Prob. 13.3-1, using the time-domain method.

13.3-12 Repeat Prob. 13.3-2, using the time-domain method.

13.3-13 Repeat Prob. 13.3-3, using the time-domain method.

13.3-14 Repeat Prob. 13.3-4, using the time-domain method.

13.3-15 Repeat Prob. 13.3-5, using the time-domain method.

13.3-16 Repeat Prob. 13.3-6, using the time-domain method.

13.3-17 Find the unit impulse response matrix $\mathbf{h}(t)$ for the system in Prob. 13.3-7, using Eq. (13.65).

13.3-18 Find the unit impulse response matrix $\mathbf{h}(t)$ for the system in Prob. 13.3-6.

13.3-19 Find the unit impulse response matrix $\mathbf{h}(t)$ for the system in Prob. 13.3-10.

13.4-1 The state equations of a certain system are given as

$$\dot{x}_1 = x_2 + 2f$$

$$\dot{x}_2 = -x_1 - x_2 + f$$

Define a new state vector $\mathbf{w}$ such that

$$w_1 = x_2$$

$$w_2 = x_2 - x_1$$

Find the state equations of the system with $\mathbf{w}$ as the state vector. Determine the characteristic roots (eigenvalues) of the matrix $\mathbf{A}$ in the original and the transformed state equations.

13.4-2 The state equations of a certain system are

$$\dot{x}_1 = x_2$$

$$\dot{x}_2 = -2x_1 - 3x_2 + 2f$$

(a) Determine a new state vector $\mathbf{w}$ (in terms of vector $\mathbf{x}$) such that the resulting state equations are in diagonalized form.
(b) If the output $\mathbf{y}$ is given by

$$\mathbf{y} = \mathbf{Cx} + \mathbf{Df}$$

where

$$\mathbf{C} = \begin{bmatrix} 1 & 1 \\ -1 & 2 \end{bmatrix} \quad \text{and} \quad \mathbf{D} = 0$$

determine the output $\mathbf{y}$ in terms of the new state vector $\mathbf{w}$.

13.4-3 Given a system

$$\dot{\mathbf{x}} = \begin{bmatrix} 0 & 1 & 0 \\ 0 & 0 & 1 \\ 0 & -2 & -3 \end{bmatrix} \mathbf{x} + \begin{bmatrix} 0 \\ 0 \\ 1 \end{bmatrix} f$$

determine a new state vector $\mathbf{w}$ such that the state equations are diagonalized.

13.4-4 The state equations of a certain system are given in diagonalized form as

$$\dot{\mathbf{x}} = \begin{bmatrix} -1 & 0 & 0 \\ 0 & -3 & 0 \\ 0 & 0 & -2 \end{bmatrix} \mathbf{x} + \begin{bmatrix} 1 \\ 1 \\ 1 \end{bmatrix} f$$

The output equation is given by

$$y = \begin{bmatrix} 1 & 3 & 1 \end{bmatrix} \mathbf{x}$$

Determine the output y if

$$\mathbf{x}(0) = \begin{bmatrix} 1 \\ 2 \\ 1 \end{bmatrix} \quad \text{and} \quad f(t) = u(t)$$

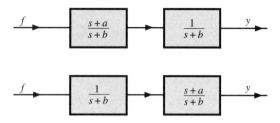

Fig. P13.5-1

13.5-1 Write the state equations for the systems depicted in Fig. P13.5-1. Determine a new state vector **w** such that the resulting state equations are in diagonalized form. Write the output **y** in terms of **w**. Determine in each case whether the system is controllable and observable.

13.6-1 An LTI discrete-time system is specified by

$$\mathbf{A} = \begin{bmatrix} 2 & 0 \\ 1 & 1 \end{bmatrix} \quad \mathbf{B} = \begin{bmatrix} 0 \\ 1 \end{bmatrix} \quad \mathbf{C} = [0 \quad 1] \quad \text{and} \quad \mathbf{D} = [1]$$

and

$$\mathbf{x}(0) = \begin{bmatrix} 2 \\ 1 \end{bmatrix} \quad \text{and} \quad f[k] = u[k]$$

(a) Find the output $y[k]$, using the time-domain method.
(b) Find the output $y[k]$, using the frequency-domain method.

13.6-2 An LTI discrete-time system is specified by the difference equation

$$y[k+2] + y[k+1] + 0.16y[k] = f[k+1] + 0.32f[k]$$

(a) Show the two canonical, series and parallel realizations of this system.
(b) Write the state and the output equations from these realizations, using the output of each delay element as a state variable.

13.6-3 Repeat Problem 13.6-1 if

$$y[k+2] + y[k+1] - 6y[k] = 2f[k+2] + f[k+1]$$

Supplementary Reading

The following books are listed for readers desiring alternate treatment of topics covered in this book.

Ambardar, Ashok, *Analog and Digital Signal Processing*, PWS Publishing, Boston, Mass, 1995.

Cadzow, J. A., and H.F. Van Landingham, *Signals and Systems*, Prentice-Hall, Englewood Cliffs, N.J., 1985.

Carlson, G.E., *Signal and Linear System Analysis*, Houghton Mifflin Co., Boston, 1992

Chen, C.T., *Systems and Signal Analysis*, Saunders College Publishing, New York, 1989.

Gabel, R.A., and R.A. Roberts, *Signals and Linear Systems*, 3rd ed., Wiley, New York, 1987.

Glisson, T.H., *Introduction to System Analysis*, McGraw-Hill, New York, 1985.

Jackson, L.B., *Signals, Systems, and Transforms*, Addison Wesley, Boston, Mass, 1991.

Kamen, E., and B.S. Heck, *Fundamentals of Signals and System*, Prentice-Hall, Englewood Cliffs, N.J., 1997.

Kwakernaak, H., and R. Sivan, *Modern Signals and Systems*, Prentice-Hall, Englewood Cliffs, N.J., 1991.

Mayhan, Robert J., *Discrete-time and Continuous-time Linear Systems*, Addison Wesley, Reading, Mass, 1984.

McGillem, C.D., and G.R. Cooper, *Continuous and Discrete Signal and System Analysis*, 3rd ed., Holt, Rinehart and Winston, New York, 1991.

Mitra, S.K., *Digital Signal Processing*, McGraw-Hill, New York, 1998

Neff, H.P., *Continuous and Discrete Linear systems*, Harper and Row, New York, 1984.

Oppenheim, A.V., A.S. Willsky, and S.H. Nawab *Signals and Systems, 2nd Ed.* Prentice-Hall, Englewood Cliffs, N.J., 1997.

Phillips, C.L., and J.M. Parr *Signals, Systems, and Transforms,* Prentice-Hall, Englewood Cliffs, N.J., 1995.

Poularikas, A., and S. Seely, *Signals and Systems,* PWS-Kent, Boston, N.J., 1985.

Proakis, J.G., and D.G. Manolakis, *Digital Signal Processing Principles, 3rd ed.* Prentice-Hall, Englewood Cliffs, N.J., 1996.

Strum, R.D., and D.E. Kirk, *First principles of Discrete Systems and Digital Signal processing,* Addison-Wesley, Reading MA, 1988.

Siebert, W.M., *Circuits, Signals, and Systems,* The MIT Press/McGraw-Hill, New York, 1986.

Soliman, S., and M. Srinath, *Continuous and Discrete Signal and Systems,* Prentice-Hall, Englewood Cliffs, N.J., 1990.

Taylor, S. J., *Principles of Signals and Systems,* McGraw-Hill, New York, 1994.

Ziemer, R.E., W.H. Tranter, and D.R. Fannin, *Signals and Systems,* Macmillan, New York, 1983.

Answers to Selected Problems

Chapter 1

1.1-1 (a), (b), and (c) π (d) 4π **1.1-4** $P_f = P_{-f} = 64/7$, $P_{2f} = 256/7$, $P_{cf} = 64c^2/7$

1.1-6 (a) 50 (b) 178 (c) 51 (d) 25 (e) 25 (f) 1/2.

1.4-2 $f_1(t) = (4t+1)u(t+1) - 6tu(t) + 3u(t) + (2t-4)u(t-2)$, $f_2(t) = t^2 u(t) - (t^2 - 2t+8)u(t-2) - (2t-8)u(t-4)$

1.4-4 (a) 0 (b) $\frac{2}{9}\delta(\omega)$ (c) $\frac{1}{2}\delta(t)$ (d) $-\frac{1}{5}\delta(t-1)$ (e) $\frac{1}{2-j3}\delta(\omega+3)$ (f) $k\delta(\omega)$

1.4-5 (a), (b) $f(t)$ (c) 1 (d) 0 (e) 20.085 (f) 5 (g) $f(-1)$ (h) -7.389

1.4-11 (a) $\pm j3$ (b) $-3 \pm j3$ (c) $2 \pm j3$ (d) -2 (e) 2 (f) 0.

1.7-1 Only (b), (f) and (h) are linear **1.7-2** Only (a) and (f) are time-invariant.

1.7-3 $y(t) = \left[e^{-(t+5)} + 2(t+5)e^{-(t+5)} + 2 \right] u(t+5) - \left[e^{-(t-5)} + 2(t-5)e^{-(t-5)} + 2 \right] u(t-5)$

1.7-7 (a) Causal (b) Noncausal (c) Noncausal (d) Noncausal.

1.8-1 $(D+3)y_1(t) = f(t)$, $(D+3)y_2(t) = Df(t)$ **1.8-3** $q_0(t) = \frac{a}{D+a}q_i(t)$, $(D+a)h(t) = \frac{1}{A}q_i(t)$

Chapter 2

2.2-1 (a) $\lambda^2 + 5\lambda + 6$, $\lambda^2 + 5\lambda + 6 = 0$, -2, -3, and e^{-2t}, e^{-3t} (b) $5e^{-2t} - 3e^{-3t}$

2.3-2 $\delta(t) + (e^{-2t} + e^{-3t})u(t)$ **2.3-4** $(2+3t)e^{-3t}u(t)$

2.4-5 (a) $tu(t)$ (b) $te^{-at}u(t)$ (c) $\frac{1}{2}t^2 u(t)$ **2.4-6** (a) $(1-\cos t)u(t)$ (b) $\sin t\, u(t)$

2.4-8 (a) $(\frac{1}{6} - \frac{2}{3}e^{-3t} + \frac{1}{2}e^{-2t})u(t)$ (b) $(e^{-2t} - e^{-3t})u(t)$ (c) $[(2-t)e^{-2t} - 2e^{-3t}]u(t)$

2.4-10 (a) $\frac{4}{\sqrt{13}}[0.555 - e^{-2t}\cos(3t+123.7°)]u(t)$ (b) $4[e^{-t} - \frac{1}{\sqrt{10}}e^{-2t}\cos(3t+71.56°)]u(t)$

2.4-11 (a) $(e^{-t} - e^{-2t})u(t)$ (b) $e^6(e^{-t} - e^{-2t})u(t)$ (c) $e^{-6}[e^{-(t-3)} - e^{-2(t-3)}]u(t-3)$

(d) $(1 - e^{-t})u(t) - [1 - e^{-(t-1)}]u(t-1)$ **2.4-13** $\tan^{-1} t + \frac{\pi}{2}$

2.4-15 $1 - \cos t$ for $0 \le t \le 2\pi$, $\cos t - 1$ for $2\pi \le t \le 4\pi$, 0 otherwise. **2.4-21** $H(s) = e^{-sT}$.

2.5-1 (a) $\frac{1}{3}e^{-3t} - \frac{1}{2}e^{-4t} + \frac{1}{6}$ (b) $\frac{1}{2}e^{-3t} - \frac{2}{3}e^{-4t} + \frac{1}{6}e^{-t}$ (c) $e^{-3t} - e^{-4t}$

2.5-3 (a) $(\frac{17}{4} + \frac{15}{2}t)e^{-2t} - 2e^{-3t}$ (b) $(\frac{9}{4} + \frac{19}{2}t)e^{-2t}$ **2.5-4** $\frac{9}{4} - \frac{1}{4}e^{-2t} + \frac{1}{2}t$ **2.5-5** $(2-t)e^{-3t} - 2e^{-4t}$

2.6-2 (a) Asympt. stable (b) marginally stable (c) unstable (d) marginally stable.

2.6-3 (a) 0 (b) marginally stable (c) BIBO unstable **2.7-1** (a) No (b) Yes

2.7-2 Width 0.6 ms, maximum rate 1667 pulses/sec. **2.7-3** (a) 10^{-4} (b) 10^4 Hz (c) 10^4 pulses/sec.

Chapter 3

3.1-2 (a) $c = 0.5$ (b) $E_e = 1/12$ **3.2-1** $c_{n_1} = c_{n_3} = 0$, $c_{n_2} = -1$, $c_{n_4} = \sqrt{2}/\pi$

3.4-1 (a) $f(t) = \frac{1}{3} + \frac{4}{\pi^2}\sum_{n=1}^{\infty}\frac{(-1)^n}{n^2}\cos n\pi t$

3.4-3 (a) $\omega_0 = \pi/2$, $a_0 = 0$, $b_n = 0$ and $a_n = \frac{4}{n\pi}\sin(n\pi/2)$ (c) $\omega_0 = 1$, $a_0 = 0.5$, $a_n = 0$ and $b_n = -\frac{1}{n\pi}$

(f) $\omega_0 = \pi/3$, $a_0 = 0.5$, $b_n = 0$ and $a_n = \frac{6}{n^2\pi^2}(\cos(\frac{n\pi}{3}) - \cos(\frac{2n\pi}{3}))$

3.4-4 (a) $\omega_0 = 2$, $a_0 = 0.504$, $a_n = 0.504(\frac{2}{1+16n^2})$ and $b_n = -0.504(\frac{8n}{1+16n^2})$

3.4-7 (a) $\omega_0 = \pi/4$, $a_0 = 0$, $a_n = \frac{4}{n^2\pi^2}(\frac{n\pi}{2}\sin\frac{n\pi}{2} - 1)$ and $b_n = \frac{4}{n^2\pi^2}\sin\frac{n\pi}{2}$

3.4-9 All except (c) and (e) are periodic. Periods (a) 2π (b) 2π (d) 2 (f) 140π (g) $8\pi/3$ (h) 2π (i) π.

3.4-11 $f(t) \simeq \frac{1}{2}x_0(t) - \frac{1}{4}x_1(t) - \frac{1}{8}x_3(t) - \frac{1}{16}x_7(t)$ **3.4-12** $f(t) = -\frac{3}{2}t + \frac{7}{8}\left(\frac{5}{2}t^3 - \frac{3}{2}t\right) + \cdots$

3.5-1   (b) $\omega_0 = 1/5$, $D_n = \frac{1}{n\pi}\sin(n\pi/5)$ (d) $\omega_0 = 2$, $D_0 = 0$, $D_n = -\frac{j}{n\pi}(\frac{2}{n\pi}\sin(\frac{n\pi}{2}) - \cos(\frac{n\pi}{2}))$

(e) $\omega_0 = 2\pi/3$, $D_0 = 1/6$, and $D_n = \frac{3}{2n^2\pi^2}[e^{-j2\pi n/3}(\frac{j2n\pi}{3} + 1) - 1]$

3.5-5 $N = 3$ **3.5-6** $N = 6$ **3.5-7** $N = 2$ **3.6-1** $y(t) = \sum_{-\infty}^{\infty} \frac{j1.08n}{(1+j4n)(\omega^2+3+j3\omega)} e^{j2nt}$

Chapter 4

4.1-5   (a) $\frac{4-2e^{-j\omega}-2e^{-j2\omega}}{j\omega}$ (b) $\frac{2}{\tau\omega^2}(\cos\omega\tau + \omega\tau\sin\omega\tau - 1)$

4.1-6   (a) $\frac{(\omega_0^2 t^2 - 2)\sin\omega_0 t + 2\omega_0 t\cos\omega_0 t}{\pi t^3}$ (b) $\frac{\sin t + \sin 2t}{\pi t}$ **4.2-4**   (a) $\frac{\omega_0}{\pi}\mathrm{sinc}[\omega_0(t - t_0)]$ (b) $\frac{1-\cos\omega_0 t}{\pi t}$

4.3-2 $F_2(\omega) = \frac{2e^{-j\omega}(\cos\omega + \omega\sin\omega - 1)}{\omega^2}$, $F_5(\omega) = \frac{3(1-j2\omega - e^{-j2\omega})}{4\omega^2}$

4.3-3   (a) $\frac{4j}{\omega}\sin^2(\frac{\omega T}{2})$ (b) $\frac{1+e^{-j\pi\omega}}{\omega^2-1}$ (c) $\frac{j\omega+e^{-j\pi\omega/2}}{1-\omega^2}$ (d) $\frac{1-e^{-(a+j\omega)T}}{j\omega+a}$ **4.3-5** $j4\,\mathrm{sinc}(\omega)\sin 3\omega$

4.3-6   (a) $\frac{\pi}{2}\left\{\mathrm{sinc}^2\left[\frac{\pi(\omega-10)}{2}\right] + \mathrm{sinc}^2\left[\frac{\pi(\omega+10)}{2}\right]\right\}$ (c) $\pi[\mathrm{sinc}\,\pi(\omega + 10) + \mathrm{sinc}\,\pi(\omega - 10)]e^{-j2\pi\omega}$

4.3-7   (a) $\frac{2}{\pi}\mathrm{sinc}(t)\cos 4t$ **4.4-1**   (a) $(e^{-t} - e^{-2t})u(t)$ (c) $\frac{1}{2}e^{-t}u(t) + \frac{1}{2}e^t u(-t)$ (d) $(1 - e^{-t})u(t)$

4.4-3   (d) Bandwidths of $y_1(t)$, $y_2(t)$ and $y(t)$ are 10 kHz, 5 kHz, and 15 kHz, respectively.

4.5-1 $t_0 \simeq 3\sqrt{2k}$ **4.5-2** $t_0 \simeq 40\ \mu s$ **4.5-3**   All unrealizable. (a) $t_d = 1\ \mu\mathrm{sec}$. (b) $t_d \approx 100\ \mu\mathrm{sec}$.

4.6-5 $B = \frac{0.366}{a}$ Hz.   **4.7-2**   (a)Yes (c) minimum $\omega_c = 2\pi B$ (d)No (e) Works only for even n.

4.7-5   (a) $A = 20$, (b) $A = 10$, (c) $A = 5$ **4.8-3** $B_{\mathrm{FM}} = 42$ kHz, $B_{\mathrm{PM}} = 38.06366$ kHz.

4.8-4 $\Delta\mathcal{F} = 100$ Hz, B= 2.2 kHz   **4.8-5** $\Delta\mathcal{F} = 20$ kHz, B= 42 kHz.

Chapter 5

5.1-2   (a), (b) 200 Hz (c) 120 Hz (d) 150 Hz   **5.1-3**   The signal recoverable only in cases (b) and (c).

5.1-4   (a) $h(t) = u(t) - u(t - 1)$ **5.1-5**   (b) $H(\omega) = T\,\mathrm{sinc}^2\left(\frac{\omega T}{2}\right)$ (c) Minimum delay$= T$

5.1-6 $\overline{F}(\omega) = 0.2F(\omega) + \frac{1}{5}\sum_{n=-\infty}^{\infty} \mathrm{sinc}\left(\frac{n\pi}{5}\right) F(\omega - 500\pi)$.

5.1-8   (a)   30 kHz (b)   16 bits (c)   480,000 bits/s (d)   705,600 bits/s

5.1-9   (a) 10.8 MHz (b)   10 bits (c) 108 Mbits/s   **5.2-1** $N_0 = 512$, zero padding required.

5.2-3 $N_0 = 256$, $T = 0.03125$ by both criteria.   **5.2-4**   (a) $N_0 = 16$, $T = 0.625$ (b) $N_0 = 8$, $T = 1.25$

Chapter 6

6.1-1   (a) $\frac{1}{s}(1-e^{-s})$ (c) $\frac{s^2 - \omega_0^2}{(s^2+\omega_0^2)^2}$ (e) $\frac{1}{2}[\frac{s}{s^2+(\omega_1+\omega_2)^2} + \frac{s}{s^2+(\omega_1-\omega_2)^2}]$ (g) $\frac{a}{s^2-a^2}$ (h) $\frac{(s+2)\cos\theta - 5\sin\theta}{s^2+4s+29}$

6.1-2   (a) $\frac{1}{s^2}(1 - e^{-s} - se^{-s})$ (b) $\frac{1+e^{-\pi s}}{s^2+1}$ (c) $\frac{1}{es^2}(1 - e^{-s} - se^{-s}) + \frac{1}{s+1}e^{-(s+1)}$

6.1-3   (b) $3.018e^{-2t}\cos(3t+6.34°)u(t)$ (c) $\delta(t)+(3.2e^{3t}-0.2e^{-2t})u(t)$ (g) $[e^{-t}-(1+t+\frac{t^2}{2}+\frac{t^3}{6})e^{-2t}]u(t)$

(h) $[\frac{1}{20} - \frac{1}{4}(1 - 2t)e^{-2t} + \frac{\sqrt{10}}{5}e^{-2t}\cos(t + 71.56°)]u(t)$

6.2-1   (c) $\frac{e^\tau}{s+1}$ (d) $\frac{1}{s+1}e^{-(s+1)\tau}$ (e) $\frac{1+\tau+s\tau}{(s+1)^2}e^{-(s+1)\tau}$ (g) $\frac{\omega_0\cos\omega_0\tau - s\sin\omega_0\tau}{s^2+\omega_0^2}$ (h) $\frac{\omega_0\cos\omega_0\tau + s\sin\omega_0\tau}{s^2+\omega_0^2}e^{-s\tau}$

6.2-3   (b) $\sqrt{2}e^{-(t-3)}\cos(t - 3 + \frac{\pi}{4})u(t - 3) + 2e^{-t}\sin tu(t)$ (c) $[\frac{e}{2}e^{(t-1)}\sin 2(t - 1)]u(t - 1) + \frac{3}{2}e^t\sin 2tu(t)$

6.2-4   (b) $\frac{1}{s}(\frac{1-e^{-2s}}{1-e^{-8s}})$

6.3-1   (a) $(e^{-t} - e^{-2t})u(t)$ (b) $(2 + 6t)e^{-2t}u(t)$ (c) $2 + 5.836e^{-3t}\cos(4t - 99.86°)u(t)$

6.3-2   The zi & zs comps. are: (b) $(2+5t)e^{-2t}$, te^{-2t} (c) $\sqrt{2}e^{-3t}\cos(4t-\frac{\pi}{4})$, $2+5.154e^{-3t}\cos(4t-112.83°)$

6.3-3   (a) $y_1(t) = (\frac{1}{2} - \frac{1}{3}e^{-t} - \frac{1}{6}e^{-4t})u(t)$, $y_2(t) = (\frac{1}{4} - \frac{1}{3}e^{-t} + \frac{1}{12}e^{-4t})u(t)$

6.3-4 $y_1(t) = [(8 + 17.89^{-1.5t}\cos(\frac{t}{2} - 26.56°)]u(t)$, $y_2(t) = 20\sqrt{2}e^{-1.5t}\cos(\frac{t}{2} - \frac{\pi}{4})u(t)$

6.3-5   (a) $\frac{5s+3}{s^2+11s+24}$ **6.3-6**   (a) $(D^2 + 3D + 8)y(t) = (D + 5)f(t)$

6.3-7   (iii) $(1.5e^{-2(t-5)} - 2e^{-3(t-5)} + 0.5e^{-4(t-5)})u(t - 5)$ (iv) $e^{20}(1.5e^{-2t} - 2e^{-3t} + 0.5e^{-4t})u(t)$

(v) $e^{-20}(1.5e^{-2(t-5)} - 2e^{-3(t-5)} + 0.5e^{-4(t-5)})u(t - 5)$

6.3-9 $\frac{1}{10}\left[e^{-t} + \frac{\sqrt{10}}{3}\cos(3t - 161.56°)\right]u(t)$ **6.4-1** $(te^{-t} - \frac{1}{2}e^{-t}\sin t)u(t)$ **6.4-2** $(5 - \frac{10}{3}e^{-2t/3})u(t)$

6.4-3 (a) $\frac{A}{L\omega_0{}^2}\delta(t)$ (b) $\frac{A}{L\omega_0}u(t)$ **6.4-4** $y_1(t) = (4 - \frac{3}{2}e^{-t} - \frac{1}{2}e^{-3t})u(t)$ $y_2(t) = (2 - \frac{3}{2}e^{-t} + \frac{1}{2}e^{-3t})u(t)$

6.4-5 $y_1(t) = 7.787e^{-t/6}\cos(\frac{\sqrt{11}}{6}t - 31.1°)u(t)$, $v_s(t) = \frac{20}{3}\delta(t) + [10 + 9.045e^{-t/6}\cos(\frac{\sqrt{11}}{6}t - 152.2°)]u(t)$

6.4-6 $40e^{-t/5}u(t)$ **6.4-7** $1.716e^{-2t}\cos(3t + 29°)u(t)$ **6.4-8** $(121.61e^{-6.53t} - 1.61e^{-2.8t})u(t)$

6.7-1b PO $\approx 3\%$, $t_r = 1.15$ secs., $t_s = 2.67$ secs., $e_s = 0$, $e_r = 0.75$, $e_p - \infty$.

6.7-2 $K_1 = 2$, $K_2 = 25.4$, $a = 6.128$ **6.7-3** t_s cannot be satisfied. **6.7-5** No.

6.8-1 (a) No region of convergence (b) $-1 < \mathrm{Re}\, s < 0$ (c) $\mathrm{Re}\, s = 0$ (d) $-1 < \mathrm{Re}\, s < 0$ (e) all s

6.8-2 (a) $\frac{-2}{s^2-1}$ $|\mathrm{Re}\, s| < 1$ (c) $\frac{-1}{(s-1)(s-2)}$ $1 < \mathrm{Re}\, s < 2$ (f) $\frac{-(s+\omega_0{}^2)}{(s-1)(s^2+\omega_0{}^2)}$ $0 < \mathrm{Re}\, s < 1$

6.8-3 (b) $e^{2t}u(t) - e^{3t}u(-t)$ (c) $(e^{-t} + e^{-2t})u(t)$ (d) $-(e^{-t} + e^{-2t})u(-t)$ (e) $(e^{-t} + e^{-3t})u(t) - e^{5t}u(-t)$

6.8-4 (a) $(e^{-t} - e^t + 2e^{-2t})u(t)$ (b) $(-e^{-t} + e^t - 2e^{-2t})u(-t)$

(c) $(e^{-t} + 2e^{-2t})u(t) + e^t u(-t)$ (d) $2e^{-2t}u(t) + (-e^{-t} + e^t)u(-t)$

6.8-5 (a) $[-\frac{4}{3}e^{-t} + 2e^{-t/2}]u(t) + \frac{2}{3}e^{t/2}u(-t)$ (d) $([\frac{1}{6}e^{-t} + \frac{1}{3}e^{2t}]u(t) + \frac{1}{2}e^t u(-t)$ (f) infinite

Chapter 7

7.1-1 (a) $\sqrt{2}\cos(2t - 15°)$ (b) $2\sqrt{2}\sin 2t$ (c) $2.28\cos(3t - 12.12°)$

7.1-2 (a) $7.5\,u(t)$ (b) $\frac{\sqrt{13}}{8}\cos(2t + 3.69°)u(t)$ (b) $\frac{\sqrt{18}}{13}\sin(3t - 112.62°)u(t)$ (c) $\frac{\sqrt{18}}{13}e^{j(3t-67.62°)}u(t)$

7.5-1 $H(s) = \frac{10^6}{s^3 + 200s^2 + 20,000s + 10^6}$ **7.5-3** $H(s) = \frac{1.002 \times 10^{20}}{s^4 + 2.615 \times 10^5 s^3 + 3.418 \times 10^{10} s^2 + 2.618 \times 10^5 s + 1.002 \times 10^{20}}$

7.6-2 $H(s) = \frac{4.9 \times 10^5}{s^3 + 98.8s^2 + 12380s + 491,300}$ **7.7-1** $H(s) = \frac{s^5}{s^5 + 52.4s^4 + 1375s^3 + 22,284s^2 + 223,192s + 1,117,708}$

7.7-2 $H(s) = \frac{s^3}{s^3 + 59.05s^2 + 942.81s + 19,079.42}$ **7.7-3** $H(s) = \frac{2.255 \times 10^4 s^2}{s^4 + 212.4s^3 + 72,550s^2 + 5.31 \times 10^6 s + 6.25 \times 10^8}$

7.7-4 $H(s) = \frac{2.24 \times 10^4 s^2}{s^4 + 164.7s^3 + 74,817s^2 4.117 \times 10^6 s + 6.25 \times 10^8}$ **7.7-5** $H(s) = \frac{(s^2 + 1200)^2}{s^4 + 56.53s^3 + 399.8s^2 + 67,824s + 1.44 \times 10^6}$

Chapter 8

8.2-2 (a), (b) $\left(-\frac{1}{e}\right)^k$ (c) $(-e)^k$ (d) $(-e)^k$ **8.2-3** Aperiodic: (b) & (c); Periodic: (a) $N_0 = 4$ & (d) $N_0 = 6$.

8.2-4 (c) Periodic; $N_0 = 20$. **8.2-5** Ω_f and $|\Omega_f|$ are (b) -0.8π, 0.8π (c) 0.6168, 0.6168 (e) 0.9π, 0.9π

8.2-7 (a) $e^{j(0.2\pi k + \theta)}$ (b) $e^{j0k} = 1$ (c) $e^{j4.333k}$ (d) $e^{j1.3\pi k}$ **8.2-8** (a) and (b) 2.7778 (c) $2.7778c^2$

8.2-9 (a) 19 (b) 19 (c) 252 (d) 40 **8.2-10** (a) 1 (b)1 (c) 0.5 (d) 0.5 (e) 0.5 **8.2-11** (a) $\frac{19}{6}$ (b) $\frac{7}{3}$

8.3-2 $10\cos\left(0.9\pi k - \frac{\pi}{6}\right)$ and $5\cos(0.9\pi k - \frac{\pi}{6})$ **8.3-3** $10\cos\left(\frac{\pi}{2}k\right) + \sqrt{10}\cos\left(\frac{3\pi}{4}k - 1.107\right)$

8.3-4 (i)1500 Hz (ii) 1500 Hz (iii) dc (iv) 1500 Hz (v) 2000 Hz. (vi) 600 Hz

8.4-4 (a) $f[k] = (k + 3)(u[k + 3] - u[k]) + (-k + 3)(u[k] - u[k - 4])$ **8.5-2** $p[k + 1] - 1.02p[k] = 1.01i[k]$

Chapter 9

9.1-4 $-12, 36$ and -63 **9.1-5** $-5, 7.667$, and -5.445

9.2-2 $-(3 + 2k)(-1)^k$ **9.2-3** $2(\sqrt{2})^k\cos(\frac{\pi}{4}k)$ **9.2-4** $\frac{100}{2^N - (0.5)^N}[2^N(0.5)^k - (0.5)^N(2)^k]$

9.3-1 $0.5\delta[k] - 0.5(-2)^k u[k]$ **9.3-4** $2.061(5)^k\cos(0.923k - 0.244)u[k]$

9.3-5 (b) $3\delta[k] - 5\delta[k - 1] - 2\delta[k - 3]$

9.4-1 $\frac{e}{2e+1}[e^{-(k+1)} - (-2)^{k+1}]u[k]$ **9.4-3** $9[\frac{11}{8}(3)^{k+1} - (2)^{k+1} - \frac{3}{8}(-5)^{k+1}]u[k]$

9.4-4 $\frac{18}{25}[(3)^{-k} - (2)^k + 5k(2)^k]u[k]$ **9.4-5** $\frac{1}{\sqrt{7}}[(3)^{k+1}\cos[\frac{\pi}{3}(k + 1) - 2.261] + 0.637(2)^{k+1}]u[k]$

9.4-7 (a) $\frac{2}{3}[(2)^{k+1} - (0.5)^{k+1}]u[k]$ (b) $\frac{1}{12}[(2)^{k+1} - (0.5)^{k+1}]u[k]$ (c) $\frac{8}{3}[(2^{k-1} - (0.5)^{k-1}]u[k - 2]$

9.4-8 $50000[(1.01)^{k+1} - 1]u[k] - 1500(1.01)^{k-4}u[k-4]$ **9.4-10** $N = 23$, residual pmt \$471.2

9.5-1 $\frac{1}{2e+1}[2e(-2)^k + e^{-k}]$ **9.5-2** $\frac{e}{2e+1}[-(-2)^k + e^{-k}]$ **9.5-3**   (b) $\frac{11}{4}(-1)^k - \frac{39}{5}(-2)^k + \frac{21}{20}(3)^k$

9.5-4 $(\frac{33}{16} + \frac{9}{4}k)(-1)^k - \frac{1}{16}(3)^{-k}$ **9.5-6** $0.241(0.2)^k - 0.377(0.8)^k + 0.765\cos(\frac{\pi k}{2} - 1.393)$

9.6-1   (a) Asympt. stable   (b) marginally stable   (c) unstable   (d) unstable   (e) marginally stable.

Chapter 10

10.1-1 $2e^{j0.4\pi k} + e^{j(0.8\pi k + \pi/2)} + e^{j(1.2\pi k - \pi/2)} + 2e^{j1.6\pi k}$

10.1-2 $\frac{1}{2}[e^{j0.5\pi k} + e^{j0.9\pi k} + e^{j1.1\pi k} + e^{j1.5\pi k}]$ **10.1-3** $e^{j(0.8\pi k + 1.6\pi)} + e^{j(1.2\pi k - 1.6\pi)}$

10.1-4 $\Omega_0 = \pi/3$, $N_0 = 6$, $\mathcal{D}_r = \frac{1}{6}[3 + 4\cos(\frac{\pi}{3}r) + 2\cos(\frac{2\pi}{3}r)]$

10.2-1   (a) 1   (b) $e^{-j\Omega k_0}$ (c) $\frac{a}{e^{j\Omega} - a}$ (d) $\frac{e^{j2\Omega}}{a(e^{j\Omega} - a)}$ **10.2-2**   (b) $e^{-j3\Omega}[3 + 4\cos\Omega + 2\cos 2\Omega]$

10.2-3 $\frac{1}{4}\text{sinc}^2(\frac{\pi k}{4})$ **10.3-1**   (a) $\frac{1-a^{10}e^{-j10\Omega}}{1-ae^{-j\Omega}}$ **10.3-2**   (a) $\frac{1+ae^{-j\Omega}}{(1-ae^{-j\Omega})^2}$ (b) $\frac{1-ae^{-j\Omega}\cos\Omega_0}{1-2ae^{-j\Omega}\cos\Omega_0 + a^2e^{-j2\Omega}}$

10.4-1   (a) $f[k] = \frac{1}{4}\text{sinc}^2(\frac{\pi k}{4})$ (b) $F_{[2k]}(\Omega) = \frac{1}{2}\sum_n \Delta(\frac{\Omega - 2\pi n}{2\pi})$, $F_{[4k]}(\Omega) = 1/4$ (c) $F_i(\Omega) = 2\Delta(\frac{2\Omega}{\pi})$.

10.5-1 $[2(-0.5)^k - \frac{8}{3}(-0.8)^k + \frac{2}{3}(-0.2)^k]u[k]$ **10.5-3** $[-(0.5)^k + \frac{8}{3}(0.8)^k]u[k] + \frac{8}{3}(2)^k u[-(k+1)]$

10.6-1   The 3-point DFT is $F_1 = 8$, $F_2 = e^{j\frac{\pi}{3}}$, $F_3 = e^{-j\frac{\pi}{3}}$. $F(\Omega) = 3e^{j\Omega} + 3 + 2e^{-j\Omega}$.   (c) The 8-point DFT

is 8, $6.5355 + j0.707$, $3 + j$, $-0.5355 + j0.707$, -2, $-0.5355 - j0.707$, $3 - j$, $6.5355 - j0.707$.

10.6-3   The 1-point DFT is $F_0 = 1$ and $F(\Omega) = 1$ (c) The N_0-point DFT is 1, 1, , $\cdots$, 1, 1.

10.6-4   (a) The N_0-point DFT is N_0, 0, 0, 0, $\cdots$, 0, 0, $F(\Omega) = \frac{e^{-jN_0\Omega} - 1}{e^{-j\Omega} - 1}$.

10.6-7   (a) The 5-point response is 9, 12, 16, 8 3.   (b) The Periodic Convolution is 17, 15, 16.

10.6-8   The 6-point response is 3, 8, 14, 9, 5, 3   **10.6-9**   The output $y[k]$ is 1, -1, 1, 3, -1, 1, $\cdots$.

Chapter 11

11.1-3   (b) $-\frac{2}{3}\delta[k] + [(2)^k - \frac{1}{3}(3)^k]u[k]$ (d) $[\frac{5}{2} - 7(2)^k + \frac{9}{2}(3)^k]u[k]$ (e) $[3(-1)^k - 3(2)^k + 2k(2)^k]u[k]$

(j) $[(2)^k + \sqrt{2}(5)^k\cos(\frac{\pi k}{3} - \frac{3\pi}{4})]u[k]$ (k) $[1 - 3(2)^k - \frac{1}{2}k(2)^k + \frac{1}{4}k(k-1)(2)^k]u[k]$

11.1-4   2, -1, and 4   **11.2-2** $\frac{z^8 - 2z^4 + 1}{z^7(z-1)^2}$ or $\frac{z^6 + 2z^5 + 3z^4 + 4z^3 + 3z^2 + 2z + 1}{z^7}$

11.2-3   (a) $\frac{\gamma z(z+\gamma)}{(z-\gamma)^3}$ (b) $\frac{z(z^2 + 4z + 1)}{(z-1)^4}$ (c) $\frac{z}{z-a}[1 - (\frac{a}{z})^m]$ (d) $\frac{e^{-2m}}{z(z-e^{-2})^2}(\frac{1-m}{e^2} + mz)$

11.3-3 $[\frac{1}{3} + \frac{5}{2}(2)^{-k} - \frac{4}{3}(4)^{-k}]u[k]$ **11.3-4** $[\frac{46}{3} - \frac{4}{3}(4)^{-k} - \frac{25}{2}(2)^{-k}]u[k]$

11.3-8 $[0.318(e)^k - 0.367(\sqrt{2})^k\cos(\frac{3\pi}{4}k - 0.525)]u[k]$ **11.3-9** $[1.32(e)^k - 0.186(-0.2)^k - 1.13(0.8)^k]u[k]$

11.3-11   (a) $12[4^{-k} + 3(2)^{-k} - 4(3)^{-k}]u[k]$ (b) $12[4^{-(k-2)} + 3(2)^{-(k-2)} - 4(3)^{-(k-2)}]u[k-2]$

(c) $192[4^{-k} + 3(2)^{-k} - 4(3)^{-k}]u[k]$ (d) $\frac{3}{4}[4^{-(k-2)} + 3(2)^{-(k-2)} - 4(3)^{-(k-2)}]u[k-2]$

11.3-13   (a) $\frac{z}{z+2}$ (b) $\frac{4z^2 - 3z}{2z^2 - 3z + 1}$ (c) $\frac{z}{4z^2 + 4z + 1}$ (d) $\frac{z+2}{z^2 + 2z + 2}$

11.3-14   (a) $[\frac{3}{2}\delta[k] - (-1)^k + \frac{1}{2}(-2)^k]u[k]$ (b) $(2 + 3k)(-1)^k u[k]$ (c) $5.099(\frac{1}{\sqrt{2}})^k\cos(\frac{\pi}{4}k - 1.373)u[k]$

11.6-1 $y[k] = \frac{e^{-kT} - e^{-T(2k+1)}}{1 - e^{-T}}u[k]$

11.7-1   (a) $\frac{-1.2z}{z^2 - 2.8z + 1.6}$ $0.8 < |z| < 2$ (c) $\frac{-z}{10(z^2 - 1.7z + 0.72)}$ $0.8 < |z| < 0.9$ (e) $\frac{4z(z-0.7)}{(z-0.4)(z-0.8)}$ $|z| > 0.8$

11.7-2   (a) $(e^{-2k} - 2^k)u[k]$ (b) $e^{-2k}u[k] + 2^k u[-(k+1)]$ (c) $(-e^{-2k} + 2^k)u[-(k+1)]$

11.7-3   (a) $[0.477e^k - 0.068(-0.2)^k - 0.412(0.8)^k]u[k]$ (b) $[\frac{1}{11}(-0.2)^k + \frac{2}{3}(0.8)^k]u[k] + 0.758(2)^k u[-(k+1)]$

(c)   The sum of (a) and (b)   **11.7-4** $[\frac{5}{66}(-0.2)^k + \frac{10}{3}(0.8)^k + 0.758(2)^k]u[k] + \frac{25}{6}u[-(k+1)]$

11.7-5 $y[k] = \infty$

12.1-1 **(a)** $|H[e^{j\Omega}]| = \frac{1}{\sqrt{1.16-0.8\cos\Omega}}$, $\angle H[e^{j\Omega}] = -\tan^{-1}[\frac{\sin\Omega}{\cos\Omega-0.4}]$

(c) $|H[e^{j\Omega}]| = [\frac{12.24-10.8\cos\Omega}{2.0256-2.32\cos\Omega+0.32\cos 2\Omega}]^{1/2}$ $\angle H[e^{j\Omega}] = \tan^{-1}[\frac{3\sin 2\Omega-1.8\sin\Omega}{3\cos 2\Omega-1.8\cos\Omega} -$

$\tan^{-1}[\frac{\sin 2\Omega-\sin\Omega}{\cos 2\Omega-\cos\Omega+0.16}]$ **12.1-2** **(a)** $|H[e^{j\Omega}]| = 4\cos\frac{\Omega}{2}+\cos\frac{3\Omega}{2}+2\cos\frac{5\Omega}{2}$, $\angle H[e^{j\Omega}] = -2.5\Omega$

12.1-3 **(a)** $|H[e^{j\Omega}]| = [\frac{1.64+1.6\cos\Omega}{1.25-\cos\Omega}]^{1/2}$, $\angle H[e^{j\Omega}] = \tan^{-1}(\frac{\sin\Omega}{\cos\Omega+0.8}) - \tan^{-1}(\frac{\sin\Omega}{\cos\Omega-0.5})$

(b) $2.86\cos(0.5k-1.6725)$ **12.1-5** **(a)** $\mathcal{F}_h = \frac{1}{2T} = \frac{10^6}{50\times 2} = 10$ kHz. **(b)** $\mathcal{F}_s \geq 2\mathcal{F}_h = 100$ kHz,

$T \leq \frac{1}{\mathcal{F}_s} = 10\,\mu$s. **12.2-2** **(c)** $BW = \frac{0.187}{T}$ Hz **12.2-3** $H[z] = \frac{1.707(1-a^2-\sqrt{2}a)(z^2-\sqrt{2}z+1)}{z^2-\sqrt{2}az+a^2}$

12.5-1 **(a)** For $T = \pi/350$, $H[z] = [\frac{0.031416z(z-0.97474)}{z^2-1.9383z+0.9391}]$

12.5-3 $H[z] = \frac{Tz}{z-1}$, $H[e^{j\omega T}] = \frac{T}{2|\sin\frac{\omega T}{2}|}$ $|\omega| \leq \frac{\pi}{T}$

12.5-4 **(a)** $\frac{0.5878z}{z^2-1.618z+1}$ **(b)** $\frac{0.5878Kz}{z^2-1.618z+1}$, K arbitrary **12.5-5** **(b)** $H[z] = \frac{1-e^{-\omega_c T}}{z-e^{-\omega_c T}}$ **(c)** $H[z] = \frac{T}{z-1}$

12.6-1 **(a)** $H[z] = \frac{2}{T}\frac{z-1}{z+1}$ **(b)** $|H[e^{j\omega T}]| = \frac{2}{T}|\tan\frac{\omega T}{2}|$ **(d)** $T = 25\,\mu$s.

12.6-3 $H[z] = \frac{0.09057(z+1)^2}{z^2-0.9871z+0.3493}$ **12.6-4** $H[z] = \frac{0.0512(z+1)^2}{z^2-1.355z+0.6125}$ **12.6-5** $H[z] = \frac{0.3203(z-1)^3}{z^3-0.9102z^2+0.5545z-0.0979}$

12.6-6 $H[z] = \frac{0.3619(z-1)^2}{z^3-0.4169z^2+0.4054}$ **12.6-7** $H[z] = \frac{0.0736(z^2-1)^2}{z^4-2.5503z^3+2.7876z^2-1.553z+0.394}$

12.6-8 $H[z] = \frac{0.04235(z-1)^2}{z^4-2.821z^3+3.507z^2-2.231z+0.6416}$ **12.6-9** $H[z] = \frac{0.4175(z^2-1.4712z+1)^2}{z^4-1.857z^3+1.538z^2-0.9003z+0.3498}$

12.8-2 $h[k] = \frac{1}{\pi(k-5)}[\sin 2(k-5)-\sin 1.6(k-5)]$ $0 \leq k \leq 10$.

12.8-4 $H[z] = \frac{2}{\pi}\sum_{k=-7}^{7}\frac{1}{k}z^{-k}$ (k odd), $H_{\text{Ham}}[z] = \frac{2}{\pi}\sum_{k=-7}^{7}\frac{1}{k}(0.54+0.46\cos\frac{\pi k}{7})z^{-k}$

12.8-5 $h[k] = \frac{1}{T}[-0.2885, 0.3140, -0.3779, 0.5282, -1.0137, 0, 1.0137, -0.5282, 0.3779, -0.3140, 0.2885]$.

Chapter 13

13.1-1 **(a)** $\dot{x}_1 = x_2$, $\dot{x}_2 = -2x_1 - 10x_2 + f$ **(b)** $\dot{x}_1 = x_2$, $\dot{x}_2 = -2e^{x_1}x_2 - \log x_1 + f$ (nonlinear) **(c)** $\dot{x}_1 = x_2$, $\dot{x}_2 = -\phi_1(x_1)x_2 - \phi_2(x_1)x_1 + f$ (nonlinear).

13.2-1 $\dot{x}_1 = -3x_1 - \dot{x}_2 + f$, $\dot{x}_2 = -\frac{3}{5}x_1 - \frac{6}{5}x_2 + \frac{3}{5}f$ **13.2-3** $\dot{x}_1 = -5x_1 + x_2 + 5f_1$, $\dot{x}_2 = -2x_1 - 2x_2 + 2f_1 - 2f_2$

13.2-5 $\dot{x}_1 = -x_1 + \frac{f}{3} + \frac{\dot{f}}{3}$, $y = -x_1 + f$ **13.2-6** $\dot{x}_1 = -10x_1 + 5x_2$, $\dot{x}_2 = -2x_2 - x_3 + f$, $\dot{x}_3 = x_1 - x_3$, $y = x_1$

13.2-7 $\dot{x}_1 = \lambda_1 x_1$, $\dot{x}_2 = \lambda_2 x_2 + f_1$, $\dot{x}_3 = \lambda_3 x_3 + f_2$, $\dot{x}_4 = \lambda_4 x_4 + f_2$, $y_1 = x_1 + x_2$, $y_2 = x_2 + x_3$

13.2-8 Controller Canonical: $\dot{x}_1 = x_2$, $\dot{x}_2 = -12x_1 - 7x_2 + f$, $y = 10x_1 + 3x_2$, Observer canonical: $x_1 =$

$-7x_1 + x_2 + 3f$, $\dot{x}_2 = -12x_1 + 10f$, $y = x_1$ Series: $\dot{x}_1 = -4x_1 - 9x_2 + 10x_2 + 3f$, $\dot{x}_2 = -3x_2 + f$, $y = x_1$

Parallel: $\dot{x}_1 = -4x_1 + f$, $\dot{x}_2 = -3x_1 + f$, $y = 2x_1 + x_2$

13.3-1 $\begin{bmatrix} (6e^{-t}-4e^{-2t})u(t) \\ (-3e^{-t}+4e^{-2t})u(t) \end{bmatrix}$ **13.3-2** $\begin{bmatrix} -34.02e^{-2t}+39.03e^{-3t}-0.01\cos 100t \\ 17.01e^{-2t}-13.01e^{-3t} \end{bmatrix}$

13.3-5 $(-1-2e^{-t}+3e^{-2t})u(t)$ **13.3-6** $(2+2e^{-t}\cos t-e^{-t}\sin t)u(t)$

13.3-8 $H(s) = \frac{-2}{s^2+3s+2}$ **13.3-9** $H(s) + \frac{s^2+3s+4}{s^2+2s+2}$

13.3-17 $h(t) = (e^{-3t}+2e^{-4t})u(t)$ **13.3-18** $h(t) = \delta(t) + (e^{-t}\sin t + e^{-t}\cos t)u(t)$

13.4-1 $\begin{bmatrix} \dot{w}_1 \\ \dot{w}_2 \end{bmatrix} = \begin{bmatrix} -2 & 1 \\ -3 & 1 \end{bmatrix}\begin{bmatrix} w_1 \\ w_2 \end{bmatrix} + \begin{bmatrix} 1 \\ -1 \end{bmatrix}f$ **13.4-2** $\begin{bmatrix} w_1 \\ w_2 \end{bmatrix} = \begin{bmatrix} 2 & 1 \\ 1 & 1 \end{bmatrix}\begin{bmatrix} x_1 \\ x_2 \end{bmatrix}$ $\mathbf{y} = \begin{bmatrix} 0 & 1 \\ -3 & 5 \end{bmatrix}\begin{bmatrix} w_1 \\ w_2 \end{bmatrix}$

13.4-4 $y(t) = (\frac{5}{2}+\frac{1}{2}e^{-2t}+\frac{5}{2}e^{-3t})$

Index